PRACTICAL FOUNDATION
ENGINEERING HANDBOOK

Other Books of Interest from McGraw-Hill

Breyer • DESIGN OF WOOD STRUCTURES
Brockenbrough and Merritt • STRUCTURAL STEEL DESIGNER'S HANDBOOK
Brown • DESIGN AND BEHAVIOR OF RESIDENTIAL AND LIGHT COMMERCIAL FOUNDATIONS
Brown • FOUNDATION BEHAVIOR AND REPAIR
Faherty and Williamson • WOOD ENGINEERING AND CONSTRUCTION HANDBOOK
Gaylord and Gaylord • STRUCTURAL ENGINEERING HANDBOOK
Merritt and Ricketts • BUILDING DESIGN AND CONSTRUCTION HANDBOOK
Merritt, Ricketts, and Loftin • STANDARD HANDBOOK FOR CIVIL ENGINEERS
Newman • DESIGN AND CONSTRUCTION OF WOOD FRAMED BUILDINGS
Newman • STANDARD HANDBOOK OF STRUCTURAL DETAILS FOR BUILDING CONSTRUCTION
Rollings and Rollings • GEOTECHNICAL MATERIALS IN CONSTRUCTION
Sharp • BEHAVIOR AND DESIGN OF ALUMINUM STRUCTURES

PRACTICAL FOUNDATION ENGINEERING HANDBOOK

Robert Wade Brown Editor in Chief

McGraw-Hill
New York San Francisco Washington, D.C. Auckland Bogotá
Caracas Lisbon London Madrid Mexico City Milan
Montreal New Delhi San Juan Singapore
Sydney Tokyo Toronto

Library of Congress Cataloging-in-Publication Data

Brown, Robert Wade.
    Practical foundation engineering handbook / Robert Wade Brown.
       p.   cm.
    Includes bibliographical references.
    ISBN 0-07-008194-8
    1. Foundations—Handbooks, manuals, etc. I. Title.
TA775.B674   1995
624.1′5—dc20   95-20386
                                                                        CIP

McGraw-Hill

A Division of The McGraw·Hill Companies

Copyright © 1996 by The McGraw-Hill Companies, Inc. All rights reserved. Printed in the United States of America. Except as permitted under the United States Copyright Act of 1976, no part of this publication may be reproduced or distributed in any form or by any means, or stored in a data base or retrieval system, without the prior written permission of the publisher.

1 2 3 4 5 6 7 8 9 0   AGM/AGM   9 0 0 9 8 7 6 5

ISBN 0-07-008194-8

The sponsoring editor for this book was Larry S. Hager, the editing supervisor was Frank Kotowski, Jr., and the production supervisor was Suzanne W. B. Rapcavage. It was set in Times Roman by Cynthia Lewis of McGraw-Hill's Professional Book Group composition unit.

Printed and bound by Quebecor/Martinsburg.

McGraw-Hill books are available at special quantity discounts to use as premiums and sales promotions, or for use in corporate training programs. For more information, please write to the Director of Special Sales, McGraw-Hill, 11 West 19th Street, New York, NY 10011. Or contact your local bookstore.

This book is printed on acid-free paper.

Information contained in this work has been obtained by McGraw-Hill from sources believed to be reliable. However, neither McGraw-Hill nor its authors guarantees the accuracy or completeness of any information published herein and neither McGraw-Hill nor its authors shall be responsible for any errors, omissions, or damages arising out of use of this information. This work is published with the understanding that McGraw-Hill and its authors are supplying information but are not attempting to render engineering or other professional services. If such services are required, the assistance of an appropriate professional should be sought.

CONTENTS

List of Contributors   xi
Preface   xv
Introduction   xvii

Part 1   Foundation and Civil Engineering Site Development   1.1

Section 1A. Water Behavior in Soils   1.3

1A.1.   Moisture Regimes / 1.3
1A.2.   Soil Moisture versus Water Table / 1.4
1A.3.   Soil Moisture versus Aeration Zone / 1.5
1A.4.   Permeability versus Infiltration / 1.6
1A.5.   Runoff / 1.7
1A.6.   Ground-Water Recharge / 1.8
1A.7.   Clay Soil / 1.8
1A.8.   Soil Moisture versus Root Development   1.8
1A.9.   Conclusions / 1.18
1A.10. References / 1.19

Section 1B. Site Preparation   1.21

1B.1.   Introduction / 1.21
1B.2.   Grading Plans / 1.23
1B.3.   Foundations on Filled Slopes / 1.24
1B.4.   Site Drainage / 1.24
1B.5.   Drainage Systems / 1.29
1B.6.   Landscaping / 1.30
1B.7.   Land Planning / 1.30
1B.8.   Foundation Drain Systems / 1.31
1B.9.   References / 1.33

Part 2   Soil Mechanics and Foundation Design Parameters   2.1

Section 2A. Soil Mechanics   2.3

2A.1.   Introduction / 2.3
2A.2.   Physical Condition / 2.3
2A.3.   Soil Identification and Classification / 2.5
2A.4.   Water Flow in Soils / 2.17
2A.5.   Geostatic Stresses in Soil / 2.21
2A.6.   Distribution of Applied Stresses in Soil / 2.22

2A.7. One-Dimensional Consolidation / 2.26
2A.8. Shear Strength of Soil / 2.32
2A.9. References / 2.44

Section 2B. Shallow Foundations — 2.45

2B.1. Introduction / 2.45
2B.2. Shallow Foundation Types / 2.45
2B.3. Design Parameters / 2.47
2B.4. Bearing Capacity of Shallow Foundations / 2.48
2B.5. Factor of Safety / 2.54
2B.6. Bearing Capacity of Mat Foundations / 2.54
2B.7. Settlement of Shallow Foundations / 2.55
2B.8. References / 2.67

Part 3 Fundamentals of Foundation Construction and Design — 3.1

Section 3A. Concrete — 3.3

3A.1. Constituent Materials / 3.3
3A.2. Computation of Required Average Compressive Strength f'_{cr} and Mix Proportioning / 3.18
3A.3. Quality Assurance / 3.33
3A.4. Mechanical Properties / 3.49
3A.5. Cold-Weather Concreting / 3.65
3A.6. Hot-Weather Concreting / 3.67
3A.7. Pumping of Concrete / 3.71
3A.8. References / 3.73

Section 3B. Fundamentals of Reinforced Concrete — 3.75

Notations / 3.75
3B.1. Introduction / 3.77
3B.2. Flexure Behavior / 3.77
3B.3. Shear Behavior / 3.93
3B.4. Torsion Behavior / 3.100
3B.5. Columns / 3.105
3B.6. Development of Reinforcing Bars / 3.117
3B.7. References / 3.120

Section 3C. Foundations Design — 3.121

3C.1. Introduction / 3.121
3C.2. Footing Types / 3.121
3C.3. Bearing Capacity of Soils under Shallow Foundations / 3.122
3C.4. Types of Failure of Footings / 3.124
3C.5. Rectangular Footings / 3.131
3C.6. Eccentrically Loaded Spread Footings / 3.135
3C.7. Combined Footings / 3.135
3C.8. Mat Foundations / 3.141
3C.9. Foundations in Cold Regions / 3.145
3C.10. Foundations in Earthquake Regions / 3.147
3C.11. References / 3.152

Section 3D. Pile Foundations   3.153

3D.1. Introduction / 3.153
3D.2. Allowable Stresses in Piles / 3.154
3D.3. Pile Loading Capacity / 3.160
3D.4. Pile Dynamics / 3.169
3D.5. Pile Load Test / 3.170
3D.6. Negative Skin Friction / 3.172
3D.7. Laterally Loaded Piles / 3.172
3D.8. Pile Cap Design / 3.174
3D.9. References / 3.180

Section 3E. Design Parameters and Procedures for Lightly Loaded Foundations   3.183

3E.1. Uniform-Thickness Ground-Supported Conventional or Posttensioned Slabs / 3.183
3E.2. Pier and Beam Foundations with Crawl Space / 3.184
3E.3. Basements / 3.184
3E.4. References / 3.185

Part 4   Earth Pressures and Retaining Systems   4.1

Section 4A. Earth Pressures   4.3

4A.1. Earth Pressure at Rest / 4.3
4A.2. Rankine Earth Pressures / 4.5
4A.3. Coulomb Earth Pressures / 4.13
4A.4. References / 4.17

Section 4B. Design of Rigid Retaining Walls   4.19

4B.1. Types of Retaining Walls / 4.19
4B.2. Earth Pressures Applied to Wall / 4.21
4B.3. Limit State for Retaining Walls / 4.21
4B.4. Limits of Design / 4.22
4B.5. Minimum Depth of Base / 4.22
4B.6. Simplified Method of Design / 4.23
4B.7. Theoretical Method of Design / 4.32

Part 5   Soil Improvement and Stabilization   5.1

Section 5A. Nongrouting Techniques   5.3

5A.1. Introduction / 5.3
5A.2. Overexcavation/Replacement / 5.4
5A.3. Near-Surface Compaction / 5.36
5A.4. Chemical Stabilization / 5.205
5A.5. Moisture Barriers / 5.264
5A.6. References / 5.267

Section 5B. Grouting to Improve Foundation Soil   5.277

5B.1 Introduction / 5.277
5B.2 Grouting Techniques / 5.306
5B.3 Types of Grout / 5.324
5B.4 Design and Specifications for a Grouting Program / 5.373
5B.5 Field Procedures / 5.380
5B.6 Some Final Comments / 5.398
5B.7 References / 5.398

Part 6   Foundation Failures and Repair: Residential and Light Commercial Buildings   6.1

Section 6A. Foundation Failures   6.3

6A.1. Causes for Foundation Failures / 6.3
6A.2. Settlement / 6.11
6A.3. Upheaval / 6.12
6A.4. Occurence of Settlement versus Upheaval / 6.12
6A.5. Diagnosis of Settlement versus Upheaval / 6.16
6A.6. Sliding / 6.17
6A.7. The Case For and Against Using Grade Elevations for Evaluating Foundation Distress / 6.18
6A.8. Soil Swell (Upheaval) versus Moisture Changes / 6.20
6A.9. How Is the Need for Foundation Repair Established? / 6.23
6A.10. References / 6.28

Section 6B. Foundation Repair Procedures   6.31

6B.1. Introduction / 6.31
6B.2. Pier and Beam Foundations / 6.32
6B.3. Slab Foundations / 6.48
6B.4. Slim Piers or Minipiles / 6.55
6B.5. Review of Repair Longevity (Common Repair Methods) / 6.60
6B.6. Grouting / 6.62
6B.7. Special Problems Adversely Affecting Leveling / 6.67
6B.8. Conclusions / 6.71
6B.9. Interesting Examples of Foundation Repair Techniques / 6.72
6B.10. Failures in Foundation Repair / 6.86
6B.11. References / 6.92

Part 7   Foundation Failures and Repair: High-Rise and Heavy Construction   7.1

7.1. Introduction / 7.3
7.2. Undermining of Safe Support / 7.7
7.3. Load-Transfer Failure / 7.21
7.4. Lateral Movement / 7.22
7.5. Unequal Support / 7.28

7.6. Drag Down and Heave / 7.31
7.7. Design Error / 7.35
7.8. Construction Errors / 7.44
7.9. Floatation and Water-Level Change / 7.52
7.10. Vibration Effects / 7.57
7.11. Cofferdams—14th Street Cofferdam / 7.59
7.12. Caissons and Piles / 7.60
7.13. Common Pitfalls 7.69
7.14. Checklist / 7.69
7.15. References / 7.70

Part 8 Miscellaneous Concerns 8.1

Section 8A. The Professional Engineer's (P.E.'s) Role in Foundation Analysis 8.3

8A.1. Introduction / 8.3
8A.2. Analysis and Evaluation of Distress / 8.3
8A.3. Other Factors that Might Be Assumed to Cause Foundation Distress / 8.4
8A.4. Recommended Repair or Other Remedial Action / 8.5
8A.5. Conclusion / 8.5
8A.6. References / 8.5

Section 8B. Business Risk Management 8.7

8B.1. Introduction / 8.7
8B.2. The Risks / 8.7
8B.3. Risk Management / 8.7
8B.4. Limitation of Liability / 8.7
8B.5. Business Entities / 8.9
8B.6. Business and Professional Liability Insurance / 8.9
8B.7. Alternative Dispute Resolution / 8.10
8B.8. Conclusion / 8.11

Section 8C. Structural Insured Protection 8.13

8C.1. Scope / 8.13
8C.2. Policies and Warranties / 8.14
8C.3. Examples of Major Structural Defects / 8.16
8C.4. Policy/Warranty Coverages / 8.17
8C.5. Typical Costs (1992 Dollars) / 8.19
8C.6. Settlement of Claims / 8.20

Section 8D. HUD 8.21

8D.1. Introduction / 8.21
8D.2. Scope / 8.21
8D.3. Policy Development and Research / 8.26
8D.4. Technical Suitability of Products Program / 8.27

8D.5. Lower-Cost Foundations / 8.29
8D.6. Manufactured Housing (Mobile Home) Foundations / 8.31
8D.7. Closure / 8.40
8D.8. Acknowledgments / 8.40
8D.9. References / 8.41

Index   I.1

CONTRIBUTORS

Section 3A

P. N. Balaguru is professor of civil engineering at Rutgers University in New Jersey. He has done extensive research in the areas of reinforced and prestressed concrete, construction, management, and the uses of new materials for construction. He is the chairman of ACI Committee 549 (Ferrocement and Other Thin Sheet Products) and a member of Committee 544 (Tendon Reinforcement). He has written a book titled *Fiber Reinforced Cement Composites*, published by McGraw-Hill, edited two books, and authored more than 140 publications in the area of concrete, new materials, and construction management, and has been a visiting professor at Northwestern University.

Sections 1A, 1B, 6A, 6B, 6C, 6D

Robert Wade Brown earned an M.S. degree in physical chemistry and math from the University of North Texas in 1955. In 1956 he joined The Western Co. as a Research Chemist specializing in such fields as hydraulics, rheology, chemical and physical properties of cement, as well as involved systems of oil well stimulation. In 1963 Mr. Brown started his own company specializing in foundation repair, pressure grouting, and soil stabilization. During the interval between 1955 and 1995 he has authored over 50 technical articles appearing in leading trade journals and has published four hard bound books, two of which were published by Van Nostrand-Reinhold and two by McGraw-Hill. He is a member of ASCE and the Society of Petroleum Engineers (AIME).

Section 8A

ECI Services, Inc., an engineering firm specializing in foundation condition reports, supplied much of the information included in this section. **Jim Irwin**, P.E. and President of ECI, was the principal contributor. He has since deceased. This section includes a compendium of other engineers' (P.E.s) contributed thoughts and editorial comments.

Jim Irwin graduated from Kansas University in 1946 with a degree in civil engineering. Prior to joining ECI Services, Inc., Mr. Irwin spent 36 years in various engineering pursuits, mostly specializing in concrete production, technology, material, and quality control. Mr. Irwin joined ECI Services, Inc. in 1985.

Sections 3B, 3C, 3D

Dr. A. Samer Ezeldin is an associate professor at Stevens Institute of Technology, New Jersey. He received his B.S. from Ain Shams University, Cairo, 1982, and his M.S. and Ph.D. degrees from Rutgers, The State University of New Jersey, in 1986 and 1989, respectively. He is an active member of the ACI and the ASCE. He is co-author of a textbook on materials for civil and highway engineers and has published more than 30 technical papers in the area of new construction materials and computer-aided design of structural elements. He is a registered professional engineer in Pennsylvania and has been involved in several consulting design projects.

Part 7

Dov Kaminetzky, P.E., is President of the New York–based consulting engineering firm of Feld, Kaminetzky & Cohen, P.C. Dov is a fellow of the ASCE and the ACI and has lectured extensively in the United States and abroad on the subjects of failures, restoration, and repairs of structures. He has published numerous articles and presented many papers on the above-mentioned subjects. He authored the book *Design and Construction Failures: Lessons from Forensic Investigation*, which was published by McGraw-Hill in 1991. This Part of the Handbook is largely based on material presented in that book.

Section 3E

Jerald W. Kunkel is a registered professional engineer in 30 states and is President of Jerald W. Kunkel Consulting Engineers, Inc., in Arlington, Texas. He received his Bachelor's degree in civil engineering in 1979 and his Master's Degree in structural engineering from the University of Texas, Arlington in 1994. He specializes in the design and investigation of residential structures, especially slab-on-grade applications. He has been involved in the design of slab-on-grade foundations in the expansive clay soils of north central Texas and at various locations nationwide for the past 22 years. Mr. Kunkel currently gives seminars and lectures at trade schools on residential construction, remodeling, and repair.

Section 5A

Dr. Evert C. Lawton is currently an associate professor of civil engineering at the University of Utah and has been teaching at the university level for 11 years. He has authored or co-authored over 40 technical papers and reports on topics related to foundation engineering, techniques for soil improvement and stabilization, and collapsible compacted soils. He is a co-patentee on two innovative methods for strengthening and stabilizing soils and is a co-winner of a national award for innovation in construction for one of these methods.

In addition to his academic teaching and research, Dr. Lawton has considerable practical experience, including 6 years as a structural engineer and a geotechnical engineer for the Virginia Department of Transportation. During his academic career, he has consulted on numerous projects, many of which have involved soil improvement and stabilization methods. He is currently registered as a professional engineer in five states.

Section 5B

Dr. Alex Naudts has a Masters degree in civil engineering (honors) from the State University in Ghent, Belgium (1977). He belongs to the Professional Engineering Association in Belgium (K.V.I.V) and holds membership in several North American and European professional associations and committees in grouting, and civil and mining engineering. Alex had been involved in the grouting and rehabilitation field since 1976. He has developed worldwide experience in design and execution of thousands of grouting projects through his involvement as manager of two major grouting contractors in Western Europe, and later as manager of his specialized consulting firm. These activities included specializing in the formulation, and the application, of chemical and conventional grouts and development of a reputation for troubleshooting projects in Western Europe, solving major water inflow problems quickly and effectively.

This expanded worldwide, and hundreds of leaks were successfully sealed off, and over 1000 grouting and rehabilitation projects were completed. Alex then had the chance to combine his strong theoretical background with field experience for challenging projects and was responsible for major changes in the grouting industry. A number of his patents and achievements include the introduction of the technology of using water reactive single component prepolymers to stop severe leaks, as well as the adaptation of hot bitumen grouting to the techniques of the 1980s.

Alex immigrated to Canada in 1982 and has been closely associated with the various inflows that have plagued the Saskatchewan potash mines during the 1980s. Several other projects were related to tunnelling problems, dam rehabilitation, restoration of underground structures, design and execution of underground flood bulkheads, restoration of heritage buildings and historical structures, and several research programs. He has been involved in over several thousands of grouting and restoration projects in North America, Europe, the Middle East, Latin America, and Africa.

Alex is the author of a university textbook on grouting and geotechnical engineering and has presented several technical papers in Europe and North America.

Section 8D

Dr. Richard J. Sazinski earned B.S. and M.S. degrees in civil engineering from the University of Connecticut (1970/1974) and a Ph.D. in structural engineering from Colorado State University (1978). He has served in the U.S. Army Corps of Engineers as a civil engineering assistant, has experience as a materials testing lab supervisor, and has worked as a project engineer over plant fabrication of prestressed structural elements and fabricated wood floor/roof truss components. Since 1978, he has worked at the Denver Office of the U.S. Department of Housing and Urban Development (HUD) as a structural engineer and has been involved with projects incorporating numerous types of *foundation* structural systems, from single-family homes to multistory high-rises. He has authored or been acknowledged in several articles in the areas of wood floor design, affordable housing, structural rehabilitation, and posttensioned *foundations* over expansive/compressible soils. He is a registered/licensed professional engineer (P.E.) in five states and is a member of ASCE and NSPE/PEG.

Section 8B

Thomas M. Smith graduated from the Texas Tech Law School in 1978. He is a licensed attorney in Texas and a partner in the law firm of Smith, Merrifield & Richards, L.L.P. in Dallas, Texas.

Mr. Smith is a member of many professional associations, including the American, Texas and Dallas Bar Associations, the State Bar College, The Real Estate Council, Real Estate Financial Executives Association, and is an affiliate Member of the Greater Dallas Association of Realtors. He is a member of the Education Committee of the Greater Dallas Association of Realtors working closely with its members to produce programs and materials to better inform and educate both the realtors and consumers in the real estate industry. Mr. Smith is also an active member of the Board of Directors of the Dallas Arboretum and Botanical Society.

Sections 2A, 2B, 4A, 4B

Dr. Richard W. Stephenson is an expert in geotechnical engineering. The thrust of his research has been in the general areas of measurement and analysis of behavioral properties of geotechnical materials. His expertise includes foundation design and analysis, laboratory analytical methods, soil dynamics, hydraulic conductivity of soils, geoenvironmental engineering, earth anchors and tiebacks, mine tailings disposal, and control as well as numerical methods.

Dr. Stephenson is active in professional societies, serves as a consultant to public and private organizations, and has lectured and given presentations on topics related to geotechnical engineering. He is the author of over 37 papers and reports and has directed or co-directed over $1,000,000 of externally funded research as professor of civil engineering at the University of Missouri, Rolla.

Section 3E

J. H. Williams is a 1942 graduate of Texas A&M with a B.S. in civil engineering. He became a registered professional engineer in Texas in 1945. During World War II Mr. Williams served four years on active duty in the Naval Air Corps. He was released to inactivity duty as a Lt. Sr. Grade in 1945. Presently he is in Retired Service.

Mr. Williams began his professional career in 1945 as a research engineer for Texas Research Foundation at Texas A&M. He entered private practice as a civil engineer in 1947 for George Smith in Fort Worth, Texas. He began private practice as an architect/engineer in 1949. He operates in this capacity to this date. During his 45 years as a consulting engineer, his activities have included design and construction of literally thousands ranging from residential to high rise. A good portion of this activity has involved the structural design of concrete foundations on expansive soils.

Section 8C

C. A. Windham has more than 40 years experience related to the residential construction industry, with more than 32 years in Texas alone.

Mr. Windham attended the University of South Carolina, Columbia, SC, and the University of Houston, Houston, TX.

Mr. Windham was employed by three different home builders in Texas in supervisory and technical capacities. He was with the Houston, Texas office of HUD/FHA (Housing and Urban Developing/Federal Housing Administration) for nearly 15 years, the last 3½ years of which was as the chief of their architectural/engineering branch. He was with the Texas company for neary 11 years.

PREFACE

The book represents the combination of hundreds of composite years experience and research. Each contributing author is a respected authority in his field and most, if not all, are widely published. The *Practical Foundation Engineering Handbook* is simply what the name implies, a text to which anyone with a concern or question about foundation design, behavior, or repair can refer. No previous publication has covered the subject to the degree provided by this handbook.

A focused attempt was made to restrict the coverage within this book to foundations or factors directly influencing foundations. Section 5B (grouting) deviates from this goal somewhat, specifically with regard to field applications. However, the principals, procedures, and products would similarily apply to foundation concerns.

In conclusion, let me give special thanks and recognition to each co-author contributing to this work and to Larry Hager of McGraw-Hill for his never ending support. In the same vein, I wish to extend appreciation to Brown Foundation Repair and Consulting, Inc. for making it all possible, to my daughters Candy for artwork and Cathy for transcribing several of the sections, and to my son Robert L. for editorial assistance.

Robert Wade Brown

INTRODUCTION

The most important concern of any structure is its foundation. This is true whether the property is high rise, residential, or anything in between. The foundation design must address such factors as site/soil conditions as well as those with individual structural load. Many times the site/soil parameters might not meet the requirements necessary to accommodate the intended structure. In this event specific steps must be taken to either enhance the bearing condition or alter the foundation load carrying capacity.

Part 1 discusses site conditions and control of surface or waste water.

Part 2 covers soil mechanics which enables contractors, architects, and engineers to evaluate and understand specific soil performance.

Part 3 deals with foundation construction and design, outlining options for site, soil, and load parameters.

Part 4 discusses methods for mechanically improving and stabilizing foundation bearing soils.

Part 5 offers methods which can be utilized to modify and enhance substandard soils.

Parts 6 and 7 address those instances where foundation designs fail. Part 6 focuses on lightly loaded structures, i.e., residential, and discusses cause and prevention as well as repair. Part 7 deals with foundations for high rise (or heavy) construction.

Part 8 covers the more or less tangential discipline which become engrossed in the foundation issue. Section 8A presents the professional engineer's role in evaluation of existing foundations, usually as a direct provision for owner transfer. Section 8B emphasises the legal concerns for foundation problems. Mostly the area concerns distressed preexisting foundations. The litigants are often consumer versus builder or insurer. Section 8C introduces an option available to residential property owners to "insure" their property against the possibility of foundation failure. Section 8D discusses the HUD concerns for insuring residential loans on either new or existing properties.

The authors' intents are to provide the reader with a comprehensive, consise source of information which can be used to design, understand, and, as the case me be, repair a defective foundation.

P · A · R · T · 1

FOUNDATION AND CIVIL ENGINEERING SITE DEVELOPMENT

Site hydrology and land planning are two initial factors that influence land use and foundation design. Part 1 addresses these concerns. Site hydrology involves both subsurface and surface water content and movement. Land planning develops construction techniques intended to accommodate hydrologic problems and provide best use of the parcel. Coverage of the topic will be rather cursory—as a rule, foundation engineers are not involved with the early stages of development, but an awareness of the potential problems is beneficial.

SECTION 1A
WATER BEHAVIOR IN SOILS

Robert Wade Brown

1A.1 MOISTURE REGIMES 1.3
1A.2 SOIL MOISTURE VERSUS WATER TABLE 1.4
1A.3 SOIL MOISTURE VERSUS AERATION ZONE 1.5
   1A.3.1 Transpiration 1.5
   1A.3.2 Gravity and Evaporation 1.5
1A.4 PERMEABILITY VERSUS INFILTRATION 1.6
1A.5 RUNOFF 1.7
1A.6 GROUND-WATER RECHARGE 1.8
1A.7 CLAY SOIL 1.8
1A.8 SOIL MOISTURE VERSUS ROOT DEVELOPMENT 1.8
1A.9 CONCLUSIONS 1.18
1A.10 REFERENCES 1.19

1A.1 MOISTURE REGIMES

The regime of subsurface water can be divided into two general classifications: the *aeration zone* and the *saturation zone*. The saturation zone is commonly termed the *water table* or *ground water*, and it is, of course, the deepest. The aeration zone includes the capillary fringe, intermediate belt (which may include one or more perched water zones), and, at the surface, the *soil water belt*, often referred to as the *root zone* (Fig. 1A.1). Simply stated, the soil water belt provides moisture for the vegetable and plant kingdom; the intermediate belt contains moisture essentially in dead storage—held by molecular forces; the perched ground water, if it occurs, develops essentially from water accumulation either above a relatively impermeable stratum or within an unusually permeable lens. Perched water usually occurs after a good rain and is relatively temporary; the capillary fringe contains capillary water originating from the water table.

The soil belt can contain capillary water available from rains or watering; however, unless this moisture is continually restored, the soil will eventually desiccate through the effects of gravity,

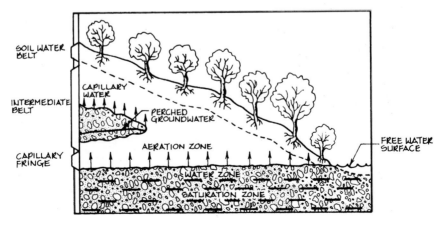

FIGURE 1A.1 Moisture regimes.

transpiration, and/or evaporation. In this process, the capillary water is lost. The soil belt is also the zone that most critically influences both foundation design and stability, as will be discussed in following sections.

As stated, the more shallow zones have the greatest influence on surface structures. Unless the water table is quite shallow, it will have little, if any, material influence on the behavior of foundations of normal residential structures. Further, the surface of the water table, the *phreatic boundary,* will not normally deflect or deform except under certain conditions, such as when it is near a producing well. Then the boundary will *draw down* or recede.

Engineers sometimes refer to a "natural" buildup of surface soil moisture beneath slab foundations that results from the lack of evaporation. This phenomenon is sometimes referred to as *center doming* or *center lift.* (Refer to Sec. 7.1.2.) If the source for this moisture is assumed to be the water table and if the water table is deeper than about 10 ft (3 m)* the boundary (as well as the capillary fringe) is not likely to "dome"; hence, no transfer of moisture to the shallow soils would be likely. The other source for moisture could involve the capillary or osmotic transfer from underlying soils to the dryer, more shallow soils. When expansive soils are involved, this intrusion of moisture can cause the soil to swell. This swell ultimately will be rather uniform over the confined area. (The expansive soil has a much greater lateral than vertical permeability.) Again no natural doming is likely to occur. Refer also to Sec. 1A.8.1.

1A.2 SOIL MOISTURE VERSUS WATER TABLE

Alway and McDole[1] conclude that deep subsoil aquifers (e.g., water table) contribute little, if any, moisture to plants and, hence, to foundations. Upward movement of water below a depth of 12 in (30 cm) was reportedly very slow at moisture contents approximating field capacity. *Field capacity* is defined as the residual amount of water held in the soil after excess gravitational water has drained and after the overall rate of downward water movement has decreased (zero capillarity). Soils at lower residual moisture content will attract water and cause it to flow at a more rapid rate. Water tends to flow from wet to dry in the same way as heat flows from hot to cold—higher energy level to lower energy level.

Rotmistrov[1] suggests that water does not move to the surface by capillarity from depths greater than 10 to 20 in (25 to 50 cm). This statement does not limit the source of water to the water table or capillary fringe. Richards[1] indicates that upward movement of water in silty loam can develop from depths as great as 24 in (60 cm). MeGee[1] postulates that 6 in (15 cm) of water can be brought to the surface annually from depths approaching 10 ft (300 cm). Again, the source of water is not restricted in origin.

The seeming disparity among these hydrologists likely results from variation in experimental conditions. Nonetheless, the obvious consensus is that the water content of the surface soil tends to remain relatively stable below very shallow depths and that the availability of soil water derived from the water table ceases when the boundary lies at a depth exceeding the limit of capillary rise for the soil. In heavy soils (e.g., clays), water availability practically ceases when the water source is deeper than 4 ft (120 cm) even though the theoretical capillary limit normally exceeds this distance. In silts, the capillary limit may approximate 10 ft (300 cm) as compared to 1 to 2 ft (30 to 60 cm) for sands. The height of capillary rise is expressed by

$$\pi \gamma_T r^2 h_c = T_{st} 2\pi r \cos \alpha \quad \text{or} \quad h_c = \frac{2T_{st}}{r\gamma_T} \cos \alpha \tag{1A.1}$$

where h_c = capillary rise, cm
T_{st} = surface tension of liquid at temperature T, g/cm

---

*The abbreviations of units of measure used in this book are listed in App. C.

γ_T = unit weight of liquid at temperature T, g/cm³
r = radius of capillary pore, cm
α = meniscus angle at wall or angle of contact

For behavior in soils, the radius r is difficult, if not impossible, to establish and depends on such factors as void ratio, impurities, grain size, and distribution and permeability. Since the capillary rise varies inversely with effective pore or capillary radius, this value is required for mathematical calculations. Accordingly, capillary rise, particularly in clays, is generally determined by experimentation. In clays, the height and rate of rise are impeded by the soil's swell (loss of permeability) upon invasion of water. Fine noncohesive soils will create a greater height of capillary rise, but the rate of rise will be slower. More information on soil moisture, particularly that dealing with clay soils, will be found in Secs. 6A and 6B.

1A.3 SOIL MOISTURE VERSUS AERATION ZONE

Water in the upper or aeration zone is removed by one or a combination of three processes: transpiration, evaporation, and gravity.

1A.3.1 Transpiration

Transpiration refers to the removal of soil moisture by vegetation. A class of plants, referred to as *phreatophytes,* obtain their moisture, often more than 4 ft (120 cm) of water per year, principally from either the water table or capillary fringe. This group includes such seemingly diverse species as reeds, mesquite, willows, and palms. The remaining two groups, mesophytes and xerophytes, obtain their moisture from the soil water zone. These include most vegetables and shrubs, along with some trees.

In all vegetation, root growth is toward soil with greater available moisture. Roots will not penetrate a dry soil to reach moisture. The absorptive area of the root is the tip, where root hairs occur. The loss of soil moisture due to transpiration follows the root pattern and is generally somewhat circular about the stem or trunk. The root system develops only to the extent necessary to supply the vegetation with required water and nutrition. Roots not accessible to water will wither and die. These factors are important to foundation stability, as will be discussed in following sections.

In many instances, transpiration accounts for greater loss of soil moisture than does *evaporation.* Another process, *interception,* is the procedure whereby precipitation is caught and held by foliage and partially evaporated from exposed surfaces. In densely planted areas, interception represents a major loss of rainfall, perhaps reaching as high as 10 to 25% of total precipitation.[1]

1A.3.2 Gravity and Evaporation

Gravity tends to draw all moisture downward from the soil within the aeration zone. Evaporation tends to draw moisture upward from the surface soil zone. Both forces are retarded by molecular, adhesive, and cohesive attraction between water and soil as well as by the soil's ability for capillary recharge. If evaporation is prevented at the surface, water will move downward under the forces of gravity until the soil is drained or equilibrium is attained with an impermeable layer or saturated layer. In either event, given time, the retained moisture within the soil will approximate the field capacity for the soil in question.

In other words, if evaporation were prevented at the soil surface, as, for example, by a foundation, an "excessive" accumulation of moisture would initially result. However, given sufficient time, even this protected soil will reach a condition of moisture equilibrium somewhere between that originally noted and that of the surrounding uncovered soil. The natural tendency of covered

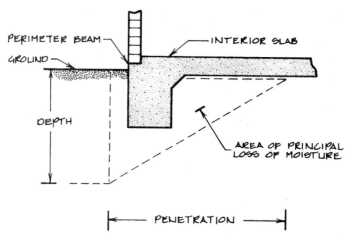

FIGURE 1A.2 Typical loss of soil moisture from beneath a slab foundation during prolonged drying cycle.

soil is to retain a moisture above that of the uncovered soil—except, of course, during periods of heavy inundation (rains) when the uncovered soil reaches a temporary state at or near saturation. In this latter instance, the moisture content decreases rapidly with the cessation of rain or other sources of water.

The loss of soil moisture from beneath a foundation caused by unabated evaporation would tend to follow a triangular configuration with one leg vertical and extending downward into the bearing soil and the other leg horizontal and extending under the foundation.[2] The relative lengths of the legs of the triangle would depend upon many factors including the particular soil characteristics, foundation design, weather, or availability of moisture (Fig. 1A.2).

In their study, Davis and Tucker reported the depth as about 5 ft (1.5 m) and the penetration approximately 10 ft (3 m).[2] In any event, the affected distances (legs of the triangle) are relatively limited. As with all cases of evaporation, the greatest effects are noted closer to the surface. In an exposed soil, evaporation forces are ever present, provided the relative humidity is less than 100%. The force of gravity is effective whether the soil is covered or exposed.

1A.4 PERMEABILITY VERSUS INFILTRATION

The infiltration feature of soil relates more directly to penetration from rain or water at the surface than to subsurface vertical movement. The exceptions are those relatively rare instances in which the ground surface is within the capillary fringe. Vertical migration or permeation of the soil by water infiltration could be approximately represented by the single-phase steady-state flow equation as postulated by Darcy[3]:

$$Q = -\left(\frac{AK}{\mu}\right)\left(\frac{\Delta P}{L} + g_c \sin \alpha \gamma\right) \qquad (1A.2)$$

where Q = rate of flow in direction L
 A = cross-sectional area of flow
 k = permeability
 μ = fluid viscosity
 $\Delta P/L$ = pressure gradient in direction L

L = direction of flow
γ = fluid density
α = angle of dip: $\alpha > 0$ if flow L is up dip
g_c = gravity constant

If $\alpha = 90°$, $\sin \alpha = 1$ and, simplified,

$$Q = -\left(\frac{AK}{\mu L}\right)(\Delta P + g_c h \gamma)$$

where $h = L \sin \alpha$
$g_c h \gamma$ = hydrostatic head

If $H = P + g_c h \gamma$ where H is the fluid-flow potential:

$$Q = -\left(\frac{Ak}{\mu}\right)\left(\frac{H}{L}\right)$$

When flow is horizontal, the gravity factor g_c drops out. Any convenient set of units may be used in Eq. (1A.2) as long as the units are consistent. Several influencing factors represented in this equation pose a difficulty for mathematical calculations. For example, the coefficient of permeability k can be determined only by experimental processes and is subject to constant variation even within the same soil. The pore sizes, water saturation, particle gradation, transportable fines, and mineral constituents all affect the effective permeability k. In the instance of expansive clays, the variation is extremely pronounced and subject to continuous change upon penetration by water. The hydraulic gradient ΔP and the distance over which it acts ΔL are also elusive values. For these reasons, permeability values are generally established by field or laboratory tests where the variables can be controlled. In the case of clean sand, the variation is not nearly as extreme, and reasonable approximations for k are often possible.

In essence, Eq. (1A.2) provides a clear understanding of factors controlling water penetration into soils but does not always permit accurate mathematical calculation. The rate of water flow does not singularly define the moisture content or capacity of the soil. The physical properties of the soil, available and residual water, each affect infiltration, as does permeability. A soil section 3 ft (90 cm) thick may have a theoretical capacity for perhaps 1.5 ft (0.46 m) of water. This is certainly more water than occurs from a serious storm; hence, the moisture-holding capacity is seldom, if ever, the limiting criterion for infiltration.

In addition to the problems of permeability, infiltration has an inverse time lag function (Fig. 1A.2). This figure is a typical, graphical representation of the relationship between infiltration and runoff with respect to time. At the onset of rain, more water infiltrates, but over time most of the water runs off, and little is added to the infiltration.

Clays will have a greater tendency for runoff, as opposed to infiltration, than will sands. The degree of the slope of the land will have a comparable effect, since steeper terrains deter infiltration. Only the water that penetrates the soil is of particular concern with respect to foundation stability. The water which fails to penetrate the soil is briefly discussed in Sec. 1A.5.

1A.5 RUNOFF

Any soil above the capillary fringe tends to lose moisture through the various forces of gravity, transpiration, and evaporation. Given sufficient lack of recharge water, the soil water belt will merge with and become identical to the intermediate belt. However, nature provides a method for replenishing the soil water through periodic rainfall. Given exposure to rain, all soils absorb water to some varying degree, depending on such factors as residual moisture content, soil composition and gradation, and time of exposure. The excess water not retained by the soil is termed *runoff* (Fig. 1A.2).

1.8 FOUNDATION AND CIVIL ENGINEERING SITE DEVELOPMENT

As would be expected, sands have a high absorption rate, and clays have a relatively low absorption rate. A rainfall of several inches over a period of a few hours might saturate the soil water belt of sands but penetrate no more than 6 in in a well-graded, high-plasticity soil. A slow, soaking rain would materially increase penetration in either case. The same comparison holds whether the source of water is rain or watering. Section 6 will develop the importance of maintaining soil moisture to aid in preventing or arresting foundation failures.

1A.6 GROUND-WATER RECHARGE

Even in arid areas, an overabundance of water can occur sporadically due, principally, to storm runoff. If these surpluses can be collected and stored, a renewable resource is developed, conserving water during periods of plenty for future use during times of shortage. In general, this storage can be in the form of surface reservoirs or recharged aquifers.[5]

Surface reservoirs suffer losses from evaporation as well as occasional flooding and are somewhat limited due to topographical demands.

Underground storage can be effected through natural ground-water recharge or artificial recharge. The obvious advantages to either form of underground storage is high capacity, simplicity, no evaporative losses, and low costs. Natural ground-water recharge occurs when aquifers are unconfined, surface soils are permeable, and vadose (aeration) zones have no layers that would restrict downward flow. When and where the foregoing conditions do not exist, artificial recharge is necessary. The latter requires that a well be drilled into the aquifer. Such wells can be used to inject water into or remove water from the aquifer or both, depending on supply and demand. The prime storage zones include limestone, sand, gravel, clayey sand, sandstone, and glacial drift aquifers. The quality of the aquifers and recharge water depend mostly on availability. Under the most adverse conditions, appropriate thought, well design, and operation procedures can produce potable water. Additional detail on this topic can be found in reference 5.

1A.7 CLAY SOIL

Preceding sections have suggested the influence of ground-water hydrology on foundation stability. This is most certainly true when the foundation-bearing soil contains an expansive clay. One complex and misunderstood aspect is the effect roots have on soil moisture. Without question, transpiration removes moisture from the soil. Exactly how much, what type, and from where represent the basic questions. If the roots take only pore (or capillary) water and/or remove the moisture from depths deeper than about 3 to 7 ft (1 to 2 m), the moisture loss is not likely to shrink soils enough to threaten foundation stability.

1A.8 SOIL MOISTURE VERSUS ROOT DEVELOPMENT

Logically, in semiarid climates (Fig. 6C.3), the root pattern tends to develop deeper. In wetter areas, the root systems are closer to the surface, and the availability of moisture allows the roots' needs to be supplied without desiccation of the soil (see Figs. 1A.3 and 1A.4 and Table 1A.1). [An explanation of the Atterberg Limits (LL, PL, and PI) is given in Sec. 2A.]

The soil in question is identified as a London clay with physical and chemical characteristics similar to many of the typical fat clays found in the United States. The London climate has a C_w*

---

*C_w is the climatic factor developed by Building Research Advisory Bulletin and used in the design of slab-on ground foundations. Refer to Fig. 6C.3.

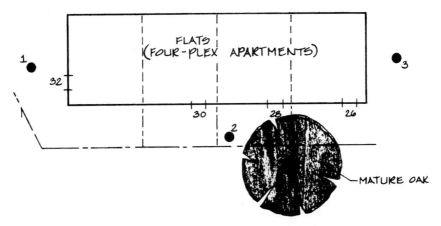

FIGURE 1A.3 Location plan.

factor in the range of 35 to 40, which is similar to that for Mississippi and Washington. Note that the soil moisture content remains constant from 2 to 5 m (6.6 to 16.4 ft) despite the close proximity of the mature oak tree (Table 1A.1). Although this observation might be surprising, it was by no means an isolated instance. The test borings provided data on the loss of soil moisture, but nothing indicated the root pattern. This is not critical but would have been interesting. Note, however, that all tests commenced below the 2-ft (0.6-m) level, which seems to be the maximum depth from which roots remove moisture in this environment. (Refer to Sec. 2A.3.1.2, "Clay Mineralogy," and Sec. 6A1.2, "Expansive Soils," for additional information concerning the water behavior in clay soils.)

In areas with more extreme climates and the same general soil, the root development pattern would more closely resemble that in Fig. 1A.5. During earlier growth stages, particularly if the tree was being conscientiously watered, the root system might be quite shallow—within the top 1 ft (30 cm) or so. Dry weather (lack of "surface" moisture) forces the roots to seek deeper soils for adequate water. The surface roots can remain dormant in a low-moisture environment for extended periods and become active again when soil moisture is restored.

Although the so-called *fat clays* are generally impermeable, thus limiting true capillary transfer of water, intrinsic fractures and fissures allow the tree or plant root systems to pull water from soil a radial distance away somewhat in excess of the normal foliage radius. As an aside, when transpiration is active, evaporation diminishes. (The shaded areas lose less moisture.) The net result is often a conservation of soil moisture.

With respect to Fig. 1A.5, Dr. Don Smith, botanist at North Texas State University, Denton, Texas, suggests certain generalities:

1. D_1 is in the range of 2 ft (0.6 m) maximum.
2. W_R is in the range of $1.25 \times W$, where W is the natural canopy diameter (unpruned).
3. When moisture is not readily available at D_1, the deeper roots D_2 increase activity to keep the tree's needs satisfied. If this is not possible, the tree wilts.
4. H has no direct correlation to W_R, D_1, or D_2 except the indirect relation that H is relative to the age of the tree.

Koslowski[6] and the National House-Building Council[7] suggest values for D_2, and the effective D_1, as shown in Table 1A.2. Note that the depth of soil moisture loss due to the near surface feeder roots is not to be confused with depth of total soil moisture loss (activity zone). The important point is that soil moisture losses from either transpiration or evaporation normally occur from rel-

FIGURE 1A.4 Borehole log.

TABLE 1A.1 Atterberg Limits and Soil Moisture for London Clay BH No. 2: Brown-Gray Mottled Silty Clay

Depth						
Meters	Feet	LL, %*	PL, %*	PI*	Natural moisture contents, %	Soil classification
2.00	6.6	93	27	66	30	CE
3.5	11.5	86	28	59	30	CV
4.5	14.8	89	28	61	30	CV
5.00	16.4	85	26	59	29	CV

*LL = liquid limit; PL = plastic limit; PI = plasticity index. The British Soil Classification uses *CV* for soils with an LL between 70 and 90 and *CE* for soils with an LL in excess of 90.

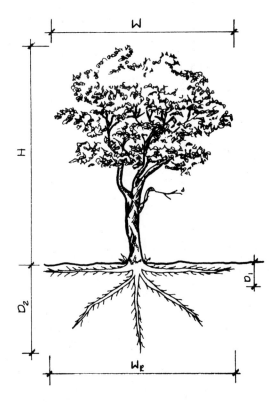

W - DIAMETER OF CANOPY (UNPRUNED) DRIP LINE
H - HEIGHT
D_1 - DEPTH OF LATERAL ROOTS
D_2 - DEPTH OF DEEP ROOTS (TAP ROOTS)
W_R - DIAMETER OF LATERAL ROOTS

FIGURE 1A.5 Root system.

TABLE 1A.2 Depth of Tree Roots, Plains Area, USA*

Name	Age (years)	D_2
Plantanus occidentalis (American Sycamore)	6	7 ft (2.1 m)
Juglans nigra (Black Walnut)	6	5 ft (1.5 m)
Quercus rubra (Red Oak)	6	5 ft (1.5 m)
Carya ovata (Shag Bark Hickory)	6	5 ft (1.5 m)
Fraxinus americana (Ash)	6	5 ft (1.5 m)
Populus deltoides (Poplar or Cottonwood)	6	6 ft (1.8 m)
Robinia pseudoacacia (Black Locust)	Unknown	24–27 ft (7–8.2 m)

*From Ted Koslowski.[6]

TABLE 1A.3 Depth of Tree Roots, London, England, (PI above 40)*

Name	Age	D_1†	H (height)
High water demand			
Elm	Mature	3.25 m (10.6 ft)	18–24 m (59–79 ft)
Oak	Mature	3.25 m (10.6 ft)	16–24 m (52–79 ft)
Willow	Mature	3.25 m (10.6 ft)	16–24 m (52–79 ft)
Moderate water demand			
Ash	Mature	2.2 m (7.2 ft)	23 m (75 ft)
Cedar	Mature	2.0 m (6.6 ft)	20 m (65.6 ft)
Pine	Mature	2.0 m (6.6 ft)	20 m (65.6 ft)
Plum	Mature	2.0 m (6.6 ft)	10 m (32.8 ft)
Sycamore	Mature	2.2 m (7.2 ft)	22 m (72 ft)
Low water demand			
Holly	Mature	1.55 m (4.9 ft)	12 m (39.4 ft)
Mulberry	Mature	1.45 m (4.7 ft)	9 m (29.5 ft)

*From National House-Building Council, UK.[7]
†Interpolation of *maximum* depth of root influence on foundation design at $D=2$ m, per reference 2. Refer also to Fig. 1A.2.

atively shallow depths. Tucker and Davis[2] and Tucker and Poor[8] report test results that 84% of total soil moisture loss occurs within the top 3 to 4 ft (1 to 1.25 m) (Fig. 1A.6). The soil involved was the EagleFord (Arlington, Texas) with a PI in the range of 42. Other scientists, such as Holland and Lawrence,[9] report similar findings. The latter publications suggest soil moisture equilibrium below about 4 ft (1.25 m) from test data involving several different clay soils in Australia with PIs ranging from about 30 to 60.

It might be interesting to note that the data accumulated by Tucker, Davis et al.[2,8,10] seem to indicate both minimal losses (if any) in soil moisture beneath the foundation and shallow losses outside the perimeter (Fig. 1A.6). Curve *B* represents moisture values taken from soil beneath the foundation. These data suggest slightly higher moisture levels than those plotted in curve *A* but

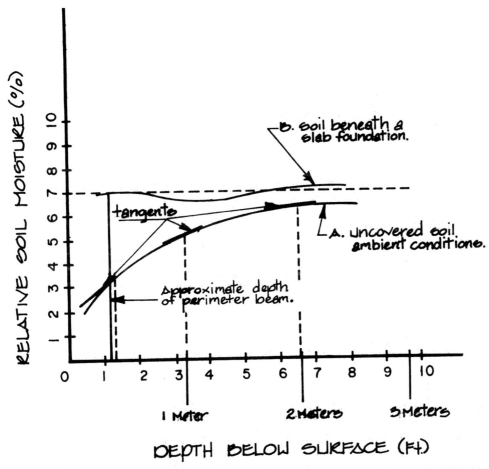

FIGURE 1A.6 Typical loss of soil moisture vs. depth during a prolonged drying cycle. The tangent lines indicate the dramatic change in comparative soil moisture vs. depth. (*From Tucker and Davis.*)

also reflect a generally uniform buildup. The data in Fig. 1A.6 show that, although soil moisture varied to a depth of perhaps 7 ft (2.14 cm), over 85% of total soil moisture change occurred within the top 3 feet or so. Data published by McKeen and Johnson[12] reflect the same general conclusion. Their data portray a relationship between the depth of the active zone, which varies with both suction (or capillary) pressure, and the number of cycles of wetting and drying that occur within the year. Nonetheless, between 80 and 90% of the total soil moisture variation occurred within the top 1.5 meter (4.5 ft). Komornik and coworkers present data on an Israel soil which presents similar results.[13] The depth of moisture change extended to 11½ ft (3.5 m), but approximately 71% of the total change occurred within the top 3.2 ft (1 m). Sowa presented data that suggest an active depth of 0.3 to 1.0 m (1 to 3.2 ft) for a Canadian soil.[14] These observations, again, would seem to support foregoing conclusions and opinions.

A source for similar information can be found in "Building Near Trees."[7] This article presents data compatible with those previously cited. Again the only question involves the issue of whether the tree height H is the important dimension describing root behavior or whether the canopy width W is the true concern, as apparently believed by most botanists.

Other authorities who agree with the statements concerning shallow feeder roots are John Haller,[15] Neil Sperry,[16] and Gerald Hall.[17] Haller states that the majority of feeder roots are found within 1 to 1 1/2 ft (30 to 45 cm) of the surface. He explains that "it is here that the soil is the richest and aeration the simplest." Both air and nutrition (water) are required by the healthy tree. Sperry and Hall concur. Deeper root systems are present, but their primary function is to provide stability to the tree. In fact, the tap roots have the principal relationship to tree height. This correlation is exploited by the Japanese, who dwarf trees by shortening the tap roots.

Many geotechnical engineers do not seem to share these views expressed by the botanists. Dr. Poor seems to feel that the radial extent of a tree's root pattern is greater (1 H to 1.5 H) and the depth of moisture loss to transpiration is deeper.[8] His beliefs seem to be based in part on data presented in Figs. 1A.5 and 1A.6. These data as interpreted by this author seem to provide a limit on root radius of 0.5 W (canopy width) and transpiration effective depth due to shallow feeder roots of less than 2.0 ft (61 cm).[11] These values are relevant to foundation stability.

The overall maximum depth of soil moisture loss (active zone) appears to be in the range of about 1 to 4 m (3.2 to 12.8 ft) depending on the proximity of trees and geographic location.[8,9,12,13,14,18] Transpiration losses at depths below 2 m (6.6 ft) may not materially influence foundation stability.[18] These conclusions are also supported by the author's experience from 1963 to the present.

The root systems for plants and shrubs would be similar to that shown in Fig. 1A.5 except on a much smaller scale. The interaction of tree root behavior to foundation failure is considered in Secs. 6A and 6C.

The following list summarizes our conclusions about soil moisture and root development:

1. Roots per se provide a benefit to soil (and foundation) stability, since their presence increases the soil's resistance to shear.[19,23] Also the plant canopies (shade) reduce evaporation and overall may conserve soil moisture.

2. Tree roots tend to remove soil moisture; hence the net result, if any, is foundation settlement. Settlement is normally slow in developing and limited in overall scope and can be arrested (or reversed) by a comprehensive maintenance program. (See Sec. 7.) Chen[20] states, "The end result of shrinkage around or beneath a covered area seldom causes structural damage and therefore is not an important item to be considered by soil engineers." Other noted authors might disagree, at least to some extent. Mike Crilly, Building Research Establishment, London England (and others within that organization[22]), presents data in Fig. 1A.7[21] collected using rods embedded in the ground. Group 1 data away from trees suggest negligible soil movement at depths below surface. (The surface loss was likely due to grass and evaporation. Refer also to point 9.) Group 2 data show vertical movement potential at the surface of 100 mm (4 in) and about 60 mm (2.4 in) at 1 m (3.3 ft), but below 2 m (6.6 ft) the movement is on the order of less than 15 mm (0.6 in). These data bring to mind two questions: (1) what would the moisture (and vertical movement) profiles look like if the data were taken from foundation slabs designed with perimeter beams and (2) would the conventional foundation design preclude damage? Others have suggested that surface soil movement can be related to the movement of slab foundations, although how the correlation might be made is not always clear.[2,8,22] For example, would tests using 1 m^2 (10.89 ft^2) pads poured on the ground surface relate to tests using larger pads, i.e., 400 m^2 (4356 ft^2) or conventional foundations?

3. While some degree of settlement is noted in most light foundations on expansive soils, that specific problem by itself is seldom sufficiently serious to demand repair. In fact, based on a random sampling of over 25,000 repairs performed (principally within the Dallas–Fort Worth area) over a period of nearly 30 years, the incidence of settlement versus upheaval (as the preponderant cause for repair) was about 1.0 to 2.3 (30 to 70%). [Three out of four foundations repaired were of slab construction (as opposed to pier and beam), and over 94% of the foundations were of steel-reinforced concrete construction.] Most of the repairs catalogued as "settlement" involved instances of: (a) shimming of interior pier caps (pier and beam foundation), (b) underpinning (raising) slab foundation wherein proper mudjacking was not included in the initial

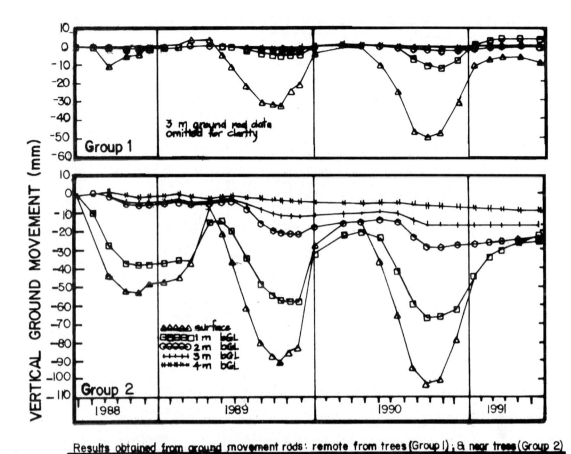

FIGURE 1A.7 Results obtained from ground movement rods: remote from trees (Group 1); and near trees (Group 2).

repairs and subsequent mudjacking of the interior slab was required, or (c) the foundations were constructed on uncompacted fill. Delete these from the settlement statistics and the incidence of settlement repairs is reduced to something like 3%.

4. Texas's shallow soils generally exist at moisture levels between the SL and PL with, as a rule, the moisture contents somewhat closer to the PL. In deeper soils the $W\%$ are sometimes higher, between the PL and LL. (For comparative purposes, C_w rating $\simeq 20$.)

5. All soil *shrinkage* ceases when $W\%$ approaches the SL (by definition) and does not commence until the moisture content is decreased below the LL. *Soil swell* in expansive soils effectively ceases at $W\%$ contents above or near the PL. (Refer to Sec. 6A.) Thus, moisture changes at levels much below the LL or much above the SL do not affect expansive soil volume (or foundation movement) to any appreciable extent.

6. Expansive soil particles tend to shrink at moisture *reductions* between something below the LL and the SL. Refer to Fig. 1A.8.[23] Those existing at a $W\%$ between the SL and PL tend to swell upon access to water. Refer to Fig. 6A.4 and Fig. 1A.8.[23] [*Nonexpansive* (or noncohesive) soils are prone to shrink when water is removed from saturation (or LL). Particle consolidation accounts for this volumetric decrease rather than particle shrinkage.]

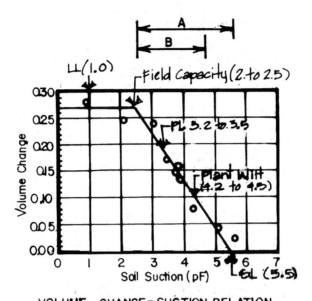

FIGURE 1A.8 Range of relative volume change: A: evaporation and transporation; B: transporation.

7. The data depicted in Fig. 1A.8 (McKeen)[24] suggest a basic relationship between soil volume change and $W\%$ expressed as pF. [pF is the logarithm to base 10 of the pressure in centimeters of water. (1 pF = 1 kPa, 2pF = 10 kPa, 3 pF = 100 kPa, etc.)] The range of volume change V_s pF decreases between the field capacity (2.2 pF) and shrinkage limit (5.5 pF). For more practical concerns the plant's removal of water (transpiration) is probably limited even further, to that level between field capacity (2.2 pF) and the point of wilt pF = 4.2 to 4.5. Note that the field capacity represents a $W\%$ less than the LL, and the point of plant wilt is well above the SL. Similar conclusions have been published by F. H. Chen.[20] Evapotranspiration, however, would transcend a wider scope. The combined effect of soil moisture withdrawal could reflect soil volume changes between the field capacity and SL—a wider range than that likely for transpiration alone. A soil can gain or lose moisture, within specific limits, without a corresponding change in volume.[20,23,24]

8. Heave of surface soils occurs only within rather confined limits as noted above (SL to proximity of PL). It would seem that removal of surface vegetation in a $C_w \simeq 20$ climate would encourage soil desiccation as opposed to net $W\%$ gain (assuming reasonable drainage). If expansive soils are properly drained, it would seem likely that $W\%$ variations largely would occur at relatively shallow depths. In climates such as London's [30 in (76 cm) annual rain distributed over about 152 days)], the in situ $W\%$ in absence of transpiration (lack of evaporation) should, in fact, increase. However, once again, this effect on soil *movement* begins to cease as the $W\%$ approaches or somewhat exceeds the PL. It would seem that $W\%$ in London, for example, would be consistently higher than in the United States. London's rainfall (though roughly equivalent to Dallas–Fort Worth's annual rainfall of 30 in) is distributed rather evenly over 152 days as opposed to the 15 days which account for 80% of the Dallas–Fort Worth precipitation. The considerably

more moderate temperature ranges would combine with the extended rain to logically produce both higher and generally more stable $W\%$. [The annual average temperature in the Dallas–Fort Worth area is about 65°F (18°C), whereas that for London is about 52°F (11°C). The relative temperature *ranges* are 15° to 105°F (−9 to 40°C) for Dallas–Fort Worth and 38° to 78°F (3° to 25°C) for London.]

9. Vegetation (transpiration) removes soil moisture only at very shallow depths.[15,16,17] The U.S. arborist community invariably recommends that trees be watered and fed at or near the drip line (extent of canopy). Further, most agree that nutritional roots are classically quite shallow—within 12 to 24 in (30 to 60 cm). The reasoning for the recommendation includes: (a) root development favors loosely compacted soil, (b) roots like oxygen, (c) roots like water, (d) roots like sunlight (to some extent), and (e) roots exert only that energy necessary for survival. Under particularly adverse conditions (such as a prolonged draught) feeder roots may develop at deeper depths. Still it is generally agreed that 90% of the tree's moisture needs are taken from 12 to 24 in (30 to 60 cm).

10. It has been well established by many research projects that foundation stability is not influenced by soil behavior below the soil active zone (SAZ). In Dallas, Texas, the preponderance (87%) of that influence on foundation stability is limited to about 3 ft (1 m), although the SAZ may extend to depths in excess of 7 ft (2.3 m). Other geographical locations report different depths for the active zone: (a) For a Canadian soil, Sowa[14] indicates the depth of SAZ to be 1 to 3 ft (0.3 to 1 m). (b) For an Israeli soil, Komornik[13] reports an SAZ as deep as 11.5 ft (3.5 m) but also reports that approximately 71% of the total moisture variation occurred within the top 3.2 ft (1 m).

11. Other factors of concern include such issues as: (a) overburden tends to suppress soil expansion. Doubling the effective overburden pressure (1000 psf to 2000 psf) can reduce swell by about one-third (F. H. Chen).[20] (b) The surcharge load on the soil diminishes with depth. (For strip footings the effect of load is in the range of only 10% at a depth of twice the width.) (c) Low soil permeabilities severely inhibit soil moisture movement, particularly in a vertical direction. [Expansive (sedimentary) soils in general have much higher lateral than vertical permeability.]

12. Without a doubt, the age and proximity of the tree (and the depth of the perimeter beam) are very important concerns to the amount of water a tree might remove from the foundation-bearing soils. Certainly younger trees tend to remove moisture at a faster and greater relative rate. Also, trees tend to require much more water during growth periods. Without the leaves or during dormancy, a tree might require a little as 1% of the "growth" amount of moisture. The influence of transpiration or foundation stability should thus be relative to season. It would seem wise that in most cases "new" trees should not be planted in close proximity to the foundation. Nonetheless, concrete evidence available to the author seems to suggest that the impact of vegetation on the stability of foundations is grossly overstated. Any *proof* to the contrary would be welcome.

13. Many engineers in the United States (and probably elsewhere as well) confuse center heave with perimeter settlement. Hence, the influence of trees is often overstated. (Refer to Sec. 6A.) Sound evidence and not wishful thinking should be the final criterion for decision making. One source for reliable data offers a history of over 25,000 actual repairs performed over 30 years. Many of these repairs were performed on structures with trees (in some cases multiple trees) located in close proximity of the foundations, sometimes as close as 1 ft (0.3 m). There is no record of the repair company suggesting or requiring the removal of any tree, bush, or other vegetation. Yet in absence of tree removal not on subsequent failure could be attributed to the presence of a tree, bush, or vegetation. (These data were collected primarily from the Dallas–Fort Worth area of Texas, but data points included other U.S. states from Arizona to Illinois and Oklahoma to Florida.) Does this seem to dispute the deleterious influence of trees on foundation stability? If the trees played a predominate part in causing the initial foundation failure, why did not the same or similar problem recur? Also, *many* other foundations within the same areas have a tree (or trees) in close proximity to the foundation, yet never suffer foundation distress. It would not stand to reason that trees are capable of preferentially selecting one address over another.

14. Again, with reference to the study mentioned above and paragraph 3, most of the repair causes were attributed to upheaval brought about by the accumulation of water beneath the slab foundations. (Once the source for water was removed, the foundation stabilized.)

There seems to be some confusion in terminology when addressing slab heave on expansive soils. An often misused term is *natural center doming,* which allegedly describes the buildup of soil moisture due to capillary and/or osmotic transfer. Proponents believe that this phenomenon occurs in most slab on-grade foundations with the net result being a central high or domed area. Research does not verify this conclusion.[9,11] For greater detail, refer to Secs. 6A and 6B.

Center lift is another term used in the Building Research Advisory Board Bulletin No. 33 and Post Tension Institute books. (Refer to Sec. 3.) This is an important design concern which relates more to *upheaval* than to center doming.

1A.9 CONCLUSIONS

What factors relating to soil moisture clearly influence foundation stability?

1. Soil moisture definitely affects foundation stability, particularly if the soil contains expansive clays.
2. The soil belt is the zone which affects or influences foundation behavior the most.
3. Constant moisture is beneficial to soil (foundation) stability.
4. The water table per se has little, if any, influence on soil moisture or foundation behavior, especially where expansive soils are involved.
5. Vegetation can remove substantial moisture from soil. Roots tend to "find" moisture. In general, transpiration occurs from relatively shallow depths.
6. Introduction of excessive (differential) amounts of water under a covered area are accumulative and threaten stability of some soils. Sources for excessive water could be subsurface aquifers (e.g., temporary perched ground water), surface water (poor drainage), and/or domestic water (leaks or improper watering). Slab foundations located on expansive soils are most susceptible to the latter. Refer to Sec. 6.
7. Assuming adequate drainage, proper watering (uniformly applied) is absolutely necessary to maintain consistent soil moisture during dry periods—both summer and winter.
8. The detrimental effects to foundations, resulting from transpiration, appear to be grossly overstated.

The homeowner can do little to affect either the design of an existing foundation or the overall subsurface moisture profile. From a logistical standpoint, about the only control the owner has is to maintain moisture around the foundation perimeter by both watering and drainage control and to preclude the introduction of domestic water under the foundation. Adequate watering will help prevent or arrest settlement of foundations on expansive soils brought about by soil shrinkage resulting from the loss of moisture.

From a careful study of the behavior of water in the aeration zone, the most significant factor contributing to distress relative to expansive soils appears to be excessive water beneath a protected surface (foundation) which causes the soil to swell (upheaval). From field data collected in a 27-year (1964 to 1991) study including over 2000 repairs, a wide majority of these instances of soil swell clearly were traceable to domestic water sources as opposed to drainage deficiencies. Further, the numerical comparison of failures due to upheaval versus settlement was estimated to be in the range of about 2 to 1. (Refer to Secs. 6A and 6B for more detailed information.) The data described were accumulated from studies within a C_w rating (climatic rating) of about 20 (see Fig. 6C.3). This describes an area with annual rainfall in the range of 30 in (75 cm) and mean temperatures of about 65°F (18°C).

1A.10 REFERENCES

1. O. E. Meinzer et al., *Hydrology,* New York, McGraw-Hill, 1942.
2. R. C. Davis and Richard Tucker, "Soil Moisture and Temperature Variation beneath a Slab Barrier on Expansive Clay," Report No. TR-3-73. Construction Research Center, UTA, May 1973.
3. S. J. Pirson, *Soil Reservoir Engineering,* New York, McGraw-Hill, 1958.
4. David B. McWhorter and Daniel K. Sunada, *Ground-Water Hydrology and Hydraulics,* Ft. Collins, Colorado, Water Resources, 1977.
5. H. Bouwer, R. A. G. Pyne, J. A. Goodwich, "Recharging Ground Water," *Civil Engineering,* June 1990.
6. T. T. Koslowski, *Water Deflicts and Plant Growth,* vol. 1, New York, Academic Press, 1968.
7. "Building Near Trees," Practice Note 3, National House-Building Council, London, 1985.
8. Richard Tucker and Arthur Poor, "Field Study of Moisture Effects on Slab Movement," *Journal of Geotechnical Engineering,* ASCE, vol. 104, April 1978.
9. John E. Holland and Charles E. Lawrence, "Seasonal Heave of Australian Clay Soils" and "The Behavior and Design of Housing Slabs on Expansive Clays," *4th International Conference on Expansive Soils,* ASCE, June 16–18, 1980.
10. Thomas M. Petry and Clyde J. Armstrong, "Geotechnical Engineering Considerations for Design of Slabs on Active Clay Soils," ACI Seminar, Dallas, February 1981.
11. Robert Wade Brown, *Foundation Behavior and Repair: Residential and Light Commercial,* New York, McGraw-Hill, 1992.
12. R. G. McKeen and L. D. Johnson, "Climate Controlled Soil Design-Parameters for Mat Foundations," J.G.E., vol. 116, no. 7, July 1990.
13. D. Komornik, et al., "Effect of Swelling Clays on Piles," Israel Institute of Technology, Haifa, Israel.
14. V. A. Sowa, "Influences of Construction Conditions on Heave of Slab-on-Grade Floors Constructed on Swelling Clays," *Theory and Practice in Foundation Engineering,* 38th Canadian Geotechnical Conference, September 1985.
15. John Haller, *Tree Care,* New York, Macmillan, 1986, p. 206.
16. Neil Sperry, *Complete Guide to Texas Gardening,* Dallas, Taylor, 1982.
17. Gerald Hall, "Garden Questions—How to Get a Fruitful Apple Tree," *Dallas Times Herald,* March 24, 1989.
18. T. J. Freeman, et al., "Seasonal Foundation Movements in London Clay," Ground Movements and Structures, Fourth International Conference, University of Wales College of Cardiff, July 1991.
19. T. H. Wu, et al., "Study of Soil-Root Interaction," *Journal of Geotechnical Engineering,* vol. 119, December 1988.
20. F. H. Chen, *Foundations on Expansive Soils,* New York, Elsevier, 1988.
21. M. S. Crilly, et al., "Seasonal Ground and Water Movement Observations from an Expansive Clay Site in the UK," Seventh International Conference on Expansive Soils, Dallas, Texas, 1992.
22. T. J. Freeman, et al., *Has Your House Got Cracks?* Institute of Civil Engineers and Building Research Establishment, London, 1994.
23. N. J. Choppin and I. G. Richards, *Use of Vegetation in Civil Engineering,* Butterworth, London, 1990.
24. R. Gordon McKeen, "A Model for Predicting Expansive Soil Behavior," *7th International Conference on Expansive Soils,* ASCE, Dallas, 1992.

SECTION 1B
SITE PREPARATION

Robert Wade Brown

1B.1 INTRODUCTION 1.21
1B.2 GRADING PLANS 1.23
1B.3 FOUNDATIONS ON FILLED SLOPES 1.24
1B.4 SITE DRAINAGE 1.24
   1B.4.1 Surface Drainage 1.25
   1B.4.2 Catch Basins and Drop Inlets 1.26
   1B.4.3 Manholes 1.26
   1B.4.4 Pipes 1.27

1B.4.5 Sanitary Sewer Systems 1.27
1B.4.6 Subsurface Drainage 1.29
1B.5 DRAINAGE SYSTEMS 1.29
1B.6 LANDSCAPING 1.30
1B.7 LAND PLANNING 1.30
1B.8 FOUNDATION DRAIN SYSTEMS 1.31
1B.9 REFERENCES 1.33

1B.1 INTRODUCTION

The foundation engineer is not generally on the scene for site preparation any more than that person is involved with site hydrology. However, a few areas in site planning can influence the engineer's decisions. The following identifies some of these areas and provides references for further study.

A concept of grading design is essential in developing the physical form of the construction site. This grading concept must strengthen the overall project rather than detract from it, as often happens. Positive drainage, providing a stable base for foundations, appearance, utilization, cost, future maintenance, and safety represent the major concerns.

Positive drainage, an important concept in grading, allows storm water runoff to flow away from the structure. When water flows away from structures toward drainage facilities, flooding is prevented, and in areas with expansive soils, expensive foundation problems can often be avoided. Studying existing topography aids design of a proper drainage system around the structure, which consequently minimizes both problems. The following terms are used in grading (refer also to Fig. 1B.1)[3]:

Batter Amount of deviation from vertical such as 2 in/ft (5 cm/30 cm) for a vertical surface such as a wall

Cut and Fill Cut is indicated when a proposed contour is moved back into an existing slope; fill is indicated when a proposed contour is moved away from an existing slope. Earthwork calculations determine if a balance exists between cut and fill or whether material will have to be added or carried away from the site.

Cross Slope or Pitch Provides for runoff on paved areas and is given in feet per 100 ft (mm/cm) or as a percentage.

Crown Provides for runoff of water on roads or walks.

Grade Percent of rise or fall or feet per 100 ft (30 m). A fair rule of thumb is ¼ in/ft (6 mm/30 cm) or 2%.

Slopes The ratio of horizontal to vertical maximum slopes

Solid rock	¼:1
Loose rock	½:1
Loose gravel	1½:1
Firm earth	1½:1
Soft earth	2:1
Mowing grass	3:1

Wash Provides for runoff on steps and is given in inches per foot, usually ⅛ to ¼ in/ft (3 to 6 mm /30 cm).

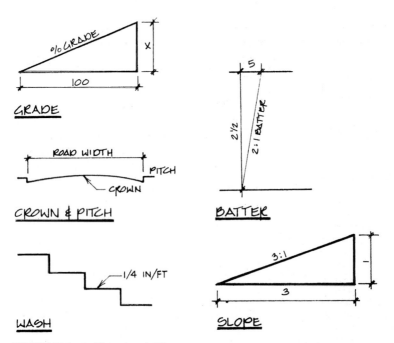

FIGURE 1B.1 An illustration of different terms used in grading. (*From Rubenstein, Ref. 3.*)

1B.2 GRADING PLANS

The grading plan is significant in the technical development of the site plan. The major consideration in a grading plan is to set elevations to achieve a positive drainage pattern. The following factors should be considered in setting preliminary elevations:

1. Setting the first-floor elevations of building so that: (a) Finished floor elevations are a minimum of 6 in (15.3 cm) above the finished landscape grade, thus ensuring positive drainage away from the foundation. (b) Drainage swales, ditches, storm inlets, and other drainage facilities can safely remove storm water from the site. (c) Should drain inlets become blocked, flooding will not reach heights above the foundation-finished floor grade.
2. Meeting existing building elevations and relating grading to adjacent properties so as not to disturb existing grade or disposal runoff.
3. Relating elevations of roads, walks, parking, and other activities to building elevations to achieve positive drainage.
4. Saving the cost of unnecessary retaining walls where other types of grading concepts may be used.
5. General balancing of cut-and-fill areas to avoid the transport of material to or from the site.

Table 1B.1 lists commonly accepted gradients for various situations.

After elevations are studied in relation to each other, preliminary contour lines are drawn on the grading plan at a chosen contour interval such as 1, 2, or 5 ft (30, 60, or 150 cm). When the proper balance of cut and fill has been obtained, final spot elevations and contour lines are set. Finished spot elevations are placed at the following locations:

1. First-floor elevations of buildings
2. All corners of buildings and door stoops or landings
3. Corners of parking areas, terraces, or other paved areas
4. Corners at the top of landings and bottom of steps
5. Top and bottom of walls, curbs, and gutters
6. Flow-line elevations in ditches, swales, and at storm water inlets
7. Points where grades change

Existing Topsoil. Two types of grading that reshape existing contours are *rough grading* (before construction of the foundation) and *finished grading* (after construction). Before rough

TABLE 1B.1 Gradients

Desirable grades	Maximum, %	Minimum, %
Streets (concrete)	8	0.50
Parking (concrete)	5	0.50
Service areas (concrete)	5	0.50
Main approach walks to buildings	4	1
Stoops or entries to buildings	2	1
Collector walks	8	1
Terraces and sitting areas	2	1
Grass recreational areas	3	2
Mowed banks of grass	3:1 slope	—
Unmowed banks	2:1 slope	—

grading is begun, existing topsoil should be stripped from the area to be graded and stockpiled away from the construction area. Topsoil of good quality can be reused in the process of finished grading. Consider the kind of soil, how it reacts in cut-and-fill situations, its bearing capacity, and expansive tendencies.

Erosion Control Plans.   Many states require erosion and sedimentation control plans during site construction. Factors to consider in developing these plans are the type of soil on the site, topographic features of the site, type of development, amount of runoff, staging of construction, temporary and permanent control facilities, and maintenance of control facilities during construction.

During construction, minimizing the area and time of exposure of disturbed soil is important. If earth-moving activities will not be completed within at least 20 days, interim stabilization measures such as temporary seeding and mulching should be carried out.

1B.3 FOUNDATION ON FILLED SLOPES

Another special problem that may be encountered after site grading is that of a foundation located on or adjacent to the filled slope (Fig. 1B.2). Note in the figure that the lack of soil on the slope side of the footing will tend to reduce the stability of the footing and favor sliding.

In this case, the overall stability of the slope should be checked for the effect of the footing load. This can be done by using slope stability charts developed by Braja M. Das (*Soil Mechanics*, Iowa State University Press, 1979) or computer programs.

If slope stability is questionable, retaining walls are often required. To prevent the future movement or sliding of fill and consequent potential foundation failure, retaining walls designed to withstand the downhill stress are sometimes used. Figure 1B.2 shows construction on a filled slope. Piers can be drilled through the fill into undisturbed soil and used to support the foundation, or a step beam that conforms to the contour of the land as illustrated in Fig. 1B.3 can be designed. Either approach will reduce or neutralize the downhill stress.

1B.4 SITE DRAINAGE

Specific site drainage is important to the design and stability of foundations. This concern includes such factors as surface and subsurface water.

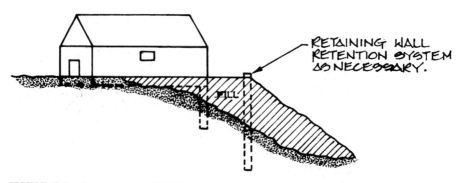

FIGURE 1B.2   Construction on a filled slope.

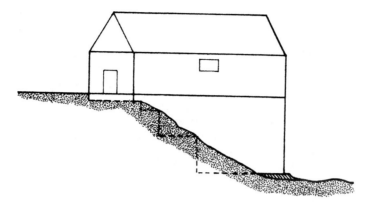

FIGURE 1B.3 Step beam construction on a slope.

1B.4.1 Surface Drainage

In residential areas, natural drainage should be used as much as possible in the design and layout of housing. Such drainage keeps the cost of the storm water system down, and drainage control can be one of the higher-cost items in site development. Natural methods often allow water to percolate into the soil and recharge the ground water system. Refer also to Sec. 1B.5.

Controlling storm water runoff is a major factor in preparing a grading plan. To prevent problems caused by flooding, the principal of positive drainage is required: Storm water is diverted from the foundation and transported from the site in a storm drainage system. Elevations are set at critical points adjacent to a building to provide drainage. Advantageous points must be chosen for placement of catch basins and their connection to existing drainage channels in the area.

Surface drainage systems, which are referred to as *storm sewers* or *storm drains*, are constructed with tight or closed joints. Surface drainage can be provided by designing slopes to allow for runoff of storm water and interception at various intervals with catch basins. This is the common conception of the accommodation of surface water (rainfall) in area development.

Diversion terraces may be constructed upgrade of a project site to convey runoff around the disturbed area. Interceptor channels may also be used within a project area to reduce the velocity of flow and thereby limit erosion.

The design of a drainage system is based on the amount of rainfall to be carried away at a given time. *Runoff* is the portion of precipitation that finds its way into natural or artificial channels either as surface flow during the storm period or as subsurface flow after the storm has subsided. Runoff is determined by calculating the volume of water discharged from a given watershed area and is measured in cubic feet of discharge water per second, or in acre-inches per hour (acre-in/h) depending upon the mathematical units utilized. To calculate runoff, the rational formula of Eq. (1B.1) can be used.

$$Q = CiA \qquad (1B.1)$$

where Q = storm water runoff from an area (acre-in/h)
C = coefficient of runoff (percentage of rainfall that runs off depending on the characteristics of the drainage area)
i = average intensity of rainfall (in/h) for a particular time of concentration, selected location, and rainfall frequency
A = area (acres)

Note that Q can be expressed in acre-cm/h if i is expressed as cm/h. The rational method can be used for drainage areas less than 5 mi² (13 km²) and is most frequently used on areas up to ½ mi² (1.3 km²).

In urban areas, the frequency of rainfall generally considered for design purposes is the greatest rainfall during the 10-year period. In residential areas, this may be reduced to a 2- to 5-year period. For the design of storm drains and inlet systems to accommodate this precipitation, inlet times of 5 to 15 min are generally used. Beyond this criterion, specific drain design varies by community. Rainfall intensity duration curves, in inches per hour, are available from the local weather bureau or city engineering office. These data can be used to determine the inlet placement and size.

John Wier published an article, "Fail-Safe Drainage Provisions in Urban Planning and Design," in the February 1990 issue of *Texas Civil Engineer*. Wier cited specific parameters which have or could have a profound influence on the capacity efficiency of drainage systems. One of the major reasons for failure in small watershed urban storm drainage systems involved misapplication of the rational formula (Eq. 1B.1). The principal misuses reportedly include:

1. Use of outdated rainfall intensity curves or ones inappropriately applied to a specific, localized, geographic area.

2. The assumption of unreasonably low runoff coefficients (C) which are not indicative of real-life densities, lot sizes, and/or imperviousness created by land development. True C values must reflect the effect of different soil types (even within the same locality), antecedent rainfall, as well as the realistic percentage of human-made surface imperviousness.

3. The assumption of unrealistically high time of concentration. Using a typical value of 15 to 20 min time concentration at the first design point (upstream end) results in predicting too small an intensity and thus too small a runoff. To compensate, at least in part, for the above factors, Wier suggests the inclusion of another constant, Ca, in the rational equation. This factor principally addresses the issue of antecedent precipitation and effectively increases the C value as well as Q. Adequate consideration of the foregoing appears to improve dramatically the usefulness of the rational equation.

Even assuming more accurate runoff, deficiencies could still prompt underdesign. These deficiencies could include: (1) undersizing the area of the inlet openings, (2) design of system based on watershed conditions which are later modified or altered, (3) economical persuasions which slant toward underdesign, and (4) obstruction or clogging of the system.

1B.4.2 Catch Basins and Drop Inlets

Catch basins intercept storm water for transfer to a storm water system of pipes. The catch basins must be cleaned periodically to remove debris. Drop inlets do not have sediment traps below the outlet line and must be designed with self-cleaning water velocities to function properly. Drop inlets often may be used in low-maintenance areas where sediment would clog improperly maintained catch basins. Both structures generally use cast-iron grates to allow water to enter the basin. Catch basins or drop inlets are generally placed 100 to 200 ft (30 to 60 m) apart on roads; however, they may be closer where swales have been created around buildings in developed areas, thus increasing the localized volume of water to be handled. Since the real problems can be most complicated, interested readers should pursue further reading.[3-8]

1B.4.3 Manholes

Manholes are used to inspect and clean sewer lines and are placed at these points:

1. Changes of direction of pipe lines
2. Changes in pipe sizes
3. Changes in pipe slope
4. Intersection of two or more pipe lines
5. Intervals not greater than 300 to 500 ft (90 to 150 m)

1B.4.4 Pipes

Pipes used in closed systems are generally concrete, vitrified clay, cast iron, galvanized corrugated metal, or, upon occasion, plastic (PVC or ADS). In some cases where corrugated-metal pipe is used, it has a paved invert for flow where the slope of the pipe is small, such as 0.5% (or 0.005 ft/ft). Minimum slopes for sewer pipes are determined by the hydraulics which will produce a minimum flow velocity of 2 ft/s (30 cm/s) during off-peak periods. In practice, this often suggests a minimum pipe slope of 1% (or 0.01 ft/ft). Pipe inverts are below frost level so that flow will not freeze in winter. Refer to following paragraphs for additional detail.

1B.4.5 Sanitary Sewer Systems

Other than the generalities touched upon in 1B.7, no attention will be given to the design of sanitary sewer systems beyond the property line. For more detailed information on this topic, refer to *Practical Manual of Land Development* by Barbara Colley,[4] or other suitable text.[3]

The use and development of any site, whether it be residential or commercial, depends on the safe disposal of sewage. In designing an adequate sewer system, many factors must be considered such as: (1) peak rates of sewage blow (estimated by determining the period of maximum consumption of water), (2) pipe size [pipe diameter affects volume and rate (velocity) of flow], (3) gradation or fall of pipe which also influences flow velocity and, (4) pipe construction material (determines friction). Sewer systems are designed to accommodate peak flows. Daily peak flows may accumulate to periods of something like 2 to 5 h.

These various items are related by the Mannings' equation:

$$V = \frac{1.49}{n}(R_H)^{2/3}(S)^{1/2} \qquad (1B.2)$$

where V = velocity, fps
n = coefficient of friction
R_H = cross-sectional area of flow divided by the wetted perimeter, ft (the hydraulic radius) (refer to Fig. 1B.4 for the mathematical relationship between factors used to establish R_H)
S = slope of pipe, ft/ft

(When metric units are used, the constant 1.49 becomes 1.00.) For most applications, the minimum slope is established at 1 ft drop per 100 ft (1%). Typical values for n would be 0.013 for concrete, vitrified, or commercial wrought iron black pipe. Polyvinyl chloride (PVC) pipe would allow a friction value (n) of 0.01. Given these constants, only R_H need be determined to permit calculation of flow velocity under the prevailing conditions. The maximum flow capacity occurs when the sewer pipe is filled to 0.8 D. Beyond this value, the increase in friction reduces the rate of flow. For simplicity, most engineers assume full pipe flow. In this instance, the hydraulic radius (R_H) equals $D/4$. This gives a somewhat conservative flow capacity since the R_H values at 0.8 pipe capacity are larger than those at full pipe capacity. Refer to example problem 1B.1 for a solution to Eqs. 1B.2 and 1B.3.

Fluid velocity and pipe diameter (area) are mathematically related to determine the flow as shown by

$$Q = VA = \frac{V\pi D^2}{4} \qquad (1B.3)$$

where Q = flow, ft/s
V = velocity ft/s
A = cross-sectonal area, ft^2
D = pipe diameter, ft

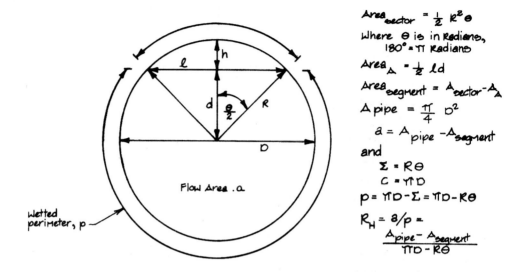

FIGURE 1B.4 The hydraulic radius, R_H.

This equation can be rearranged to allow the calculation of required pipe area (diameter) when peak rates (Q) and desired flow velocities (V) are given. Equation 1B.4 represents the rearrangement:

$$A = \frac{Q}{V} = \frac{\pi D^2}{4} \quad \text{or} \quad D^2 = \frac{4Q}{\pi V} \quad \text{or} \quad D = \sqrt{\frac{4Q}{\pi V}} \quad (1B.4)$$

One concern is to ensure an adequate flow velocity when capacities are below peak. This is important, because a scouring action is required for proper function of the system. The peak capacity, for which sanitary sewer systems are designed, is often 1.3 to 2 times the *average* flow. Assuming the factor of 2, flow velocities at average capacity will be less than half that at peak capacity.

Another concern can be the infiltration of ground water into the sanitary sewer system. The problem can often be addressed by strategic installation of check valves.

In areas where a sanitary sewer system is not available, septic systems will be required. The septic systems require a soil with a percolation rate capable of adequate dissipation of liquids. Otherwise, many of the foregoing concerns are similar.

Example Problem 1B.1 Calculate the gravity flow for a 4-in pvc pipe flowing full at a slope of 1%:

$$V = \frac{1.49}{0.01}\left(\frac{\pi D^2}{4\pi D}\right)^{2/3}(0.01)^{1/2} = \frac{1.49}{0.01}\left(\frac{D}{4}\right)^{2/3}(0.01)^{1/2}$$

$$= \frac{1.49}{0.01}\left(\frac{0.33}{4}\right)^{2/3}(0.1) = 14.9(0.083)^{2/3} = (14.9)(0.19)$$

$$= 2.80 \text{ ft/s}$$

Under the conditions shown above what volume of sewage could be transported?

$$Q = VA = 2.80 \frac{\pi D^2}{4} = 2.80(0.785)(0.11 \text{ ft}^2)$$

$$= 0.244 \text{ ft}^3/\text{s} \ (1.82 \text{ gal/s})$$

1B.4.6 Subsurface Drainage

Subsurface drainage involves the control and removal of soil moisture such as underground streams, perched water, or percolated water. It is concerned with the following:

1. Carrying water away from impervious soils, clay, and rock
2. Preventing seepage of water through the foundation walls
3. Lowering water tables for low flatland
4. Preventing unstable subgrade or frost heaving
5. Removing surface runoff in combination with underground drainage

Subsurface drainage may be accomplished by providing a horizontal passage in the subsoil that collects gravitational water and carries it to outlets. Subsurface drain lines either have open joints or use perforated pipe. Flow into subsurface drains is affected by soil permeability, depth of drain below soil surface, size and number of openings in the drain, drain spacing, and diameter.

1B.5 DRAINAGE SYSTEMS

Drainage can be accomplished by one or more of the following types of systems (See also Fig. 1B.5):

   1. *Gridiron:* System where laterals enter the main from one side. Mains and laterals may intersect at angles of less than 90°.
   2. *Interceptor:* Used near the upper edge of a wet area to drain such an area. Outlets should discharge flow without erosion and prevent flooding when they are submerged. Tile or pipe lines should be placed 2 1/2 to 5 ft (75 to 150 cm) below the soil surface or as necessary to intercept and divert the subsurface flow. In moderately permeable soils, a space approximately 24 ft (7.2 m) wide should be used for each foot (30 cm) of depth below soil surface. In general, depth varies with soil permeability and water table fluctuation.
   3. *Natural:* Used for areas that do not require complete drainage or where existing topography provides adequate drainage.
   4. *Herringbone:* Used in areas of land with a concave surface with land sloping in either direction. This system should not have angles over 45°.
   5. *French drains:* These are used to intercept and divert subsurface water. More on this subject will be presented in Sec. 6C. The slope of tile or pipe may vary from a maximum of 2 to 3% for a main to a desirable minimum of 0.2% for laterals. A minimum fluid flow velocity of 1.5 ft/s (46 cm/s) is sometimes used. Drainage tile, ADS or PVC pipe, varying in size from 4 in (10 cm) as a minimum to 6 to 8 in (15 to 20 cm) is used most frequently. The PVC pipe is perforated or slit to permit the water to enter. Generally ADS is wrapped in a geotextile material which serves as a filter. Drainage tiles may be perforated or laid with loose joints for the same purpose. The trench is often lined with screen, prior to installing the pipe and gravel, to help prevent fouling of the system.

Other drainage systems, more relative to specific construction problems, are discussed in following paragraphs.

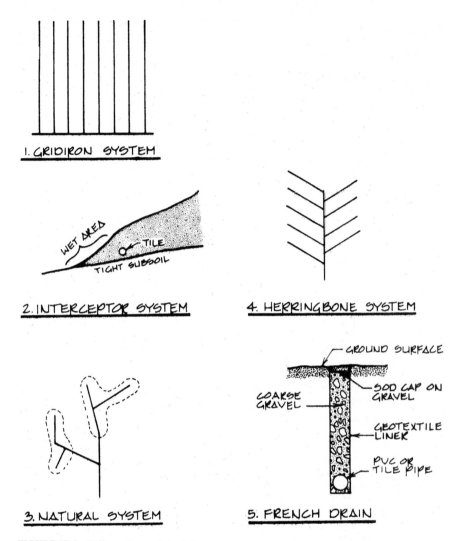

FIGURE 1B.5 Different forms of site drainage.

1B.6 LANDSCAPING

Landscaping, for purposes of discussion, involves a combination of factors such as drainage control, planting, and watering. Section 6C will provide more extensive discussion on these subjects.

1B.7 LAND PLANNING

The land use depends on architectural arrangement of a plan in terms of type of activities, linkages, and densities. Activities must be grouped so they will function in relation to each other.

When land uses have been established, the linkages between them must be evaluated. Linkage may be the movement of people, goods, waste, communication networks, or a collection of amenities, such as aesthetics or views. Land use also involves the concept of density or number of families per acre. In community development plans, density standards must be adhered to.

The type of construction will also influence the land use plan. If a plan is not economically feasible because of excessive site work, an alternative may be necessary. A particular land use may require a specific type of site—flat, rolling, or hilly. In general, the following subjects should be considered in a land planning study:

1. Visitor and other parking
2. Vehicular circulation patterns
3. Pedestrian circulation
4. Steps and ramps
5. Handicap ramps
6. Bikeways
7. Pavements
8. Walls and retaining walls
9. Sculpture
10. Fountains and pools
11. Night lighting
12. Pedestrian lighting
13. Benches
14. Seating in conjunction with raised tree planters
15. Tree planters and pots
16. Telephone booths
17. Plant material—recreational plans
18. Noise
19. Erosion control
20. View control
21. Parks and trees, shrubs, flowers

In designing residential projects, careful consideration must be given to configuration and placement of housing on the land and the relationship of the units to each other, access to public facilities, and amenities.

1B.8 FOUNDATION DRAIN SYSTEMS

Until about 1965, most foundation drains were connected to the sanitary sewers where such were available; otherwise, the foundations were served by sump pumps. With the growing demands for increased sewage treatment capacities, it became logical to eliminate as much extraneous in-flow as possible. Some municipalities have already begun to prohibit foundation drain connections into sanitary sewers, preferring connection to the storm sewer.

It is not economically feasible to size storm sewers to accommodate every possible runoff. Occasionally, rainfall is such that the storm sewer backs up to levels above the basement or lower-floor grade. Consequently, storm water backs into the foundation drains and causes flooding—the very condition it was originally designed to prevent (Fig. 1B.6).

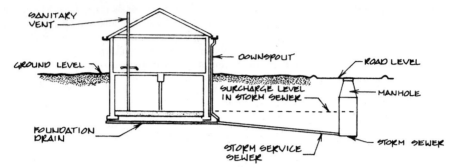

FIGURE 1B.6 Foundation gravity drains and downspouts connected to a storm sewer. (*From Ref. 7.*)

The condition becomes considerably more severe when roof water leaders are also connected to the foundation drain outlet pipe. This drastically increases the volume of water which the drain must accommodate. In addition, this practice can result in flooded basements due to the hydrostatic pressure buildup in the external soil promulgated by the drainage breakdown.

If foundation drains are connected by gravity to a storm sewer of insufficient capacity and the hydraulic grade line exceeds the basement elevation, protection against flooding of basements will not be realized. An alternative solution is a separate foundation drain collector, which could be a third pipe installed in the same trench as the sanitary sewer but with connection only to the foundation drain (Fig. 1B.7). This method has several advantages, and for many new areas, it may be the best and least expensive solution. A foundation drain collector will:

1. Eliminate the probability of hydrostatic pressure on basements due to surcharged sewers
2. Eliminate infiltration into sanitary sewers from foundation drains
3. Permit shallow storm sewers designed for lower rainfall intensity, which could reduce length of storm sewers resulting in cost savings for the system
4. Permit positive or gravity flow of both the minor and major storm drainage systems

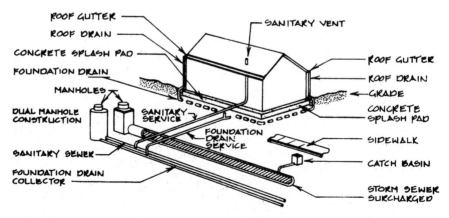

FIGURE 1B.7 Foundation drain connected to collector by gravity.

1B.9 REFERENCES

1. American Concrete Institute, *Building Code Requirements for Reinforced Concrete,* (ACI 318-71), 1975.
2. Portland Cement Association, *Design and Control of Concrete Mixtures,* 1968.
3. Harvy M. Rubenstein, *A Guide to Site and Environmental Planning,* New York, John Wiley & Sons, 1980.
4. Barbara Colley, *Practical Manual of Land Development,* New York, McGraw-Hill, 1993.
5. J. T. Adams, *The Complete Concrete Masonry and Brick Handbook,* New York, ARCP Publishing, 1979.
6. Joseph E. Bowles, *Foundation Analysis and Design,* New York, McGraw-Hill, 1977.
7. American Iron and Steel Institute, *Modern Sewer Design,* Washington, D.C., 1980.
8. Whitney Clark Huntington and Robert E. Mickadeit, *Building Construction Materials and Types of Construction,* New York, John Wiley & Sons, 1975.

PART 2

SOIL MECHANICS AND FOUNDATION DESIGN PARAMETERS

SECTION 2A
SOIL MECHANICS

Richard Stephenson

2A.1 INTRODUCTION 2.3
2A.2 PHYSICAL CONDITION 2.3
   2A.2.1 Introduction 2.3
   2A.2.2 Two-Phase Soil (Dry or Saturated) 2.3
   2A.2.3 Three-Phase Soil 2.5
2A.3 SOIL IDENTIFICATION AND CLASSIFICATION 2.5
   2A.3.1 Introduction 2.5
   2A.3.2 Soil Classification 2.13
2A.4 WATER FLOW IN SOILS 2.17
   2A.4.1 Basic Principles of One-Dimensional Fluid Flow in Soils 2.17
   2A.4.2 Determination of Permeability 2.19
2A.5 GEOSTATIC STRESSES IN SOIL 2.21
   2A.5.1 Total Stresses 2.21
   2A.5.2 Pore Stresses 2.21
   2A.5.3 Effective Stresses 2.22
2A.6 DISTRIBUTION OF APPLIED STRESSES IN SOIL 2.22
   2A.6.1 Point Load 2.22
   2A.6.2 Uniformly Loaded Strip 2.22
   2 A.6.3 Uniformly Loaded Circular Area 2.23
   2A.6.4 Uniformly Loaded Rectangular Area 2.25
   2A.6.5 Average Stress Influence Chart 2.26
2A.7 ONE-DIMENSIONAL CONSOLIDATION 2.26
   2A.7.1 One-Dimensional Laboratory Consolidation Test 2.26
2A.8 SHEAR STRENGTH OF SOIL 2.32
   2A.8.1 Direct Shear Strength of Sand 2.33
   2A.8.2 Shear Strength of Saturated Clay 2.34
2A.9 REFERENCES 2.44

2A.1 INTRODUCTION

Rock consists of an aggregate of natural minerals joined by strong and permanent cohesive bonds. *Rock mechanics* is the engineering study of rock.

   Soil is defined as natural materials consisting of individual mineral grains not joined by strong and permanent cohesive forces. Natural soils are products of the weathering of rock. *Soil mechanics* is the study of the engineering properties of soil.

2A.2 PHYSICAL CONDITION

2A.2.1 Introduction

A natural soil consists of three separate components: solids, liquids, and gases. The solids are normally natural mineral grains, although they can be human-made materials such as furnace slag or mine tailings. The liquid is usually water, and the gas is usually air. The relative amounts of each of the components in a particular soil may be expressed as a series of ratios. These ratios may be based on relative masses or weights, relative volumes, or relative mass or weight densities. These weight-volume ratios are fundamental to soil mechanics and geotechnical engineering.

2A.2.2 Two-Phase Soil (Dry or Saturated)

If a soil consists of only solids (mineral particles) and voids (either gas- or liquid-filled), then it is a two-phase soil system. Although the void spaces are interspersed throughout the mineral par-

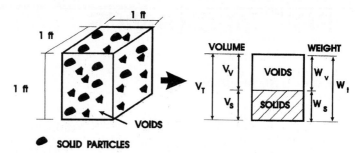

FIGURE 2A.1 Two-Phase representation of soil.

ticles, a unit volume of the soil may be viewed as in Fig. 2A.1. Using this figure, some terms can be defined. The unit weight of soil solids (γ_s) is

$$\gamma_s = \frac{W_s}{V_s} \tag{2A.1}$$

where W_s = weight of solid phase
V_s = volume of solid phase

The unit weight of total soil system (γ_t) is

$$\gamma_t = \frac{W_t}{V_t} \tag{2A.2}$$

where W_t = total weight
V_t = total volume

If the voids are filled with gas (air), then

$$W_v = 0$$

$$\gamma_t = \gamma_{dry} = \frac{W_s}{V_t} \tag{2A.3}$$

where γ_{dry} = dry density

If the voids are filled with liquid (water)

$$W_v = W_w$$

$$\gamma_t = \gamma_{sat} = \frac{W_s + W_w}{V_t} \tag{2A.4}$$

where γ_{sat} = saturated unit weight
W_w = weight of water

The buoyant unit weight is defined as:

$$\gamma' = \gamma_{sat} - \gamma_0 \tag{2A.5}$$

where γ_0 = unit weight of water
 = 62.4 pcf = 9.807 kN/m³

The voids ratio (e) is

$$e = \frac{V_v}{V_s} \tag{2A.6}$$

where V_v = volume of voids

Porosity ($n\%$) is defined

$$n(\%) = \frac{V_v}{V_T} \times 100 \tag{2A.7}$$

Water content ($w\%$) is

$$w(\%) = \frac{W_w}{W_s} \tag{2A.8}$$

Specific gravity of soil solids (G_s) is

$$G_s = \frac{\gamma_s}{\gamma_0} = \frac{W_s}{V_s \gamma_0} \tag{2A.9}$$

2A.2.3 Three-Phase Soil

Soils are not always either dry or saturated. Often the voids are only partially filled with water. The soil block then consists of a three-phase system as shown in Fig. 2A.2. Using Fig. 2A.2, some weight-volume relationships can be defined. Moist mass density is

$$\gamma_{\text{moist}} = \frac{W_t}{V_T} \tag{2A.10}$$

Water content ($w\%$) is

$$w(\%) = \frac{W_w}{W_s} \tag{2A.11}$$

Degree of saturation is

$$S_r = \frac{V_w}{V_V} \times 100 \tag{2A.12}$$

2A.3 SOIL IDENTIFICATION AND CLASSIFICATION

2A.3.1 Introduction

Soil is identified and classified using several systems. These systems include (*a*) the mechanism by which the solid particles are formed; (*b*) the size of the individual particles; and (*c*) the engineering properties of the soil.

2A.3.1.1 Soil Formation. Soil can be classified based on the origin of its constituents. The major means of soil origin is rock weathering.

Rock Weathering. Rock weathering can be either mechanical or chemical. Either will break a large rock mass into smaller particles.

MECHANICAL WEATHERING Mechanical weathering can be caused by exfoliation of large rock masses, differential thermal expansion and contraction of minerals within a rock mass, or the

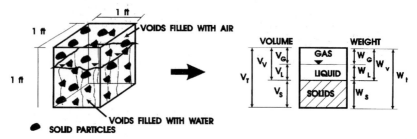

FIGURE 2A.2 Three-phase soil system.

freezing and subsequent thawing of water in minute fissures in the rock mass. In addition, mechanical weathering can be caused by the impact of running water on a rock mass, scouring of a rock mass by glacial movement, or breakdown of the rock mass by the impact of wind-blown particles. Rocks are also weathered by the expansion of roots of growing vegetation in the minute cracks and fissures in the rock.

CHEMICAL WEATHERING   Chemical weathering occurs when hard rock minerals are transformed into soft, easily erodible matter by chemical processes in nature. The primary chemical processes that weather hard rock minerals include oxidation, carbonation, hydration, and leaching.

When a rock containing iron comes into contact with moist air, $2Fe_2O_3 \cdot H_2O$ (rust) is formed. This rust is very soft and easily eroded from the rock mass.

When water is added to CO_2 in the atmosphere, a weak carbonic acid is formed. This acid decomposes minerals containing iron, calcium, magnesium, sodium, or phosphates and transforms them into the clay minerals.

Hydration is the taking up of water which is then bound chemically to form new minerals. Not only does hydration alter the mineral, often to a softer state, but it also causes a volume increase which in turn increases the decomposition of the rock mass.

When water comes into contact with salt, gypsum, feldspars, or limestone, the minerals will dissolve. The dissolved minerals are then transported and redeposited elsewhere. This is called *leaching*.

Organic Decomposition.   Because organic soils are so unique and difficult to deal with, only a relatively small fraction of the soil solids needs to be organic for the organic constituent to control its engineering behavior.

PEAT   Peat is formed by the growth and subsequent decay of plants. Peaty soils tend to be fibrous and black in color and smell like rotten eggs.

CORAL   Coral is the accumulation of fragments of inorganic skeletons or shells of organisms. These soils are easily identified visually.

2A.3.1.2 Mineralogy.

A mineral is an inorganic compound found in nature. For engineering purposes, the minerals are separated into rock and soil minerals.

Principal Rock Minerals.   The principal rock minerals are:

Quartz. Quartz is the principal mineral in granite, sands, and rock flour. The mineral is clear, transparent, and quite hard. Quartz is very resistant to chemical weathering and has a specific gravity of 2.66.

Feldspar (silicates of aluminum). Feldspars are important because these minerals chemically weather into the clay minerals. Their specific gravities are about 2.7.

Micas. Mica is the primary mineral in granites and gneiss. Mica can be split into thin, elastic sheets. It is colorless and transparent and has a specific gravity of 2.8.

Carbonates. Carbonates are derived from the chemical weathering of calcium-bearing feldspars and other calcium-bearing rocks and have been reformed into new rock masses.

Principal carbonates include calcite, dolomite, and the various limestones. The carbonates are quite soft and white to colorless and are highly susceptible to chemical weathering. Their specific gravities vary from 2.7 to 2.8.

Principal Clay Minerals. The principal clay minerals are:

Kaolinite. Kaolinite is derived from the chemical weathering of feldspars and other aluminum-bearing rocks. Its primary structure consists of a single sheet of gibbsite bound to a sheet of silicon (Fig. 2A.3). Successive two-layer sheets are bound by weaker hydrogen bonds. These hydrogen bonds can be broken relatively easily forming a basic two-layer sheet structure.

Illite. Illite takes the structure of kaolinite a further step. The hydroxyl of the octahedral layer are stripped of their hydrogen ions on both sides, and the oxygen ion is the tip of a tetrahedral layer on both sides of the octahedral layer. The octahedral layer is electrically neutral. However, the tetrahedra are not neutral. Approximately one tetrahedron in seven contains an aluminum (+3) in place of the usual silicon (+4) because of isomorphous substitution. This results in an overall charge deficiency in each of the tetrahedral layers. This charge deficiency draws potassium ions (+1) into the structure in that octagonal void that exists as discussed previously. Thus a potassium ion fits into this void and the void of the next sheet of illite; the ion is drawn by the negative charge of the tetrahedral layer and thereby hold the sheets together. This bond is obviously not as strong as the hydrogen bond of kaolinite, but it is not weak either.

Montmorillonite (Smectite). Montmorillonite is very similar to illite; montmorillonite is a three-layer mineral except the tetrahedral layer is relatively neutral with almost no substitution of aluminum for silicon. The octahedral layer has the charge deficiencies with aluminum in the vacated positions. Since the seat of the charge deficiency is in the center of the sheet, there is a weaker attraction by the negative charge to outside positive charges. Instead of strongly attracting potassium, hydrated ions, such as sodium, are weakly attracted. The sodium ions are weakly attracted to the potassium ions and to the negative faces of the sheets of montmorillonite because of its dipole nature. Thus montmorillonite surrounds its sheets with oriented water and hydrated cations. Soils with montmorillonite are known for their propensity to swell in the presence of water.

2A.3.1.3 Grain (Particle) Size Classification.

One method of classifying soils is by the size of the individual particles. The size and distribution of soil particles are determined through a grain size analysis.

Coarse-grained soils behave in nature as individual particles. They are subdivided into gravel and sand. *Gravel soils* have particle sizes coarser (larger) than about the #4 or the #10 mesh sieve

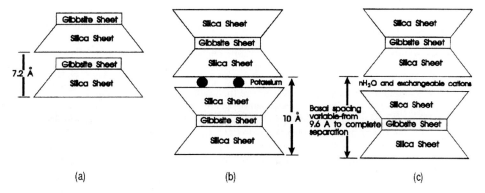

FIGURE 2A.3 Structure of principal clay minerals: (*a*) kaolinite, (*b*) illite, (*c*) montmorillonite.

opening, depending upon which particular classification system is used. *Sands* have particle sizes finer than gravel (#4 or #10 mesh) and coarser than the #200 mesh sieve. Coarse sand particles pass the #4 sieve and are retained on the #10 mesh sieve. Medium sand has a particle size that is smaller than the #10 and larger than the #40 mesh. Fine sand has particles in the #40 to the #200 mesh size.

Fine-grained soils largely behave as a mass and not as individual particles. Their particle size can be divided, however, into silt and clay. *Silt* particles are smaller than the #200 (0.0074 mm) but larger than 2 μm. Silts are derived by the mechanical weathering of rock. *Clay* particles are smaller than 2 μm. They are developed from the chemical weathering process whereby the rock minerals are chemically altered into the clay minerals.

2A.3.1.4 Grain (Particle) Size Distribution.

The particle size distribution analysis of a soil involves determining the relative amounts of particles within given size ranges in a soil mass. Different test methods are used for coarse-grained soils and fine-grained soils.

The particle size distribution of a coarse-grained soil is determined by a sieve analysis (ASTM D-422). This test uses a set of calibrated sieves, stacked in descending opening size, through which the soil is passed. The largest screen opening is several inches and the smallest size commonly used is 200 mesh (0.074 mm). Intermediate-size screens are used to separate various sizes of particles down to 200 mesh. Larger particles are retained on the upper sieves while the smaller particles pass through onto the lower sieves. The grain size distribution of coarse grained soils influences their density, permeability, shear strength, and liquefaction potential.

The grain size distribution of fine-grained soil is determined by sedimentation. The method is based on Stoke's law:

$$D = \sqrt{\frac{18\eta v}{\gamma_s - \gamma_f}} \tag{2A.13}$$

where D = diameter of sphere
η = viscosity
v = velocity of fall of sphere
γ_s = unit weight of sphere
γ_f = unit weight of fluid

A sample of the soil is mixed into a suspension in water and the suspension placed in a sedimentation cylinder. Using Stoke's law, it is possible to calculate the time t for particles of diameter D to settle a specified depth in the suspension. A hydrometer is used to measure the specific gravity of the suspension. Details of the test are given in ASTM D 422.

Although testing of the grain size distribution of fine-grained soils is often performed, the properties of these soils are more affected by grain structure, shape, and geologic origin than particle size distribution.

The amount of different particle sizes in a soil is represented by a *grain size distribution curve*. This curve is a semilogarithmic plot with the ordinates being the percentage by weight of particles smaller than the size given by the abscissa (Fig. 2A.4). The flatter the distribution curve, the larger the range of particle sizes in the soil. The steeper the distribution curve, the smaller the size range. A soil that has a relatively even distribution of particle sizes is called a *well-graded soil*. A soil that consists primarily of particles of one size is called *uniform*. Descriptive coefficients are used to quantify various characteristics of the grain size distribution curve.

The particle size corresponding to any specified value on the ordinate (percent smaller) curve can be read on the size distribution curve. For example, the diameter of particle that corresponds to the 50% smaller ordinate is the D_{50} size. Similarly, the D_{10} size is the diameter of grains that only 10% of the soil is finer than. The D_{10} is known as the effective diameter of the soil.

The uniformity coefficient is defined as

$$C_u = \frac{D_{60}}{D_{10}} \tag{2A.14}$$

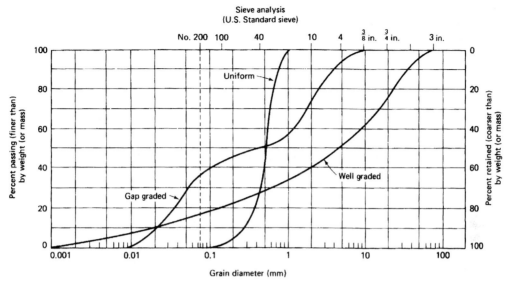

FIGURE 2A.4 Typical grain size distribution curves.

and is a measure of the general slope of the grain size distribution curve. A C_u near unity indicates a one-size (uniform) soil. However, a $C_u = 6$ indicates a well-graded soil. The coefficient of curvature (C_z) indicates the constancy of the slope of the grain size distribution curve.

$$C_z = \frac{D_{10}^2}{D_{60} \times D_{10}} \qquad (2A.15)$$

C_z of between 1 and 3 indicate a well-graded soil. C_z values outside these limits indicate either uniform or gap-graded soils.

2A.3.1.5 Grain Shape

Coarse-Grained Soils. The shape of individual coarse-grained soil particles (larger than #200 mesh) can be classified as being either bulky, platey, or needlelike in shape.

Bulky Grain. Bulky grain particles are roughly equidimensional, i.e., the length ≈ width ≈ height. This is true of most gravels, sands, and silts. Bulky grain soils are strong and relatively incompressible.

Plate-Shaped Particles. Sands that are derived from mica have shapes that are platelike, i.e., length ≈ width >> thickness. Platey particles are more elastic and prone to breakage than bulky grain particles.

Needle-Shaped Particles. Some corals and some clays have particle shapes that are like needles. These particles are resilient under low static loads but tend to break under higher loads.

Particle Angularity. Under load, angular corners crush and break but resist displacement. Smoother particles are less resistant to displacement but are less likely to crush. Particles are classified as angular, subangular, subrounded, rounded, or well-rounded.

Fine-Grained Soils. Silts are products of the mechanical weathering of rock, and therefore their shape is primarily bulky. Because the clay minerals are crystalline with an orderly, sheetlike molecular arrangement, the clay particles break down into very small (<2 μm) sheets or plates where the length ≈ width >> thickness.

2A.3.1.6 Soil Plasticity.
Soil plasticity is defined as the ability to undergo deformation without rupture. Plasticity of a soil is caused by the physical-chemical properties of the plate-shaped

clay particles in the soil. The negative charges on the surfaces of the clay platelets attract and bind polar water molecules to their surfaces. The negative end of the water dipole similarly attracts other water molecules. This phenomenon continues with the attractive forces decreasing at larger distances from the clay surface. This layer of tightly bound water around the clay particle is known as absorbed water. The absorbed water thickness will vary depending upon the strength of the surface charge as well as the presence of cations in the water phase (Fig. 2A.5).

Atterberg Limits. The clay particles are held together by a number of different forces including electrical attraction, hydrogen bonding, cation sharing, and Van Der Wall's forces. When the particles are at large spacings, there is relatively little particle-to-particle attraction. Therefore, it is relatively easy for particles to slip past each other. However, as the particle spacing reduces due to an applied force or the removal of water, the interparticle attraction will increase, and the slip potential between the particles will reduce.

As a clay slurry is reduced in volume by desiccation, it passes from a liquid phase through a viscous fluid stage through a plastic and finally to a solid state as the particle-to-particle attraction forces increase and the slip potential decreases. Atterberg divided these ranges arbitrarily into five ranges as defined by the water content of the slurry. Two important ranges for engineering work are defined by three limits:

Liquid limit. The *liquid limit* (LL) is defined as the lower limit of viscous flow. Casagrande defines the LL as the water content at which a 2-mm wide trapezoidal groove cut in moist soil held in a special cup would close 0.5 in along the bottom after 25 taps on a hard rubber plate (ASTM D-4318).

Plastic limit. The *plastic limit* (PL) is defined as the lower limit of the plastic range and is the water content at which a sample of soil begins to crumble when rolled into a thread $\frac{1}{8}$-in in diameter (ASTM D-4318).

Shrinkage limit. The *shrinkage limit* is the lower limit of volume change upon drying and is defined as the water content at which the soil is at a minimum volume as it dries out from saturation (ASTM D-4943).

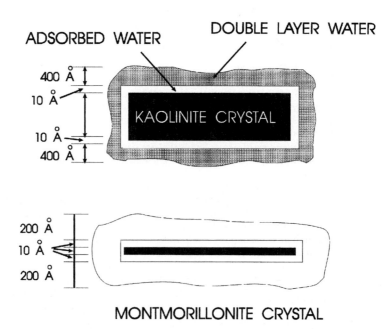

FIGURE 2A.5 Clay particle.

Index properties are used to provide relative measures of the plasticity of a soil. The *plasticity index* (PI) is defined as

$$PI = LL - PL \tag{2A.16}$$

The *liquidity index* (IL) is defined as

$$IL = \frac{W_n - PL}{PI} \tag{2A.17}$$

The *activity number* (A) is applied to plastic soils in reference to their propensity for undergoing volume change in the presence of varying moisture conditions:

$$A = \frac{PI}{\% < 0.002 \text{ mm}} \tag{2A.18}$$

Uses of the Atterberg Limits. In general, the index properties are indicative of remolded soil properties. They can be used as general indicators of clay mineralogy, volume change potential, compressibility, geologic history of the deposit, and the undrained shear strength.

CLAY MINERALOGY Since the clay content (fraction smaller than 0.002 mm) governs the Atterberg limits of a soil, the Atterberg limits are thus an indicator of the type of clay mineral. The activity (*A*) number is used to estimate the type of clay mineral in a soil (Table 2A.1).

TABLE 2A.1 Typical Values of Activity

Mineral	Activity
Kaolinite	0.2–0.4
Illite	0.5–0.9
Calcium montmorillonite	1.0–2.0
Sodium montmorillonite	4 or more

TABLE 2A.2 Swell Potential

LL	Swell potential
0–30	Slight to low
31–50	Moderate to intermediate
>50	High

VOLUME CHANGE POTENTIAL The propensity for a clay soil to undergo expansion or shrinkage with increases or decreases in moisture content can be estimated using the Atterberg limits (Table 2A.2). As the plasticity index increases and the shrinkage limit decreases, volume change potential increases (Table 2A.3).

COMPRESSIBILITY Skempton[1] developed a statistical relationship between the liquid limit and the compression index for remolded soils:

$$C_c = 0.007 \, (LL - 10) \tag{2A.19}$$

Research at Cornell[2] has justified the Terzaghi expression for undisturbed, normally consolidated soils:

$$C_c = 0.009 \, (LL - 10) \tag{2A.20}$$

These expressions apply only to normally consolidated soils and are valid to about ±30%.

Geologic history A plot of liquidity index versus depth will tend to smooth out depositional differences between soils. Deposits with a common depositional history show a smooth curve. A

TABLE 2A.3 Estimation of Volume Change Potential from Atterberg Limits

Plasticity index	Shrinkage limit	Probable expansion % total volume change dry to saturated	Degree of expansion
>35	<11	>30	Very high
25–41	7–12	20–30	High
15–28	10–16	10–20	Medium
<18	>15	<10	Low

normally consolidated deposit will show a continuous decrease in liquidity index with depth.[3] A plot of liquidity index versus vertical effective consolidation pressure provides a means of estimating whether the soil has been overconsolidated.

UNDRAINED SHEAR STRENGTH For normally consolidated soils, the curve of water content versus logarithm of effective vertical consolidation pressure, $\bar{p}$, and water content versus logarithm of undrained shear strength, s_u, are parallel. Thus, for a normally consolidated soil, the ratio of $s_u/\bar{p}$ is a constant. Skempton developed a statistical relationship between plasticity index and $s_u/\bar{p}$:

$$\frac{s_u}{\bar{p}} = 0.11 + 0.0037 \, \text{PI} \tag{2A.21}$$

This expression is valid for normally consolidated soils tested in situ by vane shear and for soils tested in unconfined compression.

2A.3.1.7 Soil Aggregate Properties

Primary Structure. Soil structure is the arrangement of individual soil grains in relation to each other. Terzaghi classified soils into three broad classes: cohesionless, cohesive, and composite soils.

COHESIONLESS SOILS *Cohesionless soils* consist of particles of gravel, sand, or silt depending upon the size of their individual particles. The structure of these soils can take two forms: single-grained or honeycombed.

Single grained
Gravel, sand, or silt particles greater than about 0.02 mm settle out of suspension in water as individual grains independent of other grains. Their weight causes the grains to settle and roll to equilibrium positions practically independent of other forces. The particles may come to equilibrium in either a loose condition, a dense condition, or anywhere in between (Fig. 2A.6). Relative density is used to measure the compactness of a single grained soil.

$$D_r = \frac{e_{max} - e}{e_{max} - e_{min}} \times 100 \tag{2A.22}$$

A soil that is in its most dense condition will have a relative density of 100 percent. Vibration can cause rapid reduction of the volume of a loose single-grained soil structure.

Honeycombed
When silt grains with diameters between 0.0002 and 0.02 mm settle out of suspension, molecular forces at the contact areas between particles may be large enough compared to the submerged unit weight of the grains to prevent the grains from rolling down immediately to positions of equilibrium among other grains already deposited. Electrostatic and other forces can cause miniature arches to form, bridging over large void spaces (Fig. 2A.7). Because the particles themselves are strong, these honeycombs can carry relatively large static loads without excessive volume change. However, if the load increases beyond the soil's critical value, large and rapid volume decrease will occur.

COHESIVE SOILS *Cohesive soils* are fine-grained soils whose particles form either a flocculated or dispersed structure.

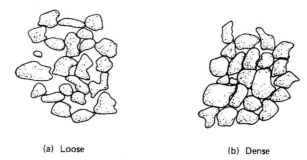

FIGURE 2A.6 Typical single-grained soil structure.

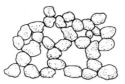

FIGURE 2A.7 Honeycombed structure.

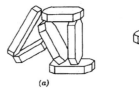

FIGURE 2A.8 Soils that are (*a*) flocculated; (*b*) dispersed.

Particles smaller than 0.0002 mm will not settle out of solution individually due to the Brownian motion. However, because of the electrical charges that exist on the surfaces and edges of clay particles, the negatively charged surfaces are attracted to the positively charged edges, and relatively large edge-to-face aggregates or flocs are formed. These flocs grow to masses great enough to settle out of suspension and form *flocculated soils* (Fig. 2A.8).

If a flocculated soil is remolded, then the edge to face structure collapses and the particles slip into nearly parallel positions. In this configuration there is very little particle-to-particle contact. This structure is called an *oriented* or *dispersed structure* (Fig. 2A.8).

COMPOSITE   In natural clays which contain a significant proportion of larger particles, the structural arrangement of the particles can be highly complex. Single grains of silt and/or sand can be interspersed within a clay platelet matrix.

Consistency and Sensitivity of Clays.   Consistency is a measure of the degree with which a clay soil will resist deformation when loaded. Consistency of a clay is measured by the unconfined compression test, which will be described later. Table 2A.4 is often used to describe the consistency of a clay.

Sensitivity describes the loss of strength of a soil upon remolding. Numerically it is defined as the unconfined compressive strength of the undisturbed soil (q_u) divided by the unconfined compressive strength of the same soil remolded at an identical water content (q_r). Table 2A.5 classifies soils according to their sensitivity.

2A.3.2 Soil Classification

Soil classification is the placing of a soil into a group of soils all of which exhibit similar characteristics.

2A.3.2.1 Classification According to Origin of Natural Deposits

All soils are either residual or transported.

Residual Soils.   *Residual soils* are formed when rock weathers faster than erosion can carry the soil particles away. In these soils, all soluble materials have been leached out. The chemical

2.14 SOIL MECHANICS AND FOUNDATION DESIGN PARAMETERS

TABLE 2A.4 Consistency of Clay Soils

Consistency	Unconfined compressive strength (tsf)
Very soft	<1/4
Soft	1/4 to 1/2
Medium	1/2 to 1
Stiff	1 to 2
Very stiff	2 to 4
Hard	>4

TABLE 2A.5 Sensitivity of Clay Soils

Sensitivity	q_u/q_r
Insensitive	<1.0
Low	1.0 to 2.0
Medium	2.0 to 4.0
Sensitive	4.0 to 8.0
Extra	8.0 to 10.0
Quick	>10.0

disintegration becomes less active with depth, and the alteration becomes less and less with depth until the parent rock is reached. These soils tend to be highly mixed grain with gravel or cobble-sized remnants of chemically resistant rock intermixed with clay particles. The particles tend to be very angular.

Transported Soils. Transported soils consist of soil particles that have been moved from their original location by various agents and redeposited. The types of deposits are classified according to their erosion and transportation methods.

Aeolian deposits are wind transported and deposited and are characterized by their high degree of sorting (all one-size particles) and the uniformity of the deposits. Principal aeolian soils include sand dunes and loess (wind-deposited silt).

Gravitational deposits are soil deposits that have collected at the base of mountains. Chief among the gravitational deposits is talus, which is the accumulation of rock and soil that builds up at the base of cliffs.

2A.3.2.2 Engineering Soil Classification

AASHTO Soil Classification. The American Association of State Highway and Transportation Officials (AASHTO) soil classification identifies kinds of soil in terms of their suitability to serve as a highway base course. This system classifies soils into one of eight groups, A-1 through A-7 (Tables 2A.6A and 2A.6B). A-1 soils consist of well-graded gravels and sands and are the best soils for highway subgrades. Organic soils that are highly unsuitable for use as a subgrade are classified as A-8. Some groupings are further subdivided. The classifications are based on the soil's grain size distribution and plasticity of the fraction passing the #40 sieve.

Unified Soil Classification. The *unified soil classification system* classifies soils into three major groups based on their predominant particle size and plasticity. Soils are coarse-grained (sand or gravel) if 50% or more of the soil particles by weight are larger than the #200 mesh sieve. Soils are fine-grained (silt or clay) if 50% or more of the soil particles by weight are smaller than the #200 mesh sieve. If the soil contains organic matter, the soil is designated either organic or peat, no matter what the size of the mineral grains. Each class of soil is further divided into subclasses depending upon either its grain size distribution (coarse-grained soils) or its plasticity (fine-grained soils). All soils are given a 2-letter designation descriptive of the soil's primary and secondary classification.

COARSE-GRAINED SOILS Coarse-grained soils are classified as gravel (G) if 50% or more of the coarse fraction is larger than the #4 mesh sieve. The coarse-grained soil is classified as sand (S) if 50% or more of the coarse fraction lies between the #4 and the #200 mesh sieve. Subclasses are either well graded (W), poorly graded (P), silty (M), or clayey (C) based on the amount, distribution, and plasticity of their particles. Details of the classification are given in Table 2A.7.

FINE-GRAINED SOILS Fine-grained soils are classified as either silt (M) or clay (C) depending upon their plasticity as measured by the Atterberg limits. To make this determination, the plasticity chart (Fig. 2A.9) is utilized. Once the liquid limit and plasticity index are known for the soil, the appropriate soil identification is made by plotting the point on the chart. Soils are high plasticity (H) if their liquid limits exceed fifty. Soils are low plasticity (L) if their liquid limit is less than 50. The *A line* separates the clays from the silts and the organic soils. Soils whose liquid

TABLE 2A.6A AASHTO Soil Classification System*

General Classification	Granular Materials (35% or less passing No. 200)							Silt-Clay Materials (More than 35% passing No. 200)			
Group Classification	A-1	A-3[A]	A-2				A-4	A-5	A-6	A-7
			A-2-4	A-2-5	A-2-6	A-2-7				
Sieve analysis, % passing:										
No. 10 (2.00 mm)	...	...								
No. 40 (425 µm)	50 max	51 min								
No. 200 (75 µm)	25 max	10 max	35 max				36 min	36 min	36 min	36 min
Characteristics of fraction passing No. 40 (425 µm):										
Liquid limit	...	...	B				40 max	41 min	40 max	41 min
Plasticity index	6 max	N.P.	B				10 max	10 max	11 min	11 min
General rating as subgrade	Excellent to Good						Fair to Poor			

*Reprinted with permission of American Association of State Highway and Transportation Officials.
[A]The piecing of A-3 before A-2 is necessary in the "left to right elimination process" and does not indicate superiority of A-3 over A-2.
[B]See Table 2A.6B for values.

TABLE 2A.6B AASHTO Soil Classification System*

General Classification	Granular Materials (35% or less passing No. 200)							Silt-Clay Materials (More than 35% passing No. 200)			
	A-1		A-3	A-2				A-4	A-5	A-6	A-7
Group Classification	A-1-a	A-1-b		A-2-4	A-2-5	A-2-6	A-2-7				A-7-5, A-7-6
Sieve analysis, % passing:											
No. 10 (2.00 mm)	50 max	...	...								
No. 40 (425 µm)	30 max	50 max	51 min								
No. 200 (75 µm)	15 max	25 max	10 max	35 max	35 max	35 max	35 max	36 min	36 min	36 min	36 min
Characteristics of fraction passing No. 40 (425 µm):											
Liquid limit			...	40 max	41 min	40 max	41 min	40 max	41 min	40 max	41 min
Plasticity index	6 max		N.P.	10 max	10 max	11 min	11 min	10 max	10 max	11 min	11 min[A]
Usual types of significant constituent materials	Stone Fragments, Gravel and Sand		Fine Sand	Silty or Clayey Gravel and Sand				Silty Soils		Clayey Soils	
General rating as subgrade	Excellent to Good							Fair to Poor			

*Reprinted with permission of American Association of State Highway and Transportation Officials.
[A]Plasticity index of A-7-5 subgroup is equal to or less than LL minus 30. Plasticity index of A-7-6 subgroup is greater than LL minus 30.

TABLE 2A.7 Unified Soil Classification Chart

				Soil classification	
Criteria for assigning group symbols and group names using laboratory tests[a]				Group symbol	Group name[b]
Coarse-grained soils—more than 50% retained on #200 sieve	Gravels—more than 50% of coarse fraction retained on #4 sieve	Clean gravels—less than 5% fines	$C_u \geq 4$ and $1 \leq C_c \leq 3$[e]	GW	Well-graded gravel[f]
			$C_u < 4$ and/or $1 > C_c > 3$[e]	GP	Poorly graded gravel[f]
		Gravels with fines—more than 12% fines	Fines classify as ML or MH	GM	Silty gravel[f,g,h]
			Fines classify as CL or CH	GC	Clayey gravel[f,g,h]
	Sands—50% or more of coarse fraction passes #4 sieve	Clean sands—less than 5% fines	$C_u \geq 6$ and $1 \leq C_c \leq 3$[e]	SW	Well-graded sand[f]
			$C_u < 6$ and/or $1 > C_c > 3$[e]	SP	Poorly graded sand[f]
		Sands with fines—more than 12% fines	Fines classify as ML or MH	SM	Silty sand[g,h,i]
			Fines classify as CL or CH	SC	Clayey sand[g,h,i]
Fine-grained soils—50% or more passes the #200 sieve	Silts and clays—liquid limit less than 50	Inorganic	PI > 7 and plots on or above A line[j]	CL	Lean clay[k,l,m]
			PI < 4 or plots below A line[j]	ML	Silt[k,l,m]
		Organic	Liquid limit—oven dried / Liquid limit—not dried < 0.75	OL	Organic clay[k,l,m,n] / Organic silt[k,l,m,o]
	Silts and clays—liquid limit 50 or more	Inorganic	PI plots on or above A line	CH	Fat clay[k,l,m]
			PI plots below A line	MH	Elastic silt[k,l,m]
		Organic	Liquid limit—oven dried / Liquid limit—not dried < 0.75	OH	Organic clay[k,l,m,p] / Organic silt[k,l,m,o]
Highly organic soils	Primarily organic matter, dark in color, and organic odor			PT	Peat

[a]Based on the material passing the 3-in (75 mm) sieve.
[b]If field sample contained cobbles or boulders, or both, add "with cobbles or boulders, or both" to group name.
[c]Gravels with 5 to 12% fines require dual symbols: GW-GM well-graded gravel with silt; GW-GC well-graded gravel with clay; GP-GM poorly graded gravel with silt; GP-GC poorly graded gravel with clay.
[d]Sands with 5 to 12% fines require dual symbols: SW-SM well-graded sand with silt; SW-SC well-graded sand with clay; SP-SM poorly graded sand with silt; SP-SC poorly graded sand with clay.
[e]$C_u = D_{60}/D_{10}$ $C_c = \dfrac{(D_{30})^2}{D_{10} \times D_{60}}$
[f]If soil contains ≥ 15% sand, add "with sand" to group name.
[g]If fines classify as CL-ML, use dual symbol GC-GM, or SC-SM.
[h]If fines are organic, add "with organic fines" to group name.
[i]If soil contains ≥ 15% gravel, add "with gravel" to group name.
[j]If Atterberg limits plot in hatched area, soil is a CL-ML, silty clay.
[k]If soil contains 15 to 29% plus #200, add "with sand" or "with gravel," whichever is predominant.
[l]If soil contains ≥ 30% plus #200, predominantly sand, add "sandy" to group name.
[m]If soil contains ≥ 30% plus #200, predominantly gravel, add "gravelly" to group name.
[n]PI ≥ 4 and plots on or above A line.
[o]PI < 4 or plots below A line.
[p]PI plots on or above A line.
[q]PI plots below A line.

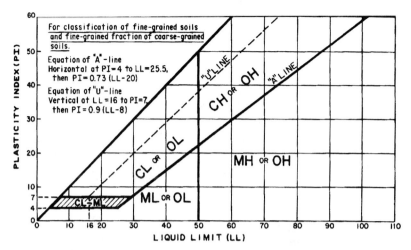

FIGURE 2A.9 Plasticity chart.

limit–plasticity index plot above the A line are classified as clays, and soils whose liquid limit–plasticity index plot below the A line are classified as silts. Organic soils plot below the A line and have organic material.

2A.4 WATER FLOW IN SOILS

Many geotechnical engineering problems involve the flow of water through soils. The study of water flow through soils has received much attention due to its importance in seepage and soil consolidation problems. Chemical transport through soils is receiving much more attention due to increased interest in ground-water pollution, waste disposal and storage, and so on.

2A.4.1 Basic Principles of One-Dimensional Fluid Flow in Soils

2A.4.1.1 Darcy's Law. Water flow is related to the corresponding driving forces according to water flow:

$$q = ki \text{ (Darcy's law)} \tag{2A.23}$$

where q = flow rate (flux) of water
  i = hydraulic gradient
  k = hydraulic conductivity

Two general assumptions govern the analysis of fluid flow through soil and rock. The first is that all the voids are interconnected. The second is that water can flow through even the densest of natural soils. Although the second assumption appears to be valid, it is generally accepted that not all the void spaces provide passageways for pore fluids.

Darcy's law [Eq. (2A.23)] can be presented as follows for the condition of Fig. 2A.10:

$$Q = kiA\Delta t \tag{2A.24}$$

where Q = quantity of flow (volumetric fluid flux, volume units)
  i = a forcing function (gradient)

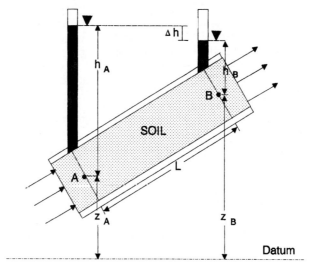

FIGURE 2A.10 Darcy's law.

A = cross-sectional area of flow
k = fluid conductivity, a coefficient of proportionality related to the properties of the flow medium *and* the fluid (often called permeability or, more correctly, *hydraulic conductivity*)
Δt = time length of flow duration

Referring to Fig. 2A.10, we get

$$q = \frac{Q}{\Delta t} = k\left[\frac{(h_A - h_B)}{L}\right]A = -kiA \tag{2A.25}$$

where h_A = the upstream head = $(h_p + h_z)_A$
h_B = the downstream head = $(h_p + h_z)_B$
h_p = pressure head
h_z = elevation head
L = the length of flow
q = specific fluid flux (flow rate)

2A.4.1.2 Flow Velocity. Since flow can only occur through the soil voids,

$$q = vA = v_s A_v \tag{2A.26}$$

where v = approach, apparent or superficial velocity = q/A = ki
v_s = seepage velocity through soil (*assumed to be in a straight line*)
A = total cross-sectional area of soil volume
A_v = volume of soil voids

solving

$$v_s = v\left[\frac{A}{A_v}\right]\left[\frac{L}{L}\right] = v\left[\frac{V}{V_v}\right] = \frac{v}{n_e} \tag{2A.27}$$

$$v_s = \frac{ki}{n_e} \tag{2A.28}$$

where n_e = effective porosity of the soil (pores available for flow)

2A.4.2 Determination of Permeability

2A.4.2.1 Laboratory Methods It is obvious that the hydraulic conductivity of a soil is, in many cases, the controlling factor in subsurface migration of hazardous wastes. Estimates of contaminant flow quantities and patterns can only be as accurate as the values of hydraulic conductivity used to make them. Subsurface fluid flow may occur under either partially saturated or fully saturated conditions. Numerous methods are available for measurement of hydraulic conductivity of soils either in situ or in the laboratory.

Soils are generally nonhomogeneous and anisotropic. Fine-grained (clay) soils are often stratified, containing root holes, fissures, and cracks. Therefore, it is desirable to test as large a volume of soil as possible. This usually means that testing should be done in the field. Field methods can usually provide more representative values than laboratory methods because they test a larger volume of material, thus integrating the effects of macrostructure and heterogeneities. However, field methods presently available to determine the conductivity of compacted fine-grained materials in reasonable times either require the tested interval to be below a water table or to be fairly thick or require excavation of the material to be tested at some point in the test. These restrictions generally lead to the requirement that the fluid conductivity of liner and cover materials be determined in the laboratory. Further, comparative tests to determine the effects of different permeants on conductivity are typically performed in the laboratory.

The most common laboratory methods of measuring the coefficient of hydraulic conductivity in the laboratory are the constant head test and the falling head test.

Constant Head Tests. In the constant head test, the hydraulic gradient, i, is maintained constant at h_w/L (Fig. 2A.11). From Darcy's law,

$$k = \frac{q}{iA} = \frac{QL}{(\Delta t h A)} \tag{2A.29}$$

where q = flow rate per unit time
Q = flow volume in elapsed time, t
L = length of soil specimen
h = head of water
A = cross-sectional area of specimen
Δt = elapsed time of test

The main advantages of the constant head test are the simplicity of interpretation of the data and the fact that the use of a constant head minimizes confusion due to changing volume of air bubbles when the soil is not saturated.

Falling Head Tests. In this test,

$$q = a\frac{dh}{dt} \tag{2A.30}$$

where a = the cross-sectional area of the burette (Fig. 2A.12). Therefore

$$k = \left[\frac{al}{At}\right] \log_e \left[\frac{h_1}{h_2}\right] \tag{2A.31}$$

The main advantage of this procedure is that small flows are easily measured using the burette. The observation time may be long, in which case corrections for water losses due to evaporation or leakage may be needed.

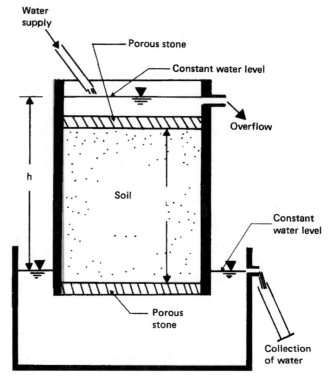

FIGURE 2A.11 Constant head test.

Triaxial Flexible Wall Permeability Tests. Two techniques are used to confine the soil sample during either the constant or falling head tests: rigid wall confining ring or a flexible membrane.

A flexible-wall-confined specimen placed in a triaxial permeameter allows us to[4]:

1. Back-pressure saturate the specimen
2. Reapply isotropic stresses to simulate field conditions
3. Insure against short-circuiting of permeant
4. Measure independently soil sample volume change

Sources of Error. The following are sources of error:

1. Use of nonrepresentative samples (an overriding source of error)
2. Voids formed during sample preparation
3. Smear zones
4. Alteration in clay chemistry
5. Air in sample
6. Growth of microorganisms
7. Menisci problems in capillary tubes
8. Temperature
9. Volume change due to stress change

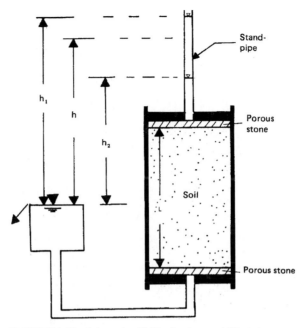

FIGURE 2A.12 Schematic of falling head permeability test.

2A.5 GEOSTATIC STRESSES IN SOIL

The three phases of soil (solid, water, and gas) will react differently to applied stresses, and thus a relationship between the phases must be established. The solid particles and water are relatively incompressible, and the gaseous phase is highly compressible. The following definitions and relationship between the various phases has been proposed by Terzaghi.[3]

2A.5.1 Total Stresses

Total stress, σ, is the stress acting at a point in a soil mass with a horizontal top surface. The total stress is computed as the total weight of a column of unit area above the point:

$$\sigma = \gamma \times z \tag{2A.32}$$

2A.5.2 Pore Stresses

The pore or neutral stress (u_w) is the stress within the water voids. Since this stress is hydrostatic, it acts equally in all directions. Under no flow conditions (static),

$$u_w = \gamma_w \times h_w \tag{2A.33}$$

where γ_w = the unit weight of water (9.8 kN/m³ = 62.4 lb/ft³)
h_w = head of water

2A.5.3 Effective Stresses

The intergranular force acting between points of contact of the solid constituents per unit area is termed the effective stress ($\bar{\sigma}$). Effective stress cannot be measured but can be calculated from the general relationship for saturated soils:

$$\bar{\sigma} = \sigma - u \qquad (2A.34)$$

Effective stress, not total stress, governs the shear and compressibility behavior of soils.

2A.6 DISTRIBUTION OF APPLIED STRESSES IN SOIL

The stress in a soil mass due to an applied load can most easily be computed from elastic theory. The soil mass is generally assumed to be semi-infinite, homogeneous, and isotropic.

2A.6.1 Point Load

Boussinesq[5] in 1885 published a relationship for the stress at any point with coordinates (x,y,z) beneath the location of a point load on the surface of the mass. His equation was

$$\sigma_z = \frac{3P}{2\pi} \frac{z^3}{(x^2 + y^2 + z^2)^{5/2}} \qquad (2A.35)$$

if $x^2 + y^2 + z^2 = R^2$

$$\sigma_z = \frac{3P}{2\pi} \frac{z^3}{R^5}$$

If $r^2 = x^2 + y^2$,

$$\sigma_z = \frac{P}{z^2} \frac{3}{2\pi} \left[\frac{1}{\left(\frac{r}{z}\right)^2 + 1} \right]^{5/2}$$

or

$$\sigma_z = \frac{P}{z^2} I \qquad (2A.36)$$

where I = influence factor = $f\left(\frac{r}{z}\right)$

If Eq. (2A.36) is plotted versus depth z for $r = 0$ (center line), Fig. 2A.13 results. As can be seen, the stress decreases rapidly as the depth beneath the point of load application increases. If Eq. (2A.36) is plotted versus r for a constant depth z, Fig. 2A.14 results. The stress decreases rapidly as the distance from the axis of the load increases.

2A.6.2 Uniformly Loaded Strip

Equation (2A.36) can be used to determine the stress beneath a flexible strip load of width B (Fig. 2A.15). Using the terms defined in Fig. 2A.12 the following equation can be developed:

$$\sigma_v = \frac{q}{\pi}[\beta + \sin\beta\cos(\beta + 2\delta)] \qquad (2A.37)$$

where $q = P/A$ = load/unit area

Table 2A.8 shows the variation of σ/q with $2z/B$ for various values of $2x/B$.

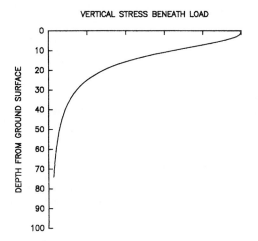

FIGURE 2A.13 Vertical stress versus depth. FIGURE 2A.14 Vertical stress versus r.

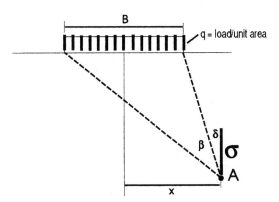

FIGURE 2A.15 Vertical stress due to a flexible strip load.

2A.6.3 Uniformly Loaded Circular Area

The Boussinesq point load equation can be used to develop an expression for the vertical stress below the center of a uniformly loaded flexible circular area:

$$\sigma_v = q\left[1 - \frac{1}{\left[\left(\frac{R}{z}\right)^2 + 1\right]^{3/2}}\right] \quad (2A.38)$$

$$= qI$$

where R = radius of loaded area
I = influence factor

Table 2A.9 gives I for various values of z/R.

TABLE 2A.8 Variation of σ/q with $2z/B$ and $2x/B$

$2x/B$	$2z/B$	σ/q	$2x/B$	$2z/B$	σ/q
0	0.0	1.0000	1.5	1.0	0.2488
	0.5	0.9594		1.5	0.2704
	1.0	0.8183		2.0	0.2876
	1.5	0.6678		2.5	0.2851
	2.0	0.5508	2.0	0.25	0.0027
	2.5	0.4617		0.5	0.0194
	3.0	0.3954		1.0	0.0776
	3.5	0.3457		1.5	0.1458
	4.0	0.3050		2.0	0.1847
0.5	0.0	1.0000		2.5	0.2045
	0.25	0.9787	2.5	0.5	0.0068
	0.5	0.9028		1.0	0.0357
	1.0	0.7352		1.5	0.0771
	1.5	0.6078		2.0	0.1139
	2.0	0.5107		2.5	0.1409
	2.5	0.4372	3.0	0.5	0.0026
1.0	0.25	0.4996		1.0	0.0171
	0.5	0.4969		1.5	0.0427
	1.0	0.4797		2.0	0.0705
	1.5	0.4480		2.5	0.0952
	2.0	0.4095		3.0	0.1139
	2.5	0.3701			
1.5	0.25	0.0177			
	0.5	0.0892			

TABLE 2A.9 I vs z/R for Circular Loaded Area

z/R	I
0.0	1.0
0.02	0.9999
0.05	0.9998
0.10	0.9990
0.2	0.9925
0.4	0.9488
0.5	0.9106
0.8	0.7562
1.0	0.6765
1.5	0.4240
2.0	0.2845
2.5	0.1996
3.0	0.1436
4.0	0.0869
5.0	0.0571

2A.6.4 Uniformly Loaded Rectangular Area

The vertical stress below a corner of a flexible rectangular load of width B and length L can be computed from

$$\sigma_v = qI_3$$

where
$$I_3 = \frac{1}{4\pi}\left[\frac{2mn\sqrt{m^2 + n^2 + 1}}{m^2 + n^2 + m^2n^2 + 1}\left(\frac{m^2 + n^2 + 2}{m^2 + n^2 + 1}\right)\right]$$
$$+ \frac{1}{4\pi}\left[\tan^{-1}\left(\frac{2mn\sqrt{m^2 + n^2 + 1}}{m^2 + n^2 - m^2n^2 + 1}\right)\right] \quad (2A.39)$$

$$m = \frac{B}{z}$$

$$n = \frac{L}{z}$$

The variation of I_3 is shown in Fig. 2A.16. The special case of the stress beneath the center of a square loaded area with side B is given in Table 2A.10.

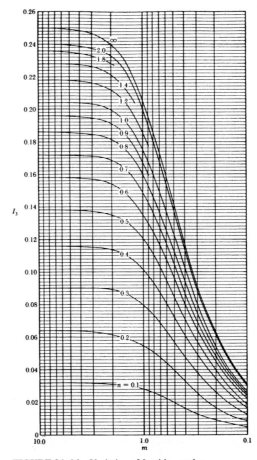

FIGURE 2A.16 Variation of I_3 with m and n.

TABLE 2A.10 Influence Factors for Stress Beneath the Center of a Square Load

B/z	I	B/z	I
0.0	1.0000	2.4	0.7832
20.0	0.9992	2.0	0.7008
16.0	0.9984	1.8	0.6476
12.0	0.9968	1.6	0.5844
10.0	0.9944	1.4	0.5108
8.0	0.9892	1.2	0.4276
6.0	0.9756	1.0	0.3360
5.0	0.9604	0.8	0.2410
4.0	0.9300	0.6	0.1494
3.6	0.9096	0.4	0.0716
3.2	0.8812	0.2	0.0188
2.8	0.8408	0.0	0.0000

2A.6.5 Average Stress Influence Chart

The point load equation has been used to compute average stresses beneath circular, square, and rectangular uniformly loaded footings. The results are presented as Fig. 2A.17. To use this chart enter the chart with the appropriate depth ratio (z/B) with B being the least dimension. Using the curve for the appropriate shape, read the influence factor on the abscissa. The average stress is the product of the influence factor times the applied stress.

2A.7 ONE-DIMENSIONAL CONSOLIDATION

Assume that the soil stratum shown in Fig. 2A.18 has been formed by sedimentation. If we assume that the element of soil is in equilibrium with its overburden, then the effective overburden stress on that soil element is $\bar{\sigma} = \gamma' z = (125 - 62.4)16 = 1000$ psf. In this state, the soil has a void ratio of e ($wG_s = 1.17$) and an element height of H_o. If a very wide fill is placed on top of the soil strata, the stress on the soil element is increased by an amount Δ_p ($\gamma_{\text{fill}} \cdot h_{\text{fill}}$). This increase in stress causes the void ratio (and the height) of the soil to decrease (settle). As you can see from the figure,

$$\frac{\Delta e}{1 + e} = \frac{\Delta h}{H_o}$$

$$\Delta h = \frac{H_o \Delta e}{1 + e} \tag{2A.40}$$

Therefore, if we can determine the relationship between the change in void ratio (Δ_e) and the applied stress, we can compute the settlement of the soil. This relationship is usually determined from a laboratory one-dimensional consolidation test.

2A.7.1 One-Dimensional Laboratory Consolidation Test

The one-dimensional consolidation test is performed using a disc-shaped sample of soil that is confined around its periphery by a rigid, impervious ring. The specimen is loaded on its flat surfaces through porous stones that allow the water from the soil voids to escape but restrain the soil particles from moving. All deformation occurs parallel to the axis of the specimen (Fig. 2A.19).

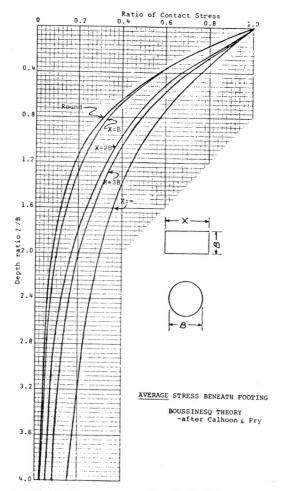

FIGURE 2A.17 Average stress beneath a loaded area.

2A.7.1.1 Performance of a Consolidation Test. The procedures used in performing one-dimensional consolidation tests vary widely depending on such factors as soil type, sampling method, and nature of the problem in the field (foundation of building, embankment, and so on). The procedures recommended in ASTM-D2435 are recommended for normal situations.

The soil sample is removed from the sampling tube to minimize further disturbance. A suitable length of soil is removed and is carefully trimmed into the ring. The cell is placed in the loading frame, and a seating load is applied, the size of which depends on the strength of the soil. The dial indicator used to measure the axial deformation of the soil is mounted in the frame, and an initial reading is taken. The consolidation cell is then filled with water. The applied pressure is adjusted continuously until the soil comes to equilibrium with the pressure at constant volume.

The first increment of consolidation pressure is then applied, and the soil begins to consolidate. A series of readings of axial deformation are made at preselected times. The consolidation pressure is maintained constant until consolidation has essentially ceased. The standard load increment duration for each pressure is 24 hours (ASTM-D2435 Method A), although shorter or longer times may be used depending on the coefficient of consolidation of the soil and the thick-

2.28 SOIL MECHANICS AND FOUNDATION DESIGN PARAMETERS

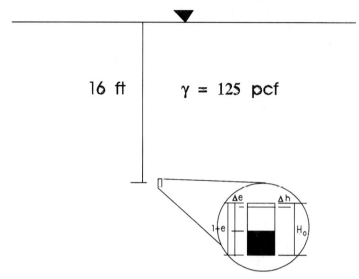

FIGURE 2A.18 Consolidation of a soil element.

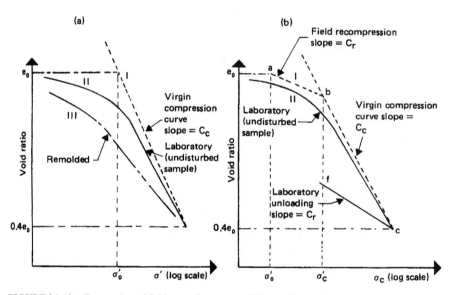

FIGURE 2A.19 Construction of field curve for overconsolidated soil.

ness of the sample. Sufficient time-deformation readings are taken to ensure that consolidation is complete. For some soils, a period of more than 24 hours may be required to reach end-of-primary consolidation. Usually, the load increment durations is some multiple of 24 hours and should be the standard duration for *all load intervals.*

When consolidation is essentially complete under the first pressure, the next increment of pressure is applied and a deformation reading taken again. The process is repeated until some preselected maximum consolidation pressure is attained. It is usual practice to double the pressure for each successive increment. Thus, for a typical consolidation test, the sequence of pressures might be 125, 250, 500, 1000, 2000, 4000, 8000, 16,000, 32,000 and 64,000 psf. The highest pressure is controlled by the capacity of the frame, economics, and time limitations, or a variety of other factors.

After the deformation has ceased under the highest load, the loads are removed in a series of decrements such that the pressure is usually reduced by four times for each decrement. The final pressure is usually of the order of 125 psf. When the specimen has equilibrated under the final pressure, the apparatus is dismantled rapidly to prevent the soil from imbibing a significant amount of water after the final pressure has been removed.

2A.7.1.2 Calculation of the Laboratory Consolidation Curve. The results of the consolidation test are normally presented as plots of void ratio, e, versus the logarithm of applied pressure, $\bar{p}$. A plot of the pressure-void ratio curve from the test is shown in Fig. 2A.19. The e-log p plot has an initially shallow slope that transitions into a steeper slope.

As you can see from the plot, the deformations are relatively small until the load approaches the in situ $\bar{p}_o$) pressure. When the load exceeds the in situ pressure, the deformation increases dramatically. The pressure where the e-log p curve increases slope is called the *maximum past consolidation pressure (preconsolidation pressure, $\bar{p}_c$)* and is the greatest stress that has ever been on the soil. For the case studied here, $\bar{p}_c = 1000$ psf $= \bar{p}_o$. The portion of the e-log p curve from the first load to $\bar{p}_c$ is called the *reload* or *recompression* curve, since the soil is being reloaded to its maximum past value. The portion of the e-log p curve for loads greater than $\bar{p}_c$ is called the *virgin consolidation* curve because each additional load is greater than any load that has ever been on the soil previously. Soils where $\bar{p}_c \approx \bar{p}_o$ are called *normally consolidated soils*.

In some (in fact, in most) cases, $\bar{p}_c >> \bar{p}_c$. This implies that the soil has been more heavily loaded in its past than the pressure that exists on it now. Soils with this stress history are called *overconsolidated soils*. Compared to normally consolidated soils, overconsolidated soils are stiffer and less compressible. As before, that portion of the e-log p curve to the $\bar{p}_c$ is the reloading branch, and that portion beyond $\bar{p}_c$ is the virgin consolidation branch.

2A.7.1.3 Reconstruction of the Field Consolidation Curve. The consolidation curve shown in Fig. 2A.19 is assumed to be typical of field consolidation curves. The recompression curves merge smoothly with the virgin consolidation curve and have a region of sharpest curvature in the vicinity of the maximum previous consolidation pressure, $\bar{p}_c$. Disturbance of soil samples during sampling, transportation, storage, and trimming, causes the laboratory curve to be displaced to lower void ratios and to have less pronounced curvature in the vicinity of $\bar{p}_c$.[5-7] A procedure is needed for reconstruction of the field consolidation curve from a slightly disturbed laboratory curve. No exact method for reconstruction of the field curve exists. The available methods are based on laboratory tests and field experience.

The first step is to obtain an estimate of the maximum previous stress under which the soil was consolidated, $\bar{p}_c$. Although several procedures have been proposed, the one suggested by Casagrande[8] seems to be the most satisfactory and most widely used. The procedure is shown in Fig. 2A.19. At the point of sharpest curvature of the laboratory curve, two lines are drawn, one tangent to the laboratory curve and the other horizontal. The angle between these lines is bisected by a third line. The intersection of the bisecting line and the extended laboratory virgin consolidation curve is taken as an approximation of the maximum previous consolidation pressure. If this pressure is approximately equal to the calculated effective overburden pressure in the field, $\bar{p}_o$, then the soil is assumed to be normally consolidated. One point on the field curve is then e_o, $\bar{p}_c$, where e_o is the initial void ratio of the sample. The field consolidation curve is drawn to pass through the point e_o, $\bar{p}_c$, and to be asymptotic to the laboratory curve at high pressures.

If the samples used in the laboratory are normally consolidated and have a maximum of disturbance, the e_o, $\bar{p}_o$ point is so near the backward extension of the laboratory virgin curve that the field curve can be drawn without use of Casagrande's construction. If the soil is badly disturbed, the laboratory consolidation curve will not have an obvious point of sharp curvature and no reasonable approximation of the field curve can be done. The construction, then, is of greatest value for slightly disturbed or overconsolidated specimens.

If Casagrande's construction indicates that $\bar{p}_c$ exceeds $\bar{p}_c$ by a significant amount, then the soil is overconsolidated. The reconstruction of the field curve then is based on the procedure recommended by Schmertmann.[9] A laboratory curve for a sample of highly plastic, overconsolidated, clay is shown in Fig. 2A-19B. Casagrande's construction is used to estimate the maximum previous consolidation pressure, the effective overburden pressure in the field is calculated, and the point $e_o, \bar{p}_o$ is plotted. For the moment we will assume that the soil is simply overconsolidated, i.e., that it was consolidated to some maximum effective stress in the field and then rebounded directly to the point $e_o, \bar{p}_o$. Further, the field rebound curve is assumed to be parallel to the laboratory rebound curve. Thus, a curve may be drawn parallel to the laboratory rebound curve back from a point on the $\bar{p}_c$ line such that the rebound curve passes through the $e_o, \bar{p}_o$ point. The point on the $\bar{p}_c$ pressure line is then one point on the field virgin consolidation curve. The actual field curve must pass through the point $e_o, \bar{p}_o$, must pass through the pressure $\bar{p}_c$ lower than the "known" point on the field virgin curve, and must remain above the laboratory curve. The reconstructed field curve is drawn in by eye and merged gradually with the laboratory virgin curve (Fig. 2A.19B). Schmertmann[9] suggested certain refinements to the method just suggested.

A number of problems arise when we attempt to apply the foregoing method. First, for many highly overconsolidated soils the reloading curve of even hand-carved samples appears to be a continuous smooth curve, and there is no apparent method for estimating the maximum previous consolidation pressure. Such curves are common for clayey glacial tills. The soil also may have rebounded to a pressure much less than the existing overburden pressure and may now be on a reloading, rather than rebounding, curve.

Based on the foregoing comments, it seems apparent that reconstruction of the field curve is based largely on the judgment of the soils engineer aided by certain constructions. The engineer can reduce uncertainty in estimating the position of the field curve by using relatively undisturbed samples.

2A.7.1.4 Calculation of the Coefficient of Consolidation.

The time required for consolidation of a soil can be computed from

$$T = \frac{c_v t}{H_d^2} \tag{2A.41}$$

where T = time factor
H_d = maximum drainage distance

For the case of a soil layer loaded very quickly and drainage allowed on both sides, the time-settlement curve can be calculated provided that the ultimate settlement, S_u, and the coefficient of consolidation, c_v, are known. The ultimate settlement can be calculated from the reconstructed field consolidation curve using the methods discussed previously. The coefficient of consolidation for use in field analyses is also usually estimated from laboratory consolidation tests.

In the laboratory initial settlement happens nearly instantaneously, which may be caused by elastic compression of the experimental apparatus, seating of the porous stones against improperly trimmed faces of the soil specimen, or compression of gas bubbles in the soil. This rapid settlement, termed *initial compression,* obviously cannot be taken into account by the theory. Thus, an adjustment must be made to the laboratory curve to remove the effects of initial compression.

Two procedures are commonly used for estimating the appropriate values of S_0 and S_{100}. They are designated *Taylor's method*[10] and *Casagrande's method.*[11]

Taylor's Method of Finding c_v. When Taylor's method is used, the settlement is plotted versus the square root of time. A square root versus time curve from a one-dimensional consolidation test is shown in Fig. 2A.20.

To find the corrected initial point for the theoretical curve, extend the linear portion of the laboratory square root of time curve (Fig. 2A.20) back to time zero.

Taylor found that there was no distinct change in the square root of time curve to show where primary and secondary compression merged. In an attempt to find S_{100}, Taylor made the assumption that secondary effects could be ignored of U less than or equal to 90%. Further, he noted that at 90% ultimate consolidation ($U = 90\%$), the abscissa of the laboratory curve would be

$$F\sqrt{t_{90}} = F\left[\frac{H^2}{C_v}\right]^{0.5}(T_{90})^{0.5} \qquad (2A.42)$$

where F is just the scale factor originally used to lay out the time scaler on the graph paper. If the linear portion of the square root of time laboratory curve is extended as a straight line to higher values of U, at U minus 90% the abscissa of this straight line is

$$\left(\frac{9}{5}\right)F\left(\frac{H^2}{C_v}\right)^{0.5}(T_{50})^{0.5} \qquad (2A.43)$$

This abscissa was found by linear extrapolation from the point at which U equals 50%. But the same abscissa is found if the extrapolation is from some other point on the linear portion of the laboratory curve. At the settlement corresponding to U equals 90%, the ratio of the abscissa of the laboratory curve to that of the extended linear curve is

$$\frac{\left[F\left(\frac{H^2}{C_v}\right)^{0.5}(0.848)^{0.5}\right]}{\left[\left(\frac{9}{5}\right)F\left(\frac{H^2}{C_v}\right)^{0.5}(0.197)^{0.5}\right]} \qquad (2A.44)$$

which is 1.15. Thus, if a line is drawn from the point S_0 with abscissa equal to 1.15 times the abscissa of the extended linear portion of the square root of time curve, this new line must intersect the laboratory curve at the point at which U equals 90% (Fig. 2A.20), provided secondary effects are in fact negligible for U less than 90%. At the point U minus 90%, $\Delta S_{90} = 0.9 \Delta S_{100}$; since ΔS_{90} is known, ΔS_{100} is easily calculated.

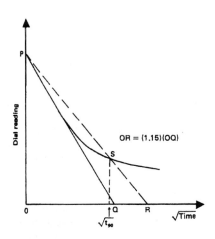

FIGURE 2A.20 Taylor's square root of time plot.

It is convenient to select the point at which U is 45% to calculate c_v because the settlement at this point is the average of the known settlements S_0 and S_{90}, thus simplifying the construction used to find the point, and because selection of a point on the linear portion of the laboratory curve ensures that the theoretical and experimental curves will coincide in the region of greatest practical interest. The time factor at 45% consolidation is 0.159 (Table 2A.11). Note that Taylor recommended use of the U equals 90% point.[9] The coefficient of consolidation is then calculated from:

$$C_v = \frac{0.159 H_d^2}{t_{45}} \qquad (2A.45)$$

where H_d is the average drainage distance during the consolidation period (half the average total thickness for double drainage) and t_{45} is the time corresponding to U equals 45%.

Casagrande's Method of Calculating c_v. When Casagrande's method is used,[11] the settlement is plotted versus the logarithm of time. Curves such as the one shown in Fig. 2A.21 (plotted using the same data previously shown in Fig. 2A.19) are typically obtained. An even spacing of points along the curve is obtained by using a geometric progression of times at which the deformations of the specimen are recorded. Typical times are 6, 15, and 30 seconds; 1, 2, 4, 8, 15, and 30 minutes; and 1, 2, 4, 8, and 24 hours, all measured from the instant of load application.

TABLE 2A.11 Time Factors

U, %	T	U, %	T
10	0.008	55	0.238
15	0.018	60	0.287
20	0.031	65	0.342
25	0.049	70	0.405
30	0.071	75	0.477
35	0.096	80	0.565
40	0.126	85	0.684
45	0.159	90	0.848
50	0.197	95	1.127

The value of S_0 is again obtained as shown on Fig. 2A.21, based on the parabolic approximation of the early part of the theoretical S-t curve:

$$\Delta S_2 = \Delta S_1 \sqrt{\frac{t_2}{t_1}} \qquad (2A.46)$$

If $t_2 = 4t_1$, then, $\Delta S_2 = 2\Delta S_1$ and $S_0 = S_1 - (\Delta S_2 - \Delta S_1)$. Because both Casagrande's method and Taylor's method for finding the corrected zero point are based on the assumption that the early part of the S-t curve is a parabola, they should yield the same corrected zero point. They differ mainly in the fact that the parabolic part of the curve is clearly visible in the square root plot but is masked in the logarithmic plot. Thus, if Casagrande's method is used, t_2 should be chosen at about t_{50} to maximize the possibility that both points will be on the parabolic part of the curve.

The construction used to locate the maximum theoretical settlement, S_{100}, is shown in Fig. 2A.21. On the semilogarithmic plot, the experimental curves do not become asymptotic to a horizontal line, as required by Terzaghi's theory but, instead, often become linear, or nearly linear, with a finite slope. The settlement S_{100} is estimated to be the settlement at the intersection of two straight lines, one drawn tangent to the sloping part of the laboratory curve and the other drawn tangent to the laboratory curve at the point of inflection.

To simplify construction and to make the theoretical and experimental curves coincide in the range of greatest field interest, c_v is calculated using t at U equals 50%.

Note that the two methods give quite different values of S_{100} and thus different values of c_v. Since Taylor's method of finding S_{100} is based on the invalid assumption that secondary effects are negligible prior to 90% consolidation and Casagrande's method has no theoretical justification at all, it is not surprising that they sometimes yield different results.

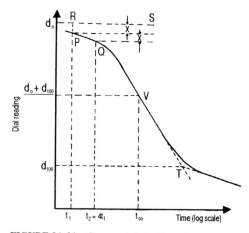

FIGURE 2A.21 Casagrande's log time plot.

2A.8 SHEAR STRENGTH OF SOIL

The shear strength of soil depends on the consolidation pressure, the drainage during shear, the volumetric history (initial relative density of sands or stress history for clays), and other factors such as disturbance, strain rate, and so on.

2A.8.1 Direct Shear Strength of Sand

A schematic section through a direct shear apparatus is shown in Fig. 2A.22. A vertical or normal load is applied. The horizontal shear force, F, is then increased until failure occurs. The average normal stress is defined as σ_n, and the average shearing stress is τ. If the measured shearing stress τ is plotted against the horizontal displacement, Fig. 2A.23 results. The volume change plot for the test specimen is given as Fig. 2A.24.

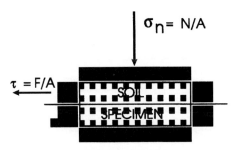

FIGURE 2A.22 Direct shear apparatus.

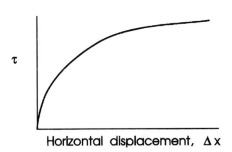

FIGURE 2A.23 Stress-strain for a loose dry sand.

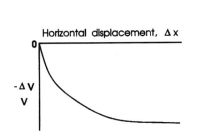

FIGURE 2A.24 Displacement versus volumetric strain for a dry loose sand.

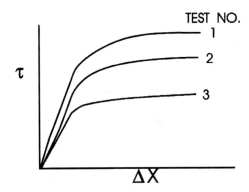

FIGURE 2A.25 Stress-strain curves for direct shear tests at differing vertical stresses.

Now a *series* of drained tests are performed on saturated or dry loose sand, with every test specimen initially at a given loose relative density. However, each test is performed with a different vertical normal stress (Fig. 2A.25). The peak shear stress for each test is called the *shear strength*. Plotting the shear strengths from each test against the normal (vertical) stresses and connecting these failure points results in a failure line, as shown in Fig. 2A.26. This failure line passes through the origin and is inclined at an angle Φ to the horizontal. Since $\tau_{max} = \sigma \tan \Phi$, the phenomenon is analogous to the friction of a block sliding on a plane, and Φ is known as the *friction angle*. The results of a direct shear test performed on a sand with a high initial relative density are shown in Fig. 2A.27 and Fig. 2A.28. The maximum shear stress (shear strength) at a given normal stress occurs at a relatively low displacement. The stress then decreases with increasing displacement. The volume first decreases during initial shearing. At a relatively low deformation, the volume of the specimen increases (dilates) until it reaches equilibrium at high deformations.

If a series of drained tests are performed on a dense sand, with each test having a different applied normal pressure, the peak stress can be plotted against the normal stress. These data can be superimposed on Fig. 2A.26 as shown in Fig. 2A.29.

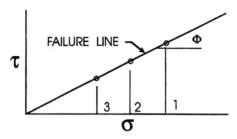

FIGURE 2A.26 Failure diagram for a dry sand tested in direct shear.

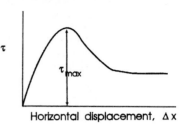

FIGURE 2A.27 Stress-displacement of drained dense sand.

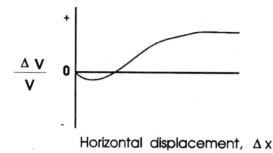

FIGURE 2A.28 Volume change-displacement of drained dense sand.

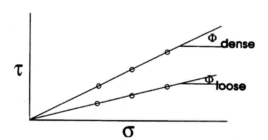

FIGURE 2A.29 Drained strength envelopes of dense and loose sand.

Figure 2A.29 shows that the drained strength of sand is proportional to the confining (normal) stress, but the coefficient of proportionality (Φ angle) is greater for an initially dense sand than for an initially loose sand.

In nature, sands are usually fully drained under static loads, since drainage occurs rapidly. Under these circumstances, confining pressures are generally effective confining pressures, since pore pressure dissipates rapidly. Full drainage occurs during shear.

2A.8.2 Shear Strength of Saturated Clay

With clays, the time to achieve drainage is important. In the direct shear test, the soil is initially loaded with a total vertical stress. The pore pressure generated by this load may be allowed to dissipate, permitting consolidation to occur. With time, substantially all the pore pressure is dissipated; the soil reaches an equilibrium volume under the total stress, which is now an effective stress, since no pore pressures exist.

The triaxial shear apparatus has proven to be superior to the direct shear box for the study of the consolidated-undrained shear strength of a clay. Preceding the discussion of consolidated undrained shear strength, Mohr's stress theory and the mechanics of the triaxial test must be discussed.

2A.8.2.1 Stresses at a Point.

Mohr's theory of stresses is fully explained in texts on mechanics of materials. However, we will review that part of the theory pertinent to soil mechanics. Compressive stress will be considered positive, since stresses in soils generally are compressive and not tensile. Shear stresses tending to cause counterclockwise rotation will be defined as positive as well.

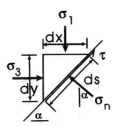

FIGURE 2A.30 Stresses on a differential element.

Consider the stresses on a small two-dimensional element shown as Fig. 2A.30: σ_n is the normal component of the resultant stress on any plane; τ is the component of the resultant stress on the plane that is parallel to the plane, i.e. a shearing stress. σ_1 is the maximum normal stress on any plane. There is no shearing stress on this plane. σ_3 is the minimum normal stress on any plane through the point under consideration. There is no shearing stress on this plane as well. σ_3 acts at right angles to σ_1. σ_2 is the normal stress acting on a plane at right angle to the planes of both σ_1 and σ_3 (out of the page). There is also no shear stress in this plane. By definition, σ_2 may not exceed the magnitude of σ_1 nor may it be less than σ_3. σ_1, σ_2, and σ_3 are called principal stresses. They are orthogonal.

If we consider Fig. 2A.30, we can determine the relationship between the normal and shear stresses, σ and τ, on a plane inclined at an angle, α, to the major principal plane, and the major and minor normal stresses, σ_1 and σ_3:

$$\tau = \frac{\sigma_1 - \sigma_3}{2} \sin 2\alpha$$

$$\sigma = \frac{\sigma_1 + \sigma_3}{2} + \frac{\sigma_1 - \sigma_3}{2} \cos 2\alpha \tag{2A.47}$$

These two equations allow the calculation of the stresses σ and τ on any plane inclined at an angle α to the plane of the major principal plane when σ_1 and σ_3 are known.

This can also be accomplished graphically using Mohr's circle (Fig. 2A.31). The ordinates represent shear stress, and the abscissa, normal stresses.

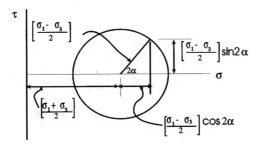

FIGURE 2A.31 Mohr's circle representation of stress on a differential element.

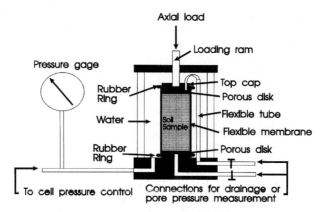

FIGURE 2A.32 Schematic of triaxial test equipment.

2A.8.2.2 Triaxial Tests. The direct shear test does not lend itself to the measurement of soil pore pressures, and the test has several other disadvantages. The triaxial test has become a popular method to determine the shear properties of a soil. The triaxial apparatus is shown in Fig. 2A.32.

Triaxial tests are performed in two stages. The first stage subjects the sample to a system of normal stresses (Fig. 2A.33). Usually this is done by application of an all-around cell pressure ($\sigma_c = \sigma_3$), in which case the stress acts in all directions. If the drainage passage from the porous stone is opened, the sample may be allowed to consolidate under the stresses applied during the first stage. Alternatively, one may want to prevent drainage during the first stage. Stage one can then be either *consolidated* or *unconsolidated*.

In the second stage an axial stress ($\Delta\sigma$) may be applied to the sample through the loading piston. Again, drainage may or may not be permitted. The second stage may then be either *drained* or *undrained*.

Assuming isotropic stresses are applied in the first stage, three types of tests are commonly performed on a soil sample:

1. *Consolidated-drained test.* In the first stage of this test, the soil is permitted to consolidate completely under the influence of the cell pressure. If the sample is saturated, the drainage connection from the porous stone may be connected directly to a burette. The progress of con-

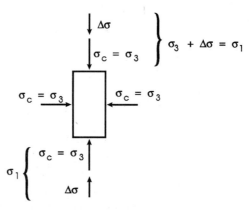

FIGURE 2A.33 Triaxial stresses on specimen.

solidation may be followed by measurement of the water outflow from (or inflow to) the sample. When consolidation is complete, with no further drainage from or into the sample, the second stage may proceed. An axial strain which causes a stress is applied so slowly that the pore pressures generated by the shear are permitted to dissipate. In the literature this test is commonly called a *consolidated-drained test* (CD test), a drained test, a slow test, or an S-test.

2. *Consolidated-undrained test.* In this test, stage one is performed identically to stage one of the preceding test. In the second stage, drainage connections are closed as the sample is sheared to failure under undrained conditions. This test is called a *consolidated-undrained test* (CU test), a *consolidated-quick test* (CQ or QC Test), or an *R-test*.

3. *Unconsolidated-undrained test.* In this test the soil is not permitted to consolidate under the cell pressure, nor is drainage permitted during the shearing stage. This test is known as an *unconsolidated-undrained test* (UU test), a *quick test,* or a *Q-test*.

2A.8.2.3 The Consolidated Drained Strength of Saturated Normally Consolidated Clay.

Since, in this test, no pore pressures are allowed to build up, the stresses in the soil specimen are effective stresses. These stresses can easily be measured or calculated from the measured forces.

In the first stage, the soil is permitted to consolidate completely under the influence of the cell pressure. Mohr's circle for both total and effective stress is a single point along the σ axis with a value equal to the cell pressure, or σ_3.

During the second stage, shear stress is slowly applied by the piston. The applied stress ($\Delta\sigma$), is equal to the diameter of Mohr's circle. The cell pressure, equal to the minor principle stress (σ_3), is not varied during the course of the shear test. The progress of shear can be shown as a series of Mohr's circles increasing in diameter, anchored at their left side on the value of the cell pressure (Fig. 2A.34).

For a single test, a plot of stress difference versus axial strain is shown in Fig. 2A.35. Points 1 through 5 on Fig. 2A.35 show values of the stress difference used to obtain the five Mohr's cir-

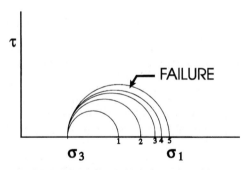

FIGURE 2A.34 Mohr's circles for increasing axial stress.

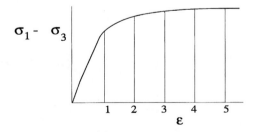

FIGURE 2A.35 Stress difference versus strain for triaxial S test on normally consolidated clay.

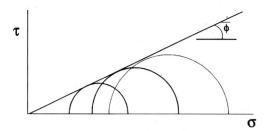

FIGURE 2A.36 S test envelope for normally consolidated clay.

cles shown in Fig. 2A.34. The stress difference obviously causes a shear stress so that the similarity of results to those obtained by direct shear is not surprising.

If a series of drained triaxial shear tests are performed on a normally consolidated soil with the consolidation (cell) pressure varied from test to test, a plot of the Mohr's circles for the maximum stress differences might appear as in Fig. 2A.36.

An envelope line, enclosing all possible Mohr's circles for this normally consolidated, saturated clay soil, will be a straight line and will pass through the origin when extended backward. The angle of inclination of this line is ϕ, approximately equal to the friction angle as measured in the drained direct shear test. The physical interpretation is that the shear strength is directly proportional to the consolidation pressure, i.e.,

$$S = t_f = \overline{\sigma} \tan \phi \tag{2A.48}$$

There is a failure on one plane in the triaxial sample, and this failure is on a plane theoretically represented by the point of tangency of Mohr's circle and the envelope.

2A.8.2.4 Consolidated-Undrained Strength of Saturated Normally Consolidated Clay.
In the consolidated undrained test, the first stage proceeds in the same manner as the consolidation phase of the consolidated drained triaxial test. The soil is consolidated to an isotropic (all around) stress equal to the cell pressure. For the sample to remain normally consolidated, the cell pressure must exceed the consolidation pressure of the soil in the ground. If plotted on a Mohr's diagram, at the end of the consolidation phase, the stresses would plot as a point circle along with the σ axis, since, as with the drained test, the progress of consolidation may be monitored. Upon completion of consolidation, the shear phase begins. To maintain an undrained condition drainage from this sample is prevented. Since no water leaves the system, the shearing occurs at constant volume.

A volume-change-versus-strain curve is not appropriate because there is zero volume change at all strains. On the Mohr's diagram, Fig. 2A.37, the diameter of the failure circle for the undrained test is about half the diameter of the failure circle for the drained case at the same consolidation stress.

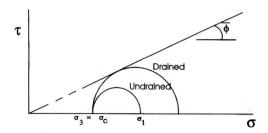

FIGURE 2A.37 Mohr's circle for undrained test compared to drained test.

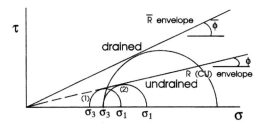

FIGURE 2A.38 Mohr's circles for increasing consolidation pressures in the undrained test.

A second sample is consolidated to a higher stress and sheared under constant volume. The corresponding Mohr's circle (Fig. 2A.38) shows a proportionately larger failure circle, but again only about one-half the diameter of the drained failure circle at the higher consolidation stress. Since this proportionality exists, one may draw a failure envelope.

This envelope is also a straight line, which can be traced backward through the origin. It is known as the *R* or *CU envelope*. The dotted line portion shows extrapolation back to the origin, at stresses below the field consolidation pressure of the soil. If, in the laboratory, a sample had been consolidated to a stress less than the field consolidation stress, the sample would no longer be normally consolidated, since stresses were at one time greater than now. Its strength would also be slightly greater than shown by the dotted line, as will be explained in the section on overconsolidated soils.

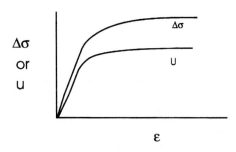

FIGURE 2A.39 Typical stress and pore pressure versus strain in undrained shear.

With only a simple adjustment to the R test, one can obtain a great deal more data. Instead of closing the drainage connection to the base of the sample, if, at the end of consolidation, a pore pressure transducer is placed in the line, one can still run a test essentially undrained during the shearing phase. But the pore pressures caused by the shearing can be constantly monitored. This test is known as the $\bar{R}$ (R-bar) *test*.

After consolidation, as the soil is sheared under the stress difference, there is a tendency for the soil to decrease in volume as with the drained test. But drainage is prevented; water cannot leave the soil. The pore water pressure gradient which existed to move the water out of the soil is now permitted to build up. A positive pore water pressure is generated and is measured by the transducer. Typical stress and porewater pressure versus strain curves are shown in Fig. 2A.39.

During drained shear it was not necessary to differentiate between total stresses and effective stresses. The pore pressures were zero; total stresses were equal to the effective stresses, and a single Mohr's circle results.

For the R test, the consolidation phase (which implies full drainage) yields a single-point Mohr's circle equal to the cell pressure. However, the moment the undrained shear begins, pore pressures are generated, and

$$\bar{\sigma}_1 = \sigma_1 - u$$
$$\bar{\sigma}_3 = \sigma_3 - u \tag{2A.49}$$

Since the cell pressure is constant and equal to σ_3 in the triaxial compression test, the total stress Mohr's circles are anchored on the left at $\sigma_3 = \sigma_c$, just as in the drained test. However, the

FIGURE 2A.40 Total and effective stress circles for undrained test.

effective stress σ_3 becomes less than the cell pressure by the magnitude of pore pressure generated. Since pore pressure constantly increases with shear, the effective stress Mohr's circles are constantly shifting to the left, i.e., σ_3 is constantly decreasing. For any given time or strain, there are two Mohr's circles, a total stress circle with $\sigma_3 = \sigma_{cell}$ and an effective stress circle with σ_3 and σ_1, displaced leftward (reduced) by the amount of the generated pore pressure at that time or at that given strain. Figure 2A.40 illustrates this.

The total stress Mohr's circle will have the same diameter as the effective stress circle:

$$\Delta\sigma = \overline{\sigma}_1 - \overline{\sigma}_3 = \text{diameter of Mohr's circle}$$
$$= (\sigma_1 - u) - (\sigma_3 - u) \qquad (2A.50)$$
$$= \sigma_1 - u - \sigma_3 + u$$
$$= \sigma_1 - \sigma_3$$

The effective stress and total stress Mohr's circles will be identical in size, and all points on the effective stress circle are simply displaced by the magnitude of the pore pressure.

In Fig. 2A.40 note that shear continues until the effective stress Mohr's circle touches the effective stress envelope. This envelope is essentially the same as the drained envelope.

In the drained direct shear test, the normal effective stress remains constant as shear stress is increased to failure. In the drained triaxial test, normal effective stress on the potential plane of failure increases slightly as shear stresses increase rapidly. In the consolidated undrained triaxial test, normal effective stresses generally decrease somewhat due to generation of pore pressures, and shear stresses increase to failure.

It should be quite clear now that effective stresses at failure govern the strength of a soil. For a given normally consolidated clay, the effective normal stress on a potential failure plane governs the shear strength of the soil, both drained and undrained. Even though this effective stress concept is relatively simple to explain, in practice we more often wish to relate the strength to stresses *before* shearing. Relating undrained strength to consolidation pressure before shearing is more useful for engineering predictions.

Previously the results of performing R tests (without pore pressure measurement) on two samples consolidated to different pressures were explained. We now perform two R tests (with pore pressure measurement) on samples consolidated to different pressures. Results are

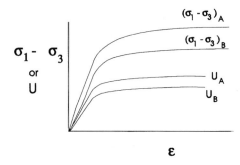

FIGURE 2A.41 Stress and pore pressure vs. strain for two tests at two consolidation pressures.

shown in Figs. 2A.41 and 2A.42. Note that the total stress circles, with σ_3 equal to the cell pressures, form the R envelope. The effective stress circles each touch the $\overline{R}$ or effective stress envelope, which is essentially the same envelope as would be obtained from drained tests. Note also that for the larger consolidation pressure, larger pore pressures are generated. With increasing consolidation pressure, Mohr's circle at failure (the shear strength) increases proportionately, as does the failure pore pressure, which is shown by the offset of the effective stress circle from its total stress circle. In each case, the offset is approximately equal to the diameter of Mohr's circle. Only the R envelope represents a failure line, where stresses become critical on a plane on which a given effective stress acts. The R envelope is simply an envelope enclosing the maximum size Mohr's circle.

Sometimes the R envelope is called a *total stress envelope*. This leads to confusion. Total stress conditions are meaningless unless they are related to effective stresses. The R envelope is an envelope of total stresses at undrained failure only if and when the σ_3 of the failure circle is the consolidation pressure of the sample. In this way total stresses bear a fixed relationship to effective stresses.

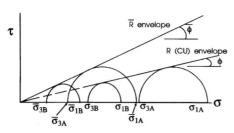

FIGURE 2A.42 Mohr's circles for two tests.

Figure 2A.42 also shows, for a given consolidation pressure, the drained strength and failure circle and both the undrained failure circles and the undrained strength. Now a main advantage of the $\overline{R}$ test becomes evident. From the $\overline{R}$ envelope, we can predict the drained strength of a soil at any given consolidation pressure. The $\overline{R}$ test gives predictions of both the drained and undrained strengths. Moreover, the drained test requires a very slow shear rate so that pore water pressures generated in the failure zone can migrate from the sample. The $\overline{R}$ test still requires time for pore pressures in the failure zone to equilibrate throughout the sample, but the time involved is substantially less than the time for full drainage.

From the foregoing we can conclude that for a normally consolidated saturated clay (constant stress history) for the drained condition, shear strength is directly proportional to the consolidation pressure. For the undrained case the same is true, but the constant of proportionality is reduced to tan Φ.

2A.8.2.5 Unconsolidated-Undrained Test and Unconfined Compression Test of Normally Consolidated Clay. The undrained strength of the soil in this test is also directly proportional to the consolidation pressure which is the consolidation pressure of the soil in the ground.

In this test no drainage is allowed, and hence no volume change in the specimen can occur. Consequently, when the sample is sheared, identical effective stress circles will develop no matter what the value of the cell pressure. At failure, the strength will also be the same for all tests. The failure total stress circle will be the same size as the effective stress circle, but displaced by the greater pore pressure (Fig. 2A.43). A Mohr's plot for total stresses for the UU tests have resulted in a series of circles of the same diameter. The UU envelope, with respect to total stresses, has a Φ of zero and cohesion intercept c (Fig. 2A.44). There may be a number of total stress circles, but only one effective stress circle and this one is governed by the common consolidation pressure.

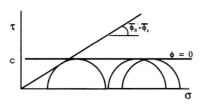

FIGURE 2A.43 Total and effective stress circles for the UU test.

Both the cost of triaxial apparatus and the requirement for highly skilled labor used in triaxial testing are large. Since the strength of the UU sam-

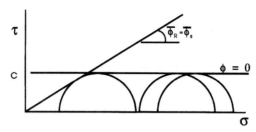

FIGURE 2A.44 UU envelope.

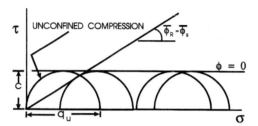

FIGURE 2A.45 UU envelope and unconfined compression Mohr's circle.

ple was dependent only on the consolidation stress in the ground and independent of the cell pressure, the use of a zero confining stress dispenses with the need for much of the triaxial apparatus and should result in a Mohr's circle of equal size by the $\Phi = 0$ principle. If we perform such a test, called the *unconfined compression test,* ideally the resulting Mohr's circle is of size equal to those failure circles produced by the unconsolidated-undrained test (Fig. 2A.45). In such a test we cannot measure pore pressures, so the position of the failure effective stress circle cannot be established. Although we have deduced the radius of the effective stress Mohr's circle (it is the same as the total stress circle radius), we cannot know what is the effective stress friction angle of the soil. The radius is also half the diameter of the circle, which is the unconfined compressive strength, q_U, which in turn is σ_1 at failure when σ_3 is zero. In general then, for undrained shear strength of cohesive soils:

$$S = \tau_f = C = \frac{q_u}{2} = \frac{\Delta\sigma}{2} \qquad (2A.51)$$

2A.8.2.6 Shear Strength of Saturated Overconsolidated Clays. A reconsolidated or overconsolidated soil at some time in its geologic history has had acting on it a consolidation pressure greater than that presently acting on it. Consolidation, of course, implies complete drainage and transfer of total stresses to the effective stresses of the soil skeleton. The causes of the preconsolidation and subsequent removal of effective stress might have been by deposition and subsequent erosion, a glacial load and melting, desiccation with subsequent more humid conditions, or any of the other causes discussed under the subject of consolidation.

Figure 2A.46 shows, on a natural scale of void ratio versus effective stress, a soil normally consolidated to point C and unloaded to point E. This plot is similar to the usual consolidation curve except that σ is shown as the effective stress on a natural scale and not log p, effective stress plotted on a logarithmic scale. What would be a relatively straight line on a logarithmic plot will appear concave upward on an arithmetic plot. This curve is a consolidation curve, even though the data were obtained from triaxial consolidation tests where $\sigma_1 = \sigma_2 = \sigma_3$.

In Fig. 2A.46 points A, B, and C are normally consolidated. Points D and E correspond to the consolidation pressures of B and A, respectively, but each of these soils has been reconsolidated to the stress level of C. Figure 2A.46 shows Mohr's failure circles for soils A–E for the consolidated drained condition. The diameter of the failure circles and the shear strength of the soil are directly proportional to the consolidation pressure, which is the σ_3 of the drained failure circles.

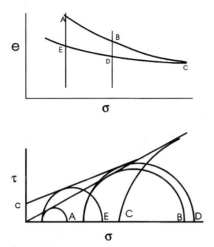

FIGURE 2A.46 Volume change and Mohr's circle plots for overconsolidated soils.

It should be expected that density expressed as void ratio and effective consolidation stress will control the strength of these soils. Soil D has been consolidated to the same effective stress as soil B. Soil D, however, is shown in Fig. 2A.46 to be denser, to have a lower void ratio than B. It would therefore be expected that D will be stronger than B. Figure 2A.46 shows that this is so.

Soil E is denser than A, and both have been consolidated to the same effective stress level. Soil E was at one time reconsolidated (with drainage, of course) to the stress level of C. In rebound it had only a slightly greater void ratio than C. The ratio of the σ_3 of E and the σ_3 of C is the overconsolidation ratio which is shown to be on the order of 8 or so. Again, soil E is stronger than A, as denoted by the larger Mohr's failure circle.

In comparing the failure circles, consolidation stresses, and void ratios, we see that the strength of these soils is controlled more by effective consolidation stresses than by void ratios.

The failure envelope through the origin and tangent to Mohr's circles *A, B,* and *C* was previously introduced as the drained strength envelope for a normally consolidated clay, or the S envelope. It indicates that the shear strength is directly proportional to the effective consolidation pressure and in particular to the consolidation pressure on the plane of failure, i.e., where the Mohr's circle touches the envelope.

The failure envelope for soils C–E, the soils which have been reconsolidated to the stress level of *C,* is not strictly a straight line; it is curved downward slightly, and does not pass through the origin. It joins the Φ_s envelope for normally consolidated soil at the point where the failure circle for *C* is tangent to the normally consolidated failure circle. When tests are performed on a reconsolidated soil, the final consolidation pressures generally are in a relatively narrow range and are generally substantially less than the preconsolidation pressure. In this range it is customary to fit a straight line to the circles obtained from the tests. Such a line will have a slope expressed as a Φ angle and will have a cohesion intercept *c*. Figure 2A.47 shows such an effective stress envelope.

The greater the preconsolidation pressure, the higher is the envelope and therefore the greater is *c* in the expression $\tau = c + \sigma \tan \Phi$. It can be seen also that *c* and Φ are not material properties of a soil.

Typical stress-strain and volume change–strain properties of heavily reconsolidated soils are shown in Figs. 2A.48 and 2A.49 respectively. The soil behaves very much the same as does a dense sand, having a peak shearing resistance at low strain and dilating or expanding in volume at strains greater than those mobilized at the peak strength.

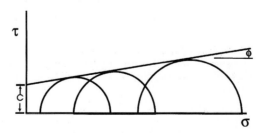

FIGURE 2A.47 Overconsolidated soil Mohr's envelope.

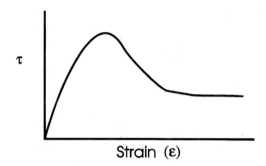

FIGURE 2A.48 Stress-strain for overconsolidated clay.

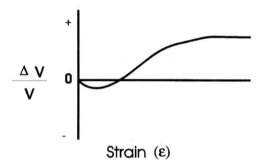

FIGURE 2A.49 Strain-volume change for overconsolidated clay.

2A.9 REFERENCES

1. A. W. Skempton, "Notes on the Compressibility of Clays," *Quarterly Journal of the Geological Society of London,* pp. 119–135 (1944).
2. Melvin I. Esrig, unpublished notes (1968).
3. Karl Terzaghi, "Influence of Geological Factors on the Engineering Properties of Sediments," *Harvard Soil Mechanics Series No. 50,* Cambridge, Massachusetts (1955).
4. G. W. Carpenter and R. W. Stephenson, "Permeability Testing in the Triaxial Cell," *ASTM Geotechnical Testing Journal,* vol. 9, no. 1, March (1986).
5. J. Boussinesq, "Application des Potentials a L'Etude de L'Equilibre et du Mouvement des Solides Elastiques," Gauthier-Villars, Paris (1885).
6. A. Casagrande, "The Structure of Clay and Its Importance in Foundation Engineering," Contributions to Soil Mechanics, BSCE 1925–1940 (1932).
7. P. C. Rutledge, *Transactions of ASCE,* vol. 109 (1944).
8. T. W. Van Zelst, in *Proceedings of the 2d International Conference on Soil Mechanics and Foundation Engineering,* vol. 7 (1948).
9. A. Casagrande, "The Determination of the Preconsolidation Load and Its Practical Significance," *Proceedings of the First International Conference on Soil Mechanics and Foundation Engineering,* Cambridge, Mass. (1936).
10. J. Schmertmann, "The Undisturbed Consolidation of Clay," *Transactions ASCE,* vol. 120 (1955).
11. D. W. Taylor, *Fundamentals of Soil Mechanics,* John Wiley & Sons, New York (1948).
12. A. Casagrande and R. E. Fadum, "Application of Soil Mechanics in Designing Building Foundations," *Transaction of ASCE,* vol. 109 (1944).

SECTION 2B
SHALLOW FOUNDATIONS

Richard W. Stephenson

2B.1 INTRODUCTION   2.45
2B.2 SHALLOW FOUNDATION TYPES   2.45
    2B.2.1 Footings   2.45
    2B.2.2 Mats   2.46
2B.3 DESIGN PARAMETERS   2.47
    2B.3.1 Structural Design Parameters   2.47
    2B.3.2 Geotechnical Design Parameters   2.47
2B.4 BEARING CAPACITY OF SHALLOW FOUNDATIONS   2.48
    2B.4.1 Development of General Bearing Capacity Equation   2.48
    2B.4.2 Choice of Unit Weight, c and Φ 2.50
2B.4.3 Influence of Water Table   2.50
2B.4.4 General Bearing Capacity Equation   2.51
2B.5 FACTOR OF SAFETY   2.54
2B.6 BEARING CAPACITY OF MAT FOUNDATIONS   2.54
2B.7 SETTLEMENT OF SHALLOW FOUNDATIONS   2.55
    2B.7.1 Allowable Settlements   2.55
    2B.7.2 Elastic (Immediate) Settlement   2.56
    2B.7.3 Consolidation Settlement of Foundations on Clays   2.67
2B.8 REFERENCES   2.68

2B.1 INTRODUCTION

A foundation is a structure built to transfer the weight of a building to the material below. This transfer must occur such that the soil below does not rupture or compress to such magnitude that the integrity of the superstructure is threatened.

Foundations are generally classified as either *deep* or *shallow*. The depth of the bearing area of shallow foundations generally is no deeper than about the width of the bearing surface. Deep foundations provide support for a structure by transferring the loads to competent soil and/or rock at some depth below the structure.

2B.2 SHALLOW FOUNDATION TYPES

Shallow foundations can be either footings or mats. They consist of reinforced concrete slabs formed directly on a prepared soil base. Footings may be spread, combined, or continuous.

2B.2.1 Footings

Spread footings support one column or load. These footings are also called *isolated* or *column* footings. They typically are 3 to 8 to 10 ft (0.9 to 2.4 to 3 m) square. Their bearing surface is typically less than 2.5 times their width ($<2.5B$) (Fig. 2B.1).

Combined footings are similar to spread footings but support two or more columns. The shape is more likely to be a rectangle or occasionally trapezoidal (Fig. 2B.2). These footings are used for column spacing that is nonuniform and for the support of exterior columns near property lines where there isn't enough room for a spread footing.

A *continuous* or *strip footing* is an elongated shallow foundation that typically supports a single row of columns or a wall or other type of strip loading. Continuous footings tie columns in

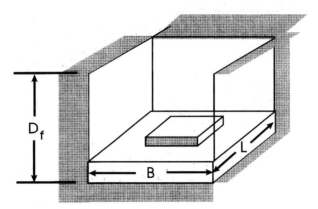

FIGURE 2B.1 Isolated shallow footing.

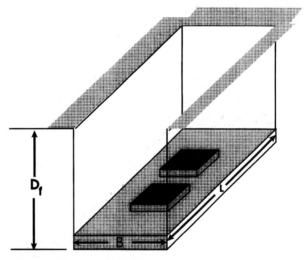

FIGURE 2B.2 Combined footing.

one direction at their base and reduce construction costs through use of appropriate equipment for trenching (Fig. 2B.3).

The advantages of shallow foundations lie primarily in their low cost and speed of construction.

2B.2.2 Mats

A mat (raft) foundation is a structural reinforced concrete slab that supports a number of columns distributed in both horizontal directions or that supports uniform pressure as from a tank. Rafts are used to bridge over soft spots if the spots are very localized and to reduce the average pressure applied to the soil (Fig. 2B.4).

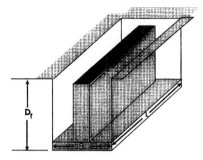

FIGURE 2B.3 Continuous footing.

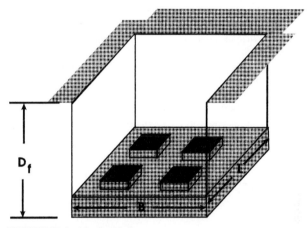

FIGURE 2B.4 Mat foundation.

2B.3 DESIGN PARAMETERS

Design parameters for shallow foundations fall into two classes: structural design parameters and geotechnical design parameters.

2B.3.1 Structural Design Parameters

Structural design parameters that influence the design of the shallow foundation include the building type and use, loading (live, dead, and uplift), column spacing, presence or absence of a basement, allowable settlement, and applicable building codes.

2B.3.2 Geotechnical Design Parameters

Geotechnical factors that influence the design include the thickness and lateral extent of bearing strata, the depth of frost penetration, the depth of seasonal volume change, and the cut/fill requirements. The strength, compressibility, and shrink-swell potential of the bearing strata are the prop-

2.48 SOIL MECHANICS AND FOUNDATION DESIGN PARAMETERS

erties of concern. In addition, the presence or absence of groundwater and its minimum and maximum elevations have an important impact on the design process.

2B.4 BEARING CAPACITY OF SHALLOW FOUNDATIONS

The design of a shallow foundation requires that the applied load does not exceed the load that would cause the soil strata beneath the foundation to rupture. The maximum load that can be applied to the foundation soil without rupture is called the *bearing capacity*.

2B.4.1 Development of General Bearing Capacity Equation

Simplifying the model of a shallow, continuous footing at impending failure, as in Fig. 2B.5, allows the problem to be treated as an earth pressure problem. When the footing is loaded, the wedge of soil beneath the footing generates lateral pressures at the wedge boundaries. These pressures cause the adjacent wedges to displace as shown in Fig. 2B.5. The slip surfaces for the wedge beneath the footing develop slip lines at $\alpha = 45 + \Phi/2$ with the horizontal. The adjacent wedge has slip line angles of $\rho = 45 - \Phi/2$ with the horizontal. If the effect of the soil above the base level of the footing is replaced with a surcharge γD_f, then

$$\bar{q} = \gamma D_f \tag{2B.1}$$

A look at the stress block on the right allows computations of the total resisting earth pressure as force P_p from Eq. (2B.2):

$$P_p = \int_0^H \sigma_1 \, (dz) = \int_0^H \left[(\gamma z + \bar{q}) \tan^2\left(45 + \frac{\phi}{2}\right) + (2c) \tan\left(45 + \frac{\phi}{2}\right) \right] dz \tag{2B.2}$$

where σ_1 is

$$\sigma_1 = \sigma_3 N_\phi + 2c N_\phi \tag{2B.3}$$

Defining K_p as in Eq. (2B.4), the result of the integration is given in Eq. (2B.5):

$$K_p = N_\phi = \tan^2\left(45 + \frac{\phi}{2}\right) \tag{2B.4}$$

$$P_p = \frac{\gamma H^2}{2} K_p + \bar{q} H K_p + 2 \, cH \sqrt{K_p} \tag{2B.5}$$

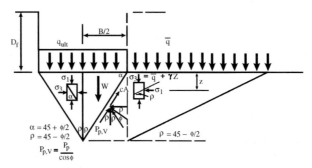

FIGURE 2B.5 Simplified bearing capacity for a c-Φ soil.

To find q_{ult}, the vertical forces are added for the half wedge to obtain

$$q_{ult}\left(\frac{B}{2}\right) + \gamma\left(\frac{B}{2}\right)\left(\frac{H}{2}\right) - (cA)\cos\rho - \frac{P_p}{(\sin\rho)(\cos\phi)} = 0 \qquad (2B.6)$$

since

$$H = \left(\frac{B}{2}\right)\tan\alpha \qquad W = \frac{1}{2}\gamma\left(\frac{B}{2}\right)\left(\frac{B}{2}\right)\tan\alpha$$

$$A = \frac{B}{2\cos\alpha} \qquad K_p = \tan^2\left(45 + \frac{\phi}{2}\right) \qquad (2B.7)$$

$$P_{p,v} = \frac{P_p}{\cos\phi} \qquad K_a = \tan^2\left(45 - \frac{\phi}{2}\right)$$

then

$$q_{ult} = C\left[\frac{2K_p}{\cos\phi} + \sqrt{K_p}\right] + \bar{q}\,\frac{\sqrt{K_p}K_p}{\cos\phi} + \frac{\gamma B}{4}\left[\frac{K_p^2}{\cos\phi} - \sqrt{K_p}\right] \qquad (2B.8)$$

This equation can be written as

$$q_{ult} = cN_c + \bar{q}N_q + \gamma B N_\gamma \qquad (2B.9)$$

where

$$N_c = \left[\frac{2K_p}{\cos\phi} + \sqrt{K_p}\right]$$

$$N_q = \frac{\sqrt{K_p}K_p}{\cos\phi} \qquad (2B.10)$$

$$N_\gamma = \frac{1}{4}\left[\frac{K_p^2}{\cos\phi} - \sqrt{K_p}\right]$$

This is the *general bearing capacity equation*. The N factors are the bearing capacity factors:

N_c = Nondimensional bearing capacity factor relating the influence of soil cohesion on bearing capacity (a function of Φ of the soil).

N_q = Nondimensional bearing capacity factor relating the influence of soil overburden on bearing capacity (a function of Φ of the soil).

N_γ = Nondimensional bearing capacity factor relating the influence of soil unit weight on bearing capacity (a function of Φ of the soil).

Equation (2B.9) generally underestimates the capacities of footings.

Various investigators have studied the bearing capacity problem. Each has made assumptions as to the character of the failure surface, the effect of the footing depth and shape, and other factors. Although almost all the investigators developed equations similar to Eq. (2B.9), the investigators computed different values of the bearing capacity factors. Some researchers included modifications to account for the footing depth, shape, and inclination of loading. Bearing capacity equations reported by several authors are given in Bowles.[1]

2B.4.2 Choice of Unit Weight, c and Φ

The choice of which soil parameters to use for a given design is primarily related to the soil's hydraulic conductivity in relation to the rate of foundation loading.

2B.4.2.1 Cohesive (Clay) Soils. For cohesive soils, the assumption is usually made that the loads are applied much more rapidly than the soil can drain. Consequently, $\Phi = 0$ and $N_c = 5.14$, $N_q = 1.00$, and $N_\gamma = 0$, and

$$q_{ult} = cN_c + \gamma D_f \tag{2B.11}$$

where γ = total unit weight

2B.4.2.2 Cohesionless (Drained) Soil. For soils that will drain rapidly, the use of effective stress parameters is suggested, i.e., use Φ and $\bar{c}$.

2B.4.3 Influence of Water Table

The presence of a water table affects Eq. (2B.9), depending upon where the water table lies with respect to the bearing surface of the footing (Fig. 2B.6). Consider the following conditions:

2B.4.3.1 Case I: $D_w = D_f$. In this case, the maximum level of the water table is at the base of the footing, i.e., $D_w = D_f$. Therefore, the soil above the footing base is at its natural moisture content, and the soil below the footing base is submerged. Therefore, the unit weight in the N_γ term should be the submerged unit weight, and the unit weight in the N_q term should be the total unit weight:

$$q_{ult} = cN_c + (\gamma_{sat} D_f)N_q + \gamma' BN_\gamma \tag{2B.12}$$

where γ' = submerged unit weight

2B.4.3.2 Case II: $D_w = 0$. In this case, the water table is at the ground surface. Therefore, the unit weight in both terms is the submerged unit weight:

$$q_{ult} = cN_c + (\gamma' D_f)N_q + \gamma' BN_\gamma \tag{2B.13}$$

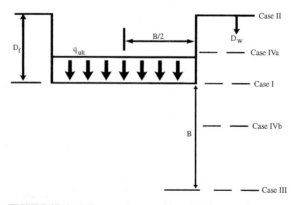

FIGURE 2B.6 Influence of water table on bearing capacity.

2B.4.3.3 Case III: $D_w = D_f + B$. In this case, the maximum height of the water table is a distance B below the base of the footing, D_f. The unit weight in both terms is the total unit weight:

$$q_{ult} = cN_c + (\gamma D_f)N_q + \gamma B N_\gamma \qquad (2B.14)$$

2B.4.3.4 Intermediate Cases. For cases where $0 < D_w < D_f$, the unit weight in N_γ term is the submerged unit weight. A weighted average is used for the unit weight in N_q terms:

$$\gamma = \gamma' + \left[\frac{D_w}{D_f}\right][\gamma - \gamma'] \qquad (2B.15)$$

For cases where $D_f < D_w < [D_f + B]$, the unit weight in N_γ terms is

$$\gamma' + \left[\frac{D_w - D_f}{B}\right][\gamma - \gamma'] \qquad (2B.16)$$

The unit weight in N_q terms is the total unit weight.

2B.4.3.5 Approximate Procedure. Peck, Hanson, and Thornburn[2] developed a procedure to adjust the bearing capacity to account for the presence of the water table. In their procedure, first solve for q_{ult} for no water table within $D_w = D_f + B$. Then set $q = C_w q_{ult}$ where

$$C_w = 0.5 + 0.5 \frac{D_w}{D_f + B} \qquad (2B.17)$$

2B.4.4 General Bearing Capacity Equation

Equation (2B.9) can be generalized to account for various geometric and loading factors:

$$q_u = cN_c s_c d_c i_c + \gamma D_f N_q s_q d_q i_q + \tfrac{1}{2} \gamma B N_\gamma s_\gamma d_\gamma i_\gamma \qquad (2B.18)$$

where s_c, s_q, s_γ = shape factors
d_c, d_q, d_γ = depth factors
i_c, i_q, i_γ = load inclination factors

These factors are discussed below.

2B.4.4.1 Bearing Capacity Factors. Various investigators have published solutions to the bearing capacity equation. Each has provided equations for the bearing capacity factors. A selection of the most useful factors is presented below:
From Meyerhof[3]:

$$N_q = \tan^2\left(45 + \frac{\phi}{2}\right) e^{\pi \tan \phi} \qquad (2B.19)$$

$$N_c = (N_q - 1) \cot \phi \qquad (2B.20)$$

$$N_\gamma = (N_q - 1) \tan(1.4 \phi) \qquad (2B.21)$$

From Hansen[4]:

$$N_\gamma = 1.5 (N_q - 1) \tan \phi \qquad (2B.22)$$

From Vesic[5]:

$$N_\gamma = 2(N_q + 1)\tan\phi \tag{2B.23}$$

2B.4.4.2 Shape Factors. Although the derivation of the bearing capacity equation was in two dimensions, the problem obviously is three-dimensional. Therefore, the relationship between the width, B, and the length, L, on the bearing capacity is of great importance. In general, this relationship is given by application of shape factors applied to the appropriate component of the general bearing capacity equation. Listed below are shape factors reported by various authors.

$$s_c = 1 + \left(\frac{B}{L}\right)\left(\frac{N_q}{N_c}\right) \quad (\phi > o) \qquad s'_c = 0.2\frac{B}{L} \quad (\phi = 0)$$

$$s_c = 1 + 0.2 K_p \frac{B}{L} \tag{2B.24}$$

$$s_c = 1.0 \text{ (strip)} = 1.3 \text{ (round)} = 1.3 \text{ (square)}$$

$$s_q = 1 + \left(\frac{B}{L}\right)\tan\phi$$

$$s_q = 1 + 0.1 K_p \frac{B}{L} \quad (\phi > 10°) \tag{2B.25}$$

$$s_q = 1 \quad (\phi = 0)$$

$$s_\gamma = 1 - 0.4\left(\frac{B}{L}\right)$$

$$S_\gamma = 1 + 0.1 K_p \frac{B}{L} \quad (\phi > 10°) \tag{2B.26}$$

$$S_\gamma = 1 \quad (\phi = 0)$$

$$S_\gamma = 1 \text{ (strip)} = 0.6 \text{ (round)} = 0.8 \text{ (square)}$$

2B.4.4.3 Depth Factors. The depth of placement of the shallow footing below the surrounding ground surface also has a significant impact on the footings' bearing capacity. In particular, this depth influences the length of the failure surface available to resist movement. These factors are usually presented in a nondimensional format similar to that of shape factors. A listing of the most useful depth factors are given below.

$$d_c = 1 + 0.4\left(\frac{D_f}{B}\right) \text{ for } \left(\frac{D_f}{B} \leq 1\right)$$

$$d_c = 1 + 0.4\left(\tan^{-1}\frac{D_f}{B}\right) \text{ for } \left(\frac{D_f}{B} > 1\right) \tag{2B.27}$$

$$d_c = 0.4\frac{D}{B} \text{ or } 0.4\left(\tan^{-1}\frac{D_f}{B}\right) \text{ for } \phi = 0$$

$$d_c = 1 + 0.2\sqrt{K_p}\frac{D_f}{B}$$

$$d_q = 1 + 2 \tan \phi (1 - \sin \phi)^2 \left(\frac{D_f}{B}\right) \text{ for } \left(\frac{D_f}{B} \leq 1\right)$$

$$d_q = 1 + 2 \tan \phi (1 - \sin \phi)^2 \left(\tan^{-1}\frac{D_f}{B}\right) \text{ for } \left(\frac{D_f}{B} > 1\right)$$

(2B.28)

$$d_q = 1 + 0.1 \sqrt{K_p} \frac{D_f}{B} \text{ for } (\phi > 10°)$$

$$d_q = 1 \text{ for } (\phi = 0)$$

$$d_\gamma = 1$$

$$d_\gamma = 1 + 0.1 \sqrt{K_p} \frac{D_f}{B} \text{ for } (\phi > 10°)$$

(2B.29)

$$d_\gamma = 1 \text{ for } (\phi = 0)$$

Note that the factor $\tan^{-1}[D_f/B]$ is in radians.

2B.4.4.4 Inclination Factors. Inclined loading significantly reduces the capacity of a shallow foundation. This reduction is determined by the use of inclination factors where β = inclination of the load on the foundation with respect to the vertical (Fig. 2B.7):

$$i_\gamma = \left(1 - \frac{\beta}{\phi}\right)^2 \quad \text{(Meyerhof)}$$

$$i_\gamma = \left(1 - \frac{0.7 H}{V + A_f c_a \cot \phi}\right)^5 \text{ for } (\eta = 0)$$

$$i_\gamma = \left(1 - \frac{(0.7 - \eta°/450) H}{V + A_f c_a \cot \phi}\right)^5 \text{ for } (\eta > 0)$$

(2B.30)

$$i_\gamma = \left(1 - \frac{H}{V + A_f c_a \cot \phi}\right)^{m+1}$$

where η = tilt angle from horizontal with (+) upward

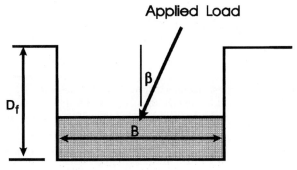

FIGURE 2B.7 Inclined loading.

$$i_c = i_q = \left(1 - \frac{\beta}{90°}\right)^2$$

$$i_c = i_q - \frac{1 - i_q}{N_q - 1}$$

$$i_c = 0.5 - 0.5\sqrt{1 - \frac{H}{A_f c_a}} \quad \text{for } (\phi = 0)$$

$$i_c = 1 - \frac{mH}{A_f c_a N_c} \quad \text{for } (\phi = 0) \tag{2B.31}$$

where $A_f = B' \times L'$
c_a = adhesion
H = horizontal component of footing load
V = total load on footing

$$m = m_B = \frac{2 + B/L}{1 + B/L} \quad \text{for } (H \text{ parallel to } B)$$

$$m = m_L = \frac{2 + L/B}{1 + L/B} \quad \text{for } (H \text{ parallel to } L)$$

2B.5 FACTOR OF SAFETY

Because of the inherent uncertainty of the bearing capacity analysis, the usual practice is to reduce the applied stress from the foundation load by some arbitrary factor. This reduction is usually presented as a factor of safety. The reduced bearing capacity is known as the allowable bearing capacity, q_{allow}.

The allowable foundation stress (q_{allow}) is

$$q_{allow} = \frac{q_{ult}}{FS} \tag{2B.32}$$

where FS is the factor of safety. In general, the factor of safety runs from 2 to 3.

2B.6 BEARING CAPACITY OF MAT FOUNDATIONS

The ultimate bearing capacity of mat foundations is calculated using Eq. (2B.18):

$$q_u = cN_c s_c d_c i_c + \gamma D_f N_q s_q d_q i_q + \tfrac{1}{2} \gamma BN_\gamma s_\gamma d_\gamma i_\gamma \tag{2B.33}$$

However, the great advantage of mat foundations is that by excavating below the ground surface for the placement of the mat, the net allowable applied load from the structure is increased. If Q is the total of the dead and live loads applied to the base of the mat, then

$$q_{net\ applied} = \frac{Q}{A} - \gamma D_f \tag{2B.34}$$

where A is the area of the mat.

It is possible to place the footing at depth $D_{critical}$ such that the net applied load is zero, i.e.:

$$q_{net\ applied} = 0 = \frac{Q}{A} - \gamma D_{critical} \tag{2B.35}$$

$$D_{critical} = \frac{Q}{A\gamma}$$

For this condition, the factor of safety is

$$FS = \frac{q_{ult}}{q_{net\ allowable}} = \frac{q_{ult}}{0} = \infty \tag{2B.36}$$

2B.7 SETTLEMENT OF SHALLOW FOUNDATIONS

Although the analysis and design of foundations usually begins with the study of the bearing capacity of the foundation-soil system, in general the settlement of the foundation controls the design.

2B.7.1 Allowable Settlements

The amount of settlement that a foundation can tolerate is called the *allowable settlement*. The magnitude of this settlement depends upon its mode.

2B.7.1.1 Uniform Settlement. A structure has undergone uniform settlement when all points within the structure have moved vertically the same amount [Fig. 2B.8(*a*)]. This type of settle-

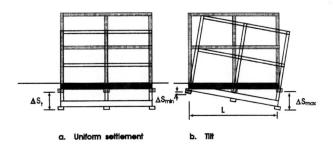

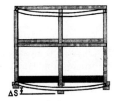

c. Differential settlement

FIGURE 2B.8 Types of foundation settlement.

ment does not result in structural damage if it is constant across the whole structure. However, there will be problems with appurtenances such as pipes and entrance ways.

2B.7.1.2 Tilt. Tilt is usually measured by its angular distortion (Fig. 2B.8):

$$\text{angular distortion} = \frac{s_{max} - s_{min}}{L} \tag{2B.37}$$

The amount of tilt that a structure can tolerate is a function of many factors, including the size and type of construction. The Leaning Tower of Pisa is currently at about 10% tilt and is still standing. However, the Campenella in Plaza San Marcos in Venice collapsed when it reached a 0.8% tilt. Tilt is visible at about $1/250 = 0.4\%$.

2B.7.1.3 Differential (Distortion) Settlement. If s_{max} is the maximum total settlement anywhere in the structure, then Δs_{max} is the maximum difference in total deformations between adjacent foundations. This is called *differential settlement*. Distortion is then defined as: $\Delta s_{max}/L$.

Field evidence indicates that architectural damage occurs when $\Delta s_{max}/L = 1/300$, and structural damage occurs when $\Delta s_{max}/L = 1/150$.

2B.7.1.4 Maximum Allowable Settlement. In general, foundations are limited to a specified amount of settlement. This settlement is called the *design* or *maximum allowable settlement*. For isolated foundations that support individual columns or small groups of columns on clay,

$$\frac{\Delta s_{max}}{L} = \frac{1}{1200} s_{max} \tag{2B.38}$$

Since $\Delta s_{max}/L = 1/300$, $s_{max} = 4$ in (10 cm).

For isolated foundations that support individual columns or small groups of columns on sand:

$$\frac{\Delta s_{max}}{L} = \frac{1}{600} s_{max} \tag{2B.39}$$

Since $\Delta s_{max}/L = 1/300$, $s_{max} = 2$ in (5 cm). Therefore we must design for total settlements of isolated foundations less than 2 to 4 in (5 to 10 cm).

2B.7.1.5 Allowable Settlement of Mat (Raft) Foundations. Few data are available that document allowable settlement for raft foundations. Therefore only minor problems probably occur.

2B.7.2 Elastic (Immediate) Settlement

Immediate settlement occurs as the load is applied to the soil. The soil particle matrix distorts, and the soil voids are compressed. If the soil voids contain air, or if the permeability of the soil matrix is high, then the volume of the voids decreases, thereby contributing to the settlement. Since sands and gravels are highly conductive, almost all the settlement of foundations on sands and gravels can be classified as immediate. Clays, however, have very low hydraulic conductivity, hence, if they are saturated, the immediate deformation is usually quite small and is limited to structural distortion of the soil fabric.

2B.7.2.1 Key Variables in Elastic Settlements. The magnitude of deformation is inversely proportional to the strength of the sand. Factors affecting the soil strength are relative density, embedment of the foundation, and the effect of groundwater. Relative density is measured in the field using the standard penetration test, the cone penetrometer, or other devices.

The relationship between settlement and footing width was described by Terzaghi and Peck.[6] For the same load on the same soil, the settlement, s, is related to the square of the footing width, B, through the settlement of a 1-ft (0.3 m) square plate, s_1, by

$$s = s_1 \left[\frac{2B}{B+1}\right]^2 \tag{2B.40}$$

The magnitude of settlement is also directly proportional to the magnitude of the applied load up to the allowable bearing pressure, with all else constant.

2B.7.2.2 Settlement Models. Many techniques are presented in the literature for predicting the settlement of shallow foundations on sand. Depending upon which method is used, this calculation can be a very simple one or can be moderately complex, and the resulting predictions can differ greatly. A recent publication by the Corps of Engineers, Waterways Experiment Station[7] reported on fifteen methods. Most of the methods can be placed within one of two categories: some are modeled after the Terzaghi and Peck[6] bearing capacity and settlement–footing width relationship, and others are modeled after elasticity methods. A few methods combine some aspects of both. The backgrounds for both the Terzaghi-based settlement methods and elastic-based settlement methods follow.

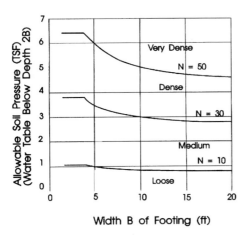

FIGURE 2B.9 Terzaghi and Peck design chart for q_{all}.

2B.7.2.2.1 Terzaghi-based Settlement. Terzaghi and Peck[6] developed the well-known design chart (Fig. 2B.9) for estimating allowable bearing pressures for shallow foundations on sand using standard penetration blowcount and footing width. These design curves correspond to a maximum footing settlement of 1 in (2.5 cm) and total differential settlement of ¾ in (1.9 cm). Data were interpreted conservatively in the development of this chart. History has proved that these values are very conservative. Modification to these values for less conservatism have been made by many. A general expression for this relation is

$$s = C\left[\frac{q}{N}\right]\left[\frac{B}{B+1}\right]^2 \tag{2B.41}$$

where s = settlement
 q = net applied load
 B = footing width
 N = blowcount
 C = empirical-constant-determined observation and/or experiment

Terzaghi's charts give $C = 8$ for footings less than 4 ft (1.2 m), and $C = 12$ for footings greater than 4 ft (1.2 m) in width.

2B.7.2.2.2 Elastic Soil Settlement. Soil is often treated as an elastic medium, linear or nonlinear, to which the elastic theory assumptions and principles of stress and strain are applied.

Settlement computations of this form use the elastic properties of Poisson's ratio and Young's modulus to represent the soil. A general expression for the elastic settlement relations is

$$s \sim \frac{qBI\mu}{E} \qquad (2B.42)$$

where μ = Poisson's ratio
E = elastic modulus
I = influence factor based on footing shape depth and extent of elastic region

One main difference between the Terzaghi model and the elastic model is the relationship between footing width and settlement. The elastic theory models a linear relation between settlement and footing width, but Terzaghi's work shows this to be a nonlinear relation. Elastic theory settlement methods can account for this nonlinear relationship through an appropriate use of the elastic or compressibility modulus.

2B.7.2.2.3 Standard Penetration Test. The relative density of the soil is the major factor that controls the settlement of foundation on cohesionless soil. The Standard Penetration Test (SPT) is the most widely used test in the United States for indirectly determining the relative density of a sand. A description of the test can be found in Bowles.[1] With all other factors being the same, the larger the blowcount, N, the less the settlement. However, a number of factors significantly affect the blowcount. The overburden pressure dramatically impacts the SPT value. In a homogeneous deposit where relative density and friction angle are constant with depth, blowcounts increase with depth due to the increasing confining pressures with depth. Therefore, each measured SPT value should be corrected for the influence of its corresponding overburden pressure. Many techniques are available to correct the SPT value for overburden pressure. In general all the techniques have the form of

$$N_c = C_N N \qquad (2B.43)$$

where N_c = corrected SPT value
C_N = correction factor based on overburden pressure

Blowcount correction factors for overburden developed by various authors are given in Table 2B.1. Each of the equations normalizes N to a standard reference overburden pressure. Typically this is 1 tsf (96 kN/m²). However, Peck and Bazarra[12] normalize to 0.75 tsf (72 kN/m²), and Teng[10] uses 40 psi (276 kN/m²). Some procedures for computing settlement do not advocate correcting the blowcount for overburden but use the blowcount values as obtained in the field. Most experiments and theories show that this correction is necessary.

TABLE 2B.1 Overburden Correction Factors

Reference	Equation for C_N	Units for p'_o
Skempton[8]	fine to medium sand: $2/(1 + p'_o)$	p'_o = effective overburden pressure
	coarse, dense sand: $3/(2 + p'_o)$	tsf
	overconsolidated fine sand: $1.7/(0.7 + p'_o)$	
Peck, Hanson, and Thornburn[2]	$0.77 \log(20/p'_o)$	tsf
Bazarra[9]	$p'_o < 1.5$ ksf: $4/(1 + 2p'_o)$	ksf
	$p'_o > 1.5$ ksf: $4/(3.25 + 0.5p'_o)$	
Teng[10]	$50/(p'_o + 10)$	psi
Liao and Whitman[11]	$(1/p'_o)^{0.5}$	tsf

Variations in the borehole diameter, rod length, and hammer type can affect the measured blowcounts for identical sands at the same overburden and relative density values. The blowcount is directly related to the driving energy of test equipment:

$$E_\epsilon = \frac{1}{2} mv^2 = \frac{1}{2} \frac{W}{g} v^2$$

$$v = (2gh)^{1/2} \quad (2B.44)$$

$$E_\epsilon = \frac{1}{2} \frac{W}{g} (2gh) = Wh$$

where W = weight or mass of hammer
h = height of fall
v = velocity of hammer

The energy ratio E_r is defined as

$$E_r = \frac{\text{Actual hammer energy to sampler, } E_a}{\text{Input energy, } E_{in}} \times 100 \quad (2B.45)$$

Bowles suggests that the energy should be adjusted to a standard energy ratio of 70 (E_{70}) and that the equation for the standard corrected blowcount be given as

$$N_{70} = C_N \times N \times \eta_1 \times \eta_2 \times \eta_3 \times \eta_4 \quad (2B.46)$$

where η_i can be found in Table 2B.2. Each of the factors correct the field blowcounts for differences in hammers, rod length, sampler differences, and borehole diameter differences.

TABLE 2B.2 Blowcount Adjustment Factors, η_i

Hammer adjustment factor, η_1
Average energy ratio E_r:

Donut		Safety	
R-P	Trip	R-P	Trip-Auto
45	—	70-80	80-100

R-P = Rope-pulley or cathead; $\eta_1 = E_r/E_{rb}$
For U.S. trip/auto w/E_r = 80: η_1 = 80/70 = 1.14

Rod length correction factor, η_2
Length >10 m $\eta_2 = 1.00$
    6–10 m   = 0.80
    4–6 m   = 0.85
    0–4 m   = 0.75

Sampler correction factor, η_3
Without liner $\eta_3 = 1.00$
With liner:
  Dense sand, clay = 0.80
  Loose sand = 0.90

Borehole diameter correction factor, η_4
Hole diameter:
  60–120 mm $\eta_4 = 1.00$
  150 mm   = 1.05
  200 mm   = 1.15

TABLE 2B.3 Embedment Correction Factors

Reference	Equation for embedment correction factor, C_D
Terzaghi and Peck[13]	$C_D = 1 - 0.25\,(D_f/B)$
Schultze and Sherif[14]	$C_D = 1/[1 + 0.4\,(D_f/B)]$
D'Appolonia et al.[15]	$C_D = 0.729 - 0.484 \log(D_f/B) - 0.224\,[\log(D_f/B)]^2$
Bowles[1]	$C_D = 1/[1 + 0.33\,(D_f/B)]1$
Teng[10]	$C_D = 1/[1 + (D_f/B)]$
Bazarra[9]	$C_D = 1 - 0.4\,(\gamma D_f/q)^{0.5}$
Schmertmann[16]	$C_D = 1 - 0.5\,[\gamma D_f/(q - D_f)]$

Terms: D_f = foundation depth, B = foundation width, q = loading pressure.

A footing placed below the ground surface will settle less than a footing at the surface. The depth correction factor reduces the calculated settlement to account for the increase in bearing capacity achieved by embedment. The embedment correction equations by the various authors are given in Table 2B.3.

2B.7.2.3 Settlement Computing Methods. Several of the more prominent methods of computing elastic settlements are presented in the following paragraphs. Other methods can be extracted from published literature.

2B.7.2.3.1 Terzaghi and Peck.[6,13] This method is based on the bearing capacity charts given in Fig. 2B.9. The equations shown below are given by Meyerhof.[3] The chart is used to determine the allowable bearing capacity for a range of footing widths and SPT blowcount values with maximum settlement not to exceed 1 in (2.5 cm) and differential settlement not to exceed ¾ in (1.9 cm). Their settlement expression is

$$s = \frac{8q}{N}(C_W C_D) \quad \text{for } B \leq 4 \text{ ft (1.2 m)}$$

$$s = \frac{12q}{N}\left[\frac{B}{B+1}\right]^2 (C_W C_D) \quad \text{for } B \geq 4 \text{ ft (1.2 m)} \quad (2B.47)$$

$$s = \frac{12q}{N}(C_W C_D) \quad \text{for rafts}$$

Their correction factors are

$$\text{Water: } C_W = 2 - \left[\frac{D_W}{2B}\right] \leq 2.0 \text{ (for surface footings)}$$

$$= 2 - 0.5\left[\frac{D_f}{B}\right] \leq 2.0 \text{ (for submerged, embedded footing; } D_W \leq D_f\text{)} \quad (2B.48)$$

$$\text{Depth: } C_D = 1 - 0.25\left[\frac{D_f}{B}\right]$$

To calculate blowcount, use the measured SPT blowcount value. If the sand is saturated, dense, and very fine or silty, correct the blowcount by

$$N_c = 15 + 0.5\,(N - 15), \text{ for } N > 15 \quad (2B.49)$$

2B.7.2.3.2 Teng.[10]

Teng's method for computing settlement is an interpretation of the Terzaghi and Peck bearing capacity chart. Teng includes corrections for depth of embedment, the presence of water, and the blowcount. The settlement expression is

$$\Delta s = \frac{q_o}{720 (N_c - 3)} \left[\frac{2B}{B+1} \right]^2 \frac{1}{(C_W)(C_D)} \quad (2B.50)$$

where q_o = net pressure in psf
Correction factors are

$$\text{Water: } C_W = 0.5 + 0.5 \left[\frac{D_W - D_f}{B} \right] \geq 0.5 \text{ for water at and below } D_f$$

$$\text{Depth: } C_D = 1 + \left[\frac{D_f}{B} \right] \leq 2.0 \quad (2B.51)$$

$$\text{Blowcount: } N_c = N \left[\frac{50}{p_o' + 10} \right]$$

where p_o' = effective overburden at median blowcount depth about $D_f + \frac{B}{2}$, in psi ($\leq$40 psi, 276 kPa)

2B.7.2.3.3 Peck, Hanson, and Thornburn.[2]

This method is based on Terzaghi and Peck settlement method:

$$\Delta s = \frac{q}{0.11 N_c C_w} \text{ for intermediate width footings } (> 2 \text{ ft, } 0.6 \text{ m})$$

$$\Delta s = \frac{q}{0.22 N_c C_w} \text{ for rafts} \quad (2B.52)$$

where q is in tsf.
Correction factors are

$$\text{Water: } C_w = 0.5 + 0.5 \left[\frac{D_w}{D_f + B} \right] \text{ for water from 0 to } D_f + B \quad (2B.53)$$

$$\text{Blowcount: } N_c = NC_n$$

$$C_n = 0.77 \log \left[\frac{20}{p'} \right] \quad (2B.54)$$

where p' = effective overburden pressure for the measured blowcount at $(D_f + B/2)$ in tsf $\geq$ 0.25 tsf (24 kPa).

2B.7.2.3.4 Bowles.[17,18]

Bowles's settlement method is based on the Terzaghi and Peck method but is modified to produce results that are not as conservative. His equations are

$$\Delta s = \frac{2.5 q_o}{N} \left[\frac{C_W}{C_D} \right] \text{ for } B \leq 4 \text{ ft}$$

$$\Delta s = \frac{4 q_o}{N} \left[\frac{B}{B+1} \right]^2 \left[\frac{C_W}{C_D} \right] \text{ for } B \geq 4 \text{ ft} \quad (2B.55)$$

$$\Delta s = \frac{4q_o}{N}\left[\frac{C_W}{C_D}\right] \text{ for mats}$$

where q is in kips/sf, N is measured in the field, and the settlement is in inches. The correction factor for water is

$$C_W = 2 - \left[\frac{D_W}{D_f + B}\right] \le 2.0 \text{ and } \ge 1.0 \tag{2B.56}$$

The correction factor for depth is

$$C_D = 1 + 0.33\left[\frac{D_f}{B}\right] \le 1.33 \tag{2B.57}$$

Therefore, the settlement can be computed as

$$\Delta s = \frac{2.5q_o}{N}\left[\frac{C_W}{C_D}\right] \text{ for } B \le 4 \text{ ft}$$

$$\Delta s = \frac{4q_o}{N}\left[\frac{B}{B+1}\right]^2\left[\frac{C_W}{C_D}\right] \text{ for } B \ge 4 \text{ ft} \tag{2B.58}$$

$$\Delta s = \frac{4q_o}{N}\left[\frac{C_W}{C_D}\right] \text{ for mats}$$

2B.7.2.3.5 Elastic Theory. Settlement computed by elastic theory uses elastic parameters to model a homogeneous, linearly elastic medium. The elastic modulus of a soil depends upon confinement and is assumed in elastic theory to be constant with depth. For uniform saturated cohesive soils, this assumption is usually valid. For cohesionless soils, elastic methods can be inappropriate because the modulus often increases with depth. However, the immediate settlement of sand is often considered to be elastic within a small strain range.

The equations are based on the theory of elasticity and are for settlement at the surface of a semi-infinite, homogeneous half-space. The equation is

$$\Delta s = \frac{q_o B'}{E_s}(1-\mu^2)\left[I_1 + \frac{(1-2\mu)}{(1-\mu)}I_2\right]I_F \tag{2B.59}$$

where

$$I_1 = \frac{1}{\pi}\left[M \ln \frac{(1+\sqrt{M^2+1})\sqrt{M^2+N^2}}{M(1+\sqrt{M^2+N^2+1})} + \ln \frac{(M+\sqrt{M^2+1})\sqrt{1+N^2}}{M+\sqrt{M^2+N^2+1}}\right] \tag{2B.60}$$

$$I_2 = \frac{N}{2\pi}\tan^{-1}\left(\frac{M}{N\sqrt{M^2+N^2+1}}\right) \quad (\tan^{-1} \text{ in rad})$$

where

$$M = \frac{L'}{B'}, \quad N = \frac{H}{B'}, \quad B' = \frac{B}{2} \tag{2B.61}$$

For the center influence factor,

$$B' = \frac{B}{2}, \quad L' = \frac{L}{2}$$

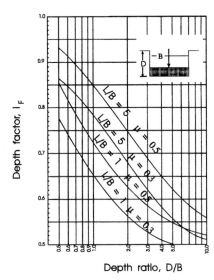

FIGURE 2B.10 Influence factors for footing at depth D. Use actual footing width and depth dimensions.[19]

For the corner influence factor,

$$B' = B, \quad L' = L$$

The correction factors are from Das[19] (Fig. 2B.10). The influence factor I_s is defined as

$$I_s = I_1 + \frac{1 - 2\mu}{1 - \mu} I_2 \quad (2B.62)$$

Therefore, Eq. (2B.10) can be written as

$$\Delta s = q_o B' \frac{1 - \mu^2}{E_s} I_s I_F \quad (2B.63)$$

For rigid footings, the value of I_s should be reduced by 7%, i.e., $I_{sr} = 0.93\, I_s$. Poisson's ratio, μ, can be determined from Table 2B.4.

Bowles suggests the following procedure:

1. Make the best estimate of q_0.
2. Convert round footings to an equivalent square.
3. Determine the point where the settlement is to be computed and divide the base so the point is at the corner or common corner of the contributing rectangles.
4. Note that the stratum depth actually causing settlement is not at $H/B \to \infty$ but is to either *a* or *b*:
   a. Depth $z = 5B$ (B = least total lateral dimension of base).
   b. Depth to where a hard stratum is encountered. Take *hard* as that where E_s in the hard layer is about $10 E_s$ of adjacent layer. Table 2B.5 can be used for approximate values. Table 2B.6 gives equations for E_s as functions of cone or standard penetration test values.
5. Compute H/B' ratio. For a depth $H = z = 5B$ and for the center of the base we have $H/B' = 5B/0.5B = 10$. For a corner $5B/B = 5$.
6. Obtain I_1 and I_2 with the best estimate for μ and compute I_s.
7. Determine I_F from Fig. 2B.10.

TABLE 2B.4 Range of Values for Poisson's Ratio*

Soil	Poisson's ratio
Loose Sand	0.2–0.4
Medium Sand	0.25–0.4
Dense Sand	0.3–0.45
Silty Sand	0.2–0.4
Soft Clay	0.15–0.25
Medium Clay	0.2–0.5

*After Das.[19]

TABLE 2B.5 Range of Elastic Modulus, E_s*

Soil	E_s psi (kPa)
Soft clay	250–500 (1725–3450)
Hard clay	850–2,000 (5860–13,800)
Loose sand	1,500–4,000 (10,350–27,600)
Dense sand	5,000–10,000 (34,000–69,000)

*After Das.[19]

TABLE 2B.6 Equations for E_s from SPT and CPT

Soil	SPT (kPa)*	CPT (units of q_c)
Sand	$E_s = 500 (N + 15)$ $E_s = 18,000 + 750N$ $E_s = (15,000$ to $22,000) \ln N$	$E_s = (2$ to $4) q_c$ $E_s = 2 (1 + D_r^2) q_c$
Clayey sand	$E_s = 320 (N + 15)$	$E_s = (3$ to $6) q_c$
Silty sand	$E_s = 300 (N + 6)$	$E_s = (1$ to $2) q_c$
Gravelly sand	$E_s = 1,200 (N+6)$	
Soft clay		$E_s = (6$ to $8) q_c$

*Divide kPa by 50 to get ksf (after Bowles).

8. Obtain the weighted average E_s in the depth $z = H$ using

$$E_{s(av)} = \frac{H_1 E_{s1} + H_2 E_{s2} + \cdots + H_n E_{sn}}{H} \tag{2B.64}$$

2B.7.2.3.6 Das's[19] Elastic Settlement of Foundations on Saturated Clay ($\mu = 0.5$). Das computes the settlement of a foundation on saturated clay using

$$S_e = A_1 A_2 \frac{q_o B}{E_s} \tag{2B.65}$$

The coefficients A_1 and A_2 are found in Figs. 2B.11 and 2B.12, respectively.

2B.7.2.3.7 Schmertmann.[16] Schmertmann proposes calculating total settlement by subdividing the compressible stratum and summing the settlements of each sublayer. The sublayer boundaries are defined by changes in the SPT or cone penetrometer (CPT) profile. His equation is

$$\Delta s = C_1 C_2 \Delta q \Sigma \frac{I_z \Delta z}{E_s} \tag{2B.66}$$

The correction factor for embedment is

$$C_1 = 1 - 0.5 \frac{\bar{q}}{q_o - \bar{q}} \tag{2B.67}$$

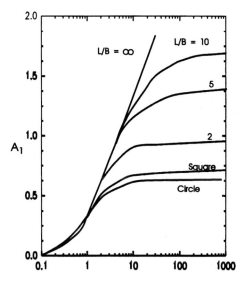

FIGURE 2B.11 Values of A_1.

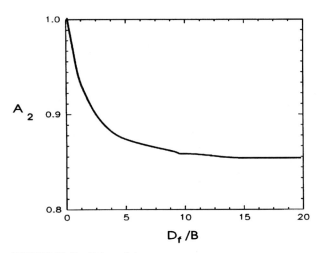

FIGURE 2B.12 Values of A_2.

where $\bar{q}$ = surcharge = γD_f
The correction factor for time is

$$C_2 = 1 + 0.2 \log \frac{t}{0.1} \qquad (2\text{B}.68)$$

where t = time (>0.1 years)

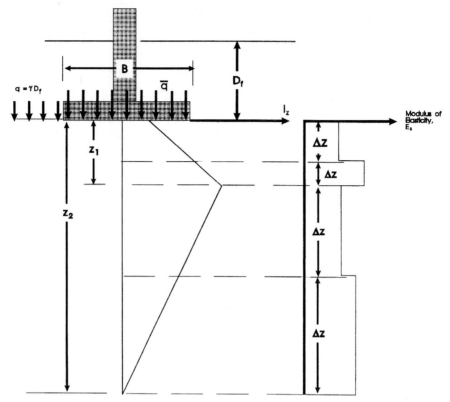

FIGURE 2B.13 Elastic settlement by the strain influence factor.

The variation of the strain influence factor is given in Fig. 2B.13. Note that for square or circular foundations,

$$I_z = 0.1 \text{ at } z = 0$$
$$= 0.5 \text{ at } z = z_1 = 0.5B$$
$$= 0 \text{ at } z = z_2 = 2B \tag{2B.69}$$

Similarly, for foundations with $L/B \geq 10$,

$$I_z = 0.2 \text{ at } z = 0$$
$$= 0.5 \text{ at } z = z_1 = B$$
$$= 0 \text{ at } z = z_2 = 4B \tag{2B.70}$$

For values of L/B between 1 and 10, necessary interpolations can be made. This profile is used to determine the elastic modulus as it changes with depth. If E_s is constant over $2B$ below the footing base, the simplified expression is

$$\Delta s = C_1 C_2 q_o \frac{0.6B}{E_s} \tag{2B.71}$$

2B.7.2.4 Proportioning Footings for Equal Settlement.
For clay soils, the usual method of proportioning footings for equal settlement is to use the equation

$$\frac{\Delta s_1}{\Delta s_2} = \frac{B'_2}{B'_1} \tag{2B.72}$$

for constant contact pressure. This has proven to work reasonably well.

For sand soils, Bowles recommends the equation

$$\frac{\Delta s_1}{\Delta s_2} = \frac{B'_2}{B'_1} \frac{I_{s_2}}{I_{s_1}} \frac{I_{f_2}}{I_{f_1}} \frac{E'_{s_2}}{E'_{s_1}} \tag{2B.73}$$

for constant contact pressure.

2B.7.3 Consolidation Settlement of Foundations on Clays

The settlement of shallow foundations located on cohesive soils is governed by consolidation theory (Sec. 2A.8).

To compute the consolidation settlement of a foundation on clay, follow these steps:

1. Determine if the soil is normally consolidated or overconsolidated.
2. Determine the thickness, H, and existing void ratio, e_o, of the consolidating soil layer.
3. Compute the average existing effective stress acting on the consolidating soil layer, $\overline{p}_0$.
4. Compute the average increase in stress $\overline{\Delta p}$ on the consolidating layer due to the addition of the foundation load.
5. Compute the consolidation settlement using these equations:
   For normally consolidated soils,

$$\Delta s = \frac{H}{1 + e_o} C_c \log \left[\frac{\overline{p}_o + \overline{\Delta p}}{\overline{p}_o}\right] \tag{2B.74}$$

For overconsolidated soils,

$$\Delta s = \frac{H}{1 + e_o} C_r \log \left[\frac{\overline{p}_o + \overline{\Delta p}}{\overline{p}_o}\right] \tag{2B.75}$$

2B.8 REFERENCES

1. Bowles, Joseph E., *Foundation Analysis and Design*, 4th ed., McGraw-Hill, New York, 1988.
2. Peck, R. B., Hansen, W. E., and Thornburn, T. H., *Foundation Engineering*, John Wiley and Sons, New York, 1974.
3. Meyerhof, G. G., "Penetration Tests and Bearing Capacity of Cohesionless Soil," *Journal of the Soil Mechanics and Foundations Division*, ASCE, vol. 82, no. SM1, 1956, pp. 1–19.
4. Hansen, J. B., "A Revised and Extended Formula for Bearing Capacity," Danish Geotechnical Institute Bull. no. 28, Copenhagen, 1970.
5. Vesic, A. S., "Analysis of Ultimate Loads of Shallow Foundations," Journal of Soil Mechanics and Foundations Division, ASCE, vol. 99, SM1, Jan., pp. 45–73, 1973.
6. Terzaghi, Karl, and Peck, Ralph B., *Soil Mechanics in Engineering Practice*, John Wiley and Sons, New York, 1948.

7. Knowles, Virginia R., "Settlement of Shallow Footings on Sand: Report and User's Guide for Computer Program CSANDSET, "Technical Report ITL-91-1," Department of the Army, Waterways Experiment Station, Corps of Engineers, Vicksburg, Mississippi, 1991.
8. Skempton, A. W., "Standard Penetration Test Procedures…," Geotechnique, vol. 36, no. 3, pp. 425–447, 1984.
9. Bazarra, A. R. S. S., "Use of the Standard Penetration Test for Estimating Settlement of Shallow Foundations on Sand," Ph.D. thesis, University of Illinois, 1967.
10. Teng, W., *Foundation Design,* Prentice-Hall, Englewood Cliffs, N.J., 1962.
11. Liao, S. S., and Whitman, R. V., "Overburden Correction Factors for Sand," *Journal of the Geotechnical Engineering Division,* ASCE, vol. 112, no. GT3, March, pp. 373–377, 1986.
12. Peck, R. B., and Bazarra, Abdel R., Discussion of "Settlement of Spread Footings on Sand," *Journal of the Soil Mechanics and Foundations Division,* ASCE, vol. 95, no. SM3, pp. 905–909, 1969.
13. Terzaghi, Karl, and Peck, Ralph B., *Soil Mechanics in Engineering Practice,* 2d ed., John Wiley and Sons, New York, 1967.
14. Schultze, E., and Sherif, G., "Prediction of Settlements from Evaluated Settlement Observations for Sand," Proceedings of the 8th International Conference on Soil Mechanics and Foundation Engineering, Moscow, pp. 225–230, 1973.
15. D'Appolonia, David J., D'Appolonia, Elio, and Brissette, Richard F., "Settlement of Spread Footings on Sand," *Journal of the Soil Mechanics and Foundations Division,* ASCE, vol. 94, no. SM3, pp. 735–760, 1968.
16. Schmertmann, John H., "Static Cone to Compute Static Settlement Over Sand," *Journal of the Soil Mechanics and Foundations Division,* ASCE, vol. 96, no. SM3, 1970, pp. 1011–1043.
17. Bowles, Joseph E., *Foundation Analysis and Design,* 2d ed., McGraw-Hill, New York, 1977.
18. Bowles, Joseph E., *Foundation Analysis and Design,* 3d ed., McGraw-Hill, New York, 1982.
19. Das, Braja M., *Principles of Foundation Engineering,* 3d ed., PWS Publishing Company, Boston, 1995.

PART 3

FUNDAMENTALS OF FOUNDATION CONSTRUCTION AND DESIGN

SECTION 3A
CONCRETE

P. Balaguru

3A.1 CONSTITUENT MATERIALS   3.3
   3A.1.1   Cement   3.4
   3A.1.2   Aggregates   3.6
   3A.1.3   Water and Water-Reducing
            Admixtures   3.11
   3A.1.4   Chemical Admixtures   3.13
   3A.1.5   Mineral Admixtures   3.16
3A.2 COMPUTATION OF REQUIRED AVERAGE
      COMPRESSIVE STRENGTH f'_{CR} AND MIX
      PROPORTIONING   3.18
   3A.2.1   Computation of Required Average
            Strength $f'cr$ 3.19
   3A.2.2   Selection of Concrete Proportions   3.25
   3A.2.3   Proportioning on the Basis of Field
            Strength Records   3.25
   3A.2.4   Proportioning on the Basis of Trial
            Mixtures   3.26
   3A.2.5   Proportioning on the Basis of Maximum
            Water-Cement Ratio   3.26
   3A.2.6   Computation of Mix Proportions   3.26
3A.3 QUALITY ASSURANCE   3.33
   3A.3.1   Tests for Fresh Concrete   3.34
   3A.3.2   Tests for Hardened Concrete   3.36
   3A.3.3   Compressive Strength Test   3.36

3A.3.4   Statistical Variation of Compressive
           Strengths and Quality Control   3.41
3A.3.5   Accelerated Strength Tests   3.43
3A.3.6   Quality Control Using Accelerated
           Strength Tests   3.44
3A.3.7   Nondestructive Tests   3.47
3A.3.8   Core Tests   3.49
3A.4 MECHANICAL PROPERTIES   3.49
   3A.4.1   Compressive Strength   3.50
   3A.4.2   Tensile Strength   3.56
   3A.4.3   Modulus of Rupture   3.57
   3A.4.4   Shear Strength   3.57
   3A.4.5   Modulus of Elasticity   3.58
   3A.4.6   Shrinkage   3.58
   3A.4.7   Creep   3.60
   3A.4.8   Estimation of Creep and Shrinkage
            Strains   3.62
   3A.4.9   Behavior under Multiaxial
            Stresses   3.62
   3A.4.10 Fatigue Loading   3.63
3A.5 COLD-WEATHER CONCRETING   3.65
3A.6 HOT-WEATHER CONCRETING   3.67
3A.7 PUMPING OF CONCRETE   3.71
3A.8 REFERENCES   3.73

Concrete is one of the basic construction materials, which finds a place in almost all structures. Even in such structures as steel bridges the deck is quite often made of concrete. Concrete is the preferred and most widely used material for foundation construction. Even if the superstructure is made of steel or wood, the foundation is usually made of concrete. In the case of slab on grade floors, whether industrial, commercial, or residential, concrete is the preferred material.

This section deals with some of the fundamental aspects of concrete. Only the basic information considered necessary for the design and construction engineer is presented. The reader can refer to the literature for more details and in-depth information on a particular aspect. This section deals only with plain concrete. Reinforced concrete is discussed in the Sec. 3B.

3A.1   CONSTITUENT MATERIALS

Concrete is a composite material made of portland cement (often simply called cement), aggregates, and water. In most cases additional constituents, called admixtures, are used to improve the properties of fresh and hardened concrete. For example, water-reducing admixtures are often used to improve the workability of fresh concrete without increasing its water content, thus maintaining the strength and durability characteristics of the hardened concrete. The admixtures can be classified broadly as chemical and mineral admixtures.

This section presents basic information with regard to the various constituent materials used in concrete. They are grouped as (1) cement, (2) aggregates, (3) water and water-reducing admixtures, (4) chemical admixtures, and (5) mineral admixtures. Even though water-reducing admixtures are chemical admixtures, they are discussed together with water because of their direct impact on the quantity of water used in the mix and their widespread use in practice.

3A.1.1 Cement

Cement, which is the binding ingredient of concrete, is produced by combining lime, silica, and alumina. A small amount of gypsum is added to control the setting time of the cement. Portland cement was first patented by Joseph Aspdin of England in 1824. David Saylor of Coplay, Pennsylvania, was the first to produce portland cement in the United States in 1871. He used vertical kilns that were similar to the ones used for burning lime. The rotary kiln was introduced in 1899. In the 1990s cement production in the United States has been running in the range of 800 million tons (725 million tonnes); worldwide it has reached 5 billion tons (4.5×10^9 tonnes).

Manufacture of Cement. The raw materials for portland cement consist primarily of limestone or some other lime-containing material such as marl, chalk, or shells, and of clay or shale or some other clayey material such as ash or slag. Sometimes other ingredients, such as high-calcium limestone, sandstone, and iron ore, are added to control the chemical composition of the final product. The manufacturing process can be briefly described as follows.

The raw materials are ground into impalpable powder and thoroughly mixed. In the dry process, blending and grinding operations are done in the dry form, and the mixing is primarily accomplished during the grinding phase. In the wet process, water is used to form a slurry. The slurry is often mixed in large vats to obtain a thorough mixing, even though the ingredients have already been mixed during the grinding process. The wet process, which requires about 15% more energy than the dry process, is often chosen for environmental and technological reasons. Continuous quality control measures are used to ascertain the proper chemical composition of the raw material so that the chemical ingredients of the final product will be within the limits specified.

In most cases the slurry is fed into the upper end of a slightly inclined rotary kiln. In some instances part of the water is removed from the slurry before feeding it into the kiln. The length and the diameter of the kilns vary between 60 and 500 ft (18 and 150 m) and between 6 and 15 ft (1.8 and 4.5 m), respectively. The kilns, set at an inclination of about 0.5 in/ft (40 mm/m), rotate between 30 and 90 revolutions per hour, moving the material toward the lower (discharge) end. Heating is usually done by using powdered coal and air. In some instances oil or gas is used instead of coal. The temperature varies along the kiln, reaching a maximum in the range of 2300 to 3450°F (1250 to 1900°C).

As the mix passes through the kiln, various reactions take place, including (1) evaporation of free water, (2) dehydroxylation of clay minerals, (3) crystallization of the products of clay mineral dehydroxylation, (4) decomposition of $CaCO_3$, (5) reaction between $CaCO_3$ (CaO) and aluminosilicates, and (6) liquefaction and formation of cement compounds. The temperature variations are controlled in such a way as to keep the compounds in the molten stage to a minimum. The molten liquid agglomerates into nodules. The nodules, ranging in size from 0.125 to 2 in (3 to 50 mm), are called cement clinkers. These clinkers are dropped off from the kiln.

The clinkers are cooled and ground to a fine powder. About 3 to 5% of gypsum ($CaSO_4 \cdot 2H_2O$) is added during the grinding process to control the setting time of the cement. Addition of gypsum retards the hydration of cement, or increases its setting time. After grinding, the cement is stored in silos.

In the United States, cement can be bought in bulk or in bags containing 94 lb (42.5 kg). It is common to designate concrete mixes as 5, 6, or 7 bag mix, and hence it is useful to remember the weight of the cement in a bag.

Composition. Compounds of four oxides containing lime, silica, alumina, and iron constitute about 95% of the portland cement clinkers. The other 5% could include magnesia, sodium and potassium oxide, titania, sulfur, phosphorous, and manganese oxide. The major components, namely, tricalcium silicate (C_3S), dicalcium silicate (C_2S), tricalcium aluminate (C_3A), and tetracalcium aluminoferrate (C_4AF), play important roles in the rate of strength development, the heat of hydration, and the ultimate cementing value. For example, the early strength of hydrated portland cement is higher if the percentage of tricalcium silicate is higher, whereas long-term strengths will be higher with higher percentages of dicalcium silicate.

Types. Various types of cement are produced to suit the various applications. The American Society for Testing and Materials (ASTM) recognizes the following five main types:

Type I	For general use in construction
Type II	For use that requires moderate heat of hydration and exposure to moderate sulfate action
Type III	For use where high early strength is needed
Type IV	For use that requires low heat of hydration
Type V	For use that requires high sulfate resistance

Types I, II, and III can be obtained with air-entraining agents. These are then designated types IA, IIA, and IIIA. Some standard blended portland cements that are available are called portland blast-furnace slag cement and portland pozzolan cement.

Typical composition values for the various compounds of the five cement types are shown in Table 3A.1. These numbers are mean values, and there is a specified minimum and maximum for each compound.

Fineness. The term fineness refers to the average size of the cement particles. The fineness of the cement determines the rate of reaction because finer particles have more surface area and, hence, generate more reactivity when water is added. Type III high-early-strength cement has more fine particles than type I cement. Finer cement bleeds less than coarser cement. In addition, finer cement contributes to better workability and produces less autoclave expansion. But the finer cement is more expensive to produce, and if the particles are overly fine, they could lead to increased shrinkage, higher water demand, strong reaction with alkali-reactive aggregates, and poor stability.

The particle size usually varies from 1 to 200 μm. Fineness of cement can be expressed using the Blaine specific surface area. The cement is considered overly fine if the Blaine specific surface area is greater than 2440 ft^2/lb (5000 cm^2/g). The specific surface areas, measured using the air permeability method, vary from 1220 to 1760 ft^2/lb (2500 to 3600 cm^2/g) and from 1760 to 2200 ft^2/lb (3600 to 4500 cm^2/g) for type I and III cements, respectively.

Cements with small particle sizes, known as microcements, are available for special purposes such as grouting. The fine particles facilitate the grouting of soils containing small pore sizes.

TABLE 3A.1 Percentage Composition of Portland Cements

	Type of cement	Component, %							General characteristics
		C_3S	C_2S	C_3A	C_4AF	$CaSO_4$	CaO	MgO	
I	Normal	49	25	12	8	2.9	0.8	2.4	All-purpose cement
II	Modified	45	29	6	12	2.8	0.6	3.0	Comparative low heat liberation; used in large structures
III	High early strength	56	15	12	8	3.9	1.4	2.6	High strength in 3 days
IV	Low heat	30	46	5	13	2.9	0.3	2.7	Used in mass concrete dams
V	Sulfate resisting	43	36	4	12	2.7	0.4	1.6	Used in sewers and structures exposed to sulfates

Testing Methods. Typically cement is tested at periodic intervals for chemical composition and certain physical properties to satisfy quality control requirements. Special tests may be needed for particular cases such as the determination of compatibility with certain admixtures. ASTM specifications exist for most of the tests. The following list contains commonly used tests and the corresponding ASTM specifications.

- *Chemical compositions:* Chemical analysis of portland cement (ASTM C114).
- *Fineness:* Sieve analysis (ASTM C184), Wagner turbidimeter method (ASTM C115), Blaine air-permeability method (ASTM C204).
- *Normal consistency* (ASTM C187): This test is used to determine the amount of water to be used in making samples to be tested for soundness, time of setting, and strength. The amount of water needed for normal consistency varies between 22 and 28% for portland cement.
- *Time of setting:* Time of setting includes both initial and final setting times. The time needed for the paste to start stiffening is called the initial setting time, whereas the final setting time represents the end of plasticity. The ASTM standards are C191 for Vicat apparatus and C266 for Gillmore needles. There is also a term called false set. This represents a stage in which the mix that is stiff can be remixed without adding water to restore plasticity. ASTM specifications require that initial setting time be at least 45 min by Vicat apparatus and 60 min by Gillmore needles. The corresponding final setting times are 8 and 10 h, respectively.
- *Soundness:* The test for soundness involves the measurement of the expansion of hardened cement paste. One of the popular tests is the autoclave test (ASTM C151).
- *Strength:* Strength is probably the most important property sought after. Both the magnitude of strength and the rate of strength development are important. The basic strength tests are compression (ASTM, C109), tension (ASTM C190), and flexure (ASTM C348 or C349).
- *Heat of hydration* (*ASTM C186*): This property, which provides an indication of the amount of heat generated during hydration, is extremely important for most concrete construction such as dams, thick slabs, and pile caps.
- *Other tests:* Other tests that are used infrequently include shrinkage or expansion tests and tests for measuring specific gravity, alkali reactivity, sulfate resistance, air entrainment, bleeding, and efflorescence.

3A.1.2 Aggregates

Aggregates are much less expensive than the cementing material and could constitute up to 90% of the volume of the concrete. They are typically considered as inert filler material, even though some aggregates do react minimally with the cement paste. Aggregates can be classified as coarse or fine based on their size; normal weight, lightweight, or heavyweight based on their bulk densities; and natural mineral or synthetic based on their type of production. Aggregate characteristics that affect the final product (namely, the concrete) include porosity, grading and size distribution, shape, surface texture, crushing strength, elastic modulus, moisture absorption, and type and amount of deleterious substances present. The primary concerns are the quality of the aggregate and its grading.

Coarse Aggregates. If the particle size is greater than 0.25 in (6 mm), the aggregates are classified as coarse aggregates. A list of the common coarse aggregate types follows.

- *Natural gravel:* About 50% of the coarse aggregates used in the United States consists of gravel. Natural cobbles and gravel are produced by weathering action. They are usually round in shape with a smooth surface. Hence these aggregates provide better workability. When they are made of siliceous rocks and uncontaminated with clay or silt, they make strong and durable aggregates. Some of the very high strength concretes with a compressive strength in the range of 20,000 psi (140 MPa) are made using gravel aggregates.

- *Natural crushed stone:* Crushed stone aggregate is produced by crushing the rocks and grading them. About two-thirds of the crushed aggregate in the United States is made of carbonate rocks (limestone, dolomite). The remainder is made of sandstone, granite, diorite, gabbro, and basalt. Carbonate rocks are softer than siliceous sedimentary rocks. The characteristics such as strength, porosity, and durability could vary considerably. Hence care should be taken to avoid the rocks that are not suitable for aggregates. Crushed stone aggregates are typically angular in shape and thus less workable than gravel under similar conditions.

- *Lightweight aggregates:* Aggregates with a bulk density of less than 70 lb/ft^3 (1120 kg/m^3) are normally considered lightweight aggregates. However, there is a whole spectrum of lightweight aggregates weighing from 5 to 55 lb/ft^3 (80 to 900 kg/m^3), as shown in Fig. 3A.1.[1] Natural lightweight aggregates are made by processing naturally occurring lightweight rock formations such as pumice, scoria, and tuff. But most of the lightweight aggregates used for structural concrete are made by expanding or thermally treating a variety of materials such as clay, shale, slate, diatomite, perlite, or vermiculite. Industrial by-products such as blast-furnace slag and fly ash are also used to manufacture lightweight aggregates. The lightweight aggregates can be grouped into three categories based on their end use: structural concrete, production of masonry units, and insulating concrete.

- *Heavyweight aggregates:* The bulk density of heavyweight aggregates ranges from 145 to 280 lb/ft^3 (2320 to 4480 kg/m^3). The primary use of these aggregates is for nuclear radiation shields. Natural rocks suitable for heavyweight aggregates may contain barium minerals, iron ores, or titanium ores. The aggregate types are witherite (BaCO$_3$), barite, (BaSO$_4$), magnetite (Fe$_3$O$_4$), hematite (Fe$_2$O$_3$), hydrous iron ores, ilmenite (FeTiO$_3$), ferrophosphorus (Fe$_3$P, Fe$_2$P, FeP), and steel aggregate (Fe). Ferrophosphorus aggregates when used with portland cement might generate flammable (and possibly) toxic gases and, hence, should be used with caution. Boron and hydrogen are very effective for neutron attenuation, and hence the aggregates containing these

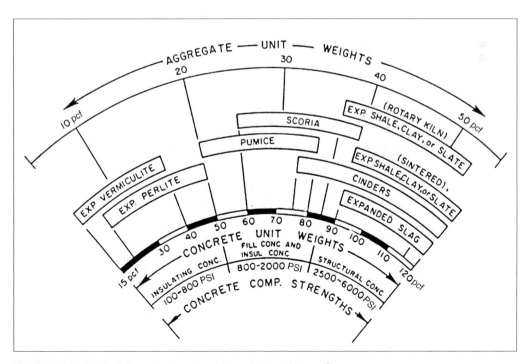

FIGURE 3A.1 Lightweight aggregate spectrum. (*From Litvin and Fiorato.*[1])

compounds are more useful for shielding. Heavyweight aggregates tend to produce more segregation. Sometimes preplaced-aggregate techniques are employed to make heavyweight concrete in order to avoid segregation.

- *Aggregates made from recycled concrete and other waste products:* Rubble from demolished buildings could be used as aggregates. The best source is the recycled concrete that contains less hydrolized cement paste and gypsum. Concrete made with these aggregates has about two-thirds of the compressive strength of concrete made with stone aggregates. One of the major projects built using recycled aggregate was a repavement project in Michigan. The work was carried out by the Michigan Department of Transportation and involved about 125,000 yd^3 (96,000 m^3) of concrete made using recycled concrete pavements.

Investigations have also been conducted for utilizing aggregates made with municipal wastes and incinerator ashes. The results are not very positive. The presence of glass particles tends to reduce workability and long-term durability. Metals such as aluminum react with alkaline materials in concrete and expand, causing deterioration. Organic wastes and paper interfere with the setting of concrete.

Fine Aggregates. Fine aggregates are usually made of natural or crushed sand. Their size ranges from 0.25 to 0.01 in (6 to 0.25 mm). The fine aggregates should be free of organic materials, clay, or other deleterious materials. Minimum particle sizes should be not less than 0.01 in (0.25 mm) because these fine particles tend to increase the water demand and reduce strength. For all-lightweight concrete, lightweight sand made of expanded shale or clay materials is used. Similarly, for heavyweight concrete, regular sand is replaced with fine steel shot, iron ore, or other high-density material.

Production of Aggregates. The natural sand and gravel aggregates are the most economical aggregates to produce. They only need cleaning and grading. Cleaning can be accomplished by either screening or washing. Washing is more effective for removing clay and silt particles, but the process is more expensive.

When aggregates are produced by crushing rock, the type of crushing equipment depends on the type of rock to be crushed. The major sequential operations are crushing, cleaning, separation of various sizes, and blending of various sizes to obtain the required gradation.

TABLE 3A.2 Grading Requirements for Coarse Aggregates

Size number	Nominal size (sieves with square openings)	Amounts finer than laboratory sieve (square openings), wt %				
		100 mm (4 in)	90 mm (3½ in)	75 mm (3 in)	63 mm (2½ in)	50 mm (2 in)
1	90–37.5 mm (3½–1½ in)	100	90–100	—	25–60	—
2	63–37.5 mm (2½–1½ in)	—	—	100	90–100	35–70
3	50–25.0 mm (2–1 in)	—	—	—	100	90–100
357	50–4.75 mm (2 in–No. 4)	—	—	—	100	95–100
4	37.5–19.0 mm (1½–¾ in)	—	—	—	—	100
467	37.5–4.75 mm (1½ in–No. 4)	—	—	—	—	100
5	25.0–12.5 mm (1–½ in)	—	—	—	—	—
56	25.0–9.5 mm (1–⅜ in)	—	—	—	—	—
57	25.0–4.75 mm (1 in–No. 4)	—	—	—	—	—
6	19.0–9.5 mm (¾–⅜ in)	—	—	—	—	—
67	19.0–4.75 mm (¾ in–No. 4)	—	—	—	—	—
7	12.5–4.75 mm (½ in–No. 4)	—	—	—	—	—
8	9.5–2.36 mm (⅜ in–No. 4)	—	—	—	—	—

Source: From *Annual Book of ASTM Standards.*[2]

In the case of synthetic lightweight aggregates, an extra step of expanding is needed. Generally, crushed or ground and pelletized raw material (clay, slate, or shale) is heated to about 2012°F (1100°C). At this temperature, the material expands because of the entrapped gases. The expanded material is processed to obtain the required grading. Lightweight aggregates are typically more expensive because of the extra effort and energy needed to produce them.

Size and Grading. The maximum size of the aggregates depends on the type of structure for which the concrete is to be used. For example, in dams and large foundations, maximum sizes greater than 2 in (50 mm) are very common, whereas for beams and columns containing extensive reinforcement, the maximum size is often restricted to 0.75 in (19 mm). The maximum size of the aggregate should be smaller than one-fifth of the narrowest dimension of the form and three-fourths of the maximum clear distance between reinforcements. For high-strength concrete the maximum size is generally limited to 0.75 in (19 mm).

As mentioned earlier, aggregates constitute a combination of different-size particles. The distribution of these particle sizes is called grading. The aim is to attain a grading (size distribution) that will produce a concrete with the required strength using minimum cement. At the same time, the mix should be workable in the plastic stage. Theoretically the best combination of particle sizes is the one that produces minimum voids and minimum total surface area. ASTM has established grading requirements for both coarse and fine aggregates so as to obtain workable concrete mixtures. Tables 3A.2 and 3A.3 present these requirements for normal-weight concrete. Similar requirements for lightweight (C320) and heavyweight (C637) concrete can be found in the *Annual Book of ASTM Standards,*, vol. 04.02.

The grading requirement for fine aggregate is straightforward (Table 3A.3). The distribution of particles has to follow a certain pattern, with particles of different sizes limited to a certain range. For the coarse aggregate, first a maximum size should be chosen. The size distribution of the other particles depends on this maximum size. For example, if the maximum size is 1.0 in (25 mm), size numbers 5, 56, or 57 can be chosen (Table 3A.2). If size 5 is chosen, 100% of the aggregates should pass through a 1.5-in (37.5-mm) sieve. Reading horizontally, 90–100, 20–55, 0–10, and 0–5% of material should pass through 1-, 0.75-, 0.5-, and 0.375-in (25-, 19-, 12.5-, and 9.5-mm) sieves, respectively. The sieve size represents the dimensions of the square openings. For example, a 1-in (25-mm) sieve has a mesh with a 1-in (25-mm) square opening. The reader should refer to ASTM C33 for other restrictions, such as fineness modulus.

TABLE 3A.2 (*Cont.*)

Amounts finer than laboratory sieve (square openings), wt %							
37.5 mm (1½ in)	25.0 mm (1 in)	19.0 mm (¾ in)	12.5 mm (½ in)	9.5 mm (⅜ in)	4.75 mm (No. 4)	2.36 mm (No. 8)	1.18 mm (No. 16)
0–15	—	0–5	—	—	—	—	—
0–15	—	0–5	—	—	—	—	—
35–70	0–15	—	0–5	—	—	—	—
—	35–70	—	10–30	—	0–5	—	—
90–100	20–55	0–15	—	0–5	—	—	—
95–100	—	35–70	—	10–30	0–5	—	—
100	90–100	20–55	0–10	0–5	—	—	—
100	90–100	40–85	10–40	0–15	0–5	—	—
100	95–100	—	25–60	—	0–10	0–5	—
—	100	90–100	20–55	0–15	0–5	—	—
—	100	90–100	—	20–55	0–10	0–5	—
—	—	100	90–100	40–70	0–15	0–5	—
—	—	—	100	85–100	10–30	0–10	0–5

TABLE 3A.3 Grading Requirements for Fine Aggregates

Sieve (Specification E11)	Percent passing
9.5 mm (3/8 in)	100
4.75 mm (No. 4)	95–100
2.36 mm (No. 8)	80–100
1.18 mm (No. 16)	50–85
600 μm (No. 30)	25–60
300 μm (No. 50)	10–30
150 μm (No. 100)	2–10

Source: From *Annual Book of ASTM Standards.*[2]

Most of the time the aggregates have a uniform gradation. However, gap-graded aggregates are shown to provide better strength characteristics. In gap grading, the sizes of aggregates do not decrease uniformly. Certain size segments are left out in order to obtain a better packing and, hence, a more efficient utilization of cement.

Other Aggregate Characteristics. Grading is the most influential parameter that affects the behavior of concrete. Other factors influencing the quality of concrete are density and apparent specific gravity, absorption and surface moisture, crushing strength, elastic modulus, abrasion resistance, soundness, shape and surface texture, and the presence of deleterious substances. The significance of some of these parameters is listed in Table 3A.4. The table also presents the ASTM test methods that can be used for the evaluation of these characteristics.

Selection of Aggregates. The ideal aggregate consists of particles that are strong, durable, clean, do not flake when wetted and dried, have somewhat rough surface texture, and contain no constituents that interfere with cement hydration. The grading of these particles should be done so that the concrete has good workability and the cement is utilized to its maximum efficiency. It is seldom possible to obtain ideal aggregates because of economical constraints.

In practical situations, aggregate selection should be made based on field conditions and end use. For example, a foundation built on sulfate-containing soil should not have aggregates vulnerable to sulfate attack. If the structure is going to be exposed to freezing and thawing cycles, durability of the aggregate plays a major role. The gradation requirements should be chosen not only for the maximum utilization of cement, but also to produce a workable concrete mixture. The availability of aggregates in close proximity plays an important economical role. If past records regarding the performance of the potential aggregate source based on existing structures are not available, suitable tests should be run to evaluate their properties. It should be noted that some of

TABLE 3A.4 Aggregate Characteristics and Their Significance

Characteristic	Effect on concrete	ASTM test method
Gradation	Economy, workability, long-term performance	C117, C136
Bulk unit weight	Mix proportioning calculations	C29
Absorption and surface moisture	Quality control of concrete	C70, C127, C128, C566
Abrasion resistance	Wear resistance of floors and pavements	C131, C295, C535
Resistance to freezing and thawing	Surface scaling, durability	C295, C666, C682
Particle shape and surface texture	Workability of fresh concrete	C295, C398
Elastic modulus	Elastic modulus of concrete	C469

the disadvantages of the aggregate could be overcome by making minor modifications using admixtures. For example, if workability is the problem because of angular surfaces, water-reducing admixtures could be used to overcome this problem. The only constraint is the economy. The additional cost of the admixture should be justifiable.

3A.1.3 Water and Water-Reducing Admixtures

Water is one of the primary ingredients of concrete. Water used during the mixing, called mixing water, performs the basic functions of hydration and lubrication. It also provides space for expanding hydration products. The lubrication action influences the workability of fresh concrete for placing, compaction, and finishing operations. The hydration, or the chemical reaction between water and cement, results in the hardening of concrete. Water is also needed for curing and sometimes for washing the aggregates. The amount of water needed for adequate workability is always greater than that needed for hydration. In addition, complete hydration of cement does not produce the highest strength. Therefore a number of admixtures were developed to improve the workability of concrete containing a limited amount of water. The most notable admixture is called high-range water-reducing admixture because of its very high efficiency. This section describes the requirements of water quality and the properties of water-reducing admixtures.

Water. Typically the water that is good for drinking is good for making concrete. Certain mineral waters that are potable may not be suitable for concrete. The water should be free of a particular taste, color, and odor, and should not foam or fizz when shaken. If in doubt, the water should be tested for suitability by evaluating the setting time of the cement paste, compressive strength, and durability. In most cases the setting time test alone may be sufficient.

The harmful contaminates not permitted in the water used for concrete are sugar, tannic acid, vegetable matter, oil, humic acid, alkali salts, free carbonic acid, sulfates, and water containing effluents from paint and fertilizer factories and sewage treatment plants. The following can be considered the general maximum limits for impurities:

Acidity	$0.1N$ NaOH; 2 mL maximum to neutralize 200-mL sample
Alkalinity	$0.1N$ HCl; 10 mL maximum to neutralize 200-mL sample; pH in the range of 6 to 9
Organic solids	$\not> 0.02\%$
Inorganic solids	$\not> 0.30\%$
Sulfuric anhydride	$\not> 0.04\%$
Sodium chloride	$\not> 0.10\%$
Turbidity	$\not> 2000$ ppm

Water containing more impurities than the limits mentioned, such as seawater, has been used successfully. However, special tests should be conducted for the particular application prior to approval. In most cases a strength reduction of 15% can be expected if the mixing water contains salts. Decreased durability and corrosion of reinforcement can also be anticipated.

The requirements for curing water are less stringent because it comes in contact with the concrete on the surfaces only and that for too short a duration. Nevertheless, water with excessive impurities should not be used because it could cause surface discoloration. Under the worst conditions curing water could cause surface deterioration.

With regard to the water used for washing the aggregates and for concrete mixing and placing equipment the primary concern is the deposit of minerals on aggregate particles and equipment. The water should be clean enough not to leave any deposit. Washing the aggregate is not usually recommended because the disposal of contaminated water presents a problem. It should be noted that the disposal of contaminated water into natural streams and drains is not permitted in the United States.

Water-Reducing Admixtures. The water-cement ratio is the most influential parameter controlling the strength. Typically, a lower water content results in higher strength. However, a certain amount of water is needed to obtain workability so that the fresh concrete can be placed in position and compacted. Water-reducing admixtures improve the workability, and thus workable concrete can be obtained without increasing the water content. These admixtures can be used either to improve workability for the same water content, as the reduction of water without losing workability results in higher compressive strength, or to reduce the cement content, maintaining workability and strength.

For example, consider a reference concrete with 500 lb/yd^3 (300 kg/m^3) cement and a water-cement ratio of 0.62. This concrete had a slump of 2 in (50 mm) and a 28-day compressive strength of 5.3 ksi (37 MPa). Addition of the admixture resulted in an increase of slump (workability) to 4 in (100 mm). The compressive strength was 5.4 ksi (38 MPa).

When the admixture was used to reduce the water-cement ratio to 0.56, maintaining a slump of 2 in (50 mm), the compressive strength increased to 6.8 ksi (46 MPa). If the slump and strength levels were kept at the reference levels of 2 in (50 mm) and 5.3 ksi (37 MPa), the cement content could be reduced to 450 lb/yd^3 (270 kg/m^3). The 10% reduction in cement provides not only economy but also other benefits such as low heat of hydration and reduced shrinkage.

The principal active ingredients in water-reducing admixtures are salts, modifications and derivatives of lignosulfonic acid, hydroxylated carboxylic acids, and polyhydroxy compounds. Typically these admixtures provide better dispersion of cement particles and, hence, could provide an increase in the early-age strengths. Larger amounts of admixtures could retard the setting time by preventing the flocculation of hydrated particles. Some commercial formulations may contain accelerating admixtures to overcome this effect. Admixtures derived from lignin products also tend to entrain considerable air. This effect is nullified by adding air-detraining agents. The period of effectiveness varies with the formulations. However, in all cases the improved workability will be at least partially lost as the cement starts to hydrate.

A special type of water-reducing admixture called high-range water-reducing admixture or superplasticizer was developed in the 1970s and is commonly used now. As the name implies, this admixture provides substantial improvement in the workability. A water reduction of 20 to 25% is possible, as compared to 5 to 10% for normal water-reducing admixtures. Introduction of the superplasticizer is also responsible for the use of high-volume fly ash and silica fume in concrete because these mineral admixtures (at high volume fractions) have a high water demand and could not be used without the aid of a superplasticizer.

Superplasticizers consist of long-chain, high-molecular-weight anionic surfactants with a large number of polar groups in the hydrocarbon chain. These compounds are adsorbed on the cement particles during the mixing and impart a strong negative charge, helping to lower the surface tension of the surrounding water. This results in a uniform distribution of the cement particles and increased fluidity. Because of the better dispersion of cement particles, concrete made with superplasticizers tends to have higher 1-, 3-, and 7-day strengths as compared to reference concrete having the same water-cement ratio. The negative effect of better cement distribution is a rapid loss of workability because of accelerated setting. Hence set-retarding admixtures are typically added to control the setting time.

The four basic types of superplasticizers available in the market are as follows:

1. Sulfonated melamine formaldehyde condensates
2. Sulfonated naphthalene formaldehyde condensates
3. Modified lignosulfonates
4. Sulfonic acid esters or other carbohydrate esters

Typical dosage of superplasticizers is in the range of 1 to 2.5%, even though dosages as high as 4% have been used successfully. One of the major problems encountered in using superplasticizers is the loss of workability with time. Some of the original versions lost their effectiveness in less than 1 h. The currently available formulations are effective for longer durations. The individual commercial product should be checked for its effective duration. Multiple dosages or retempering with superplasticizer was also found to be effective. It was found that the mix can be retempered three times without affecting the mechanical properties adversely.

The water-reducing admixtures are covered in ASTM C494. High-range water-reducing admixtures are called type F, and the regular admixtures are called type A water-reducing admixtures.

3A.1.4 Chemical Admixtures

As mentioned earlier, the primary ingredients of concrete are cement, aggregates, and water. Any other ingredient added to concrete can be classified as an admixture. The functions performed by admixtures include improved workability, acceleration or retardation of setting time, control of strength development, improved freeze-thaw durability, and enhanced resistance to water permeation, frost action, thermal cracking, aggressive chemicals, and alkali-aggregate expansion. The admixtures are also used to improve economy and save energy. In some countries up to 80% of all concrete produced contains some kind of admixtures.

The admixtures can be broadly classified as chemical and mineral admixtures. The mineral admixtures are discussed in Sec. 3A.1.5. The most frequently used chemical admixtures are (1) accelerators or retarders, (2) water reducers, and (3) air-entraining admixtures. The water-reducing admixtures were discussed in Sec. 3A.1.3. This section deals with the other admixtures.

Accelerating Admixtures. Accelerating admixtures, or accelerators, classified as type C admixtures in ASTM C494, are used in concrete to reduce the time of setting or to enhance early strength development, or both. It should be noted that increased early-strength development could lead to a reduction in strength at later ages. In any case, improvement in long-term strength should not be expected. The chemical components used in accelerating admixtures include soluble chlorides, carbonates, silicates, fluorosilicates, hydroxides, bromides, and organic compounds.

The higher early strength can be used to achieve the following benefits in construction:

- Early finishing of surfaces
- Early removal of forms
- Early opening of construction for service
- More efficient plugging of leaks against hydraulic pressure
- Partial or complete compensation for effects of low temperature
- Reduction of the time required for curing and protection against cold weather

The most common accelerator used for concrete is calcium chloride ($CaCl_2$). Calcium chloride can be safely used up to 2% by weight of cement. The influence of this chemical on setting time and strength development is presented in Fig. 3A.2. From this figure it can be seen that (1) the initial and final setting times can be reduced by as much as 50 and 70%, respectively, (2) 1- and 3-day strengths can be increased significantly, and (3) as the curing temperature decreases, the effectiveness of the admixture increases.

The influence of calcium chloride on various properties of concrete is shown in Table 3A.5. Almost all of the mechanical properties at early age are improved. The most detrimental effect is on the corrosion of metals. Corrosion becomes a major problem only in locations where there is a steep gradient in the chloride ion concentration. However, even a uniform concentration beyond 2% is viewed with suspicion. The addition of calcium chloride also increases creep and shrinkage and aggravates alkali-aggregate reaction.

A number of nonchloride accelerating admixtures have been developed during the late 1980s. Some of these admixtures are almost as effective as calcium chloride.

Air-Entraining Admixtures. Air-entraining admixtures are used to entrain small spherical air bubbles about 10 to 1000 μm in diameter. Entrained air significantly increases the frost resistance

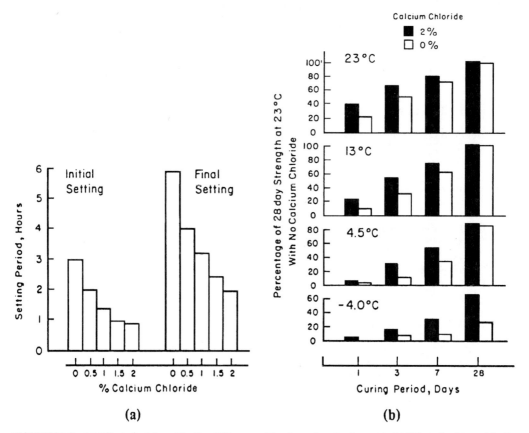

FIGURE 3A.2 (*a*) Effect of calcium chloride addition on setting time of portland cement. (*b*) Effect of calcium chloride addition on strength at various curing temperatures. (*From Ramachandran.*[3])

and durability under freezing and thawing conditions. Most specifications mandate air entrainment for exposed structures. About 9% by volume of mortar is recommended for proper freeze-thaw durability. The air voids should be distributed uniformly with low spacing factors (distance between bubbles). Smaller bubbles are more effective than larger ones. Entrapped air, which is the result of incomplete compaction, is not effective in improving durability.

The usual dosage of air-entraining mixtures is in the range of 0.02 to 0.06% by weight of cement. Higher dosages may be required for mixes containing type III cement, pozzolan cements, fly ash, or other finely divided powders. The volumetric air content in typical air-entrained concrete varies from 4 to 10%. The presence of air improves the workability of fresh concrete because the bubbles increase the spacing of the solids, resulting in decreased dilatancy. The air also reduces segregation and bleeding. Entrained air is particularly helpful in lightweight concrete because of the unfavorable shape and surface texture of the fine fraction of most lightweight aggregates. The air reduces the harshness of the mix and the bleeding rate in addition to providing improved workability. Entrained air typically reduces the compressive strength. Concrete containing 8 vol % air can be expected to register about 15% reduction in compressive strength as compared to control concrete with no entrained air and comparable workability. Note that to obtain comparable workability, the water content of the control concrete has to be increased.

TABLE 3A.5 Some Properties Influenced by Calcium Chloride Admixture in Concrete

Property	Effect	Remarks
Setting	Reduces both initial and final setting times	ASTM standard requires that initial and final setting times occur at least 1 h earlier with respect to reference concrete.
Compressive strength	Significant increase at 3 days (gain may be about 30–100%)	ASTM requires an increase of at least 125% over control concrete at 3 days.
Tensile strength	Slight decrease at 28 days	
Flexural strength	Decrease of about 10% at 7 days	This figure may vary depending on starting materials and method of curing. The decrease may be more at 28 days.
Heat of hydration	Increase of about 30% in 24 h	Total amount of heat at longer times is almost the same as that evolved by reference concrete.
Resistance to sulfate attack	Reduced	Can be overcome by using type V cement with adequate air entrainment.
Alkali-aggregate reaction	Aggravated	Can be controlled by using low-alkali cement or pozzolana.
Corrosion	Causes no problems in normal reinforced concrete if adequate precautions taken (dosage should not exceed 1.5% $CaCl_2$ and adequate cover given)	Calcium chloride admixture should not be used in prestressed concrete or in a concrete containing a combination of dissimilar metals. Some specifications do not allow use of $CaCl_2$ in reinforced concrete.
Shrinkage and creep	Increased	
Volume change	Increase of 0–15% reported	
Resistance to damage by freezing and thawing	Early resistance improved	At later ages may be less resistant to frost attack.
Watertightness	Improved at early ages	
Modulus of elasticity	Increased at early ages	At longer periods almost same with respect to reference concrete.
Bleeding	Reduced	

Source: From Ramachandran.[3]

Most of the commercially available air-entraining admixtures are in liquid form; a few are available in powder, flake, or semisolid form. The ingredients used for manufacturing admixtures include salts of wood resins, synthetic detergents, salts of sulfonated lignin, salts of petroleum acids, salts of proteinaceous materials, fatty and resinous acids and their salts, and organic salts of sulfonated hydrocarbons. The air-entraining admixture is typically added to the mixing water. Air-entrained concrete can also be made using air-entraining portland cement. The former method is preferable because the amount of air can be controlled easily.

The amount of air, or air content, is the primary factor that influences freeze-thaw durability and frost resistance. Other factors that are important include size and distribution of air voids, specific surface, and spacing of bubbles. These factors are influenced not only by the amount of air-entraining agent but also by the nature and proportions of the other ingredients of the concrete, including other admixtures, water-cement ratio and consistency of the mix, type and duration of mixing, temperature, and the type and degree of compaction employed. Based on the constituent materials and the type of mixing, placing, and compaction, the amount of air-entraining agent should be adjusted to obtain the desired air content in the final product.

Set-Controlling Admixtures. Set-controlling admixtures are typically used in conjunction with water-reducing admixtures. As mentioned in Sec. 3A.1.3, retarders have to be used in conjunction with some water-reducing admixtures in order to extend the life, or effectiveness, of the water-reducing admixtures. Materials, such as lignosulfonic acids and their salts and hydroxylated carboxylic acids and their salts, can act as water-reducing, set-retarding admixtures.

Set-retarding admixtures are used by themselves to offset the accelerating effects of high temperatures and to extend the workable period for proper placing. Set retarders are also used to keep the concrete workable for a longer duration. The longer duration may be needed to avoid cold joints in large constructions or to prevent cracking in beams, bridge decks, and composite construction because of the deflections caused when the adjacent spans are loaded.

The amount of retardation obtained depends on the chemical composition of the admixture, its dosage, type and amount of cement, temperature, mixing sequence, and other job conditions. The quantity of admixture required should be determined carefully. Excessive retardation can damage the setting and hardening process, resulting in long-term detrimental effects.

Other Chemical Admixtures. Chemical admixtures other than the ones discussed so far include admixtures used for (1) air detraining, (2) gas forming, (3) producing expansion, (4) damp proofing, (5) bonding, (6) reducing alkali-aggregate expansion, (7) inhibiting corrosion, (8) flocculating, (9) coloring, and (10) fungicidal, germicidal, and insecticidal purposes. In addition, admixtures are available for water thickening and for reducing friction in pumping concrete.

Polymer and latex-modified concretes are also being used, especially for repair and restoration. Polymers are used as bonding agents and surface coatings to reduce permeability of the surface layer. Impervious surface layers typically improve the long-term durability.

3A.1.5 Mineral Admixtures

Mineral admixtures in concrete provide improved resistance to thermal cracking because of reduced heat of hydration, reduce the permeability by reducing pore sizes, increase strength, and improve resistance against chemicals such as sulfate water and alkali-aggregate expansion. Most of the mineral admixtures have some pozzolanic property. A pozzolan is defined as a siliceous or siliceous and aluminous material that will chemically react with calcium hydroxide at normal temperatures to form compounds possessing cementitious properties. The material has to be in a finely divided form, and the reaction will take place only in the presence of moisture. Pozzolans possess little or no cementitious value by themselves. Typically, mineral admixtures are used in large volume fractions, generally in the range of 20 to 100% by weight of cement.

Even though a number of naturally occurring pozzolans exist and can be used as admixtures, most of the admixtures used in concrete, especially by industrialized nations, are industrial by-products. The most commonly used industrial by-product is coal fly ash produced by power plants. Blast-furnace slag and silica fume are the other major industrial by-products used in concrete. Silica fume, which contains much finer particles, is typically used for high-strength and impermeable concrete. These three admixtures are discussed next, followed by other mineral admixtures.

Fly Ash. In modern power plants, coal is fed into furnaces in powder form to improve thermal efficiency. As the coal powder passes through the high-temperature zone in the furnace, the volatile matter and carbon burn off, providing heat generation. The impurities, such as clay, feldspar, and quartz, melt and fuse. When the fused matter moves through zones of lower temperature, it solidifies as glassy spheres. The glass contents in these spheres range from 60 to 85%. Some of these spheres are hollow and very light. The spheres get blown out with the flue gas stream. These particles, collected with special equipment such as electrostatic precipitators, are called fly ash. Special equipment is needed for collecting most of the tiny particles so that the amount of ash discharged into the atmosphere is at an absolute minimum.

Based on the amount of calcium content, fly ash is classified as low-calcium or high-calcium. ASTM classifies high-calcium fly ash (CaO content 15 to 35%) as type C and low-calcium fly ash (CaO less than 10%) as type F fly ash. High-calcium fly ash is more reactive because the calcium occurs in the form of crystalline compounds. The chemical that is considered to be harmful to concrete is carbon. In most commercial fly ashes the carbon content is limited to 2%, and it rarely exceeds 5%. If the carbon content is high, the fly ash should not be used in concrete. A higher carbon content typically increases water demand and interferes with air-entraining admixtures.

The sizes of fly ash particles vary from <1 to 100 μm. The particle size distribution is shown in Fig. 3A.3. This figure shows the particle distributions of type C and F fly ash, cement, and silica fume. It can be seen that the fly ash particle sizes are about the same as the cement particle sizes. About 50% of the particles are smaller than 20 μm. The particle size distribution influences the workability of fresh concrete and the rate of strength development.

It is well established that the addition of fly ash in the range of 10 to 20% by weight of cement improves the workability. Fly ash containing finer particles is more effective in improving workability. As much as 7% water reduction was reported with an addition of 30% fly ash.

One of the primary reasons for using fly ash in concrete is to reduce the heat of hydration during placement and the early curing period. Fly ash could reduce the rise in temperature almost in direct proportion to the amount of cement it replaces. Fly ash has been used in concrete as early as the 1930s. In most of the earlier applications the amount of cement replaced was limited to about 30%. Recently equal amounts of cement and fly ash have been used successfully for mass concrete construction. As mentioned earlier, the advent of superplasticizers was responsible for the use of high volume fractions of fly ash. At lower dosages, fly ash improves the workability. When large volume fractions are used, more water is needed to wet the fly ash particles. The use of superplasticizers allows for the least increase in water content, and hence high strengths can be achieved even with large volume fractions of fly ash.

The pozzolanic activity of fly ash refines the pores, and thus concrete containing fly ash is less permeable. The 90-day permeability of the cement can be reduced by almost an order of magnitude by replacing 10 to 30% of cement with fly ash. This reduced permeability enhances the durability against chemical attacks significantly. Alkali-aggregate expansion can also be reduced by adding fly ash.

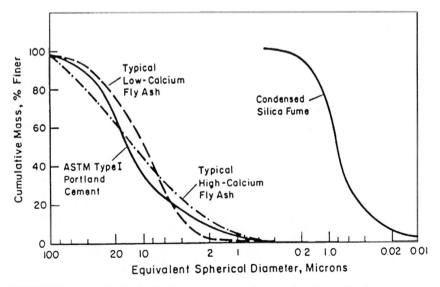

FIGURE 3A.3 Particle-size distributions—comparison of cement, fly ash, and silica fume.

Replacement of cement with fly ash normally results in a reduction of the 28-day strength. But the long-term (56- or 90-day) strengths are always higher for fly ash concrete. High-strength concretes with compressive strengths greater than 8000 psi (55 MPa) always contain some form of mineral admixture.

Blast-Furnace Slag. Blast-furnace slag is a by-product of cast iron production. The chemical components present in the slag do not react with water at normal temperatures. The slag is ground to fine particles and used as mineral admixtures. In some cases the liquid slag is cooled rapidly to produce sandlike and pelletlike particles, called granulated and pelletized slag, respectively.

The properties of concrete containing slag are almost the same as the properties of concrete containing fly ash, in both the plastic and the hardened forms. The major difference is that slag particles smaller than 10 μm were found to contribute to an increase in the 28-day strength.

Silica Fume. Silica fume, also called condensed silica fume, volatilized silica, or microsilica, is a by-product of silicon metal and ferrosilicon alloy industries. The reduction of quartz to silicon at temperatures of 3632°F (2000°C) produces SiO vapors. These vapors oxidize and condense to form small spheres with diameters in the range of 0.01 to 0.2 μm (Fig. 3A.3). Silica fume particles are about two orders of magnitude smaller than cement particles. Because of its extremely small particle size, silica fume is highly pozzolanic. But its higher surface area increases the water demand. Hence the use of silica fume was almost impossible until the advent of superplasticizers. The use of silica fume for high-strength concrete became a common occurrence after the superplasticizers were introduced.

The silica fume content in concrete ranges from 5 to 30% by weight of the cement. The addition of silica fume typically results in denser concrete with low permeability. Silica fume is more effective in reducing permeability than fly ash. The strength increase provided by silica fume is also more substantial than that obtained by the addition of fly ash. Because of the collecting and processing expense, silica fume is much more expensive than fly ash. It is available in powder and slurry form. Silica fume concrete typically has a higher Young's modulus and is more brittle than concrete of comparable strength that contains no silica fume.

Rice-Husk Ash. Rice-husk ash can be considered a natural product, even though it is produced by controlled combustion of rice husks. Rice husks are the product of a dehusking operation in which outer shells are removed from rice. Husk constitutes about 20% of paddy by weight. The combustion process reduces the weight about fivefold. Ash produced by controlled combustion was found to produce nonreactive silica minerals. These ashes have to be finely ground before using in concrete.

Naturally Occurring Mineral Admixtures. Naturally occurring pozzolans are typically mined and processed. Processing normally involves the steps of crushing, grinding, and size separation. In some cases thermal activation may be needed.

Natural pozzolans are derived from volcanic rocks and minerals and from diatomaceous earth. Diatomaceous earth consists of hydrated silica derived from skeletons of diatoms. Diatoms are tiny water plants whose walls are composed of silica shells. Pozzolans were formed during volcanic eruption because of the quick cooling of magma-containing aluminosilicates.

Based on their major chemical components, natural pozzolans can be classified as (1) volcanic glasses, (2) volcanic tuffs, (3) calcined clays or shales, and (4) diatomaceous earth.

3A.2 COMPUTATION OF REQUIRED AVERAGE COMPRESSIVE STRENGTH f'_{cr} AND MIX PROPORTIONING

The structural engineer who designs the components specifies the minimum compressive strength required for the concrete to be used. This strength is called specified compressive strength f'_c. In most cases the specified strength is measured at the age of 28 days. In some cases 56- or 90-day strengths are also specified. Compressive strength tests are normally conducted using cylinders, cubes, or prisms. It is the job of the construction professionals to ensure that the concrete placed in position satisfies the specified strength requirement. Since concrete is a composite, cast in the field using a number of constituent materials and various casting and curing procedures, there is always a variation in strength. The major parameters that influence the strength include the following:

1. Amount of water used in the mix, or water-cement ratio
2. Aggregate-cement ratio
3. Quality of cement
4. Strength, shape, texture, cleanliness, and moisture content of aggregates
5. Type and amount of mineral and chemical admixtures
6. Mixing procedure and adequacy of mixing
7. Placing, compacting, and finishing techniques used during construction
8. Curing conditions and type of curing method
9. Test procedures

Because of variations in any of these parameters, or other factors such as temperature and humidity in the field during the construction, there is always a variation in compressive strength. Typically, quality control tests are run in both the field and the laboratory to monitor the variations and take corrective measures if needed. American Concrete Institute (ACI) code 318-92 specifies the following acceptance criteria for concrete[4]:

1. The average of all sets of three consecutive strength tests must equal or exceed f'_c, and
2. No individual strength test (average of two cylinders) must fall below f'_c by more than 500 psi (3.4 MPa)

The code also provides guidelines for achieving the acceptable concrete.

Stated in simple terms, the code requires that the concrete be proportioned to obtain an average compressive strength f'_{cr} that is higher than the specified strength f'_c. The magnitude of overdesign, that is, the difference between f'_{cr} and f'_c, depends on the rigorousness and success of the quality control measures used on the job site. The concrete should be proportioned to obtain the average compressive strength f'_{cr} and not the specified strength f'_c.

This section deals with the computation of f'_{cr}, which is also known as required average strength, and the mix proportioning procedures. The computation of f'_{cr} is based on ACI code guidelines. It should be noted that other codes may require different procedures for the computation of the required average strength.

3A.2.1 Computation of Required Average Strength, f'_{cr}

If the concrete production facility has test records from previous projects, these records can be used to establish the variability and the mix proportions. In this case the computation of f'_{cr} is based on previous records. If the records are not available, a certain variability is assumed. As the project progresses, the data from the project can be used to establish the variability.

If field data are available, the computation of f'_{cr} requires determination of the standard deviation and the number of acceptable low tests. It is assumed that the strength variation of concrete (proportioned to obtain the same compressive strength) follows a normal distribution. If a large number of samples is collected, they were found to follow the normal distribution curve. The ACI code specifies a minimum of 30 samples. If 30 or more samples are available, their average and their standard deviation are assumed to be the same as the average and the standard deviation of a large population.

Once the strength variation is assumed to follow a normal distribution, the properties of the normal distribution curve can be used to predict the probability that a given sample will have a strength less than f'_c. For example, if the average is $\bar{x}$, and the standard deviation is s, it can be said that:

- Half the samples will have strengths less than $\bar{x}$, and hence the probability that a certain sample will have a strength less than $\bar{x}$ is 50%.
- About 68.27% of all samples will have strengths greater than $(\bar{x} - s)$ and less than $(\bar{x} + s)$, or 15.87% of samples will have strengths less than $(\bar{x} - s)$, or the probability that a given sample will have strength less than $(\bar{x} - s)$ is 15.87%.
- About 95.45% of all samples will have strengths between $(\bar{x} - 2s)$ and $(\bar{x} + 2s)$, or the probability that an individual sample will have strength less than $(\bar{x} - 2s)$ is 2.25%.
- About 0.13% of all samples, or 1 in 741 samples, will have a strength less than $(\bar{x} - 3s)$.

The aforementioned postulations were derived using the property of the normal or bell curve. It can be seen that the two factors that control the prediction are the sample average $\bar{x}$ and the standard deviation s.

If the average of a given set of strength results $\bar{x}$, is assumed to be the required average strength f'_{cr}, then the mix proportions used to obtain the strengths are good enough for a specified strength f'_c subjected to the condition

$$f'_{cr} = f'_c + tS \tag{3A.1}$$

where t is a statistic which depends on the number of tests permitted to fall below f'_c. If 50% of the tests are permitted to fall below, then $t = 0$, or f'_{cr} and f'_c are the same. Naturally, in the actual construction such a large number of low tests cannot be permitted. For structural concrete, the ACI code specifies the following two equations:

$$f'_{cr} \geq f'_c + 1.34S \tag{3A.2}$$

$$f'_{cr} \geq f'_c + 2.33S - 500 \text{ psi} \quad (1000 \text{ psi} = 6.89 \text{ MPa}) \tag{3A.3}$$

Equation (3A.2) results in a probability that not more than 1 in 100 averages of three consecutive strength tests (each being the average of two cylinders) will be less than f'_c. The t value for a probability of 1 in 100 is 2.33, and the number 1.34 is obtained by dividing 2.33 by $\sqrt{3}$. The division by $\sqrt{3}$ for three consecutive tests is based on theorems in statistics. Equation (3A.3) results in the probability that not more than 1 individual strength in 100 (average of two cylinders) falls below $f'_c - 500$ psi ($f'_c - 3.4$ MPa). Note that these two equations are consistent with both acceptance criteria presented previously. Overall ACI code recommendations for the computation of f'_{cr} are based on the acceptance of 1 low test in 100. It should be noted that this probability is used only for the computation of f'_{cr}, and if the low strengths do occur, correction measures should be taken to increase the average strengths.

For the construction of facilities such as footpaths, a higher number of low tests could be acceptable. In this case the t value is reduced. For example, if 1 low test in 10 is acceptable, then the t value is 1.28. More information on the t values for various probabilities and the recommended low tests for various facilities can be found in the report of ACI Committee 214.[5]

Once Eqs. (3A.2) and (3A.3) are accepted as the basis for the computation of f'_{cr}, the process becomes a set of mathematical steps. The ACI code also establishes procedures for accepting data sets with less than 30 tests. In addition there are certain restrictions to be satisfied for using the data from previous projects. The following is the gist of the various provisions of the ACI code. The restrictions and their interdependence are presented as a flowchart in Fig. 3A.4.

If a concrete production facility has test records, establish the standard deviation using its records, provided that the material quality control procedures and conditions are similar to the proposed project and the f'_c for the proposed work is within 1000 psi (6.89 MPa) of the f'_c for which the records exist.

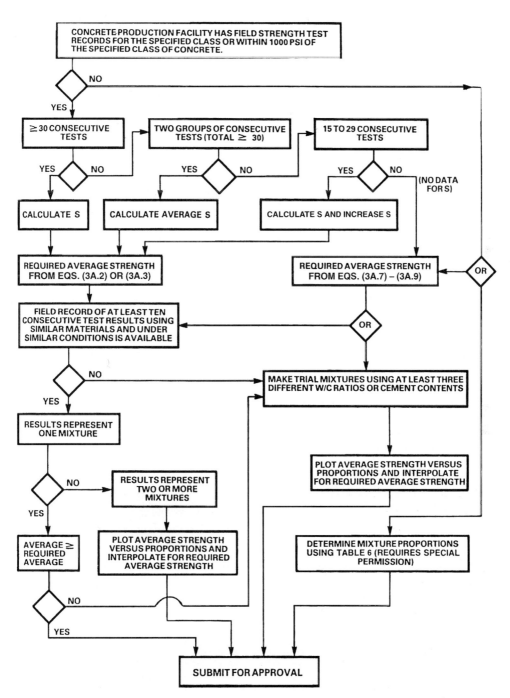

FIGURE 3A.4 Flowchart for selection and documentation of concrete proportions. (*From ACI 318-92.*[4])

If the records contain 30 or more consecutive tests, compute the standard deviations using the equation

$$S = \left[\frac{(x_i - \bar{x})^2}{n - 1} \right]^{1/2} \quad (3A.4)$$

where n = number of consecutive strength tests
x_i = individual strength tests (average of two cylinders at 28 days or at designated test age for the determination of f'_c)

$$\bar{x} = \frac{x_i}{n} \quad (3A.5)$$

If the records contain two groups of consecutive tests totaling at least 30 tests, compute the standard deviations S_1 and S_2 for the two sets using Eqs. (3A.4) and (3A.5). Compute the statistical average of S_1 and S_2 using

$$S = \left[\frac{(n_1 - 1) S_1^2 + (n_2 - 1) S_2^2}{n_1 + n_2 - 2} \right]^{1/2} \quad (3A.6)$$

where S = statistical average standard deviation where two test records are used to estimate standard deviation
S_1, S_2 = standard deviations of sets 1 and 2
n_1, n_2 = number of tests in respective test record

If the available number of tests is less than 30 but greater than 25, multiply the estimated standard deviation by a factor of 1.03 and use this value in Eqs. (3A.2) and (3A.3). The corresponding factors for data sets with a minimum of 20 and 15 tests are 1.08 and 1.16, respectively.

Using the estimated value of S, compute the required average strength f'_{cr} using Eqs. (3A.2) and (3A.3).

If data are not available for estimating the standard deviation,

$$f'_{cr} = f'_c + 1000 \text{ psi } (6.89 \text{ MPa}) \qquad f'_c < 3000 \text{ psi } (20.7 \text{ MPa}) \quad (3A.7)$$

$$= f'_c + 1200 \text{ psi } (8.27 \text{ MPa}) \qquad 3000 \leq f'_c \leq 5000 \text{ psi } (20.7 \leq f'_c \leq 34.5 \text{ MPa}) \quad (3A.8)$$

$$= f'_c + 1400 \text{ psi } (9.64 \text{ MPa}) \qquad f'_c > 5000 \text{ psi } (34.5 \text{ MPa}) \quad (3A.9)$$

Equations (3A.7) to (3A.9) are based on conservative (or overestimated) estimates of S values.

The following numerical examples further illustrate the procedure used for the computation of f'_{cr} under various conditions.

Example 3A.1 Compute the required average compressive strength f'_{cr} for the following cases if the specified compressive strength is 4000 psi (27.6 MPa). Assume similar materials and conditions for all cases. The available test results (average of two cylinders) are as follows:

(a)	(b)	(c)	(d)	(e)
4260	5200	First set:	4700	None
4760	5220			
4120	4750			
3650	4440	3100	4050	
4310	4750	4450	3500	
4960	4720	3900	3250	
4350	4650	4900	4100	
3980	5020	3750	4350	
4450	4920	4100	3750	

(a)	(b)	(c)	(d)	(e)
4430	5780	4400	3600	
4240	5350	4500	3550	
4400	5420	5100	3900	
3420	5070	3500	3850	
4760	5220	3700	4150	
4620	4200	3250	4200	
4260	5070	3750	4600	
3860	5600	4250	4300	
4290	5780	Second set:		
5120	4900	4700		
4870	5980	4750		
4150	5200	4850		
4170	4320	3750		
3740	5170	3050		
4180	5030	3000		
4860	5700	3900		
3890	5050	3850		
4720	5200	4100		
4880	4130	4150		
4980	4450	4350		
4240	5300	4400		
	4600	4950		
	4850	4050		
	6140	3450		
	5100	3200		
	5070	3250		
	4850			

1000 psi = 6.895 MPa.

Solution In case (a) the number of tests $\geq$ 30. Hence the single set can be used:

$$\bar{x} = \frac{x_i}{n}$$

$$= \frac{4260 + 4760 + \cdots + 4980}{30}$$

$$= 4364 \text{ psi (30 MPa)}$$

The average is within 1000 psi of f'_c. Thus

$$S = \left[\frac{(x_i - x)^2}{n-1}\right]^{1/2}$$

$$= \left[\frac{(4364 - 4260)^2 + \cdots + (4364 - 4240)^2}{29}\right]^{1/2}$$

$$= 426 \text{ psi (2.94 MPa)}$$

$$f'_{cr} > 4000 + 1.34 \times 426 = 4571 \text{ psi (31.5 MPa)}$$

$$> 4000 + 2.33 \times 426 - 500 = 4493 \text{ psi (31.0 MPa)}$$

Hence f'_{cr} = 4571 psi, or 4600 psi (32 MPa).

3.24 FUNDAMENTALS OF FOUNDATION CONSTRUCTION AND DESIGN

In case (b) the number of tests is greater than 30. Hence the single set can be used:

$$\bar{x} = \frac{5200 + \cdots + 4850}{36}$$

$$= 5061 \text{ psi } (34.9 \text{ MPa})$$

The difference between the average strength and the specified strength of 4000 psi is greater than 1000 psi. Hence the data set cannot be used. The problem has to be treated as though data were not available. Equation (3A.8) controls,

$$f'_{cr} = f'_c + 1200 \text{ psi}$$

$$= 4000 + 1200 \text{ psi} = 5200 \text{ psi } (36 \text{ MPa})$$

In case (c) the average for the first data set $\bar{x}_1$ is 4046 psi, $n_1 = 14$. The average for the second data set $\bar{x}_2 = 3985$ psi, $n_2 = 17$. Both averages are within 1000 psi of f'_c and can be used for the computation of f'_{cr}:

$$S_1 = 590 \text{ psi } (4.07 \text{ MPa})$$

$$S_2 = 637 \text{ psi } (4.39 \text{ MPa})$$

$$S = \left[\frac{13 \times 590^2 + 16 \times 637^2}{14 + 17 - 2} \right]^{1/2}$$

$$= 616 \text{ psi } (4.25 \text{ MPa})$$

$$f'_{cr} > 4000 + 1.34 \times 616 = 4825 \text{ psi } (33.3 \text{ MPa})$$

$$f'_{cr} > 4000 + 2.33 \times 616 - 500 = 4935 \text{ psi } (34.0 \text{ MPa})$$

Therefore,

$$f'_{cr} = 4935 \text{ psi, or } 4950 \text{ psi } (34 \text{ MPa})$$

In case (d) 15 tests are available:

$$\bar{X} = 3990 \text{ psi}$$

$\bar{X}$ is within 1000 psi of f'_c.

$$S = 415 \text{ psi } (2.86 \text{ MPa})$$

The multiplying factor is 1.16. Therefore S is used for the computation:

$$f'_{cr} = 1.16 \times 415 = 481 \text{ psi } (3.32 \text{ MPa})$$

$$f'_{cr} > 4000 + 1.34 \times 481 = 4645 \text{ psi } (32.0 \text{ MPa})$$

$$f'_{cr} > 4000 + 2.33 \times 481 - 500 = 4621 \text{ psi } (31.9 \text{ MPa})$$

Therefore,

$$f'_{cr} = 4645 \text{ psi, or } 4650 \text{ psi } (32 \text{ MPa})$$

For case (e) data are not available. Therefore,

$$f'_{cr} = 4000 + 1200 \text{ psi} = 5200 \text{ psi } (36 \text{ MPa})$$

which is the same as case (b).

3A.2.2 Selection of Concrete Proportions

Once the required average compressive strength f'_{cr} is established, the mix proportions that can produce average strengths equal to or greater than f'_{cr} can be chosen based on (1) field records or (2) trial mixes. It should be noted that the required average strength f'_{cr} is only one of the parameters to be considered in mix proportioning. Other primary requirements are workability, consistency, and resistance to special exposure conditions. The final mix proportion should also be the most economical solution for the given set of conditions.

Workability of fresh concrete determines the ease with which the concrete can be transported, placed in position, consolidated, and finished. A mix that is too difficult to handle or consolidate may result in a final product that has honeycombs. Poorly consolidated concrete will not only have poor strength but will deteriorate quickly. Similarly, a mix with excess water may lead to segregation and bleeding, again resulting in a poor final product. The consistency of the mix should be just sufficient for proper placing, consolidation, and finishing operations. Slump is the most widely used indicator of workability. Section 3A.3.1 deals with the slump and its measurement. For general applications, concrete with sufficient strength is automatically assumed to be durable. However, special precautions should be taken when proportioning concrete that is to be exposed to severe environments. The following are the requirements for certain exposure conditions. The list is not comprehensive. Each structure should be treated carefully based on the exposure conditions.

- Normal-weight and lightweight concrete exposed to freezing and thawing or deicer chemicals should be air-entrained. The ACI code specifies 3.5 to 6% and 4.5 to 7.5% air content for moderate and severe exposure conditions, respectively. Higher air content is needed for mixes with smaller-size aggregates. For concretes with strengths higher than 5000 psi (35 MPa) the air-content requirement is slightly relaxed because these concretes made with lower water-cement ratios are generally less permeable. In addition to air entrainment, the water-cement ratio should be limited to 0.45 for concrete subjected to freezing and thawing.
- If the concrete is to be used for reinforced structural components exposed to salts, seawater, or brackish water, the water-cement ratio should be limited to 0.40.
- Water-cement ratio restrictions (preferably not more than 0.45) should also be used for concrete exposed to sulfates.
- For concrete exposed to freezing and thawing in the presence of deicing salts, the ACI code also stipulates a minimum cement content of 520 lb/yd^3 (310 kg/m^3).

The importance of the economics, or the cost-optimum solution, cannot be overemphasized. Typically, cement is the most expensive ingredient in concrete. However, because of the large quantities of materials involved, even a few cents per ton of aggregate may translate into millions of dollars. When selecting ingredients, availability and transportation costs should be considered carefully. In certain instances, locally available materials could be used in conjunction with admixtures rather than transporting materials that do not need admixtures for long distances. Generally the least amount of cement that is needed for obtaining the required strength and durability provides the best economical solution. The least amount of cement also provides some technical advantages, such as lower heat of hydration and less shrinkage. The use of mineral admixtures such as fly ash also provides cost savings. The use of industrial by-products produces savings in disposal costs and better utilization of resources. These aspects are important for both industrialized countries that have limited disposal space and developing countries that have limited resources.

3A.2.3 Proportioning on the Basis of Field Strength Records

If field strength records are available for concrete with an average compressive strength in the range of f'_{cr}, the mix proportions used for this concrete can be used for the new project provided they satisfy the workability and durability requirements outlined in the previous section. The field test records should satisfy the following conditions:

- The available test records shall represent materials and conditions similar to those expected. Quality control, in terms of uniformity of materials, conditions, and proportions, shall be the same or better for the proposed work as compared to the project from which the records are taken. If field records were used for the computation of f'_{cr}, the same records can be used for the selection of mix proportions.
- It is preferable to have 30 or more consecutive test records. However, 10 consecutive test results may be used if the records encompass a period of time not less than 45 days.
- Mix proportions can also be chosen by interpolation using records that resulted in compressive strengths higher and lower than f'_{cr}.

3A.2.4 Proportioning on the Basis of Trial Mixtures

When acceptable field records are not available, mix proportions can be established using trial mixtures. Trial mixes should be made using at least three different water-cement ratios or cement contents to establish a relation between compressive strength and water-cement ratio or between strength and cement content. The relationship of strength versus water-cement ratio can be used to establish the maximum water-cement ratio that can produce the required f'_{cr}. On the other hand, if the strength versus cement content relationship is obtained, it can be used to establish the minimum cement content. In either case the chosen mix proportion should satisfy the requirements for workability and durability. The trial mixes should also meet the following restrictions:

- The combination of the materials used should be the same as that of the materials to be used for the proposed work.
- Extrapolations should not be used. The trial mixes should have strengths both smaller and higher than f'_{cr}.
- The slump of the trial mix should be within 0.75 in (19 mm) of the permitted slump of the proposed work. Similarly, the air content should be within 0.5% for air-entrained concrete.
- For each test variable, at least three cylinders should be tested at 28 days or the age designated for the determination of f'_c.

The actual proportioning of the constituent materials is explained in Sec. 3A.2.6.

3A.2.5 Proportioning on the Basis of Maximum Water-Cement Ratio

The ACI code also allows proportioning using a maximum permissible water-cement ratio for the chosen f'_{cr}. The code recommends the maximum water-cement ratio that can be used for either air-entrained or non-air-entrained concrete (Table 3A.6). These water-cement ratios cannot be used for lightweight concrete or concrete containing admixtures. The chosen water-cement ratio should also satisfy the durability requirements.

3A.2.6 Computation of Mix Proportions

This section deals with the actual computation of the amounts of the various constituent materials that will result in a concrete with the required strength and durability. A number of procedures are available for proportioning normal-weight, lightweight, and heavyweight concretes. The method proposed by ACI Committee 211 for normal-weight concrete[7] is explained here. The other popular method used in the United States is the PCA method. This method is explained in a

TABLE 3A.6 Maximum Permissible Water-Cement Ratio for Concrete When Strength Data from Field Experience or Trial Mixtures Are Not Available

	Absolute water-cement ratio by weight	
Specified compressive strength f'_c, psi* (MPa)	Non-air-entrained concrete	Air-entrained concrete
2500 (17)	0.67	0.54
3000 (21)	0.58	0.46
3500 (24)	0.51	0.40
4000 (28)	0.44	0.35
4500 (31)	0.38	†
5000 (34)	†	†

*28-day strength.
†For strengths above 4500 psi (31 MPa) (non-air-entrained concrete) and 4000 psi (28 MPa) (air-entrained concrete), concrete proportions shall be established by using trial mixes.
Source: From ACI Building Code.[6]

manual published by Portland Cement Association.[8] The details of the procedure used in Britain, which is similar to the methods used in western Europe, Australia, and Asia, can be found in Neville.[9] For the sake of brevity, lightweight and heavyweight concretes are not discussed here. The details can be found in reports published by ACI Committee 211.[7,10]

The water-cement ratio is the most influential factor that affects strength. Hence the design charts and tables developed for mix proportioning are geared toward obtaining the minimum water-cement ratio that would produce a workable concrete. The following is the step-by-step procedure to estimate the quantities of various ingredients.

1. *Collection of background information:* The following information should be collected before starting the computations:
   - Sieve analysis data for fine and coarse aggregate including fineness modulus
   - Bulk specific gravity of aggregates, cement, and admixtures in solid (powder) form
   - Dry-rodded unit weight of coarse aggregate
   - Moisture content of fine and coarse aggregates
   - Ratio of solid-to-liquid contents of liquid (or slurry) admixtures
   - Special conditions such as permissible maximum water-cement ratio, minimum cement content, minimum air content, minimum slump, maximum size of aggregate, and strength requirements at early age

2. *Selection of slump:* If the slump is not specified, choose an appropriate value from Table 3A.7. A minimum possible value should be chosen within the specified range.

TABLE 3A.7 Recommended Slump for Various Types of Construction

Type of construction	Maximum slump*	Minimum slump
Reinforced foundation walls and footings	3 in (75 mm)	1 in (25 mm)
Plain footings, caissons, and substructure walls	3 in (75 mm)	1 in (25 mm)
Beams and reinforced walls	4 in (100 mm)	1 in (25 mm)
Building columns	4 in (100 mm)	1 in (25 mm)
Pavements and slabs	3 in (75 mm)	1 in (25 mm)
Mass concrete	2 in (50 mm)	1 in (25 mm)

*May be increased 1 in (25 mm) for methods of consolidation other than vibration.
Source: From ACI Committee 211.[7]

3. *Selection of maximum size for aggregate:* For the same volume fraction a large maximum size of well-graded aggregate provides the least void space, requiring the least amount of mortar content. Hence the maximum possible aggregate size, consistent with the type of application, should be chosen. The maximum size should satisfy the following restrictions:

   ≯ $\frac{1}{5}$ narrowest dimension between sides of form

   ≯ $\frac{3}{4}$ of minimum clear spacing between reinforcing bars

   ≯ $\frac{1}{3}$ depth of slab

4. *Estimation of amount of water and air:* The amount of water required to produce a given slump depends on the aggregate properties, the concrete temperature, and the amount of entrained air. The aggregate properties that influence the slump are maximum size, shape, and grading. The admixture can also influence the slump, the most notable being a high-range water-reducing admixture. Cement content, within the normal range, does not influence the slump. Table 3A.8 can be used to estimate the approximate amount of water needed. It provides guidelines for both air-entrained and non-air-entrained concrete. Note that the table gives only approximate values, and the influence of any admixtures other than air-entraining admixtures is not considered. Only trial batches can establish the actual water and the corresponding air content.

5. *Selection of water-cement, or water-cementitious-materials, ratio.* Water-cement ratio is a classical term. When mineral admixtures such as fly ash or silica fume are used, they could be considered as part of the cementitious materials. In this case the water is computed based on the weight of the cementitious (cement + fly ash or silica fume) materials, rather than just the weight of cement. As mentioned earlier, the amount of water used should satisfy both strength and durability requirements. Information given in Table 3A.9 can be used as an approximate first step to establish the water-cement ratio. Since the strength is also affected by factors such as aggregate and cement types and by the properties of other cementitious materials, the values shown in Table 3A.9 should be used only as a guideline. It is highly desirable to develop a water-cement (cementitious materials) ratio for the particular type of materials to be used in the proposed work.

   If the structure to be built is going to be exposed to severe environmental conditions, water-cement (cementitious materials) ratios should be limited to the values shown in Table 3A.10. As mentioned earlier, a lower water content typically reduces permeability and improves overall durability.

6. *Computation of cement content:* Once the amount of water and the water-cement ratio are established, the computation of the amount of cement becomes a simple division of water content by water-cement ratio. The cement content should also satisfy any special minimum cement content stipulated in the specification.

7. *Estimation of coarse aggregate content:* Typically the use of more coarse aggregate per unit volume of concrete leads to better economy. The larger the size of the particles in coarse aggregate and the finer the sand, the more volume fraction of coarse aggregate can be incorporated without sacrificing workability. The volume fractions of coarse aggregate that will produce a workable mix for various maximum aggregate sizes and fineness moduli of sand are shown in Table 3A.11.

   For the chosen maximum aggregate size, say 1 in (25 mm), and the fineness of the sand to be used, say 2.6, the table provides the volume fraction of coarse aggregate in the dry-rodded form, 0.69. For 1 yd^3 (0.76 m^3) of concrete, the volume of coarse aggregate is 0.69 × 27, or 18.63 ft^3 (0.52 m^3). The corresponding weight is obtained by multiplying 18.63 by the dry-rodded unit weight. The values shown in Table 3A.11 can be reduced by up to 10% to improve the workability for special circumstances, such as pumping or concreting members with congested reinforcement.

8. *Estimation of fine aggregate content:* At the completion of step 7, the amounts of all ingredients, except fine aggregate, have been estimated. Hence if the unit weight of fresh concrete is known, the weight of fine aggregate can be estimated by subtracting the total weight of all

TABLE 3A.8 Approximate Mixing Water and Air Content Requirements for Different Slumps and Nominal Maximum Sizes of Aggregates

Slump, in	Water, lb/yd³ of concrete for nominal maximum aggregate sizes							
	³⁄₈ in[a]	½ in[a]	¾ in[a]	1 in[a]	1½ in[a]	2 in[a,b]	3 in[b,c]	6 in[b,c]
	Non-air-entrained concrete							
1–2	350	335	315	300	275	260	220	190
3–4	385	365	340	325	300	285	245	210
6–7	410	385	360	340	315	300	270	—
>7[a]	—	—	—	—	—	—	—	—
Approximate amount of entrapped air in non-air-entrained concrete, %	3	2.5	2	1.5	1	0.5	0.3	0.2
	Air-entrained concrete							
1–2	305	295	280	270	250	240	205	180
3–4	340	325	305	295	275	265	225	200
6–7	365	345	325	310	290	280	260	—
>7[a]	—	—	—	—	—	—	—	—
Recommended average[d] total air content for level of exposure, %								
Mild exposure	4.5	4.0	3.5	3.0	2.5	2.0	1.5[e,f]	1.0[e,f]
Moderate exposure	6.0	5.5	5.0	4.5	4.5	4.0	3.5[e,f]	3.0[e,f]
Severe exposure[g]	7.5	7.0	6.0	6.0	5.5	5.0	4.5[e,f]	4.0[e,f]

[a]The quantities of mixing water given for air-entrained concrete are based on typical total air content requirements as shown for moderate exposure. These quantities of mixing water are for use in computing cement contents for trial batches at 68–77°F. They are maximum for reasonably well-shaped angular aggregates graded within limits of accepted specifications. Rounded aggregate will generally require 30 lb less water for non-air-entrained and 25 lb less for air-entrained concretes. The use of water producing chemical admixtures (ASTM C494) may also reduce mixing water by 5% or more. The volume of the liquid admixtures is included as part of the total volume of the mixing water. The slump values of more than 7 in are only obtained through the use of water-reducing chemical admixture; they are for concrete containing nominal maximum-size aggregate not larger than 1 in.

[b]The slump values for concrete containing aggregate larger than 1½ in are based on slump tests made after removal of particles larger than 1½ in by wet screening.

[c]These quantities of mixing water are for use in computing cement factors for trial batches when 3- or 6-in nominal maximum-size aggregate is used. They are average for reasonably well-shaped coarse aggregate, well-graded from coarse to fine.

[d]Additional recommendations for air content and necessary tolerances on air content for control in the field are given in a number of ACI documents, including ACI 201, 345, 318, 301, and 302. ASTM C94 for ready-mixed concrete also gives air-content limits. The requirements in other documents may not always agree exactly, so in proportioning concrete consideration must be given to selecting an air content that will meet the needs of the job and also the applicable specifications.

[e]For concrete containing large aggregates that will be wet-screened over the 1½-in sieve prior to testing for air content, the percentage of air expected in the 1½-in material should be as tabulated in the 1½-in column. However, initial proportioning calculations should include the air content as a percent of the whole.

[f]When using large aggregate in low cement factor concrete, air entrainment need not be detrimental to strength. In most cases the mixing water requirement is reduced sufficiently to improve the water-cement ratio and to thus compensate for the strength-reducing effect of air-entrained concrete. Generally, therefore, for these large nominal maximum sizes of aggregate, air contents recommended for extreme exposure should be considered even though there may be little or no exposure to moisture and freezing.

[g]These values are based on the criteria that 9% air is needed in the mortar phase of the concrete. If the mortar volume will be substantially different from that determined in this recommended practice, it may be desirable to calculate the needed air content by taking 9% of the actual mortar volume.

Note: 1 in = 25.4 mm; 1 lb/yd³ = 0.59 kg/m³.

Source: From ACI Committee 211.

TABLE 3A.9 Relationships between Water-Cement Ratio and Compressive Strength of Concrete

Compressive strength at 28 days, psi* (MPa)	Water-cement ratio, by weight	
	Non-air-entrained concrete	Air-entrained concrete
6000 (41)	0.41	—
5000 (35)	0.48	0.40
4000 (28)	0.57	0.48
3000 (21)	0.68	0.59
2000 (14)	0.82	0.74

*Values are estimated average strengths for concrete containing not more than the percentage of air shown in Table 3A.8. For a constant water-cement ratio, the strength of concrete is reduced as the air content is increased. Strength is based on 6- by 12-in cylinders moist cured 28 days at 73.4 ± 3°F (23 ± 1.7°C) in accordance with Sec. 9(b) of ASTM C31, *Making and Curing Concrete Compression and Flexure Test Specimens in the Field.*
Source: From ACI Committee 211.[7]

TABLE 3A.10 Maximum Permissible Water-Cement Ratios for Concrete in Severe Exposures

Type of structure	Structure wet continuously or frequently and exposed to freezing and thawing*	Structure exposed to seawater or sulfates†
Thin sections (railings, curbs, sills, ledges, ornamental work) and sections with less than 1-in (25 mm) cover over steel	0.45	0.40
All other structures	0.50	0.45

*Concrete should also be air-entrained.
†If sulfate-resisting cement (type II or type V of ASTM C150) is used, the permissible water-cement ratio may be increased by 0.05.
Source: From ACI Committee 211.[7]

TABLE 3A.11 Volume of Dry-Rodded Coarse Aggregate* per Unit Volume of Concrete for Different Fineness Moduli of Sand

Maximum size of aggregate, in (mm)	Fineness modulus of sand			
	2.40	2.60	2.80	3.00
3/8 (9)	0.50	0.48	0.46	0.44
1/2 (13)	0.59	0.57	0.55	0.53
3/4 (19)	0.66	0.64	0.62	0.60
1 (25)	0.71	0.69	0.67	0.65
1 1/2 (38)	0.75	0.73	0.71	0.69
2 (50)	0.78	0.76	0.74	0.72
3 (76)	0.82	0.80	0.78	0.76
6 (152)	0.87	0.85	0.83	0.81

*Volumes are based on aggregates in dry-rodded condition as described in ASTM C29, "Unit Weight of Aggregate." These volumes are selected from empirical relationships to produce concrete with a degree of workability suitable for usual reinforced construction. For less workable concrete such as required for concrete pavement construction the volume may be increased by about 100%. For more workable concrete, such as may sometimes be required when placement is to be by pumping, it may be reduced by up to 10%.
Source: From ACI Committee 211.[7]

TABLE 3A.12 First Estimate of Weight of Fresh Concrete

Maximum size of aggregate, in (mm)	Concrete weight,* lb/yd³	
	Non-air-entrained concrete	Air-entrained concrete
³/₈ (9)	3840 (2266)	3710 (2189)
½ (13)	3890 (2295)	3760 (2218)
¾ (19)	3960 (2336)	3840 (2266)
1 (25)	4010 (2366)	3850 (2272)
1½ (38)	4070 (2401)	3910 (2307)
2 (50)	4120 (2431)	3950 (2331)
3 (76)	4200 (2478)	4040 (2384)
6 (152)	4260 (2513)	4110 (2425)

*Values calculated for concrete of medium richness (550 lb of cement per cubic yard or 325 kg per cubic meter) and medium slump with aggregate specific gravity of 2.7. Water requirements based on values for 3 to 4 in (75 to 100 mm) of slump are given in Table 3A.8. If desired, the estimated weight may be refined as follows when necessary information is available. For each 10-lb (4.5-kg) difference in mixing water from Table 3A.8 values for 3 to 4 in (75 to 100 mm) of slump, correct the weight per cubic yard by 15 lb (6.8 kg) in the opposite direction; for each 100-lb (45.4-kg) difference in cement content from 550 lb (250 kg), correct the weight per cubic yard by 15 lb (6.8 kg) in the same direction; for each 0.1 by which aggregate specific gravity deviates from 2.7, correct the concrete weight by 100 lb (45.4 kg) in the same direction.

Source: From ACI Committee 211.[7]

other ingredients from the weight of fresh concrete. This type of computation is called the weight method. In the absence of previous experience, a first estimate of the unit weight of concrete can be obtained using Table 3A.12. The table covers both air-entrained and non-air-entrained concrete. Medium-rich concrete and a coarse aggregate specific gravity of about 2.7 have been assumed for developing Table 3A.12. Even rough estimates of unit weight were found to provide satisfactory results for trial mixes.

There is another procedure, called the absolute volume method, in which the volume of fine aggregate is computed by subtracting the volumes of all other ingredients from the unit volume of fresh concrete. This method is considered more accurate, but the specific gravity of all ingredients is needed prior to the computation.

9. *Adjustments for aggregate moisture:* In most cases the stock aggregates retain some moisture. The computations of aggregate weights in steps 7 and 8 are based on saturated surface-dry conditions. Hence the weight of moisture present in the aggregate should be accounted for. The easiest way to make the correction is to adjust the weight of aggregates for moisture. For example, if the moisture content of coarse aggregate is 2%, the amount of coarse aggregate needed for the batch should be increased by 2%. The actual amount of water, which is 2% of the weight of the coarse aggregate, should be subtracted from the water to be used for the mix. The reduction in water is necessary to maintain the water-cement ratio chosen in step 5. Similar adjustments should also be made for fine aggregate.

10. *Trial batch adjustments:* Since the estimation of the various ingredients is only approximate, adjustments are needed to obtain the mix that satisfies the workability and strength requirements. Fresh concrete should be tested for slump, segregation of aggregates, air content, and unit weight. The hardened concrete, cured under standard conditions, should be tested for strength at the specified age. The test methods are described in Sec. 3A.3. In some instances it may take several trials to obtain a satisfactory mix. The following guidelines may be used for the adjustment of ingredients. The recommended numerical values are for 1 ft³ (0.028 m³) of concrete.

- If the slump of the trial mix is not correct, increase or decrease the estimated water by 10 lb (4.5 kg) for each 1-in (25-mm) required increase or decrease of slump.

- If the desired air content is not achieved, adjust the admixture content. Since the amount of air content influences the slump, the water content should also be changed with the change in air-entraining admixture. Change the water content by 5 lb (2.3 kg) for each 1% of air.
- Adjust the yield by using the unit weight of fresh concrete.

The following example further illustrates the mix proportioning process.

Example 3A.2 To compute the mix proportions of normal-weight concrete, we use these specifications:

Required (28-day) average compressive strength f'_{cr}	4100 psi (28 MPa)
Type of construction	Reinforced concrete footing
Slump	Minimum 3 in (75 mm)
Exposure condition	Below ground; no freezing, no exposure to chemicals
Maximum size of aggregate	1.5 in (38 mm)

Solution

1. *Background information on properties of constituent materials:*

Cement	ASTM type I; bulk density 196 lb/ft^3 (3136 kg/m^3)
Coarse aggregate	Maximum size 1.5 in (38 mm)
	Bulk density 168 lb/ft^3 (2690 kg/m^3)
	Dry-rodded unit weight 100 lb/ft^3 (1600 kg/m^3)
	Moisture content 2.0% over SSD condition
Fine aggregate	Bulk density 160 lb/ft^3 (2560 kg/m^3)
	Fineness modulus 2.8
	Moisture content 3.0% over SSD condition

2. *Selection of slump:*

$$\text{Specified minimum} = 3 \text{ in } (75 \text{ mm})$$

The specified minimum value is consistent with the 3 in (75 mm) recommended for reinforced concrete foundation walls and footings in Table 3A.7.

3. *Maximum aggregate size:*

$$\text{Specified maximum} = 1.5 \text{ in } (38 \text{ mm})$$

4. *Estimation of amount of water and air:* Since there is no freezing or exposure to chemicals, non-air-entrained concrete is assumed to be adequate. Using Table 3A.8, the amount of water for 1 yd^3 (0.76 m^3) = 300 lb for 3- to 4-in slump (136 kg^3 for 75- to 100-mm slump).

5. *Selection of water-cement ratio:* Using Table 3A.9, the water-cement ratio for the required average strength is 0.56. The value of 0.56 is obtained by interpolating linearly between 4000 and 5000 psi (27 and 34 MPa). Note that this value is only a first estimate.

6. *Cement content:*

$$\text{Cement content} = \frac{\text{amount of water}}{\text{water-cement ratio}}$$

$$= \frac{300}{0.56} = 536 \text{ lb/yd}^3 \text{ (316 kg/m}^3\text{)}$$

7. *Coarse aggregate content:* Using Table 3A.11:

Volume fraction of dry-rodded aggregate for 1.5-in maximum-size aggregate and a sand fineness modulus of 2.8	0.71
Volume of coarse aggregate per cubic yard of concrete	0.71×27 ft^3 = 19.2 ft^3 (0.54 m^3)
Weight of coarse aggregate (since dry-rodded unit weight is 100 lb/ft^3 or 1600 kg/m^3)	19.2×100 = 1920 lb/yd^3 (1133 kg/m^3)

8. *Estimation of fine aggregate:*

Estimated unit weight of concrete, from Table 3A.12	4070 lb/yd^3 (2401 kg/m^3)
Weight of cement (step 6)	536 lb/yd^3 (316 kg/m^3)
Weight of coarse aggregate (step 7)	1920 lb/yd^3 (1133 kg/m^3)
Weight of water (step 4)	300 lb/yd^3 (177 kg/m^3)
Weight of fine aggregate	$4070 - (536 + 1920 + 300)$ = 1314 lb/yd^3 (775 kg/m^3)

9. *Adjustments for aggregate moisture:* The moisture content of coarse aggregate above SSD is 2%. In order to obtain 1920 lb of dry-rodded weight, we have to use 1.02×1920 lb of stock sample.

Weight of stock sample of coarse aggregate	1.02×1920 lb/yd^3 = 1958 lb/yd^3 (1155 kg/m^3)
Weight of stock sample of fine aggregate	1.03×1314 lb/yd^3 = 1353 lb/yd^3 (798 kg/m^3)
Free water from coarse aggregate	$1958 - 1920$ = 38 lb/yd^3 (22 kg/m^3)
Free water from fine aggregate	$1353 - 1314$ = 39 lb/yd^3 (23 kg/m^3)
Water to be added	$300 - 39 - 38$ = 223 lb/yd^3 (132 kg/m^3)

The final weights of the constituent materials per cubic yard (0.76 m^3) are:

Cement	536 lb (243 kg)
Fine aggregate	1353 lb (614 kg)
Coarse aggregate	1958 lb (888 kg)
Water	223 lb (101 kg)
Total	4070 lb (1846 kg)

Note that the unit weight of concrete is the same even after the corrections are made for the moisture contents of aggregates.

10. *Trial batches:* Trial batches should be made using about 0.1 yd^3 (0.076 m^3) of concrete to determine the properties of fresh and hardened concrete. If the slump is too high or too low, corrections can be made immediately. Since it takes 28 days to obtain strength results, it may be more efficient to make two or three trial mixes with various water contents rather than waiting for the results from a single mix.

3A.3 QUALITY ASSURANCE

Quality assurance procedures are needed in order to ascertain that the concrete placed in the actual structures satisfies the required specifications. As mentioned earlier, the quality of the concrete is influenced by a large number of variables. Hence continuous monitoring of properties is needed. The quality and physical properties of the constituent materials, namely, cement, aggregates, water, and admixtures, should also be checked periodically. ASTM standards are available for the test procedures needed to evaluate the properties of constituent materials and concrete. The properties of constituent materials were discussed in Sec. 3A.1. This section deals with the test methods for fresh and hardened concrete and the quality assurance procedures. The frequency of testing depends primarily on the type of structure. For buildings, samples should be taken at least once a day or once for every 150 yd^3 (115 m^3) of concrete. If large surface areas are being constructed, the ACI code recommends at least one test for each 5000 ft^2 (464 m^2). For structures such as nuclear containment buildings where failure could result in disastrous consequences, more stringent quality control is needed.

3A.3.1 Tests for Fresh Concrete

The most universally used test for fresh concrete is the slump test. It measures only the consistency of concrete. However, this test is used as an indicator of workability. The slump test is also used to assure the uniformity from batch to batch. The other two tests used for fresh concrete are the V-B test and the compaction factor test. These two tests are used primarily in the laboratory environment, whereas the slump test is used both in the laboratory and in the field.

Slump Test. The slump test is covered in ASTM C143. In this test a sample of freshly mixed concrete is placed inside a mold, which has the shape of the frustrum of a cone. The concrete is compacted using a standard procedure and the mold is raised to allow the concrete to slump. The amount by which the concrete slumps is measured in inches or millimeters and is called the slump value. The following are the pertinent details of this test.

The test equipment consists of (1) mold (Fig. 3A.5), (2) a tamping rod, (3) a ruler with a reading accuracy of at least 0.125 in (3 mm), and (4) a nonabsorbent rigid pan. The mold shown in Fig. 3A.5 should be clean and free of any projections such as rivets or dents on the inside surface. The top and bottom faces should be parallel to each other and perpendicular to the vertical axis. The tamping rod consists of a $\frac{5}{8}$-in (16-mm)-diameter and 24-in (610-mm)-long steel bar with hemispherically rounded ends.

The test procedure consists of the following steps:

1. Dampen the mold and place it on a rigid, flat surface. The surface should be moist and nonabsorbent. Stand on the two foot pieces in order to hold the mold in place during the filling operation.
2. Pour concrete until the mold is filled to one-third of the volume. One third of the volume is reached when the mold is filled to a height of 2$\frac{5}{8}$ in (66 mm). Rod this bottom layer of concrete

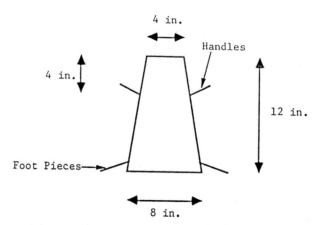

FIGURE 3A.5 Salient features and dimensions of slump cone.

with 25 strokes of the tamping rod. Apply approximately half of the strokes near the perimeter and progress spirally toward the center. The bottom layer should be rodded throughout its depth.

3. Pour concrete to fill the mold to two-thirds of its volume. To reach this volume, the mold should be filled to a height of $6\frac{1}{8}$ in (155 mm) from the bottom. Rod this layer using 25 strokes. The strokes should just penetrate into the underlying bottom layer.

4. Fill the remaining part of the mold, which forms the top layer. Again rod this top layer with 25 uniformly distributed strokes. The top level of concrete should stay slightly above the top surface of the mold. When concrete subsides below the top surface because of compaction, add more concrete. The strokes of the tamping rod should just penetrate into the second layer.

5. Strike off the excess concrete on the top using the tamping rod. A screeding and rolling motion should be used to obtain a relatively smooth top surface.

6. Remove the mold by lifting it in a vertical direction. Lateral or torsional motion should be avoided. The lifting operation should be complete in 5 ± 2 s. The entire operation of filling and lifting the mold should be completed without interruption in 2.5 min.

7. When the mold is removed, the concrete will slump down. If the concrete is stiff, it might move only a fraction of an inch. If the concrete is of flowing consistency, the whole mass will collapse. Measure the difference between the original 12 in (300 mm) and the height of the slumped concrete and record it to the nearest 0.25 in (6 mm). This value is called the slump of the concrete. If the concrete slides to one side, the slump value should be disregarded and the test repeated.

V-B Test. The equipment for the V-B test consists of a vibration table, a cylindrical pan, and a glass or plastic disk, Fig. 3A.6. The glass or plastic disk is attached to a free-moving rod which serves as a reference end point. The cone is placed in the cylindrical pan and filled with concrete. The cone is removed, the disk is brought into position on the top of the concrete, and the vibrating table is set into motion. When the table starts vibrating, the concrete in the conical shape remolds itself into a cylinder. The time required for remolding the concrete into cylindrical shape, until the disk is completely covered with concrete, is recorded in seconds and reported as V-B time. Since the V-B test is conducted using vibration, the V-B time is a better indicator of workability when in the actual construction the concrete is compacted using vibration. However, this test is not easily adoptable for testing on the construction site.

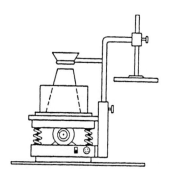

FIGURE 3A.6 V-B apparatus.

Compaction Factor Test. This test measures the amount of compaction achieved when the concrete mixture is subjected to a standard amount of compacting work. The result is expressed as a compacting factor, which is a ratio of the density actually achieved in the test to the density of the same concrete under fully compacted conditions. This test, which was developed in Britain, is not as popular as the other two tests.

The three tests mentioned primarily provide an indication of the workability of concrete. In the case of air-entrained concrete the air content of fresh concrete has to be measured to assure that the concrete has the specified amount. Other properties measured using fresh concrete include unit weight, setting time, and concrete temperature. The air content can be determined using either the pressure or the volumetric method. These tests are described in ASTM C231 and C173, respectively. The gravimetric method is covered in ASTM C138. Only the volumetric method should be used for lightweight concrete. The unit weight, setting time, and temperature measurements are covered in ASTM C138, C403, and C1064, respectively.

3A.3.2 Tests for Hardened Concrete

Hardened concrete should meet the minimum-strength requirements and should be durable. The compressive strength test is the most widely used quality control test. Various types of theoretical and empirical relationships have been developed to relate the compressive strength to other properties of concrete so that other tests need not be conducted for every situation. However, test methods exist for determining strength at various modes of loading, such as tension, flexure, shear, and torsion, and to ascertain durability under freeze-thaw conditions. The most commonly used tests and the corresponding ASTM standards are as follows:

Compressive strength of cylindrical concrete specimens (ASTM C39)
Flexural strength of concrete (ASTM C78, C293)
Splitting tensile strength (ASTM C496)
Modulus of elasticity and Poisson's ratio (ASTM C469)
Length change of hardened paste or mortar (ASTM C157)
Creep under compression (ASTM C512)
Air void parameter (ASTM C457)
Resistance of concrete to rapid freezing and thawing (ASTM C666)
Resistance to scaling (ASTM C672)

Only the compressive strength test is described here. The details for the other tests can be found in the appropriate ASTM standards.

3A.3.3 Compressive Strength Test

The compressive strength test can be conducted using cylinders, cubes, or prisms. In the United States, 6- by 12-in (150- by 300-mm) cylinders are the most popular test specimens for obtaining the compressive strength. In recent years 4- by 8-in (100- by 200-mm) cylinders have also been used, especially for high-strength concrete. In special circumstances smaller [3- by 6-in (75- by 150-mm)] and larger [(up to 24- by 48-in (600- by 1200-mm)] cylinders are also being be used. The cylinders are typically made in the laboratory for trial mixes. The cylinders made for quality control are usually made on the construction site. In either case, standard procedures should be followed for the casting, curing, capping, and testing of cylinders. This section presents the salient features of the procedure to be used for the preparation and testing of cylinders in the laboratory. For field cast and cured specimens appropriate ASTM standards should be followed for the sampling, making, and curing of cylinders.

Mixing, Molding, and Curing of Concrete Test Specimens. The concrete can be mixed either by hand or by machine. Hand mixing should be avoided as much as possible because it is very difficult to achieve uniform mixing. At least 10% more concrete than needed for molding should be mixed. Hand mixing, which is not to be used for air-entrained and no-slump concrete, and quantities exceeding 0.25 ft^3 (0.007 m^3) can be achieved by using the following steps:

1. Mix cement and fine aggregate in a watertight, clean, and damp metallic pan until the contents are thoroughly blended.
2. Add the coarse aggregate to the cement-sand mix and mix until the coarse aggregates are uniformly dispersed.
3. Add water and the admixtures, if any, and mix the contents until they become a homogeneous mass.

The mixing sequence for machine mixing is as follows. It is assumed that a drum-type mixer is used for mixing.

1. Place coarse aggregate and half the mixing water in the mixer and mix the contents for 30 s.
2. Add fine aggregate, cement, and the remaining water and mix for 3 min.
3. Stop the mixer and rest the mixture for 3 min.
4. Mix again for 2 min longer.

If admixtures are used, they should be mixed with water before starting the mixing process. The open end of the mixer should be covered with a pan to avoid evaporation. Typically, some concrete sticks to the sides of the drum. This concrete contains more mortar than the discharged concrete. Hence the concrete used for making samples could contain less mortar. This is particularly true when small quantities are mixed. This could be avoided by mixing a similar batch of concrete and disposing of the contents. The test batch is then mixed without cleaning the mixer. This process is called buttering the mixer. If the exact amount of mortar that sticks to the sides can be established, the mix can be adjusted to contain this excess mortar. This process is more difficult because it is not easy to establish the accurate amount of mortar that will stick to the sides.

Molding of specimens should be done in a place that is very near to the storing place where the specimens will be kept for the first 24 h. Molds should be made of a material which is nonabsorbent and nonreactive with concrete. They should be dimensionally stable and watertight. Reusable molds should be coated lightly with mineral oil for easy removal of the specimens.

The mixed concrete is placed in the mold in layers and compacted to form the test specimens. For a cylinder height of 12 in (300 mm) or smaller, casting should be done in three equal layers if compaction is done by rodding. If compaction is done by vibration, two layers should be used. For larger cylinders, more layers may be needed. If the slump is greater than 3 in (75 mm), compaction by rodding is preferable. If the slump is less than 1 in (25 mm), vibration should be used for compaction. If the slump is in the range of 1 to 3 in (25 to 75 mm), either rodding or vibration can be used. If the cylinder diameter is 4 in (100 mm) or less, only external vibration should be used. For larger cylinders, either internal or external vibration can be used.

The following points should be observed when compaction is done by rodding.

For 3- by 6- or 4- by 8-in (75- by 150- or 100- by 200-mm) cylinders, use a $\frac{3}{8}$-in (9-mm)-diameter 12-in (300-mm)-long metal rod with hemispherical ends. For 6- by 12-in (150- by 300-mm) cylinders use $\frac{5}{8}$-in (15-mm)-diameter 12-in (300-mm)-long metal rod with hemispherical ends.

Each layer should be rodded 25 strokes, uniformly distributed over the cross section. The bottom layer should be rodded throughout the depth. While rodding the upper layers, allow the rod to penetrate about $\frac{1}{2}$ in (12 mm) into the underlying layer for 3- by 6- and 4- by 8-in (75- by 150- and 100- by 200-mm) cylinders and about 1 in (25 mm) for 6- by 12-in (150- by 300-mm) cylinders.

After rodding each layer, tap the outside of the mold about 15 times with a rubber mallet to close any holes left by the tamping rod and to release large entrained air bubbles. Spade the top of the concrete lightly before placing the subsequent layer.

If the compaction is done by vibration, fill the mold in a number of layers of equal height and vibrate them. Place all the concrete for each layer before starting the vibrating equipment. When adding the final layer, do not overfill more than 0.25 in (6 mm). The duration of vibration required depends on the workability of the concrete and the effectiveness of the vibration. Compaction can be assumed to be complete if the top surface is smooth. Overvibration should be avoided. Each layer should be vibrated to the same extent. When using an internal vibrator, use three insertions for each layer. Allow the vibrator to penetrate through the layer being vibrated and approximately 1 in (25 mm) into the underlying layer. The vibrator should be pulled out slowly so as to avoid air pockets. After vibrating each layer, tap the side 10 to 15 times with a rubber mallet to release entrapped air bubbles. When external vibration is used, ensure that the mold is held securely against the vibrating surface.

At the end of consolidation, strike off the top surface with a trowel. Flatten the top surface such that it is level with the rim of the mold and has no depressions or projections larger than 0.125 in (3 mm). The top surface of the freshly made cylinder may be capped with a stiff cement paste.

After finishing, cover the specimens with a nonabsorptive and nonreactive plate or an impervious plastic sheet. Remove the specimens from the mold 24 ± 8 h after casting. Cure the specimens at 73 ± 3°F (23 ± 2°C) from the time of removal until testing. Curing can be done by immersing the specimens in lime-saturated water or in a room maintained at 100% relative humidity by using moist sprays.

Capping of Cylindrical Specimens. The cylinders should be capped before testing to assure two flat surfaces that are perpendicular to the axis of the cylinders (ASTM C617). The capping material should be at least as strong as the concrete being tested. Freshly made specimens can be capped with cement paste. This is not usually done because it is very difficult to achieve the required accuracy. Normally capping is done after the cylinders are cured. Common capping materials are high-strength gypsum cement or molten sulfur. Standard equipment is available for melting the sulfur and for the alignment of caps so that they are perpendicular to the axis of the cylinder.

For high-strength concrete, with compressive strength greater than 12,000 psi (80 MPa), grinding of the ends is recommended. Grinding prevents the interference of the capping materials.

The tolerance for level surfaces is 0.002 in (0.05 mm). The caps should be about 0.125 in (3 mm) thick. They should not be more than 0.31 in.(8 mm) thick. Gypsum plaster should cure at least for 4 h prior to testing. The minimum curing period for sulfur caps is 2 h.

Test Apparatus. The test apparatus consists of a testing machine, scale, and calipers. If the stress-strain relationship is measured, then a compressometer setup is also needed. Most of the testing machines are driven by hydraulic fluid pressure. These machines can be operated so as to apply the load at a constant rate at a certain number of psi per minute or at a constant displacement rate. Some of the machines with servo control mechanisms can be run under different controls such as displacement or strain control.

A compressometer setup can be used to measure the deformation and, hence, the strain at a given load. This device is needed to obtain the stress-strain relationship and the Young's modulus of elasticity. The setup consists of two yokes (rings) (Fig. 3A.7). The bottom ring (yoke) is attached to the cylinder using three screws. The top ring is attached to the cylinder using two screws placed at diametrically opposite points. This ring can rotate about the two screws. One end of the rotatable ring is connected to the bottom ring using a pivot rod. The pivot rod does not allow any translation but allows the ring to rotate. A dial gauge or other deformation-measuring instrument, such as an LVDT, is attached on the other side (Fig. 3A.7). When the cylinder deforms, the two points where the screws of the top ring are attached move downward. Since one end cannot move due to the pivot rod, the ring rotates. Due to symmetry, if the deformation along the gauge length is S, the dial gauge will record $2S$, hence improving the accuracy of the measurements, specially for small deformations. The rings are placed in proper position using spacing bars. Normally the gauge length for 6- by 12-in (150- by 300-mm) cylinders is 6 in (150 mm).

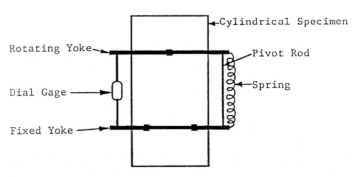

FIGURE 3A.7 Compressometer arrangement.

Specimen Preparation. As mentioned earlier, the cylinders should be capped so that the ends are flat and perpendicular to the axis of the cylinder. The diameter of one cylinder should not vary by more than 2% of that of the companion cylinder. If the difference is more than 2%, the cylinders should be discarded. The diameter should be measured to the nearest of 0.01 in (0.25 mm) by averaging two diameters measured at right angles to each other at about midheight of the specimen. If the length-to-diameter ratio is less than 1.8 or more than 2.2, or if the height is used to compute the volume of the cylinder, the height should be measured to the nearest of "0.05 diameter." The test specimen should be in saturated surface-dry condition. The specimens should be tested at the specified age. The following are the ASTM permissible tolerances:

Test age, days	Permissible tolerance, hours
1	±0.5
3	±2.0
7	±6.0
28	±20.0
90	±48.0

Test Procedure

1. Clean the upper and lower bearing blocks and place the specimen on the lower block. Carefully align the axis of the cylinder with the center of thrust of the spherically seated upper block. Most testing machines have concentric circles marked on the bearing plates, and hence centering is not a difficult task.

2. Start the machine and raise the lower block so that the top block comes in contact with the cylinder. Once the top and bottom plates are touching the cylinder, lock the top plate to prevent its rotation.

3. Start applying load without shock. The loading should be a continuous process. For hydraulically operated machines the loading rate should be in the range of 20 to 50 psi/s (138 to 344 kPa/s). The loading rate should be constant. Adjustments should not be made, even when the specimen begins to fail.

4. Record the maximum load, type of failure, and appearance of the concrete. Five types of failures shown in Fig. 3A.8, cover most of the failure modes.

5. Compute the compressive strength by dividing the maximum load by the area of cross section. If the length-to-diameter ratio is less than 1.8, apply the following correction factors. Cylinders with lower length-to-diameter ratios resist more loads due to a different strain distribution along the length.

Length/diameter	1.75	1.50	1.25	1.00
Correction factor	0.98	0.96	0.93	0.87

The values for other length-to-diameter ratios can be interpolated. These correction factors are applicable for normal-strength concrete with strengths in the range of 2000 to 6000 psi (14 to 42 MPa) and lightweight concrete with densities in the range of 100 to 120 lb/ft^3 (1600 to 1920 kg/m^3). Report the compressive strength to the nearest 10 psi (0.1 MPa).

To obtain the modulus of elasticity the following additional steps are needed.

6. Prior to testing for the modulus of elasticity determine the compressive strength of concrete.

7. Attach the compressometer to the cylinder and place the specimen in the machine.

8. Load the specimen at a rate of 35 ± 5 psi/s (240 ± 34 kPa/s) to approximately 40% of the ultimate load. If the companion cylinders are not available, load the cylinders until the longitu-

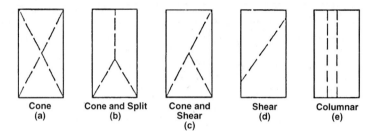

FIGURE 3A.8 Types of fractures.

TABLE 3A.13 Maximum Strain Values

		Strain at age indicated × 10^{-6}	
Unit weight at time of test, lb/ft³ (kg/m³)		7 days or more	Less than 7 days
205	(3282) and over	300	200
165–204	(2642–3266)	375	250
135–164	(2161–2626)	450	300
115–134	(1841–2145)	525	350
105–114	(1681–1825)	600	400
95–104	(1521–1665)	675	450
85–94	(1361–1505)	750	500
75–84	(1201–1345)	825	550

dinal strain reaches a value shown in Table 3A.13. Immediately upon reaching the designated load, reduce the load to zero at the same rate at which it was applied.

9. Reload the specimen and record loads and deformations at predefined intervals. A sufficient number of readings should be taken to establish the stress-strain relationship. Load the specimen up to about 50% of its capacity. Reduce the load to zero and repeat the measurements. If the two sets of measurements are the same, then proceed to compute the modulus E_c. If there is a significant difference between the two sets, an additional set of readings should be taken.
10. After obtaining a consistent set of load-deformation information, remove the compressometer and test the cylinder to failure.
11. Compute the stress by dividing the load by the area and the strain by dividing the deformation by the gauge length. Plot the stress-strain curve. The curve should be approximately linear, at least up to 40% of the failure load.
12. Compute E_c using the equation

$$E_c = \frac{S_2 - S_1}{\epsilon_2 - 0.00005} \tag{3A.10}$$

where E_c = modulus of elasticity
S_1 = stress corresponding to a longitudinal strain of 0.00005
S_2 = stress corresponding to 40% of ultimate load
ϵ_2 = longitudinal strain corresponding to S_2

Report the modulus of elasticity to the nearest 50,000 psi (0.5 GPa).

3A.3.4 Statistical Variation of Compressive Strengths and Quality Control

Due to the influence of various factors, there is always a variation in the strengths of concrete. The variations occur in all modes of loading, including compression, splitting tension, flexure, shear, and torsion. In most construction practices the variations of compressive strength are used for quality control. Hence only compressive strength variations and their tracking procedures are discussed here. In some instances involving pavements, flexural strengths are measured and used for quality assurance. The statistical procedure used for compressive strength can also be used for flexural strength.

As mentioned in Sec. 3A.2.1, the variation of compressive strength is usually assumed to follow a normal distribution. The properties of the normal curve, sample average, and sample standard deviation are used to compute the required average compressive strength to satisfy the specified strength requirements. Hence at the beginning of a project, the strength level of the concrete being produced is based on the calculation of the required average strength. This hypothetical production strength is based on the assumption that the variables affecting the strength of concrete will be the same in the future as they have been in the past. As the data become available, the average of the actual production strength will replace the hypothetical average strength and the standard deviation. If the average and the standard deviation obtained during the project are about the same as the values used in the computation, the project average strength should be carefully maintained.

If the project average strength is smaller than the required average strength while the standard deviation is the same, the percentage of tests below the specified strength will be greater than the acceptable value, and steps must be taken to increase the average strength of the concrete. The average strength should also be increased if the standard deviation of the project is greater than the assumed standard deviation. On the other hand, if the project average is higher, or if the standard deviation is lower; the average strength could be reduced.

Continuous evaluation of the data is necessary in order to assure that the concrete being placed satisfies the specified strength requirements. An updated determination of the average strength and the standard deviation will provide an indication of how well the quality control procedures are working. At any given time the approximate percentage of tests falling below the specified strength can be calculated using the equation

$$p = \frac{\bar{x} - f'_c}{S} \tag{3A.11}$$

where p = probability factor
$\bar{x}$ = average strength
f'_c = specified strength
S = standard deviation

Note that Eq. (3A.11) is a transformed version of Eq. (3A.1). The variables have been changed to permit using the project average strength $\bar{x}$ and the project standard deviation S. In Eq. (3A.11), if $p = 2.33$, the probability of the cylinder strength (average of two cylinders) falling below f'_c is 1%. Probabilities for other values can be estimated using statistical tables in statistics books or tables provided in the report of ACI Committee 214.[5]

In order to satisfy the equations of ACI code 318-92 [Eqs. (3A.2) and (3A.3)], the project average and the standard deviation should satisfy the following two inequalities:

$$1.34 < \frac{\bar{x} - f'_c}{S} \quad \text{or} \quad \frac{\bar{x} - f'_c}{S} \geq 1.34 \tag{3A.12}$$

$$2.33 < \frac{\bar{x} - (f'_c - 500)}{S} \quad \text{or} \quad \frac{\bar{x} - (f'_c - 500)}{S} \geq 2.33 \tag{3A.13}$$

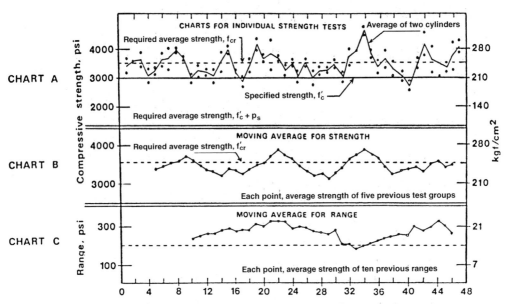

FIGURE 3A.9 Quality control charts for concrete production and evaluation. (A) Individual strength tests. (B) Moving average for strength. (C) Moving average for range. (*From ACI Committee 214.*[11])

Quality control charts are quite often used for a visual picture of concrete performance. Three typical quality control charts used in the industry are shown in Fig. 3A.9. Various forms of other charts are also being used. With the advent of tabletop computers it is very easy to develop, maintain, and update these charts. The charts can also be transferred from location to location using phone lines.

Figure 3A.9(*a*) shows the variations of (1) the individual cylinder strength, (2) the average of two cylinders, and (3) specified and required average strengths. The number of low tests can be easily picked out from this chart. Note that the number of low tests is computed using the average of two cylinders (solid line). If the volume of concrete produced requires more than one test per day, the average of all the tests (instead of two) can be plotted for that day. The charts can also be plotted using calendar dates.

Figure 3A.9(*b*) and (*c*) is plotted using the values of Fig. 3A.9(*a*). Each point in Fig. 3A.9(*b*) represents the average of the previous five tests. The number of tests used to calculate this moving average depends on the type of job and the number of tests per day. In Fig. 3A.9(*b*) some of the high variabilities of individual tests are suppressed. This chart can be used to identify the influence of major factors such as seasonal changes and changes in materials. Figure 3A.9(*c*) shows the moving average range of the previous 10 groups of cylinders. Considerable change in this chart is an indication of high variability.

The control charts are valuable tools not only for the current project, but also for future projects. As discussed earlier, good records can be used for the computation of f'_{cr} and mix proportions instead of trial mixes, thus saving a considerable amount of time and effort.

The variability caused by the test procedure is always a concern in quality control. It is always advisable to separate the variability caused by the testing procedure from variabilities caused by other factors such as change in material properties because the variability in testing does not represent a variability in the strength of the concrete used in the actual construction. The following procedure can be used to estimate the magnitude of variation due to testing.

A test consists of all the cylinders made under identical conditions. The cylinders should be made using the same sample of concrete, cured at the same conditions, and tested at the same age. If it is assumed that two or more test cylinders made from the same sample of concrete and tested at the same age should have the same strength, variations in the strengths of these cylinders can be attributed to the testing procedures. However, since differences in casting and curing could also make a difference, only a major part (and not 100%) of the variation can be attributed to testing. Differences between cylinders cast from the same sample are called within-test variations. The within-test standard deviation S_{wt} can be calculated using the equation

$$S_{wt} = \frac{R}{d_2} \tag{3A.14}$$

where R = average range for all tests of a class of concrete
d_2 = factor based on number of cylinders within test

The values of d_2 are 1.128, 1.693, and 2.059 for two, three, and four cylinders, respectively. The range is the difference between the highest and lowest values of strength.

The within-test coefficient of variation V_{wt} can be computed using the equation

$$V_{wt} = \frac{S_{wt}}{\bar{x}} \times 100 \tag{3A.15}$$

where $\bar{x}$ is the average strength for the class of concrete.

If V_{wt} is less than 1.5%, the field control testing can be considered excellent. If the value is greater than 4%, the within-test variation should be considered as being poor, and errors in testing may be a major contributing factor to strength variation. If V_{wt} is between 1.5 and 2.0, 2 and 3, or 3 and 4, the testing performance is considered very good, good, or fair, respectively.

3A.3.5 Accelerated Strength Tests

In modern-day construction, large volumes of concrete are placed in a single day. In some cases, such as slip-formed construction, it is possible to complete a substantial portion of a structure in a single day. For example, in the case of the CN Communication Tower in Toronto, Canada, the slip-formed construction procedure was used to complete almost 20 ft/day (6 m/day). Therefore one cannot wait 28 days to ascertain the strength. If the strength were found to be unsatisfactory after 28 days, hundreds of feet of structure would have to be taken down. Accelerated strength test procedures were developed for use in such situations. Using these procedures potential 28-day strengths can be estimated in 1 or 2 days. The four procedures, procedure A (warm-water method), procedure B (boiling-water method), procedure C (autogenous curing method), and procedure D (high-temperature and -pressure method), are recognized in ASTM C 684. This section deals with brief descriptions of these methods and their use in quality assurance.

Warm-Water Method (Procedure A). In this method the cylinders are placed in warm water right after casting. The specimens are cured in their molds in the water maintained at 95 ± 5°F (35 ± 3°C) for a period of 23.5 ± 0.5 h. After this curing period of about 24 h the cylinders are capped and tested to determine their compressive strengths. An extensive study conducted by the U.S. Corps of Engineers established that this is a reliable method for routine quality control of concrete. The primary limitation of this method is that the strength gain is not substantial as compared to 24-h moist cured samples.

Boiling-Water Method (Procedure B). In this method the cylinders are stored at 70 ± 10°F (21 ± 5°C) for the first 23 ± 0.25 h and then placed in boiling water for a period of 3.5 h ±5 min. The specimens are allowed to cool for 1 h and tested. The strength increase provided by this type of accelerated curing is much higher than for the warm-water method, and hence the specimen can be transported to the laboratory site without being damaged. This method is the most commonly used one among the four methods.

Autogenous Method (Procedure C). In this procedure the accelerated curing effect is obtained by using the heat of hydration. The cylinders are placed in an insulated container right after casting to retain the heat generated by hydration. The cylinders are tested after curing for 48 ± 0.25 h and a rest period of 30 min at room temperature. The strength gain obtained in this method is lower than that obtained by the boiling-water method. This procedure was found to be less accurate than procedures A and B. However, it was used successfully in the CN tower in Toronto, Canada. The project, which was completed in 1974, involved the placement of about 51,000 yd^3 (39,000 m^3) of concrete.

High-Temperature and -Pressure Method (Procedure D). This procedure is limited to concrete containing aggregates smaller than 1 in (25 mm). Wet sieving can be used for concrete containing larger aggregates. Sealed 3- by 6-in (75- by 150-mm) cylinders are cured at a temperature of 300 ± 5°F (149 ± 3°C) and a pressure of 1500 ± 25 psi (10.3 ± 17 MPa) for a period of 5 h ± 5 min. The curing process starts right after casting. In most cases capping is not required because of the presence of end plates and external pressure. Hence the specimens can be tested within 15 min after curing. For specimens that need capping, testing is done after 30 min. This procedure needs sophisticated equipment and hence is more expensive compared to the three other procedures.

A summary of all four procedures is presented in Table 3A.14. This table shows the curing medium, the temperature and duration of curing, and the age at testing for all four procedures.

3A.3.6 Quality Control Using Accelerated Strength Tests

The most important use of accelerated test data is quality control. These tests permit rapid adjustment of batching and mixing. In the current practice, the accelerated strength results are used to estimate 28-day strengths because of the traditional use of 28-day strength for design purposes. A correlation between the chosen accelerated strength and the 28-day strength should be established before starting the project. The mix proportions and materials used for developing the correlation equation should be the same as the materials and mix proportions to be used for the project.

ACI Committee 214[12] recommends a minimum of 30 data sets for establishing the correlation between the 28-day strength and the accelerated strength. The 28-day strength range should include the specified strength and should not fall below 75% of the specified strength. The 28-day strength is normally expressed as a linear function of accelerated strength using the equation

$$y = ax + b \tag{3A.16}$$

where y = 28-day strength
x = accelerated strength
a, b = constants

TABLE 3A.14 Accelerated Curing Procedures

	Procedure	Molds	Accelerated curing medium	Curing begins	Duration of curing	Age at testing
A	Warm water	Reusable or single use	Warm water, 95°F (35°C)	Immediately after casting	23½ h ± 30 min	24 h ± 15 min
B	Boiling water	Reusable or single use	Boiling water	23 h ± 15 min after casting	3½ h ± 5 min	28½ h ± 15 min
C	Autogenous	Single use	Heat of hydration	Immediately after casting	48 h ± 15 min	49 h ± 15 min
D	High temperature and pressure	Reusable	External heat and pressure, 300°F (149°C)	Immediately after casting	5 h ± 5 min	5¼ h ± 5 min

Source: From ASTM C684, 1993.[2]

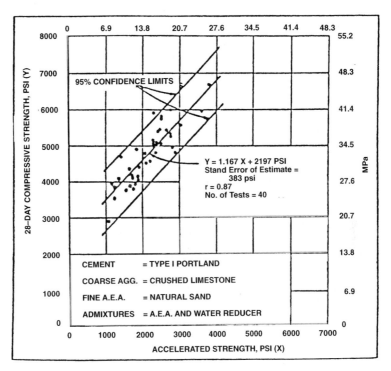

FIGURE 3A.10 Relationship between accelerated and 28-day compressive strengths of concrete—Data obtained using boiling-water method (ASTM procedure B). (*From ACI Committee 214.*[12])

The constants a and b are obtained using statistical correlation. The correlation coefficient should be at least 0.8. A typical correlation curve is shown in Fig. 3A.10. The 95% confidence limits in Fig. 3A.10 show the variations of 28-day strength that can be expected 95% of the time. For example, if the accelerated strength is 3000 psi (21 MPa), the expected 28-day strength is about 5700 psi (40 MPa). Of the 28-day strengths 95% can be expected to be in the range of 4800 to 6500 psi (33 to 45 MPa). An interpretation of the accelerated strength test results and their use for quality assurance follows.

Interpretation of Test Results. Accelerated strength test results can be interpreted using the same procedures as those used for 28-day strength results (Sec. 3A.3.4). The required average strength f'_{cr} can still be computed using the equation

$$f'_{cr} = f'_c + t\sigma \tag{3A.17}$$

where f'_c = specified design strength
t = constant depending on proportion of tests that may fall below f'_c (Table 3A.15).
σ = standard deviation of data set used for prediction of f'_{cr}

Table 3A.15 presents the t values for a number of low test values ranging from 1 in 2 to 1 in 1000. Permissible low tests should be chosen primarily based on the type of structure, as explained earlier.

The required average strength can be established based on either specified accelerated strength or specified 28-day strength and the correlation equation between accelerated and 28-day strengths. The following examples illustrate the computation procedure.

TABLE 3A.15 Values of t for the Equation $f'_{cr} = f'_c + t$

Number	Likelihood of low test results, %	t
1 in 1000	0.1	3.09
1 in 500	0.2	2.88
1 in 100	1.0	2.33
1 in 50	2.0	2.06
1 in 25	4.0	1.75
1 in 20	5.0	1.65
1 in 10	10.0	1.28
1 in 5	20.0	0.84
1 in 2	50.0	0.00

Source: From ACI Committee 214.[12]

Example 3A.3 The specifications require an accelerated strength of 2000 psi (13.8 MPa). Compute the required average (accelerated) strength f'_{cr} if the acceptable number of low tests is 1 in 100. The standard deviation from past records for accelerated strength tests is 500 psi (3.4 MPa).

Solution From Table 3A.15, the value of t for 1 in 100 low tests is 2.33

$$f'_{cr} = f'_c + t\sigma$$
$$= 2000 + 2.33(500)$$
$$= 3165 \text{ psi } (21.8 \text{ MPa})$$

Note that f'_{cr}, f'_c, and σ correspond to accelerated strengths.

Example 3A.4 The specifications require a 28-day compressive strength of 4500 psi (31 MPa). Compute the required accelerated average strength using the following information. The relationship between 28-day strength y and accelerated strength x is

$$y = 1.167x + 2197$$

The acceptable number of low tests is 1 in 10, and the standard deviation for accelerated strengths is 410 psi (2.83 MPa).

Solution The specified 28-day strength is 4500 psi (31 MPa). Using the correlation equation, the corresponding accelerated strength is

$$x = \frac{y - 2197}{1.167}$$

$$= \frac{4500 - 2197}{1.167} = 1973 \text{ psi } (13.6 \text{ MPa})$$

For 1 low test in 10, $t = 1.28$,

$$(f'_{cr})_{\text{accelerated}} = 1973 + 1.28(410) = 2498 \text{ psi}$$

Hence the required average accelerated strength is 2500 psi (17 MPa).

If the number of pairs of data used for the regression line relating accelerated and 28-day strengths is less than 30, special statistical procedures can be used for the prediction of f'_{cr}. The procedure can be found in the report of ACI Committee 214.[12] The number of pairs should be at least 10.

3A.3.7 Nondestructive Tests

Nondestructive tests are valuable tools for evaluating the properties of in situ concrete. These methods can be used to estimate the strength, durability, or elastic properties of concrete. In addition they can also be used to estimate the location and condition of reinforcement, and for locating cracks, large voids, and moisture content. The test methods can be classified as:

Surface hardness methods
Penetration resistance techniques
Pull-out tests
Ultrasonic pulse velocity method
Maturity concepts
Electromagnetic methods
Acoustical methods

These methods are described briefly in this section.

Surface Hardness Methods. In surface hardness methods the hardness of the surface measured, using the size of the indentation or the amount of rebound, is taken as an indicator of the strength of concrete. The most popular method in this category is the rebound hammer test. This test is also known as Schmidt rebound hammer, impact hammer, or sclerometer test.

The rebound hammer, shown in Fig. 3A.11, consists of a spring-loaded mass and a plunger. When the plunger is pressed against the concrete, it retracts against the force imparted by a spring. When the spring is retracted to a certain position, it releases automatically. Upon release the mass rebounds, taking a rider with it along a guide scale. The distance traveled by the mass, expressed as a percentage of the initial extension of the spring, is called the rebound number. ASTM C805 covers the procedure for conducting the rebound hammer test.

The rebound number, which is a measure of the hardness of the concrete surface, can be empirically related to the compressive strength of concrete. However, in certain circumstances the surface hardness may not represent the strength of the concrete inside the structure. For example, the presence of a large aggregate immediately underneath the plunger would result in a large rebound number. A large void underneath the plunger, on the other hand, would provide an unusually low rebound number. Other factors that influence the rebound number include the type of aggregate, smoothness of the surface, moisture condition, size and age of the specimen, degree of carbonation, and position of the hammer (vertical versus inclined or horizontal). The hammer should be used only against a smooth surface. Troweled surfaces should be rubbed smooth using a carborundum stone. If the concrete being tested is not part of a large mass, it has to be supported so that the specimen does not move during the impact. Concrete in the dry state tends to record a

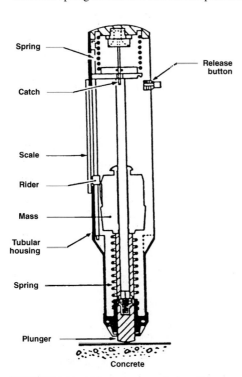

FIGURE 3A.11 Major components of rebound hammer.

higher rebound number. Because of gravity, the rebound number of floor concrete would be smaller than that of the soffit concrete, even though both concretes are similar. The rebound number of inclined and vertical surfaces would fall somewhere in between.

The rebound hammer is best utilized for checking the uniformity of concrete in a large structure or for comparing the quality of concrete in similar structural components such as precast beams. It can also be used for estimating the strength of concrete that is being cured for the purpose of removing formwork. If the rebound numbers have to be used for estimating strength, correlation should be established between the rebound number and the compressive strength for each type of concrete used on a site. If mix proportions or constituent materials are changed, then a new set of tests should be conducted and a new correlation equation obtained.

Typically there is a large variation in rebound numbers. At least 10 to 12 readings should be taken in each location. ASTM C805 provides guidelines for averaging the rebound numbers. In certain cases some individual readings might have to be omitted from the average. If proper calibration is used, the accuracy of prediction of concrete strength is about ±20% for laboratory specimens and ±25% for in situ concrete.

Penetration Resistance Techniques. In this method the penetration resistance of concrete is used as the indicator of its strength. The most commonly used test is the Windsor probe test. In this test a hardened alloy probe is fired (or driven) by a driver using a standard charge of powder. The exposed length of the probe is taken as a measure of penetration resistance. It is assumed that the compressive strength of concrete is proportional to its penetration resistance. Here again the hardness of the aggregates plays an important role. A correlation has to be developed for a particular concrete if this method is to be used for predicting strengths.

The probes are driven in sets of three in close vicinity, and the average value is used for estimating the strength. The test procedure is covered in ASTM C803. This test is more expensive than the rebound hammer test, but much less expensive than core tests. This test, which is considered to be more accurate than the rebound hammer test because the measurement is not made just on the surface, is also an excellent tool for determining the uniformity of concrete and the relative rate of strength gain at early age for the purpose of removing formwork.

Pull-Out Tests. In a standard pull-out test the concrete strength is estimated using the force required to pull out a specially shaped steel insert whose enlarged end has been cast into the fresh concrete. The specifications of the test are covered in ASTM C900. Because of the shape of the insert, a lump of concrete, in the shape of a frustrum of a cone, is pulled out along with the insert. In most cases the fracture occurs at about a 45° angle. Even though the failure occurs due to tension and shear, the strength computed using an idealized area of the frustrum was found to be approximately equal to the shear strength of concrete. An approximate linear correlation seems to exist between pull-out strength and compressive strength. The ratio of pull-out strength to compressive strength decreases slightly for concrete with higher compressive strengths.

The pull-out test is more accurate than the penetration or the rebound hammer test because it is based on the actual failure load. However, this test, which is more involved, has to be planned in advance, and the damaged area has to be repaired. In some cases, such as estimation of strength for form removal, pulling out the assembly may not be necessary. The test can be stopped when a predetermined force is reached, assuring sufficient strength for removing the forms.

Other forms of tests similar to the standard pull-out test are still being developed. Those test methods include the pull-out of inserts placed in drilled holes, pulling out a wedge anchor using torque, and break-off tests. In the break-off test the flexural strength of concrete is determined by applying a transverse force on the cylinder created by inserting a tube in the fresh concrete. Inserts placed in drilled holes can be used for existing structures. But their success is still to be established.

Ultrasonic Pulse Velocity Method. In this method the longitudinal wave velocity in concrete is used to estimate the compressive strength and check the uniformity of concrete. An exciter is used to initiate a pulse which is picked up at another designated location. The distance between the two locations divided by the time required for the travel is the pulse velocity. Correlation relationships have been developed between pulse velocity, compressive strength, and modulus of elasticity. The relationships are affected by a number of variables, such as the moisture condition of the specimen, type and volume fraction of the aggregate, water-cement ratio, and age of specimen. In general, pulse velocity alone cannot be used to estimate the strength, unless a correlation exists for the particular type of concrete being tested. But the method is an excellent tool for quality control measures. The test method is covered in ASTM C597. The pulse velocity method can be used for both laboratory and field tests. In field tests, the presence of reinforcement and the vibration of elements being tested (such as piers under a roadway in use) might pose problems. A combination of rebound hammer and pulse velocity methods is sometimes used to evaluate both the surface characteristics and the quality of concrete located well inside the surface.

Maturity Concepts. It is well established that the strength development of concrete depends on duration, curing temperature, and pressure. If moist curing is done at atmospheric pressure, then the combination of duration and curing temperature could be used to estimate the maturity and hence the compressive strength. A number of relationships relating strength and maturity exist in the published literature. Maturity meters are also available for use in the field and the laboratory. This concept can be used to estimate the early strength of structural members for form removal. The correlation between maturity and strength should be established before starting the actual project.

Electromagnetic and Acoustical Methods. Various forms of magnetic, electrical, and acoustical techniques have been tried to determine the properties of concrete such as dynamic modulus, presence of cracks, honeycombing, and measurement of cover to reinforcement. These methods have not attained common acceptance so far.

3.A.3.8 Core Tests

If there is a reason to believe that the concrete in place may not have the specified strength, nondestructive tests discussed in the previous section can be used to determine the uniformity. If the location in question behaves very similar to other locations, the quality of the concrete could be satisfactory. However, if the tests indicate variability, cores might have to be taken to determine the strength. The procedure for evaluation using cores is covered in ASTM C42.

Typically, cores are drilled using diamond drills. A number of factors should be considered in evaluating core strengths. The following are some of the important points:

- Typically core strengths are lower than standard cylinder strengths. The differences could be more significant for high-strength concrete.
- The strength of the core could depend on its position in the structure. Cores taken near the top of a structural element are typically weaker than cores taken from the bottom.
- Cores taken from very thick sections could contain microcracks due to excessive heat of hydration and hence register lower strengths.
- The presence of large aggregates in small cores could result in erroneous strengths.

3A.4 MECHANICAL PROPERTIES

Strength, stiffness, and dimensional stability constitute the core of the mechanical properties. Strength can be measured under various modes of loading such as compression, tension, flexure, shear, and torsion. This section deals with these basic mechanical properties of concrete. Properties such as the durability of concrete exposed to various chemicals or to freezing and thawing, and permeability are also very important for some structures. These properties are not covered in this book for lack of space. The reader is referred to the literature.

3A.4.1 Compressive Strength

Compressive strength is the most commonly used design parameter for concrete. In most cases the concrete is specified using its compressive strength measured at 28 days. In some instances 56-day strength or minimum early strength is specified. Up to around 1960, the compressive strength of concrete was limited to about 6000 psi (42 MPa). The advent of new admixtures as well as mixing, placing, and compacting techniques led to the development of higher strengths. Concrete with a specified compressive strength of 12,000 psi (84 MPa) was used in Water Tower Place in Chicago, which was topped off in 1972. The development of high-range water-reducing admixtures (superplasticizers) resulted in routine use of high-strength concrete. Concrete used in a Seattle building in the late 1980s had an average strength in excess of 20,000 psi (135 MPa). This section deals with the various factors that affect the compressive strength of concrete. An understanding of the influence of the various factors is needed for effectively proportioning, making, and casting concrete in the field.

The major factors that influence compressive strength are water-cement ratio; aggregate-cement ratio; maximum size of aggregate; grading, surface texture, shape, strength, and stiffness of aggregate particles; degree of compaction; curing conditions; and testing parameters. The admixtures used can also influence the strength by improving workability and through better compaction.

Water-Cement Ratio. The relation between compressive strength and water-cement ratio was established in 1918 by Duff Abrams. For a fully compacted concrete, he found that the strength f'_c can be expressed as

$$f'_c = \frac{k_1}{k_2 w/c} \tag{3A.18}$$

where w/c = water-cement ratio
k_1, k_2 = empirical constants

If the amount of air voids does not exceed 1% by volume, the concrete can be considered as fully compacted concrete. Typical variations of compressive strength and the influence of compaction are shown in Fig. 3A.12. From this figure it can be seen that compaction plays an impor-

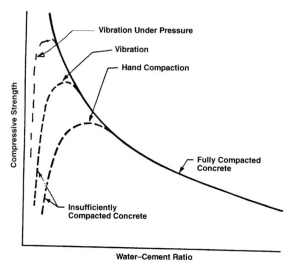

FIGURE 3A.12 Relationship between strength and water-cement ratio of concrete.

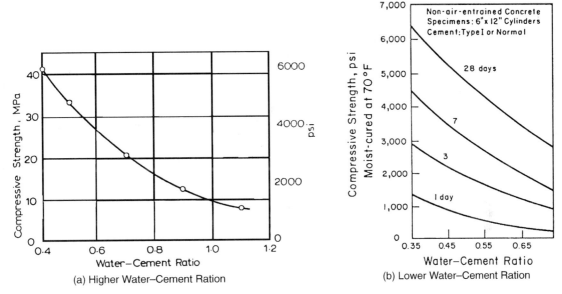

FIGURE 3A.13 Influence of water-cement ratio and moist curing on concrete strengths. (*a*) Higher water-cement ratio. (*b*) Lower water-cement ratio. (*From PCA.*[8])

tant role. The use of admixtures that improved workability led to better compaction, resulting in higher and higher strengths in the 1980s. Typical variations of compressive strength for water-cement ratios ranging from 0.35 to 1.2 are shown in Fig. 3A.13. In field construction the commonly used water-cement ratios vary from 0.25 to 0.65. As mentioned in Sec. 3A.2, the water-cement ratio should be restricted to 0.4 for obtaining durable concrete. If the water-cement ratio is less than 0.4, in most cases some form of admixture is needed for obtaining a workable concrete.

The amount of voids present in concrete controls its strength. In concrete with a high water-cement ratio, the excess water results in more voids and hence lower strength. Improperly compacted concrete also has higher voids, resulting in low strength. Hence a balanced approach should be used in mix proportioning. A water-cement ratio of about 0.4 is needed for complete hydration of the cement. However, complete hydration of cement does not produce the highest strength. The presence of unhydrated cement as inert particles was found to provide better strength. In most cases the lower limit for the water-cement ratio is 0.28. However, water-cement ratios lower than 0.28 have been used with superplasticizers and other admixtures for producing very high-strength concrete.

Aggregate-Cement Ratio. The aggregate-cement ratio affects the strength of concrete if the strength is about 5000 psi (35 MPa) or more. The influence of the aggregate-cement ratio is not as significant as that of the water-cement ratio, but it has been found that for a constant water-cement ratio, leaner mixes provide higher strengths, as shown in Fig. 3A.14. The increase in strength could be due to absorption of water by the aggregate and hence a lower effective water-cement ratio. In addition leaner mixes have lower amounts of total water and paste content and hence lower amounts of voids.

A more recent study indicates that the strength increases with an increase in the cement content if the volume of aggregates is less than 40%. But the trend reverses in the aggregate volume ratios of 40 to 80% (Fig. 3A.15).

Cement Type and Age. The degree of cement hydration determines the porosity of the hydrated cement paste and hence the compressive strength. Under standard curing conditions type III

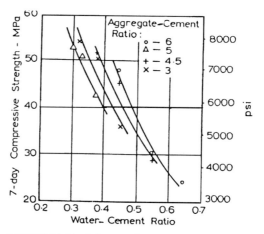

FIGURE 3A.14 Influence of aggregate-cement ratio on strength of concrete. (*From Singh.*[13])

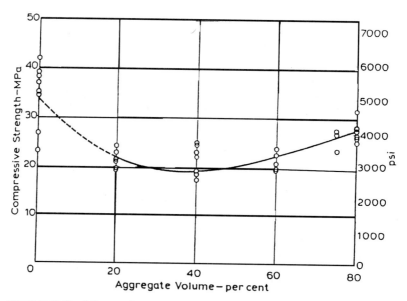

FIGURE 3A.15 Influence of aggregate content on strength. (*From Stock et al.*[14])

cement hydrates faster than type I cement. Hence at early ages type III cement provides higher strengths. The variations in strength, for type I and III cements at 1, 3, 7, and 28 days are shown in Fig. 3A.16. The bands shown in this figure cover the majority of the data obtained in the laboratories. Relative strength gains for three water-cement ratios are shown in Fig. 3A.17. From this figure it can be seen that early strength gain is higher for lower water-cement ratios. Empirical relationships are available in the literature for predicting 28-day strengths based on the results of 1-, 3-, or 7-day strengths and vice versa.

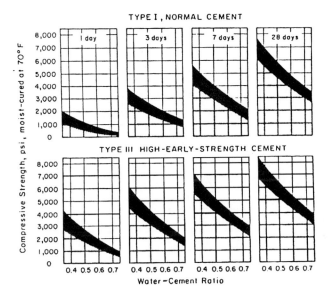

FIGURE 3A.16 Influence of water-cement ratio, duration of moist curing, and cement type on strength. (*From PCA.*[8])

At normal temperatures ASTM type II, IV, and V portland cements hydrate at a slower rate, and hence concrete containing these cements will have lower early strengths. The results presented in Table 3A.16 show typical relative strengths at 1, 7, and 28 days for various cements. After 90 days the variation in strength between cement types is negligible.

Coarse Aggregate. Typically, aggregates are stronger than the matrix and hence do not control the fracture strength. However, factors such as aggregate size, shape, surface texture, and mineralogy can influence the workability, degree of compaction, and formation of gel around the aggregate. Consequently it can affect the compressive strength.

The maximum size of aggregate has considerable influence, especially at low water-cement ratios, as shown in Fig. 3A.18. Concrete with large-size aggregates requires less water, and hence the same water-cement ratio provides better compaction, resulting in higher strength. On the other hand, larger sizes provide weaker transition zones around the aggregate, resulting in lower strength. These opposing influences are water-cement ratio dependent and provide more pronounced effects at lower water-cement ratios. In most cases the strength can be expected to go down with an increase in the maximum size of the aggregate.

If water-cement ratio and maximum size of aggregate are kept constant, aggregate grading influences the consistency of the concrete and hence the strength. An increase in fines typically increases the water demand, and if the amount of water is not increased, consistency decreases. The decrease in strength was found to be as high as 12%.

Aggregates with rough texture were found to result in early high strengths as compared to aggregates with smooth textures. The rough surface provides a better bond between matrix and aggregate, especially at early stages when the hydration is not complete. At later stages the influence diminishes. Aggregates with smooth surfaces are easier to work with and hence provide a better final product.

The mineralogical composition of aggregates was also found to influence the concrete strength. Calcareous aggregates tend to provide better strength than siliceous aggregates. Fig. 3A.19 shows the compressive strengths obtained using various types of aggregates. It can be seen that the strength variation could be as high as 50%.

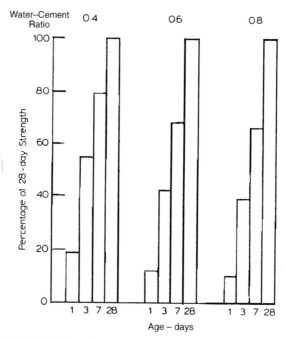

FIGURE 3A.17 Relative strength gain of concrete with time for different water-cement ratios.

TABLE 3A.16 Approximate Relative Strength of Concrete as Affected by Type of Cement

	Type of portland cement	Compressive strength, % of strength of type I*		
		1 day	7 days	28 days
I	Normal or general-purpose	100	100	100
II	Moderate heat of hydration and moderate sulfate resisting	75	85	90
III	High early strength	190	120	110
IV	Low heat of hydration	55	65	75
V	Sulfate resisting	65	75	85

*Compressive strength is 100% for all cements at 90 days.
Source: From PCA.[8]

Air Content. Air is entrained in concrete to improve its durability. The air bubbles tend to improve the workability of fresh concrete and hence improve the compaction. But the presence of air bubbles in the hardened concrete increases its porosity and reduces the density of the composite. Hence air entrainment leads to a decrease in compressive strength.

The decrease in strength due to air entrainment was found to depend on both the water-cement ratio and the cement content. As the water-cement ratio decreases, the strength loss increases. Hence the strength loss is considerable for high-strength concrete. When the cement content is reduced, the influence of the air content also decreases. In fact the air content improves the strength slightly if the cement content is very low. In the normal strength range, about a 2% loss in compressive strength can be expected for each 1% increase in air content.

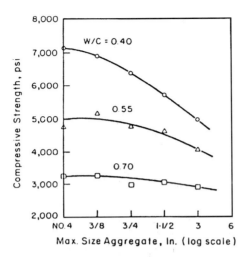

FIGURE 3A.18 Influence of aggregate size and water-cement ratio on strength. (*From Cordon and Gillespie.*[15])

The air content was found to improve the workability of lightweight concrete. This is particularly true for mixes of low paste content.

Curing Conditions. Concrete should be moist cured for at least 28 days to obtain best results. Premature drying can reduce the strength by more than 50%. In terms of curing, the major factors are time, humidity, temperature, and pressure.

A longer curing time under moist conditions always provides better results. A 100% relative humidity is the best condition. This can be achieved by pooling water, placing wet burlaps, or making other arrangements. Figure 3A.20 shows the influence of moist curing. It can be seen that 3 days of moist curing provides 50% improvement over air-dried samples. It is preferable to moist cure for 28 days. In any circumstance, the moist curing should be done for at least 7 days.

Typically higher temperatures provide faster curing. At about 12°F (−11°C) the cement stops hydrating. Hence at this temperature there may not be any increase in strength, even after long periods of time. Early age strength increase can be accelerated by using warm water. Temperatures higher than 100°F (38°C) are not normally used because the increase in acceleration beyond this temperature is not high enough for economic reasons and the initial high temperature could result in lower long-term strength. However, precast elements are sometimes steam cured in order to reduce turnaround time and speed up reuse of forms.

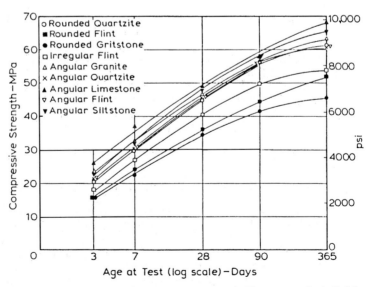

FIGURE 3A.19 Influence of aggregate type on strength. (*From a report by the Building Research Station, London, 1969.*)

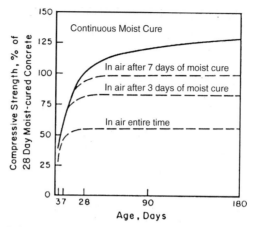

FIGURE 3A.20 Influence of curing conditions on strength. (*From PCA.*[8])

Higher pressure typically accelerates the curing. But the increase is not significant for commercial applications. Hence pressure curing is used only for steam curing.

Test Conditions. In North America 6- by 12-in (150- by 300-mm) cylinders are the standard test specimens, even though 4- by 8-in (100- by 200-mm) cylinders are sometimes used for high-strength concrete. Other types of specimens include cubes and prisms. In the case of cylinders and prisms, the minimum length-to-diameter ratio (or longer dimension versus lateral dimension) should be at least 2 to avoid end effects. If specimens with lower length-to-diameter ratios are used, correction factors should be applied for determining the compressive strength. ASTM specifications provide guidelines, as explained in Sec. 3A.2. Cylinders tend to record lower strengths than cubes. The ratio of cylinder to cube strength is about 0.85. Larger samples tend to record lower strengths as compared to smaller samples. Other test conditions that influence strength include end conditions, rate of loading, moisture condition of the specimen, and condition of the plattens of the machine.

Cylinders should be capped properly, making sure that the ends are plane, parallel to each other, and perpendicular to the longitudinal axis. The testing machine should have capability to allow slight rotation of the ends. The machine should also be calibrated periodically to avoid errors in the measurement of loads.

3A.4.2 Tensile Strength

The tensile strength can be measured using uniaxial tension specimens or cylinders. Uniaxial tension specimens are very difficult to test because special gripping devices are needed. The most common practice is to test a 6- by 12-in (150- by 300-mm) cylinder along the longitudinal axis. This test is called the splitting tension test, and the strength value obtained is the splitting tensile strength.

There is a relationship between compressive and tensile strength. But the relationship is not well established. Tensile strength is about 10% of compressive strength for normal-strength concrete. As the compressive strength increases, the ratio decreases. Table 3A.17 presents typical tensile strength values for compressive strengths varying from 1000 to 9000 psi (7 to 60 MPa). A number of factors, including type and shape of coarse aggregate, properties of fine aggregate, grading of aggregate, air entrainment, and age at testing, affect the ratio between tensile and com-

TABLE 3A.17 Relation between Compressive, Flexural, and Tensile Strengths of Concrete

Strength of concrete, psi (MPa)			Ratio, %		
Compressive strength	Modulus of rupture	Splitting tensile strength	Modulus of rupture to compressive strength	Splitting tensile strength to compressive strength	Splitting tensile strength to modulus of rupture
1000 (6.9)	230 (1.6)	110 (0.8)	23.0	11.0	48
2000 (13.8)	375 (2.6)	200 (1.4)	18.8	10.0	53
3000 (20.7)	485 (3.3)	275 (1.9)	16.2	9.2	57
4000 (27.6)	580 (4.0)	340 (2.3)	14.5	8.5	59
5000 (34.5)	675 (4.7)	400 (2.8)	13.5	8.0	59
6000 (41.3)	765 (5.3)	460 (3.2)	12.8	7.7	60
7000 (48.2)	855 (5.9)	520 (3.6)	12.2	7.4	61
8000 (55.1)	930 (6.4)	580 (4.0)	11.6	7.2	62
9000 (62.0)	1010 (7.0)	630 (4.3)	11.2	7.0	63

Source: From Price.[16]

pressive strengths. A number of empirical relations are available for estimating the tensile strength. Most of these equations are of the form

$$f_t = k (f'_c)^n \tag{3A.19}$$

where f'_c = compressive strength
f_t = tensile strength
k, n = empirical constants

Concretes containing lightweight aggregate typically have lower tensile strengths.

3A.4.3 Modulus of Rupture

The flexural strength of concrete is called modulus of rupture. The strength values are determined using 4- by 4- by 14- or 6- by 6-by 20-in (100- by 100- by 350- or 150- by 150- by 550-mm) prisms subjected to four-point loading. When analyzing reinforced concrete beams and slabs for flexural loading, the modulus of rupture is used to compute cracking load and deflection. Since the beams are in the flexure mode, the modulus of rupture is more representative than splitting or direct tensile strengths. Typical values of flexural strength are presented in Table 3A.17.

For design purposes, the modulus of rupture f_r can be estimated using the following equation:

$$f_r = 7.5\sqrt{f'_c} \tag{3A.20}$$

In most cases this equation provides a conservative estimate. For high-strength concrete, other forms of equations have been proposed. Nevertheless, Eq. (3A.20) provides a good estimate for design purposes. For lightweight concrete a constant smaller than 7.5 should be used.

Most factors that influence the ratio of tensile to compressive strength also influence the modulus of rupture. The magnitude of influence is slightly lower for flexural strength.

3A.4.4 Shear Strength

Concrete is seldom subjected to pure shear. But when structural members are loaded under various modes, the concrete in those members could be subjected to a shear force. It is extremely difficult to measure direct shear strength. Torsion or deep beam specimens can be used to measure shear strength indirectly.

Equations for the prediction of shear strength are not well established. Most researchers agree that the shear strength is proportional to the square root of compressive strength. But the constant of proportionality is not established. For beams subjected to bending and shear, the ACI code allows a shear stress of $4\sqrt{f'_c}$ for uncracked sections. However, this value is useful only for beams subjected to shear. A number of researchers have tested specimens under torsion to establish the shear strength. But the results are still inconclusive in terms of having a single equation for predicting shear strength.

3A.4.5 Modulus of Elasticity

The modulus of elasticity of a material is the slope of the initial linear portion of the stress-strain curve. Since the stress-strain curve of concrete is nonlinear, there are three types of moduli, namely, the tangent, the secant, and the chord modulus. The secant modulus is the most commonly used parameter for design purposes. It is defined as the slope of the line joining the origin and the point on the stress-strain curve corresponding to 40% of compressive strength. For high-strength concrete the initial portion of the curve is almost linear, and hence the tangent and the secant moduli are same. Since measurement of the modulus is a more involved process, its value is normally estimated using the compressive strength and the unit weight of concrete.

The ACI code recommends the following equations for normal-strength concrete[4]:

$$E_c = 33 W_c^{1.5} \sqrt{f'_c} \tag{3A.21}$$

where E_c = secant modulus
W_c = unit weight of concrete
f'_c = compressive strength of concrete

For normal-weight concrete which has a unit weight of about 145 lb/ft³ (2321 kg/m³) E_c can be computed using the following equation:

$$E_c = 57,000 \sqrt{f'_c} \tag{3A.22}$$

For high-strength concrete the aforementioned equations were found to overestimate the value of the modulus. The equation recommended for high-strength concrete is

$$E_c = (40,000\sqrt{f'_c} + 10^6)\left(\frac{W_c}{145}\right)^{1.5} \tag{3A.23}$$

A number of other equations are also available in the literature. The primary factors that influence the modulus of elasticity are aggregate type, amount of paste, mineral admixtures such as fly ash and silica fume, and moisture conditions of the specimen. Tests should be conducted carefully using ASTM procedures. The modulus value is very sensitive to test conditions such as rate of loading.

3A.4.6 Shrinkage

Shrinkage is a phenomenon in which concrete shrinks under no load due to the movement and evaporation of water. There are two types of shrinkage known, plastic and drying shrinkage. Plastic shrinkage occurs during the initial and final setting times, extending to several hours after the initial placement of concrete. Major factors that affect plastic shrinkage are area of exposed surface, surface and concrete temperature, and surface air velocity.

Drying shrinkage occurs in hardened concrete, and it can continue for up to 2 years. It occurs due to the evaporation of water and the movement of water within, resulting in a more compact matrix.

If concrete is immersed in water, a small amount of expansion known as swelling can be observed. However, the expansion cannot completely reverse the shrinkage. Shrinkage is a time-dependent process. In the first few months it is much larger than at later ages. A number of factors influence the shrinkage. These factors are briefly discussed in the following. More details can be found in the literature dealing with concrete.

Aggregate Type and Volume Fraction. The aggregates restrain the shrinkage of cement paste because they do not shrink. Hence the concrete with lower cement content has less shrinkage. In addition the elastic properties and other characteristics of the aggregate also influence the shrinkage because the amount of restraint prodivided by the aggregate depends on its elastic modulus and how much force it can transmit along the interface. Typical shrinkage values for different aggregate-cement and water-cement ratios are shown in Table 3A.18. From this table it can be seen that shrinkage can be reduced by as much as four times by increasing the aggregate content.

In terms of type of aggregates, quartz provides the least amount of shrinkage. Aggregates that produce more shrinkage in progressive order are limestone, granite, basalt, gravel, and sandstone. Lightweight aggregate concretes typically shrink more than normal-weight concrete.

Water-Cement Ratio. As the water-cement ratio increases, shrinkage increases, as shown in Table 3A.18. This should be expected because more water leads to more evaporation and movement. Water-cement ratios higher than 0.7 could lead to excessive shrinkage.

Exposure Conditions. The relative humidity, temperature, and air movement to which the element is exposed affect both the rate of shrinkage and the total (ultimate) shrinkage. Higher relative humidity, lower temperature, and low air velocity decrease shrinkage.

Size of Member. As the size of the member increases, the rate of evaporation decreases and hence the rate of shrinkage is lower. As the concrete matures, the amount of water lost to evaporation also decreases, because water cannot move freely in matured concrete. This results in a decrease in ultimate shrinkage.

Type of Cement. Cements that hydrate faster tend to produce more shrinkage. Special cements, called shrinkage-compensating cement, are available for reducing shrinkage. ASTM type K cement is called expansive cement, as it provides increases in volume rather than a decrease, or shrinkage.

Admixtures. As mentioned in Sec. 3A.1, a number of mineral and chemical admixtures are used in concrete to obtain certain properties in the fresh and hardened states. Accelerating admixtures tend to increase the rate of shrinkage whereas retarding admixtures will decrease the rate of shrinkage. Since concrete with water-reducing admixtures tends to have lower water-cement ratios, both the rate and the ultimate shrinkage for this concrete decreases. The effect of mineral

TABLE 3A.18 Typical Values ($\times 10^{-6}$) of Shrinkage after 6 Months for Mortar and Concrete Specimens*

Aggregate-cement ratio	Water-cement ratio			
	0.4	0.5	0.6	0.7
3	800	1200	—	—
4	550	850	1050	—
5	400	600	750	850
6	300	400	550	650
7	200	300	400	500

*Based on 5- by 5-in (125- by 125-mm) prisms, stored at a relative humidity of 50% and a temperature of 70°F (21°C).
Source: From Lea.[17]

admixtures depends on the type and volume fraction. Air-entraining agents were found to have little effect on shrinkage.

Carbonation. Carbonation, which occurs due to a reaction between the carbon dioxide present in the atmosphere and the cement paste, tends to produce shrinkage which is called carbonation shrinkage. This phenomenon occurs only near the exposed surface. Carbonation tends to occur over a longer period of time than drying shrinkage. At humidities less than 50%, carbonation decreases drastically whereas drying shrinkage accelerates.

External Restraints. External restraints reduce shrinkage. The most common restraints are the reinforcing bars. Reinforcement tends to resist movement, and hence the total magnitude of shrinkage decreases. Because of the restriction in movement, a stress state occurs in reinforced concrete elements due to shrinkage. Typically reinforcement is subjected to compression and the surrounding concrete is subjected to tension. If the concrete is weak and its cross section is not sufficient to withstand the tensile forces, it might crack.

Mixing, Placing, Consolidation, and Curing. The way the concrete is mixed, placed in position, compacted, finished, and cured affects the quality of the concrete and hence the shrinkage. If these operations are not done properly, the resulting concrete could be porous, sustaining higher shrinkage strains. Curing is very important because the presence of moisture on the exposed surface reduces the evaporation loss considerably, resulting in reduced shrinkage.

Reduced shrinkage strain is better for the integrity of the concrete elements. Shrinkage increases the long-term deflections of beams and slabs, as well as the crack widths. In prestressed concrete shrinkage produces a loss in prestress. In composite construction and indeterminate structures, shrinkage produces a redistribution of stresses. These effects should be considered in the structural design.

3A.4.7 Creep

The phenomenon of creep has a number of similarities with shrinkage. Most factors that affect shrinkage also affect creep. Shrinkage occurs under no load whereas creep occurs under a state of sustained stress. Typical variations of creep and shrinkage strains with time are shown in Fig. 3A.21. Figure 3A.21(*a*) shows the variation of shrinkage strain obtained using an unloaded specimen. If the specimen is sealed and subjected to sustained stress, the variation of strain with respect to time is shown in Fig. 3A.21(*b*). Since the specimen is sealed, the shrinkage is essentially eliminated. As soon as the specimen is loaded, there is an elastic response producing an elastic or instantaneous strain. The strain continues to increase under the sustained stress. This additional strain is called creep strain. If the unsealed specimen is kept under sustained stress, both creep and shrinkage phenomenon occur. In addition, more shrinkage occurs because of the stress. The stress aids the movement of water, resulting in additional shrinkage strain.

As mentioned earlier, the factors that affect shrinkage also affect creep. The influences are similar in most cases. The factors that have less effect on creep than on shrinkage are type of cement and carbonation. The following additional factors influence the creep strain.

Level of Stress. There is a proportionality between the magnitudes of sustained stress and creep if the level of stress is less than 50% of compressive strength. Hence the creep strain is proportional to the elastic strain under normal working load conditions. However, if the level of stress increases beyond 70% of compressive strength, excessive creep strain occurs, leading to failure.

Time of Loading. The time of loading influences the creep strain because the strength of concrete increases with time. For example, if the concrete is subjected to sustained load at 7 days, it undergoes more creep strain as compared to concrete loaded at 28 days. It should be noted that the level of stress should be the same for both loading conditions. The influence of the time at loading decreases after 28 days and becomes almost insignificant if the time at loading exceeds

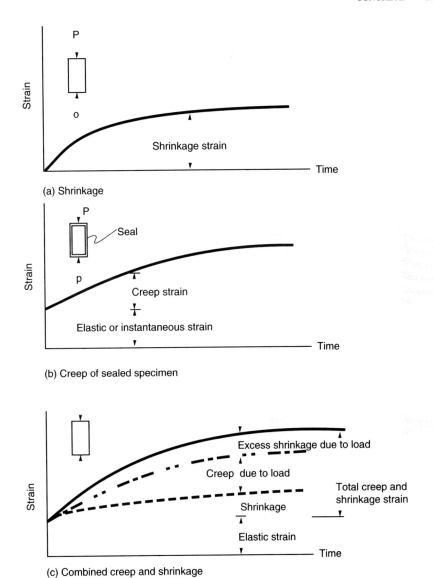

FIGURE 3A.21 Typical variations of creep and shrinkage strains with time. (*a*) Shrinkage. (*b*) Creep of sealed specimens. (*c*) Combined creep and shrinkage.

56 days. This should be expected because the change in compressive strength after 56 days is negligible for normal concrete.

As is the case of shrinkage strain, creep strain increases deflections and crack widths, causes loss of prestress, and results in a redistribution of stresses in indeterminate structures and composite members. In the case of reinforced concrete columns, the redistribution of stresses could result in yielding of steel and buckling of eccentrically loaded columns.

Unloading typically leads to elastic and creep recovery. As in the case of shrinkage, creep recovery is not complete, leading to some amount of permanent deformation.

3A.4.8 Estimation of Creep and Shrinkage Strains

A number of rheological models have been developed to simulate creep and shrinkage. These models consisting of Kelvin solid and Maxwell fluid can be used to predict the behavior of concrete subjected to various modes of loading such as constant stress, constant strain, and loading for a certain amount of time. These models can predict both creep and creep recovery. However, most designers prefer to use simple empirical equations for the prediction of creep and shrinkage strains.

A number of empirical equations exist for the prediction shrinkage and creep at a given time under load. The equations recommended by ACI Committee 209[18] are

$$(\epsilon_{sh})_t = \frac{t}{35 + t}(\epsilon_{sh})_u \tag{3A.24}$$

$$\gamma_t = \frac{t}{10 + t^{0.6}}(C_u) \tag{3A.25}$$

where $(\epsilon_{sh})_t$ = shrinkage strain at time t
$(\epsilon_{sh})_u$ = ultimate shrinkage strain
γ_t = creep coefficient
$(C)_u$ = ultimate creep coefficient

For a given concrete the ultimate creep coefficient and ultimate shrinkage strain have to be assumed. These values can also be obtained using short-term readings taken at, say, 14, 28, or 56 days.

The equations recommended by the Comité Euro-International du Béton (CEB), Paris, France,[19] involve the use of coefficients for various exposure conditions and hence are more accurate. The shrinkage creep strains are computed using the following equation:

$$\epsilon_{sh} = \epsilon_c K_b K_t K_e \tag{3A.26}$$

where

$$\epsilon_{cr} = \phi \text{ (elastic strain)} \tag{3A.27}$$

$$\phi = K_c K_b K_d K'_t K'_e \tag{3A.28}$$

where ϵ_c, K_b, K_e, K_t, K_c, K_d, K'_t, and K'_e are coefficients. The values for these coefficients can be obtained using Figs. 3A.22 and 3A.23.

3A.4.9 Behavior under Multiaxial Stresses

In some structures such as two-way slabs and nuclear reactor containment vessels the concrete is subjected to biaxial and triaxial stresses. Another example for triaxial state of stress is off-shore oil platforms. Concrete located near the bottom of the sea is subjected to water pressure and external load, resulting in a triaxial state of stress.

In the case of biaxial loading, the strength of concrete increases by about 20% under biaxial compression. Figure 3A.24 shows the failure envelope for concrete subjected to biaxial loading in tension and compression. The increase in strength under biaxial loading is normally neglected in design computations. Hence structural components subjected to biaxial loading can be designed using uniaxial strengths.

Concrete subjected to triaxial loading can withstand much higher stresses as compared to uniaxial strength, as shown in Fig. 3A.25. In general, triaxial compression improves the compressive strength of leaner or low-strength concrete more than that of a stronger or high-strength concrete. Failure theories have been developed for the triaxial state of stress. The most popular ones are known as octahedral shear stress theory and Mohr-Coulomb failure theory. Some researchers have also developed empirical relationships between minor axial stress σ_3 and major axial stress

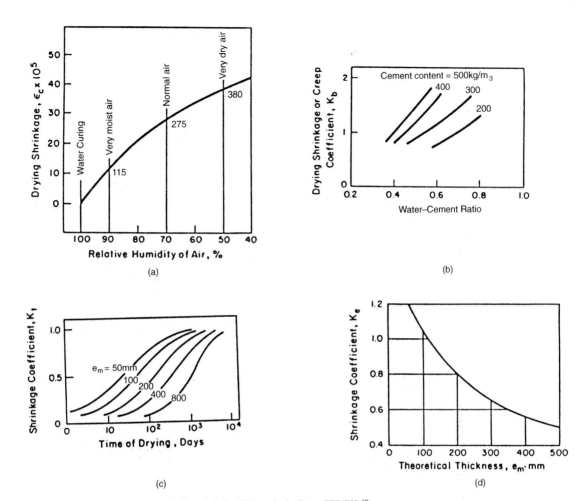

FIGURE 3A.22 Coefficients ϵ_c, K_b, K_t, and K_e for CEB method. (*From CEB/FIP.*[19])

σ_1, which creates failure. Under the uniaxial state of stress σ_3 is zero, and hence σ_1 is the compressive strength f'_c. One such empirical equation is

$$\sigma_1 = f'_c + 4.8\,\sigma_3 \tag{3A.29}$$

where f'_c is the compressive strength.

For example, concrete with a compressive strength of 4000 psi (28 MPa) can be expected to withstand 8800 psi (61 MPa) under a lateral pressure of 1000 psi (6.89 MPa) ($\sigma_1 = 4000 + 4.8 \times 1000$). More detailed information regarding various theories and stress-strain behavior can be found in the literature.

3A.4.10 Fatigue Loading

In fatigue loading, the structural components are subjected to varying stresses. Typical examples are bridge beams, offshore structures subjected to wave loads, and machine foundations. Extensive

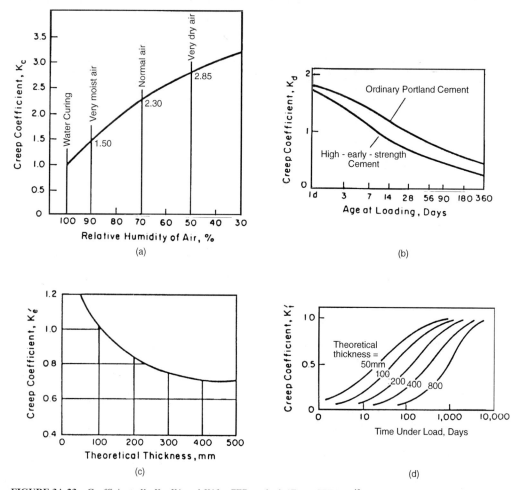

FIGURE 3A.23 Coefficients K_c, K_d, K'_e, and K'_t for CEB method. (*From CEB/FIP.*[19])

research have been conducted on the behavior of concrete subjected to fatigue loading. The following major conclusions, arrived at by the various investigators, may be useful for the designer.

- Concrete can withstand 10 million cycles if the stress range (difference between maximum and minimum stresses) is less than 55% of compressive strength and the minimum stress is about zero.
- The fatigue strength decreases with an increase in minimum strength.
- Components subjected to different intensities of fatigue loading can be designed using Minor's hypothesis.
- Frequency of loading of between 70 and 900 cycles has little effect on fatigue strength, provided the stress range is less than 70% of compressive strength.
- The variables of mix proportion, such as water-cement ratio and cement content, affect fatigue strength in the same way as static compressive strength. Behavior in tension and flexure is about the same as behavior in compression. For example, the flexural fatigue strength at a stress range of 55% of static flexural strength (modulus of rupture) is about 10 million cycles.

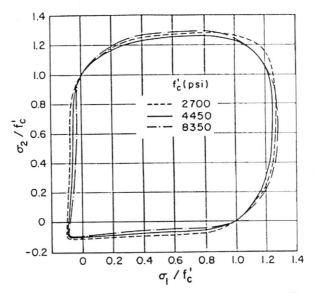

FIGURE 3A.24 Biaxial stress interaction curves. (*From Kupfel et al.*[20])

- The stress gradient increases the fatigue life.
- Rest periods do not affect the fatigue life significantly.
- The creep strain under fatigue loading is higher than the creep strain under the static load that corresponds to the maximum fatigue load. In other words, even though the average stress is lower for fatigue load conditions the creep strain is higher.

3A.5 COLD-WEATHER CONCRETING

When the temperature falls below 20°F (−6°C), the hydration of cement becomes extremely slow. Therefore concrete placed at low temperatures should be protected until it gains sufficient strength to resist freezing action. Normally the minimum recommended compressive strength is 500 psi (4 MPa). Hydration of cement also reduces the degree of saturation because the chemical action consumes the water.

The practice recommended by ACI Committee 306 for cold-weather concreting is shown in Table 3A.19. The committee recommends that the concrete be maintained at a certain minimum temperature for 1 to 3 days, depending on whether it is the conventional or the high-early-strength type. For moderately and fully stressed members longer durations are recommended. The recommended minimum temperature depends on the exposure temperature and the thickness of the specimen. Since thicker specimens dissipate the heat of hydration more slowly, they could be maintained at a lower temperature than thin sections. Non-air-entrained concrete should be protected for at least twice the number of days because it is much more susceptible to freeze-thaw damage than air-entrained concrete. In most cases the concrete could be maintained without using external heat sources if proper care is taken to maintain the temperature of the ingredient materials and the insulation is properly placed. The effect of frozen ground, reinforcing bars, and formwork should be considered in computing the temperature requirements of the ingredients. For very thin sections external heat may be needed to maintain the recommended concrete temperature. Temperatures higher than 70°F (21°C) are not normally recommended.

3.66 FUNDAMENTALS OF FOUNDATION CONSTRUCTION AND DESIGN

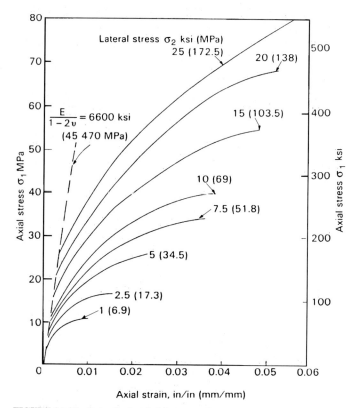

FIGURE 3A.25 Behavior in triaxial compression.

TABLE 3A.19 Recommended Concrete Temperatures for Cold-Weather Construction, Air-Entrained Concrete

	Sections < 12 in (0.3 m) thick		Sections 12–36 in (0.3–0.9 m) thick		Sections 36–72 in (0.9–1.8 m) thick		Sections > 72 in (1.8 m) thick	
	°F	°C	°F	°C	°F	°C	°F	°C
Minimum temperature for fresh concrete *as mixed* in weather indicated								
Above 30°F (−1°C)	60	16	55	13	50	10	45	7
0 to 30°F (−18 to −1°C)	65	18	60	16	55	13	50	10
Below 0°F (−18°C)	70	21	65	18	60	16	55	13
Minimum temperature for fresh concrete *as placed* and *maintained*	55	13	50	10	45	7	40	5
Maximum allowable *gradual* drop in temperature in first 24 h after end of protection	50	28	40	22	30	17	20	11

Source: From ACI 306 R-88.[21]

Most specifications are written based on minimum compressive strength. It is assumed that if the concrete has the specified compressive strength, it is durable. This may not be true in all cases, particularly for concrete exposed to cold weather in its fresh state. The frost action could impart considerable internal damage, making the concrete less durable. If durability is a main consideration the concrete should be protected for longer periods than recommended in Table 3A.19.

The temperature of the fresh concrete can be controlled by controlling the temperature of the constituent materials, namely, cement, aggregates, and water. Since the specific heat of water is 1.0 as compared to 0.22 for cement and aggregates, it is more efficient to heat the water. In addition it is easier to heat the water than the other ingredients used in concrete.

If the outside temperature is above freezing, aggregates are not usually heated. In temperatures below freezing, heating of fine aggregates could be sufficient. Coarse aggregate is heated only as a last resort because it is more difficult to heat loosely packed materials. Fine aggregates are generally heated by circulating hot air or steam through pipes that are embedded in them. In any case, the final temperature of the freshly mixed concrete should be maintained at the specified level. The temperature of fresh concrete T can be estimated using the following equation:

$$T = \frac{0.22(T_a W_a + T_c W_c) + T_w W_w + T_{wa} W_{wa}}{0.22(W_a + W_c) + W_w + W_{wa}} \tag{3A.30}$$

where T_a, T_c, T_w, and T_{wa} are the temperatures of aggregate, cement, water, and free moisture in aggregates, respectively, and W_a, W_c, W_w, and W_{wa} are the weights of aggregate, cement, water, and free moisture in aggregates, respectively. All the temperatures are in °F and the weights are in pounds. For SI units the temperatures are expressed in °C and the weights in kilograms.

The temperatures of concrete should be checked using thermometers. Both mercury and bimetallic thermometers are available for that purpose.

3A.6 HOT-WEATHER CONCRETING

Concreting in hot weather leads to a rapid hydration rate and evaporation loss, resulting in microcracks and inferior final product. Since lower humidity and higher wind velocity also lead to rapid evaporation, hot weather for concreting purposes is taken as a combination of temperature, relative humidity, and wind velocity. If the relative humidity is high, such as near sea shores, special precautions may not be needed up to about 85°F (29°C). If the humidity is very low, such as in desert conditions, precautions might be needed even at temperatures lower than 80°F (27°C). The combined effect of temperature, relative humidity, and wind velocity can be judged using the amount of water that evaporates from fresh concrete. Nomograms are available for estimating the rate of evaporation for a given site condition.

Solar radiation can also affect the properties of fresh concrete in a number of ways. If the ingredients are stored outside, they can absorb heat during transport or during the waiting period. Heating of reinforcement and formwork can further aggravate the situation.

Fresh Concrete. Hot weather adversely affects workability. For the same workability, or to improve the workability, more water is needed at higher temperatures. Unfortunately the excess water added results in reduced strength and less durable concrete. Hence it is advisable to use water-reducing admixtures such as superplasticizers rather than excess water. Researchers have shown that superplasticizers can be used effectively to increase workability without adversely affecting strength and durability.

Concrete stiffens much faster at higher temperatures. The rate of hydration approximately doubles for every 18°F (10°C) increase in temperature. Hence the workability decreases much faster at high temperatures. The change in workability can be best represented by the loss in slump values. Figure 3A.26 shows the variation of slump at two different temperatures. It can be seen that

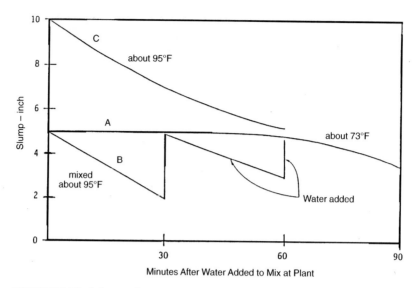

FIGURE 3A.26 Influence of temperature on slump variation with time. (*From Shilstone.*[22])

slump loss is much more rapid at 95°F (35°C) than at 73°F (23°C). Set-retarding admixtures may be used to delay the stiffening of concrete.

When the concrete is placed in position, vibrated, and the top surface finished to the desired texture, a certain amount of water rises up to the surface. This water, called bleed water, should be kept to a minimum. The bleed water does help to prevent shrinkage cracking. However, at high temperatures the bleed water may evaporate rapidly, causing excessive shrinkage cracks.

Hardened Concrete. Concrete placed and cured at higher temperatures develops higher strengths at the early ages of maturity. But the final strength, measured after 28 days, decreases with an increase in placing and curing temperature. The reduction in strength occurs both in the compression and the flexure mode. Typical variations of compressive strengths at various temperatures are shown in Fig. 3A.27.

Concrete cast during hot weather develops more shrinkage cracks. This results in more permeable and less durable concrete. In general, concrete placed at high temperatures without taking special precautions can be considered weak under both thermal (freeze-thaw) and chemical attacks.

In general, creep of concrete increases with the increase in temperature. Concrete exposed to temperatures higher than 90°F (32°C) could undergo considerably more creep strains. In certain cases the higher creep strains should be taken into consideration in structural design.

Precautionary Measures for Hot-Weather Concreting. The best approach is to cast concrete at temperatures lower than 85°F (29°C). In some instances concrete could be cast during the evening or night hours rather than the daytime. If concrete has to be placed at higher temperatures, the best way to avoid problems is to keep the concrete temperature low. This can be achieved by cooling the aggregates or the water.

Using ice in the mixing water is the most efficient way to lower the concrete temperature. The approximate temperature of the concrete can be calculated using Eq. (3A.30). If ice is used as part of the mixing water, the equation can be modified as follows:

$$T = \frac{0.22(T_a W_a + T_c W_c) + T_w W_w + T_{wa} W_{wa} - 112 I}{0.22(W_a + W_c) + W_w + W_{wa} + I} \qquad (3A.31)$$

where I is the weight of ice in pounds.

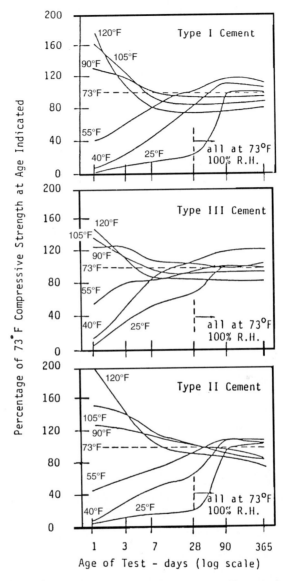

FIGURE 3A.27 Effect of temperature on compressive strength for different types of cement. (*From Kliegar.*[23])

The amount of ice needed to obtain a certain concrete temperature can be computed by the following modified form of Eq. (3A.31):

$$I = \frac{0.22[W_a(T_a - T) + W_c(T_c - T)] + W_{wa}(T_a - T) + W_{wa}(T_w - T)}{112 + T} \quad (3A.32)$$

Shaved ice can be added to the mixer as part of the mixing water. If block ice is used, it should be crushed before adding. In either case, ice must be completely melted at the end of the mixing cycle. Water from melted ice must be considered as part of the total mixing water, keeping the water-cement ratio the same.

Under certain circumstances, liquid nitrogen has been shown to be economical and practical, especially if low concrete temperatures are needed. However, this method can be used only when nitrogen manufacturing facilities are available locally. Liquid nitrogen can be used to cool the aggregates, the water, or the concrete mix. Experience shows that a combination provides a repeatable and quality mix.

Some of the proven techniques helpful for hot-weather concreting are listed here. The recommendations are not elaborated upon for lack of space.

- Keep the concrete temperature low, preferably below 85°F (29°C).
- Provide shade over stockpiles of coarse and fine aggregates.
- Use chilled water or partly replace mixing water with ice for mixing concrete, or use liquid nitrogen to cool the concrete.
- Keep the cement content to the minimum required for strength and durability.
- Use appropriate water-reducing, superplasticizing, and set-retarding admixtures after establishing, under site conditions, dosage and compatibility with the cement used.
- Paint all mixing and conveying equipment for concreting with reflective or light-colored paint.
- Precool the forms, reinforcement, and surroundings prior to concrete placement.
- Place concrete at night.
- Place concrete in layers of optimum thickness for efficient compaction and avoidance of cold joints.
- Wet or moist curing and membrane protection are necessary after the concrete has hardened.

Retempering of Concrete in Hot Weather. Normally, retempering of concrete to improve its workability is considered a bad option. But recent research shows that the retempering technique can be used successfully without adverse effects. A study conducted using water and superplasticizer for retempering concretes, mixed at temperatures up to 140°F (60°C), led to the following conclusions.

- Additional water and cement are needed (maintaining the same water-cement ratio) at higher ambient temperatures to achieve the same slump. The additional water demand is very high for concretes with low water-to-cement ratios (0.4) at temperatures higher than 104°F (40°C). However, for concretes with higher water-cement ratios (0.5 and 0.6) there is only a slight increase in the water required to maintain the same slump for a temperature range of 86 to 140°F (30 to 60°C).
- For all concretes the quantity of retempering water required to restore their initial slumps, after an elapse of a 30-min period, increases with an increase in the ambient temperature for both first and second retemperings. The quantity of water needed for second retempering is significantly higher than that for first retempering at all temperatures. Concretes with lower water-cement ratios (0.4) need considerably higher quantities of retempering water at higher temperatures.
- The cohesiveness and finishability of concrete seems to be better after retempering than after initial mixing.
- Slump loss is considerably higher for a concrete with a water-cement ratio of 0.4 than for concretes with water-cement ratios of 0.5 and 0.6. After retempering the rate of slump loss is higher for all concretes. The rate of slump loss is not significantly higher at higher temperatures.
- No appreciable change in the unit weight of fresh concrete occurs after first and second retemperings at all temperatures tested. For concretes with low water-cement ratios (0.4) an increase in the ambient temperature causes a decrease in the plastic unit weight.

- There is no apparent change in the entrapped air either due to a temperature increase or due to retemperings.
- All hardened concrete properties (compressive strength, splitting tensile strength, flexural strength, static modulus of elasticity, dynamic modulus, pulse velocity, and dry unit weight) are affected similarly by an increase in the ambient temperature from 86 to 140°F (30 to 60°C) and due to first and second retemperings. There is a successive, though not significant, reduction in the strength and modulus values after first and second retemperings. There is a slight reduction (less than 5%) as the temperature increases from 86 to 140°F (30 to 60°C). The concretes most affected by temperature increase are those with a water-cement ratio of 0.4.
- There is no positively recognizable change in the relationships between the various properties of hardened concrete either due to an increase in temperature from 86 to 140°F (30 to 60°C) or due to two retemperings.
- There is no significant difference in the properties, particularly in the case of compressive strength, of concretes mixed, cast, and cured under identical conditions and with two different agents, namely, superplasticizer and water. The observed difference between the two is less than 15%. Moreover the variation is not consistent. A statistical analysis conducted using regression equations relating the compressive strength and other properties, such as splitting tensile strength, flexural strength, pulse velocity, static modulus, dynamic modulus, and dry unit weight, confirms the observation that the two different retempering agents have the same influence on the properties of concrete.

3A.7 PUMPING OF CONCRETE

In modern construction, pumping of concrete has become quite common. Pumping has a number of advantages, such as providing a continuous supply of concrete, access to hard to reach places, and economy. The pumping system consists of a hopper, a concrete pump, and pipes that can be connected and dismantled easily.

Two typical pumps are shown in Fig. 3A.28. Direct-acting horizontal piston pumps shown in Fig. 3A.28(*a*) are more common. The semirotary valves allow the passage of coarse aggregate particles. Concrete, which is fed into the hopper from the mixer, gets sucked into the pipes by gravity and the vacuum action created during the suction stroke. These pumps can pump up to 1500 ft (457 m) horizontally and 140 ft (43 m) vertically.

The squeeze-type pumps shown in Fig. 3A.28(*b*) are normally smaller and truck-mounted. In this pump concrete placed in the collecting hopper is fed by rotating blades into a pliable pipe located in the pumping chamber. The pumping chamber, which can maintain a vacuum of about 26 in (660 mm) of mercury, supplies continuous feed to the delivery pipe. Delivery is normally done using folding boom consisting of 3- and 4-in (75- and 100-mm) pipes. Squeeze pumps can pump up to 300 ft (91 m) horizontally and 100 ft (30 m) vertically.

Different pump and pipe sizes are available. Squeeze pumps can deliver up to 25 yd^3 (19 m^3) per hour whereas piston pumps can deliver up to 80 yd^3 (61 m^3) per hour. Piston pumps work with pipes up to 9 in (230 mm) in diameter. The pipe diameter should be at least three times the maximum size of the aggregate. Hence even squeeze pumps with 3-in (75-mm)-diameter pipes can handle most concrete used for buildings.

Two types of blockages can occur in pumping. The first occurs due to segregation, the second due to excessive friction. When the concrete has too much water, the solid particles consisting of aggregates cannot be carried through by the liquid medium because the water escapes through the mix. Since water is the only medium pumpable in its natural state, the aggregates stay behind and get clogged. If the mix is cohesive, then the water will carry the solid particles with it.

When the mix is very cohesive, the friction between the walls and the mix becomes high and the pump cannot overcome this friction, which will result in a blockage. This type of failure is

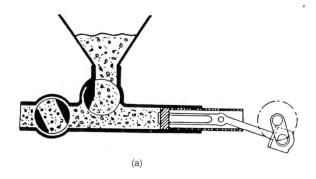

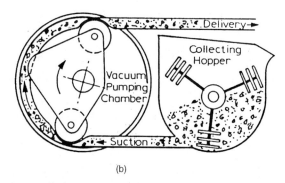

FIGURE 3A.28 Concrete pumps. (*a*) Direct-acting horizontal piston pump. (*b*) Squeeze-type pump.

more common in high-strength concrete mixes and in mixes containing a high proportion of very fine material such as crushed dust or fly ash.

The optimum mix is the one that produces maximum internal frictional resistance within the ingredients and minimum frictional resistance against the pipe walls. Void sizes should also be minimum. For concrete containing 0.75-in (19-mm) maximum size aggregate, fine aggregate should be in the range of 35 to 40%. Of these 15 to 20% should pass through ASTM sieve 50, or finer than 300 μm.

Concrete should be mixed well before feeding it into the hopper. In some cases additional mixing is done in the hopper using stirrers. The mix cannot be too harsh, too dry, too wet, or too sticky. A slump of 1.5 to 4 in (38 to 100 mm) normally produces satisfactory results. Since pumping provides some compaction, the mix at the delivery point could have a slump lowered by as much as 1 in (25 mm). When the concrete is at the correct consistency, a thin lubricating film forms near the surface of the pipes, allowing smooth flow of concrete.

The following are some of the additional factors to be considered in pumping concrete.

- Pumping is economical only if it can be used over long uninterrupted periods because considerable effort is needed for lubricating and cleaning at the beginning and end.
- A short piece of flexible hose near the end makes the placement easier. But the flexible hose could increase the friction loss.
- Bends should be kept to a minimum.
- Aluminum pipes should not be used because they react with alkalis in cement and generate hydrogen bubbles, weakening the hardened concrete.
- The shape of the aggregate influences the pumpability of the mix. Natural sands are preferable to crushed sands because of their spherical shape and continuous uniform grading.

- The presence of entrained air increases the pumping effort. If large amounts of entrained air are present, the entire movement of the piston could be wasted on compressing the air bubbles, resulting in no flow. In general air-entrained concrete can be pumped only shorter distances as compared to non-air-entrained concrete.
- Typically pumping lightweight aggregate concrete needs more effort. Sealing of the surface of the aggregates may be necessary, or special admixtures may be needed to pump lightweight concrete. Otherwise aggregates can absorb more water under pressure, making the mix stiffer and hence more difficult to pump. Some of the aggregate may also crush during the pumping process.
- Concrete with unsatisfactory conditions in the fresh state cannot be pumped. Pumpable concrete, in almost all cases, has the right consistency for placing and finishing.

3A.8 REFERENCES

1. A. Litvin and A. E. Fiorato, *Concrete Int.,* vol. 3, no. 3, p. 49, 1981.
2. ASTM, *Annual Book of ASTM Standards,* vol. 04.02, *Concrete and Aggregates,* American Society of Testing and Materials, Philadelphia, Pa, 1993.
3. V. S. Ramachandran, in V. M. Malhotra (Ed.), *Progress in Concrete Technology,* CANMET, Ottowa, Ont., Canada, 1980, pp. 421–450.
4. ACI Committee 318, "Building Code Requirements for Reinforced Concrete and Commentary," ACI 318-92, ACI 318R-92, American Concrete Institute, Detroit, Mich., 1992.
5. ACI Committee 214, "Recommended Practice for Evaluation of Strength Test Results of Concrete," American Concrete Institute, Detroit, Mich., 1983.
6. ACI Building Code/Commentary, American Concrete Institute, Detroit, Mich., 1989.
7. ACI Committee 211, "Standard Practice for Selecting Proportions for Normal, Heavyweight, and Mass Concrete," American Concrete Institute, Detroit, Mich., 1990.
8. PCA, *Design and Control of Concrete Mixtures,* 12th ed., Portland Cement Association, Skokie, Ill., 1979.
9. A. M. Neville, *Properties of Concrete,* Pitman, Marshfield, Mass., 1981.
10. ACI Committee 211, "Standard Practice for Selecting Proportions for Structural Lightweight Concrete," American Concrete Institute, Detroit, Mich., 1990.
11. ACI Committee 214, "Simplified Version of the Recommended Practice for Evaluation of Strength Results," American Concrete Institute, Detroit, Mich., 1989.
12. ACI Committee 214, "Use of Accelerated Strength Testing," American Concrete Institute, Detroit, Mich., 1981.
13. B. G. Singh, "Specific Surface of Aggregates Related to Compressive and Flexural Strength of Concrete," *J. ACI,* vol. 54, pp. 897–907, 1958.
14. A. F. Stock, D. J. Hannant, and R. I. T. Williams, "The Effect of Aggregate Concentration upon the Strengths and Modulus of Elasticity of Concrete," *Mag. Concrete Res.,* vol. 31, no. 109, pp. 225–234, 1979.
15. W. A. Cordon and H. A. Gillespie, *J. ACI,* vol. 60, no. 8, 1963.
16. W. H. Price, *J. ACI,* vol. 47, p. 429, 1951.
17. F. M. Lea, *The Chemistry of Cement and Concrete,* Arnold, London, 1970.
18. ACI Committee 209, "Prediction of Creep, Shrinkage, and Temperature Effects in Concrete Structures," ACI SP-76, American Concrete Institute, Detroit, Mich., 1982, pp. 193–300.
19. CEB/FIP, "International Recommendations for the Design and Construction of Concrete Structures," Paris, France, 1970.
20. H. Kupfel, H. K. Hilsdorf, and H. Rusch, *J. ACI,* vol. 66, pp. 662–663, 1969.
21. ACI Committee 306R-88, "Cold Weather Concreting," Detroit Mich., 1989.
22. J. W. Shilstone, "Concrete Strength Loss and Slump Loss in Summer," *Concrete Constr.,* vol. 27, pp. 429–432, 1982.
23. P. Kliegar, "Effect of Mixing and Curing Temperature on Concrete Strength," *J. ACI,* vol. 54, pp. 1063–1082, 1958.

SECTION 3B
FUNDAMENTALS OF REINFORCED CONCRETE

A. Samer Ezeldin

NOTATIONS   3.57
3B.1 INTRODUCTION   3.77
3B.2 FLEXURE BEHAVIOR   3.77
   3B.2.1 Uncracked, Elastic Range   3.78
   3B.2.2 Cracked, Elastic Range   3.81
   3B.2.3 Cracked, Inelastic Range, and Flexural Strength   3.82
   3B.2.4 Rectangular Sections with Tension Reinforcement Only   3.83
   3B.2.5 Doubly Reinforced Sections   3.86
   3B.2.6 Flanged Section   3.90
3B.3 SHEAR BEHAVIOR   3.93
   3B.3.1 Plain Concrete   3.93
   3B.3.2 Reinforced Concrete Beams   3.94
3B.4 TORSION BEHAVIOR   3.100
   3B.4.1 Torsion Strength of Reinforced Concrete Beams   3.100
   3B.4.2 Combined Shear, Bending, and Torsion   3.102
3B.5 COLUMNS   3.105
   3B.5.1 General   3.105
   3B.5.2 Axially Loaded Columns   3.106
   3B.5.3 Uniaxial Bending and Compression   3.106
   3B.5.4 Load-Moment Interaction Diagrams   3.109
   3B.5.5 Slender Columns (Buckling Effect)   3.110
   3B.5.6 Biaxial Bending   3.116
   3B.5.7 ACI Code Requirements   3.117
3B.6 DEVELOPMENT OF REINFORCING BARS   3.117
   3B.6.1 General   3.117
   3B.6.2 Tension Development Length   3.118
   3B.6.3 Compression Development Length   3.120
3B.7 REFERENCES   3.120

NOTATIONS

a = depth of equivalent rectangular stress block as defined in Sec. 3B.2.4.1
A_c = area of core of spirally reinforced compression member measured to outside diameter of spiral
A_g = gross area of concrete section
A_s = area of tension steel reinforcement
     area of steel per unit width of slab or plate
A_s' = area of compression steel reinforcement
A_{st} = total area of longitudinal reinforcement
b = width of cross section
b_w = web of cross section
c = distance from extreme compression fiber to neutral axis
d = distance from extreme compression fiber to centroid of tension reinforcement
d' = distance from extreme compression fiber to centroid of compression reinforcement
d_c = thickness of concrete cover measured from extreme tension fiber to center of bar or wire located closest thereto
E_c = modulus of elasticity of concrete
E_s = Young's modulus of steel
EI = flexural stiffness of compression member
f_c = compressive strength in concrete

3.76 FUNDAMENTALS OF FOUNDATION CONSTRUCTION AND DESIGN

f'_c = cylinder compressive strength of concrete
f_r = modulus of rupture of concrete
f_s = calculated stress in reinforcement at service loads
f'_t = tensile splitting strength of concrete
f_y = specified yield strength of nonprestressed reinforcement
h = overall thickness of member
I_{cr} = moment of inertia of cracked section
I_e = effective moment of inertia for computation of deflection
I_g = gross moment of inertia
k = effective length factor for compression members
L = clear long span length
l = effective beam span
l_d = development length
l_e = equivalent embedded length of a hook
M = positive moment
M' = negative moment
M_a = moment of maximum service load in span
M_{cr} = cracking moment
M_n = nominal moment strength at section
P_b = nominal axial load strength at balanced strain conditions
P_c = critical load
P_n = nominal axial load strength at given eccentricity
P_0 = nominal axial load strength at zero eccentricity
P_u = factored axial load at given eccentricity; $\leq \phi P_n$, ϕ being a reduction factor
s = spacing of stirrups or ties
V_c = nominal shear strength provided by concrete
V_s = nominal shear strength provided by steel reinforcement
W = total uniform load per unit area
x = shorter overall dimension of rectangular part of cross section
x_1 = shorter center-to-center dimension of closed rectangular stirrup
y = longer overall dimension of rectangular part of cross section
y_t = distance from centroidal axis of gross section, neglecting reinforcement, to extreme fiber in tension
y_1 = longer center-to-center dimension of closed rectangular stirrup
β_1 = factor, varying from 0.85 for f'_c = 4000 psi to 0.65 minimum; it decreases at a rate of 0.05 per 1000-psi strength above 4000 psi (27.6 MPa)
δ = moment magnification factor
ε_s = unit strain in reinforcement steel
ν = Poisson's ratio
ρ = ratio of nonprestressed tension reinforcement; A_s/bd for beam and $A_s/12d$ for slab
$\bar{\rho}_b$ = reinforcement ratio producing balanced strain conditions
ρ_s = ratio of volume of spiral reinforcement to total volume of core (out-to-out of spirals) of spirally reinforced compressive member

ρ_t = active steel ratio; = A_s/A_t
ω = reinforcement index; = $\rho f_y/f'_c$
Σ^0 = sum of circumferences of reinforcing elements
$\Sigma x^2 y$ = torsional section properties

3B.1 INTRODUCTION

Concrete is obtained by mixing cement, water, fine aggregate, coarse aggregate, and frequently other additives in specified proportions. The hardened concrete is strong in compression but weak in tension, making it vulnerable to cracking. Also, concrete is brittle and fails without warning. In order to overcome the negative implications of these two main weaknesses, steel bars are added to reinforce the concrete. Hence reinforced concrete, when designed properly, can be used as an economically strong and ductile construction material.

In many civil engineering applications, reinforced concrete is used extensively as a construction material for structures and foundations. This chapter provides a basic knowledge of reinforced concrete elements (beams, columns, and slabs) subjected to flexure, shear, and torsion. Once the response of an individual element is understood, the designer will have the necessary background to analyze and design reinforced concrete systems composed of these elements, such as foundations and buildings.

Two popular methods are available for analyzing and designing the strength of reinforced concrete members. The first is referred to as the working stress design method. This method is based on limiting the computed stresses in members as they are subjected to service loads up to the allowable stresses. The second design method, called the strength method, is based on predicting the maximum resistance of a member rather than predicting stresses under service loads.

The strength method is the design method recommended by the current edition of the ACI code.[1] Members are designed for factored loads that are greater than the service loads. The factored loads are obtained by multiplying the service load by load factors greater than 1. Table 3B.1 gives the load factors for various types of load. Using factored loads, the designer performs an elastic analysis to obtain the required strength of the members. Members are designed so that their design strength is equal to or greater than the required strength,[2–4]

$$\text{Required strength} \leq \text{design strength} \tag{3B.1}$$

The design strength is used to express the nominal capacity of a member reduced by a strength reduction factor ϕ. The nominal capacity is evaluated in accordance with provisions and assumptions specified by the ACI code. Reduction factors for different loading conditions are presented in Table 3B.2. The design criteria just presented provide for a margin of safety in two ways. First, the required strength is computed by increasing service loads by load factors. Also, the design strength is computed by reducing the nominal strength by a strength reduction factor. This design criterion applies to all possible states of stress, namely, bending, shear, torsion, and axial stresses.

3B.2 FLEXURE BEHAVIOR

Four basic assumptions are made when deriving a general theory for flexure behavior of reinforced concrete members.

1. Plane sections before bending remain plane after bending.
2. The stress-strain relationships for the steel reinforcement and the concrete are known.
3. The tensile strength of the concrete is neglected after cracking.
4. Perfect bonding exists between the concrete and the steel reinforcement.

TABLE 3B.1 Load Factors for Ultimate-Strength Design Method

Condition	U
1. Dead load D + live load L	$1.4D + 1.7L$
2. Dead + live + wind load W when additive	$0.75(1.4D + 7L + 1.7W)$
3. Same as item 2 when gravity counteracts wind-load effects	$0.9D + 1.3W$
4. In structures designed for earthquake loads or forces E, replace W by $1.1E$ in items 2 and 3	
5. When lateral earth pressure H acts in addition to gravity forces when effects are additive	$1.4D + 1.7L + 1.7H$
6. Same as item 5 when gravity counteracts earth-pressure effects	$0.9D + 1.7H$
7. When lateral liquid pressure F acts in addition to gravity loads, replace $1.7H$ by $1.4F$ in items 5 and 6	
8. Vertical liquid pressures shall be considered as dead loads D	
9. Impact effects, if any, shall be included in live loads L	
10. When effects T of settlement, creep, shrinkage, or temperature change are significant and additive	$0.75(1.4D + 1.4T + 1.7L)$
11. Same as item 10 when gravity counteracts T	$1.4(D + T)$
12. In no case shall U be less than given by item 1	

Source: From ACI.[1]

TABLE 3B.2 Reduction Factors for Ultimate-Strength Design Method

Kind of strength	Strength reduction factor ϕ
Flexural, with or without axial tension	0.90
Axial tension	0.90
Axial compression, with or without flexure:	
Members with spiral reinforcement	0.75
Other reinforced members	0.70
Exception: for low values of axial load, ϕ may be increased in accordance with the following:	
For members in which f_y does not exceed 60,000 psi, with symmetrical reinforcement, and with $(h - d' - d_s)/h$ not less than 0.70, ϕ may be increased linearly to 0.90 as ϕP_n decreases from $0.10 f'_c A_s$ to zero.	
For other reinforced members, ϕ may be increased linearly to 0.90 as ϕP_n decreases from $0.10 f'_c A_g$ or ϕP_{nb}, whichever is smaller, to zero.	
Shear and torsion	0.85
Bearing on concrete	0.70
Flexure in plain concrete	0.65

Source: From ACI.[1]

The simple reinforced concrete beam shown in Fig. 3B.1(*a*) exhibits the basic characteristics of flexural behavior. Depending on the magnitude of the bending moment, the beam's response will be in the elastic range, either uncracked or cracked, or in the inelastic range and cracked.

3B.2.1 Uncracked, Elastic Range

When the bending moment is smaller than the cracking moment, the strain and stress distributions at the maximum moment section are as shown in Fig. 3B.1(*b*). Stresses are related to strain by the

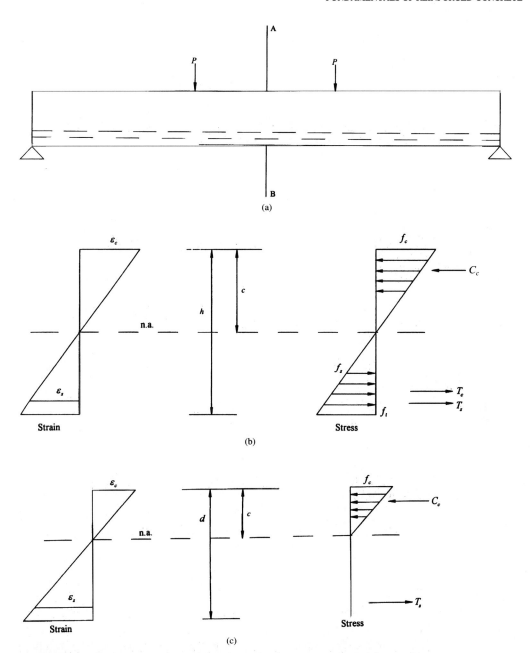

FIGURE 3B.1 Flexural behavior of simple reinforced concrete beam. (*a*) Reinforced concrete beam. (*b*) Uncracked, elastic stage. (*c*) Cracked, elastic stage.

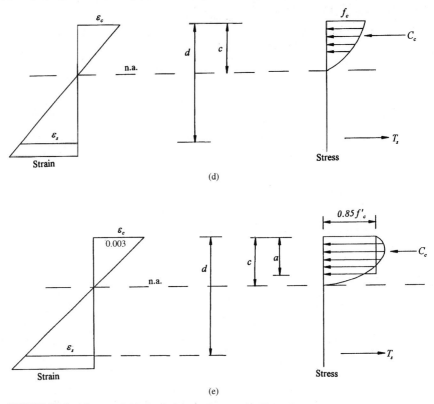

FIGURE 3B.1 (*Continued*) (*d*) Cracked, inelastic stage. (*e*) Flexural strength.

modulus of elasticity of concrete E_c. The reinforcement in concrete makes only a minor contribution at this stage, because its strain is too small to achieve an appreciable resisting stress. The internal force couple formed by the tension and compression forces in concrete provides the required resistance to the externally applied moment. The principle of elastic theory and the transformed area concept are employed for computing stresses in concrete and in steel reinforcement.

The stress in concrete with the extreme fiber in tension is given by

$$f_t = \frac{M(h - c)}{I_t} \tag{3B.2}$$

and the stress in the steel reinforcement by

$$f_s = \frac{M(d - c)}{I_t} n \tag{3B.3}$$

where

$$n = \frac{E_s}{E_c}$$

and I_t is the moment of inertia of the transformed section.

3B.2.2 Cracked, Elastic Range

As the applied load is increased, the maximum moment reaches higher values, resulting in the maximum tension in concrete approaching the modulus of rupture. At this level, the tension cracks start forming on the tension face of the concrete section. The cracking moment can be computed using the equation

$$M_{cr} = \frac{f_r I_t}{y_t} \tag{3B.4}$$

Here I_t can be replaced by the gross moment of inertia I_g, neglecting the contribution of the steel reinforcement without appreciable error.

Under increasing loads and progressive cracking, the neutral axis shifts upward. Since concrete cannot resist the developed tensile stresses, the reinforcing steel is called upon to resist the entire tension force. Up to a compressive stress of about $0.5 f'_c$, the linear relationship is still valid. Figure 3B.1(c) shows the response of a reinforced concrete section at this stage. The internal resisting couple is provided by a concrete compressive force C_c and a steel tensile force T_s. The moment equilibrium equation is given by

$$M = T_s \left(d - \frac{c}{3} \right) = C_c \left(d - \frac{c}{3} \right) \tag{3B.5}$$

The concept of transformed section can still be used to compute the stress in concrete and in steel reinforcement. However, at this stage the cracked concrete is assumed to make no contribution to I_t.

Example 3B.1 Calculate the bending stresses in the beam shown in Fig. X.1, using the transformed area method.

Given:
$f'_c = 4000$ psi (27.6 MPa)
$M = 900,000$ in · lb (101.7 kN · m)
$E_s = 29,000,000$ psi (199,810 MPa)
$b = 14.0$ in (356 mm)
$h = 20.0$ in (508 mm)
$d = 17.0$ in (432 mm)
$A_s = 3$ no. 9 bars [3 in² (1935 mm²)]

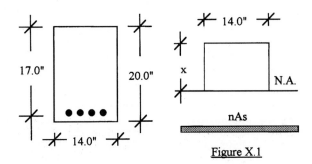

Figure X.1

Solution

$$M_{cr} = \frac{f_r I_t}{y_t}$$

$$= \frac{(7.5\sqrt{4000})(14 \times 20^3/12)}{10} = 442{,}717 \text{ in} \cdot \text{lb} < 900{,}000 \text{ in} \cdot \text{lb} \quad \therefore \text{ Section has cracked.}$$

$$n = \frac{E_s}{E_c} = \frac{29 \times 10^6}{57{,}000\sqrt{f'_c}} = \frac{29 \times 10^6}{4 \times 10^6} = 7.2$$

Taking moments about the neutral axis,

$$14x\left(\frac{x}{2}\right) = nA_s(d - x)$$

$$7x^2 = 367.2 - 21.6x$$

Solving the quadratic equation we get

$$x = \frac{-21.6 \pm \sqrt{21.6^2 + 4 \times 7 \times 367.2}}{2 \times 7} = 5.86 \text{ in (148.8 mm)}$$

The moment of inertia of the transformed section is

$$I_t = \frac{bx^3}{3} + nA_s(d - x)^2$$

$$= \frac{14 \times 5.86^3}{3} + 7.2 \times 3(17 - 5.86)^2 = 3620 \text{ in}^4 \ (1506 \times 10^6 \text{ mm}^4)$$

The concrete top-fiber compression stress is

$$f_c = \frac{Mc}{I_t} = \frac{900{,}000 \times 5.86}{3620} = 1457 \text{ psi (10 MPa)}$$

The steel tension stress is

$$f_t = n\frac{Mc}{I_t} = \frac{7.2 \times 900{,}000(17 - 5.86)}{3620} = 19{,}941 \text{ psi (137.4 MPa)}$$

3B.2.3 Cracked, Inelastic Range, and Flexural Strength

With increasing loads the stresses in concrete exceed the $0.5f'_c$ value. The proportionality of stresses and strains ceases to exist, and nonlinear characteristics are observed, as shown in Fig. 3B.1(d).

It is a common practice to utilize an elastic-plastic idealization for the stress-strain relationship of steel. This implies that stresses and strains are related with the steel modulus of elasticity E_s up to the yield stress and its corresponding strain ε_y. At higher strain values, stresses in steel are taken equal to the yield stress f_y irrespective of the strain magnitude. However, the inelastic performance of concrete under load will result in a parabolic stress distribution. The diagram of the area of stress as well as the line of action of the resulting force have to be obtained in order to compute the stresses acting on the section.

The section reaches its flexural strength (nominal strength) when the extreme compressive fiber of concrete reaches its maximum usable strain. The ACI code recommends a maximum usable strain value of 0.003. If the properties of the compressive stress block just prior to failure are defined, the flexural strength can be computed [Fig. 3B.1(e)]. The ACI code allows the use of a simplified equivalent rectangular stress block to represent the stress distribution of the concrete. The rectangular block has a mean stress of 0.85 f'_c and a depth a, where $a/c = \beta_1 = 0.85$ for $f'_c \le$ 4000 psi. β_1 is reduced incrementally by 0.05 for each 1000 psi of strength in excess of 4000 psi (27.6 MPa), provided that it does not go below 0.65. The reduction of β_1 is mainly due to less favorable properties of the stress-strain relationship for higher-strength concrete.

When the ultimate flexural capacity is reached, two types of failure can occur, depending on the amount of steel reinforcement. If a relatively low percentage of steel is used, the strain on the tension face will be beyond the yield strain ε_y. This triggers excessive deflections and wide cracks, providing a warning of imminent failure. This type of ductile failure is a desirable mode for flexural members. On the other hand, if a relatively high percentage of steel reinforcement is incorporated in the section, failure will occur prior to yielding of the reinforcement. Hence violent concrete crushing occurs in a sudden brittle manner. Because of the nonductile behavior of this mode of failure, it should be avoided. When the extreme compressive fiber of concrete reaches its maximum usable strain simultaneously with the steel reaching its yielding strain, the section is defined as a balanced section. The balanced section is used as a datum to identify ductile and nonductile reinforced concrete sections.

3B.2.4 Rectangular Sections with Tension Reinforcement Only

3B.2.4.1 Moment Capacity. From Fig. 3B.1(e), equating the horizontal forces C and T and solving for the depth of the compression block a, we obtain

$$0.85 f'_c ba = A_s f_s \tag{3B.6}$$

$$a = \beta_1 c = \frac{A_s f_s}{0.85 f'_c b} \tag{3B.7}$$

Hence the nominal flexural strength is

$$M_n = A_s f_s \left(d - \frac{a}{2} \right) \tag{3B.8}$$

3B.2.4.2 Balanced Condition. From Fig. 3B.1(e), the linear strain distribution gives

$$\frac{c}{d} = \frac{\varepsilon_c}{\varepsilon_c + \varepsilon_y} = \frac{0.003}{0.003 + f_y/29 \times 10^6} = \frac{87{,}000}{87{,}000 + f_y} \tag{3B.9}$$

From Eqs. (3B.7) and (3B.9), we can write

$$c = \frac{a}{\beta_1} = \left(\frac{87{,}000}{87{,}000 + f_y} \right) d = \frac{A_s f_y}{0.85 f'_c b \beta_1}$$

Defining $\rho_b = A_s/bd$ we obtain

$$\rho_b = \left(\frac{87{,}000}{87{,}000 + f_y} \right) \left(\frac{0.85 f'_c}{f_y} \right) \beta_1 \tag{3B.10}$$

Equation (3B.10) gives the balanced section reinforcement ratio. To ensure ductile failure, the ACI code limits the maximum tension reinforcement ratio to 75% of ρ_b. Reinforcement ratios higher than ρ_b produce nonductile failure, with steel reinforcement not yielding prior to the crushing of concrete.

3B.2.4.3 Ductile Tension Failure.
For this condition Eqs. (3B.7) and (3B.8) can be written as follows:

$$a = \frac{A_s f_y}{0.85 f'_c b a} \quad (3B.11)$$

$$M_n = A_s f_y \left(d - \frac{a}{2}\right) \quad (3B.12)$$

3B.2.4.4 Nonductile Failure.
The nominal flexural strength is obtained from

$$M_n = A_s f_s \left(d - \frac{a}{2}\right) \quad (3B.13)$$

where f_s is obtained from the following quadratic equation:

$$A_s(f_s)^2 + (87{,}000 A_s) f_s - (0.85 f'_c b) \times (87{,}000 \beta_1 d) = 0 \quad (3B.14)$$

$$a = \frac{A_s f_s}{0.85 f'_c b} \quad (3B.15)$$

3B.2.4.5 Minimum Percentage of Steel.
If the nominal flexural strength of the section is less than its cracking moment, the section will fail immediately when a crack occurs. This type of failure occurs in very lightly reinforced beams without warning.

In order to avoid such a failure, the ACI code requires a minimum steel percentage ρ_{min} equal to

$$\rho_{min} = \frac{200}{f_y} \quad (3B.16)$$

Example 3B.2 Determine the nominal flexural strength M_n of the rectangular section shown in Fig. X.2.

Given:
$f'_c = 4000$ psi (27.6 MPa)
$f_y = 60{,}000$ psi (413.4 MPa)
$b = 14.0$ in (356 mm)
$h = 24.0$ in (610 mm)
$d = 21.5$ in (546 mm)
$A_s = 4$ no. 10 bars [5.08 in² (3277 mm²)]

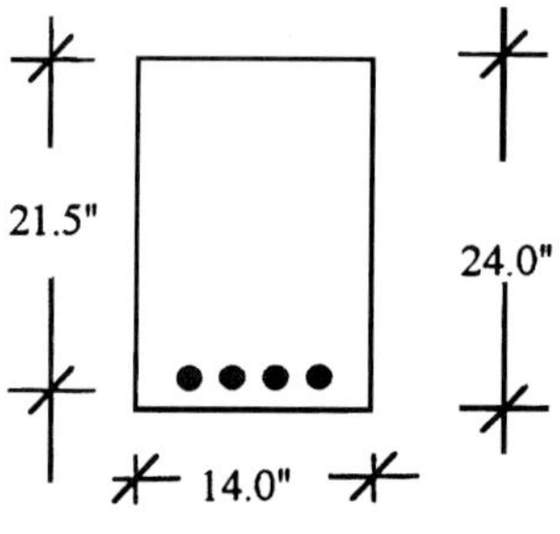

Figure X.2

Solution

$$\rho_b = \left(\frac{87{,}000}{87{,}000 + f_y}\right)\left(\frac{0.85 f'_c}{f_y}\right)\beta_1 = \left(\frac{87{,}000}{87{,}000 + 60{,}000}\right)\left(\frac{0.85 \times 4000}{60{,}000}\right) 0.85 = 0.0285$$

$$\rho_{max} = 0.75 \rho_b = 0.75 \times 0.0285 = 0.0213$$

$$\rho = \frac{A_s}{bd} = \frac{5.08}{14 \times 21.5} = 0.0169 < 0.0213 \quad \therefore \text{Ductile failure.}$$

$$a = \frac{A_s f_y}{0.85 f'_c b} = \frac{5.08 \times 60{,}000}{0.85 \times 4000 \times 14} = 6.4 \text{ in } (16.3 \text{ mm})$$

$$M_n = A_s f_y \left(d - \frac{a}{2}\right)$$

$$= 5.08 \times 60{,}000 \left(21.5 - \frac{6.4}{2}\right) = 5.58 \times 10^6 \text{ in} \cdot \text{lb } (630.63 \text{ kN} \cdot \text{m})$$

Example 3B.3 Determine the flexural strength of the rectangular section shown in Fig. X.3.

Given:
$f'_c = 5000$ psi (34.45 MPa)
$f_y = 60{,}000$ psi (413.4 MPa)
$b = 14.0$ in (356 mm)
$d = 20.0$ in (508 mm)
$h = 24.0$ in (610 mm)
$A_s = 8$ no. 10 bars [10.16 in² (6553.2 mm²)]

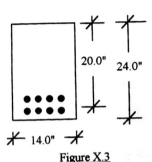

Figure X.3

Solution

$$\rho_b = \left(\frac{87{,}000}{87{,}000 + f_y}\right)\left(\frac{0.85 f'_c}{f_y}\right)\beta_1 = \left(\frac{87{,}000}{87{,}000 + 60{,}000}\right)\left(\frac{0.85 \times 5000}{60{,}000}\right) 0.8 = 0.0335$$

$$\rho = \frac{A_s}{bd} = \frac{10.16}{14 \times 20.0} = 0.036 > \rho_b \qquad \therefore \text{ Nonductile failure}$$

(The section does not satisfy ACI code requirements for ductility.)

Hence $f_s < f_y$. To compute f_s,

$$(A_s)f_s^2 + (87{,}000 A_s)f_s - (0.85 f'_c b \times 87{,}000 \beta_1 d) = 0$$

$$10.16 f_s^2 + (87{,}000 \times 10.16)f_s - (0.85 \times 5000 \times 14 \times 87{,}000 \times 0.8 \times 20) = 0$$

$$10.16 f_s^2 + 883{,}920 f_s - 8.28 \times 10^{10} = 0$$

Hence

$$f_s = \frac{-883{,}920 \pm \sqrt{883{,}920^2 + 4 \times 10.16 \times 8.28 \times 10^{10}}}{2 \times 10.16}$$

$$= 56{,}720 \text{ psi } (390 \text{ MPa}) < 60{,}000 \text{ psi } (413.4 \text{ MPa})$$

$$a = \frac{A_s f_y}{0.85 f'_c b} = \frac{10.16 \times 56{,}720}{0.85 \times 5000 \times 14} = 9.68 \text{ in } (24.59 \text{ mm})$$

$$M_n = A_s f_s \left(d - \frac{a}{2}\right)$$

$$= 10.16 \times 56{,}720 \left(20 - \frac{9.68}{2}\right) = 8.73 \times 10^6 \text{ in} \cdot \text{lb } (986.63 \text{ kN} \cdot \text{m})$$

3B.2.5 Doubly Reinforced Sections

3B.2.5.1 Moment Capacity.
Doubly reinforced sections contain reinforcement on the tension side as well as on the compression side of the cross section. These sections become necessary when the size of the rectangular section is restricted due to architectural or mechanical limitations such that the required moment is larger than the resisting design moment of singly reinforced sections.

The analysis of a doubly reinforced section is carried out by theoretically dividing the cross section into two parts, as shown in Fig. 3B.2. Beam 1 is comprised of compression reinforcement at the top and sufficient steel at the bottom to have $T_1 = C_1$. Beam 2 consists of the concrete web and the remaining tensile reinforcement.

The nominal strength of part 1 can be obtained by taking the moment about the tension steel,

$$M_{n1} = A_{s1}f_y(d - d') = A'_s f_y(d - d') \tag{3B.17}$$

The nominal strength of part 2 is obtained by taking the moment about the compression force,

$$M_{n2} = (A_s - A_{s1})f_y\left(d - \frac{a}{2}\right) = (A_s - A'_s)f_y\left(d - \frac{a}{2}\right) \tag{3B.18}$$

where

$$a = \frac{(A_s - A_{s1})f_y}{0.85 f'_c b} \tag{3B.19}$$

Hence the nominal strength is

$$M_n = M_{n1} + M_{n2}$$

or

$$M_n = A'_s f_y(d - d') + (A_s - A'_s)f_y\left(d - \frac{a}{2}\right) \tag{3B.20}$$

This equation is only valid when A'_s reaches the yield stress prior to concrete crushing. This condition is satisfied if

$$\rho - \rho' \geq \left(\frac{0.85\beta_1 f'_c d'}{f_y d}\right)\left(\frac{87,000}{87,000 - f_y}\right) \tag{3B.21}$$

Otherwise the nominal strength equation is written as

$$M_n = A'_s f'_s(d - d') + (A_s f_y - A'_s f'_s)\left(d - \frac{a}{2}\right) \tag{3B.22}$$

where

$$a = \frac{A_s f_y - A'_s f'_s}{f'_c b} \tag{3B.23}$$

and $f'_s < f_y$.

The following iterative procedure can be followed to obtain f'_s:

1. For the first trial assume

$$f'_s = 87,000\left[1 - \frac{0.85\beta_1 f'_c}{(\rho - \rho')f_y}\frac{d'}{d}\right]$$

2. Obtain the depth of the compression block,

$$a = \frac{A_s f_y - A'_s f'_s}{0.85 f'_c b}$$

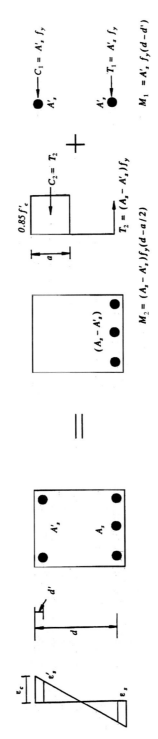

FIGURE 3B.2 Flexural behavior of a doubly reinforced concrete section.

3. Calculate the depth of the neutral axis, $c = a/\beta_1$.
4. Using similar triangles as in Fig. 3B.2, compute the strain ε'_s and then the adjusted compression stress of the steel, $f'_s = \varepsilon'_s E_s$.
5. Repeat steps 2 to 4 until an acceptable convergence is reached. Usually one trial will give acceptable results.

3B.2.5.2 Ductile Failure. To ensure ductile failure, the tension steel reinforcement ρ should be limited to

$$\rho \leq 0.75\bar{\rho}_b + \rho'\frac{f'_s}{f_y} \quad (3B.24)$$

where $\bar{\rho}_b$ is the balanced steel ratio for a singly reinforced beam with a tension steel area of $A_{s2} = A_s - A'_s$.

Example 3B.4 Determine the nominal strength of the rectangular section shown in Fig. X.4.

Given:
$f'_c = 5000$ psi (34.45 MPa)
$f_y = 60,000$ psi (413.4 MPa)
$b = 14.0$ in (356 mm)
$h = 30.0$ in (762 mm)
$d = 26.0$ in (660 mm)
$d' = 2.0$ in (51 mm)
$A_s = 8$ no. 9 bars [8 in² (5160 mm²)]
$A'_s = 2$ no. 8 bars [1.58 in² (1019 mm²)]

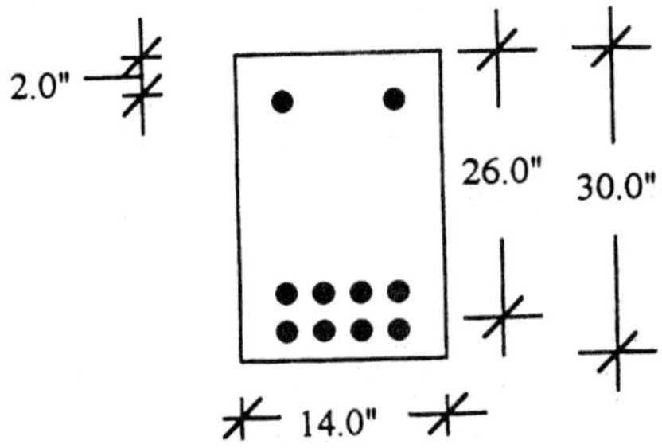

Figure X.4

Solution

$$\rho = \frac{A_s}{bd} = \frac{8.0}{14 \times 26.0} = 0.022$$

$$\rho' = \frac{A'_s}{bd} = \frac{1.58}{14 \times 26.0} = 0.0043$$

$$\bar{\rho}_b = \left(\frac{87,000}{87,000 + f_y}\right)\left(\frac{0.85 f'_c}{f_y}\right)\beta_1 = \left(\frac{87,000}{87,000 + 60,000}\right)\left(\frac{0.85 \times 5000}{60,000}\right)0.8 = 0.0336$$

Check for yielding of steel in compression:

$$\rho - \rho' = 0.022 - 0.0043 = 0.0177$$

$$\left(\frac{0.85\beta_1 f'_c d'}{f_y d}\right)\left(\frac{87,000}{87,000 - f_y}\right) = \left(\frac{0.85 \times 0.8 \times 5000 \times 2}{60,000 \times 26}\right)\left(\frac{87,000}{87,000 - 60,000}\right) = 0.014$$

$$\rho - \rho' = 0.0177 > 0.014 \quad \therefore \text{ Steel is yielding in compression.}$$

$$f'_s = f_y = 60,000 \text{ psi (413.4 MPa)}$$

Check for yielding of steel in tension:

$$0.75\bar{\rho}_b + \rho'\frac{f'_s}{f_y} = 0.75 \times 0.0336 + 0.0043 = 0.0295$$

$\rho = 0.022 < 0.0295$ ∴ Steel is yielding in tension.

$$a = \frac{(A_s - A_s')f_y}{0.85f_c'b} = \frac{(8 - 1.58)60{,}000}{0.85 \times 5000 \times 14} = 6.47 \text{ in } (16.4 \text{ mm})$$

$$M_{n1} = A_s'f_y(d - d') = 1.58 \times 60{,}000(26 - 2) = 2.27 \times 10^6 \text{ in} \cdot \text{lb } (256.5 \text{ kN} \cdot \text{m})$$

$$M_{n2} = (A_s - A_s')f_y\left(d - \frac{a}{2}\right)$$

$$= (8 - 1.58) \times 60{,}000\left(26 - \frac{6.47}{2}\right) = 8.76 \times 10^6 \text{ in} \cdot \text{lb } (990.02 \text{ kN} \cdot \text{m})$$

$$M_n = (2.27 + 8.76)10^6 = 11 \times 10^6 \text{ in} \cdot \text{lb } (1246.52 \text{ kN} \cdot \text{m})$$

Example 3B.5 Determine the nominal strength M_n of the rectangular section shown in Fig. X.5.

Given:
$f_c' = 5000$ psi (34.45 MPa)
$f_y = 60{,}000$ psi (413.4 MPa)
$b = 14.0$ in (356 mm)
$h = 24.0$ in (610 mm)
$d = 21.0$ in (533 mm)
$d' = 2.0$ in (51 mm)
$A_s = 4$ no. 10 bars [5.08 in² (3277 mm²)]
$A_s' = 3$ no. 7 bars [1.8 in² (1161 mm²)]

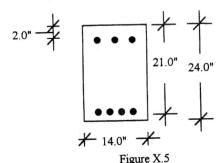

Figure X.5

Solution

$$\rho = \frac{A_s}{bd} = \frac{5.08}{14 \times 21.0} = 0.0173$$

$$\rho' = \frac{A_s'}{bd} = \frac{1.8}{14 \times 21.0} = 0.0061$$

$$\overline{\rho_b} = 0.0336$$

Check for yielding of steel in compression:

$\rho - \rho' = 0.0173 - 0.0061 = 0.0112 < 0.0173$ ∴ Steel did not yield in compression.

Assume

$$f_s' = 87{,}000\left[1 - \frac{0.85\beta_1 f_c'd'}{(\rho - \rho')f_y d}\right]$$

$$= 87{,}000\left[1 - \frac{0.85 \times 0.8 \times 5000 \times 2}{0.0112 \times 60{,}000 \times 21}\right] = 45{,}000 \text{ psi } (310 \text{ MPa})$$

Start the iteration cycle,

$$a = \frac{A_s f_y - A_s' f_s'}{0.85 f_c' b} = \frac{5.08 \times 60{,}000 - 1.8 \times 45{,}000}{0.85 \times 5000 \times 14} = 3.76 \text{ in } (95.5 \text{ mm})$$

$$(f_s')\text{adjusted} = 29 \times 10^6 \times 0.003\left(\frac{3.76/0.8 - 2}{3.76/0.8}\right) = 50{,}000 \text{ psi } (344.5 \text{ MPa})$$

Take $f_s' = 50{,}000$ psi,

$$a = \frac{A_s f_y - A_s' f_s'}{0.85 f_c' b} = \frac{5.08 \times 60{,}000 - 1.8 \times 50{,}000}{0.85 \times 5000 \times 14} = 3.6 \text{ in } (91.4 \text{ mm})$$

$$M_n = (A_s f_y - A_s' f_s')\left(d - \frac{a}{2}\right) + A_s' f_s'(d - d')$$

$$= (5.08 \times 60{,}000 - 1.8 \times 50{,}000)\left(21 - \frac{3.6}{2}\right) + 1.8 \times 50{,}000(21 - 2)$$

$$= 5.83 \times 10^6 \text{ in} \cdot \text{lb } (658.88 \text{ kN} \cdot \text{m})$$

3B.2.6 Flanged Section

3B.2.6.1 Moment Capacity. Because rectangular beams are generally cast monolithically with concrete slabs, a full composite action between the slab and the beam is obtained. In the positive moment diagram, the slab is in compression and, hence, contributes to the moment strength of the section. The effective cross section of the beam has a T shape or an L shape, consisting of the rectangular beam as the web and a portion of the slab as the flange (Fig. 3B.3). The effective width of the slab contributing to the section strength has to satisfy the following requirements.

For the T shape (interior beam),

$$b \leq 16h_f + b_w$$
$$\leq b_w + l_c$$
$$\leq l_n/4 \quad \quad (3\text{B}.25)$$

For the L shape (edge beam),

$$b \leq 6h_f + b_w$$
$$\leq b_w + l_c/2$$
$$\leq l_n/12 \quad \quad (3\text{B}.26)$$

Flanged beams possess a high compression capacity because of the large contribution of the concrete on the compression face. Hence the neutral axis generally lies in the flange. When this situation occurs, the section behaves as a rectangular singly reinforced section having a width b equal to the effective width of the slab. The flexural strength of this section is

$$M_n = A_s f_y \left(d - \frac{a}{2}\right) \quad \quad (3\text{B}.27)$$

where

$$a = \frac{A_s f_y}{0.85 f_c' b}$$

The neutral axis will fall below the flange if the tension force $A_s f_y$ is greater than the compression force capacity of the flange $0.85 f_c' b h_f$,

$$A_s f_y > 0.85 f_c' b h_f$$

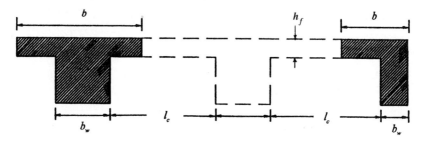

FIGURE 3B.3 Effective cross section of a flanged concrete beam. l_n = beam span length.

Hence

$$a = \frac{A_s f_y}{0.85 f'_c b} > h_f \tag{3B.28}$$

In this case the analysis can be conducted by considering the resistance provided by the overhanging flanges and that provided by the remaining rectangular beam, as shown in Fig. 3B.4. Beam 1 consists of the overhanging flange area A_f, stressed to $0.85 f'_c$, giving a compressive force C_f, which acts at the centroid of the area of the overhanging flanges. To maintain equilibrium, beam 1 has a tensile steel area A_{sf} chosen such that

$$A_{sf} f_y = A_f (0.85 f'_c)$$

This steel area A_{sf} is a portion of the total area A_s and is assumed to be at the same centroid. The moment capacity of beam 1 is obtained by taking the moment about the tensile steel area A_{sf},

$$M_{n1} = 0.85 f'_c (b - b_w) h_f \left(d - \frac{h_f}{2} \right) \tag{3B.29}$$

Beam 2 is a rectangular beam having a width b_w. The compressive force of the beam, $C_2 = 0.85 f'_c b_w a$, acts through the centroid of the compression area. Equilibrium is maintained by utilizing the remaining tensile steel area, $A_s - A_{sf} = A_{sw}$. The moment capacity is obtained by taking the moment about the compression force C_2,

$$M_{n2} = A_{sw} f_y \left(d - \frac{a}{2} \right) \tag{3B.30}$$

where

$$A_{sw} = A_s - A_{sf} = A_s - \frac{0.85 f'_c (b - b_w) h_f}{f_y} \tag{3B.31}$$

and

$$a = \frac{(A_s - A_{sf}) f_y}{0.85 f'_c b_w} \tag{3B.32}$$

Hence the total nominal strength is

$$M_n = M_{n1} + M_{n2} = 0.85 f'_c (b - b_w) h_f \left(d - \frac{h_f}{2} \right) + A_{sw} f_y \left(d - \frac{a}{2} \right) \tag{3B.33}$$

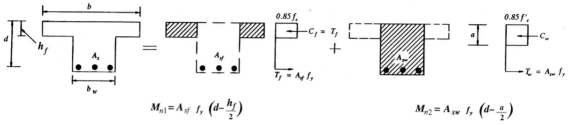

FIGURE 3B.4 Flexural behavior of concrete beam.

3B.2.6.2 Ductile Failure. To ensure ductile failure, the tension steel reinforcement ratio ρ should be limited to

$$\rho \leq 0.75 \frac{b_w}{b}(\bar{\rho}_b + \rho_f) \qquad (3B.34)$$

where $\rho = A_s/bd$
$\bar{\rho}_b$ = balanced steel ratio for a rectangular section (b_w and h) with tension reinforcement, $A_{sw} = A_s - A_{sf}$

$$\rho_f = \frac{0.85 f'_c (b - b_w) h_f}{f_y b_w d} = \frac{A_{sf}}{b_w d}$$

Example 3B.6 Determine the nominal flexural strength M_n of the precast T beam shown in Fig. X.6.

Given:
$f'_c = 5000$ psi (34.45 MPa)
$f_y = 60,000$ psi (413.4 MPa)
$b = 36.0$ in (914 mm)
$bw = 12.0$ in (305 mm)
$h = 20.0$ in (508 mm)
$h_f = 2.0$ in (51 mm)
$d = 17.0$ in (432 mm)
$A_s = 6$ no. 9 bars [6 in² (3870 mm²)]

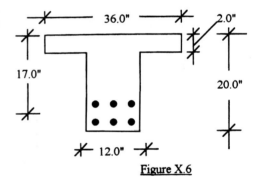

Figure X.6

Solution Check whether the tension force is greater than the compressive force,

$$a = \frac{A_s f_y}{0.85 f'_c b} = \frac{6 \times 60,000}{0.85 \times 5000 \times 36} = 2.35 \text{ in (59.7 mm)} > h_f = 2 \text{ in (51 mm)}$$

Hence the section will act as a T beam.

$$A_{sf} = \frac{0.85 f'_c (b - b_w) h_f}{f_y} = \frac{0.85 \times 5000 (36 - 12) 2}{60,000} = 3.4 \text{ in}^2 \text{ (2193 mm}^2\text{)}$$

$$a = \frac{(A_s - A_{sf}) f_y}{0.85 f'_c b_w} = \frac{(6 - 3.4) 60,000}{0.85 \times 5000 \times 12} = 3.06 \text{ in (77.7 mm)}$$

$$M_{n_1} = 0.85 f'_c(b - b_w)h_f\left(d - \frac{h_f}{2}\right)$$

$$= 0.85 \times 5000(36 - 12)2\left(17 - \frac{2}{2}\right) = 3.26 \times 10^6 \text{ in} \cdot \text{lb } (368.43 \text{ kN} \cdot \text{m})$$

$$M_{n_2} = (A_s - A_{sf})f_y\left(d - \frac{a}{2}\right)$$

$$= (6 - 3.4)60{,}000\left(17 - \frac{3.06}{2}\right) = 2.4 \times 10^6 \text{ in} \cdot \text{lb } (271.24 \text{ kN} \cdot \text{m})$$

Hence the total nominal strength is

$$M_n = M_{n_1} + M_{n_2} = (3.26 + 2.41)10^6 = 5.67 \times 10^6 \text{ in} \cdot \text{lb } (639.67 \text{ kN} \cdot \text{m})$$

3B.3 SHEAR BEHAVIOR

3B.3.1 Plain Concrete

Because structural members are usually subjected to shear stresses combined with axial, flexure, and tension forces rather than to pure shear stresses, the behavior of concrete under pure shear forces is not of major importance. Furthermore, even if pure shear is encountered in a member, a principal stress of equal magnitude will be produced on another inclined plane, leading to the failure of concrete in tension before its shearing strength can be reached.

Consider a small element A at the neutral axis of the beam presented in Fig. 3B.5(a). It can be shown that the case of pure shear [Fig. 3B.5(b)] is equivalent to a set of normal tension and compression stresses σ_1 and σ_2, respectively. Cracking of concrete will occur if the tension stress referred to as diagonal tension exceeds the tension strength of the concrete. As stated earlier, shear stresses are usually combined with other stresses. Considering a small element B located below the

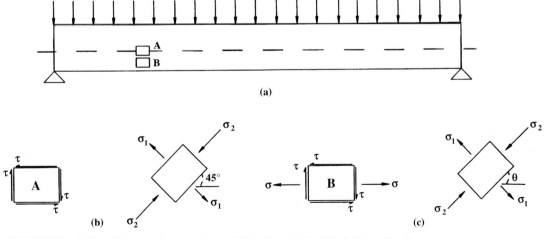

FIGURE 3B.5 (a) Shear behavior of concrete beam. (b) Pure shear. (c) Combined shear and bending.

neutral axis of the beam in Fig. 3B.5(a), two types of stresses occur, bending stresses and shear stresses. If the beam is behaving in the elastic range, these stresses can be obtained as follows:

$$\sigma = \frac{Mc}{I} \tag{3B.35}$$

$$\tau = \frac{VQ}{Ib} \tag{3B.36}$$

This element can be rotated at an angle θ to obtain the principal normal stresses σ_1 and σ_2. The magnitude of the principal stresses and their orientations [Fig. 3B.5(c)] are determined from the following expressions:

$$\sigma_1 = \frac{\sigma}{2} + \sqrt{\left(\frac{\sigma}{2}\right)^2 + \tau^2} \tag{3B.37}$$

$$\sigma_2 = \frac{\sigma}{2} - \sqrt{\left(\frac{\sigma}{2}\right)^2 + \tau^2} \tag{3B.38}$$

$$\tan 2\theta = \frac{\tau}{\sigma/2} \tag{3B.39}$$

It should be clear at this stage that, depending on the relative values of the bending moments and shear forces, the magnitudes of the principal stresses and their orientations will vary. When bending moments are relatively predominant compared to shearing forces, flexure cracks are observed. However, if the shear stresses are sufficiently higher than the bending stresses, inclined cracks of the web propagating from the neutral axis are to be expected.

3B.3.2 Reinforced Concrete Beams

The shear strength of concrete beams reinforced with steel bars resisting flexural loading is denoted by V_c. The failure plane shown in Fig. 3B.6 indicates that V_c is mainly provided by three sources—shear resistance of the uncracked concrete V_{uc}, resistance by aggregate interlock V_a, and dowel action provided by the longitudinal flexural steel reinforcements V_d. These three terms cannot be determined individually. Their combined total effect is evaluated empirically based on a large number of available test results. The ACI code suggests the following conservative equation to determine V_c:

$$V_c = 2\sqrt{f'_c} b_w d \tag{3B.40}$$

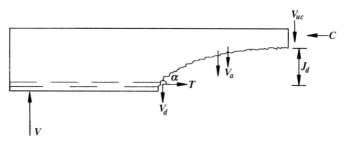

FIGURE 3B.6 Shear-resisting mechanism in reinforced concrete beams.

A less conservative expression, which takes into account the effects of the longitudinal reinforcement and the moment-to-shear ratio, may also be used:

$$V_c = \left(1.9\sqrt{f'_c} + 2500\rho_w \frac{V_u d}{M_u}\right) b_w d \leq 3.5\sqrt{f'_c} b_w d \quad (3B.41)$$

In this expression $V_u d/M_u$ may not be taken as greater than 1.0. In spite of the fact that Eq. (3B.41) is less conservative, its complex form makes its use justifiable in cases of large numbers of similar members. If the shear strength is to be determined for lightweight concrete members, the term $\sqrt{f'_c}$ should be replaced with $f_{ct}/6.7 \leq f'_c$, where f_{ct} is the split cylinder strength of concrete. If the f_{ct} value is not available, then the term $\sqrt{f'_c}$ is to be multiplied by 0.75 for all lightweight concrete and by 0.85 for sand lightweight concrete.

When axial compression exists, the ACI code permits the use of the following equation:

$$V_c = 2\left(1 + \frac{N_u}{2000 A_g}\right)\sqrt{f'_c} b_w d \quad (3B.42)$$

A more conservative equation is adopted by the ACI code for the case of axial tension,

$$V_c = 2\left(1 + \frac{N_u}{500 A_g}\right)\sqrt{f'_c} b_w d \quad (3B.43)$$

where N_u is negative for tension and the ratio N_u/A_g is expressed in pounds per square inch.

When the factored shear force V_u is relatively higher compared to V_c, an additional web reinforcement is provided, as shown in Fig. 3B.7. Several theories have been presented to explain the behavior of web reinforcement. The truss analogy theory is being used widely to illustrate the contribution of web reinforcement to the shear strength of the reinforced concrete beams. According to this theory, the behavior of a reinforced concrete beam with shear reinforcement is analogous to that of a statically determinate paralleled chord truss with pinned joints.

If it is conservatively assumed that the horizontal projection of the crack is equal to the effective depth of the section d, it can be shown that the shear contribution of the vertical stirrups is

$$V_s = \frac{A_v f_y d}{s} \quad (3B.44)$$

If inclined stirrups are used, their contribution to the section shear strength is

$$V_s = \frac{A_v f_y d}{s}(\sin \alpha + \cos \alpha) \quad (3B.45)$$

where α is the angle between the stirrups and the longitudinal axis of the member.

The total nominal shear strength of a section is therefore

$$V_n = V_c + V_s \quad (3B.46)$$

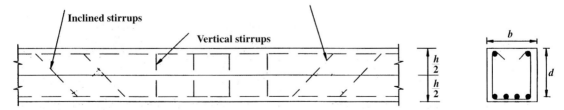

FIGURE 3B.7 Web reinforcement for shear resistance.

3.96 FUNDAMENTALS OF FOUNDATION CONSTRUCTION AND DESIGN

The ACI code limits the maximum vertical stirrup spacing to $d/2 \le 24$ in (305 mm). If the shear resistance V_s of the web reinforcement exceeds $4\sqrt{f'_c}b_w d$, the maximum spacing limit is reduced by one-half to $d/4 \le 12$ in (305 mm). A minimum practical spacing that could be adopted is approximately 3 to 4 in (75 to 100 mm).

The code also provides maximum and minimum limits for the area of shear reinforcement. To avoid concrete crushing prior to the yielding of shear reinforcement, a maximum limit is set. This is provided by limiting the contribution of V_s to the shear resistance to

$$V_s \le 8\sqrt{f'_c}b_w d \qquad (3B.47)$$

If a safe design calls for a higher V_s contribution than the limit set in Eq. (3B.47), the concrete section needs to be enlarged to increase the contribution of concrete to the shear resistance V_c.

The ACI code also requires a minimum shear reinforcement $(A_v)_{min}$ if V_u exceeds $\phi V_c/2$ such that

$$(A_v)_{min} = \frac{50 b_w s}{f_y} \qquad (3B.48)$$

This requirement is necessary to avoid possible brittle failure after the formation of early-stage diagonal cracking. The yield strength of stirrups is limited by the code to 60,000 psi (413.4 MPa) to control the crack width and provide for aggregate interlock availability.

Example 3B.7 The rectangular cross section shown in Fig. X.7 is subjected to the following factored loads:

$$V_u = 70{,}000 \text{ lb } (311.5 \text{ kN})$$

$$M_u = 1 \times 10^6 \text{ in} \cdot \text{lb } (113 \text{ kN} \cdot \text{m})$$

Given:
$f'_c = 4000$ psi (27.56 MPa)
f_y(stirrups) = 60,000 psi (413.4 MPa)
$b = b_w = 12.0$ in (305 mm)
$h = 28.0$ in (711 mm)
$d = 25.0$ in (635 mm)
$A_s = 3$ no. 9 bars (3 in² (1935 mm²)]

Determine the required shear reinforcement.

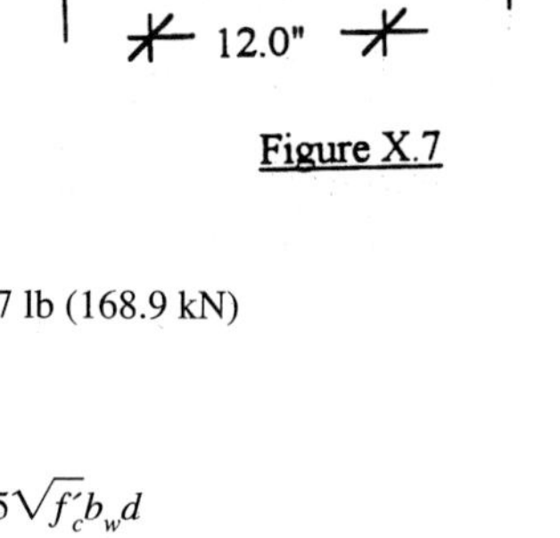

Figure X.7

Solution Determine the shear resistance V_c provided by concrete:

1. Using the simplified expression in Eq. (3B.40),

$$V_c = 2\sqrt{f'_c}b_w d = 2\sqrt{4000} \times 12 \times 25 = 37{,}947 \text{ lb } (168.9 \text{ kN})$$

2. Using the detailed expression in Eq. (3B.41),

$$V_c = \left(1.9\sqrt{f'_c} + 2500\rho_w \frac{V_u d}{M_u}\right) b_w d \le 3.5\sqrt{f'_c}b_w d$$

where $\rho_w = \dfrac{A_s}{b_w d} = \dfrac{3}{12 \times 25} = 0.01$

$\dfrac{V_u d}{M_u} = \dfrac{70{,}000 \times 25}{10^6} = 1.75 > 1.0 \qquad \therefore$ Use 1.0.

$$V_c = (1.9\sqrt{4000} + 2500 \times 0.01 \times 1)12 \times 25 \le 3.5\sqrt{4000} \times 12 \times 25$$
$$= 43{,}550 \text{ lb } (193.8 \text{ kN}) \le 66{,}407 \text{ lb } (295.5 \text{ kN})$$

Thus $V_c = 43{,}550$ lb (193.8 kN), compared to 37,947 lb (168.9 kN) obtained using the simplified expression. Use $V_c = 43{,}550$ lb (193.8 kN) for the rest of the example.

$$V_s = V_n - V_c = \frac{V_u}{\phi} - V_c = \frac{70{,}000}{0.85} - 43{,}550 = 38{,}803 \text{ lb } (172.7 \text{ kN})$$

$V_s = 38{,}803$ lb (172.7 kN) $\le 8\sqrt{f'_c}b_w d = 151{,}788$ lb (675.5 kN) ∴ No need to enlarge section.

$V_s = 38{,}803$ lb (172.7 kN) $\le 4\sqrt{f'_c}b_w d = 75{,}894$ lb (337.7 kN)

The maximum stirrup spacing is

$$s = \frac{d}{2} = \frac{25}{2} = 12.5 \text{ in } (318 \text{ mm}) \le 24 \text{ in } (610 \text{ mm})$$

Hence a maximum spacing of 12.5 in (318 mm) should be used.

Use a U-shape no. 3 bars for stirrups. Hence $A_v = 0.11 \times 2 = 0.22$ in² (142 mm²). The required spacing is

$$s = \frac{A_v f_y d}{V_s} = \frac{0.22 \times 60{,}000 \times 25}{38{,}803} = 8.5 \text{ in } (216 \text{ mm}) < 12.5 \text{ in } (318 \text{ mm}) \quad \text{O.K.}$$

Check the minimum reinforcement:

$$(A_v)_{min} = \frac{50 b_w s}{f_y} = \frac{50 \times 12 \times 8.5}{60{,}000} = 0.085 \text{ in}^2 \text{ (55 mm}^2\text{)} < 0.22 \text{ in}^2 \text{ (142 mm}^2\text{)} \quad \text{O.K.}$$

Therefore use U-shape no. 3 bars with spacing $s = 8.0$ in (200 mm).

Example 3B.8 Provide shear reinforcement for the beam shown in Fig. X.8.

Given:
$b_w = b = 14$ in (356 mm)
$d = 23$ in (584 mm)
$h = 26$ in (660 mm)
$f'_c = 5000$ psi (34.45 MPa)
$f_y = 60{,}000$ psi (413.4 MPa)
$w_D = 3000$ lb/ft (437.5 N/m)
$w_L = 5000$ lb/ft (729.2 N/m)

Solution Critical section:

$$W_u = 1.4 \times 3000 + 1.7 \times 5000 = 12{,}700 \text{ lb/ft } (4710 \text{ N/m})$$

$$V_u \text{ at face of support} = 12{,}700\left(\frac{15}{2}\right) = 95{,}250 \text{ lb } (423.9 \text{ kN})$$

$$V_u \text{ at distance } d = 23 \text{ in from face of support} = 95{,}250\left(\frac{90 - 23}{90}\right) = 70{,}908 \text{ lb } (315.5 \text{ kN})$$

$$V_c \text{ using simplified expression} = 2\sqrt{f'_c}b_w d = 2\sqrt{5000} \times 14 \times 23 = 45{,}537 \text{ lb } (202.6 \text{ kN})$$

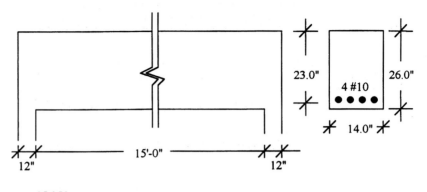

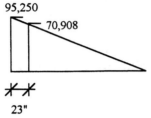

Figure X.8

$$V_s \text{ required} = \frac{V_u}{\phi} - V_c = \frac{70{,}908}{0.85} - 45{,}537 = 37{,}884 \text{ lb } (168.6 \text{ kN})$$

$V_s = 37{,}884 \leq 8\sqrt{f'_c}b_w d = 182{,}148 \text{ lb } (810.6 \text{ kN})$ ∴ No need to enlarge section.

$V_s = 37{,}884 \leq 4\sqrt{f'_c}b_w d = 91{,}075 \text{ lb } (405.3 \text{ kN})$

The maximum stirrup spacing is

$$s = \frac{d}{2} = \frac{23}{2} = 11.5 \text{ in } (292 \text{ mm}) \leq 24 \text{ in } (610 \text{ mm})$$

Hence a maximum spacing of 11.5 in (292 mm) should be used.
Use U-shape no. 3 bars for stirrups, and we get

$$s = \frac{A_v f_y d}{V_s} = \frac{0.22 \times 60{,}000 \times 23}{37{,}884} = 8.01 \text{ in } (203 \text{ mm})$$

Therefore use $s = 8$ in (200 mm) $< s_{max} = 11.5$ in (292 mm).   O.K.
Check the minimum reinforcement:

$$(A_v)_{min} = \frac{50 b_w s}{f_y} = \frac{50 \times 14 \times 8}{60{,}000} = 0.093 \text{ in}^2 \text{ (60 mm}^2\text{)} < 0.22 \text{ in}^2 \text{ (142 mm}^2\text{)} \quad \text{O.K.}$$

Section at which s_{max} can be used:

$$V_s \text{ with } s = 11.5 \text{ in} = \frac{A_v f_y d}{s} = \frac{0.22 \times 60{,}000 \times 23}{11.5} = 26{,}400 \text{ lb } (117.5 \text{ kN})$$

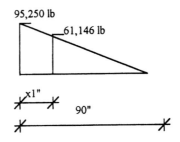

$$V_n = \frac{V_u}{\phi} = V_s + V_c = 26{,}400 + 45{,}537 = 71{,}937 \text{ lb } (320.1 \text{ kN})$$

$$V_u = 71{,}937 \times 0.85 = 61{,}146 \text{ lb } (272.1 \text{ kN})$$

This value is obtained at distance x_1 from the face of the support, such that

$$90 - x_1 = 90\left(\frac{61{,}146}{95{,}250}\right)$$

$$x_1 = 32 \text{ in } (813 \text{ mm})$$

Section at which stirrups can be omitted: In order not to use stirrups, the ACI code requires that

$$V_u \leq \frac{\phi V_c}{2} = \frac{0.85 \times 45{,}537}{2} = 19{,}353 \text{ lb } (86.1 \text{ kN})$$

This value is obtained at distance x_2 from the face of the support,

$$90 - x_2 = 90\left(\frac{19{,}353}{95{,}250}\right)$$

$$X_2 = 72 \text{ in } (1829 \text{ mm})$$

Shear reinforcement summary: This is illustrated by the following figure.

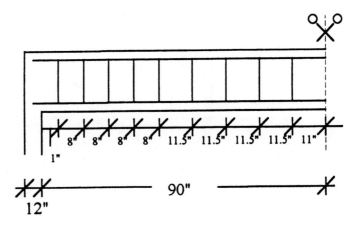

3B.4 TORSION BEHAVIOR

3B.4.1 Torsion Strength of Reinforced Concrete Beams

Torsion occurs in many structures, such as main girders of bridges, edge beams in buildings, spiral stairways, and balcony girders. Torsion, however, usually occurs in combination with shear and bending.

If a plain concrete member is subjected to pure torsion, it will first crack, then fail along 45° spiral lines. This failure pattern is due to the diagonal tension corresponding to the torsional stresses, which have opposite signs on both sides of the member (Fig. 3B.8).

The maximum torsional stresses v_{max} caused in a rectangular cross section can be calculated theoretically using the following expression:

$$v_{max} = \alpha_1 \frac{T_{cr}}{x^2 y} \tag{3B.49}$$

where α is a constant varying from 3 to 5, according to the y/x ratio.

Using the lowest value of α (namely, 3) and equating the maximum torsional stress that concrete can resist without cracking to $6\sqrt{f'_c}$, the torsional moment at which diagonal tension cracking occurs can be estimated using the following expression:

$$T_{cr} = 2\sqrt{f'_c} x^2 y \tag{3B.50}$$

If the member cross section is T- or L-shaped, the following expression can be used satisfactorily:

$$T_{cr} = 2\sqrt{f'_c} \Sigma x^2 y \tag{3B.51}$$

In this case the section is divided into a set of rectangles, each resisting part of the twisting moment in proportion to its torsional rigidity. The ACI code limits the length of the overhanging flange to be considered effective in torsional rigidity computations to three times its thickness.

Due to the combined effect of bending and torsion, a portion of the diagonal cracks will fail in the compression zone on one side of the beam. As a result, even though diagonal torsion cracks have developed on part of the beam, the other part continues to resist some torsion. The ACI conservatively limits the torsional resistance of the cracked section to 40% of the uncracked section. Hence,

$$T_c = 2 \times 0.4\sqrt{f'_c} \Sigma x^2 y = 0.8\sqrt{f'_c} \Sigma x^2 y \tag{3B.52}$$

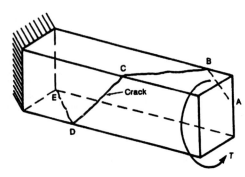

FIGURE 3B.8 Cracking pattern due to torsion.

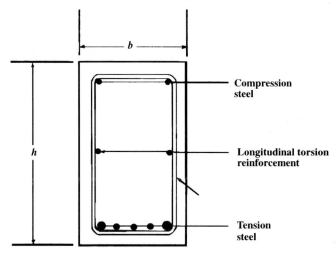

FIGURE 3B.9 Reinforcement for torsion resistance.

If the factored torsional moment T_u exceeds the resistance of the reinforced concrete section without web reinforcement, additional torsional reinforcement needs to be provided such that

$$T_u \leq \phi(T_c + T_s) \tag{3B.53}$$

The torsion reinforcement is provided using closed vertical stirrups to form a continuous loop and additional longitudinal reinforcing bars, as shown in Fig. 3B.9. The longitudinal bars should be no smaller than no. 3, and they should be 12 in (0.3 m) apart.

According to the ACI code, the total twisting moment T_s resisted by the vertical closed stirrups and the longitudinal reinforcing bars can be estimated using the following expression:

$$T_s = \frac{A_t \alpha_t x_1 y_1 f_y}{s} \tag{3B.54}$$

where A_t = area of one leg of a closed stirrup spaced at a distance s
x_1 = shorter dimension of stirrup
y_1 = longer dimension of stirrup
α_t = empirical coefficient; $= 0.66 + 0.33 y_1/x_1 \leq 1.5$.

The area of the longitudinal bars required for torsional resistance is obtained as the larger of the following two equations:

$$A_l = 2 A_t \left(\frac{x_1 + y_1}{s} \right) \tag{3B.55}$$

or

$$A_l = \left[\frac{400 b_w s}{f_y} \left(\frac{T_u}{T_u + V_u/3C_t} \right) - 2 A_t \right] \left(\frac{x_1 + y_1}{s} \right) \tag{3B.56}$$

where

$$C_t = \frac{b_w d}{\Sigma x^2 y}$$

In Eq. (3B.56) the value of A_t need not exceed the value obtained if $50b_w s/f_y$ is substituted in place of $2A_t$.

The spacing of the closed stirrups selected to provide torsion resistance should not be greater than $(x_1 + y_1)/4$ or 12 in (0.3 m). Also the code limits the amount of torsion reinforcement in order to ensure ductility. Hence the torsion resistance obtained for steel reinforcement is limited to

$$T_s \leq 4T_c \tag{3B.57}$$

Furthermore, the steel reinforcement yield stress is limited to 60,000 psi (413.4 MPa).

3B.4.2 Combined Shear, Bending, and Torsion

Due to the combined action of shear, bending, and torsion the following modified expressions are provided in the ACI code to obtain V_c and T_c, respectively:

$$V_c = \frac{2\sqrt{f'_c} b_w d}{\sqrt{1 + (2.5 C_t T_u/V_u)^2}} \tag{3B.58}$$

$$T_c = \frac{0.8\sqrt{f'_c} \Sigma x^2 y}{\sqrt{1 + (0.4 V_u/C_t T_u)^2}} \tag{3B.59}$$

Equations (3B.44) and (3B.54) could be used to provide shear and torsion reinforcement, respectively. For reasons of ductility, the code limits the reinforcement contribution as follows:

$$V_s \leq 8\sqrt{f'_c} b_w d \tag{3B.60}$$

$$T_s \leq 4T_c \tag{3B.61}$$

The minimum area of steel reinforcement may not be less than $50b_w s/f_y$. Hence

$$A_v + 2A_t \leq \frac{50 b_w s}{f_y} \tag{3B.62}$$

The minimum reinforcement needs to be provided if T_u exceeds $0.5\sqrt{f'_c} \Sigma x^2 y$. Otherwise torsional effects can be neglected.

Example 3B.9 A 6-in (152-mm) slab cantilevers 6 ft (1829 mm) from the face of a 14 × 24 in (356×610 mm) simple beam, as shown in Fig. X.9. The beam spans 30 ft (9144 mm). It carries a uniform service live load of 25 psf (1.02 kPa) on the cantilever.

Given:
$f'_c = 4000$ psi (27.56 MPa)
$f_y = 60,000$ psi (413.4 MPa)
$A_s = 3$ in^2 (1935 mm^2)
Column dimensions = 14 × 14 in (356 × 356 mm)

Determine the required shear and torsion reinforcement at the critical section.

Solution

$$w_D = \frac{6}{12} \times 150 = 75 \text{ psf } (3.59 \text{ kPa})$$

$$w_u = 1.4 \times 75 + 1.7 \times 25 = 148 \text{ psf } (7.08 \text{ kPa})$$

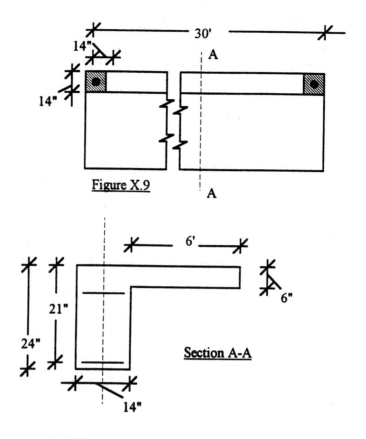

Figure X.9

Section A-A

At the centerline of the column,

$$V_u = \frac{148 \times 7.17 \times 30}{2} + 1.4\left[\frac{14(24-6)}{144} \times 150\right]\frac{30}{2} = 21{,}430 \text{ lb } (95.36 \text{ kN})$$

$$T_u = \frac{148 \times 30 \times 6}{2}\left(3 + \frac{7}{12}\right) = 47{,}730 \text{ ft} \cdot \text{lb } (64.72 \text{ kN} \cdot \text{m})$$

At a distance $d = 21$ in (533 mm) from the face of the support (critical section),

$$V_u = 21{,}430\left[\frac{15 - \left(\dfrac{7+21}{12}\right)}{15}\right] = 18{,}096 \text{ lb } (80.53 \text{ kN})$$

$$T_u = 47{,}730\left[\frac{15 - \left(\dfrac{7+21}{12}\right)}{15}\right] = 40{,}305 \text{ ft} \cdot \text{lb } (54.65 \text{ kN} \cdot \text{m})$$

$$\Sigma x^2 y = 14^2 \times 24 + 6^2 \times 24 = 5352 \text{ in}^3 \ (8.8 \times 10^7 \text{ mm}^3)$$

$$\phi(0.5\sqrt{f'_c}\,\Sigma x^2 y) = 0.85(0.5\sqrt{4000} \times 5352) = 143{,}858 \text{ in} \cdot \text{lb } (16.26 \text{ kN} \cdot \text{m})$$

$< T_u = 483{,}660$ in · lb (54.65 kN · m)   ∴ Torsional effects need to be considered.

$$C_t = \frac{b_w d}{\Sigma x^2 y} = \frac{14 \times 21}{5352} = 0.055$$

$$V_c = \frac{2\sqrt{f'_c}\, b_w d}{\sqrt{1 + \left(\dfrac{2.5 C_t T_u}{V_u}\right)^2}} = \frac{2\sqrt{4000} \times 14 \times 21}{\sqrt{1 + \left(\dfrac{2.5 \times 0.055 \times 483{,}660}{18{,}096}\right)^2}} = 9763 \text{ lb (43.4 kN)}$$

$$T_c = \frac{0.8\sqrt{f'_c}\, \Sigma x^2 y}{\sqrt{1 + \left(\dfrac{0.4 V_u}{C_t T_u}\right)^2}} = \frac{0.8\sqrt{4000} \times 5352}{\sqrt{1 + \left(\dfrac{0.4 \times 18{,}096}{0.055 \times 483{,}660}\right)^2}} = 261{,}291 \text{ in · lb (29.53 kN · m)}$$

$$V_s = \frac{V_u}{\phi} - V_c = \frac{18{,}096}{0.85} - 9763 = 11{,}526 \text{ lb (51.29 kN)} < 8\sqrt{f'_c}\, b_w d = 148{,}753 \text{ lb (662 kN)}$$

∴ No need to enlarge section.

$$T_s = \frac{T_u}{\phi} - T_c = \frac{483{,}660}{0.85} - 261{,}291 = 307{,}721 \text{ in · lb (34.77 kN · m)} < 4T_c = 1{,}045{,}164 \text{ in · lb}$$
$$(118.1 \text{ kN · m}) \quad \therefore \text{No need to enlarge section.}$$

$$\frac{A_v}{s} = \frac{V_s}{f_y d} = \frac{11{,}526}{60{,}000 \times 21} = 0.0091 \text{ in}^2/\text{in spacing for two legs}$$

Assume 1.5 in (38 mm) clear cover and no. 4 closed stirrup,

$$x_1 = 14 - 2(1.5 + 0.25) = 10.5 \text{ in (267 mm)}$$
$$y_1 = 24 - 2(1.5 + 0.25) = 20.5 \text{ in (521 mm)}$$
$$\alpha_t = 0.66 + 0.33\left(\frac{y_1}{x_1}\right) = 0.66 + 0.33\left(\frac{20.5}{10.5}\right) = 1.3 \leq 1.5$$

Hence

$$\frac{A_t}{s} = \frac{T_s}{f_y \alpha_t x_1 y_1} = \frac{307{,}721}{60{,}000 \times 1.3 \times 10.5 \times 20.5} = 0.0183 \text{ in}^2/\text{in spacing for two legs}$$

Stirrup for combined shear and torsion:

$$\frac{A_v}{s} + \frac{2A_t}{s} = 0.0091 + 2 \times 0.0183 = 0.046 \text{ in}^2/\text{in spacing for two legs}$$

$$\frac{50 b_w}{f_y} = \frac{50 \times 14}{60{,}000} = 0.0117 < 0.046 \quad \text{O.K.}$$

Using no. 4 and area = $2 \times 0.2 = 0.4$ in² (258 mm²), we get

$$s = \frac{\text{area}}{A_v/s + 2A_t/s} = \frac{0.4}{0.046} = 8.7 \text{ in (221 mm)}$$

The maximum allowable spacing is

$$s_{max} = \frac{x_1 + y_1}{4} = \frac{10.5 + 20.5}{4} = 7.75 \text{ in } (197 \text{ mm})$$

Therefore use no. 4 closed stirrups at spacing $s = 7.75$ in (197 mm).
Longitudinal torsional steel:

$$A_l = 2A_t\left(\frac{x_1 + y_1}{s}\right) = 2 \times 0.0183(10.5 + 20.5) = 1.13 \text{ in}^2 \text{ (728.9 mm}^2\text{)}$$

or

$$A_l = \frac{400xs}{f_y}\left[\left(\frac{T_u}{T_u + V_u/3C_t}\right) - 2A_t\right]\left(\frac{x_1 + y_1}{s}\right)$$

Substituting $50b_w d/f_y$ for $2A_s$ if it is larger than $2A_t$, we get

$$\frac{50b_w d}{f_y} = \frac{50 \times 14 \times 7.5}{60,000} = 0.875 < 2A_t = 2 \times 0.0183 \times 7.5 = 2.745$$

Hence we use $2A_t = 0.2745$,

$$A_l = \frac{400 \times 14 \times 7.5}{60,000}\left[\left(\frac{40,305}{40,305 + 18,096/(3 \times 0.055)}\right) - 0.274\,5\right]\left(\frac{10.5 + 20.5}{7.5}\right)$$

$$= -0.017 \text{ in}^2 \, (-10.96 \text{ mm}^2)$$

Hence we use $A_l = 1.13$ in² (728.9 mm²), and add $A_l/4 = 0.3$ in² (193.5 mm²) on each face of the cross section. Thus we use 2 no. 4 bars on each vertical side of the cross section and on the top face [= 0.4 in² (258 mm²)]. The reinforcement on the bottom face becomes $A_s = 3$ in² + 0.3 in² = 3.3 in², and we use 2 no. 9 + 2 no. 8 bars [= 3.56 in² (2296 mm²)]. The final design is shown in Fig. X.10.

2 # 4

2 # 4

2 # 4

4 @ 7.5" closed stirrups

2 # 9 + 2 # 8   **Figure X.10**

3B.5 COLUMNS

3B.5.1 General

Columns are vertical members subjected to axial loads or a combination of axial loads and bending moments. They can be divided into three categories, depending on their structural behavior. Short compression blocks or pedestals are members with a height less than three times the least lateral dimensions. They may be designed with plain concrete with a maximum stress of $0.85\phi f'_c$, where $\phi = 0.7$. For higher stresses the pedestal should be designed with reinforced concrete. Short reinforced concrete columns have low slenderness ratios, resulting in transverse deformations that will not affect the ultimate strength. Slender reinforced concrete columns have slenderness ratios that exceed the limits given for short columns. For this case the secondary moments due to transverse deformations reduce the ultimate strength of the member.

Sections of reinforced concrete columns are usually square or rectangular shapes for lower construction costs. Longitudinal steel bars are added to increase the load-carrying capacity. A substantial strength increase is obtained by providing lateral bracing for the longitudinal bars. If bracing is provided with separate closed ties, the column is referred to as a tied column. If a continuous helical spiral is used to contain the longitudinal bars, the columns are referred to as spiral columns. Commonly, a circular shaped cross section is used for spiral columns. Composite columns consist of structural steel shapes encased in concrete. The concrete may or may not be reinforced with longitudinal steel bars.

3B.5.2 Axially Loaded Columns

The theoretical nominal strength of an axially loaded short column can be determined by the following expression:

$$P_n = 0.85 f'_c (A_g - A_{st}) + f_y A_{st} \quad (3B.63)$$

where A_g = gross concrete area
A_{st} = total cross-sectional area of longitudinal reinforcement

In actual construction situations there are no perfect axially loaded columns. Minimum moments occur even though no calculated moments are present. To account for these minimum moments, the ACI code requires that the theoretical nominal strength obtained from Eq. (3B.63) be multiplied by a reduction factor α. This reduction factor is equal to 0.85 for spiral columns and 0.80 for tied columns. The use of Eq. (3B.63) and of the reduction factor α is applicable for small moments where the eccentricity e is less than $0.1h$ for tied columns and less than $0.05h$ for spiral columns. For higher eccentricity values the procedures presented in the next section should be used.

3B.5.3 Uniaxial Bending and Compression

The ultimate-strength behavior of sections under combined bending and axial compression is presented in Fig. 3B.10. Depending on the magnitude of the strain in the tension steel, the section

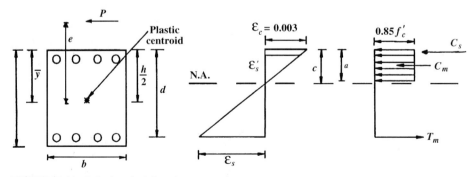

FIGURE 3B.10 Behavior of reinforced concrete columns.

would fail either in tension or in compression. Tension failure is characterized by initial yielding of steel preceding crushing of concrete. Compression failure is due to concrete crushing before yielding of steel. If the tension steel yields at the same time that the concrete crushes, this condition is termed balanced condition and is defined by the following expressions:

$$\varepsilon_s = \varepsilon'_s = \varepsilon_y = \frac{f_y}{E_s} \tag{3B.64}$$

$$c_b = d\left(\frac{0.003}{0.003 + \varepsilon_y}\right) \tag{3B.65}$$

$$a_b = B_1 c_b \tag{3B.66}$$

$$P_{nb} = 0.85 f'_c b a_b + A'_s f_y - A_s f_y \tag{3B.67}$$

$$M_{nb} = 0.85 f'_c b a_b \left(\bar{y} - \frac{a}{2}\right) + A'_s f_y (\bar{y} - d') + A_s f_y (d - \bar{y}) \tag{3B.68}$$

$$e_b = \frac{M_{nb}}{P_{nb}} \tag{3B.69}$$

Tension failure is obtained if $P_n < P_b$ or $e > e_b$, whereas compression failure is obtained if $P_n > P_b$ or $e < e_b$. For both types of failure the strain compatibility and equilibrium relationships have to be maintained.

The following procedure is used to obtain P_n and M for a given section and a known eccentricity e:

1. Assume a depth for neutral axis c. Then obtain $a = c\beta_1$.
2. Compute the compression strain of steel ε'_s and the tension strain of steel ε_s,

$$\varepsilon'_s = 0.003\left(\frac{c - d'}{c}\right) \tag{3B.70}$$

$$\varepsilon_s = 0.003\left(\frac{d - c}{d}\right) \tag{3B.71}$$

3. Compute the compression stress of steel f'_s and the tension stress of steel f_s,

$$f'_s = \varepsilon'_s E_s \leq f_y \tag{3B.72}$$

$$f_s = \varepsilon_s E_s \leq f_y \tag{3B.73}$$

4. Compute the value of P_n and M_n,

$$P_n = 0.85 f'_c b a + A'_s f'_s - A_s f_s \tag{3B.74}$$

$$M_n = 0.85 f'_c b a \left(\bar{y} - \frac{a}{2}\right) + A'_s f'_s (\bar{y} - d') + A_s f_s (d - \bar{y}) \tag{3B.75}$$

5. Compute $e^* = M_n/P_n$.
6. Compare e^* with the known eccentricity e. If equal, the values of P_n and M_n represent the nominal strength of the cross section. If different, repeat steps 1 to 6 with a different c value.

Example 3B.10 A 12 × 20 in (305 × 508 mm) tied column is carrying a vertical load with an eccentricity $e = 8$ in (203 mm), as illustrated in Fig. X.11.

Given:

$f'_c = 4000$ psi (27.56 MPa)
$f_y = 60,000$ psi (413.4 MPa)
$A_s = A'_s = 4$ no. 7 bars $= 2.4$ in² (1548 mm²)

Find the nominal load P_n

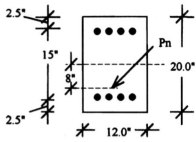

Figure X.11

Solution To obtain the balanced condition,

$$\varepsilon_s = \varepsilon'_s = \frac{60,000}{29 \times 10^6} = 0.0021$$

$$c_b = d\left(\frac{0.003}{0.003 + \varepsilon_y}\right) = 17.5\left(\frac{0.003}{0.003 + 0.0021}\right) = 10.29 \text{ in (261.4 mm)}$$

$$a_b = \beta_1 c_b = 0.85 \times 10.29 = 8.75 \text{ in (222.3 mm)}$$

$$P_{nb} = 0.85 f'_c b a_b + A'_s f_y - A_s f_y = 0.85 \times 4000 \times 12 \times 8.75 = 357,000 \text{ lb (1588.7 kN)}$$

$$M_{nb} = 0.85 f'_c b a_b \left(\bar{y} - \frac{a}{2}\right) + A'_s f_y (\bar{y} - d') + A_s f_y (d - \bar{y})$$

$$= 0.85 \times 4000 \times 12 \times 8.75\left(10 - \frac{8.75}{2}\right) + 2.4 \times 60,000(10 - 2.5) + 2.4 \times 60,000(17.5 - 10)$$

$$= 4.17 \times 10^6 \text{ in} \cdot \text{lb (471.21 kN} \cdot \text{m)}$$

$$e_b = \frac{M_{nb}}{P_{nb}} = \frac{4.17 \times 10^6}{357,000} = 11.7 \text{ in (297.2 mm)}$$

$e = 8$ in (203 mm) $< e_b = 11.7$ in (297.2 mm)   ∴ Column will fail in compression.

First trial

1. Assume $c = 15$ in (381 mm).

    $a = 15 \times 0.85 = 12.75$ in (323.9 mm)

2. $\varepsilon'_s = 0.003\left(\frac{c - d'}{c}\right) = 0.003\left(\frac{15 - 2.5}{15}\right) = 0.0025$

    $\varepsilon_s = 0.003\left(\frac{d - c}{c}\right) = 0.003\left(\frac{17.5 - 15}{15}\right) = 0.0005$

3. $f'_s = 0.0025 \times 29 \times 10^6 = 72,500$; take $f'_s = 60,000$ psi (413.4 MPa)

    $f_s = 0.0005 \times 29 \times 10^6 = 14,500$ psi (99.9 MPa)

4. $P_n = 0.85 f'_c b a + A' f'_s - A_s f_s$

    $= 0.85 \times 4000 \times 12 \times 12.75 + 2.4 \times 60,000 - 2.4 \times 14,500 = 629,400$ lb (2800.8 kN)

$$M_n = 0.85 f'_c ba\left(\bar{y} - \frac{a}{2}\right) + A'_s s_y(\bar{y} - d') + A_s s_y(d - \bar{y})$$

$$= 0.85 \times 4000 \times 12 \times 12.75\left(10 - \frac{12.75}{2}\right) + 2.4 \times 60{,}000(10 - 2.5) + 2.4 \times 14{,}500(17.5 - 10)$$

$$= 3.2 \times 10^6 \text{ in} \cdot \text{lb } (361.6 \text{ kN} \cdot \text{m})$$

5. $e^* = \dfrac{M_n}{P_n} = \dfrac{3.2 \times 10^6}{629{,}400} = 5.12$ in (130.05 mm)

Second trial
Assume $c = 11$ in (279.4 mm), and by repeating the same procedure we get

$$P_n = 465{,}000 \text{ lb } (2069.3 \text{ kN})$$

$$M_n = 3.8 \times 10^6 \text{ in} \cdot \text{lb } (429.4 \text{ kN} \cdot \text{m})$$

$$e^* = \frac{3.8 \times 10^6}{465{,}000} = 8.1 \text{ in} \cong 8 \text{ in given}$$

Hence the nominal force P_n that can be applied on this section with an eccentricity $e = 8$ in (203 mm) is 465,000 lb (2069.3 kN).

3B.5.4 Load-Moment Interaction Diagrams

The strength of a concrete section subjected to combined axial and bending loads can be conveniently determined with the help of interaction diagrams. An interaction diagram gives a relationship between the nominal axial load P_n and the nominal moment capacity M_n of a given sec-

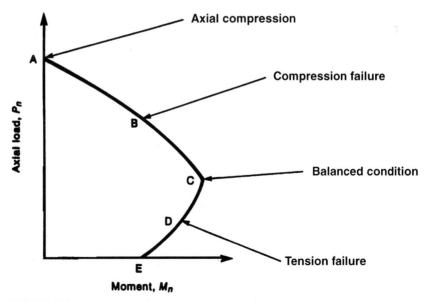

FIGURE 3B.11 Load-moment interaction diagrams for columns.

tion, as shown in Fig. 3B.11. Each point on the diagram represents one possible combination of nominal axial load and nominal axial moment. The interaction diagram is divided into the tension control region and the compression control region by the balanced condition point (P_{nb}, M_{nb}). These diagrams are available as design aids for different column sections with different reinforcement percentages and arrangements. Typical diagrams are presented in Fig. 3B.12.

Example 3B.11 Solve Example 3B.10 using the moment interaction diagrams.

Solution

$$\gamma_h = 20 - 2.5 = 15 \text{ in (381 mm)}$$

$$\gamma = \frac{\gamma_h}{h} = \frac{15}{20} = 0.75$$

$$\rho_{total} = \frac{2 \times 2.4}{20 \times 12} = 0.02$$

$$\frac{e}{h} = \frac{8}{20} = 0.4$$

From load-moment interaction diagrams,

$$\frac{\phi P_n}{A_g} = 1.35$$

Hence

$$P_n = \frac{1.35 \times 20 \times 12}{0.7} = 463{,}000 \text{ lb (2060.3 kN)}$$

3B.5.5 Slender Columns (Buckling Effect)

The strength of slender columns is affected by the secondary stresses due to the slenderness effect. The slenderness is usually expressed in terms of the slenderness ratio KL_u/r, where K depends on the column's end conditions, L_u is the column length, and r is the radius of gyration ($= \sqrt{I/A}$). The lower limits for neglecting the slenderness effect are

$$\text{Braced frames} \quad \frac{KL_u}{r} < 34 - 12\frac{M_1}{M_2} \quad (3\text{B}.76)$$

$$\text{Unbraced frames} \quad \frac{KL_u}{r} < 22 \quad (3\text{B}.77)$$

where M_1 and M_2 are the smaller and larger moments at the opposite ends of the compression member, respectively. This ratio is positive if the column is bent in single curvature, negative if it is bent in double curvature.

Two methods can be used to analyze slender columns.

1. The moment magnification method, where the member is designed for a magnified moment δM, with $\delta \geq 1.0$ and M the nominal moment based on an analysis neglecting the slenderness effect. This method is presented in the ACI code and is applicable to compression members with slenderness ratios lower than 100.

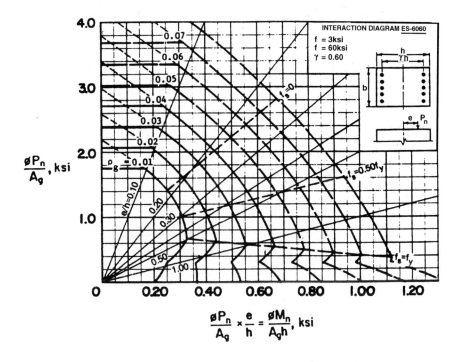

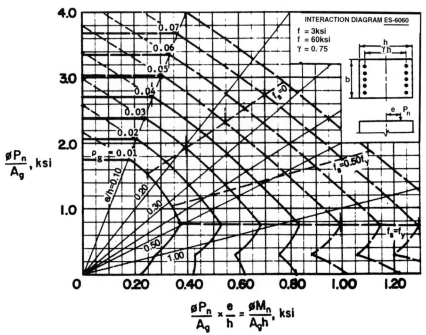

FIGURE 3B.12 Typical load-moment interaction diagrams.

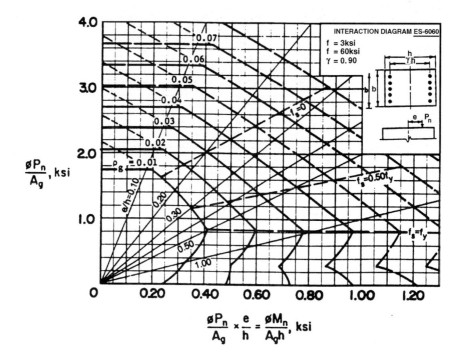
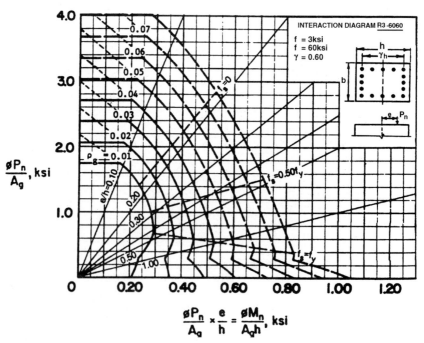

FIGURE 3B.12 (Continued)

FUNDAMENTALS OF REINFORCED CONCRETE **3.113**

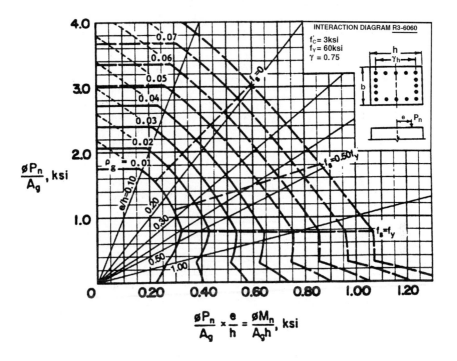

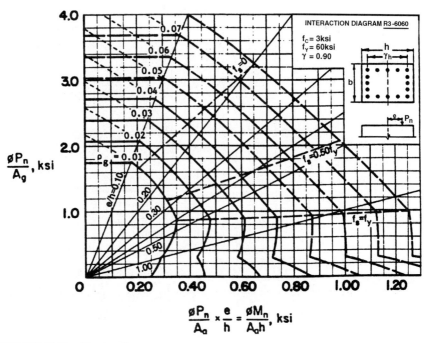

FIGURE 3B.12 (*Continued*)

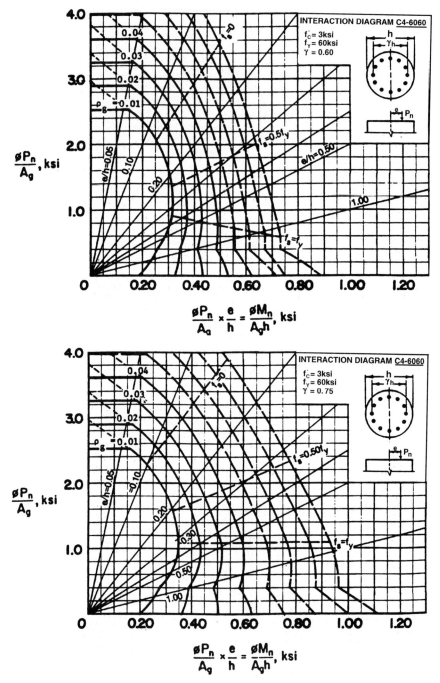

FIGURE 3B.12 (*Continued*)

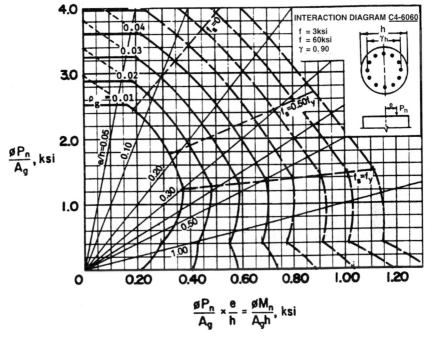

FIGURE 3B.12 *(Continued)*

2. The second-order analysis takes into consideration the effect of deflection, change in stiffness, sustained load effects, and stability. This type of analysis is usually done with the aid of computers and is only required by the code for slenderness ratios greater than 100. It should be noted that the majority of reinforced concrete columns does not require such an analysis because in most cases the slenderness ratio is below 100.

In the moment magnification method, the magnified moment M_c can be determined using the following equation:

$$M_c = \delta_b M_{2b} + \delta_s M_{2s} \tag{3B.78}$$

where the subscripts b and s refer to the moments due to gravity loads and lateral loads, respectively. Also,

$$\delta_b = \frac{C_m}{1 - P_u/\phi P_c} \geq 1.0 \tag{3B.79}$$

$$\delta_s = \frac{1}{1 - \Sigma P_u/\phi \Sigma P_c} \geq 1.0 \tag{3B.80}$$

$$P_c = \frac{\pi^2 EI}{(KL_u)^2} \tag{3B.81}$$

and $C_m = 0.6 + 0.4(M_1/M_2) \geq 0.4$ for columns braced against side sway and not exposed to transverse loads between supports. For all other cases $C_m = 1.0$. EI in Eq. (3B.81) must account for the effects of cracking, creep, and nonlinearity of concrete. The ACI code provides the following two equations. For a heavily reinforced member,

$$EI = \frac{E_c I_g/5 + E_s I_s}{1 + \beta_d} \quad (3B.82)$$

and for a lightly reinforced member,

$$EI = \frac{E_c I_g/2.5}{1 + \beta_d} \quad (3B.83)$$

where $\quad \beta_d = \dfrac{\text{design dead-load moment}}{\text{design total moment}}$

In practical situations, columns have end conditions that are partially restricted by adjoining members. Therefore the factor K will vary with the ratio of column stiffness to flexure member stiffness, which provides restraint at the column ends. The value of K can be determined from charts given in Fig. 3B.13. The ACI code recommends the use of $0.5I_g$ for flexural members and I_g for compression members to compute the stiffness parameter ψ in Fig. 3B.13.

3B.5.6 Biaxial Bending

Many reinforced concrete columns are subjected to biaxial bending, or bending about both axes. Corner columns in buildings where beams and girders frame into the columns from both direc-

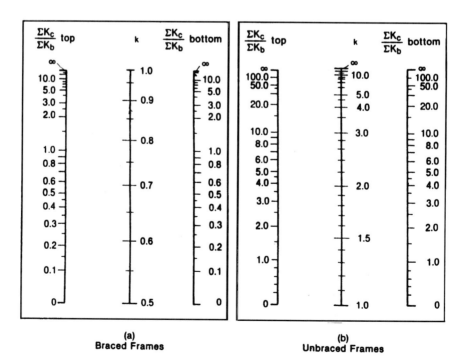

FIGURE 3B.13 Column stiffness factors ψ. (*a*) Braced frames. (*b*) Unbraced frames.

tions are typical examples of biaxial bending. An approximate procedure developed by Bresler[5] has been found to provide satisfactory results. For a given column section, the nominal axial capacity under biaxial bending P_n can be computed using the following equation:

$$\frac{1}{P_n} = \frac{1}{P_{n_x}} + \frac{1}{P_{n_y}} - \frac{1}{P_{n_0}} \tag{3B.84}$$

where P_{n_x} = axial load capacity when load is placed at eccentricity e_x with $e_y = 0$
P_{n_y} = axial load capacity when load is placed at eccentricity e_y with $e_x = 0$
P_{n_0} = capacity for axially loaded case

Bresler's equation produces reliable results if the axial load P_n is larger than $0.1\,P_{n_0}$. For lower P_n values it is satisfactory to neglect the axial force and design the section as a member subjected to biaxial bending.

3B.5.7 ACI Code Requirements

The ACI code requires the following limitations on dimensions, reinforcing steel, and lateral restraint:

1. The percentage of longitudinal reinforcement should not be less than 1% nor greater than 8% of the gross cross-sectional area of the column. Usually for practical considerations the percentage of reinforcement does not exceed 4%.
2. Ties provided shall not be less than no. 3 for longitudinal bars no. 10 or smaller. They shall not be less than no. 4 for longitudinal bars larger than no. 10 and for bar bundles.
3. Tie spacing is restricted to the least of the following three values:
   a. Least lateral column dimension
   b. 16 times the diameter of the longitudinal bars
   c. 48 times the diameter of the tie
4. The spiral reinforcement size is determined from

$$\rho_s = \frac{\text{volume of spiral in one loop}}{\text{volume of concrete core for pitch}} = \frac{a_s \pi (D_c - d_b)}{\frac{1}{4}\pi \Delta_s^2 s}$$

where a_s = cross section of spiral bar
D_c = diameter of concrete core
d_b = diameter of spiral

$$(\rho_s)_{\min} = 0.45 \left(\frac{A_g}{A_c} - 1 \right) \frac{f'_c}{f_y}$$

where A_g = cross-sectional area
A_c = core cross section

3B.6 DEVELOPMENT OF REINFORCING BARS

3B.6.1 General

Reinforced concrete is a composite material where the compressive stresses are resisted by concrete and the tensile stresses are resisted by steel reinforcement. For this mechanism to work properly, a bond or a force transfer must exist between the two materials. The bond strength is controlled by several factors: (1) adhesion between concrete and steel reinforcement, (2) frictional resistance between steel and surrounding concrete, (3) shear interlock between bar deformations and concrete, (4) concrete strength in tension and compression, and (5) the geometrical characteristics of the steel bar—diameter, deformation spacing, and deformation height. The actual bond stress along the length of a steel bar embedded in concrete and subjected to tension varies with its location and the crack pattern. For this reason the ACI code uses the concept of development length rather than bond stress to ensure an adequate anchorage. Development length is defined as the minimum length of a bar in which the bar stress can increase from zero to the yield stress f_y. Hence a shorter embedded length will result in the bar pulling out of the concrete. The ACI code specifies different development length for steel bars in tension and in compression.

3B.6.2 Tension Development Length

A basic development length l_{db} is determined according to the following ACI code equations. Its value, however, should not be less than 12 in (0.3 m).

- No. 11 bars and smaller and deformed wire,

$$l_{db} = \frac{0.04 A_b f_y}{\sqrt{f'_c}} \tag{3B.85}$$

- No. 14 bars,

$$l_{db} = \frac{0.085 f_y}{\sqrt{f'_c}} \tag{3B.86}$$

- No. 18 bars,

$$l_{db} = \frac{0.125 f_y}{\sqrt{f'_c}} \tag{3B.87}$$

where A_b is the cross-sectional area of the bar, and $\sqrt{f'_c}$ shall not be taken greater than 100. Values of l_{db} are given in Table 3B.3.[6]

The basic development length is multiplied by a series of multipliers given in the ACI code, secs. 12.2.3, 12.2.4, and 12.2.5, to obtain the necessary development length l_d. These multipliers account for bar spacing, amount of cover, transverse reinforcement, reinforcement location, concrete unit weight, coating of steel reinforcement, and amount of reinforcement.

For development length multipliers

1. Compute $l_d = \lambda_d l_{db}$, where λ_d is from *a*, *b*, or *c*:
   a. $\lambda_d = 1$ if (1) clear spacing $s = 3d_b$ or more and stirrups are used with minimum ACI code cover requirements, or (2) bars in inner layer of slab or wall with clear spacing $s \geq 3d_b$, or (3) bars with cover $\geq 2d_b$ and clear spacing $\geq 3d_b$.
   b. $\lambda_d = 2$ for bars with cover $< d_b$ or clear spacing $< 2d_b$.
   c. $\lambda_d = 1.4$ for bars not covered by *a* or *b*.
2. Reduce the multiplier λ_d by multiplying it by
   a. 0.8 for no. 11 bars and smaller if clear spacing $\geq 5d_b$ and cover $\geq 2.5\, d_b$, and
   b. 0.75 for reinforcement enclosed within spiral reinforcement of diameter $\geq \tfrac{1}{4}$ in and pitch ≤ 4 in, or ties of no. 4 or more with spacing ≤ 4 in.
3. The resulting development length as modified in **1** and **2** should not be taken less than $0.03 d_b f_y / \sqrt{f'_c}$, with the value of $\sqrt{f'_c}$ not to exceed 100.
4. The following additional multipliers λ_{dd} are applied for special conditions to obtain the development length $l_d = \lambda_d \lambda_{dd} l_{db}$:

TABLE 3B.3 Basic Tension Development Length*

Bar no.	f'_c = 3000 psi (20.7 MPa)			f'_c = 3750 psi (25.9 MPa)			f'_c = 4000 psi (27.6 MPa)			f'_c = 5000 psi (34.5 MPa)			f'_c = 6000 psi (41.4 MPa)		
	Bottom bar	Top bar	Lower limit†	Bottom bar	Top bar	Lower limit†	Bottom bar	Top bar	Lower limit†	Bottom bar	Top bar	Lower limit†	Bottom bar	Top bar	Lower limit†
						f_y = 60,000 psi, normal = weight concrete									
3	4.8	6.3	12.3	4.3	5.6	11	4.2	5.4	11	3.7	4.9	9.5	3.4	4.4	8.7
4	8.8	11.4	16.4	7.8	10.2	15	7.6	9.9	14	6.8	8.8	13	6.2	8.1	12
5	14	18	21	12	16	18	12	15.5	18	10.5	14	16	9.6	12.5	15
6	19	25	25	17	22	22	18	23	21	15	20	19	14	18	18
7	26	34	29	24	31	26	23	30	25	20	27	22	19	24	20
8	35	45	33	31	40	29	30	39	29	27	35	25	25	32	23
9	44	57	37	39	51	33	38	49	32	34	44	29	31	40	26
10	56	72	42	50	65	37	48	62	36	43	56	32	39	51	30
11	68	89	46	61	80	46	59	77	44	53	69	36	48	63	33
14	93	121	56	83	108	50	81	105	48	72	94	43	66	86	39
18	137	178	74	122	159	66	119	154	64	106	138	57	97	126	52
						f_y = 40,000 psi, normal-weight concrete									
3	3.2	4.2	8.2	2.9	3.7	7.4	2.8	3.6	7.1	2.5	3.2	6.4	2.3	3	5.8
4	5.8	7.6	11	5.2	6.8	9.8	5.1	6.6	9.5	4.5	5.9	8.5	4.1	5.4	7.7
5	9	12	14	8.1	10.5	12	7.8	10.2	12	7.0	9.1	10.6	6.4	8.3	9.7
6	13	17	17	11.5	15	15	11.1	14.5	14.5	10	13	13	9.1	11.8	12

*$l_d = l_{db} \times$ factors in ACI Sec. 12.2.3 but not less than lower limit $\times$ factors in Secs. 12.2.4 and 12.2.5 but not less than 12 in (0.3 m).
Source: From MacGregor.[6]

a. $\lambda_{dd} = 1.3$ for top reinforcement, that is, horizontal reinforcement with more than 12 in of concrete cast below the bars.
   b. $\lambda_{dd} = 1.3$ for lightweight concrete. When f_{ct} is specified, use $6.7\sqrt{f'_c}/f_{ct}$, where f_{ct} is the splitting tensile strength of concrete.
   c. For epoxy-coated reinforcement (1) when cover $<3d_b$ or clear spacing between bars $<6d_b$, use $\lambda_{dd} = 1.5$; (2) for other conditions use $\lambda_{dd} = 1.2$.
   d. Multiply by the excess reinforcement ratio,

   $$\lambda_{dd} = \frac{A_s(\text{required})}{A_s(\text{provided})}$$

3B.6.3 Compression Development Length

The basic development length l_{db} is computed according to the ACI code from

$$l_{db} = \frac{0.02 d_b f_y}{\sqrt{f'_c}} \geq 0.0003 d_b f_y \tag{3B.85}$$

Then the following multipliers are applied to obtain $l_d = \lambda_d l_{db}$:

1. For excess reinforcement, $\lambda_d = A_s(\text{required})/A_s(\text{provided})$.
2. For spirally enclosed reinforcement, $\lambda_d = 0.75$.

The minimum total development length should be greater than 8 in (200 m).

3B.7 REFERENCES

1. ACI Committee 318, "Building Code Requirements for Reinforced Concrete," ACI 318-89; also "Commentary on Building Code Requirements for Reinforced Concrete," American Concrete Institute, Detroit, Mich., 1989.
2. J. C. McCormac, *Design of Reinforced Concrete,* 3d ed., HarperCollins, New York, 1993.
3. E. G. Nawy, *Reinforced Concrete: a Fundamental Approach,* 2d ed., Prentice-Hall, Englewood Cliffs N.J., 1990.
4. C. K. Wang and C. G. Salmon, *Reinforced Concrete Design,* 5th ed., HarperCollins, New York, 1992.
5. B. Bresler, "Design Criteria for Reinforced Concrete Columns under Axial Load and Biaxial Bending," *Journal of the American Concrete Institute,* vol. 57, pp 481–490, November 1960.
6. J. G. MacGregor, *Reinforced Concrete: Mechanics and Design,* 2d ed., Prentice-Hall, Englewood Cliffs, N.J., 1992.

SECTION 3C
FOUNDATIONS DESIGN

A. Samer Ezeldin

3C.1 INTRODUCTION 3.121
3C.2 FOOTING TYPES 3.121
3C.3 BEARING CAPACITY OF SOILS UNDER SHALLOW FOUNDATIONS 3.122
3C.4 TYPES OF FAILURE OF FOOTINGS 3.124
   3C.4.1 Diagonal Tension Failure 3.124
   3C.4.2 One-Way Shear Failure 3.124
   3C.4.3 Flexure Failure 3.124
   3C.4.4 Additional Design Aspects 3.126
3C.5 RECTANGULAR FOOTINGS 3.131
3C.6 ECCENTRICALLY LOADED SPREAD FOOTINGS 3.135
3C.7 COMBINED FOOTINGS 3.135
3C.8 MAT FOUNDATIONS 3.141
   3C.8.1 Introduction 3.141
   3C.8.2 Types of Mat Foundations 3.142
   3C.8.3 Design Methods 3.142
3C.9 FOUNDATIONS IN COLD REGIONS 3.145
3C.10 FOUNDATIONS IN EARTHQUAKE REGIONS 3.147
   3C.10.1 General 3.147
   3C.10.2 Dynamic Properties of Soils 3.148
   3C.10.3 Design Considerations 3.150
3C.11 REFERENCES 3.152

3C.1 INTRODUCTION

Footings are structural elements that transfer the loads from a structure above ground surface (superstructure) to the underlying soil. The soil-carrying capacity is in general much lower than the high stress intensities carried by the columns and walls in the superstructure. Hence the footings (substructure or foundations) can be considered as interface elements that spread the high-intensity stresses in the supporting elements to much lower stress levels along the weaker soil. This section will be limited to the design of foundations at a shallow depth. Design considerations for foundations in cold regions and in earthquake regions will be also presented.

3C.2 FOOTING TYPES

The most common types of footings are illustrated in Fig. 3C.1.

- Isolated spread footings are used beneath individual columns. They can be square or rectangular in shape. They spread the load of the column to the soil in two perpendicular directions.
- Strip footings or wall footings support bearing walls essentially in a one-dimensional action by cantilevering out on both sides of the wall.
- Combined footings are used to support two or more columns. Usually they have a rectangular or trapezoidal plan. Such footings are often used when a column is close to a property line.
- Pile caps are used to transmit the loads of columns or bearing walls to a series of piles. These piles transfer the loads from the upper poor soil layers to deeper and stronger soil layers.
- A mat or raft foundation is one large footing carrying the loads of all the columns of the structure. This type of foundation is used when weak soil layers are present but piles are not used.

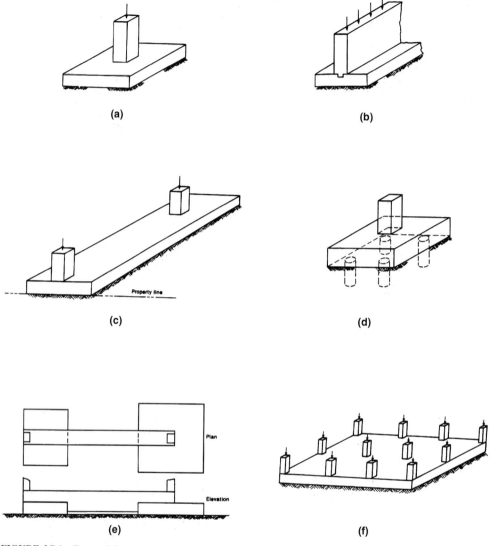

FIGURE 3C.1 Types of footings. (*a*) Spread footing. (*b*) Strip or wall footing. (*c*) Combined footing. (*d*) Pile cap. (*e*) Strap footing. (*f*) Mat or raft footing.

3C.3 BEARING CAPACITY OF SOILS UNDER SHALLOW FOUNDATIONS

In order to avoid a bearing failure of the footing, in which the soil beneath the footing moves downward and outward from under the footing, the service load stress under the footing must be limited. This limitation is provided by ensuring that the service load stress q_s is less than or equal to an allowable bearing capacity q_a,

$$q_s \leq q_a = \frac{q_{\text{ult}}}{FS} \tag{3C.1}$$

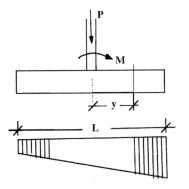

FIGURE 3C.2 Pressure distribution under footing.

where q_{ult} is the ultimate bearing capacity corresponding to failure of the footing, and FS is a factor of safety, usually taken to be 2.0 to 3.0. Soil mechanics principles are relied upon to establish the ultimate bearing capacity, which depends on the shape of the footing, its depth, the surcharge on top of the footing, the position of the underground water table, and the soil type. The allowable bearing capacity may vary from 15,000 psf for rock to 2000 psf for clay.

The soil beneath a footing is assumed to be subjected to a linearly elastic compression action. The service pressure distribution, shown in Fig. 3C.2, is obtained by the equation

$$q_s = \frac{P}{A} \pm \frac{My}{I} \qquad (3C.2)$$

where P = vertical load, positive in compression
A = area of contact surface between soil and footing (length of footing L × width of footing B)
I = moment of inertia of area A
M = moment about centroidal axis of area
y = distance from centroidal axis to point where stress is being calculated

In general tensile stresses are not acceptable underneath concrete footing.

The gross soil pressure is considered the pressure caused by the total load applied on a footing, including dead loads (structure, footing, and surcharge) and live loads. In Fig. 3C.3 the gross soil pressure is

$$q_{gross} = (h_f - h_c)\gamma_s + h_c\gamma_c + \frac{P}{A} \qquad (3C.3)$$

The gross soil pressure must not exceed the allowable bearing capacity q_a in order to avoid failure of the footing.

The net soil pressure is taken as the pressure that will cause internal forces in the footing. Considering Fig. 3C.3, the net soil pressure is

$$q_{net} = h_c(\gamma_c - \gamma_s) + \frac{P}{A} \qquad (3C.4)$$

The net soil pressure is used to calculate the flexural reinforcement and the shear strength of the concrete footing.

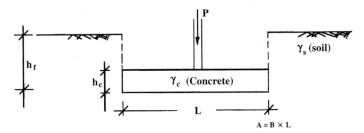

FIGURE 3C.3 Gross and net soil pressures.

3C.4 TYPES OF FAILURE OF FOOTINGS

Three different types of failure may occur in a concrete footing subjected to a concentrated load (Fintel, 1985; Winterkorn and Fang, 1975).

3C.4.1 Diagonal Tension Failure

This type of failure is also referred to as punching shear failure(Fig. 3C.4). The footing fails due to the formation of inclined cracks around the perimeter of the column. Test results have indicated that the critical section can be taken at $d/2$ from the face of the column. To avoid such a failure, the upward ultimate shearing force V_u increased by applying the strength reduction factor ϕ must be lower than the nominal punching shear strength V_c.

$$\frac{V_u}{\phi} \leq V_c \tag{3C.5}$$

V_u, acting on the tributary area shown in Fig. 3C.4, is computed with the load factors applied (see Table 3B.1) and ϕ taken as 0.85. V_c is taken as the smallest of

$$V_c = \left(2 + \frac{4}{\beta_c}\right)\sqrt{f'_c}\,b_0 d \tag{3C.6}$$

$$V_c = \left(\frac{\alpha_s}{b_0/d} + 2\right)\sqrt{f'_c}\,b_0 d \tag{3C.7}$$

$$V_c = 4\sqrt{f'_c}\,b_0 d \tag{3C.8}$$

where b_0 = perimeter of critical section taken at $d/2$ from face of column
d = depth at which tension steel reinforcement is placed
β_c = ratio of long side to short side of column section
α_s = 40 for interior columns, 30 for edge columns, and 20 for corner columns

3C.4.2 One-Way Shear Failure

The footing fails due to the formation of inclined cracks that intercept the bottom of the slab at a distance d from the face of the column (Fig. 3C.5). For footings carrying columns with steel base plates, the distance d is measured from a line halfway between the face of the column and the edge of the base plate.

In order to avoid such a failure, Eq. (3C.5) must be satisfied. V_u is the upward ultimate shearing force acting on the tributary area shown in Fig. 3C.5 and ϕ is taken to be 0.85. V_c is taken in accordance with the ACI code as

$$V_c = 2\sqrt{f'_c}\,Bd \tag{3C.9}$$

3C.4.3 Flexure Failure

A moment M_u/ϕ is acting at the face of column, as shown in Fig. 3C.6, where

$$M_u = (q_{n_u} BX)\frac{X}{2} \tag{3C.10}$$

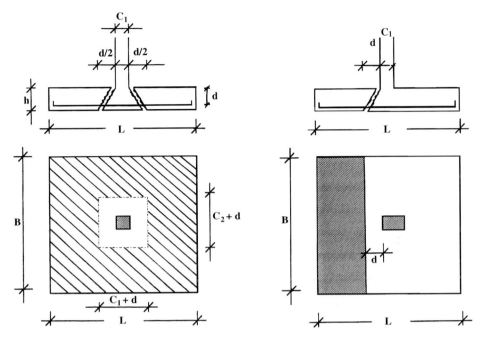

FIGURE 3C.4 Diagonal tension failure.

FIGURE 3C.5 One-way shear failure.

and $\phi = 0.9$. The factored net soil pressure q_{nu} is obtained by dividing the factored applied loads on the footing by its area.

The moment M_u/ϕ must be lower than or equal to the nominal strength of the concrete section having an effective depth d, a width b, and reinforced with tension steel A_s. Thus

$$\frac{M_u}{\phi} \leq M_n \tag{3C.11}$$

where

$$M_n = A_s f_y \left(d - \frac{a}{2} \right) \tag{3C.12}$$

$$a = \frac{A_s f_y}{0.85 f'_c B} \tag{3C.13}$$

In a similar manner, the moment M_u/ϕ acting at the perpendicular face of the column must be resisted by the tension reinforcement layer placed orthogonally, resulting in two layers of steel, one in each direction. ACI requires the minimum steel reinforcement placed in structural slabs of uniform thickness to be

$$(A_s)_{min} = 0.002bh, \quad \text{for } f_y = 40 \text{ or } 50 \text{ ksi (276 or 345 MPa)} \tag{3C.14}$$

$$(A_s)_{min} = 0.0018bh, \quad \text{for } f_y = 60 \text{ ksi (414 MPa)} \tag{3C.15}$$

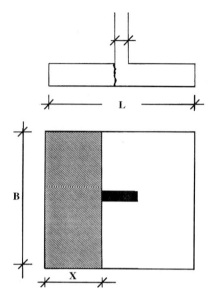

FIGURE 3C.6 Flexure failure.

3C.4.4 Additional Design Aspects

3C.4.4.1 Development of Reinforcement. The flexural reinforcement is provided in the footing with the assumption that the reinforcement stress reaches the yield stress f_y at the face of the column. In order to ensure that, the reinforcement must be extended beyond the critical section to develop this stress. This implies that a development length l_d must be provided from the critical section. The ACI code development length requirements for different bar diameters were presented in Sec. 3B.6.

3C.4.4.2 Load Transfer from Column to Footing. The ACI code requires that the forces acting on the column be safely transmitted to the footing. Dowels in steel connection are used to transfer any tension forces whereas the compression forces are transferred by bearing.

The bearing capacity of the column is checked by

$$\frac{P_u}{\phi} \leq 0.85 f'_c A_1 \tag{3C.16}$$

where $\phi = 0.7$
 P_u = ultimate load applied on column
 A_1 = the column area

The bearing capacity of the concrete footing is checked by

$$\frac{P_u}{\phi} \leq 0.85 f'_c A_1 \sqrt{\frac{A_2}{A_1}} \tag{3C.17}$$

where A_2 is the maximum area of the supporting surface that is geometrically similar and concentric with A_1. The value of $\sqrt{A_2/A_1}$ should not be greater than 2.

Example 3C.1: Design of Square Spread Footing Design an interior spread footing to carry a service load of 500 kips (2225 kN) and a service live load of 350 kips (1558 kN) from a 20-in (508-mm)-

square tied column containing no. 11 bars [1.56 in² (960 mm²)] as the principal column steel. The top of the footing will be covered with 12 in (305 mm) of fill having a density of 110 lb/ft³ (1337 kg/m³) and a 6-in (152-mm) basement floor. The basement floor loading is 100 psf (4.78 kPa). The allowable bearing pressure on the soil q_a is 7000 psf (335 kPa). Use f'_c = 5000 psi (34.45 MPa) and f_y = 60,000 psi (413.4 MPa).

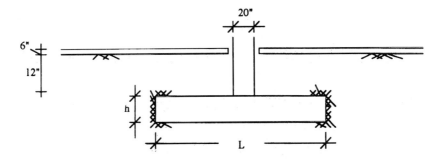

Solution

1. Estimate the thickness of the footing as between one and two times the width of the column, say $h = 36$ in (914 mm). The allowable net soil pressure is

$$q_{net} = 7 \text{ ksf} - (\text{weight of footing} + \text{soil} + \text{floor} + \text{floor load})$$

$$= 7 - \left(\frac{36}{12} \times 0.15 + 1 \times 0.11 + 0.5 \times 0.15 + 0.1\right) = 6.265 \text{ ksf (299.8 kPa)}$$

2. Required area = $\dfrac{P_D + P_L}{q_{net}} = \dfrac{500 + 350}{6.265} = 135.7 \text{ ft}^2 \text{ (12.9 m}^2\text{)}$

   Try a 12-ft (3.6-m) square by 36-in (914-mm)-thick footing.

3. The factored net soil pressure is obtained from

$$q_{n_u} = \frac{1.4 \times 500 + 1.7 \times 350}{12^2} = 9 \text{ ksf (430.6 kPa)}$$

4. The two-way shear check is performed on the critical section at the distance $d/2$ from the face of the column,

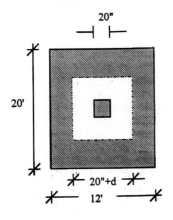

$d = h -$ concrete cover $-$ bar diameter

$= 36 \text{ in} - 3 \text{ in} - 1 \text{ in} = 32 \text{ in (813 mm)}$

$V_u = q_{n_u}(\text{tributary area}) = 9\left[12^2 - \left(\dfrac{20 + 32}{12}\right)^2\right]$

$= 1127$ kips (5015 kN)

$\dfrac{V_u}{\phi} = \dfrac{1127}{0.85} = 1326$ kips (5900.7 kN)

V_c is the smallest of

$$\left(\frac{2+4}{\beta_c}\right)\sqrt{f'_c}b_o d = \frac{(2+4/1)\sqrt{5000}\times 52\times 4\times 32}{1000} = 2824 \text{ kips (12,567 kN)}$$

$$\left(\frac{\alpha_s}{b_o/d}+2\right)\sqrt{f'_c}b_o d = \frac{[40/(52\times 4/32)+2]\sqrt{5000}\times 52\times 4\times 32}{1000} = 3836 \text{ kips (17,070 kN)}$$

$$4\sqrt{f'_c}b_o d = \frac{4\sqrt{5000}\times 52\times 4\times 32}{1000} = 1883 \text{ kips (8379 kN)} \quad \therefore \text{ Controls design.}$$

Hence

$$V_c = 1883 \text{ kips (8379 kN)} > \frac{V_u}{\phi} = 1326 \text{ kips (5900.7 kN)}$$

The thickness of the footing is adequate to prevent two-way shear failure.

5. The one-way shear is performed on a critical section at the distance d from the face of the column. The width of the tributary area is

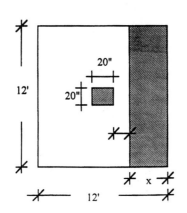

$$x = \frac{144-20}{2} - 32 = 30 \text{ in (762 mm)}$$

$$V_u = q_{n_u}(\text{tributary area}) = 9\left(12\times \frac{30}{12}\right) = 270 \text{ kips (1201 kN)}$$

$$\frac{V_u}{\phi} = \frac{270}{0.85} = 318 \text{ kips (1415 kN)}$$

$$V_c = 2\sqrt{f'_c}bd = \frac{2\sqrt{5000}\times 12\times 12\times 32}{1000}$$

$$= 652 \text{ kips (2901 kN)}$$

Hence

$$V_c = 652 \text{ kips (2901 kN)} > \frac{V_u}{\phi} = 318 \text{ kips (1415 kN)}$$

The thickness of the footing is capable of preventing one-way shear failure.

6. Design for flexure reinforcement. The width of the tributary area is

$$y = \left(\frac{144-20}{2}\right) = 62 \text{ in (1575 mm)}$$

$$M_u = 9\left(12\times \frac{62}{12}\times \frac{62}{2\times 12}\right) = 1442 \text{ ft}\cdot\text{kip (1955 kN}\cdot\text{m)}$$

$$\frac{M_u}{\phi} = \frac{1442}{0.9} = 1602 \text{ ft}\cdot\text{kips (2172.6 kN}\cdot\text{m)}$$

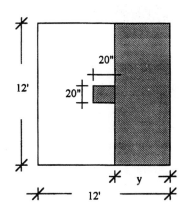

$$M_n = A_s f_y \left(d - \frac{a}{2}\right)$$

Assuming $(d - a/2) = 0.9d$,

$$(A_s)_{req} = \frac{M_u/\phi}{f_y(0.9d)} = \frac{1602 \times 12{,}000}{60{,}000(0.9 \times 32)} = 11 \text{ in}^2 \text{ (7095 mm}^2\text{)}$$

$$(A_s)_{min} = 0.0018bh = 0.0018(144 \times 36) = 9.3 \text{ in}^2 \text{ (5998 mm}^2\text{)}$$

Choose 11 no. 9 each way; then $A_s = 11 \text{ in}^2 > (A_s)_{min}$.
Check the chosen area of the reinforcement:

$$a = \frac{A_s f_y}{0.85 f'_c b} = \frac{11 \times 60{,}000}{0.85 \times 5000 \times 144} = 1.08 \text{ in (27.4 mm)}$$

Then

$$M_n = A_s f_y \left(d - \frac{a}{2}\right) = 11 \times 60{,}000 \left(32 - \frac{1.08}{2}\right) = 20.8 \times 10^6 \text{ in} \cdot \text{lb (2350 kN} \cdot \text{m)}$$

$$= 1730 \text{ ft} \cdot \text{kips (2350 kN} \cdot \text{m)} > \frac{M_u}{\phi} = 1602 \text{ ft} \cdot \text{kips (2172.6 kN} \cdot \text{m)}$$

7. Check the development length:

$$l_{db} = 0.04 \frac{A_b f_y}{\sqrt{f'_c}} = 0.04 \frac{1.0 \times 60{,}000}{\sqrt{5000}} = 33.94 \text{ in} \cong 34 \text{ in (864 mm)}$$

No increase in the basic development length l_d is needed to account for the effects of bar spacing, cover, stirrup confinement, and reinforcement location.

$$(l_d)_{min} = 0.03 d_b \frac{f_y}{\sqrt{f'_c}} = 0.03 \times 1.125 \frac{60{,}000}{\sqrt{5000}} = 28.7 \text{ in (729 mm)}$$

Hence choose $l_d = 34$ in (864 mm).
The bar length available from the location of the maximum moment on each side is

$$y - \text{concrete cover} = 62 - 3 = 59 \text{ in (1499 mm)} > 34 \text{ in (864 mm)}$$

Therefore the development length is provided.

8. Check the bearing at the column-footing interface:

$$P_u = 1.4 \times 500 + 1.7 \times 350 = 1295 \text{ kips (5763 kN)}$$

$$\frac{P_u}{\phi} = \frac{1295}{0.7} = 1850 \text{ kips (8233 kN)}$$

Footing capacity:

$$P_n = 0.85 f'_c A_1 \sqrt{\frac{A_2}{A_1}}$$

FUNDAMENTALS OF FOUNDATION CONSTRUCTION AND DESIGN

$$\sqrt{\frac{A_2}{A_1}} = \sqrt{\frac{144 \times 144}{20 \times 20}} = 7.2 \quad \therefore \text{Use } \sqrt{\frac{A_2}{A_1}} = 2.$$

Then

$$P_n = \frac{0.85 \times 5000 \times 20 \times 20 \times 2}{1000} = 3400 \text{ kips } (15{,}130 \text{ kN})$$

$$P_n = 3400 \text{ kips } (15{,}130 \text{ kN}) > \frac{P_u}{\phi} = 1850 \text{ kips } (8233 \text{ kN}) \quad \text{O.K.}$$

Column capacity:

$$P_n = 0.85 f'_c A_1 = \frac{0.85 \times 5000 \times 20 \times 20}{1000} = 1700 \text{ kips } (7565 \text{ kN})$$

$$P_n < \frac{P_u}{\phi}$$

Hence dowels are needed to transfer the excess load.

$$\text{Area of dowel required} = \frac{1850 - 1700}{f_y} = \frac{150}{60} = 2.5 \text{ in}^2 \ (1613 \text{ mm}^2)$$

The area of the dowel must be higher than the minimum specified by the ACI code,

$$(\text{Area of dowel})_{\min} = 0.005 A_g = 0.005 \times 20 \times 20 = 2 \text{ in}^2 \ (1290 \text{ mm}^2) < 2.5 \text{ in}^2 \ (1613 \text{ mm}^2)$$

Hence the value of 2.5 in² controls. Choose 4 no. 8 bars [3.16 in² (2038 mm²)]. The dowels must extend at least the compression development length of the 8 bar into the footing,

$$l_{d_b} = 0.02 d_b \frac{f_y}{\sqrt{f'_c}} \geq 0.0003 d_b f_y$$

$$l_{d_b} = 15 \text{ in } (381 \text{ mm}) < 16 \text{ in } (406 \text{ mm})$$

Hence extend 4 no. 8 dowels at least 16 in (406 mm) into the footing.

The complete design is detailed here.

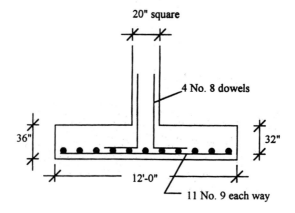

3C.5 RECTANGULAR FOOTINGS

Rectangular footings are usually employed as spread footings when the space is inadequate for a square footing. The design procedures for these footings are basically similar to those of square footings, except that the one-way shear and bending moments have to be checked in both principal directions. Also in such footings the flexural reinforcement in the short direction has to be distributed in three regions with more concentration in the region beneath the column (Fig. 3C.7). The total required reinforcement A_s is obtained such that the bending moment at the column face (section A–A) is resisted. The reinforcement in the central region under the column shall be $A_s[2/(ß + 1)]$, where ß is the ratio of the long side of the footing to the short side. The remaining reinforcement is distributed equally between the two outer regions of the footing.

Example 3C.2: Design of Rectangular Footing Redesign the footing of Example 3C.1, given that the maximum width of the footing cannot exceed 10 ft (3 m).

Solution

1. From the solution of Example 3C.1, take $h = 36$ in (914 mm) and $d = 32$ in (813 mm). Then

$$q_{net} = 6.265 \text{ ksf } (299.8 \text{ kPa})$$

2.
$$\text{Required area} = \frac{P_D + P_L}{q_{net}} = \frac{500 + 350}{6.265} = 135.7 \text{ ft}^2 \text{ (12.9 m}^2\text{)}$$

$$\text{Required length} = \frac{135.7}{10} = 13.57 \text{ ft (4.2 m)} \quad \text{Take } L = 14 \text{ ft (4.25 m)}.$$

Try a footing 10 ft (3 m) wide by 14 ft (4.25 m) long by 36 in (914 mm) thick.

3. The factored net soil pressure is

$$q_{n_u} = \frac{1.4 \times 500 + 1.7 \times 350}{10 \times 14} = 9.25 \text{ ksf } (442.6 \text{ kPa})$$

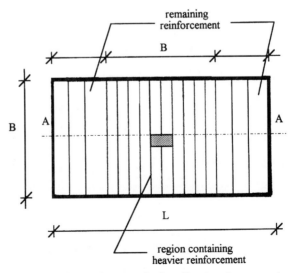

FIGURE 3C.7 Reinforcement in short direction of a rectangular footing.

4. Two-way shear analysis:

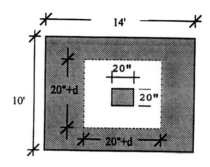

$$V_u = q_{n_u} \text{(tributary area)} = 9.25\left[140 - \left(\frac{20+32}{12}\right)^2\right] = 1121 \text{ kips (4988 kN)}$$

$$\frac{V_u}{\phi} = \frac{1121}{0.85} = 1319 \text{ kips (5869 kN)}$$

$$V_c = 4\sqrt{f'_c}b_0d = \frac{4\sqrt{5000} \times 52 \times 4 \times 32}{1000} = 1883 \text{ kips (8379 kN)}$$

Hence

$$V_c = 1883 \text{ kips (8379 kN)} > \frac{V_u}{\phi} = 1319 \text{ kips (5869 kN)}$$

The thickness of the footing is adequate to prevent two-way shear failure.

5. The one-way shear is performed along two sections.
   a. Section A–A:

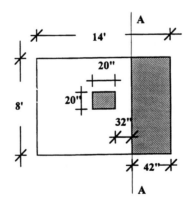

$$V_u = 9.25\left(8 \times \frac{42}{12}\right) = 259 \text{ kips (1153 kN)}$$

$$\frac{V_u}{\phi} = \frac{259}{0.85} = 305 \text{ kips (1356 kN)}$$

$$V_c = 2\sqrt{f'_c}\,bd = \frac{2\sqrt{5000} \times 8 \times 12 \times 32}{1000} = 434 \text{ kips (1931 kN)}$$

Hence

$$V_c = 434 \text{ kips (1931 kN)} > \frac{V_u}{\phi}\, 305 \text{ kips (1356 kN)}$$

b. Section B–B:

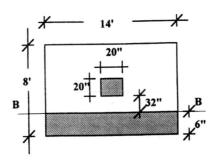

$$V_u = 9.25\left(14 \times \frac{6}{12}\right) = 64.75 \text{ kips (289 kN)}$$

$$\frac{V_u}{\phi} = \frac{64.75}{0.85} = 76 \text{ kips (340 kN)}$$

$$V_c = \frac{2\sqrt{5000} \times 14 \times 12 \times 32}{1000} = 760 \text{ kips (3382 kN)}$$

The thickness of the footings is capable of preventing two-way shear failure in both directions.

6. Design for flexure reinforcement

a. Section A–A (Long direction):

$$M_u = 9.25\left(10 \times \frac{74}{12} \times \frac{74}{2 \times 12}\right) = 1759 \text{ ft} \cdot \text{kips (2385 kN} \cdot \text{m)}$$

$$\frac{M_u}{\phi} = \frac{1759}{0.9} = 1954 \text{ ft} \cdot \text{kips (2650.2 kN} \cdot \text{m)}$$

$$(A_s)_{\text{req}} = 13.6 \text{ in}^2 \text{ (8772 mm}^2\text{)}$$

$$(A_s)_{\text{min}} = 0.0018bh = 0.0018 \times 10 \times 12 \times 36 = 7.8 \text{ in}^2 \text{ (5031 mm}^2\text{)} < (A_s)_{\text{req}}$$

Choose 14 no. 9 [(14 in²)(9030 mm²)] in the long direction.

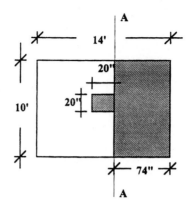

b. Section B–B (short direction):

$$M_u = 9.25\left(14 \times \frac{50}{12} \times \frac{50}{2 \times 12}\right) = 1124 \text{ ft} \cdot \text{kip } (1524 \text{ kN} \cdot \text{m})$$

$$\frac{M_u}{\phi} = \frac{1124}{0.9} = 1249 \text{ ft} \cdot \text{kips } (1693.5 \text{ kN} \cdot \text{m})$$

$(A_s)_{req} = 8.8 \text{ in}^2 \ (5676 \text{ mm}^2)$

$(A_s)_{min} = 0.0018 bh = 0.0018 \times 14 \times 12 \times 36 = 10.9 \text{ in}^2 \ (7031 \text{ mm}^2)$ ∴ Controls.

Choose $A_s = 10.9 \text{ in}^2 \ (7031 \text{ mm}^2)$ in the short direction.

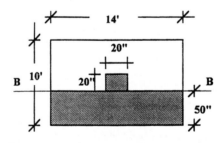

In the 10-ft inner region provide

$$10.9\left(\frac{2}{\beta + 1}\right) = 10.9\left(\frac{2}{14/10 + 1}\right) = 9.08 \text{ in}^2 \ (5857 \text{ mm}^2) \quad (12 \text{ no. } 8)$$

In the 2-ft outer regions provide

$$\frac{10.9 - 9.08}{2} = 0.91 \text{ in}^2 \ (587 \text{ mm}^2) \text{ on each side} \quad (2 \text{ no. } 8)$$

The checks for the development length and the bearing at the column-footing interface are similar to those in Example 3C.1. The details of the final design are shown here.

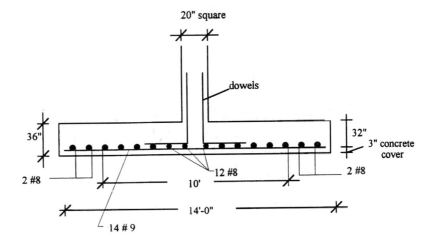

3C.6 ECCENTRICALLY LOADED SPREAD FOOTINGS

In some cases, due to a moment at the column base or an eccentrically applied load, the bearing pressure beneath the footing will deviate from the uniform distribution shown in Fig. 3C.2. The design of such a footing can be performed in a manner similar to that of a square or rectangular footing with the following conditions satisfied:

1. Tensile stresses are not generated beneath the footing under extreme loading conditions.
2. The difference in compressive stresses between the two edges of the footing is not extremely high in order to avoid tilting settlement of the footing.
3. The designs for one-way shear, two-way shear, and bending moment are performed using the actual pressures under the footing resulting from critical loading conditions that might occur.

3C.7 COMBINED FOOTINGS

A combined footing is usually used when an exterior column is close to a property line, preventing the use of an isolated spread footing (see Fig. 3C.1). Thus a combined footing is used to support the exterior column along with an interior column. The shape of a combined footing is usually rectangular or trapezoidal. That shape is carefully designed in order to have the centroid of the footing coincide with the resultant of the column loads applied to the footing. For cases where the load is lower on the exterior column P_{ext} than on the interior column P_{int} a rectangular combined footing is considered an economical solution. In cases when $0.5 < P_{int}/P_{ext} < 1$ a trapezoidal footing is preferred. However, when $P_{int}/P_{ext} < 0.5$, a strip or cantilever footing should be considered. In a strip or cantilever footing, the overturning of the exterior footing is prevented by connecting it with an adjacent interior footing using a strip beam. The exterior footing is designed for one-way bending whereas the interior footing is designed for two-way bending, as in isolated footings. The strip beam is subjected to a constant shear force and a linearly decreasing negative moment. This behavior is similar to a cantilever beam. It is preferable that all three elements, namely, the exterior footing, the interior footing, and the strip beam, have the same thickness. This thickness is chosen such that the shear requirement for the footings and the shear and flexure requirements for the strip beam are satisfied.

Example 3C.3: Design of a Combined Footing Design a combined rectangular footing to support two columns. The exterior column is 20 in (508 mm) square, carrying service loads of 150 kips (667.5 kN) dead load and 120 kips (534 kN) live load. The interior column is 22 in (559 mm) square carrying service loads of 200 kips (890 kN) dead load and 180 kips (801 kN) live load. The distance between the columns is 15 ft (4.6 m) centerline to centerline. The top of the footing is 3 ft (914 mm) below grade and the fill above the footing is 120 lb/ft³ (1459 kg/m³). Use $f'_c = 4000$ psi (27.56 MPa) and $f_y = 60,000$ psi (413.4 MPa).

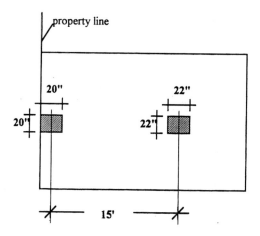

Solution

1. Estimate the depth of the footing to be one to two times the column dimension. Take $h = 36$ in (914 mm). Hence

$$d = h - \text{cover} - \text{bar diameter} = 36 - 3 - 1 = 32 \text{ in (813 mm)}$$

The allowable net soil pressure is

$$q_{net} = 5 - (\text{weight of footing} + \text{soil}) = 5 - (3 \times 0.15) - (3 \times 0.12) = 4.19 \text{ ksf (200.5 kPa)}$$

2. $$\text{Required area} = \frac{P_D + P_L}{q_{net}} = \frac{(150 + 200) + (120 + 180)}{4.19} = 155 \text{ ft}^2 \text{ (14.73 m}^2\text{)}$$

The distance of the center of gravity of loads from the exterior column is

$$\frac{(150 + 120)0 + (200 + 180)15}{(150 + 120) + (200 + 180)} = 8.77 \text{ ft (2.67 m)}$$

The distance from the property line to the center of gravity is

$$\frac{10 \text{ in}}{12} + 8.77 \text{ ft} = 9.6 \text{ ft (2.93 m)}$$

Length of footing $= 2 \times 9.6$ ft $= 19.2$ ft (5.85 m) ∴ Say 19.5 ft (5.9 m).

Width of footing $= \dfrac{155}{19.5} = 7.95$ ft (24 m) ∴ Say 8 ft (2.5 m).

Try a 19.5 × 8 ft (5.9 × 2.5 m) rectangular footing with 36-in (914-mm) thickness.

3. The factored net soil pressure is

$$q_{n_u} = \frac{1.4(150 + 200) + 1.7(120 + 180)}{19.5 \times 8} = 6.4 \text{ ksf (306.25 kPa)}$$

$$= 6.4 \times 8 = 51.2 \text{ kips/ft (7.47 kN/m)}$$

4. Using q_{n_u}/ft, determine the factored bending moment and shearing force diagrams for the footing. These diagrams are plotted here for the full 8-ft (2.5-m) width of the footing.

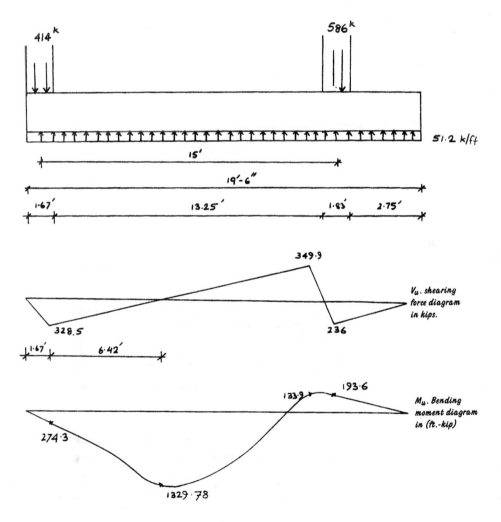

5. Check one-way shear: V_u at a distance d from the interior face of the inner column is

$$V_u = 349.9 - \frac{32}{12} \, 51.2 = 213.3 \text{ kips (949.5 kN)}$$

$$\frac{V_u}{\phi} = \frac{213.37}{0.85} = 251 \text{ kips } (1117 \text{ kN})$$

$$V_c = 2\sqrt{f'_c}\,bd = 2\sqrt{4000} \times 8 \times 12 \times 32 = 388.6 \text{ kips } (1729.3 \text{ kN})$$

Hence,

$$V_c = 388.6 \text{ kips } (1729.3 \text{ kN}) > \frac{V_u}{\phi} = 251 \text{ kips } (1117 \text{ kN})$$

The thickness of the footing is adequate to prevent one-way shear failure.

6. Check of two-way shear
   a. Exterior column:

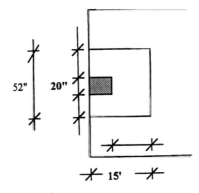

$$V_u = 414 - q_{nu}\left(\frac{52}{12}\right)\left(\frac{36}{12}\right) = 414 - 6.4 \times 4.33 \times 3 = 330.9 \text{ kips } (1472.5 \text{ kN})$$

$$\frac{V_u}{\phi} = \frac{330.9}{0.85} = 389.3 \text{ kips } (1732 \text{ kN})$$

$$b_0 = 2 \times 36 + 52 = 124 \text{ in } (3150 \text{ mm})$$

V_c is the smallest of

$$\frac{(2+4)\sqrt{4000} \times 124 \times 32}{1000} = 1506 \text{ kips } (6702 \text{ kN})$$

$$\left(\frac{30}{124/32} + 2\right)\sqrt{4000} \times 124 \times 32 = 2444 \text{ kips } (10876 \text{ kN})$$

$$4\sqrt{4000} \times 124 \times 32 = 1003 \text{ kips } (4463 \text{ kN}) \quad \therefore \text{ Controls design.}$$

Hence

$$\frac{V_u}{\phi} = 389.3 \text{ kips } (1732 \text{ kN}) < 1003 \text{ kips } (4463 \text{ kN})$$

The thickness is adequate for the exterior column.

b. Interior column:

$$V_u = 586 - 6.4\left(\frac{54}{12}\right)\left(\frac{54}{12}\right) = 456.4 \text{ kips (2031 kN)}$$

$$\frac{V_u}{\phi} = \frac{456.4}{0.85} = 537 \text{ kips (2389 kN)}$$

$$b_0 = 4 \times 54 = 216 \text{ in (5486 mm)}$$

$$V_c = 4\sqrt{4000} \times 216 \times 32 = 1749 \text{ kips (7783 kN)}$$

Hence

$$\frac{V_u}{\phi} = 537 \text{ kips (2389 kN)} < 1749 \text{ kips (7783 kN)}$$

The thickness is adequate for the interior column.

7. **Design for flexure reinforcement**
   a. Midspan negative moment:

$$\frac{M_u}{\phi} = \frac{1329.78}{0.9} = 1477.5 \text{ ft} \cdot \text{kips (2003.5 kN} \cdot \text{m)}$$

$$M_n = A_s f_y \left(d - \frac{a}{2}\right)$$

Assuming $(d - a/2) = 0.9d$,

$$(A_s)_{req} = \frac{M_u/\phi}{f_y(0.9d)} = \frac{1477.5 \times 12,000}{60,000 \times 0.9 \times 32} = 10.26 \text{ in}^2 \text{ (6618 mm}^2\text{)}$$

$$(A_s)_{min} = 0.0018bh = 0.0018 \times 8 \times 12 \times 36 = 6.22 \text{ in}^2 \text{ (4012 mm}^2\text{)}$$

Choose 11 no. 9 $\therefore A_s = 11 \text{ in}^2 \text{ (7095 mm}^2\text{)} > (A_s)_{min}$
Checking the area of the reinforcement,

$$a = \frac{A_s f_y}{0.85 f'_c b} = \frac{11 \times 60,000}{0.85 \times 4000 \times 8 \times 12} = 2.0 \text{ in (50.8 mm)}$$

$$M_n = 11 \times 60,000 \left(32 - \frac{2}{2}\right) = 20.45 \times 10^6 \text{ in} \cdot \text{lb}$$

$$= 1704 \text{ ft} \cdot \text{kips (2311 kN} \cdot \text{m)} > 1477.5 \text{ ft} \cdot \text{kips (2003.49 kN} \cdot \text{m)}$$

Use 11 no. 9 top bars at midspan.

b. Interior column positive moment:

$$\frac{M_u}{\phi} = \frac{193.6}{0.9} = 215 \text{ ft} \cdot \text{kips (291.54 kN} \cdot \text{m)}$$

This would require $A_s = 2 \text{ in}^2 \text{ (1290 mm}^2\text{)}$, which is less than $(A_s)_{min} = 6.22 \text{ in}^2 \text{ (4012 mm}^2\text{)}$.
Use 7 no. 9 bottom bars for the interior column.

3.140 FUNDAMENTALS OF FOUNDATION CONSTRUCTION AND DESIGN

8. Design for transverse beams under columns. It is assumed that transverse beams under each column transmit the load from the longitudinal direction into the columns. The width of the transverse beam is taken to be the width of the column plus an extension d/2 on each side of the column.

   a. Transverse steel under interior column:

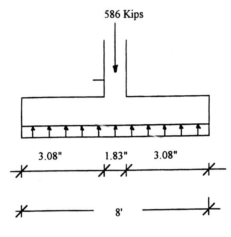

$$\text{Beam width} = 22 + 2\left(\frac{32}{2}\right) = 54 \text{ in } (1372 \text{ mm})$$

$$q_{n_u} = \frac{586}{8} = 73.25 \text{ kips/ft } (10.68 \text{ kN/m})$$

$$M_u = 73.25 \, \frac{3.08^2}{2} = 347.44 \text{ ft} \cdot \text{kips } (471.1 \text{ kN} \cdot \text{m})$$

$$\frac{M_u}{\phi} = \frac{347.44}{0.9} = 386.04 \text{ ft} \cdot \text{kips } (523.47 \text{ kN} \cdot \text{m})$$

This would require $A_s = 2.5 \text{ in}^2 \, (1613 \text{ mm}^2)$.

$$(A_s)_{\min} = 0.0018bh = 0.0018 \times 54 \times 36 = 3.5 \text{ in}^2 \, (2258 \text{ mm}^2) \qquad \therefore \text{ Controls.}$$

Select 6 no. 7 [3.6 in² (2322 mm²)]

b. Transverse steel under exterior column:

$$\text{Beam width} = 20 + \frac{32}{2} = 36 \text{ in } (914.4 \text{ mm})$$

$$q_{n_u} = \frac{414}{8} = 51.75 \text{ kips/ft } (7.55 \text{ kN/m})$$

$$M_u = 51.75 \, \frac{3.17^2}{2} = 260 \text{ ft} \cdot \text{kips } (352.6 \text{ kN} \cdot \text{m})$$

$$\frac{M_u}{\phi} = 289 \text{ ft} \cdot \text{kips } (392 \text{ kN} \cdot \text{m})$$

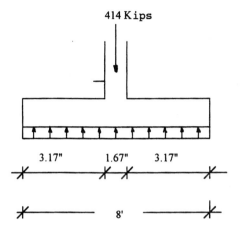

This would require $A_s = 1.9$ in² (1226 mm²).

$$(A_s)_{min} = 0.0018bh = 0.0018 \times 36 \times 36 = 2.33 \text{ in}^2 \text{ (1503 mm}^2\text{)}$$

Select 4 no. 7 [2.4 in² (1548 mm²)].

The checks for the development length and the bearing at the column-footing interface are similar to those in Example 3C.1 and will not be repeated here. The final details of the design are shown here.

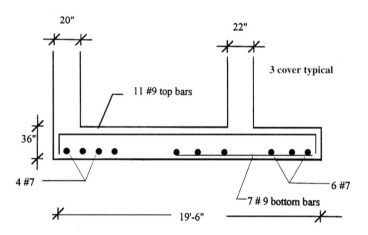

3C.8 MAT FOUNDATIONS

3C.8.1 Introduction

A mat foundation consists of a large concrete slab that supports the column of the entire structure (see Fig. 3C.1). It is generally used when the underlying soil has a low bearing capacity. The advantages of using mat foundations are (1) the applied pressure on the supporting soil is reduced because a larger area is used and (2) the bearing capacity of the supporting soil is increased because of the larger foundation depth. Mat foundations can also be used on rock exhibiting irregular compositions, creating weak regions. To overcome the differential settlements that could result from such nonhomogeneous behavior, the mat foundation presents a practical solution. Mat foundations also present an attractive solution to support structures and machinery sensitive to differential settlements.

3C.8.2 Types of Mat Foundations

The most common type of mat foundation is a flat concrete slab of uniform thickness (see Fig. 3C.1). This type provides an economical solution for structures with moderate column loads and uniform and small column spacings. For large column loads the slab thickness is increased beneath the columns to resist resulting shear stresses. If the column spacing becomes large, thickened beams may be used along the column lines. For structures requiring foundations with large flexural rigidity, box structures made of rigid frames or cellular construction are used.

3C.8.3 Design Methods

Different methods for designing mat foundations can be used, depending on the assumptions pertaining to the structure.

3C.8.3.1 Rigid Method. If the mat is rigid enough compared to the subsoil, flexural deflections of the mat will not vary the contact pressure. Hence the contact pressure can be assumed to vary linearly. The line of action of the resultant for the column loads coincides with the centroid of the contact pressure. This assumption is justifiable when the following conditions apply:

1. The column load does not vary by more than 20% compared to adjacent columns.
2. The column spacing is less than $1.75/\lambda$. The coefficient λ is defined as

$$\lambda = \sqrt[4]{\frac{K_b b}{4 E_c I}} \qquad (3C.18)$$

where K_b = coefficient of subgrade reaction
b = width of a strip of mat between centers of adjacent bays
E_c = modulus of elasticity of concrete
I = moment of inertia of strip of width b

On soft soils the actual contact pressure distribution is close to being linear. Hence it is commonly acceptable to design a mat on soft clay or organic soils using the rigid method.

The resultant force of all column loads and its location are first determined. Then the contact pressure q can be calculated using the principles of the strength of materials [Fig. 3C.8(a)],

$$q = \frac{\Sigma Q}{A} \pm \frac{(\Sigma Q \cdot e_y) x}{I_y} \pm \frac{(\Sigma Q \cdot e_x) y}{I_x} \qquad (3C.19)$$

where ΣQ = resultant force of all column loads
A = total area of mat
e_x, e_y = coordinates determining location of resultant force
x, y = coordinates for a given point under mat
I_x, I_y = moments of inertia of mat with regard to x and y axes

The mat could then be analyzed in each of the two perpendicular directions. As an example, the total shear force acting on section a–a is equal to the algebraic sum of the column loads P_1, P_2, and P_3 and the contact pressure reaction on the tributary area $R_{a\text{-}a}$ [Fig. 3C.8(b)],

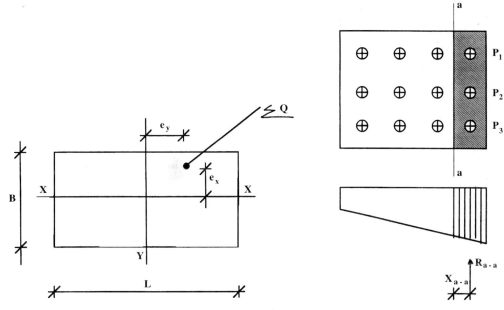

FIGURE 3C.8 Analysis of mat foundations using the rigid method.

$$V = P_1 + P_2 + P_3 - R_{a-a} \tag{3C.20}$$

Similarly the total bending moment acting on section a–a is equal to

$$M = \sum_{i=1}^{n} P_i x_i - (R_{a-a} x_{a-a}) \tag{3C.21}$$

The rigid mat should be checked for shear and bending failures. The designer must calculate the shear force at each section and the punching shear under each column and provide an adequate mat thickness. Flexural reinforcement is provided on the top and bottom of the raft foundation in order to guarantee adequate resistance to applied moments.

3C.8.3.2 Elastic Method. This method is based on the theory of plates on elastic foundations (Hetenyi, 1946). For a typical mat on stiff or compact soils it has been found that the effect of a concentrated load damped out quickly. By determining the effect of a column load on the surrounding area, and by superimposing the effects of all the column loads within the influence area, the total effect at any point can be determined. The influence area is usually considered no more than two bays in all directions. The use of polar coordinates is necessary when applying this method since the effect of loads is transferred through the mat to the soil in a radial direction. The application of this method involves extensive mathematical manipulations. Tables have been developed to speed up the solutions. However, for cases involving variable moments of inertia of the mat and possibly variable coefficients of subgrade reaction, the work to be performed remains tedious.

ACI Committee 436 (1966), based on the theory of plates on elastic foundations, recommends the following procedure to design mat foundations of constant moment of inertia and constant coefficient of subgrade reaction.

1. The mat thickness h is chosen such that shear at critical sections is adequately resisted.
2. The coefficient K of subgrade reaction is determined.

3. The flexural rigidity of the mat foundation is calculated using

$$D = \frac{Eh^3}{12(1-\mu^2)} \qquad (3C.22)$$

where E = modulus of elasticity of concrete
μ = Poisson's ratio of concrete

4. The radius of effective stiffness l is determined using

$$l = \sqrt[4]{\frac{D}{K_b}} \qquad (3C.23)$$

where K_b is the coefficient of subgrade reaction adjusted for mat size.

5. Radial moment M_r, tangential moment M_t, and deflection Δ at any point are calculated:

$$M_r = -\frac{P}{4}\left[Z_4\left(\frac{r}{l}\right) - (1-\mu)\frac{Z'_3(r/l)}{(r/l)}\right] \qquad (3C.24)$$

$$M_t = -\frac{P}{4}\left[\mu Z_4\left(\frac{r}{l}\right) - (1-\mu)\frac{Z'_3(r/l)}{(r/l)}\right] \qquad (3C.25)$$

$$\Delta = -\frac{Pl^2}{4D}Z_3\left(\frac{r}{l}\right) \qquad (3C.26)$$

where P = column load
r = distance of point of interest from column load along radius l
$Z_3\left(\frac{r}{l}\right), Z'_3\left(\frac{r}{l}\right), Z_4\left(\frac{r}{l}\right)$ = functions for moments and deflections (Fig. 3C.9)

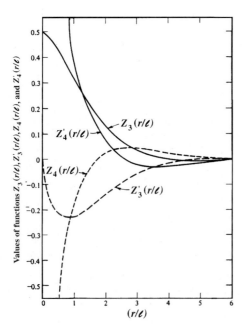

FIGURE 3C.9 Functions for mat foundation design using the elastic method. (*From Hetenyi, 1946.*)

6. The radial and tangential moments are transformed to rectangular coordinates,

$$M_x = M_r \cos^2\phi + M_t \sin^2\phi \tag{3C.27}$$

$$M_y = M_r \sin^2\phi + M_t \cos^2\phi \tag{3C.28}$$

where $\tan \phi = y/x$.

7. The shear force Q for a unit width of the mat foundation is determined by

$$Q = -\frac{P}{4l} Z'_4\left(\frac{l}{r}\right) \tag{3C.29}$$

where $Z'_4(l/r)$ is the function for shear (Fig. 3C.10).

8. The moments and shear forces computed for each column are superimposed to obtain the total moment on shear design values.

3C.8.3.3 Numerical Methods. With the increasing use of computers in design applications, numerical methods capable of handling cases of variable moments of inertia and variable coefficients of subgrade reaction are becoming more and more attractive. The methods of finite difference and of finite elements are among the mostly used numerical techniques.

The finite-difference method is based on the assumption that the effect of the underlying soil can be represented by uniformly distributed elastic springs. These springs have an elastic constant K equal to the subgrade reaction. The differential equation of such a mat foundation is

$$\frac{\delta^4\Delta}{\delta^4 x} + \frac{2\delta^4\Delta}{\delta^2 x \delta^2 y} + \frac{\delta^4\Delta}{\delta^4 y} = \frac{q - Kw}{D} \tag{3C.30}$$

where q = subgrade reaction per unit area of mat
K = coefficient of subgrade reaction
w = deflection
D = rigidity of mat defined in Eq. (3C.22)

The deflection of any point can be related to the deflection at the adjacent points to the right, left, top, and bottom using a numerical difference equation. The mat foundation is divided into a network of points. The difference equations for these points are formulated and rapidly solved for the deflections with a programmed computer. With the knowledge of deflections, the bending moments and shear forces can be determined from the theory of elasticity. The accuracy of the results obtained with the finite-difference method largely depends on the size and number of networks used.

The finite-element method uses the concept of matrix structural analysis to address the problem of plates on elastic foundations. The mat foundation is idealized as a mesh of plates (finite elements) interconnected only at the nodes, where isolated springs are used to model the soil reactions. A more detailed discussion of this method and its applications for foundation design can be found in Weaver and Johnston (1984).

3C.9 FOUNDATIONS IN COLD REGIONS

A locality, city, or state that spends a large amount of its financial resources to maintain a program for continuous social and economical operations under cold weather conditions and snow storms is considered located in a cold region. Seasonal and permanently frozen grounds are characteristics of cold regions and require special attention from the foundations designer.

In areas of seasonal frost during winter months, the foundation depth is carefully taken below the frost line (Fig. 3C.10). This is a necessary measure to prevent heaving of the structure due to

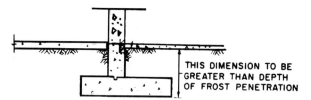

FIGURE 3C.10 Depth of frost penetration.

freezing of the underlying soil. Heave is a phenomenon caused by the formation and growth of ice particles in the soil. If a foundation is placed at or above the frost line, it will move upward as the underlying soil freezes and expands. Later it will suddenly settle when thawing occurs. An additional problem is encountered in the case of fine-grained soils, namely, the decrease in the soil shear strength when it thaws after being frozen. This loss of strength is due to thawing, liberating moisture that had been soaked up by the soil particles during freezing. Thus the moisture content of the soil is increased compared to conditions prior to freezing. Such a loss of shear strength could result in a foundation failure.

An estimate of the depth of the frost line in different regions can be obtained from data supplied by the U.S. Weather Bureau (see Fig. 3C.11). The depth values obtained from such charts are only approximate. They should be corrected to account for several factors, such as susceptibility of the soil type to frost, location of the footing (interior versus exterior), and local experience (local regulations and adjacent buildings).

For frost action to occur, the following conditions must apply:

1. *Presence of frost-susceptible soil.* These are soils with enough fine pores to initiate and enhance the mechanism of ice formation and growth. Several criteria have been proposed, based on the particle-size distribution of the soil. One of the most widely known of these criteria was proposed by Casagrande (1932):

   > Under natural conditions and with sufficient water supply, one should expect considerable ice segregation in uniform soils containing more than 3% of grains smaller than 0.2 mm and in very uniform soils containing more than 10% smaller than 0.02 mm. No ice segregation was observed in soils containing less than 1% of grains smaller than 0.02 mm, even if the groundwater level was as high as the frost line.

   A definite distinction between soils that are frost-susceptible and those that are not is not available. Thus soils that are borderline should be used with caution.

2. *Availability of water.* For the ice particles to grow, water in the liquid phase must move in the soil to the frost line. This movement is carried by the capillary action and by suction due to supercooling at the frost front.

3. *Freezing conditions.* These conditions are determined by air temperature, solar radiation, snow cover, and exposure to wind.

In permanently frozen regions (permafrost areas), the loads of the structure are transmitted to the frozen soil with utmost attention to maintaining the frozen state. This is usually performed by insulating and ventilating between the building and the frozen ground such that the presence of the building will not alter the temperature of the ground. Another possible solution is to excavate the soil down to foundation depth and then replace it with soil that is not susceptible to frost action. Thus the foundations will not be affected by the freezing and thawing cycles. In some cases foundations are allowed to bear on frozen ground with a source of artificial refrigeration provided to keep the soil under the footings permanently frozen. This approach is, however, used rarely because of its high cost.

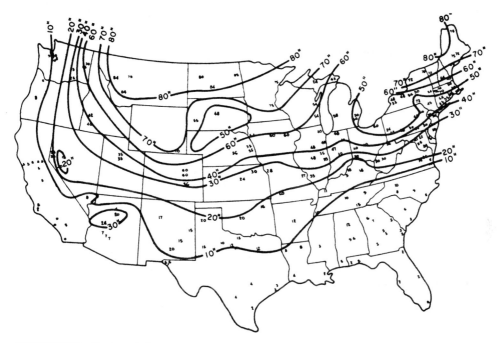

FIGURE 3C.11 Frost penetration map.

In general the same foundation types used in moderate regions can be used in cold regions, such as spread footings, mat foundations, piles, and caissons. The selection of a specific type of foundation will depend on the particular site conditions, particularly soil type, temperature characteristics, and structural loads. Detailed discussions of the mechanical properties of frozen soil and its bearing capacity are presented in the *Canadian Foundation Engineering Manual,* Andersland and Anderson (1978), and Sodhi (1991). The design of foundations in cold regions rarely requires higher-strength materials to resist the stresses induced from the frost-susceptible soils. What is necessary instead are techniques to avoid problems of frost heaving.

3C.10 FOUNDATIONS IN EARTHQUAKE REGIONS

3C.10.1 General

Earthquakes can produce extensive damage to foundations and structures supported on them. This damage could be related to a gross instability of the soil or to ground movement developing high-intensity stress on the structural systems. Instability of the soil can occur in loose dry sand deposits which are compacted by the ground vibrations of earthquakes, leading to large settlements and differential settlements of the ground surface. The settlements are larger for sands with smaller relative density. In cases where the soil consists of saturated loose sand, the compaction by ground vibrations could increase the hydrostatic pressure to a sufficient magnitude to cause "liquefaction" of the soil. Liquefaction is a phenomenon whereby saturated loose granular soil loses its shear strength due to the earthquake motion. Reports on many earthquakes refer to such liquefaction causing large settlements, tilting, and overturning of structures. Sudden increases in pore water pressures due to ground vibration in deposits of soft clay and sands have been the cause for major landslides in earthquake regions.

Liquefaction is likely to occur under the following soil conditions (Oshaki, 1970):

1. The sand layer is within 45 to 60 ft (15 to 20 m) of the ground level and is not subjected to high overburden pressure.
2. The sand deposits consist of uniform medium-size particles and are below the groundwater level (saturated).
3. The standard penetration test is below a certain value.

To reduce the possibility of liquefaction, several measures can be taken:

1. Increasing the sand relative density by compaction
2. Replacing the sand with another soil having better characteristics to withstand liquefaction
3. Lowering the ground water level or installing drainage equipment

3C.10.2 Dynamic Properties of Soils

In order to perform a seismic design for foundations in an earthquake region, the dynamic soil characteristics must be determined. The following is a brief description of tests used to obtain such data (Wakabayashi, 1986).

3C.10.2.1 Particle Size Distribution.
The soil particle size distribution is related to the liquefaction of saturated cohesionless soils. Figure 3C.12 indicates a liquefaction potential zone based on the performance of cohesionless soils in previous earthquakes.

3C.10.2.2 Relative Density Test.
This test indicates the degree of soil compaction. It gives helpful information in determining the possibility of excessive settlement for dry sands and the potential of liquefaction for saturated sands in earthquake regions. The relative density is obtained from one of the equations

$$D_r = \frac{e_{max} - e}{e_{max} - e_{min}}$$

$$D_r = \frac{\rho_{max}(\rho - \rho_{min})}{\rho(\rho_{max} - \rho_{min})} \tag{3C.31}$$

where e_{max}, e_{min} = maximum and minimum void ratios
ρ_{max}, ρ_{min} = maximum and minimum unit mass
e = in situ void ratio
ρ = in situ unit mass

3C.10.2.3 Cyclic Triaxial Test.
This test is performed to determine the shear modulus and damping of cohesive and cohesionless soils. The shear modulus can be obtained from the compressive modulus of elasticity E using

$$G = \frac{E}{2(1 + \nu)} \tag{3C.32}$$

where ν is Poisson's ratio.

It can also be obtained directly by performing cyclic shear tests to obtain stress-strain relationships. The shear modulus is strain-dependent. Hence the level at which G is determined must be defined. Average relationships of shear modulus to strain for clay and sand are shown in

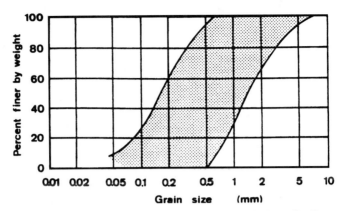

FIGURE 3C.12 Zone of liquefaction potential for cohesionless soils. (*From Ohsaki, 1970.*)

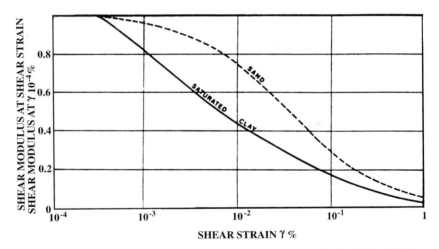

FIGURE 3C.13 Average relationships of shear modulus to strain. (*From Seed and Idriss, 1970.*)

Fig. 3C.13. During earthquakes, developed shear strains may range from 10^{-3} to 10^{-1}%, with a different maximum strain at each cycle. For this reason it has been suggested to use a value of two-thirds the shear modulus measured at the maximum strain developed for earthquake design purposes. In the field, the shear modulus of soil can be estimated from a shear wave velocity test. An explosive charge or a vibration source is used to initiate waves in the soil. The velocity of these waves is measured and the following relationship is used to determine the shear modulus of elasticity:

$$G = \rho v_s^2 \tag{3C.33}$$

where ρ = mass density of soil
v_s = shear wave velocity

The second chief dynamic parameter for soils is damping. Two different damping phenomena are related to soils—material damping and radiation damping. Material damping takes place when

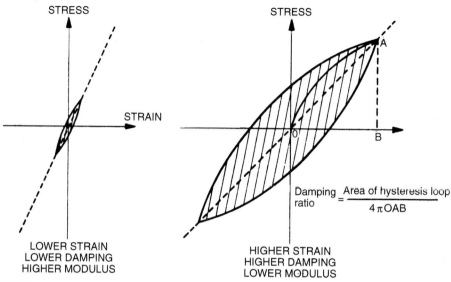

FIGURE 3C.14 Calculation of material damping ratio. (*From Seed and Idriss, 1970.*)

any vibration wave travels through the soil. It is related to the loss of vibration energy resulting from hysteresis in the soil. Damping is generally expressed as a fraction of critical damping and thus referred to as damping ratio. The damping ratio is expressed as (Fig. 3C.14)

$$\epsilon = \frac{W}{4\pi \Delta W} \qquad (3C.34)$$

where W = energy loss per cycle (area of hysteresis loop)
ΔW = strain energy stored in equivalent elastic material (area OAB in Fig. 3C.14)

Typical material damping ratios, representing average values of laboratory test results on sands and saturated clays, are give in Fig. 3C.15.

Radiation damping is a measure of the loss of energy through the radiation of waves from the structure. It is related to the geometrical properties of the foundation. The theory for the elastic half-space has been used to provide estimates for radiation damping. Figure 3C.16 shows values of radiation damping for circular footings of machinery obtained by Whitman and Richart (1967).

3C.10.3 Design Considerations

Additional design considerations for foundations in earthquake regions include (1) transmission of horizontal base shear forces from the structure to the soil, (2) resisting the earthquake-induced overturning moments, and (3) differential settlements and liquefaction of the subsoil (Wakabayashi, 1986; Dorwick, 1987).

Generally in earthquake design practice two separate stress systems are considered—seismic vertical stress resulting from overturning moments and seismic horizontal stresses caused by the base shear on the structure. Unless it is very slender, overturning moments are not a design problem for the structure as a whole. However, they can drastically impact on individual footings. Hence the foundation should be proportioned so as to maintain the maximum bearing pressures

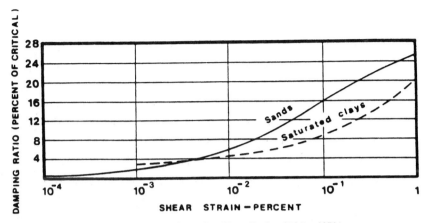

FIGURE 3C.15 Typical material damping ratios. (*From Seed and Idriss, 1970.*)

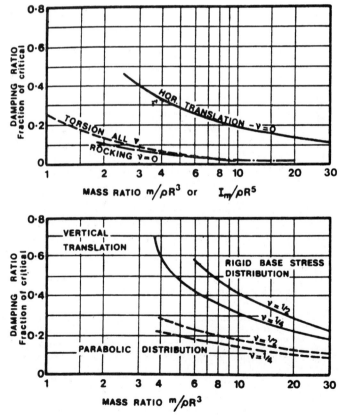

FIGURE 3C.16 Radiation damping values. (*From Whitman and Richart, 1967.*)

caused by the overturning moments and gravity loads within the allowable seismic bearing capacity of the present soil. Safe seismic bearing pressures vary from one location to another, and local codes should be used for guidelines. In general most soils are capable of resisting higher short-term loads than long-term loads. Some sensitive clays that lose strength under dynamic loading are an exception.

With shallow foundations the base shear is assumed to be resisted by friction on the bottom surfaces of footings. The total resistance to horizontal displacement of a structure is taken to be equal to the product of the dead load of the structure and the coefficient of friction between soil and footings. Some codes recommend the use of 75% of the standard friction coefficients. Additional horizontal resistance can be obtained from the passive soil pressures developed against footing surfaces. However, if this resistance is to be relied upon, reducing the computed total resistance becomes necessary. This can be done by reducing either the frictional force or the passive resistance by about 50%. Also, careful compaction of the backfill against the sides of the footing must be performed in order to rely on the passive restraint of the soil.

To avoid or minimize damage to the foundation structure in earthquake regions due to differential settlements, it is recommended to provide ties or beams between column footings. These ties should be designed to withstand a prescribed differential movement between the connected footings.

3C.11 REFERENCES

ACI Committee 436, "Suggested Design Procedures for Combined Footings and Mats," *ACI J.,* Oct. 1966.

Andersland, O. B., and D. M. Anderson (Eds.), *Geotechnical Engineering for Cold Regions,* McGraw-Hill, New York, 1978.

Canadian Foundation Engineering Manual, 2d ed., Canadian Geotechnical Society, 1985.

Casagrande, A., "A New Theory on Frost Heaving," *Highway Res. Board Proc.,* no. 11, pp. 168–172, 1932.

Dorwick, D. J., *Earthquake Resistant Design for Engineers and Architects,* 2d ed., Wiley, New York, 1987.

Fintel, M. (Ed.), *Handbook of Concrete Engineering,* 2d ed., Van Nostrand Reinhold, Princeton, N.J., 1985.

Hetenyi, M., *Beams on Elastic Foundation,* University of Michigan Press, Ann Arbor, Mich., 1946.

Ohsaki, Y., "Effects of Sand Compaction on Liquefaction during the Tokachi-Oki Earthquake," *Soil Foundations,* vol. 10, no. 2, pp. 112–128, 1970.

Seed, H. B., and I. M. Idriss, "Soil Moduli and Damping Factors for Dynamic Response Analysis," Report EERC 70-10, Earthquake Engineering Research Center, University of California, Berkeley, Calif., 1970.

Sodhi, D. S. (Ed.), *Cold Regions Engineering, Proc. 6th Int. Specialty Conf.,* American Society of Civil Engineers, West Lebanon, N.H., 1991.

Wakabayashi, M., *Design of Earthquake-Resistant Buildings,* McGraw-Hill, New York, 1986.

Weaver, W., Jr., and P. R. Johnston, *Finite Elements for Structural Analysis,* Prentice-Hall, Englewood Cliffs, N.J., 1984.

Whitman, R. V., and F. E. Richart, "Design Procedures for Dynamically Loaded Foundations," *J. Soil Mech. Found. Div., ASCE,* vol. 93, no. SM6, pp. 169–91, 1967.

Winterkorn, H. E., and H. Y. Fang (Eds.), *Foundation Engineering Handbook,* Van Nostrand Reinhold, Princeton, N.J., 1975.

SECTION 3D
PILE FOUNDATIONS

A. Samer Ezeldin

3D.1 INTRODUCTION   3.153
3D.2 ALLOWABLE STRESSES IN PILES   3.154
    3D.2.1 Concrete Piles   3.154
    3D.2.2 Steel Piles   3.157
    3D.2.3 Timber Piles   3.159
3D.3 PILE LOADING CAPACITY   3.160
    3D.3.1 General   3.160
    3D.3.2 Axial Single-Pile Capacity   3.160
    3D.3.3 Pile Group Capacity   3.167
3D.4 PILE DYNAMICS   3.169
    3D.4.1 Dynamics Equations for Pile Capacity   3.169
    3D.4.2 Dynamics Considerations   3.170
3D.5 PILE LOAD TEST   3.170
3D.6 NEGATIVE SKIN FRICTION   3.172
3D.7 LATERALLY LOADED PILES   3.172
3D.8 PILE CAPS DESIGN   3.174
3D.9 REFERENCES   3.180

3D.1 INTRODUCTION

Piles are vertical or slightly inclined members used to transmit the loads of the superstructure to lower layers in the soil mass. The load transfer mechanism relies either on the skin resistance occurring along the surface contact of the pile with the soil or on the end bearing on a dense or firm layer. The design of some piles can also be based on the utilization of both the skin resistance and the end bearing to carry the applied load jointly. In general, pile foundations are relied upon to transfer the load acting on the superstructures in situations where the use of shallow foundations becomes inadequate or unreliable. Such situations include (1) the top soil layers have a weak bearing capacity, with the soil layer at greater depth possessing a high bearing capacity; (2) large values of concentrated loads are to be transmitted from the superstructure to the foundation; and (3) the structure to be designed is very sensitive to unequal settlements.

Materials usually used to make piles are concrete, steel, and timber. The upper part of the pile connected to the superstructure is referred to as the pile head. The middle part is called the shaft, the lower is the pile tip. The pile cross section can either be maintained throughout the length of the pile or it can be tapered to a rather pointed pile. The cross section can be circular, octagonal, hexagonal, square, triangular, or H-shaped. Figure 3D.1 illustrates typical pile shapes and various cross sections.

Piles can be classified into two types—displacement piles and nondisplacement piles. Displacement piles are those which displace the soil to allow for the pile penetration. These piles can be of solid cross section, driven into the ground, and left in position. Timber, steel, prestressed concrete piles, and precast concrete piles are of this type. Displacement piles are also obtained by driving shell (hollow) piles by means of an internal steel mandrel onto which the shell is threaded. After the mandrel is pulled out, the shell pile is filled with concrete internally. The Raymond pile is a mandrel-driven steel-shell pile; the Western pile is a mandrel-driven concrete-shell pile. Another method for obtaining displacement piles is driving a pilelike body into the ground and withdrawing it while filling the void with concrete (Franki pile). Nondisplacement piles are those in which the soil is removed to accommodate the pile. Typically a borehole is formed in the ground, then concrete is cast in place in the hole.

3.154 FUNDAMENTALS OF FOUNDATION CONSTRUCTION AND DESIGN

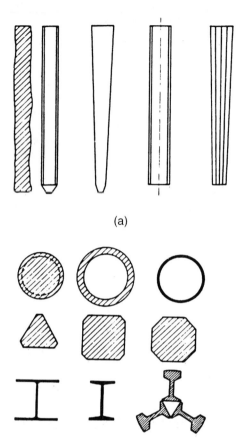

FIGURE 3D.1 Typical pile shapes and cross sections. (*a*) Piles shapes. (*b*) Various pile cross sections.

3D.2 ALLOWABLE STRESSES IN PILES

In pile foundation design it is necessary to determine the required number of piles, their cross section, and their length. This will require knowledge about the pile loading capacity as well as their allowable stresses. This section will present the allowable design stresses for service loads as adopted in the design guides of the U.S. Army Corps of Engineers and published by the American Society of Civil Engineers (ASCE, 1993). The pile loading capacity is discussed in Sec. 3D.3. The allowable stresses presented in this section may be increased by one-third to account for unusual loading such as maintenance, infrequent floods, barge impact, construction, or hurricanes.

3D.2.1 Concrete Piles

Concrete piles can be prestressed, precast-reinforced, cast in place, or mandrel-driven.

3D.2.1.1 Prestressed Concrete Piles. Prestressed concrete piles are formed by tensioning high-strength steel cable having an ultimate stress f_{pu} of 250 to 270 ksi with a prestress of about 0.5 to $0.7 f_{pu}$ before casting the concrete. After casting the concrete, and only when it develops adequate strength, the prestress cables are cut. Due to the bond between steel and concrete, the cables

TABLE 3D.1 Allowable Concrete Stresses for Prestressed Concrete Piles

Uniform axial tension	0
Bending (extreme fiber)	
Compression	$0.40 f'_c$
Tension	0

Source: ASCE, 1993.

will apply a compressive stress on the concrete pile as they attempt to return to their original length. When designing prestressed concrete piles, both strength and serviceability requirements must be satisfied. Strength design should be conducted in accordance with the American Concrete Institute code (ACI, 1989), except that a strength reduction factor ϕ of 0.7 is to be used for all failure modes and a load factor of 1.9 for both dead and live loads. The use of these factors will result in a factor of safety of 2.7 for all dead and live load combinations. The axial strength to be used in design is the least of: (1) 80% of the concentric axial strength or (2) the axial strength corresponding to an eccentricity equal to 10% of the pile diameter or width. Cracking control is achieved by limiting the actual concrete compressive and tensile stresses resulting from working conditions to the values presented in Table 3D.1. For the combined condition of axial force and bending, the concrete stresses should satisfy the following:

$$f_a + f_b + f_{pc} \leq 0.4 f'_c \tag{3D.1a}$$

$$f_a - f_b + f_{pc} \geq 0 \tag{3D.1b}$$

where f_a = computed axial stress (tensions negative)
f_b = computed bending stress (tensions negative)
f_{pc} = effective prestress
f'_c = concrete compressive strength

The allowable stresses for hydraulic structures are limited to 0.85% of the values recommended by ACI Committee 543 for improved serviceability (ACI, 1986). Permissible stresses in the prestressing steel cables should be in accordance with the ACI code requirements (ACI, 1989). In cases where the pile is free-standing or when the soil is too weak to provide a reliable lateral support, the pile capacity should be reduced due to slenderness effects. The moment magnification method of ACI as modified by PCI can be used to perform such design (PCI, 1988). Figure 3D.2 illustrates typical prestressed concrete piles.

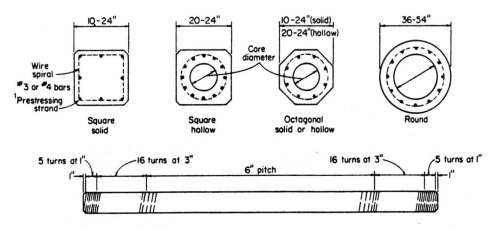

[1] Strand: 1/2 - 7/16-in diam., f_u = 270 ksi

FIGURE 3D.2 Typical prestressed concrete pile. (*From Bowles, 1982.*)

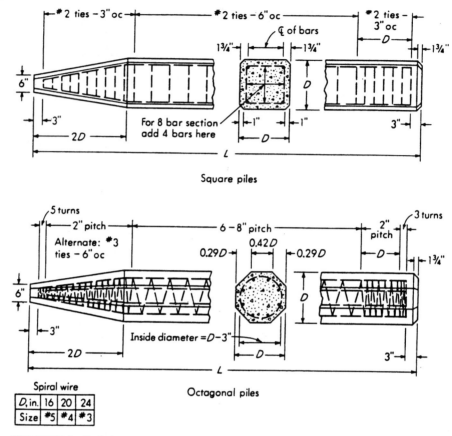

FIGURE 3D.3 Typical precast concrete pile. (*From Bowles, 1982.*)

3D.2.1.2 Precast-Reinforced Concrete Piles. Precast-reinforced concrete piles are designed in accordance with the ACI code (ACI, 1989). For hydraulic structures, the ultimate load is to be increased by a hydraulic load factor H_f. The hydraulic load factor is taken as (1) 1.30 for reinforcement calculations in flexure or compression, (2) 1.65 for reinforcement in direct tension, and (3) 1.30 for reinforcement in shear. When performing shear reinforcement design, the calculations should exclude the shear carried by the concrete prior to application of the hydraulic load factor. The axial strength limitations are taken as in the case for prestressed piles. The slenderness effects are accounted for according to the ACI code moment magnification method (ACI, 1989). Figure 3D.3 shows typical precast-reinforced concrete piles.

3.D.2.1.3 Cast-in-Place and Mandrel-Driven Piles. Figure 3D.4 illustrates various types of cast-in-place piles. The depths indicated are for the usual ranges for the different piles. These piles are mostly used when continuous lateral support is present. Cast-in-place and mandrel-driven piles are designed such that working stresses are limited to the allowable stresses shown in Table 3D.2. In case of axial load combined with bending, the concrete stresses are such that

$$\left| \frac{f_a}{F_a} + \frac{f_b}{F_b} \right| \leq 1.0 \tag{3D.2}$$

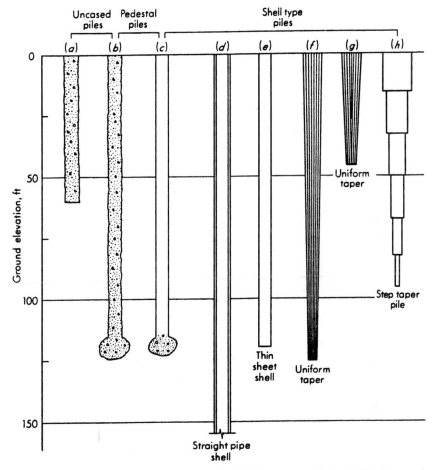

FIGURE 3D.4 Typical cast-in-place concrete pile. (*a*) Western uncased pile. (*b*) Franki uncased-pedestal pile. (*c*) Franki cased-pedestal pile. (*d*) Welded or seamless pile. (*e*) Western cased pile. (*f*) Union or Monotube pile. (*g*) Raymond standard. (*h*) Raymond step-taper pile. (*From Bowles, 1982.*)

where f_a = computed axial stress
F_a = allowable axial stress
f_b = computed bending stress
F_b = allowable bending stress

3D.2.2 Steel Piles

Steel piles are usually rolled, H-shaped, or pipe piles. The lower region of these piles could be subjected to damage during driving. This is why the U.S. Army Corps of Engineers uses allowable stresses with a high factor of safety for that region, as shown in Fig. 3D.5. Pile shoes are usually used when driving piles in the dense layer. Table 3D.3 shows the allowable stresses for fully supported piles when using pile shoes. The upper portion of the pile is designed as a beam-column where the lateral support conditions are accounted for. Bending moments are, however, negligible in the lower portion of the pile. The moment diagram along the pile is shown in Fig. 3D.5.

TABLE 3D.2 Allowable Concrete Stresses for Cast-in-Place and Mandrel-Driven Piles

Uniform axial compression	
Confined	$0.33f'_c$
Unconfined	$0.27f'_c$
Uniform axial tension	0
Bending (extreme fiber)	
Compression	$0.40f'_c$
Tension	0

Source: ASCE, 1993.

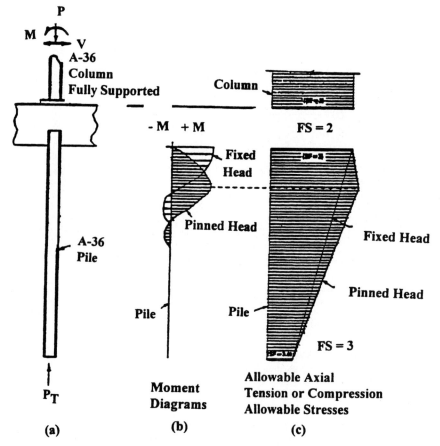

FIGURE 3D.5 Behavior of steel piles. (*From ASCE, 1993.*)

TABLE 3D.3 Allowable Stresses in Lower Pile Region for Steel Piles

Concentric axial tension or compression only 10 ksi ($1/3 \times F_y \times 5/6$)	10 ksi for A-36 material
Concentric axial tension or compression only with driving shoes ($1/3 \times F_y$)	12 ksi for A-36 material
Concentric axial tension or compression only with driving shoes, at least one axial load test and use of a pile driving analyzer to verify pile capacity and integrity ($1/2.5 \times F_y$)	14.5 ksi for A-36 material

Source: ASCE, 1993.

For laterally unsupported piles the allowable stress should be five-sixths of the values the American Institute of Steel Construction code gives for beam columns (AISC, 1989). For combined axial compression and bending conditions, the stress should be

$$\left| \frac{f_a}{F_a} + \frac{f_{bx}}{F_b} + \frac{f_{by}}{F_b} \right| \leq 1.0 \tag{3D.3}$$

where f_a = computed axial stress
F_a = allowable axial stress, = $0.5 F_y$
f_{bx}, f_{by} = computed bending stress
F_b = allowable bending stress, $0.5 F_y$ for noncompact section and $5/9 F_y$ for compact section

3.D.2.3 Timber Piles

Timber piles are cut from tree trunks and driven with the smaller cross section down. Representative allowable stress values for pressure-treated round timber piles are presented in Table 3D.4. These stresses have been adjusted to account for treatment. For untreated piles, or piles that were either air- or kiln-dried before pressure treatment, the allowable stress shown in Table 3D.4 should be increased by dividing each value by 0.9 for Pacific Coast Douglas fir and by 0.85 for southern pine. To account for combined axial load and bending moment effects, the stresses should satisfy

$$\left| \frac{f_a}{F_a} + \frac{f_b}{F_b} \right| \leq 1.0 \tag{3D.4}$$

where f_a = computed axial stress
F_a = allowable axial stress
f_b = computed bending stress
F_b = allowable bending stress

TABLE 3D.4 Allowable Stresses for Pressure-Treated Round Timber Piles

Species	Compression parallel to grain, psi F_a	Bending, psi F_b	Horizontal shear, psi	Compression perpendicular to grain, psi	Modulus of elasticity, psi
Pacific Coast Douglas fir	875	1700	95	190	1,500,000
Southern pine	825	1650	90	205	1,500,000

Source: ASCE, 1993.

3D.3 PILE LOADING CAPACITY

3D.3.1 General

The mechanism of the load transfer from piles to the soil layer is illustrated in Fig. 3D.6. The horizontal earth pressures act on the shaft surface area, creating vertical frictional reactions that increase with depth. If enough displacement occurs, adhesion could also contribute to these reactions. The sum of these reactions is referred to as the mantle friction or skin resistance. In addition vertical reactions occur at the tip of the pile, mobilizing tip-bearing resistance. The ratio of the mantle friction to the tip-bearing resistance varies according to the physical properties and profile of the soil, the pile dimensions, and the method of installation.

3D.3.2 Axial Single-Pile Capacity

The pile loading capacity consists of the sum of the skin resistance and the tip-bearing resistance. Hence it may be represented by the equation

$$Q_{ult} = Q_s + Q_t = f_s A_s + q A_t \tag{3D.5}$$

where Q_{ult} = ultimate pile capacity
Q_s = shaft resistance of pile due to skin friction
Q_t = tip-bearing resistance of pile
f_s = average skin resistance stress
A_s = surface area of shaft in contact with soil
q = tip-bearing stress
A_t = effective area at tip of pile in contact with soil

3D.3.2.1 Piles in Cohesionless Soil. The skin resistance of piles in cohesionless soil is assumed to increase linearly up to a critical depth d_c. For design purposes, the critical depth is taken as $10B$ for loose sand, $15B$ for medium dense sand, and $20B$ for dense sand, where B is the pile diameter or width. Below the critical depth, the skin resistance is taken to be a constant value

FIGURE 3D.6 Load transfer mechanism using skin friction resistance and tip-bearing resistance.

TABLE 3D.5 Typical K values* for Piles in Compression and in Tension

Soil type	K_c	K_t
Sand	1.00 to 2.00	0.50 to 0.70
Silt	1.00	0.50 to 0.70
Clay	1.00	0.70 to 1.00

*Values do not apply to piles that are prebored, jetted, or installed with a vibratory hammer. Picking K values at the upper end of these ranges should be based on local experience. K, δ, and N_q values back-calculated from load tests may be used.
Source: ASCE, 1993.

(equal to the critical-depth skin resistance stress). The average skin resistance stress at a particular depth may be calculated using the equation

$$f_s = K\sigma'_v \tan \delta \qquad (3D.6)$$

where K = coefficient for lateral earth pressure (see Table 3D.5 for K values for piles in compression and tension)
σ'_v = effective overburden pressure at particular depth d
  = $\gamma' d$ for $d < d_c$
  = $\gamma' d_c$ for $d \geq d_c$ using γ' as the effective unit weight for soil
δ = friction angle between soil and pile material (see Table 3D.6 for typical δ values)

TABLE 3D.6 Typical δ Angles in terms of ϕ

Pile material	δ
Steel	0.67 ϕ to 0.83 ϕ
Concrete	0.90 ϕ to 1.0 ϕ
Timber	0.80 ϕ to 1.0 ϕ

Source: ASCE, 1993.

It must be emphasized that the K and δ values presented should be selected based on experience and site conditions and could be replaced with better representative values if such are available to the designer. When using steel H piles, the value of δ should be the average friction angle of steel against soil and soil against soil (ϕ value). Also, the value of A_s for steel H piles is to be taken as the block perimeter of the pile.

The tip-bearing stress, q can be determined from the expression

$$q = \sigma'_v N_q \qquad (3D.7)$$

where σ'_v is as defined earlier, and the bearing capacity factor N_q is obtained from Fig. 3D.7. When using steel H piles, the area A_t is taken as the area included within the block perimeter. The pile tension capacity in cohesionless soil is obtained by solely calculating the shaft resistance of the pile due to skin friction Q_s using the corresponding K values in Table 3D.5.

Example 3D.1 Find the allowable compression capacity of a 12-in-diameter (305 mm) reinforced concrete pile with a total length of 45 ft (13.7 mm) driven in medium dense sand. The K and δ values are found to be 1.5 and 0.9ϕ, respectively. The soil profile is shown in Fig. 3D.X.1. The soil angle of friction ϕ is 30°. Use a factor of safety of 3.0.

Solution The critical depth is

$$d_c = 15B = 15\left(\frac{12}{12}\right) = 15 \text{ ft } (4.57 \text{ m})$$

The effective overburden pressure σ'_v at water table level is

$$\sigma'_v = 110 \times 10 = 1100 \text{ psf } (53.647 \text{ kPa})$$

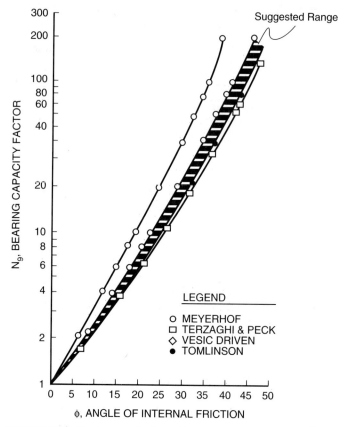

FIGURE 3D.7 Bearing capacity factor N_q versus angle of internal friction ϕ. (*From Terzaghi and Peck, 1967.*)

The effective overburden pressure σ'_v at the critical depth d_c is

$$\sigma'_{v(-15)} = 1100 + 5(125 - 62.4) = 1413 \text{ psf (68.91 kPa)}$$

The shaft resistance due to skin friction from level (0.00) to (−10.00) (3 m) is

$$\left[1.5\left(\frac{1100 + 0}{2}\right)(\tan 0.9 \times 30°)\right]\left[\pi\left(\frac{12}{12}\right)10\right] = 13{,}185 \text{ lb (58.65 kN)}$$

The shaft resistance due to skin friction from level (−10.00) (3 m) to (−15.00) (4.6 m) is

$$\left[1.5\left(\frac{1100 + 1413}{2}\right)(\tan 0.9 \times 30°)\right]\left[\pi\left(\frac{12}{12}\right)5\right] = 15{,}061 \text{ lb (66.99 kN)}$$

The shaft resistance due to skin friction from level (−15.00) (−4.6 m) to (−45.00) (13.7 m) is

$$[1.5 \times 1413(\tan 0.9 \times 30°)]\left[\pi\left(\frac{12}{12}\right)30\right] = 101{,}625 \text{ lb (452 kN)}$$

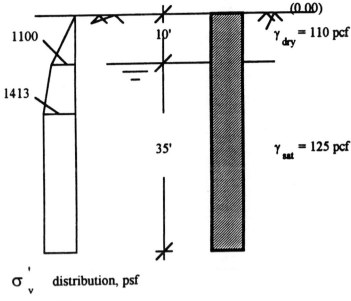

FIGURE 3D.X.1

The total shaft resistance due to skin friction is

$$Q_s = 13{,}185 + 15{,}061 + 101{,}625 = 129{,}871 \text{ lb } (582 \text{ kN})$$

The tip-bearing resistance is, using $\bar{N}_q = 18$,

$$Q_t = \sigma'_v N_q A_t$$

$$= 1413 \times 18 \times \frac{\pi \times 1^2}{4} = 19{,}965 \text{ lb } (88.8 \text{ kN})$$

The allowable compression capacity is

$$Q_{\text{all}} = \frac{Q_{\text{ult}}}{FS} = \frac{129{,}871 + 19{,}965}{3} = 49{,}945 \text{ lb} \approx 50 \text{ kips } (222 \text{ kN})$$

3D.3.2.2 Piles in Cohesive Soil. The skin resistance is due to adhesion of the cohesive soil to the pile shaft. The following expression can be used to estimate the average skin resistance stress:

$$f_s = \alpha c \tag{3D.8}$$

where c = undrained shear strength of soil from a Q test
α = adhesion factor (see Fig. 3D.8 for values of α in terms of undrained shear strength)

An alternate method proposed by Semple and Rigden (1984) is especially applicable for long piles. It consists of obtaining the adhesion factor α as the product of two factors α_1 and α_2. These two factors can be obtained using Fig. 3D.9.

3.164 FUNDAMENTALS OF FOUNDATION CONSTRUCTION AND DESIGN

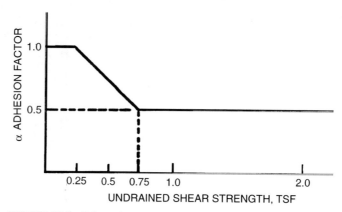

FIGURE 3D.8 Values of α versus undrained shear strength. (*From ASCE, 1993.*)

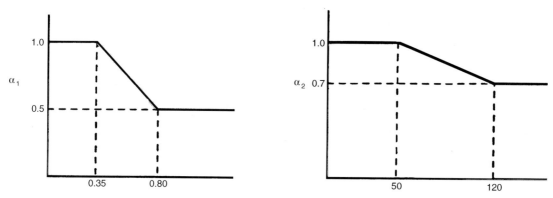

FIGURE 3D.9 Values of α_1 and α_2 applicable for long piles. (*From Semple and Rigden, 1984.*)

The tip-bearing stress q is obtained from the expression

$$q = 9c \tag{3D.9}$$

To develop such tip-bearing stress the required pile movement may have to be larger than that necessary to mobilize skin resistance. The pile tension capacity in cohesive soil can be calculated using only the shaft resistance due to skin friction.

Example 3D.2 Find the allowable compression capacity of a 12-in-diameter (305 mm) reinforced concrete pile with a total length of 45 ft (13.7 m) driven in clay layers as shown in Fig. 3D.X.2. Use a factor of safety of 2.5.

Solution The shaft resistance due to skin friction from level (0.00) to (−10.00) [3 m] is

$$1 \times 400 \left[\pi \left(\frac{12}{12} \right) 10 \right] = 12{,}560 \text{ lb } (55.9 \text{ kN})$$

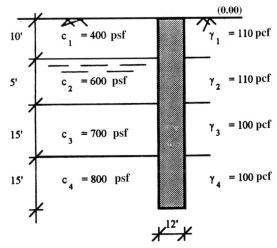

FIGURE 3D.X.2

The shaft resistance due to skin friction from level (-10.00) $[-3$ mm$]$ to (-15.00) $[-4.5$ m$]$ is

$$0.95 \times 600 \left[\pi \left(\frac{12}{12} \right) 5 \right] = 8949 \text{ lb } (39.8 \text{ kN})$$

The shaft resistance due to skin friction from level (-15.00) $[-4.5$ m$]$ to (-30.00) $[-9$ m$]$ is

$$0.9 \times 700 \left[\pi \left(\frac{12}{12} \right) 15 \right] = 29{,}673 \text{ lb } (131.98 \text{ kN})$$

The shaft resistance due to skin friction from level (-30.00) $[-9$ m$]$ to (-45.00) $[-13.7$ m$]$ is

$$0.85 \times 800 \left[\pi \left(\frac{12}{12} \right) 15 \right] = 32{,}028 \text{ lb } (143.5 \text{ kN})$$

The total shaft resistance due to skin friction is

$$Q_s = 12{,}560 + 8949 + 29{,}673 + 32{,}028 = 83{,}210 \text{ lb } (370.1 \text{ kN})$$

The tip-bearing resistance is

$$Q_r = 9c \frac{\pi(d)^2}{4} = 9 \times 800 \frac{\pi(1)^2}{4} = 5652 \text{ lb } (25.14 \text{ kN})$$

The allowable compression capacity is

$$Q_{all} = \frac{Q_{ult}}{FS} = \frac{83{,}210 + 5652}{2.5} = 35{,}545 \text{ lb} \approx 36 \text{ kips } (158.1 \text{ kN})$$

3D.3.2.3 Piles in Silt. The skin resistance of piles in silt is generated from two sources—friction along the pile shaft and adhesion of soil to the pile shaft. The portion of the friction resistance increases linearly up to the critical depth d_c, below which the frictional resistance remains constant. In design, the critical depth is assumed as $10B$ for loose silts, $15B$ for medium silts, and $20B$

for dense silts, where B is the pile diameter or width. The portion of the adhesion is controlled by the undrained shear strength of the soil. The combined average skin resistance stress f_s can be determined using the equation

$$f_s = K\sigma'_v \tan \delta + \alpha c \tag{3D.10}$$

where all variables are as defined in Secs. 3D.3.2.1 and 3D.3.2.2.

The tip-bearing stress q can be calculated from Eq. (3D.7). The pile tension capacity in silt soil is obtained by excluding the tip-bearing stress calculation and including only the effect of the skin resistance stress f_s along the pile shaft.

Example 3D. 3 Determine the allowable compression capacity of a 12-in-diameter (304 mm) reinforced concrete pile with a total length of 50 ft (15 m) driven in medium dense silt. The K and δ values are found to be 1.0 and 1.0ϕ, respectively. The soil angle of friction ϕ is 20°, and its undrained shear strength c is 200 psf (9.5 kPa). The soil profile is shown in Fig. 3D.X.3. Use a factor of safety of 3.0.

Solution The critical depth is

$$dc = 15B = 15\left(\frac{12}{12}\right) = 15 \text{ ft (4.57 m)}$$

The effective overburden pressure σ'_v at water table level is

$$\sigma'_{v(-b)} = 110 \times 10 = 1100 \text{ psf (52.67 kPa)}$$

The effective overburden pressure σ'_v at the critical depth d_c is

$$\sigma'_{v(-15)} = 1100 + 5(110 - 62.4) = 1334 \text{ psf (63.87 kPa)}$$

The shaft resistance due to skin friction from level (0.00) to (−10.00) [3 m] is

$$\left[\left(\frac{1100 + 0}{2}\right)\tan 20° + 200\right]\left[\pi\left(\frac{12}{12}\right)10\right] = 12{,}566 \text{ lb (55.89 kN)}$$

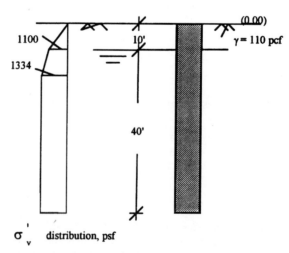

FIGURE 3D.X.3

The shaft resistance due to skin friction from level (-10.00) $[-3$ m$]$ to (-15.00) $[-4.57$ m$]$ is

$$\left[\left(\frac{1100+1334}{2}\right)\tan 20° + 200\right]\left[\pi\left(\frac{12}{12}\right)5\right] = 10{,}095 \text{ lb } (44.9 \text{ kN})$$

The shaft resistance due to skin friction from level (-15.00) $[-4.57$ m$]$ to (-50.00) $[-15$ m$]$ is

$$[1334(\tan 20°) + 200]\left[\pi\left(\frac{12}{12}\right)35\right] = 75{,}340 \text{ lb } (335.1 \text{ kN})$$

The total shaft resistance due to skin friction is

$$Q_s = 12{,}566 + 10{,}095 + 75{,}340 = 98{,}001 \text{ lb } (435.9 \text{ kN})$$

The tip-bearing resistance is, using $N_q = 8$,

$$Q_t = \sigma'_v N_q A_t$$

$$= 1334 \times 8 \frac{\pi(1)^2}{4} = 8378 \text{ lb } (37.26 \text{ kN})$$

The allowable compression capacity is

$$Q_{all} = \frac{Q_{ult}}{FS} = \frac{98{,}001 + 8378}{3} = 35{,}460 \text{ lb} \approx 35 \text{ kips } (157.7 \text{ kN})$$

3D.3.3 Pile Group Capacity

In actual construction applications it is rare to encounter a foundation consisting of a single pile. Rather, a group of piles is used to transmit the load of the superstructure to the soil mass. Figure 3D.10 presents some typical pile groupings. When several piles are placed with a distance s from each other, it is expected that both the skin resistance stress and the tip-bearing stresses developed in the soil will overlap. Due to this overlap it is reasonable to expect that the pile group capacity could be less than the sum of the individual pile capacities. The pile group efficiency η is the ratio of the pile group capacity to the sum of the individual pile capacities. Several equations have been proposed to determine the numerical value of η. Of these equations, the Converse-Labarre equation seems to be one of the most accepted. According to this equation, the ratio η is equal to

$$\eta = 1 - \frac{\theta}{90°}\frac{(n-1)m + (m-1)n}{mn} \tag{3D.11}$$

where m = number of rows
n = number of piles in a row
θ = arctan (d/s)
d = pile diameter
s = center-to-center spacing of piles

The drawback of using the η equation is neglecting the beneficial effects of the pile-driving operations on increasing the relative density and the friction angle of the sand. Another approach to determine the pile group capacity was presented by Terzaghi and Peck (1967). They assumed the group of piles with the enclosing soil to form a rigid pier which behaves as a unit

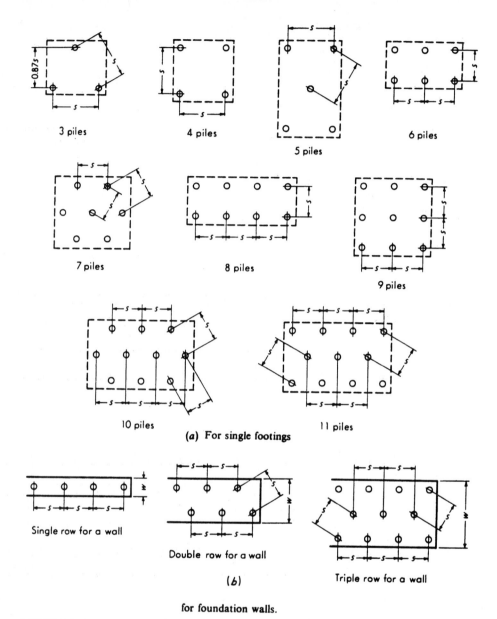

FIGURE 3D.10 Typical pile group patterns. (*a*) Single footings. (*b*) Foundation walls. (*From Bowles, 1982.*)

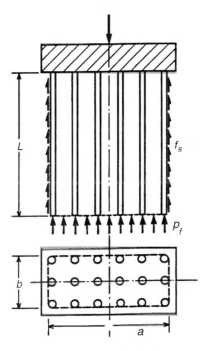

FIGURE 3D.11 Load capacity of a pile group. (*From Terzaghi and Peck, 1967.*)

(see Fig. 3D.11). The bearing capacity of the pier is calculated as the sum of the tip-bearing resistance of the area $a \times b$ and the skin resistance on the perimeter of the pier. Thus,

$$Q_{\text{ult(group)}} = f_s[2(a \times b)L] + q(a \times b) \quad (3\text{D}.12)$$

where f_s and q are obtained as presented in Sec. 3D.3.2 for different soil types.

For design purposes the pile group capacity of driven piles in sand not underlain by a weak layer is to be taken as the sum of the single-pile capacities. For other conditions the pile group capacity is the least of either the sum of the single-pile capacities or the group capacity as determined from Eq. (3D.12). The pile spacing is generally taken not less than three times the pile diameter on centers for bearing piles, and a minimum of three to five times the pile diameter on centers for friction piles, depending on the characteristics of the soil and the piles.

3D.4 PILE DYNAMICS

3D.4.1 Dynamics Equations for Pile Capacity

It is well known that the pile load capacity is affected by the method of installation. There are several installation methods based on dynamic processes such as vibration and ramming. In order to estimate the capacity of the pile while it is being driven at the site, many driving formulas have been proposed. These formulas are all based on the following relation:

$$\text{Energy input} = \text{energy used} + \text{energy lost} \quad (3\text{D}.13)$$

Unfortunately these equations are not consistently reliable. The main reasons for this are:

1. The pile elastic compression is calculated using a static approach.
2. A portion of the input energy is used in displacing the soil laterally.

3. The total resistance is assumed acting on the pile tip.
4. The effects of driving velocity and duration of intermissions on the dynamic penetration are neglected.

In spite of these limitations, application of the driving formula could be beneficial to compare the dynamic resistance of piles driven in a site. This would give the engineer a way to judge the uniformity of the site subsoil. The most commonly used driving formulas are the *Engineering News* formula,

$$R = \frac{1.25 e_h E_h}{S + 0.1} \frac{W_r + (n \cdot n) W_p}{W_r + W_p} \quad (3D.14)$$

and the Danish formula

$$R = \frac{e_h E_h}{S + C_1} \quad (3D.15)$$

where R = load capacity of pile (just after driving)
e_h = hammer efficiency
E_h = hammer energy rating
S = amount of point penetration per blow
W_r = weight of ram
W_p = weight of pile
n = coefficient of restitution
$C_1 = (e_h E_h L/2AE)^{1/2}$ with AE and L being the pile cross section, modulus of elasticity, and length.

Currently the best means for estimating the pile capacity dynamically consists of recording the pile-driving history and then load testing the pile. It would be a reasonable assumption to expect that piles with similar driving history will develop the same load capacity.

3D.4.2 Dynamics Considerations

The vibration attenuation and the liquefaction potential due to construction dynamic loading should be investigated. The U.S. Army Corps of Engineers requires the removal or densification of liquefiable soil (ASCE, 1993). Also the first few natural frequencies of the structure-foundation assemblage should be determined and compared to the construction operations frequencies in order to avoid resonance.

3D.5 PILE LOAD TEST

As mentioned previously, load testing a pile is considered the most dependable way of determining its carrying capacity. The pile load test consists of applying a series of increasing load values and measuring the corresponding settlements to obtain a pile load-settlement curve (Fig. 3D.12). Many empirical methods have been proposed to determine the pile capacity from the pile load-settlement data. Table 3D.7 includes a list of most of these methods. The methods used by the U.S. Army Corps of Engineers consists of taking the average of three load values obtained from the load test data as the pile load capacity. These three load values are

1. The load causing a movement of 0.25 in on the net settlement curve
2. The load corresponding to the point on the curve with a significant change in slope
3. The load matching the point on the curve that has a slope of 0.01 in per ton

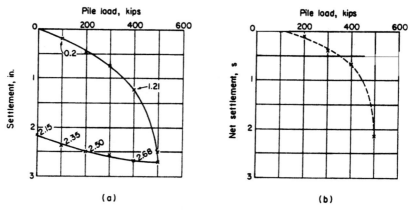

FIGURE 3D.12 Typical pile load test data. (*a*) Total settlement curve. (*b*) Net settlement curve.

TABLE 3D.7 Methods of Pile Load Test Interpretation

1. Limiting total butt settlement
    a. 1.0 in (Holland)
    b. 10% of tip diameter (United Kingdom)
    c. Elastic settlement+D/30 (Canada)
2. Limiting plastic settlement
    a. 0.25 in (AASHTO, N.Y. State, Louisiana)
    b. 0.5 in (Boston) [complete relaxation of pile assumed]
3. Limiting ratio: plastic/elastic settlement
    1.5 (Christiani and Nielson of Denmark)
4. Limiting ratio: settlement/unit load
    a. Total   0.01 in/ton (California, Chicago)
    b. Incremental   0.03 in/ton (Ohio)
                     0.05 in/ton (Raymond International)
5. Limiting ratio: plastic settlement/unit load
    a. Total   0.01 in/ton (N.Y. City)
    b. Incremental   0.003 in/ton (Raymond International)
6. Load-settlement curve interpretation
    a. Maximum curvature: Plot log total settlement versus log load; choose point of maximum curvature.
    b. Tangents: Plot tangents to general slopes of upper and lower portions of curves; observe point of intersection.
    c. Break point: Observe point at which plastic settlement curve breaks sharply; observe point at which gross settlement curve breaks sharply (Los Angeles).
7. Plunge   Find loading at which pile "plunges" (i.e., the load increment could not be maintained after pile penetration was greater than 0.2B).
8. Texas quick load   Construct tangent to initial slope of load versus gross settlement curve; construct tangent to lower portion of the load versus gross settlement curve at 0.05 in/ton slope. The intersection of the two tangent lines is the ultimate bearing capacity.

Source: ASCE, 1993.

3D.6 NEGATIVE SKIN FRICTION

In some cases, piles are driven into compressible soil before its consolidation is complete. If a fill is placed on this compressible soil, the soil will move downward against the pile. This relative movement creates a skin friction between the pile and the moving soil that increases the axial load acting on the pile. This mechanism is known as negative skin friction (Fig 3D.13). In excessive soil consolidation cases a gap may form between the bottom of the pile cap and the fill. This results in the full cap weight being transferred directly to the piles and could alternate the stresses in the pile cap. The value of the negative skin friction can be computed as follows:

$$Q_{NF} = cLP' \tag{3D.16}$$

where c = shear strength of soil
L = length of pile in contact with compressible layers
P' = perimeter of pile

Thus the total applied load on the pile becomes

$$Q_{TOTAL} = Q + Q_{NF} \tag{3D.17}$$

where Q is the load transmitted from the superstructure.

If piles are spaced a small distance apart, the developed negative skin friction may also be represented as the dragging force acting on the perimeter of the pier formed by the group of piles and the enclosed soil. In these situations, two modes of negative skin friction require investigation:

1.
$$Q_{NF} = N(Q_{NF}/\text{pile}) \tag{3D.18a}$$

where Q_{NF}/pile is as given as by Eq. (3D.16) and N is the number of piles.

2.
$$Q_{NF} = AL\gamma + cLP \tag{3D.18b}$$

where A = area bounded by pile group
L = length of pile in contact with compressible layers
γ = unit weight of compressible layer
c = shear strength of soil
P = perimeter of area A.

In the presence of expansive soils the negative skin friction phenomenon can generate upward tension stresses in the pile. These stresses could be larger if no or an insufficient gap is left between the expansive soil and the bottom of the pile cap.

3D.7 LATERALLY LOADED PILES

Piles are slender vertical members that have only limited capability to resist nonvertical loads. Therefore batter piles are used to resist large inclined or horizontal loads when acting on a structure. Beresantsev et al. (1961) suggested the following practice to transmit inclined loads in terms of the angle α, where α is the angle of the force to be transmitted with the vertical (Fig. 3D.14):

1. Vertical piles $\alpha < 5°$
2. Batter piles $5° \leq \alpha < 15°$
3. Deadman $\alpha < 15°$

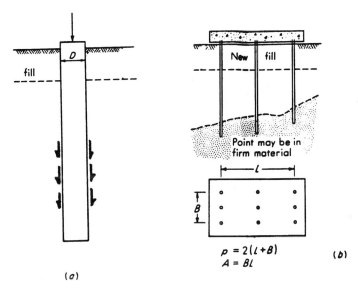

FIGURE 3D.13 Negative skin friction. (*a*) Single pile. (*b*) Pile-group effect.

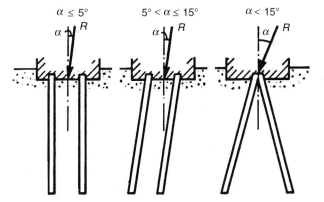

FIGURE 3D.14 Recommended practice to transmit inclined loads to soil mass. (*From Beresantsev et al., 1961.*)

Brooms (1965) presented charts that give the limit lateral load to act on a vertical pile versus the ratio of the pile embedment length to its diameter. These diagrams, applied to short piles, are presented in Fig. 3D.15 for cohesive and cohesionless soils. The term "short piles" refers to rigid piles where the lateral capacity is dependent mainly on the soil resistance. Long piles are those whose lateral capacity is primarily dependent on the yield moment of the pile itself. Figure 3D.16 shows the relationship between the limit lateral load and the yield bending moment of the pile for cohesive and cohesionless soils. In these figures the dashed lines represent the case of a fixed pile head, whereas the full lines indicate different e/l ratios, where e is the height of the line of action of the force P above ground surface and l is the length of pile in the ground.

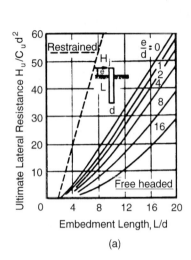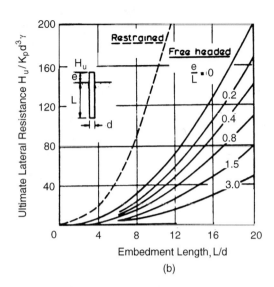

FIGURE 3D.15 Ultimate lateral resistance of short piles. (*a*) Cohesive soils. (*b*) Cohesionless soils. (*From Brooms, 1965.*)

The symbols used in these figures are

d = diameter of pile
γ = soil bulk unit weight
k_p = coefficient of passive earth pressure
H_u = limit value of horizontal load
c_u = undrained shear strength
L = pile embedment length

Very few test results are available for inclined forces acting on vertical piles or on a batter pile. Petrasovits and Awad (1968) conducted model tests on piles having a length of 50 cm and a diameter of 1.3 to 3.5 cm embedded in a soil with an angle of internal friction $\phi = 37.5°$. Figure 3D.17 gives the percentage of increase for the applied load for different cases when the inclination angle β of the pile is varied.

3D.8 PILE CAP DESIGN

Pile caps are used to distribute the loads and moments acting on the column to all of the piles in the group. The pile cap is usually made of reinforced concrete and rests directly on the ground, except when the soil underneath is expansive. The design considers the effects of a number of concentrated reactions due to the column load, surcharge load, fill weight, and pile cap weight. For the design of a rigid pile cap it is usual to assume that each pile carries an equal amount of concentric axial load and that a planar stress distribution is valid for nonconcentric loading. This assumption is justified when the pile cap is resting on the ground, the piles are vertical, the load is applied at the center of the pile group, and the pile group is symmetrical. The structural design of a reinforced concrete pile cap requires consideration of the following critical conditions:

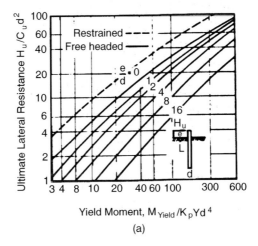
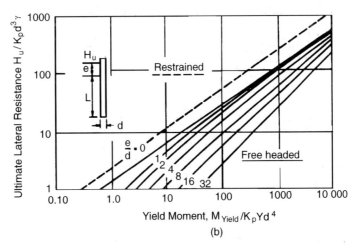

FIGURE 3D.16 Ultimate lateral resistance of long piles. (*a*) Cohesive soils. (*b*) Cohesionless soils. (*From Brooms, 1965.*)

1. Punching shear failure at sections located at a distance $d/2$ from the face of the column and around each pile
2. Beam shear failure at sections at a distance d from the face of the column
3. Bending failure at the sections located at the face of the column

where d is the effective depth at which the steel layer is placed.

The rules followed during the design are essentially the same as the ones used for spread footings. When deciding on the piles causing shear, attention is drawn to Chap. 15 of ACI 318 (1989), which states that

> Computation of shear on any section through a footing supported on piles shall be in accordance with the following: a) Entire reaction from any pile whose center is located $d_p/2$ (d_p is the pile diameter at the upper end) or more outside the section shall be assumed as producing shear on the section,

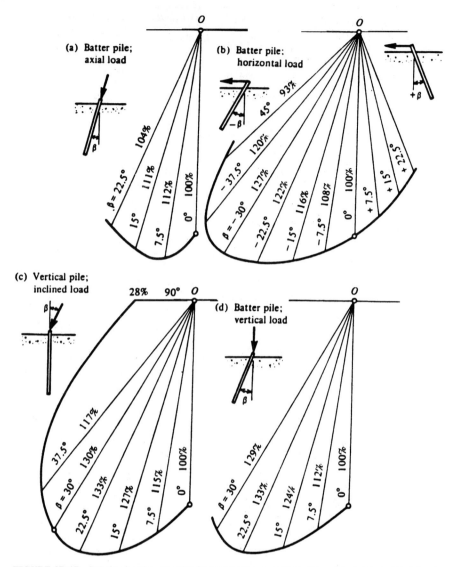

FIGURE 3D.17 Relative bearing capacity for batter piles or vertical piles subjected to inclined forces. (*From Petrasovits and Awad, 1968.*)

b) reaction from any pile whose center is located $d_p/2$ or more inside the section shall be assumed as producing no shear on the section, c) for intermediate positions of the pile center, the portion of the pile reaction to be assumed as producing shear on the section shall be based on straight line interpolation between full value at $d_p/2$ outside the section and zero at $d_p/2$ inside the section.

The designer is urged to keep the pile cap design on the safe side because the individual actual pile reaction may differ from the value used in design due to group action and possible differences between layout on drawings and driven piles.

Example 3D.4 A 28-in-square (710-mm²) column carries the following loads: P_D = 500 kips (2224 kN), P_L = 700 kips (3114 kN), M_D = 200 ft · kips (271 kN · m), and M_L = 300 ft · kips (406.8 kN · m). The column is to rest on a 4-ft-thick (1.2-m) cap supported by piles having an allowable load capacity of 100 kips (445 kN) and a diameter of 12 in (304 mm). The cap is topped with 12 in (304 mm) of fill having a unit weight of 120 lb/ft³ (1922 kg/m³) and 6-in concrete (152-mm) slab carrying a surcharge load of 100 psf (4788 Pa) (see Fig. 3D.X.4a). Design the pile cap using f'_c = 4000 psi (27.6 MPa) and f_y = 60,000 psi (414 MPa).

Solution The total vertical load is

$$P_{total} = 500 + 700 = 1200 \text{ kips (5338 kN)}$$

To account for the bending effects as well as the surcharge and cap weight, choose a total number of 15 piles spaced at 3 ft on centers (Fig. 3D.X.4b).

The surcharge load per pile is

$$P_s/\text{pile} = 3^2(0.5 \times 150 + 100 + 1 \times 120 + 4 \times 150) = 8055 \text{ lb} \approx 8.1 \text{ kips (36 kN)}$$

$$\Sigma x^2 = 3(3^2 + 6^2 + 3^2 + 6^2)144 = 38{,}880 \text{ in}^2 \ (250 \times 10^3 \text{ cm}^2)$$

The net allowable capacity of a pile is

$$P_{net} = 100 - 8.1 = 91.9 \text{ kips (408.8 kN)}$$

The maximum axial force of a pile is

$$\frac{1200}{15} + \frac{500 \times 12 \times 72}{38{,}880} = 91.1 \text{ kips (405.2 kN)} < 91.9 \text{ kips (408.8 kN)} \quad \text{O.K.}$$

The factored load on piles p_1 through p_5 is

$$P_1 = \frac{500 \times 1.4 + 700 \times 1.7}{15} + \frac{(200 \times 12 \times 1.4 + 300 \times 12 \times 1.7)72}{38{,}880} = 143.6 \text{ kips (638.7 kN)}$$

$$P_2 = \frac{500 \times 1.4 + 700 \times 1.7}{15} + \frac{(200 \times 12 \times 1.4 + 300 \times 12 \times 1.7)36}{38{,}880} = 134.8 \text{ kips (599.6 kN)}$$

$$P_3 = \frac{500 \times 1.4 + 700 \times 1.7}{15} = 126 \text{ kips (560.41 kN)}$$

$$P_4 = \frac{500 \times 1.4 + 700 \times 1.7}{15} - \frac{(200 \times 12 \times 1.4 + 300 \times 12 \times 1.7)36}{38{,}880} = 117.2 \text{ kips (521.3 kN)}$$

$$P_5 = \frac{500 \times 1.4 + 700 \times 1.7}{15} - \frac{(200 \times 12 \times 1.4 + 300 \times 12 \times 1.7)72}{38{,}880} = 108.4 \text{ kips (482.16 kN)}$$

Pile punching shear check:

$$V_u = 143.6 \text{ kips (639 kN)}$$

$$V_c = 4\sqrt{f'_c} b_0 d = \frac{4\sqrt{4000} \times \pi \times 36 \times 44)}{1000} = 1258 \text{ kips}$$

3.178 FUNDAMENTALS OF FOUNDATION CONSTRUCTION AND DESIGN

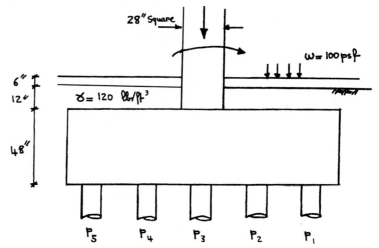

FIGURE 3D.X.4a

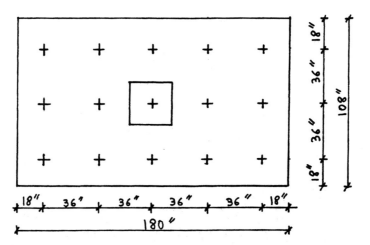

FIGURE 3D.X.4b

$$\frac{V_u}{\phi} = \frac{143.6}{0.85} = 169 < 1258 \text{ kips } (752 < 5596 \text{ kN}) \qquad \text{O.K.}$$

Two-way shear check (using a conservative approach for computation of applied shear):

$$V_u = 3 \times 143.6 + 3 \times 134.8 + 2 \times 126 + 3 \times 117.2 + 3 \times 108.4 = 1764 \text{ kips } (7846 \text{ kN})$$

$$V_c = 4\sqrt{f'_c}\, b_0\, d = \frac{4\sqrt{4000} \times 72 \times 4 \times 44}{1000} = 3206 \text{ kips } (14{,}260 \text{ kN})$$

$$\frac{V_u}{\phi} = \frac{1764}{0.85} = 2075 < 3206 \text{ kips } (9229 < 14{,}260 \text{ kN}) \qquad \text{O.K.}$$

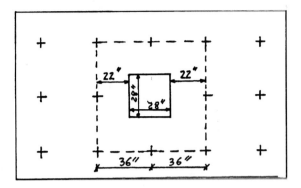

Beam shear check:

$$V_u = 3 \times 143.6 = 430.8 \text{ kips (1916 kN)}$$

$$V_c = 2\sqrt{f'_c}\, b_0\, d = \frac{2\sqrt{4000} \times 108 \times 44}{1000} = 601 \text{ kips (2673 kN)}$$

$$\frac{V_u}{\phi} = \frac{430.8}{0.85} = 507 < 601 \text{ kips} \qquad \text{O.K.}$$

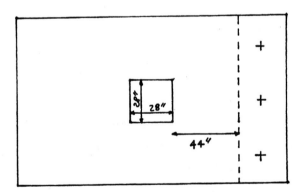

Flexure bending design: Check Sec. 3B.

1. Section A–A:

$$M_u = (3 \times 143.6 \times 1000 \times 58 + 3 \times 134.8 \times 1000 \times 22 = 33.9 \times 10^6 \text{ lb} \cdot \text{in (3831 kN} \cdot \text{m)}$$

$$A_s = \frac{M_u}{\phi f_y (0.9d)} = \frac{33.9}{0.9 \times 60{,}000 \times 0.9 \times 44} \times 10^6 = 15.85 \text{ in}^2 \text{ (102.2 cm}^2\text{)}$$

$$(A_s)_{\min} = \left(\frac{200}{f_y}\right) bd = 0.0033 \times 108 \times 44 = 15.68 \text{ in}^2 \text{ (101.2 cm}^2\text{)}$$

3.180 FUNDAMENTALS OF FOUNDATION CONSTRUCTION AND DESIGN

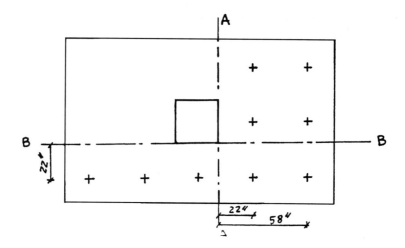

Choose $A_s = 15.85$ in² (102.2 cm²)

$$a = \frac{15.85 \times 60,000}{0.85 \times 4000 \times 108} = 2.59 \text{ in (65.8 mm)}$$

$$M_n = 15.85 \times 60,000\left(44 - \frac{2.59}{2}\right) = 40.6 \times 10^6 > \frac{33.9 \times 10^6}{0.9} = 37.7 \times 10^6 \text{ lb} \cdot \text{in (4260 kN} \cdot \text{m)}$$

Choose $A_s = 15.85$ in² (101.2 cm²) in the x direction.

2. Section B–B:

$$M_u = (143.6 + 134.8 + 126 + 117.2 + 108.4)1000 \times 22 = 11 \times 10^6 \text{ lb} \cdot \text{in (1243 kN} \cdot \text{m)}$$

$$A_s = \frac{M_u}{\phi f_y(0.9d)} = \frac{11}{0.9 \times 60,000 \times 0.9 \times 44} \times 10^6 = 5.18 \text{ in}^2 \text{ (33.42 cm}^2\text{)}$$

$$(A_s)_{min} = \left(\frac{200}{f_y}\right)bd = 0.0033 \times 180 \times 44 = 26.1 \text{ in}^2 \text{ (168.39 cm}^2\text{)}$$

Choose $A_s = 26.1$ in² (168.39 cm²) in the y direction.

The selection and distribution of the bars as well as the development length checks are performed as presented in Sec. 3B.

3D.9 REFERENCES

ACI Committee 318, "Building Code Requirements for Reinforced Concrete," ACI 318-89; and "Commentary on Building Code Requirements for Reinforced Concrete," American Concrete Institute, Detroit, Mich., 1989.

ACI Committee 543, "Recommendations for Design, Manufacture and Installation of Concrete Piles," American Concrete Institute, Detroit, Mich., 1986.

"Design of Pile Foundations" (Technical Engineering and Design Guides as Adapted from U.S. Army Corps of Engineers, no.1), American Society of Civil Engineers, ASCE, New York, 1993.

Beresantsev, V. G., V. S. Khristoforov, and V. N. Golubkov, "Load Bearing Capacity and Deformation of Pile Foundations," in *Proc., 5th Conf. on Soil Mechanics and Foundation Engineering,* Paris, France, 1961, pp. 11–15.

Bowles, J. E., *Foundation Analysis and Design,* 3d ed., 1982, pp. 575–703.

Brooms, B. B., "Design of Laterally Loaded Piles," *Proc. ASCE,* pp. 79–99, May 1965.

Das, B. M., *Principles of Foundation Engineering,* PWS-Kent Publishing, Boston, Mass., 1984, pp. 330–415.

Fintel, M. (Ed.), *Handbook of Concrete Engineering,* Van Nostrand Reinhold, Princeton, N.J., 1985, chap. 5, Footings, pp. 139–168.

Manual of Steel Construction, 9th ed., American Institute of Steel Construction, AISC, New York, 1989.

PCI Committee on Prestressed Concrete Columns, "Recommended Practice for the Design of Prestressed Concrete Columns and Walls," vol. 33, pp. 56–95, July-Aug. 1988.

Petrasovits, G., and A. Awad, "Considerations of the Bearing Capacity of Vertical and Batter Piles Subjected to Forces Acting in Different Directions," in *Proc. 3d Conference on Soil Mechanics and Foundation Engineering, Budapest, 1968,* pp. 483–497.

Poulos, H. G., and E. H. Davis, *Pile Foundation Analysis and Design,* Wiley, New York, 1980.

Prakash, S., and H. Sharma, *Pile Foundations in Engineering Practice,* Wiley, New York, 1990.

Semple, R. M., and W. J. Rigden, "Shaft Capacity of Driven Pile in Clay," in *Analysis and Design of Pile Foundations,* J. R. Meyer (Ed.), American Society of Civil Engineers, New York, 1984, pp. 59–79.

Terzaghi, K., and R. B. Peck, *Soil Mechanics in Engineering Practice,* 2d ed., Wiley, New York, 1967.

Whitaker, T., *The Design of Piled Foundation,* 2d ed., Pergamon, New York, 1976.

Winterkorn, H. E., and H. Y. Fang, (Eds.), *Foundation Engineering Handbook,* Van Nostrand Reinhold, Princeton, N.J., 1975.

SECTION 3E
DESIGN PARAMETERS AND PROCEDURES FOR LIGHTLY LOADED FOUNDATIONS

Jim Williams and Jerry Kunkel

3E.1 UNIFORM-THICKNESS GROUND-SUPPORTED CONVENTIONAL OR POSTTENSIONED SLABS   3.183
   3E.1.2 Nonstructural Slabs   3.183
   3E.1.3 Structural Slabs   3.183

3E.2 PIER AND BEAM FOUNDATIONS WITH CRAWL SPACE   3.184
3E.3 BASEMENTS   3.184
3E.4 REFERENCES   3.185

3E.1 UNIFORM-THICKNESS GROUND-SUPPORTED CONVENTIONAL OR POSTTENSIONED SLABS

Foundations in this category can be divided into two types depending on soil activity or expected movement. They can be broadly referred to as nonstructural and structural.

3E.1.2 Nonstructural Slabs

These slabs are for sites where little or no volume change in the supporting soils is expected, that is, when plasticity indexes (PI) are less than 20 and the potential vertical rise (PVR) is less than 1 in. Slabs would normally be between 4 and 8 in thick, with a turned-down perimeter beam at least 12 in deep. Reinforcing steel should be furnished for crack control. The percentage of reinforcing steel should be equal to or greater than 0.0014 times the cross-sectional area of the slab. All steel is placed in the slab at middepth.

3E.1.3 Structural Slabs

These slabs are for sites where soil volume changes are expected, that is, PIs are greater than 20 and PVRs greater than 1 in. Slabs are sometimes more than 6 in thick, with a perimeter turned-down beam at least 12 in deep. The design will meet the requirements for the site as furnished by the geotechnical engineer. The structural design will be in accordance with a commonly employed rational method that has provided satisfactory results for similar sites.[1, 2]

1. Building Research Advisory Board (BRAB), "Criteria for Selection and Design of Residential Slabs-on-Ground," 1968.[1]
2. Post-Tensioning Institute, *Design and Construction of Post-Tensioned Slabs-on-Ground,* 1980.[2] As an alternate, posttensioning furnishing a minimum of 50 psi prestress force at midslab may be used.
3. Wire Reinforcement Institute and Concrete Reinforcing Steel Institute, "Design of Slab-on-Ground Foundations."

Minimum reinforcing specified in ACI 318[3] does not apply to ground-supported slabs. Minimum reinforcement is 0.14% for rebar or mesh, based on the cross-sectional area being considered. Reinforcement is placed at middepth.

3E.1.3.1 Ribbed (Waffle) Slabs-on-Ground. Foundations in this category can be used for either active or nonactive soils. All slabs will be considered structural and designed to meet minimum requirements of the following design methods. The references provide typical design illustrations.[1,2]

3E.1.3.2 Ribbed (Waffle) Slabs-on-Ground, Pier-Supported. Foundations in this category are normally used for very active soils or where significant settlements are expected. Two types are used, depending on whether or not settlement or heave is anticipated. In both cases piers designed in accordance with the geotechnical engineer's recommendations are placed at the perimeter and at intersections of interior beams.

Where settlement is expected, the soil is a form for the concrete, and the final design assumes no soil support for the slab. Beams are designed to carry the imposed loads, and slabs are designed to carry their dead load and minimal uniform live load. Beams are usually connected to the pier with dowels or by extending the pier reinforcement.

For sites that have expansive soils, void forms are usually placed at the bottom of the beams to allow soils under the bottom of the beam to heave. The slabs are usually in direct contact with the soil. Beams are designed to carry the imposed loads, and slabs are designed to carry their dead load and minimal uniform live loads. The soil provides additional support for the slab. Beams are usually not connected to the piers, and piers to resist uplift should have reinforcing based on the gross concrete area. Minimum reinforcement specified in ACI 318 is generally not used.

3E.1.3.3 Ribbed Slab with Void Space. Pier-supported foundations in this category are used for very active soils where significant heave is anticipated. The void space under the beams and slab needs to be adequate so that the soil can heave to its maximum potential without coming into contact with the beams or slab.

Beams are supported with piers, and slabs span between the beams. The design of beams and slabs should comply with requirements of ACI 318 and carry full code loading. Piers are usually connected to grade beams with dowels or extended pier reinforcing.

3E.2 PIER AND BEAM FOUNDATIONS WITH CRAWL SPACE

Foundations in this category are generally used for sites of moderate to very active soils, or where wood floors are desired. This foundation is generally accepted as the most positive method to prevent damage from active soils. This design requires a pier system capable of safely supporting the applied structural weight coupled with a continuous beam to distribute the load to the piers. The floor systems are generally composed of wood systems which span between exterior concrete-grade beams and interior girders. Unlike the slab foundations, there is no single reference for "typical" pier and beam foundation design.

Girders are supported as required by piers and are designed in accordance with applicable codes and standards. Exterior beams are generally concrete, and interior girders can be concrete, steel, or wood. Void spaces can be provided under concrete beams to allow for soil movement, and piers are generally connected to the concrete beams with dowels or extended pier reinforcing. A minimum clearance space of 18 in under joists and 12 in under girders needs to be provided for wood members. Proper ventilation is required for wood structures.

3E.3 BASEMENTS

Foundations in this category are generally used on all types of soils. Depth of frost lines and expensive lot costs influence this practice. The floor system is wood or concrete which spans between exterior concrete walls. (Wood and masonry walls are also used in some areas.) Walls and basement slabs are supported by the soils on footings or concrete foundations in areas of low to moderate soil activity. Walls and basement floor slabs may be structurally suspended in areas of very active soils. Site-specific geotechnical reports should provide guidance as to which type of system is needed. Structural systems should be designed in accordance with applicable codes and standards.

3E.4 REFERENCES

1. FHA, "Criteria for Selection and Design of Residential Slab-on-Ground," BRAB Report 33, National Academy of Science, 1968.
2. *Design and Construction of Post-Tensioned Slabs-on-Ground,* 2d ed., Post-Tensioning Institute, 1995.
3. ACI, "Building Codes Requirements for Reinforced Concrete," Report 318, American Concrete Institute, Box 19150, Redford, Mich. 48219.

PART 4

EARTH PRESSURES AND RETAINING SYSTEMS

SECTION 4A
EARTH PRESSURES

Richard W. Stephenson

4A.1 EARTH PRESSURE AT REST 4.3
   4A.1.1 Calculation of At-Rest Earth Thrust 4.4
4A.2 RANKINE EARTH PRESSURES 4.5
   4A.2.1 Rankine Active Pressure 4.6
   4A.2.2 Rankine Passive Pressure 4.9

4A.2.3 Deformation and Boundary Conditions 4.12
4A.3 COULOMB EARTH PRESSURES 4.13
   4A.3.1 Coulomb Active Pressure 4.13
4A.4 REFERENCES 4.17

The vertical pressures that exist within a semi-infinite soil mass are easy to compute using the following equation:

$$\sigma_v = \gamma z \quad (4A.1)$$

However, the lateral stresses within the soil are not as easily determined. Since the lateral dimension is large, there is little reason for significant lateral compression to occur. Consequently it is reasonable to assume that the vertical locked in effective stresses, $\overline{\sigma}_v$ would be larger than the effective lateral stresses, $\overline{\sigma}_h$ at the same point. The ratio of horizontal to lateral stresses is usually presented as

$$\overline{K} = \frac{\overline{\sigma}_h}{\overline{\sigma}_v} \quad (4A.2)$$

where $\overline{K}$ is the earth pressure ratio.

4A.1 EARTH PRESSURE AT REST

For conditions where no lateral deformation occurs, $\overline{K}$ is termed $\overline{K}_0$, the coefficient of earth pressure at rest. In discussing coefficients of earth pressure in real soils, care must be taken to show whether total or effective stresses are used. The two are clearly not equal because the total stresses are obtained by adding a constant (the pore water pressure) to the effective vertical and horizontal stresses.

Measurements of $\overline{K}_0$ were reported by Terzaghi as early as 1920. Many other studies have suggested that the coefficient depends primarily on the effective friction angle $\overline{\phi}$ and stress history for sands, and on the plasticity and stress history for clays.

Jaky[1] suggested that for sands and normally consolidated clays,

$$\overline{K}_0 = 1 - \sin \overline{\phi} \quad (4A.3)$$

A study of the Brooker and Ireland[2] data suggests that the best relationship might be

$$\overline{K}_0 = (0.95 - \sin \overline{\phi}) + 0.15 \quad (4A.4)$$

Alpan[3] gave the following equation for overconsolidated soils:

$$\overline{K}_{0(oc)} = \overline{K}_{0(nc)}\sqrt{OCR} \tag{4A.5}$$

where $\overline{K}_{0(oc)}$ = coefficient of overconsolidated earth pressure at rest
$\overline{K}_{0(nc)}$ = coefficient of normally consolidated earth pressure
OCR = overconsolidation ratio, $= \overline{p}_c/\overline{p}_0$

Alpan also suggested that the following empirical equation could be used for normally consolidated clays:

$$\overline{K}_0 = 0.19 + 0.233 \log (PI) \tag{4A.6}$$

where PI is the plasticity index.

Brooker and Ireland[2] further suggested that

$$\overline{K}_{0(oc)} = \overline{K}_{0(nc)}(OCR)\, e^{\sin(1.2\overline{\phi})} \tag{4A.7}$$

Mayne[4] reported that

$$\overline{K}_{0(oc)} = \overline{K}_{0(nc)}\left(A + \frac{s_u}{\overline{p}_0}\right) \tag{4A.8}$$

where s_u = undrained strength
$\overline{p}_0$ = effective vertical pressure at depth z
A = 0.8 for K_0 consolidated undrained triaxial tests
= 0.7 for isotropically consolidated triaxial tests
= 1.0 for K_0 consolidated drained simple shear tests

s_u can be estimated according to Bjerrum[5] as

$$\frac{s_u}{\overline{p}_0} = 0.45(PI)^{0.5} \quad \text{for PI} > 5$$

$$\frac{s_u}{\overline{p}_0} = 0.18(LI)^{0.5} \quad \text{for LI} > 0.5 \tag{4A.9}$$

where LI = liquidity index = $\dfrac{W - PL}{PI}$
W = water content

4A.1.1 Calculation of At-Rest Earth Thrust

Assume that a wall is placed without disturbing the soil (Fig. 4A.1). The thrust is computed as the area of the at-rest stress versus depth,

$$\overline{P}_0 = \tfrac{1}{2}H\, \overline{\sigma}_h = \tfrac{1}{2}H\, \overline{K}_0 \gamma\, H = \tfrac{1}{2}\overline{K}_0 \gamma\, H^2 \tag{4A.10}$$

If the water table is at the ground surface (Fig. 4A.2),

$$\overline{\sigma}_v = \gamma\, z - u_z$$

Then

$$\overline{P}_0 = \tfrac{1}{2}\overline{K}_0\, \gamma\, H^2 \tag{4A.11}$$

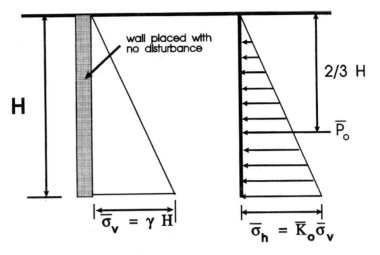

FIGURE 4A.1 Thrust of earth pressure at rest.

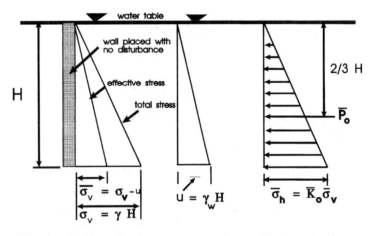

FIGURE 4A.2 Thrust of earth pressure at rest with water table at ground surface.

If
$$u_z = H(\gamma_w)$$

$$U = \tfrac{1}{2} H(\gamma_w H)$$

Then
$$P_0 = \overline{P}_0 + U \qquad (4A.12)$$

4A.2 RANKINE EARTH PRESSURES

In 1857 Rankine[6] investigated the plastic equilibrium of soil. The following discussion presents Rankine's solutions for the active and passive soil stress states.

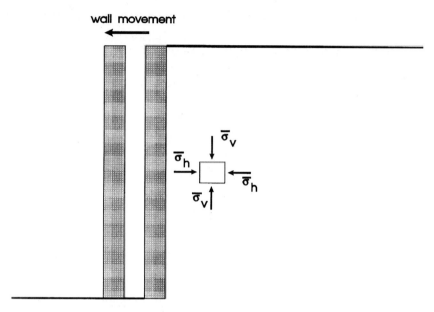

FIGURE 4A.3 Rankine active stress state.

4A.2.1 Rankine Active Pressure

If an unyielding (rigid) frictionless wall in the at-rest condition is allowed to move away from the soil, each soil element next to the wall can expand laterally as σ_h is reduced (Fig. 4A.3). The horizontal stresses on the element of soil behind the wall are reduced from the at-rest state such that, in a cohesionless or normally consolidated and drained clay soil, a Mohr's failure circle develops and the failure condition occurs (Fig. 4A.4). The stress is presented as

$$\overline{\sigma}_{h_f} = \overline{\sigma}_a = \overline{K}_a \overline{\sigma}_v \tag{4A.13}$$

Hence
$$\overline{K}_a = \frac{\overline{\sigma}_a}{\overline{\sigma}_v} \tag{4A.14}$$

where $\overline{K}_a$ is the coefficient of active earth pressure. From Fig. 4A.4,

$$\frac{\overline{\sigma}_a}{\overline{\sigma}_v} = \tan^2\left(45° - \frac{\overline{\phi}}{2}\right) \tag{4A.15}$$

Hence
$$\overline{K}_a = \tan^2\left(45° - \frac{\overline{\phi}}{2}\right) = \frac{1 - \sin\overline{\phi}}{1 + \sin\overline{\phi}} \tag{4A.16}$$

$$= \frac{1}{N_\phi}$$

where
$$N_\phi = \tan^2\left(45° + \frac{\overline{\phi}}{2}\right) \tag{4A.17}$$

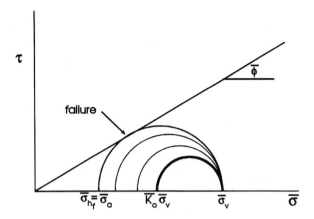

FIGURE 4A.4 Mohr's circle for Rankine active stress state.

The active thrust is computed from the area of the stress diagram as

$$\overline{P}_a = \tfrac{1}{2} H \overline{\sigma}_a = \tfrac{1}{2} H \overline{K}_a \gamma H = \tfrac{1}{2} \gamma H^2 \overline{K}_a \tag{4A.18}$$

For those cases where the groundwater table is at the ground surface

$$\overline{P}_a = \tfrac{1}{2}(\gamma H - u) \overline{K}_a H = \overline{P}_a + U \tag{4A.19}$$

If the soil behind the wall is a saturated clay, then the Mohr's circle relationships are as shown in Fig. 4A.5. Note that for the undrained case, the stresses are total stresses. The active stress can be computed from

$$\sigma_a = \sigma_v - 2c = \gamma z - 2c \tag{4A.20}$$

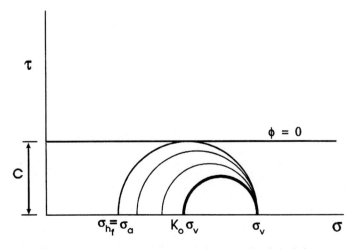

FIGURE 4A.5 Active Rankine failure circle for saturated undrained clay.

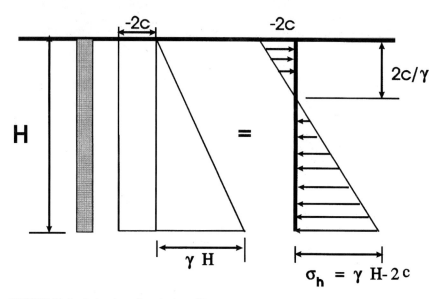

FIGURE 4A.6 Active thrust for cohesive soils.

The active thrust for this case can then be determined using Fig. 4A.6,

$$P_a = \tfrac{1}{2}\gamma H^2 - 2cH \tag{4A.21}$$

Setting $P_a = 0$ and solving,

$$2cH = \frac{\gamma H^2}{2} \tag{4A.22}$$

$$H = \frac{4c}{\gamma}$$

Therefore for depths less than $4c/\gamma$, the soil is in tension.

For partly saturated soils that possess both c and ϕ, Fig. 4A.7 shows the development of the active Mohr's circle. The equation for the active pressure σ_a is

$$\sigma_a = \sigma_v \tan^2\left(45° - \frac{\phi}{2}\right) - 2c\tan\left(45° - \frac{\phi}{2}\right) \tag{4A.23}$$

$$= \gamma z \tan^2\left(45° - \frac{\phi}{2}\right) - 2c\tan\left(45° - \frac{\phi}{2}\right)$$

The active thrust P_a is computed as

$$P_a = \tfrac{1}{2}\gamma H^2 \tan^2\left(45° - \frac{\phi}{2}\right) - 2cH\tan\left(45° - \frac{\phi}{2}\right) \tag{4A.24}$$

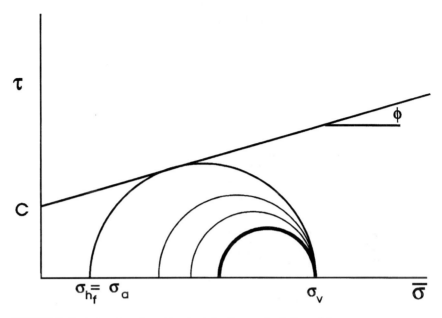

FIGURE 4A.7 Active stress for partly saturated soil processing both c and ϕ.

4A.2.2 Rankine Passive Pressure

If the wall of Fig. 4A.3 moves into the retained backfill instead of away from the backfill, passive stresses are generated (Fig. 4A.8). The development of the Mohr's circle at failure for a cohesionless backfill ($c = 0$) is shown in Fig. 4A.9. At failure, $\overline{\sigma}_h > \overline{\sigma}_v$. Therefore

$$\overline{\sigma}_h = \overline{\sigma}_p = \overline{\sigma}_v \tan^2\left(45° + \frac{\phi}{2}\right) = \overline{\sigma}_v \overline{K}_p \qquad (4A.25)$$

Hence
$$\overline{K}_p = \tan^2\left(45° + \frac{\phi}{2}\right) \qquad (4A.26)$$

The passive thrust, $\overline{P}_p$ is computed as the area of the corresponding stress diagram (Fig. 4A.10),

$$\overline{P}_p = \tfrac{1}{2}\gamma H^2 \overline{K}_p \qquad (4A.27)$$

If the water table is at the surface of the backfill,

$$\sigma_p = (\gamma_b z)\, \overline{K}_p \qquad (4A.28)$$

The total thrust P_p is then

$$\overline{P}_p = \tfrac{1}{2}\gamma_b H^2 \overline{K}_p \qquad (4A.29)$$

$$P_p = \overline{P}_p + U = \tfrac{1}{2}\gamma_b H^2 \overline{K}_p + \tfrac{1}{2}\gamma_w H^2 \qquad (4A.30)$$

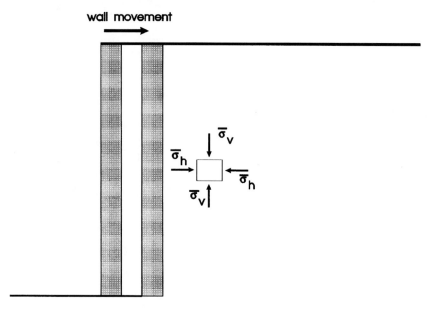

FIGURE 4A.8 Passive stress conditions.

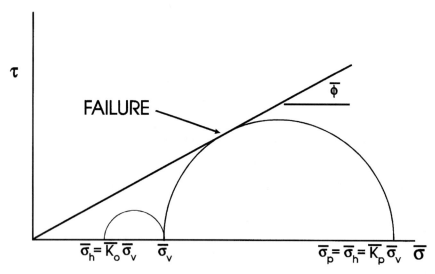

FIGURE 4A.9 Mohr's failure circle for passive state of stress.

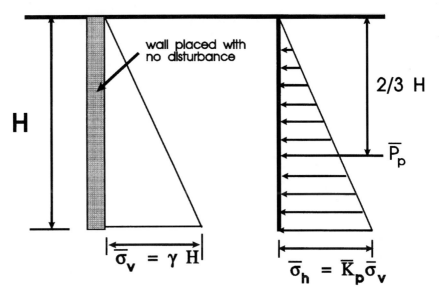

FIGURE 4A.10 Passive thrust.

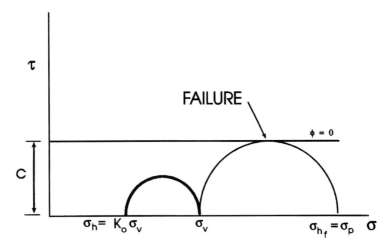

FIGURE 4A.11 Mohr's failure circle for passive stress in clay soil.

If the soil behind the wall is a saturated clay, then the Mohr's circle relationships are as shown in Fig. 4A.11. Note that for the undrained case, the stresses are total stresses. The passive stress can be computed from

$$\sigma_p = \sigma_v + 2c \tag{4A.31}$$

Therefore the passive thrust is

$$P_p = \tfrac{1}{2}\gamma H^2 + 2cH \tag{4A.32}$$

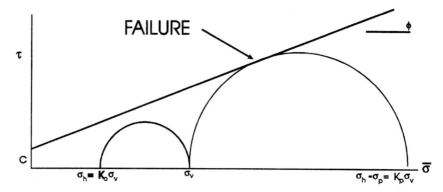

FIGURE 4A.12 Mohr's failure circle for passive stress in partly saturated soil.

For a soil possessing both c and ϕ (partly saturated soil), the failure Mohr's circles are shown in Fig. 4A.12. The corresponding passive stress is

$$\sigma_p = \sigma_v \tan^2\left(45° + \frac{\phi}{2}\right) + 2c \tan\left(45° + \frac{\phi}{2}\right) \tag{4A.33}$$

The passive thrust can be found as

$$P_p = P_c + P_\phi = \frac{1}{2}\gamma H^2 \tan^2\left(45° + \frac{\phi}{2}\right) + 2cH \tan\left(45° + \frac{\phi}{2}\right) \tag{4A.34}$$

4A.2.3 Deformation and Boundary Conditions

In both the active and the passive states the zones of soil adjacent to the frictionless wall that are in a state of impending shear failure or plastic equilibrium form plane wedges. The failure plane from Mohr's circles is $45° + \phi/2$ from the plane of $\sigma_{1\ max}$. The failure planes developed for the active case are shown in Fig. 4A.13 and for the passive case in Fig. 4A.14. The amount of hori-

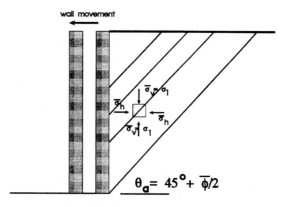

FIGURE 4A.13 Active-case failure planes.

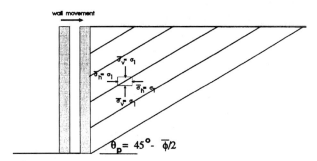

FIGURE 4A.14 Passive-case failure planes.

TABLE 4A.1 Movement Required to Generate Active and Passive Stresses

Soil	Typical minimum movement	
	Active	Passive
Dense, cohesionless	0.0005H	0.005H
Loose, cohesionless	0.0020H	0.010H
Stiff, cohesive	0.0100H	0.020H
Soft, cohesive	0.0200H	0.040H

zontal movement of any point of the wall that is necessary to produce either the active or the passive state depends on the width of the shear zone at that point. Typical minimum movements necessary to generate the active and passive states are given in Table 4A.1. Figure 4A.15 shows schematically the amount of movement necessary to generate active and passive stresses.

4A.3 COULOMB EARTH PRESSURES

The Rankine earth pressure theory assumes a general failure in a soil mass. The stresses in the mass are determined by assuming plasticity and then developing stresses in terms of overburden stresses.

In order for the Rankine theory to be valid, both the deformation conditions and the boundary conditions must be valid. In Fig. 4A.16 it is apparent that the boundary of the active Rankine zone, which is inclined at $45° + \phi/2$ with the horizontal, will intersect the retaining wall. Thus there will be a restraint on the deformation of the soil, invalidating the Rankine assumptions. Because of this, other methods are based on the evaluation of the forces on the active wedge. The simplest and oldest of these methods was developed by Coulomb.[6]

4A.3.1 Coulomb Active Pressure

If it is assumed that the failure surface is planar and that it intersects the ground surface, then a free-body diagram can be drawn, as shown in Fig. 4A.17. The force W is the weight of the failure wedge. The normal force N is the resultant of all the normal stresses along the assumed failure plane AB. The shear force S is the sum of the unit shearing stress along the failure plane multi-

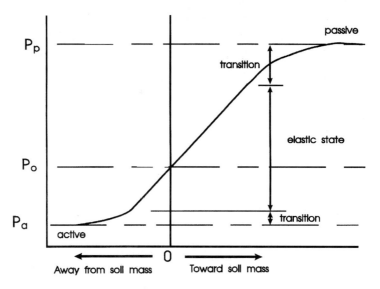

FIGURE 4A.15 Movement required to generate active and passive stresses.

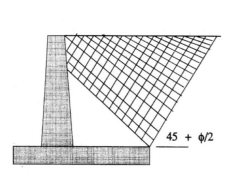

FIGURE 4A.16 Retaining wall with active Rankine zone.

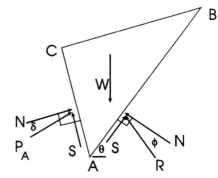

FIGURE 4A.17 Coulomb active wedge.

plied by the length of the plane. The unit failure stress is defined by the Mohr-Coulomb shear strength equation,

$$s = \tau_f = \bar{c} + \bar{\sigma} \tan \bar{\phi}$$

For a cohesionless soil ($c = 0$) the shear force S at failure, is the normal force N times $\tan \bar{\phi}$. The single resultant R is inclined at an angle $\bar{\phi}$ with the normal force.

If the soil is cohesive ($\bar{\phi} = 0$), then the shear force is simply c times the length of the failure plane $\overline{AB}$. R is then the resultant of $c\overline{AB}$ and N.

The retaining wall exerts forces on the wedge also. N is the normal to the boundary $\overline{AC}$. At the point of impending failure, a shear force is acting upward along the retaining wall. The angle of friction δ between soil and wall is equal to or less than that between soil and soil, $\bar{\phi}$. The angle of friction between wall and soil depends primarily on the smoothness of the wall material. For

FIGURE 4A.18 Vector solution.

retaining walls δ is usually assumed to be two-thirds of φ. P_A, the resultant force exerted between the retaining wall and the soil behind it, is inclined at an angle δ below the normal to the wall-soil boundary.

Since the magnitude and direction of W are known as well as the directions of R and P_A, the system of forces can be solved (Fig. 4A.18). However, this solution depends on the selected failure plane θ. Obviously, the selected plane may not be the most critical failure plane. Consequently several wedges should be investigated, and the failure wedge through the base of the wall that generates the highest P_A is the governing wedge.

4A.3.1.1 Trial Wedge Method. There are many graphic methods to determine the Coulomb active earth pressure. The simplest to understand is the trial wedge method. Its advantage is that forces on the diagram act in the directions that one normally expects them to act, that is, a force acting downward is a vector directed to the bottom of the page. The trial wedge is a force polygon in which several trial failure wedges are investigated, with their weight vectors acting from the origin.

The method is best illustrated by an example. A retaining wall is 24 ft high with a backfill slope *i* of 20° (1:2.75 slope) and a wall batter of 8 ft in 24 ft. The backfill consists of sand with a unit weight γ = 125 pcf and φ = 36°. The wall friction angle δ is assumed to be ⅔ φ = 24°. Various assumed trail failure surfaces, AV, AC, etc., are shown in Fig. 4A.19.

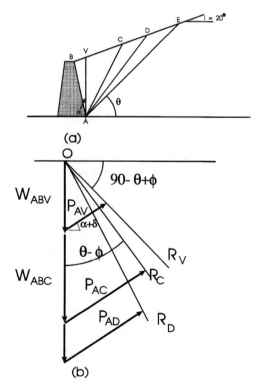

FIGURE 4A.19 Coulomb trial wedge solution.

The solution proceeds as follows:

1. The area of the first trial wedge, area $\overline{ABV}$, is determined. The magnitude of vector W_{ABV} is then area $\overline{ABV}$ times the unit weight of the backfill soil γ. This vector is drawn to scale vertically downward from the origin O.
2. The second vector, R_V, acts at an angle ϕ from the normal to the failure surface $\overline{AV}$. Simple geometry shows that R_V then acts at an angle $90 - \theta + \phi$ clockwise from the horizontal. This vector is drawn through the origin, O.
3. The direction of P_{AV} and all of the active force vectors will be at an angle δ below the normal to the wall-wedge interface and will be the same for all trials. Simple geometry shows that this angle is $\omega + \delta$ from the horizontal, where ω is the back-face batter angle of the wall. P_{AV} is drawn from the head of vector W_{ABV} to vector R_V.
4. The magnitude of P_{AV} is then scaled from the diagram.
5. The Coulomb active thrust is the largest vector found using trial wedges.

In actuality, graphic methods have given way to a general formula that can be solved easily,

$$P_A = \tfrac{1}{2} \gamma H^2 \overline{K_A} \qquad (4A.35)$$

$$\overline{K_A} = \frac{\cos^2(\phi - \omega)}{\cos^2 \omega \cos(\omega + \delta)\left[1 + \sqrt{\dfrac{\sin(\phi + \delta)\sin(\phi - i)}{\cos(\omega + \delta)\cos(\omega - i)}}\right]^2} \qquad (4A.36)$$

where i = slope backfill angle
ω = angle of batter between wall and vertical
ϕ = friction angle of soil
δ = angle of friction between soil and wall

4A.3.1.2 Point of Application of Earth Pressure. Simple statics show that the resultant R acts at the one-third point of the trial failure plane (Fig. 4A.20). As can be seen, the vectors are not concurrent. Consequently moment equilibrium is not obeyed. This is the result of the assumption that the failure plane is straight. It is in actuality curved very slightly convex downward in the active case. The magnitude of the error introduced is negligible. However, the curvature is much greater for the passive Coulomb case, and the difference accounts for the gross inaccuracy of the Coulomb method for passive earth pressure. Unfortunately the passive Coulomb method results in an overestimation of the passive earth pressure and therefore an error on the unsafe side. The Rankine method, although not theoretically correct either, gives more conservative and usable values.

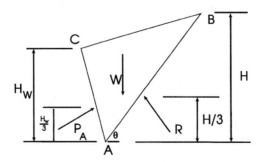

FIGURE 4A.20 Coulomb active thrust vector locations.

4A.4 REFERENCES

1. J. Jaky, "The Coefficient of Earth Pressure at Rest," *J. Soc. Hungar. Architects and Engineers,* vol. 7, pp. 355–358, 1944.
2. E. W. Brooker and H. O. Ireland, "Earth Pressure at Rest Related to Stress History," *Can. Geotech. J.,* vol. 2, pp. 1–15, 1965.
3. I. Alpan, "The Empirical Evaluation of the Coefficients K_0 and K_{0r}," *Soils and Foundations,* vol. 7, p. 31, 1967.
4. P. W. Mayne, "$K_0 - c_u/\sigma_{v0}$ trends for Overconsolidated Clays," *J. Geotech. Eng., ASCE,* vol. 110, pp. 1511–1516, Oct. 1984.
5. L. Bjerrum, "Theoretical and Experimental Investigations on the Shear Strength of Soils," Norwegian Geotechnical Institute Publ. no. 50510, 1954.
6. W. M. J. Rankine, "On Stability of Loose Earth," *Phil. Trans. R. Soc. London,* p. I, pp. 9–27, 1857.
7. C. A. Coulomb, "Essai sur une Application des Règles de Maximis et Minimis à quelques Problèmes de Statique, Relatifs à l'Architecture," *Mem. R. Soc. Sci.,* Paris, vol. 3, p. 38, 1776.

SECTION 4B
DESIGN OF RIGID RETAINING WALLS

Richard W. Stephenson

4B.1 TYPES OF RETAINING WALLS 4.19
   4B.1.1 Foundation Walls 4.21
   4B.1.2 Gravity Walls 4.21
   4B.1.3 Cantilever Walls 4.21
   4B.1.4 Counterfort Walls 4.21
   4B.1.5 Buttressed Walls 4.21
4B.2 EARTH PRESSURES APPLIED TO WALL 4.21
4B.3 LIMIT STATE FOR RETAINING WALLS 4.21
4B.4 LIMITS OF DESIGN 4.22
4B.5 MINIMUM DEPTH OF BASE 4.22
4B.6 SIMPLIFIED METHOD OF DESIGN 4.23
   4B.6.1 Estimating of Stresses on Nonyielding Walls 4.23
   4B.6.2 Estimating Applied Forces under Active Conditions 4.24
   4B.6.3 Stability 4.29
4B.7 THEORETICAL METHOD OF DESIGN 4.32
   4B.7.1 Unyielding Walls 4.32
   4B.7.2 Active Earth Pressure on Solid Gravity Walls 4.33
   4B.7.3 Active Forces on Cantilever, Counterfort, and Buttressed Walls 4.33
   4B.7.4 Passive Earth Pressures 4.35
   4B.7.5 Stability 4.35

A retaining wall is used to support an earthen backfill. Typically it is constructed of concrete with reinforcing steel, unreinforced concrete, and other materials. A retaining wall may also support loads directly, such as a bridge abutment, or indirectly, as in the support of structures placed on or in the backfill.

Common uses for a retaining wall include:

- Bridge abutments
- Retaining soil in highway and railroad cuts
- Retaining backfill
- Retaining piles of ore, coal, sand, and so on
- Retaining water as a floodwall or canal wall

4B.1 TYPES OF RETAINING WALLS

In general there are five types of traditional retaining walls (Fig. 4B.1):

1. Foundation walls
2. Gravity walls
3. Cantilever walls
4. Counterfort walls
5. Buttressed walls

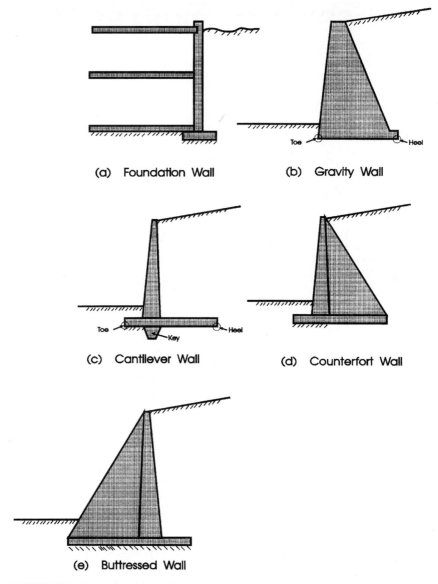

FIGURE 4B.1 Types of retaining walls.

4B.1.1 Foundation Walls

Foundation walls are basement walls of buildings. These walls are typically supported by the additional floor slabs and floor joists.

4B.1.2 Gravity Walls

Gravity walls are massive walls that rely on their trapezoidal shape and great weight to minimize tensile stresses in the concrete. No reinforcing steel is used. These walls have been constructed as high as 130 ft (39 m). However, they are used mainly for low walls. The base width is typically 0.4 to 0.7 of the wall height H. The outer face is normally battered 0.75 to 1.0 in/ft (6.3 to 8.4 cm/m).

4B.1.3 Cantilever Walls

Cantilever walls are reinforced concrete vertical walls (stems) supported on a slab (base). The wall is usually tied into the slab by reinforcing steel that runs the entire height of the wall. The wall may be keyed into the ground to resist lateral movement. The width of the base is typically 0.4 to 0.7H. The thickness of the base and stem is typically $H/12$ to $H/10$. The projection of the toe beyond the stem is usually $B/3$. The main advantage of cantilever walls over gravity walls is a considerable savings in concrete.

4B.1.4 Counterfort Walls

Counterfort walls are cantilever walls that are equipped with reinforced concrete ribs on the inner (earth) side of the wall. Less reinforcing steel is required for counterfort walls than for cantilever walls, but more forming is necessary. The base width is typically 0.4 to 0.7H and the slab thickness is $B/14$ to $B/12$. The ribs are normally placed 0.3 to 0.6H apart.

4B.1.5 Buttressed Walls

Buttressed walls are cantilever walls that are equipped with reinforced concrete ribs on the outer side of the wall. The proportions are about the same as for counterfort walls. These types of walls are sometimes used for the support of slopes. A main disadvantage is that the buttresses protrude into the usable space in front of the wall.

4B.2 EARTH PRESSURES APPLIED TO WALL

If the wall is fixed in place, such as a foundation wall supported by additional horizontal bracing, then the pressures K_0 applied to the wall are at rest. If the wall has moved or is moving away from the backfill, then active pressures K_A are being applied to the wall. If the wall has been or is being forced into the backfill, then passive pressures K_P are being applied to the wall. A schematic of the applied pressures is given in Fig. 4B.2.

4B.3 LIMIT STATE FOR RETAINING WALLS

Retaining walls can fail either by complete collapse of the wall itself or by allowing so much movement that the wall no longer serves its design purpose.

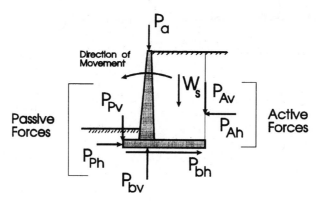

FIGURE 4B.2 Forces typically applied to retaining walls.

Complete collapse of the wall is usually limited to the breaking of the stems of cantilever, counterfort, or buttressed walls due to excessive earth pressures or inadequate structural design.

Excessive movement of the wall with the wall intact can occur in four ways:

1. Overturning by rotation about its toe
2. Sliding of wall in its base
3. Bearing capacity failure of soil beneath the wall
4. General shear failure of entire wall-backfill unit where backfill, subsoil, and wall fail together

Movements may occur rapidly during or shortly after construction, or they may build up gradually with time.

4B.4 LIMITS OF DESIGN

The design of a retaining wall is usually done at one of two levels. Simplified methods are generally used for low walls [$H \leq$ about 20 ft (6 m)]. For these simplified methods, only generalized characteristics of the backfill must be known and only minor backfill placement control is necessary. The factors of safety are such that adequate performance is assured.

Higher retaining walls [$H >$ 20 ft (6 m)] require a more theoretical design approach using earth pressure theory. These approaches, coupled with a high level of backfill control, can reduce the cost of these walls significantly.

4B.5 MINIMUM DEPTH OF BASE

The base of the retaining wall should be below the depth of frost penetration. This is normally 1 to 3 ft (0.3 to 0.9 m), depending on the climate.

Although it would be best to place the wall base below the depth of seasonal variations in moisture content, it generally is not practical to do so. Consequently the use of fully softened soil properties is recommended for the design unless positive moisture control measures are used.

It is further necessary to make sure that the wall base is placed below the maximum depth of scour.

4B.6 SIMPLIFIED METHOD OF DESIGN

4B.6.1 Estimating of Stresses on Nonyielding Walls

Pile-supported walls for abutments with tops constrained by bridge decks, basement walls supported by floor slabs, the insides of U-shaped abutments, and box culverts are relatively fixed. The horizontal stresses produced on such structures depend on the soil properties, backfill placement procedures, and a variety of other factors.

For relatively dry, loosely dumped cohesionless gravel to sandy silty soils, the lateral soil pressure σ_h may be estimated using

$$\sigma_h = \gamma_e z \tag{4B.1}$$

where γ_e = equivalent unit weight
z = depth below a horizontal ground surface

The equivalent unit weight is normally taken as 60 pcf (961 kg/m³) for this case. For any soil beneath the water table γ_e should be taken as 100 pcf (1602 kg/m³). Thus if the depth to the water table is z_w, the total horizontal stress below the water table (Fig. 4B.3) is

$$\sigma_h = 60 z_w + 100 (z - z_w) \tag{4B.2}$$

For loosely dumped soft clay or organic soil γ_e = 100 pcf (1602 kg/m³) whatever the position of the water table.

Compacted backfill will generate larger horizontal stresses than loosely dumped backfill. The magnitude of these stresses will depend to a large part on the method of compaction. Cohesionless soils that are compacted in a moist to air-dry condition with hand tampers or small vibratory compactors will have a γ_e of 50 pcf (800 kg/m³). If the water table subsequently rises, 90 pcf (1441 kg/m³) should be used below the water table and 50 pcf (800 kg/m³) above. The equivalent unit weight for clays that are compacted using hand-operated pneumatic tampers is 120 pcf (1922 kg/m³).

Compacted soil placed next to a nonyielding wall with heavy rollers will induce larger stresses than those indicated. The value of γ_e may exceed 500 pcf (8000 kg/m³) at a few feet below the surface, but will diminish with depth.

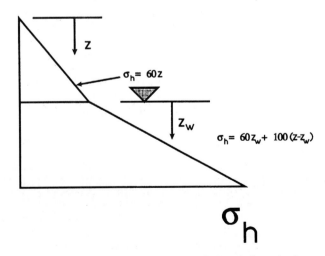

FIGURE 4B.3 At-rest horizontal pressures for loosely dumped soils.

4.24 EARTH PRESSURES AND RETAINING SYSTEMS

TABLE 4B.1 Types of Backfill for Retaining Walls

1. Coarse-grained soil without admixture of fine soil particles, very permeable (clean sand or gravel)
2. Coarse-grained soil of low permeability due to admixture of particles of silt size
3. Residual soil with stones, fine silty sand, and granular materials with conspicuous clay content
4. Very soft or soft clay, organic silts, or silty clays
5. Medium or stiff clay, deposited in chunks and protected in such a way that a negligible amount of water enters the spaces between chunks during floods or heavy rains

4B.6.2 Estimating Applied Forces under Active Conditions

To carry out these computations, the backfill is first classified into one of the five categories listed in Table 4B.1. The forces applied to the wall are then estimated using charts for any one of the following four cases:

1. The backfill slopes upward away from the wall, at an angle i (Fig. 4B.4), to a distance well removed from the wall. No significant loads are applied to the ground surface.
2. The backfill slopes upward away from the wall to a point a and then becomes horizontal. The ground surface is not subjected to significant loads.
3. The backfill is horizontal, but is subjected to a uniform surface pressure.
4. The backfill is horizontal, but is subjected to a uniformly distributed line load parallel to the wall.

The forces are calculated on vertical planes through the heel of the wall (Fig. 4B.4). The vertical force is P_{Av} and the horizontal force is P_{Ah}. For case 1 the forces on the wall are given by

$$P_{Av} = \tfrac{1}{2} K_v H^2 \qquad (4B.3)$$

$$P_{Ah} = \tfrac{1}{2} K_h H^2 \qquad (4B.4)$$

The values of K_v are given in Fig. 4B.5 and the values of K_h in Fig. 4B.6. H is the height of the vertical plane extending from the heel to the ground surface, not the height of the wall. For type

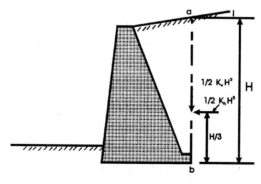

FIGURE 4B.4 Active thrust.

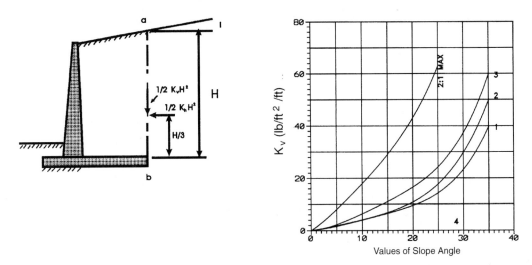

FIGURE 4B.5 Vertical earth pressure coefficients for retaining walls with plane surface backfills.

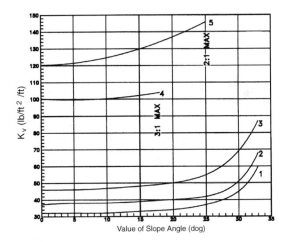

FIGURE 4B.6 Horizontal earth pressure coefficients for planar backfill.

5 backfill, the computations may be performed using a height 4 ft (1.2 m) less than H. The stresses applied to the vertical plane are assumed to increase linearly with depth. Thus the resultant forces P_{Av} and P_{Ah} act at the lower one-third point ($H/3$) above the heel. This point of application is used even for type 5 backfill.

For case 2 the forces on the vertical plane through the heel are again calculated using Eqs. (4B.3) and (4B.4), but the coefficients K_v and K_h are taken from Figs. 4B.7 to 4B.11. The pressure distributions and points of application are determined as in case 1. The forces for case 2 are shown in Fig. 4B.12.

4.26 EARTH PRESSURES AND RETAINING SYSTEMS

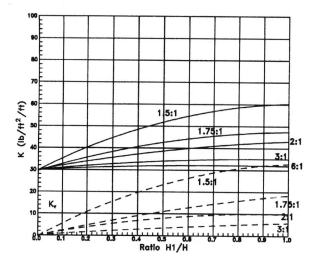

FIGURE 4B.7 Case 2, K values for soil type 1.

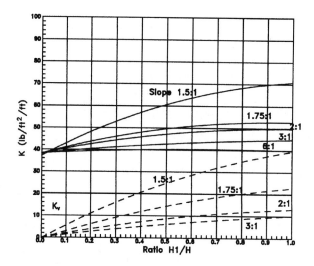

FIGURE 4B.8 Case 2, K values for soil type 2.

For case 3 the application of a uniform pressure over the ground surface behind a retaining wall causes a uniform increase in the horizontal pressure on the vertical plane through the heel by an amount P_q,

$$P_q = Cq \tag{4B.5}$$

The coefficient C is determined from Table 4B.2. The application of a line load at the surface parallel to the wall is assumed to produce a horizontal line load on the vertical plane through the heel. The point of application is found by drawing a line at an angle of 40° to the horizontal (Fig. 4B.13)

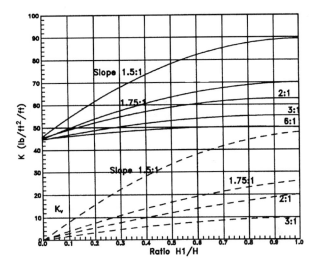

FIGURE 4B.9 Case 2, K values for soil type 3.

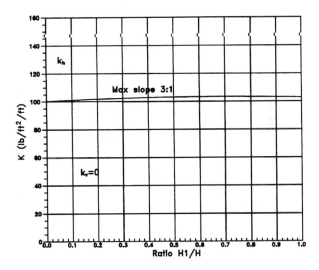

FIGURE 4B.10 Case 2, K values for soil type 4.

from the surface line load to a point d_1 on the back of the wall. The line load on the vertical plane through the heel is assumed to act at the same height as the point d_1 and is computed as

$$P'_q = Cq' \tag{4B.6}$$

where q' is the surface line load as a force per unit of length and C is given in Table 4B.2.

The line load also causes an increase in the vertical stress on the top of the base of a cantilever, counterfort, or buttressed wall and on the top of any heel projection of a solid gravity wall (Fig. 4B.13),

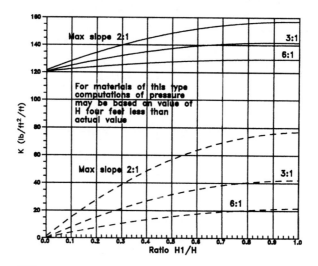

FIGURE 4B.11 Case 2, K values for soil type 5.

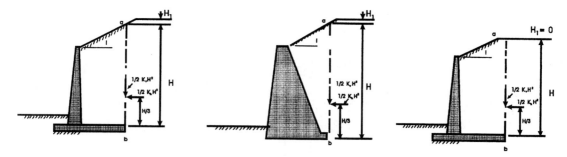

FIGURE 4B.12 Forces for case 2—backfill with limit height.

$$p_q'' = \frac{q'}{ef} \tag{4B.7}$$

It generally is not a good idea to use expansive soil as backfill. However, if it must be used, then classify the soil as type 5 and make the following adjustments.

1. When analyzing stability against overturning and sliding, increase P_{Ah} by 50% and set $P_{Av} = 0$.
2. When analyzing for bearing capacity failure, increase P_{Av} by 50% and leave P_{Ah} unaltered.

Backfills of silts and sandy silts which have access to moisture from a high water table or seepage from natural soil behind the wall and are not protected by suitable drains may have moisture drawn into them near the wall which will form ice lenses. When these lenses expand, they will push the wall outward. Subsequent thawing will soften the backfill. For soils of this type, assume a type 4 backfill.

It is important that adequate drainage be provided in the backfill against the wall. There should be no buildup of pore pressures pushing against the backfill wall face. Measures taken should include the use of freely draining backfill, impervious cover, and the use of weep holes 2 to 4 in

TABLE 4B.2 Case 3

Backfill type	C
1	0.3
2	0.4
3	0.4
4	1.0
5	1.0

(5 to 10 cm) in diameter, 5 ft (1.5 m) on center, located just above the base. However, if no positive measures are taken to control water behind the wall, then the earth pressure coefficients K_h in the previous figures must be increased by 50 lb/ft²/lin ft (81 kg/m²/m) to account for the additional pressures caused by the presence of the unbalanced water forces.

Passive pressures at the toe (P_{Ph} and P_{Pv}; Fig. 4B.14) are generally assumed to be zero because of their high potential for removal due to excavation, erosion, or scour. However, if a key is used, substantial passive forces can be developed. The passive force should be computed only over the key height, $K_v = 0$ and K_h is given in Table 4B.3.

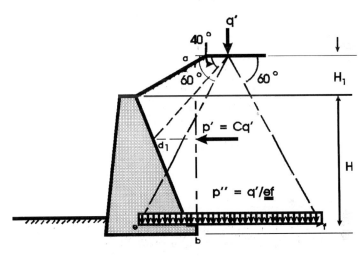

FIGURE 4B.13 Forces on wall due to line load at surface.

4B.6.3 Stability

Retaining walls must have an adequate factor of safety against overturning, bearing capacity failure, sliding, and general failure. The selected factors of safety generally allow for small but not negligible movement of the walls. The normally selected factors of safety are listed in Table 4B.4. The forces and distances for the computations are given in Fig. 4B.15.

4B.6.3.1 Overturning. In this mode of failure the entire wall rotates away from the backfill, with the center of rotation at the toe. The factor of safety F_{so} is then

$$F_{so} = \frac{\Sigma M_r}{\Sigma M_d} = \frac{\text{resisting moment}}{\text{driving moment}} \tag{4B.8}$$

For a solid gravity wall and neglecting passive pressures,

$$\Sigma M_r = W_c l_c + W_s l_s \tag{4B.9}$$

and

$$\Sigma M_d = P_{Ah} l_{Ah} - P_{Av} l_{Av} \tag{4B.10}$$

4.30 EARTH PRESSURES AND RETAINING SYSTEMS

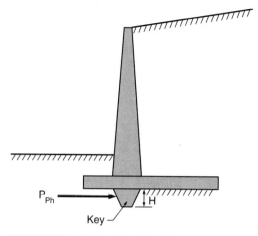

FIGURE 4B.14 Passive force due to key.

TABLE 4B.3 Passive Pressure Coefficients for Keys

Soil type	K_h, lb/ft²/ft
1	400
2	350
3	220
4	100
5	100

TABLE 4B.4 Factors of Safety for Wall Stability Analysis

Mode of failure	Wall supports a structure	Wall does not support a structure
Overturning	3.0	1.5
Sliding	3.0	1.5
Bearing capacity	3.0	1.5
General	3.0	1.5

For cantilever, counterfort, and buttressed walls,

$$\Sigma M_r = W_s l_s + W_{cs} l_{cs} + W_{cb} l_{cb} \tag{4B.11}$$

4B.6.3.2 Bearing Capacity Failure. A retaining wall must be evaluated for the bearing capacity of its base. The factor of safety against bearing capacity failure is

$$FS_b = \frac{Q_f}{Q_b} \tag{4B.12}$$

where Q_f = maximum total force that can be applied to soil beneath wall
Q_b = actual applied force

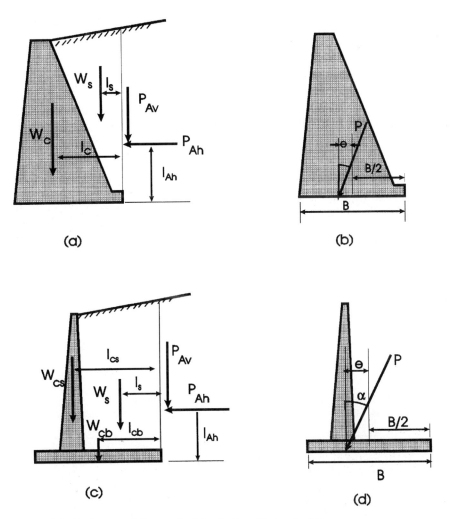

FIGURE 4B.15 Forces and distances for determination of factor of safety.

Q_f must be defined for the force with the same eccentricity and inclination as Q_b. The eccentricity and inclination of Q_b are found by satisfying the equation of static equilibrium.

For a solid gravity wall,

$$Q_b = [P_{Ah}^2 + (W_c + W_s + P_{Av})^2]^{0.5} \tag{4B.13}$$

$$i = \tan^{-1}\left(\frac{P_{Ah}}{W_{cs} + W_{cb} + W_s + P_{Av}}\right) \tag{4B.14}$$

$$e = \left(\frac{W_{cs}l_{cs} + W_{cb}l_{cb} + W_s l_s + P_{Ah}l_{Ah}}{W_{cs} + W_{cb} + W_s + P_{Av}}\right) - \frac{1}{2}B \tag{4B.15}$$

For a cantilever wall,

$$Q_b = [(W_{cs} + W_{cb} + W_s + P_{Av})^2 + P_{Ah}^2]^{0.5} \tag{4B.16}$$

4.32 EARTH PRESSURES AND RETAINING SYSTEMS

$$i = \tan^{-1}\left(\frac{P_{Ah}}{W_{cs} + W_{cb} + W_s + P_{Av}}\right) \qquad (4\text{B}.17)$$

$$e = \left(\frac{W_{cs}l_{cs} + W_{cb}l_{cb} + W_s l_s + P_{Ah}l_{Ah}}{W_{cs} + W_{cb} + W_s + P_{Av}}\right) - \tfrac{1}{2}B \qquad (4\text{B}.18)$$

In all cases e must be less than $\tfrac{1}{6}B$. If a larger value is obtained, the wall should be redesigned or a theoretical analysis performed.

4B.6.3.3 Sliding Failure. Sliding failure occurs when the wall moves laterally away from the backfill, such as horizontal displacement without rotation. The factor of safety is

$$FS_s = \frac{(W_w + W_s + P_{Av})\mu + A}{P_{Ah}} \qquad (4\text{B}.19)$$

where W_w = total weight of wall
μ = coefficient of friction between base of wall and subsoil
A = adhesive force between base of wall and subsoil, $= a \times B$
a = adhesion between wall and subsoil
B = wall base width

For a freely draining granular subsoil, $A = 0$ and $\mu = 0.60$. For a subsoil of clay or silt that is fully drained, use Table 4B.5. For fully undrained loading, use $\mu = 0$ and $a = c$ (undrained cohesion).

If a relatively thin layer of cohesionless soil fill is placed between the base and a cohesive subbase, the factor of safety should be calculated for sliding in both the cohesionless fill and the cohesive subsoil. The lower value should be used for the design.

TABLE 4B.5 Friction and Adhesion for Drained Clay or Silt

	A	μ
LL*<50	0	0.40
50<LL<100	0	0.20
LL>100	0	0.10

*LL = liquid limit.

4B.7 THEORETICAL METHOD OF DESIGN

Theoretical methods should be used for high walls [$H > 20$ ft (6 m) or so] and for critical walls that have unusual or important functions.

4B.7.1 Unyielding Walls

Walls that are generally assumed to be unyielding include basement walls braced by floor slabs before placement of the backfill and rigid walls placed on unyielding bases such as rock or unyielding batter piles.

The flexure of the stem of a cantilever wall is enough to develop active stresses except in the region near the base, as shown in Fig. 4B.16, where either insufficient rotation of the stem or simultaneous rotation of both the stem and the base leads to an at-rest condition.

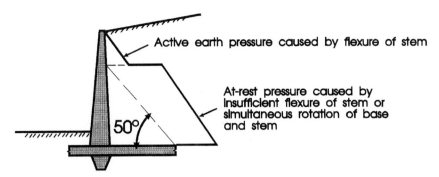

FIGURE 4B.16 Pressure distribution on unyielding wall.

Earth pressures against walls that are unyielding are indeterminate because the pressures are a function of the method of placement, the amount of loading after placement, soil properties, drainage conditions, and special conditions such as frost, or expansive soil.

At-rest pressures are computed by determining the value of $\overline{K}_0$,

$$\overline{K}_0 = \frac{\overline{\sigma}_h}{\overline{\sigma}_v} \tag{4B.20}$$

The minimum value of $\overline{K}_0$ is normally determined using the Jaky equation,

$$\overline{K_0 min} = 1 - \sin \overline{\phi} \tag{4B.21}$$

This equation is used for loosely dumped or lightly compacted backfill and for cohesionless backfill compacted using vibratory methods.

The use of heavy rollers leads to values of $\overline{K}_0$ as high as 5 for soil near the surface. However, the value at any given depth decreases rapidly as additional fill is added. Heavy rollers should not be used for compaction near the wall to avoid excessive pressures.

4B.7.2 Active Earth Pressure on Solid Gravity Walls

The active thrust on solid gravity walls should be determined using Coulomb's trial wedge analysis procedures (Fig. 4B.17). If the backfill used is well-compacted granular soil, then $\overline{\phi}$ and $\overline{\phi}_w = \overline{\phi}/2$ should be used in the equations. If the backfill is cohesive, then both drained and undrained analysis should be conducted and the most critical analysis used.

If the backfill is sloping, then the following equation for $\overline{K}_A$ should be used:

$$\overline{K}_A = \left[\frac{\csc \beta \sin (\beta - \phi)}{\sqrt{\sin (\beta + \phi_w)} + \sqrt{\frac{\sin (\phi + \phi_w) \sin (\phi - i)}{\sin (\beta - i)}}} \right]^2 \tag{4B.22}$$

4B.7.3 Active Forces on Cantilever, Counterfort, and Buttressed Walls

For this case, the active force P_{AR} should be calculated using Rankine methods on a vertical plane through the heel. For cohesionless soils ($c = 0$),

$$\overline{\sigma}_h = \gamma \, z \overline{K}_{AR} \tag{4B.23}$$

4.34 EARTH PRESSURES AND RETAINING SYSTEMS

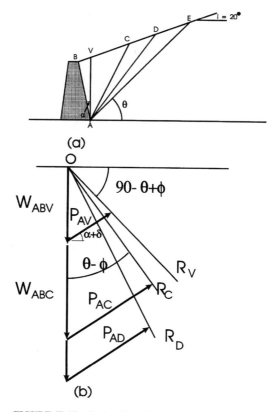

FIGURE 4B.17 Coulomb's wedge analysis.

$$\overline{K_{AR}} = \cos i \; \frac{\cos i - \sqrt{\cos^2 i - \cos^2 \phi}}{\cos i + \sqrt{\cos^2 i - \cos^2 \phi}} \tag{4B.24}$$

$$\overline{P_{AR}} = \tfrac{1}{2} \gamma H^2 \overline{K_{AR}} \tag{4B.25}$$

$\overline{P_{AR}}$ is parallel to the ground surface.

For cohesive backfill with $i = 0$ (no slope) the pressures are computed as follows:

$$\sigma_h = \gamma z \tan^2\left(45° - \frac{\phi}{2}\right) - 2c \tan\left(45° - \frac{\phi}{2}\right) \tag{4B.26}$$

The earth pressure is zero to a depth:

$$H_c = 2 \frac{c}{\gamma} \tan\left(45° - \frac{\phi}{2}\right) \tag{4B.27}$$

$$P_{AR} = P_c + P_\gamma \tag{4B.28}$$

P_c acts at $(H - H_c)/2$ above the base and PV acts at $(H - H_c)/3$ above the base:

$$P_c = -2c (H - H_c) \tan\left(45° - \frac{\phi}{2}\right) \tag{4B.29}$$

$$P_\gamma = \tfrac{1}{2}\gamma\,(H^2 - H_c^2)\tan^2\left(45° - \frac{\phi}{2}\right) \qquad (4\text{B}.30)$$

For uniform surcharge q,

$$\sigma_q = q\tan^2\left(45° - \frac{\phi}{2}\right) \qquad (4\text{B}.31)$$

$$P_q = qH\tan^2\left(45° - \frac{\phi}{2}\right) \qquad (4\text{B}.32)$$

4B.7.4 Passive Earth Pressures

The Rankine analysis procedures should be used for passive earth pressures.

4B.7.4.1 Horizontal Backfill Ground Surfaces.
The following equations should be used to compute the passive thrusts for walls with a horizontal backfill ground surface:

$$\sigma_h = \gamma z \tan^2\left(45° + \frac{\phi}{2}\right) + q\tan^2\left(45° + \frac{\phi}{2}\right) + 2c\tan\left(45° + \frac{\phi}{2}\right) \qquad (4\text{B}.33)$$

$$P_{PR} = P_{P\gamma} + P_{Pq} + P_{Pc} \qquad (4\text{B}.34)$$

$$P_{P\gamma} = \frac{1}{2}\gamma H^2 \tan^2\left(45° + \frac{\phi}{2}\right) \qquad (4\text{B}.35)$$

$$P_{Pq} = qH\tan^2\left(45° + \frac{\phi}{2}\right) \qquad (4\text{B}.36)$$

$$P_{Pc} = 2cH\tan\left(45° + \frac{\phi}{2}\right) \qquad (4\text{B}.37)$$

4B.7.4.2 Cohesionless Backfill Soil with Sloping Ground Surfaces.
For cohesionless backfill soils with a sloping ground surface,

$$P_{PR} = \frac{1}{2}\gamma H^2 K_{PR} \qquad (4\text{B}.38)$$

$$K_{PR} = \cos i\,\frac{\cos i + \sqrt{\cos^2 i - \cos^2 \phi}}{\cos i - \sqrt{\cos^2 i - \cos^2 \phi}} \qquad (4\text{B}.39)$$

4B.7.5 Stability

4B.7.5.1 Overturning of Solid Gravity Walls.
The factor of safety against overturning of solid gravity walls is computed using

$$FS_o = \frac{\Sigma M_r}{\Sigma M_d} \qquad (4\text{B}.40)$$

The forces and moment arms are shown in Fig. 4B.18. The moments are computed as follows:

$$\Sigma M_r = l_c W_c \qquad (4B.41)$$

$$\Sigma M_d = P_{Ac} \cos(\phi_w + \omega) l_h - P_{Ac} \sin(\phi_w + \omega) l_v \qquad (4B.42)$$

The passive forces are assumed to be negligible.

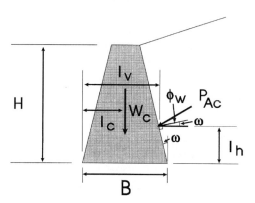

FIGURE 4B.18 Forces and moment arms for overturning of solid gravity walls.

4B.7.5.2 Bearing Capacity Failure of Solid Gravity Walls. The forces used for the computation of the factor of safety against bearing capacity failure are shown in Fig. 4B.19. The equations are

$$Q_b = \frac{P_{Ah}}{\sin i} = \frac{P_A \cos(\phi_w + \omega)}{\sin i} \qquad (4B.43)$$

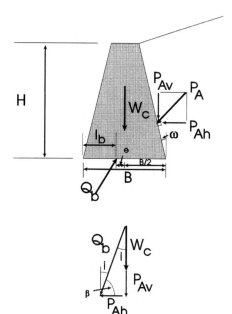

FIGURE 4B.19 Forces and locations for bearing capacity of solid gravity walls.

TABLE 4B.6 Sliding Friction for Gravity Walls

Type	μ
Clean sound rock	0.70
Clean gravel, sand-gravel mixtures, coarse sand	0.55–0.65
Clean fine to medium sand, silty coarse sand, silty or clayey gravel	0.45–0.55
Silty or clayey fine to medium sand	0.35–0.45
Silt	0.30

$$i = \tan^{-1}\left[\frac{P_A \cos(\phi_w + \omega)}{W_c + P_A \sin(\phi_w + \omega)}\right] \qquad (4B.44)$$

$$e = \frac{(W_c/P_A)(B - l_c) + l_v[\cos(\phi_w + \omega) + \sin(\phi_w + \omega)\tan\omega]}{\cos(\phi_w + \omega)\cot\beta} - \frac{B}{2} \qquad (4B.45)$$

4B.7.5.3 Sliding of Solid Gravity Walls. The factor of safety against sliding of solid gravity walls is computed using the equation

$$FS_s = \frac{[W_c + P_A \sin(\phi_w + \omega)]\mu + A}{P_A \cos(\phi_w + \omega)} \qquad (4B.46)$$

For a granular base beneath the wall, the coefficient of base friction μ may be taken as the full coefficient of friction of the soil itself ϕ and $A = 0$.

For a cohesive soil an analysis should be performed for both the undrained and the drained cases unless construction proceeds so slowly that the undrained case cannot develop. For the undrained case, $\mu = 0$ and A is calculated using the full undrained strength of the subsoil c. For the drained case, $A = 0$ and μ is the fully drained coefficient of friction of the subsoil ϕ. When measurements are not available, use μ from Table 4B.6.

PART · 5

SOIL IMPROVEMENT AND STABILIZATION

SECTION 5A
NONGROUTING TECHNIQUES

Evert C. Lawton

5A.1 INTRODUCTION 5.3
5A.2 OVEREXCAVATION/REPLACEMENT 5.4
   5A.2.1 Ultimate Bearing Capacity 5.5
   5A.2.2 Settlement 5.14
   5A.2.3 Usefulness and Limitations of the Finite Element Method 5.30
   5A.2.4 Limitations of the Overexcavation/Replacement Method 5.30
5A.3 NEAR-SURFACE COMPACTION 5.36
   5A.3.1 Equipment 5.39
   5A.3.2 Moisture-Density Relationships 5.54
   5A.3.3 Laboratory Compaction and Testing 5.59
   5A.3.4 Engineering Properties and Behavior of Compacted Soils 5.72
   5A.3.5 Quality Control and Assurance 5.173
5A.4 CHEMICAL STABILIZATION 5.205
   5A.4.1 Chemicals and Reactions 5.205
   5A.4.2 Engineering Properties and Behavior of Chemically Stabilized Soils 5.224
   5A.4.3 Construction Methods 5.252
5A.5 MOISTURE BARRIERS 5.264
   5A.5.1 Horizontal Barriers 5.264
   5A.5.2 Vertical Barriers 5.265
5A.6 REFERENCES 5.267

5A.1 INTRODUCTION

Increasing urbanization continues to occur throughout much of the world, and at this time this trend appears likely to continue for at least the near future. As the population grows in any metropolitan area, a variety of additional facilities are needed to serve these people. Many of these facilities are in the form of structures—such as houses, apartment buildings, restaurants, and office buildings—that occupy areas within the metropolis. The additional space required for these structures is generally obtained in three ways: (*a*) Existing structures are torn down to make room for new structures; (*b*) new structures are built on land within the existing metropolitan boundaries that were previously "unimproved" (relative term); and (*c*) the boundaries of the metropolis are expanded to provide additional land for development.

One of the primary criteria used to select a site for development is the suitability of the ground for supporting the structure to be built. In most urban areas, the best sites were developed first, and, as urbanization continues, when a previously undeveloped site is purchased, the engineering properties of the existing near-surface materials are often such that the structure cannot be supported by shallow foundations. The traditional solution for these situations is to support the structures on deep foundations—typically piles or drilled piers—where a small portion of the load is transmitted to the poorer near surface materials and a large portion of the load is transmitted to better bearing materials deeper within the ground. However, even with the development and use of more economical types of deep foundations such as auger-cast piles, the demand has increased for more economical solutions to the unsuitability of near-surface soils for shallow foundations.

To meet this demand, numerous soil stabilization and improvement techniques (also commonly called *ground modification techniques*) have been developed within the past 25 years or so. These techniques involve modifying the engineering properties and behavior of the near-surface soils at a site so that shallow foundations can be used where they previously could not or in some instances so that more economical shallow foundations can be used. The state of the art in this field is currently changing so fast that it is difficult, if not impossible, for any one person to keep up with it. It is likely that significant new technologies or modifications to existing technologies will have occurred between the time of writing and the time of publication of this handbook. Therefore, it is incumbent upon the reader to review the literature frequently to keep up with developments and changes in ground modification techniques.

Many of the techniques described in this section have proprietary restrictions to their use. In many instances, the restrictions are not in the use of the method itself but in other areas such as in the manufacture of a material or product used in the technique or in equipment used to perform the modifications to the soil or to install a particular product. Because patents associated with ground improvement techniques are continuously expiring and new patents are being obtained, it would be onerous and unfruitful to attempt to describe all these restrictions. Few patents are worldwide, and certain restrictions that apply in one country may not apply in the country where the project is being undertaken. Therefore, anyone who wishes to use or recommend the use of a particular soil improvement technique for a project should first perform a thorough examination of all potential proprietary restrictions associated with that technique.

Although many techniques are currently available, and more are likely to become available, not all techniques are appropriate for every project. The primary responsibilities of the foundation engineer, therefore, are to (*a*) determine which techniques can be used to safely support the structure; (*b*) determine which of the suitable methods is most economical for the project; (*c*) design or supervise the design of the details for the technique that best meets criteria (*a*) and (*b*); and (*d*) ensure that the actual modifications produce the desired result. In some instances, especially where proprietary restrictions are involved, the company or individuals who developed a particular method may require that they perform the design themselves, so that the project foundation engineer's responsibility is to ensure that the design prepared by the company meets criteria (*a*) and (*b*) and that the final product achieves the desired objectives stated in the specifications.

In keeping with the goals of this handbook, soil improvement and stabilization techniques that are applicable to buildings are discussed in this chapter. The chapter is divided into two main sections—nongrouting and grouting techniques. These two sections are organized according to type of technique. In building applications, the most important static properties for bearing soils are strength, compressibility, drainage characteristics (permeability), and potential for wetting- and drying-induced volume changes. In seismically active regions, liquefaction potential of saturated silty sands and sands is extremely important. The discussions in this chapter are primarily aimed at showing how each technique can be used to improve one or more of these properties.

5A.2 OVEREXCAVATION/REPLACEMENT

The technique of overexcavation/replacement is one of the oldest, most intuitive, and simplest methods for modifying bearing materials to increase support for shallow foundations. The method consists of excavating poor or inadequate bearing material and replacing it with a stiffer and stronger material (see Fig. 5A.1). As long as the in-place replacement material is stiffer and stronger than the excavated bearing soil, the settlement that the foundation undergoes when loaded is reduced and the factor of safety against ultimate bearing capacity failure is increased. The greater the stiffness and strength of the replacement material, the greater the reduction in settlement and increase in ultimate bearing capacity.

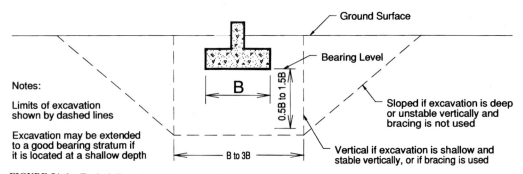

FIGURE 5A.1 Typical dimensions of excavation for overexcavation/replacement.

Overexcavation/replacement is most commonly used when the bearing soils are very weak and highly compressible. The replacement material can be the excavated material that has been modified in some way, or it can be borrow material (obtained from another location on or off the site). The replacement material is usually sand, gravel, or a sand-gravel mixture, especially in situations where the ground-water table is high or when it is desirable to have a free-draining bearing material. To obtain the stiffest and strongest material possible, the replaced sand or gravel is usually compacted in lifts during the replacement process. If good-quality sands and gravels are not readily or economically available, the excavated soils can be chemically stabilized and used as the replacement material. However, chemically stabilized soils are generally not free-draining (they have relatively low permeabilities), so their use as a replacement material may change the local or regional ground-water seepage and precipitation infiltration patterns, which may be an environmental consideration in some instances. It is also possible to use the excavated soil as the replacement material without chemically modifying it, although this is seldom done. If the inadequate bearing soil is cohesionless, several techniques are available for densifying in situ soil that are more economical than removing, drying or wetting, replacing, and compacting it. Cohesive soils can be excavated, dried, and compacted at an appropriate water content to produce a material that is initially stiff and strong, but these soils may be susceptible to wetting-induced volume changes and reductions in stiffness and strength from wetting (see Sec. 5A.3.4).

The excavation is deeper and often wider than is needed to place the foundation, hence the term *overexcavation*. Typical dimensions for an overexcavation are shown in Fig. 5A.1. The width of the bottom of the excavation typically varies from one to three times the width of the foundation (B to $3B$ in Fig. 5A.1), and the depth below the bearing level is generally about ½ to 1½ times the foundation width ($0.5B$ to $1.5B$). If a good bearing stratum (medium dense sand, dense sand, gravel, or bedrock) exists close to the bearing level, the excavation is usually taken to the top of the bearing stratum or a shallow depth into it.

The replacement material is usually compacted so that the replaced zone is as stiff and strong as possible. A variety of compaction procedures can be used. If the excavation is narrow and shallow, hand-operated compaction equipment appropriate for confined areas is used, including rammers, tampers, vibrating plates, and small rollers (see Sec. 5A.3.1). For deep, narrow excavations, backhoes or hydraulic excavators with special compaction attachments can be used. Other techniques include pounding the material with the bucket of a backhoe and dropping a weight from a crane. For wide excavations, full-size compaction rollers are normally used.

5A.2.1 Ultimate Bearing Capacity

When the overexcavation/replacement process is properly designed and implemented, it results in a composite bearing zone that is stronger and stiffer than the unreinforced material. An increase in ultimate bearing capacity occurs because the potential failure surface must pass either through or around the stronger replaced zone. Three possible failure methods are shown in Fig. 5A.2 for a foundation bearing on a strong replaced zone of finite width and depth. It is possible (but unlikely) that the most critical failure surface is a general bearing failure surface that develops through the replaced zone and into the in situ soil [Fig. 5A.2(a)]; a theoretical solution for this case where the foundation bears on a long (continuous) granular trench within a soft saturated clay matrix immediately after loading (the clay is undrained) has been given by Madhav and Vitkar (1978). Their solution is presented in the form of charts from which modified bearing capacity factors (N_{ct}, N_{qt}, and $N_{\gamma t}$) can be obtained (Fig. 5A.3). These factors are valid for the following conditions:

1. The width of the granular trench (W) varies from zero (no trench) to twice the foundation width ($2B$).
2. The friction angle for the granular trench material (ϕ_1) varies from 20° to 50°, and the cohesion intercept (c_1) varies from zero to the same value as for the clay matrix (c_2).
3. The clay matrix is undrained ($\phi_2 = 0$, $c_2 =$ undrained shear strength $= s_{u2}$).
4. The unit weights of the granular trench material and the clay matrix are the same ($\gamma_1 = \gamma_2$).

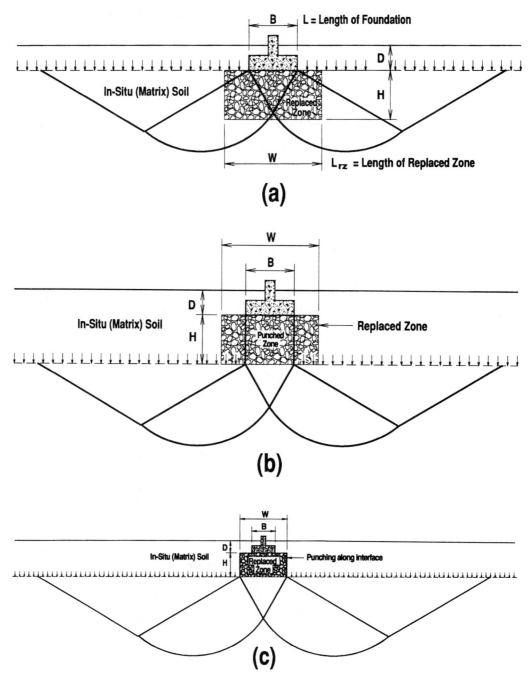

FIGURE 5A.2 Potential bearing shear failure mechanisms for foundation supported by overexcavated and replaced zone: (a) general shear failure through replaced zone and in situ soil; (b) punching failure through replaced zone; (c) punching failure of replaced zone through in situ soil.

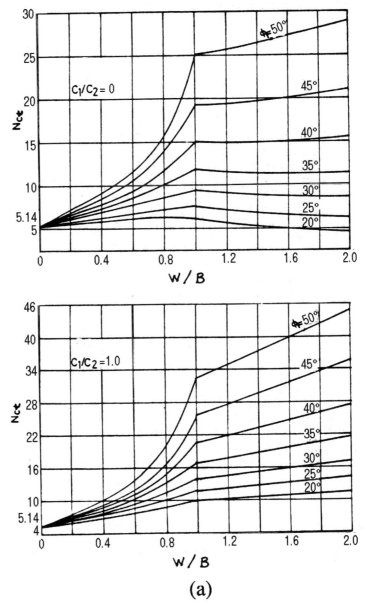

FIGURE 5A.3 Bearing capacity factors for general shear failure in a weak clay stabilized with a granular trench (from Madhav and Vitkar, 1978): (a) N_{ct}, (b) N_{qt}, (c) $N_{\gamma t}$.

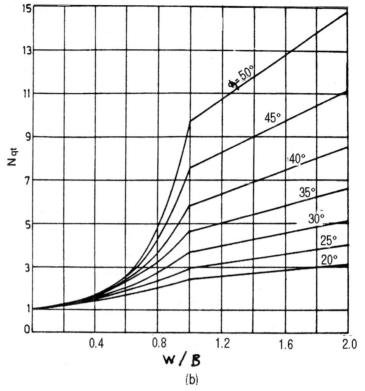

FIGURE 5A.3 (Continued)

With these factors, the ultimate bearing capacity (q_{ult}) can be predicted from the following equation, which has the same form as the general bearing capacity equations for homogeneous bearing soil:

$$q_{ult} = c_2 N_{ct} + q N_{qt} + 0.5 \gamma_2 B N_{\gamma t} \tag{5A.1}$$

Hamed and coworkers (1986) conducted laboratory model tests to determine the variation in q_{ult} for a continuous foundation bearing on a granular trench ($W = B$) constructed within a soft clay matrix. By comparing their experimental values of q_{ult} with values predicted from Eq. (5A.1) (see Fig. 5A.4), Hamed and colleagues concluded that Madhav and Vitkar's theory overpredicts q_{ult}. In addition, they found that the minimum height of the granular trench necessary to obtain the maximum value of q_{ult} is about $3B$. Hamed and colleagues also developed their own theory for q_{ult} based on the following assumptions:

1. Failure occurs by bulging at the bearing level along the interface between the trench and matrix materials.
2. Undrained conditions exist at failure in the saturated clay matrix.
3. The principal planes at the failure point are horizontal and vertical in both the trench and matrix materials.
4. The horizontal stress (σ_h) in the trench material is the minor principal stress (σ_3); σ_h in the matrix material is the major principal stress (σ_1). These two values of σ_h are equal.

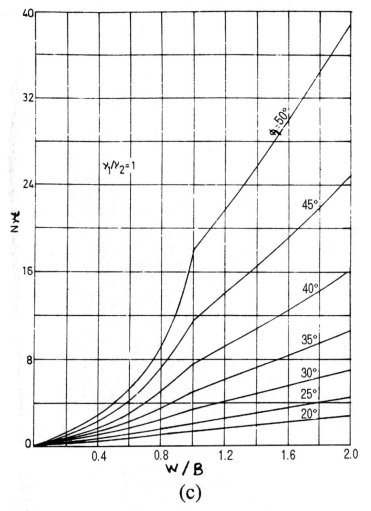

FIGURE 5A.3 *(Continued)*

From these assumptions, the following equation for q_{ult} was derived:

$$q_{ult} = (\gamma_2 D + 2s_{u2}) \cdot \tan^2\left(45° + \frac{\phi_1}{2}\right) \tag{5A.2}$$

A comparison of the maximum experimental values of q_{ult} with values calculated from Eq. (5A.2) (Fig. 5A.4) suggests that this theory provides a reasonable (but probably somewhat conservative) estimate of q_{ult} for $W = B$ and $H/B \geq 3$.

If the in situ soil is a moderate or stiff clay or a granular soil, it is likely that failure would occur by punching through or around the perimeter of the replaced zone combined with general shear failure in the underlying in situ soil [Fig. 5A.2(*b*) and (*c*)]. The wider the replaced zone is, the more likely that punching failure would occur through it rather than around it. If it is not apparent which punching mechanism is more likely to occur, estimates of q_{ult} for both types can be determined and the lesser value used for design. It is also possible that general shear failure would

5.10 SOIL IMPROVEMENT AND STABILIZATION

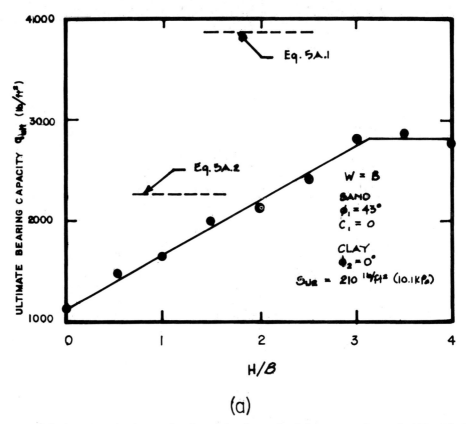

FIGURE 5A.4 Ultimate bearing capacity of a model continuous foundation on a granular trench within a soft clay matrix (from Hamed and coworkers, 1986): (*a*) trench material is dense sand.

occur completely within the replaced zone (q_{rz}), but this is not likely unless the replaced zone is deep and wide. It is prudent, however, to calculate q_{rz} in all cases to ensure that the computed values of q_{ult} for the two punching cases are less than q_{rz}.

For punching through the replaced zone [Fig. 5A.2(*b*)], the following equation based on Meyerhof and coworkers (Meyerhof and Hanna, 1978; Valsangkar and Meyerhof 1979) and Bowles (1988) can be used to estimate q_{ult}:

$$q_{ult} = q_b + \frac{pP_h \tan \phi_1 + pHc_1 - W_{pz}}{A_f} \leq q_{rz} \tag{5A.3a}$$

where q_b = ultimate bearing capacity of the in situ soil beneath the replaced zone based on the dimensions of the foundation

p = perimeter length of the punched zone = $2(B + L)$ for a rectangular foundation or πd_0 for a circular foundation (where d_0 = diameter)

P_h = lateral earth pressure thrust (force per unit horizontal length) acting along the perimeter surface of the punched zone at failure = $\int_0^H K_s \sigma_v dH$

K_s = lateral earth pressure coefficient along the perimeter surface of the punched zone at failure

W_{pz} = weight of the material in the punched zone

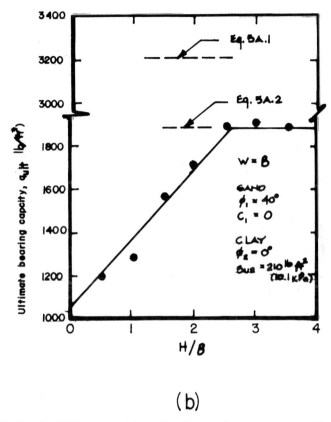

FIGURE 5A.4 (*Continued*) (*b*) Trench material is medium dense sand.

A_f = area of the foundation = BL for rectangular and $0.25\pi d_0^2$ for circular
q_{rz} = ultimate bearing capacity for general shear failure within the replaced zone

If the ground-water table is below the lowest point on the potential failure surface, the following equations apply:

$$q_b = c_2 N_{c2} S_{c2} D_{c2} + \gamma_2(D + H)N_{q2} S_{q2} D_{q2} + 0.5\gamma_2 B N_{\gamma2} S_{\gamma2} D_{\gamma2} \tag{5A.3b}$$

where $N_{c2}, N_{q2}, N_{\gamma2}$ are bearing capacity factors (see Sec. 2B) for general shear failure in the in situ soil beneath the punched zone based on ϕ_2 and $S_{c2}, S_{q2}, S_{\gamma2}$ are shape factors and $D_{c2}, D_{q2}, D_{\gamma2}$ are depth factors (see Sec. 2B) based on the dimensions of the foundation and an embedded depth of $(D + H)$.

The use of γ_2 for the N_{q2} term in Eq. (5A.3b) is conservative if W is greater than B, because a portion of the surcharge soil for the general failure surface beneath the punched zone consists of the replacement material (γ_1), which in most cases is denser than the in situ soil ($\gamma_1 > \gamma_2$). The wider the replaced zone, the more conservative is the use of γ_2 in the N_{q2} term. If desired, an equivalent unit weight can be calculated that accounts for the relative portions of the replacement material and in situ soil that are in the surcharge zone, but this refinement is probably not necessary in most instances.

$$q_{rz} = c_1 N_{c1} S_{c1} D_{c1} + \gamma_2 D N_{q1} S_{q1} D_{q1} + 0.5\gamma_1 B N_{\gamma1} S_{\gamma1} D_{\gamma1} \tag{5A.3c}$$

5.12 **SOIL IMPROVEMENT AND STABILIZATION**

where N_{c1}, N_{q1}, $N_{\gamma1}$ are bearing capacity factors for general shear failure within the replaced zone based on ϕ_1 and S_{c1}, S_{q1}, $S_{\gamma1}$ are shape factors and D_{c1}, D_{q1}, $D_{\gamma1}$ are depth factors based on the dimensions of the foundation and an embedded depth of D. In addition,

$$W_{pz} = BLH\gamma_1 \quad \text{for a rectangular foundation} \quad (5A.3d.1)$$

$$W_{pz} = 0.25\pi d_0^2 H\gamma_1 \quad \text{for a circular foundation} \quad (5A.3d.2)$$

If K_s is assumed to be constant at all depths along the perimeter surface of the punched zone, the following equation applies for P_h:

$$P_h = K_s(\gamma_2 DH + 0.5\gamma_1 H^2) \quad (5A.3e)$$

If the ground-water table is within the potential failure zone, its influence on the effective stresses and unit weights (and hence q_b, q_{rz}, W_{pz}, and P_h) must be considered, which may alter Eqs. (5A.3b), (5A.3c), (5A.3d), and (5A.3e).

Selection of a value for K_s is not necessarily simple. K_s probably varies along the depth of the punched zone, so if one value is selected, it constitutes an average or equivalent value. Meyerhof and Hanna (1978) provided values of K_s as a function of ϕ_1 and q_2/q_1 for the case of a continuous foundation bearing on a strong layer of infinite horizontal extent overlying a weak layer (Fig. 5A.5); q_1 and q_2 are the ultimate bearing capacities for a continuous foundation of width B under a vertical centric load on the surfaces of homogeneous thick deposits of the strong and weak soils and can be calculated from the following equations:

$$q_1 = c_1 N_{c1} + 0.5\gamma_1 BN_{\gamma1} \quad (5A.4a)$$

$$q_2 = c_2 N_{c2} + 0.5\gamma_2 BN_{\gamma2} \quad (5A.4b)$$

Thus, the term q_2/q_1 represents the relative strengths of the weaker and stronger materials, with $q_2/q_1 = 1$ representing soils of the same strength and $q_2/q_1 = 0$ corresponding to a strong soil that is infinitely stronger than the weak soil.

Values for K_s can be selected from Fig. 5A.5 but should be used with caution, since these values do not strictly apply to either finite length foundations or a replaced zone of finite width. The reader should also note that K_s does not equal Rankine's $K_p = \tan^2(45° + \phi_1/2)$ for two reasons: (a) Shear stresses develop along the vertical punched perimeter, and therefore σ_h and σ_v cannot be principal stresses; and (b) values of σ_v along the punched perimeter are likely to be higher than calculated from the weights of the overburden material owing to the application of the foundation load and the shearing action along the punched perimeter. Item (a) tends to reduce the value of K_s, whereas item (b) tends to raise it. Hence, K_s may be either greater or lesser than K_p when K_p is based on σ_v calculated from the weights of the overburden materials [as in Eq. (5A.3e)]; for example, the values of K_s for $\phi_1 = 30°$ in Fig. 5A.5 range from about 1.0 for $q_2/q_1 = 0$ to about 5.6 for $q_2/q_1 = 1$, while $K_p = 3.0$. If a conservative value is desired, K_s can be assumed equal to the at-rest coefficient for normally consolidated soils (K_{0-nc}):

$$K_s = K_{0-nc} = 1 - \sin\phi_1 \quad (5A.5)$$

It is unlikely that K_s would ever be less than K_{0-nc}. For a design situation, a better estimate of K_s can be obtained from Fig. 5A.5 than from Eq. (5A.5) so long as a reasonable factor of safety is applied to q_{ult}.

If the replaced zone punches through the matrix soil [Fig. 5A.2(c)], punching resistance will develop within the in situ soil adjacent to the vertical perimeter surface of the replaced zone. The following equation applies to this case:

$$q_{ult} = q_b + \frac{pP_h\tan\phi_2 + pHc_2 - W_{rz}}{A_{rz}} \leq q_{rz} \quad (5A.6a)$$

where q_b = ultimate bearing capacity of the in situ soil beneath the replaced zone based on the dimensions of the replaced zone

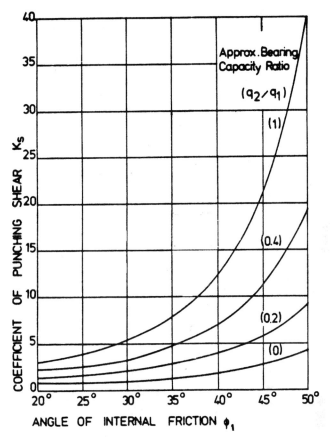

FIGURE 5A.5 Coefficients of punching shear resistance under vertical load (from Meyerhof and Hanna, 1978).

p = perimeter length of the replaced zone = $2(W + L_{rz})$ for a rectangular replaced zone (where L_{rz} is the length of the replaced zone) and πd_{rz} for a circular replaced zone (where d_{rz} is the diameter of the replaced zone)

P_h = lateral earth pressure thrust (force per unit length) acting along the perimeter surface of the replaced zone at failure $\int_0^H K_s \sigma_v dH$

K_s = lateral earth pressure coefficient along the perimeter surface of the replaced zone at failure

W_{rz} = weight of the material in the replaced zone

A_{rz} = area of the replaced zone = WL_{rz} for a rectangular zone and $0.25\pi d_{rz}^2$ for a circular zone

q_{rz} = ultimate bearing capacity for general shear failure within the replaced zone [from Eq. (5A.3c)]

If the ground-water table is below the lowest point on the potential failure surface, the following equations apply:

$$q_b = c_2 N_{c2} S_{c2} D_{c2} + \gamma_2(D + H) N_{q2} S_{q2} D_{q2} + 0.5\gamma_2 W N_{\gamma2} S_{\gamma2} D_{\gamma2} \quad (5A.6b)$$

where N_{c2}, N_{q2}, $N_{\gamma 2}$ are bearing capacity factors for general shear failure in the in situ soil beneath the replaced zone based on ϕ_2 and S_{c2}, S_{q2}, $S_{\gamma 2}$ are shape factors and D_{c2}, D_{q2}, $D_{\gamma 2}$ are depth factors based on the dimensions of the replaced zone and an embedded depth of $(D + H)$. In addition,

$$W_{rz} = WL_{rz}H\gamma_1 \quad \text{for a rectangular replaced zone} \tag{5A.6c.1}$$

$$W_{rz} = 0.25\pi d_{rz}^2 H\gamma_1 \quad \text{for a circular replaced zone} \tag{5A.6c.2}$$

If K_s is assumed to be constant at all depths along the perimeter surface of the replaced zone, the following equation applies for P_h:

$$P_h = K_s(\gamma_2 DH + 0.5\gamma_2 H^2) = K_s\gamma_2(DH + 0.5H^2) \tag{5A.6d}$$

Values for K_s can be obtained from Fig. 5A.5 for $q_1 = q_2$ (since passive resistance develops within the in situ soil) or by assuming $K_s = K_{0-nc}$ and using the following equation:

$$K_s = K_{0-nc} = 1 - \sin \phi_2 \tag{5A.7}$$

The method employed to obtain the friction angles used in Eqs. (5A.3) through (5A.7) should model the type of failure assumed in each equation. Values of ϕ_1 in Eq. (5A.3a) and ϕ_2 in Eq. (5A.6a) obtained from direct shear tests would be appropriate. Values of ϕ_1 or ϕ_2 used to calculate factors in the other equations should be axisymmetric (triaxial) values if the foundation or replaced zone is square or circular or plane-strain if rectangular and L/B or L_{rz}/W is greater than about 5. Most values for ϕ found in tables based on relative density or Standard Penetration Test (SPT) blowcounts are correlated to triaxial values. Conservative values for plane-strain friction angle (ϕ_{ps}) can be estimated from triaxial values (ϕ_{tx}) using the following equations (Lade and Lee, 1976):

$$\phi_{ps} = \phi_{tx} \quad \text{for } \phi_{tx} \leq 34° \tag{5A.8a}$$

$$\phi_{ps} = 1.5\phi_{tx} - 17° \quad \text{for } \phi_{tx} > 34° \tag{5A.8b}$$

Values of ϕ from direct shear tests are usually about 1° or 2° greater than ϕ_{tx} for the same range of confining stresses.

5A.2.2 Settlement

When properly designed and implemented, overexcavation and replacement results in reduced settlement owing to one or both of the following factors: (*a*) The replaced zone is stiffer so it settles less than the replaced in situ soil would have, and (*b*) the vertical stresses induced in the in situ soil beneath the replaced zone may be less than without the replaced zone, so the settlement of this underlying soil may also be less. As will be discussed subsequently, the vertical stresses induced in the underlying in situ soil may be greater with a replaced zone than without it, so care must be exercised when using overexcavation/replacement where saturated fine-grained soils are within the zone of influence for settlement.

Several solutions are available in the literature to calculate settlement for a uniform, circular load on the surface of two- and three-layer elastic systems (e.g. Burmister, 1945; Burmister, 1962; Thenn de Barros, 1966; Ueshita and Meyerhof, 1967). The solutions for two-layer systems take the following general form:

$$S_i = \frac{q_0 d_0}{E_2} I_s \tag{5A.9}$$

where q_0 = uniform stress applied to the surface of the uppermost layer
d_0 = diameter of the loaded area
E_2 = stress-strain modulus for the lower layer
I_s = influence factor for settlement beneath the center of the loaded area

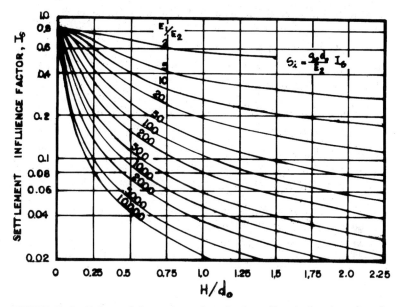

FIGURE 5A.6 Settlement influence factors for a circular, uniform load on the surface of a two-layer elastic system ($v_1 = 0.2$, $v_2 = 0.4$) (from Burmister, 1962).

The magnitude of I_s depends on the height of the stiff layer relative to the diameter of the loaded area (H/d_0), the modulus ratio for the two layers (E_2/E_1), and the Poisson's ratios for the two layers (v_1, v_2). Values of I_s for a two-layer system with $v_1 = 0.2$ and $v_2 = 0.4$ are given in Fig. 5A.6 (Burmister, 1962), from which it can be seen that settlement decreases as E_1/E_2 increases or H/d_0 increases.

The applicability of charts such as Fig. 5A.6 for estimating the settlement of a foundation bearing on a replaced zone is limited in the following ways:

1. The charts are not applicable to replaced zones of finite width unless the replaced zone is wide enough to get the full reduction in settlement.
2. Embedment of the foundation, which results in reduced settlement, is not considered.
3. The possible existence of a rigid base (such as bedrock) within the zone of influence, which also reduces settlement, is not considered.
4. The charts apply only to foundations with circular or square shapes. (Square foundations can be considered by converting to an equivalent circle with the same area.)

No solutions are available in the literature to estimate settlement for a replaced zone of finite width or for infinitely wide layered soils when the foundation is either embedded or rectangular. Solutions are available for calculating immediate settlement for semi-infinite, homogeneous soils that consider the effect of depth of embedment for both circular and rectangular foundations [Nishida (1966) for circular and E. Fox (1948) for rectangular]. Settlement of a circular foundation bearing on the *surface* of a two-layer system underlain by a rough, rigid base can be estimated from the influence chart or tables given in Uzan and coworkers (1980). An extensive collection of elastic solutions for stresses and displacements of geologic materials can be found in Poulos and Davis (1974).

The induced vertical stresses ($\Delta\sigma_z$) in a two-layer elastic system subjected to a uniform, circular stress on the surface of the upper layer are shown in Fig. 5A.7 for $v_1 = v_2 = 0.5$ and the fol-

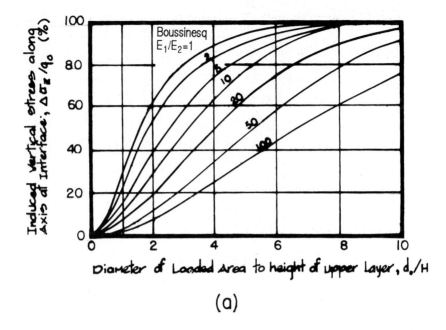

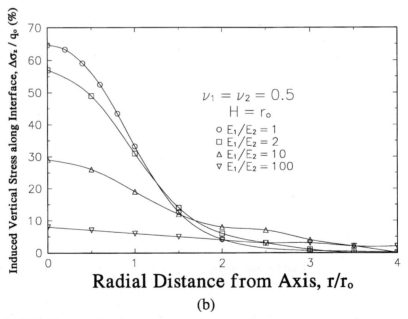

FIGURE 5A.7 Induced stresses along the interface of a two-layer elastic system subjected to a circular, uniform load on the surface ($v_1 = v_2 = 0.5$): (a) along axis for varying d_0/H and E_1/E_2 (from Burmister, 1958); (b) as a function of r/r_0 and E_1/E_2 for $H = r_0$ (data from L. Fox, 1948).

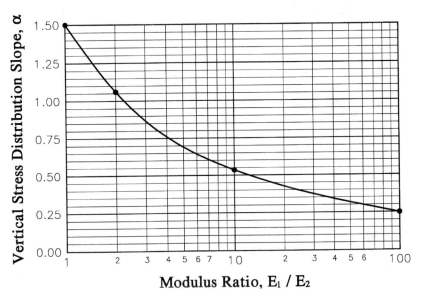

FIGURE 5A.8 Vertical stress distribution slope for a two-layer elastic system.

lowing two distributions: (a) Along the axis at the interface between the two layers as a function of d_0/H and E_1/E_2, and (b) horizontal distribution along the interface for $H = r_0$ and varying E_1/E_2. Figure 5A.7 shows that increasing either the height or the stiffness of the upper layer distributes the load over a larger area and reduces the maximum $\Delta\sigma_z$ (along the axis of the loaded area). To obtain an estimate of the approximate slope of the vertical stress distribution [α(vertical):1(horizontal)] as a function of E_1/E_2, the curves in Fig. 5A.7 were integrated numerically by the author to determine the radial distance within which 95% of the applied load is distributed, with the results shown in Fig. 5A.8. This technique is similar in concept to the 2:1 or 60° stress distribution methods. Thus, an estimate of the minimum dimensions of the replaced zone needed to obtain the full reduction in settlement can be determined with the aid of Fig. 5A.8 and the following equations. For a rectangular foundation use

$$W \geq B + \frac{2H}{\alpha} \qquad (5A.10a.1)$$

$$L_{rz} \geq L + \frac{2H}{\alpha} \qquad (5A.10a.2)$$

For a circular foundation use

$$d_{rz} \geq d_0 + \frac{2H}{\alpha} \qquad (5A.10b)$$

In many cases where overexcavation/replacement is used, the replaced zone is not wide enough to develop the full reduction in settlement. As an example, consider the case of a circular foundation bearing on a cylindrical replaced zone with $H = d_0$, and $E_1/E_2 = 20$. From Fig. 5A.8:

$$\alpha \cong 0.42$$

From Eq. (5A.10b):

$$d_{rz} \geq d_0 + \frac{2d_0}{0.42} = 5.8d_0$$

d_{rz} is typically in the range of d_0 to $3d_0$, which is less than the minimum diameter of about $5.8d_0$ required for full reduction in settlement. Estimating S_i from Fig. 5A.6 and Eq. (5A.9) for this case would therefore tend to underestimate the settlement.

To determine the influences of depth of embedment and width, height, and stiffness of the replaced zone on S_i, a parametric finite element study was performed by the author for a flexible, uniform, circular stress applied to the surface of a cylindrical replaced zone ($v_1 = 0.2$) within an otherwise homogeneous, isotropic, elastic soil mass ($v_2 = 0.4$). In this investigation, the following values of parameters were used:

$$\frac{E_1}{E_2} = 5, 10, 20, 50, \text{ and } 100$$

$$\frac{D}{d_0} = 0, 0.5, 1.0, \text{ and } 1.5$$

$$\frac{H}{d_0} = 0.5, 1.0, \text{ and } 1.5$$

$$\frac{d_{rz}}{d_0} = 0.25, 0.5, 0.75, 1, 2, 3, 4, 6, 8, \text{ and } 10$$

The results are presented in Table 5A.1 in the form of settlement influence factors (I_s), which can be used in Eq. (5A.9) to estimate S_i for an embedded circular or square foundation bearing on a finite-width replaced zone within a relatively homogeneous in situ soil. Values of I_s are given for settlement at the center of the loaded area and the mean settlement over the entire loaded area. The values of mean I_s represent a better estimate for real foundations, which are neither perfectly flexible nor perfectly rigid. The values of I_s at the center of the loaded area for $d_{rz}/d_0 = 10$ in Table 5A.1 match the corresponding theoretical values for an infinitely wide upper layer obtained from Fig. 5A.6. In addition, the values of central I_s for a homogeneous soil ($E_1/E_2 = 1$) are within 2% of the theoretical values obtained from Nishida's (1966) equations for a uniform, circular stress embedded within a homogeneous half-space. These comparisons indicate that the finite element results are reasonable.

The values of $v_1 = 0.2$ and $v_2 = 0.4$ used in the parametric study were selected as reasonable estimates for most overexcavation/replacement cases. For other values of v_1 and v_2, adjustments may be needed to the values shown in Table 5A.1. A limited investigation has indicated that v_2 has a significant influence on S_i but that v_1 has very little effect. This can be illustrated by the following finite element results for $E_1/E_2 = 20$, $D/d_0 = 1$, $H/d_0 = 1$, and $d_{rz}/d_0 = 2$:

v_1	v_2	I_s center	I_s mean
0.2	0.4	0.24	0.23
0.3	0.4	0.23	0.23
0.4	0.4	0.23	0.23
0.2	0.3	0.29	0.28
0.3	0.3	0.28	0.28
0.4	0.3	0.28	0.28
0.2	0.2	0.31	0.31
0.3	0.2	0.31	0.31
0.4	0.2	0.31	0.30

Therefore, values of I_s from the above table can be used for $v_2 = 0.4$ and any reasonable value of v_1. Preliminary results suggest that for $v_2 \neq 0.4$, a *rough* estimate for I_s can be obtained from the following equation:

$$I_{s(v_2 \neq 0.4)} = \frac{1 - v_2}{1 - 0.4} \cdot I_{s(v_2 = 0.4)} \tag{5A.11}$$

For homogeneous soils, I_s is proportional to $(1 - v^2)$, so the immediate settlement for overexcavated and replaced soils appears to be affected more by the Poisson's ratio of the matrix soil than are homogeneous soils.

I_s also depends on the shape of the loaded area. Reasonable values of I_s for square foundations and rectangular foundations with L/B less than about 1.5 can be obtained by converting to an equivalent circle with the same area. However, this method is not applicable to higher L/B ratios. To illustrate the effect that shape of the loaded area has on I_s, results from plane-strain finite element analyses (strip loading on an infinitely long replaced zone) are compared with axisymmetric results for the case of $v_1 = 0.2$, $v_2 = 0.4$, $E_1/E_2 = 20$, $D/d_0 = D/B = 1$, and $H/d_0 = H/B = 1$ as follows:

Axisymmetric [$S_i = (q_0 d_0/E_2) I_s$]			Plane-Strain [$S_i = (q_0 B/E_2) I_s$]		
d_{rz}/d_0	I_s center	I_s average	W/B	I_s center	I_s average
0.25	0.41	0.44	0.25	0.16	0.16
0.5	0.38	0.40	0.5	0.15	0.15
0.75	0.35	0.35	0.75	0.15	0.15
1	0.32	0.32	1	0.14	0.14
1.5	0.27	0.26	1.5	0.14	0.14
2	0.24	0.23	2	0.13	0.13
3	0.21	0.20	3	0.13	0.13
4	0.20	0.19	4	0.12	0.12
6	0.19	0.18	6	0.12	0.12
8	0.19	0.18	8	0.12	0.12
10	0.18	0.18	10	0.12	0.12

Therefore, for comparable conditions, a long rectangular foundation will settle substantially less than a circular foundation with the same width ($d_0 = B$). No additional information is currently available for estimating S_i for rectangular foundations of finite length bearing on replaced zones of finite extent, so engineering judgment is required for these cases. A very conservative estimate for I_s could be obtained by using Table 5A.1 with $D/d_0 = D/B$, $H/d_0 = H/B$, and $d_{rz}/d_0 = W/B$. One could also perform plane-strain finite element analyses for comparable conditions to obtain a lower-bound value for S_i, with S_i from the axisymmetric case as an upper-bound value, and use engineering judgment to estimate S_i for the actual L/B ratio. A better method would be to perform a three-dimensional finite element analysis for the actual conditions; unfortunately, few engineers have ready access to a three-dimensional finite element code suited to this type of problem.

If the in situ bearing soils within the depth of influence for settlement are highly stratified with widely varying stiffnesses, or if there is one or more layers of saturated clay within the zone of influence, more involved techniques are needed. The increased stiffness of a replaced zone that is sufficiently wide to develop the full reduction in settlement [Eq. (5A.10)] can be readily incorporated into most methods for calculating immediate settlement (e.g., Bowles, 1987; Schmertmann, 1970 and Schmertmann et al., 1978; Burland and Burbidge, 1985; Meyerhof, 1956, 1965). If either the Bowles or Schmertmann method is used, a value of equivalent stress-strain modulus (E) for the replaced soil is needed and can be estimated in a variety of ways (see Sec. 2B.4). If a method based on SPT blowcounts (N) is used (Burland and Burbidge or Meyerhof), SPT tests can be conducted within the replaced zone to obtain N values for estimating S_i.

TABLE 5A.1 Settlement Influence Factors for a Flexible, Uniform, Circular Stress Applied to the Surface of a Cylindrical Replaced Zone within an Otherwise Homogeneous, Isotropic, Elastic Half-Space ($v_1 = 0.2$, $v_2 = 0.4$)

$$S_i = \frac{q_0 d_0}{E_2} I_s$$

E_1/E_2	D/d_0	H/d_0	d_{rz}/d_0	I_s center	I_s mean
1	0	0	0	0.84	0.75
1	0.5	0	0	0.65	0.59
1	1	0	0	0.54	0.50
1	1.5	0	0	0.50	0.46
5	0	0.5	0.25	0.71	0.69
5	0	0.5	0.5	0.65	0.65
5	0	0.5	0.75	0.61	0.59
5	0	0.5	1	0.57	0.54
5	0	0.5	1.5	0.53	0.49
5	0	0.5	2	0.51	0.47
5	0	0.5	3	0.50	0.46
5	0	0.5	4	0.50	0.46
5	0	0.5	6	0.50	0.46
5	0	0.5	8	0.50	0.46
5	0	0.5	10	0.50	0.46
5	0	1	0.25	0.67	0.67
5	0	1	0.5	0.60	0.60
5	0	1	0.75	0.54	0.53
5	0	1	1	0.49	0.47
5	0	1	1.5	0.43	0.41
5	0	1	2	0.39	0.37
5	0	1	3	0.37	0.35
5	0	1	4	0.37	0.34
5	0	1	6	0.36	0.34
5	0	1	8	0.36	0.34
5	0	1	10	0.36	0.33
5	0	1.5	0.25	0.66	0.66
5	0	1.5	0.5	0.57	0.58
5	0	1.5	0.75	0.51	0.50
5	0	1.5	1	0.46	0.44
5	0	1.5	1.5	0.40	0.38
5	0	1.5	2	0.36	0.34
5	0	1.5	3	0.33	0.31
5	0	1.5	4	0.32	0.30
5	0	1.5	6	0.31	0.29
5	0	1.5	8	0.31	0.29
5	0	1.5	10	0.31	0.28
5	0.5	0.5	0.25	0.57	0.56
5	0.5	0.5	0.5	0.54	0.53
5	0.5	0.5	0.75	0.51	0.49
5	0.5	0.5	1	0.48	0.46
5	0.5	0.5	1.5	0.44	0.42
5	0.5	0.5	2	0.43	0.40
5	0.5	0.5	3	0.42	0.39
5	0.5	0.5	4	0.42	0.39
5	0.5	0.5	6	0.42	0.39
5	0.5	0.5	8	0.42	0.39
5	0.5	0.5	10	0.42	0.39
5	0.5	1	0.25	0.55	0.55
5	0.5	1	0.5	0.51	0.50
5	0.5	1	0.75	0.46	0.45
5	0.5	1	1	0.43	0.41
5	0.5	1	1.5	0.38	0.36
5	0.5	1	2	0.35	0.33
5	0.5	1	3	0.33	0.31
5	0.5	1	4	0.32	0.30
5	0.5	1	6	0.32	0.30
5	0.5	1	8	0.32	0.30
5	0.5	1	10	0.32	0.30
5	0.5	1.5	0.25	0.55	0.55
5	0.5	1.5	0.5	0.49	0.49
5	0.5	1.5	0.75	0.44	0.44
5	0.5	1.5	1	0.41	0.39
5	0.5	1.5	1.5	0.35	0.34
5	0.5	1.5	2	0.33	0.31
5	0.5	1.5	3	0.30	0.28
5	0.5	1.5	4	0.29	0.27
5	0.5	1.5	6	0.28	0.26
5	0.5	1.5	8	0.28	0.26
5	0.5	1.5	10	0.28	0.26
5	1	0.5	0.25	0.49	0.48
5	1	0.5	0.5	0.46	0.46
5	1	0.5	0.75	0.45	0.43
5	1	0.5	1	0.42	0.40
5	1	0.5	1.5	0.40	0.37
5	1	0.5	2	0.39	0.36
5	1	0.5	3	0.38	0.35
5	1	0.5	4	0.38	0.35
5	1	0.5	6	0.38	0.35
5	1	0.5	8	0.38	0.35
5	1	0.5	10	0.38	0.35
5	1	1	0.25	0.47	0.47
5	1	1	0.5	0.44	0.44
5	1	1	0.75	0.41	0.40
5	1	1	1	0.38	0.37
5	1	1	1.5	0.35	0.33
5	1	1	2	0.32	0.31
5	1	1	3	0.31	0.29
5	1	1	4	0.30	0.28
5	1	1	6	0.30	0.28
5	1	1	8	0.30	0.28
5	1	1	10	0.29	0.28

TABLE 5A.1 (*Continued*)

E_1/E_2	D/d_0	H/d_0	d_{rz}/d_0	I_s center	I_s mean	E_1/E_2	D/d_0	H/d_0	d_{rz}/d_0	I_s center	I_s mean
5	1	1.5	0.25	0.47	0.47	10	0	1	0.25	0.61	0.64
5	1	1.5	0.5	0.43	0.43	10	0	1	0.5	0.53	0.55
5	1	1.5	0.75	0.39	0.39	10	0	1	0.75	0.48	0.48
5	1	1.5	1	0.37	0.35	10	0	1	1	0.43	0.42
5	1	1.5	1.5	0.33	0.31	10	0	1	1.5	0.35	0.34
5	1	1.5	2	0.30	0.29	10	0	1	2	0.32	0.30
5	1	1.5	3	0.28	0.26	10	0	1	3	0.28	0.27
5	1	1.5	4	0.27	0.25	10	0	1	4	0.27	0.26
5	1	1.5	6	0.26	0.25	10	0	1	6	0.27	0.25
5	1	1.5	8	0.26	0.25	10	0	1	8	0.26	0.25
5	1	1.5	10	0.26	0.24	10	0	1	10	0.26	0.25
5	1.5	0.5	0.25	0.45	0.45	10	0	1.5	0.25	0.59	0.62
5	1.5	0.5	0.5	0.43	0.43	10	0	1.5	0.5	0.49	0.52
5	1.5	0.5	0.75	0.42	0.41	10	0	1.5	0.75	0.43	0.44
5	1.5	0.5	1	0.40	0.38	10	0	1.5	1	0.39	0.38
5	1.5	0.5	1.5	0.37	0.35	10	0	1.5	1.5	0.32	0.31
5	1.5	0.5	2	0.36	0.34	10	0	1.5	2	0.28	0.27
5	1.5	0.5	3	0.36	0.34	10	0	1.5	3	0.24	0.23
5	1.5	0.5	4	0.36	0.33	10	0	1.5	4	0.23	0.22
5	1.5	0.5	6	0.36	0.33	10	0	1.5	6	0.22	0.21
5	1.5	0.5	8	0.36	0.33	10	0	1.5	8	0.21	0.20
5	1.5	0.5	10	0.36	0.33	10	0	1.5	10	0.21	0.20
5	1.5	1	0.25	0.44	0.44	10	0.5	0.5	0.25	0.55	0.55
5	1.5	1	0.5	0.41	0.41	10	0.5	0.5	0.5	0.51	0.51
5	1.5	1	0.75	0.38	0.38	10	0.5	0.5	0.75	0.48	0.47
5	1.5	1	1	0.36	0.35	10	0.5	0.5	1	0.44	0.42
5	1.5	1	1.5	0.33	0.31	10	0.5	0.5	1.5	0.39	0.37
5	1.5	1	2	0.31	0.29	10	0.5	0.5	2	0.37	0.35
5	1.5	1	3	0.29	0.28	10	0.5	0.5	3	0.36	0.34
5	1.5	1	4	0.29	0.27	10	0.5	0.5	4	0.36	0.34
5	1.5	1	6	0.29	0.27	10	0.5	0.5	6	0.35	0.33
5	1.5	1	8	0.28	0.27	10	0.5	0.5	8	0.35	0.33
5	1.5	1	10	0.28	0.27	10	0.5	0.5	10	0.35	0.33
5	1.5	1.5	0.25	0.44	0.44	10	0.5	1	0.25	0.51	0.53
5	1.5	1.5	0.5	0.40	0.40	10	0.5	1	0.5	0.46	0.47
5	1.5	1.5	0.75	0.37	0.37	10	0.5	1	0.75	0.42	0.41
5	1.5	1.5	1	0.35	0.33	10	0.5	1	1	0.38	0.37
5	1.5	1.5	1.5	0.31	0.29	10	0.5	1	1.5	0.32	0.31
5	1.5	1.5	2	0.29	0.27	10	0.5	1	2	0.29	0.28
5	1.5	1.5	3	0.27	0.26	10	0.5	1	3	0.26	0.25
5	1.5	1.5	4	0.26	0.25	10	0.5	1	4	0.25	0.24
5	1.5	1.5	6	0.26	0.24	10	0.5	1	6	0.25	0.24
5	1.5	1.5	8	0.25	0.24	10	0.5	1	8	0.24	0.23
5	1.5	1.5	10	0.25	0.24	10	0.5	1	10	0.24	0.23
10	0	0.5	0.25	0.68	0.68	10	0.5	1.5	0.25	0.50	0.52
10	0	0.5	0.5	0.61	0.62	10	0.5	1.5	0.5	0.43	0.45
10	0	0.5	0.75	0.57	0.56	10	0.5	1.5	0.75	0.39	0.39
10	0	0.5	1	0.51	0.50	10	0.5	1.5	1	0.35	0.34
10	0	0.5	1.5	0.45	0.43	10	0.5	1.5	1.5	0.30	0.29
10	0	0.5	2	0.42	0.40	10	0.5	1.5	2	0.26	0.25
10	0	0.5	3	0.41	0.39	10	0.5	1.5	3	0.23	0.22
10	0	0.5	4	0.40	0.38	10	0.5	1.5	4	0.22	0.21
10	0	0.5	6	0.40	0.38	10	0.5	1.5	6	0.21	0.20
10	0	0.5	8	0.40	0.38	10	0.5	1.5	8	0.20	0.19
10	0	0.5	10	0.40	0.38	10	0.5	1.5	10	0.20	0.19

TABLE 5A.1 (*Continued*)

E_1/E_2	D/d_0	H/d_0	d_{rz}/d_0	I_s center	I_s mean	E_1/E_2	D/d_0	H/d_0	d_{rz}/d_0	I_s center	I_s mean
10	1	0.5	0.25	0.47	0.47	10	1.5	1.5	0.25	0.40	0.42
10	1	0.5	0.5	0.44	0.45	10	1.5	1.5	0.5	0.36	0.38
10	1	0.5	0.75	0.42	0.41	10	1.5	1.5	0.75	0.33	0.33
10	1	0.5	1	0.39	0.38	10	1.5	1.5	1	0.30	0.30
10	1	0.5	1.5	0.35	0.34	10	1.5	1.5	1.5	0.26	0.25
10	1	0.5	2	0.34	0.32	10	1.5	1.5	2	0.24	0.23
10	1	0.5	3	0.33	0.31	10	1.5	1.5	3	0.21	0.20
10	1	0.5	4	0.32	0.31	10	1.5	1.5	4	0.20	0.19
10	1	0.5	6	0.32	0.30	10	1.5	1.5	6	0.19	0.18
10	1	0.5	8	0.32	0.30	10	1.5	1.5	8	0.19	0.18
10	1	0.5	10	0.32	0.30	10	1.5	1.5	10	0.19	0.18
10	1	1	0.25	0.44	0.46	20	0	0.5	0.25	0.65	0.66
10	1	1	0.5	0.40	0.42	20	0	0.5	0.5	0.59	0.61
10	1	1	0.75	0.37	0.37	20	0	0.5	0.75	0.54	0.53
10	1	1	1	0.34	0.34	20	0	0.5	1	0.48	0.47
10	1	1	1.5	0.30	0.29	20	0	0.5	1.5	0.40	0.38
10	1	1	2	0.27	0.26	20	0	0.5	2	0.36	0.35
10	1	1	3	0.25	0.24	20	0	0.5	3	0.34	0.33
10	1	1	4	0.24	0.23	20	0	0.5	4	0.33	0.32
10	1	1	6	0.23	0.22	20	0	0.5	6	0.33	0.31
10	1	1	8	0.23	0.22	20	0	0.5	8	0.32	0.31
10	1	1	10	0.23	0.22	20	0	0.5	10	0.32	0.31
10	1	1.5	0.25	0.43	0.45	20	0	1	0.25	0.56	0.61
10	1	1.5	0.5	0.38	0.40	20	0	1	0.5	0.49	0.52
10	1	1.5	0.75	0.35	0.35	20	0	1	0.75	0.44	0.44
10	1	1.5	1	0.32	0.31	20	0	1	1	0.39	0.39
10	1	1.5	1.5	0.27	0.27	20	0	1	1.5	0.31	0.31
10	1	1.5	2	0.25	0.24	20	0	1	2	0.27	0.26
10	1	1.5	3	0.22	0.21	20	0	1	3	0.23	0.22
10	1	1.5	4	0.21	0.20	20	0	1	4	0.22	0.21
10	1	1.5	6	0.20	0.19	20	0	1	6	0.21	0.20
10	1	1.5	8	0.20	0.19	20	0	1	8	0.20	0.20
10	1	1.5	10	0.19	0.18	20	0	1	10	0.20	0.19
10	1.5	0.5	0.25	0.44	0.44	20	0	1.5	0.25	0.52	0.58
10	1.5	0.5	0.5	0.41	0.42	20	0	1.5	0.5	0.44	0.48
10	1.5	0.5	0.75	0.39	0.39	20	0	1.5	0.75	0.39	0.39
10	1.5	0.5	1	0.37	0.35	20	0	1.5	1	0.35	0.34
10	1.5	0.5	1.5	0.33	0.32	20	0	1.5	1.5	0.28	0.27
10	1.5	0.5	2	0.32	0.30	20	0	1.5	2	0.24	0.23
10	1.5	0.5	3	0.31	0.29	20	0	1.5	3	0.20	0.19
10	1.5	0.5	4	0.31	0.29	20	0	1.5	4	0.18	0.17
10	1.5	0.5	6	0.31	0.29	20	0	1.5	6	0.17	0.16
10	1.5	0.5	8	0.31	0.29	20	0	1.5	8	0.16	0.16
10	1.5	0.5	10	0.31	0.29	20	0	1.5	10	0.15	0.15
10	1.5	1	0.25	0.41	0.43	20	0.5	0.5	0.25	0.53	0.54
10	1.5	1	0.5	0.38	0.39	20	0.5	0.5	0.5	0.50	0.50
10	1.5	1	0.75	0.35	0.35	20	0.5	0.5	0.75	0.46	0.45
10	1.5	1	1	0.32	0.32	20	0.5	0.5	1	0.41	0.41
10	1.5	1	1.5	0.28	0.27	20	0.5	0.5	1.5	0.35	0.34
10	1.5	1	2	0.26	0.25	20	0.5	0.5	2	0.33	0.31
10	1.5	1	3	0.24	0.23	20	0.5	0.5	3	0.31	0.30
10	1.5	1	4	0.23	0.22	20	0.5	0.5	4	0.30	0.29
10	1.5	1	6	0.23	0.21	20	0.5	0.5	6	0.30	0.29
10	1.5	1	8	0.22	0.21	20	0.5	0.5	8	0.30	0.29
10	1.5	1	10	0.22	0.21	20	0.5	0.5	10	0.30	0.28

TABLE 5A.1 (Continued)

E_1/E_2	D/d_0	H/d_0	d_{rz}/d_0	I_s center	I_s mean	E_1/E_2	D/d_0	H/d_0	d_{rz}/d_0	I_s center	I_s mean
20	0.5	1	0.25	0.48	0.50	20	1.5	0.5	0.25	0.43	0.44
20	0.5	1	0.5	0.43	0.45	20	1.5	0.5	0.5	0.40	0.41
20	0.5	1	0.75	0.39	0.39	20	1.5	0.5	0.75	0.38	0.38
20	0.5	1	1	0.35	0.35	20	1.5	0.5	1	0.35	0.34
20	0.5	1	1.5	0.29	0.28	20	1.5	0.5	1.5	0.30	0.29
20	0.5	1	2	0.25	0.25	20	1.5	0.5	2	0.28	0.27
20	0.5	1	3	0.22	0.21	20	1.5	0.5	3	0.27	0.26
20	0.5	1	4	0.21	0.20	20	1.5	0.5	4	0.27	0.25
20	0.5	1	6	0.20	0.19	20	1.5	0.5	6	0.26	0.25
20	0.5	1	8	0.19	0.19	20	1.5	0.5	8	0.26	0.25
20	0.5	1	10	0.19	0.18	20	1.5	0.5	10	0.26	0.25
20	0.5	1.5	0.25	0.45	0.49	20	1.5	1	0.25	0.38	0.41
20	0.5	1.5	0.5	0.39	0.41	20	1.5	1	0.5	0.36	0.37
20	0.5	1.5	0.75	0.35	0.35	20	1.5	1	0.75	0.33	0.33
20	0.5	1.5	1	0.32	0.31	20	1.5	1	1	0.30	0.30
20	0.5	1.5	1.5	0.26	0.26	20	1.5	1	1.5	0.25	0.25
20	0.5	1.5	2	0.23	0.22	20	1.5	1	2	0.23	0.22
20	0.5	1.5	3	0.19	0.19	20	1.5	1	3	0.20	0.19
20	0.5	1.5	4	0.17	0.17	20	1.5	1	4	0.19	0.18
20	0.5	1.5	6	0.16	0.16	20	1.5	1	6	0.18	0.18
20	0.5	1.5	8	0.16	0.15	20	1.5	1	8	0.18	0.17
20	0.5	1.5	10	0.15	0.15	20	1.5	1	10	0.18	0.17
20	1	0.5	0.25	0.46	0.47	20	1.5	1.5	0.25	0.37	0.40
20	1	0.5	0.5	0.43	0.44	20	1.5	1.5	0.5	0.33	0.35
20	1	0.5	0.75	0.41	0.40	20	1.5	1.5	0.75	0.30	0.30
20	1	0.5	1	0.37	0.36	20	1.5	1.5	1	0.28	0.27
20	1	0.5	1.5	0.32	0.31	20	1.5	1.5	1.5	0.23	0.23
20	1	0.5	2	0.30	0.29	20	1.5	1.5	2	0.21	0.20
20	1	0.5	3	0.29	0.27	20	1.5	1.5	3	0.18	0.17
20	1	0.5	4	0.28	0.27	20	1.5	1.5	4	0.16	0.16
20	1	0.5	6	0.28	0.27	20	1.5	1.5	6	0.15	0.15
20	1	0.5	8	0.28	0.26	20	1.5	1.5	8	0.15	0.15
20	1	0.5	10	0.28	0.26	20	1.5	1.5	10	0.15	0.14
20	1	1	0.25	0.41	0.44	50	0	0.5	0.25	0.63	0.65
20	1	1	0.5	0.38	0.40	50	0	0.5	0.5	0.58	0.60
20	1	1	0.75	0.35	0.35	50	0	0.5	0.75	0.52	0.52
20	1	1	1	0.32	0.32	50	0	0.5	1	0.45	0.45
20	1	1	1.5	0.27	0.26	50	0	0.5	1.5	0.36	0.35
20	1	1	2	0.24	0.23	50	0	0.5	2	0.31	0.30
20	1	1	3	0.21	0.20	50	0	0.5	3	0.27	0.26
20	1	1	4	0.20	0.19	50	0	0.5	4	0.26	0.25
20	1	1	6	0.19	0.18	50	0	0.5	6	0.25	0.25
20	1	1	8	0.19	0.18	50	0	0.5	8	0.25	0.24
20	1	1	10	0.18	0.18	50	0	0.5	10	0.24	0.24
20	1	1.5	0.25	0.39	0.43	50	0	1	0.25	0.52	0.58
20	1	1.5	0.5	0.35	0.37	50	0	1	0.5	0.47	0.50
20	1	1.5	0.75	0.32	0.32	50	0	1	0.75	0.42	0.42
20	1	1.5	1	0.29	0.29	50	0	1	1	0.37	0.37
20	1	1.5	1.5	0.24	0.24	50	0	1	1.5	0.29	0.29
20	1	1.5	2	0.21	0.21	50	0	1	2	0.24	0.24
20	1	1.5	3	0.18	0.18	50	0	1	3	0.19	0.19
20	1	1.5	4	0.17	0.16	50	0	1	4	0.17	0.17
20	1	1.5	6	0.16	0.15	50	0	1	6	0.16	0.15
20	1	1.5	8	0.15	0.15	50	0	1	8	0.15	0.15
20	1	1.5	10	0.15	0.14	50	0	1	10	0.14	0.14

TABLE 5A.1 (Continued)

E_1/E_2	D/d_0	H/d_0	d_{rz}/d_0	I_s center	I_s mean	E_1/E_2	D/d_0	H/d_0	d_{rz}/d_0	I_s center	I_s mean
50	0	1.5	0.25	0.46	0.54	50	1	1	0.25	0.39	0.43
50	0	1.5	0.5	0.40	0.45	50	1	1	0.5	0.36	0.39
50	0	1.5	0.75	0.36	0.36	50	1	1	0.75	0.34	0.34
50	0	1.5	1	0.32	0.32	50	1	1	1	0.30	0.30
50	0	1.5	1.5	0.25	0.25	50	1	1	1.5	0.25	0.24
50	0	1.5	2	0.21	0.21	50	1	1	2	0.21	0.21
50	0	1.5	3	0.17	0.17	50	1	1	3	0.18	0.17
50	0	1.5	4	0.15	0.14	50	1	1	4	0.16	0.16
50	0	1.5	6	0.13	0.13	50	1	1	6	0.15	0.15
50	0	1.5	8	0.12	0.12	50	1	1	8	0.14	0.14
50	0	1.5	10	0.12	0.11	50	1	1	10	0.14	0.13
50	0.5	0.5	0.25	0.52	0.53	50	1	1.5	0.25	0.36	0.40
50	0.5	0.5	0.5	0.49	0.49	50	1	1.5	0.5	0.33	0.35
50	0.5	0.5	0.75	0.45	0.44	50	1	1.5	0.75	0.30	0.30
50	0.5	0.5	1	0.39	0.39	50	1	1.5	1	0.27	0.27
50	0.5	0.5	1.5	0.32	0.32	50	1	1.5	1.5	0.22	0.22
50	0.5	0.5	2	0.28	0.28	50	1	1.5	2	0.19	0.19
50	0.5	0.5	3	0.26	0.25	50	1	1.5	3	0.16	0.16
50	0.5	0.5	4	0.25	0.24	50	1	1.5	4	0.14	0.14
50	0.5	0.5	6	0.24	0.23	50	1	1.5	6	0.12	0.12
50	0.5	0.5	8	0.24	0.23	50	1	1.5	8	0.12	0.12
50	0.5	0.5	10	0.23	0.23	50	1	1.5	10	0.11	0.11
50	0.5	1	0.25	0.45	0.48	50	1.5	0.5	0.25	0.42	0.43
50	0.5	1	0.5	0.41	0.43	50	1.5	0.5	0.5	0.40	0.41
50	0.5	1	0.75	0.37	0.37	50	1.5	0.5	0.75	0.37	0.37
50	0.5	1	1	0.33	0.33	50	1.5	0.5	1	0.33	0.33
50	0.5	1	1.5	0.27	0.26	50	1.5	0.5	1.5	0.27	0.27
50	0.5	1	2	0.23	0.22	50	1.5	0.5	2	0.25	0.24
50	0.5	1	3	0.19	0.18	50	1.5	0.5	3	0.23	0.22
50	0.5	1	4	0.17	0.16	50	1.5	0.5	4	0.22	0.21
50	0.5	1	6	0.15	0.15	50	1.5	0.5	6	0.22	0.21
50	0.5	1	8	0.15	0.15	50	1.5	0.5	8	0.21	0.21
50	0.5	1	10	0.14	0.14	50	1.5	0.5	10	0.21	0.21
50	0.5	1.5	0.25	0.41	0.46	50	1.5	1	0.25	0.36	0.40
50	0.5	1.5	0.5	0.36	0.39	50	1.5	1	0.5	0.34	0.36
50	0.5	1.5	0.75	0.33	0.33	50	1.5	1	0.75	0.32	0.32
50	0.5	1.5	1	0.29	0.29	50	1.5	1	1	0.29	0.28
50	0.5	1.5	1.5	0.24	0.24	50	1.5	1	1.5	0.23	0.23
50	0.5	1.5	2	0.20	0.20	50	1.5	1	2	0.20	0.20
50	0.5	1.5	3	0.16	0.16	50	1.5	1	3	0.17	0.17
50	0.5	1.5	4	0.14	0.14	50	1.5	1	4	0.16	0.15
50	0.5	1.5	6	0.13	0.12	50	1.5	1	6	0.15	0.14
50	0.5	1.5	8	0.12	0.12	50	1.5	1	8	0.14	0.14
50	0.5	1.5	10	0.12	0.11	50	1.5	1	10	0.14	0.13
50	1	0.5	0.25	0.45	0.46	50	1.5	1.5	0.25	0.33	0.38
50	1	0.5	0.5	0.42	0.43	50	1.5	1.5	0.5	0.31	0.33
50	1	0.5	0.75	0.40	0.39	50	1.5	1.5	0.75	0.29	0.29
50	1	0.5	1	0.35	0.35	50	1.5	1.5	1	0.26	0.26
50	1	0.5	1.5	0.29	0.29	50	1.5	1.5	1.5	0.21	0.21
50	1	0.5	2	0.26	0.25	50	1.5	1.5	2	0.18	0.18
50	1	0.5	3	0.24	0.23	50	1.5	1.5	3	0.15	0.15
50	1	0.5	4	0.23	0.22	50	1.5	1.5	4	0.14	0.13
50	1	0.5	6	0.23	0.22	50	1.5	1.5	6	0.12	0.12
50	1	0.5	8	0.22	0.22	50	1.5	1.5	8	0.12	0.12
50	1	0.5	10	0.22	0.21	50	1.5	1.5	10	0.11	0.11

TABLE 5A.1 (*Continued*)

E_1/E_2	D/d_0	H/d_0	d_{rz}/d_0	I_s center	I_s mean	E_1/E_2	D/d_0	H/d_0	d_{rz}/d_0	I_s center	I_s mean
100	0	0.5	0.25	0.63	0.65	100	0.5	1.5	0.25	0.38	0.44
100	0	0.5	0.5	0.58	0.59	100	0.5	1.5	0.5	0.35	0.38
100	0	0.5	0.75	0.52	0.51	100	0.5	1.5	0.75	0.32	0.32
100	0	0.5	1	0.44	0.44	100	0.5	1.5	1	0.28	0.28
100	0	0.5	1.5	0.34	0.34	100	0.5	1.5	1.5	0.23	0.23
100	0	0.5	2	0.28	0.28	100	0.5	1.5	2	0.19	0.19
100	0	0.5	3	0.23	0.23	100	0.5	1.5	3	0.15	0.15
100	0	0.5	4	0.22	0.21	100	0.5	1.5	4	0.13	0.13
100	0	0.5	6	0.21	0.21	100	0.5	1.5	6	0.11	0.11
100	0	0.5	8	0.21	0.20	100	0.5	1.5	8	0.11	0.10
100	0	0.5	10	0.20	0.19	100	0.5	1.5	10	0.10	0.10
100	0	1	0.25	0.50	0.57	100	1	0.5	0.25	0.44	0.46
100	0	1	0.5	0.46	0.49	100	1	0.5	0.5	0.42	0.43
100	0	1	0.75	0.41	0.41	100	1	0.5	0.75	0.39	0.39
100	0	1	1	0.36	0.36	100	1	0.5	1	0.35	0.35
100	0	1	1.5	0.28	0.28	100	1	0.5	1.5	0.28	0.28
100	0	1	2	0.23	0.23	100	1	0.5	2	0.24	0.24
100	0	1	3	0.18	0.18	100	1	0.5	3	0.21	0.21
100	0	1	4	0.15	0.15	100	1	0.5	4	0.20	0.20
100	0	1	6	0.13	0.13	100	1	0.5	6	0.19	0.19
100	0	1	8	0.13	0.13	100	1	0.5	8	0.19	0.19
100	0	1	10	0.12	0.12	100	1	0.5	10	0.18	0.18
100	0	1.5	0.25	0.43	0.52	100	1	1	0.25	0.38	0.42
100	0	1.5	0.5	0.39	0.43	100	1	1	0.5	0.36	0.38
100	0	1.5	0.75	0.35	0.35	100	1	1	0.75	0.33	0.33
100	0	1.5	1	0.31	0.31	100	1	1	1	0.30	0.30
100	0	1.5	1.5	0.25	0.24	100	1	1	1.5	0.24	0.24
100	0	1.5	2	0.20	0.20	100	1	1	2	0.20	0.20
100	0	1.5	3	0.16	0.16	100	1	1	3	0.16	0.16
100	0	1.5	4	0.13	0.13	100	1	1	4	0.14	0.14
100	0	1.5	6	0.11	0.11	100	1	1	6	0.13	0.13
100	0	1.5	8	0.11	0.11	100	1	1	8	0.12	0.12
100	0	1.5	10	0.10	0.10	100	1	1	10	0.12	0.12
100	0.5	0.5	0.25	0.52	0.53	100	1	1.5	0.25	0.34	0.39
100	0.5	0.5	0.5	0.48	0.49	100	1	1.5	0.5	0.32	0.34
100	0.5	0.5	0.75	0.44	0.44	100	1	1.5	0.75	0.29	0.29
100	0.5	0.5	1	0.39	0.39	100	1	1.5	1	0.26	0.26
100	0.5	0.5	1.5	0.31	0.30	100	1	1.5	1.5	0.22	0.21
100	0.5	0.5	2	0.26	0.26	100	1	1.5	2	0.18	0.18
100	0.5	0.5	3	0.22	0.22	100	1	1.5	3	0.15	0.15
100	0.5	0.5	4	0.21	0.21	100	1	1.5	4	0.13	0.13
100	0.5	0.5	6	0.20	0.20	100	1	1.5	6	0.11	0.11
100	0.5	0.5	8	0.20	0.19	100	1	1.5	8	0.11	0.10
100	0.5	0.5	10	0.19	0.19	100	1	1.5	10	0.10	0.10
100	0.5	1	0.25	0.43	0.47	100	1.5	0.5	0.25	0.41	0.43
100	0.5	1	0.5	0.40	0.42	100	1.5	0.5	0.5	0.39	0.41
100	0.5	1	0.75	0.37	0.37	100	1.5	0.5	0.75	0.37	0.37
100	0.5	1	1	0.33	0.32	100	1.5	0.5	1	0.33	0.32
100	0.5	1	1.5	0.26	0.26	100	1.5	0.5	1.5	0.26	0.26
100	0.5	1	2	0.22	0.22	100	1.5	0.5	2	0.23	0.22
100	0.5	1	3	0.17	0.17	100	1.5	0.5	3	0.20	0.20
100	0.5	1	4	0.15	0.15	100	1.5	0.5	4	0.19	0.19
100	0.5	1	6	0.13	0.13	100	1.5	0.5	6	0.19	0.18
100	0.5	1	8	0.13	0.12	100	1.5	0.5	8	0.18	0.18
100	0.5	1	10	0.12	0.12	100	1.5	0.5	10	0.18	0.17

TABLE 5A.1 (Continued)

E_1/E_2	D/d_0	H/d_0	d_{rz}/d_0	I_s center	I_s mean	E_1/E_2	D/d_0	H/d_0	d_{rz}/d_0	I_s center	I_s mean
100	1.5	1	0.25	0.35	0.39	100	1.5	1.5	0.25	0.32	0.37
100	1.5	1	0.5	0.34	0.36	100	1.5	1.5	0.5	0.30	0.33
100	1.5	1	0.75	0.31	0.31	100	1.5	1.5	0.75	0.28	0.28
100	1.5	1	1	0.28	0.28	100	1.5	1.5	1	0.25	0.25
100	1.5	1	1.5	0.23	0.23	100	1.5	1.5	1.5	0.21	0.20
100	1.5	1	2	0.19	0.19	100	1.5	1.5	2	0.18	0.18
100	1.5	1	3	0.16	0.16	100	1.5	1.5	3	0.14	0.14
100	1.5	1	4	0.14	0.14	100	1.5	1.5	4	0.13	0.12
100	1.5	1	6	0.13	0.12	100	1.5	1.5	6	0.11	0.11
100	1.5	1	8	0.12	0.12	100	1.5	1.5	8	0.11	0.10
100	1.5	1	10	0.12	0.11	100	1.5	1.5	10	0.10	0.10

For a replacement zone of finite width within highly stratified matrix soils, Eq. (5A.9) and Table 5A.1 can be used to estimate S_i if an average or equivalent value for E_2 is calculated for the settlement influence zone. A weighted average modulus based on the thicknesses of the layers may be appropriate and can be calculated from the following equation (Bowles, 1987):

$$E_{2-av} = \frac{\sum_{i=1}^{i=n} H_i \cdot E_i}{\sum_{i=1}^{i=n} H_i} \quad (5A.12)$$

where n = number of layers within the settlement influence zone
H_i = thickness of layer i
E_i = stress-strain modulus for layer i

If certain strata are deemed to be more critical than others, additional weighting factors can be applied. For example, if a stratum is considered to be twice as critical as the other strata, its height can be multiplied by two, with the weighted height used in both the numerator and denominator of Eq. (5A.12).

When a saturated clay layer is within the depth of influence for settlement, S_i for a wide replaced zone can be calculated using the undrained modulus for the clay and any appropriate method based on elastic theory. An estimate of $\Delta\sigma_z$ throughout the height of the clay layer is needed to estimate primary consolidation settlement (S_c). Many engineers use Boussinesq-type analyses based on a uniform stress applied to the surface of a homogeneous half-space to estimate $\Delta\sigma_z$ in the clay layer; however, if the replaced zone is much stiffer than the undrained clay, or if the foundation bears below the ground surface, Boussinesq-type analyses are overly conservative and should not be used. Thus, it is better to use a method that considers the effects of the stiffer upper layer and the depth of embedment. A simple method for a wide replaced zone is to determine an equivalent uniform stress induced at the top of the lower zone (q_{LZ}) and dimensions of the loaded area (B_{LZ} and L_{LZ} or d_{LZ}) using the following equations:

$$B_{LZ} = B + \frac{2H}{\alpha} \quad \text{for a rectangular foundation} \tag{5A.13a.1}$$

$$L_{LZ} = L + \frac{2H}{\alpha} \quad \text{for a rectangular foundation} \tag{5A.13a.2}$$

$$d_{LZ} = d_0 + \frac{2H}{\alpha} \quad \text{for a circular foundation} \tag{5A.13b}$$

$$q_{LZ} = q \cdot \frac{BL}{B_{LZ}L_{LZ}} \quad \text{for a rectangular foundation} \tag{5A.14a}$$

$$q_{LZ} = q \cdot \frac{d_0^2}{d_{LZ}^2} \quad \text{for a circular foundation} \tag{5A.14b}$$

Methods for estimating $\Delta\sigma_z$ for foundations embedded within homogeneous, semi-infinite soils [Nishida (1966) for circular; Skopek (1961) for rectangular] can be used with Eqs. (5A.13) and (5A.14) to estimate $\Delta\sigma_z$ beneath the replaced zone.

An alternate method for a wide replaced zone is to estimate $\Delta\sigma_z$ using the average bearing stress applied to surface of the replaced zone and solutions for layered elastic systems. All available methods for layered systems are for a circular load on the ground surface, so the effect of embedment is ignored, and square foundations must be converted to equivalent circles. If the replaced zone bears directly on a saturated clay layer and the clay layer is thick (extends to the bottom of the settlement influence zone or deeper), a two-layer analysis is appropriate, and Table 5A.2 can be used to estimate $\Delta\sigma_z$ beneath the center of the loaded area at various depths within the clay layer. If the replaced zone is underlain by a thin clay layer, or if the clay layer is the second layer beneath the replaced zone, a three-layer analysis is appropriate, and $\Delta\sigma_z$ can be estimated from the tables given in Jones (1962) or the charts provided in Peattie (1962).

If the width of the replaced zone is less than the minimum required for full reduction in settlement, the stresses induced within the underlying in situ soil may be either lesser or greater than for the same foundation in a homogeneous soil. The finite element results provided in Fig. 5A.9 illustrate this concept for a cylindrical replaced zone of varying diameter with $E_1/E_2 = 20$, $D/d_0 = 1$, and $H/d_0 = 1$. Along a vertical profile near the axis [Fig. 5A.9(a)], the induced vertical stress near the interface between the replaced zone and the underlying in situ soil decreases as the diameter of the replaced zone increases. For a replaced zone with $d_{rz}/d_0 \leq 1$, the induced vertical stresses near the interface are greater than for the homogeneous case ($d_{rz}/d_0 = 0$), and the induced stresses are lesser for $d_{rz}/d_0 \geq 1.5$ than for the homogeneous case. At greater depths, the trends are different. The induced vertical stresses do not change much for $d_{rz}/d_0 \geq 6$; this result is consistent with the approximate value of $d_{rz}/d_0 = 5.8$ calculated previously for minimum diameter required for full reduction in settlement when $H/d_0 = 1$ and $E_1/E_2 = 20$.

The induced vertical stresses along a horizontal profile located a short distance below the bottom of the replaced zone are shown in Fig. 5A.9(b). For any replaced zone with $d_{rz}/d_0 \leq 2$, $\Delta\sigma_z/q_0$ increases from the axis to the edge of the footprint for the replaced zone and then decreases with distance from the footprint. For $d_{rz}/d_0 \geq 3$, $\Delta\sigma_z/q_0$ decreases gradually with distance from the axis. The induced vertical stresses for a homogeneous soil fall within the range for replaced zones with $0.25 \leq d_{rz}/d_0 \leq 10$ at all distances from the axis. To obtain an estimate of the relative magnitudes of settlement that would occur in the soil underlying the replaced zone, an estimate of the force transmitted to the underlying soil within the footprint of the loaded area (Q_{tr}) was obtained for each d_{rz}/d_0 by numerically integrating the curves for $\Delta\sigma_z/q_0$ in Fig. 5A.9 from $r/r_0 = 0$ to $r/r_0 =$

TABLE 5A.2 $\Delta\sigma_z/q_0$ beneath the Center of a Uniform Circular Stress Applied to the Surface of a Two-Layer System with a Perfectly Rough Interface and $v_1 = v_2 = 0.5$

		$\Delta\sigma_z/q_0$ (%)			
H/d_0	z_i/H	$E_1/E_2 = 1$	$E_1/E_2 = 10$	$E_1/E_2 = 100$	$E_1/E_2 = 1000$
0.25	0	91.1	64.4	24.6	7.10
	1	64.6	48.0	20.5	6.06
	2	42.4	34.0	16.5	5.42
	3	28.4	24.4	13.3	4.80
	4	20.0	18.1	10.8	4.28
0.5	0	64.6	29.2	8.1	1.85
	1	28.4	16.8	6.0	1.62
	2	14.5	10.5	4.6	1.43
	3	8.70	7.0	3.6	1.24
	4	5.70	5.0	2.9	1.10
1	0	28.4	10.1	2.38	0.51
	1	8.70	4.70	1.58	0.42
	2	4.03	2.78	1.17	0.35
	3	2.30	1.84	0.91	0.31
	4	1.48	1.29	0.74	0.28

Note: z_i = depth below the interface between the two layers
Source: After L. Fox (1948).

1. Values of Q_{tr} are shown as a percentage of the applied load in the table below, along with Q_{tr} for each d_{rz}/d_0 divided by Q_{tr} for a homogeneous soil:

$\dfrac{d_{rz}}{d_0}$	$Q_{tr}(\%)$	$\dfrac{Q_{tr}}{Q_{tr(d_{rz}/d_0 = 0)}}$
0	10.9	1.00
0.25	14.5	1.34
0.50	18.3	1.68
0.75	21.1	1.94
1.00	18.4	1.69
1.50	10.3	0.95
2.00	7.5	0.69
3.00	5.8	0.53
4.00	5.3	0.49
6.00	5.1	0.47

Thus, the settlement in the underlying soil would be greater for a replaced zone with d_{rz}/d_0 less than about 1.5 than for the homogeneous case, and is a maximum for $d_{rz}/d_0 \cong 0.75$. Where the induced vertical stresses in the underlying clay matrix are greater for overexcavation/replacement than without it, either a net increase or decrease in settlement could occur depending on whether the increase in primary consolidation settlement is greater or lesser than the decrease in immediate settlement. Thus, more involved analyses such as these are warranted when overexcavation/replacement is being considered for foundations underlain by one or more layers of saturated clay. For cases where a granular replaced zone is constructed within a saturated clay matrix, the replaced zone will act as a drainage well, and the rate of consolidation will be increased owing to increased horizontal drainage.

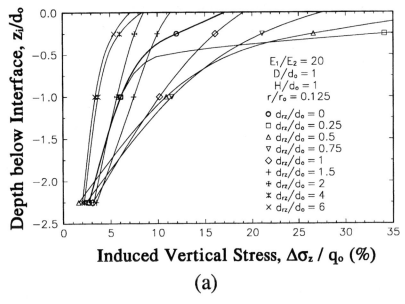

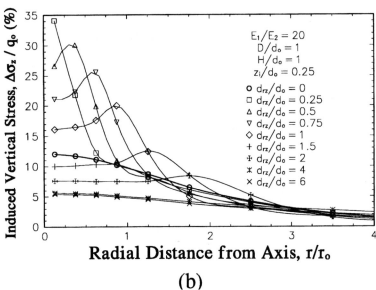

FIGURE 5A.9 Vertical stresses induced in underlying matrix soil by a uniform circular stress applied to the surface of an embedded replaced zone of varying diameter within an otherwise homogeneous half-space as predicted by finite element analysis: (a) vertical profile near axis; (b) horizontal profile near interface.

5A.2.3 Usefulness and Limitations of the Finite Element Method

On many projects where one or more of the ground modification techniques described in this handbook are used to increase bearing support for building foundations, the availability of reliable theoretical or empirical methods for estimating settlement or bearing capacity for the improved ground are either limited or nonexistent. In these cases, the finite element method—if used properly—can provide valuable insight into how ground modification affects settlement and bearing capacity. There are currently several axisymmetric and plane-strain finite element software packages available at relatively low cost that run on microcomputers and are capable of modeling the stress-strain-strength behavior of soils. Thus, the finite element technique is now readily available to many practicing engineers and should be used more frequently in foundation engineering practice than it is. Although it is not a cure-all, the finite element method is a powerful tool that all foundation engineers should consider using when available theoretical or empirical techniques are inadequate.

If the results of finite element analyses are to be meaningful, sufficient data are needed to determine reliable stress-strain-strength parameters for the bearing soils (replaced zone material and in situ soils for overexcavation/replacement). However, even under the best of circumstances, finite element analyses for soils seldom give reliable quantitative predictions for settlement or bearing capacity unless the input parameters for the stress-strain-strength characteristics of the soils are adjusted until the finite element results match the known or expected values. Therefore, an appropriate procedure for using the finite element method to estimate the bearing capacity or settlement of a foundation bearing on a replaced zone is as follows:

1. Calculate the estimated settlement or ultimate bearing capacity for the foundation with and without the replaced zone using the theories given in this section and Sec. 2B.4.
2. Obtain reliable values for the stress-strain-strength parameters for the replacement material and the in situ soils within the zone of influence. For the hyperbolic finite element model (Duncan and Chang, 1970), which is the most frequently used model for soils, the stress-strain-strength parameters are obtained from triaxial tests conducted on specimens of the soils or are estimated from published data on similar soils (e.g., Duncan et al., 1980).
3. Perform finite element (FE) analyses for the foundation without the replaced zone (in situ geologic profile). Compare the FE values for settlement or q_{ult} with those obtained from theoretical analyses (step 1). If the FE results are comparable to the theoretical results, proceed to step 4. If not, adjust the input soil parameters until the FE results are comparable.
4. Conduct finite element analyses for the foundation with the replaced zone. The input parameters for the replaced zone should be adjusted in a manner similar to that used for the in situ soils. These FEM results should provide reasonable estimates for settlement or ultimate bearing capacity.

In many projects, settlement must be estimated from limited soil data. In the United States it is fairly common to have only SPT blowcounts from which to estimate the settlement of granular soils, and the use of nonlinear stress-strain parameters for the finite element analyses is not justified in these cases. Therefore, it is better to estimate a reasonable range in values of elastic stress-strain modulus for the replacement material and each in situ soil and to perform a parametric study to estimate the probable percentage reduction in settlement which would occur from overexcavation/replacement. To provide an estimate of the settlement that would occur with the replaced zone, the range of values for percentage reduction in settlement can be applied to estimated values of settlement for the same foundation without the replaced zone.

5A.2.4 Limitations of the Overexcavation/Replacement Method

Although the overexcavation/replacement method has been used for centuries to improve bearing soils, it has the following limitations (Lawton et al., 1994):

1. Expensive bracing or sloping of the excavation is required if it is deep.
2. The excavation may cause settlement of adjacent existing structures.
3. High ground water may cause instability of the excavation and difficulties in compacting the replacement material. Therefore, bracing may be required to support the sides of the excavation, and pumping may be needed to lower the ground-water level within the excavation.
4. Sufficient high-quality replacement material may not be readily or economically available.
5. The densification that can be achieved in cohesionless replacement soils is sometimes limited to low levels, owing to lack of confinement in wide excavations and limited access in narrow excavations.

Example Problem 5A.1 A 1-m (3.28-ft) diameter footing will support a centric vertical load of 471 kN (106 kips) and will bear at a depth of 1 m (3.28 ft) below the ground surface. The in situ soil consists of a thick deposit of fairly homogeneous sandy micaceous silt. The overexcavation/replacement technique will be used to improve the bearing soils. A replacement zone 1 m (3.28 ft) in height and 2 m (6.56 ft) in diameter is being considered. The replacement material will be compacted gravelly sand. Relevant properties of the in situ and replacement soils are given below:

Replacement soil	In situ soil
Gravelly sand	Sandy micaceous silt
$\phi_1 = 45°$	$\phi_2 = 28°$
$c_1 = 0$	$c_2 = 0$
$E_1 = 100$ MPa (2090 ksf)	$E_2 = 5$ MPa (104 ksf)
$\nu_1 = 0.2$	$\nu_1 = 0.4$
$\gamma_1 = 22$ kN/m³ (140 pcf)	$\gamma_2 = 16.5$ kN/m³ (105 pcf)

Estimate the ultimate bearing capacity, factor of safety against bearing capacity failure, immediate settlement, and vertical stresses induced in the underlying in situ soil. Compare results with those for the same footing without the replaced zone.

Solution: Ultimate bearing capacity without replaced zone Use Meyerhof's (1951, 1963) equations and an equivalent square foundation:

$$B_{eq} = L_{eq} = (0.25\pi d_0^2)^{1/2} = [0.25\pi(1)^2]^{1/2} = 0.8862 \text{ m (2.91 ft)}$$

For $c_2 = 0$,

$$q_{ult} = \gamma_2 D N_{q2} S_{q2} D_{q2} + 0.5\gamma_2 B_{eq} N_{\gamma2} S_{\gamma2} D_{\gamma2}$$

$$S_{q2} = S_{\gamma2} = 1 + 0.1 \tan^2\left(45° + \frac{\phi_2}{2}\right) \cdot \frac{B_{eq}}{L_{eq}}$$

$$= 1 + 0.1 \tan^2(59°)(1) = 1.277$$

$$D_{q2} = D_{\gamma2} = 1 + 0.1 \tan\left(45° + \frac{\phi_2}{2}\right) \cdot \frac{D}{B_{eq}}$$

$$= 1 + 0.1 \tan(59°) \frac{1}{0.8862} = 1.188$$

$$N_{q2} = e^{\pi \tan \phi_2} \cdot \tan^2\left(45° + \frac{\phi_2}{2}\right) = e^{\pi \tan(28°)} \cdot \tan^2(59°) = 14.72$$

$$N_{\gamma2} = (N_{q2} - 1) \tan(1.4\phi_2) = (14.72 - 1) \tan(39.2°) = 11.19$$

$$q_{ult} = (16.5)(1)(14.72)(1.277)(1.188) + 0.5(16.5)(0.8862)(11.19)(1.277)(1.188) = 493 \text{ kPa (10.3 ksf)}$$

Calculate the average applied stress at bearing level:

$$q_0 = \frac{471}{0.25\pi(1)^2} = 600 \text{ kPa}$$

$$F_s \text{ (bearing capacity failure)} = \frac{493}{600} = 0.82$$

Solution: Ultimate bearing capacity with replaced zone—punching through replaced zone q_b is calculated for B_{eq}, L_{eq}, and an embedded depth of 2 m. N_{q2}, $N_{\gamma 2}$, S_{q2}, and $S_{\gamma 2}$ are the same as before.

$$D_{q2} = D_{\gamma 2} = 1 + 0.1 \tan(59°) = \frac{2}{0.8862} = 1.376$$

From Eq. (5A.3b):

$$q_b = (16.5)(2)(14.72)(1.277)(1.376) + 0.5(16.5)(0.8862)(11.19)(1.277)(1.376)$$

$$= 997 \text{ kPa } (20.8 \text{ ksf})$$

Calculate the perimeter length based on a circular rather than a square punching area, because the perimeter length for an equivalent square based on area is not the same as for the circle.

$$p = \pi d = \pi(1) = 3.142 \text{ m}$$

From Eq. (5A.4) with $c_1 = c_2 = 0$:

$$\frac{q_2}{q_1} = \left(\frac{\gamma_2}{\gamma_1}\right)\left(\frac{N_{\gamma 2}}{N_{\gamma 1}}\right)$$

$$N_{q1} = e^{\pi \tan(45°)} \cdot \tan^2(67.5°) = 134.9$$

$$N_{\gamma 1}(134.9 - 1) \tan(63°) = 262.7$$

$$\frac{q_2}{q_1} = \left(\frac{16.5}{22}\right)\left(\frac{11.19}{262.7}\right) = 0.032$$

From Fig. 5A.5 for $q_2/q_1 = 0.032$ and $\phi_1 = 45°$

$$K_s \cong 3.2$$

From Eq. (5A.3e):

$$P_h = (3.2)[(16.5)(1)(1) + 0.5(22)(1)^2]$$

$$= 88.0 \text{ kN/m } (6.03 \text{ kips/ft})$$

From Eq. (5A.3d):

$$W_{pz} = 0.25\pi(1)^2(1)(22) = 17.28 \text{ kN } (3.88 \text{ kips})$$

$$A_f = \pi \frac{d_0^2}{4} = \pi \frac{(1)^2}{4} = 0.7854 \text{ m}^2 \text{ } (8.454 \text{ ft}^2)$$

From Eq. (5A.3a):

$$q_{ult} = 997 + \frac{(3.142)(88.0)\tan(45°) + 0 - 17.28}{0.7854} = 1327 \text{ kPa } (27.7 \text{ ksf})$$

Solution: Ultimate bearing capacity with replaced zone—punching of replaced zone through in situ soil

$$W_{eq} = [0.25\pi(2)^2]^{1/2} = 1.772 \text{ m (5.82 ft)}$$

From Eq. (5A.6b):

$$q_b = (16.5)(2)(14.72)(1.277)(1.376) + 0.5(16.5)(1.772)(11.19)(1.277)(1.376) = 1141 \text{ kPa (23.8 ksf)}$$

$$p = \pi d_{rz} = \pi(2) = 6.283 \text{ m (20.6 ft)}$$

From Fig. 5A.5 for $q_2/q_1 = 1$ and $\phi_1 = \phi_2 = 28°$:

$$K_s \cong 4.5$$

From Eq. (5A.6d):

$$P_h = 4.5(16.5)[(1)(1) + (1)^2/2] = 111.4 \text{ kN/m (7.63 kips/ft)}$$

From Eq. (5A.6c):

$$W_{rz} = 0.25\pi(2)^2(1)(22) = 69.12 \text{ kN (15.5 kips)}$$

$$A_{rz} = 0.25\pi(2)^2 = 3.142 \text{ m}^2 \text{ (33.8 ft}^2\text{)}$$

From Eq. (5A.6a):

$$q_{ult} = 1141 + \frac{(6.283)(111.4)\tan(28°) + 0 - 69.12}{3.142} = 1237 \text{ kPa (25.8 ksf)}$$

Solution: Ultimate bearing capacity with replaced zone—general shear failure within replaced zone
Because of the limited horizontal extent of the replaced zone, it is unlikely that general shear failure would occur within this zone. However, q_{rz} is calculated here to ensure that it is greater than either of the two values of q_{ult} computed for the punching cases. N_{q1} and $N_{\gamma 1}$ were calculated previously.

$$S_{q1} = S_{\gamma 1} = 1 + 0.1 \tan^2(67.5°)(1) = 1.583$$

$$D_{q1} = D_{\gamma 1} = 1 + 0.1 \tan(67.5°)(1/0.8862) = 1.272$$

From Eq. (5A.3c):

$$q_{rz} = (16.5)(1)(134.9)(1.583)(1.272) + 0.5(22)(0.8862)(262.7)(1.583)(1.272) = 9638 \text{ kPa (201 ksf)}$$

Comparing q_{rz} with q_{ult} for the two punching cases, it is seen that the value of q_{ult} for punching of the replaced zone through the in situ soil controls. Now calculate the factor of safety against bearing capacity failure:

$$F_s = \frac{q_{ult}}{q_0} = \frac{1237}{600} = 2.1$$

Most engineers would prefer a higher factor of safety, preferably in the range of 3 to 4. According to Vesic (1975), a minimum factor of safety of 2.0 is acceptable for apartment and office buildings if a thorough and complete soil exploration has been completed, but higher factors of safety are required for other structures and conditions. Thus, for most situations $F_s = 2.1$ is unacceptable. The ultimate bearing capacity could be increased for this case by widening the replacement zone (because punching of the replaced zone controlled) or by deepening the replacement zone (which increases q_b and punching resistance for both punching cases) or both. If punching through the replaced zone controls, a stronger replacement material could be used or the replacement material could be chemically stabilized before placement (see Sec. 5A.4).

Solution: Settlement without replaced zone From Table 5A.1 for $E_1/E_2 = 1$ and $D/d_0 = 1$, mean $I_s = 0.50$. From Eq. (5A.9):

$$S_i = \frac{(600)(1)}{5000} \cdot 0.50 = 0.060 \text{ m} = 60 \text{ mm (2.4 in)}$$

Solution: Settlement with replaced zone From Fig. 5A.9 for $E_1/E_2 = 20$, $\alpha \cong 0.42$. The minimum diameter required to obtain the full reduction in settlement can be estimated from Eq. (5A.10b):

$$d_{rz}(\min) = \frac{1 + (2)(1)}{0.42} = 5.8 \text{ m (19 ft)}$$

Since the actual d_{rz} is less than $d_{rz}(\min)$, the full reduction in settlement will not be achieved. From Table 5A.1 for $E_1/E_2 = 20$, $D/d_0 = 1$, $H/d_0 = 1$, and $d_{rz}/d_0 = 2.0$, the mean $I_s \cong 0.23$. From Eq. (5A.9):

$$S_i = \frac{(600)(1)}{5000}(0.23) = 0.028 \text{ m} = 28 \text{ mm (1.1 in)}$$

The effect of the replaced zone is to reduce the estimated immediate settlement from 60 mm (2.4 in) to 28 mm (1.1 in). A maximum tolerable settlement of 25 mm (1.0 in) is frequently specified for buildings by the structural engineer or architect, so this value of immediate settlement may not be acceptable. Since there may be some additional long-term settlement in granular soils (see Burland and Burbidge 1985), a lower estimated value for S_i may be preferable, perhaps in the range of 13 to 19 mm (0.5 to 0.75 in). In this situation the settlement could be reduced by deepening or widening the replaced zone, using a stiffer replacement material, chemically stabilizing the replacement material before placement (see Sec. 5A.4), increasing the diameter of the footing, increasing the depth of embedment, or by some combination of these factors.

A design value of $S_i = 19$ mm (0.75 in) will be used. A number of changes could be made to reduce estimated S_i to the design value, including the following.

If footing size and depth of embedment remain the same:

$$d_0 = D = 1 \text{ m} = 3.3 \text{ ft} \Rightarrow \frac{D}{d_0} = 1$$

The required value of I_s can be calculated by rearranging Eq. (5A.9) as follows:

$$I_s \leq \frac{S_i E_2}{q_0 d_0} = \frac{(0.019)(5000)}{(600)(1)} = 0.16$$

For the same replacement material, $E_1/E_2 = 20$

$$H = 1.5 \text{ m (4.9 ft)}, d_{rz} = 4 \text{ m (13.1 ft)} \Rightarrow \frac{H}{d_0} = 1.5, \frac{d_{rz}}{d_0} = 4, \text{ and } I_s = 0.16.$$

If the replacement material is chemically stabilized to give $E_1 = 250$ MPa (5,220 ksf) $\Rightarrow E_1/E_2 = 50$:

$$H = 1 \text{ m (3.3 ft)}, d_{rz} = 4 \text{ m (13.1 ft)} \Rightarrow \frac{H}{d_0} = 1, \frac{d_{rz}}{d_0} = 4, \text{ and } I_s = 0.16$$

$$H = 1.5 \text{ m (4.9 ft)}, d_{rz} = 3 \text{ m (9.8 ft)} \Rightarrow \frac{H}{d_0} = 1.5, \frac{d_{rz}}{d_0} = 3, \text{ and } I_s = 0.16$$

If the diameter of the footing is increased to 2 m (6.6 ft) with the same $D = 1$ m (3.3 ft) $\Rightarrow D/d_0 = 0.5$:

$$q_0 = \frac{471}{\frac{\pi}{4}(2)^2} = 150 \text{ kPa (3.1 ksf)}$$

$$I_s \le \frac{S_i E_2}{q_0 d_0} = \frac{(0.019)(5000)}{(150)(2)} = 0.32$$

For the same replacement material ($E_1/E_2 = 20$), the following dimensions of the replaced zone will work:

$$H = 1 \text{ m (3.3 ft)}, d_{rz} = 3.6 \text{ m (11.8 ft)} \Rightarrow \frac{H}{d_0} = 0.5, \frac{d_{rz}}{d_0} = 1.8, \text{ and } I_s \cong 0.32$$

$$H = 2 \text{ m (6.6 ft)}, d_{rz} = 2.4 \text{ m (7.9 ft)} \Rightarrow \frac{H}{d_0} = 1, \frac{d_{rz}}{d_0} = 1.2, \text{ and } I_s \cong 0.32$$

Solution: Stress distribution, in situ soil beneath replaced zone Although not needed to estimate settlement in this example problem, the stresses along the axis of the applied load are calculated for various depths below the interface to illustrate the procedure. The following four methods are used to calculate values of $\Delta\sigma_z$:

1. Values of $\Delta\sigma_z$ for an infinitely wide replaced zone are calculated from Table 5A.2 for $H/d_0 = 1$. Manual curve fitting is used to interpolate values of $\Delta\sigma_z/q_0$ for $E_1/E_2 = 20$ at each level of Z_i/H.
2. Values of $\Delta\sigma_z$ for $d_{rz}/d_0 = 2$ are obtained by interpolation from a curve of $\Delta\sigma_z/q_0$ versus z_i/d_0 [similar to Fig. 5A.9(a)] drawn from results of finite element analyses.
3. Values of $\Delta\sigma_z$ for an embedded load within a homogeneous half-space are obtained from the equation given in Nishida (1966).
4. Values of $\Delta\sigma_z$ for a surface-loaded homogeneous half-space (ignoring the embedment) are obtained from the equation given in Foster and Ahlvin (1954).

The results are summarized in the following table:

		Induced vertical stress beneath axis, $\Delta\sigma_z$							
		Wide replaced zone* $E_1/E_2 = 20$		$d_{rz}/d_0 = 2$† $E_1/E_2 = 20$		Homogeneous considering D‡ $E_1/E_2 = 1$		Homogeneous ignoring D§ $E_1/E_2 = 1$	
z_i (m)	z_i/H	kPa	ksf	kPa	ksf	kPa	ksf	kPa	ksf
0	0	41	0.86	50	1.0	98	2.0	171	3.6
1	1	20	0.42	34	0.71	35	0.73	52	1.1
2	2	14	0.29	21	0.44	18	0.38	24	0.50
3	3	9.6	0.20	15	0.31	11	0.23	14	0.29
4	4	7.0	0.15	11	0.23	7.3	0.15	8.9	0.19

*Interpolated by manual curve fitting from Table 5A.2.
†From results of finite element analyses.
‡Nishida (1966).
§Foster and Ahlvin (1954).

The induced vertical stresses for an infinitely wide replaced zone are less than for the case of $d_{rz}/d_0 = 2$ by an average of 33%. The induced stresses for the embedded homogeneous case are greater by 96%

at the interface and lesser by 34% at $z_i/H = 4$. For the homogeneous case where embedment is ignored, the values of $\Delta\sigma_z$ are greater by 242% at the interface and less by 19% at $z_i/H = 4$.

5A.3 NEAR-SURFACE COMPACTION

The density of a soil is one of the primary factors that influences its compressibility and strength. So long as all other characteristics of the soil remain unchanged—such as state of stress, moisture condition, fabric, and physical and chemical nature of the soil particles and pore fluid—the densification of a soil will generally result in a stronger, less compressible material. However, it is generally neither desirable nor possible to densify a soil without changing one or more of these other characteristics. Therefore, the process of densifying a soil to achieve desired engineering characteristics is more complicated than it seems because many other factors will affect the engineering properties of the soil after densification. In addition, these other factors generally vary with time, which not only may affect the density of the soil (resulting in settlement or heave of the soil) but also may substantially alter the strength and compressibility of the soil.

Compaction is the process by which mechanical energy is applied to a soil to increase its density. At the time of compaction, the void spaces of the soil are occupied by air, water, and various chemicals and substances, or some combination thereof. For simplicity, it will be assumed in this discussion that the voids are occupied by air, water, or both. The soil is said to be *dry* if the voids are completely filled with air (water content, w, is 0 and degree of saturation, S_r, is 0) and *saturated* (also called *wet* by some engineers) if entirely occupied by water (S_r is 100%). In the more general case where both water and air occupy the voids ($0 \leq S_r \leq 100\%$), the moisture condition of the soil is described as *partially saturated* or *moist*. In reality soils are never dry, so the term *unsaturated* is sometimes used for soils that are not completely saturated ($S_r < 100\%$).

The total volume of the soil is decreased during densification; because the volume of the soil solids remains essentially unchanged, the volume of the voids decreases by the same amount that the total volume decreases. The result is that air, water, or both are expelled from the void spaces of the soil. The soil is usually partially saturated when compaction is initiated. Because the air permeability of a soil is typically much greater than the water permeability, air is expelled from the voids of a partially saturated soil during compaction. If the soil is initially saturated or becomes saturated or essentially saturated during the compaction process, water will be expelled.

Numerous methods and types of equipment are used to apply mechanical energy to soil during compaction. The methods and equipment used on any particular project depend on several factors, including cost, availability of equipment, the contractor's experience and preference, regional practices, volume and types of soil to be compacted, and conditions under which the compaction occurs (such as weather conditions, proximity to structures and utilities, and availability of water). In the following sections, the discussion of compaction is organized according to the equipment and method used to effect the densification of the soil.

The most common type of compaction, and what most engineers and contractors think of when the term *compaction* is used, is *near-surface* or *shallow compaction*. In open areas where the size of the equipment is not a consideration, near-surface compaction is generally accomplished using self-propelled or towed compaction rollers, although rubber-tired and tracked construction vehicles, sheep, elephants (Meehan, 1967), and human feet (Kyulule, 1983) have also been used. Hand-held compactors, tampers, or rammers are used in confined areas such as narrow trenches and small excavations, around pipes, behind retaining and basement walls, and adjacent to buildings, bridges, and other structures.

Near-surface compaction can be performed on unexcavated site soils or on fill materials placed in thin layers called *lifts* (Fig. 5A.10). Near-surface compaction of unexcavated site soils is usually done to increase the stiffness and strength of granular soils for support of lightly loaded (residential and light commercial) shallow foundations. Two common types of fills—embankments and fills produced from cut-and-fill operations—are illustrated in parts (*b*) and (*c*) of Fig. 5A.10. Embankments may be used to impound water (earth dams, levees, and so on) or to support struc-

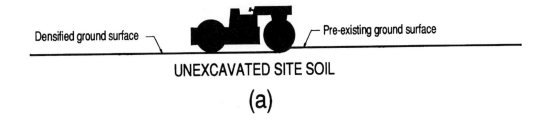

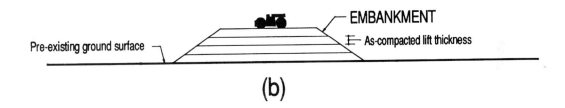

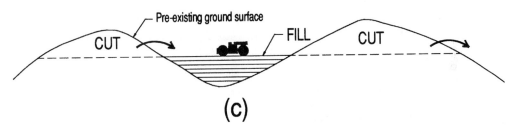

FIGURE 5A.10 Common types of near-surface compaction: (*a*) compaction of unexcavated site soil; (*b*) construction of embankment; and (*c*) cut-and-fill operations.

tures such as highways, buildings, and abutments for bridges. Cut-and-fill operations are conducted in hilly areas to provide relatively flat grades primarily for highways and residential developments.

Fill materials can consist of on-site soils that have been excavated and replaced in the same location (called *overexcavation/replacement*—see Sec. 5A.2) or borrow materials obtained from either on-site or off-site borrow pits. Soils to be used in fills are frequently moisture conditioned, pulverized, chemically modified, blended with other materials, or otherwise modified before compaction. The maximum depth to which soil can be densified using rollers or hand-held equipment is about 6 ft (2 m) and usually is much less (hence the term *near-surface*), depending on the compaction equipment used and the type of soil compacted. Substantial densification generally occurs only to a depth of about 6 to 24 in (150 to 610 mm). The degree of densification of a dune sand with depth as a function of the number of passes of a vibratory roller is illustrated in Fig. 5A.11 (D'Appolonia et al., 1969). Because of the shallow depth to which substantial densi-

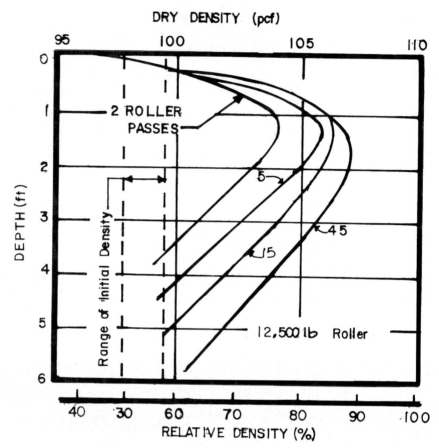

FIGURE 5A.11 Relationship between as-compacted dry density and depth for an increasing number of passes from a 12,500-lb (56-KN) vibratory smooth-drum roller on an 8-ft (2.4-m) lift height of a dune sand (from D'Appolonia et al., 1969).

fication occurs, compaction of unexcavated site soils is generally limited to situations where either the applied loads are light or the depth of influence of the applied load is shallow (the width of the loaded area is small). As a rule of thumb, the depth of substantial densification decreases as the cohesiveness of the soil increases. Lift thicknesses in fill operations range from about 3 to 36 in (75 to 910 mm), with the smallest values for highly plastic (fat) clays and the largest values for select coarse-grained soils.

Engineering properties of the soil that can be affected by compaction and that can be controlled to some extent during the compaction process include the following:

1. Density
2. Strength
3. Compressibility
4. Potential for volume changes caused by changes in moisture condition (swell, collapse, shrinkage)
5. Potential for frost heave
6. Permeability

5A.3.1 Equipment

A variety of construction equipment is used in a near-surface compaction process. For borrow materials, equipment is needed to excavate the material at the borrow location, transport it to the site, place it in lifts, and compact it. If in situ soils are to be compacted in place, only compaction equipment is required. A detailed discussion of the wide variety of methods and equipment used on compaction projects is beyond the scope of this book, so only a general overview is given here.

5A.3.1.1 Excavation, Hauling, Spreading. Borrow materials are usually excavated using some combination of power shovels, draglines, scrapers (pans), loaders, backhoes, and hydraulic excavators (see Figs. 5A.12 to 5A.15). Scrapers are multipurpose vehicles that can also be used to transport the soil and spread it in lifts on the fill area. If scrapers are not used, borrow materials are usually transported to the fill area in haul trucks (Fig. 5A.13) and spread in lifts using graders (Fig. 5A.16), loaders (Fig. 5A.13), or bulldozers (Fig. 5A.17). Some compaction rollers have a blade on the front that allows the soil to be spread and compacted in the same process. If the borrow material consists of dry, cohesive soils, it may need to be pulverized using a pulverizer or pugmill prior to spreading. Depending on the application, it may also be necessary to remove cobbles, boulders, undecomposed organic parts, or other undesirable components prior to hauling or spreading.

Moisture conditioning is frequently required to bring the soil to the desired compaction water content. This process consists of either drying or wetting the soil, reworking (mixing) the soil, and allowing the soil sufficient time to achieve relatively uniform moisture distribution. This can be accomplished either after the soil is spread in a loose lift using a water truck and discing equip-

FIGURE 5A.12 Scraper (courtesy of Caterpillar Inc., Peoria, IL).

5.40 SOIL IMPROVEMENT AND STABILIZATION

FIGURE 5A.13 Loader dumping soil into a haul truck (courtesy of Caterpillar Inc., Peoria, IL).

ment or in a stockpile located in a designated processing area. The conditioning time required to achieve moisture equilibrium in a soil after either wetting or drying depends on the soil type, varying from a few minutes for clean coarse sands and gravels to several days or weeks for highly plastic clays. When dry cohesive soils are wetted, the required moisture conditioning time depends primarily on the time required for the water to penetrate the hard clods and is therefore a function of clod size—larger clods need more conditioning time. At least 24 to 72 h are generally needed for proper moisture conditioning of cohesive soils, and more time usually is desirable.

5A.3.1.2 Compaction. Near-surface compaction is accomplished using a variety of equipment that can be classified into two primary categories depending on whether the general shape of the portion of the equipment that contacts the soil is round or flat. The major types of compactors are listed in Fig. 5A.18.

The types of equipment used on any compaction project depend on the grain-size distribution and mineralogy of the soils being compacted, the engineering properties that need to be controlled during the compaction process, and the equipment readily available to the compaction contractor. In some instances the type of equipment to be used is designated in the compaction specifications (usually just general type but sometimes more specifically, including type, weight, length of feet, vibration frequencies, and so on), but more commonly the choice of equipment is left to the compaction contractor. This will be discussed in more detail in Sec. 5A.3.5.

FIGURE 5A.14   Backhoe (courtesy of Deere & Company, Moline, IL).

Smooth-drum or *smooth-wheel rollers* (Fig. 5A.19) have smooth steel drums that provide 100% areal coverage over the width of the drum, and compaction occurs primarily from static pressures generated by the weight of the roller. A conceptual drawing illustrating the forces generated by a moving smooth-drum roller is given in Fig. 5A.20. The peripheral force (U) generated by the rotating drum acts horizontally but opposite in direction to the tractive force (Z) generated by the momentum of the roller. Because the peripheral and tractive forces tend to negate each other (although a small net horizontal force may result in either direction), the primary force generated upon the soil results from the weight of the roller (W).

The vertical contact pressure generated by a roller depends on W and the magnitude of the contact area. The shape of the contact area can be approximated by a rectangle whose length is equal to the length of the drum and whose width depends on the magnitude of vertical deflection, which in turn depends on the magnitude of W and the stiffness of the soil. During the initial pass of the roller on a loose lift, the vertical deflection (and hence the width of the contact area) is greatest; during subsequent passes of the roller, the soil is stiffer and the contact area is less, which produces higher contact stresses.

Smooth drum rollers have been used historically to compact soils of all types, but their usefulness in current practice is generally limited to proofrolling (smoothing) the surface of lifts compacted by other types of rollers and compacting granular soils and asphaltic pavements. Their usefulness for compacting relatively soft cohesive soils is limited owing to poor traction and a "plowing" effect in which soil is displaced laterally without significant compaction (Hausmann 1990). Smooth drum rollers are also typically very slow compared to newer types of rollers.

A wide variety of *rubber-tired* or *pneumatic rollers* (Fig. 5A.21) are available. Most contain a series of tires closely spaced at regular intervals across the width of the roller. The areal coverage is typically about 80%, and tire pressures up to about 100 psi (700 kPa) are used (Holtz and Kovacs, 1981). Compaction occurs by a combination of static pressure (approximately equal to the inflation pressure of the tires) and kneading action of the tires. *Wobble-wheel rollers* have the wheels mounted such that the tires wobble laterally as the roller is towed forward, which imparts

FIGURE 5A.15 Hydraulic excavator (courtesy of Deere & Company, Moline, IL).

additional kneading action to the soil that facilitates compaction in cohesive soils (Spangler and Handy, 1982). The factors that influence the compactive effort include gross weight of the roller; wheel diameter and load; and tire width, size, and inflation pressure (Hausmann, 1990). Rubber-tired rollers are effective compactors of a wide range of soil types. Large-size tires are desirable in cohesionless soils to avoid shear and rutting. In general, low inflation pressures of 20 to 40 psi (140 to 275 kPa) are desirable for clean sands and gravelly sands with low gravel content, whereas inflation pressures greater than 65 psi (425 kPa) are desirable for cohesive soils and very gravelly soils (Spangler and Handy, 1982). Although not technically classified as compaction equipment, standard rubber-tired and tracked construction vehicles are sometimes used for compaction but are inefficient compared to rubber-tired or other rollers designed specifically for compaction.

In *sheepsfoot rollers* (Fig. 5A.17) and *tamping foot* or *padfoot rollers* (Fig. 5A.22), protrusions or "feet" of various lengths, sizes, and shapes (Fig. 5A.23) extend outward from a steel drum. Although tamping foot rollers are considered by some to be sheepsfoot rollers, they are categorized separately here because the effectiveness of each type can vary for different cohesive soils and compaction conditions. Sheepsfoot and tamping foot rollers are most effective in cohesive soils and are the best rollers currently available for compacting most cohesive soils. The results from model studies simulating the action of a sheepsfoot in clay and in sand are shown in Fig. 5A.24 (Spangler and Handy, 1982). These results were obtained by photographing the soil movements in

FIGURE 5A.16 Grader (courtesy of Caterpillar Inc., Peoria, IL).

FIGURE 5A.17 Tracked bulldozer towing a sheepsfoot roller (courtesy of Case Corporation, Racine, WI).

5.44 SOIL IMPROVEMENT AND STABILIZATION

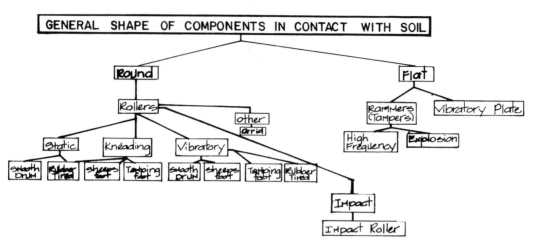

FIGURE 5A.18 Classification of near-surface compaction equipment (modified after Poesch and Ikes, 1975).

FIGURE 5A.19 Vibratory smooth-drum roller (courtesy of Vibromax 2000 USA Inc., Racine, WI).

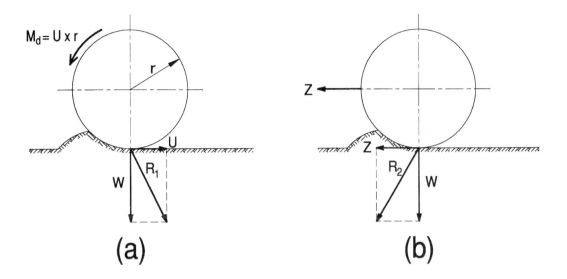

Note: The shifts in the points of application of the horizontal forces during travel of the roller are neglected in this analysis.

FIGURE 5A.20 Forces applied to the ground by a static smooth-drum roller (after Poesch and Ikes, 1975): (*a*) drum rotating in place; (*b*) drum pulled by a tractive force (*Z*).

FIGURE 5A.21 Rubber-tired roller (courtesy of Caterpillar Inc., Peoria, IL).

FIGURE 5A.22 Self-propelled tamping foot roller with leveling blade on front (courtesy of Caterpillar Inc., Peoria, IL).

a Plexiglas-fronted box caused by vertical penetration of a spherically tipped device and indicate a much larger zone of compaction in clay than in sand.

Sheepsfoot rollers generally have feet about 6 to 10 in (150 to 250 mm) long with tip areas from 5 to 12 in^2 (30 to 80 cm^2), provide an areal coverage of 8 to 12%, with contact pressures ranging from about 200 to 1000 psi (1400 to 7000 kPa) (Holtz and Kovacs, 1981). Compaction occurs as a combination of static pressure from the weight of the roller and kneading action of the feet as they penetrate the soil and are rotated through it. In cohesive soils, the high contact pressures permit crushing or breaking of dry clods and remolding of wet clods into a denser soil mass. If the lift thickness is smaller than the length of the feet, densification first occurs along the interface between the previously compacted lift and the new lift as the loose soil along the bottom of the new lift is pushed into and molded with the upper portion of the underlying lift. In this manner a good bond is developed between lifts, which helps prevent sloughing along lift interfaces in fills constructed on slopes and reduces preferential drainage paths for contaminant migration in compacted clay. It is therefore important to compact cohesive soils in thin lifts to promote good interfacial bonding. The upper surface of a newly completed lift is sometimes scarified using discing or other equipment before the next lift is placed to promote better bonding between the lifts. As compaction continues, the soil in the new lift is densified from the bottom upward, and the roller is said to "walk out" of the soil as the upper portion of the lift becomes densified.

The feet on tamping foot rollers are typically shorter, less than 6 in (150 mm), but have greater contact area [20 to 30 in^2 (130 to 190 cm^2)] than those on sheepsfoot rollers. The areal coverage is also much greater, typically about 40%, with contact pressures ranging from about 200 to 1200 psi (1400 to 8400 kPa) (Holtz and Kovacs, 1981). Some tamping foot rollers have hinged feet for greater kneading action. Like sheepsfoot rollers, tamping foot rollers are most effective in com-

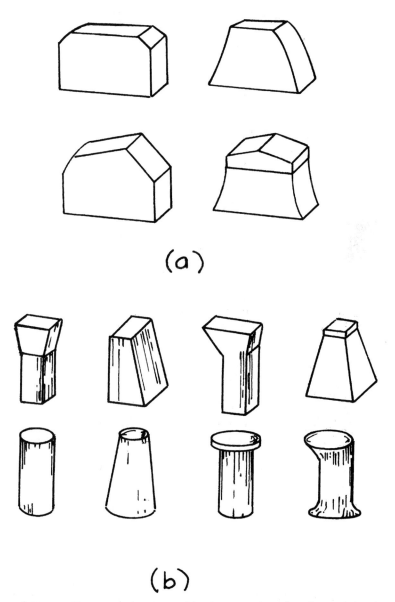

FIGURE 5A.23 Some typical shapes of feet for compaction rollers (from Poesch and Ikes, 1975): (*a*) tamping foot; and (*b*) sheepsfoot.

pacting cohesive soils. Tamping foot rollers are generally more effective than sheepsfoot rollers in compacting softer, wetter cohesive soils in which the clods need to be remolded rather than crushed or broken; the larger foot contact area and areal coverage of the tamping foot roller provides better and more uniform remolding of the clods. For the same lift thickness, however, the shorter tamping feet are less effective than sheepsfeet in producing interfacial bonding of lifts; therefore, thinner lift thicknesses or scarification of the lift surfaces is needed when using tamping foot rollers if good bonding between lifts is required.

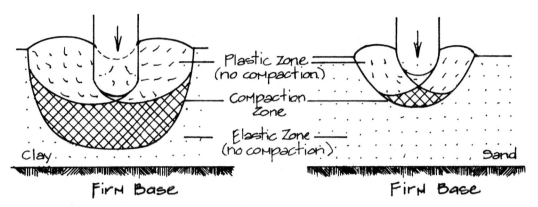

FIGURE 5A.24 Results of model studies simulating the action of a sheepsfoot roller in clay and sand (from Spangler and Handy, 1982; after Butt et al., 1968).

The drums on *grid rollers* consist of heavy steel mesh that provides about 50% coverage and contact pressures from about 200 to 900 psi (1400 to 6200 kPa) (Holtz and Kovacs, 1981). The moderate areal coverage of the mesh allows high contact pressures while helping to prevent excessive shear deformation responsible for producing plastic waves in front of the roller (Hausmann, 1990). Grid rollers are effective in breaking and rearranging gravel and cobble-sized particles and are primarily used in weathered rocks and rocky soils containing large sand, gravel, and cobble fractions. Cohesive soils tend to clog the mesh, rendering it ineffective by essentially changing the roller into a smooth drum roller. By varying the operating speed of the grid roller, more effective compaction can be obtained; faster speeds can be used in the initial passes to break down the material, and lower speeds can be used in the final passes to produce greater densification (Hausmann, 1990).

Smooth-drum, rubber-tired, sheepsfoot, and tamping foot rollers can have vibrators attached to make them *vibratory rollers*. The vibration is typically produced by rotating eccentric weights. Vibratory rollers are the most effective method for near-surface compaction of cohesionless soils. The vibration produces cyclic deformation of the soil, which can result in significant densification beyond that provided by the static weight of the roller, as shown in Fig. 5A.25. Vibration can, in some instances, produce additional densification in soils with some cohesion (Selig and Yoo, 1977).

The effectiveness of vibratory compaction on any given project is a function of the characteristics of the compactor, the properties of the soil, and the construction procedures used. The variables that influence vibratory compaction include the following (Forssblad, 1981; Holtz and Kovacs, 1981):

1. Characteristics of the compactor
   a. Static weight
   b. Vibration frequency and amplitude
   c. Ratio between frame mass and drum mass
   d. Drum diameter
2. Properties of the soil
   a. Initial density (density prior to compaction)
   b. Grain-size distribution
   c. Shapes of particles
   d. Compaction (molding) water content
3. Construction procedures
   a. Number of passes of the roller per lift
   b. Lift thickness
   c. Speed at which the roller is operated

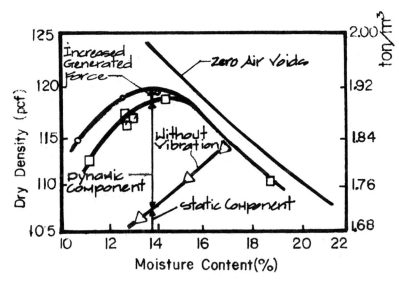

FIGURE 5A.25 Compaction results on 12-in (305-mm) layers of silty sand with and without vibration using a 17,000-lb (76-kN) towed vibratory roller (after Parsons et al. 1962 as cited in Selig and Yoo, 1977).

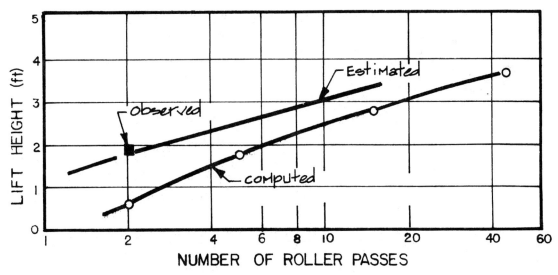

FIGURE 5A.26 Relationship between lift height and number of passes of a 12,500-lb (56-kN) vibratory smooth-drum roller required to achieve a minimum relative density of 75% in a dune sand (from D'Appolonia et al., 1969).

The relationship between lift thickness and number of passes for a given roller and desired soil density is illustrated in Fig. 5A.26. By decreasing the lift thickness, the number of passes needed to achieve the desired density is reduced; conversely, a thicker lift can be used if the number of passes is increased.

The vibration frequency of the compactor has been shown to affect the as-compacted dry density of a wide variety of soil types. As indicated in Fig. 5A.27, in most soils a peak dry density is

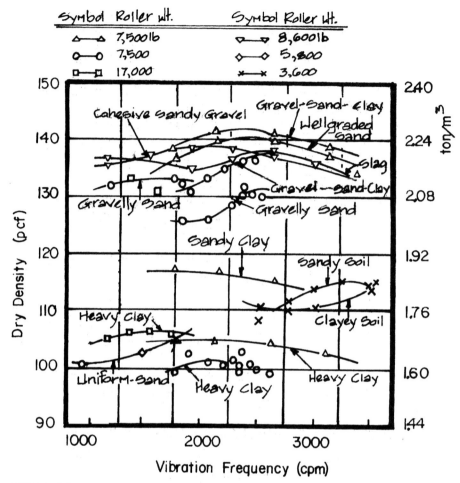

FIGURE 5A.27 Effect of vibration frequency on the as-compacted dry density of different soils (from Selig and Yoo, 1977).

achieved at an optimum vibration frequency. However, the differences in dry density achieved at various frequencies are relatively small. Better compaction (higher density and greater depth of influence) results from a combination of a large amplitude and a frequency just over the resonance frequency (usually about 25 Hz) than a combination of high frequency and small amplitude (Forssblad, 1981).

The effect of roller speed and number of passes on the as-compacted dry density of a well-graded sand and a highly plastic clay are shown in Fig. 5A.28. It is apparent from this figure that the dry density increases with increasing number of passes up to a certain point, beyond which little additional densification is achieved from more passes. The influence of roller speed is also obvious; for comparable conditions, reducing the travel speed results in higher as-compacted dry density.

The *impact roller* was developed in the 1950s in South Africa for in situ densification of collapsible sands (see Sec. 5A.3.4.2) (Clifford, 1980). An impact roller typically consists of a four-faced rolling mass that delivers four impact blows per revolution and is towed at an ideal speed of about 6 to 9 mi/h (10 to 14 km/h). The typical shape of an impact roller is shown in Fig. 5A.29

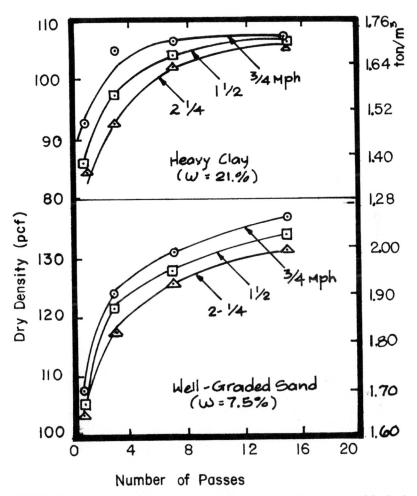

FIGURE 5A.28 Influence of travel speed and number of passes on the as-compacted dry densities of a clay and a sand (from Parsons et al., 1962 as cited in Selig and Yoo, 1977).

and includes rounded corners to simplify transition from one lift and fall cycle to the next. A groove is incorporated behind each rolling corner to allow the roller mass to revolve freely during impact even in soft ground when the corner penetrates some distance into the surface of the soil. The impact faces of the roller are rounded so that as the soil densifies, the impact blow is dissipated over a smaller and smaller area, effecting compaction to greater depth. The high energy delivered by an impact roller compacts materials up to about 10 ft (3 m) thick. A corrugated surface is left by the impact roller, which has been found to produce good interfacial bonding of lifts and to reduce potential interfacial shear movements. If a smooth final surface is desired, the top lift must be proofrolled with a smooth-drum roller.

Most *vibratory plate compactors* (Fig. 5A.30) are self-propelled and hand-guided and are used primarily for base and subbase compaction for streets, sidewalks, and the like; street repairs; fills behind bridge abutments, retaining walls, basement walls, and so on; fills below floors; and trench fills (Broms and Forssblad, 1969). Tractor-mounted and crane-mounted models are also available. Compaction is achieved by the application of high-frequency (10 to 80 Hz), low-amplitude vibra-

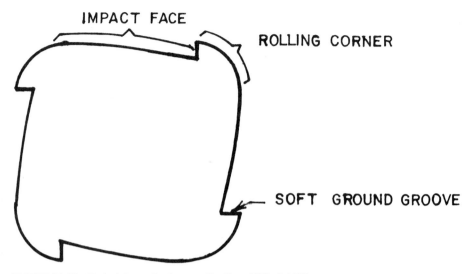

FIGURE 5A.29 Typical shape of an impact roller (from Clifford, 1980).

FIGURE 5A.30 Vibratory plate compactor (courtesy of Wacker Corporation, Menomonee Falls, WI).

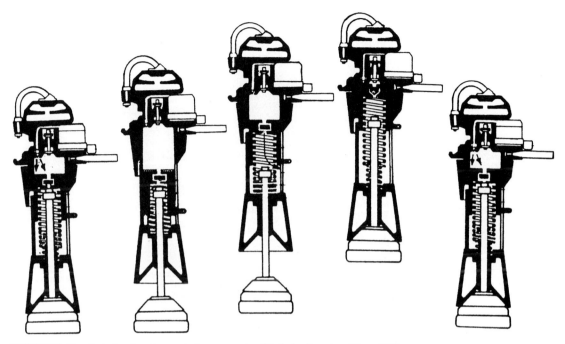

FIGURE 5A.31 Typical action of an explosion rammer (modified after Poesch and Ikes, 1975).

tions (Wacker, 1987). Vibratory plates are usually powered by gasoline or diesel engines and are used mainly for compacting granular and low-cohesion soils. The hand-guided vibratory plate compactors generally weigh between 100 and 6000 lb (0.4 to 27 kN) and come in a wide variety of plate sizes.

Hand-guided *rammers* or *tampers* densify the soil by a combination of impact and vibratory compaction. The most common applications for rammers are for street repairs; fill behind bridge abutments, retaining walls, basement walls, and so on; and trench fills (Broms and Forssblad, 1969). In *explosion* or *combustion rammers* (Fig. 5A.31), a mixture of either diesel fuel or gasoline and air is ignited in an internal chamber that simultaneously forces the rammer against the ground and lifts the casing off the ground. After the casing has moved upward a short distance independent of the rammer, the piston rod (which is attached to the rammer) is pulled upward with the casing until the maximum trajectory reached. The casing and rammer then fall freely to the ground, where a second impact occurs. Thus, two impacts are produced from one explosion. Explosion rammers typically weigh about 200 to 2000 lb (1 to 10 kN), with impact surface areas of about 80 to 930 in^2 (500 to 6000 cm^2) (Poesch and Ikes, 1975). *High-frequency rammers* (Fig. 5A.32) are usually powered by gasoline or diesel engines and produce high-frequency impacts through a piston/spring system connected to the engine via a gear transmission system (Wacker, 1987). Typical characteristics of high-frequency rammers are as follows (Broms and Forssblad, 1969; Wacker, 1987):

Frequency: 6 to 14 Hz
Weight: 100 to 300 lb (450 to 1300 N)
Impact area of ramming shoe: 100 to 500 in^2 (650 to 3200 cm^2)
Maximum height rammer comes off ground: 1 to 3 in. (25 to 75 mm)

High-frequency rammers can be used on granular and cohesive soils.

FIGURE 5A.32 High-frequency rammer (courtesy of Wacker Corporation, Menomonee Falls, WI).

5A.3.2 Moisture-Density Relationships

The quality of a compacted soil is frequently correlated with the dry density of the as-compacted soil; that is, a greater dry density is commonly assumed to mean a better soil. Although this correlation is generally valid for many soils in terms of their strength and compressibility, other parameters (especially water content) may have an important effect, especially in cohesive soils. In addition, some engineering properties of certain soils are only slightly influenced by dry density, and other parameters such as moisture condition have a greater effect. These relationships will be discussed in more detail in Sec. 5A.3.4. Despite these other considerations, the fact remains that dry density is the property used most often to determine whether a compacted soil will exhibit the desired engineering behavior.

The primary factors that influence the value of dry density obtained by compacting a specific soil are as follows:

1. Moisture condition of the soil during compaction (molding or compaction water content)
2. Compactive energy or effort per unit volume of soil (lift thickness and number of passes, blows, or tamps of the equipment)
3. Method of compaction (compaction equipment)
4. Rate of compaction (rate of travel, frequency of impact, or frequency of vibration of the equipment)

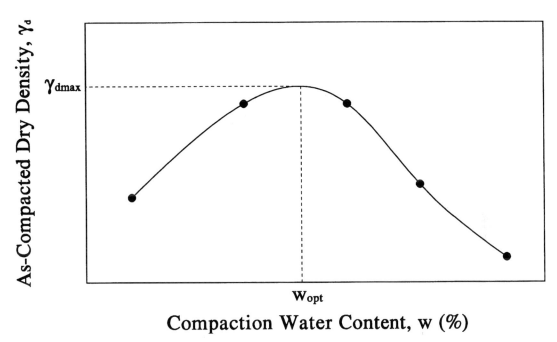

FIGURE 5A.33 Typical moisture-density relationship for compacted soil.

For a given situation in which factors 2, 3, and 4 are held constant, the as-compacted dry density of the soil may vary significantly depending on the molding water content. For many soils, the moisture-density relationship for a given method of compaction and compactive effort can be described by a single peak curve similar to that shown in Fig. 5A.33. The water content and dry density corresponding to the highest peak in the moisture-density curve for one method of compaction and one level of compactive effort are referred to as the *optimum* water content and *maximum* dry density. If the method of compaction or the compactive effort is varied, the moisture-density curve for the same soil will be different. For example, the effect of varying the compactive effort for the same soil and the same method of compaction is illustrated in Fig. 5A.34; similarly, the influence of method of compaction is shown in Fig. 5A.35. Therefore, for a specific soil there are an infinite number of optimum water contents and maximum dry densities, and the terms *optimum* and *maximum* are relative rather than absolute and should be used in proper contexts.

For a given method of compaction, the optimum moisture condition for achieving the maximum dry density is represented by the locus of points corresponding to the peak points for various compactive efforts (Fig. 5A.34). This locus of points is known as the *line of optimums* or the *locus of optimums*. Analysis of the general trend of the line of optimums shows that the maximum dry density increases and the optimum water content decreases as the compactive effort is increased. This trend is not without bounds, however, as a point is reached where applying additional compactive effort produces no measurable increase in dry density and in some instances may produce a decrease in dry density.

The line of optimums is generally approximately parallel to the zero air voids line (ZAVL), which represents a degree of saturation of 100%, and as such the line of optimums is approximately a line of constant degree of saturation. In the author's experience, the degree of saturation represented by the line of optimums varies from about 60 to 95%, with the lower values for clean sands and higher values for highly plastic clays; for most soils this value ranges between 70 and 85%. This correlation is important with respect to the engineering behavior of many soils (to be discussed in Sec. 5A.3.4) and suggests that the moisture condition of a soil may, in some circum-

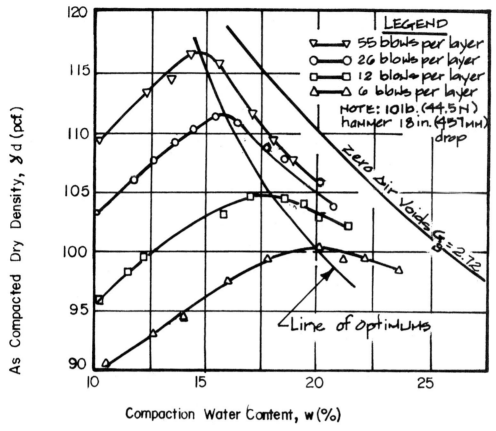

FIGURE 5A.34 Effect of compactive effort on the moisture-density relationship of an impact-compacted soil (from Foster, 1962).

stances, be better represented by degree of saturation rather than by water content. The approximate degree of saturation represented by the line of optimums is referred to as optimum saturation (S_{opt}). The line of optimums may vary for a given soil depending on the method of compaction, as shown in Figure 5A.36 for a lean clay. The range in degree of saturation for each of the four lines of optimums shown in Fig. 5A.36 are summarized as follows (assuming $G_s = 2.70$):

Method of Compaction	Approximate Range of S_{opt}
Sheepsfoot roller	73–83%
Laboratory impact	77–86%
Laboratory static	80–81%
Rubber-tired roller	89–90%

For each method of compaction, S_{opt} varies within a fairly narrow range ($\leq 10\%$). Although the lines of optimums vary somewhat from each other, the line of optimums from the laboratory impact (Proctor) tests provides a reasonable approximation of the line of optimums for the two field-compacted soils, and this is generally true for most soils.

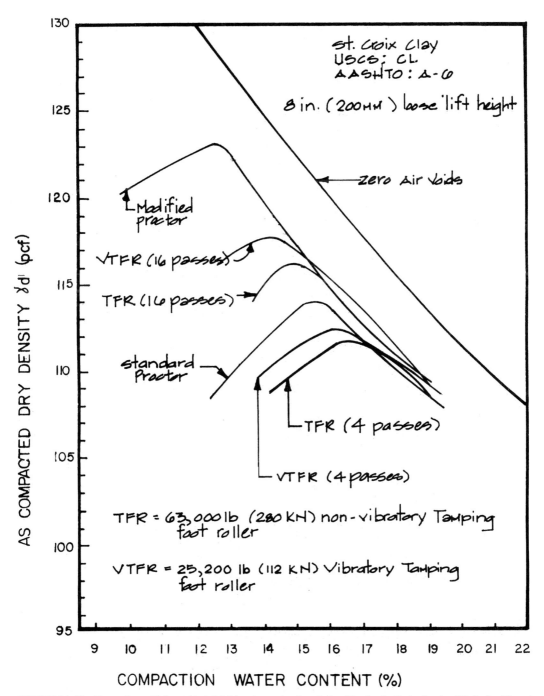

FIGURE 5A.35 Comparison of laboratory and field moisture-density relationships for St. Croix clay (modified after Lin and Lovell, 1981; data from Terdich, 1981).

5.58 SOIL IMPROVEMENT AND STABILIZATION

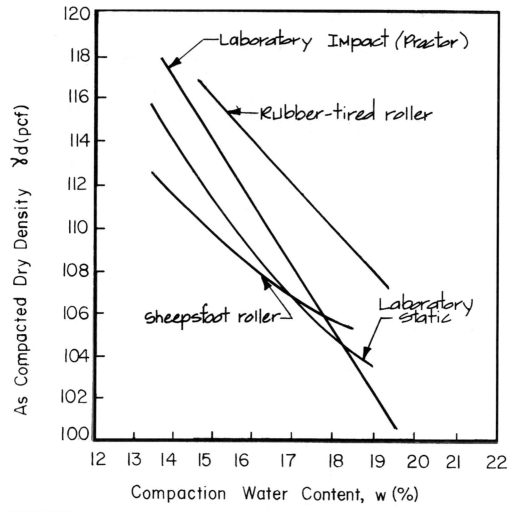

FIGURE 5A.36 Comparison of lines of optimums for four methods of compaction (Foster, 1962).

The ZAVL varies as a function of specific gravity of the soil solids, and the ZAVL can be found from the following equation by setting $S_r = 100\%$:

$$w = \left(\frac{\gamma_w}{\gamma_d} - \frac{1}{G_s}\right) \cdot S_r \tag{5A.15}$$

where γ_d = dry density (by mass or weight) of the soil
γ_w = density (by mass or weight) of water
w = water content
S_r = degree of saturation
G_s = average specific gravity of the soil solids

Any line of constant saturation (e.g., 60%, 70%, 80%) can be found using Eq. (5A.15).

Although the single peak curve is most common, an extensive laboratory study by Lee and Suedkamp (1972) showed that three irregularly shaped curves can also occur, as illustrated in Fig. 5A.37. The results from their research indicate that (1) soils with a liquid limit between 30 and 70 typically exhibited a single-peak moisture-density curve; (2) both double-peak and oddly shaped curves were present in soils with a liquid limit greater than 70; (3) soils with a liquid limit less than 30 usually produced either a double-peak or a one-and-one-half peak curve; and (4) the length of the moisture conditioning period substantially affected the moisture-density relationships in high-plasticity (high-liquid-limit) soils but had little influence on low-plasticity or nonplastic soils.

The moisture-density curves for free-draining sands and sandy gravels typically have $1\frac{1}{2}$ peaks, as illustrated in Figure 5A.38 (Foster, 1962). The maximum dry density for these soils is obtained in either the air-dried or saturated condition. In the partially saturated state, capillary suction increases the effective stresses in the soil and produces a resistance to compaction called *bulking*. Because it is difficult to keep large quantities of borrow materials in the air-dried condition, sands and sandy gravels are usually placed saturated. Keeping free-draining soils saturated is not easy and generally requires continuous wetting of the soil in front of the compaction equipment.

The temperature of the soil during compaction has a noticeable effect on as-compacted dry density when the temperatures are above freezing—the same compactive effort produces a lower dry density at the same compaction water content (Waidelich, 1990). Thus, more compactive effort is needed to obtain the same value of dry density at lower temperatures. If the temperature of the soil is below freezing, it is extremely difficult, uneconomical, and virtually impossible under some circumstances to obtain the densities required for proper performance of a compacted fill. The influence of temperatures below freezing on the standard and modified Proctor moisture-density relationships of a sand with a trace of silt is illustrated in Fig. 5A.39. At a temperature of 30°F (-1°C) standard Proctor γ_{dmax} was 94% of the value at 74°F (23°C), and increasing the compactive effort to modified Proctor (4.5 times that of standard Proctor) at 30°F (-1°C) only increased γ_{dmax} to 98% of standard Proctor γ_{dmax} at 74°F (23°C). An additional problem with fills compacted at temperatures below freezing is that the interior portions of the fill may take several years to thaw, depending on the type of fill and its dimensions. For these reasons, it is highly recommended that compaction not be conducted during freezing weather; in fact, many government agencies located in freezing climates that engage in earthwork construction do not permit the construction of fills during winter months.

5A.3.3 Laboratory Compaction and Testing

The best method for predicting the engineering behavior of compacted soil is to conduct full-scale field testing on the soil that simulates as closely as possible the actual service conditions. For example, if the compacted soil is to support shallow foundations for a building, full-scale load tests can be conducted on all footings prior to any construction of the superstructure to ensure that the compacted soil can support the footings without excessive settlement or failure. This method is rarely used in practice for obvious reasons—the cost is extremely high, and a long time is required for testing. For a compacted fill, a test pad is sometimes constructed using the actual soil and compaction equipment, and appropriate tests (field, laboratory, or both) are conducted on the test pad to determine the important engineering properties. For example, many state and federal regulatory agencies currently require that a test pad be constructed to simulate a compacted clay liner for waste containment systems. In situ permeability tests (usually double-ring infiltration tests, ASTM D5093) are then conducted on the test pad to verify that the permeability of the test pad is less than the maximum allowable value (usually 1.0×10^{-7} cm/s = 2.8×10^{-4} ft/day). However, laboratory tests are used on many projects as the primary method for predicting, controlling, and verifying the engineering properties and behavior of compacted soil because laboratory tests generally cost much less than comparable field tests.

It is important to model field conditions as closely as possible when conducting moisture-density tests and when preparing samples to be used in other types of laboratory tests. For example, recent research on compacted soils has shown that clod size and percentage of large-size particles

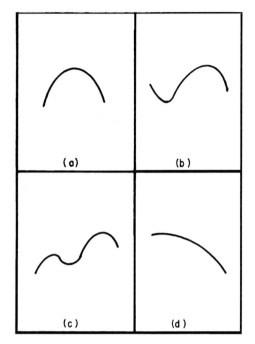

FIGURE 5A.37 Types of moisture-density curves: (a) single peak; (b) one-and-one-half peaks; (c) double peak; and (d) oddly shaped (Lee and Suedkamp, 1972).

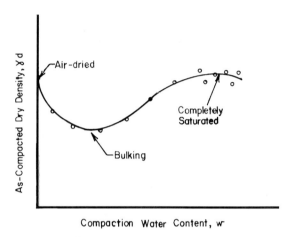

FIGURE 5A.38 Typical compaction curve for free-draining sands and sandy gravel (Foster, 1962).

(gravel and larger) have an important effect on the engineering properties of compacted soils (e.g., Benson and Daniel, 1990; Day, 1989, 1991; Houston and Randeni, 1992; Houston and Walsh, 1993; Larson et al., 1993; Shelley and Daniel, 1993). A general rule of thumb for laboratory testing is that the maximum particle or clod size should be less than one-sixth to one-tenth the smallest dimension of the sample. It is therefore important to test samples that are large enough to obtain results that are representative of field behavior. Several methods are available for correcting for oversize particles, where the term *oversize* refers to particles that are too large to be included in a particular testing apparatus (see ASTM D698, D1557, D4253, and D4718; Hausmann, 1990; Houston and Walsh, 1993; Fragaszy et al., 1992; Siddiqi et al., 1987).

When preparing soils for moisture-density determination or when making specimens for laboratory testing, the soil should be brought to the desired water content and double-sealed in airtight containers for a sufficient time to allow the soil to come to moisture equilibrium (called the *moisture conditioning period*). For proper modeling of field compaction techniques, this moisture conditioning period should duplicate that to be used in the field. As noted previously, for cohesive soils a minimum of 24 h is desirable, and for highly plastic soils 72 h or more may be needed for proper moisture conditioning. When preparing samples for testing at specific values of water content and density, a method of compaction should be used that simulates as closely as possible the field compaction method. A comparison of the most common laboratory and field compaction methods is provided in Table 5A.3, and additional details on laboratory compaction methods are given below.

TABLE 5A.3. Comparison of Common Laboratory and Field compaction Methods

Type of Compaction	Laboratory methods	Field methods
Static	Loading plate with outer diameter slightly smaller than inner diameter of mold is pushed onto soil using a speed-regulated compression machine, a hydraulic jack within a loading frame, or a dead weight loading system (Fig. 5A.40).	Smooth drum roller (nonvibratory) Rubber-tired roller*
Impact	Impact hammer is used to drop a rammer with a diameter about one-third to one-half the diameter of the mold from a given height onto the soil (Fig. 5A.43). In some setups, the diameter of the rammer is slightly smaller than the diameter of the mold.	Impact roller Hand-guided rammers[†] Dynamic compaction
Kneading	Harvard miniature tamper is pushed into the soil until the spring deflects (Fig. 5A.44). California kneading compactor is used to push a compaction foot into the soil at constant pressure for a given time (Fig. 5A.45).	Sheepsfoot roller Tamping foot roller Rubber-tired roller*
Vibratory	Mold containing soil plus surcharge weight on top of the soil is placed on vibrating table and vibrated (Fig. 5A.47). High-frequency rammer compacts the soil within a mold (Fig. 5A.48). The ramming plate has a diameter slightly smaller than the diameter of the mold.	Vibratory rollers[‡] Vibrating plates Hand-guided rammers[†]

*Combination static and kneading compaction.
[†]Combination impact and vibratory compaction.
[‡]Some vibratory rollers, such as vibratory sheepsfoot and vibratory rubber-tired rollers, compact by using a combination of vibration and kneading action, and require more sophisticated tools to model.

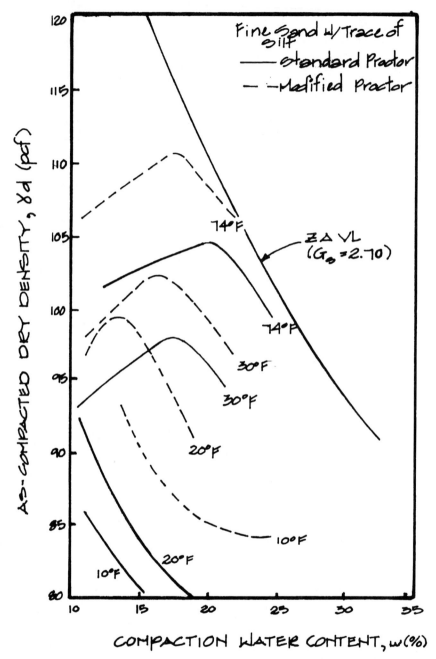

FIGURE 5A.39 Influence of temperature on the moisture-density relationships of a sand with a trace of silt (from Waidelich, 1990).

5A.3.3.1 Static Compaction. For all laboratory compaction methods, the soil is placed loosely into a compaction mold (usually cylindrical with a base plate and generally made of steel or aluminum), and the surface of the soil is leveled as well as possible using a spatula or similar device. In static compaction, a loading plate slightly smaller than the inner diameter of the mold (usually with an attached piston) is placed gently onto the surface of the loose soil and a compressive force is applied to the soil via the loading plate in one of several ways (see Fig. 5A.40). The most common methods are (1) slowly pushing the loading plate using a speed-regulated compression machine; (2) jacking against a reaction frame using a hydraulic jack; or (3) applying constant force using a dead-weight reaction frame. The compressive stress is increased until the desired density has been achieved or a predetermined energy has been applied. A diagram of a mold used by the writer to prepare statically compacted oedometer specimens is shown in Fig. 5A.41. Compactive energy can be determined by numerically integrating the load-deformation curve (Fig. 5A.42). It is difficult to apply a predetermined magnitude of compactive effort, especially if a speed-regulated compression machine is used, as the energy must be calculated immediately after each set of load and deformation readings and the test stopped when the desired energy level has been reached.

Static compaction is a relatively inefficient laboratory compaction method, because for any given type of soil and compaction water content, the same density can be achieved with less effort using another compaction method. Most samples need to be compacted in thin lifts because the force required to obtain a given density can be enormous. For example, the author once attempted to use static compaction to prepare a 6-in (152-mm) diameter sample of clayey sand in 2-in (50.8-mm) thick lifts to modified Proctor maximum dry density at the modified Proctor optimum water content. Using a 250,000-lb (1.1-MN) capacity compression machine normally used for structural engineering testing, the capacity of the machine was reached before the desired density was achieved. However, static compaction is commonly used in parametric studies because the reproducibility of nominally identical fine-grained specimens is greater than for the other methods. Because the method of compaction can have a significant influence on the as-compacted fabric of soils—and hence their engineering properties (see Sec. 5A.3.4)—static compaction should be used with caution.

5A.3.3.2 Impact Compaction. Impact compaction, usually in the form of Proctor tests, is the most commonly used method for estimating the moisture-density relationships of compacted soil. Of all the types of compaction rollers described in Sec. 5A.3.1, only the impact roller can be clas-

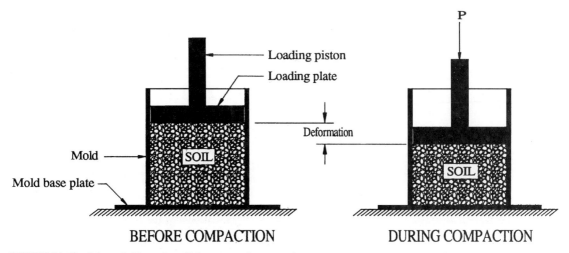

FIGURE 5A.40 Schematic illustration of laboratory static compaction.

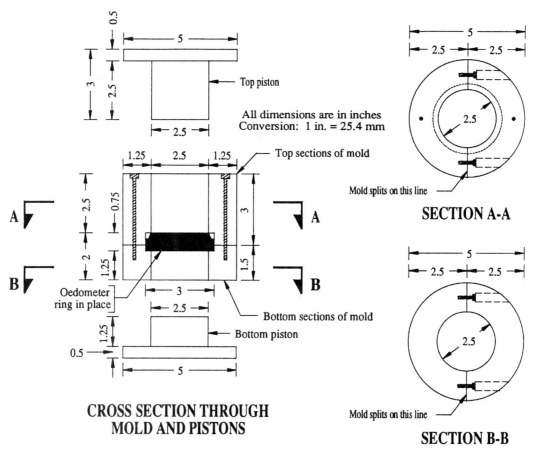

FIGURE 5A.41 Schematic diagram of mold for laboratory static compaction of oedometer specimens.

sified as an impact method. Therefore, the use of impact tests to predict field behavior of most compaction rollers may create some problems because of differences in moisture-density relationships between the laboratory and field compaction methods. A comparison of the differences in moisture density curves for two compaction rollers and laboratory impact compaction was given previously (Fig. 5A.35). Furthermore, differences in compaction method can have an important effect on the fabric of fine-grained soils and therefore the engineering behavior of compacted soils. Impact methods are used so frequently primarily because of their relatively low cost and the ease and accuracy in calculating compactive effort. However, the potential differences in moisture-density relationships and other engineering characteristics must be considered when deciding which compaction method to use for laboratory and field testing of compacted soils.

The standard test for determining the moisture-density relationships of compacted soils are impact tests in which rammers of a given weight are dropped from a constant height onto soil contained within a mold (Fig. 5A.43). Either hand-held or automatic hammers may be used. These tests are routinely conducted in the laboratory and the field and are commonly referred to as the *Proctor tests* in honor of R. R. Proctor, who in 1933 introduced the principles of modern compaction in a series of four articles in the *Engineering News-Record* (Proctor, 1933). Two variations of the Proctor test are used—the *standard* and *modified* tests (ASTM D698 and D1557). The basic procedures for the two tests are the same, but the compactive efforts are considerably dif-

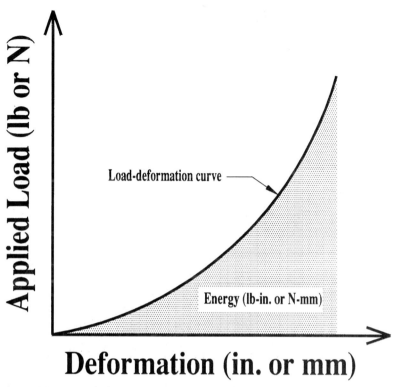

FIGURE 5A.42 Typical load-deformation curve from laboratory static compaction illustrating energy of compaction.

ferent. The variation in details for the two tests are outlined in Table 5A.4. The original (standard) Proctor test was developed to model the compactive effort of light rollers in use at that time. The modified Proctor test was developed later to simulate the compactive effort of heavier compaction equipment that subsequently became common.

Two types of procedures can be used to prepare impact-compacted specimens for oedometric, triaxial, and other types of testing. One method involves compacting oversize specimens using a Proctor hammer and mold and trimming the specimens to the desired shape and dimensions. This method is applicable only to cohesive soils and is often difficult to perform on compacted specimens because they do not trim easily at low to moderate water contents (owing to the stiff, brittle nature of the compacted soil). A second procedure involves compacting a specimen in a mold or ring with the desired shape and dimensions using a properly sized impact hammer. For some applications, stock or special order molds and hammers are available from various geotechnical equipment suppliers. If the impact face of the hammer has a diameter that is smaller than the diameter of the confining ring or mold, it is impossible to get the surface of the specimen perfectly flush with the top of the confining ring or exactly perpendicular to the axis of the specimen at the desired height by compaction with the hammer alone. To compensate for this unevenness, the surface of the specimen must be left slightly above the top of the ring or the desired height and brought to the appropriate level in one of the following ways:

1. If the specimen is contained within a confining ring, the specimen can be trimmed level with the top of the ring using a stiff, sharpened, metal trimming bar. For cohesive specimens that will be removed from the mold—such as for unconfined compression or triaxial testing—the

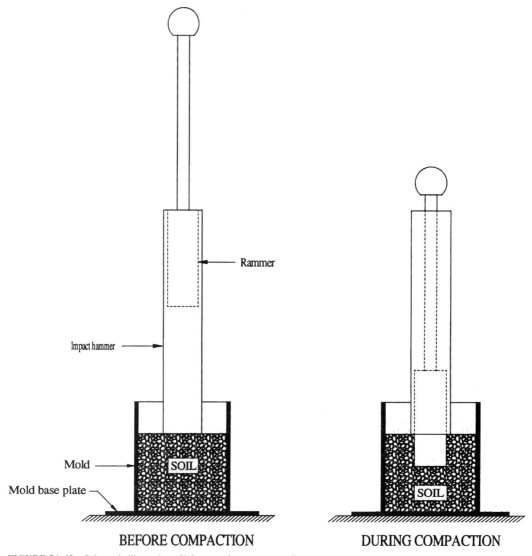

FIGURE 5A.43 Schematic illustration of laboratory impact compaction.

    specimens can be cut to the proper height using a miter box and wire saw. However, if the specimen is brittle, as is the case for compacted specimens at low to moderate compaction water contents, this method will not work because the specimen will crumble or chunks will break off the specimen, leaving the end uneven and unsuitable for testing.

2. The preferred method is to complete the specimen using static compaction, which involves removing the impact hammer and replacing it with a loading plate that has a diameter slightly smaller than the inside diameter of the ring or mold. The loading plate is then pushed to the top of the ring or to the desired height of specimen via a loading piston (Fig. 5A.40).

TABLE 5A.4 Comparison of Standard and Modified Proctor Tests

Detail	Standard Proctor	Modified Proctor
Rammer		
Diameter*	2.0 in (50.8 mm)*	2.0 in (50.8 mm)*
Drop height	12.0 in (304.8 mm)	18.0 in (457.2 mm)
Weight (mass)	5.5 lb (2.49 kg)	10.0 lb (4.54 kg)
Mold		
Option 1		
Diameter	4.0 in (101.6 mm)	4.0 in (101.6 mm)
Height	4.584 in (116.4 mm)	4.584 in (116.4 mm)
Volume	1/30 ft^3 (944 cm^3)	1/30 ft^3 (994 cm^3)
Option 2		
Diameter	6.0 in (152.4 mm)	6.0 in (152.4 mm)
Height	4.584 in (116.4 mm)	4.584 in (116.4 mm)
Volume	1/13.333 ft^3 (2124 cm^3)	1/13.333 ft^3 (2124 cm^3)
No. of soil layers	3	5
Blows per layer		
Mold Option 1	25	25
Mold Option 2	56	56
Compactive effort	12,375 ft-lb/ft^3 (593 kJ/m^3)	56,250 ft-lb/ft^3 (2693 kJ/m^3)

*Diameter shown is for a handheld hammer or an automatic compactor used with a 4.0-in (101.6-mm) mold. If an automatic compactor is used with a 6.0-in (152.4-mm) mold, a foot in the shape of a sector of a circle with a radius of 2.90 in (73.7 mm) is used.

5A.3.3.3 Kneading Compaction. Two types of kneading apparatuses are commonly used to compact soils in the laboratory—the Harvard miniature tamper (Wilson, 1970) and the California kneading compactor (ASTM D1561, D2844). The Harvard miniature tamper consists of a 0.5-in (12.7-mm) diameter piston attached to a spring contained within an otherwise hollow handle (Fig. 5A.44). An adjusting nut is used to preset the compression of the spring to a specified force, commonly 40 lb (178 N). Springs of different stiffnesses can be used to provide various values of preset force. The tamper is inserted into the mold and pushed firmly into the soil until the spring just begins to compress; at this point, a force equal to the preset value has been applied to the soil. The tamper is then pulled from the soil, and one "tamp" has been completed.

Specimens for testing can be made in layers within a mold by applying a sufficient number of tamps to each layer to obtain the desired density. Moisture-density relationships can be obtained by varying the water content of the soil and applying a predetermined level of compactive effort. Compactive effort is controlled by the volume of soil in the layer and the number of tamps per layer. A standard procedure for conducting moisture-density tests using the Harvard miniature tamper is given in Wilson (1970).

With the California kneading compactor, the soil is compacted within a 4-in (102-mm) diameter mold by applying a constant pressure to the soil via a pie-shaped tamper foot (see Figs. 5A.45 and 5A.46). When each tamp is completed, the mold is rotated an equal angle (usually 5 to 7 tamps per revolution), and another tamp is applied. This process is continued until a given number of tamps have been applied or until additional tamps produce no additional compaction in the lift. The compactive effort applied to the soil can be changed by varying the pressure of application, the lift height, or both. Specimens that will be subjected to additional testing are usually trimmed from the 4-in (102-mm) diameter specimen created during the compaction process.

The top surface of specimens prepared using either the Harvard miniature tamper or the California kneading compactor are generally neither level nor square (perpendicular to the axis). If the specimens are to be tested without being trimmed, a final application of static compaction

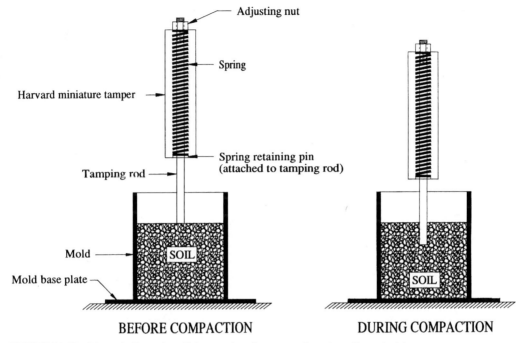

FIGURE 5A.44 Schematic illustration of laboratory kneading compaction using a Harvard miniature tamper.

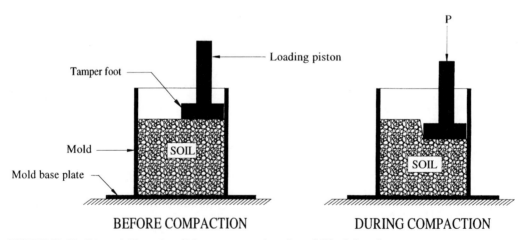

FIGURE 5A.45 Schematic illustration of laboratory compaction using a California kneading compactor.

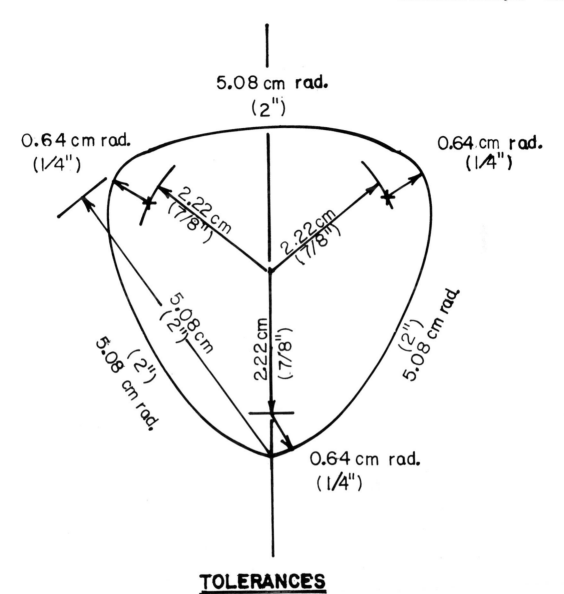

FIGURE 5A.46 Characteristics of pie-shaped tamper foot for California kneading compactor (from ASTM D2844).

5.70 SOIL IMPROVEMENT AND STABILIZATION

is needed to provide a level and square surface. Procedures similar to those described for impact-compacted specimens in Sec. 5A.3.3.2 can be used.

5A.3.3.4 Vibratory Compaction. In the United States, vibratory compaction of cohesionless soils is most commonly conducted in the laboratory using a vibratory table (Fig. 5A.47). To simulate field compaction, soil at an appropriate water content is placed loosely in a mold, a surcharge weight corresponding to the expected surcharge pressure in the field is placed gently on top of the loose soil, and the mold containing the soil and surcharge weights is secured to a vibratory table and vibrated at a given amplitude and frequency for a period. The soil to be densified can be dry, wet, or moist depending on the field conditions to be modeled. Vibration is continued until the desired density is achieved. If no further densification can be achieved but the desired density has not been obtained, the amplitude of vibration can, in some instances, be varied to achieve a greater densification.

Standard test methods are also available for determining the maximum index density of free-draining soils using a vibratory table (ASTM D4253). Because capillary tension in partially saturated soils creates a resistance to compaction called *bulking* (see Fig. 5A.38), the maximum index density test is performed on either oven-dried or saturated soil. A surcharge weight corresponding to a pressure of 2 psi (14 kPa) is applied to the soil contained within a mold, and the mold, soil, and surcharge are vibrated vertically at a double amplitude of vertical vibration of 0.013 in (0.33 mm) for 8 min at 60 Hz or at 0.019 in (0.48 mm) for 10 min at 50 Hz. Because the peak density for various soil types and gradations can vary as a function of double amplitude of vibration, ASTM D4253 permits the use of amplitudes other than those given above in special circumstances. It is important to note that an absolute maximum density is not necessarily obtained by these test procedures, rather they provide a standard method for obtaining reproducible values of "maximum" density. The relative density of a compacted soil can be calculated from the actual density of the soil in relation to the minimum (ASTM D4254) and maximum index densities.

Other vibratory methods can be used in the laboratory to compact soils and to determine the maximum index density (Forssblad, 1981). These methods include a vibrating rammer used in

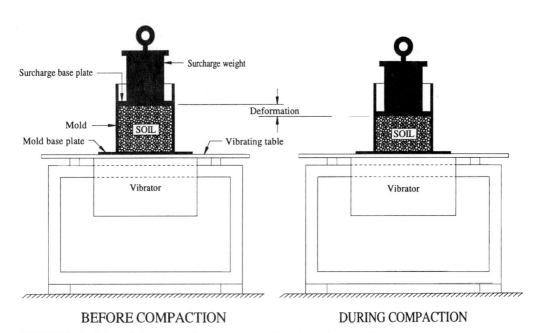

FIGURE 5A.47 Schematic illustration of laboratory compaction using a vibratory table.

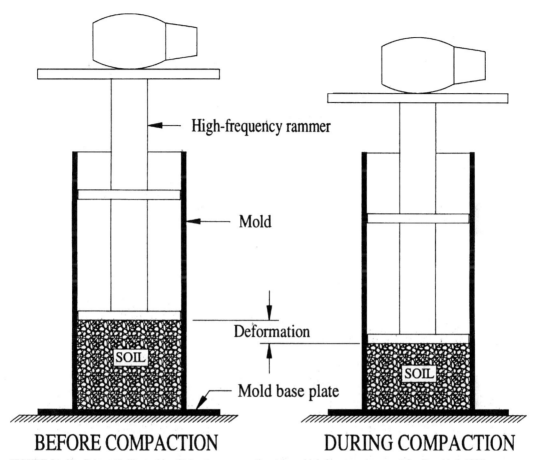

FIGURE 5A.48 Schematic illustration of laboratory compaction using a high-frequency rammer (after Forssblad, 1981)

Sweden (Fig. 5A.48), a vibrating hammer used in England, and a manually operated steel fork used in Germany.

5A.3.3.5 Special Considerations for Confined-Ring Specimens. When tests are to be conducted on specimens within small confining rings, such as for one-dimensional compression or wetting-induced volume change tests, the samples can be prepared in four ways:

1. The soil can be compacted directly into the confining ring using any of the four laboratory methods of compaction discussed above.
2. A sample can be compacted into an oversize mold in the laboratory using static, impact, or kneading compaction, and the specimen can be trimmed into the ring from the oversize specimen.
3. The ring can be pushed into the field-compacted soil, trimmed, sealed in an air-tight container, and then brought to the laboratory for testing.
4. A chunk sample of the field-compacted soil can be brought into the laboratory and a specimen trimmed into the ring from the chunk sample.

5.72 SOIL IMPROVEMENT AND STABILIZATION

For laboratory-compacted specimens, compaction directly into the ring is preferable to trimming an oversize sample because disturbance to the specimen caused during the trimming procedure is eliminated. Typical procedures for compacting specimens directly into confining rings using different methods of compaction can be found in Lawton (1986), Lawton and coworkers (1993), and Booth (1976).

Confined cohesive specimens obtained from the field-compacted soil (methods 3 and 4) are theoretically better than those prepared by either of the two laboratory methods (1 and 2) because the specimens ideally have the same fabric, density, and water content as the soil being modeled (that is, the field-compacted soil). Sometimes, however, the field-compacted soil cannot be trimmed into a confining ring without substantial disturbance, especially if the soil is dry to moderately dry or contains a significant fraction of oversize particles. In addition, a long time may elapse from removal in the field to testing in the laboratory, during which time the specimen may either dry out or bond to the ring. In these instances, laboratory compaction directly into the ring may produce a specimen more representative of the field-compacted soil if a laboratory compaction technique is used that closely simulates the field compaction.

5A.3.4 Engineering Properties and Behavior of Compacted Soils

The engineering behavior and properties of compacted soils are discussed here according to type of soil. For convenience, the soil types are divided into three general categories that differentiate the behavior of most compacted soils: clean coarse-grained soils (those containing less than about 2% fines), granular (cohesionless) soils with fines, and cohesive soils. The reader should bear in mind, however, that deviations from the typical engineering behavior described in the following sections may occur for each type of soil, and it is incumbent upon the engineer for a particular project to determine the relevant engineering properties of the soils to be compacted.

5A.3.4.1 Granular (Cohesionless) Soils. In building applications, the most important static properties for the bearing soil are strength, compressibility, permeability, and potential for volumetric changes induced by variations in available moisture. The engineering behavior of granular soils depends mainly on the relative density of the soil, although cementation of the particles upon aging after compaction may also play an important role. In seismically active regions, liquefaction potential of saturated silty sands and sands is extremely important.

Static Stress-Strain-Strength Behavior. The static settlement of a building to be founded on granular soils can be substantially reduced by compacting the soil to a greater density (lower void ratio) before constructing the building. The influence of relative density on the one-dimensional and isotropic compressibility of sand is shown in Figs. 5A.49 and 5A.50. For example, according to the data shown in Fig. 5A.49, the one-dimensional loading-induced vertical strain for Platte River sand (a clean, medium sand) at an applied stress of 25 psi (172 kPa) was about 58% less for $D_r = 73\%$ ($\varepsilon_{VL} = 0.8\%$) than for $D_r = 39\%$ ($\varepsilon_{VL} = 1.9\%$). Similar trends are shown in Fig. 5A.50 for the isotropic compressibility of Sacramento River sand (fine, uniform sand). The $\varepsilon_{vol}-\sigma_3$ plots for all four relative densities are reasonably straight for σ_3 greater than about 350 psi (2400 kPa), so a measure of the increased stiffness of the soil with increased density can be obtained by calculating values for bulk compressive modulus ($M_B = -\Delta\sigma_3/\Delta\varepsilon_{vol}$), with the following results:

D_r, %	Bulk modulus, M_B	
	ksi	MPa
38	13.9	96
60	17.4	120
78	21.2	146
100	32.5	224

Thus, M_B for $D_r = 100\%$ is more than twice the value of M_B for $D_r = 38\%$.

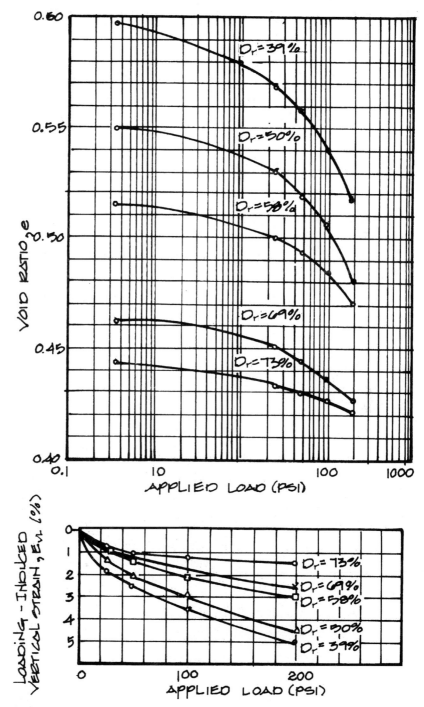

FIGURE 5A.49 Compressibility of a clean medium sand at various relative densities (from Hilf, 1991).

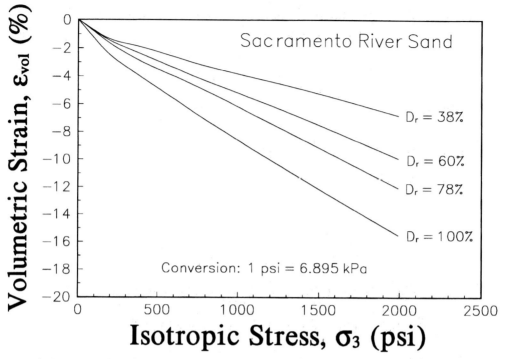

FIGURE 5A.50 Influence of density on the isotropic compressibility of Sacramento River sand (data from Lee and Seed, 1967b).

For foundation conditions in which the width of the foundation is small compared to the total thickness of the compressible strata beneath the bearing level, the strains induced by loading will be three-dimensional, and the magnitudes of the horizontal stresses within the zone of influence for settlement are important. Compaction increases both the density and horizontal stresses within the compacted soil, and both effects increase the stiffness of the soil, as shown by the results in Fig. 5A.51 for drained triaxial tests on Sacramento River sand. Thus, compaction is especially effective in reducing settlement for situations in which horizontal strains contribute significantly to vertical settlement.

The strength and hence ultimate bearing capacity of a granular bearing stratum can be substantially enhanced by a reduction in void ratio produced by compaction. The variation in angle of internal friction between the loose and dense conditions for most naturally occurring granular soils is about 4 to 10°. The influence of density on the drained shearing strength of Sacramento River sand is illustrated in Fig. 5A.52, for which the angle of internal friction at low confining pressures increased from 34° in the loose condition (D_r = 38%) to 41° in the dense condition (D_r = 100%). The magnitude of potential increase in ultimate bearing capacity can be measured by the change in the bearing capacity factors [using Meyerhof's (1951, 1963) factors]:

Condition	D_r, %	ϕ', degrees	N_q	N_γ
Loose	38	34	29.4	31.1
Dense	100	41	73.9	114.0

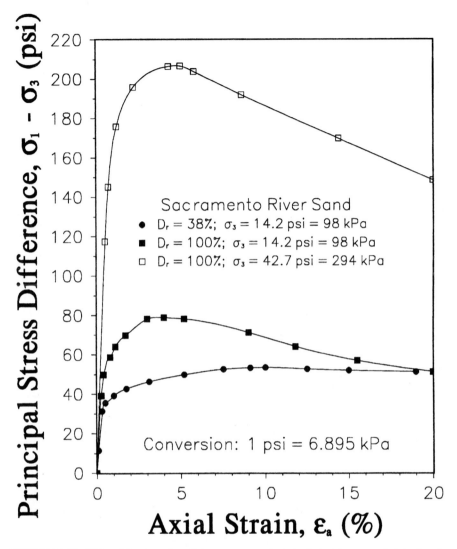

FIGURE 5A.51 Effect of density and confining pressure on the drained triaxial stress-strain behavior of Sacramento River sand (data from Lee and Seed, 1967b).

It can be seen that N_q and N_γ are 151 and 267% greater, respectively, in the dense condition than in the loose condition, indicating that the increase in ultimate bearing capacity would be within this range.

Contrary to common perception, the fabric of coarse-grained soils may have an important effect on compressibility and strength. Thus, method of compaction may affect the fabric of a compacted coarse-grained soil and hence its stress-strain-strength behavior. The effects of fabric anisotropy are greater in coarse-grained soils with elongated grains than in those with more spherical grains (Mitchell, 1993). Fabric anisotropy produces differences in volume change (dilatancy) tendencies for different directions of loading, which results in differences in stress-deformation

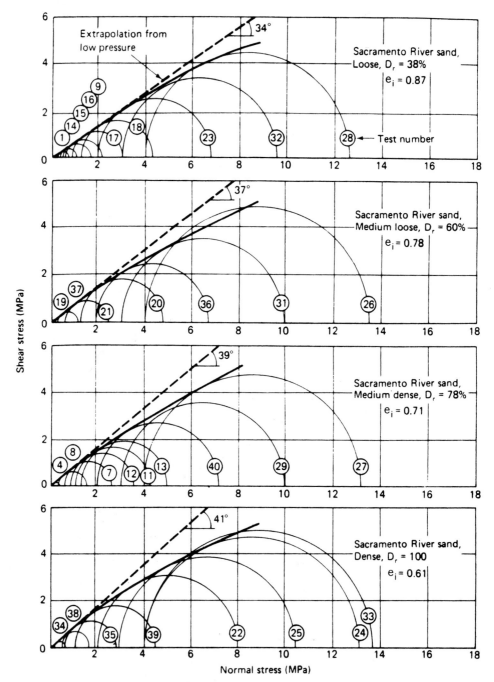

FIGURE 5A.52 Influence of relative density on the drained triaxial shear strength of Sacramento River sand (from Holtz and Kovacs, 1981; data from Lee, 1965 and Lee and Seed, 1967b).

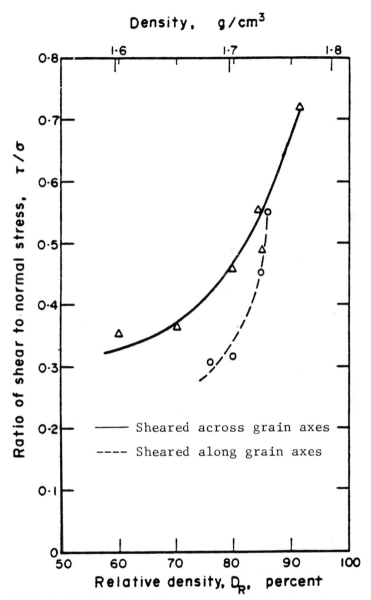

FIGURE 5A.53 Effect of shear direction on the strength of specimens of crushed basalt prepared by pouring into a shear box (from Mahmood & Mitchell 1974).

and strength behavior. This effect is illustrated in Fig. 5A.53 for specimens of crushed basalt tested in direct shear (Mahmood and Mitchell, 1974). At relative densities less than about 90%, specimens sheared along the plane of preferred orientation (along the long axis of most particles) were weaker than comparable specimens sheared across the plane of preferred orientation. This difference in strength decreased with increasing density, and the strengths were about the same for relative densities above 90%. These results were consistent with the finding that the intensity of preferred orientation decreased as relative density increased.

The influence of laboratory method of compaction on the triaxial stress-strain and volume-change behavior of a uniform sand with rounded to subrounded grains and a mean axial length ratio of 1.45 is shown in Fig. 5A.54. In these tests, oven-dried soil was poured into a cylindrical mold and then compacted to a dense void ratio of 0.64 by two methods: (1) plunging a hand-held tamper into the sand, and (2) tapping the side of the mold with the hand tamper (vibration). The method of compaction had a significant influence on both the volume-change and stress-deformation characteristics. The specimen compacted by tapping dilated more and was substantially stiffer than the specimen compacted by tamping. These differences in behavior were attributed to differences in fabric—the specimens prepared by tapping tended toward some preferred orientation of long axes parallel to the horizontal plane, and the intensity of orientation increased slightly during deformation; specimens prepared by tamping initially had weak preferred orientation in the vertical direction, and this disappeared with deformation. These results illustrate the importance of using a method of compaction in the laboratory that closely simulates the method to be used in the field, as discussed in Sec. 5A3.3.

The influence of fabric anisotropy on stress-strain-strength properties of compacted granular soils can be summarized as follows (Mitchell, 1993):

1. Anisotropic fabric and anisotropic mechanical properties are likely to occur in compacted soils.
2. The magnitude of strength and modulus anisotropy depends on density and the extent to which particles are platy and elongated. When axial ratios of particles are 1.6 or greater, differences in peak strength on the order of 10 to 15% may exist. Stress-strain moduli in different directions may vary by a factor of two or three.

Effect of Moisture on Compressibility and Strength

CLEAN COARSE-GRAINED SOILS Moisture has little effect on the static stress-strain-strength behavior of most coarse-grained soils, although clean, very fine sands are generally slightly stiffer and stronger in the partially saturated condition than in either the dry or saturated condition owing to matric suction (all other factors being the same). The compressibility and strength of some coarse-grained soils, however, are substantially affected by moisture condition. For example, Bishop and Eldin (1953) showed that the angle of internal friction of a fine to medium clean sand tested in drained triaxial compression was consistently higher in the dry condition than in the saturated condition—about 5° higher for dense condition and 2° higher for loose condition. Triaxial compression tests conducted by Lee and Seed (1967c) on a fine uniform sand indicated that the friction angle, stress-strain behavior, and creep behavior were substantially affected by moisture condition. This sand was stronger, less compressible, and exhibited less creep in the oven-dried condition than in the saturated condition, with air-dried behavior intermediate between the two extreme states. Increased moisture also caused a decrease in dilatant volume change behavior and an increase in particle crushing during shearing. The moisture sensitivity was therefore attributed to the presence of cracks in some of the particles, into which water infiltrated and weakened those particles so that they crushed during shearing. Moisture sensitivity has also been found in some gravels and rockfills (e.g., Zeller and Wulliman, 1957).

Based on evidence found in the literature and their own work, Lee and Seed (1967c) came to the following conclusions regarding moisture sensitivity of granular soils:

> From the available information it would appear that moisture sensitivity is likely to be greatest in granular soils whose particles either contain cracks or are susceptible to the formation of fine cracks as a result of loading. Such soils would be those derived from weathered rock, and angular material such as broken rock in which low contact areas would lead to high contact stresses and the formation of minute cracks at the contact points. Furthermore, it would also appear that water sensitivity may be greater at high pressures since fracturing of particles will be facilitated under these conditions.
>
> In light of the available evidence, therefore, it would seem to be desirable to check the nature of the component particles in investigating the strength of cohesionless soils and ascertain their sensitivity to moisture changes.

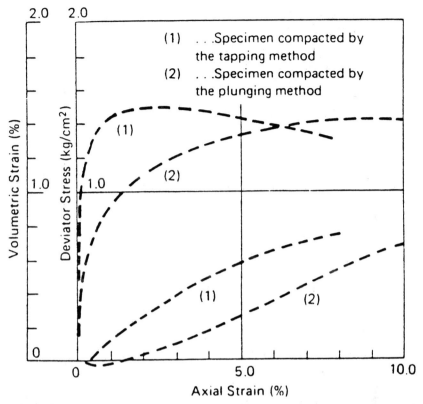

FIGURE 5A.54 Influence of compaction method on the triaxial stress-strain and volume change behavior of a uniform sand (from Mitchell, 1993; data from Oda, 1972).

GRANULAR SOILS WITH FINES   Any granular soil containing particles that are cracked or that develop cracks when loaded may be susceptible to loss of strength and stiffness with increasing moisture content, as discussed in the previous section. Granular soils with a substantial nonplastic fines fraction may also be susceptible to loss of strength and stiffness with increasing moisture owing to a reduction in suction. The total suction in an unsaturated soil can be subdivided into the osmotic suction and the matric suction. *Osmotic suction* arises from differences in ionic concentrations at different locations in the soil water within a soil. Osmotic suction is of significance primarily in cohesive soils where the cations in the soil water are electrochemically attracted to the clay particles, with the result that the concentration of cations is greatest near the surface of the clay particles and decreases with distance from the particles. *Matric suction* is the difference between the pore air pressure and the pore water pressure ($u_a - u_w$) in a partially saturated soil and reflects both the capillary forces and the electrochemical forces between the soil particles and the water molecules in the soil water (adsorptive forces).

In granular soils, the adsorptive forces are small, and thus the matric suction is predominately associated with the capillary menisci that occur in the partially saturated condition owing to *surface tension* that develops along the air-water interface (also called the *contractile skin*). Surface tension is a tensile force per unit length that is tangential to the contractile skin and results from a difference in intermolecular attractions between water molecules along the air-water interface and those within the interior of the water. Within the interior of the water, the water molecules are

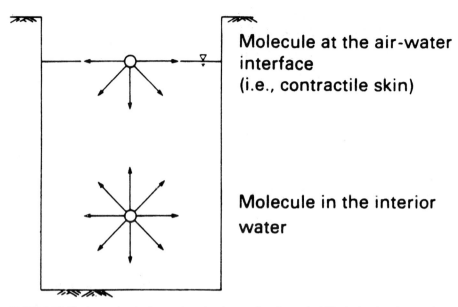

FIGURE 5A.55 Intermolecular forces along the air-water interface and within the interior of water (from Fredlund and Rahardjo, 1993).

attracted to neighboring water molecules, and there is no net force acting on any of the water molecules because the molecules are distributed equally in all directions (see Fig. 5A.55). The water molecules along the air-water interface are attracted to neighboring molecules within the interior and those along the interface, which results in a net attraction into the bulk of the water. Thus, the surface layer of water molecules is pulled toward the interior until the contractile skin assumes a spherical shape, which is the most stable configuration from an energy standpoint because the surface area is minimal (Davis et al., 1984). Surface tension develops, therefore, to overcome the net inward intermolecular force and to provide equilibrium to the contractile skin.

In partially saturated granular soils, u_a is usually zero (atmospheric) or compressive (positive), and u_w is tensile (negative). Kelvin's capillary model equation can be used to relate the matric suction to the curvature of a spherical contractile skin and the surface tension:

$$(u_a - u_w) = \frac{2T_s}{r_m} \qquad (5A.16)$$

where T_s = surface tension
r_m = radius of the meniscus (contractile skin)

Theoretical values of $(u_a - u_w)$ from Eq. (5A.16) for a contact angle of zero are shown in Fig. 5A.56 for saturated soils as a function of pore radius, such as might occur at the top of a saturated capillary zone or in a saturated soil as it dries.

The influence of moisture condition on matric suction is illustrated in Fig. 5A.57 using the capillary model. In drier soils (lower degree of saturation), the capillary menisci have smaller radii and higher values of $(u_a - u_w)$. Note, though, that the contact area between the pore water and the soil particles is less for drier soils, as indicated by the parameter χ in the following equation for effective stress in unsaturated soils proposed by Bishop (1959):

$$\sigma' = (\sigma - u_a) + \chi(u_a - u_w) \qquad (5A.17)$$

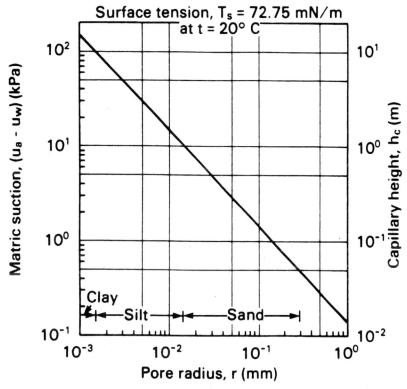

FIGURE 5A.56 Relationships among pore radius, matric suction, and capillary height for saturated soils (from Fredlund and Rahardjo, 1993).

χ relates the portion of $(u_a - u_w)$ that is transmitted to the soil in terms of a change in effective stress. $\chi = 0$ for a completely dry soil and $\chi = 1$ for a saturated soil, and the relationship between χ and degree of saturation for a silt is shown in Fig. 5A.58. In granular soils, χ depends primarily on the ratio of area of pore water in contact with the soil particles to the total surface area of the particles but also on factors such as the fabric and stress history of the soil. χ has also been found to differ when determined for volume change behavior and shear strength for the same soil under otherwise identical conditions (Morgenstern 1979).

The net effect of a lower water content may be either to decrease or increase the effective stress in a soil depending upon the combined effect of increased suction and decreased contact area—that is, how the combined factor $\chi(u_a - u_w)$ varies as a function of water content. The relation between matric suction and change in effective stress is shown in Fig. 5A.59 for a clayey sand. This concept is also illustrated for three types of soils in Fig. 5A.60, which shows the unconfined compressive strength as a function of water content. It is clear from Fig. 5A.60 that even perfectly cohesionless soils (e.g., very fine clean sand) can have an unconfined compressive strength greater than zero owing to matric suction, which provides some effective stress in the soil even at essentially zero total stress. The effective stress induced by matric suction in cohesionless soils is sometimes referred to as *apparent cohesion.*

Because the strength of finer-grained soils is not negligible at very low water contents (silty sand and clay in Fig. 5A.60), the capillary model obviously cannot be used by itself to describe the changes in matric suction that result from changes in moisture condition. Thus, the influence

5.82 SOIL IMPROVEMENT AND STABILIZATION

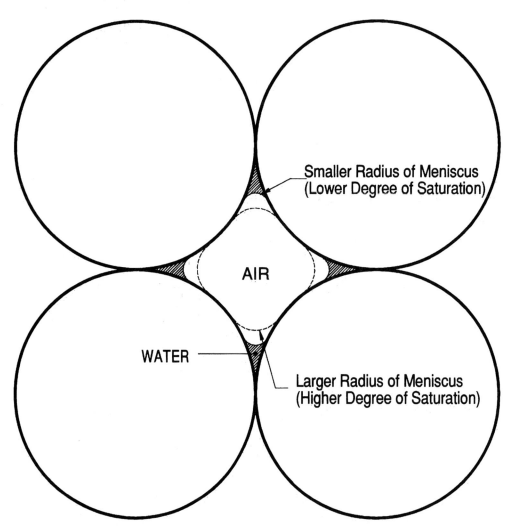

FIGURE 5A.57 Influence of moisture condition on the capillary menisci in an unsaturated soil according to the capillary model.

of adsorptive forces on matric suction is significant for cohesive soils and will be discussed in more detail in Sec. 5A.3.4.2.

In most granular soils the values of matric suction are small, and the stress-strain-strength behavior of these soils is generally little affected by changes in moisture condition. In current "standard" geotechnical engineering practice, the effect of partial saturation on granular soils is generally ignored for three primary reasons:

1. Ignoring the effect is simpler.
2. The effect is nearly always conservative.
3. There is always a chance the soil will become saturated at some later point.

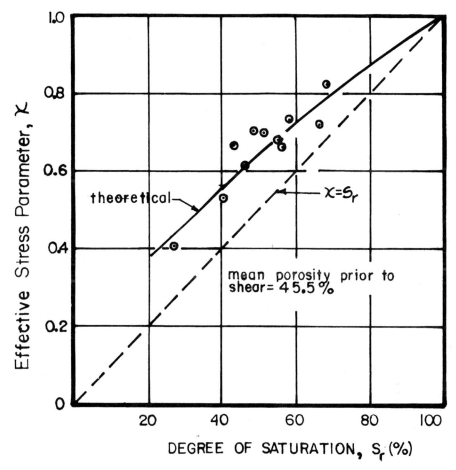

FIGURE 5A.58 Relationship between effective stress parameter χ and degree of saturation for Braehead silt (from Bishop et al., 1960).

Techniques for including the effect of partial saturation are considered to be too difficult to implement in practice.

Thus, most laboratory tests on granular soils are conducted on either dry or saturated specimens, and stress-strain-strength analyses are performed on the basis of total stresses, which are assumed to be equal to effective stresses. In a similar manner, the results of field tests are used to determine appropriate stress-strain-strength parameters based on total stresses, assuming that the total stresses are equal to the effective stresses.

Although ignoring the effects of partial saturation usually has no practical consequence in granular soils, in some instances doing so is uneconomical. Bishop's effective stress equation for unsaturated soils—Eq. (5A.17)—has proved to have little practical use primarily because of difficulties in obtaining values of χ that adequately predict field behavior. These difficulties stem from the fact that stress-strain-strength behavior cannot be uniquely defined in terms of a single effective stress variable (Morgenstern, 1979). Thus, modern methods of predicting the stress-strain-strength behavior of unsaturated soils use two independent stress variables, usually $(\sigma - u_a)$

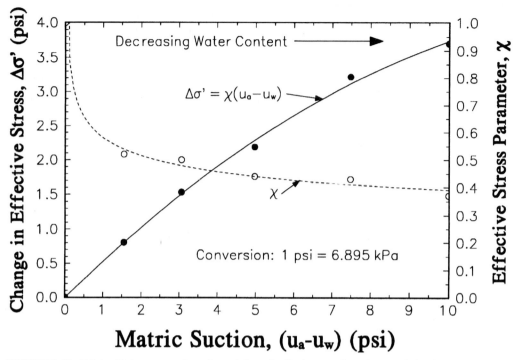

FIGURE 5A.59 Relationship between matric suction and change in effective stress for a clayey sand (data from Blight, 1965).

and $(u_a - u_w)$. Although the use of these more sophisticated techniques is slowly becoming more common in geotechnical engineering practice, most geotechnical engineers prefer to use traditional total stress analyses. Because these techniques have limited application in granular soils, they will not be discussed any further here but will be discussed in more detail in Sec. 5A.3.4.2.

Permeability. Although the terms *permeability* and *hydraulic conductivity* are often used interchangeably in engineering practice, they have different meanings. The *intrinsic permeability* of a soil is independent of the fluid which may flow through it, and the relationship between intrinsic permeability (K) and coefficient of permeability (k) can be described by the following equation (Das, 1983):

$$k = \frac{K \cdot \gamma_f}{\mu_f} \tag{5A.18}$$

where γ_f = weight density (unit weight) of the fluid
μ_f = absolute viscosity of the fluid

Thus, the coefficient of permeability depends on the density and viscosity of the fluid flowing through the soil. The terms *hydraulic conductivity* and *water permeability* describe the same characteristic of a soil—the relative ease with which water can flow through the voids in the soil. In soils, the two most common fluids that flow through the soil in foundation engineering contexts are water and air. For simplicity, the term *permeability* will be used to mean water permeability or hydraulic conductivity, and the permeability with respect to other types of fluids will be differentiated as required.

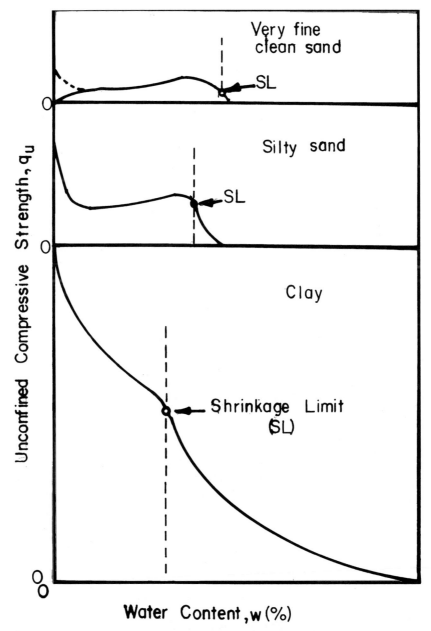

FIGURE 5A.60 Unconfined compressive strength as a function of water content for three types of soils (from Terzaghi and Peck, 1967).

The coefficient of permeability of clean, coarse-grained soils increases as void ratio (e) increases. Several correlations have been proposed to describe this relationship, including the following (Das, 1983):

$$k \propto \frac{e^3}{1+e} \qquad (5A.19a)$$

$$k \propto \frac{e^2}{1+e} \qquad (5A.19b)$$

$$k \propto e^2 \qquad (5A.19c)$$

Permeability is important in granular soils with respect to their drainage characteristics. It is advantageous in most instances for granular bearing soils to be free-draining, that is, any significant accumulation or introduction of moisture should drain rapidly through the granular layer. Although compaction results in somewhat lower permeability, the permeability of even dense coarse-grained soils is generally still high enough to permit free drainage under most conditions.

k decreases with decreasing grain size, as illustrated by the typical values for various types of granular soils given in Table 5A.5. Casagrande (1938) has established the boundary between good

TABLE 5A.5 Typical Values of Coefficient of Permeability for Granular Soils

Type of soil	k cm/s	k ft/day
Clean gravel	1×10^0 to 1×10^2	3×10^3 to 3×10^5
Clean sand	1×10^{-3} to 1×10^0	3×10^0 to 3×10^3
Silty sand/sandy silt	1×10^{-4} to 1×10^{-2}	3×10^{-1} to 3×10^1
Silt	1×10^{-6} to 1×10^{-4}	3×10^{-3} to 3×10^{-1}

drainage and poor drainage as 1×10^{-4} cm/s (3×10^{-1} ft/day). Thus, granular soils with a large percentage of fines may not be free-draining soils and may require a perimeter or interior drainage system to prevent unwanted buildup of moisture in situations where moisture accumulation may have a deleterious effect on the load-carrying capacity of the foundation soil.

Wetting-Induced Volume Changes

CLEAN COARSE-GRAINED SOILS   Coarse-grained soils will not swell (increase in volume) when wetted because they contain no expansive clay minerals. However, all compacted soils are susceptible to *wetting-induced collapse* or *hydrodensification* (decrease in volume when wetted) under some conditions (Lawton et al., 1992). The magnitude of collapse depends on the state of the soil at the *time of wetting*; in general, collapse increases with increasing mean normal total stress, decreasing water content or degree of saturation, and decreasing dry density (Lawton et al., 1989, 1991). Therefore, one must consider the potential changes in these properties that may occur between the end of compaction and the time when wetting may occur. Examples of changes that may occur include the placement of additional fill or structures above a given lift, which increases the stresses acting on that lift and densifies the soil (thereby changing both the state of stress and density); drying of the exposed surfaces of the soil from evaporation or transpiration (which changes the moisture condition); and redistribution of moisture throughout the partially saturated soil as it tries to achieve hydraulic, thermal, chemical, and electrical equilibrium.

The addition of water necessary for collapse in compacted soil may occur in a variety of ways, including precipitation, irrigation, regional ground-water buildup, broken water pipes, moisture buildup beneath covered areas, ponding, inadequate or ineffective drainage, and flooding (Lawton et al., 1992). Some of these methods are difficult to predict or foresee, so it is prudent to consider the wetting-induced collapse potential for all compacted soils.

Although some notable exceptions have occurred, compacted coarse-grained soils will not collapse significantly upon wetting under most circumstances. Some examples of clean coarse-grained soils that have collapsed are provided below to illustrate the potential problems.

Jaky (1948) reported that apartment houses and other structures along the banks of the Danube suffered extensive damage and cracking when the level of the Danube rose 2 m (6.2 ft) above the flood stage. The results from one-dimensional collapse tests on specimens made from six different granular soils indicated that the collapse potential could be high for all the soils under certain conditions. Grain-size distribution properties and USCS classification for the six soils studied are summarized in Table 5A.6, and the effect of relative density on measured wetting-induced collapse strains is shown in Fig. 5A.61. As would be expected, the collapse potential decreased as the relative density of each soil increased. The optimal or critical relative density (the minimum relative density for which the collapse was zero) varied from 69 to 97%.

Collapse settlements have also been observed in rockfill dams. Sowers and colleagues (1965) reported that the Dix River dam was accidentally flooded during construction, with a wetting-induced vertical strain of about 0.5%. They also conducted one-dimensional wetting tests on specimens of freshly broken rocks of various types and found that additional (collapse) settlements occurred rapidly when the dry specimens were wetted. These collapse settlements were attributed to water entering microfissures in the particles that were produced at highly stressed contact points, with the water causing local increases in stress and additional crushing of particles at the contact points. Collapse settlements were also observed during filling of the El Infiernillo dam (Marsal and Ramizeq de Arellano, 1967).

The results of triaxial collapse tests on specimens of Antioch sand and Ottawa sand were reported by Lee and Seed (1967c). In these tests, oven-dried or air-dried dense ($D_r = 100\%$) specimens of both soil types were first loaded to an axisymmetric state of stress with the sustained deviator stress ($\sigma_1 - \sigma_3$) equal to 76 or 83% of the maximum deviator stress for Antioch sand and 90% of the maximum deviator stress for Ottawa sand. To ensure that the pore water pressures were maintained at a very low level during wetting, water was introduced into the specimens by gravity flow through the bottom drainage line under a head of less than 1.0 in (25 mm). Upon wetting, the Antioch sand specimens first underwent substantial collapse strains before culminating in a sudden large deformation along a well-developed shear plane. In contrast, the addition of water to the Ottawa sand specimens produced virtually no collapse. The differences in collapse behavior for these two types of soils were attributed to differences in the soundness of the particles—a significant number of the Antioch sand particles contained small cracks, whereas the Ottawa sand grains were essentially sound.

TABLE 5A.6 Grain Size Properties and Classification of Six Granular Soils Investigated by Jaky (1948)

Soil	Grain size distribution by weight, %*					USCS Classification*	
	Gravel	Sand	Silt	D_{50}, mm	D_{10}, mm	Symbol	Descriptive name
Fine sand	0	97	3	0.14	0.10	SP	Poorly graded sand
Coarse sand	0	100	0	0.28	0.15	SP	Poorly graded sand
Danube gravel 1	43	57	0	3.0	0.19	SW	Well-graded sand with gravel
Danube gravel 2	31	52	17	2.2	0.041	SM	Silty sand with gravel
Pea gravel	20	80	0	1.6	0.29	SW	Well-graded sand with gravel
Slag	N.A.	N.A.	N.A.	N.A.	N.A.	—	—

*ASTM D-2487

N.A. = not available

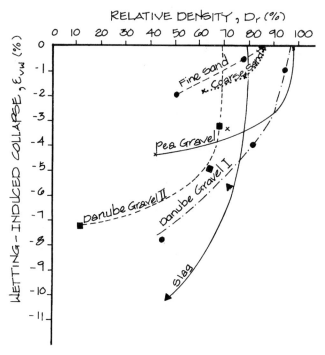

FIGURE 5A.61 Influence of relative density on the one-dimensional collapse of six granular soils (from Jaky, 1948).

Hellweg conducted more than 300 one-dimensional collapse tests on specimens of various sands from northern Germany with different densities, moisture contents, and applied loads (Rizkallah and Hellweg, 1980). The results from these tests indicated that uniform fine sands can only be compacted with extraordinary difficulties and that such soils collapse about 8% when wetted. Based on these results, it was concluded that compacting uniform fine sands to a high relative compaction based on the standard Proctor maximum dry density does not necessarily ensure low collapse potential.

It has been recognized for a long time that fill materials containing soft rock fragments—such as shales, mudstones, siltstones, chalks, and badly weathered igneous and metamorphic rocks—can deform significantly upon wetting when used in embankment dams and should be avoided where possible (Sherard et al. 1963). That excessive wetting-induced settlement can also occur in buildings founded on compacted fills containing soft rock fragments has been reported by Wimberley and coworkers (1994). In this project, the excessive settlement (>12 in = 30 cm) of a one-story building founded on a 50-ft (15-m) deep compacted fill occurred. The compacted fill consisted of fine-grained fill with rock fragments (siltstone, limestone, claystone, and sandy shale) and some rocky layers. Much of the settlement was believed to relate to the introduction of moisture into the fill by precipitation and watering of plants, which caused softening of highly stressed contact points between coarser particles and degradable coarse particles (e.g., poorer shales).

In summary, the collapse potential of compacted clean coarse-grained soils under most conditions is insignificant. However, substantial collapse potentials can exist if one or more of the following factors are present in the compacted soil: (1) The compaction is poor or insufficient; (2) defects exist within some of the particles; (3) some of the particles are water soluble or degradable; or (4) the soil is subjected to very high stress levels. The types of granular soils that have defects sufficient to cause potentially high wetting-induced collapse potentials include those derived from highly weathered rock and blasting. Angular particles with low contact areas and high applied stresses seem to be more susceptible to the formation of minute cracks at contact points. Compacted soils containing substantial amounts of soft rock fragments may also be susceptible to wetting-induced collapse. Structures founded on deep fills are especially susceptible to damage resulting from collapse settlements because of the high stresses generated at lower portions of the fill from the weight of the overlying material.

Because potential problems from wetting-induced collapse can be detected by performing one-dimensional collapse tests on representative compacted specimens, it is recommended that collapse tests be conducted in all situations where collapse is *possible*. Granular soils containing medium and fine sand particles only can be tested in standard consolidometers; soils containing larger particles require special testing equipment that can be found in some universities and other research institutes. For example, San Diego State University has the facilities to test compacted specimens 4 ft (1.2 m) in diameter (Noorany, 1987).

GRANULAR SOILS WITH FINES There are several cases described in the literature of damage resulting from wetting-induced collapse of compacted granular soils containing fines. Examples include severe cracking of two earth dams in California (Leonards and Narain, 1963), and deformation, cracking, and failure of numerous highway embankments in South Africa (Booth, 1977). The earth dams were constructed of sand/silt mixtures, and silty sands and wetting occurred during filling of the reservoirs. The highway embankments were composed of silty sands, with wetting generally resulting from infiltration of water following periods of heavy rainfall.

Collapse settlements in compacted granular soils generally occur because the soils are too dry, not dense enough, or both. For Woodcrest dam (Leonards and Narain, 1963), which suffered severe cracking, the average compaction water content was 1.7% below standard Proctor optimum, the average degree of saturation was 30% below optimum saturation, and the average standard Proctor relative compaction was 94%. The highway embankments reported by Booth (1977), which suffered extensive damage from collapse settlements, were loosely compacted (modified Proctor relative compaction, $R_m = 80\%$) and were very dry (estimated degree of saturation, $S_r <$ 10% in some instances). It was not known whether the embankments were compacted dry (no compaction water content records were available) or whether substantial moisture loss occurred between compaction and wetting.

Recent research (Alwail et al. 1992) has shown that the amount of silt and shape of the silt particles has an important influence on the collapse potential of granular soils. These effects are shown in Fig. 5A.62 for compacted mixtures of uniform silica sand and two different silts compacted to $R_m = 90\%$ at water contents 3% dry of modified Proctor optimum water content (w_{om}). For both silts, the maximum one-dimensional collapse potential for applied vertical stresses ranging from 0.52 to 33.4 ksf (25 to 1600 kPa) increased as the percentage of silt was increased. It is also evident that the group 2 soils had substantially higher collapse potentials than the group 1 soils. Scanning electron micrographs taken on samples of the two groups of soils showed that the group 1 silt grains were rounded to subrounded, whereas the group 2 silt grains were flaky and angular. Thus, the greater collapse potential for the group 2 soils was attributed to a more open structure resulting from the angular and flaky shapes of the silt grains.

Shrinkage. Shrinkage (decrease in volume induced by drying) is not a problem in granular soils because by definition granular soils are cohesionless, and thus they contain no appreciable amount of expansive minerals.

Liquefaction Potential. Liquefaction occurs in a saturated sandy soil when deformation during undrained or partially drained conditions produces a buildup in pore water pressure to a level that approaches or equals the total confining pressure. At this point the effective confining pressure is zero or near zero, and the soil will continue to deform until enough water has been squeezed from the soil to dissipate a substantial portion of the excess pore water pressure. Liquefaction may occur from either statical strains induced from monotonic loading (e.g., Casagrande 1936a) or from cyclic strains induced by dynamic or vibratory loading (e.g., Lee and Seed, 1967a). Damage to buildings caused by static liquefaction is rare, so only liquefaction caused by cyclic loadings (also called *cyclic mobility*) will be considered herein.

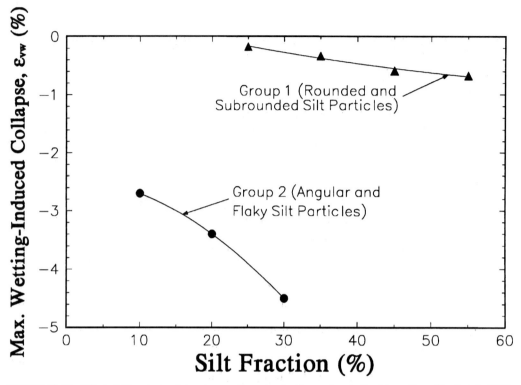

FIGURE 5A.62 Effect of silt content and shape on one-dimensional collapse of sand-silt mixtures (from Alwail et al., 1992).

Extensive damage can occur to buildings and other engineering structures during earthquakes owing to liquefaction in saturated sandy soils beneath the structures. For example, hundreds of buildings were severely damaged as a result of liquefaction during the 1964 earthquake in Nigata, Japan (Seed and Idriss, 1967). Many structures settled more than 3 ft (0.9 m), and the settlement was frequently accompanied by severe tilting (up to 80°). Lateral ground displacements generated during earthquakes by liquefaction-induced lateral spreads and flows have also caused severe damage to structures and their appurtenances (Youd, 1993). Furthermore, liquefaction can occur from lower levels of excitation produced by other phenomena, including blasting, pile driving, construction equipment, road and train traffic, and dynamic compaction (Carter and Seed, 1988). Thus all structures underlain by saturated sand deposits may be susceptible to liquefaction-induced damage, and it is imperative that liquefaction potential be considered for these cases even when the site is not located in a seismically active region.

Three definitions differentiate the possible states of liquefaction (Lee and Seed, 1967a):

Complete liquefaction—when a soil exhibits no resistance or negligible resistance over a wide range of strains (e.g., a double amplitude of 20%).

Partial liquefaction—when a soil exhibits no resistance to deformation over a strain range less than that considered to constitute failure.

Initial liquefaction—when a soil first exhibits any degree of partial liquefaction during cyclic loading.

The following characteristics are known with respect to liquefaction in saturated sandy soils (Lee and Seed, 1967a; Seed, 1986):

1. Application of cyclic stresses or strains will induce liquefaction or partial liquefaction over a considerable range of densities, even in moderately dense sands.
2. If the generated pore pressures do not exceed about 60% of the confining pressure, liquefaction will not be triggered in the soil, and there is no serious deformation problem.
3. The smaller the cyclic stress or strain to which the soil is subjected, the higher is the number of stress cycles required to induce liquefaction (Fig. 5A.63).
4. The higher the confining pressure acting on a soil, the higher the cyclic stresses, strains, or number of cycles required to induce liquefaction (Fig. 5A.64).
5. The higher the density of a soil, the higher the cyclic stresses, strains, or number of cycles required to induce liquefaction (Figs. 5A.63 and 5A.64).
6. When loose sandy soils liquefy under cyclic stresses of constant amplitude, deformations immediately become very large (complete liquefaction occurs quickly—see Fig. 5A.65).
7. When dense sandy soils liquefy under cyclic stresses of constant amplitude, deformations are initially small (partial liquefaction occurs) and gradually increase with increasing number of cycles (Fig. 5A.66). Even when the pore water pressure becomes equal to the confining pressure in dense sands, it is often of no practical significance because the strains required to eliminate the condition are very small. Thus, dense cohesionless soils do not normally present problems with regard to the design of foundations.
8. Numerous structures built on liquefiable soils have stood for hundreds of years without liquefaction occurring, simply because there has been no triggering mechanism strong enough to induce liquefaction. Therefore, liquefaction can be prevented by designing to keep the pore pressures well below the confining stresses.

It is evident from the preceding information that the liquefaction potential of any saturated sandy soil depends on both the properties of the soil (density, permeability, and lengths of drainage paths) and the characteristics of the cyclic loading (magnitude of stress and strain, duration). Three possible methods have been used to various extents for estimating liquefaction potential (Youd, 1993): (1) analytical methods, (2) physical modeling, and (3) empirical procedures. Because it is difficult to model the soil conditions at liquefiable sites either analytically or physically, empirical procedures are commonly used in routine engineering practice. A complete discussion of methods for evaluating liquefaction potential for all types of cyclic loading is beyond the scope of this presentation, so details are given below only for earthquake-induced liquefaction. Methods for evaluating liquefaction for cyclic loads from other sources (blasting, pile driving, construction equipment, road and train traffic, and dynamic compaction) are given in Carter and Seed (1988).

The factor of safety against the occurrence of earthquake-induced liquefaction is commonly defined as the available soil resistance to liquefaction (expressed in terms of the cyclic stresses required to cause liquefaction) divided by the cyclic stresses generated by the design event (Youd, 1993). Both factors are usually normalized with respect to the preevent effective overburden stress at the depth being analyzed. In equation form, the factor of safety is defined as

$$F_s = \frac{\text{CSRL}}{\text{CSRE}} \quad (5A.20)$$

where CSRL = cyclic stress ratio required to generate liquefaction
CSRE = cyclic stress ratio generated by the design earthquake = τ_{av}/σ'_{v0}
τ_{av} = average earthquake-induced cyclic shear stress
σ'_{v0} = preearthquake effective overburden stress at the depth under consideration

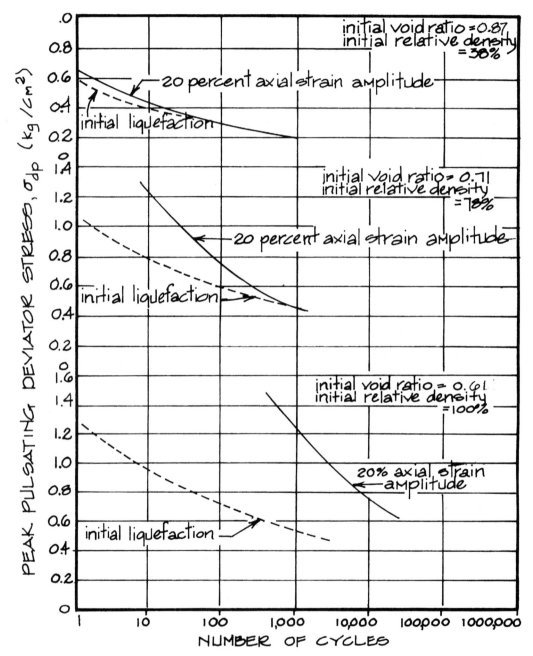

FIGURE 5A.63 Effect of density, number of cycles, and failure criterion on the cyclic stress required to cause liquefaction of Sacramento River sand (from Lee and Seed, 1967a).

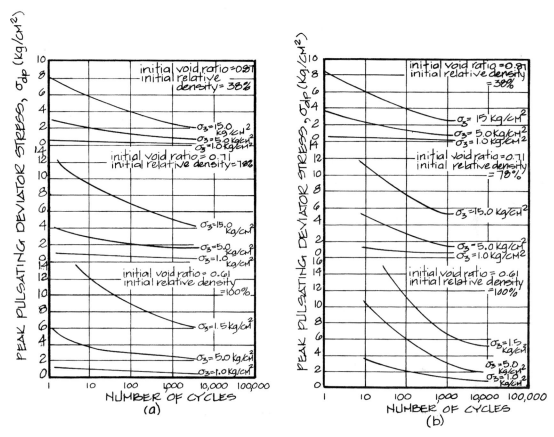

FIGURE 5A.64 Influence of confining pressure, density, and number of cycles on the cyclic stress required to cause (*a*) initial liquefaction and (*b*) 20% strain in Sacramento River sand (from Lee and Seed, 1967a).

CSRE can evaluated either by using a computer code (such as SHAKE or DESRA) to estimate τ_{av} or by estimating it directly from the following equation (Seed & Idriss 1971):

$$\text{CSRE} = 0.65 \cdot \frac{a_{max}}{g} \cdot \frac{\sigma_{v0}}{\sigma'_{v0}} \cdot r_d \tag{5A.21}$$

where a_{max} = maximum acceleration at the ground surface that would occur in the absence of liquefaction
g = acceleration caused by gravity
σ_{v0} = total overburden stress at the depth under consideration
r_d = depth-related stress reduction factor that varies from a value of 1.0 at the ground surface to about 0.9 at a depth of 35 ft (10.7 m)

Estimated values for a_{max} are usually determined using one of the following three methods (Youd, 1993): (1) standard peak acceleration attenuation curves that are valid for comparable soil conditions; (2) standard peak acceleration attenuation curves for bedrock sites, with correction for local site effects using standard site amplification curves or computerized site response analysis;

5.94 SOIL IMPROVEMENT AND STABILIZATION

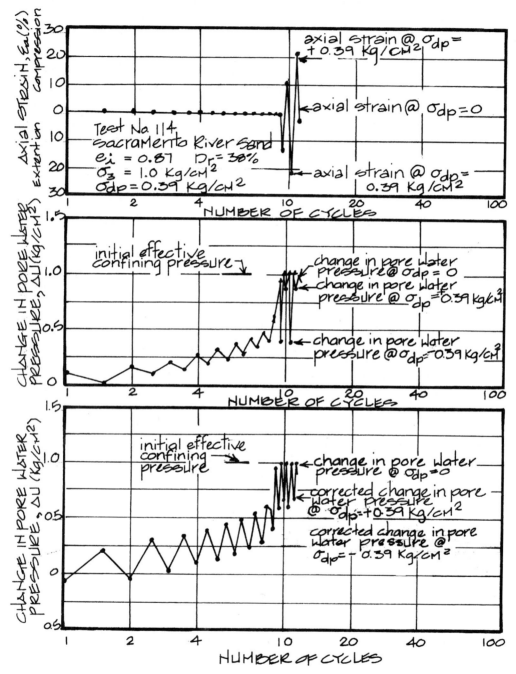

FIGURE 5A.65 Results from pulsating load test on loose Sacramento River sand (from Seed and Lee, 1966).

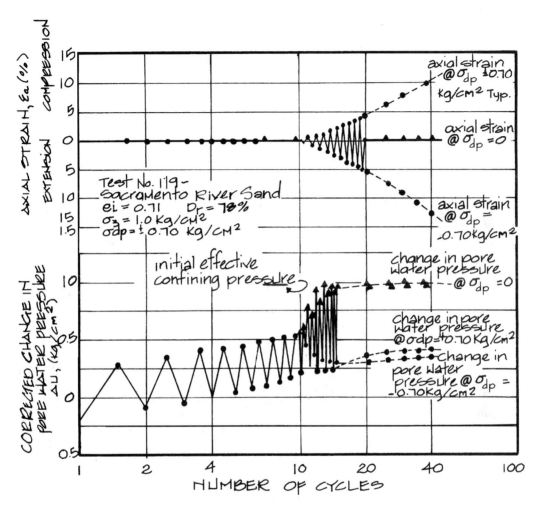

FIGURE 5A.66 Results from pulsating load test on dense Sacramento River sand (from Seed and Lee, 1966).

and (3) probabilistic maps of a_{max}, with or without correction for site attenuation or amplification depending on the rock or soil conditions used to generate the map.

CSRL is most commonly estimated from empirical correlations with corrected SPT blowcount $[(N_1)_{60}]$ such as those shown in Fig. 5A.67, which are valid for magnitude 7.5 earthquakes and sands and silty sands with up to 35% fines. Open symbols represent sites where surface liquefaction did not develop, and filled symbols are from sites where surface liquefaction did occur. Curves are given in the figure for varying percentage of fines that represent the boundary between the liquefaction and no liquefaction zones. Youd (1993) recommends using a factor of safety of 1.2 for engineering design based on this chart because it is possible that liquefaction may have occurred at some sites but was not detected at the ground surface. One should exercise care when using the curves in Fig. 5A.67 for soils containing substantial gravel fractions because verified corrections for gravel content have not been established (Youd, 1993).

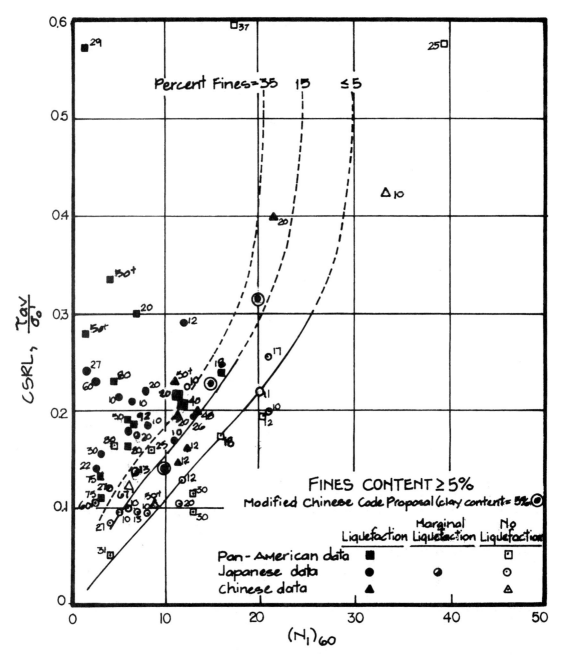

FIGURE 5A.67 Relationship between CSRL and $(N_1)_{60}$ values for silty sands and magnitude 7.5 earthquakes (from Seed et al., 1985).

$(N_1)_{60}$ is the field SPT blowcount corrected for overburden pressure (reference stress = 1 tsf = 95.8 kPa), energy efficiency, length of drill rod, and characteristics of sampling barrel (Seed et al., 1985) and can be represented by the following equation:

$$(N_1)_{60} = N_{field} \cdot C_E \cdot C_N \cdot C_L \cdot C_S \tag{5A.22}$$

where N_{field} = SPT blowcount measured in the field
C_E = correction factor for energy ratio
C_N = correction factor for overburden pressure
C_L = correction factor for length of drill rod
C_S = correction factor for sampling barrel

$(N_1)_{60}$ is standardized to an energy ratio (actual energy divided by theoretical free-fall energy) of 60%, so the energy ratio correction can be calculated as

$$C_E = \frac{E_r}{60} \tag{5A.23}$$

where E_r = energy ratio for the equipment being used

E_r can be either measured during use of the actual equipment (preferred) or estimated from the type of equipment (e.g., Table 5A.7). C_N can be obtained from Fig. 5A.68 or the following equation suggested by Liao and Whitman (NRC, 1985):

$$C_N = \frac{2}{\sigma'_{v0}} \quad \text{for } \sigma'_{v0} \text{ in units of ksf} \tag{5A.24a}$$

$$C_N = \frac{95.76}{\sigma'_{v0}} \quad \text{for } \sigma'_{v0} \text{ in units of kPa} \tag{5A.24b}$$

C_L is 0.75 for drill rod shorter than 10 ft (3.05 m) and 1.0 for drill rod longer than 10 ft (3.05 m). C_S is 1.0 for a constant-diameter barrel (i.e., with liner); if a liner is not used, C_S is 1.1 for loose sands and 1.25 to 1.30 for dense sands.

The following SPT procedure is recommended by Seed and colleagues (1985) for use in liquefaction correlations:

1. *Borehole.* 4- to 5-in (102- to 127-mm) diameter rotary borehole with bentonitic drilling mud to stabilize borehole.
2. *Drill bit.* Upward deflection of drilling mud (tricone or baffled drag bit) to prevent erosion of soil below the cutting edge of the bit.
3. *Sampler.* 2.00-in (50.8-mm) outer diameter, 1.38-in (35.0-mm) constant inner diameter.
4. *Drill rods.* A or AW for depths less than 50 ft (15 m); N or NW for greater depths.
5. *Energy ratio.* 60% of theoretical free-fall energy.
6. *Blowcount rate.* 30 to 40 blows per minute.
7. *Penetration resistance count.* Measured over depth of 6 to 18 in (152 to 457 mm) below the bottom of the borehole.

If the design earthquake has a magnitude other than 7.5, the value of CSRL obtained from Fig. 5A.67 must be multiplied by a scaling factor to obtain the adjusted CSRL. A list of commonly used correction factors is given in Table 5A.8. It is preferable to use moment magnitude in liquefaction analyses, but surface-wave magnitude or local (Richter) magnitude may also be used for magnitudes less than 7.5 (Youd, 1993).

TABLE 5A.7 Estimated Energy Ratios for SPT Procedures (from Seed et al., 1985)

Country	Hammer type	Hammer release	Estimated rod energy, %	Correction factor for 60% rod energy
Japan	Donut	Free-fall	78	1.30
	Donut	Rope and pulley with special throw release	67	1.12
United States	Safety	Rope and pulley	60	1.00
	Donut	Rope and pulley	45	0.75
Argentina	Donut	Rope and pulley	45	0.75
China	Donut	Free-fall	60	1.00
	Donut	Rope and pulley	50	0.83

Several empirical methods for estimating liquefaction potential from the results of cone penetration tests (CPTs) are also available. The primary advantages of the CPT over the SPT for liquefaction analyses are that the CPT provides data much more rapidly than does the SPT, the CPT provides a continuous record of penetration resistance in any borehole, and the CPT is less vulnerable to operator error than the SPT (Seed et al., 1983). Unfortunately, there is a limited database to provide correlation between soil liquefaction characteristics and CPT values. Douglas and coworkers (1981) recommended performing preliminary studies at each site to establish a correlation between CPT and SPT data, converting CPT data to SPT N value, and then using the correlations between liquefaction potential and SPT N value to determine the liquefaction potential of a deposit. Seed and colleagues (1983) prepared charts of CSRL versus modified cone penetration resistance that were based on the CPT-SPT correlations proposed by Schmertmann (1978). Robertson and Campanella (1985) also presented a chart of CSRL versus modified cone penetration resistance.

The previous discussions clearly establish that density plays an important role in the liquefaction potential of a saturated sandy soil. Therefore, the liquefaction potential of a compacted granular fill that may become saturated can be reasonably controlled in many instances by ensuring that the fill soil is highly densified. Reducing the liquefaction potential of an existing saturated granular soil deposit can be more difficult. Near-surface compaction is not a viable alternative in many cases because it is effective only to a maximum depth of about 5 to 10 ft (1.5 to 3.0 m) from the ground surface, and many potentially liquefiable deposits extend to much greater depths. In these instances, several types of deep compaction techniques have been successfully used. Liquefaction potential can also be controlled by shortening the drainage path within the liquefiable deposit, which prevents the excess pore pressure from building to a level where liquefaction can occur. This can be accomplished by the inclusion of granular columns (usually gravel or stone columns) with much higher permeability than the existing deposit.

Effects of Cementation. Many soils contain carbonates, iron oxide, alumina, and organic matter that may precipitate at interparticle contacts and act as cementing agents (Mitchell, 1993). The effect of cementation is to increase the stiffness and strength of the soil while reducing the permeability and liquefaction potential somewhat. For highly cemented coarse-grained soils, the increase in stiffness and strength and the decrease in liquefaction potential can be significant, whereas the reduction in permeability is generally small, so that the overall effect is beneficial for bearing soils. Cemented granular soils that are densified in place by near-surface or dynamic compaction may lose most of or all their natural cementation during the compaction process. The net effect may be either an increase or a decrease in stiffness and strength depending on which mech-

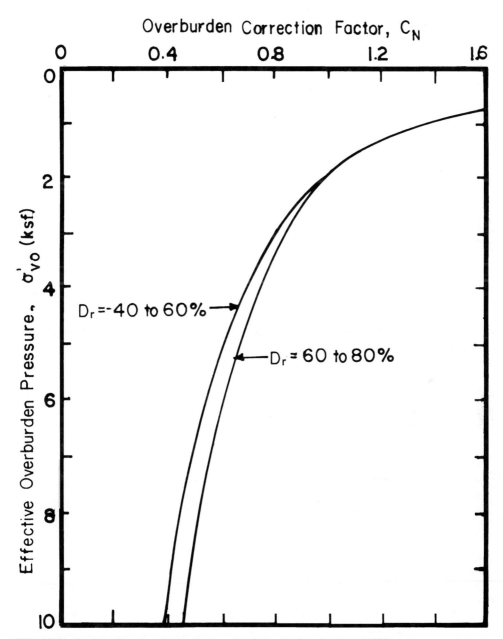

FIGURE 5A.68 Chart for values overburden correction factor, C_N (from Seed et al., 1985).

TABLE 5A.8 Correction Factors for Influence of Earthquake Magnitude on Liquefaction Resistance

Earthquake magnitude, M	Number of representative cycles at $0.65\tau_{max}$	$\dfrac{(\tau_{av}/\sigma'_{vo})_l \text{ for } M = M}{(\tau_{av}/\sigma'_{vo})_l \text{ for } M = 7.5}$
8.5	26	0.89
7.5	15	1.0
6.75	10	1.13
6	5 to 6	1.32
5.25	2 to 3	1.5

anism predominates—the increase produced by densification or the decrease resulting from loss of cementation. Therefore, compaction should be used with caution in naturally cemented soils.

Any cementation of borrow materials will also likely be substantially destroyed during the excavation and compaction procedures. However, borrow soils that were cemented before excavation contain minerals that will cement, and therefore it is likely they will develop new cementation with time after compaction. The effect of this postcompaction cementation on the engineering properties of compacted soils is commonly ignored in foundation engineering practice primarily because it is difficult or impractical to determine both the rate at which the cementation will occur and what effect this cementation will have at critical stages of construction and loading of the structure being built. In addition, since cementation has a positive effect on compressibility, strength, and liquefaction potential, and little effect on the permeability, it is conservative to use the as-compacted properties in design and analysis.

In some instances neglecting the postcompaction cementation of fills may be highly uneconomical. For example, in southeastern Florida a common borrow material is carbonate sand that either is a residual soil from the oolitic limestone found near the ground surface or is excavated and crushed from rock deposits of the same material. It is common knowledge among engineers in the area that recementing of the carbonate sand occurs after compaction so that the resulting fill behaves more like a soft rock than a soil. A significant improvement in the postcompaction stress-strain-strength characteristics of the compacted fills often occurs, so that the stiffness and strength of the fills commonly continue to increase during their service life. This improvement is sometimes included in a qualitative way in the design of the foundations for structures constructed on fills composed of these carbonate materials, but more in-depth studies of this phenomenon would surely result in more economical foundation designs. The effects of aging on the engineering properties and behavior of compacted soils are discussed in more detail in Sec. 5A.3.4.3.

5A.3.4.2 Cohesive Soils. The engineering characteristics of compacted cohesive soils are strongly influenced by the as-compacted structure, moisture condition, and density of the soil; by the method used to compact the soil; and by any postcompaction changes in structure, moisture condition, and density that may occur.

Fabric and Structure. Although the terms *fabric* and *structure* are sometimes used interchangeably, it is preferable to define *fabric* as the arrangement of particles, particle groups, and pore spaces within a soil, and to use *structure* to refer to the combined effects of fabric, composition, and interparticle forces. The structure of cohesive soils is extremely complex, and the reader is referred to Mitchell (1993) for a comprehensive discussion. Therefore, only fabric will be discussed here. The fabric of a soil can be considered to consist of the following three components (Mitchell, 1993):

1. *Microfabric.* The microfabric consists of the regular aggregations of particles and the very small pores between them. Typical fabric units are up to a few tens of micrometers across.

2. *Minifabric.* The minifabric contains the aggregations of the microfabric and the interassemblage pores between them. Minifabric units may be a few hundred micrometers in size.
3. *Macrofabric.* The macrofabric may contain cracks, fissures, root holes, laminations, and so on.

For a given cohesive borrow material, the fabric of the as-compacted soil can vary widely depending on the compaction water content and the method of compaction. Conventional descriptions of the fabric of compacted cohesive soils assume that the soil is composed entirely of plate-shaped clay particles and are based on the microfabric of the soil, that is, how the individual clay particles are arranged (Lambe, 1958). If most of the particles are oriented parallel to each other, the fabric is said to be *dispersed* or *oriented* (Fig. 5A.69). In clays where most of the particles are arranged perpendicular to each other, the fabric is considered to be *flocculated* or *random.* Compaction dry of optimum or with low effort tends to produce a more flocculated fabric, whereas compaction wet of optimum or with high effort generally yields a more dispersed fabric (Fig. 5A.70).

Although the engineering behavior of pure clay samples compacted in the laboratory can be adequately explained on the basis of the microfabric, it is difficult to justify its use for field compacted cohesive soils for two reasons: (1) Most of the soils contain a variety of particles, including both coarse-grained and fine-grained particles; and (2) the minifabric and macrofabric have a predominant influence in many cases on the engineering characteristics. Excellent reviews of models for compacted soil behavior can be found in Hausmann (1990) and Hilf (1991). A practical theory that considers both microfabric and minifabric is the *clod model,* a slightly modified version of the model proposed by Hodek and Lovell (1979), which built upon the basic concepts developed by others (e.g., Barden and Sides, 1970; Mitchell et al., 1965; Olsen, 1962). In this model, the borrow soil consists of aggregations of particles held together by cohesive (clay) particles and separate granular particles. These aggregations (called *clods*), may contain particles as coarse as gravel-sized, but their behavior during compaction is dominated by the clay particles. The granular particles are generally brittle, but the clods may be either brittle or plastic depending on their moisture condition. The total void space within the soil consists of *intraclod pores* (voids within the clods) and *interclod pores* (void spaces between bulky units, either clods or separate granular particles). The interclod pores are considerably larger than the intraclod pores.

Densification is achieved during compaction by one or more of the following three mechanisms: (1) Pushing the clods and granular particles closer together; (2) breaking brittle clods; and

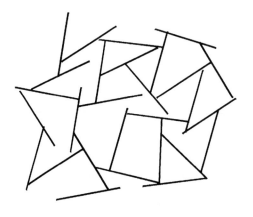
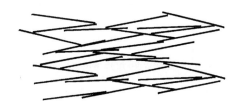

FIGURE 5A.69 Microfabrics for individual particle interaction in pure clays.

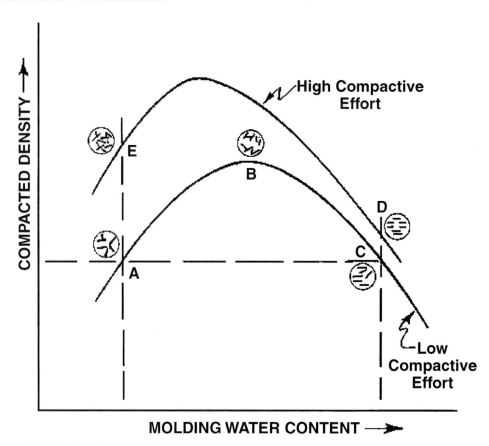

FIGURE 5A.70 Effects of molding water content and compactive effort on the microfabric of pure clays (from Lambe, 1958).

(3) deforming plastic clods. The moisture condition of the clods during compaction is the key factor that determines the density, fabric, and behavior of the as-compacted soil. The granular particles are generally little affected by moisture within the soil. At relatively low compaction water contents, the capillary (matric) suction within the clods is very large, and the clods are dry, strong, and brittle and behave as discrete units. During compaction these clods may be either broken or moved as a unit, but they deform very little, and thus behave essentially as granular particles. Compaction reduces the interclod space but does not affect the intraclod space, which is relatively small because the clods are shrunken (Fig. 5A.71). Thus, the as-compacted fabric of dry cohesive soil essentially consists of discrete, hard clods. Sheepsfoot and tamping foot rollers are generally more effective in densifying dry cohesive soils because the higher contact pressures break down more of the hard clods.

At relatively high water contents, the suction is small, and the clods are plastic and may be remolded into almost any shape. As the clods are deformed during compaction, the interclod space can be nearly eliminated without affecting the intraclod space, which is relatively large because the aggregates are swollen. Depending on the amount of remolding of clods that occurs during compaction, the as-compacted fabric may vary. Owing to their high contact pressures and large induced shearing strains, sheepsfoot rollers tend to remold the clods into a relatively homogeneous fabric. Smooth-drum rollers remold the clods primarily by compressive forces, resulting in a fabric that retains to some extent the identity of the individual clods.

Dry Side of Optimum

Individual gravel, sand, and silt grains (G), as well as cohesive clods, are brittle; clods are shrunken

Compaction reduces the interclod space without affecting the intraclod space, which is small

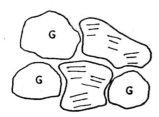

Wet Side of Optimum

Cohesive clods are plastic and swollen

Compaction can almost eliminate the interclod space without affecting the intraclod space, which is large

FIGURE 5A.71 Schematics of compaction according to the clod model (modified from Hodek and Lovell, 1979).

The engineering behavior and properties of compacted cohesive soils often vary with time as changes occur in moisture condition, stress state, and fabric. The as-compacted characteristics of cohesive soils, and how these properties may change as other parameters change, is discussed in the following sections.

Stress-Strain-Strength Behavior. The discussion in this section on the stress-strain-strength behavior of compacted cohesive soils is divided into two primary classifications based on the moisture condition of the soil at the time of loading: (1) Soils that remain at their as-compacted water content up to and throughout the loading, and (2) soils that become soaked prior to loading. These two moisture conditions represent the limiting cases usually considered for compacted cohesive soils. Although some compacted cohesive soils dry out during their service life, the drying generally results in a stiffer and stronger soil (unless substantial cracking occurs), so it is usually conservative to ignore the effects of drying and use the compressibility and strength characteristics of the soil at its as-compacted moisture content. However, drying of the soil may result in shrinkage and settlement of any structures founded on that soil. In addition, the loss of moisture increases the potential for subsequent wetting-induced volume changes (swelling or collapse), and this increased potential for volume changes should also be considered (see Sec. 5A.3.4.2).

The term "as-compacted" will be used throughout the next section to refer to a compacted cohesive soil that has remained at its compaction water content up to the time of loading. Thus, *as-compacted* refers to no change in moisture content but does not preclude changes in void ratio, for example, owing to compression under the weight of fill placed on top of the soil prior to the inducement of shearing stresses by a structure constructed on or within the fill. This definition of *as-compacted* is somewhat different than that given by others (for example, Seed et al., 1960), who have defined as-compacted soils as those in which no change in either moisture condition or void ratio occurs prior to or during loading.

As-compacted moisture content

Compactive Prestress The as-compacted compressibility and strength of compacted cohesive soils are influenced by the moisture condition and density of the soil, as well as method and effort of compaction. The cyclic application and removal of mechanical energy during compaction prestresses the soil, and the stress-strain behavior of an as-compacted cohesive soil is similar to that of an overconsolidated, saturated clay. A typical strain-stress plot for one-dimensional compression of an as-compacted cohesive soil is shown in log $\sigma_v - \varepsilon_{VL}$ space in Fig. 5A.72, in which the recompression and virgin compression portions of the loading curve are evident. The vertical stress at which the behavior changes from recompression to virgin compression is called the *compactive prestress* (σ_{vc}) and is analogous to the effective preconsolidation pressure (σ_p') in saturated clays. Casagrande's (1936b) graphical procedure for estimating σ_p' of saturated clays can also be used to approximate σ_{vc} for compacted cohesive soils. σ_{vc} for a particular soil increases with increasing dry density and decreasing water content (Lin and Lovell, 1983). Method of com-

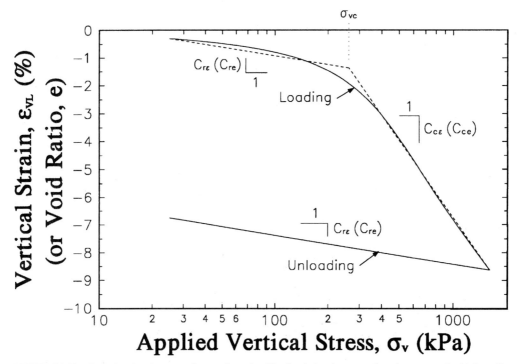

FIGURE 5A.72 Typical strain-stress curve for one-dimensional loading-induced compression of a compacted cohesive soil.

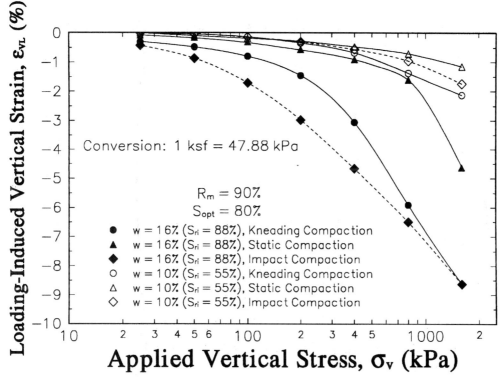

FIGURE 5A.73 Effect of compaction method on one-dimensional compressibility of compacted clayey sand (data from Lawton, 1986).

paction also has a significant influence on σ_{vc}, as indicated in Fig. 5A.73 for specimens of a clayey sand compacted at two water contents to the same dry density using kneading, static, and impact compaction. For both compaction water contents, σ_{vc} for static compaction is greater than for kneading compaction because static compaction is much less efficient in cohesive soils than kneading compaction. Thus, for the same compaction water content, higher pressures are needed in static compaction to produce the same density. σ_{vc} for impact compaction is less than for static compaction at both water contents but is greater than for kneading compaction at $w = 10\%$ and less than for kneading compaction at $w = 16\%$. This suggests that impact compaction is less efficient than kneading compaction for compaction below optimum moisture condition (as-compacted $S_r < S_{opt}$) but more efficient above optimum moisture condition (as-compacted $S_r > S_{opt}$). The apparent anomalies in the behavior of impact-compacted cohesive soils will be discussed in more detail in a subsequent section.

Influence of Moisture Condition Moisture condition has a dramatic influence on the compressibility and strength of compacted cohesive soils owing to changes in suction. As discussed previously (Sec. 5A.3.4.1), the total suction in an unsaturated soil consists of two components—osmotic suction and matric suction. Values of total, matric, and osmotic suction are shown in Fig. 5A.74 for specimens of a glacial till compacted with modified Proctor effort at various water contents. Matric suction is the largest component of suction in unsaturated soils. Because osmotic suction varies little for the normal variation in water contents for most compacted soils, changes in suction are primarily related to changes in matric suction, which can be substantial. For example, for the glacial till shown in

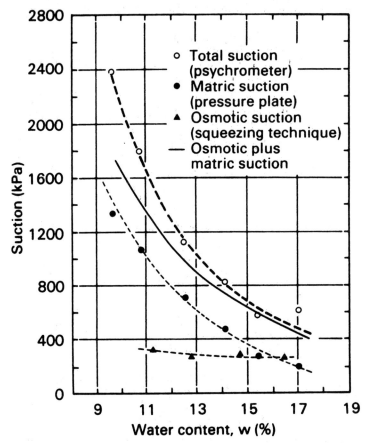

FIGURE 5A.74 Total, matric, and osmotic suctions for glacial till (Fredlund and Rahardjo, 1993, data from Krahn and Fredlund, 1972).

Fig. 5A.74, the matric suction decreased from about 900 kPa (131 psi) at $w = 11\%$ to about 200 kPa (29 psi) at $w = 17\%$, a 78% reduction in matric suction for a 6% increase in water content.

The effect of moisture condition on the one-dimensional compressibility of a compacted clayey sand is illustrated in Fig. 5A.75. At low water contents the clods are hard and the compacted soil is relatively stiff. Well-compacted, highly plastic, dry soils can be nearly incompressible under stress levels encountered in some projects. For the same method of compaction and as-compacted density, the compressibility of the soil increases as the water content increases, owing to the reduction in suction described previously. Similar trends are seen in Fig. 5A.76 for triaxial stress-strain-strength behavior. An increase in compaction moisture content at constant energy level results in a weaker, more compressible soil. Note also the reduction in brittleness as water content is increased.

The shear strength of an unsaturated soil is commonly formulated in terms of the stress-state variables net normal stress $(\sigma - u_a)$ and matric suction $(u_a - u_w)$. The resulting shear strength equation, which takes the form of a three-dimensional extension of the Mohr-Coulomb failure criterion (see Fig. 5A.77), is as follows (Fredlund et al., 1978):

$$\tau_{ff} = c' + (\sigma_f - u_a)_f \cdot \tan \phi' + (u_a - u_w)_f \cdot \tan \phi^b \tag{5A.25}$$

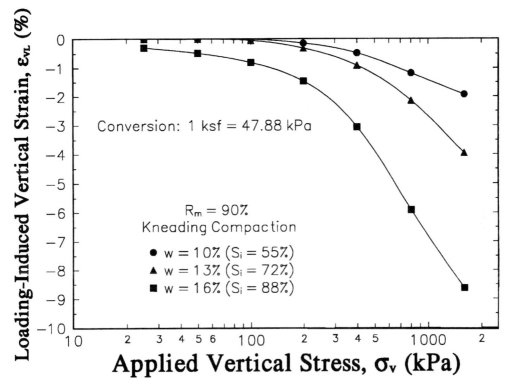

FIGURE 5A.75 Influence of moisture condition on one-dimensional compressibility of compacted clayey sand (data from Lawton, 1986).

where τ_{ff} = shear stress on the failure plane at failure (shear strength)
c' = intercept of the extended Mohr-Coulomb failure envelope on the shear stress axis where the net normal stress and the matric suction at failure are equal to zero (commonly referred to as the *effective cohesion intercept*)
$(\sigma_f - u_a)_f$ = net normal stress state on the failure plane at failure
u_{af} = pore air pressure on the failure plane at failure
ϕ' = angle of internal friction associated with the net normal stress variable
$(u_a - u_w)_f$ = matric suction on the failure plane at failure
ϕ^b = angle indicating the rate of increase in shear strength relative to the matric suction, $(u_a - u_w)_f$

The effect of changes in matric suction on the shear strength of an unsaturated soil is illustrated in Fig. 5A.78(a). The projections of the three-dimensional failure envelope on the shear stress–net normal stress plane are shown in Fig. 5A.78(b). Thus, the effect of wetting an unsaturated cohesive soil is to reduce its shear strength owing to a reduction in matric suction. This loss of shear strength upon wetting can be considered to result from a reduction in cohesion intercept, where the cohesion intercept is defined as

$$c = c' + (u_a - u_w)_f \cdot \tan \phi^b \qquad (5A.26)$$

Values of ϕ' and c' can be measured using saturated soil specimens and conventional triaxial and direct shear testing equipment. Tests on unsaturated specimens using specialized triaxial or

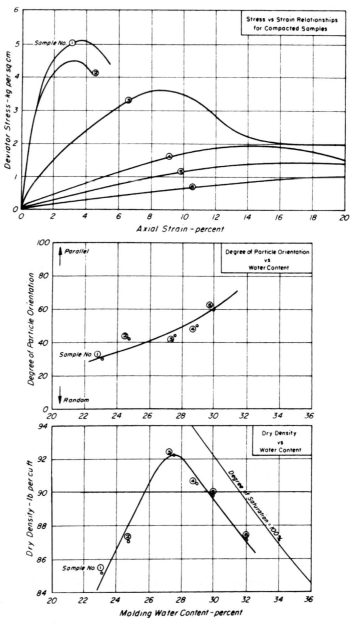

FIGURE 5A.76 Influence of molding water content on structure and triaxial stress-strain relationship for as-compacted samples of Kaolinite (from Seed and Chan, 1959a).

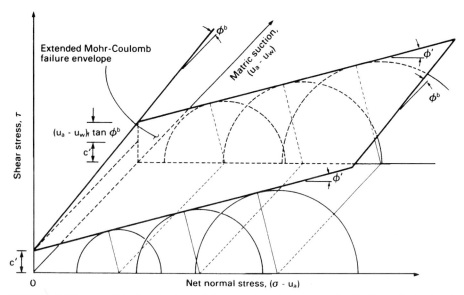

FIGURE 5A.77 Extended Mohr-Coulomb failure envelope for unsaturated soils (from Fredlund and Rahardjo, 1993).

direct shear testing apparatuses are needed to obtain values of ϕ^b, and the reader is referred to Fredlund and Rahardjo (1993) for details of these tests and equipment. Experimentally determined values of ϕ^b have ranged from about 7 to 26°, with most values in the range of about 13 to 18°.

Effect of Density In general, an increase in density of a compacted cohesive soil for the same method of compaction and water content results in an increase in stiffness and strength. The decrease in as-compacted one-dimensional compressibility of a compacted clayey sand with increasing density at the same water content is shown in Fig. 5A.79. Note also that the compactive prestress increases with increasing density, as would be expected because more energy is required, at the same compaction water content, to achieve a greater density.

The effects of density on the unconsolidated, undrained triaxial compressive strength of two cohesive soils at small and large strains are shown in Figs. 5A.80 and 5A.81. In these graphs, the strength is defined as the maximum stress that the sample sustained to reach the designated value of strain and thus is not necessarily the point on the stress-strain graph corresponding to that value of strain. The strength at large strain (20 or 25%) is applicable to situations where large deformations are tolerable, such as a flexible (wet) core of a high earth dam. The strength at small strain (5%) is appropriate for bearing soils where only small deformations are tolerable, such as for a building founded on a compacted fill, or where the soil is brittle (dry) and may crack at small strains.

At large strains, the strength of the sandy clay and the silty clay either increased or remained constant as the dry density increased. At low strains, the strength of both soils increased with increased density at low water contents, but at high water contents the strength peaks at a given density and then decreases with increased density beyond the peak. This loss of strength at high water contents is especially pronounced in the silty clay; for example, at a water content of 16% the strength peaks at about 5.2 kg/cm² (10.7 ksf = 510 kPa) for γ_d = 112.5 pcf (17.7 kN/m³) but drops to 2.0 kg/cm² (4.1 ksf = 200 kPa) for γ_d = 114.9 pcf (18.0 kN/m³), a reduction of 62%. The rapid decrease in strength at low strains near optimum water content can be seen in the left side of Fig. 5A.81 for each of the three compactive energy levels. At the same water content of 16%, the sample compacted with lower energy [γ_d = 112.5 pcf (17.7 kN/m³)] is dry of optimum

5.110 SOIL IMPROVEMENT AND STABILIZATION

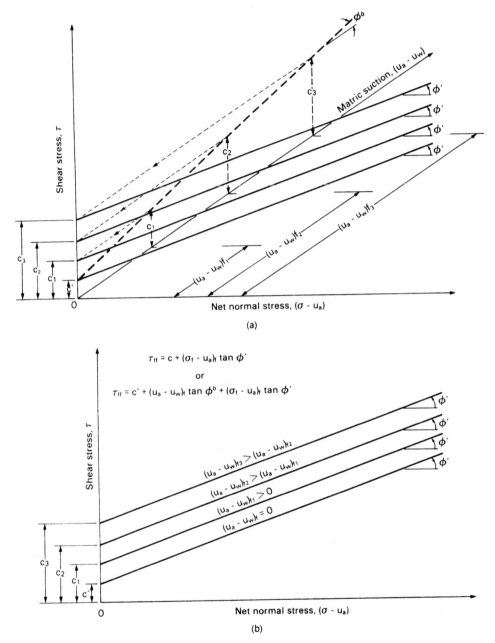

FIGURE 5A.78 Horizontal projection of the failure envelope onto the τ versus $(\sigma - u_a)$ plane, viewed parallel to the $(u_a - u_w)$ axis (from Fredlund and Rahardjo, 1993): (*a*) failure envelope projections onto the τ versus $(\sigma - u_a)$ plane; and (*b*) contour lines of the failure envelope onto the τ versus $(\sigma - u_a)$ plane.

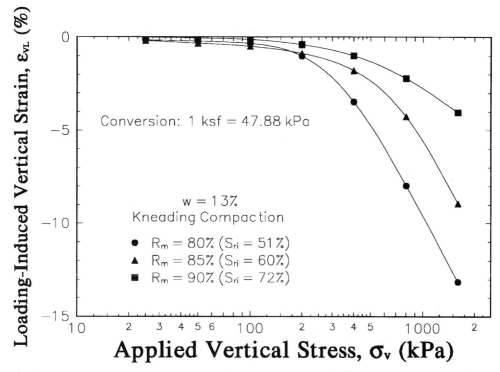

FIGURE 5A.79 Effect of density on as-compacted one-dimensional compressibility of a compacted clayey sand (data from Lawton, 1986).

($S_r < S_{opt}$) and therefore stronger than the sample compacted with higher energy [$\gamma_d = 114.9$ pcf (18.0 kN/m³)] that is denser but wet of optimum ($S_r > S_{opt}$). The same trend can be seen in the right side of Fig. 5A.81, where the author has added the line of optimums. Seed and coworkers (1960) attributed this marked decrease in strength near optimum moisture condition to differences in fabric—higher pore water pressures develop in undrained cohesive soils compacted wet of optimum owing to a greater degree of dispersion of the microfabric. An alternative interpretation of this phenomenon can be given in terms of the air/water phases and their pressures. For $S_r < S_{opt}$, the air phase is continuous, and the matric suction likely remains positive even under undrained loading conditions because the air is somewhat compressible, and therefore only small values of positive air pressure are likely to develop compared to the negative water pressures. This suction adds to the strength and stiffness of the soil. As S_r approaches S_{opt}, there is a transition from a continuous to an occluded air phase, which changes the pore water pressures from negative to positive during undrained loading, with a corresponding reduction in strength and stiffness of the soil.

Method of Compaction Method of compaction can have an important influence on the fabric of as-compacted cohesive soils, which in turn can have a substantial effect on their stress-strain-strength characteristics. For dry cohesive borrow materials, the clods are hard, and densification occurs either by rearrangement of the clods, which retain their basic shape, or by breaking of the clods into smaller (but still hard) pieces. During static compaction (smooth-drum roller), the induced stresses are primarily compressive and of relatively low magnitude compared to other types of rollers of comparable weight, so little breakage of dry clods occurs, and densification

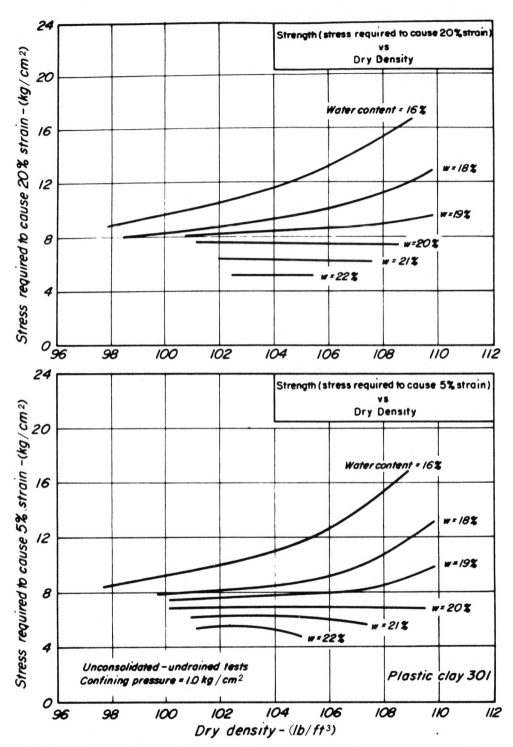

FIGURE 5A.80 Relationships among dry density, water content, and triaxial strength for as-compacted specimens of highly plastic clay prepared by kneading compaction (from Seed et al., 1960).

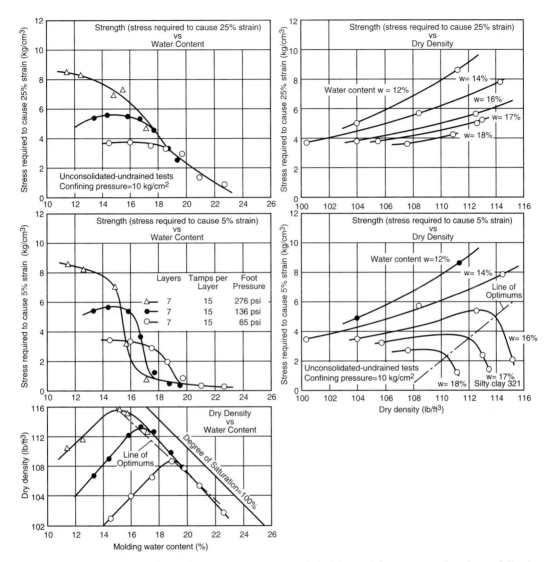

FIGURE 5A.81 Relationships among dry density, water content, and triaxial strength for as-compacted specimens of silty clay prepared by kneading compaction (from Seed et al., 1960).

results mainly from rearrangement of clods. In contrast, kneading compaction (sheepsfoot and padfoot rollers) induces both shear stresses and compressive stresses in the soil that are of relatively high magnitude owing to the smaller contact areas of the feet, which results in densification from both rearrangement and breakage of particles. Thus, the microfabric of dry clods for both types of compaction is the same because it is essentially unaffected by the compaction, but the minifabric of the compacted soil is different—larger clods and larger intraclod pores for statically compacted soils, and smaller clods and smaller intraclod pores for soils compacted by kneading compaction. If the clods are very hard or the compactors are light, little breakage of clods will occur even for kneading compactors, and the method of compaction used will have little effect on the fabric of the compacted soil. Whether breakage occurs during compaction has little effect on the strength and stiffness of the as-compacted soil for the applied stress levels normally encountered in engineering practice because both characteristics depend on the strength and stiffness of the clods, which are independent of their size. At very high applied stress levels where some breakage might occur, a soil with a fabric consisting of small, hard clods might be stronger and stiffer than one at the same density but with larger clods because of the higher number of interclod contacts, which would result in less force being carried per clod and therefore less breakage of particles. This tendency toward greater stiffness may be offset to some degree by the possibility of more cracks being induced in the smaller clods during compaction, which would make the smaller clods more susceptible to breakage for the same applied force.

Wet cohesive borrow materials are soft, and clods are easily remolded during compaction. Static compaction tends to flatten the clods from the applied compressive stresses, but little change occurs to the microfabric of the clods. The resulting compacted soil mass tends to consist of relatively large interclod pore spaces and clods with flocculated microfabrics that retain to some extent their individual identity. Kneading compaction results in a more homogeneous fabric owing to the shearing stresses induced during kneading, and the microfabric tends to be more dispersed than for static compaction. For the same moisture content and density, a cohesive soil with a flocculated fabric is less compressible and stronger than the same soil with a dispersed fabric. Thus, static compaction at high water contents produces a stronger and stiffer soil than does kneading compaction. The reasons for this difference in strength and stiffness for differing fabric are quite complex (see Lambe, 1958 for details). Simplified explanations for differences in strength and stiffness for the same soil, water content, and density but different methods of compaction can be given as follows:

Compression. Compression from an applied load tends to align the particles in a parallel array perpendicular to the direction of the applied load. Applying a load to particles that are already parallel merely brings them closer together. Applying the same load to particles that are randomly oriented requires that a portion of the load be spent reorienting the nonaligned particles to a more parallel arrangement before any reduction in average void ratio occurs.

Shear Strength. For the same water content and average spacing between clay particles (same void ratio), increasing the angle between particles increases the net attractive force between the particles, and hence increases the shear stress required to cause sliding of the particles relative to one another.

It is unlikely that field compaction would be undertaken on a cohesive borrow material so dry that no remolding would occur during compaction. Therefore, it is reasonable that method of compaction will have some effect on the compressibility and strength of all field-compacted cohesive soils.

Typical results from a series of laboratory one-dimensional compression tests conducted on specimens of a clayey sand compacted directly into the confining rings using static, impact, and kneading methods are shown in Fig. 5A.73. Two sets of results are provided—those for $w = 16\%$ and $R_m = 90\%$ corresponding to a wet of optimum moisture condition (as-compacted degree of saturation, S_{ri}, greater than optimum saturation, S_{opt}), and those for $w = 10\%$ and $R_m = 90\%$ corresponding to a dry of optimum moisture condition ($S_{ri} < S_{opt}$). In both cases, the statically compacted specimens are less compressible than the kneading-compacted specimens. For the dry of optimum specimens at low stresses [$\sigma_v < 200$ kPa (4.2 ksf)], there is little discernible difference in the results. Isograms of loading-induced vertical strain (ε_{VL}) for kneading and statically compacted specimens loaded to $\sigma_v = 400$ kPa (8.4 ksf) are shown in dry density-water content space in Fig. 5A.82. Note that ε_{VL} is greater for kneading compaction than for static compaction for the entire ranges of dry densities and water contents studied. It is interesting that the isograms of ε_{VL} for kneading compaction and S_r greater than about 70% are approximately horizontal, indicating that the compressibility of the specimens at these higher values of S_r is essentially independent of the water content and suggesting that the air phases become occluded in this soil after kneading compaction at about $S_r = 70$ to 80% for $R_m = 80$ to 90%. In contrast, the isograms ε_{VL} for the statically compacted specimens in the same region indicate that compressibility is a function of both water content and density, which suggests that occlusion of the air phases has not occurred.

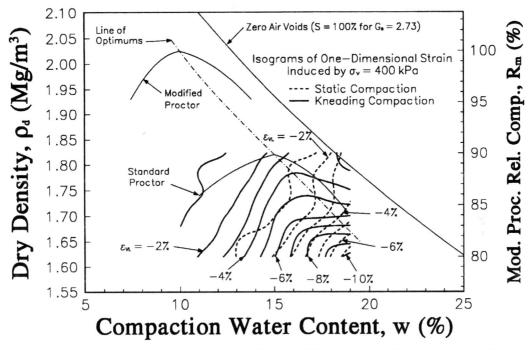

FIGURE 5A.82 Isograms of one-dimensional as-compacted compressibility of clayey sand for specimens prepared by static and kneading compaction (data from Lawton, 1986).

One-dimensional compression results are also shown in Fig. 5A.73 for specimens compacted directly into oedometer rings using a specially designed impact hammer. At both water contents, the impact-compacted soil was more compressible than the statically compacted soil. However, the comparison between the effects of kneading and impact compaction is not as straightforward—impact compaction produced a less compressible soil at $w = 10\%$ but a more compressible soil at $w = 16\%$. To provide additional insight into the differences in compressibility produced by impact compaction compared to static and kneading compaction, isograms of ε_{vL} at $\sigma_v = 400$ kPa for the same clayey sand are given in Fig. 5A.83(a) for static and impact compaction, and Fig. 5A.83(b) for kneading and impact compaction. It can be seen in Fig. 5A.83(a) that static compaction produced a stiffer soil than kneading compaction at all values of ρ_d and w studied. In Fig. 5A.83(b), the equivalent isograms for kneading and impact compaction intersect, with the kneading-compacted soil tending to be more compressible than the impact-compacted soil at low densities and less compressible at high densities. Also note that the isograms for the impact-compacted soil are nearly vertical, suggesting that the as-compacted compressibility of the impact-compacted soil is highly dependent on water content and nearly independent of density for the ranges studied.

Similar results have been found with respect to the effect of method of compaction on the unconsolidated-undrained shear strength of a silty clay (Seed and Chan, 1959a). Values of relative unconsolidated, undrained triaxial strength at 5% strain, based on the strength for kneading compaction, are shown in Fig. 5A.84 for specimens prepared to nominally identical values of (γ_d, w) at various points along the standard Proctor moisture-density curve. The method of compaction had little influence on the strength dry of optimum but had a substantial influence on the strength wet of optimum. Note that the strength of the statically compacted soil was equal to or greater than the strength of the kneading-compacted and impact-compacted soil at all water contents. Impact compaction produced a weaker soil than kneading compaction at low water contents (<18.5%)

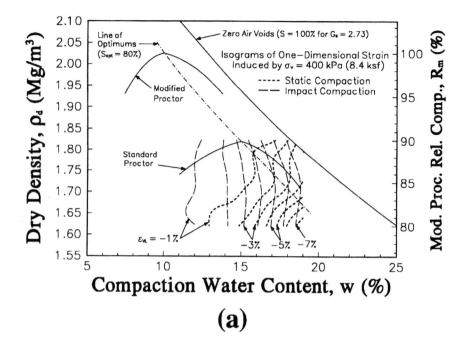
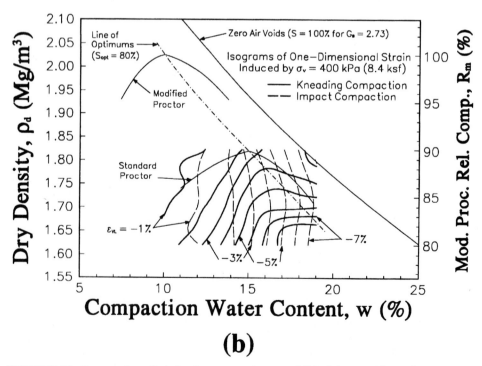

FIGURE 5A.83 Isograms of one-dimensional as-compacted compressibility of clayey sand for specimens prepared by (*a*) static and impact compaction; and (*b*) kneading and impact compaction (data from Lawton, 1986).

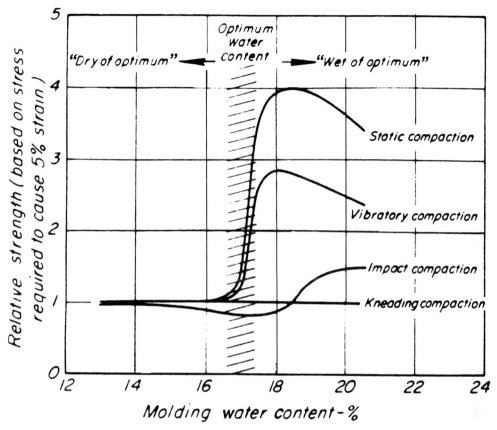

FIGURE 5A.84 Influence of compaction method on the triaxial strength of silty clay (from Seed and Chan, 1959a).

and a stronger soil at high water contents (>18.5%). For purposes of further analysis, the strength results shown in Fig. 5A.84 for the silty clay are subdivided into three regions:

1. At the low end of the compaction water contents (13 to 14%), the soil was very dry, and the strengths of the soil for both impact and static compaction were about the same.
2. Around optimum water content (15.5 to 17.5%), the specimens were dense (standard Proctor relative compaction, $R_s \cong 99$ to 100%), and impact-compacted specimens were about 80 to 90% as strong as the kneading-compacted specimens.
3. At the upper end of the compaction water content range for the silty clay (19.5 to 20.5%), the densities are lower ($R_s \cong 94$ to 96%), and impact compaction produced soil that was about 1.4 to 1.5 times stronger than the soil prepared by kneading compaction.

The reasons for the differences in stress-strain-strength behavior of impact-compacted and kneading-compacted cohesive soils are not completely understood, and further research is needed in this area. A preliminary explanation for these differences can be given as follows (modified from Seed and Chan, 1959a):

1. For cohesive borrow materials that are very dry, no method of compaction produces any appreciable shear deformation, and consequently the method of compaction has no appreciable effect on the fabric and hence the stress-strain-strength characteristics of the as-compacted soil.

2. For the water contents at which cohesive borrow soils are usually compacted, impact compaction causes somewhat less shearing strain during compaction than does kneading compaction. At the same density, the as-compacted fabric for impact compaction tends to have larger but fewer intraclod voids and less dispersion within the clods than for kneading compaction. The larger intraclod voids tend to decrease the strength and increase the compressibility of the impact-compacted soil, but the lesser degree of dispersion tends to increase the strength and decrease the compressibility. Whether the impact-compacted soil is stronger and stiffer than the kneading-compacted soil or vice versa depends on which characteristic dominates the stress-strain-strength behavior—larger intraclod voids or decreased dispersion. At high densities, the fabric of impact-compacted cohesive soil apparently has a degree of dispersion that is close to that of kneading-compacted soil but still has larger intraclod voids. In this case, the larger intraclod voids dominate, resulting in a more compressible soil. At low densities, the increased stiffness produced by the more flocculated fabric of the impact-compacted soil is greater than the reduction in stiffness resulting from the larger intraclod voids.

The results presented in this section provide salient evidence to substantiate that the methods and procedures used to prepare laboratory-compacted specimens for testing should simulate as closely possible the methods and procedures used in the field compaction process. It is also apparent that specimens obtained for stress-strain-strength testing by trimming Proctor samples—a routine practice in some testing laboratories—are unlikely to have stress-strain-strength characteristics representative of the soil as compacted in the field using most types of rollers. The following procedure is the best way to ensure that the fabrics of cohesive specimens to be tested in the laboratory are similar to those of the field-compacted soil:

1. Construct several lifts of the borrow soil in a test pad in which the actual compaction procedure is simulated, that is, using the same compaction roller, lift thicknesses, and construction techniques that will be employed in the field compaction process.
2. Obtain samples of the compacted soil from the test pad that are several times larger than the sizes of the specimens to be tested in the laboratory.
3. Carefully trim laboratory specimens to the appropriate sizes for testing from within the interior portions of the larger samples obtained from the field.

Unfortunately, this procedure is not economically feasible for many projects, and thus test specimens are commonly prepared using laboratory compaction procedures.

Time Rate of Settlement As with saturated naturally deposited cohesive soils, the settlement of compacted cohesive soils can be considered to consist of three parts: immediate settlement, primary compression settlement, and secondary compression settlement. *Primary compression* is defined herein for compacted cohesive soils as the loading-induced compression that occurs until either the matric suction reaches a steady-state value under the applied load (unsaturated soil where the air phase is continuous), or the pore water pressure reaches a steady-state value under the applied load (soil is unsaturated but the air phase is occluded or becomes occluded during compression). Primary compression in unsaturated compacted cohesive soils generally occurs much more quickly than does primary consolidation in the same cohesive soil that is saturated (all other factors being the same), owing to differences in water permeability and air permeability. Coefficients of air permeability (k_a) and water permeability (k_w) for a compacted cohesive soil (10% clay) as a function of water content are shown in Fig. 5A.85 (Barden and Pavlakis, 1971). At degrees of saturation less than about S_{opt}, the air phase is continuous, and loading-induced compression expels air from the voids rather than water. For $S_r \ll S_{opt}$, the portion of the void through which air can flow (the area not occupied by water) is large and primary compression occurs very quickly because of the high value of k_a. As S_r increases, k_a decreases owing to a decrease in the size of the air phase, with a concomitant decrease in the rate of primary compression. For $S_r < S_{opt}$, the air phase becomes occluded, water rather than air is expelled during compression, and the rate of primary compression (consolidation) is controlled by k_w.

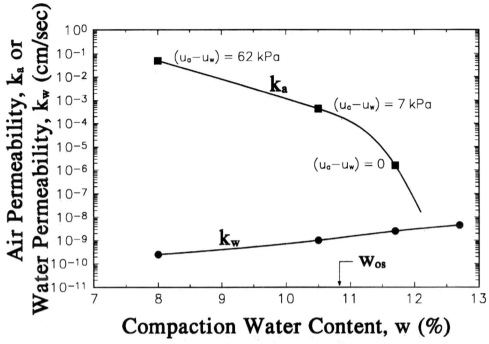

FIGURE 5A.85 Effect of compaction water content on air and water permeability of Westwater soil (from Barden and Pavlakis, 1971).

The effect of moisture condition on the time rate of one-dimensional compression in a compacted clayey sand is illustrated in Fig. 5A.86. The total compressive strains are shown in Fig. 5A.86(a), and the time-dependent strains (after the first reading at 0.25 min) are plotted in Fig. 5A.86(b). In these tests, oedometer specimens were compacted at different water contents to the same nominal dry density ($R_m = 80\%$) and then loaded incrementally in a standard one-dimensional compression test. The results shown in Fig. 5A.86 are for the loading increment from 200 to 400 kPa (4.2 to 8.4 ksf). During the previous loading increments, the magnitudes of the strains produced in the specimens were different, so the void ratios of the specimens at the beginning of the 200 to 400 kPa loading increment (e_{200}) varied somewhat, but the influence of these small differences in void ratio on the loading-induced compression was negligible compared to the effect produced by the differences in degree of saturation (S_{r-200}). It can be seen in Fig. 5A.86(b) that the magnitude of the time-dependent strains increased substantially with increasing degree of saturation. It is also apparent that the strain-log time plot for the wettest sample ($S_{r-200} > S_{opt}$) has the classical "backward S" shape associated with saturated cohesive soils; in contrast, the plot for the driest sample ($S_{r-200} \ll S_{opt}$) is essentially linear, suggesting that primary compression was completed prior to the first reading (0.25 min).

Differentiating between primary and secondary compression in compacted cohesive soils is not easy unless tests are conducted in which the pore air and pore water pressures are simultaneously measured during compression, which requires special testing equipment. Tests of this type are rarely conducted in geotechnical engineering practice. Details of equipment and methods that can be used to measure or control u_a and u_w in unsaturated samples are given in Fredlund and Rahardjo (1993). Simultaneous measurement of u_a and u_w is further complicated by the fact that the air phase in the compacted specimen may change from continuous to occluded during the compression. For example, the specimen with $w = 16\%$ in Fig. 5A.86 was loaded incrementally

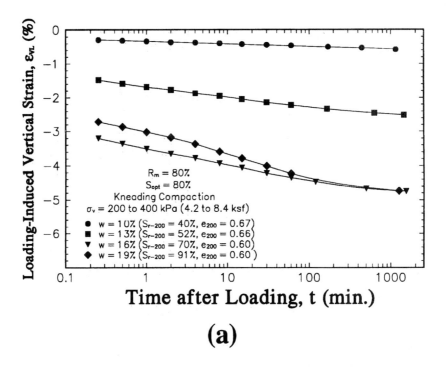

(a)

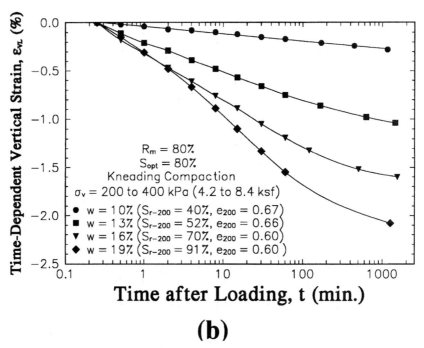

(b)

FIGURE 5A.86 Effect of moisture condition on the time rate of one-dimensional loading-induced compression of a compacted clayey sand (data from Lawton, 1986): (*a*) total vertical strain; and (*b*) time-dependent vertical strain.

from the as-compacted state ($\sigma_v \cong 0$) to $\sigma_v = 1600$ kPa (33.4 ksf), which increased S_r from 63% (17% below S_{opt}) to nearly 100%, and it is likely that the air phase changed from continuous to occluded. [Note that S_r increased from 70 to 81% during the loading increment from 200 to 400 kPa (4.2 to 8.4 ksf) shown in Fig. 5A.86.] Although the air phase was continuous, the pore air pressures were either positive or zero during loading, whereas the pore water pressures were negative [$u_w \cong -80$ kPa (-1.7 ksf)] in the as-compacted state determined from suction measurements. After the air phase became occluded, the pore water pressures were positive (perhaps as much as 800 kPa during the 800–1600-kPa loading increment) or zero, and the pore air pressures in the occluded bubbles could not have been measured but probably had little effect on the behavior during compression. To have measured both the pore air and pore water pressures throughout this test would have required the following setup:

1. A sealed sample with a low air entry porous disk at the top of the sample (for measuring or controlling u_a) and a high air entry porous disk at the bottom of the sample (for measuring u_w).
2. A pressure transducer for measuring positive air pressures up to perhaps 800 kPa (16.7 ksf = 116 psi).
3. A pressure transducer for measuring u_w from about -80 kPa (-1.7 ksf = -12 psi) to perhaps $+800$ kPa (16.7 ksf = 116 psi).

Because high air entry ceramics usually have low permeabilities, the response time is generally slower than desired for rapidly changing pore water pressures so that there may be somewhat of a difference between measured and actual pore water pressures at any time.

Owing to the difficulty and expense associated with measuring u_a and u_w during compression tests on compacted cohesive soils, the delineation between primary and secondary compression is usually accomplished by visually examining the strain-log time curves in the same manner as for saturated cohesive soils. The following guidelines are provided for conducting an immediate/primary/secondary compression analysis for a compacted cohesive soil:

1. The values for immediate strain can be calculated from the first readings taken after applying any loading increment (usually 6 or 15 s). From the data in Fig. 5A.86(*a*), the immediate strains for the 200–400-kPa (4.2–8.4-ksf) loading increment would be 0.29, 1.5, 3.2, and 2.7% for $w = 10, 13, 16$, and 19%, respectively.

2. Primary and secondary (creep) compression occur during the phase called *primary compression,* but there is no method currently available to separate the two components other than to assume that the secondary compression that occurs during primary compression is the same as (or similar to) that measured after primary compression has ended. In typical geotechnical engineering practice, the secondary compression component is not differentiated from the primary compression component, and the two components together are considered to constitute primary compression.

3. When the strain-log time plot has the classical backward S shape [e.g., $w = 19\%$ in Fig. 5A.86(*b*)], the time to the end of primary compression is estimated to occur at the point of intersection of a straight line extended forward from the central (steepest) portion of the backward S (some judgment involved) and a best-fit straight line extended backward from the tail of the backward S. Secondary compression is assumed to begin when primary compression ends. The secondary compression index, $C_{\alpha\varepsilon}$ (or $C_{\alpha e}$ from void ratio-log time plot), can be determined from the slope of the best-fit straight line through the tail, and settlement estimated using the following equations:

$$C_{\alpha\varepsilon} = -\frac{d\varepsilon_v}{d(\log t)} \quad (5A.27a)$$

$$C_{\alpha e} = -\frac{de}{d(\log t)} \quad (5A.27b)$$

$$S_s = \int_{z=0}^{z=H} \int_{t=t_p}^{t=t_{design}} C_{\alpha\varepsilon} \cdot d(\log t) \cdot dz \tag{5A.28a}$$

$$S_s = \int_{z=0}^{z=H} \int_{t=t_p}^{t=t_{design}} \frac{C_{\alpha e}}{1+e_p} \cdot d(\log t) \cdot dz \tag{5A.28b}$$

where ε_v = vertical strain
e = void ratio
z = depth from bearing level
H = height of the compacted soil layer
e_p = void ratio at the end of primary compression
t_{design} = design life of the structure

If $C_{\alpha\varepsilon}$ (or $C_{\alpha e}$) is assumed to be independent of time and stress level, and e_p is assumed to be approximately constant with depth, Eq. (5A.28) reduces to the form more commonly seen in the literature:

$$S_s = C_{\alpha\varepsilon} \cdot H_0 \cdot \log \frac{t_{design}}{t_p} \tag{5A.29a}$$

$$S_s = \frac{C_{\alpha e}}{1+e_p} \cdot H_p \cdot \log \frac{t_{design}}{t_p} \tag{5A.29b}$$

The height of the compacted soil layer is shown as H_0 in Eq. (5A.29a) because strain is generally plotted as engineering strain ($\Delta H/H_0$). In Eq. (5A.29b) e_p and H_p are used because secondary compression is assumed to begin when primary compression ends (t_p). However, for one-dimensional compression the ratio $H/(1+e)$ is a constant, so e_0 with H_0 can also be used, which is the form more commonly seen in the literature, but the equation as shown is the strictly correct form. If the soil is highly layered or if $C_{\alpha\varepsilon}$ (or $C_{\alpha e}$) varies significantly as a function of stress level, the depth integral in Eq. (5A.28) can be approximated by a summation:

$$S_s = \sum_{i=1}^{i=n} C_{\alpha\varepsilon n} \cdot \log \frac{t_{design}}{t_p} \cdot \Delta z_n \tag{5A.30a}$$

$$S_s = \sum_{i=1}^{i=n} \frac{C_{\alpha e n}}{1+e_{pn}} \cdot \log \frac{t_{design}}{t_p} \cdot \Delta z_n \tag{5A.30b}$$

The coefficient of primary compression, c_v, can be calculated from the same techniques used for saturated cohesive soils. A more direct method for extrapolating one-dimensional laboratory results to field time scale is to assume that the time rate of compression is proportional to H_{dp}^2, where H_{dp} is the height of the longest drainage path. For example, ε_{VL} after 1 h in a 1-in (25-mm) high, doubly drained laboratory specimen can be extrapolated to field time scale for a 20-ft (6.1-m) thick, singly drained compacted fill in the following manner:

$$H_{dp-lab} = 0.5 \text{ in } (13 \text{ mm})$$

$$H_{dp-field} = 20 \text{ ft } (6.1 \text{ m}) = 240 \text{ in}$$

$$t_{field} = t_{lab} \cdot \frac{H_{dp-field}^2}{H_{dp-lab}^2} = (1 \text{ h}) \cdot \frac{(240 \text{ in})^2}{(0.5 \text{ in})^2} = 230{,}400 \text{ h} = 9600 \text{ day} = 26.3 \text{ yr}$$

Although this method is crude, it is similar in concept to the methods commonly used for saturated cohesive soils and is simple enough to use in routine practice. More sophisticated methods are available for estimating the time rate of compression for unsaturated soils (see Fredlund and Rahardjo, 1993), but these techniques are beyond the scope of this handbook. An alternative method for establishing the effect of H_{dp} on the time rate of compression is to conduct a series of one-dimensional compression tests on specimens compacted at the same water content to the same density but of varying thicknesses, and then comparing the values of strain for the same loading increments and time after loading.

   4. If the strain-log time plot can be approximated quite closely with a straight line [e.g., plot for $w = 10\%$ in Fig. 5A.86(*b*)], it can be assumed that primary compression was complete and secondary compression began before the first reading was taken. Thus, the strain at the first reading can be conservatively used as the immediate plus primary compression strain. In this case, the approach for estimating settlement is similar to that used for sands wherein the primary compression settlement occurs so quickly that it cannot be separated from the immediate (distortion) settlement, so both components are included in a single component called *immediate settlement.*

   5. For time-rate plots that are neither straight lines nor backward *S* shapes [e.g., $w = 13$ and 16% in Fig. 5A.86(*b*)], the time to the end of primary compression can be estimated as the point of intersection between a straight line extended forward for the initial (steeper) portion of the plot and a straight line extended backward from the tail (flatter) portion of the plot. Some judgment is commonly required to draw these straight lines, especially the line drawn through the initial portion of the plot, because this part is generally curved.

   6. For loading and field conditions that cannot be reasonably approximated by one-dimensional compression tests, more sophisticated laboratory testing techniques—such as the stress path method (Lambe and Whitman, 1979)—are needed to model the field compression in a reasonable way.

Total Settlement It is difficult to differentiate between immediate compression and secondary compression in the laboratory and even more difficult to extrapolate this difference to field time scale. Using a traditional dead-weight consolidometer setup, the first reading commonly taken in a one-dimensional compression test is either 6 s or 15 s. Using the method given in the previous section, these two laboratory times are extrapolated to field scale for a 1.0-in (25-mm) high, doubly drained laboratory specimen and various singly drained fill thicknesses in the following table:

Singly drained field thickness		$\dfrac{H_{dp-\text{field}}}{H_{dp-\text{lab}}}$	t_{field} (days)	
ft	m		$t_{\text{lab}} = 6$ s	$t_{\text{lab}} = 15$ s
1	0.3	24	0.04	0.1
5	1.5	120	1	2.5
10	3.0	240	4	10
20	6.1	480	16	40
50	15.2	1200	100	250
100	30.5	2400	400	1000
200	61.0	4800	1600	4000

Thus, it is clear that extrapolating laboratory immediate settlement to field time scale depends on the relative values of $H_{dp-\text{lab}}$ and $H_{dp-\text{field}}$ and is not easy to do for large values of $H_{dp-\text{lab}}/H_{dp-\text{field}}$. Because of these difficulties, it is common in geotechnical engineering practice to perform a combined settlement analysis for the immediate and primary compression components, S_{ic}. It should be recognized, though, that usually a significant portion of the settlement that occurs at the first reading is immediate settlement that would occur regardless of when the first reading is taken. Thus, the author recommends taking readings as quickly as possible and assigning the value of strain at the first reading to immediate strain, with the time-dependent strains assumed to occur after the first reading.

5.124 SOIL IMPROVEMENT AND STABILIZATION

The results from one-dimensional compression tests are commonly plotted as either a log $\sigma_v - \varepsilon_v$ or log $\sigma_v - e$ plot, as shown in Fig. 5A.72. In either case, the plot can be approximated by two straight lines (dashed lines in Fig. 5A.72) as is commonly done for saturated clays, with the slopes of the recompression and virgin compression lines designated $C_{r\varepsilon}$ and $C_{c\varepsilon}$, respectively (or C_{re} and C_{ce} if plotted in log $\sigma_v - e$ space). Because the initial portion of the strain-stress plot is curved, some judgment is needed to establish $C_{r\varepsilon}$. The method preferred by the author is to use the average slope for the rebound curve obtained by unloading the specimen after the loading portion is completed, as shown in Fig. 5A.72. $C_{c\varepsilon}$ is obtained by approximating the virgin compression curve with a straight line. The intersection of the two straight lines representing initial recompression and virgin compression establishes the value of σ_{vc}. With the calculated values for $C_{r\varepsilon}$, $C_{c\varepsilon}$, and C_{vc}, one-dimensional settlement for a shallow foundation can be estimated from the following equation:

$$S_{ic} = \int_{z=0}^{z=\infty} \varepsilon_{VL} dz \tag{5A.31}$$

where $z = 0$ corresponds to the bearing level, with z increasing with depth below the bearing level.

The symbol for settlement in Eq. (5A.31) is designated S_{ic} because the results from one-dimensional compression tests include both the immediate and primary compression components of the settlement. The loading-induced, one-dimensional vertical strain for any point within a compacted soil mass can be expressed by one of the following equations:

For $\sigma_{v1} \leq \sigma_{vc}$:

$$\varepsilon_{VL} = C_{r\varepsilon} \cdot \log \frac{\sigma_{v1}}{\sigma_{v0}} \tag{5A.32a}$$

For $\sigma_{v1} > \sigma_{vc}$:

$$\varepsilon_{VL} = C_{r\varepsilon} \cdot \log \frac{\sigma_{vc}}{\sigma_{v0}} + C_{c\varepsilon} \cdot \log \frac{\sigma_{v1}}{\sigma_{vc}} \tag{5A.32b}$$

where σ_{v0} = initial total vertical stress
σ_{v1} = total vertical stress after loading

If true one-dimensional compression is closely approximated—such as a wide foundation bearing on a relatively thin and homogeneous compacted soil layer overlying a thick, incompressible layer—and $\sigma_{v1} \leq \sigma_{vc}$ at all depths, Eq. (5A.31) can be used directly to estimate settlement:

$$S_{ic} = \int_{z=0}^{z=H} C_{r\varepsilon} \cdot \log \frac{\sigma_{v1}}{\sigma_{v0}} \cdot dz \tag{5A.33}$$

where H = the height from the bearing level to the bottom of the compacted soil layer
$\sigma_{v0} = \gamma D + z$
$\sigma_{v1} = \sigma_{v0} + q_0$
γ = average unit weight of the compacted soil
D = depth of embedment of the foundation
q_0 = applied uniform, centric bearing stress

Assuming that $C_{r\varepsilon}$ is constant throughout H, $C_{r\varepsilon}$ can be pulled out of the integrand, and the solution for S_{ic} is as follows:

$$S_{ic} = C_{r\varepsilon}\left\{\frac{\gamma(H+D)+q_0}{\gamma} \cdot \log[(\gamma(H+D)+q_0)]\right.$$

$$\left. - (H+D) \cdot \log[\gamma(H+D)] - \frac{\gamma D + q_0}{\gamma} \cdot \log(\gamma D + q_0) + D \cdot \log(\gamma D)\right\} \tag{5A.34}$$

A similar equation can be derived for $\sigma_{v1} > \sigma_{vc}$. In Eqs. (5A.33) and (5A.34), the settlement term S_{ic} is used because C_{ce} and C_{re} are commonly calculated based on laboratory data that include both immediate and primary compression strains. If the time rate of settlement is an important consideration, a preferred alternative is to separate the immediate and primary compression components, with values of C_{ce} and C_{re} determined separately for immediate compression and primary compression (that is, C_{cei} and C_{rei} for immediate compression and C_{cec} and C_{rec} for primary compression).

Equation (5A.34) is valid if the excavation is backfilled after construction of the foundation, and if no rebound occurs from the time of excavation to the completion of backfilling (which depends on time rate of rebound for the soil versus the elapsed time between excavation and construction of the foundation). If the foundation is not backfilled, or if the soil rebounds completely before the backfill is placed, the initial stress at the bearing level is zero, and S_{ic} should be calculated based on $\sigma_{v0} = \gamma z$. For this case, Eq. (5A.34) simplifies to

$$S_{ic} = C_{re}\left\{\frac{\gamma(H+D)+q_0}{\gamma} \cdot \log[(\gamma(H+D)+q_0)] - H \cdot \log(\gamma H) \right.$$

$$\left. - \frac{\gamma D + q_0}{\gamma} \cdot \log(\gamma D + q_0)\right\} \quad (5A.35)$$

It is assumed in Eq. (5A.35) that the total weight of the foundation, the portion of the wall or stem below the ground surface, and the backfilled soil above the foundation is about the same as the weight of the excavated soil within the footprint of the foundation.

In many situations the width of the foundation is small in comparison to H, which means that the equation for σ_{v1} is much more complex than that given in Eq. (5A.33), and the $\log(\sigma_{v1}/\sigma_{v0})$ term in Eq. (5A.33) cannot be easily integrated. For these cases, the integration must be approximated with a summation, and the appropriate equations for S_{ic} using the one-dimensional compression indices are

$$S_{ic} = \sum_{i=1}^{i=n} S_{ici} \quad (5A.36a)$$

For $\sigma_{v1} \leq \sigma_{vc}$:

$$S_{ici} = C_{rei} \cdot \log \frac{\sigma_{v1i}}{\sigma_{v0i}} \cdot \Delta z_i \quad (5A.36b)$$

For $\sigma_{v1} > \sigma_{vc}$:

$$S_{ic} = \left(C_{rei} \cdot \log \frac{\sigma_{vci}}{\sigma_{v0i}} + C_{cei} \cdot \log \frac{\sigma_{v1i}}{\sigma_{vci}}\right) \cdot \Delta z_i \quad (5A.36c)$$

where n = the number of sublayers into which the bearing soil is subdivided for purposes of settlement analysis
i = subscript added to each parameter indicating the value of that parameter for the ith sublayer
$\sigma_{v1} = \sigma_{v0} + \Delta\sigma_v$
$\Delta\sigma_v$ = increase in total vertical stress caused by the foundation load

The greater the value of n, the more closely the integration is approximated. Values of σ_{v0} and σ_{v1} at the center of each sublayer are used as "average" values. $\Delta\sigma_v$ can be estimated for each sublayer using an appropriate stress distribution method. If the soil within the settlement influence zone is relatively homogeneous, the Boussinesq-type equations can be used for foundations bearing on the ground surface. For foundations embedded within a relatively homogeneous soil, Skopek's (1961) and Nishida's (1966) solutions can be used for rectangular and circular foundations, respectively. For layered soils and other conditions, refer to the discussion in Sec. 5A.2.2. Spreadsheet programs are ideally suited for estimating settlement in this manner.

If one-dimensional strain conditions are closely approximated but the soil within the settlement influence zone is layered, the settlement can be calculated by integrating the strains for each layer using Eq. (5A.33) (or a comparable equation for $\sigma_{v1} > \sigma_{vc}$) and the appropriate value of H for each layer. As an alternative, the settlement can be estimated using Eq. (5A.36) and subdividing each layer into a number of sublayers.

Three-Dimensional Settlement It is important to note that for small-width foundations relative to the thickness of the compressible zone (a common situation), the use of Eqs. (5A.32) to (5A.36) would be inconsistent, because one-dimensional compression indices ($C_{c\varepsilon}$ and $C_{r\varepsilon}$) would be used to estimate settlement for conditions of three-dimensional strain. The influence of principal stress ratio on the compression characteristics of a clayey sand compacted at w_{om} to $R_m = 85\%$ is shown in Fig. 5A.87. These results were determined from a series of tests conducted in a triaxial cell in which an initial vertical stress of 25 kPa (3.6 psi) was applied to four identically compacted specimens, and the initial radial (horizontal) stress was varied to provide ratios of $\sigma_r/\sigma_v = \frac{1}{3}, \frac{1}{2}, \frac{2}{3}$, and 1. The vertical strain induced in each sample by the application of the initial stresses was very small and could not be accurately measured, and thus are shown as zero in Fig. 5A.87. The stresses on the specimens were then increased incrementally, while maintaining the appropriate value of lateral stress coefficient (σ_r/σ_v), up to the controlling capacities of the equipment (a maximum cell pressure of 690 kPa = 100 psi or a maximum vertical stress of 1600 kPa = 232 psi). Therefore, the maximum value of σ_v for each test varied.

The results from these triaxial compression tests are compared in Fig. 5A.87 with those from one-dimensional compression tests conducted in an oedometer on specimens compacted directly into the oedometer ring at the same water content using the same compaction method to the same relative compaction. At low stresses the strains are small and the effect of σ_r/σ_v is small, and it is

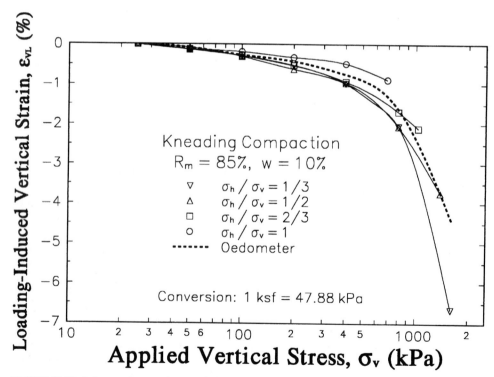

FIGURE 5A.87 Influence of stress ratio on vertical compressibility of compacted clayey sand (data from Lawton, 1986).

difficult to determine a consistent relationship. At the higher stresses, it is clear that the stress ratio has a significant influence on the loading-induced strain, with ε_{VL} increasing with decreasing σ_r/σ_v. Thus, the lower the confining pressure, the greater the induced vertical strain. Comparing the position of the curve for the oedometer test with those from the triaxial test indicates that the value of σ_r/σ_v in the oedometer sample was consistently between ⅔ and 1 for stresses below the compactive prestress ($\sigma_{vc} \cong 600$ kPa = 87 psi) but decreased at stresses above σ_{vc} to a value of about 0.5 at σ_v = 1600 kPa (232 psi).

These results suggest that the results of oedometer tests can be used to estimate three-dimensional settlements for dry, stiff compacted cohesive soils at low applied stresses but that more sophisticated techniques are needed for higher stresses or wet, soft soils. Some methods that might be used include the following:

1. The Skempton and Bjerrum (1957) method for correcting one-dimensional settlement of saturated cohesive soils for three-dimensional conditions might be used for soils with $S_r > S_{opt}$ because the induced pore pressures in the compacted soil may be similar to those for saturated soil. However, as noted by Lambe and Whitman (1979), Skempton and Bjerrum's method assumes a discontinuous—and therefore incorrect—stress path and tends to underestimate settlement.

2. The stress-path method (Lambe, 1967) could be employed and would involve conducting stress-path triaxial tests on a number of samples that duplicate the expected stress paths in the soil at various depths beneath the foundation. The major problem with using the stress-path method is that the induced changes in vertical and horizontal stress must be predicted, and the predicted values may vary substantially from the actual changes in stress.

3. A series of plate-load tests using at least three sizes of plates could be conducted in the field-compacted soil at the expected bearing level, and the results could be extrapolated to full field scale.

4. A finite element settlement analysis could be performed (see Sec. 5A.2.3 for recommended procedures).

Ultimate Bearing Capacity Analyses Any appropriate method for estimate ultimate bearing capacity can be used (see Sec. 2B). It is important that the strength parameters (friction angle and cohesion intercept) be consistent with the desired analysis, that is, undrained strength parameters for short-term bearing capacity and drained strength parameters for long-term bearing capacity.

Final Comments One might conclude from the previous discussions that it is best to compact cohesive bearing soils at low water contents to maximize strength and minimize loading-induced settlement. Although dry cohesive soils are stronger and stiffer than wet (all other factors being the same), the potential for wetting-induced and drying-induced volume changes must also be considered, as will be discussed. Thus, a settlement analysis for a foundation bearing on a compacted cohesive soil must consider the potential for settlement or heave from loading, wetting, and drying throughout the design life of the structure.

Soaked

In some instances it is highly likely that compacted cohesive soils will become wetter at some time during the design life of the structure founded on these soils. Wetting may occur from obvious sources such as precipitation (rainfall and snowmelt) and changes in regional or local groundwater conditions (rising of the ground-water table, diversion of runoff, and so on). Wetting may also result from less obvious sources, including broken or leaking water pipes, landscape irrigation, and clogged drainage systems. Therefore, it is nearly impossible to ensure that wetting will not occur in cohesive bearing soils, which prompts many engineers to estimate settlement and ultimate bearing capacity based on soaked compressibility and strength characteristics of the soil.

Compressibility Wetting of a compacted cohesive soil generally results in some volume change (swelling or collapse, as will be discussed), which must be considered in the overall settlement/heave analysis, and which also changes the compressibility and strength of the soil. The changes in compressibility and strength depend on the prewetting moisture condition and the

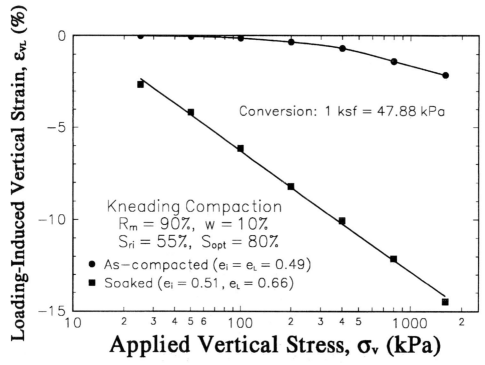

FIGURE 5A.88 Effect of soaking on one-dimensional compressibility of dry compacted clayey sand (data from Lawton, 1986).

magnitude and sign of the wetting-induced volume change (if any). For example, a soil that swells when wetted becomes more compressible owing to both the reduction in suction and the increase in void ratio. The significant increase in one-dimensional compressibility caused by wetting a high-density (R_m = 90%), dry (S_r = 55%) compacted clayey sand is illustrated in Fig. 5A.88. After inundation at a very low applied stress (σ_v = 1 kPa = 0.02 ksf), the specimen swelled 14.4%, which changed the void ratio from 0.51 in the as-compacted condition (e_i) to 0.66 at the beginning of the loading process (e_L). In addition, the matric suction in the specimen was reduced by soaking from about 800 kPa (16.7 ksf) to essentially zero. At an applied stress of 1600 kPa (33.4 ksf), the soaked sample compressed 14.5%—about 7 times as much as the as-compacted sample (2.1%). Soaking the specimen eliminated the prestressing induced during compaction, as indicated by the essentially linear strain-log stress relationship for the soaked sample.

Cohesive soils compacted to high degrees of saturation will undergo little or no wetting-induced volume change and their compressibility will not be appreciably affected by soaking. This is illustrated in Fig. 5A.89 for a compacted clayey sand with $S_r \cong S_{opt}$ that swelled only 0.4% after soaking. The compressibility of this soaked specimen did not differ significantly from that of the as-compacted specimen. In addition, the soaking had no noticeable effect on the compactive prestress.

Soaking can also induce a decrease in void ratio (collapse) in the soil. This decrease in void ratio tends to reduce the compressibility, but in most instances this reduction is more than offset by the increase in compressibility produced by the reduction in matric suction.

Strength The soaked strength of compacted cohesive soils depends on many factors, including the following:

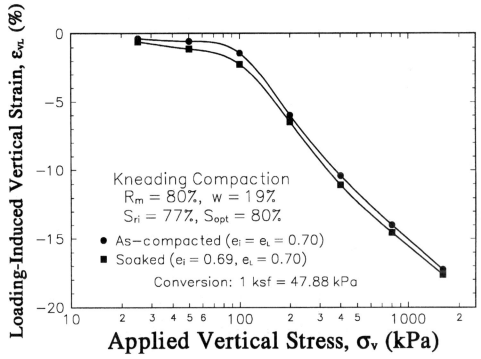

FIGURE 5A.89 Influence of soaking on one-dimensional compressibility of wet compacted clayey sand (data from Lawton, 1986).

1. Fabric (microfabric, minifabric, and macrofabric)
2. Method of compaction
3. Compaction water content
4. Soaked water content and degree of saturation
5. Volume change induced by the soaking
6. Dry density or void ratio after soaking
7. Magnitude and sign of the pore water pressures that develop during loading

Most of these factors are interdependent. Because these interrelationships are quite complicated, only general discussions will be given below. Further details can be found in Seed and Chan (1959b) and Yi (1991).

Effect of fabric The fabric of the soaked soil at the initiation of loading can have a substantial effect on the strength. It is generally assumed that the fabric of the soaked soil is essentially the same as that of the as-compacted soil, but some changes in fabric may occur because of soaking, and further study is needed of this issue. There is no doubt, however, that the soaked fabric depends to some extent on the as-compacted fabric.

The fabric affects the undrained strength of soaked soil primarily in terms of the developed pore pressures. The results from consolidated-undrained triaxial compression tests on two specimens of compacted silty clay are given in Fig. 5A.90. The two specimens were prepared by kneading compaction to different as-compacted states—one dry of optimum and one wet of optimum, but after soaking the specimens had the same composition (γ_d, w). The dry-of-optimum specimen

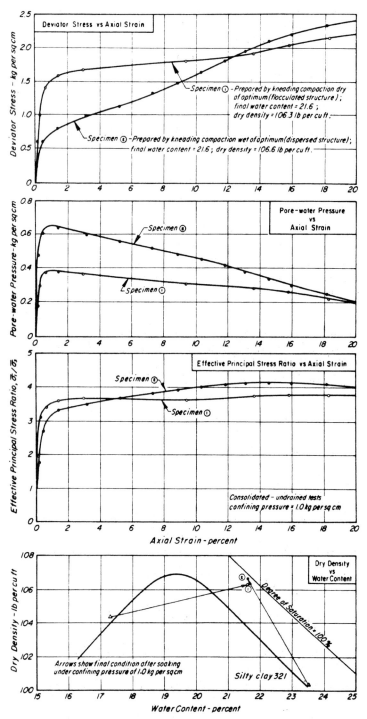

FIGURE 5A.90 Effect of microfabric on pore water pressure and strength from triaxial consolidated-undrained tests on a silty clay (from Seed et al., 1960).

had an initially flocculated microfabric and developed a high strength at low strain ($\cong 2\%$) and thereafter showed only a small increase in strength with additional deformation. The wet-of-optimum specimen had an initially dispersed microfabric, and at strains less than about 12.5%, appreciably larger pore water pressures were developed than in the dry-of-optimum specimen, with the result that the wet-of-optimum specimen was weaker. These differences in developed pore water pressures result from the different shapes of the pore spaces in the microfabric. The magnitude of developed pore water pressure in any void space is inversely related to the permeability of that void space, and the permeability of a void space depends to a great extent on its smallest dimension. Thus, for two void spaces with the same volume and identical loading conditions, greater pore water pressures will be induced in a void space with a long, narrow shape (dispersed microfabric) than in a more equidimensionally shaped void space (flocculated microfabric). At strains greater than 12.5%, the differences in pore water pressures were small and the strengths of the two specimens were nearly the same, suggesting that the microfabric in the zones of the failure planes for the two specimens were essentially the same (dispersed). The effective principal stress ratios for both specimens were nearly the same at all strains, suggesting that the effective stress strengths were similar for the two specimens so that the differences in total stress strengths (load-carrying capacity) were primarily the result of differences in developed pore water pressures.

Yi (1991) conducted a similar study on soils that collapsed when wetted. The results of consolidated-undrained simple shear tests conducted on specimens of a clayey sand that were compacted to two different as-compacted states and then soaked to the same γ_d and w are shown in Fig. 5A.91. The dry-of-optimum specimens collapsed on wetting, whereas the wet-of-optimum specimens underwent no volume change. The results clearly show that at all three vertical confining pressures, the dry-of-optimum specimens developed higher pore water pressures and were weaker than the wet-of-optimum specimens. Similar results were obtained for three other soils. These results cannot be explained in terms of flocculated versus dispersed microfabric and are opposite to those found by Seed and Chan (1959b) for the silty clay discussed above. It is not known if these differences in results are related to differences in types of soils, type of wetting-induced volume change (swelling versus collapse), or perhaps differences in sample preparation (although kneading compaction was used in both studies). Scanning electron micrographs of the dry-of-optimum clayey sand specimens showed that the effect of collapse on fabric was to reduce the size of the interclod pores without any apparent change to the microfabric. Unfortunately, no micrographs of the wet-of-optimum specimens were taken, so the characteristics of the fabric for these specimens are not known. Yi assumed both that the soaked fabrics of the wet- and dry-of-optimum specimens were similar, with the wet-of-optimum specimens containing larger interclod pores and therefore denser microfabric, and that the strengths of these specimens were governed by the density of their microfabric. The author believes that the soaked fabrics of the wet- and dry-of-optimum specimens were probably not as similar as assumed by Yi, so that the explanation for the differences in strength is probably quite complex and cannot be determined from the available information.

Clearly, fabric can play an important role in the soaked strength of compacted cohesive soils. However, with our present knowledge and the uncertainty regarding the relative importance of microfabric and minifabric, additional research is needed before a reasonable explanation can be given for the influence of fabric on soaked strength.

Influence of wetting-induced volume change The volume changes that occur from soaking a compacted cohesive soil can be summarized as (*a*) no change in volume, (*b*) an increase in volume (swelling), and (*c*) a decrease in volume (collapse). Compacted cohesive soils that undergo no volume change after soaking are a special case and may be of two types—nonexpansive soils that are soaked under a low surcharge pressure and expansive soils soaked under a surcharge pressure sufficiently high to prevent swelling.

The undrained strength of a compacted cohesive soil that has been soaked depends on the characteristics of the soil in the soaked condition—dry density, water content, and fabric. These soaked characteristics depend on the as-compacted characteristics of the soil (γ_d, w, and fabric) as well as the stress state of the soil at the time of soaking, as discussed previously.

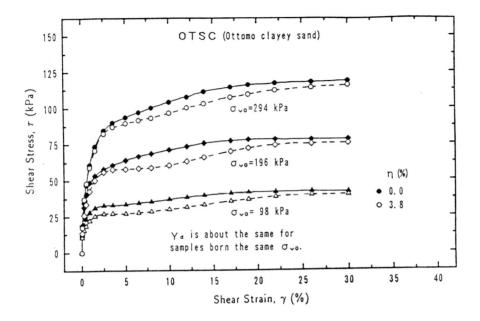

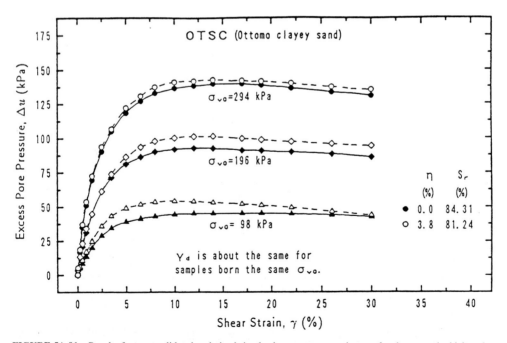

FIGURE 5A.91 Results from consolidated-undrained simple shear tests on specimens of a clayey sand which underwent different amounts of collapse (from Yi, 1991).

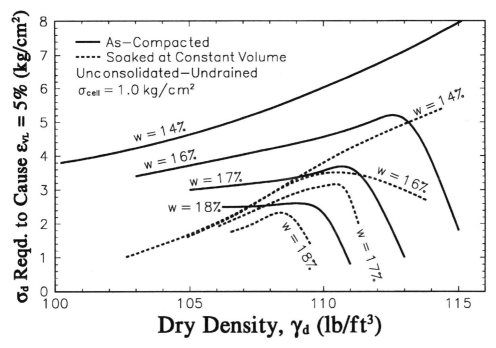

FIGURE 5A.92 Influence of soaking on the triaxial compressive strength of kneading-compacted silty clay (data from Seed and Chan, 1959a,b).

The influence of soaking at constant volume on the unconsolidated-undrained (UU) triaxial compressive strength at low strain (ε_{VL} = 5%) of a silty clay prepared by kneading compaction is shown in Fig. 5A.92. At all densities and water contents, the soaked soil is weaker than the comparable as-compacted soil, with the differences decreasing with increasing compaction water content.

Isograms of triaxial UU triaxial compressive strength at low strain are shown in Fig. 5A.93 for kneading-compacted specimens of sandy clay soaked under two different surcharge loads. Under the low surcharge pressure (1 psi = 7 kPa), the soaked strength depended significantly on the compaction water content owing to the differences in swelling that occurred upon soaking. At lower water contents the soil swelled more than at higher water contents, and thus for the same as-compacted dry density, the dry density after soaking was less and the water content was more for the initially dryer soil. Thus, for the same dry density, the soaked strength increased significantly with increasing compaction water content. Note that this relationship is the opposite of that for as-compacted strength, which decreases with increasing compaction water content at the same dry density (see Fig. 5A.92).

The high-surcharge specimens underwent little swelling, and the corresponding soaked-strength isograms are nearly horizontal, indicating that the strength depended primarily on as-compacted density and was nearly independent of compaction water content. Dry of optimum ($S_r < S_{opt}$), the strength at constant as-compacted dry density increases slightly with increasing compaction water content owing to slightly more swelling at the lower compaction water contents. Note, though, that the strength decreases slightly with increasing water content wet of optimum ($S_r > S_{opt}$) because the microfabric became increasingly more dispersed but no swelling occurred.

There is no information available with regard to the influence of the magnitude of collapse on the soaked strength of compacted cohesive soils. For a given composition and fabric, increasing the applied stresses causes greater wetting-induced collapse and thus increases the density of the

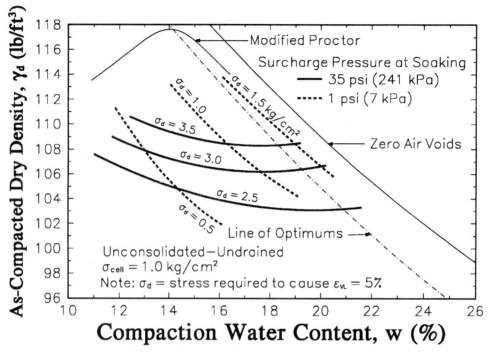

FIGURE 5A.93 Influence of surcharge pressure at soaking on the triaxial compressive strength at low strain for kneading-compacted sandy clay (data from Seed and Chan, 1959a,b).

soil, which likely increases the soaked strength of the soil. However, differences in magnitude of collapse may produce variations in fabric, which in turn may affect the soaked strength of the soil.

Method of compaction For a cohesive soil compacted to the same composition, varying the method of compaction may alter the soaked strength in two ways: (*a*) The as-compacted fabric (and hence the soaked fabric) may be different, and (*b*) the wetting-induced volume change may vary. As noted previously, flocculated microfabrics are stronger than dispersed microfabrics, so, other than very dry of optimum where compaction method has little effect on fabric, static compaction would be expected to produce a stronger soil than kneading compaction. However, the wetting-induced volume change also affects the soaked strength. Static compaction induces a higher swelling potential or a higher collapse potential in the soil than does kneading compaction, depending on the applied stress level (see discussion below). As discussed above, greater swelling reduces the soaked strength of the soil while greater collapse likely increases it. Thus, for conditions where swelling is likely to occur (highly expansive soils and low stresses), static compaction may result in greater or lesser soaked strength than kneading compaction depending on which mechanism controls. For nonexpansive soils or high surcharge pressures, either no volume change or collapse is likely to occur from wetting, and static compaction will probably produce a stronger soaked soil.

Wetting-Induced Volume Changes. *Swelling*, which is a decrease in total volume of a soil upon wetting, and *collapse*, which is an increase in total volume upon wetting, are considered separate and distinct phenomena by many engineers and researchers. Recent research has shown, however, that any unsaturated cohesive soil may either collapse, swell, or not change volume when wetted, depending on the condition of the soil at the time of wetting (Lawton et al., 1996). Thus, swelling and collapse in unsaturated cohesive soils are associated aspects of the general phenomenon of wetting-induced volume changes.

The magnitude and sign of the volume change that occurs when an unsaturated cohesive soil is wetted are primarily related to changes in matric suction. As suction decreases during wetting, the change in pore water pressure is positive (the pore water pressure becomes less negative), which reduces both the effective stresses acting on the soil and the stiffness of the soil structure. The decrease in effective stress results in an increase in volume (swelling), whereas the reduction in stiffness produces a decrease in volume (collapse). The net change in volume may be positive, negative, or zero depending on the condition of the soil at the time of wetting.

The primary factors that affect the wetting-induced volume change in a compacted cohesive soil are as follows (Lawton et al., 1992, 1993):

1. Gradation of the soil and mineralogy of the clay particles
2. State of stress at the time of wetting
3. Void ratio or dry density at the time of wetting
4. Moisture condition at the time of wetting
5. Degree of wetting
6. Fabric of the soil at the time of wetting
7. Chemistry of the water that is introduced into the soil

Soil Type The effect of clay content on the one-dimensional wetting-induced volume change of compacted silt-clay and sand-clay mixtures was studied by El Sohby and Rabbaa (1984), with the results shown in Fig. 5A.94. The clay was highly expansive, with calcium montmorillonite the predominant mineral (45% by weight). For any given soil type (same silt or sand and clay fraction), the net collapse increased or the net swelling decreased with increasing applied vertical stress. For the applied stress levels studied (σ_v = 400, 800, or 1600 kPa or 8.4, 16.7, or 33.4 ksf), the specimens containing low clay fractions collapsed and those containing high clay fractions

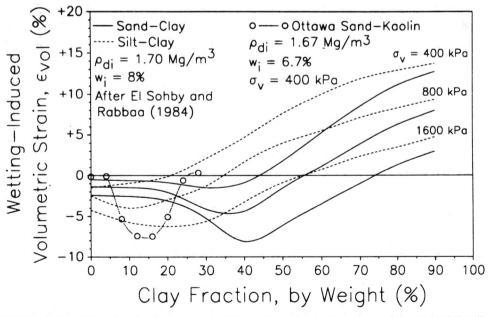

FIGURE 5A.94 Influence of clay fraction on potential for wetting-induced volume change of clay-sand and clay-silt mixtures (from Lawton et al., 1992).

swelled. The magnitude of swelling at high clay fractions increased with increasing clay content. At low clay contents, the silt-clay collapsed more than the sand-clay; at moderate to high clay contents, the silt-clay collapsed less or swelled more than the sandy clay. The maximum collapse occurred at clay contents of 10 to 20% for the silt-clay and 30 to 40% for the sand-clay.

A similar study was conducted by the author to determine the influence of clay content on the wetting-induced volume change of mixtures of uniform sand and slightly expansive clay (kaolin). The results from this study are also shown in Fig. 5A.94 and can be summarized as follows:

1. At very low clay contents (<5%), little or no wetting-induced volume change occurred.
2. At clay contents from between 5 and 25%, collapse occurred, reaching a maximum between 12 and 16%.
3. At clay contents greater than 25%, a small amount of swelling occurred.

The effects of soil type on the potential for wetting-induced volume change in cohesive soils can be summarized as follows (Lawton et al., 1992):

1. *Very low clay contents.* The clay acts as a binder between silt and sand particles. When wetted, the clay binders soften and lubricate the intergranular contacts, thereby facilitating collapse at high applied stresses. If the clay is highly expansive, there may be some swelling at low pressures and high densities, but the magnitude will likely be small because the clay content is too low for the soil to have appreciable swelling. Silts tend to collapse more than sands.

2. *Low to moderate clay contents.* With increasing clay content, more clods are present in the borrow material, and for the compaction water contents at which significant collapse and swelling potentials exist, the as-compacted soil retains to some extent the identities of the individual clods. Upon wetting, the clods swell but also soften and distort under the applied load, so that the net volume change depends on the stress state and the expansiveness of the clay minerals. Swelling tends to occur at low stresses and for expansive clay minerals, and collapse tends to occur at high stresses and for nonexpansive clay minerals.

3. *Moderate to high clay contents.* Although some collapse occurs during wetting, swelling generally predominates over collapse, especially for highly expansive clay minerals and low stresses. For highly expansive clay minerals and high clay contents, swelling can occur at overburden pressures as great as 1600 kPa (33.4 ksf).

Density and Moisture Condition The effects of prewetting density and prewetting moisture condition on the one-dimensional wetting-induced volume change of a compacted clayey sand subjected to an applied total vertical stress (σ_v) of 400 kPa (8.35 ksf) are shown in Fig. 5A.95. The prewetting dry density is represented as a percentage of modified Proctor maximum dry density, R_{mw}, and the prewetting moisture condition is expressed as degree of saturation, S_{rw}. The values of wetting-induced volumetric strain, ε_{vw}, shown in Fig. 5A.95 are for complete inundation with water and therefore represent the maximum potential for wetting-induced volume change. For the same S_{rw}, ε_{vw} is more positive (greater swelling) or less negative (less collapse) as R_{mw} increases because the stiffness of the soil increases with increasing dry density. Thus, for the same moisture condition and total vertical stress, the swelling resulting from the reduction in effective stresses would be the same, although the collapse resulting from the reduction in stiffness would be less, with a net increase in swelling or decrease in collapse. For the same R_{mw}, the matric suction in the soil decreases with increasing moisture, with the result that the net swelling or net collapse decreases as S_{rw} increases up to the value of S_{rw} at which the suction is zero (typically in the range of 70 to 90%). Therefore, for the same dry density and total vertical stress, the net wetting-induced volume change—whether positive or negative—will be less for an initially wetter soil because the magnitude of the dominating mechanism—reduction in effective stress for net swelling or reduction in stiffness for net collapse—will be less for lower values of prewetting suction.

For the same prewetting moisture condition (S_{rw}), an increase in dry density either reduces the collapse potential or increases the swelling potential. Collapse in most compacted cohesive soils

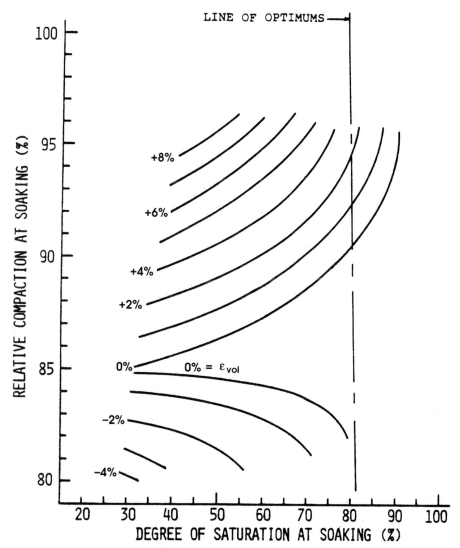

FIGURE 5A.95 Isograms of one-dimensional wetting-induced vertical strain for an applied vertical stress of 200 kPa (4.2 ksf) as a function of prewetting relative compaction and prewetting degree of saturation (from Lawton, 1986).

seems to occur primarily as a reduction in volume of the interclod voids (Yi, 1991). In a denser soil, the interclod voids are smaller. After wetting, the soil will come to an equilibrium density based on the state of stress. Therefore, less reduction in volume will occur for a denser soil with the same prewetting moisture condition. Swelling is a complex phenomenon that cannot be easily characterized. A greatly simplified explanation for the influence of density on swelling potential can be given as follows:

1. For a given saturated cohesive soil with the same soil fabric and water chemistry, there is a unique relationship between void ratio and state of stress.

2. For the same fabric and moisture condition of the prewetting soil, when soaked the soil will come to an equilibrium void ratio that depends on the state of stress acting on the soil but is independent of the prewetting density. Therefore, the increase in void ratio caused by soaking will be greater for an initially denser soil.

The wetting-induced volume change in compacted cohesive soils can also be zero. In compacted cohesive soils, this can occur in three ways: (a) The swelling and collapsing components are equal and therefore offset each other; (b) the prewetting suction in the soil is small enough that no significant reduction in effective stress or stiffness will occur upon wetting; or (c) cementation of particles is strong enough to preclude any wetting-induced volume change. Cementation of particles may exist in clods prior to compaction and may survive the compaction process to some extent. Cementation may also occur naturally with time after compaction or may be intentionally induced in the soil by chemical stabilization prior to compaction. These latter two possibilities are discussed in more detail in subsequent sections.

For a cohesive soil loaded and wetted shortly after compaction (before any significant postcompaction cementation occurs), there is a range of prewetting dry densities and degrees of saturation for which there is no wetting-induced volume change, as illustrated by the zone in Fig. 5A.95 bounded by the two isograms for $\varepsilon_{vw} = 0$. No wetting-induced volume change occurs at low degrees of saturation because the swelling and collapsing components offset each other, and in Fig. 5A.95 this occurs at $R_{mw} \cong 85\%$. At degrees of saturation equal to or greater than a critical value (S_{rc}), the prewetting suction is small, and $\varepsilon_{vw} \cong 0$. This phenomenon is indicated in Fig. 5A.95 by the increase in distance between the two isograms for $\varepsilon_{vw} = 0$ at higher values of S_{rw}. S_{rc} varies as a function of soil type, prewetting dry density, and prewetting state of stress. For a given soil, the influence of density and state of stress is small, and S_{rc} can be approximated as S_{opt} obtained from Proctor tests or can be determined more accurately by conducting a series of one-dimensional wetting-induced volume change tests.

Another important characteristic illustrated by the isograms of Fig. 5A.95 is that for any soil and any value of σ_v, there is range in values of R_{mw} for which $\varepsilon_{vw} = 0$ regardless of moisture condition. The value of R_{mw} at the middle of this range is called the *critical prewetting relative compaction* (R_{mwc}), and for a given soil R_{mwc} increases as σ_v increases (Fig. 5A.96).

STATE OF STRESS   The effect of increased stress for the same prewetting density, moisture condition, and fabric is to reduce the swelling potential or increase the collapse potential. This relationship is illustrated in Fig. 5A.97, where isograms of one-dimensional wetting-induced strain are given for a compacted clayey sand at two levels of overburden pressure. At each point in $R_{mw} - S_{rw}$ space, ε is less positive or more negative at $\sigma_v = 800$ kPa than at $\sigma_v = 400$ kPa.

For three-dimensional strain conditions, the total wetting-induced volume change is uniquely related to the mean normal total stress (Lawton et al. 1991) and thus is independent of the principal stress ratio, as illustrated in Fig. 5A.98 for wetting tests on kneading-compacted specimens of a compacted clayey sand conducted under axisymmetric loading conditions. However, the individual components of wetting-induced volume change depend significantly on the stress ratios. For a given magnitude of stress in a direction of interest, the swelling increases or the collapse decreases in that direction as the perpendicular stresses increase. This concept is illustrated in Fig. 5A.99, where vertical and radial wetting-induced strains are plotted as a function of vertical stress. An increase in lateral stress ratio (σ_r/σ_v) for any given value of vertical stress results in an increase in vertical strain (increase in swelling or decrease in collapse) and a decrease in radial strain (decrease in swelling or increase in collapse).

Results from wetting tests on oedometer specimens compacted at the same water content to the same density using the same method of compaction are also shown in Fig. 5A.99(a) for comparison with the triaxial results. This comparison clearly indicates that the one-dimensional wetting tests may not give reliable predictions for situations where three-dimensional strains occur. It is also apparent that for the range of lateral stress ratios studied ($\sigma_r/\sigma_v = 1/3$ to 1), collapse was affected more by the stress ratio than swelling. In addition, the results appear to indicate that the oedometer tests significantly overestimate the swelling for three-dimensional volume change. However, the process of application and removal of loads during compaction can result in significant lateral earth pressures, which can be many times greater than the theoretical at-rest values

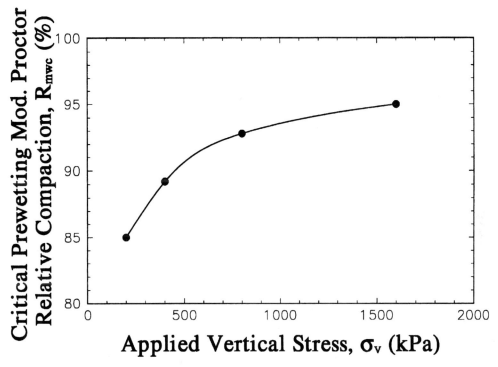

FIGURE 5A.96 Effect of vertical stress on critical prewetting density (data from Lawton, 1986).

and may approach limiting passive earth pressure values (Duncan and Seed, 1986). Thus, the values of σ_r/σ_v for as-compacted soils may be much greater than the maximum value of 1.0 shown in Fig. 5A.99. In addition, for a constant vertical stress, the horizontal stresses may vary substantially as the soil undergoes wetting-induced volume change. Preliminary research indicates that the horizontal stress in a compacted soil increases with increasing vertical collapse and decreases with increasing vertical swelling, that the limiting passive horizontal pressure may be approached for large vertical collapse, and that limiting active horizontal pressure may be achieved for large vertical swelling (Lawton et al., 1991). Thus, laboratory tests in which σ_r/σ_v is kept constant do not accurately model the stress path of soils in the field that undergo three-dimensional wetting-induced volume change, and additional research is needed to clarify the interdependence of horizontal stress and wetting-induced volume change of soil.

METHOD OF COMPACTION The influence of compaction method on the one-dimensional wetting-induced volume change of a clayey sand is shown in Fig. 5A.100 for three different as-compacted compositions and can be summarized as follows:

1. For the same water content, density, and state of stress, the statically compacted specimens swelled more than the kneading-compacted specimens.
2. Within the lower range of stresses at which collapse occurred, the magnitude of collapse was less for the statically compacted specimens than for the kneading-compacted specimens.
3. At stresses near or greater than the stresses at which maximum collapse would occur for any given composition, the statically compacted specimens collapsed more than the kneading-compacted specimens. This trend is not observed for $R_m = 90\%$ and $w = 10\%$ because the maximum stress on the specimens was less than the stress at which maximum collapse would occur.

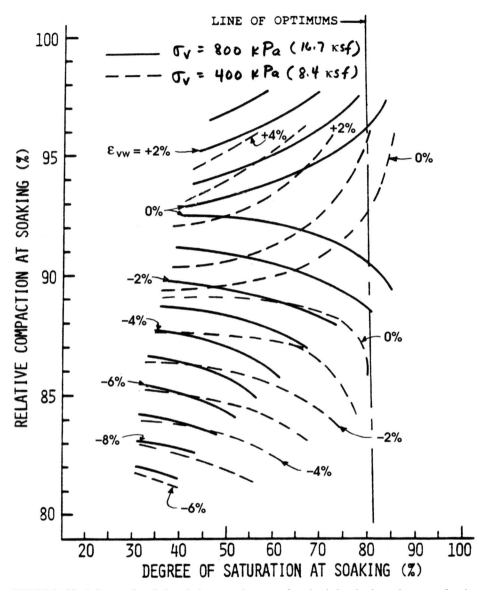

FIGURE 5A.97 Influence of applied vertical stress on isograms of wetting-induced volume change as a function of prewetting relative compaction and prewetting degree of saturation for a compacted clayey sand (data from Lawton, 1986).

Statically compacted cohesive soil can be either more or less collapsible than the comparable kneading-compacted soil because static compaction tends to produce a fabric that is susceptible to both more swelling and more collapse. Recall from previous discussions that both swelling and collapse occur when an expansive soil is wetted. At low stresses, the swelling component clearly dominates, and swelling is greater for static compaction than for kneading compaction. At high stresses, the collapsing component dominates and the statically compacted soil collapses more than the kneading-compacted soil. There is a transition zone for moderate stresses wherein the net wetting-

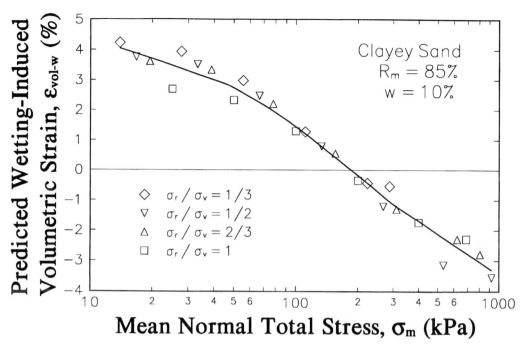

FIGURE 5A.98 Predicted wetting-induced volumetric strain as a function mean normal total stress from double triaxial tests on kneading-compacted specimens of a clayey sand (from Lawton et al., 1991).

induced volume change is negative (collapse) but the tendency for greater swelling in the statically-compacted soil results in less collapse than for the kneading-compacted soil.

Although method of compaction can have a measurable effect on ε_{vw}, it is apparent from the results in Fig. 5A.100 that w and σ_v influence ε_{vw} to a much greater degree than method of compaction. To illustrate this point, examples of changes in ε_{vw} produced by changes in w, σ_v, R_m, and method of compaction using $\varepsilon_{vw} = +8.6\%$ (static compaction, $R_m = 90\%$, $w = 10\%$, $\sigma_v = 25$ kPa) as the base are as follows:

Change	ε_{vw}, %	$\Delta\varepsilon_{vw}/\varepsilon_{vw-base}$, %
Increase w to 16%	+1.4	−84
Increase σ_v to 200 kPa	+3.1	−64
Decrease R_m to 80%	+4.4	−49
Change to kneading compaction	+7.0	−19

This discussion is not meant to imply that method of compaction is unimportant; rather that with respect to wetting-induced volume change in cohesive soils, it is a relatively minor factor compared to w and σ_v. In addition, it should be borne in mind that specimens which will be tested in the laboratory should always be prepared with a method of compaction that simulates as closely as possible the method of compaction to be used in the field compaction process.

DEGREE OF WETTING The degree of wetting has a substantial influence on the wetting-induced volume change; that is, the results of wetting tests in which the specimen is inundated with water represent the maximum wetting-induced volume change for the conditions under which the specimen was tested. Partial wetting reduces the matric suction somewhat but to a lesser degree than full wetting. Thus, the magnitude of wetting-induced volume change, which is significantly influ-

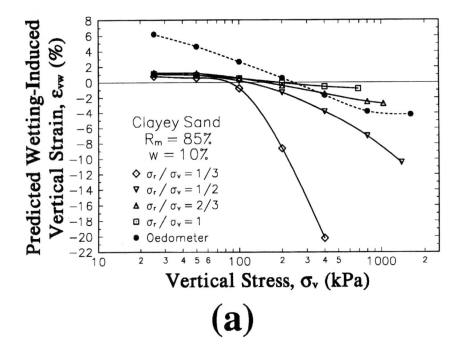

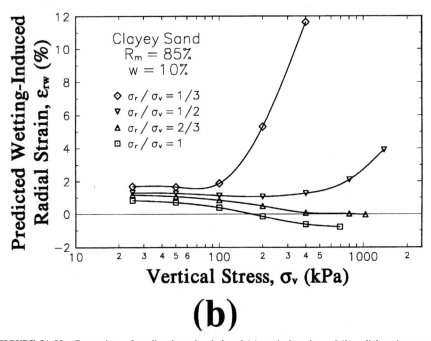

FIGURE 5A.99 Comparison of predicted wetting-induced (*a*) vertical strain, and (*b*) radial strain versus vertical stress curves from double triaxial and double oedometer tests on kneading-compacted specimens of a clayey sand (from Lawton et al., 1991).

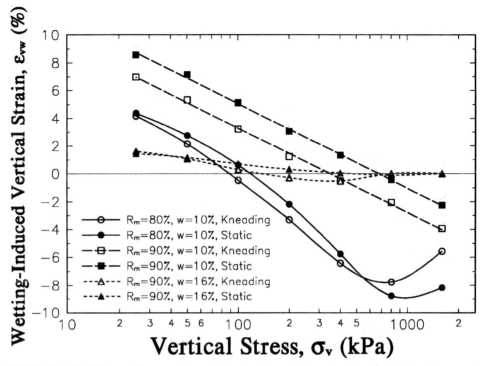

FIGURE 5A.100 Influence of compaction method on one-dimensional wetting-induced strain for a clayey sand (data from Lawton, 1986).

enced by the changes in suction, will be less for partial wetting than for full wetting. This is illustrated in Fig. 5A.101 for collapse of three naturally deposited soils. Note that the collapse for the three soils shown in Fig. 5A.101 is essentially completed at degrees of saturation less than 100%, typically about 80%. Similar trends have been noted by the author for the wetting-induced volume change (both collapse and swelling) of compacted cohesive soils. In general, little or no wetting-induced volume change occurs in compacted cohesive soils for additional wetting above S_{opt}.

OVERSIZE PARTICLES When tests are conducted in the laboratory to estimate the potential for wetting-induced volume changes in field-compacted soils, the specimens used for testing are generally small, and the maximum particle size that can be used in a specimen is limited to approximately one-sixth the least dimension of the specimen. In this context, the term *oversize* refers to sizes of particles that are larger than the maximum size that can be used in a specimen for any particular test, and the maximum size will therefore vary depending on the test being performed and the size of the sample being tested. In general, the oversize particles are gravel size and larger. As with other types of tests, the results from laboratory wetting-induced volume change tests should be corrected to account for the effects of the oversize particles that are present in the field-compacted soil but not in the specimens tested in the laboratory.

A common method of correcting for the lack of oversize particles in laboratory wetting tests is based on the following two assumptions:

1. The oversize particles are inert and therefore do not change volume when wetted.
2. The presence of the oversize particles in the borrow material during field compaction does not influence the potential for wetting-induced volume change within the matrix soil (the soil exclusive of the oversize particles).

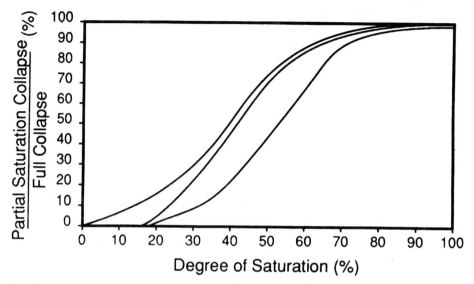

FIGURE 5A.101 Partial collapse curves for three collapsible silts (from Walsh et al., 1993).

With these two assumptions, the results from laboratory tests on specimens prepared at the same compaction water content as the field matrix soil to the same dry density as the field matrix soil using the same (or similar) method of compaction and loaded to the same state of stress can be used to predict the potential for wetting-induced volume change of the field-compacted soil. The dry density and water content of the field matrix soil (γ_{dm} and w_m) can be calculated from the following equations, which are derived from phase relations:

$$\gamma_{dm} = \frac{(1 - P_{ow}) \cdot G_{so} \cdot \gamma_w \cdot \gamma_d}{G_{so} \cdot \gamma_w - P_{ow} \cdot \gamma_d(1 + G_{so} \cdot w_o)} \tag{5A.37}$$

$$w_m = \frac{w - w_o \cdot P_{ow}}{1 - P_{ow}} \tag{5A.38}$$

where w = water content of the field-compacted soil
 γ_d = dry density of the field-compacted soil
 P_{ow} = fraction of the soil that is oversized particles, by dry weight, expressed as a decimal
 G_{so} = specific gravity of the oversize particle solids
 w_{os} = water content of the oversize particle fraction
 γ_w = density of water

w_{os} can be estimated by removing the oversize particles from a sample of the field-compacted soil and scraping the surfaces of the oversize particles to remove any cohesive material that is adhered to them. This can be a cumbersome task, and a value for w_{os} is usually estimated in the range of 0.5 to 2%. The following example illustrates the calculation of γ_{dm} and w_m for the preparation of laboratory specimens for wetting tests (after Noorany and Stanley, 1994): Given that $w = 11.7\%$, $\gamma_d = 112.1$ pcf (17.61 kN/m³), $P_{ow} = 15\%$, $G_{so} = 2.66$, and w_{os} assumed to be 1%, calculate values of γ_{dm} and w_m for the field-compacted soil:

$$\gamma_{dm} = \frac{(1 - 0.15)(2.66)(62.43)(112.1)}{(2.66)(62.43) - (0.15)(112.1)[1 + (2.66)(0.01)]} = 106 \text{ pcf} \left(16.7 \frac{\text{kN}}{\text{m}^3}\right)$$

$$w_m = \frac{0.117 - (0.01)(0.15)}{1-0.15} \cdot 100 = 13.6\%$$

Some preliminary research has been conducted to determine the validity of this method for predicting wetting-induced volume change of field-compacted soil. Day (1991) studied the effect of gravel content on the magnitude of one-dimensional swelling of specimens with an expansive silty clay matrix. The wetting tests were conducted at low stresses—$\sigma_v = 0.1$ to 0.2 psi (0.7 to 1.4 kPa)—and swelling occurred in all tests. The results from these tests are given in Fig. 5A.102 in the form of the wetting-induced strain for the specimens containing gravel normalized with respect to the wetting-induced strain for the comparable specimen with no gravel at the same applied stress ($\varepsilon_{vw}/\varepsilon_{vw0}$). Also shown in Fig. 5A.102 is the theoretical relationship for $\varepsilon_{vw}/\varepsilon_{vw0}$ obtained from the following equation, which was derived from the same assumptions discussed above except that the gravel was assumed to be dry:

$$\frac{\varepsilon_{vw}}{\varepsilon_{vw0}} = \frac{(1 - P_{ow}) \cdot G_{so} \cdot \gamma_w}{(1 - P_{ow}) \cdot G_{so} \cdot \gamma_w + P_{ow} \cdot \gamma_{dm}} \qquad (5A.39)$$

A comparison of the experimental and theoretical results suggests that this method provides a reasonable estimate for predicting the one-dimensional swelling potential of compacted cohesive soils.

A similar study was conducted by the author and his associates (Larson et al., 1993) with regard to the influence of oversize particles on the one-dimensional collapse of compacted clayey sands. Experimental values of $\varepsilon_{vw}/\varepsilon_{vw0}$ from this study are plotted in Fig. 5A.103 as a function of percentage of oversize particles by volume (P_{ov}). The experimental results are compared with the

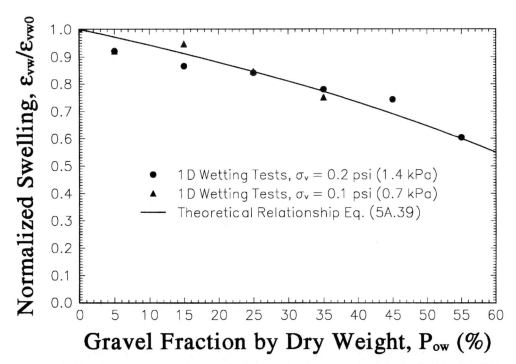

FIGURE 5A.102 Effect of gravel content on one-dimensional swelling of a compacted silty clay (modified from Day, 1991).

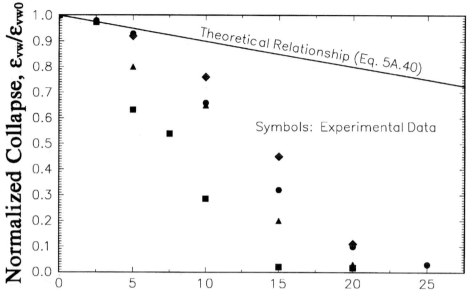

FIGURE 5A.103 Influence of oversize particles on the collapse of compacted clayey sands (from Larson et al., 1993).

following theoretical equation, which is equivalent to Eq. (5A.39) except that $\varepsilon_{vw}/\varepsilon_{vw0}$ is given in terms of P_{ov} rather than P_{ow}:

$$\frac{\varepsilon_{vw}}{\varepsilon_{vw0}} = 1 - P_{ov} \qquad (5A.40)$$

For the soils and conditions studied, the theoretical relationship is valid only for small oversize fractions ($P_{ov} \cong 2.5$ to 5%; $P_{ow} \cong 4$ to 8%), thus indicating that at higher contents of oversize particles, the presence of the oversize particles changes the fabric of the matrix soil, the manner in which collapse occurs, or both. The magnitudes of collapse for the five specimens with $P_{ov} \geq 20\%$ ($P_{ow} \geq 28\%$) were less than 11% of the values for the comparable specimens without any oversize particles.

Why this method of correcting for oversize particles apparently works for swelling but not for collapse is not well understood. We know that the oversize particles affect the fabric of the as-compacted soil (e.g., Siddiqi et al., 1987; Fragaszy et al., 1990). In soils with P_{ow} less than about 40%, the oversize particles "float" in the matrix soil, as shown in Fig. 5A.104; therefore, there is little or no interaction between oversize particles, and the behavior of the soil is controlled by the matrix soil. Soils with the same overall matrix density and water content but varying amounts of oversize particles will have different fabrics, with the matrix soil adjacent to the surfaces of the oversize particles (near-field matrix) having higher void ratios than the matrix soil that is distant from the oversize particles (far-field matrix soil). For conditions where net swelling occurs, the amount of swelling in the near and far-field matrices is different, but the overall magnitude of swelling for the soil is about the same. Because the oversize particles move farther apart as the swelling occurs, they do not interact, and their presence does not appreciably affect the swelling process. The influence of the oversize particles on the collapse process is less well understood. For the results shown in Fig. 5A.103, as little as 5% oversize particles (by volume) decreased the collapse by as much as 37%. At this low-oversize particle content, appreciable interaction among the oversize particles is unlikely, so the reduction in collapse must be primarily related to changes

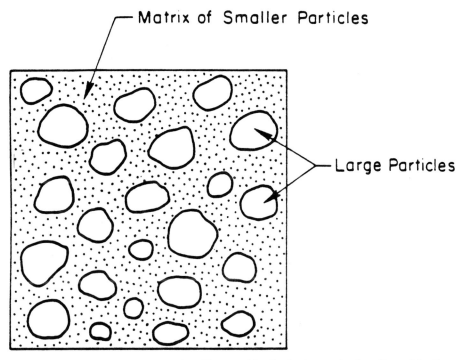

FIGURE 5A.104 Schematic representation of large particles floating in a matrix of smaller particles (from Siddiqi et al., 1987).

in fabric. Apparently little additional collapse occurs within the looser near-field matrix compared to the soil without any oversize particles, and collapse within the denser far-field matrix is substantially reduced. Some of the reduction may occur from the oversize particles' bridging the near-field matrix zones that underlie the particles so that even if collapse occurs in these areas, it may not decrease total volume of the soil.

TIME RATE OF WETTING-INDUCED VOLUME CHANGE The time rate of one-dimensional wetting-induced volume change for two specimens of a clayey sand with the same as-compacted condition but loaded to a low and a high vertical stress and then soaked are shown in Fig. 5A.105. The specimen loaded to $\sigma_v = 1$ kPa swelled, and the specimen loaded to $\sigma_v = 400$ kPa collapsed. When plotted versus the log of time, the wetting-induced strain curves are similar in shape to those for loading-induced strain of saturated fine-grained soils. In Fig. 5A.105, three distinct regions of wetting-induced strain are evident. Each curve is initially flat as the water flows through the porous stones and enters the specimen. As water infiltrates the specimen, the majority of the wetting-induced strain occurs and the curve is relatively steep. After the matric suction in the specimen is eliminated, the wetting-induced strain is nearly complete, and the curve again becomes flat. At some time after wetting, the effective stress in the specimen becomes constant, but secondary compression will continue to occur. This region of secondary compression is apparent in the specimen that collapsed but not in the specimen that swelled because the magnitude of the secondary compression for the low applied stress level is too small to be measured.

Wetting-induced volume changes occur quickly in oedometer specimens inundated in the laboratory, usually within a few hours or less, which suggests that the wetting-induced volume change at any location in a field-compacted soil takes place within a few hours of sustained wetting (Lawton et al., 1992). Thus, the time required for complete wetting-induced volume change to occur in a field-compacted soil depends on (*a*) the method by which wetting occurs, and (*b*) the rate of water infiltration, which is a function of the permeability of the soil.

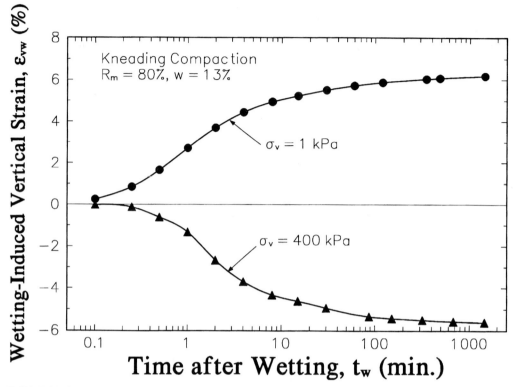

FIGURE 5A.105 Time rate of one-dimensional wetting-induced collapse for compacted clayey sand (data from Lawton, 1986).

CONTROL OF WETTING-INDUCED VOLUME CHANGE   The potential for wetting-induced volume change can be maintained within acceptable limits (a) before compaction by changing the engineering properties of the borrow material, (b) during design and construction by properly written compaction specifications and adequate supervision and inspection of the compaction process, or (c) after the compaction process by preventing water from infiltrating the soil or by modifying the properties of the soil. Preventing the infiltration of water into a compacted soil over a long period is nearly impossible and is not a viable alternative. Modifying the properties of the soil after compaction is generally difficult and expensive. Thus, the preferred alternatives are a and b, and they will be discussed in detail below.

From the data shown in Fig. 5A.97, there are two ways in which the potential for wetting-induced volume changes in a fill consisting of expansive compacted soil can be theoretically eliminated or substantially reduced during the compaction process without modifying the borrow soil: (a) by compacting the soil to a degree of saturation greater than the optimum saturation or (b) by compacting the soil at any depth within the fill to a dry density corresponding to the appropriate critical relative compaction based on the anticipated value of σ_v at that level in the fill. However, both methods have practical limitations that may preclude their use on any project. If the borrow materials are highly variable—and they usually are on large fills—determining appropriate values of S_{rwc} and R_{mwc} for the different materials would be difficult or very expensive. In addition, the quality control procedures required to ensure that the compacted soil has the desired properties are much more extensive and stringent than typically used for most projects. Some potential drawbacks to compacting to a high degree of saturation include:

1. The degree of saturation changes from the as-compacted condition owing to compression caused by placement of additional fill or structures built on or within the fill. However, compression causes an increase in S_r, so that a fill constructed with $S_r \geq S_{rwc}$ will meet the requirements as long as the soil does not dry out so much that S_r drops below S_{rwc}.

2. Postcompaction wetting or drying of the soil may occur, especially for shallow depths along the exposed surfaces of the fill. Continuous or intermittent wetting wherein the moisture content increases gradually over a long period constitutes part of the wetting process and will not result in any significant volume change if the as-compacted soil is properly compacted ($S_r \geq S_{rwc}$). Cyclic wetting and drying can be troublesome, especially near the top of the fill, where the stress levels are low and cyclic swelling/shrinkage may occur. Swelling and shrinkage are primarily elastic phenomena and will continue to occur so long as the wetting/drying cycles occur, although recent research suggests that the magnitude of the swelling and shrinkage may decrease somewhat for the first few cycles and then remain constant thereafter. Wetting-drying cycles at depths within the fill where the stress levels are high enough to result in collapse are unusual, but these cycles may occur near the bottom of moderately deep to deep embankments along the exposed surfaces. Wetting-drying cycles are generally less of a problem for collapsible conditions than for swelling conditions because collapse is primarily plastic, and any shrinkage caused by drying is small and decreases the potential for further collapse. Drying of the soil produces an increase in density owing to shrinkage and an increase in suction, which can be a major problem at low stress levels in an expansive fill because both phenomena increase the swelling potential. At high stress levels, the increase in density decreases the collapse potential, but this decrease is usually more than offset by the increase in collapse potential resulting from the reduction in suction. Thus, postcompaction drying at all depths within a compacted cohesive soil can be important, because it increases either the collapse potential or the swelling potential depending on the stress level. Postcompaction drying commonly occurs in embankments located in regions with semiarid and arid climates.

3. Compacted cohesive soils with high degrees of saturation have more time-dependent compression than the same soil at low degrees of saturation (for the same stress level and dry density), as discussed previously. For $S_r > S_{opt}$, the air phase is probably occluded and the time-dependent compression behavior may be similar to that for the same soil in the saturated condition.

Swelling of the bearing soils for a foundation can be eliminated by ensuring that the applied bearing stress from the foundation is greater than the *swelling pressure* of the bearing soil. *Swelling pressure* is defined as the pressure needed to prevent swelling of a soil as it is takes on moisture from its in situ condition to its stable moisture level. Swelling pressure is usually determined in the laboratory on a specimen contained within a cylindrical mold that is subjected to water and is nominally prevented from swelling. A calibrated steel bar or a strain gauge is commonly used to measure the force that develops in the specimen because it is prevented from expanding. However, in both types of devices, the developed force is determined from a calibration factor or curve that is based on that amount of deformation that occurs in the bar or the gauge. Thus, some upward movement of the specimen does occur, with the magnitude of movement depending on the stiffness of the force measuring device, and the calculated swelling pressure (measured force divided by the area of the specimen) is less than the actual swelling pressure by some unknown amount. An alternative method for determining swelling pressure is to perform a series of oedometric wetting tests on nominally identical specimens subjected to increasing applied stresses and to determine graphically the swelling pressure from a plot of ε_{vw} versus σ_v as the value of σ_v at which the wetting-induced volume change is zero. The results from such a series of tests would be similar to any one of the curves shown in Fig. 5A.100. For example, the swelling pressure is about 340 kPa (7.1 ksf) for the specimen in Fig. 5A.100 prepared by kneading compaction at $w = 10\%$ to $R_m = 90\%$. One must make sure, however, that the applied bearing stress is not so high that an unacceptable potential for wetting-induced collapse exists.

If the borrow soil is nonexpansive or slightly expansive, swelling potential is not a concern, and collapse potential can be controlled by compacting the soil to a high density. The critical relative compaction above which little or no collapse occurs for any reasonable compaction water content varies considerably from soil to soil but is typically in the range of R_m = 85 to 100% (R_s = 90 to 110%). The value for any soil can be estimated from a series of one-dimensional laboratory tests performed on specimens prepared at the driest water content to be used in the field compaction to various densities.

Chemical stabilization of expansive borrow soils prior to compaction has been used successfully for many years to control the swelling potential of the as-compacted soil. The most commonly used stabilizing agents are Portland cement, hydrated lime, and fly ash. Chemical stabilization can also be used to control the collapse potential of compacted soils but to date has been little used for this purpose. The primary advantage to using chemical stabilization to control wetting-induced volume change is that the potential for swelling or collapse is unaffected by post-compaction drying of the soil. Other advantages include increased stiffness and strength of the as-compacted soil and little or no reduction in stiffness and strength if the soil becomes soaked. Potential disadvantages include increased brittleness and decreased permeability.

As shown previously, the addition of gravel to a cohesive borrow material may substantially reduce the potential for collapse but not swelling. Thus, for nonexpansive or slightly expansive soils, blending gravel with the cohesive borrow material may be an economically viable method for controlling collapse potential if sufficient quantities of inexpensive gravel are available.

Shrinkage. Expansive soils are also susceptible to shrinkage (decrease in total volume) caused by drying, which is also known as *desiccation shrinkage* to differentiate it from other phenomena that produce a decrease in volume. During drying, the matric suction increases, which in turn increases the effective stresses acting on the soil, resulting in a decrease in volume that may be accompanied by cracking. In general, the trends of factors that affect shrinkage potential are opposite to the trends of the same factors for swelling, that is, if increasing a certain factor tends to increase the swelling potential of a compacted soil, increasing that same factor tends to decrease the shrinkage potential. The most important factors that affect the shrinkage potential of compacted cohesive soils are as follows:

1. *Soil type.* The more expansive a soil is, the more susceptible it is to shrinkage. Soils that have high shrinkage potentials generally contain some highly active clay minerals such as montmorillonite.
2. *Moisture condition.* The drier a soil is prior to desiccation, the greater its matric suction and the lesser the increase in effective stress for complete drying. Thus, shrinkage potential decreases with decreasing moisture condition (see Fig. 5A.106).
3. *Dry density.* For the same moisture condition and fabric, denser soils shrink less than looser soils.
4. *Fabric.* The influence of fabric on the axial shrinkage of a silty clay is shown in Fig. 5A.107 for three pairs of specimens where each pair had the same density and water content but different fabrics when shrinkage was initiated. For each pair of specimens, one specimen was compacted dry of optimum and the other was compacted wet of optimum, and then both specimens were soaked to the same composition. In each case the dry-compacted specimen (flocculated microfabric) shrank substantially less than the wet-compacted specimen (dispersed microfabric). The greater resistance to shrinkage for the flocculated microfabric can be attributed to two factors: (*a*) The flocculated microfabric contains more interparticle contacts that resist shrinkage, and (*b*) a portion of the load must be spent reorienting the nonaligned particles in the flocculated fabric to a more parallel arrangement before any reduction in average void ratio occurs.

During drying of compacted expansive soils with occluded air phases, the increase in suction resulting from drying acts on the entire surface area of all soil particles and thus produces an increase in effective stress equal to the increase in suction. When some of the pore air spaces become continuous through adjacent void spaces, the suction no longer acts on the total surface area of the particles, and not all the increase in suction is transmitted to the soil as effective stress. As drying continues to occur, the portion of the increase in suction that is transmitted to the soil

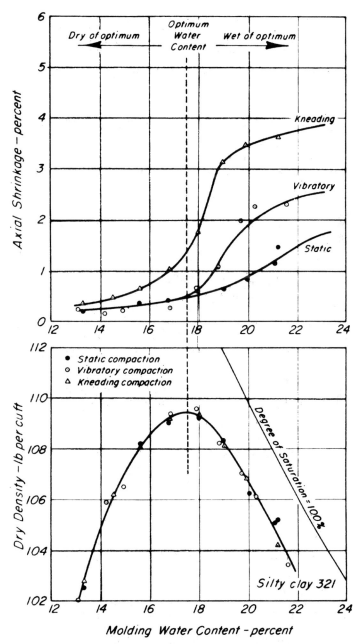

FIGURE 5A.106 Influence of method of compaction and molding water content on axial shrinkage of a silty clay (from Seed et al., 1960).

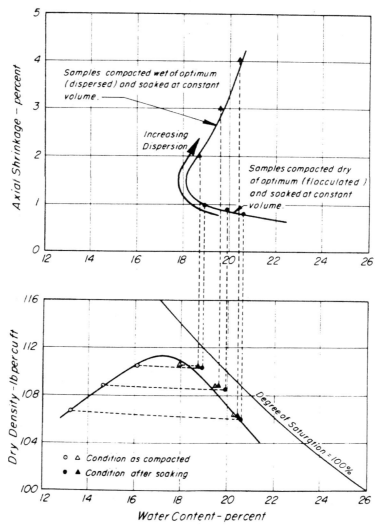

FIGURE 5A.107 Effect of fabric on the axial shrinkage of a silty clay (from Seed and Chan, 1959a).

as effective stress decreases (see Fig. 5A.59). Shrinkage also produces an increase in stiffness of the soil owing to both the increase in density and the increase in suction. Shrinkage ceases when the increase in effective stress resulting from desiccation is no longer sufficient to produce a measurable decrease in volume. The water content of the soil when the shrinkage stops is called the *shrinkage limit,* which generally decreases with increasing plasticity of the soil. Partial drying of the soil to a water content above the shrinkage limit results in less shrinkage than drying to or below the shrinkage limit. The shrinkage limit of the soil as it dries from an as-compacted, partially saturated state, may be a somewhat different value than that determined by the standard shrinkage limit test (ASTM D427) where the soil is dried from a loose, saturated state. The shrinkage limit for compacted cohesive soils may also vary somewhat depending on the density and water content at which drying initiates.

It is important to note that shrinkage of an expansive soil may also occur from moisture use by vegetation (transpiration). However, according to Brown (1992), the potential for foundation distress relative to transpiration is probably much less than normally assumed, and as long as the moisture loss is confined to capillary water, the soil itself is not particularly affected. However, at some point other moisture may be pulled from the clay and may result in shrinkage and potential foundation settlement. The greatest potential for transpiration-induced settlement is from trees located close to shallow foundations. However, tree roots can be beneficial to foundation stability, since they tend to act as soil reinforcement and increase the soil's stiffness and shearing resistance. Thus, the practical result of a tree located close to a shallow foundation may be to impede foundation settlement, even though the roots remove moisture from the soil (refer to Sec. 1A).

Available data indicate that significant structural distress occurs much more frequently from swelling-induced heave than from shrinkage-induced settlement. Of the 502 foundation inspections in Dallas County, Texas, reviewed by Brown (1992) during a particular study period, 216 showed some movement but did not warrant repairs. Of the 286 cases that did warrant foundation repairs, 87 resulted from settlement and 199 resulted from heave. Because it is likely that nearly all of the heave cases were caused by swelling, and the settlement cases could have resulted from a number of phenomena other than shrinkage (e.g., loading-induced compression, marginal bearing capacity, or erosion of supporting soil), it is clear that structural distress is much more likely to occur from swelling than from shrinkage.

Cyclic Swelling and Shrinkage. Cyclic swelling and shrinkage of expansive bearing soils is a serious problem that introduces two issues that are not a concern for a single cycle of either swelling or shrinkage: (*a*) Cyclic swelling and shrinkage may cause fatigue cracking or failure in the foundation or structure for situations where the same amount of movement in one cycle of swelling or shrinkage would not, and (*b*) the magnitude of cyclic swelling and shrinkage cannot be controlled by compaction alone. Cyclic swelling and shrinkage is of most concern in regions where expansive surficial soils are prevalent and therefore might commonly be used as borrow material, where the summers are hot and dry, and where there is a substantial rainy season in the winter (e.g., southern California and Texas).

As discussed in the two previous sections, the potential for either swelling or shrinkage can be kept to acceptable levels by proper compaction of the soil—compaction wet of optimum to control swelling, and compaction dry of optimum to control shrinkage. However, both swelling and shrinkage potential cannot be controlled by compaction moisture content alone because reducing the potential for either phenomenon increases the potential for the other. In addition, the beneficial effect of compaction moisture control is eliminated by subsequent cycles of wetting and drying. For example, Day (1994) showed that the one-dimensional swelling of a silty clay compacted wet of optimum ($R_m = 92\%$, $S_{ri} = 89\% > S_{opt} = 80\%$) was only 1.0% for the first cycle of wetting (after an aging period of 21 days) but increased to 8.0% for the second and subsequent cycles of wetting after drying.

Because problems from cyclic swelling and shrinkage cannot be eliminated by compaction moisture control alone, alternative methods must be used. It is theoretically possible to reduce substantially postcompaction wetting or drying of the bearing soil, but this is difficult to accomplish in practice over a long period of time. Blending gravel with the borrow material may effectively eliminate the shrinkage potential if a sufficient amount is used, but swelling potential will only be reduced in proportion to the volume of gravel added; thus, this method is unlikely to be economical in many instances. Therefore, chemical stabilization is the preferred alternative on most projects.

Permeability. Compacted cohesive soils have low permeabilities, with values of coefficient of permeability (k) for small specimens tested in the laboratory typically within the range of 1×10^{-4} cm/s to 1×10^{-11} cm/s (3×10^{-1} to 3×10^{-8} ft/day). The low permeability of compacted cohesive soils generally causes two types of concerns in applications for building foundations:

1. Rainfall and snowmelt tend to pond on the surface or run off the surface rather than infiltrate the compacted soil. Thus, more extensive drainage systems are required around buildings founded in compacted cohesive soils than are needed for more permeable granular soils.

2. Local or regional ground-water aquifers and flow patterns may be affected by compacting an existing cohesive soil or by constructing a compacted cohesive fill. This factor may require extensive (and expensive) measures to mitigate potential problems or to compensate adjacent property owners who may be adversely affected. Potential problems include reduction in available water supplies for nearby landowners and damage from diversion of additional water onto adjacent properties because of erosion or wetting-induced volume changes.

The low permeability nature of compacted cohesive soils can be put to productive engineering use in the form of compacted clay liners, which are used to impound water, hazardous waste, and other liquids. Some information on compacted clay liners is given below, but because they are used infrequently in building projects, an in-depth discussion of this topic is beyond the scope of this book. A good reference for additional details on compacted clay liners is Daniel (1993).

The in situ permeability of a compacted cohesive soil is generally controlled by the macrofabric, which may include macroscopic characteristics such as cracks, fissures, root holes, animal burrow holes, and flow channels between lifts that increase the permeability of the in situ soil. Because laboratory specimens are typically quite small, these macroscopic features are generally not included in laboratory specimens. Therefore, laboratory specimens contain only the microfabric and minifabric features of the in situ soil, so that values of k obtained from laboratory testing represent lower-bound values. To obtain a realistic estimate of the in situ permeability, field tests—such as the sealed double ring infiltration test (see Sec. 5A.3.5)—should be conducted over an area of the compacted soil that is large enough to represent the macrofabric of the soil.

The following factors can have a major influence on the field permeability of a compacted cohesive soil:

1. *Soil gradation and mineralogy.* The permeability of any soil depends on the size distribution and shape of the portions of the voids through which water can flow. In cohesionless soils, the water essentially can flow through the entire voids. In cohesive soils, the adsorbed (double-layer) water is strongly attracted to the clay particles, and the flow of water is limited to the portions of the voids outside the boundaries of the double layers. Thus, the permeability of a cohesive soil depends on the clay content as well as the size distribution and activity of the clay particles. In general, the permeability of a cohesive soil decreases with increasing clay content and activity and decreasing size of the clay particles. Because activity of the clay minerals is inversely related to particle size, plasticity index is a general indicator of relative permeability, with k decreasing with increasing plasticity of the soil. For clay contents greater than about 30% by weight, the clay fraction dominates the permeability to the extent that the gradation of the rest of the soil has little influence on the permeability unless there is a high percentage of particles gravel size or larger. Laboratory tests on uniformly blended mixtures of gravel and cohesive soils have shown that gravel contents less than about 60% have little effect on permeability (Fig. 5A.108) because the gravel floats in the finer matrix (see Fig. 5A.104). At higher gravel contents, the finer particles exist within the void spaces of a skeleton formed by the gravel particles, and the finer soil may not completely fill the voids of the skeleton. Thus, the permeability of the soil decreases rapidly with increasing gravel content above 60%. In the field, the borrow materials cannot be blended as uniformly as in the laboratory so that some segregation of the gravel particles occurs, with the result that gravel skeletons may form at gravel contents as small as 30%. For compacted clay liners where very low permeability is required, it is recommended that the gravel content of the borrow soil be limited to a maximum of 30%.

Large particles (coarse gravel, cobbles, and boulders) inhibit the compaction of the near-field matrix soil (finer soil close to the particles), whereas the far-field matrix soil is unaffected by the presence of the large particles. This produces zones of near-field matrix soil that may be significantly less dense and much more permeable than the far-field matrix soil. These relatively high permeability zones provide a preferential path for water flow through the soil and hence substantially increase its overall permeability. Thus, for low-permeability applications, all particles larger than about 1 to 2 in (25 to 50 mm) should be removed from the borrow material prior to moisture conditioning and compaction.

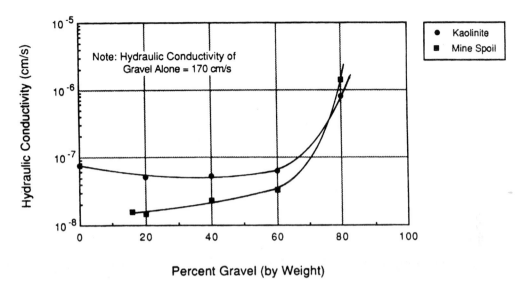

FIGURE 5A.108 Influence of gravel content on hydraulic conductivity of kaolinite/gravel and mine spoil gravel mixtures (from Shelley and Daniel, 1993).

2. *Fabric.* The ease with which water can flow through any void in a given soil depends on the smallest dimension of the void in the direction perpendicular to the flow. The voids in flocculated microfabrics may be irregular in shape but tend to be equidimensional, whereas the voids in dispersed microfabrics are long and narrow. Therefore, for the same composition and density of the microfabric, flocculated fabrics are more permeable than dispersed fabrics. At the minifabric scale, a fabric wherein the clods have been remolded into a relatively homogeneous material is much less permeable than one in which the clods retain some of their individual identity and thus have relatively large interclod pores.

3. *As-compacted moisture condition and density.* Wetter cohesive soils have softer clods that are more easily remolded. For otherwise comparable compaction conditions, wetter compacted cohesive soils tend to have more dispersed microfabric and more homogeneous minifabric and therefore lower permeability. For the same method of compaction and compaction water content, increased density tends to reduce the permeability in two ways: (*a*) smaller interclod and intraclod void spaces, and (*b*) more dispersed microfabric. Isograms of k for laboratory specimens of silty clay prepared by kneading compaction are shown in water content-density space in Fig. 5A.109; k for this soil varied by over four orders of magnitude depending on the compaction water content and as-compacted density. (Note: Not all permeability data for the silty clay is reflected in Fig. 5A.109; see Mitchell et al., 1965 for additional data.) Wet of optimum ($S_r > S_{opt}$), the permeability of the soil was primarily a function of degree of saturation and was little affected by density. Note also the rapid decrease in k with increasing S_r. Density played a more substantial role dry of optimum, but S_r was still the primary factor affecting k. To achieve low permeability in compacted clay liners, the soil should be compacted wet of the line of optimums. Other engineering characteristics of the soil—such as shear strength and shrinkage potential—may also be important, so a moisture-density zone should be specified that meets all the important criteria (Fig. 5A.110).

4. *Method of compaction.* Very dry of optimum, the method of compaction generally has little influence on the fabric of the compacted soil. Wet of optimum, increasing the shearing strain induced in the soil during compaction tends to produce a more dispersed microfabric and greater remolding of the clods. Therefore, kneading compaction results in lower permeability than does

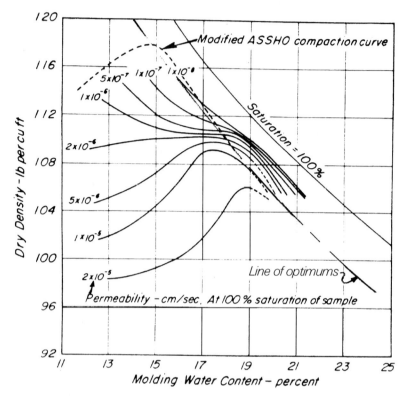

FIGURE 5A.109 Isograms of coefficient of permeability for specimens of silty clay prepared by kneading compaction (from Mitchell et al., 1965).

static compaction for otherwise comparable conditions, and impact compaction produces an intermediate permeability. A comparison of permeability for kneading and statically compacted specimens of a silty clay is shown in Fig. 5A.111. All kneading rollers (sheepsfoot, tamping foot, and rubber-tired) can remold a wet cohesive borrow material into a relatively homogeneous mass for low permeability applications. However, unless good bonding is achieved between lifts, hydraulic defects in adjacent lifts may be hydraulically connected to each other. To produce good interfacial bonding in compacted clay liners, sheepsfoot rollers with feet longer than the lift thickness are usually specified. Heavier rollers are needed in drier soils to break down the clods; in wet soils, lighter rollers should be used that do not become bogged down in the soft soil.

5. *Compactive effort.* Increased compactive effort generally results in higher density, greater dispersion of the microfabric, and increased remolding of clods, all of which result in lower permeability. At some point, additional passes of the compaction equipment will produce no significant changes in the compacted soil. In some soft cohesive soils, additional passes beyond a certain number may produce loosening and weakening of the soil and a concomitant increase in permeability.

6. *Compaction water content and moisture conditioning period.* The water content of the soil and the distribution of moisture within the clods determine to a great extent the amount of remolding that will occur during compaction for a given method of compaction. Very dry clods can be extremely hard and brittle and will not remold but may be broken during compaction, and the permeability of dry-compacted cohesive soils is relatively high owing to flocculated microfabrics and large interclod pores. Wet clods are soft and are easily remolded, and wet-compacted soils

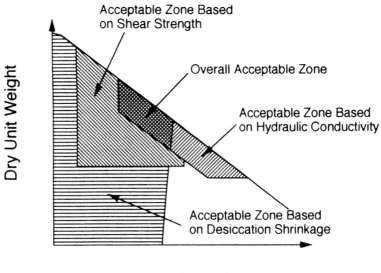

FIGURE 5A.110 Overall acceptable moisture-density zone based on hydraulic conductivity, shear strength, and shrinkage upon desiccation (from Daniel, 1993).

have relatively low permeabilities because of dispersed microfabrics and small interclod pores. The distribution of moisture within the clods also influences the as-compacted fabric. Long periods of time—up to several weeks in some cases—may be required to achieve moisture equilibrium within highly plastic clods, with the required length of time increasing with the size of the clods. If an adequate moisture conditioning time is not used, the outer portions of the clods will be wet and soft, and the inner portions will remain dry and hard. For this reason—and because of potential problems with postcompaction shrinkage and swelling (see number 9 below)—highly plastic borrow materials are not recommended for use in compacted clay liners. The water content of the borrow material for compacted clay liners should be wet and soft enough to allow easy remolding of the clods but not so wet that the compaction and construction equipment have inadequate traction to function properly.

   7. *Clod size.* Laboratory studies have shown that the size of the clods may affect k by several orders of magnitude (Benson and Daniel, 1990; Daniel, 1984; Houston and Randeni, 1992) and that k tends to increase with increasing size of clods, as illustrated in Fig. 5A.112. The influence of clod size on k decreases with increasing compaction water content, as shown in Fig. 5A.113. Because of this effect, a maximum clod size of about 0.75 in (19 mm) should be specified for compacted clay liners.

   8. *Lift thickness.* The lift thicknesses for cohesive soils are usually in the range of 4 to 12 in (0.1 to 0.3 m). If the lift thickness is too large for the borrow soil and compaction equipment being used, the lower portions of the lift will not be properly compacted, resulting in a lift with lower k near the top and higher k near the bottom.

   9. *Postcompaction moisture changes.* Postcompacted wetting and drying of the compacted soil may result in changes in permeability, especially in highly plastic soils. Wetting may result in swelling and increased k owing to decreased density. Drying may produce cracks that can penetrate compacted cohesive soils to a depth of several inches in just a few hours, with a substantial increase in k (Boynton and Daniel, 1985). In low-permeability applications, it is especially important that swelling and shrinkage be prevented to the extent possible. Shrinkage cracking can

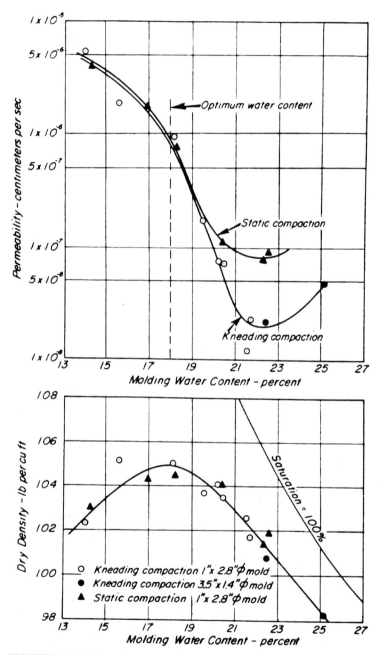

FIGURE 5A.111 Effect of method of compaction on the permeability of silty clay (from Mitchell et al., 1965).

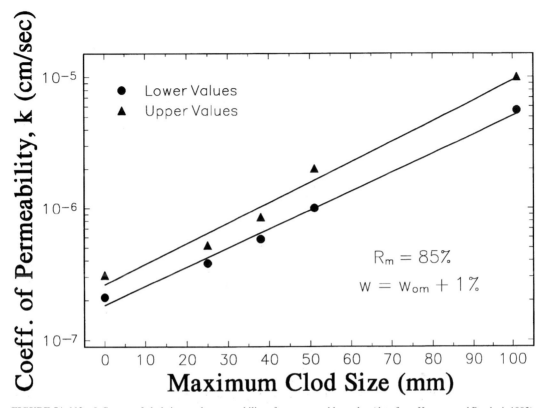

FIGURE 5A.112 Influence of clod size on the permeability of a compacted lean clay (data from Houston and Randeni, 1992).

be prevented by lightly spraying the surface of each compacted lift with water as needed to prevent drying, but care should be taken not to add so much water that swelling is induced. The compacted soil can also be protected from shrinkage by covering it with a light-color plastic liner. If heavy rain is expected, the soil should be covered with a plastic liner to prevent swelling and erosion. To prevent postcompaction drying of the compacted clay liner in composite liner systems, the geomembrane liner should be placed as quickly as possible after proofrolling the final lift of the clay liner.

10. *Postcompaction freeze-thaw cycles.* Postcompaction freezing and thawing of compacted cohesive soils may produce cracking and an increase in permeability of about 5 to 1000 times that of the as-compacted soil. The effect of freeze-thaw on four compacted clays is shown in Fig. 5A.114. The change in permeability caused by freezing and thawing appears to increase with increasing compaction water content. From the results of permeability tests conducted on a glacial clay (Fig. 5A.115), it can be seen that five freeze-thaw cycles increased the permeability by about 2 to 6 times for specimens compacted dry of optimum and about 100 to 250 times for specimens compacted wet of optimum (Kim and Daniel, 1992). Field tests conducted on a test pad constructed of compacted clay in which the uninsulated portion of the pad underwent up to 10 cycles of freeze-thaw showed that the permeability of the soil within the depth of frost penetration (0.5 m = 1.6 ft) increased by approximately 50 to 300 times compared to its as-compacted condition (Benson et al. 1995). However, because the frost penetrated only about 30% of the total thickness (1.5 m = 4.9 ft) of the test pad, the overall hydraulic conductivity of the liner was not affected by the freeze-

5.160 SOIL IMPROVEMENT AND STABILIZATION

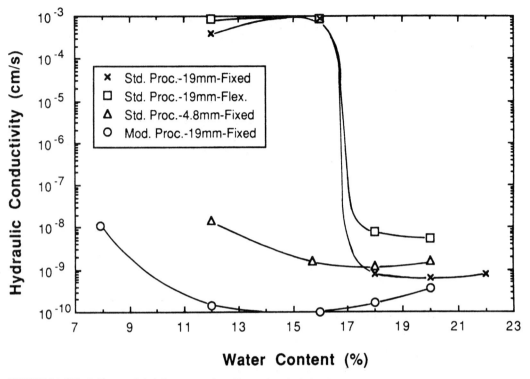

FIGURE 5A.113 Influence of clod size, compactive effort, and method of testing on the hydraulic conductivity of highly plastic, compacted clay (from Benson and Daniel, 1990).

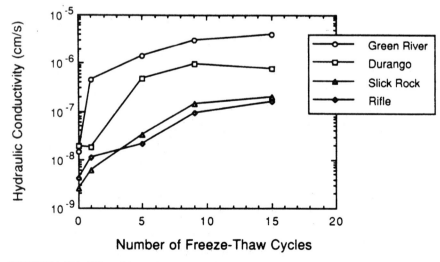

FIGURE 5A.114 Effect of freeze-thaw on the hydraulic conductivity of four compacted clays (from Kim and Daniel, 1992, data from Chamberlain et al., 1990).

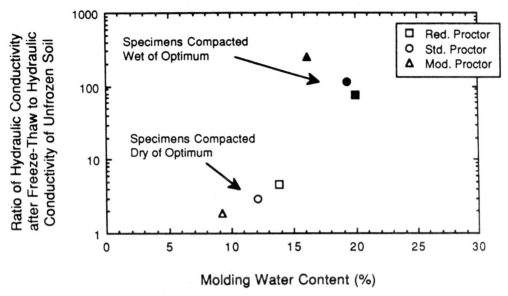

FIGURE 5A.115 Ratio of average hydraulic conductivity after five cycles of freeze-thaw to hydraulic conductivity of unfrozen soil for a glacial clay (from Kim and Daniel, 1992).

thaw cycles. Compacted clay liners in cold climates should be insulated to protect them from the potential effects of freeze-thaw.

11. *Macrofabric.* It is obvious from the preceding discussions that cracks produced by such phenomena as desiccation shrinkage and freeze-thaw cycles can substantially increase the permeability of a compacted cohesive soil. Macroscopic hydraulic channels can also be opened by other phenomena, including burrowing animals, erosion by wind or water, penetration of plant roots, dissolution of water-soluble coarse-grained particles (e.g., gypsum), and chemical dissolution of soil particles.

12. *Degree of saturation during flow.* The influence of degree of saturation during flow on the permeability of a compacted silty clay is shown in Fig. 5A.116. The increase in permeability with increasing degree of saturation is expected because of the increased pore volume through which water can flow (Mitchell et al., 1965). Theory and experimental results suggest that the permeability varies approximately as the cube of degree of saturation ($k \propto S_r^3$).

13. *Stress level.* The permeability of compacted cohesive soils decreases with increasing compressive stress level owing to an increase in density and perhaps some increase in dispersion of the microfabric, as depicted in Fig. 5A.117 (Boynton & Daniel 1985). Because field permeability tests are usually conducted on test pads at essentially zero overburden stress and therefore overestimate the in situ permeability, the results from field permeability tests should be corrected for stress level based on the results of laboratory permeability tests performed over a range of compressive stresses, as shown in Fig. 5A.118 (Daniel, 1993). Higher stress levels also tend to "heal" desiccation cracks, as illustrated in Fig. 5A.117. The data indicate that the cracks began to close at confining pressures of about 4 psi (28 kPa) and that the cracks were essentially closed at confining pressures greater than about 8 psi (55 kPa).

14. *Aging.* The effect of 21 days of aging on specimens of a compacted silty clay stored at constant water content and density is shown in Fig. 5A.119. This increase in permeability with aging was ascribed to an increase in flocculation of clay particles at the microscale (Mitchell et al., 1965). Boynton and Daniel (1985) reported that the permeability of the fire clay they studied

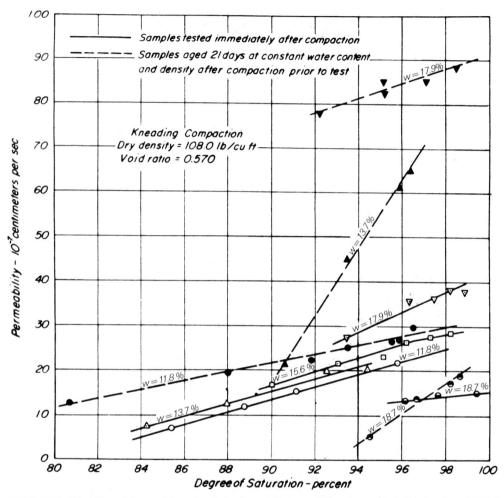

FIGURE 5A.116 Effect of degree of saturation during flow on the permeability of compacted silty clay (from Mitchell et al., 1965).

increased for storage times up to 2 months and then decreased up to a storage time of 6 months. Thus, the effect of aging on the permeability of cohesive soils is not completely understood. Additional details on the effects of aging on the engineering characteristics of soils can be found in Sec. 5A.3.4.3.

Liquefaction. Both laboratory tests and field performance data indicate that the great majority of clayey soils will not liquefy during earthquakes (Seed et al., 1983). However, clayey soils with the following characteristics may be vulnerable to severe loss of strength from earthquake shaking:

Percent finer than 0.005 mm $< 15\%$

Liquid limit < 35

Water content $> 0.9 \times$ liquid limit

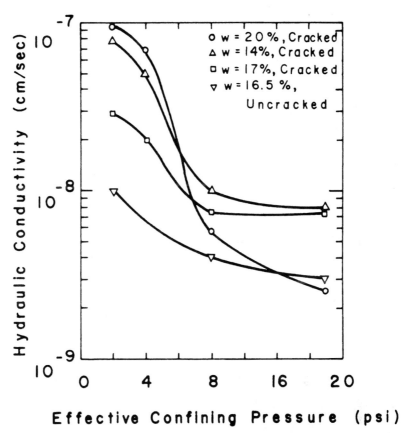

FIGURE 5A.117 Hydraulic conductivity versus effective confining pressure for uncracked and desiccation-cracked specimens of fire clay compacted at various molding water contents (from Boynton and Daniel, 1985).

A rough estimate of the liquefaction potential for these soils may be obtained using the procedures given previously for sands and silty sands, but the best means of determining their cyclic load characteristics is by test.

5A.3.4.3 Aging, Thixotropy, and Cementation. Changes occur in the engineering properties of many compacted soils as the soils age. In general, the strength, stiffness, and permeability of compacted soils increase with aging, whereas the potentials for liquefaction, swelling, and shrinkage typically decrease. Examples of these changes are given in Figs. 5A.119 to 5A.123. Although the reasons for these changes are not completely understood, the following phenomena have been suggested as likely contributors: thixotropy, secondary compression, particle interference, clay dispersion, and cementation. Each of these phenomena is described in limited detail below. The reader is referred to Mitchell (1960) for additional details on thixotropy and to Schmertmann (1991) for further information about mechanical aging.

Thixotropy. There is no universally accepted definition of *thixotropy* in soil mechanics. Mitchell (1960) defines *thixotropy* as an isothermal, reversible, time-dependent process occurring under conditions of constant composition and volume whereby a cohesive soil stiffens while at rest and softens upon remolding. Simply stated, thixotropy is an increase in flocculation (ran-

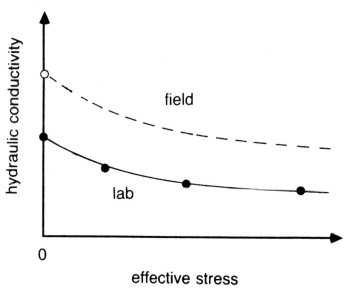

FIGURE 5A.118 Procedure for adjusting the hydraulic conductivity measured in the field on a test pad for the influence of compressive stress (from Daniel, 1993).

domness) of a cohesive soil fabric in undrained (constant-volume) conditions. The properties of a purely thixotropic material are given in Fig. 5A.124.

The energy that drives thixotropy comes from the forces of attraction and repulsion between clay particles. A cohesive soil will exhibit thixotropy if the following two conditions exist (Mitchell 1960):

1. The net interparticle force balance is such that the clay particles will flocculate if given the chance.
2. The flocculation is not so strong, however, that the particles cannot be dispersed by applied shearing strains.

Mitchell (1960) has postulated that the mechanism of thixotropy may occur in the following manner:

1. When a thixotropic soil is compacted, a portion of the compactive shearing energy acts with the repulsive forces between platy particles to produce a more parallel arrangement. During compaction, the energy of interaction between particles is at a level commensurate with the externally applied compactive forces, and the adsorbed water layers and ions are distributed according to this high energy level. The net result is a structure similar to that shown schematically in Fig. 5A.125(*a*).
2. When compaction ceases, the externally applied energy drops to zero, and the net repulsive force decreases; that is, the attractive forces exceed the repulsive forces for the particular arrangement of particles and distribution of water, and the structure attempts to adjust itself to a lower energy condition. The dissipation of energy may be accompanied by changes in particle arrangements, adsorbed water structure, and distribution of ions. These structural changes are time-dependent, because physical movement of particles, water, and ions must take place. Since the process occurs at constant volume, the particle movements are probably small and of a rotational nature. A schematic diagram of the structure at some intermediate time after compaction is given in Fig. 5A.125(*b*).

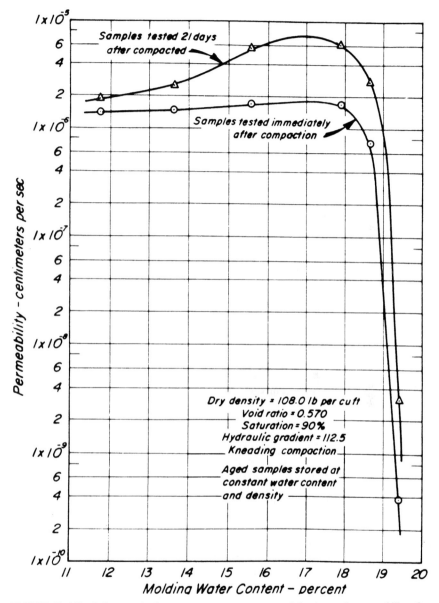

FIGURE 5A.119 Influence of aging at constant water content and density on the permeability of a compacted silty clay (from Mitchell et al., 1965).

3. After some time, the soil will achieve an equilibrium structure, as depicted in Fig. 5A.125(*c*). The time required to reach equilibrium is related to such factors as water content, particle size distribution, particle shape, the ease of displacement of adsorbed water molecules, pore water chemistry, and the magnitude of effective stress during aging.

Nalezny and Li (1967) suggested that thixotropic flocculation may result from Brownian motion that brings some clay particles close enough to flocculate.

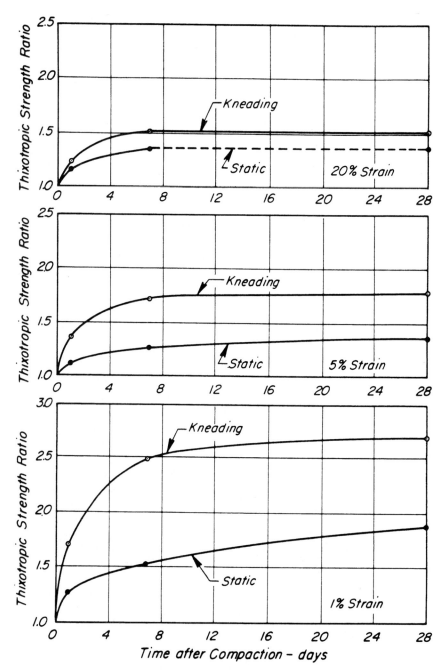

FIGURE 5A.120 Effect of method of compaction on the thixotropic strength ratio for a compacted silty clay (from Mitchell, 1960).

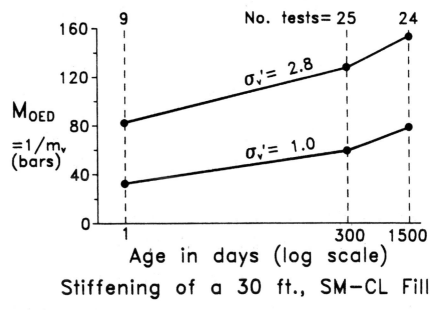

FIGURE 5A.121 Aging-induced increase in oedometer modulus of a compacted, cohesive fill (from Schmertmann, 1991).

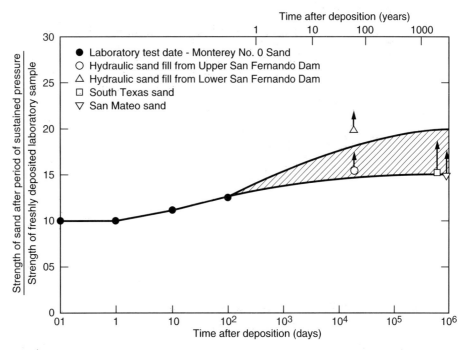

FIGURE 5A.122 Influence of aging under sustained pressure on the stress ratio causing liquefaction of sand (from Seed, 1979).

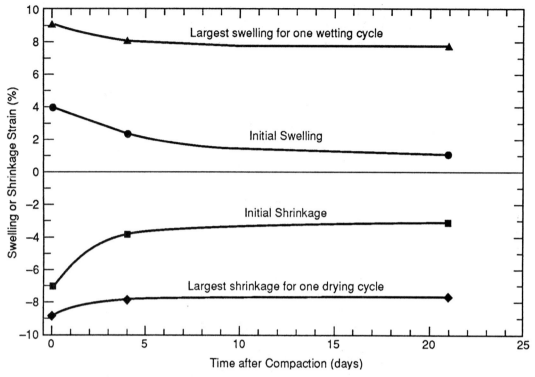

FIGURE 5A.123 Effect of aging on the swelling and shrinkage of a compacted silty clay (data from Day, 1994).

Available evidence suggests that all compacted cohesive soils may undergo thixotropy under certain conditions. The following factors may affect the magnitude of the gain in strength and stiffness that occurs during thixotropy (Mitchell, 1960; Seed et al., 1960):

1. *As-compacted fabric.* Little or no thixotropic hardening occurs in soils with an initially flocculated fabric, such as those produced by compaction to dry-of-optimum saturation. Thixotropic hardening increases with increasing dispersion (all other factors being the same).

2. *Water content.* The influence of compaction water content on thixotropic hardening of a kneading-compacted silty clay after 6 days of aging is shown in Fig. 5A.126 (Mitchell 1960). The thixotropic strength ratio is defined as the strength of the aged soil to the strength of the as-compacted soil. Dry of optimum ($w < 17.7\%$), the fabric is essentially flocculated, and the thixotropic strength ratio is small. Wet of optimum, the thixotropic strength ratio increases with increasing water content up to about 28 to 29%, beyond which the strength ratio decreases with increasing water content. The disparity between the induced and equilibrium structures is apparently as great as it can be for this soil and method of compaction at a water content of about 28 to 29%. The decrease in thixotropic effect at higher water contents may be attributable to the natural dispersing tendency of the soil at these water contents.

3. *Magnitude of strain.* The thixotropic increase in strength is greater at low strains and decreases with increasing strain (Fig. 5A.126) and is attributable to progressive destruction during shearing of the thixotropically formed structure.

4. *Method of compaction.* The effect of method of compaction on thixotropic strength ratio for the same silty clay is shown in Fig. 5A.120. As expected, the kneading-compacted soil is more

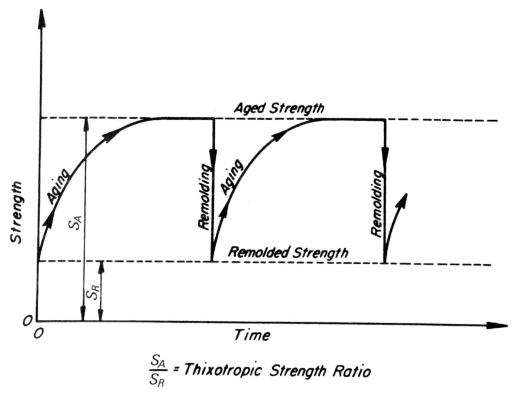

FIGURE 5A.124 Properties of a purely thixotropic material (from Mitchell, 1960).

thixotropic than the statically compacted soil because of the greater dispersion—and hence greater tendency toward thixotropic flocculation—produced by kneading compaction.

Secondary Compression. Secondary compression, which is a creep-type rearrangement of particles into a slightly denser fabric, is closely associated with aging effects (Schmertmann 1991). The driving energy for secondary compression is supplied by the in situ effective stresses acting on the soil. Although some engineers believe that aging effects can be fully explained in terms of secondary compression, the magnitude of aging-induced changes in engineering behavior cannot be entirely attributed to the small changes in density that occur from secondary compression. Thus, secondary compression is one of several phenomena that contribute to aging effects in soils.

Particle Interference. The term *particle interference* is used to describe the tendency of aged soils to show increased dilation during shearing (see Fig. 5A.127). This increased dilatancy probably results from small slippages of particles during secondary compression, which produce additional particle-to-particle interlocking (Schmertmann, 1991).

Clay Dispersion and Internal Arching. The drained movements associated with secondary compression tend to disperse plate-shaped particles in cohesive soils, which leads to an increased basic frictional capability of the soil. Schmertmann (1991) has presented two possible mechanisms of internal arching by which this increased frictional resistance may occur:

1. As parts of the fabric of the soil stiffen owing to dispersion under drained conditions during aging, the stresses may arch to these stiffer parts.

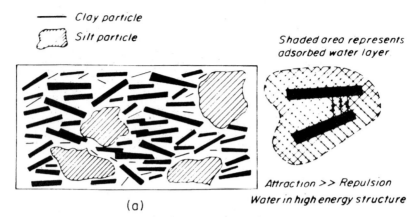
(a) Structure Immediately after Remolding or Compaction

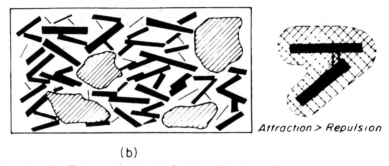
(b) Structure after Thixotropic Hardening Partially Complete.

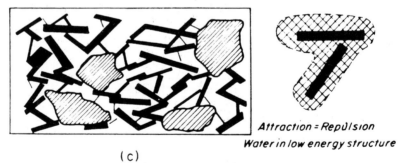
(c) Final Structure at End of Thixotropic Hardening

FIGURE 5A.125 Schematic diagram of thixotropic change in structure for a fine-grained soil (from Mitchell, 1960).

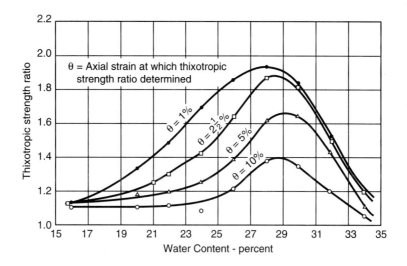

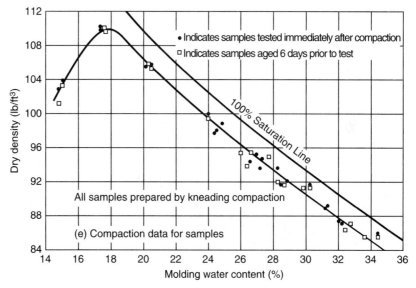

FIGURE 5A.126 Thixotropic strength ratio for a silty clay as a function of molding water content and axial strain (from Mitchell, 1960).

2. The dispersive movements may occur primarily in the weaker, softer parts of the fabric and cause an arching stress transfer to the stiffer parts.

In either case the arching is assumed to occur internally at the minifabric level, with the average effective stress remaining constant.

Cementation and Bonding. Although Schmertmann (1991) has shown that substantial increases in stiffness and strength can occur with aging in the absence of any measurable cemen-

5.172 SOIL IMPROVEMENT AND STABILIZATION

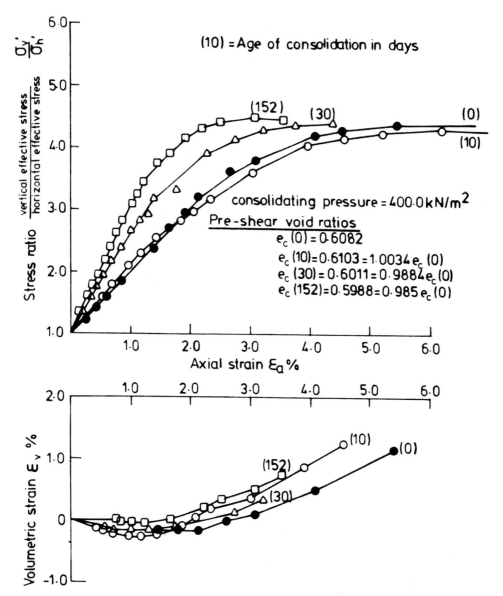

FIGURE 5A.127 Effect of aging on the triaxial stress-strain and volume change characteristics of Ham River sand (from Daramola, 1980).

tation, postcompaction cementation may play a significant role in aging effects for some soils, especially those containing carbonates, iron oxide, alumina, and organic matter. The decrease in swelling potential with aging shown in Fig. 5A.123 cannot be explained by any combination of factors that excludes cementation or some other form of bonding. Day (1994) suggested that the decrease in swelling potential for the silty clay he studied may be partially related to bonds that develop during aging. These bonds were manifested in the form of an aging-induced increase in effective cohesion intercept at zero effective stress, with $c'(\sigma'_v = 0)$ increasing from zero for no aging to 45 psf (2.2 kPa) after 21 days of aging (see Day, 1992). The nature of this bonding was not determined. Nalezny and Li (1967) also reported experimental evidence that showed swelling potential decreased with aging in the highly expansive clay they studied, which they attributed to an increase in cross-linking of clay particles. This additional cross-linking—which results from Coulombic and van der Waals attractive forces that develop at edge-to-face contacts—is assumed to occur with aging as Brownian motion brings some clay particles close enough to flocculate. Note, however, that increasing flocculation in freshly compacted soils tends to increase the swelling potential (compare the results for the kneading and statically compacted specimens shown in Fig 5A.100). Therefore, if the aging-induced decrease in swelling potential is produced entirely by cross-linking, the reduction in swelling potential caused by the cross-linking must be greater than the increase in swelling potential resulting from increased flocculation. Whatever its nature, some form of bonding occurs in cohesive soils with aging that contributes not only to a reduction in swelling potential but also to the other aging-induced changes in engineering behavior.

Summary. Aging effects have been reported in virtually all types of remolded and compacted soils. Aging effects in sands appear to result from increased density and particle interlocking as a consequence of secondary compression. Cementation may also be a contributing factor in some cases. In clays, aging effects may also be associated with thixotropy, dispersion, and some form of bonding. Schmertmann (1991) estimates that for most soils, the aging-induced improvement in many key soil properties is on the order of 50 to 100% during engineering times. Unfortunately, the mechanisms by which these improvements occur are not completely understood, and the beneficial effects of these improvements are seldom considered in engineering practice. However, because the potential degree of improvement is significant, a substantial savings in foundation costs may be realized on some projects involving compaction of bearing soils by performing laboratory and field tests to determine the extent of this improvement for important properties of the borrow soils or in situ soils to be compacted.

5A.3.5 Quality Control and Assurance

To ensure that a compacted soil has the desired engineering characteristics, a two-step quality control and assurance process is commonly employed that consists of compaction specifications and postcompaction testing of the compacted soil. The compaction specifications are usually written by a geotechnical engineer and become part of the legal construction documents. The primary purpose of the compaction specifications is to provide guidelines that assist the compaction contractor in achieving the desired engineering product. The postcompaction testing program is used to verify that compacted soil with appropriate engineering properties has been constructed.

5A.3.5.1 Compaction Specifications. Most compaction specifications fall into one of two general categories—*end product specifications* or *method specifications.* In end product specifications, the desired characteristics of the compacted soil are specified, and the compaction contractor may use any equipment and procedures to produce compacted soil that meets the requirements. In method specifications, the methods and equipment that the contractor must use to compact the soil are prescribed. In some instances, combination end product and method specifications are provided.

End Product Specifications. The desired engineering characteristics for a compacted soil depend significantly on the use to which the soil will be put and therefore may vary considerably from project to project and at different locations within the same project. For the bearing soils beneath a building, the most important properties (in decreasing order of importance for most projects) are (*a*) volume change characteristics, (*b*) strength, and (*c*) permeability. Section 5A.3.4 provided a detailed discussion of these characteristics for compacted soils.

In many end product specifications, either a criterion for dry density or criteria for density and compaction water content are furnished. These specifications are based on the assumption that the desired engineering characteristics—usually low compressibility and high strength—are related to water content and/or density. Density and water content specifications are commonly used primarily for economic reasons, that is, it is much faster and cheaper to verify that density and water content specifications have been met than to verify, for example, that compressibility and strength specifications have been met. However, the engineer writing the specifications should remember that in some instances the desired characteristics may not be achieved even if the density and water content criteria are met.

DENSITY SPECIFICATIONS Dry density specifications are generally given in terms of *relative compaction* (or *percent compaction*), R, defined as

$$R(\%) = \frac{\gamma_d}{\gamma_{dmax}} \cdot 100 \qquad (5A.41)$$

where γ_d = dry density of the compacted soil
γ_{dmax} = "maximum" dry density of the same soil

Relative compaction specifications are used rather than density specifications in most fills because of the inherent variability of the borrow soils, for which small changes in size and gradation may result in significant changes in absolute values of maximum dry density and optimum water content (Hilf, 1991). The test method from which γ_{dmax} is to be determined should be specified because the value of γ_{dmax} for any soil depends on the method of compaction and compactive effort, as discussed in Sec. 5A.3.2. The most common tests used to determine γ_{dmax} are the standard and modified Proctor tests (ASTM D698 and D1557, or equivalent). The same dry density specifications can be given using γ_{dmax} from either test, and in the United States the preferred test varies from region to region. Poulos (1988) recommends using the modified Proctor test because the test errors are likely to be smaller than for the standard Proctor test.

The importance of specifying the method from which γ_{dmax} is to be determined cannot be overemphasized. If the required method is not given in the specifications, the compaction contractor can legally use any reasonable method she or he wishes. The author knows of several projects in which the method was not specified, and the contractor assumed standard Proctor γ_{dmax} (to the contractor's advantage) when the specifying engineer intended modified Proctor γ_{dmax}. In these cases, either the contractor was paid more than his or her bid price to compact the soil according to modified Proctor γ_{dmax}, or the soil was compacted using standard Proctor γ_{dmax} and the compacted soil did not meet the desired criteria (which resulted in either expensive litigation or removal and replacement of the unacceptable material). The net result in each case was that more money than necessary was paid to compact the soil, an expense that could have been avoided if the engineer writing the compaction specifications had simply specified the method for determining γ_{dmax} rather than assume that everyone would understand what was meant.

Density requirements for clean granular soils are sometimes given in terms of *relative density* (D_r) rather than relative compaction. Relative density can be calculated in terms of either void ratios or dry densities using one of the following equations:

$$D_r(\%) = \frac{e_{max} - e}{e_{max} - e_{min}} \cdot 100 \qquad (5A.42a)$$

$$D_r(\%) = \frac{\gamma_{dmax}}{\gamma_d} \cdot \frac{\gamma_d - \gamma_{dmin}}{\gamma_{dmax} - \gamma_{dmin}} \cdot 100 \qquad (5A.42b)$$

where e = void ratio of the compacted soil
e_{min} = "minimum" void ratio for the same soil
e_{max} = "maximum" void ratio for the same soil
γ_{dmin} = "minimum" dry density for the same soil
γ_{dmax} = "maximum" dry density for the same soil

e_{min} and e_{max} (or γ_{dmin} and γ_{dmax}) are commonly determined from laboratory tests as described in Sec. 5A.3.3.4, and the method on which they are to be based should be specified (usually ASTM D4253 and D4254). Nominal values of relative density vary from 0% for a soil in its "loosest" condition to 100% for a soil in its "densest" state. However, note that, because γ_{dmin} and γ_{dmax} are generally based on laboratory tests that do not exactly model field compaction, the actual values of relative density for naturally deposited and compacted soils may be less than 0 or greater than 100%. Compacted soils never have D_r less than 0% but may have D_r greater than 100% depending on the method of compaction and the procedure used to determine e_{min} or γ_{dmax} (refer also to Sec. 2A).

A statistical investigation by Lee and Singh (1971) of data found in the literature for 47 granular soils has shown that values for relative density and relative compaction for any soil are related. By assuming that the values of γ_{dmax} in Eqs. (5A.41) and (5A.42) are the same, they obtained the following equation relating D_r to R:

$$R = \frac{R_0}{1 - D_r(1 - R_0)} \tag{5A.43}$$

where

$$R_0 = \frac{\gamma_{dmin}}{\gamma_{dmax}} \tag{5A.44}$$

The term R_0 has the physical meaning of being the relative compaction at zero relative density. The mean value of R_0 for the soils studied was 81.8. Inserting this value of R_0 in Eq. (5A.43) yields

$$R = \frac{0.818}{1 - 0.182 D_r} \tag{5A.45}$$

Sixty-seven percent of the data fell within ± 1 standard deviation. Unfortunately, the values for γ_{dmax} and γ_{dmin} from which the mean value of R_0 was calculated were obtained using a wide variety of test methods, and frequently the procedures used were not adequately described in the literature. Thus, Eq. (5A.45) is of limited practical use because the bases for calculating R and D_r are not standardized to particular methods.

According to the obsolete standard ASTM D2049 (replaced by D4253 and D4254), relative density rather than relative compaction should be used if the soil contains less than 12% fines. Lee and Singh (1971) recommend using relative density for granular soils because small deviations in field density appear as large numbers (a change in relative compaction of one percentage point is approximately equivalent to a change in relative density of five percentage points) and thus tend to convey the severity of the differences to the compaction contractor. However, there appears to be a growing trend among engineers toward using relative compaction for all types of soils. For example, Poulos (1988) recommends against the use of relative density for compaction control of granular soils for the following reasons:

1. Both γ_{dmin} and γ_{dmax} are functions of soil composition (gradation, particle shape, mineralogy, etc.) and are linearly related (with some scatter) for most granular soils. Therefore, only one of the two factors is needed for controlling the compaction of granular soils.
2. γ_{dmin} is not of great interest in engineering practice.
3. Relative compaction appears to be a slightly better index of engineering properties than relative density.

In most density specifications, only a minimum acceptable relative compaction or relative density is prescribed, which results in an allowable moisture-density range as shown in Fig. 5A.128. It may also be prudent in some instances to include a maximum acceptable relative compaction specification, as indicated in Fig. 5A.129. Typical values of relative compaction specification for different applications are summarized in Table 5A.9 and are provided only as general guidelines. Actual specifications should be based on the desired engineering characteristics and economics.

MOISTURE CONDITION SPECIFICATIONS The moisture condition of the soil during and after compaction determines, to a great extent, the maximum density that can be achieved with a given compaction roller, as well as the engineering characteristics of the as-compacted soil. The moisture condition of soils can be described in terms of either water content or degree of saturation. In many projects, therefore, it is imperative that the compaction specifications include some requirements as to the moisture condition of the soil during compaction. The moisture condition can be

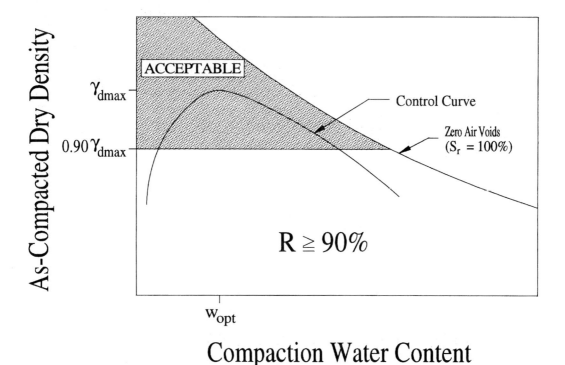

FIGURE 5A.128 Acceptable moisture-density region for specified minimum relative compaction.

specified in terms of water content, degree of saturation, or both. In addition, because of the potential for volume changes owing to changes in the availability of free moisture, especially in compacted cohesive soils, it may be necessary to specify methods for preventing excessive drying or wetting of the compacted soil either prior to compaction of the next lift or after the compaction process is completed.

Compaction moisture condition is nearly always specified as a water content range, usually referenced to either standard or modified Proctor optimum water content. Typical water content specifications for various applications are provided in Table 5A.9. When water content specifications are combined with relative compaction specifications, the resultant acceptable moisture-density zone is depicted in Fig. 5A.130. However, because of the confusion regarding optimum water content and optimum moisture condition (line of optimums) discussed in Sec. 5A.3.2, moisture condition specifications given in terms of water content only may not result in compacted soil with the desired engineering properties. For example, the permeability of a compacted clay may be several orders of magnitude lower when compacted wet of optimum than when compacted dry of optimum (see Fig. 5A.109). Optimum moisture condition refers to the line of optimums rather than one value of optimum water content referenced to some standard test. An example is given in Fig. 5A.131 showing the permeabilities that would be achieved in a compacted silty clay by specifying moisture condition as a water content range (Benson and Daniel, 1990). Because the specified water content range is above "optimum" water content, some engineers would believe that the given specifications would result in permeabilities less than the specified maximum value of 1×10^{-7} cm/s (3×10^{-4} ft/day). As can be seen in Fig. 5A.131, the actual permeabilities within some portions of the specified moisture-density zone are more than 100 times the desired maximum value.

In most compacted soils it is desirable to control more than one property. For example, in a compacted clay liner, low permeability is the dominant consideration, but moderate strength and

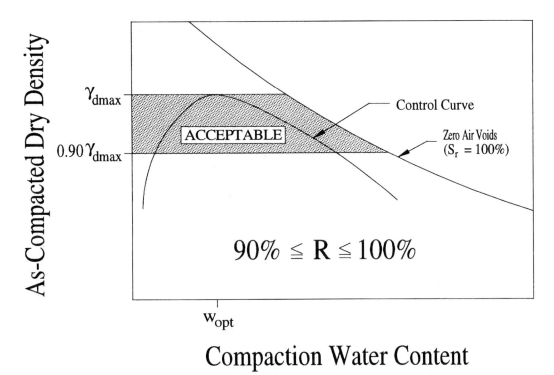

FIGURE 5A.129 Acceptable moisture-density region for specified minimum and maximum relative compaction.

low shrinkage potential are also desirable characteristics. An acceptable moisture-density range that would result in a compacted soil with all the desired characteristics can be established by delineating acceptable zones separately for each characteristic, as illustrated in Fig. 5A.110. Moisture-density specifications that meet the desired objectives in this type of situation can be given in terms of relative compaction and degree of saturation, as shown in Fig. 5A.132. Since many compaction personnel are familiar with the concept of water content but not degree of saturation, a water content range can also be specified to alleviate somewhat any concern they may have because of their unfamiliarity with degree of saturation, but the specifications must be clearly written to emphasize that the requirement for degree of saturation must also be met. It may also be prudent when working with cohesive soils to specify a maximum water content so that the soil remains dry enough that the hauling, spreading, and compaction equipment have sufficient traction to operate efficiently. The engineering properties of many compacted cohesive soils are better correlated with degree of saturation than water content. Hence, the desired engineering behavior frequently will be better achieved by specifying the as-compacted moisture condition in terms of degree of saturation rather than compaction water content.

Many compaction contractors are resistant to specifications that include degree of saturation because it cannot be measured directly but must be calculated indirectly from phase relations using an equation of the following form:

$$S_r = \frac{w}{(\gamma_w/\gamma_d) - (1/G_s)} \tag{5A.46}$$

Therefore, to calculate degree of saturation, γ_d and w must be measured, and G_s either estimated or determined from laboratory tests. Since γ_d and w are normally measured during the construc-

TABLE 5A.9 Typical Density, Water Content, and Lift Thickness Specifications*

Use for compacted soil	Relative compaction†		Water content range†‡	Lift thickness, in (mm)
	standard	modified		
Bearing soils for structures	98–100	92–95	−2 to +2	6–10 (152–254)
Lining for canal or reservoir	95	90	−2 to +2	6 (152)
Low earth dam	95	90	−1 to +3	6–12 (152–305)
High earth dam	98	93	−1 to +2	6–12 (152–305)
Highway or airfield	95	90	−2 to +2	6–12 (152–305)
Backfill surrounding structures	95–98	90–93	−2 to +2	6–10 (152–254)
Backfill in pipe or utility trenches	95–98	90–93	−2 to +2	6–8 (152–203)
Drainage blanket or filter	98	93	Thoroughly wetted	10 (254)
Subgrade of excavation for structure	98	93	−1 to +2	—
Rock fill	—	—	Thoroughly wetted	24–36 (610–914)

*After NAVDOCKS (1961).
†All values for relative compaction and water content in percent.
‡Referenced to optimum water content (either standard or modified).

tion control process anyway, the only additional factor needed is G_s, which can be estimated or measured fairly easily. Because each of these three terms has an inherent variability associated with measuring it, the calculated value of S_r has a greater variability associated with it than any of the three terms from which it was calculated. This inherent variability in calculated values for degree of saturation must be considered when performing quality control and assurance checks on the compacted soil (Schmertmann, 1989). For example, this variability will result in a certain percentage of calculated values for S_r plotting above the zero air voids line, indicating that $S_r > 100\%$, which is physically impossible. Routinely rejecting moisture-density tests simply because they plot above the zero air voids line may result in a fill that has a higher S_r than intended and the fill may therefore be weaker, more compressible, and more susceptible to developing high pore water pressures during construction. A detailed procedure for using test data with $S_r > 100\%$ is given by Schmertmann (1989) (refer also to Sec. 2A).

Other end product specifications Although end product specifications are generally given in terms of dry density and water content, it is also possible—and sometimes more effective—to specify other criteria for quality control and assurance of compacted soil. The types of tests that may be used include the following:

- Proctor penetrometer (ASTM D1558)
- Plate load test (ASTM D1194, D1195, D1196)
- California bearing ratio (CBR) test (ASTM D4429)
- Dynamic penetration test

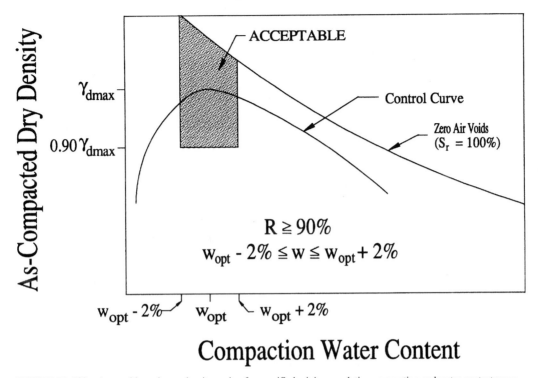

FIGURE 5A.130 Acceptable moisture-density region for specified minimum relative compaction and water content range.

- Unconfined compression test (ASTM D2166)
- One-dimensional compression test (ASTM D2435 and D4186)
- Triaxial compression test (ASTM D2850 and D4767)
- Standard penetration test (ASTM D1586)
- Cone penetration test (ASTM D3441)
- Sealed double ring infiltration test (ASTM D5093)
- One-dimensional swelling/collapse test (ASTM D4546)
- Laboratory permeability test (ASTM D2434 and D5084)
- Suction measured using tensiometer

Although many of these tests are more expensive to conduct than density/water content tests, they frequently will provide better correlations with the desired engineering characteristics. For example, if settlement of a building is the primary consideration, results from plate load tests will give a better indication of the compressibility of the compacted soil than density/water content tests. Hausmann (1990) has reported that in Europe it is not uncommon to have acceptability of compacted soil specified in terms of the reload modulus determined by the plate load test. For compacted clay liners for waste containment facilities, it is often required that a sealed double ring infiltration test be performed on a field test pad constructed using the same equipment and procedures to be used on the actual liner. Laboratory permeability tests on samples taken from the actual liner are also typically required. Therefore, it is the responsibility of the engineer writing the specifications to ensure that the quality control/assurance testing procedures will result in a compacted soil that has the desired engineering characteristics. This cannot always be accom-

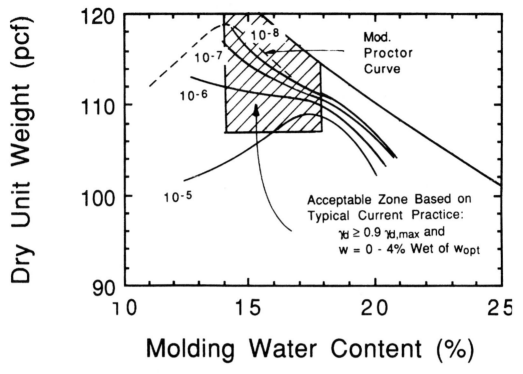

FIGURE 5A.131 Acceptable moisture-density zone based on typical current practice superimposed on isograms of coefficient of permeability for a silty clay prepared by kneading compaction (from Daniel and Benson, 1990).

plished with density/water content criteria only, and in those instances the engineer should specify supplemental testing criteria that will assist in reliably assessing the appropriate engineering characteristics of the compacted soil.

Testing variability and statistical Specifications In traditional compaction specifications wherein only a minimum acceptable value or a range of acceptable values is specified for each parameter (density, water content, and so on), the acceptability of any compacted lift or portion of a lift is judged strictly by whether all measured values of that parameter fall within the acceptable range. If not, that lift or portion of the lift is deemed unsuitable and must be either recompacted in place to a greater density (if the water content is acceptable) or removed, moisture conditioned, and recompacted to meet the specifications. However, these traditional specifications do not consider the inherent variability of the soil and the variation between measured values and true values. Therefore, in recent years it has become more common to base compaction specifications and quality control on statistical considerations of soil and testing variability.

Compacted soil masses, whether compacted fills or natural soil deposits compacted at their original location, often vary considerably in terms of their composition and engineering characteristics. In addition, all tests conducted on the field-compacted soil or on samples taken therefrom all have some inherent inaccuracies, that is, the measured values will always vary from the "true" value by some amount. The more accurate the test, the less deviation between measured and true values. The accuracy of a particular test depends on random errors and systematic errors related to the particular equipment being used and the operator (Hausmann, 1990). Because of the many variables associated in testing compacted soils, the results of water content and density tests tend to be normally distributed (Turnbull et al., 1966; Hausmann, 1990). As an example, the dis-

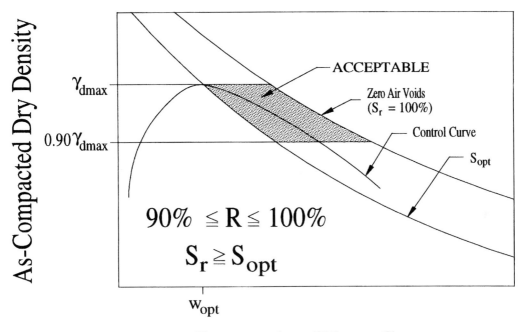

FIGURE 5A.132 Acceptable moisture-density region for specified minimum and maximum relative compaction and minimum degree of saturation.

tribution of water content variation from optimum and dry density variation from maximum dry density for the impervious fill of Ferrells Bridge dam are shown in Fig. 5A.133.

In a statistical approach to specification and control, it is recognized that measured values of density and water content will vary from the desired values. If the measured values are normally distributed about the desired value, every undervalue is balanced by an overvalue, and if the standard deviation is relatively small, the overall behavior of the compacted soil is not likely to be much different than if the entire soil had been compacted to the desired value. Therefore, some deviation from the desired value may be acceptable, so long as there is no large, contiguous zone of material for which the minimum acceptable value is not met.

The following abbreviated specifications for Cutter Dam illustrate the use of statistical control of compaction based on frequency distribution concepts (from Hilf, 1991). Note that these specifications are referenced to optimum water content and maximum dry density from the modified Proctor test.

1. *Moisture control.* The moisture content of the earthfill prior to and during compaction shall be distributed uniformly throughout each layer of the material. In addition, the moisture content of the compacted earthfill determined by testing shall be within the following limits.
    a. If $w > w_{opt} - 3.5\%$ or $w < w_{opt} + 1.0\%$, the material shall be removed or reworked until the water content is between these limits.
    b. No more than 20% of the samples shall be drier than $w_{opt} - 3\%$, and no more than 20% of the samples shall be wetter than $w_{opt} + 0.5\%$.
    c. The average water content of all accepted embankment material shall be between $w_{opt} - 1.0\%$ and $w_{opt} - 0.5\%$.

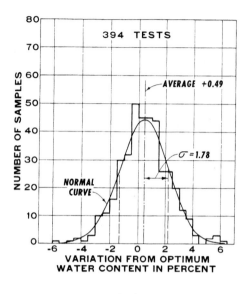 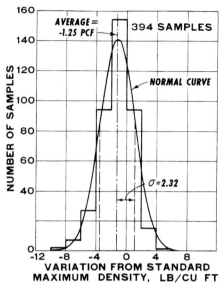

FIGURE 5A.133 Variation in (*a*) compaction water content and (*b*) as-compacted dry density for the impervious fill of Ferrells Bridge dam (from Turnbull et al., 1966).

    d. As far as practicable, the material shall be brought to the proper water content in the borrow pit before excavation. Supplementary water, if required, shall be added to the material by sprinkling on the earthfill on the embankment, and each layer of the earthfill shall be conditioned so that the moisture is uniform throughout the layer.
 2. *Density control.* The dry density of the compacted material shall conform to the following limits:
    a. Material will be rejected if $R < 96.0\%$. Rejected material shall be rerolled until $R \geq 96\%$.
    b. No more than 20% of the samples shall have $R < 97.0\%$.
    c. The average density of all material shall be greater than or equal to $R = 100\%$.

Method Specifications. Method specifications typically prescribe the type of equipment to be used, the thickness of the lift to be compacted, and the number of passes of the equipment per lift (Hilf, 1991). Moisture requirements and a maximum size of material may also be specified. The most important advantages and disadvantages of method specifications are summarized as follows (Holtz and Kovacs, 1981):

1. An expensive field testing program generally must be conducted to determine the most efficient and economical methods and equipment for obtaining a compacted soil with the desired characteristics. Method specifications, therefore, are normally used only on major compaction projects such as earth dams.
2. A major part of the uncertainty associated with compaction will be eliminated for the contractor. Therefore, the contractor should be able to estimate the construction costs more accurately, and a substantial savings in earthwork costs should be realized.
3. The owner or owner's engineer has the major portion of the responsibility for the quality of the earthwork rather than the contractor. If the contractor follows the specifications but the compacted material does not meet the desired characteristics, the contractor will be paid extra for any work needed to make the material meet the requirements.

5A.3.5.2 Postcompaction Testing and Verification

Moisture-Density Tests. The most common type of postcompaction testing and verification program consists of density and water content tests conducted at random locations within the compacted material. In many fills, at least several tests are conducted on each lift. The frequency with which moisture-density tests are conducted in compacted fills varies considerably, with typical values as follows (after Hausmann 1990 and Hilf 1991):

Type of fill	Volume of fill per test, yd^3 or m^3
Embankment	500–4000
Impermeable liner	200–1000
Subgrade	500–1500
Base course	500–1000
Backfill	100–500

Additional tests should be conducted when the borrow material changes significantly. If the surface of the most recent lift is uneven (e.g., sheepsfoot or tamping foot roller) or some densification may occur during compaction of overlying lifts (e.g., sand), testing should be done one or two lifts below the top lift. To avoid potential bias of testing personnel, the locations for moisture-density testing should be randomly selected (that is, each location should have the same chance for being chosen). This can be accomplished using a table of random numbers such as that shown in Table 5A.10 or by generating random numbers on a computer or handheld calculator. The following example is given to illustrate the procedure for randomly selecting locations for testing (from Sherman et al. 1967):

A compacted lift is 30,000 ft (9,144 m) long and 26 ft (7.92 m) wide, and 50 moisture-density tests will be conducted. Starting at any point in the random number table and proceeding up or down (but not omitting any numbers), 50 *pairs* of numbers are read. For example, starting at the top of column 4, the following pairs of numbers are read: (.732, .721); (.153, .508); (.009, .420);...(.698, .539). The first or A number in each pair is multiplied by the length, and the second or B number in each pair is multiplied by the width to establish the locations for testing. Using the first pair of numbers, a moisture-density test would be conducted at (21,960 ft, 19 ft) or (6693 m, 5.7 m). The locations of the 50 tests are then plotted and numbered in the order in which they will be performed. Should two locations be so close together that they both could not be tested properly, the second one is discarded and the next pair of numbers in the table is substituted.

SAND CONE AND RUBBER BALLOON TESTS Moisture-density tests can be either destructive (fill material is excavated and removed) or nondestructive (density and water content are determined indirectly). The most common destructive tests are the sand cone test (ASTM D1556) and the rubber balloon method (ASTM D2167) (Fig. 5A.134). The following steps are used in these tests (after Holtz and Kovacs, 1981):

1. A hole is excavated in the compacted fill at the desired elevation. The size of the hole depends on the maximum size of included particle in the excavated material (ASTM D1556 and D2167):

Sand cone				Rubber balloon			
Maximum particle size		Minimum test hole volume		Maximum particle size		Minimum test hole volume	
in	mm	ft^3	cm^3	in	mm	ft^3	cm^3
½	12.5	0.05	1420	No. 4 sieve	4.75	0.04	1130
1	25	0.075	2120	¾	19.0	0.06	1700
2	50	0.1	2830	1½	37.5	0.10	2840

5.184 SOIL IMPROVEMENT AND STABILIZATION

TABLE 5A.10 Random Numbers for Selecting Locations of Field Testing (from Sherman et al., 1967)

1		2		3		4		5	
A	B	A	B	A	B	A	B	A	B
.576	.730	.430	.754	.271	.870	.732	.721	.998	.239
.892	.948	.858	.025	.935	.114	.153	.508	.749	.291
.669	.726	.501	.402	.231	.505	.009	.420	.517	.858
.609	.482	.809	.140	.396	.025	.937	.310	.253	.761
.971	.824	.902	.470	.997	.392	.892	.957	.640	.463
.053	.899	.554	.627	.427	.760	.470	.040	.904	.993
.810	.159	.225	.163	.549	.405	.285	.542	.231	.919
.081	.277	.035	.039	.860	.507	.081	.538	.986	.501
.982	.468	.334	.921	.690	.806	.879	.414	.106	.031
.095	.801	.576	.417	.251	.884	.522	.235	.398	.222
.509	.025	.794	.850	.917	.887	.751	.608	.698	.683
.371	.059	.164	.838	.289	.169	.569	.977	.796	.996
.165	.996	.356	.375	.654	.979	.815	.592	.348	.743
.477	.535	.137	.155	.767	.187	.579	.787	.358	.595
.788	.101	.434	.638	.021	.894	.324	.871	.698	.539
.566	.815	.622	.548	.947	.169	.817	.472	.864	.466
.901	.342	.873	.964	.942	.985	.123	.086	.335	.212
.470	.682	.412	.064	.150	.962	.925	.355	.909	.019
.068	.242	.667	.356	.195	.313	.396	.460	.740	.247
.874	.420	.127	.284	.448	.215	.833	.652	.601	.326
.897	.877	.209	.862	.428	.117	.100	.259	.425	.284
.875	.969	.109	.843	.759	.239	.890	.317	.428	.802
.190	.696	.757	.283	.666	.491	.523	.665	.919	.146
.341	.688	.587	.908	.865	.333	.928	.404	.892	.696
.846	.355	.831	.218	.945	.364	.673	.305	.195	.887
.882	.227	.552	.077	.454	.731	.716	.265	.058	.075
.464	.658	.629	.269	.069	.998	.917	.217	.220	.659
.123	.791	.503	.447	.659	.463	.994	.307	.631	.422
.116	.120	.721	.137	.263	.176	.798	.879	.432	.391
.836	.206	.914	.574	.870	.390	.104	.755	.082	.939
.636	.195	.614	.486	.629	.663	.619	.007	.296	.456
.630	.673	.665	.666	.399	.592	.441	.649	.270	.612
.804	.112	.331	.606	.551	.928	.830	.841	.602	.183
.360	.193	.181	.399	.564	.772	.890	.062	.919	.875
.183	.651	.157	.150	.800	.875	.205	.446	.648	.685

The mass or weight of the excavated material is determined in the field using a scale or balance.
2. The water content is determined from either the total excavated material or a representative sample taken from the excavated material. The standard water content procedure consists of oven drying the sample at 110°C (230°F) (ASTM D2216). If the total excavated material is not used, requirements for minimum mass of the representative sample are given in ASTM D2216. The material is dried in the oven until a constant mass is reached. The time required to obtain constant mass depends on such factors as type of material, size of specimen, and oven type and capacity. As a general rule of thumb, sands will dry to constant mass in about 4 h, whereas highly plastic soils may require 16 h or more. For most compaction projects, even 4 h is unacceptable for proper control of compacted fills, and one of the following faster but approximate methods are commonly used (after Hilf, 1991):
   a. The Proctor penetration needle, used to determine the field-compacted water content by comparing the penetration resistance of the field-compacted soil with a calibration curve of

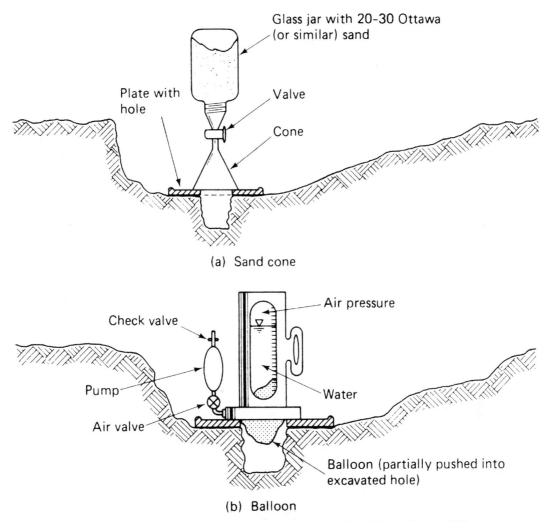

FIGURE 5A.134 Schematic illustrations of sand cone and rubber balloon tests (from Holtz and Kovacs, 1981).

penetration resistance water content from tests on Proctor specimens. (Additional details are given below.)
- *b.* Alcohol burning method (HRB, 1952).
- *c.* Bouyoucos alcohol method using a hydrometer (Bouyoucos 1931).
- *d.* A moisture meter using calcium carbide to generate acetylene in a closed container connected to a pressure gauge (Reinhold 1955; ASTM D4944).
- *e.* Drying the soil in a container using any of a variety of types of direct heating equipment, such as a hotplate, stove, or blowtorch (ASTM D4959).
- *f.* Drying the soil in a microwave oven (ASTM D4643).

If one or more of these approximate methods are used, they must somehow be correlated with the oven-drying method. Usually this involves testing duplicate samples at selected intervals using the oven-drying method, and comparing the result with the result from the approximate method being used.

3. The original volume of the excavated material is determined by measuring the volume of the hole.
   a. *Sand cone.* Clean, dry, uniform sand of known "loose" density is allowed to flow through a cone-shaped pouring device into the hole. The volume of the hole is calculated from the weight of the sand in the hole (by weighing the sand container before and after placing the sand) and the calibrated loose dry density of the sand. Care must be taken to ensure that the loose sand is not densified by vibrations from any source, such as construction vehicles, nearby pile driving or blasting operations, or personnel walking near the hole. In soft soils, volume changes of the hole may occur by deformations induced by personnel (including those performing the test) who walk or stand close to the hole. In dry, cohesionless soils, raveling of the sides of the hole may occur by personnel or vibrations.
   b. *Rubber balloon.* The volume of the hole is determined by expanding the balloon directly into the hole and reading the change in volume of the water in the apparatus directly from the calibration marks. As with the sand cone test, care must be exercised to avoid deformation of the hole in soft soils or raveling of the sides of the hole in loose, dry, cohesionless soils.
4. Compute the total density and dry density of the compacted soil (γ_{field} and $\gamma_{d\text{-}field}$) from the total mass or weight of material excavated from the hole, the volume of the hole, and the water content.
5. Compare $\gamma_{d\text{-}field}$ with γ_{dmax} and calculate relative compaction (R).

The following destructive methods are also sometimes used.

DRIVE-CYLINDER METHOD (ASTM D2937)   This method involves driving a short, thin-walled cylinder into the soil using a special drop-hammer sampler. The cylinder is dug from the ground, excess soil is removed from the sides of the cylinder, and the ends of the sample are trimmed flush with the ends of the cylinder using a sharpened straightedge. The volume of the sample is equal to the inner volume of the cylinder. The total mass is determined by separately weighing the cylinder only (usually before sampling) and the cylinder with the trimmed sample. Water content tests are also performed, from which the dry density is calculated. This method is not appropriate for organic soils, very hard soils, soils of low plasticity that will not readily stay in the cylinder, or soils which contain appreciable amounts of coarse material.

SLEEVE METHOD (ASTM D4564)   The sleeve method is used primarily for soils that are predominantly fine gravel size, with a maximum of about 5% fines, and a maximum particle size of 0.75 in (19 mm). The density is obtained by rotating a metal sleeve into the soil, removing the soil from the sleeve, and calculating the dry mass of soil removed per linear inch (per linear 25 mm). The mass per inch is correlated with the dry density of the in-place material using a calibration equation that has been predetermined for the soil being tested. Details can be found in ASTM D4564.

TEST PIT METHODS (ASTM D4914 AND D5030)   In these methods, a test pit is excavated in materials containing particles larger than about 3 in (75 mm). Test pit methods are generally limited to unsaturated materials and are not recommended for materials that are soft or friable or where water will seep into the excavated hole. Two methods are used which differ primarily by the manner in which the volume of the test pit is determined. In the *sand replacement method,* calibrated sand is poured into the excavated hole, and the mass of sand within the pit is determined. The volume of the hole is determined from the calibrated density of the sand and the total mass of sand within the hole in the same manner as for the sand cone test. In the *water replacement method,* an impervious liner is placed on the surface of the test pit, water is poured into the pit up to the appropriate level, and the mass or volume of the water within the pit is measured. Comparisons for the two tests are given below:

	Sand replacement	Water replacement
Maximum particle size	3–5 in (75–125 mm)	>5 in (>125 mm)
Excavated volume	1–6 ft³ (0.03–0.17 m³)	3–100 ft³ (0.08–2.83 m³)

NUCLEAR METHODS   The use of nondestructive nuclear methods for moisture density testing of soils (ASTM D2922) has become increasingly popular in recent years because both density and water content readings can be obtained within a few minutes, which allows rapid feedback on the quality of the compacted soil. Density is determined by the emission of electromagnetic radiation (gamma rays or photons) from a radioactive source (usually either radium or cesium), as shown in Fig. 5A.135. Some of the gamma rays are absorbed by the soil particles while others are reflected. The number of photons reaching the detector (usually a Geiger-Mueller tube) is an indication of the total density of the soil; a lesser number of photons reaching the detector indicates a denser soil. Three modes of operation can be used. In the *backscatter* mode, the source and detector are on the bottom of the gauge and only about the top 2 to 3 in (50 to 75 mm) of the soil is tested. The backscatter technique is very sensitive to surface roughness and quality of site preparation (U.S. Army, 1985) and is not recommended for general use in soils. The gamma source is lowered into the ground through a hole punched into the soil in the *direct transmission* mode, which is therefore a pseudonondestructive technique. The depth of measurement can be varied in 1- or 2-in (25- or 50-mm) increments on most gauges, allowing a better measurement of the average density in a lift. The *air gap* mode, in which an open space is maintained between the bottom of the gauge and the ground surface, may also be used. Density calibration is achieved using standard blocks consisting of materials of known density. Best calibration occurs using blocks of uniform rock because these materials best simulate those found in soils (Hausmann, 1990). Limestone and granite blocks are recommended in ASTM D2922. The gauge should be calibrated every day prior to testing. Sand cone or rubber balloon tests are commonly conducted at selected intervals of nuclear testing to provide comparison with the density and moisture results obtained from the nuclear gauges.

The measurement of moisture is nondestructive and is accomplished in the backscatter mode by emitting fast neutrons from a radioactive source (usually americium mixed with beryllium) located in the base of the gauge. The detector (usually a boron trifluoride or helium 3 tube) can only detect slow (thermal) neutrons and is also located in the base at some distance from the source. Hydrogen is the only atom normally found in soils that can slow down the neutrons to a level where they can be detected, and thus the moisture detector is really a hydrogen analyzer (U.S. Army, 1985). The moisture per unit volume of the soil is related to the total hydrogen content of the soil, and the water content (per unit mass of dry solids) can be calculated if the density is known. If hydrogen is present in the soil in any form other than free water, errors in measuring the water content will occur. Such errors may occur from the presence of various forms of hydrogen, including the following (U.S. Army, 1985; Hausmann, 1990):

1. Water of hydration in the mineral matrix
2. Organic matter
3. Oil or bitumen
4. Geosynthetics which contain hydrogen

Another factor which affects the accuracy of water content determined from a nuclear gauge is absorption interaction (Ballard and Gardner, 1965). Certain elements—particularly boron and cadmium, and to a lesser extent chlorine and iron—will absorb thermal neutrons to a very high degree. Thus, the presence of these elements in significant quantities may substantially affect the accuracy of the measured water content.

The major advantages and disadvantages of using nuclear gauges for moisture-density determination are summarized as follows (after Holtz and Kovacs, 1981 and Hausmann, 1990). Advantages of nuclear gauges are:

1. Tests can be performed quickly and results obtained within minutes. If necessary, corrective action can be taken before much additional fill has been placed.
2. Better statistical control of the fill is obtained because more tests can be conducted in the same amount of time.
3. The use of nuclear gauges on large projects or over extended periods is often more economical than other alternatives.

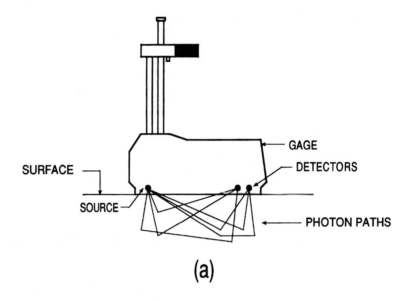

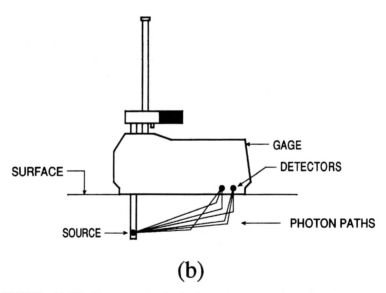

FIGURE 5A.135 Measurement of density using a nuclear density meter: (*a*) backscatter; (*b*) direct transmission (after Troxler Electronic Laboratories, Inc., Research Triangle Park, NC).

Disadvantages of nuclear gauges include:

1. The use of nuclear gauges is regulated by governmental agencies concerned with radiation safety. A radiation license must be obtained by the organization using the nuclear equipment.
2. Personnel using and supervising the use of nuclear gauges must be licensed and trained.
3. Operators are exposed to very small amounts of radiation and must wear film badges that monitor their exposure to radiation.
4. Extraordinary care must be taken during storage and transport of the gauges. If the seal containing the radioactive source is broken or is possibly broken, such as by a construction accident, a large area surrounding the gauge must be evacuated and the appropriate governmental agency notified. This may result in the shutdown of the entire construction project—and possible evacuation of nearby buildings and shutdown of nearby roads—for several hours or days while the status of the radioactive source is determined and necessary remedial measures taken.
5. Excessive concern or fear of the uninformed public and some transportation authorities when the radiation warning labels are sighted.
6. The initial cost of a nuclear gauge is relatively high compared to other moisture-density testing equipment.
7. There are differences in measured values of density and water content obtained using nuclear gauges versus the traditional tests (sand cone or rubber balloon for density, oven drying for moisture). Since some engineers consider the values obtained from the traditional tests to be "correct" even though those tests are also subject to inaccuracies, some engineers and governmental agencies have prohibited the use of nuclear gauges on some projects.
8. As discussed previously, there are some problems associated with determining density and moisture content indirectly using nuclear methods for certain soils and conditions. In some instances, therefore, the nuclear gauges simply will not give reliable or correctable answers. However, in most cases, if comparisons are made at selected intervals with the method specified as the standard, reliable values for density and water content can be obtained directly from the gauges or by applying corrections based on comparison with the values obtained from the specified standard test.

HILF'S RAPID METHOD Because of the problems associated with determining the water content of field-compacted cohesive soils—the long time required for oven drying and the differences in values obtained by the faster methods compared to those obtained by oven drying—Hilf (1959) developed a method that rapidly gives an exact value for standard Proctor relative compaction (R_s) and a close approximation of the field water content relative to the standard Proctor optimum water content (w_{os}). This method has been used satisfactorily on about 150 earth dams since 1957 and has been adopted for control of all compacted cohesive soils on U.S. Bureau of Reclamation projects (Hilf, 1991). The theoretical background and procedure for using this rapid method of field control is given as follows (after Hilf, 1991):

1. Excavate a field density hole in the compacted soil, weigh the excavated material, and determine the volume of the hole. Calculate the total density of the field-compacted soil (γ_{tf}) from the total weight and total volume. The amount of material removed should be enough to perform two reliable oven-dried water content tests and at least three standard Proctor tests without reusing any of the excavated material.
2. Perform a water content test by placing the excavated material in an oven at the site immediately after weighing it and drying the soil in the oven for a minimum of 16 h at 110°C (230°F) (preferred), or by placing the soil in an airtight container and transferring it to a laboratory for oven drying.
3. Conduct a minimum of three Proctor tests on the excavated material—one at the field water content and at least two others at different water contents.

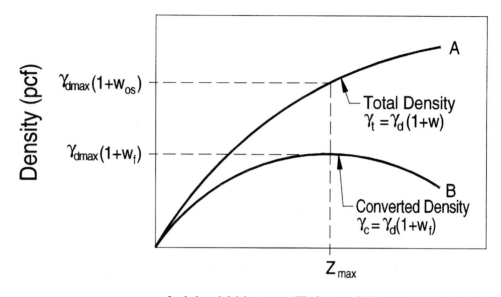

FIGURE 5A.136 Compaction curves for Hilf's rapid method of control (after Hilf, 1990).

4. Plot the results from the Proctor tests as total density (γ_t) versus added water in percentage of the total weight of the soil (Z), as shown by curve A in Fig. 5A.136. Z may be either positive or negative; that is, the water content of the soil during the Proctor tests can be either greater than or less than the water content of the field-compacted soil. Note that $\gamma_t = \gamma_d(1 + w_f)$, where w_f is the water content of the field-compacted soil. A relationship between Z and w (the water content of the soil for any Proctor test) can be developed as follows:

$$\Delta W_w = W_s \cdot \Delta w = W_s \cdot (w - w_f) \tag{5A.47}$$

$$W_{tf} = W_s \cdot (1 + w_f) \tag{5A.48}$$

$$Z = \frac{\Delta W_w}{W_{tf}} = \frac{W_s \cdot (w - w_f)}{W_s \cdot (1 + w_f)} = \frac{w - w_f}{1 + w_f} \tag{5A.49}$$

where ΔW_w = weight of water added or removed from soil
W_s = weight of the solids (dry weight of the soil)
Δw = change in water content of the soil
w_f = water content of the field-compacted soil
W_{tf} = total weight of the soils at w_f

5. Calculate the converted density (γ_c) for each of the Proctor tests as follows:

$$\gamma_c = \frac{\gamma_t}{1 + Z} = \frac{\gamma_d \cdot (1 + w)}{1 + \frac{w - w_f}{1 + w_f}} = \frac{\gamma_d \cdot (1 + w)}{\frac{1 + w}{1 + w_f}} = \gamma_d \cdot (1 + w_f) \tag{5A.50}$$

Plot a curve of γ_c versus Z (curve B in Fig. 5A.136). Curve B can be drawn with only three data points by assuming that the curve is parabolic and using the method shown in Fig. 5A.137.

6. Obtain Z_{max} as the abscissa at the peak point of curve B. Since w_f is a constant for every point on curve B, the maximum ordinate of curve B must have the value of $\gamma_{dmax}(1+w_f)$. Dividing this value into γ_{tf} gives the desired value of relative compaction as follows:

$$R_s = \frac{\gamma_{tf}}{\gamma_{dmax} \cdot (1 + w_f)} = \frac{\gamma_{df} \cdot (1 + w_f)}{\gamma_{dmax} \cdot (1 + w_f)} = \frac{\gamma_{df}}{\gamma_{dmax}} \quad (5A.51)$$

7. Estimate w_{os} from Fig. 5A.138, which shows a best-fit curve of γ_t versus w_{os} based on data from 1300 specimens. Estimate $(w_{os} - w_f)$ and w_{os} from the following equation:

$$w_{os} - w_f = Z_{max} \cdot (1 + w_f) = \frac{Z_{max} \cdot (1 + w_{os})}{1 + Z_{max}} \quad (5A.52)$$

$$w_{os} = w_f + (w_{os} - w_f) \quad (5A.53)$$

The estimate of $(w_{os} - w_f)$ obtained in this manner is usually very close to the actual value and can be used to determine the acceptability of the field-compacted soil in terms of moisture condition only when the moisture condition is specified in terms of a range of water contents referenced to w_{os}. The value of w_{os} obtained from Eqs. (5A.52) and (5A.53) is an approximation only and may vary significantly from the actual value of w_{os}.

8. After the oven-dried water content has been determined, calculate the actual values of R_s and $(w_{os} - w_f)$ and compare with the estimated values determined previously. Both values of R_s should be the same, and the actual value of $(w_{os} - w_f)$ should be close to the estimated value.

The following is an abbreviated example given to illustrate the procedure (from Hilf, 1991): From a field density test we have $\gamma_{tf} = 127.5$ pcf (20.03 kN/m³), and from standard Proctor tests on excavated material we have these values:

Point	Z, %	γ_t pcf	γ_t kN/m³	γ_c pcf	γ_c kN/m³
1	0	123.4	19.39	123.4	19.39
2	2	128.6	20.20	126.1	19.81
3	4	124.6	19.57	119.8	18.82

By the parabolic method:

$$Z_{max} = 1.6\%$$

$$\gamma_{dmax}(1 + w_f) = 126.3 \text{ pcf (kN/m}^3)$$

From Eq. (5A.51):

$$R_s = \frac{127.5}{126.3} \cdot 100 = 101.0\%$$

From Fig. 5A.138:

$$w_{os} \cong 16.0\%$$

From Eq. (5A.52):

$$w_{os} - w_f = \frac{1.6 \cdot (1 + 0.160)}{1 + 0.16} = +1.6\% \text{ (dry of } w_{os})$$

Parabola Method

Graphical solution for vertex, O, of a parabola whose axis is vertical, given three points A, B, and C. If more than three points are available, use the three closest to optimum.

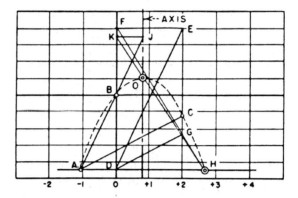

1. Draw horizontal base line through the left point, A, and draw vertical lines through points B and C.
2. Draw line DE parallel to AB, point E lies on the vertical line through point C, project E horizontally to establish point F on the vertical line through B.
3. Draw line DG parallel to AC, point G lies on the vertical line through point C.
4. Line FG intersects the base line at H. Axis of parabola bisects AH, draw the axis.
5. Intersection of line AB with the axis is at J, project J horizontally to K, which lies on the vertical line through point B.
6. Line KH intersects the axis at O, the vertex.

NOTE: If points A, B, and C are equally spaced horizontally (this is true when 2 points are obtained by adding water or when soil is dried exactly 2 percent) steps 2 and 3 above are eliminated. Point F coincides with point B and point G is halfway between the base line and point C. Hence, point H is obtained by drawing BG and point O is obtained by steps 5 and 6 as usual. See graph below.

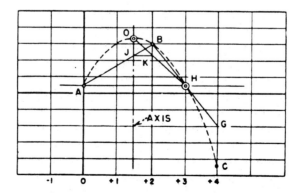

FIGURE 5A.137 Hilf's parabolic method for obtaining the peak point for the curve of converted density versus added water based on three data points (from Hilf, 1991).

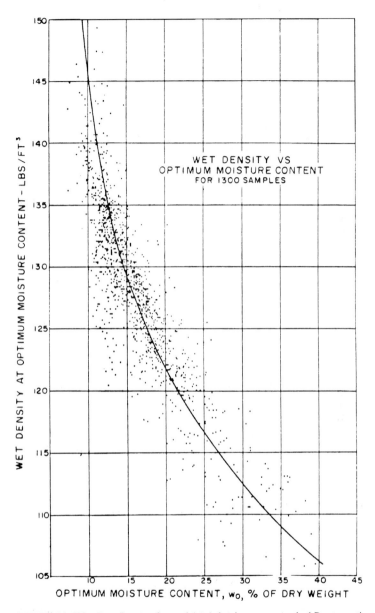

FIGURE 5A.138 Best-fit curve for total (wet) density versus standard Proctor optimum water content based on data from 1300 specimens (from Hilf, 1991).

After oven-drying water content specimens, the actual values are calculated as follows:

$$w_f = 15.0\%$$

$$\gamma_{df} = \frac{\gamma_{tf}}{1 + w_f} = \frac{127.5}{1 + 0.150} = 110.9 \text{ pcf } (17.42 \text{ kN/m}^3)$$

$$\gamma_{dmax} = \frac{\gamma_{dmax} \cdot (1 + w_f)}{1 + w_f} = \frac{126.3}{1 + 0.150} = 109.8 \text{ pcf } (17.25 \text{ kN/m}^3)$$

$$R_s = \frac{\gamma_{df}}{\gamma_{dmax}} = \frac{110.9}{109.8} \cdot 100 = 101.0\% \text{ (checks exactly with previous value)}$$

$$w_{os} - w_f = Z_{max} \cdot (1 + w_f) = 1.6 \cdot (1 + 0.150) = 1.8\% \text{ (compared to estimated 1.6\%)}$$

$$w_{os} = w_f + (w_{os} - w_f) = 15.0 + 1.8 = 16.8\%$$

Other Verification Tests. As noted in Section 5A.3.5.1, many different types of tests may be used to verify that the compacted soil will be acceptable from an engineering standpoint. Some of these tests are discussed in more detail here.

PROCTOR PENETROMETER *(ASTM D1558)* Proctor (1933) developed the soil plasticity needle—now more commonly known as the *Proctor penetrometer*—to assist in evaluating the suitability of compacted soils for use in earth dam construction. A modern version of the penetrometer is depicted in Fig. 5A.139 and can be used in the field and the laboratory. It consists of a special spring dynamometer with a pressure-indicating stem on the handle that is graduated to 90 lb in 2-lb divisions with a line encircling the stem at each 10-lb interval (or 45 kg in 1-kg divisions with a line at each 5-kg interval). The needle is pushed into the field-compacted soil or laboratory-compacted specimen at a rate of 0.5 in/s (13 mm/s) for a distance of at least 3 in (76 mm). As the spring deflects under load, a sliding ring on the stem indicates the developed force and remains at the maximum value as the force decreases during additional penetration beyond the point where the maximum force is developed. Interchangeable needles with heads varying in area from 0.025 in^2 (0.16 cm^2) to 1 in^2 (6.45 cm^2) are available so that the maximum force in any soil can be measured within a consistent range (40 to 80 lb or 10 to 20 kg according to ASTM D1558). The maximum penetration resistance is calculated as the maximum force divided by the area of the head.

The procedure used to determine the acceptability of field-compacted soils is as follows:

1. Establish the relationship between maximum penetration resistance and compaction water by conducting tests on Proctor specimens compacted in the laboratory or field, or on field-compacted test pad sections, for the soil compacted at various water contents with the same method and energy. A minimum of three tests at different water contents is specified by ASTM D1558, but at least five tests are recommended by the author.

2. Plot the results of the penetrometer tests on the same graph with the moisture-density results, as illustrated in Fig. 5A.140. Conduct laboratory or field tests to correlate penetration resistance with the desired engineering properties. Establish an acceptable range of penetration resistances. For bearing situations where strength and settlement are the most important engineering characteristics, a minimum value of penetration resistance is usually prescribed. The specifications can also be in the form of a range of values.

3. Conduct penetrometer tests at random locations within each compacted lift using a procedure for selecting testing locations similar to that given previously for density-water content testing. Compare the measured penetration resistance with the allowable criteria to determine the acceptability or unacceptability of the compacted soil.

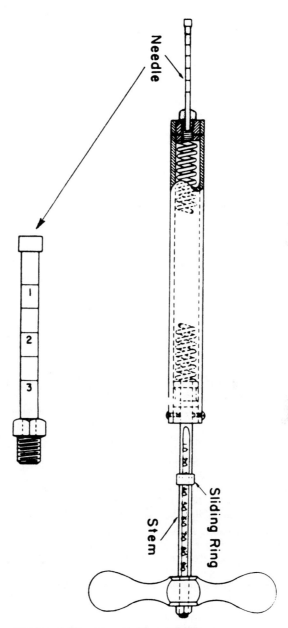

FIGURE 5A.139 Schematic illustration of Proctor penetrometer (from ASTM D1558).

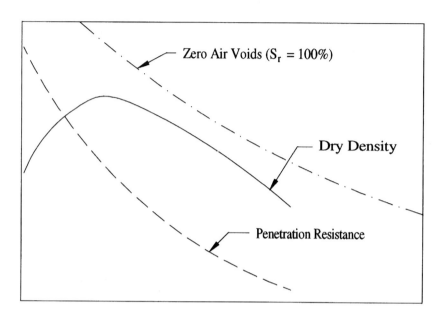

FIGURE 5A.140 Proctor penetration resistance and as-compacted dry density as a function of compaction water content.

PLATE LOAD TEST (ASTM D1194, D1195, D1196) The plate load test can be used to determine the stiffness and/or bearing capacity of a compacted soil. The steps performed in a plate load test are as follows (see Fig. 5A.141):

1. A steel plate [commonly 12 in (305 mm) in diameter] or a small concrete footing is placed at the desired depth within the compacted soil.
2. Vertical load is applied to the plate or footing by jacking against a reaction beam, a dead weight loading frame, or an axle of a heavy construction vehicle. The magnitude of the load is usually determined from a pressure gauge located on a hydraulic jack that has been calibrated to indicate applied force or is read directly from an electronic load cell. The load is increased until a predetermined load or the maximum load for the setup has been reached or until failure has occurred.
3. Vertical settlement of the plate is measured at various times after the application of each loading increment. Settlement readings are taken at one or more locations on the plate or footing using a dial gauge or electronic distance transducer attached to an independent reference beam.

The results from a plate load test are plotted as settlement, S, versus average applied pressure, q_0 (Fig. 5A.142), from which the subgrade modulus ($k_s = \Delta q_0/\Delta S$) and ultimate bearing capacity (q_{ult}, if loaded to failure) can be calculated. Because soils have nonlinear stress-deformation characteristics, a variety of methods can be used to calculate k_s, so the method to be used must be stated in the compaction specifications. Common methods for calculating k_s are shown in Fig. 5A.142 and include the initial tangent modulus (line A), secant modulus between zero and one-

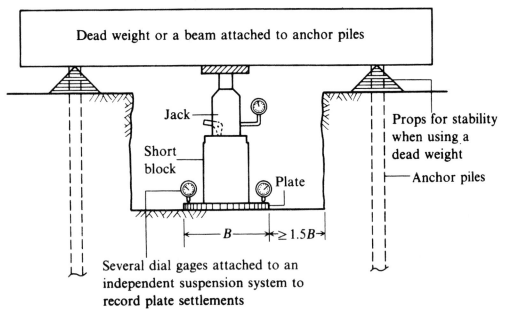

FIGURE 5A.141 Schematic diagram of a typical setup for a plate load test (from Bowles, 1988).

half the maximum or ultimate applied pressure (line *B*), tangent modulus at one-half the maximum or ultimate applied pressure (line *C*), or reload secant modulus from zero to the maximum or ultimate applied pressure (line *D*). The reload curve is often deemed to give better prediction of the stiffness of the field-compacted soil than the curve for first loading because problems associated with bedding errors are reduced or eliminated on reloading.

Because the subgrade modulus is a function of the width and shape of the loaded area, a scaling relationship is needed to extrapolate the results of plate load tests to full field scale (e.g., the width and shape of the footings for a building). It is best to establish this scaling relationship at the site for the actual soils and conditions. This can be accomplished by conducting a series of plate load tests, wherein the size of the plates or small footings are varied but are of the same shape as the actual foundation. For example, if the actual footings will be square, a series of tests using square plates with widths of 12, 18, and 24 in (305, 407, and 610 mm) could be conducted at stress levels comparable to those expected in the actual foundation. From the results of these tests, k_s can be plotted as a function of plate width and the curve extrapolated to obtain an estimate of k_s for the actual foundation width. It is desirable from an engineering standpoint to conduct tests at full field scale, but this is seldom done because of the cost and time needed to perform full-scale tests.

In the absence of data obtained for the particular site and soil, the scaling laws established by Terzaghi (1955) are often used. Terzaghi considered two types of soils: (1) Those whose stiffness is proportional to the effective confining pressure (e.g., sands and most partially saturated soils of all types), and (2) those whose stiffness is independent of the confining pressure (e.g., saturated, undrained clays). The following relationships were developed for estimating subgrade modulus for a full-scale footing based on the results of plate load tests on 1 ft (0.3 m) wide square plates.

For sands and square or rectangular foundations:

$$k_S = k_{S1} \cdot \left(\frac{B+1}{2B}\right)^2 \quad \text{for B in units of ft} \tag{5A.54a}$$

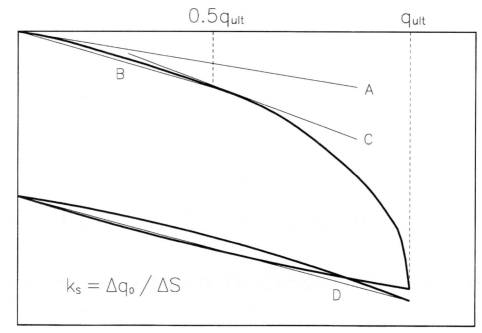

FIGURE 5A.142 Typical results from a plate load test.

$$k_S = k_{S1} \cdot \left(\frac{B + 0.3}{2B}\right)^2 \text{ for B in units of m} \quad (5A.54b)$$

where k_S = subgrade modulus for a foundation of width B
k_{S1} = subgrade modulus for a 1-ft (0.3-m) *square* plate
B = width of actual foundation

For saturated, undrained clays and rectangular foundations:

$$k_S = \frac{k_{S1}}{B} \cdot \frac{n + 0.5}{1.5n} \text{ for B in ft} \quad (5A.55a)$$

$$k_S = \frac{0.3 k_{S1}}{B} \cdot \frac{n + 0.5}{1.5n} \text{ for B in m} \quad (5A.55b)$$

where $n = L/B$
L = length of actual foundation

For a square foundation ($n = 1$), Eq. (5A.55) reduces to

$$k_S = \frac{k_{S1}}{B} \text{ for B in units of ft} \quad (5A.56a)$$

$$k_S = \frac{0.3 k_{S1}}{B} \quad \text{for B in units of m} \tag{5A.56b}$$

and for a strip or continuous foundation ($n \approx \infty$), Eq. (5A.55) reduces to

$$k_S \cong \frac{2}{3} \cdot \frac{k_S}{B} \quad \text{for B in units of ft} \tag{5A.57a}$$

$$k_S \cong \frac{1}{5} \cdot \frac{k_S}{B} \quad \text{for B in units of m} \tag{5A.57b}$$

These relationships are valid for applied stresses less than one-half the ultimate bearing capacity of the actual footing.

These scaling laws were developed for relatively homogeneous soils, and one should exercise caution when the depth of influence for the actual foundation (approximately $2B$) extends into a stratum or strata that are not influenced during the plate load test. This concept is illustrated in Fig. 5A.143 for the case of a dense sand overlying a soft clay layer wherein the depth of influence for the plate load test is entirely within the sand layer, while the depth of influence for the actual footing extends into the soft clay layer. Therefore, extrapolating the results from the plate load test to field scale using Terzaghi's scaling relationships would substantially overestimate the subgrade modulus, which would not be conservative.

Some engineers prefer to use stress-strain modulus (E_s) for the soil rather than k_s. An average value of E_s for the actual foundation can be estimated from the equations for settlement based on elastic theory for a uniformly loaded foundation on a semi-infinite, homogeneous, linearly elastic material. These equations for settlement are of the following general form:

$$S = \frac{q_0 \cdot B \cdot (1 - \nu^2)}{E_s} \cdot I_s \cdot I_d \tag{5A.58}$$

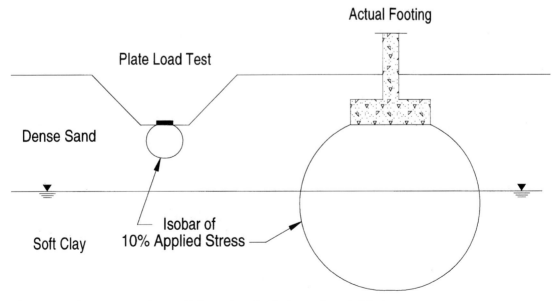

FIGURE 5A.143 Comparison of depth of influence for a plate load test and an actual footing.

where q_0 = applied pressure
B = width or diameter of loaded area
v = Poisson's ratio
E_s = stress-strain modulus for the soil
I_s = influence factor that accounts for the shape and rigidity of the foundation, and the location where settlement is calculated
I_d = correction factor for depth of embedment (D)

Using the secant value of subgrade modulus ($k_s = q_0/S$), Eq. (58) can be rearranged to solve for $E_s = f(k_s)$ as follows:

$$E_s = k_s B (1 - v^2) I_s I_d \qquad (5A.59)$$

When calculating E_s for the actual foundation, the value of k_s input into Eq. (5A.59) should be the field scale k_s. Equations or values of I_s for either rigid or flexible foundations of circular or rectangular shape can be found in a number of reference books (e.g., Poulos and Davis, 1974; Das, 1983). Selected values of I_s for circular and square foundations—the most common shapes used in plate load tests—are given in Table 5A.11. The plates in most plate load tests are quite rigid. If the plate is wide, usually a stack of increasingly narrower plates is placed on the bearing plate to increase the rigidity of the system. Thus, values of I_s for rigid foundations generally should be

TABLE 5A.11 Selected Values of Settlement Influence Factor I_s for Circular and Square Foundations

| | | | | I_s | |
| | | | | | Flexible |
Shape	v	H_c/B	Rigid	Center	Edge
Circular	0 to 0.5	∞	0.79	1.00	0.64
	0.45	3		0.80	0.47
	0.45	2		0.77	0.43
	0.45	1		0.61	0.29
	0.3	3		0.85	0.50
	0.3	2		0.81	0.46
	0.3	1		0.67	0.34
	0.15	3		0.86	0.51
	0.15	2		0.83	0.47
	0.15	1		0.70	0.36
Square	0 to 0.5	∞	0.89	1.12	0.56
	0.5	5		1.00	0.43
	0.5	2	0.77	0.82	0.29
	0.5	1	0.57	0.57	0.14
	0.4	5		1.01	0.45
	0.4	2		0.84	0.31
	0.4	1		0.61	0.17
	0.3	5		1.01	0.45
	0.3	2		0.86	0.32
	0.3	1		0.64	0.19
	0.2	5		1.02	0.46
	0.2	2		0.87	0.33
	0.2	1		0.67	0.20

used. For example, $I_s = 0.79$ for a rigid, circular foundation on a homogeneous half-space, and this value is commonly used for circular plate load tests. If the compressible layer or layers being tested are underlain by a hard layer, a value of I_s for a finite height of compressible layer (H_c) should be used. E. Fox (1948) presented equations to calculate I_d for rectangular foundations embedded in a homogeneous half-space, and charts for estimating I_d based on these equations are given in Fig. 5A.144. Equations for calculating I_d for circular foundations can be found in Nishida (1966). Figure 5A.144 can also be used for circular foundations with little loss of accuracy by converting to an equivalent square based on area. Note that the values of I_d in Fig. 5A.144 are not strictly applicable to finite-height compressible layers but may be used as approximations. In tests conducted below the ground surface, the bottom of the excavated pit is typically several times wider than the width of the plate, so the use of a correction factor for embedment in these cases is probably not justified. ν is usually estimated based on the soil type and moisture condition, but also can be determined from triaxial testing either directly by measuring radial deformations using a Poisson's ratio clamp or Hall effect transducers or indirectly by measuring the changes in height and total volume of the specimen and assuming that the specimen retains the shape of a right circular cylinder during the testing.

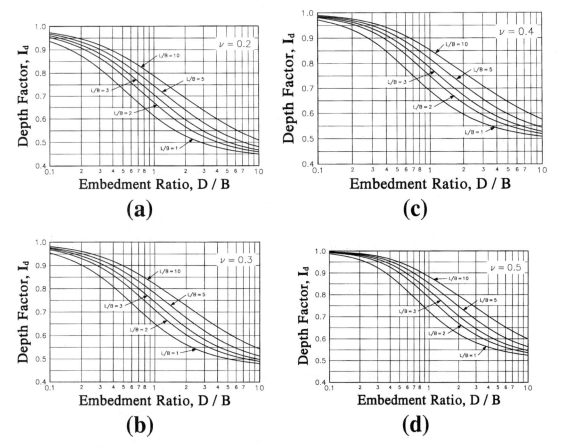

FIGURE 5A.144 Charts for estimating settlement correction factor depth of embedment (I_d) for rectangular foundations based on E. Fox's (1948) equations: (a) $\nu = 0.2$; (b) $\nu = 0.3$; (c) $\nu = 0.4$; and (d) $\nu = 0.5$ (format adapted from Bowles, 1988).

The advantages of the plate load test over moisture-density testing can be summarized as follows (after Hausmann, 1990):

1. The stiffness and strength of the compacted soil are evaluated directly.
2. The results are available immediately.
3. No laboratory testing (e.g., compaction tests) is needed.
4. The test is suitable for a wide range of soil types and maximum particle sizes.
5. Once established as a routine test, the plate load test can be conducted rapidly at low cost.

CALIFORNIA BEARING RATIO (CBR) TEST (ASTM D4429)   In the CBR test, a 1.95 in (49.6 mm) diameter piston with a flat head is slowly pushed into the soil, and the resisting force developed at various depths of penetration is measured. CBR value is calculated at penetrations of 0.1 and 0.2 in (2.54 and 5.08 mm) by dividing the calculated bearing stress by the appropriate standard stress (which varies as a function of penetration), and multiplying by 100. The higher of the two values is reported as the CBR value. If the piston is penetrated deeply enough, failure occurs in the soil beneath the piston, and the test is essentially a miniature bearing capacity test.

CBR value is an indication of the stiffness and strength of the soil and is widely used in the design and control of subsoils beneath pavements. By correlating CBR value with the desired strength and compressibility requirements for a particular compacted soil through laboratory testing or empirical data on similar soils, the CBR test can, in some circumstances, provide a rapid and effective method for assessing the suitability of compacted soils.

The CBR test is similar to the plate load test in the type of results (penetration or settlement versus stress) that are obtained. In fact, one can easily calculate subgrade modulus from the CBR data. The primary difference in the two tests is the width of the loaded area—approximately 2 in (51 mm) in the CBR test and commonly 12 in (305 mm) in the plate load test—with a depth of influence of about 4 in (102 mm) in the CBR test and about 24 in (610 mm) in the plate load test. Therefore, the plate load test is suitable for a wider range of soils. However, for thin lifts (no more than about 8 in = 203 mm) and soils for which particles larger than medium sand size are only a small fraction of the soil, CBR tests often can be conducted faster and cheaper in many instances than can plate load tests, usually with equivalent success.

STANDARD PENETRATION TEST (ASTM D1586)   The standard penetration test (SPT) is seldom used for quality control of compacted fills for the following reasons: (*a*) The test is conducted over a depth of 12 in (305 mm), so that it would be valid only for this lift thickness or greater; (*b*) the setup time and cost for performing one test per location is excessive; and (*c*) the results are reliable only for granular soils with little or no particles gravel-size or larger. The main application for the SPT in compacted soils is for checking the degree of improvement in compacted in situ soils (primarily deep methods such as dynamic compaction and vibrocompaction).

CONE PENETRATION TEST (ASTM D3441)   The cone penetration test (CPT) has limited application in near-surface compacted soils primarily because of setup time and cost. The CPT is not applicable in stiff or hard clays or soils that contain appreciable amounts of gravel or larger particles and is primarily applicable for establishing the improvement in both deeply compacted sands and silty sands.

SEALED DOUBLE RING INFILTRATION TEST (ASTM D5093)   The sealed double-ring infiltration (SDRI) test has become increasingly popular in recent years for determining the field coefficient of permeability (hydraulic conductivity) of compacted clay liners. It was developed to eliminate the two major problems associated with using other field hydraulic conductivity tests on low permeability soils—difficulties in measuring small changes in elevation of the water surface (caused by low flow rates and evaporation) and separating lateral flow from vertical flow (Trautwein 1993).

The SDRI consists of an open outer ring and a sealed inner ring (Fig. 5A.145). The rings may be either circular or square, but square rings are usually used because they make it easier to excavate straight trenches in the soil. The most common sizes for the rings are 12 ft (3.7 m) and 5 ft (1.5 m) square for the outer and inner rings, respectively.

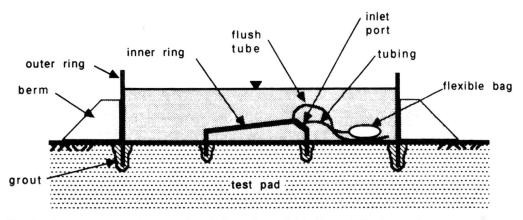

FIGURE 5A.145 Schematic illustration of a typical setup for a sealed double ring infiltration test (from ASTM D5093).

The general procedure for conducting an SDRI test is summarized as follows (ASTM D5093; Trautwein, 1993):

1. The rings are embedded in trenches excavated in the soil and the trenches are sealed with grout to prevent water loss.
2. The outer ring is filled with water to a level such that the inner ring is completely submerged.
3. The rate of flow is measured by connecting a flexible bag filled with a known weight of water to a port on the inner ring. As water from the inner ring infiltrates into the ground, an equal amount of water flows from the flexible bag into the inner ring. The flexible bag is removed after a known interval of time and weighed. The volume of water that has infiltrated the ground is determined from the weight loss of the flexible bag.
4. The infiltration rate I is calculated from the following equation:

$$I = \frac{Q}{At} \quad (5A.60)$$

where Q = volume of infiltrated water
A = area of the inner ring
t = time interval over which Q is determined

5. The process is repeated, and a plot is constructed of infiltration rate versus time. The test is continued until the infiltration rate becomes steady or until it becomes less than or equal to a specified value.
6. The coefficient of permeability k is calculated as follows:

$$k = \frac{Q}{iAt} = \frac{I}{i} \quad (5A.61)$$

where i = hydraulic gradient = $\Delta H/\Delta L$
ΔH = total head loss
ΔL = length of flow path for which ΔH is measured

This equation for k assumes steady state, vertical, one-dimensional flow under saturated conditions.

5.204 SOIL IMPROVEMENT AND STABILIZATION

Although Eq. (5A.61) seems simple, the calculation of i is complex because the flow is transient and the soil is usually unsaturated at the start of testing. The problem of transient flow is overcome by waiting for quasi-steady-state conditions to be reached, which unfortunately results in long testing times. The unsaturated flow makes it difficult to determine the gradient, but this can be alleviated by making some simplifying assumptions. Three methods are commonly used for calculating i (refer to Fig. 5A.146 for illustration of parameters).

Apparent Hydraulic Conductivity Method This is the simplest method, and it yields the most conservative (highest) value of k. It is assumed that the wetting front has passed completely through the compacted liner, which means that the depth of the wetting front is equal to the thickness of the liner ($D_W = H_L$) and that the suction head at the bottom of the liner is zero ($H_s = 0$). This results in $\Delta H = H_W + H_L$, and $\Delta L = H_L$. The advantage of this method is that neither the location of the wetting front nor the suction at the wetting front need be known. The disadvantage is that the correct value of k is calculated only if the wetting front has passed through the test zone; otherwise, k is overestimated. Therefore, this method can be used to determine compliance with the specified maximum acceptable value of k ($k \leq k_{max}$) but cannot be used to determine noncompliance ($k > k_{max}$).

Suction Head Method In this method it is assumed that H_s is equal to the ambient suction in the soil below the wetting front, and the location of the wetting front and the ambient soil suction must be known. For this method, $\Delta H = H_W + D_W + H_s$, and $\Delta L = D_W$. Unfortunately, the basic assumption appears to be incorrect. Although the suction below the wetting front may have some impact on infiltration rate, this impact is apparently offset by low hydraulic conductivity in the unsaturated transition zone between the wetting front and the unaffected unsaturated soil, which restricts downward flow. This method is not recommended because it can yield highly unconservative (low) values of k.

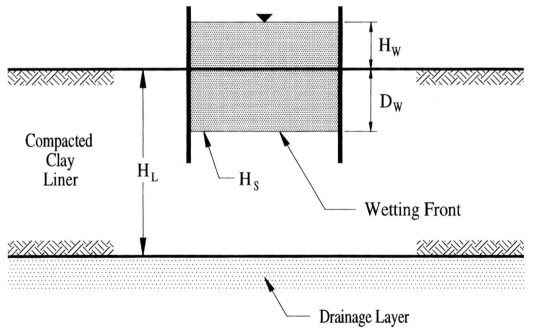

FIGURE 5A.146 Parameters for calculating hydraulic gradient for the sealed double-ring infiltration test (after Trautwein, 1993).

Wetting Front Method This method is based on the assumption that the suction head at the wetting front is zero ($H_s = 0$) and requires that the location of the wetting front be known. With this assumption and a known D_w, $\Delta H = H_w + D_w$ and $\Delta L = D_w$. Because ambient suction may have a small effect on infiltration (and hence k), this method is considered to be conservative. Recent research suggests that the wetting front method is the best of the three methods, and it is the recommended procedure.

The location of the wetting front and changes in suction within the compacted soil as the water advances can be monitored by installing tensiometers in the compacted soil. Typically nine tensiometers are used per test, three at each depth of 6, 12, and 18 in (152, 305, 457 mm). If desired, tensiometers can also be installed in the native soil to measure and monitor ambient suctions.

The SDRI test is the best test currently available for determining the hydraulic conductivity of low-permeability in situ soils. Many governmental agencies now require the SDRI test to establish the acceptability of compacted clay liners for containing waste and hazardous materials. The disadvantages of SDRI testing include lack of overburden, long test times (a few weeks to several months), and high cost ($8000 to $15,000 per test).

5A.4 CHEMICAL STABILIZATION*

The mechanical properties of a soil can be improved by chemical stabilization, which involves the injection or mixing of a chemical agent or agents into the soil. Chemical stabilization can be used on in situ soils or on borrow or excavated materials prior to compaction. The two most common types of chemical reactions used to stabilize soils are cation or base exchange reactions with clay particles and cementitious or pozzolanic reactions. Commonly used chemical agents for cementitious and pozzolanic stabilization include Portland cement, lime, fly ash, sodium silicate, polyacrylamides, and bituminous emulsions. Chemicals used for base exchange include inorganic cations such as Ca^{2+} and organic products such as polyquarternaryamines. Engineering properties of a soil that may be intentionally altered during chemical stabilization, or that may occur as a byproduct of the stabilization process, include the following:

1. Strength
2. Compressibility
3. Permeability
4. Expansiveness
5. Collapsibility
6. Potential for liquefaction
7. Erodability

Because of environmental concerns and regulations, an engineer must consider the potentially adverse impact that chemical stabilization may have on the soil and local groundwater. Before any chemical stabilization is conducted, the appropriate local, state, and federal environmental regulatory agencies should be consulted to ensure that the chemicals to be used are not prohibited.

5A.4.1 Chemicals and Reactions

This section discusses chemicals that are commonly used to stabilize soils, along with the reactions that may occur and the manner in which the chemicals stabilize the soil.

---

*Coauthored by Robert Wade Brown.

5A.4.1.1 Cation and Base Exchange with Clay Particles

Clay Mineralogy, Clay-Water Interaction, and Cation Exchange. The near-surface clay minerals are basically composed of various hydrated oxides of silicon, aluminum, iron, and to a lesser extent potassium, sodium, calcium, and magnesium (Brown, 1992). Since clays are produced from the weathering of certain rocks, the particular origin of a clay determines its nature and properties. Chemical elements present in a clay are aligned or combined in a specific geometric pattern referred to as *structural* or *crystalline lattice*, which is generally sheetlike in appearance. This structure, coupled with ionic substitution, accounts principally for both the various clay classifications and their specific physical and chemical behavior.

Owing to their loose crystalline structure, most clays exhibit the properties of ion exchange and moisture absorption and adsorption. Among the more common clays, the activity of the clays—defined in general terms as the capacity for electrochemical interaction with water and chemicals in the soil—decreases in the following order: montmorillonite, illite, attapulgite, chlorite, and kaolinite. A specific clay may adsorb and absorb water to varying degrees—from a single layer to six or more layers—depending on such factors as its structural lattice, presence of exchange ions, temperature, and environment. The moisture taken on by a clay may be described as one of three basic forms: interstitial or pore water, surface adsorbed water, or crystalline interlayer water. This combined moisture accounts for the differential movement (e.g., swelling and shrinkage) problems encountered with expansive soils. If this movement is to be controlled, each of these forms of moisture must be controlled and stabilized.

Interstitial water and surface adsorbed water both occur within the soil mass external to individual soil grains and are generally accepted as capillary moisture. The interstitial water is held by interfacial tension, whereas the surface adsorbed water is held by molecular attraction between the clay particles and the dipolar water molecules. Variations in this combined moisture are believed to account for the principal volume change potential of the soil (see also Sec. 1).

Interlayer moisture is the water situated within the crystalline layers of the clay. The amount of this water that can be accommodated by a particular clay depends on three primary factors: crystalline spacing, elements present in the crystalline structure, and presence of exchange ions. For example, bentonite (sodium montmorillonite) will swell approximately thirteen times its original volume when saturated in fresh water. If the same clay is added to water containing sodium chloride (NaCl), the expansion is reduced to about threefold. If the bentonite is added to a calcium hydroxide [$Ca(OH)_2$] solution, the expansion is suppressed even further, to less than twofold. This reduction in swelling is produced principally by ion exchange within the crystalline lattice of the clay. The absorbed sodium ions (Na^+) or calcium ions (Ca^{2+}) limit the space available to the water and cause the clay lattice to collapse and further decrease the capacity.

As a rule (and as indicated by the preceding example), the divalent cations such as Ca^{2+} produce a greater collapse of the lattice than the monovalent ions such as Na^+. Exceptions to this rule may be found with potassium (K^+) and hydrogen (H^+) ions. The potassium ion, owing to its atomic size, is believed to fit almost exactly within the cavity on the oxygen layer. Consequently, the structural layers of the clay are held more closely and more firmly together. As a result, the K^+ becomes abnormally difficult to replace by other exchange ions.

The hydrogen ion, for the most part, behaves like a divalent or trivalent ion, probably because of its relatively high bonding energy. Therefore, in most cases the presence of H^+ interferes with the cation exchange capacity of most clays. This has been verified by several authorities. Orcutt and coworkers (1955) showed that sorption of Ca^{2+} by halloysite was increased by a factor of 9 as the pH was increased from 2 to 7. Although these data are limited and qualitative, they are sufficient to establish a trend. Grim (1953, 1962) indicated that this trend would be expected to continue to a pH range of 10 or higher. Stabilization of clays with cement or lime represents one condition in which clays are subjected to Ca^{2+} at high pH (see Secs. 5A.4.1.2 and 5A.4.1.3). Under any conditions, as the concentration of exchanged ions within the clay increases, the capacity of a soil for ion exchange decreases, and hence the potential for moisture sorption and desorption (swelling/shrinkage) also decreases.

The preceding discussions have referred to changes in the potential for volumetric expansion caused by induced cation exchange. In nature, various degrees of exchange preexist, giving rise to widely variant soil behavior even among soils containing the same type and amount of clay. For example, soils containing Na^+ substituted montmorillonite will be more volatile (expansive) than will soils containing montmorillonite with equivalent substitution of Ca^{2+} or Fe^{3+}.

To this point, the discussion has been limited to inorganic ion exchange. However, available data suggest that organic ion adsorption might have even more practical importance to construction problems (McLaughlin et al., 1976). The exchange mechanism for organic ions is basically identical to that discussed above except that, in all probability, more organic sorption occurs on the surface of the clays than in the interlayers, and once attached is more difficult to exchange. Gieseking (1939) reported that montmorillonitic clays lost or reduced their tendency to swell in water when treated with several selected organic cations. The surface that surrounds the clay platelets can be removed or reduced by certain organic chemicals. When this layer shrinks, the clay particles tend to pull closer together (flocculate) and create macropores or shrinkage cracks (fractures in the macrofabric). The effect of this cracking is to increase the permeability of the soil (Foreman and Daniel, 1986), which can be helpful to reduce runoff and ponding and to facilitate penetration by chemicals for stabilization. The extent of these benefits would depend on the performance of the specific chemical product. From a broad viewpoint, certain chemical qualities—such as high pH, high OH substitution, low ionic radii, high molecular size, high polarity, high valence (cationic), and highly polar vehicles—tend to help stabilize active clays. Examples of organic chemicals that possess a combination of these features include polyacrylamides, polyvinylalcohols, polyglycolethers, polyamines, polyquarternaryamines, pyridine, collidine, and certain salts of each.

Because none of the organic chemicals listed above possess all the desired qualities, they are often blended with other additives to enhance their performance. For example, the desired pH can be attained by the addition of hydrated lime $[Ca(OH)_2]$, hydrochloric acid (HCl), or acetic acid (C_2H_3OOH); the polar vehicle is generally satisfied by dilution with water; high molecular size can be accomplished by polymerization; and surfactants can be used to improve penetration of the chemical through the soil. Additional details concerning the chemical reactions of base exchange in montmorillonite are given in the next two sections.

Organic chemicals generally can be formulated to be far superior to lime with respect to stabilization of clay. The organic products can be rendered soluble in water for easy introduction into the soil. Chemical characteristics can be more finitely controlled, and the stabilization process is more nearly permanent. About the only advantages lime has over specific organics, at present time, are lower treatment cost, more widespread usage (general knowledge), and greater availability.

Chemistry of Cation Exchange. The chemistry of cation exchange and moisture capacity within a particular clay is neither exact nor predictable. A given clay under seemingly identical circumstances will often indicate variations in cation exchange, as well as the extent to which a particular cation is exchanged. The specifics with this exchange capacity dictate the affinity for water and bonding to the structure of the clay.

Chemical Bonding in Cation Exchange. The principal conditions that create the affinity of a clay to cation bonding include the following:

1. *Broken bonds around the edges of the silica-alumina units.* These broken bonds give rise to unsatisfied negative charges that can be balanced by adsorbed cations. In montmorillonite, this accounts for about 20% of the cation exchange capacity.
2. *Substitutions of cations within the crystalline lattice.* Examples include Al^{3+} for Si^{4+} in the tetrahedral sheet, and ions of lower valence, particularly Mg^{2+} for Al^{3+}, in the octahedral sheet. The latter substitution results in unbalanced negative charges within the structural units of the clay (see Fig. 5A.147). These exchanges account for about 80% of the total cation exchange capacity of montmorillonite.
3. *Hydrogens of exposed hydroxyl groups that are replaced by exchangeable cations.* There is some doubt that this occurrence is substantial in the cation exchange reaction, since it seems that the hydrogen bond to the hydroxyl group would resist the exchange reaction (Grim 1962).

A simplified relationship between the exchanged cation and the substituted cation, as well as that relationship to adsorbed water, is also shown in Fig. 5A.147. Water held directly on the surface

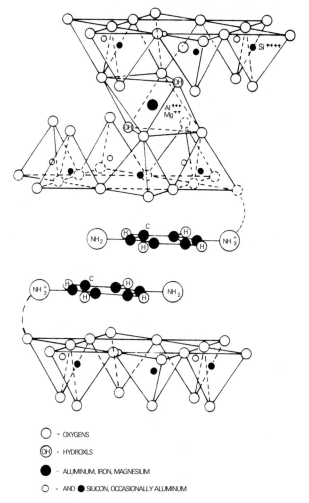

FIGURE 5A.147 Cationic substitution within clay crystalline structure (from Brown, 1992).

of the unit is in a physical state different from liquid water. Generally, the thickness of the nonliquid layer is restricted to 3 to 10 molecular layers of interlayer water (Brown, 1989).

The manner in which adsorbed ions on the surface of clay affect the adsorption of water can be described in simple terms as follows:

1. Adsorbed ions may serve as a bond to hold the clay mineral units together or limit the distance over which they can be separated. Either phenomenon would resist the intrusion of water.
2. The adsorbed cations may become hydrated (that is, develop an envelope of water molecules), and then might either interfere with or influence the configuration of adjacent adsorbed water molecules.
3. The size and geometry of the adsorbed ion may influence the manner in which it fits into the configuration of the adsorbed water molecules. This would thereby influence the nature of the configuration of the adsorbed water molecules and the extent to which the configuration could develop.

Hence it follows that the type of bonding (or the energy required to break this bond) suggests the ease by which water can be removed or exchanged from the clay. Temperature is generally the source of that energy. Water lost at various levels of temperature can be categorized as (*a*) water in the pores (interstitial), on the surface, and on the surface and around the edges of the clay mineral particles (surface adsorbed); (*b*) interlayer water between the unit-cell layers of the minerals; and (*c*) the water that occurs within the tubular openings between elongated structural units (sepiolite, attapulgite, polygorskite). Type *a* water is easily removed, substantially in its entirety, with little energy such as drying at slightly above room temperature. Types *b* and *c* water require definite energy for their removal. Specifically, in the case of montmorillonite, temperatures in the range of about 100°C (212°F) are required for the removal of interlayer water. Some clay minerals—montmorillonite in particular—rehydrate with great difficulty if the dehydration is absolutely complete but with great ease if only trace amounts of water remain between unit layers.

The amount of energy required to liberate a water molecule depends on the strength of the bond between that molecule and the clay particle. These bonds typically involve intermolecular electrostatic forces rather than the more conventional ionic or covalent bonds. [Ionic bonds (typical of salts) result from the complete transfer of electrons. Covalent bonds result from the sharing of a pair of electrons, one supplied by each atom forming the bond. (Shared electrons must be of opposite spin.)] The electrostatic (coulombic) forces generally develop from some variation of hydrogen bonding, ion-dipole attraction, or dipole-dipole attraction. A dipole is said to exist when two equal and opposite charges are separated in space. In a pictorial representation (see Fig. 5A.148), one end of the molecule bears a partial negative charge whereas the other end has a positive charge of equal value. Molecular shapes plays an important role in determining polarity, as bond polarity alone does not necessarily lead to a polar molecule. It is entirely possible for a molecule to contain polar bonds but for the entire molecule to be nonpolar, as is the case for CO_2. The two C–O bond dipoles in the linear CO_2 molecule cancel each other, and therefore their combined effects cancel. Hydrogen bonding is an exceptionally strong dipole-dipole attraction and results from the situation where a hydrogen atom serves as a bridge between two electronegative atoms. The hydrogen is held by a rather strong covalent bond (about 75 to 100 kcal/mol depending on the electronegative atom) and by a hydrogen bond (4 to 10 kcal/mol) of purely electrostatic forces. For hydrogen bonding to become predominate, both electronegative atoms must come from the group F, O, or N (Fig. 5A.148).

FIGURE 5A.148 Dipole moments and hydrogen bonding (from Brown, 1992).

In the absence of a true dipole moment, electrons in constant movement can, at any instant, exist in an unequal distribution, thus creating in effect a net dipole. The electrostatic forces so produced are referred to as van der Waals forces. The larger the molecule, the stronger the van der Waals forces. This, at least in part, gives rise to the organic treatment or modification of clay.

Organic Modification of Clay. Many clay units have a natural attraction to specific organic ions through van der Waals and coulombic forces. Larger organic ions are difficult, if not impossible, to replace by smaller ions owing to the intensity of the van der Waals forces. Furthermore, whereas smaller ions are absorbed only up to the exchange capacity, larger ions may be absorbed to a greater extent and are not disassociated.

The dipolar nature of water (plus its capacity for hydrogen bonding) makes it an excellent vehicle for the implementation of cation exchange or organic modification. The most important reactions for modification of clays include absorption, cation exchange, and interlocation (Evans and Pancoski, 1989). Interlocation refers to the process of absorption owing to the pillaring of clay. Pillaring is the process whereby the clay platelets are forced apart, thereby creating greater porosity. Both effects provide greater access to the surface for organic cations to replace exchangeable inorganic cations (see Fig. 5A.149). The organic cations are more restrictive to the inclusion of water than are the inorganic cations. In fact, the organically modified clays become organophilic (hydrophobic) rather than hydrophilic when treated with long-chain organic amines. The structure of smectite modified by an organic amine is depicted in Fig. 5A.149.

In addition to the foregoing, many organic chemicals tend to shrink the double diffuse layer that surrounds the clay particles, causing the clay particles to flocculate and the soil skeleton to shrink. The net result is the formation of cracks referred to as syneresis cracks. The combination of these effects, coupled with attendant desiccation, increases the permeability of the clay

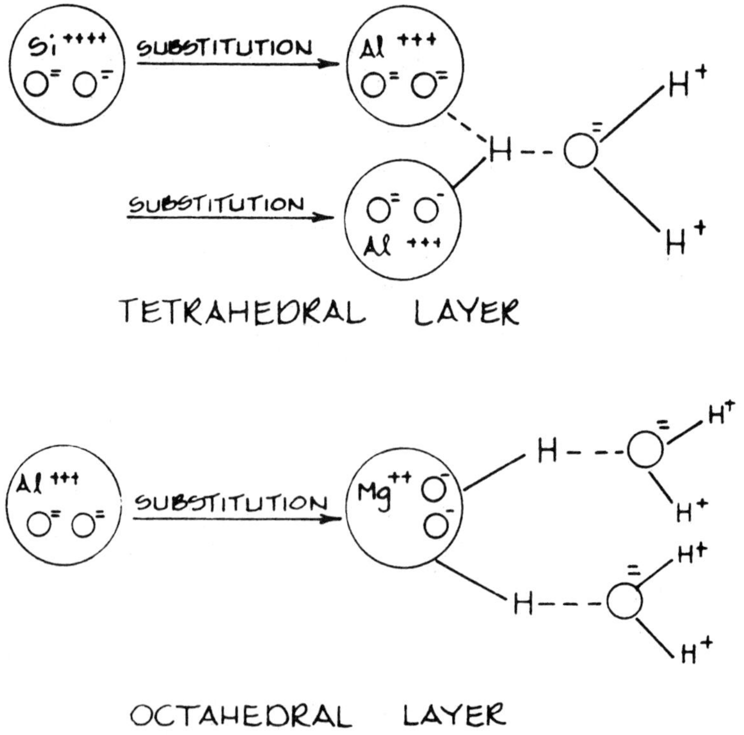

FIGURE 5A.149 Structure of organically modified smectite layers (from Grim, 1968).

(Broderick and Daniel, 1990). The exposure of greater surface area may further facilitate the base exchange of certain organic molecules (Jordan, 1949).

5A.4.1.2 Cement. The most common application of cement-stabilized soils is for roadway and airfield pavement systems, but cement stabilization has also been used successfully in many other types of applications, including the following:

1. Strengthening and stiffening bearing soils to support structures by the overexcavation/replacement method (Sec. 5A.2)
2. Reducing the potential for swelling and shrinkage in expansive soils
3. Reducing the potential for liquefaction in liquefiable sand and silty sand deposits
4. Slope protection (erosion control) for embankments, dams, canals, reservoirs, etc.
5. Strengthening of embankment materials to allow steeper side slopes
6. Repairs of failed slopes

Except for grouting techniques, which are discussed in Sec. 5B, cement stabilization is achieved by wetting or drying the soil to an appropriate moisture condition, mixing the cement into the soil, and compacting the stabilized soil into place. The soil to be stabilized may be either an in situ soil that is excavated and replaced in the same location or a borrow material that is used in a fill.

Cement can be used to stabilize all types of soils except organic soils containing more than about 1 to 2% organic matter (Ingles and Metcalf, 1973). Fat (highly plastic) clays are difficult to stabilize with cement for two reasons: (*a*) It is difficult to mix the cement into the clay in the field, and (*b*) high percentages of cement are required to effect a significant change in engineering properties. Pretreatment of fat clays with a small percentage (2 to 3%) of either hydrated lime or cement is a common method used to overcome these difficulties. The pretreatment, also called *modification*, reduces the plasticity of the soil and renders the soil more workable. After a curing period of 1 to 3 days, the modified soil is stabilized with cement in the usual way. Certain salt solutions, especially sulfates, can disrupt the structure of the cement and thus reduce the effectiveness of the cement. This problem can be overcome in most cases by increasing the cement content. Some salts, if present in the soil-cement mixture at the time the cementitious reactions occur, may have a beneficial effect (e.g., Lambe et al., 1960).

Portland cement, because it is readily available and is relatively inexpensive, is the most widely used type of cement for stabilizing soils, but any type of cement may be used. Ordinary Portland cement (type I) and air-entraining cement (type IA) have been used extensively and have given about the same results (Little et al., 1987). Type II cement, because of its greater resistance to sulfates, has been used more frequently in recent years. High early strength cement (type III) has produced higher strengths in some soils. Subsequent discussions about cement will be limited to Portland cement, because it is used predominantly in stabilizing soils.

Portland cements are hydraulic cements—that is, they set and harden by reacting with water (hydration). The four principal compounds in Portland cement are the following (PCA, 1968):

Name of compound	Chemical formula	Abbreviation
Tricalcium silicate	$3CaO \cdot SiO_2$	C_3S
Dicalcium silicate	$2CaO \cdot SiO_2$	C_2S
Tricalcium aluminate	$3CaO \cdot Al_2O_3$	C_3A
Tetracalcium aluminoferrite	$4CaO \cdot Al_2O_3 \cdot Fe_2O_3$	C_4AF

The primary reactions that occur when water is added to Portland cement can be summarized as follows:

$$2(3Ca \cdot SiO_2) + 6H_2O \rightarrow 3CaO \cdot 2SiO_2 \cdot 3H_2O + 3Ca(OH)_2$$

$$2(2Ca \cdot SiO_2) + 4H_2O \rightarrow 3CaO \cdot 2SiO_2 \cdot 3H_2O + Ca(OH)_2$$

$$3\text{CaO} \cdot \text{Al}_2\text{O}_3 + 12\text{H}_2\text{O} + \text{Ca(OH)}_2 \rightarrow 3\text{CaO} \cdot \text{Al}_2\text{O}_3 \cdot \text{Ca(OH)}_2 \cdot 12\text{H}_2\text{O}$$

$$4\text{CaO} \cdot \text{Al}_2\text{O}_3 \cdot \text{Fe}_2\text{O}_3 + 10\text{H}_2\text{O} + 2\text{Ca(OH)}_2 \rightarrow 6\text{CaO} \cdot \text{Al}_2\text{O}_3 \cdot \text{Fe}_2\text{O}_3 \cdot 12\text{H}_2\text{O}$$

where $\text{CaO}_2 \cdot \text{SiO}_2 \cdot 3\text{H}_2\text{O}$ = calcium silicate hydrate (tobermorite gel)
Ca(OH)_2 = calcium hydroxide (hydrated lime)
$\text{CaO} \cdot \text{Al}_2\text{O}_3 \cdot \text{Ca(OH)}_2 \cdot 12\text{H}_2\text{O}$ = calcium aluminate hydrate
$\text{CaO} \cdot \text{Al}_2\text{O}_3 \cdot \text{Fe}_2\text{O}_3 \cdot 12\text{H}_2\text{O}$ = calcium aluminoferrite hydrate

C_3S hardens quickly and is primarily responsible for initial set and early strength. C_2S hardens slowly and contributes mainly to strength increase at ages beyond 1 week. C_3A liberates a large amount of heat during the first few days of hardening and contributes slightly to early strength development. Sulfate-resistant cements have less than 8% of C_3A. The formation of C_4AF during manufacturing reduces the clinkering temperature and thus assists in the manufacturing process. C_4AF hydrates rather rapidly but contributes very little to strength. A detailed discussion on the chemistry, preparation, and properties of concrete can be found in Sec. 3A.

Unhydrated Portland cements contain particles in the range of 0.5 to 100 μm, with a mean of about 20 μm (Ingles and Metcalf, 1973). Calcium silicate hydrate (CSH), also known as *tobermorite,* is the predominant cementing compound in hydrated Portland cement. Although the specific surface of dry portland cement powder is only about 0.3 m^2/g (1460 ft^2/lb), the tobermorite gel after hydration has a specific surface of about 300 m^2/g (1.46 × 10^6 ft^2/lb) (Little et al., 1987). In predominately coarse-grained soils, this large specific surface is responsible for the cementing action of cement pastes owing to adhesion forces to adjacent surfaces of particles. In fine-grained soils, the clay phase may also contribute to the stabilization through solution in the high-pH environment and reaction with free lime from the cement to form additional CSH. The crystallization structure formed by the hardened cement is primarily extraneous to the soil particles. This structure can be disrupted by subsequent swelling of soil particle groups (clods) if insufficient cement is used.

The amount of cement needed for treatment depends on the type of soil being treated and the intended use and desired engineering characteristics of the treated soil. The extent of treatment can be classified by two general categories—modification and stabilization. *Modification* occurs when a low percentage of cement is incorporated within a soil that improves its workability owing to a reduction in plasticity and an agglomeration of particles. If a high percentage of cement is mixed with the soil, the surface molecular properties of the soil are changed and the grains are cemented together. Typical values of cement content used to stabilize soils according to their AASHTO and USCS group symbols are given in Table 5A.12, which can be used as a general guideline to estimate the amount of cement that may be required to stabilize a particular soil.

TABLE 5A.12 Typical Cement Requirements According to AASHTO and USCS Soil Groups

AASHTO group symbol	Corresponding USCS group symbols*	Usual range in cement requirement†	
		Percent by volume	Percent by weight
A-1-a	GW, GP, GM, SW, SP, SM	5 to 7	3 to 5
A-1-b	GM, GP, SM, SP	7 to 9	5 to 8
A-2	GM, GC, SM, SC	7 to 10	5 to 9
A-3	SP	8 to 12	7 to 11
A-4	ML, CL-ML	8 to 12	7 to 12
A-5	ML, MH, CH	8 to 12	8 to 13
A-6	CL, CH	10 to 14	9 to 15
A-7	OH, MH, CH	10 to 14	10 to 16

*Based on correlation presented by Air Force.

†For most *A* horizon soils, the cement content should be increased by 4 percentage points if the soil is dark gray to gray, and by 6 percentage points if the soil is black.

As a general rule, the cement requirement increases as the fine-grained fraction increases. The one exception to this rule is that poorly graded (uniform) sands require more cement than sandy soils containing some silt and clay.

The actual proportion of cement to be used on any particular project should be determined from an appropriate mixture design evaluation. The mixture design procedure typically consists of Proctor tests on the cement-stabilized soil to determine the approximate optimum water content and maximum dry density for either standard or modified energy, followed by a series of laboratory tests on cement-stabilized specimens with varying cement contents to determine the minimum cement content that will achieve the desired engineering properties. For example, if strength is the primary requirement, the general procedures for a typical mixture design would be as follows:

1. Establish a minimum desired strength based on rational engineering procedures. In soil-cement applications, the minimum strength is typically specified as unconfined compressive strength. In applications where the cement-stabilized soil will be subjected to significant confining pressures, it is more appropriate to specify strength in terms of friction angle and cohesion intercept determined by triaxial or other appropriate strength tests.
2. Estimate the required cement content based on personal experience or guidelines provided by others (such as Table 5A.12).
3. Conduct a series of Proctor tests on the cement-stabilized soil at three cement contents—the estimated value from step 2 and one value above and one value below the estimated value. It is common practice to use 1% increments for estimated cement contents of 5% or less and 2% increments for estimated cement contents of 6% or more. As an illustration, if the estimated cement content is 8%, the Proctor tests typically would be conducted at cement contents of 6, 8, and 10%. Determine the optimum water content and maximum dry density for each of the three cement contents.
4. Perform a series of unconfined compression tests on specimens compacted at the corresponding optimum water content to the corresponding maximum dry density for each of the three cement contents used in step 3. Cement-stabilized specimens are generally stored (cured) at room temperature in a relative humidity of approximately 100% for periods of 2, 7, and 28 days and then immersed in water for 4 h prior to testing (see ASTM D1632 and D1633). Usually at least three replicate specimens are tested for each cement content and curing period.
5. Three situations are possible depending on how the measured values of strength obtained in step 4 compare with the specified minimum value developed in step 1.
    a. The measured values of unconfined compressive strength at the three cement contents are both above and below the specified minimum value. The actual (design) cement content to be used can be determined as the lowest cement content that yields strengths greater than or equal to the minimum desired value. In cases where a small difference in cement content would significantly affect the economics of the project, steps 2 and 3 could be repeated at intermediate cement contents to better establish the most economical cement content for the desired strength characteristics.
    b. The measured values of strength at all three cement contents are less than the specified minimum value. Repeat steps 2 to 4 at three cement contents higher than the previous maximum value.
    c. The measured values of strength at all three cement contents are greater than the specified minimum value. Repeat steps 2 to 4 at three cement contents lower than the previous minimum value.
6. Additional testing can be conducted to determine if other important engineering characteristics are met for the preliminary design cement content established in steps 2 through 5. If the other criteria are met, the preliminary design cement content is the final design cement content. If not, the final design cement content is determined based on the results of the additional tests, ensuring that the strength requirement is also met. For example, in soil-cement applications in pavement systems, wet-dry and freeze-thaw tests (ASTM D559 and D560) are also conducted. In applications for building foundations, compressibility and liquefaction resistance are also important engineering properties.

The presence of organic material or sulfates in a soil may inhibit the proper hydration of the cement. Organics with low molecular weights, such as nucleic acid and dextrose, retard hydration and reduce strength (Clare and Sherwood, 1954). Certain types of organics, such as undecomposed vegetation, may not adversely affect cement stabilization. For situations where organics are present in the soil, Little and coworkers (1987) recommend conducting a pH test on a 10:1 mixture (by weight) of soil and cement. If the pH of the soil-cement 15 min after mixing is at least 12.1, the organic content probably will not interfere with normal hardening. One method by which organics repress the hydration of cement is absorption of calcium ions liberated during the hydrolysis of the calcium silicates and aluminates in the cement grains (Clare and Sherwood, 1954, 1956). These ions are subsequently not available to form the compounds constituting the set cement matrix. The addition of a material with readily available calcium, such as calcium chloride ($CaCl_2$) or hydrated lime, may satisfy the adsorptive capacity of the organic matter before hydration of the bulk of the cement has begun, thus enabling the soil to be successfully stabilized with cement. For the calcium additive to be effective, it must be present in a sufficient quantity to completely satisfy the adsorptive capacity of the organic matter, as illustrated by the following data from Clare and Sherwood (1956) for an otherwise clean silica sand stabilized with 10% ordinary Portland cement:

Additive (by weight of dry soil)	Unconfined compressive strength	
	psi	kPa
None (clean silica sand)	355	2450
0.01% tartaric acid	285	1970
0.02% tartaric acid	190	1310
0.03% tartaric acid	150	1030
0.04% tartaric acid	25	172
0.05% tartaric acid	10	69
0.10% tartaric acid + 0.25% calcium chloride	10	69
0.10% tartaric acid + 0.50% calcium chloride	10	69
0.10% tartaric acid + 1.00% calcium chloride	260	1793

Research by Sherwood (1962) has shown that the presence of sulfates can significantly affect the durability, integrity, and strength of cement-stabilized cohesive soils but has relatively little influence on cement-stabilized granular soils. These trends are illustrated in Fig. 5A.150, which shows how immersion in magnesium sulfate solutions of different concentrations affected the unconfined compressive strength of cement-stabilized specimens of London clay, silica sand, and various mixtures of the two soils. The presence of magnesium sulfate in the immersion water reduced the strength of all specimens containing London clay except one. In contrast, the strengths of the 100% silica sand specimens were somewhat greater when immersed in magnesium sulfate solutions than when immersed in water without magnesium sulfate. Note that 0.5% magnesium sulfate was sufficient to cause complete disintegration of the stabilized London clay. Similar trends were found for the same soils immersed in calcium sulfate solutions. Owing to the problems that sulfates can cause in cement-stabilized cohesive soils, Little and colleagues (1987) suggest that the use of cement be avoided for fine-grained soils containing more than about 1% sulfates.

5A.4.1.3 Lime. Two types of lime are commonly used to stabilize cohesive soils—CaO (calcium oxide, also known as *quicklime*) and $Ca(OH)_2$ (calcium hydroxide, also referred to as *hydrated lime* or *construction lime*). Hydrated lime is the most frequently used lime product for soil stabilization in the United States, and quicklime is used more often in Europe (Bell, 1988). Both quicklime and hydrated lime are produced by burning limestone ($CaCO_3$, calcium carbonate), as illustrated by the following reactions (Fox, 1968):

$$CaCO_3 + Heat \rightarrow CaO + CO_2$$

$$CaO + H_2O \rightarrow Ca(OH)_2 + Heat$$

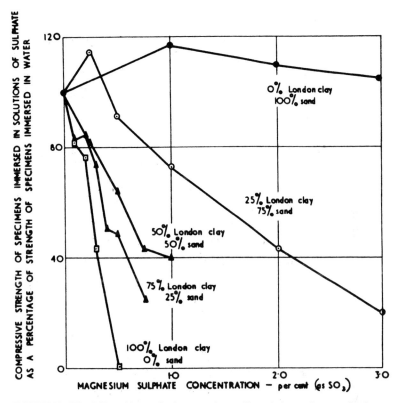

FIGURE 5A.150 Effect of immersion in magnesium sulfate solution on the unconfined compressive strength of London clay-silica sand mixtures stabilized with 10% ordinary Portland cement (from Sherwood, 1962).

Less pure forms of lime are also sometimes used to stabilize soils. These alternate forms include monohydrated dolomitic lime [Ca(OH)$_2$ · MgO], dolomitic quicklime (CaO · MgO), and by-product limes (Little et al., 1987). Ground limestone (CaCO$_3$), also known as *lime dust* or *agricultural lime,* is ineffective in chemically stabilizing soils but is sometimes used as a filler to increase the fines content of a soil, which modifies the gradation of the soil for better compaction and sometimes enhances stabilization by other chemicals.

Treatment with lime is generally effective in soils with plasticity indices in the range of 10 to 50 (Bell, 1988). Lime stabilization is typically ineffective in soils with little or no clay content because the primary improvements in engineering behavior are produced by reactions between the lime and the clay minerals. All clay minerals react with lime, with the strength of the reaction generally increasing in proportion to the amount of available silica. Thus, the three-layer clay minerals (e.g., montmorillonite, illite, chlorite) are more reactive than the two-layer clay minerals (e.g., kaolinite). The availability of the silica in the clay minerals is also important. Hence, illite and chlorite are much less reactive than montmorillonite because their silicas are bound to other silicas by ions that are not readily exchangeable.

Four basic types of reactions occur in cohesive soils treated with lime: (*a*) carbonation, (*b*) cation exchange, (*c*) flocculation-agglomeration, and (*d*) pozzolanic cementation. Carbonation occurs if carbon dioxide from the air or stagnant water gains entry into the lime-soil matrix and converts the lime back to calcium carbonate. Calcium carbonate is a weak cement because it is soluble in acidic water. Carbonation is undesirable because it reduces that amount of lime available to produce the primary cementitious reactions, which are pozzolanic. Carbonation can be prevented by using appropriate construction techniques.

Lime mixed with water results in free calcium cations that may replace other cations within the exchange complex of the soil. In some instances the exchange complex may be saturated with Ca^{2+} before the addition of lime, yet cation exchange may still occur, since the capacity for cation exchange increases as the pH increases (TRB, 1987). Cation exchange is at least partially responsible for the flocculation and agglomeration of clay particles that occur in lime-treated soils (Herzog and Mitchell, 1963). The rapid formation of calcium aluminate hydrate cementing materials probably significantly assists in the flocculation-agglomeration process (Diamond & Kinter 1965). The net result of flocculation and agglomeration is a change in texture of the soil as the clay particles clump together into larger clods.

The pozzolanic reactions that occur in lime-treated soils, although not completely understood, are similar to those that occur in cement-treated soil (see Sec. 5A.4.1.2). We know that the lime and water react with silica and alumina in the soil to form various cementitious compounds (TRB, 1987). Typical sources of alumina and silica in soils include clay minerals, quartz, feldspars, micas, and similar silicate or alumino-silicate materials (either crystalline or amorphous). The addition of lime also raises the pH of the soil, which is beneficial because it increases the solubility of the silica and alumina present in the soil. If a significant amount of lime is added to the soil, the pH may reach 12.4, which is the pH of saturated lime water. The following is an oversimplification of the reactions that occur in a lime-treated soil (TRB, 1987):

$$Ca(OH)_2 \rightarrow Ca^{2+} + 2(OH)^-$$

$$Ca^{2+} + 2(OH)^- + SiO_2 \text{ (clay silica)} \rightarrow CSH$$

$$Ca^{2+} + 2(OH)^- + Al_2O_3 \text{ (clay alumina)} \rightarrow CAH$$

where $C = CaO$
$S = SiO_2$
$A = Al_2O_3$
$H = H_2O$

These reactions occur only if water is present and able to bring Ca^{2+} and $(OH)^-$ to the surfaces of the clay particles (that is, while the pH is still high) (Bell, 1988). Hence, the reactions will not occur in dry soils and will cease in a previously wet soil if it dries out. The primary characteristics of a soil that determine the effectiveness of lime treatment include the following (TRB, 1987):

1. Type and amount of clay minerals
2. Silica-alumina ratio
3. Silica-sesquioxide ratio
4. pH
5. Organic carbon content
6. Natural drainage
7. Degree of weathering
8. Presence of carbonates, sulfates, or both
9. Extractable iron

Caution is needed when one is considering lime stabilization of soils that contain sulfates and carbonates. Sherwood (1962) provided experimental results showing that lime-stabilized soils can be deleteriously affected by sulfates in a similar manner to cement-stabilized soils. A first set of experiments was conducted in which specimens of London clay were stabilized with 10% lime and cured at constant water content for 7 days. The specimens were then immersed in solutions of either sodium sulfate or magnesium sulfate at concentrations ranging from 0 to 1.5%. In all cases, the specimens immersed in the sulfate solutions cracked and swelled to the extent that they were too weak to be tested in unconfined compression. A second series of experiments was then performed in which varying amounts of calcium, sodium, and magnesium sulfates were mixed with the soil before adding 10% lime. Two sets of specimens were made for each sulfate concentration and were cured at constant moisture content. After 7 days of curing, one set of specimens was immersed in water while the other set was allowed to continue curing at constant moisture content. At the end of a second 7-day period, both sets of specimens were tested in unconfined compression. The following trends can be observed from the results of these tests, which are given in Fig. 5A.151:

1. For the specimens cured at constant moisture condition for 14 days, the calcium sulfate had a substantial beneficial effect, and the magnesium sulfate and sodium sulfate had no significant influence.
2. All the immersed specimens containing sulfates swelled and cracked, resulting in a loss in strength, with a general trend of decreased strength with increased sulfate content. At the higher sulfate concentrations ($\geq 2\%$), the losses in strength were considerable, ranging from about 65 to 85%.

Poststrength testing analyses of the 2% calcium sulfate specimens indicated that the calcium sulfate had reacted with the clay and that ettringite $\{Ca_6 \cdot [Al(OH)_6]_2 \cdot (SO_4)_3 \cdot 26H_2O\}$ had been formed as a result of the reaction. Ettringite is a very expansive material that is one of the materials responsible for the deterioration of concrete by sulfate attack.

Mitchell (1986) presented a case history in which the presence of sodium sulfate in the soil contributed to the failure of lime-stabilized subbases in Las Vegas, Nevada. The results of postmortem investigations that showed the effect of the sulfates can be summarized as follows:

1. The untreated soil contained up to 1.5% soluble sodium sulfate.
2. The lime-treated soil in the failed zones had swelled more than untreated soils also exposed to water.
3. Calcium silicate hydrates (CSH), the cementitious material produced during successful lime stabilization, were not found in specimens from the failed zones. This lack of cementation was attributed to the low pH of the treated soils (8 to 10.5), well below the minimum pH of 12.4 necessary for the pozzolanic reactions to occur.
4. Significant amounts of ettringite and thaumasite $\{Ca_6 \cdot [Si(OH)_6]_2 \cdot (SO_4)_2 \cdot (CO_3)_2 \cdot 24H_2O\}$ were found in the failed zones. Thaumasite also is very expansive and is known to contribute to the deterioration of concrete by sulfate attack.
5. Negligible amounts of ettringite and thaumasite were found in treated samples from locations where failure had not occurred.

Owing to problems such as these, Mitchell (1986) recommended that lime stabilization not be used for soils containing more than 5000 ppm (0.5%) of soluble sulfates. Petry and Little (1992) indicated that the minimum level of sulfates that may represent potential problematic situations depends on the method used to extract the sulfates from the soil and may be as low as 500 ppm (0.05%) for 1:1 soil-to-water ratios to 2000 ppm (0.2%) for 1:10 soil-to-water ratios. Additional case histories in which sulfates in lime-treated soils resulted in swelling and structural damage to pavements related to the formation of ettringite and thaumasite can be found in Hunter (1988), Grubbs (1990), and Perrin (1992). Hunter (1988) also provided a geochemical model for the formation of ettringite and thaumasite.

Note that lime stabilization in expansive soils with high sulfate contents is possible. This is confirmed by the fact that lime has been used successfully to stabilize sulfate-bearing soils in Texas for over 30 years (Petry and Little, 1992). Unfortunately, the construction and application techniques responsible for successful stabilization have yet to be identified. The following meth-

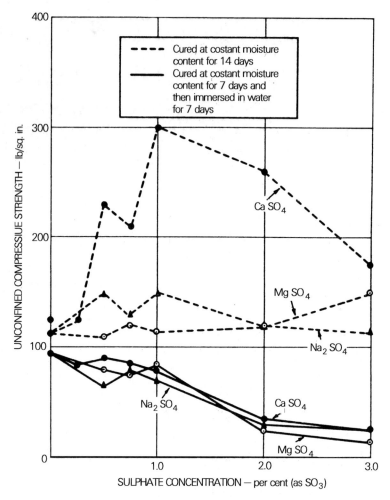

FIGURE 5A.151 Influence of sulfates on the unconfined compressive strength of London clay stabilized with 10% hydrated lime (from Sherwood, 1962).

ods have been studied or are currently being evaluated as potential techniques to overcome the detrimental aspects of sulfates in lime-treated soils:

- Pretreatment of the soils with compounds of barium (Little and Deuel, 1989; Ferris et al., 1991).
- Double applications of lime
- Use of compounds to sustain high pH values and thus promote pozzolanic reactions
- Addition of materials to enhance pozzolanic reactions

The amount of lime needed for treatment depends on the characteristics of the soil being treated and the intended use and desired engineering characteristics of the treated soil. As with cement stabilization, the treatment of soils with lime can be divided into two general classes (Ingles and Metcalf, 1973): (*a*) modification, which reduces plasticity, improves workability, increases resistance to deflocculation and erosion, and produces a rapid increase in strength as a construction expedient and (*b*) stabilization, which provides permanent increases in strength and stiffness to increase bearing capacity, reduce settlement, and so on (Table 5A.13).

Several lime-soil mixture design methods have been developed. TRB (1987) summarizes eight procedures. These procedures are of two general types—those relating to soil modification and those pertaining to soil stabilization. The Illinois procedure will be summarized below to illustrate the basic concepts for both modification and stabilization (from TRB, 1987).

Modification. The mixture design procedure for modification is based on the effect that the lime has on the plasticity index (PI) of the soil. CBR testing can also be performed but is optional.

1. The liquid limit (LL), plastic limit (PL), and PI of the soil treated with various percentages of lime are determined. The soil-lime-water mixtures are loose-cured (conditioned) for 1 h before the LL and PL tests are conducted.
2. A plot is prepared of PI versus lime content. The design lime content may be based on one of the following criteria: (*a*) The lime content above which no additional appreciable reduction in PI occurs; or (*b*) the minimum lime content that produces an acceptable reduction in the PI.
3. If desired, CBR tests may also be performed to assess the stability or swelling potential of the soil, or both. Depending upon the objectives of the lime treatment, curing and soaking the CBR specimens before testing is optional. If the results of the CBR tests indicate that a higher lime content is required for successful modification, the design lime content is changed accordingly.
4. The design lime content for field construction is increased by 0.5 to 1% to offset the effects of variability in soil properties and construction procedures.

Stabilization. The mixture design procedure for stabilization is based on the results of unconfined compressive strength tests:

1. Standard Proctor tests (AASHTO T-99, ASTM D698) are conducted on the natural soil and soil-lime mixtures at various lime contents to determine the optimum water content and maximum dry density for the natural soil and each of the soil-lime mixtures.
2. Specimens 2 in (51 mm) in diameter and 4 in (102 mm) in height of the natural soil and soil-lime mixtures are prepared at the corresponding optimum water content and corresponding maximum dry density. The soil-lime mixtures are cured at 120°F (48.9°C) for 48 h before testing.
3. The unconfined compressive strength of the soil-lime mixture containing 3% lime must be at least 50 psi (345 kPa) greater than that of the natural soil. If not, the soil is considered unsuitable for lime stabilization. If the criterion for 3% lime is met, the design lime content is the value above which additional lime does not produce a significant increase in strength. Minimum strength requirements are 100 psi (690 kPa) for use as a subbase and 150 psi (1030 kPa) for use as a base course.
4. The design lime content for field construction is increased by 0.5 to 1% to offset the effects of variability in soil properties and construction procedures.

5A.4.1.4 Fly Ash. Two waste products are produced by the burning of powdered coal—fly ash and bottom ash (boiler slag). Bottom ash collects at the base of the furnace, and fly ash is collected from the flue gases by mechanical and/or electrostatic precipitation. The particles in bottom ash range in size from fine sand to gravel, and bottom ash serves well as structural fill in road construction (Hausmann, 1990). Fly ash is powdery and of predominantly silt size and is used for a variety of purposes, including (*a*) as a stand-alone cement, (*b*) as a partial replacement for cement in concrete, (*c*) in combination with lime, cement, or both in soil stabilization, (*d*) as a fill material in the form of lime–fly ash-aggregate (LFA), cement-fly ash-aggregate (CFA), or lime-cement-fly ash-aggregate (LCFA) mixtures, and (*e*) as a drying agent to facilitate compaction of overly wet soils.

The chemical and physical properties of fly ash vary widely and depend primarily on the mineralogy and purity of the coal and the process and equipment used to burn the coal and recover the fly ash. Owing to variations in the properties of the coal burned at any power plant, fly ash within

TABLE 5A.13 Typical Range of Hydrated Lime Content for Modification and Stabilization of Various Types of Soils (from Ingles and Metcalf, 1973)

	Typical range of hydrated lime content, % by dry weight of soil	
Type of soil	Modification	Stabilization
Fine crushed rock*	2 to 4	Not recommended
Well-graded clayey gravels	1 to 3	Approximately 3
Sands†	Not recommended	Not recommended
Sandy clay	Not recommended	Approximately 5
Silty clay	1 to 3	2 to 4
Heavy clay	1 to 3	3 to 8
Very heavy clay	1 to 3	3 to 8
Organic soils	Not recommended	Not recommended

*Lime only effective if fines are plastic.
†Lime used in bitumen stabilization for promoting adhesion; quicklime in loessial materials.

TABLE 5A.14 Typical Ranges in Values for the Chemical Composition of Fly Ash (from NCHRP, 1976)

Principal constituent	Content by weight, %
SiO_2	28 to 52
Al_2O_3	15 to 34
Fe_2O_3	3 to 26
CaO	1 to 40
MgO	0 to 10
SO_3	0 to 4

the same stockpiles can be quite heterogeneous. The primary chemical components of fly ash are silica (SiO_2), alumina (Al_2O_3), ferric oxide (Fe_2O_3), and calcium oxide (CaO), and the secondary constituents include magnesium oxide (MgO), titanium oxide (TiO_2), alkalies (Na_2O and K_2O), sulfur trioxide (SO_3), phosphorous oxide (P_2O_5), and carbon (Hausmann, 1990). Typical chemical compositions and physical properties of fly ashes are given in Tables 5A.14 and 5A.15. Fly ash is classified as either class F (pozzolanic properties only) or class C (pozzolanic and some cementitious properties) depending primarily on its CaO content (ASTM C618). Type F fly ashes are generally produced by burning anthracite or bituminous coal, and type C fly ashes are normally derived from the combustion of subbituminous or lignite coal. In the United States, characteristics of the fly ashes from the major coal-producing regions can be broadly summarized as follows (Little et al., 1987; refer to Fig. 5A.152):

- Fly ashes from the bituminous coals of the Appalachian region generally behave as true pozzolans, with little or no inherent cementitious properties.
- Fly ashes from the midcontinent coals, which are mostly bituminous, have some inherent cementitious properties owing to the CaO present in the ash.

TABLE 5A.15 Typical Physical Properties for Fly Ash (from NCHRP, 1976)

Property	Description or typical range in values
Color	Usually light gray, but can vary from light tan through black
Shape	Spherical, solid, or hollow
Glass content (amorphous material)	71 to 88%
Size	1 to 80 μm (3.9×10^{-5} to 3.1×10^{-3} in)
Specific surface	0.2 to 0.8 m²/g (8800 to 35,000 in²/oz)

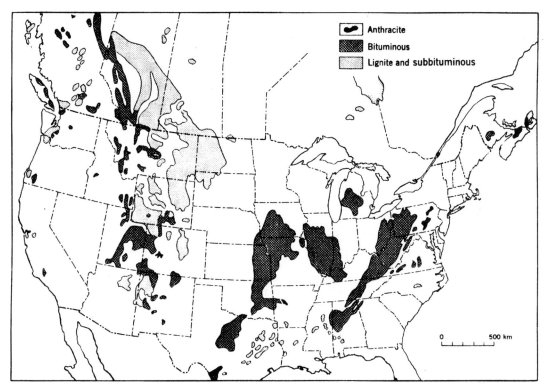

FIGURE 5A.152 Coal fields of the United States and southern Canada (from Skinner and Porter, 1987).

- Fly ashes from the subbituminous and lignite coals of the northern and western plains states have a high natural CaO content and are highly cementitious even without the addition of lime.

Several kinds of chemical reactions occur in naturally cementitious fly ash and lime-fly ash mixtures. These reactions are complex, owing in part to the heterogeneous nature of fly ash and have not been completely defined. The principal cementitious reactions are similar to those that take place in cement and lime-treated soils (see Secs. 5A.4.1.2 and 5A.4.1.3): Alumina and silica in the fly ash react with lime (either naturally occurring in the fly ash or added quicklime or hydrated lime), in the presence of water, to produce calcium silicate hydrates and calcium aluminosilicate hydrates. An illustrative list of *some* reactions that may occur is given as follows (Minnick, 1967):

$$CaO \xrightarrow{H_2O} Ca(OH)_2$$

$$CaO \xrightarrow{H_2;CO_2} CaCO_3 + H_2O$$

$$Ca(OH)_2 \xrightarrow{CO_2} CaCO_3 + H_2O$$

$$Ca(OH)_2 + SiO_2 \xrightarrow{H_2O} xCaO \cdot ySiO_2 \cdot zH_2O$$

$$Ca(OH)_2 + Al_2O_3 \xrightarrow{H_2O} xCaO \cdot yAl_2O_3 \cdot zH_2O$$

$$\text{Ca(OH)}_2 + \text{SiO}_2 + \text{Al}_2\text{O}_3 \xrightarrow{\text{H}_2\text{O}} x\text{CaO} \cdot y\text{SiO}_2 \cdot z\text{Al}_2\text{O}_3 \cdot w\text{H}_2\text{O}$$

$$\text{Ca(OH)}_2 + (\text{SO}_3)^{2-} + \text{Al}_2\text{O}_3 \xrightarrow{\text{H}_2\text{O}} x\text{CaO} \cdot y\text{Al}_2\text{O}_3 \cdot z\text{CaSO}_4 \cdot w\text{H}_2\text{O}$$

Note that Mg^{2+} may be substituted for Ca^{2+} in all equations.

Not all fly ashes are reactive even when lime is added, so the procedures provided in ASTM C593 can be used to evaluate the suitability of a fly ash for use in lime–fly ash mixtures. Low temperatures, high organic content, and high sulfate content can have deleterious effects on the cementitious reactions. These reactions are retarded at low temperatures and are almost nonexistent at temperatures below about 40°F (14°C) (Leonard and Davidson, 1959; Little et al., 1987). The organics probably affect the cementitious reactions in fly ash in a similar manner as in Portland cement, primarily by absorbing calcium ions. If high organic content is suspected, loss-on-ignition tests (ASTM C114) can be conducted to determine carbon content. Fly ashes having sulfate contents of 5 to 10% appear to have lower rates of initial hydration (Ferguson 1993). For high sulfate contents, high initial strengths have been observed, but durability appears to be substantially reduced. The formation of ettringite, which is highly expansive, is also a potential problem when using fly ashes with high sulfate content.

The amount of fly ash used in any application depends on a variety of factors such as the desired engineering characteristics of the stabilized material, and the properties of the fly ash, the additives (if any), and the soil to be stabilized. For applications where self-cementing fly ash is used without additives, more fly ash is required to achieve the same stabilizing effect as that produced by Portland cement, with typical fly ash requirements on the order of 15 to 25%. In LFA, CFA, and LCFA applications, the stabilizer is defined by designating the total amount of stabilizer (fly ash plus additives) and the lime-to-fly-ash ratio or the cement-to-fly-ash ratio (or both). For example, an LFA mixture of 5% lime, 15% fly ash, and 80% aggregate would be specified as 20% lime plus fly ash with a 1:3 lime-to-fly-ash ratio. Lime-plus-fly-ash contents generally range from 12 to 30%, with fine-grained soils requiring higher percentages and well-graded granular soils requiring lower percentages (NCHRP 1976). The lime-plus-fly-ash content generally increases with increasing angularity and roughness of the particles. Lime-to-fly-ash ratios normally range from 1:10 to 1:2, with ratios of 1:3 to 1:4 being common. Factors that tend to increase the lime requirement are greater fines content, increased PI, and increased pozzolanic reactivity of the fly ash.

Because of the unique characteristics of each stockpile of fly ash and the complex chemical reactions that occur in fly ash-stabilized soils, mixture proportions cannot be established from separate analyses of the components (fly ash, soil, additives). Proportions are properly determined using laboratory or field-based design procedures with representative samples of the component materials to be incorporated into the field mixture. Mixture design procedures for fly ash stabilization are similar to those used for cement or lime-treated soils (Secs. 5A.4.1.2 and 5A.4.1.3) but are somewhat more complicated when additives are used, because the optimal content for the fly ash and the additive(s) must be determined. Figure 5A.153 illustrates a typical mixture design procedure for LFA when strength and durability are the most important engineering properties.

5A.4.1.5 Emulsions, Sodium Silicates, and Other Chemicals. A variety of other chemicals—including emulsions, sodium silicates, polyacrylamides, phosphoric acid, and iron and aluminum oxides—have been successfully used to stabilize soils. The use of these other chemicals in foundation applications has been limited and thus will be only briefly discussed here. A discussion of some of the less frequently used chemicals can be found in ASCE (1987).

An emulsion is a fluid formed by the suspension of a very finely divided oily or resinous liquid in another liquid. Various emulsions have been used to stabilize soils, of which the most common types are bituminous. The principal component of bituminous emulsions is *bitumen*, which refers to a class of cementitious substances composed mainly of hydrocarbons with high molecular weights. Included in the class of bitumens are asphalts, tars, pitches, and asphaltites. Bituminous emulsions are typically either anionic or cationic, depending on the predominance of the type of charges on the discontinuous phase, and are incorporated into the soil using either hot

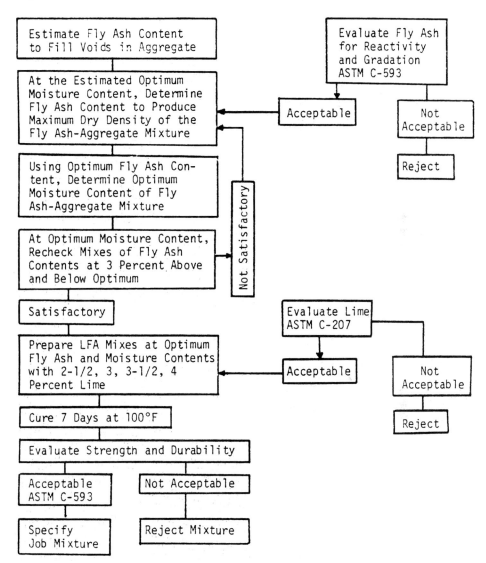

FIGURE 5A.153 Mixture design flow chart for lime–fly-ash-aggregate (LFA) mixtures (from Little et al., 1987).

or cold mixing methods. A number of nonbituminous emulsions have been developed to stabilize soils, many of which are proprietary. The primary use for emulsions in treating soils has been in pavement applications (stabilization of soils for use in subgrade, subbase, and base courses). Detailed discussions of bituminous stabilization of soils are given in Ingles and Metcalf (1973) and Little and coworkers (1987), among others.

Details on the use of sodium silicates and acrylamides in grouting applications are given in Secs. 5B.3.2.3 and 5B.3.2.4. The difference between the usage of these materials for grouting and chemical stabilization is the nature of the chemical reactions. In solution grouting, the basic chemical is combined with a hardener that, upon setting, produces a solid material that fills the pore spaces of the soil. Thus, the chemical reaction in grouting is external to the soil particles; that is, the reaction is between the primary component and the hardener. In contrast, when used in chemical stabilization, sodium silicates and polyacrylamides react chemically with the soil particles to produce the characteristics of stabilization (reduced swelling potential, increased strength and stiffness, and so on). Sodium silicates and polyacrylamides are injected as solutions into the soil.

The cementitious reactions that occur between sodium silicate and the clay fraction of a soil can be summarized as follows (Sokolovich and Semkin, 1984).

Adsorbed calcium cations are exchanged with sodium cations from the introduced reagent:

$$Ca^{2+} + 2NaOH \rightarrow 2Na^+ + Ca(OH)_2$$

Hydrated alumina interact with liberated calcium hydroxide:

$$Al(OH)_3 + NaOH \rightarrow NaAlO_2 + 2H_2O$$

$$2NaAlO_2 + Ca(OH)_2 \rightarrow Na_2O \cdot CaO \cdot Al_2O_3 \cdot 2H_2O$$

as well as

$$2Al(OH)_3 + 3Ca(OH)_2 \rightarrow Ca_3[Al(OH)_6]_2$$

Interaction of hydrated silica with calcium hydroxide and sodium aluminate is as follows:

$$Si(OH)_4 + Ca(OH)_2 \rightarrow CaO \cdot SiO_2 \cdot 3H_2O$$

$$2Si(OH)_4 + 2NaAlO_2 \rightarrow Na_2O \cdot Al_2O_3 \cdot 2SiO_2 + 4H_2O$$

Because the permanence of sodium silicate grouts has been questioned (see Sec. 5B.3.2.3), caution should be exercised when considering the use of sodium silicate for soil stabilization.

5A.4.2 Engineering Properties and Behavior of Chemically Stabilized Soils

In many instances, chemical stabilization of a soil produces significant and often dramatic changes in the engineering properties and behavior of the soil compared to its natural condition. Although the magnitude of the changes produced in any soil depend on the type and amount of chemicals used, the qualitative features of these changes in successfully stabilized soils are often similar for all the types of chemicals commonly used. In this section, results illustrating the nature and magnitude of these changes will be presented and discussed according to the different engineering properties that may be affected by the stabilization process. Most of the data and results presented will be for soils treated with cement, lime, and fly ash because of the vast amount of information publicly available for these stabilizers and the dearth of data available for most other chemicals.

5A.4.2.1 Plasticity and Workability. As discussed in Sec. 5A.4.1, chemical modification using admixtures involves mixing a relatively small amount of a chemical stabilizer into a cohesive soil to improve its workability from a construction standpoint—primarily resulting in a reduction in plasticity and an agglomeration of particles into larger clods. The net result of successful modification is a friable soil with a silty texture. These changes are visually illustrated in Fig. 5A.154 in which a highly plastic clay is shown in its natural state and after modification with hydrated lime.

Quantitative changes in the liquid limit (LL), plastic limit (PL), plasticity index (PI), and shrinkage limit (SL) are shown in Figs. 5A.155 and 5A.156 for lime and cement-treated soils, respectively, as a function of stabilizer content. In general, chemical modification of a cohesive soil reduces the liquid limit and increases the plastic limit, with both changes producing a reduction in the plasticity of the soil as indicated by a decrease in plasticity index. The reduction in plasticity can be extensive; for example, the PI of Porterville Clay (Fig. 5A.155) was reduced from 47 to 3 with the addition of 8% lime. The reduction in plasticity increases with increasing stabilizer con-

FIGURE 5A.154 Modification of highly plastic clay with lime in Vietnam in 1967, illustrating the reduction in plasticity and agglomeration of particles into clods (courtesy of Dr. Nathaniel S. Fox, Geopier Foundation Company, Atlanta, Georgia).

tent up to a certain content, above which little or no additional effect occurs. Many, but not all, soils can be made nonplastic through chemical stabilization (at contents typically used in practice), and generally the amount of a particular stabilizer required to effect a nonplastic state increases with increasing PI of the natural soil. The changes in plasticity are also time-dependent, as depicted in Fig. 5A.157. In most cases, a considerable reduction in PI occurs within the first 24 h, and most of the reduction occurs within the first week. However, the process may continue for years.

5A.4.2.2 Swelling. Areas of the United States that have a general and local abundance of high-clay, expansive soils are shown in Fig. 5A.158. The darker areas indicate those regions suffering most seriously from problems related to expansive soils. These data indicate that eight states have extensive, highly active soils and nine others have sufficient distribution and content to be considered serious. An additional 10 to 12 states have problems that are generally viewed as relatively limited. As a rule, the 17 "problem" states have soil containing montmorillonite, which is, of course, the most expansive clay mineral. The 10 to 12 states with so-called limited problems (represented by the lighter shading) generally have soils which contain clays of lesser volatility such as illite and/or attapulgite, or montmorillonite in lesser abundance.

Chemical stabilization is the most effective and permanent method available for controlling the swelling potential of expansive soils. In new construction where controlled, intimate mixing with the expansive soil is feasible, hydrated lime [$Ca(OH)_2$] is the most commonly used stabilizer for controlling swelling. Data illustrating the effectiveness of lime in reducing the swelling potential of a number of soils are given in Table 5A.16. Other stabilizers such as cement, fly ash, lime–fly ash, and organic chemicals can also be used but are generally more expensive than lime. In remedial applications, such as stabilizing the soil beneath an existing structure, lime has inherent shortcomings. Lime is difficult to introduce into the soil matrix with any degree of uniformity, penetration, or saturation largely because (*a*) lime is sparsely soluble in water and (*b*) the expansive soil needing stabilization is both impermeable and heterogeneous. In addition, the presence

5.226 SOIL IMPROVEMENT AND STABILIZATION

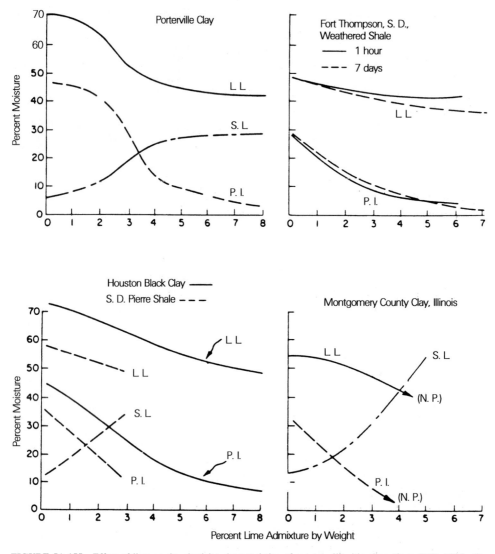

FIGURE 5A.155 Effect of lime on the plasticity characteristics of montmorillonitic clays (from TRB, 1987; after Holtz, 1969).

of sulfates or organics in the soil may produce detrimental side effects when lime, cement, fly ash, or lime–fly ash are used (see Secs. 5A.4.1.2, 5A.4.1.3, and 5A.4.1.4).

For the reasons cited in the previous paragraph, organic chemicals are the preferred stabilizer for remedial applications. As early as 1965, certain surface-active organic chemicals were evaluated and utilized with some degree of success. One chemical used in the late 1960s and early 1970s was unquestionably successful in stabilizing the swell potential of montmorillonitic clay. The chemical was relatively inexpensive and easily introduced into the soil. However, the product maintained a "nearly permanent" offensive aroma which chemists were never able to mask. Generally, this product was a halide salt of the pyridine-collidine-pyrillidene family. In the late 1970s, the quest began to focus more on the potential use of polyamines, polyethanol glyco-

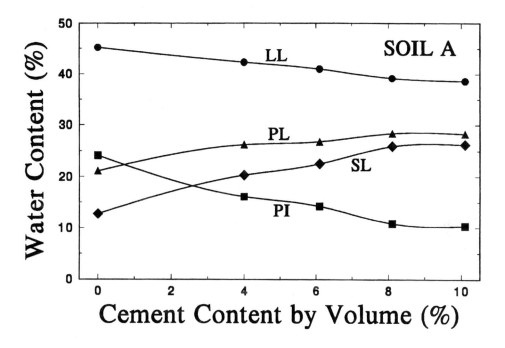
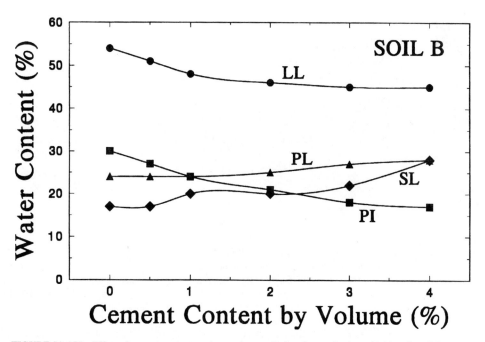

FIGURE 5A.156 Effect of cement content on the consistency limits of two cohesive soils (data from PCA, 1971).

5.228 SOIL IMPROVEMENT AND STABILIZATION

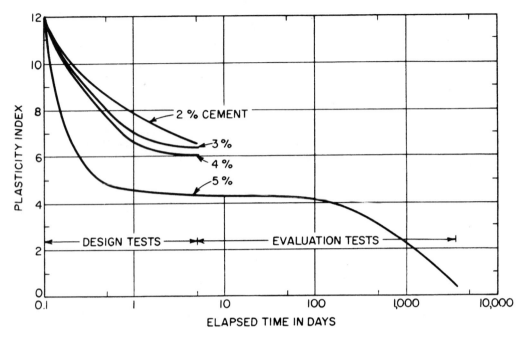

FIGURE 5A.157 Influence of aging on the plasticity index of cement-treated soil mixes from Hot Springs, Arkansas (from Redus, 1958).

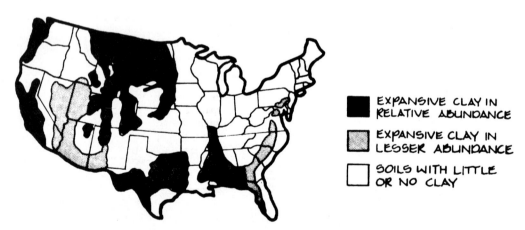

FIGURE 5A.158 Areas in the United States with expansive soils (from Godfrey, 1978).

TABLE 5A.16 Values of Swell for Selected Soils and Soil-Lime Mixtures (data from Thompson 1969)

		Swell, %		
			Soil-lime mixtures	
Name of soil	Natural soil	No curing*	48-h curing at 120°F (48.9°C)	Lime content, %
Accretion Gley 2	2.1	0.1	0.0	5
Accretion Gley 3	1.4	0.0	0.1	5
Bryce B	5.6	0.2	0.0	3
Champaign Co. till	0.2	0.5	0.1	3
Cisne B	0.1	0.1	0.1	5
Cowden B	1.4	—	0.0	3
Cowden B	2.9	0.1	0.1	5
Cowden C	0.8	0.0	0.0	3
Darwin B	8.8	1.9	0.1	5
East St. Louis clay	7.4	2.0	0.1	5
Fayette C	0.0	0.0	0.1	5
Illinoian B	1.8	0.0	0.0	3
Illinoian till	0.3	0.1	0.0	3
Illinoian till	0.3	0.9	0.1	3
Sable B	4.2	0.2	0.0	3
Fayette B	1.1	0.0	0.0	3
Miami B	0.8	0.0	0.0	3
Tama B	2.0	0.2	0.1	3

*Specimens were placed in 96-h soak immediately after compaction.

lethers, polyacrylamides, and so on, generally blended and containing surface-active agents to enhance soil penetration (Williams and Undercover, 1980). Certain combinations of chemicals seemed to be synergistic in behavior (the combined product achieved superior results to those noted for any of its constituents). Such a product is Soil Sta (proprietary to Robert Brown); similar laboratory and field data are not publicly available for any other organic stabilizer. Organic stabilizers function differently from lime, and no standardized testing procedures existed for the evaluation of these type products. The following discussions and data should prove beneficial to others wishing to evaluate an organic chemical stabilizer. All descriptive information was supplied through the courtesy of Brown Foundation Repair & Consulting, Inc., Dr. Cecil Smith, Professor of Civil Engineering, Southern Methodist University, Dr. Tom Petry, Professor of Civil Engineering, University of Texas, Arlington, all of Dallas, Texas, and Dr. Malcom Reeves, Soil Survey of England and Wales, London, England.

Soil Sta is basically a mixture of surfactant, buffer, inorganic cation source, and polyquaternaryamine in a polar vehicle. By virtue of its chemical nature, Soil Sta would be expected to have a lesser influence on kaolinite or illite than on the more expansive clays such as montmorillonite. Prior research had also indicated that soils exhibiting LL less than 35 or PI less than 23 (montmorillonite content less than about 10% by weight) would not swell appreciably (Sherif et al., 1982). Hence, the soils used in the laboratory tests and field applications contained at least 10% montmorillonite.

Soil Sta was first subjected to laboratory evaluation in 1982, and field testing commenced in mid-1983. The laboratory tests indicated that Soil Sta:

- Reduced the free swell potential of montmorillonitic clay (see Fig. 5A.159)
- Appeared stable to repeated cycles of weathering (a simulated period of 50 years)
- Increased shear strengths in some soils by 2-fold
- Increased soil permeabilities up to 40-fold
- Reduced soil shrinkage by amounts varying from 11 to 50% (Petry and Brown, 1987; Brown, 1989)

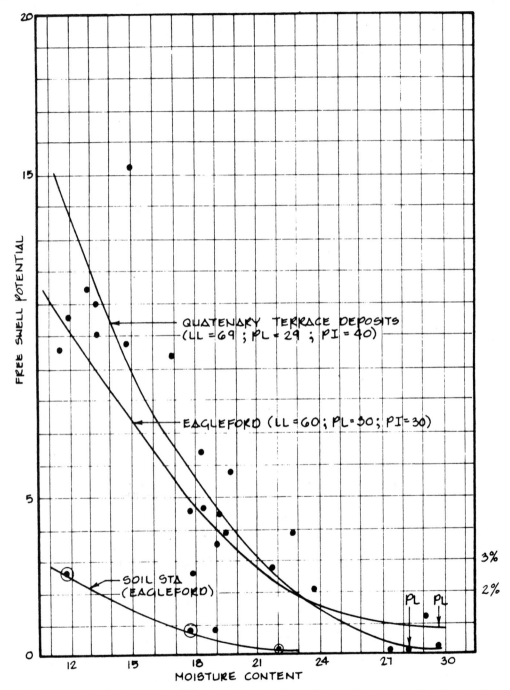

FIGURE 5A.159 Effectiveness of Soil Sta in reducing the free swell potential of montmorillonitic clays (from Brown, 1992).

By 1991 Soil Sta had been subjected to literally thousands of field applications with few, if any, failures. That is, less than 1% of the foundations treated with the chemical experienced recurrent movement, and, in a goodly proportion of those, there was a serious question as to the cause.

Some interesting points are illustrated in Fig. 5A.159:

1. At moisture contents above about 10.5%, the chemical treatment of these particular montmorillonitic soils reduced the free swell potential to levels often considered as "tolerable." Free swell potential was essentially nil above in situ moisture contents of about 18%.
2. The untreated soil is somewhat moisture-stabilized (against swell) at natural moisture contents above about 21% and would be totally stabilized against swell at natural moisture contents above about 27%.
3. The PL of this soil was 30%; hence, the moisture stabilization (against swell) occurs several percentage points below the PL.
4. Even at natural moisture contents below 10.5%, the effective abatement of swell is still profound, nearly fourfold. However, in the "real world," the natural soil moisture in foundation bearing soils (expansive in nature) is seldom found to be materially less than 10.5%.

Again, the principal intended function for this particular product (Soil Sta) is to abate swell. Under certain conditions, recurrent settlement could be a persistent problem. For example, Soil Sta is not designed to prevent soil shrinkage resulting from either soil moisture losses (evaporation or transpiration), compaction, or consolidation.

5A.4.2.3 Shrinkage. Successful chemical stabilization of expansive soils results in reduced shrinkage potential. This reduction in shrinkage potential is illustrated in Figs. 5A.155 and 5A.156 by the increase in the SL for the stabilized soils compared to the natural soils and by the data for linear shrinkage of three cement-stabilized cohesive soils provided in Fig. 5A.160. Natural granular (cohesionless) soils do not shrink because of a lack of clay content, but cement-stabilized granular soils will shrink because the cement shrinks owing to increased cohesion associated with the cementing action.

Shrinkage of properly stabilized soils is usually of little or no significance in most foundation applications. The magnitude of the shrinkage is generally so small that little or no settlement will occur in foundations bearing on chemically stabilized soils. For example, data from Nakayama and Handy (1965) showed that the total shrinkage was less than 1% for four soils (two sands, a silt, and a clay) treated with various levels of cement and tested under a variety of curing and drying conditions. The cement-stabilized sands shrank very little (<0.1%). Shrinkage may produce minor cracking of the stabilized soil, which may result in some small but probably insignificant reduction in strength. Perhaps the most important potential problem related to shrinkage-induced cracking is that it may subject the stabilized soil to greater attack from sulfates that are leached from adjacent sulfate-bearing soils by providing easier access into the interior of the stabilized mass.

5A.4.2.4 Collapse. The preconstruction collapse potential of both naturally deposited and compacted soils can be substantially reduced or eliminated by chemical stabilization. Naturally deposited collapsible soils can be chemically treated either by adding the chemical in an aqueous solution to the soil via either injection or ponding or by excavating the soil, mixing lime or cement with the soil at an appropriate water content, and recompacting the soil in place (Clemence and Finbarr, 1981). The collapse potential of compacted fills can be controlled by mixing the stabilizer with the borrow soil prior to compaction (Lawton et al., 1993). To the authors' knowledge, there are no methods of chemical stabilization currently available for remedial applications in collapsible soils.

The stabilization of naturally deposited collapsible soils using sodium silicate has been demonstrated by laboratory and field tests conducted in the former Soviet Union and in the United States. Sokolovich and Semkin (1984) reported the results of laboratory collapse tests conducted on specimens of loess that had been prewetted in the field by injecting either water or a 2% solution of sodium silicate in water. The prewetted soil was allowed to dry for several months before

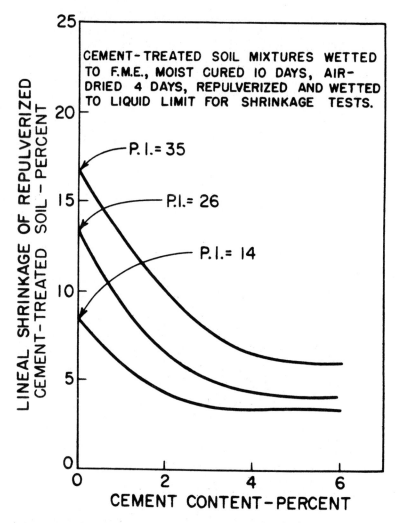

FIGURE 5A.160 Influence of cement content on the linear shrinkage of repulverized cement-treated soil (from HRB, 1961).

samples were taken from the field at depths of 4 m and 6 m (13 ft and 20 ft). Upon wetting in the laboratory (applied stress not given), the specimens prewetted with water collapsed an average of 4.1%, whereas those prewetted with the sodium silicate solution collapsed an average of 0.26%. Rollins and Rogers (1994) conducted a series of full-scale field load tests to evaluate the effectiveness of different treatment methods in reducing the collapse potential of a deposit of clayey sandy silt located in an alluvial fan. One of the test sections was prewetted with a 2% sodium silicate solution that was ponded on the ground surface and allowed to percolate into the ground. After periods of 1.5 months for curing and drying and two weeks for sampling, a 1.5 m square reinforced concrete footing was built and loaded to an average bearing stress of 85 kPa (1.8 ksf). The site was then wetted with water. The resulting collapse settlement was less than 27 mm (1.1 in), compared to an average settlement of 280 mm (11 in) for the untreated soil. Furthermore, the treatment with sodium silicate reduced the creep settlement after 6 months from 12 mm (0.47 in) to 9 mm (0.35 in).

Laboratory results provided by Lawton and coworkers (1993) have demonstrated the effectiveness of lime and cement in substantially reducing the collapse potential of compacted cohesive soils. One-dimensional collapse tests were conducted on comparable untreated and treated specimens of compacted clayey sand wherein the dry density and water content of the soil exclusive of the stabilizer were the same. By comparing the results of the collapse tests on these comparable untreated and treated specimens, which are provided in Table 5A.17, the fundamental influence of the stabilizer on the collapse potential can be determined. In all instances where there was sufficient free water in the soil for complete hydration of the stabilizer (a compaction water content of at least 8.8% in these tests), the reduction in collapse was significant, and the magnitude of collapse was small ($\leq 0.08\%$). The most dramatic example of reduction in collapse was from 17.04 to 0.02% by the addition of 5% cement (by weight) at a dry density of 1.50 Mg/m^3 (93.6 pcf), a water content of 8.8%, and an applied stress of 400 kPa (8.4 ksf). For a compaction water content $\leq 7.5\%$, the hydration of the stabilizer was incomplete, and, although collapse was generally reduced, the magnitude of collapse was still quite high ($\geq 4.04\%$) and unacceptable from a design standpoint.

The addition of chemical stabilizers also substantially reduced the time to the end of collapse (t_c), as illustrated in Fig. 5A.161. For $\gamma_d = 1.60$ Mg/m^3 (100 pcf), $w = 7.5\%$, and $\sigma_v = 400$ kPa (8.4 ksf), t_c for the 5% lime specimen was 4 min compared to 60 min for the untreated specimen. Similarly, for $\gamma_d = 1.67$ Mg/m^3 (104 pcf), $w = 7.0\%$, and $\sigma_v = 800$ kPa (16.7 ksf), the addition of 2% cement reduced t_c from 240 min. to 60 min. Similar reductions in t_c occurred in the other treated specimens.

The chemically treated soils also exhibited significantly less postcollapse creep strain than the comparable untreated soils. For example, for the same specimens discussed in the preceding paragraph, the modified secondary compression index [$C_{\alpha\varepsilon} = -\Delta\varepsilon_v/\Delta(\log t)$] decreased from 0.094 to

TABLE 5A.17 Results from Single Oedometer Collapse Tests on Comparable Untreated and Chemically Stabilized Specimens Consisting of 70% Silica Sand and 30% Kaolin (from Lawton et al., 1993)*

| Stabilizer† | γ_d*‡ | | w‡, % | S_r‡, % | σ_v | | ε_{vw}, % |
	Mg/m^3	pcf			kPa	ksf	
None	1.50	94	8.8	30	400	8.4	−17.04
5% cement	1.50	94	8.8	30	400	8.4	−0.02
None	1.60	100	12.5	50	400	8.4	−0.60
2% cement	1.60	100	12.5	50	400	8.4	−0.08
None	1.60	100	7.5	30	400	8.4	−13.47
5% lime	1.60	100	7.5	30	400	8.4	−4.04
None	1.67	104	10.0	45	400	8.4	−0.31
0.5% cement	1.67	104	10.0	45	400	8.4	−0.01
None	1.67	104	10.0	45	800	16.7	−0.06
0.5% cement	1.67	104	10.0	45	800	16.7	−0.01
None	1.67	104	7.0	31	400	8.4	−13.06
2% cement	1.67§	104§	7.0§	31§	400	8.4	−13.51
None	1.67	104	7.0	31	800	16.7	−11.28
2% cement	1.67	104	7.0	31	800	16.7	−9.56

*γ_d = dry density of soil (exclusive of stabilizer); w = compaction water content of soil (exclusive of stabilizer); S_r = degree of saturation of soil (exclusive of stabilizer); σ_v = applied (total) vertical stress; ε_{vw} = wetting-induced strain.
†Cement = Portland cement; lime = hydrated lime [Ca(OH)$_2$].
‡Values given are nominal. Actual values for all specimens, except as noted below, were close to their nominal values.
§Actual γ_d = 1.60 Mg/m^3 = 100 pcf; actual w = 6.4%; actual S_r = 25%

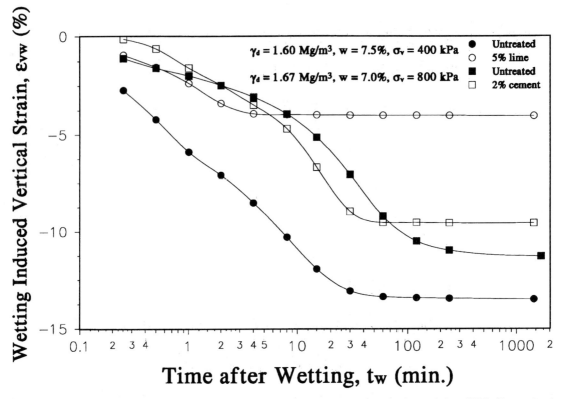

FIGURE 5A.161 Effect of lime and cement on the time rate of collapse of compacted specimens consisting of 70% silica sand and 30% kaolin (from Lawton et al., 1993).

0.025% for the lime-treated specimen and decreased from 0.36 to 0.014% for the cement-treated specimen. This trend was consistent in all treated specimens.

5A.4.2.5 Stress-Strain-Strength Behavior. The general effect of chemical stabilization on soils is to increase both the strength and stiffness of the soil while making the soil more brittle. Where the stabilization occurs primarily by cation and base exchange with clay particles, the changes in these characteristics are generally modest. For example, as noted in Sec. 5A.4.2.2, the increases in shear strength of montmorillonitic soils produced by Soil Sta, an organic stabilizer designed to control swelling potential, are typically a maximum of about 100% (that is, it doubles the shear strength of the natural soil). In contrast, cementitious or pozzolanic stabilizers may produce 50-fold or greater increases in stiffness and strength. To illustrate the substantial changes in stiffness and strength that may occur from chemical stabilization, the results of triaxial compression tests on untreated and cement-treated specimens of a micaceous silty sand are presented in Fig. 5A.162. At a confining pressure (σ_3) of 30 psi (207 kPa), the maximum deviator stress increased from 78.7 psi (543 kPa) for the untreated soil to 971 psi (6.70 MPa) for the same soil treated with 10% cement (by dry weight), an increase of more than 12-fold. The stiffness of the cemented-treated specimen, as measured by the secant modulus at 50% of the maximum deviator stress (E_{50}), was about 37 times that of the natural soil: 73.9 ksi (510 MPa) for the cement-stabilized soil compared to 2.0 ksi (13.8 MPa) for the untreated soil. Note also that the stress-strain behavior changed from ductile for the untreated soil to brittle for the cement-treated soil. The

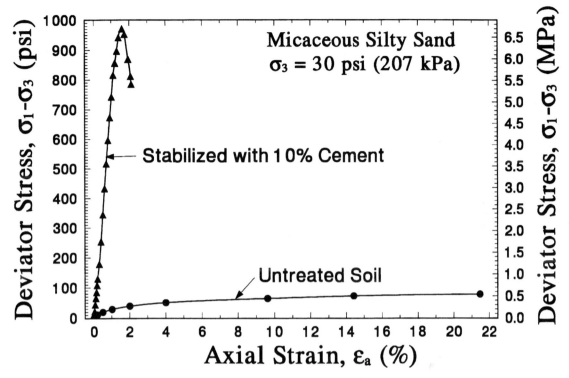

FIGURE 5A.162 Influence of cement stabilization on the triaxial stress-strain characteristics of a micaceous silty sand.

cement stabilization also produced substantial increases in the strength parameters. For σ_3 ranging from 5 to 30 psi (34 to 207 kPa), the friction angle (ϕ) increased from 33° to 64°, and the cohesion intercept increased from 1.4 psi (9.7 kPa) to 60 psi (410 kPa).

In general, an increase in stabilizer content will produce an additional increase in stiffness and strength so long as the additional stabilizer produces further cementitious or pozzolanic reactions in the soil. Thus, cement-stabilized soils will continue to increase in strength and stiffness with increased cement content so long as there is sufficient water for hydration of the cement (up to the point where the cement constitutes a vast majority of the material and the strength of the mixtures is essentially equal to the strength of the cement). In contrast, the maximum strength and stiffness of lime-treated soils is achieved at a specific lime content that depends primarily on the amount of clay available in the soil for pozzolanic reactions. The addition of lime beyond this point produces no further pozzolanic reactions and may result in a decrease in strength and stiffness. The minimum lime content that produces the maximum benefit in terms of strength and stiffness is about 8% for most clayey soils (Ingles and Metcalf, 1973). Typical characteristics of compressive strength versus stabilizer content for both cement and lime are given in Fig. 5A.163.

Because the cementitious and pozzolanic reactions are time-dependent, the strength and stiffness of stabilized soils generally increases with time for at least several years. Typical results showing the effect of aging on the strength and stiffness of cement-stabilized soils are provided in Figs. 5A.164 and 5A.165. Aging can have a significant effect on both strength and stiffness. For example, for Soil 2 treated with 14% cement in Figs. 5A.164 and 5A.165, the unconfined compressive strength increased from 1400 psi (kPa) after 7 days of curing to 2300 psi (kPa) after 90 days of curing, while the compressive secant modulus at one-third of the ultimate load

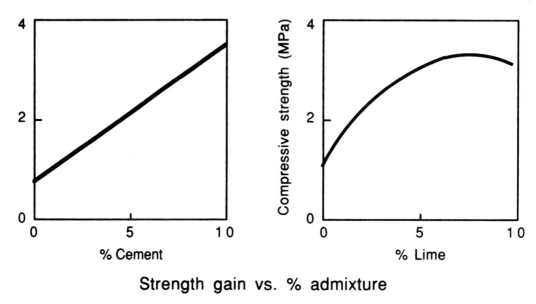

FIGURE 5A.163 Typical characteristics of gain in strength versus admixture content for cement and lime (from Hausmann, 1990).

increased from 2000 ksi (13,800 MPa) to 2,800 ksi (19,300 MPa) during the same period of time. Similar effects of aging on the stress-strain-strength characteristics of a lime-treated silty clay are shown in Fig. 5A.166. That the increase in strength and stiffness may occur over long periods of aging is demonstrated in Fig. 5A.167 for two cement-treated soils.

When chemical stabilizers are mixed with the soil, a delay between the time of mixing and the time of compaction may result in a substantial decrease in strength and stiffness of the treated soil. Thus, many specifications require that compaction be completed within a short time after mixing, usually less than 1 or 2 h. A delay in compaction reduces the strength and stiffness in two ways:

1. Cementitious products (clods) are formed that harden with time. These clods must be broken down or remolded if the material is to be densified. For the same compactive effort, the density will decrease with increasing delay in compaction, as shown in Fig. 5A.168 for a soil stabilized with self-cementing fly ash.

2. As clods form and harden with increasing delay, the effective area of contact between clods in the compacted soil decreases. The result is that a smaller volume of stabilized soil is cemented together because cementation between clods occurs only along the interclod contact areas. In contrast, a well-mixed, chemically treated soil that is compacted immediately after compaction will be relatively homogeneous with cementation occurring throughout the treated mass. Thus, a delay in compaction means that a portion of the stabilizer is lost to the clod-forming process and is unavailable for cementation of the entire mass.

The loss in strength resulting from a delay in compaction can be significant. For example, a 2-h delay in compaction of the fly-ash-treated material shown in Fig. 5A.168 reduced the maximum strength for a given compactive effort from about 2300 kPa (330 psi) to about 700 kPa (100 psi)—a decrease of 70%. This magnitude of reduction in strength is typical for class C (cementitious) fly ashes, which "flash-set" when water is added. To counteract this characteristic, set retarders often must be added to the mixture to delay the cementitious reactions for about 2 h to permit hauling, placing, and compaction of the treated material (ASCE, 1987). A delay in compaction may also be a significant problem in cement-treated soils, as illustrated in Fig. 5A.169,

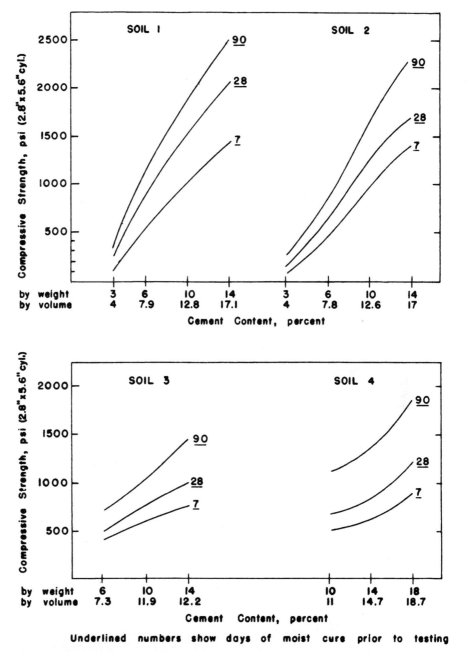

FIGURE 5A.164 Influence of aging and cement content on unconfined compressive strength (from HRB, 1961; after Felt and Abrams, 1957).

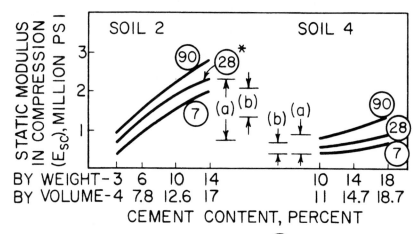

FIGURE 5A.165 Effect of aging and cement content on static unconfined compressive modulus (from HRB, 1961; data from Felt and Abrams, 1957).

but is generally not a significant problem in soils treated with pozzolanic stabilizers (typically lime and lime–fly ash).

Temperature affects the rate at which the cementitious or pozzolanic reactions take place and thus the strength and stiffness of the soil. In general, higher temperatures produce faster gains in strength and stiffness, as illustrated in Fig. 5A.170 for various soils treated with lime and cement, and in Fig. 5A.171 for an LFA mixture. At temperatures below about 40°F (4°C), the cementitious or pozzolanic reactions are severely retarded. Thus, during the winter, the cementitious reactions will be minimal when temperatures drop below 40°F (4°C) but will resume when the weather warms, as illustrated in Fig. 5A.172. Low temperatures, including long periods of freezing, seem to have no permanent effect on the strength that will eventually be achieved.

Some typical guidelines for estimating the stress-strain-strength relationships of soils stabilized with lime, cement, and fly ash have been developed. The typical shape of a stress-strain plot for these types of chemically treated soils is as shown in Fig. 5A.173. In general, the stress-strain relationship is approximately linear up to about 50 to 70% of the peak strength, beyond which the relationship is nonlinear (strain-softening) up to failure.

Typical relationships between unconfined compressive strength (q_u) and stabilizer content have been established for cement-stabilized soils. Data illustrating these relationships are provided in Fig. 5A.174, from which the following approximate equations have been developed for estimating q_u based on cement content (C in percent by dry weight) for coarse-grained and fine-grained soils. For coarse-grained soils, the equations are

$$q_u(\text{psi}) = (80 \text{ to } 150) \cdot C \tag{5A.62a}$$

$$q_u(\text{kPa}) = (550 \text{ to } 1030) \cdot C \tag{5A.62b}$$

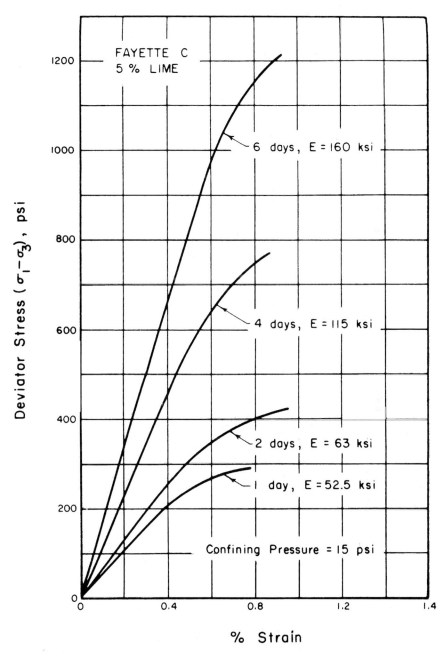

FIGURE 5A.166 Influence of aging on the stress-strain properties of a lime-treated cohesive soil (from Thompson, 1966).

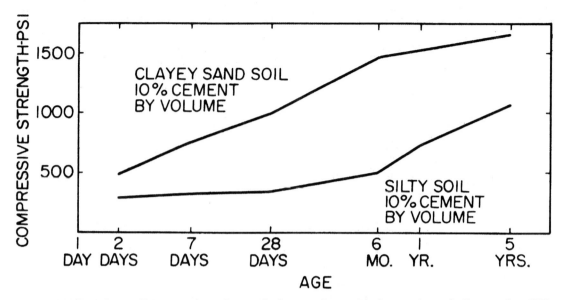

FIGURE 5A.167 Influence of long-term aging on the unconfined compressive strength of two specimens of soil-cement (from HRB, 1961; after Leadabrand, 1956).

For fine-grained soils the equations are

$$q_u(\text{psi}) = (40 \text{ to } 80) \cdot C \tag{5A.63a}$$

$$q_u(\text{kPa}) = (275 \text{ to } 550) \cdot C \tag{5A.63b}$$

Note that Eqs. (5A.62) and (5A.63)—as well as the rest of the equations given in this section—should be used only to estimate values, not to obtain values for design or analysis.

Typical values of q_u for lime–fly-ash-stabilized materials are given in Table 5A.18 according to general soil type. An equation for estimating q_u for cement-stabilized soils at a later age (d = number of days) based on a known value of q_u at a certain age (d_0) is given as follows (Little et al., 1987; Mitchell, 1981):

$$(q_u)_d = (q_u)_{d_0} + F \cdot \log\left(\frac{d}{d_0}\right) \tag{5A.64}$$

where $F = 70C$ for coarse-grained soils and q_u in psi
$F = 480C$ for coarse-grained soils and q_u in kPa
$F = 10C$ for fine-grained soils and q_u in psi
$F = 70C$ for fine-grained soils and q_u in kPa

The stiffness and strength of a chemically stabilized soil are obviously interrelated. The following equation gives an approximate relationship between compressive modulus (E_c) at a confining pressure of 15 psi (100 kPa) and unconfined compressive strength for lime-treated soils (Thompson, 1966):

$$E_c(\text{ksi}) = 10 + 0.124 \cdot q_u(\text{psi}) \tag{5A.65a}$$

$$E_c(\text{MPa}) = 70 + 0.124 \cdot q_u(\text{kPa}) \tag{5A.65b}$$

Values of E_c for cement-stabilized soils are typically within the range of 1000 to 5000 ksi (7000 to 35,000 MPa) for coarse-grained soils and 100 to 1000 ksi (700 to 7000 MPa) for fine-grained

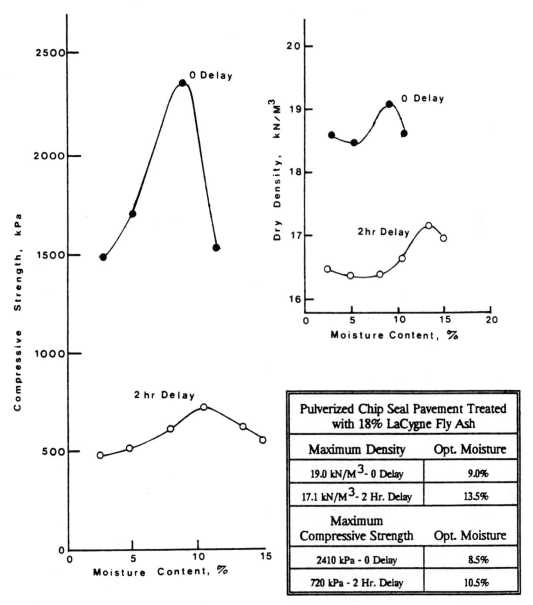

FIGURE 5A.168 Moisture-density and moisture-strength relationships for a fly-ash-treated material (from Ferguson, 1993).

soils. E_c depends on the confining pressure and can be expressed in the following form based on an assumed hyperbolic shape of the stress-strain relationship (Duncan et al., 1980):

$$E_{tc} = K \cdot P_a \cdot \left(\frac{\sigma_3}{P_a}\right)^n \cdot \left[1 - \frac{0.75(1 - \sin\phi)(\sigma_1 - \sigma_3)}{2(c \cdot \cos\phi + \sigma_3 \cdot \sin\phi)}\right]^2 \qquad (5A.66)$$

where E_{tc} = tangent compressive modulus at a given stress level $(\sigma_1 - \sigma_3)$
ϕ = angle of internal friction

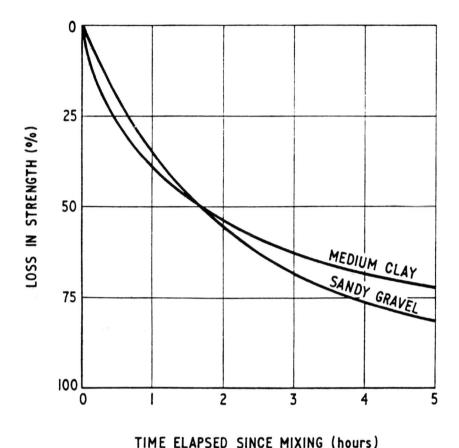

FIGURE 5A.169 Loss in strength caused by delay in compaction for two soils stabilized with 10% cement (standard Proctor compaction) (from Ingles and Metcalf, 1973; after West, 1959).

c = cohesion intercept
P_a = atmospheric pressure
n = modulus exponent (typically 0.1 to 0.5 for cement-stabilized soils)
K = modulus number (typically 1000 to 10,000 for cement-stabilized soils)

E_c for lime–fly-ash-stabilized soils generally varies from about 500 to 2500 ksi (3500 to 17,500 MPa) (Little et al., 1987).

Values of the Mohr-Coulomb strength parameters, angle of internal friction (ϕ) and cohesion intercept (c), vary considerably for chemically stabilized soils depending primarily upon the type of soil, the amount of stabilizer, and the amount and type of curing. Typical values of ϕ for soils treated with cement, lime, and lime–fly ash are given in Table 5A.19. For cement-stabilized soil, c may be as high as a few hundred psi (a few thousand kPa) (FHWA, 1979; Mitchell, 1981). The following equation can be used to estimate c based on q_u for lime-stabilized soils (Thompson, 1966):

$$c \text{ (psi)} = 9.3 + 0.292 \cdot q_u \text{ (psi)} \tag{5A.67a}$$

$$c \text{ (kPa)} = 64 + 0.292 \cdot q_u \text{ (kPa)} \tag{5A.67b}$$

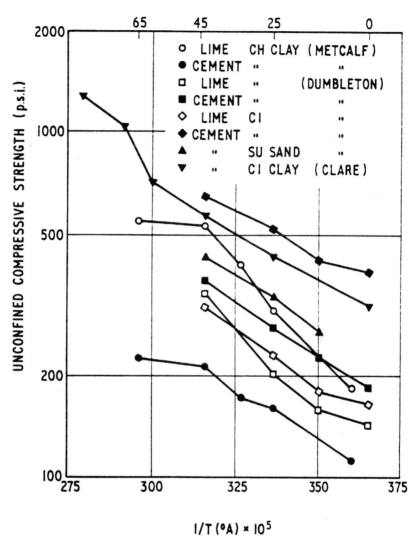

FIGURE 5A.170 Relationship between the rate of gain in strength and curing temperature for lime and cement-stabilized soils (from Ingles and Metcalf, 1973).

For the lime–fly ash-stabilized gravels evaluated by Hollon and Marks (1962), c varied from 55 to 128 psi (380 to 880 kPa).

Although few data in the literature address values of Poisson's ratio (v) for chemically stabilized soils, some general guidelines can be given for soils stabilized with cement, lime, and fly ash. Values of static v for chemically stabilized soils are typically smaller than for natural soils. In general, v is relatively constant within the linear stress-strain range for a given stabilized soil and confining pressure, as shown in Fig. 5A.175 for a lime–fly ash-stabilized gravel. Values of

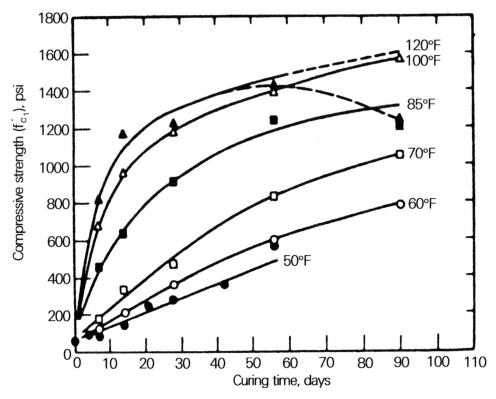

FIGURE 5A.171 Effect of curing time and temperature on the strength development of a lime–fly ash-aggregate mixture (from NCHRP, 1976; after MacMurdo and Barenberg, 1973).

ν within this range of stresses typically vary from about 0.08 to 0.15 for granular soils and 0.15 to 0.20 for cohesive soils. Thus, these values are typical for the normal range of stresses that a stabilized soil would be subjected to during its service life. At higher stresses levels, ν increases with increasing stress up to values at failure of about 0.20 to 0.30 for granular soils and 0.25 to 0.35 for cohesive soils. Values of dynamic Poisson's ratio appear to be somewhat higher. In a study of four soils stabilized with cement (Felt and Abrams, 1957), values of dynamic ν were found to range from 0.22 to 0.27 for two granular soils, 0.24 to 0.31 for a silty soil, and 0.30 to 0.36 for a clayey soil.

5A.4.2.6 Compaction Moisture-Density Relationships. The addition of a chemical stabilizer usually alters the compaction moisture-density relationships of a soil. The moisture-density curves for a loess treated with varying amounts of quicklime compacted 24 h after mixing are shown in Fig. 5A.176. For nearly all cohesive soils, the addition of lime produces the same types of changes to the moisture-density relationships illustrated in Fig. 5A.176—a decrease in maximum dry density (γ_{dmax}) and an increase in optimum water content (w_{opt}) for the same method of compaction and compactive effort. In addition, γ_{dmax} typically decreases and w_{opt} normally increases with increasing stabilizer content. These changes can be explained in simplified terms as follows:

1. Some cementitious or pozzolanic reactions occur immediately upon hydration. This cementing effect produces a resistance to compaction that results in a decrease in γ_{dmax}.

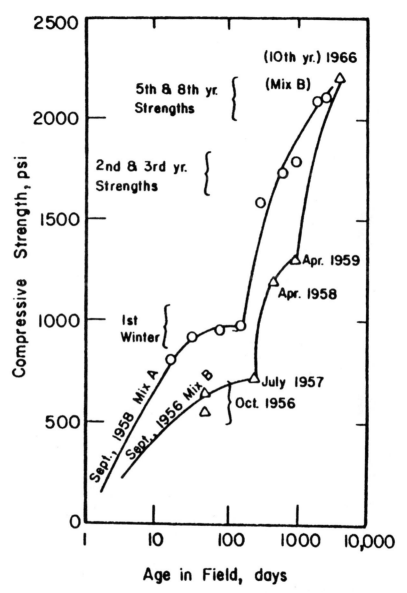

FIGURE 5A.172 Age-strength relationships of lime–fly ash-aggregate mixtures (from Little et al., 1987; after FHWA, 1979).

2. Extra water (in addition to that needed to facilitate rearrangement of the soil particles during compaction) is required to hydrate the lime and produce the pozzolanic reactions. Hence, w_{opt} is increased.

In soils treated with cement, γ_{dmax} and w_{opt} may either increase or decrease (HRB, 1961). Increases in γ_{dmax} usually occur for sands and sandy soils and sometimes to a small degree for heavy clays. Little or no change in γ_{dmax} generally occurs for light to medium clays. Decreases in

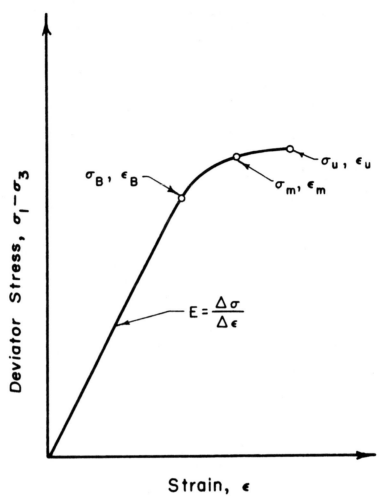

FIGURE 5A.173 Idealized stress-strain curve for lime-stabilized soils (from Thompson, 1966).

γ_{dmax} may occur in silts; w_{opt} normally decreases for clays, increases for silts, and changes little for sands and sandy soils. Typical values for changes in w_{opt} and γ_{dmax} compared to the values for the comparably compacted untreated soil, are summarized in Table 5A.20 according to type of soil for the normal range of cement contents. The reasons for the wide variation in the changes in w_{opt} and γ_{dmax} that occur in soils because of cement stabilization are not completely understood. The diversity of changes in γ_{dmax} may relate to the opposing trends caused by the high specific gravity of Portland cement ($G_s \cong 3.15$) compared to that for the soil particles (typical average $G_s \cong 2.65$ to 2.75), which tends to increase the density, and the resistance to compaction produced by the cementitious reactions, which tends to decrease the density.

As the length of the delay between mixing and compaction increases, the characteristics of the treated soil, including the moisture-density relationships, continuously changes as the cementitious or pozzolanic reactions occur with time. The effect of a 2-h delay on the moisture-density

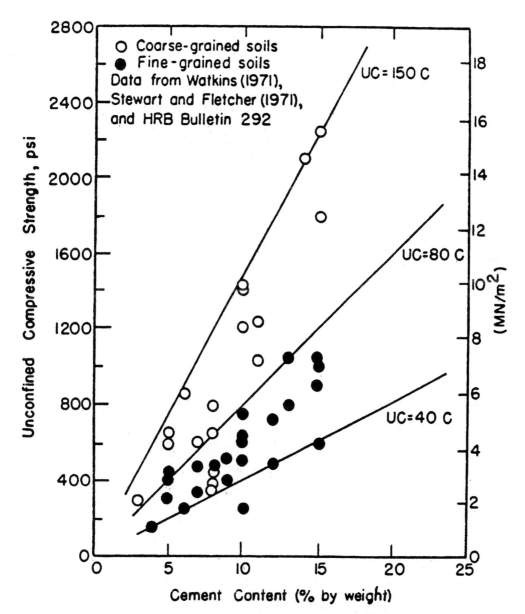

FIGURE 5A.174 Relationship between cement content and unconfined compressive strength for soil-cement mixtures (from Little et al., 1987; after FHWA, 1979).

TABLE 5A.18 Ranges of Unconfined Compressive Strength for Lime–Fly-Ash-Stabilized Materials (from NCHRP, 1976).

Type of soil	28-day immersed unconfined compressive strength	
	psi	kPa
Gravels	400 to 1300	2800 to 9000
Sands	300 to 700	2100 to 4800
Silts	300 to 700	2100 to 4800
Clays	200 to 500	1400 to 3400
Crushed stones and slag	1400 to 2000	10,000 to 14,000

TABLE 5A.19 Typical Ranges of Values for Angle of Internal Friction for Soil Stabilized with Cement, Lime, and Lime–Fly Ash

Stabilizer	Typical values of ϕ (degrees)	
	Coarse-grained	Fine-grained
Cement	40 to 60	30 to 40
Lime	N.A.*	25 to 35
Lime–fly ash	49 to 53†	N.A.*

*N.A. indicates not available or not applicable.
†Based on limited data for gravel (Hollon and Marks, 1962).

relationships of a pulverized chip seal pavement stabilized with a cementitious fly ash is illustrated in Fig. 5A.168. In general, γ_{dmax} decreases and w_{opt} increases with increasing length of the delay. The effect of delay in compaction is usually much less important in pozzolanic stabilizers (lime and lime–fly ash) than in self-cementing stabilizers (cement and class C fly ash), as illustrated in Fig. 5A.177. Because the length of the delay may significantly affect the engineering characteristics of the stabilized soil (for example, see the reduction in strength shown in Fig. 5A.168), it is essential that when preparing laboratory specimens for testing, the delay between mixing and compaction match the delay expected during the actual field construction. Sometimes retarders are used with cement or class C fly ash to inhibit the cementitious reactions during the anticipated delay between mixing and compaction.

5A.4.2.7 Permeability. The changes in permeability that occur because of chemical stabilization depend on the type of soil, the amount of stabilizer, the process used to add the stabilizer (injection versus mixing), and the density achieved during compaction (admixture-stabilized soil). Unfortunately, the data available in the literature regarding the permeability of chemically stabilized soils is rather limited, with most of the information pertaining to soils treated with cement and lime. Data reflecting changes in permeability produced by the chemical stabilization of particular soils can be found in HRB (1961), Townsend and Klym (1966), Ingles and Metcalf (1973), Poran and Ahtchi-Ali (1989), Bowders and colleagues (1990), and Taki and Yang (1991), among others.

In general, chemical stabilization increases the permeability of soils with very low natural permeabilities and reduces the permeability of soils with moderate to high natural permeabilities. Examples illustrating these two general trends are given as follows: (*a*) Soil Sta, an organic chemical stabilizer that is injected into expansive soils to control swelling potential (see Sec. 5A.4.2.2), increases the permeability of montmorillonitic soils by as much as 40-fold; and as shown in Fig. 5A.178, the treatment of two fines sands with cement reduced their permeabilities by more than two orders of magnitude. Ingles and Metcalf (1973) have indicated the following

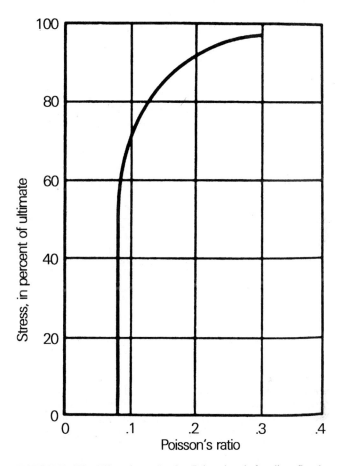

FIGURE 5A.175 Effect of stress level on Poisson's ratio for a lime–fly ash-gravel mixture (from NCHRP, 1976; after Ahlberg and Barenberg, 1965).

general trends for cement stabilization regarding how the permeability changes for different types of soils:

Permeability is generally decreased	Permeability is generally increased
Well-graded gravel	Silt
Well-graded sand	Silty clay
Well-graded gravel-sand-clay	Very poorly graded soil
Sand and gravel	Heavy clay
Poorly graded sand	Organic and sulfate-rich soil
Silty sand	
Sandy clay	
Silty-sandy clay	

Deviations from these general trends may occur, particularly for the borderline soil types (sandy clay, silty-sandy clay, silt, and silty clay), but deviations for the extreme soil types (clean coarse-

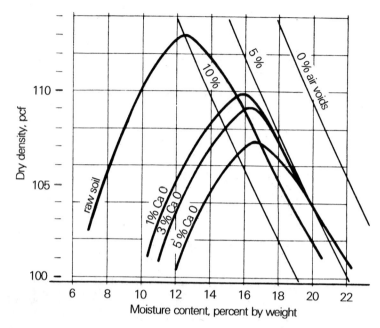

FIGURE 5A.176 Influence of quicklime content on the standard Proctor moisture-density relationship of a loess (from Brand and Schoenberg, 1959).

TABLE 5A.20 Maximum Dry Densities and Optimum Water Contents for Cement-Treated Soils Compared to the Corresponding Values for Untreated Soils (from HRB, 1961)

Type of soil	Change in γ_{dmax}		Change in w_{opt}, %
	pcf	kN/m^3	
Sandy loams	0 to +3	0 to +0.5	−1 to +1
Sands	0 to +6	0 to +0.9	0 to −1
Silts and loams	0 to −6	0 to −0.9	0 to +3
Silts	−3 to +1	−0.5 to +0.2	0 to −3
Medium clays	0 to +1	0 to +0.2	0 to −2
Heavy clays	−1 to +2	−0.2 to +0.3	0 to −4

grained soils and highly plastic clays) are unlikely. These trends are also applicable to other chemical admixtures (lime, lime–fly ash, and fly ash). The effect of the stabilizer on permeability usually increases with increasing stabilizer content, as illustrated in Fig. 5A.178, up to a limiting value of stabilizer content above which little or no additional change occurs.

5A.4.2.8 Liquefaction Potential. For many years it has been recognized that cementation in saturated granular soils, whether occurring naturally or induced artificially, tends to inhibit liquefaction when the soil is subjected to seismic loading. In a laboratory study conducted by Clough and coworkers (1989), the following results and conclusions were determined regarding the liquefaction resistance of sands treated with cement:

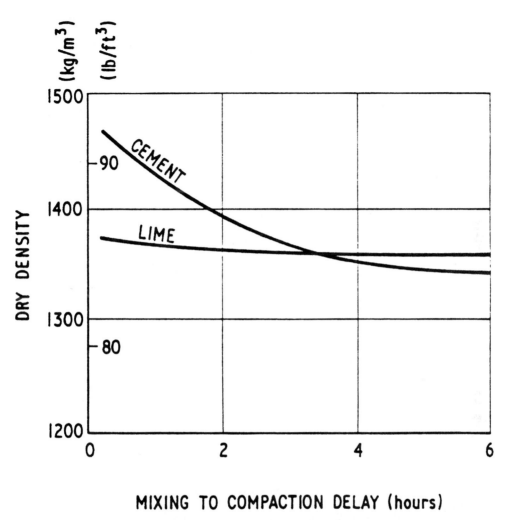

FIGURE 5A.177 Effect of delay in compaction on the as-compacted density of a heavy clay treated with 10% stabilizer (from Ingles and Metcalf, 1973; after Dumbleton, 1962).

1. For the sands studied, when the unconfined compressive strength (q_u) exceeded 100 kPa (14.5 psi), the soil was, for practical purposes, not liquefiable.
2. When q_u for the cemented sand was greater than 60 kPa (8.7 psi), differences in unit weight had little influence on the resistance to liquefaction.
3. The addition of small amounts of cement (1 and 2%) to the sands substantially reduced the potential for liquefaction (see Fig. 5A.179)

Although the use of chemical stabilization to control liquefaction potential is technically viable, it has been used only infrequently in practice. No case histories have been found in the literature in which an entire liquefiable mass beneath a structure has been chemically stabilized. In two recent projects, in situ soil mixing techniques (see Sec. 5A.4.3) were used to reduce the poten-

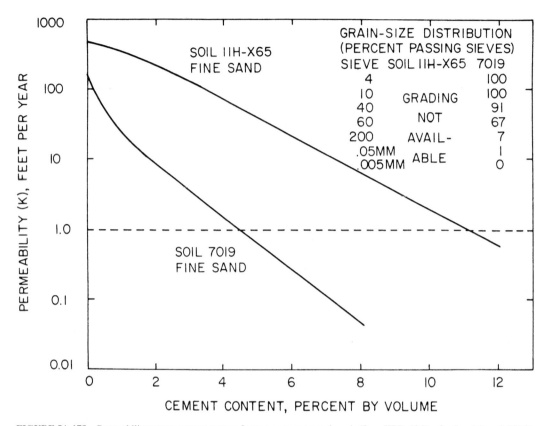

FIGURE 5A.178 Permeability versus cement content for two cement-treated sands (from HRB, 1961; after Leadabrand, 1956).

tial for damage from liquefaction by containing the liquefiable soil within cells formed by cement-stabilized walls. In the Jackson Lake Dam modification project, cement-stabilized walls arranged in a hexagonal shape (Fig. 5A.180) were built within the liquefiable soils beneath the earth dam (Taki and Yang, 1991). Research has shown that soil-cement walls constructed in a cellular arrangement can reduce seismically generated pore pressures within loose granular soils by as much as 90%. In a similar application, soil-cement walls were constructed in a circular pattern beneath the walls of two spill tanks for a paper mill (Broomhead and Jasperse, 1992). The soil-cement walls were intersected by cross-drains at regular intervals to allow water generated by the dissipation of excess pore water pressures to drain to the exterior of the tanks (Fig. 5A.181). The cross-drains were connected by a continuous perimeter drain inside the wall footing, which in turn was connected to a gravel drainage blanket beneath the entire floor slab.

5A.4.3 Construction Methods

Numerous methods can be used to incorporate chemical stabilizers within soils. These methods can be divided into two general categories—injection and mixing. The information provided in this section is meant not as a complete treatment of the subject but as a summary of the principles, techniques, and equipment used to effect chemical stabilization of soils. Additional details on this

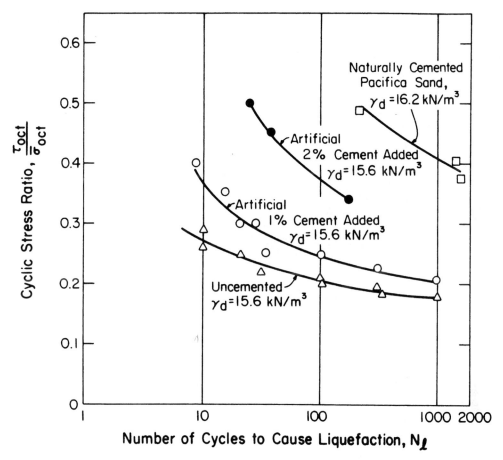

FIGURE 5A.179 Influence of cementation on the number of cycles required to cause liquefaction of cemented sands (from Clough et al., 1989).

topic can be found in NCHRP (1976), PCA (1979), Boynton and Blacklock (1985), Little and coworkers (1987), TRB (1987), and Broms (1993).

5A.4.3.1 Injection. Chemical stabilizers contained within solutions or suspensions (slurries) can be injected into the soil through the openings (void spaces and cracks). Many techniques have been used to inject chemical stabilizers into soils, some of which are similar or the same as those used for grouting (see Sec. 5B.5.1). In the following discussions, a technique is described to inject Soil Sta, a proprietary organic chemical stabilizer (refer to Sec. 5A.4.2.2). Bear in mind that Soil Sta is used almost exclusively in remedial applications to stabilize expansive soils beneath existing foundations, and the following discussions are necessarily aimed toward that application.

In special cases in the United States and for most applications within the United Kingdom, chemical injection is performed through a specially designed stem. In the system utilized, Soil Sta is injected under pressure to some depth—usually 4 to 6 ft (1.2 to 1.8 m). Penetration of the stem is accomplished by pumping through the core and literally washing the tool down. Once positioned, the hand valves are switched to close the core and divert flow into the annular space and out the injection ports (Fig. 5A.182).

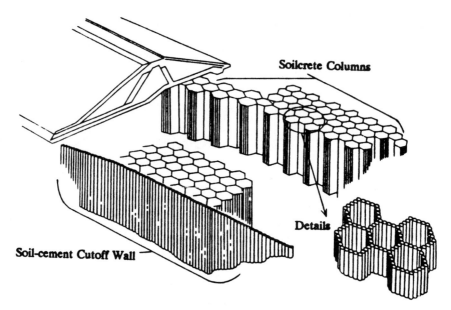

FIGURE 5A.180 Schematic illustration of the soil-cement walls used to control liquefaction potential in the Jackson Lake Dam modification project (from Taki and Yang, 1991).

In some instances, a particular soil might tend to resist penetration by Soil Sta. Both the rate of penetration and the volume of chemical placed can be enhanced by utilizing hydraulic pulsation (high pressure of short duration) during the injection phase. Alternatively, the stem can be equipped with a packer assembly to isolate selective zones (Fig. 5A.183). This equipment permits higher injection pressures and also allows some zone selectivity.

From a practical viewpoint, minimal concern should be given to the exact volume of chemical injected into a specific hole. The primary intent is to distribute the stabilizer volume reasonably uniformly over the area to be treated. Time (days, weeks, or months—depending on the specific site conditions) will produce a nearly equal distribution.

A similar analysis would be true for depth interval injection. Within a short depth (that is, 6 ft ≅ 2 m or less), chemical penetration can be fairly uniform given time, although placement of the stinger might vary a foot or so one way or the other. This condition could change if two factors coexisted: (*a*) The depth of penetration is increased substantially, that is, 15 to 20 ft (4.6 to 6 m), and (*b*) packers or other positive seal methods are used to isolate each zone to be injected. Even in the latter example, true penetration of the zone may not occur if the placement pressure or specific soil characteristics favor communication between zones. No matter what precautions might be taken, the normal heterogeneous and fractured nature of the soil would tend to preclude exact placement of a specified volume.

At a pressure differential of about 3.5 psi (24 kPa), the system in Fig. 5A.182 would theoretically place about 12 gal/min (45 L/min) *neglecting line friction.* Hence, 10 s would theoretically place 2 gal (7.6 L) of chemical through the stinger perforations. [This volume equates to $\frac{1}{8}$ gal/ft^2 (5 L/m^2) on a 4-ft (1.2-m) spacing pattern.] By timing the period of injection and maintaining a reasonably constant supply pressure, an acceptably uniform treatment spread would result. Carelessness in either timing or pressure would not be disastrous so long as it was not excessive. [In field practice, a pressure differential of about 60 psi (414 kPa) delivered 2 gal (7.6 L) of chemical in 30 s through the stinger and approximately 60 ft (18.3 m) of ½-in (12.7-mm) ID hose.] The following equations are used to calculate velocities and pressure differentials:

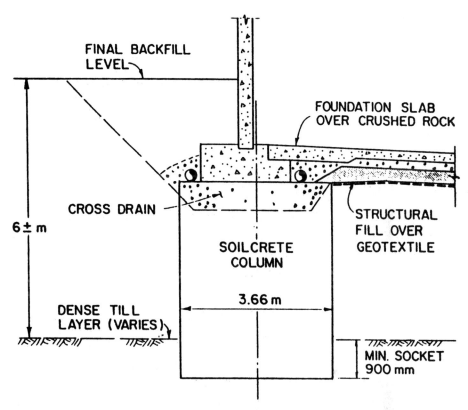

FIGURE 5A.181 Foundation details showing soil-cement walls used to control liquefaction potential beneath two spill tanks (Broomhead and Jasperse, 1992).

Annular velocity is

$$V_a = \frac{Q}{A} \tag{5A.68}$$

which for an annular area, A, = 0.175 in² (113 mm²) reduces, to

$$V_a \text{ (ft/s)} = 1.83\, Q \text{ for } Q \text{ in gal/min} \tag{5A.69a}$$

$$V_a \text{ (m/s)} = 0.148\, Q \text{ for } Q \text{ in L/min} \tag{5A.69b}$$

Port velocity

$$V_p = \frac{Q}{A_p C_D} = \frac{4Q}{D_p^2 \pi n C_D} \tag{5A.70}$$

which for D_p = 3/16 in (4.8 mm), the number of ports, n, = 6, and the discharge coefficient C_D = 0.8 (Brown and Gilbert, 1957), reduces to

$$V_p \text{ (ft/s)} = 2.42\, Q \text{ for } Q \text{ in gal/min} \tag{5A.71a}$$

$$V_p \text{ (m/s)} = 0.195\, Q \text{ for } Q \text{ in L/min} \tag{5A.71b}$$

The pressure differential is represented as

$$\Delta P = \frac{\gamma_f(V_p^2 - V_a^2)}{2g} = \frac{\rho_f(V_p^2 - V_a^2)}{2} \tag{5A.72}$$

where γ_f = unit weight (weight density) of the fluid (= 62.43 pcf = 9.807 kN/m³ for water)
ρ_f = mass density of the fluid (= 1.0 Mg/m³ = 1000 kg/m³ for water)
g = acceleration owing to gravity = 32.17 ft/s² = 9.807 m/s²

For V_p and V_a in ft/s [from Eqs. (5A.69a) and (5A.71a), Eq. (5A.72) can be represented as

$$\Delta P \text{ (psi)} = \frac{G_f(V_p^2 - V_a^2)}{148} \tag{5A.73a}$$

where G_f is the specific gravity of the fluid (= 1.0 for water).
For V_p and V_a in m/s [from Eqs. (5A.69b) and (5A.71b), Eq. (5A.72) can be represented as

$$\Delta P \text{ (kPa)} = \frac{G_f(V_p^2 - V_a^2)}{2} \tag{5A.73b}$$

Force developed from the hydraulic pressure is

$$F = PA \tag{5A.74}$$

where F = force
P = pressure
A = area

Equation (5A.74) illustrates the factors that create a lifting force on a foundation. Generally, pressure injection of chemical soil stabilizers does not involve lifting, as mudjacking or pressure grouting would. In fact, as a rule, chemical injection pressures are limited to prevent lifting. The safe operating pressure could then be estimated from the rearrangement of Eq. (5A.74) as follows:

$$P = \frac{F}{A} \tag{5A.75}$$

5A.4.3.2 Mixing. The primary objective of all mixing techniques is to obtain a thorough mixture of soil and stabilizer at an appropriate moisture content. There are three main types of mixing methods, broadly categorized as off-site, on-site, and deep in situ mixing. A general overview of these mixing methods will be given in the following subsections.

Off-Site Mixing. In off-site mixing techniques, excavated or borrow material is brought to the mixing site, stabilizers and water (if required) are thoroughly mixed with the soil, and the mixture is then hauled to the site for spreading and compaction. Depending upon preference or performance, the soil can be brought to the compaction moisture content either during the off-site mixing process or on-site with the soil remixed to distribute thoroughly the additional water. If the haul distances are very long or if there is a long delay between the off-site mixing and on-site compaction of the material, the soil should be brought to the compaction water content on-site to avoid potentially substantial losses in strength and stiffness of the stabilized material (see Sec. 5A.4.2.5).

Off-site mixing can be performed either in pads or in a central plant. When prepared in a mixing pad, the soil is spread on the ground surface and the stabilizer is added in either a dry or wet condition. If applied wet, the stabilizer is contained within either a thick suspension (slurry) or a solution depending on the chemical nature of the stabilizer and the primary fluid (usually water). Mixing is accomplished using a disc harrow, a grader, or a traveling mixing machine (see below for discussion of different types of mixing equipment).

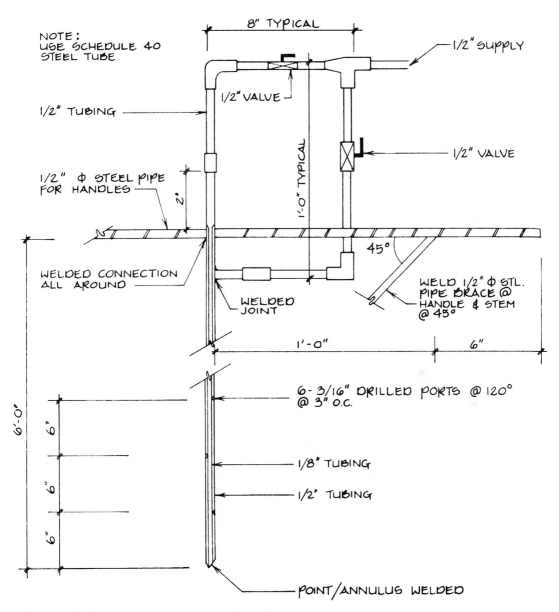

FIGURE 5A.182 Pressure injection stem (from Brown, 1992).

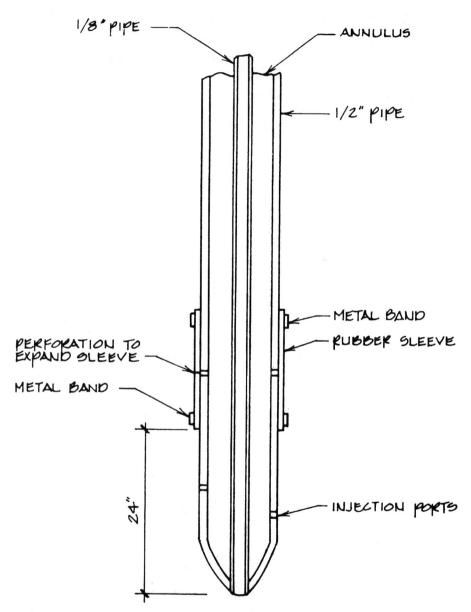

FIGURE 5A.183 Injection stinger with pack-off (from Brown, 1992).

The most common type of off-site mixing is conducted in a central plant. A typical layout of a plant for lime–fly ash mixtures is shown schematically in Fig. 5A.184, which is also typical of the plants used for lime and cement stabilization except only one stabilizer is added instead of two. The main components of a mixing plant are as follows (NCHRP, 1976):

1. Conveyor system
2. Aggregate hopper with belt feeder
3. Hoppers for stabilizers with feed control devices
4. Water storage tank with calibrated pump
5. Surge hopper for temporary storage of soil-stabilizer mixture

As soon as possible after mixing, the soil-stabilizer mixture is hauled to the site in open-bed or bottom dump trucks. If the haul distances are long, or if drying or scattering of the material en route is a problem, the trucks should be covered with tarpaulins or other suitable covers.

Some soils, particularly highly plastic clays, cannot be adequately pulverized and mixed in central mixing plants unless the soils are modified beforehand with a small amount of stabilizer to reduce plasticity and increase agglomeration. If feasible, however, central plant mixing accomplishes the most thorough and uniform mixing and reduces safety problems associated with the chemicals (for example, health problems for workers caused by the causticity of lime).

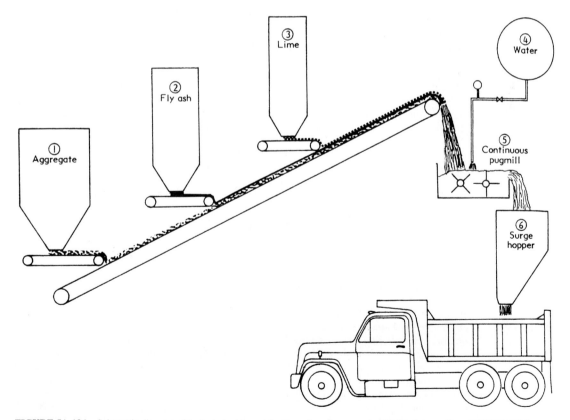

FIGURE 5A.184 Schematic diagram of typical plant layout for lime–fly ash-aggregate (LFA) mixtures (from NCHRP, 1976).

On-Site Mixing. On-site mixing can be accomplished in test pads, in a plant, or in portable mechanical mixers or through use of mixed-in-place techniques. Details on mixing in test pads or a central plant are given in the previous section. Mixing in pads is usually conducted off-site rather than on-site owing to a lack of sufficient on-site open space in most cases. On-site mixing plants are similar in concept and equipment to off-site plants but are usually smaller and hence cannot produce mixing at the same volumetric rate. The portable mechanical mixers used to mix soil and stabilizer are the same as or similar to the portable mixers used for concrete.

Mixed-in-place methods can be used on excavated materials, borrow materials, or in situ surficial soils. When borrow or excavated materials are mixed in place, the soil is first spread in either uniform layers or windrows. Wet or dry stabilizer is then added to the soil. Dry stabilizer can be added either in bags (lime or cement) or in bulk from pneumatic tanker trucks or mechanical spreaders. Bags are typically spotted by hand by dropping them at previously marked locations from a slowly moving truck. The bags are then slit open and the contents dumped into piles or transverse windrows across the area to be stabilized. The cement or lime is leveled by hand raking, or using a spike-tooth harrow, a nail drag, or a length of chain-link fence pulled by a tractor or truck. When both lime and fly ash are used, they can be spread separately, or if dry they can be pre-blended before being added to the soil. After lime (but not cement or lime–fly ash) is placed, the surface is usually sprinkled with water to reduce dusting. Either hydrated lime or quicklime can be added to the soil as a slurry. Lime and water are mixed in a central mixing tank, jet mixer, or tank truck and spread by either gravity or pressure spraying from a tank truck through spray bars.

Many different types of equipment can be used to mix soil and stabilizer in place. Pugmill travel plants can be used on excavated or borrow materials only. When hopper-type pugmill travel plants are used, soil is dumped into a hopper from a dump truck, and stabilizer and water (if desired) are added prior to mixing. Water can also be added after the soil-stabilizer mixture has been placed. For windrow-type pugmill travel plants, soil is placed in windrows and stabilizer added using one of the methods described in the previous paragraph. Rotary mixers, graders, and disc harrows can be used on excavated, borrow, or in situ surficial soils.

The difficulty of in-place mixing increases with increasing fineness and plasticity of the soils being treated. A double application of lime (modification of the soil with a small amount of lime followed by a suitable period for curing and then a second treatment for stabilization) or modification with lime prior to stabilization with cement or lime–fly ash, is often required when highly plastic clays (PI $\geq$ 50) are to be stabilized. In general, compaction should begin as soon as possible after mixing is completed.

After mixing, accomplished by any of the off-site or on-site methods, and placement, the stabilized soil must be compacted. The same equipment and methods used for near-surface compaction (see Sec. 5A.3.1 for details) can also be used for chemically stabilized soils. For cohesive soils, tamping foot or sheepsfoot rollers are typically used. Static or vibratory rubber-tired or smooth-drum rollers are generally used to compact granular soils or chemically modified cohesive soils that behave essentially as granular materials. When a relatively smooth surface finish is desired, smooth drum rollers can be used to proof-roll the material.

Compaction should commence as soon as possible after uniform mixing of soil, stabilizer, and water has taken place. It is common to specify that the chemically treated materials be compacted within 2 h after mixing. Because the reactions associated with lime stabilization are slow compacted to those that occur in cement and lime-fly ash stabilization, additional time may be available between mixing and compaction for lime-treated soils.

A typical density specification for a chemically treated soil is a minimum relative compaction of 95% based on standard Proctor maximum dry density. However, the actual specification given on any project should be based on criteria that will yield the desired engineering characteristics. The reference Proctor tests should be conducted on samples of the soil treated with the same percentage of stabilizer used in the field and allowing the same delay between mixing and compaction as the average (or maximum) delay that occurs in the field. The specifications for compaction moisture condition are critical to successful stabilization. If the soil is too dry when compacted, incomplete hydration of the stabilizer may occur, resulting in engineering properties that may not meet the requirements for the project. In general, the compaction water content should not be less than standard Proctor optimum water content, which may be several percentage points higher than modified Proctor optimum water content. The specifications for compaction water content ideally should be based on the results of laboratory or field tests conducted to establish the range in water contents that will produce the desired engineering characteristics.

Deep in Situ Mixing. Although deep in situ soil mixing techniques originated in the United States in the 1950s, the major development of these methods has occurred over the last two decades in Japan and Sweden (Broomhead and Jasperse, 1992; Hausmann, 1990). Although the techniques were developed almost concurrently in Japan and Sweden and are similar in concept, they are generally referred to by different names—*deep soil mixing* for the Japanese method and *lime columns* for the Swedish method. In general, unslaked lime (quicklime, CaO) is used as the primary stabilizer in the Swedish method, and cement is used most frequently in the Japanese method. Other stabilizers and additives that have been used include fly ash, furnace slag, gypsum, hydroxyaluminum, potassium chloride, and bentonite (Broomhead and Jasperse, 1992; Broms, 1993).

Lime columns are used only in soft clays and silts. The mixing tool, which is shaped like a giant dough mixer (see Fig. 5A.185), is rotated down to the depth that corresponds to the desired length of the columns (Broms 1993). The direction of rotation is then reversed, and the mixing tool is slowly withdrawn as quicklime is forced into the soil by compressed air pushing through holes located just above the horizontal blades of the mixing tool. The blades are inclined so that the stabilized soil will be compacted during the withdrawal of the tool. The rotary table and kelly are on a mast, which is mounted on either a standard front wheel loader or on a special off-road vehicle. The lime-stabilized columns are typically 0.5 to 0.6 m (1.6 to 2.0 ft) in diameter and can be constructed up to about 30 m (100 ft) in length.

Deep soil mixing can be used in soils ranging from soft clays to gravels (Broomhead and Jasperse, 1992; Taki and Yang, 1991). The equipment used in deep soil mixing consists of a crane-supported set of leads that guide a series of one to eight hydraulically driven mixing augers, as illustrated in Fig. 5A.186. The augers are typically 0.45 to 1.0 m (1.5 to 3.3 ft) in diameter, and columns of more than 60 m (200 ft) in length can be constructed. Each auger shaft is mounted with discontinuous auger flights and mixing paddles. The purposes of the discontinuous auger flights are to provide some vertical displacement for soil mixing and to prevent the transportation of soil to the surface. As penetration by the auger flights occurs, the soil is broken loose and the cement slurry is injected into the soil through the tip of the hollow-stemmed augers. The auger flights lift the soil and slurry to the mixing paddles where the materials are blended. As the auger continues to advance, the soil and slurry are remixed by additional paddles attached to the shaft. When multiple augers are used, overlapped soil-cement columns are created that can be extended to form any geometric shape or pattern.

Lime columns have been used extensively in Sweden, Norway, and Finland, with some applications in the United States, for the following purposes (Broms, 1993):

- To improve the total and differential settlements of light structures
- To increase the settlement rate and control the settlements of relatively heavy structures
- To improve the stability of embankments, slopes, trenches, and deep cuts
- To reduce the vibrations from, for example, traffic, blasting, pile driving

More than 3,000,000 m (10,000,000 ft) of lime columns have been installed in Sweden alone since 1975, with a production rate (1990) of about 600,000 m/yr (2,000,000 ft/yr).

Deep soil mixing has been used on more than 1500 projects in Japan and the Far East and was reintroduced in the United States in 1986 (Broomhead and Jasperse, 1992; Taki and Yang, 1991). Soil-cement walls constructed by deep soil mixing have been used in the following applications:

- Cutoff walls for ground-water control
- Walls for excavation support
- Stabilization of soft or liquefiable soils

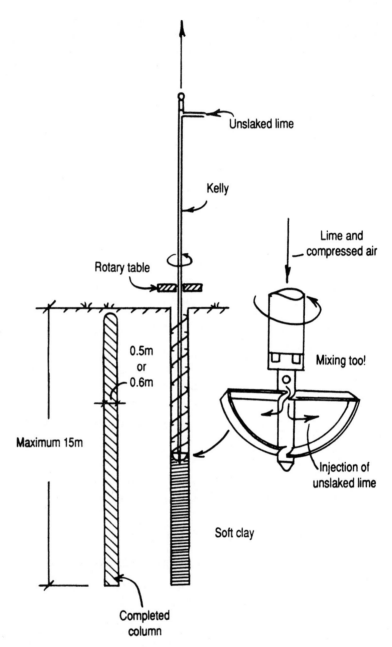

FIGURE 5A.185 Typical equipment used to manufacture lime columns (from Broms, 1993).

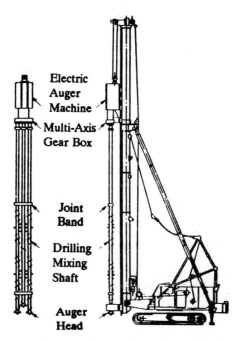

FIGURE 5A.186 Typical equipment used to manufacture deep soil-cement mixed columns (from Taki and Yang, 1991).

A spinoff technology to deep soil mixing has recently been developed and has been dubbed *shallow soil mixing* (Broomhead and Jasperse, 1992). This process uses a single auger 1.0 to 3.7 m (3 to 12 ft) in diameter and can be used to stabilize soils to a depth of 10 m (33 ft) or less. Shallow soil mixing has been used for both geotechnical and geoenvironmental applications.

Curing. Three conditions are needed for proper curing of soils stabilized with cementitious or pozzolanic admixtures:

1. *Favorable temperatures.* Temperatures greater than 40°F (4°C) are needed for the cementitious or pozzolanic reactions to occur at a reasonable rate. Therefore, construction of admixture stabilized soils should not commence if the temperatures during the month or so following the completion of the construction will drop below the minimum required temperature for extended periods.

2. *Passage of time.* Usually at least a 3- to 7-day undisturbed curing period is specified before any appreciable load can be applied.

3. *Moist conditions.* The primary purposes for requiring the soil to be kept moist are to ensure that adequate moisture is available for hydration of the stabilizer and to keep the shrinkage cracking to a minimum. In lime-stabilized soils, moist conditions are also needed to prevent carbonation. If the stabilized soil is backfilled immediately after construction, no special measures are required for adequate curing. If the stabilized soil is exposed to the atmosphere for more than a few hours following construction, appropriate procedures should be taken to ensure that complete curing will occur. The procedures used are of two general types—moist or asphaltic curing. In *moist curing,* the surface is sprinkled with water to keep it damp. On small jobs, the stabilized soil may also be covered with plastic sheets. Light compaction with small smooth drum rollers or vibratory plate compactors may be used, as needed, to keep the surface knitted together. *Asphaltic curing* involves sealing the surface with a coat of cutback asphalt or bituminous emulsion.

Quality Control and Assurance. For admixture-stabilized soils prepared by off-site or on-site mixing methods, the procedures for quality control and assurance are similar to those for compacted untreated soils (Sec. 5A.3.5), with the additional requirements that the stabilizer content and uniformity of mixing be checked. The control of stabilizer content is a two-step process involving monitoring and spot-checking the amount of stabilizer that is added to the soil and conducting tests on samples obtained from the material immediately after mixing is completed. Standard chemical analyses for determining the cement or lime content of freshly mixed, treated soils are given in ASTM D2901 and D3155, respectively.

The uniformity of cement-stabilized soils can be checked by visually inspecting the exposed soil within a trench dug across the width of the treated area and to the desired depth of treatment. A satisfactory mixture will exhibit a uniform color throughout, and a streaked appearance indicates a nonuniform mixture. Checking the uniformity of lime-soil mixtures is more difficult because the visual appearances of the untreated soil and the treated soil are the same.

Phenolphthalein, a color-sensitive indicator solution, can be used to check that the minimum lime content required for soil treatment is present. When the solution is sprayed on the soil, the soil will turn a reddish pink color if free lime is available (pH = 12.5). Short-cut strength methods can also be used.

The quality control and assurance procedures used for deep in situ mixing are necessarily complex because the soil is stabilized to substantial depths. Details regarding the methods used for lime columns are given in Broms (1993).

5A.5 Moisture Barriers*

As implied by their names, the intended purpose of horizontal or vertical moisture barriers is to prevent the seepage of moisture into or out of an expansive soil located beneath a foundation. Moisture barriers have been used as both preventive and remedial measures. Unfortunately, moisture barriers are frequently ineffective in controlling swelling in expansive soils for the following reasons (Jones and Jones, 1987):

1. Moisture may enter the protected area from below the barrier, such as by upward migration of water vapor from an underlying ground-water table.
2. Variations in moisture content of the protected soil will still occur in covered soils owing to changes in soil temperature (called *hydrogenesis*). Even in dry climates the air has some moisture, which enters the subgrade soil. As the ground temperatures cool at shallow depths, the moisture in the air condenses and accumulates in areas that are covered.
3. Moisture barriers cannot prevent moisture flow through shrinkage cracks, slickensides, or permeable inclusions below the barriers.

The use of moisture barriers is not a viable preventative or remedial measure for expansive soils because the money spent could be better spent on an alternative solution that is both more effective and more permanent.

5A.5.1 Horizontal Barriers

Horizontal moisture barriers are intended to prevent or substantially reduce the moisture loss from an area of soil resulting primarily from evaporation. A typical design of a horizontal barrier using a polyethylene membrane overlain by a thin layer of gravel is depicted in Fig. 5A.187. Membranes made from polyvinyl chloride (PVC) and polypropylene have also been used. These plastic membranes typically range from about 4 to 20 mils (0.2 to 0.8 mm) in thickness. Other materials that have been used to make the impervious barrier include concrete and asphalt. Regardless of the type of barrier material used, construction techniques are important to the effectiveness of the barriers (Nelson and Miller, 1992). Care should be taken to seal joints, seams, rips, or holes in the barrier.

Jones and Jones (1987) indicate that the effectiveness of horizontal barriers can be improved in the following ways:

1. The barriers should be placed at least 2 ft (0.6 m) below the ground surface because at this depth, the daily variations in soil temperature are negligible.
2. The preconstruction moisture condition of the bearing soil and the soil underlying the barrier should be controlled to maintain a high water content that is uniform throughout the soil.

---

*Coauthored by Robert Wade Brown.

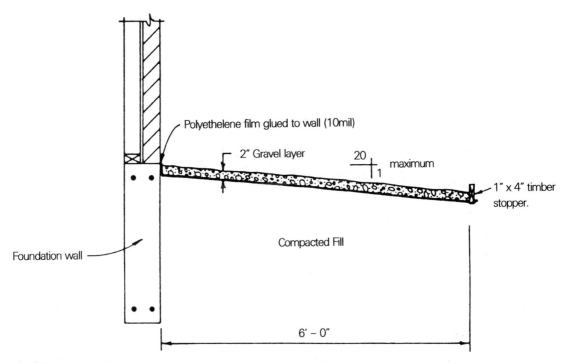

FIGURE 5A.187 Horizontal moisture barrier consisting of a polyethylene membrane overlain by a thin gravel layer (from Chen, 1988).

3. Open areas surrounding the barriers should be irrigated regularly using a sprinkler system to sustain a consistently high water content.

Even if the measures outlined above are taken, the behavior of a horizontal barrier is not reliably predictable. Discussions of data reflecting the success or benefit in using horizontal barriers can be found in Chen (1988) and Nelson and Miller (1992). Both sources seem to agree that a horizontal barrier does little to prevent or lessen the increase in moisture (heave) within a foundation soil over several years. However, the presence of a barrier does seem to cause the buildup in moisture to develop more slowly and to be more uniform. The pertinent question is whether these benefits justify the cost.

5A.5.2 Vertical Barriers

Vertical moisture barriers are intended to avoid or substantially limit the lateral transfer of water between the foundation bearing soils and those outside the perimeter. The theory is that if expansive soils are maintained at a constant level of moisture, no volumetric changes occur. The vertical barrier can consist of an excavated trench lined or filled with any impermeable material such as polyethylene membrane, fiberglass, asphalt, concrete, semihardening slurries, or sometimes even tar paper (see Fig. 5A.188). If polyethylene membrane is used, it should be sufficiently thick and durable to resist puncturing and tearing during installation. The trench should ideally extend as deep as the zone of seasonal changes in moisture (active zone). Sometimes a horizontal barrier is used with a vertical barrier to prevent wetting of the soil between the vertical barrier and the building.

The published data dealing with the effectiveness of impermeable vertical barriers are similar to those described for horizontal barriers. Vertical barriers tend to reduce increases in soil mois-

5.266 SOIL IMPROVEMENT AND STABILIZATION

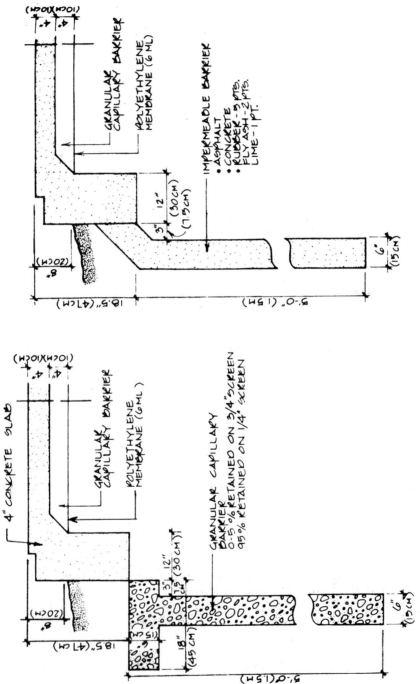

FIGURE 5A.188 Typical permeable and impermeable vertical moisture barriers (from Brown, 1992).

ture (heave) during the first year or so; however, after 4 or 5 years, the results become nearly the same with or without the barrier. As with horizontal barriers, the rate of increase in moisture is generally slower and more uniform when a vertical barrier is used. Again the question arises: Does the result justify the cost? As stated by Chen (1988), "...in view of the high cost involved in the installation of a vertical moisture barrier, especially where great depth is required, it is doubtful that such an installation is of sufficient merit to warrant the expense."

A variation of this design is the permeable, vertical soil barrier, also shown in Fig. 5A.188. This technique is primarily intended to prevent or minimize external water from entering the bearing soil matrix, somewhat similar to a French drain. Published information on the performance of permeable vertical barriers is not readily available.

References

Ahlberg, H. L., and Barenberg, E. J. (1965). "Pozzolanic Pavements." *Bulletin 473,* Engineering Experiment Station, University of Illinois, Urbana.

Alwail, T. A., Ho, C. L., and Fragaszy, R. J. (1992). "Collapse Mechanisms of Low Cohesion Compacted Soils." *Bulletin of the Association of Engineering Geologists,* vol. 29, no. 4, pp. 345–353.

ASCE (1987). *Soil Improvement—A Ten Year Update,* Geotechnical Special Publication no. 12, edited by J. P. Welsh.

ASTM (1991). *Annual Book of Standards, Volume 04.08: Soil and Rock; Dimension Stone; Geosynthetics,* American Society for Testing and Materials, Philadelphia.

Ballard, L. F., and Gardner, R. P. (1965). "Density and Moisture Content Measurements by Nuclear Methods." *National Cooperative Highway Research Program Report No. 14,* Highway Research Board, Washington, D.C.

Barden, L., and Pavlakis, G. (1971). "Air and Water Permeability of Compacted Unsaturated Cohesive Soil." *Journal of Soil Science,* vol. 22, no. 3, pp. 302–317.

Barden, L., and Sides, G. R. (1970). "Engineering Behavior and Structure of Compacted Clay." *Journal of the Soil Mechanics and Foundations Division,* ASCE, vol. 96, no. SM4, July, pp. 1171–1200.

Bell, F. G. (1988). "Stabilization and Treatment of Clay Soils with Lime: Part 1—Basic Principles." *Ground Engineering,* British Geotechnical Society, London, vol. 21, no. 1, January, pp. 10–15.

Benson, C. H., Abichou, T. H., Olson, M. A., and Bosscher, P. J. (1995). "Winter Effects on Hydraulic Conductivity of Compacted Clay." *Journal of Geotechnical Engineering,* ASCE, vol. 121, no. 1, January, pp. 69–79.

Benson, C. H., and Daniel, D. E. (1990). "Influence of Clods on the Hydraulic Conductivity of Compacted Clay." *Journal of Geotechnical Engineering,* ASCE, vol. 116, no. 8, August, pp. 1231–1248.

Bishop, A. W. (1959). "The Principle of Effective Stress." *Teknisk Ukeblad,* vol. 39, pp. 859–863.

Bishop, A. W., Alpan, I., Blight, G. E., and Donald, I. B. (1960). "Factors Controlling the Strength of Partly Saturated Cohesive Soils." *Proceedings of the Research Conference on Shear Strength of Cohesive Soils,* ASCE, Boulder, Colorado, pp. 503–532.

Bishop, A. W., and Eldin, A. K. G. (1953). "The Effect of Stress History on the Relation between f and Porosity in Sand." *Proceedings of the Third International Conference on Soil Mechanics and Foundation Engineering,* Zurich, Switzerland, vol. 1, pp. 100–105.

Blight, G. E. (1965). "A Study of Effective Stresses for Volume Change." *Moisture Equilibria and Moisture Changes in Soils beneath Covered Areas,* edited by G. D. Aitchison, Butterworths, Sydney, Australia.

Booth, A. R. (1976). "Compaction and Preparation of Soil Specimens for Oedometer Testing." *Soil Specimen Preparation for Laboratory Testing, ASTM STP 599,* American Society for Testing and Materials, Philadelphia, Pennsylvania, pp. 216–228.

Booth, A. R. (1977). "Collapse Settlement in Compacted Soils." *CSIR Research Report 324,* Council for Scientific and Industrial Research, Pretoria, South Africa.

Bouyoucos, G. J. (1931). "The Alcohol Method for Determining the Water Content for Soils." *Soil Scientist,* vol. 32, pp. 173–179.

Bowders, J. J., Jr., Gidley, J. S., and Usmen, M. A. (1990). "Permeability and Leachate Characteristics of Stabilized Class F Fly Ash." *Transportation Research Record no. 1288: Geotechnical Engineering 1990,* Transportation Research Board, Washington, D.C., pp. 70–77.

Bowles, J. E. (1987). "Elastic Foundation Settlements on Sand Deposits." *Journal of Geotechnical Engineering,* ASCE, vol. 113, no. 8, pp. 846–860.

Bowles, J. E. (1988). *Foundation Analysis and Design,* 4th ed., McGraw-Hill, New York.

Boynton, R. S., and Blacklock, J. R. (1985). "Lime Slurry Pressure Injection Bulletin." *Bulletin No. 331,* National Lime Association, Arlington, Virginia.

Boynton, S. S., and Daniel, D. E. (1985). "Hydraulic Conductivity Tests on Compacted Clay." *Journal of Geotechnical Engineering,* ASCE, vol. 111, no. 4, April, pp. 465–478.

Brand, W., and Schoenberg, W. (1959). "Impact of Stabilization of Less with Quicklime on Highway Construction." *Highway Research Board Bulletin No. 231: Lime and Lime-Flyash as Soil Stabilizers,* Washington, D.C., pp. 18–23.

Broderick, G. P., and Daniel, D. E. (1990). "Stabilizing Compacted Clay against Chemical Attack." *Journal of Geotechnical Engineering,* ASCE, vol. 116, no. 10, October, pp. 1549–1567.

Broms, B. B. (1993). "Lime Stabilization." Chapter 4 in *Ground Improvement,* edited by M. P. Moseley, CRC Press, Boca Raton, Florida, pp. 65–99.

Broms, B. B., and Forssblad, L. (1969). "Vibratory Compaction of Cohesionless Soils." *Proceedings of the Seventh International Conference on Soil Mechanics and Foundation Engineering,* Specialty Session no. 2, pp. 101–118.

Broomhead, D., and Jasperse, B. H. (1992). "Shallow Soil Mixing—a Case History." *Grouting, Soil Improvement and Geosynthetics,* edited by R. H. Borden, R. D. Holtz, and I. Juran, ASCE Geotechnical Special Publication no. 30, vol. 1, pp. 564–576.

Brown, R. W. (1989). *Design and Repair of Residential and Light Commercial Foundations,* McGraw-Hill, New York.

Brown, R. W. (1992). *Foundation Behavior and Repair: Residential and Light Commercial,* 2d ed., McGraw-Hill, New York.

Brown, R. W., and Gilbert, B. (1957). "Pressure Drop Perforations Can Be Computed." *Petroleum Engineer,* September.

Bureau of Reclamation (1968). *Earth Manual,* rev. ed., U.S. Bureau of Reclamation, Denver, Colorado.

Burland, J. B., and Burbidge, M. C. (1985). "Settlement of Foundations on Sand and Gravel." *Proceedings of the Institution of Civil Engineers,* part 1, vol. 78, December, pp. 1325–1381.

Burmister, D. M. (1945). "The General Theory of Stresses and Displacements in Layered Soil Systems." *Journal of Applied Physics,* vol. 16, no. 2, pp. 89–96; no. 3, pp. 126–127; no. 5, pp. 296–302.

Burmister, D. M. (1958). "Evaluation of Pavement Systems of the WASHO Road Test by Layered System Methods." *Highway Research Board Bulletin 177,* Washington, D.C., pp. 26–54.

Burmister, D. M. (1962). "Application of Layered System Concepts and Principles to Interpretations and Evaluations of Asphalt Pavement Performances and to Design and Construction." *Proceedings of the International Conference on Structural Design of Pavements,* University of Michigan, Ann Arbor, pp. 441–453.

Butt, G. S., Demirel, T., and Handy, R. L. (1968). "Soil Bearing Tests Using a Spherical Penetration Device." *Highway Research Record No. 243: Soil Properties from in Situ Measurements,* Highway Research Board, Washington, D.C., pp. 62–74.

Carter, D. P., and Seed, H. B. (1988). "Liquefaction Potential of Sand Deposits under Low Levels of Excitation." *Report No. UCB/EERC–88/11,* University of California at Berkeley, College of Engineering.

Casagrande, A. (1936a). "Characteristics of Cohesionless Soils Affecting the Stability of Slopes and Earth Fills." *Journal of the Boston Society of Civil Engineers,* January; reprinted in *Contributions to Soil Mechanics 1925–1940,* BSCE, pp. 257–276.

Casagrande, A. (1936b). "The Determination of the Preconsolidation Load and Its Practical Significance." *Proceedings of the First International Conference on Soil Mechanics and Foundation Engineering,* Cambridge, vol. 3, pp. 60–64.

Casagrande, A. (1938). "Notes on Soil Mechanics—First Semester." Harvard University (unpublished), cited in Holtz, R. D. and Kovacs, W. D. (1981).

Chamberlain, E. J., Iskander, I., and Hunsicker, S. E. (1990). "Effect of Freeze-Thaw Cycles on the Permeability and Macrostructure of Clays." *Proceedings of the International Symposium on Frozen Soil*

Impacts on Agricultural, Range, and Forest Lands, CRREL Special Report 90–1, U.S. Army Corps of Engineers Cold Regions Research and Engineering Laboratory, Hanover, New Hampshire, pp. 145–155.

Chen, F. H. (1988). *Foundations on Expansive Soils,* Elsevier Science Publishers, New York.

Clare, K. E., and Sherwood, P. T. (1954). "The Effect of Organic Matter on the Setting of Soil-Cement Mixtures." *Journal of Applied Chemistry,* Society of Chemical Industry, London, vol. 4, November, pp. 625–630.

Clare, K. E., and Sherwood, P. T. (1956). "Further Studies on the Effect of Organic Matter on the Setting of Soil-Cement Mixtures." *Journal of Applied Chemistry,* Society of Chemical Industry, London, vol. 6, August, pp. 317–324.

Clemence, S. P., and Finbarr, A. O. (1981). "Design Considerations for Collapsible Soils." *Journal of the Geotechnical Engineering Division,* ASCE, vol. 107, no. GT3, March, pp. 305–317.

Clifford, J. M. (1980). "An Introduction to Impact Rollers." *Proceedings of the International Conference on Compaction,* Paris, April, pp. 621–626.

Clough, G. W., Iwabuchi, J., Rad, N. S., and Kuppusamy, T. (1989). "Influence of Cementation on Liquefaction of Sands." *Journal of Geotechnical Engineering,* ASCE, vol. 115, no. 8, August, pp. 1102–1117.

Daniel, D. E. (1984). "Predicting Hydraulic Conductivity of Clay Liners." *Journal of Geotechnical Engineering,* ASCE, vol. 110, no. 2, February, pp. 285–300.

Daniel, D. E. (1993). "Clay Liners." Chapter 7 in *Geotechnical Practice for Waste Disposal,* edited by D. E. Daniel, Chapman & Hall, New York, pp. 137–163.

Daniel, D. E., and Benson, C. H. (1990). "Water Content-Density Criteria for Compacted Soil Liners." *Journal of Geotechnical Engineering,* ASCE, vol. 116, no. 12, December, pp. 1811–1830.

D'Appolonia, D. J., Whitman, R. V., and D'Appolonia, E. D. (1969). "Sand compaction with vibratory rollers." *Journal of the Soil Mechanics and Foundations Division,* ASCE, vol. 95, no. SM1, January, pp. 263–284.

Daramola, O. (1980). "Effect of Consolidation Age on Stiffness of Sand." *Geotechnique,* vol. 30, no. 2, June, pp. 213–216.

Das, B. M. (1983). *Advanced Soil Mechanics,* Hemisphere Publishing, New York.

Davis, R. E., Gailey, K. D., and Whitten, K. W. (1984). *Principles of Chemistry,* Saunders College Publishing, Philadelphia.

Day, R. W. (1989). "Relative Compaction of Fill Having Oversize Particles." *Journal of Geotechnical Engineering,* ASCE, vol. 115, no. 10, October, pp. 1487–1491.

Day, R. W. (1991). "Expansion of Compacted Gravelly Clay." *Journal of Geotechnical Engineering,* ASCE, vol. 117, no. 6, June, pp. 968–972.

Day, R. W. (1992). "Effective Cohesion for Compacted Clay." *Journal of Geotechnical Engineering,* ASCE, vol. 118, no. 4, April, pp. 611–619.

Day, R. W. (1994). "Swell-Shrink Behavior of Compacted Clay." *Journal of Geotechnical Engineering,* ASCE, vol. 120, no. 3, March, pp. 618–623.

Diamond, S., and Kinter, E. B. (1965). "Mechanisms of Soil-Lime Stabilization, an Interpretive Review." *Highway Research Record No. 92: Lime Stabilization,* Highway Research Board, Washington, D.C., pp. 83–102.

Douglas, B. J., Olsen, R. S., and Martin G. R. (1981). "Evaluation of Cone Penetrometer Test for SPT Liquefaction Assessment." *In Situ Testing to Evaluate Liquefaction Susceptibility,* Preprint 81-544, ASCE.

Dumbleton, M. J. (1962). "Lime Stabilized Soil for Road Construction in Great Britain—A Laboratory Investigation." *Roads and Road Construction,* vol. 40, November, pp. 321–325.

Duncan, J. M., Byrne, P., Wong, K. S., and Mabry, P. (1980). "Strength, Stress-Strain, and Bulk Modulus Parameters for Finite Element Analyses of Stresses and Movements in Soil Masses." *Report No. UCB/GT/80–01,* University of California, Berkeley, August.

Duncan, J. M., and Chang, C.-Y. (1970). "Nonlinear Analysis of Stress and Strain in Soils." *Journal of the Soil Mechanics and Foundations Division,* ASCE, vol. 96, no. SM5, September, pp. 1629–1653.

Duncan, J. M., and Seed, R. B. (1986). "Compaction-Induced Earth Pressures under K_0-Conditions." *Journal of Geotechnical Engineering,* ASCE, vol. 112, no. 1, January, pp. 1–22.

Evans, J. C., and Pancoski, S. E. (1989). "Organically Modified Clays." *Paper No. 880587,* Transportation Research Board, Washington, D.C., January.

Felt, E. J., and Abrams, M. S. (1957). "Strength and Elastic Properties of Compacted Soil-Cement Mixtures." *Special Technical Publication No. 206,* ASTM.

Ferguson, G. (1993). "Use of Self-Cementing Fly Ashes as a Soil Stabilization Agent." *Fly Ash for Soil Improvement,* ASCE Geotechnical Special Publication no. 36, pp. 1–14.

Ferris, G. A., Eades, J. L, McClellan, G. H., and Graves, R. E. (1991). "Improved Characteristics in Sulfate Soils Pretreated with Barium Compounds Prior to Lime Stabilization." Paper presented at the 1991 Annual Meeting of the Transportation Research Board, Washington, D.C., January.

FHWA (1979). Soil Stabilization in Pavement Structures—a User's Manual, Report no. FHWA-IP–80–2, Federal Highway Administration, Washington, D.C., October.

Foreman, D. E., and Daniel, D. E. (1986). "Permeation of Compacted Clay with Organic Chemicals." *Journal of Geotechnical Engineering,* ASCE, vol. 112, no. 7, July, pp. 669–681.

Forssblad, L. (1981). *Vibratory Soil and Rock Fill Compaction.* Dynapac Maskin AB, Stockholm, Sweden.

Foster, C. R. (1962). "Field Problems: Compaction." Chapter 12 in *Foundation Engineering,* edited by G. A. Leonards, McGraw-Hill, New York, pp. 1000–1024.

Foster, C. R., and Ahlvin, R. G. (1954). "Stresses and Deflections Induced by a Uniform Circular Load." *Proceedings of the Highway Research Board*, Washington, D.C., vol. 33, pp. 467–470.

Fox, E. N. (1948). "The Mean Elastic Settlement of a Uniformly Loaded Area at a Depth below the Ground Surface." *Proceedings, Second International Conference on Soil Mechanics and Foundation Engineering,* Rotterdam, Netherlands, vol. 1, pp. 129–132.

Fox, L. (1948). "Computation of Traffic Stresses in a Simple Road Structure." *Proceedings of the Second International Conference on Soil Mechanics and Foundation Engineering,* vol. 2, pp. 236–246.

Fox, N. S. (1968). *Clay Stabilization in the Mekong Delta,* U.S. Army Report, 579th Engineer Detachment.

Fragaszy, R. J., Su, J., and Siddiqi, F. H. (1992). "Effects of Oversize Particles on Density of Clean Granular Soils." *Geotechnical Testing Journal,* vol. 12, no. 2, June, pp. 106–114.

Fragaszy, R. J., Su, J., Siddiqi, F. H., and Ho, C. L. (1992). "Modeling Strength of Sandy Gravel." *Journal of Geotechnical Engineering,* ASCE, vol. 118, no. 6, June, pp. 920–935.

Fredlund, D. G., and Rahardjo, H. (1993). *Soil Mechanics for Unsaturated Soils,* John Wiley & Sons, New York.

Fredlund, D. G., Morgenstern, N. R., and Widger, R. A. (1978). "The Shear Strength of Unsaturated Soils." *Canadian Geotechnical Journal,* vol. 15, no. 3, pp. 313–321.

Gieseking, J. E. (1939). "Mechanics of Cation Exchange in the Montmorillonite-Beidilite-Nontronite Type of Clay Mineral." *Soil Science,* vol. 4, pp. 1–14.

Godfrey, K. A. (1978). "Expansive and Shrinking Soils—Building Design Problems Being Attacked." *Civil Engineering,* ASCE, October, pp. 87–91.

Grim, R. E. (1953). *Clay Mineralogy,* McGraw-Hill, New York.

Grim, R. E. (1962). *Applied Clay Mineralogy,* McGraw-Hill, New York.

Grubbs, B. R. (1990). "Lime-Induced Heave of Eagle Ford Clays in Southeast Dallas County." Presented at Texas Section ASCE Fall Meeting, El Paso, October 13.

Hamed, J. T., Das, B. M., and Echelberger, W. F. (1986). "Bearing Capacity of a Strip Foundation on Granular Trench in Soft Clay." *Civil Engineering for Practicing and Design Engineers,* Pergamon Press, vol. 5, pp. 359–376.

Hanna, A. M. (1981). "Foundations on Strong Sand Overlying Weak Sand." *Journal of the Geotechnical Engineering Division,* ASCE, vol. 107, no. GT7, pp. 915–927.

Hausmann, M. R. (1990). *Engineering Principles of Ground Modification,* McGraw-Hill, New York.

Herzog, A., and Mitchell, J. K. (1963). "Reactions Accompanying Stabilization of Clay with Cement." *Highway Research Record No. 36,* Highway Research Board, Washington, D.C., pp. 146–171.

Hilf, J. W. (1959). *A Rapid Method of Construction Control for Embankments of Cohesive Soil,* Engineering Monograph no. 26, U.S. Department of the Interior, Bureau of Reclamation, Denver, Colorado.

Hilf, J. W. (1991). "Compacted Fill." Chapter 8 in *Foundation Engineering Handbook,* 2d ed., edited by H.-Y. Fang, Van Nostrand Reinhold, New York, pp. 249–316.

Hodek, R. J., and Lovell, C. W. (1979). "A New Look at Compaction Processes in Fills." *Bulletin of the Association of Engineering Geologists,* vol. 16, no. 4, pp. 487–499.

Hollon, G. W., and Marks, B. A. (1962). "A Correlation of Published Data on Lime-Pozzolan-Aggregate Mixtures for Highway Base Course Construction," *Circular No. 72,* Engineering Experiments Station, University of Illinois, Urbana.

Holtz, R. D., and Kovacs, W. D. (1981). *An Introduction to Geotechnical Engineering,* Prentice-Hall, Inc., Englewood Cliffs, New Jersey.

Holtz, W. G. (1969). "Volume Change in Expansive Clay Soils and Control by Lime Treatment." Presented at the Second International Research and Engineering Conference on Expansive Clay Soils, Texas A & M University, College Station, August.

Houston, S. L., and Randeni, J. S. (1992). "Effect of Clod Size on Hydraulic Conductivity of Compacted Clay." *Geotechnical Testing Journal,* ASTM, vol. 15, no. 2, June, pp. 123–128.

Houston, S. L., and Walsh, K. D. (1993). "Comparison of Rock Correction Methods for Compaction of Clayey Soils." *Journal of Geotechnical Engineering,* ASCE, vol. 119, no. 4, April, pp. 763–778.

HRB (1952). *Highway Research Board Bulletin No. 58: Compaction of Embankments, Subgrades, and Bases,* Washington, D.C.

HRB (1961). *Highway Research Board Bulletin No. 292: Soil Stabilization with Portland Cement.*

Hunter, D. (1988). "Lime-Induced Heave in Sulfate-Bearing Clay Soils." *Journal of Geotechnical Engineering,* ASCE, vol. 114, no. 2, February, pp. 150–167.

Ingles, O. G., and Metcalf, J. B. (1973). *Soil Stabilization: Principles and Practice,* John Wiley & Sons, New York.

Jaky, J. (1948). "Influence of Ground Water Level Oscillation on Subsidence of Structures." *Proceedings of the Second International Conference on Soil Mechanics and Foundation Engineering,* Rotterdam, pp. 148–150.

Jones, A. (1962). "Tables of Stresses in Three-Layer Elastic Systems." *Highway Research Board Bulletin 342: Stress Distribution in Earth Masses,* Washington, D.C., pp. 176–214.

Jones, D. E. Jr., and Jones, K. A. (1987). "Treating Expansive Soils." *Civil Engineering,* ASCE, vol. 57, no. 8, August, pp. 62–65.

Jordan, J. W. (1949). "Alteration of the Properties of Bentonite by Reaction with Amines." *Journal of Mineralogical Society,* London, June.

Kim, W.-H., and Daniel, D. E. (1992). "Effects of Freezing on Hydraulic Conductivity of Compacted Clay." *Journal of Geotechnical Engineering,* ASCE, vol. 118, no. 7, July, pp. 1083–1097.

Kyulule, A. L. (1983). "Additional Considerations on Compaction of Soils in Developing Countries." *No. 121, Mitteilungen des Institutes fur Grundbau und Bodenmechanik Eidgenossische Technische Hochschule,* Zurich.

Lade, P. V., and Lee, K. L. (1976). "Engineering Properties of Soils." *Report No. UCLA-ENG–7652,* University of California at Los Angeles.

Lambe, T. W. (1958). "The Structure of Compacted Clay." *Journal of the Soil Mechanics and Foundations Division,* Proceedings of ASCE, no. SM2, May, pp. 1654-1 to 1654–34.

Lambe, T. W. (1967). "Stress Path Method." *Journal of the Soil Mechanics and Foundations Division,* Proceedings of ASCE, no. SM6, November, pp. 309–331.

Lambe, T. W., Michaels, A. S., and Moh, Z. C. (1960). "Improvements of Soil-Cement with Alkali Metal Compounds." *Highway Research Board Bulletin No. 241: Soil Stabilization with Asphalt, Portland Cement, Lime and Chemicals,* Washington, D.C., pp. 67–103.

Lambe, T. W., and Whitman, R. V. (1979). *Soil Mechanics, SI Version,* John Wiley & Sons, New York.

Larson, C. T., Lawton, E. C., Bravo, A., and Perez, Y. (1993). "Influence of Oversize Particles on Wetting-Induced Collapse of Compacted Clayey Sand." *Proceedings of the 29th Symposium on Engineering Geology and Geotechnical Engineering,* Reno, Nevada, March, pp. 449–459.

Lawton, E. C. (1986). "Wetting-Induced Collapse in Compacted Soil." Ph.D. thesis, Washington State University, Pullman, Washington.

Lawton, E. C., Chen, J., Larson, C. T., Perez, Y., and Bravo, A. (1993). "Influence of Chemical Admixtures and Pore Fluid Chemistry on Wetting Induced Collapse of Compacted Soil." *Proceedings of the 29th Symposium on Engineering Geology and Geotechnical Engineering,* Reno, Nevada, March, pp. 425–448.

Lawton, E. C., Chen, J., and Roadifer, J. (1996). "One-Dimensional Wetting-Induced Volume Change of Unsaturated Clayey Sand." *Journal of Geotechnical Engineering,* ASCE (in review).

Lawton, E. C., Fox, N. S., and Handy, R. L. (1994). "Control of Settlement and Uplift of Structures Using Short Aggregate Piers." *In-Situ Deep Soil Improvement,* ASCE Geotechnical Special Publication No. 45, pp. 121–132.

Lawton, E. C., Fragaszy, R. J., and Hardcastle, J. H. (1989). "Collapse of Compacted Clayey Sand." *Journal of Geotechnical Engineering,* ASCE, vol. 115, no. 9, September, pp. 1252–1267.

Lawton, E. C., Fragaszy, R. J., and Hardcastle, J. H. (1991). "Stress Ratio Effects on Collapse of Compacted Clayey Sand." *Journal of Geotechnical Engineering,* ASCE, vol. 117, no. 5, May, pp. 714–730.

Lawton, E. C., Fragaszy, R. J., and Hetherington, M. D. (1992). "Review of Wetting-Induced Collapse in Compacted Soil." *Journal of Geotechnical Engineering,* ASCE, vol. 115, no. 9, September, pp. 1376–1394.

Leadabrand, J. A. (1956). "Some Engineering Aspects of Soil-Cement Mixtures." Mid-South Section, ASCE, April 27.

Lee, K. L. (1965). "Triaxial Compressive Strength of Saturated Sands under Seismic Loading Conditions." Ph.D. dissertation, University of California at Berkeley.

Lee, K. L., and Seed, H. B. (1967*a*). "Cyclic Stress Conditions Causing Liquefaction of Sand." *Journal of the Soil Mechanics and Foundations Division,* ASCE, vol. 93, no. 1, January, pp. 47–70.

Lee, K. L., and Seed, H. B. (1967*b*). "Drained Strength Characteristics of Sands." *Journal of the Soil Mechanics and Foundations Division,* ASCE, vol. 93, no. SM6, November, pp. 117–141.

Lee, K. L., and Seed, H. B. (1967*c*). "Effect of Moisture on the Strength of a Clean Sand." *Journal of the Soil Mechanics and Foundations Division,* ASCE, vol. 93, no. SM6, November, pp. 17–40.

Lee, K. L., and Singh, A. (1971). "Compaction of Granular Soils." *Proceedings of the Ninth Annual Symposium on Engineering Geology and Soils Engineering,* Boise, Idaho, pp. 161–174.

Lee, P. Y., and Suedkamp, R. J. (1972). "Characteristics of Irregularly Shaped Compaction Curves of Soils." *Highway Research Record No. 381: Compaction and Stabilization,* Highway Research Board, Washington, D.C., pp. 1–9.

Leonard, R. J., and Davidson, D. T. (1959). "Pozzolanic Reactivity Study of Flyash." *Highway Research Board Bulletin No. 231: Lime and Lime-Flyash as Soil Stabilizers,* Washington, D.C., pp. 1–14.

Leonards, G. A., and Narain, J. (1963). "Flexibility of Clay and Cracking of Earth Dams." *Journal of the Soil Mechanics and Foundations Division,* ASCE, vol. 89, no. 2, pp. 47–98.

Lin, P. S., and Lovell, C. W. (1981). "Compressibility of Field Compacted Clay." *Joint Highway Research Project Report no. 81-14,* Purdue University, West Lafayette, Indiana.

Lin, P. S., and Lovell, C. W. (1983). "Compressibility of Field Compacted Clay." *Transportation Research Record no. 897, Transportation Research Board,* Washington, D.C., pp. 51–60.

Little, D. N., and Deuel, L. (1989). "Evaluation of Sulfate-Induced Heave at Joe Pool Lake." Report Prepared for Chemical Lime Company, June.

Little, D. N., Thompson, M. R., Terrell, R. L., Epps, J. A., and Barenberg, E. J. (1987). "Soil Stabilization for Roadways and Airfields." *Report No. ESL-TR-86-19,* Air Force Engineering and Services Center, Tyndall AFB, Florida, July.

MacMurdo, F. D., and Barenberg, E. J. (1973). "Determination of Realistic Cutoff Dates for Late-Season Construction with Lime-Fly Ash and Lime-Cement-Fly Ash Mixtures." *Highway Research Record No. 442,* Highway Research Board, Washington, D.C., pp. 92–101.

Madhav, M. R., and Vitkar, P. P. (1978). "Strip Footing on Weak Clay Stabilized with a Granular Trench or Pile." *Canadian Geotechnical Journal,* vol. 15, pp. 605–609.

Mahmood, A., and Mitchell, J. K. (1974). "Fabric-Property Relationships in Fine Granular Soils." *Clays and Clay Minerals,* vol. 22, nos. 5/6, pp. 397–408.

Marsal, R. J., and Ramigez de Arellano, L. (1967). "Performance of El Infiernillo Dam, 1963–1966." *Journal of the Soil Mechanics and Foundations Division,* ASCE, vol. 93, no. SM4, July, pp. 265–298.

McLaughlin, H. C., et al. (1976). "Aqueous Polymers for Treating Clays." *SPE Paper No. 6008,* Society of Petroleum Engineers, Bakersfield, California.

Meehan, R. L. (1967). "The Uselessness of Elephants in Compacting Soil." *Canadian Geotechnical Journal,* vol. 4, no. 3, September, pp. 358–360.

Meyerhof, G. G. (1951). "The Ultimate Bearing Capacity of Foundations." *Geotechnique,* vol. 2, no. 4, pp. 301–331.

Meyerhof, G. G. (1956). "Penetration Tests and Bearing Capacity of Cohesionlesss Soils." *Journal of the Soil Mechanics and Foundations Division,* ASCE, vol. 82, no. SM1, pp. 1–19.

Meyerhof, G. G. (1963). "Some Recent Research on the Bearing Capacity of Foundations." *Canadian Geotechnical Journal,* vol. 1, no. 1, September, pp. 16–26.

Meyerhof, G. G. (1965). "Shallow Foundations." *Journal of the Soil Mechanics and Foundations Division,* ASCE, vol. 91, no. SM2, pp. 21–31.

Meyerhof, G. G., and Hanna, A. M. (1978). "Ultimate Bearing Capacity of Foundations on Layered Soils under Inclined Load." *Canadian Geotechnical Journal,* vol. 15, no. 4, pp. 565–572.

Minnick, L. J. (1967). "Reactions of Hydrated Lime with Pulverized Coal Fly Ash." *Proceedings of the Fly Ash Utilization Conference, Bureau of Mines Information Circular 8348.*

Mitchell, J. K (1960). "Fundamental Aspects of Thixotropy in Soils." *Journal of the Soil Mechanics and Foundations Division,* ASCE, vol. 86, no. SM3, pp. 19–52.

Mitchell, J. K. (1981). "Soil Improvement: State-of-the-Art." *Proceedings of the Tenth International Conference on Soil Mechanics and Foundation Engineering,* Stockholm, Sweden, vol. 4, pp. 509–565.

Mitchell, J. K. (1986). "Practical Problems from Surprising Soil Behavior." *Journal of Geotechnical Engineering,* ASCE, vol. 112, no. 3, March, pp. 259–289.

Mitchell, J. K. (1993). *Fundamentals of Soil Behavior,* 2d ed., John Wiley & Sons, New York.

Mitchell, J. K., Hooper, D. R., and Campanella, R. G. (1965). "Permeability of Compacted Clay." *Journal of the Soil Mechanics and Foundations Division,* ASCE, vol. 91, no. SM4, July, pp. 41–65.

Morgenstern, N. R. (1979). "Properties of Compacted Soils." *Proceedings of the Sixth Pan-American Conference on Soil Mechanics and Foundation Engineering,* Lima, Peru, vol. 3, pp. 349–354.

Nakayama, H., and Handy, R. L. (1965). "Factors Influencing Shrinkage of Soil-Cement." *Highway Research Record No. 86: Soil-Cement Stabilization,* Highway Research Board, Washington, D.C., pp. 15–27.

Nalezny, C. L., and Li, M. C. (1967). "Effect of Soil Structure and Thixotropic Hardening on the Swelling Behavior of Compacted Clay Soils." *Highway Research Record No. 209: Physicochemical Properties of Soils,* Highway Research Board, Washington, D.C., pp. 1–19.

NAVDOCKS (1961). "Design Manual: Soil Mechanics, Foundations, and Earth Structures." *NAVDOCKS DM-7,* Department of the Navy, Washington, D.C.

NCHRP (1976). *Lime-Fly Ash-Stabilized Bases and Subbases,* National Cooperative Highway Research Program Synthesis of Highway Practice No. 37, Transportation Research Board, Washington, D.C.

Nelson, J. D., and Miller, D. J. (1992). *Expansive Soils: Problems and Practice in Foundation and Pavement Engineering,* John Wiley & Sons, New York.

Nishida, Y. (1966). "Vertical Stress and Vertical Deformation of Ground under a Deep Circular Uniform Pressure in the Semi-Infinite." *Proceedings of the First Congress of the International Society of Rock Mechanics,* Lisbon, Portugal, vol. 2, pp. 493–497.

Noorany, I. (1987). "Precision and Accuracy of Field Density Tests in Compacted Soils." *Research Report,* Department of Civil Engineering, San Diego State University, San Diego, California.

Noorany, I., and Stanley, J. V. (1994). "Settlement of Compacted Fills Caused by Wetting." *Vertical and Horizontal Deformations of Foundations and Embankments,* ASCE Geotechnical Special Publication no. 40, vol. 2, pp. 1516–1530.

NRC (1985). *Liquefaction of Soils during Earthquakes,* National Research Council, National Academy Press, Washington, D.C.

Oda, M. (1972). "Initial Fabrics and Their Relations to Mechanical Properties of Granular Material." *Soils and Foundations,* vol. 12, no. 1, March, pp. 17–36.

Olsen, H. W. (1962). "Hydraulic Flow through Saturated Clays." *Clays and Clay Minerals,* vol. 9, no. 2, pp. 131–161.

Orcutt, R. G., et al. (1955). "The Movement of Radioactive Strontium through Naturally Porous Media." *Atomic Energy Commission Report,* November 1.

PCA (1968). *Design and Control of Concrete Mixtures,* 11th ed., Portland Cement Association, Skokie, Illinois, July.

PCA (1971). *Soil-Cement Laboratory Handbook,* Portland Cement Association, Skokie, Illinois.

PCA (1979). *Soil-Cement Construction Handbook,* Portland Cement Association, Skokie, Illinois.

Peattie, K. R. (1962). "Stress and Strain Factors for Three-Layer Elastic Systems." *Highway Research Board Bulletin 342: Stress Distribution in Earth Masses,* Washington, D.C., pp. 215–253.

Perrin, L. L. (1992). "Expansion of Lime-Treated Clays Containing Sulfates." *Proceedings of the Seventh International Conference on Expansive Soils,* Dallas, Texas, August, vol. 1, pp. 409–414.

Petry, T. M., and Brown, R. W. (1987). "Laboratory and Field Experiences Using Soil Sta Chemical Soil Stabilizer." ASCE Paper Presented at San Antonio, Texas, October 3, 1986. *Texas Civil Engineer,* February 1987, pp. 15–19.

Petry, T. M., and Little, D. N. (1992). "Update on Sulfate Induced Heave in Treated Clays; Problematic Sulfate Levels." *Transportation Research Record No. 1362: Aggregate and Pavement-Related Research,* Transportation Research Board, Washington, D.C., pp. 51–55..

Poesch, H., and Ikes, W. (1975). *Verdichtungstechnik and Verdichtungsgerate in Erdbau (Compaction Techniques and Compaction Equipment in Earth Construction)*, in German, Wilhelm Ernst & Sohn, Berlin.

Poran, C. J., and Ahtchi-Ali, F. (1989). "Properties of Solid Waste Incinerator Fly Ash." *Journal of Geotechnical Engineering,* ASCE, vol. 115, no. 8, August, pp. 1118–1133.

Poulos, H. G., and Davis, E. H. (1974). *Elastic Solutions for Soil and Rock Mechanics,* John Wiley and Sons, New York.

Poulos, S. J. (1988). "Compaction Control and the Index Unit Weight." *Geotechnical Testing Journal,* ASTM, Philadelphia, vol. 11, no. 2, June, pp. 100–108.

Proctor, R. R. (1933). "Fundamental Principles of Soil Compaction." *Engineering News-Record,* vol. 111, Nos. 9, 10, 12, and 13.

Redus, J. F. (1958). "Study of Soil-Cement Base Courses on Military Airfields." *Highway Research Board Bulletin No. 198: Cement-Soil Stabilization,* Washington, D.C., pp. 13–19.

Reinhold, F. (1955). "Soil-Cement Methods for Residential Streets in Western Germany." *World Construction,* March-April, p. 41.

Rizkallah, V., and Hellweg, V. (1980). "Compaction of Non Cohesive Soils by Watering." *Proceedings of the International Conference on Compaction,* Paris, April, pp. 357–361.

Robertson, P. K., and Campanella, R. G. (1985). "Liquefaction Potential of Sands Using the CPT." *Research Report R69-15,* Department of Civil Engineering, Massachusetts Institute of Technology, Cambridge, Massachusetts.

Rollins, K. M., and Rogers, G. W. (1994). "Mitigation Measures for Small Structures on Collapsible Alluvial Soils." *Journal of Geotechnical Engineering,* ASCE, vol. 120, no. 9, September, pp. 1533–1553.

Schmertmann, J. H. (1970). "Static Cone to Compute Settlement over Sand." *Journal of the Soil Mechanics and Foundations Division,* ASCE, vol. 96, no. SM3, pp. 1011–1043.

Schmertmann, J. H. (1978). "Guidelines for Cone Penetration Test Performance and Design." *Report No. FHWA-TS-78-209,* U.S. Department of Transportation, Federal Highway Administration, Washington, D.C., July.

Schmertmann, J. H. (1989). "Density Tests above Zero Air Voids Line." *Journal of Geotechnical Engineering,* ASCE, vol. 115, no. 7, July, pp. 1003–1018.

Schmertmann, J. H. (1991). "The Mechanical Aging of Soils." *Journal of Geotechnical Engineering,* ASCE, vol. 117, no. 9, September, pp. 1288–1330.

Schmertmann, J. H., Hartman, J. P., and Brown, P. R. (1978). "Improved Strain Influence Factor Diagrams." *Journal of the Geotechnical Engineering Division,* ASCE, vol. 104, no. GT8, pp. 1131–1135.

Seed, H. B. (1979). "Soil Liquefaction and Cyclic Mobility Evaluation for Level Ground during Earthquakes." *Journal of the Geotechnical Engineering Division,* ASCE, vol. 105, no. GT2, February, pp. 201–255.

Seed, H. B. (1986). "Design Problems in Soil Liquefaction." *Report No. UCB/EERC-86/02,* University of California at Berkeley, February.

Seed, H. B., and Chan, C. K. (1959a). "Structure and Strength Characteristics of Compacted Clays." *Journal of the Soil Mechanics and Foundations Division,* ASCE, vol. 85, no. SM5, October, pp. 87–128.

Seed, H. B., and Chan, C. K. (1959b). "Undrained Strength of Compacted Clays after Soaking." *Journal of the Soil Mechanics and Foundations Division,* ASCE, vol. 85, no. SM6, December, pp. 31–47.

Seed, H. B., and Idriss, I. M. (1967). "Analysis of Soil Liquefaction: Nigata Earthquake." *Journal of the Soil Mechanics and Foundations Division,* ASCE, vol. 93, no. SM3, May, pp. 83–108.

Seed, H. B., and Idriss, I. M. (1971). "Simplified Procedure for Evaluating Soil Liquefaction Potential." *Journal of the Soil Mechanics and Foundations Division,* ASCE, vol. 97, no. SM9, September, p. 1249.

Seed, H. B., Idriss, I. M., and Arango, I. (1983). "Evaluation of Liquefaction Potential Using Field Performance Data." *Journal of Geotechnical Engineering,* vol. 109, no. 3, March, pp. 458–482.

Seed, H. B., and Lee, K. L. (1966). "Liquefaction of Saturated Sands during Cyclic Loading." *Journal of the Soil Mechanics and Foundations Division,* ASCE, vol. 92, no. SM6, November, pp. 105–134.

Seed, H. B., Mitchell, J. K., and Chan, C. K. (1960). "The Strength of Compacted Cohesive Soil." *Proceedings of the Research Conference on Shear Strength of Cohesive Soils,* ASCE, Boulder, Colorado, June.

Seed, H. B., Tokimatsu, K., Harder, L. F., and Chung, R. M. (1985). "Influence of SPT Procedures in Soil Liquefaction Resistance Evaluations." *Journal of Geotechnical Engineering,* ASCE, vol. 111, no. 12, December, pp. 1425–1445.

Selig, E. T., and Yoo, T.-S. (1977). "Fundamentals of Vibratory Roller Behavior." *Proceedings of the Ninth International Conference on Soil Mechanics and Foundation Engineering,* Tokyo, vol. 2, pp. 375–380.

Shelley, T. L., and Daniel, D. E. (1993). "Effect of Gravel on Hydraulic Conductivity of Compacted Soil Liners." *Journal of Geotechnical Engineering,* ASCE, vol. 119, no. 1, January, pp. 54–68.

Sherard, J. L., Woodward, R. J., Gizienski, S. F., and Clevenger, W. A. (1963). *Earth and Earth-Rock Dams,* John Wiley & Sons, New York.

Sherif, M. A., et al. (1982). "Swelling of Wyoming Montmorillonite and Sand Mixtures." *Journal of Geotechnical Engineering,* ASCE, vol. 108, no. GT1, January, pp. 33–45.

Sherman, G. B., Watkins, R. O, and Prysock, R. H. (1967). "A Statistical Analysis of Embankment Compaction." *Highway Research Record No. 177: Symposium on Compaction of Earthwork and Granular Bases,* Highway Research Board, Washington, D.C., pp. 157–185.

Sherwood, P. T. (1962). "Effect of Sulfates on Cement- and Lime-Stabilized Soils." *Highway Research Board Bulletin No. 353: Stabilization of Soils with Portland Cement,* Washington, D.C., pp. 98–107. Also in *Roads and Road Construction,* vol. 40, February, pp. 34–40.

Siddiqi, F. H., Seed, R. B., Chan, C. K., Seed, H. B., and Pyke, R. M. (1987). "Strength Evaluation of Coarse-Grained Soils." *Report No. UCB/EERC-87/22,* University of California, Berkeley.

Skempton, A. W., and Bjerrum, L. (1957). "A Contribution to the Settlement Analysis of Foundations on Clay." *Geotechnique,* vol. 7, no. 4, pp. 168–178.

Skinner, B. J., and Porter, S. C. (1987). *Physical Geology,* John Wiley & Sons, New York.

Skopek, J. (1961). "The Influence of Foundation Depth on Stress Distribution." *Proceedings of the Fifth International Conference on Soil Mechanics and Foundation Engineering,* Paris, France, vol. 1, p. 815.

Sokolovich, V. E., and Semkin, V. V. (1984). "Chemical Stabilization of Loess Soils." *Soil Mechanics and Foundation Engineering,* vol. 21, no. 4, July–August, pp. 8–11.

Sowers, G. F., Williams, R. C., and Wallace, T. S. (1965). *Proceedings of the Sixth International Conference on Soil Mechanics and Foundation Engineering*, Montreal, vol. 2, pp. 561–565.

Spangler, M. G., and Handy, R. L. (1982). *Soil Engineering,* 4th ed., Harper & Row, New York.

Taki, O., and Yang, D. S. (1991). "Soil-Cement Mixed Wall Technique." *Geotechnical Engineering Congress 1991,* edited by F. G. McLean, D. A. Campbell, and D. W. Harris, ASCE Geotechnical Special Publication no. 27, vol. 1, pp. 298–309.

Terdich, G. M. (1981). "Prediction and Control of Field Swell Pressures of Compacted Medium Plastic Clay." *Joint Highway Research Project Report No. 81-4,* Purdue University, West Lafayette, Indiana.

Terzaghi, K. (1955). "Evaluation of Coefficients of Subgrade Reaction." *Geotechnique,* vol. 5, no. 4, December, pp. 297–326.

Terzaghi, K, and Peck, R. B. (1967). *Soil Mechanics In Engineering Practice,* 2d ed., John Wiley & Sons, New York.

Thenn de Barros, S. (1966). "Deflection Factor Charts for Two and Three-Layer Elastic Systems." *Highway Research Record No. 145,* Highway Research Board, Washington, D.C., pp. 83–108.

Thompson, M. R. (1966). "Shear Strength and Elastic Properties of Lime-Soil Mixtures." *Highway Research Record No. 139: Behavior Characteristics of Lime-Soil Mixtures,* Highway Research Board, Washington, D.C., pp. 1–14.

Thompson, M. R. (1969). "Engineering Properties of Soil-Lime Mixtures." *Journal of Materials,* ASTM, vol. 4, no. 4, December.

Townsend, D. L., and Klym, T. W. (1966). "Durability of Lime-Stabilized Soils." *Highway Research Record No. 139: Behavior of Lime-Soil Mixtures,* Highway Research Board, Washington, D.C., pp. 25–41.

Trautwein, S. J. (1993). "Field Hydraulic Conductivity Testing." *Permeability of Soil Liners for Waste Containment Facilities,* Notes for Short Course, Reno, Nevada, March.

TRB (1987). *Lime Stabilization: Reactions, Properties, Design, and Construction,* State of the Art Report 5, Transportation Research Board, Washington, D.C.

Turnbull, W. J., Compton, J. R., and Ahlvin, R. G. (1966). "Quality Control of Compacted Earthwork." *Journal of the Soil Mechanics and Foundations Division,* ASCE, vol. 92, no. SM1, January, pp. 93–103.

Ueshita, K., and Meyerhof, G. G. (1967). "Deflection of Multilayer Soil Systems." *Journal of the Soil Mechanics and Foundations Division,* ASCE, vol. 93, no. SM5, pp. 257–282.

U.S. Army (1985). "Operator's and Organizational Maintenance Manual for Tester, Density and Moisture (Soil and Asphalt), Nuclear Method." *Technical Manual (TM) 5-6635-386-12&P,* U.S. Department of the Army.

Uzan, J., Ishai, I, and Hoffman, M. S. (1980). "Surface Deflexion in a Two-Layer Elastic Medium Underlain by a Rough Rigid Base." *Geotechnique,* vol. 30, no. 1, pp. 39–47.

Valsangkar, A. J., and Meyerhof, G. G. (1979). "Experimental Study of Punching Coefficients and Shape Factor for Two-Layered Soils." *Canadian Geotechnical Journal,* vol. 16, no. 4, November, pp. 802–805.

Vesic, A. S. (1975). "Bearing Capacity of Shallow Foundations." Chapter 3 in *Foundation Engineering Handbook,* edited by H. F. Winterkorn and H.-Y. Fang, Van Nostrand Reinhold, New York, pp. 121–147.

Wacker (1987). *Soil Compaction and Equipment for Confined Areas,* Wacker Corporation, Menomonee Falls, Wisconsin.

Waidelich, W. C. (1990). "Earthwork Construction." Chapter 4 in *Guide to Earthwork Construction,* State of the Art Report 8, Transportation Research Board, Washington, D.C., pp. 25–48.

Walsh, K. D., Houston, W. N., and Houston, S. L. (1993). "Evaluation of in-Place Wetting Using Soil Suction Measurements." *Journal of Geotechnical Engineering,* ASCE, vol. 119, no. 5, May, pp. 862–873.

West, G. (1959). "A Laboratory Investigation into the Effect of Elapsed Time after Mixing on the Compaction and Strength of Soil-Cement." *Geotechnique,* vol. 9, pp. 22–28.

Williams, L. H., and Undercover, D. R. (1980). "New Polymer Offers Effective, Permanent Clay Stabilization Treatment." *Society of Petroleum Engineers Paper No. 8797,* January.

Wilson, S. D. (1970). "Suggested Method of Test for Moisture-Density Relations of Soils Using Harvard Compaction Apparatus." *STP No. 479,* ASTM, pp. 101–103.

Wimberley, P. L. III, Mazzella, S. G., and Newman, F. B. (1994). "Settlement of a 15-Meter Deep Fill Below a Building." *Vertical and Horizontal Deformations of Foundations and Embankments,* ASCE Geotechnical Special Publication no. 40, edited by A. T. Yeung and G. Y. Felio, vol. 1, pp. 398–416.

Yi, F. (1991). "Behavior of Compacted Collapsible Soils Subjected to Water Infiltration." Doctor of Engineering thesis, University of Tokyo, Tokyo, Japan.

Youd, T. L. (1993). "Liquefaction-Induced Lateral Spread Displacement." *NCEL Technical Note,* Naval Civil Engineering Laboratory, Port Hueneme, California.

Zeller, J., and Wullimann, R. (1957). "The Shear Strength of Shell Materials for the Goschenalp Dam, Switzerland." *Proceedings of the Fourth International Conference on Soil Mechanics and Foundation Engineering,* London, England, vol. 2, pp. 399–404.

SECTION 5B
GROUTING TO IMPROVE FOUNDATION SOIL

Alex A. Naudts

5B.1 INTRODUCTION   5.277
  5B.1.1 Increasing the Bearing Capacity of the Soil   5.278
  5B.1.2 Slope or Excavation Stabilizing Technique   5.289
  5B.1.3 Pile or Pier Enhancement   5.295
  5B.1.4 Hazardous Chemical or Waste Control   5.298
  5B.1.5 Ground Water and Erosion Control   5.299
  5B.1.6 Remedial Grouting   5.303
5B.2 GROUTING TECHNIQUES   5.306
  5B.2.1 Permeation Grouting   5.306
  5B.2.2 Hydrofracture Grouting   5.310
  5B.2.3 Compaction Grouting   5.311
  5B.2.4 Jet Grouting   5.312
  5B.2.5 Unusual Approaches   5.322
5B.3 TYPES OF GROUT   5.324
  5B.3.1 Cement-Based Suspension Grouts   5.324
  5B.3.2 Solution Grouts   5.337
5B.4 DESIGN AND SPECIFICATIONS FOR A GROUTING PROGRAM   5.373
  5B.4.1 Introduction   5.373
  5B.4.2 Contractual Arrangements   5.374
  5B.4.3 Pay Items   5.375
  5B.4.4 Design of a Soil-Grouting Program   5.376
  5B.4.5 Drilling for Soil-Grouting Projects   5.379
5B.5 FIELD PROCEDURES   5.380
  5B.5.1 Grouting Equipment   5.380
  5B.5.2 Engineering Supervision of a Grouting Program   5.384
  5B.5.3 Evaluation of the Grouting Operation and Nondestructive Quality Control   5.397
5B.6 SOME FINAL COMMENTS   5.398
5B.7 REFERENCES   5.398

5B.1 INTRODUCTION

Soil grouting is often used as a "last resort" technique, a step taken by the desperate to correct a problem not created by design but because of an accident or negligence or as a result of extreme conditions.

The grouting gospel has been preached by the real and the false prophets and has been practiced as an applied science by some, and as "art" or "magic" by others. The overall success rate of grouting technology in achieving its intended goals is probably less than 50%, because of a fundamental lack of understanding and knowledge coupled to utter ignorance and laziness by the majority of the people involved in grouting.

The pragmatic and "rigid" approach by some engineering firms to grouting and grouting-related technologies is often counterproductive, especially if a good (engineering) grouting contractor is on site. However, the sloppy performance by a number of established grouting contractors has done serious harm to the credibility of the industry.

The intent of this section is to cut through some of the mystique and to lift the smokescreen which has been created by some in this industry to obfuscate the real issues at hand.

Soil grouting is an applied engineering technology, which is rooted in the following sciences and applied sciences:

Soil mechanics

Structural engineering

Chemistry

Fluid mechanics

Mathematics

Geology

Soil grouting is only one of the several fields of grouting. Grouting is also used for:

General seepage control in underground structures (tunnels, sewers, parkades, mines, etc.)

Erosion control and reforestation

Structural rehabilitation of damaged (fire, explosion, accident) or deteriorated concrete, masonry, wood, and plastics

Cemented backfill in mining

The installation of water stops or to glue structural members together, using the injectable tube technique

An engineered cutoff system for environmental purposes

Creating a low-permeability curtain below dams and water-retaining structures

Restoration of historical structures (wood, masonry, concrete)

Engineered, high-performance soil and rock anchors

Creation of pin piles and soil nails

Structural concrete elements, using the preplaced aggregate grouting technique

The above applications will be discussed only very briefly—if at all—in this section. The applications for grouting are virtually unlimited. Grouting jobs are rarely designed or found, they are "created" by innovative individuals looking for more economical and durable solutions to engineering challenges.

Section 5B discusses several applications in detail to familiarize the reader with the numerous aspects of a grouting project. A number of general applications are highlighted followed by an analysis of a field application in which the author was directly involved.

5B.1.1 Increasing the Bearing Capacity of the Soil

One of the most popular applications is remedial soil grouting to restore or increase the bearing capacity of soils below foundation systems. Note that there is a distinct difference between *absolute bearing capacity* of the soil (i.e., when exceeded, the foundation system punches through the soil), and the *deformation bearing capacity* (when this capacity is exceeded, substantial settlement occurs, to the extent that it causes excessive stresses and potentially damages the structure supported by the foundation system). Grouting is predominantly used to prevent or reduce excessive settlement.

Often, prior to the pouring of the concrete for footings, the soil gets disturbed. Granular soil "boils" under artesian pressure. This occurs predominantly in areas with a high ground-water table and insufficient lowering of the ground-water table. This phenomenon occurs even more often when deep foundation systems (caissons, auger piles) are used (see Section 5B.1.3).

When a load is applied to a footing, the vertical stress in the soils below the footings increases, and the soil undergoes compression: settlement (however minor) occurs. The compression of soils under a given load is governed by its compressibility coefficient C. The compressibility coefficient C of disturbed soil is very low, and considerable compression of this soil will take place (i.e., substantial settlement) before a stable condition is established.

This settlement is several magnitudes higher than the engineered estimate obtained with Terzaghi's settlement equations. As a result, a properly engineered footing will not perform as designed if the founding soils are disturbed and may cause major problems in the structure supported by those footings.

A grouting operation dramatically increases the C value and significantly reduces its compression under a given load. Permeation grouting of the soil below footings has been successfully used in a number of instances to reverse the disturbance below footings, often applied in conjunction with a compaction and/or hydrofracturing grouting program. A well-engineered grouting program focuses on increasing the compressibility coefficient of the soils subjected to the greatest increase in vertical stress, induced by the load applied to the footing.

Permeation grouting, of course, is not always applicable and depends on the injectability of the soils. Permeation grouting might have to be replaced by a "permeation in conjunction with hydrofracturing exercise" or merely a hydrofracturing program or by jet grouting. As will be further explained in detail, the depth to which the grouting has to be executed can be engineered and is not an arbitrary process, as is commonly executed by the self-proclaimed "grouting experts."

The analysis of the following field applications provides an insight into the systematic engineered approach to the various aspects of foundation grouting.

5B.1.1.1 Field Application: Rogers' Reservoir (Ontario, Canada).

See Figs. 5B.1–5B.3 for illustrations of this field application.

History. In 1980, a new bridge was constructed over the Holland River in East-G., Ontario (Canada). The north end of the bridge was founded on piles, and the south end was founded on the massive concrete walls of an old lock, in turn supported by very dense "silty sand to fine to medium sand" founding soils.

In the fall of 1989, after several days of heavy rain, during which substantial amounts of water passed over the spillway, adjacent to the lock structure, large slabs of concrete dropped out of the lock wall. Scour activity apparently had undermined a large section of the lock wall, and unsupported slabs of concrete dropped in the space created by the scour action.

This resulted in the lock wall acting as an arch to support the bridge and its own weight. Minor rotation of the wall toward the river was also evident. The structural calculations indicated that both the lock wall and the bridge were in an imminent state of collapse. Emergency measures were immediately taken to stop the scour activity by constructing a sand/cement berm alongside the lock-wall. Concrete infilling beneath the wall and between the concrete slabs of the lock wall, including infilling of the arch, was then undertaken to temporarily stabilize the most unstable section of the lock wall. Over 40 m^3 (53 yd^3) of a flowable sand cement grout were gravity-fed below the lock wall and cut back on the outward thrust on the abutments of the "arch" in the lock wall.

Field Investigation. The investigation that followed focused on determining the depth and extent of the scour activity and developing remedial measures. An appropriate number of boreholes were executed, and routine SPT testing and soil sampling were performed. A sleeve pipe was installed, through the hollow stem of the auger in each borehole, properly embedded in casing grout.

In situ hydraulic conductivity tests (using a double packer), were performed in each pipe, at each elevation, and a permeability profile for the various soil horizons was developed.

The tests were done by pressure grouting water at the various locations, recording the pressures and takes via an X-Y recorder. The head-losses curve (line losses + friction losses to squeeze fluid past the rubber sleeves) was established for various flows prior to testing, and the hydrostatic head was also taken into account to calculate the effective testing pressures and the permeability coefficients.

The in situ hydraulic conductivity (k) values calculated from these tests were then compared with the calculated k values determined from grain size distribution curves using Hazen's formula. Hazen's formula is given by

$$k \text{ (cm/s)} = 116 \, (0.7 + 0.034t) \, d_{10}^2$$

where d_{10} = the grain size in cm at which 10% of the particles by weight are smaller
t = the temperature in °C
k = the permeability coefficient

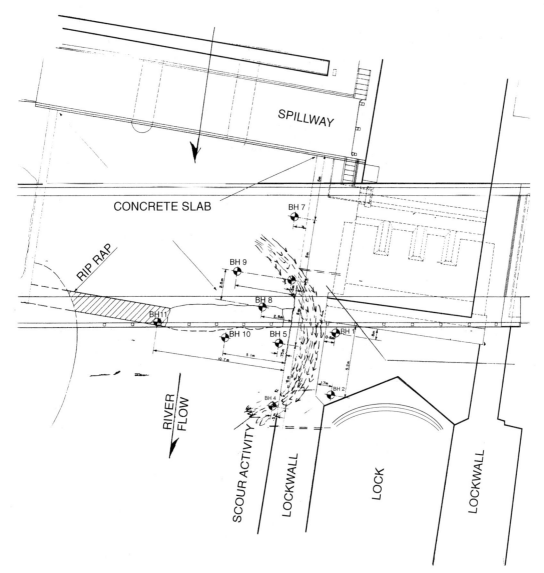

FIGURE 5B.1 Site layout—borehole locations.

In the disturbed zones or horizons, the in situ measured hydraulic conductivities were several orders of magnitude higher than the theoretical value derived from the sieve curves. In the undisturbed zones, there was a good correlation between the value derived from the sieve curves and the ones established via the hydraulic conductivity testing. In the disturbed areas, the in situ permeability values were much higher than the theoretical values.

As an example, we look at the results of two test intervals, 1 m long. For borehole 5 the location test interval was -1.75 to -2.75 m (5.7 to 9 ft); the average water take was 46 L/min (stable) = 768.2 cm^3/s (47 in^3/s); P_{gage} = 25 psi (172 kPa), $P_{effective}$ = 1725−1587 cm WC = 138 cm WC (54 in); and

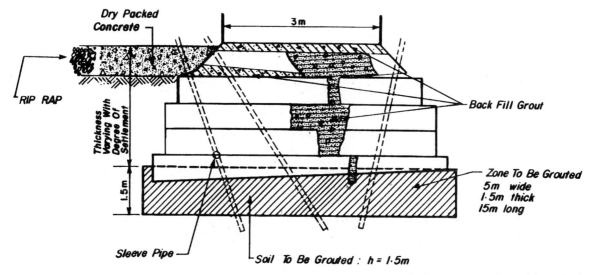

FIGURE 5B.2 Sleeve-pipe placement. The first row of sleeve pipes (the ones closest to the river) are to extend to a minimum depth of 5 m, regardless of soil conditions.

$$k = \frac{768}{4(3.14)\frac{(7.5)}{2}138} \text{ cm/s} = 1.2 \times 10^{-1} \text{ cm/s } (5 \times 10^{-2} \text{ in/s})$$

The k value derived from the grain-size distribution obtained from samples in that borehole in the test location is

$$k_{\text{Hazen}} = 10^{-3} \text{ cm/s } (3.9 \times 10^{-4} \text{ in/s})$$

The in situ hydraulic conductivity was two orders of magnitude higher than the one derived from the sieve curves, therefore it was concluded that the disturbance in the upper layers was responsible for the discrepancy.

For borehole 7 the location test interval was -3 to -3.5 m (-9.8 to -11.5 ft); the average take: 18 liters/min (stable) or 300 cm^3/s (18.3 in^3/s); $P_{\text{gage}} = 20$ psi (138 kPa), and $P_{\text{effective}} = 1380 - 483 + 50$ cm WC or 947 cm WC (373 in)

$$k = \frac{300}{15(3.14)947} \text{ cm/s or } 6.7\ 10^{-3} \text{ cm/s } (2.6 \times 10^{-3} \text{ in/s})$$

$$k_{\text{Hazen}} = 2 \times 10^{-3} \text{ cm/s } (7.9 \times 10^{-4} \text{ in/s})$$

The in situ hydraulic conductivity on this project, as derived from the field test even in undisturbed material, tended to be a factor of 3 higher than that obtained using Hazen's equation for this type and density of soil. Therefore, the conclusion was reached that in the above test interval, the soil had not been disturbed.

It is recommended, especially in case of disputes, to perform two or three pilot holes as close to the area to be grouted as possible, in undisturbed soils, and to obtain a correlation between the in situ hydraulic conductivity values and the ones determined with Hazen's equation for the various soil horizons. This eliminates further controversial debate.

The author used this methodology to provide irrefutable evidence that disturbance below the footing of a heavy concrete structure had occurred (at the Mannheim water treatment plant in

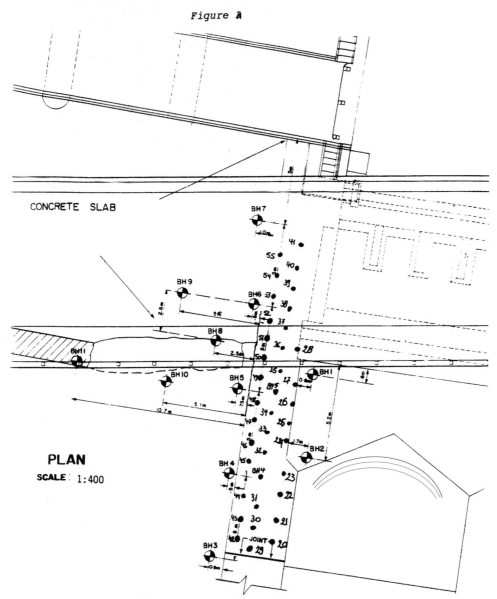

FIGURE 5B.3 Layout of the grout holes.

1991). The in situ permeability results (as well as the SPT numbers) were compared with the calculated values (Hazen) from two pilot holes, outside the disputed area. The in situ hydraulic conductivity values in the disputed area were much higher than the ones obtained using Hazen's equation. This information, combined with the low blowcounts, erased the last doubts about the existence of a disturbed soil horizon. The general contractor accepted the verdict and paid for the remedial grouting work, which was similar in approach to the one described.

Assessment of the Data and Design of the Grouting Program. The in situ hydraulic conductivity values, combined with the standard penetration data, allowed the development of a model for the disturbed area. Based on this model, and the measured in situ permeabilities, the appropriate grout formulations were selected and the grout-hole pattern was engineered.

Grout selection Microfine cement-based suspension grouts were selected for the following reasons:

1. Properly formulated microfine cement-based suspension grouts have a low effective viscosity: $v_{\text{water}}/v_i = 1.08$.
2. Average grain size of the best commercially available microfine cements was 4 μm, with less than 5% in the 9- to 12-μm range. The d_{95} of the selected grout after adding silica-fume was approximately 8 μm. For the type and nature of the disturbed soil, grouting with microfine cement formulations could achieve a grout spread of approximately 50–90 cm (19.7 to 35.4 in) (depending on the permeability and thickness of the soil horizons).
3. Microfine cement cures slowly, making it possible to grout each sleeve pipe two to four times by breaking through the curing grout cylinders from previous phases.
4. The disturbed soil mass did not have to be 100% impregnated. Minor fracturing (claquage) and soil encapsulation were acceptable.
5. This is a durable product—no documented cases exist about long-term deterioration. The author's long experience with this type of product also played a role.
6. This product was more economical than any other type of "chemical grout." Microfine cement-based formulations were one of the few types of chemical grouts acceptable in this environment.
7. The finest commercially available regular cement-based suspension grouts, properly balanced with all the appropriate ingredients, cannot permeate soils (in a predictable fashion) with a hydraulic conductivity of less than 6 to 8×10^{-2} cm/s (2.4 to 3.1×10^{-2} in/s). Therefore regular cement-based cement grouts were only to be used to perform the initial void filling. Microfine cement-based suspension grouts are therefore more suitable to saturate the soils by impregnation.
8. Other grouting methods, such as hydrofracturing and compaction grouting, were not particularly suitable, and jet grouting was not economical, given the size of the project.

Because there was some fear of washout of grout, due to the potential for fast-changing site conditions, provisions were made to use water reactive hydrophobic polyurethane prepolymers to create a barrier and prevent washout of the suspension grout.

The intent was to vary the formulations from location to location, depending on the in situ permeability and on the reduction of the apparent permeability during the grouting.

The following *ingredient range* was eventually selected for the cement-based suspension grouts:

Water: 100 kg (220 lb)

Bentonite: 1 kg (2.2 lb)

Naphthalene sulphonate: between 200 and 400 g (0.44 and 0.88 lb)

Silica fume: between 2 and 5 kg (4.4 and 11 lb)

Microfine cement (MC 500): between 35 and 60 kg (77 and 132 lb)

The plan was to lower the solids content of the grout during the second and third stage while grouting the sleeve pipes.

Spacing of the grout holes The spacing of the grout pipes was estimated with the equations of Cambefort (see to section 5B.4) for radial spread. The theoretical grout quantities were calculated. The grout spread was defined as

$$r_{\text{grout}} = \sqrt[3]{\frac{3kv}{n_i v_i} r_0 p_{\text{effective}} t + r_0^3}$$

5.284 SOIL IMPROVEMENT AND STABILIZATION

or, by substituting the proper values in the above equation, we found that the anticipated spread for a horizon with a hydraulic conductivity of 10^{-3} cm/s would be in the order of 52 cm (21 in)

$$r_{grout} = \sqrt[3]{\frac{3 \times 10^{-3}}{0.3} \text{ cm/s} \times \frac{1}{1.08} \times 2068 \text{ cm} \times 2000 \text{ s} \times 3.75 \text{ cm} + 3.75^3 \text{ cm}^3}$$

or
$$r_{grout} = 52 \text{ cm (20.5 in)}$$

Where the hydraulic conductivity of the soil was a factor of 10 or higher, the anticipated grout spread was 107 cm (43 in). In reality, the hydraulic conductivity in most disturbed horizons was predominantly higher than 10^{-2} cm/s (4×10^{-3} in/s) and about 10^{-3} cm/s (4×10^{-4} in/s) in the undisturbed areas. As a result, a conservative value of 65 cm (25.6 in) for the grout spread was selected to work out the grout hole spacing.

The estimated take for the operation was calculated under the assumption of 75% accessible pore volume (based on the silt encountered). The pore volume was estimated to be 40% of the soil matrix. The volume of grout per lineal meter of sleeve pipe was therefore estimated to be

$$v_{grout} = 3.14 \, (0.65)^2 \, 1 \, (0.75) \, (0.4) = 397 \text{ L (105 gal)}$$

Once the number of sleeve pipes was known, calculating the estimated grout quantities was easy. The intent was to inject the theoretical quantities during the first grouting phase and to bring the various horizons to a proper refusal during subsequent grouting stages.

A cost estimate for a professional grouting operation was worked out.

Various other remedial methods, including caissons, piles, auger cast piles, and "dewatering and underpinning" were thoroughly investigated, and cost estimates were obtained.

The grouting solution was by far the most economical solution for restoring the necessary support to the lock wall. As an added benefit, the grouting operation would cause the least hinderance to traffic and was the most environmentally friendly method in a very sensitive area. The grouting operation involved the permeation grouting of a zone 20 m (22 yd) long and varying in width between 1 and 4 m (3.2 and 13 ft) and in thickness between 1 and 3 m (3.2 and 9 ft).

The estimated quantity of the most expensive ingredient, MC 500, was 16,000 kg (25,273 lb).

Execution. The project was tendered, and the low bidder, one of the few professional grouting contractors in North America, executed the project for less than the tendered amount. An *X-Y* recorder was used to continuously check the "apparent Lugeon value" and the grouting trend (for apparent Lugeon value see Sec. 5B.3.1.6). All appropriate mixing (colloidal) and pumping (mohno) equipment was used.

The operation in the field was under the "hands-on" direction of a qualified grouting engineer. Allowable grouting pressures were adjusted in the field. The grouting engineer monitored hydrofracturing phenomena, and the theoretically calculated maximum pressures were adjusted in the field. The actual pressures [45 to 75 psi (310 to 517 kPa) for the first stage and 75 to 125 psi (512 to 862 kPa) in subsequent stages] were considerably higher than the theoretical values. Minor hydrofracturing occurred from time to time, but the operation went according to the script. The grouting mode was compatible with permeation grouting induced by microfracturing.

The theoretical grout quantities proved to be accurate within 15%. Most sleeve pipes were grouted three times. All zones were brought to a proper refusal. A threefold double packer (covering 1 m at the time) was used to perform the sleeve-pipe grouting.

The total cost for the grouting operation was less than $100,000.

5B.1.1.2 Field Application: The "Pl." Case—an Ill-Engineered Grouting Project. Nobody likes to talk about projects that did not work out as planned or were a complete disaster. It is often dangerous to name the projects for fear of lawsuits. Nobody likes to be associated with failures, but, unfortunately, there is probably one bad grouting project for every good one that is performed. The reason is simple: Most grouting contractors and engineers are *not* knowledgeable about the applied science, called *grouting*.

History. When the author was 24 years old, he designed a grouting project to provide support to footings which were going to be loaded in excess of the "deformation bearing capacity" of the soil. This was only noticed when the construction was 60% complete, when settlement was starting to take place. The engineer, looking after both structural and geotechnical aspects for the construction of a high-rise building had made an error when calculating the load to be transferred via the columns to the footings.

The author was asked to provide the design for a grouting program below the cylindrical footings, including detailed calculations and execution techniques. Contrary to North American geotechnical engineering practice, in most (continental) European countries, the engineer has to provide calculations of bearing capacity and anticipated settlement of footings in full detail before receiving a construction permit. The design of the grouting plan, as explained further, was sound, but the author made two mistakes: The selected grouting technique was flawed and so was the product selection.

Grouting Plan Design. The objective was to limit the *total* settlement of each cylindrical footing to 35 mm (1.4 in), under the full load. Some of the initial settlement had already occurred, and the different loads applied to the various footings had already induced additional vertical stresses dissipating with depth, below the footing. Therefore, the remedial grouting program could not provoke additional settlement. In other words, drilling below footing level was not allowed. Therefore regular sleeve pipes could not be used. The grout pipes had to be jacked in place.

The design had to provide the size and the depth of the stabilized soil in order that the ungrouted soil below as well as the treated soil conglomerate would not compress more than the allowable 35 mm (1.4 in). In other words, a settlement calculation had to be submitted, identifying the anticipated compression of each soil horizon below the footing under the additional vertical stress (reducing with depth).

The method used consisted of the "singular points" and applied Frohlich's equations to determine the additional vertical stress (σ_z) on the vertical line through the singular points in each footing. (Frohlich's equations are the same as Boussinesq's for most soils, but the former take the sequence of softer and denser layers into account).

The settlement equations of Terzaghi were in turn used to calculate the compression of each ungrouted layer under the final load and under the "present" loads (*present* means loads applied at that point—prior to grouting).

The compressibility coefficient C was derived from the static penetrometer tests carried out prior to the construction of the building. Because of the depth of the footings below the original surface, it was safe to use the following equation to calculate the value of C:

$$C = \frac{3C_{kd}}{2p_0}$$

where C = compressibility factor
C_{kd} = cone resistance in the penetrometer test
p_0 = the vertical stresses in the soil prior to applying the load

The static penetrometer data of the subsoils on this site are no longer available. To demonstrate the mathematical process to determine grouting depth, the author has created a similar set of static penetrometer data, which for all intents and purposes are comparable with the actual in situ values (Fig. 5B.4). The corresponding C values of the soil layers are derived from the static penetrometer records.

Figure 5B.5 depicts the vertical stress curves as a function of the depth. In this semilogarithmic diagram, the original vertical stresses p_0 as well as the additional vertical stresses superposed on the original vertical stresses ($p_0 + \sigma_z$) are displayed, as a function of the depth z.

The anticipated compression of each layer under the design load is calculated in Fig. 5B.5. The total compression of each layer under the load is equal to the surface area between the two curves

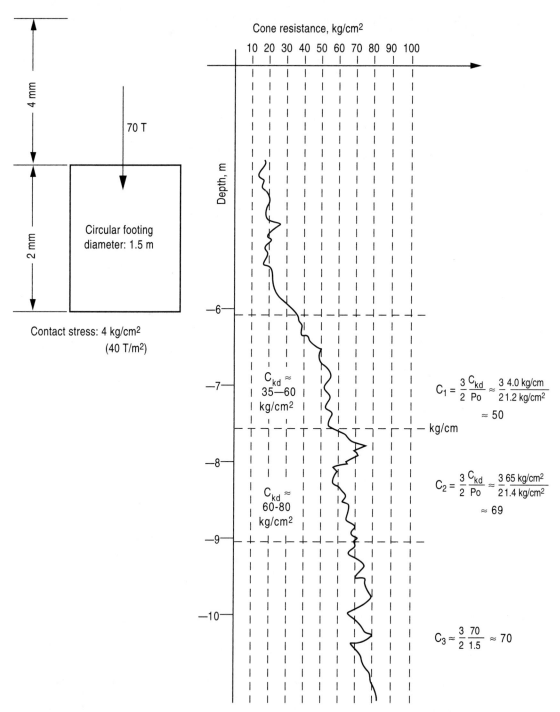

FIGURE 5B.4 Static penetrometer data.

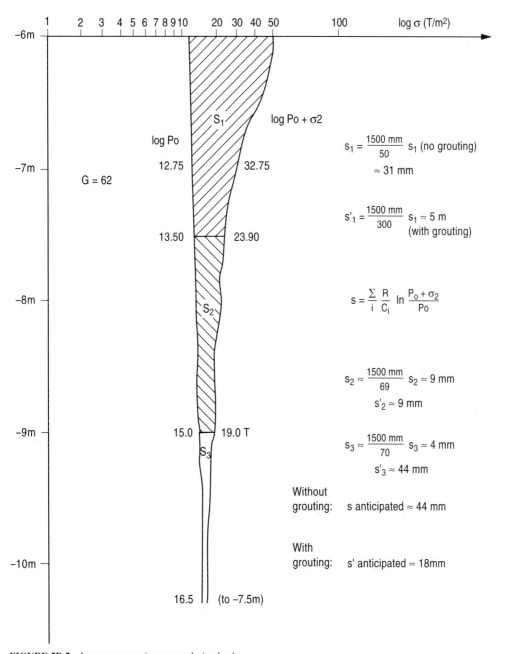

FIGURE 5B.5 Layer compression versus design load.

in Fig. 5B.5 multiplied by the scale factors and the thickness of the layer, divided by the compressibility factor for each layer:

$$s = \Sigma_i \frac{h_i}{C_i} [\ln(p_0 + \sigma_z) - \ln p_0]$$

A permeation grouting program will increase the C factor several orders of magnitude. It is the author's experience that, depending on the grain-size distribution of the granular soils, the C factor of grouted soils varies between 300 to well over 1000 using a hard gel. Especially when "rigid," water-reactive, hydrophobic polyurethane prepolymers are used, the latter value is often exceeded. The compressibility of the grouted soil layers subjected to the highest additional vertical stress therefore is reduced to a fraction of its original anticipated deformation.

It is also obvious that the total compression of the deeper (ungrouted) soil layers under the applied design load is not substantial. The surface area between the two vertical stress curves (Fig. 5B.5) for the lower layers is rather insignificant compared to the one in the layers in which there is a substantial increase in vertical stress (i.e., the layers directly below the footing). These layers are providing the largest contribution to the total settlement, especially if their compressibility factor is rather low. The grouting operation therefore has to focus on the layers that would provide the bulk of the settlement. The grouting depth is selected in such a fashion, that the total contribution of the ungrouted layers (plus a small contribution by the grouted layers) to the total settlement is within the tolerated limits.

The complicating factor often is that there is not enough injectable soil below the footing that can be treated with permeation grouting. In those cases, the goal of the grouting will be to sufficiently widen the footing (using permeation grouting) to limit the average additional vertical stress transferred by the grouted soil to the ungrouted layers, to keep the settlement within design criteria.

In the studied field application, there was approximately 2.5 to 3 m (8.2 to 10 ft) of properly injectable soil below the footing, and the grouting did not need to be extended laterally.

The grouting depth is easily determined. In the example, the ungrouted soil below the footing would undergo a compression of 44 mm (1.7 in) (almost 50% higher than the tolerated 30 mm), under the total load, applied to the footing.

By executing a proper permeation grouting program, we were of the opinion that the C value could easily be increased to 500. Using $C = 300$, therefore, was a conservative approach. By grouting the first 1.50 m (5 ft) below the footing, the anticipated settlement would be reduced to 18 mm (0.7 in) which met the design criteria.

Details of the Design for the Practical Execution. The conceptual design was accepted by the engineer for the insurance company, after he was assured that the grout pipes were going to be 2-in (5-cm) perforated steel pipes, driven in place with a jackhammer. A massive steel point was welded on the bottom of each "strainer-pipe."

The grout selected was a sodium-silicate-based solution. This solution grout had triggered a grouting scandal in Nurnberg (Germany), a few years earlier. The opinion prevailed that, using a formulated organic hardener instead of the ethyl acetate used in the Nurnberg catastrophe would avoid the problem. The Durcisseur 600B, manufactured by Rhone Poulenq, was selected based, at least in part, on the belief that the problems of deterioration in time due to syneresis had been resolved. Especially since the leading French grouting contractors at that time were using sodium silicate solutions for permanent applications, our concern about the durability was eliminated. What a mistake that was!

The Execution. The execution was a complete fiasco. The sleeve pipes were driven (through cored holes) to the required depth. The pipes filled up immediately with fine sand (which was causing some additional disturbance below the footing). Initially, the ability to develop a system to prevent the "quick sand" from entering the grout pipes (through the perforations) posed a problem. Various techniques were tried, including condoms and rubber sleeves, but each tore or rolled up while the pipe was vibrated into the soil. The result was the installation of a perforated pipe, through which only one grout stage was possible (in contrast to sleeve pipes through which several grouting "passes" can be made). Still it seemed likely that the soil could be grouted "in one shot" because of the demonstrated ability to alleviate (at least temporarily) one of the most significant problems: the flow of the sand into the grout pipe. This was done by filling the pipes with a very weak silicate gel, which, in turn was squeezed out during the grouting operation.

The first pipe grouted appeared to go well, until grout started to show up, around the footing, as fast as the grout was placed. The operation had to be halted, causing the sodium silicate to gel in the pipes. Sand, mixed with sodium silicate, had entered the grout pipe, through the perforations. The hard gel prevented removal and resulted in the loss of the grout pipe.

Basically the same story was repeated in adjacent pipes: The pathway alongside the footing was the path of least resistance. Packers were attempted, but forcing the packers down caused too much gel to enter the packer rods, and the gel could not be squeezed out. To add insult to injury, grout-saturated soil entered through the upper holes of the grout pipe and migrated along the packer rods to the top of the footing, where it was blocked, more or less successfully, using a hot patch of quick-setting cement. By the time the second "sleeve" had been finished (counting from the bottom of the strainer pipe), the packer was grouted in place.

This methodology was not providing the grouted soil-conglomerate in the desired and designated locations and the project was aborted. The client resorted to screw piles to solve the problem. This was one of the rare moments where the author wished never to have been in this type of business. Similar amateurish grouting approaches are more the norm than the exception in North America more than 15 years later.

Shortly after this "project" the author developed a (patented) steel sleeve pipe which does not have the problems of rubbers rolling off the pipes, because each rubber, covering the perforations, is protected by a short perforated steel sleeve, welded on the sleeve pipe, protecting the rubber.

5B.1.1.3 Field Application: Ryerson Polytechnical Institute (Toronto) and Construction of a Condominium Complex in Ghent. The presence of substantial piles of rubble (below grade) on the construction site, caused by demolition of old buildings, filling old basements, and overlying several foundations, presented an unwelcome problem on both sites. The removal of the debris would have been a time-consuming and very expensive operation, especially since environmental regulations were strictly applied.

In both applications, the engineer decided to leave the rubble above the old footings in place, and stabilized the very pervious rubble backfill with cement-based suspension grout to form the foundation for the new construction. Sleeve pipes were installed through the brick and concrete rubble and grouted several times, until complete refusal was obtained, with a stable, cementitious grout. Multiple-phase, multiple-hole grouting was selected to ensure complete filling of the voids in an economical way. After grouting, plate tests were performed to verify the bearing capacity.

5B.1.2 Slope or Excavation Stabilizing Technique

An important application field for grouting, very often "design-oriented," involves the stabilization of the soil to enable excavation under safe conditions. Hundreds of case histories pertaining to this technique involve soil grouting to enable excavation through cohesionless soils, above or below the water table, to prevent settlement of roads or structures, or to provide necessary safety to the workers. Other applications involve the underpinning of existing buildings, or structures, using grouting technology, to enable excavation and construction adjacent to and below the existing footings, without causing structural damage to these structures.

Some extensive grouting projects were executed for shaft sinking well below the water table to prevent ground and water inflows into the excavation. For these applications, grouting often offers an alternative to the freezing technique, especially when dewatering (deep wells) cannot be executed.

Grouting has often been used in conjunction with other retaining systems, to enable excavating below the water table to prevent settlement to adjacent structures caused by the lowering of the water table.

Applied grouting techniques such as soil nailing and soil anchoring are used extensively to stabilize slopes and allow deep, semivertical excavations. Excellent publications on soil nailing and soil anchoring by Bruce[1,2] provide in-depth information about these techniques, which are growing in popularity.

Grouting the (leaking) corners of sheet-pile cofferdams is another frequently applied grouting procedure, especially when loss of soil has to be prevented. Grouting below sheet piles, driven through saturated overburden to bedrock, to extend the cutoff wall into hard and fissured rock is another application.

A typical example pertains to a cut-and-cover tunnel section during the construction of the subway in Charleroi (1980–1982), involving sheet piles driven to fissured bedrock. The large cracks in the bedrock brought water and fine soils into the excavation, which resulted in subsidence of the adjacent buildings. A grout curtain adjacent to and below the sheet piles was installed to prevent further settlement to the buildings. As a preventive measure, prior to the excavation of the soils between the two rows of sheet piles, the grout curtain was extended over the entire length of the cut-and-cover section of the tunnel. One-component water reactive hydrophobic polyurethane prepolymers were used for this project.

Gularte[3] reports on the grouting for the construction of the subway in Los Angeles. The sections constructed under the protection of a grout cover did not collapse during a major fire in the late 1980s; classic sodium silicate solutions were used for this project.

5B.1.2.1 Field Application: Grouting for European Subways

History. Pregrouting the soils below the structures within the influence zone of tunnels or major sewers, prior to their construction has been a routine technique to prevent settlement.

In most European major cities, tunnels cannot follow the trajectory of the winding narrow streets and therefore often pass underneath old buildings and structures. To reduce settlement, the soils above the tunnel are pregrouted to form a "cap" or arch, reducing the settlement to tolerable limits.

One of the complicating factors in most old European towns is the existence of centuries-old, wooden pile foundations and a high ground-water table. The water table cannot be lowered, since this would cause irreparable damage to the wooden foundation system of the old structures.

Traditional "umbrella grouting" from shafts adjacent to the tunnel trajectory, fanning out below footings, has been performed for a number of years (Figs. 5B.6 and 5B.7). Classic sodium silicate–based grouts were the most commonly used grouts for this type of application in the fine granular soils. While the tunnel is constructed, backfill grouting from within the tunnel provides permanent support for the grouted soil and indirectly for the footings.

One of the most infamous sodium silicate grouting applications of the century took place in Nurnberg (Germany) during the mid-1970s, but most of the grouting work in other cities has proved to be successful, especially in Paris, where gigantic soil-grouting projects using sodium silicate–based grouts have been executed during the last three decades. The quality control program applied for the construction of most grout umbrellas consists of coring grouted soil and performing unconfined compressive strength tests on the grouted soil.

The soil-freezing method was most often used in soils that were not injectable. Since the introduction of jet grouting, the freezing method has lost popularity and even soils treatable with classic permeation grouting techniques are now routinely jet-grouted. Classic sodium silicate–based grouts are gradually being replaced by microfine cement-based suspension grouts, especially under pressure of environmental groups.

Note however that there is a continuous evolution in the grouting techniques and approaches for this type of work. In Antwerp, it was noticed that the drilling for the installation of the sleeve pipes, prior to tunneling, caused more settlement than when tunneling took place without any grouting. As a result, after the execution for one part of the subway line, under the protection of grouted soil-arches, tunneling projects below structures since the late 1970s have been predominantly executed without grouting.

Since the leading grouting contractors always try to be innovative, "frac-grouting" to control settlement, made its appearance in tunneling below buildings. Pototschnik[4] reports on a "new and

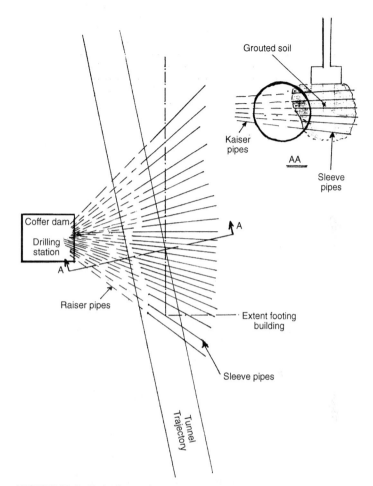

FIGURE 5B.6 Umbrella grouting.

remarkable compensation grouting method" to minimize and continuously offset settlements, used when tunneling for the Vienna subway underneath a five-story building.

The same approach (compensation grouting) was followed for the construction of a new railway tunnel (Fall 1993), through clay horizons, passing underneath a research facility in Sarnia (Ontario). The cost of this type of grouting is extremely high. Other types of grouting and underpinning should be evaluated first before resorting to this method.

5B.1.2.2 Field Application: Small-Diameter Shaft-Sinking Operation through Water-Bearing Fine Sand for the Deep Sewer Network in the Paris Region—Alfortville, France, 1984.
A manhole for a sewer tunnel had to be installed to a total depth (below surface) of 8 m (26 ft), 4 m (13 ft) below the ground-water table, through predominantly uniform fine sand with approximately 10% silt. The contractor's engineering design involved sinking of the shaft through a concrete "guidance slab" and excavation under water, to an elevation of approximately 800 mm below the required depth, followed by the installation of a concrete floor, using the preplaced aggregate grouting method.

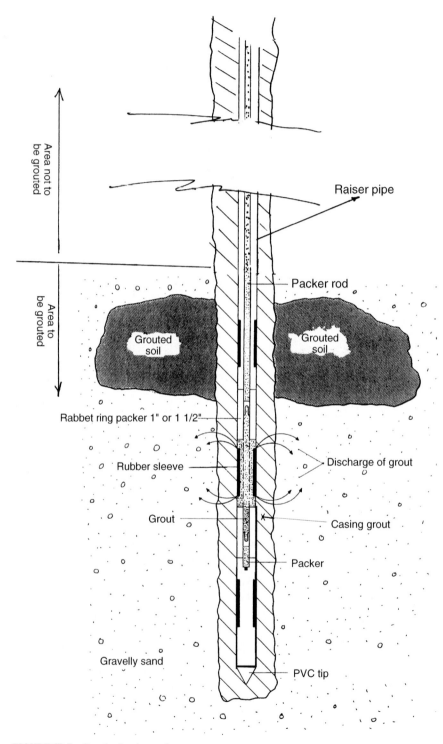

FIGURE 5B.7 Grouting by sleeve pipe.

The shaft did not sink uniformly and got stuck, approximately 2 m (6.5 ft) above the initial target elevation.

The following solution in five steps was provided for the contractor:

1. Install a grout curtain over a length of 3.5 m (11.5 ft), overlapping 1 m with the shaft and reaching to an elevation of 2.5 m (8.2 ft) below the initial target depth. The grout to be used above the shaft shoe would be a weak gel. Below the shoe only a temporary soil modification would be necessary. Because of the low permeability of the soil, a classic sodium silicate solution, (utilizing the commercial hardener Durcisseur 1000) was selected. It should be noted that the chemical engineer of the manufacturer of the Durcisseur, "guaranteed" the stability of the produced silica gel for "one week only," even at high (but workable) hardener concentrations.

2. Excavate below the level at which the shaft was stuck, using a sump pump to remove the incoming water coming through the grout curtain, and excavate below the shaft shoe. Failing this, sink a slightly smaller shaft within the large shaft, to the required depth under the protection of the grout curtain. A structural and watertight connection between the two shafts (using injectable tubes in the cold joint) was designed in order to transfer the loads from the upper shaft to the foundation of the new shaft.

3. Vertical and oblique sleeve pipes (*tubes à manchettes*) were installed in cased drill holes, in order to form a grout curtain around the shaft, as well as a 2-m (6.5-ft) thick plug below the future bottom of the shaft.

4. The initial calculations for the grout spread were made, based on permeability coefficients derived from sieve curves on disturbed samples from the shaft bottom, using Hazen's equation. As soon as the casing grout around the first sleeve pipes had stiffened, in situ hydraulic conductivity tests were conducted over the entire stretch to be grouted, using a double packer within the sleeve pipe, starting from the bottom, to obtain a permeability profile for the strata to be impregnated. Some silty sand layers were discovered with hydraulic conductivities in the order of 10^{-3} cm/s (4×10^{-4} in/s). The grout spacing was recalculated, and the only practical alternative was to double up the number of sleeve pipes.

5. The theoretical grout quantity per sleeve was determined, and the grouting operation was executed sleeve per sleeve. To reduce vertical migration of solution grout alongside the shaft, bentonite slurry was injected in the upper portions of the sleeve pipes (hydrofracturing technique) to create a "bentonite sheet" around the shaft extending to the surface. In total, each sleeve pipe was injected three times, systematically increasing grouting pressures, to break through the treated ground. Approximately 30% more grout than anticipated was necessary due to grout-migration via the badly disturbed areas into the shaft.

The results obtained were the following: The shaft was dewatered, and the shaft-sinking operation was brought back on (vertical) track by undercutting on the high side below the shaft shoe. The shaft excavation was continued to the desired depth in fairly wet circumstances, but no piping conditions were observed, and a reinforced concrete slab was poured in dry conditions, using a sump which was subsequently tremie-grouted, once the shaft was submerged. The total duration for the shaft grouting, including design, was 1 week. The total cost for the grouting, including engineering and supervision, was less than $25,000.

This shaft-sinking problem and its solution has striking similarities with cases involving large-diameter caisson foundations, predominantly installed from within the basements of existing buildings.

5B.1.2.3 Field Application: Underpinning of Foundations with Jet Grouting. The jet-grouting technique has made it possible to create structural retaining walls (similar to caisson walls). Numerous applications have been documented by several authors such as Bruce,[5] Gallavresi,[6] Kauschinger,[7] Welsh,[8] Wolosick[9] and others.

Applications often involve the following: Existing structures and buildings are underpinned by installing a wall of jet-grouted vertical columns, through the footings of existing buildings, immediately alongside the wall to be underpinned, creating a new foundation system for the old building. Excavation alongside the building to a depth well below the foundation level of the adjacent structure can be performed. Often the "underpinning retaining wall" needs to be anchored, to take up the lateral thrust on the jet-grouted wall. The jet-grouting technique was explained in detail in Sec. 5B.2.4).

5B.1.2.4 Field Application: Grouting of Loose Rubble—Roosevelt Square, Antwerp. During the construction of the pilot tunnels for the subway in Antwerp, below the Roosevelt Square, in 1978, a substantial rubble zone was intersected. The rubble zone extended from the tunnel horizon to the surface. To make matters worse, this rubble zone extended laterally all the way to the Schelde River. Each time the high tide came up, the tunnel was plagued with a severe in-rush of water (and fine soils). It was discovered that an old canal, extending through the Roosevelt Square, had been backfilled with construction debris during the eighteenth century.

A major grouting program from surface was undertaken to stabilize the rubble in and adjacent to the trajectory of the tunnels. Stable cement-based suspension grouts were injected via plastic sleeve pipes. No engineering was done, and the pipes were spaced 3 m apart in an arbitrarily selected triangular pattern. This proved to be overkill in certain locations and totally inadequate in other areas.

During the subsequent excavation of the tunnels, the rubble was well cemented, but in some locations the stabilization did not reach to the roof of the pilot gallery, and additional grouting from within the tunnel had to be performed.

The formulation (per m^3 or 1.3 yd^3) consisted of approximately

450 kg (992 lb) cement

150 kg (331 lb) trass (natural pozzolan-clay phyllo-silicate 1-1 structures)

700 kg (1543 lb or 185 gal) water

This grout was only partly stable and generated approximately 5% bleeding. The unconfined compressive strength of the grout exceeded 38 Mpa (5511 psi) after 1 year.

The subway authorities had no understanding about grouting and were lead by the grouting contractor, who had no interest in limiting the grout spread. Over 2,000,000 L (528,346 gal) of this grout were injected in a rather primitive fashion. The general contractor claimed considerable extras for delays. The tunneling contractor wanted to use a weaker grout which could have been done easily by introducing bentonite and (to a lesser extent) type F fly-ash into the mix.

Note that trass is still used as an important additive in suspension grouts in Belgium and Germany to increase the stability of the grout and to enhance its durability.

5B.1.2.5 Field Application: Slope Stabilization of Embankments and Excavations. Applied grouting technology is also used for temporary and permanent slope stabilization and erosion control of fairly steep excavations under particular circumstances. The application is a shotcretelike material in which the cement has been replaced by a hydrophilic polyurethane prepolymer (introduced into the nozzle); grass and tree seeds are added to the sand aggregate. The urethane reacts very fast and "loosely" glues the particles together, creating a pervious blanket.

This allows grass and vegetation to grow within and through the stabilized crust, eventually taking over the role of the stabilized blanket, providing erosion protection for the slope. This application was especially developed for sites where aesthetics are a major issue.

To prevent gully erosion or worse—slope failure—potentially causing undermining of a high-pressure pipeline installed on a freshly excavated slope toward a canal, the decision was made to create a flexible soil conglomerate by "shotcreting" a 200-mm (7.9-in) thick grout-sand blanket on the slope.

Regular (dry-mix) shotcreting equipment was used to convey sand to the nozzle in which water and a hydrophilic polyurethane grout were introduced via a dose pump. The very reactive polyurethane acted as the glue for the soil particles. The water and polyurethane introduced in the nozzle amounted to less than 20% of the volume of sand. The water-urethane mixing ratio was

3-1. The intent was to form a pervious blanket that would allow ground water to percolate through the slope without causing sloughing or even triggering a failure. Rain predominantly ran off, but some of it percolated through the grout-sand blanket without causing gully erosion or triggering a slope failure.

This rather unusual application was executed in 1981 for the N.V. Distrigaz during the installation of the 90-km (56-mi) pipeline from Zeebruges to Antwerp.

Taking this application one step further, grass seeds and pellet fertilizer were premixed with the sand and shotcreted in the same fashion as the above application, for stabilization of rugged slopes along highways and even on acidic tailings piles, to prevent wind erosion.

5B.1.3 Pile or Pier Enhancement

A variety of techniques have been developed to enhance the performance of in situ cast piles, using grouting techniques. Toe grouting, or base grouting, has been successfully applied in several occasions to improve the pile base stiffness by reversing the disturbance caused by base heaving or boiling during the installation. Grouting reduces settlement of large-diameter piles before full bearing capacity is mobilized. There are, however, a number of inventive ways to enhance the bearing capacity dramatically, but only a few engineering contractors have explored the potential of "in-depth" base grouting to take advantage of the evolution in grouting technology and new product development to increase the end-bearing capacity or, by locking in increased horizontal stresses, to enhance the lateral bearing capacity. Most of the improvement in caisson performance at present is due to increased lateral bearing capacity, which is the result of shaft grouting.

More emphasis should be placed on methods enhancing the end-bearing capacity by base grouting, including

1. Dual-phase toe grouting (compaction grouting followed by permeation grouting or even jet grouting)
2. Multiple-stage permeation grouting (or, alternatively, jet grouting, depending on soil permeability) to a considerable depth below the pile in conjunction with an "artificial widening" of the pile base, which goes far beyond reversing the base disturbance

The end-bearing capacity can be quadrupled by toe grouting and by proper product selection and grouting methodology.

In 1989, Yeats and O'Riordan[10] and Sherwood and Mitchell[11] published papers on the influence of base grouting on the performance of caissons in the Thanet sands in London. The grouting approach they described contrasted sharply with the sparkling and enthusiastic reporting by Gouvenot and Gabaix[12] in 1975. The latter authors presented a sound case about the effects of grouting on the lateral bearing capacity of caissons, indicating an increase of 200 to 300% in ultimate load on caissons in sands and clays. They also provided a direct relationship between ultimate load (as a result of improved shaft friction) and volume of injected grout.

The authors of the papers about the base grouting on the caissons in the Thanet sands in London, describe a horizontal sleeve-pipe arrangement placed at the pile base. The reports indicated that the clock of progress was turned back. Only limited quantities of an unstable cement grout were used to carry out a crude base grouting program, whereby the sleeve pipes were grouted several times. The grouting procedure created a substantial increase in horizontal stresses at the pile base, but only succeeded in reversing most of the disturbance created by the boiling of the soil. A substantial amount of grout was "lost" in thin hydrofracturing planes. These planes break when minor settlement occurs and do not contribute to an increase in bearing capacity except at very low base stresses.

In Holland and Germany, geotextile bags filled with gravel have been installed at the base of caissons. The "preplaced aggregate" is grouted after the pile is cast. The geotextile is used to prevent hydrofracturing at elevated pressures.

The common practice is to use rather high pressures in order to lift the pile to mobilize the lateral friction and to lock in the horizontal stresses at the toe. Oosterbos[12a] documented a case history whereby this type of toe grouting technology was implemented below deep tubular piles in the bed of the Oude Maas River near Dordrecht (The Netherlands). Bandimere[13] describes a compaction grouting program below caissons at the IBM building in Dallas, Texas, in 1989. Settlement was halted by performing a classic compaction grouting program through the toe of a number of caissons.

Bruce[14] points out that to fully mobilize shaft friction, displacements in the order of 0.5 to 1% of the caisson diameter are required but to mobilize the end bearing (without toe grouting), displacements of 10 to 15% of the diameter are needed. This clearly demonstrates the need for improving the end bearing of the pile. In an excellent and comprehensive study about "enhancing the performance of large diameter piles," Bruce describes a number of case histories in which pile base grouting has been applied.

Although Gouvenot and Gabaix[12] only concentrated their efforts on the improvement of the pile skin friction, they demonstrated that it was possible to increase it 200 to 500%! Their method, in essence, involved the placement of a grout tube (a kind of homemade sleeve pipe) in the middle of the caisson, which was gravity grouted during the placement of the concrete. Subsequent grouting in two more phases produced remarkable results.

Stocker[15] produced a useful contribution to shaft grouting (vertical sleeve pipes close to pile perimeter) in conjunction with base grouting utilizing a flat jack at the pile base.

There are only a few documented cases in which deep base grouting has been applied. Littlejohn[16] and Bruce discuss cases in which solution grout has been used to consolidate the soil below and adjacent to the pile base. The authors describe the successful deep base grouting exercise below caissons at the Corniche Centre in Jeddah. Although the product selection is most debatable, (resorcin formaldehyde was used), an enlarged base of strengthened soil was created to prevent excessive settlement. Classic sleeve pipes were installed to approximately 1.5 to 2 m (5 to 6.6 ft) below the pile base. A very fine silty sand with a hydraulic conductivity between 1×10^{-3} to 5×10^{-5} cm/s (4×10^{-4} to 2×10^{-5} in/s) was grouted with diluted resorcin formaldehyde.

Endo[17] documented a case of base grouting whereby the end bearing was doubled. A hard water reactive prepolymer (Tacss 25) was used in this application.

In personal conversations, Heenan has reported a successful base grouting project in the Toronto area using microfine cement-based suspension grouts. A zone with a thickness of over four times the pile diameter was grouted using multiple-phase sleeve pipe grouting.

It is obvious that crude base grouting programs, using cementitious grout, have resulted in considerable loss of grout along the pile shaft or due to soil fracturing. Only when geotextile bags or similar arrangements are used, containing the grout at the caisson base, will base grouting effectively increase the ultimate bearing capacity. Repeated grouting operations at the base result in higher locked-in stresses.

The author[18] proposed the following methodology to obtain the best of both worlds (increased shaft friction and increased end bearing) to increase the total bearing capacity of piles by at least 400%:

1. Drive a number of steel sleeve pipes (1 to 4, depending on the nature and permeability of the soil and the size of the pile) through the pile base after excavating the pile and before pouring the concrete. The boiling of the caisson floors should be partly offset by placing a bentonite suspension in the caisson. The "pointed" sleeve-pipes are vibrated into the soil to a depth of at least $2D$ (D = pile diameter) into sufficiently dense, injectable soil horizon.

2. The minimum required quantity of casing grout is injected through the lowest sleeve of each sleeve pipe, to obtain sleeve pipes properly enrobed in casing grout. If sleeve pipes cannot be driven, casings should be installed in the pile, and sleeve pipes are installed after the concrete (of the pile) has been poured. If the soil below the pile base is not injectable, the casings are used to maintain access to the pile base and perform a jet-grouting program to widen the pile base by creating jet-grouted soil columns to the required depth, below the base of the pile. The lower section of the casings are constructed as "homemade sleeve pipes."

3. Have an oversized geotextile sock installed around the lower section of the caisson. This sock should cover the zone in which compaction grouting is to take place, and should be properly attached to the reinforcing cages.

4. Immediately after the concrete has been poured, all the sleeves above the pile base are "cracked" (using water) for future shaft grouting. The sleeve pipes, which should be attached to the reinforcing cage to maintain constant spacing, could also be used to test pile integrity prior to the start of the grouting program.

5. The goal of the first grouting program is to increase the lateral bearing capacity of the piles. A balanced, stable cementitious grout with medium resistance against pressure filtration is injected above the pile base. This causes the geotextile sock to expand, widening the shaft, and the geotextile prevents hydrofracturing of the soils. The horizontal stresses, induced by the grouting pressure, are effectively locked in. The additional lateral friction in the grouted zone can be estimated with Littlejohn's equation:

$$F = D10\,\phi p$$

where F = total lateral friction
D = ultimate pile diameter
ϕ = angle of internal friction of the surrounding soil
p = applied grouting pressure

An extensive testing program (scale model and small-diameter pipe testing), indicated that the lateral friction increases easily by 200 to 300% through geotextile-contained shaft grouting compared to the ungrouted piles. The tests also indicated that Littlejohn's equation provides a reasonable (and even conservative) estimate for the actual shaft friction, mobilized by grouting.

6. The second stage consists of executing a base grouting program. Grout selection is very important. Microfine cement-based suspension grouts should be used if possible. The grouting should be performed several times, to obtain a competent soil conglomerate below the pile base. Solution grouts, especially "rigid" water reactive prepolymers, are the most suitable products to permeate soils which are marginally injectable by permeation grouting. Under no circumstances should "classic" sodium silicate solutions be used.

A method to calculate the end bearing of the piles is explained in detail in a paper by Naudts.[18] The disturbance at the pile base can be reversed and the end bearing can be increased by approximately 200 to 400%.

During a 2-year intensive and thorough research program, the author, in conjunction with a specialized grouting contractor, developed an engineered method to install and calculate bearing capacity (and pull-out load) of special mini–grout piles and grout anchors. The patented method in essence involves geotextile membranes, strapped around the sleeve pipe containing the grout that is injected. The geotextile prevents hydrofracturing and guarantees that the applied grouting pressures are effectively locked in, permanently increasing the in situ horizontal stresses. The geotextile also prevents washout of grout where anchors are installed under flow conditions (dams) and further enhances the compressive strength of the grout injected into the geotextile as a result of the pressure filtration. The problem of installing piles through aquifers under artesian pressure is also resolved with this system: The inflated grout bags form an excellent seal in the confining layers above (and below) the aquifer.

The pull-out strength or lateral bearing capacity has empirically been established and is reflected in the following equation (equation by Heenan and Naudts):

$$LB = LB_0 + (\Omega_1 - \Omega_0)\,p10\phi$$

where LB = lateral bearing capacity
LB_0 = lateral bearing capacity of straight shaft anchor, in the same formation

Ω_1, Ω_0 = the bond surface ($\pi \times$ diameter $\times$ bond length) for the pressure-grouted and ungrouted piles
ϕ = angle of internal friction

The second term in the equation pertains to the predictable improvement that is accomplished by using this technique.

The equation can be used for the design, to determine the bond length of the anchor or pin pile, to obtain the desired capacity.

The grout formulation is rather critical: A stable, balanced suspension grout with reasonable resistance against pressure filtration turned out to be the most suitable. Because of pressure filtration, very high compressive strengths are obtained in a very short time [over 50 MPa (7252 psi) in less than 72 h]. The advantage is that a very flowable grout with a viscosity of less than 45 s can be used. Special hardware is manufactured to accommodate multiple-strand anchors, rock anchors, pin piles, and soil anchors.

5B.1.4 Hazardous Chemical or Waste Control

Grouting techniques are often used to create cutoff systems in fractured rock, around tailings ponds and land fills. In soils, slurry walls or sheet-pile walls are usually selected, because they are considered to be positive cutoff systems.

Rock grouting with cement-based suspension grout has been the main topic in textbooks by Houlsby[19] and Weaver.[20] Both textbooks provide a wealth of information. However, the technology does not stop evolving and their approach to formulating cement-based suspension grouts is outdated.

For environmental applications, the hydraulic conductivity has to be reduced to much lower levels than what used to be the acceptable norm in dam grouting. The grout curtains have to act as an effective cutoff system. Average Lugeon values in the order of 0.1 Lu are therefore necessary and can easily be accomplished by combining a professionally executed cement-based suspension grouting program in conjunction with some solution grouting (not classic sodium silicate formulations).

The amenability theory guides the engineer in selecting and changing formulations, in optimizing spacing, and in determining the limitations of the cement-based suspension grouting exercise.

In Manitouwadge (northern Ontario, Canada), a grout curtain through fissured granite, approximately 1200 m (3937 ft) long and 15 m (49 ft) deep was executed to create a cutoff wall around an acidic tailings pond, during the summer of 1991. Only regular and microfine cement-based suspension grouts were utilized. The average hydraulic conductivity was well over 100 Lu. After the execution of a single-row grout curtain, the average hydraulic conductivity was reduced to less than 0.5 Lu. The spacing was adjusted according to the amenability of the formation for a particular grout formulation, so that in certain zones only secondary holes were selected, and tertiary and quaternary holes were used in areas with poor amenability. The engineering and hands-on direction of a rock-grouting program are briefly discussed in the section on cement-based suspension grouts.

Grouting in conjunction with other cutoff systems is often used in soils for containment of regular and nuclear waste. The so-called positive cutoff systems have an Achilles heel—the joints. Grouting is necessary to provide protection for these weak spots.

For sheet piles, the author developed a system whereby steel sleeve pipes were welded to the sheet piles, adjacent to the slots. (Fig. 5B.8) When the sheet piles are driven, the steel sleeve pipe is only millimeters away from the joint. A properly executed grouting program easily takes care of creating a grout sheet over the joint, even if the soil is not injectable. When driving the sheet piles, there is always a pathway of loosened soil or even a minor gap (in cohesive soils). During grouting, the loosened soil as well as the gap are filled with grout.

A grouted soil conglomerate or grout sheet can be established on the upstream side of the joint if the soil is injectable. This method is particularly interesting for "constructing" tight corners in sheet-pile pits and for making tight connections between sheet piles and other constructions.

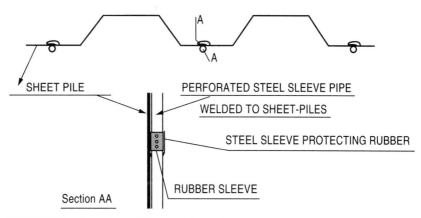

FIGURE 5B.8 Sheet piles with integrated sleeve pipes.

For slurry walls, a number of specialized contractors have developed systems whereby sleeve pipes are attached to the guidance cylinders (forming the joints), which are in turn left in place when the guidance cylinders were pulled. The sleeve pipes are usually pressure-flushed, to remove bentonite from the joints, prior to grouting.

Jet grouting is more often used to make positive cutoff walls for environmental purposes. GKN Keller reports on applications in Germany, whereby the contaminated sludge, produced by the jet grouting through contaminated areas, was treated in an on-site treatment plant.

Some field applications in the following section on ground water and erosion control provide more details about application techniques.

5B.1.5 Ground Water and Erosion Control

Grouting competes with other technologies but often is the only reasonable approach to deal with ground water and (internal) erosion problems. The following four field applications illustrate the potential for grouting in dealing with ground water and erosion. Later in Sec. 5B, more case histories document use of grouting in ground-water control. The majority of the projects in this category fall under the heading of *crisis management*.

5B.1.5.1 Field Application: Careless Tunneling below Railway Embankment. During the winter of 1990, a tunnel-jacking operation took place near Newcastle in Ontario, Canada, below the main railway line linking Montreal and Toronto. The tunnel was located in sandy till with boulders approximately 3 m (10 ft) below the ground-water table. Loss of ground went unnoticed until a near derailment occurred. A sinkhole had formed in the crest of the (previously) marginally stable railway embankment. It was postulated that the water running from the tunnel face had carried at least 15 m^3 (18 yd^3) of soil into the shaft. Voids had formed above the ground-water level, which in turn collapsed, due to the vibration caused by train traffic, leading to the formation of a sinkhole at the crest. A very large disturbed area, at least 13 m (43 ft) high and 30 m (98 ft) in diameter had formed above the tunnel, in the railway embankment. The extremely cold temperatures—below $-20°C$ ($-4°F$)—most likely prevented a major slope failure.

The railway line could not be taken out of service. Emergency measures included dumping several hundred liters of water-reactive hydrophobic, polyurethane prepolymers into the sinkhole. The face of the tunnel was barricaded with jute bags filled with sand and water reactive prepolymer (very slow set time). Sleeve pipes (horizontal and inclined) were installed from the embankment (one heading toward the tunnel) in a fan shape pattern, via cased drill holes. Several cubic meters of casing grout were necessary to fill the boreholes, during the installation of the sleeve pipes.

Grouting was performed around the clock; initially a stable cement-based suspension grout with high solids content was used to fill the voids and to perform minor compaction grouting. Inclinometers, installed through the slopes of the embankment in several locations, "informed" the engineers that the grouting could easily trigger a slope failure, and effective pressures had to be limited to the weight of the overburden.

Because of the nature of the problem, fast-curing, permanent, hard-solution grouts had to be used to impregnate the disturbed soils in the most critical areas. Water-reactive hydrophobic polyurethane prepolymers were injected via the sleeve pipes into the most critical areas. The selected set time was 30 min. Once the inclinometers showed signs that the "worst was over," microfine cement-based suspension grouts were injected in the remainder of the disturbed areas. Microfine cement is slow curing and in its initial stages "liquifies" disturbed soils. Therefore the grouting program had to be carefully controlled and monitored to avoid destabilizing too much soil at one time. Tests indicated that it took between 24 and 48 h for microfine-treated soils to gain increased shear resistance and cohesion at these low soil temperatures. Measurable compressive strength on grouted soil was only obtained after approximately 6 to 8 days.

The curing process of soils stabilized with microfine cement-based suspension grout, at low temperatures has been the subject of research by this author, during the grouting of the tremie concrete and gravel base for the Detroit Windsor tunnel.[21] One conclusion was that balanced stable microfine cement grouts (even if they contain, amongst other additives, 2% bentonite) cure considerably faster than the neat water-cement microfine cement suspension grouts with the same water-cement ratio.

The problem was that the tunnel still had to be completed. The railway authorities no longer accepted a "tunnel—wait and see" approach. A "Roman arch" was grouted in the soil over the area in which the new tunnel had to be installed to enable tunneling without destabilizing the embankment. Even though the soils in the last portion of the tunnel trajectory were only marginally injectable with microfine cement-based suspension grouts, the multiple-phase grouting program, via horizontal sleeve pipes, was successful in preventing disturbance during the final tunneling phase. Tunneling was performed from the other side toward the "abandoned tunnel." An oversized pipe was used as liner, and a water-tight mechanical connection system was created between the existing tunnel.

For several months, the embankment was monitored and the inclinometers provided evidence that the embankment was permanently stable. Over 50 m³ (60 yd³) of microfine cement-based grout were injected after the void filling had been completed, using regular cement-based formulations.

5B.1.5.2 Field Application: Grout Curtain in Rubble Backfill to Prevent Flooding of Building Site for a High-Rise Complex near Lake Ontario

History. An unrestricted inflow of water into an excavation for a high-rise development in metropolitan Toronto occurred in the early spring of 1989. The site, located on the shore of Lake Ontario, is a fill deposit largely comprised of silty to clayey soil with occasional sand beds. The excavation was near to completion, to a point about 0.9 m below lake level, when at its nearest point to the lake (30 m) a layer of highly permeable brick rubble, producing an uncontrollable inflow, was encountered. The area was diked, and pumping was attempted. It was concluded that a direct link between the lake and the excavation, via the rubble backfill, existed. Several permanent water control systems were evaluated. Based on time and economic considerations, the grout-curtain solution was found to be the most suitable system.

Execution. A plan view of the double-row grout curtain is depicted in Fig. 5B.9. Because of the very large openings in the rubble zone, a double grout curtain was deemed necessary to ensure complete filling of the very pervious formation.

Sleeve pipes were installed through the stem of the auger drill. The casing grout disappeared immediately into the open formation. The decision was made to use a slow-curing, cohesive, and very viscous bentonite cement suspension grout to limit the sagging of the grout. During each grouting pass, the viscosity was gradually increased.

The initial viscosity was approximately 60 s (marsh cone). The bentonite content never exceeded 15% of the weight of cement to make sure curing of the slurry occurred. Ultimately the

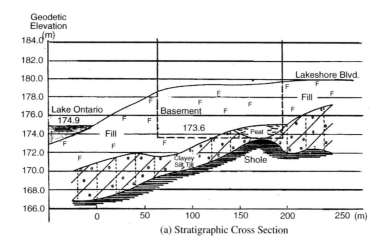

(a) Stratigraphic Cross Section

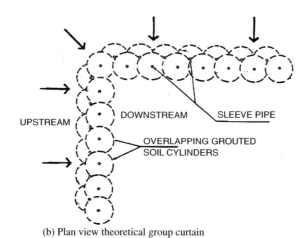

(b) Plan view theoretical group curtain

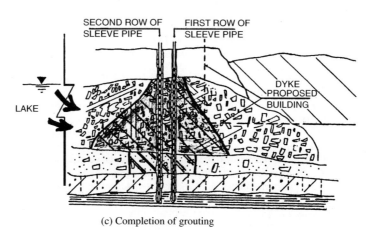

(c) Completion of grouting

FIGURE 5B.9 Plan view of double-row grout curtain: (*a*) stratigraphic cross-section; (*b*) plan view of theoretical grout curtain; (*c*) completion of grouting.

viscosity had to be increased to a toothpastelike consistency (no flow through marsh funnel). Because of the high cohesion of the grout at that point, permeation occurred more through "extrusion" than through flow. The slow curing was necessary to make it possible to grout each sleeve pipe three to five times.

The silty sand overlying the bedrock located below the rubble was also grouted using a microfine cement-based suspension grout. Each horizon in each sleeve pipe was brought to satisfactory refusal, and the excavation was dewatered. A total residual inflow of approximately 60 L(16 gal) per min was recorded (predominantly through minor fissures in the bedrock). The flow stabilized around the 100 L/min (26 gal/min) mark and has been stable since.

The grout curtain was used as foundation for an addition, and footings were placed on it. The curtain has been performing satisfactorily as a cutoff wall and as a foundation system.

The use of bentonite cement-based suspension grouts is not without risks. A prominent grouting specialist directing a grouting program to arrest settlement below the footings of an industrial building in Nova Scotia appeared to be unaware that the bentonite content cannot exceed 15% of the weight of cement when placed in relatively small voids and pores. He directed the injection of an improperly formulated bentonite cement suspension grout, most likely causing internal erosion of finer particles from a very course (boulder) matrix, causing further subsidence to the building. When the grouting program was finally halted, the additional settlement stopped. Moreover, the engineering fees for this bad grouting project virtually equalled the cost of the execution. Another sad chapter in the history of grouting had been written.

Especially in colder climates, a number of issues have to be considered when formulating cement-based suspension grouts. The dissipation of the exotherm into the surrounding soil, rock, or structure can cause complete grout failure.

5B.1.5.3 *Field Application: Internal Erosion Caused by Driving Piles through Aquifers at the Credit River (Brampton, Ontario)*

Problem. Driving concrete or steel piles through a deep artesian aquifer can cause major problems. Even the improper sealing of investigation holes intersecting aquifers under pressure can destroy a structure at a later date.

While driving the deepest piles for a new bridge over the Credit River, the piles first perforated dense clays before intersecting an aquifer under substantial pressure. Fine sand and silt eroded a hole alongside the pile, and silty water started to boil up to the surface. The worst part of the problem was that a large area partly below and adjacent to more shallow piles was disturbed as well.

On a bridge project in Nepal, a similar problem occurred. It was mainly caused by a number of investigation holes intersecting an aquifer under considerable artesian pressure sandwiched between two rather impervious layers. The holes were not sealed and eroding fine material was allowed to run from within the aquifer. Although the pile foundation did not go as deep as the aquifer, the ongoing internal erosion process eliminated the support for the pile foundation. Settlement in the order of 1 m occurred to the pier, closest to the unsealed borehole after the bridge had been built.

Solution. The geotechnical consultants for the Credit River bridge recommended dewatering. This was attempted but failed due to the high silt content in the aquifer. The engineering contractor recommended grouting, which was executed at a cost lower than the aborted dewatering program.

Two sleeve pipes were installed adjacent to each one of the piles causing the erosion. Geotextile bags were strapped around the sleeve pipes and were inflated (with cement grout) to create a tight barrier in the clay horizon, above the aquifer. Hydrophobic, water-reactive polyurethane prepolymers were used to stop the flow around the piles. At that point, a permeation grouting program (microfine cement-based suspension grout) followed by a hydrofracturing and compaction grouting program (using regular cement-based grouts) were executed to reverse the disturbance below and around the other piles.

The engineers of the Nepal bridge abandoned the bridge and constructed a new one a few hundred meters away from the old one, redirecting the trajectory of the highway. This cost several million dollars.

5B.1.5.4 Field Application: Tunnel for the SIAAP in Paris. During the installation of a storm collector (box-culvert type), part of the gigantic sewer system of Paris, certain sections were installed with the cut-and-cover method. Sheet piles were driven to retain the soil. Near one of the old arms of the Seine River, granular soil layers were encountered, making the dewatering of the wide sheet-pile trench very difficult. When obvious signs appeared of settlement to the huge natural gas tanks, located less than 50 m (55 yd) from the culvert trajectory, a quick investigation identified that an internal soil erosion process was taking place as a direct result of the dewatering.

The Parisian grouting contractors proposed to create a grout curtain alongside the sheet-pile wall to a sufficient depth to dissipate the flow energy (utilizing the criteria of Bligh and Lane). The solution consisted of a preplaced aggregate grouting program:

1. The 220-m (240-yd) long trench was flooded (using pumps to create an overpressure inside the cofferdamlike trench)
2. The trench was deepened (excavation under water, using clamshells and air-lifting techniques, to remove the soil close to the sheet piles)
3. Slotted (horizontal) grout pipes, connected to vertical feeder pipes, were assembled to reinforcing cages. A geotextile was used to prevent sinking of the cages and grout pipes into the mud.
4. Once the grouting and reinforcing steel assembly had been lowered, 1.5 m (5 ft) of clean coarse aggregate were dumped and uniformly spread (under water) over the entire length of the excavation.
5. A multiple-hole, cementitious grouting program was launched, injecting more than 175 m^3 (229 yd^3) of cement-based suspension grout, in one day. The grout was prepared in concrete mixers and was a thixotropic stable clay-phyllo-silicate-cement (CPSC) grout. The clay-phyllo-silicates were available at a "negative price," since they were a waste product from the petroleum cracking industry. The formulation was tested prior to grouting, and its water-repellant characteristics were confirmed. The grout displaced the water from the bottom up. The interesting feature was that, during the entire grouting operation, the same grout remained in contact with the water.

Initially, the first injected grout gets partly diluted by the water it is displacing. Once the horizontal grout tubes are fully covered with this diluted grout, the new grout displaces the (lighter) diluted grout. The poor-quality grout therefore floats above the good grout and ends up at the surface of the preplaced aggregate concrete. The water does not interfere with the grouting, and the quality of the preplaced aggregate concrete is the same as if it were executed in dry circumstances.

Seven days after the concrete had been installed, the trench was dewatered. The operation was a complete success.

5B.1.6 Remedial Grouting

Grouting to restore the integrity of old, damaged, or deteriorated buildings and structures offers an economical alternative to rebuilding and other renovation techniques. Remedial grouting is one of the most important application fields for grouting. Unfortunately, far too often only the symptoms are treated without eliminating the cause of the deterioration, cracking, or bulging.

Foundation problems, water infiltration, fire, accidents, and the "elements" are the most commonly noted causes for decay of structures.

Jones[22] describes the grouting of churches with specially formulated cement-based suspension grouts to prevent the formation of calcium salts on the facing stones. Heenan & Naudts[23] describe structural (stitch) grouting in conjunction with the installation of pin piles and preplaced aggregate grouting for the restoration of a badly cracked concrete culvert. Naudts[24] describes the restoration of old, historical lock structures on the Rideau Canal, using regular and microfine cement-based suspension grouts. Hundreds of small contractors are performing crack injection with epoxy resins, polyurethane elastomers, methacrylate grouts, and vinyl esters to seal or restore structural integrity to buildings, bridges, and so on.

5B.1.6.1 Field Application: Underpinning of an Historical Building in Ontario

History and Program Design. During construction of a new office building, the general excavation took place at the same level of the base of an old rubble wall, acting as a party wall. Caissons, located less than 3 m (10 ft) away from the party wall, extended through cohesionless fine to medium (slightly silty) sand, approximately 2 m (6.6 ft) below the base of this old party wall. Cracks developed in the existing building besides the excavation. The building was condemned, and construction of the office building was suspended by court order.

Several methods were evaluated to perform the safe construction of the remaining caissons and to restore the inflicted damage. After several meetings, the following solution was accepted:

1. Provide temporary lateral support (bracing) to the (leaning) party wall.
2. Stitch-grout the crumbly party (rubble) wall, using the steel grout pipes as anchors. Only the section below grade was to be done. Use shotcrete to create confinement for the grout.
3. Execute a permeation grouting program in the upper 2 m (6.6 ft) of sandy soil (below the party wall) to reverse the disturbance and to prevent further loosening of the soil below the party wall. This operation was to be done only after the structural grouting had been completed.

The in situ hydraulic conductivity was to be determined on site (using sleeve pipes), and calculation of the spacing was to be performed with the equations of Naudts-Cambefort. Based on soil samples from test holes, and applying Hazens' equation, it appeared that the hydraulic conductivity was around 10^{-2} cm/s. Microfine cement-based suspension grouts were selected for the permeation grouting program, to be performed in three stages, via 1.5-in (diameter) (3.8-cm) sleeve pipes. An artificial overburden of 1.5 m (5 ft) was installed to provide additional confinement for the grout.

Execution. The project was executed by a professional grouting contractor. The first phase of the grouting exercise involved the structural rehabilitation of the rubble wall. The filling ratio was exceptionally high: The amount of grout injected, amounted to over 24% of the volume of the wall. The formulation consisted of

Cement: 140 kg (308 lb)

Silica fume: 10 kg (22 lb)

Naphthalene sulphonate: 1 kg (2.2 lb)

Water: 100 kg (220 lb)

The grout travel was excellent. The wall was grouted in two lifts to prevent fracturing as a result of internal hydrostatic pressures in this virtually cohesionless rubble wall.

The second phase of the operation involved the installation of sleeve pipes, for which a rotary drill was used. Sleeve pipes were installed inside the casing (drill rods), filled up with casing grout. The grouting took place, typically, 48 h after the installation of the sleeve pipes. The in situ hydraulic conductivity tests were performed in each sleeve pipe at each elevation to obtain a profile of the disturbed soil. During the first stage, each sleeve received the theoretical quantity of grout in the undisturbed areas, and up to 50% more in the disturbed areas. Within 24 h a second and a third grouting stage took place, in each sleeve pipe at each elevation. The entire operation was recorded and evaluated on a continuous basis.

In a few locations, back-flush to surface occurred. The operation was halted, and a grout-cap was installed around the grout-pipe, at the base level of the rubble wall. In most instances, the grouting operation was successfully completed 24 h later.

The grouting project was completed in less than 3 weeks, to the entire satisfaction of the parties involved. The excavation activities, including the installation of the remainder of the caissons, were resumed, and no additional movement to the old building took place.

5B.1.6.2 Field Application: Restoration of Historical Locks

History. The Rideau Canal links the St. Lawrence River with Ottawa. It was constructed more than 160 years ago, and more than 40 locks are still in service. During the last 4 years, 6 of the locks have been restored using state-of-the-art grouting techniques.

Over time, water has washed out fine mortar particles from within the massive limestone block walls and floors, creating channels for internal erosion, in turn causing damage to adjacent structures and substantial freeze-thaw damage to the lock structure. Some of the lock walls are bulging over 300 mm (12 in) and are a *wahrheitszeichen* (sign of truth) about what the harsh Canadian climate does to this type of structure. In the past, these structures were, at best, rebuilt using the old authentic stones. Now, grouting technology is used to restore the structural integrity of these structures. Not only does grouting cost a fraction of classic repair methods, it is a superior technique creating far less disturbance to the authentics of the industrial patrimonium of the country.

A full-size testing program in the Fall of 1989 at Kingston Mills Locks, followed by the grouting of the locks at Hartwells, demonstrated that properly repointed locks can be grouted to obtain a final hydraulic conductivity comparable to the one of concrete (i.e., 10^{-8} m/s) using cement-based suspension grouts (regular and microfine ones).

Methodology. Vertical sleeve pipes with barrier bags (MPSP pipes) are placed in cored drill holes through the wall (Fig. 5B.10). After flushing the holes, the barriers on the MPSP pipes are

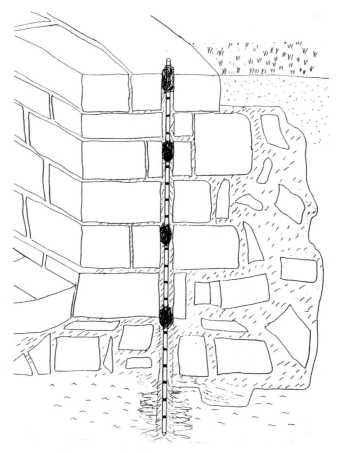

FIGURE 5B.10 Pictorial representation of MPSP system.

inflated with cement-based suspension grout. The interior of the wall consists of rubble and mortar. The sleeve pipe with barriers keeps the hole accessible and allows the grouting of the various zones, separated by the inflated barriers. The various zones often have different hydraulic conductivities and are often treated with different formulations. Moreover, in the lower zones higher pressures can be used, which is often beneficial to obtain the required spread. Via sensitive check valves, the simple and combined Lugeon tests are carried out in each zone [typically 1.5 m (5 ft) high] in each drill hole.

At the beginning of the grouting, in each zone, the amenability (see Sec. 5B.3.1) of the formation is continuously calculated, and the formulations are adjusted to obtain the highest possible amenability with the most competent grout possible. During the grouting operation, the apparent Lugeon values are continuously calculated, and all zones in each hole are brought to an absolute refusal. Multiple-hole grouting in zones with similar hydraulic conductivities is necessary to conduct an economical grouting program.

As a result of its excellent amenability for cement-based suspension grouts, the grouting at Hartwells Locks was a great success. At the Ottawa Locks, however, amenability was usually in the order of 30 to 50%, even with microfine-based formulations. Debris had accumulated in the joints over the years, and the concrete lining on the inside of the locks had prevented natural flushing. Only routine flushing prior to grouting could be performed on the walls. Pressure washing and acidization could not be conducted because of the time constraints and perceived environmental problems. (These were political, not real, since complete neutralization of the phosphoric acid would take place in the limestone environment.)

The microfine cement-based grouts were silicafume-enhanced and typically contained 1 to 2% bentonite and 0.5% naphthalene sulphonate. The stable cement-based grouts contained the same additives, plus small amounts of type C fly ash.

5B.2 GROUTING TECHNIQUES

5B.2.1 Permeation Grouting

Permeation grouting involves the pressurizing of a grout into the matrix or the pores of the medium to be impregnated. The grout will travel through pores and cracks within this medium and will eventually gel or solidify, filling the travel channels. Soil grouting will create a conglomerate, in which the soil particles are glued together without being dislodged. Grout will travel as long as the driving force is greater than the resistance (friction loss) to which it is subjected as it travels through the pores and capillary channels.

The *hydraulic conductivity* of the soil, which is directly related to grain-size distribution and percentage of *fines* (i.e., particles smaller than 70 μm), governs the treatability of this soil using permeation-grouting techniques. The permeability coefficient of a soil deposit is usually different in vertical direction than in horizontal direction. Usually, and especially in layered soil deposits, the hydraulic conductivity is much higher in the horizontal direction than in the vertical direction. Soil is rarely a homogeneous mass, which further complicates permeation grouting.

5B.2.1.1 Solution Grouts. How do we determine the injectability of a particular soil horizon? Injectability and grout spread in a soil with a given hydraulic conductivity are determined by the grout characteristics. For solution grouts, the injectability and the grout spread of a given soil are determined by the kinematic viscosity, of the grout, viscosity, and initial gelation time. For the analysis, a completely homogeneous soil has been assumed, which is fully injectable with a solution grout, with a pump time t and a kinematic viscosity γ_i (Fig. 5B.11).

The lateral grout spread from a sleeve pipe can be estimated using the following characteristics of the horizon:

GROUTING TO IMPROVE FOUNDATION SOIL

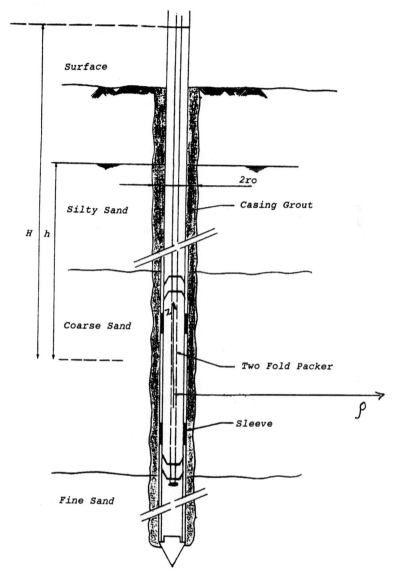

FIGURE 5B.11 Down-hole setup for solution grouting.

Hydraulic conductivity horizon for the injection product = k_i
Thickness horizon = e
Kinematic viscosity grout = v_i
Specific weight grout = σ_i
Diameter borehole = $2r_0$

When an effective pressure of $p = \delta_i H$ is exerted over the thickness of the layer (H is height of fluid column), assuming the permeation of the much lower permeability horizons above and below the injected layer is negligible, we can express that

SOIL IMPROVEMENT AND STABILIZATION

$$\text{flow } Q = 2\pi\rho e k_i \left(-\frac{dh}{d\rho}\right)$$

$$-dh = \frac{Q}{2\pi k_i e} \frac{d\rho}{\rho}$$

$$h_1 - h_2 = \frac{Q}{2\pi k_i e} \ln \frac{R}{r_0} \tag{5B.1}$$

R is the distance from the middle of the sleeve pipe to where the influence of the grouting pressure is not being felt. R cannot be calculated and has to be estimated or empirically determined. The following table provides a good indication for R.

Type of soil	R (m)
Gravel	4–6
Coarse sand	6–10
Medium sand	4–6
Fine sand	2–4
Silty sand	1–2

Since $kv = k_i v_i$ (k and v pertain to the hydraulic conductivity of the horizon and the kinematic viscosity of water, respectively), the Eq. (5B.1) can be rewritten as

$$h_1 - h_2 = \frac{Qv_i}{2\pi kve} \ln \frac{R}{r_0} \tag{5B.2}$$

Since

$$h_1 = Z_1 + \frac{P_1}{\delta_i} = Z_1 + \frac{\delta_i H}{\delta_i} = Z_1 + H$$

and

$$h_2 = Z_2 + \frac{P_2}{\sigma_{water}} = Z_2 + \frac{\delta_w h}{\delta_w} = Z_2 + h$$

and, since in a horizontal plane $Z_1 = Z_2$

$$h_1 - h_2 = H - h$$

Equation (5B.2) can be rewritten as

$$H - h = \frac{Qv_i}{2\pi kve} \ln \frac{R}{r_0}$$

or

$$\delta_i(H - h) = \frac{Q\delta_i v_i}{2\pi kve} \ln \frac{R}{r_0}$$

or

$$P_{\text{effective}} = \frac{Qv_i \delta_i}{2\pi kve} \ln \frac{R}{r_0} \tag{5B.3}$$

This is the so-called layer equation, reflecting the significance of the thickness of the soil layers during a grouting process.

The effective injection pressure required to inject a flow Q increases when the layers are thinner and the viscosity is higher. The volume of grout, dV, injected in the interval dt is given by

$$dV = Q\, dt = \frac{2\pi k v e\, P_{\text{eff}}\, dt}{v i\, \delta_i \ln \dfrac{R}{r_0}} = 2\pi\rho e n_i\, d\rho$$

or

$$\frac{2\pi k v e P_{\text{eff}}\, t}{v_i\, \delta_i \ln \dfrac{R}{r_0}} = 2\pi e n_i\, \frac{r^2 - r_0^2}{2}$$

or

$$r^2 = \frac{2 k v P_{\text{eff}}\, t}{v_i\, \delta_i \ln \dfrac{R}{r_0} n_i} + r_0^2 \qquad (5B.4)$$

Here r is the distance the grout will travel from the sleeve pipe through the soil or other porous medium, during a time t. The spread is theoretically limited by the set time t_{set} of the grout (t_{set} in reality is the pump time of the grout). As a result, the theoretical spread with sleeve-pipe grouting is given by the equation of Cambefort and Naudts

$$r_{\text{spread}} = \sqrt{\frac{2 k v P_{\text{eff}}\, t_{\text{set}}}{v_i\, \delta_i \ln \dfrac{R}{r_0} n_i} + r_0^2} \qquad (5B.5)$$

The in situ hydraulic conductivity of a horizon is determined in the lab and compared with the in situ measured value (Lugeon test). When the head losses to squeeze the grout through the matrix of the soil or structure, at the maximum tolerable injection pressure, become closer to this value, the flow reduces and eventually drops below uneconomical pump values. Because the reduced flow also causes the grout to stiffen up slowly, thereby increasing in viscosity, the hole will refuse to accept grout.

In practical terms, uniform (homogeneous) soils with a hydraulic conductivity greater than 8×10^{-4} cm/s (3×10^{-4} in/s) are injectable in a predictable fashion, using solution grouts, provided the grouting takes place in multiple stages, applying gradually increased grouting pressures.

5B.2.1.2 Soil Grouting with Suspension Grouts. The injectability of soils, using a particulate or suspension grout, is more difficult to predict. As a rule of thumb, only very coarse soils (coarse sand and up) can be grouted in a predictable fashion using a stable, balanced regular cement-based suspension grout. The minimum permeability of the soil has to exceed 0.1 cm/s (0.04 in/s).

As a rule of thumb, balanced, microfine cement-based suspension grouts are suitable for the grouting of soils with less than 5% fines and a hydraulic conductivity greater than 10^{-3} cm/s (0.4×10^{-3} in/s). Note that very dense, homogeneous soils with lower permeability but with virtually no fines can be grouted as well. In a full-scale field test, Heenan and Naudts demonstrated that a well-compacted sand blanket, with an average hydraulic conductivity of 8×10^{-4} cm/s (3×10^{-4} in/s) was perfectly injectable, creating uniform grout cylinders.

The equations to calculate the grout spread as established for solution grouts could be used to estimate the grout spread for microfine cement-based suspension grouts, but some interpretation as well as verification and adjustments in the field are absolutely necessary. A small grout-test program, if at all possible, is very much in order. Grouting should be followed by excavation to verify the grout spread and the uniformity of the grout cylinders. Section 5B.3.1 provides specific information regarding the various factors governing the injectability of soils with regular and microfine cement-based suspension grouts.

5B.2.1.3 Maximum Pressures and Limitations of Permeation Grouting.
Vertical hydrofracturing (claquage) usually occurs if the grouting pressures exceed the horizontal stresses, predominantly induced by the overburden. *Controlled hydrofracturing* (fracturing in conjunction with permeation grouting) is appropriate, provided the smallest movement of the structure is immediately detected and the soil is injectable with the suspension grout (but at lower pump flows). A multiple-phase soil grouting program also must be executed to obtain a uniformally grouted soil.

Theoretical considerations by French researchers have produced equations providing the pressure at which hydrofracturing occurs. Since the soil is virtually never homogeneous, the author's experience is that these equations are of no practical use in the field. It is much better to perform and monitor (X-Y recorder) a controlled fracturing test to determine the effective fracturing values. Controlled microfracturing is a natural part of a proper permeation-grouting exercise. Section 5B.3.1 provides more detailed information about the injectability and behavior of suspension grouts.

If no properly applied grouting techniques are utilized, soil permeation grouting is not feasible. Sleeve pipes embedded in casing grout are necessary to construct a quality product in a controllable fashion. Too often, contractors resort to makeshift sleeve pipes or even dare to use perforated pipes. In North America some contractors still resort to "upstage soil grouting," a totally uncontrollable method leading to unpredictable results which should not be accepted under any circumstances. Papers by Abidi, Barnes, and Young[25] and Aziz and Levay[26] remind us that there are different views on grouting methods.

5B.2.2 Hydrofracture Grouting

With this process, the ground to be treated is deliberately split by injecting predominantly suspension grouts, either at higher flows (and, consequently, pressures) than the soil channels can accept or into capillaries too fine to establish a grout flow (Fig. 5B.12).

Grout "wings" or lenses are formed from the injection point, fracturing the soil. Those grout sheets can extend a long way from the pressure source. The lenses consist of pure grout and are thickest near the grout pipe. Hydrofracture grouting supposedly creates some kind of a compres-

FIGURE 5B.12 Hydrofracture grouting.

sion of the soils between the grout blades. If the process is repeated several times, new lenses are formed between the old wings, and some kind of soil improvement can be achieved, especially if the subsequent loading is not excessive. Otherwise, the "wings" shear off, and the benefits of the program are virtually lost.

Hydrofracture grouting definitely is suitable to reverse disturbance, to fill unconnected voids, but the extra consolidation of the soil is difficult to predict or to control from an engineering standpoint. The potential danger of damaging (uplifting) structures should not be underestimated. Also, when grouting near walls (of underground structures such as tunnels or underground parking), the high pressures could potentially cause excessive local stresses in these structures.

Hydrofracture grouting has been practiced predominantly by the charlatans in the grouting industry, although some reputable French contractors toyed with this technique during the 1960s. It is no longer justifiable to exploit this technique deliberately, except in some uncommon situations.

It should be noted that, during a regular permeation grouting program, soil hydrofracturing often takes place. When grouting coarse strata with regular cement-based suspension grouts, minor fracturing will occur when the in situ hydraulic conductivity is below but near 0.1 cm/s (0.04 in/s). When grouting microfine cement-based suspension grouts, in media which are marginally injectable by permeation grouting, micro hydrofracturing occurs all the time. Permeation grouting in conjunction with micro hydrofracturing is a generally accepted process and works very well if the grouting is executed in several stages or "passes" (i.e., the same sleeve pipes are grouted several times).

The initial hydrofracturing occurs via semivertical planes and is not harmful, especially if it results in feeding a larger surface area of soil, in which a regular permeation grouting process can take place. The only way to assess permeation in conjunction with hydrofracturing is by continuously analyzing the grouting data as they appear on the X-Y (pressure-flow) recorder. A hydrofracturing phenomenon is characterized by a sudden and significant increase in "apparent Lugeon value" (i.e., drop in pressure at constant or increased flow rate). If permeation grouting occurs along the grout wings, the pressure at a constant grout flow will only gradually recover. The consequent decrease in apparent Lugeon value should be virtually the same after each occurrence. If this is not the case, more extensive "mere hydrofracturing" is taking place. Grouting requires continuous assessment and decision making. This technique should not be practiced by the "journeyman" grouters (most engineers fall in this category), because the chances of not accomplishing the intended goals are very high.

One of the latest developments in hydrofracturing grouting is the so-called controlled fracture grouting. Even synthetic fibers are added to the grout, resulting in higher shear resistance of the grout as well as added tensile strength. "Frac-grouting" has been described briefly earlier in this section. Only time will tell if this is just one more scam or if this technology will evolve from its infant stage into an economical, reliable technique. For a field application see Fig. 5B.12.

5B.2.3 Compaction Grouting

This technique has most often been described as a "uniquely American" process, but in fact it has been used in a number of countries, before and after the term *compaction grouting* was invented. Compaction grouting is one of the crudest grouting techniques and is rooted in the concrete pumping technology.

Instead of pumping concrete to make slabs in high-rise buildings or bridge decks, people have resorted to pumping concrete as crude emergency measures to support footings and to prevent settlement. Sophistication as well as some engineering principles have continuously been added since the first basic applications. Compaction grouting has been well promoted by the people in the business.

Classic compaction grouting consists of pumping very stiff concrete mixes at high pressures, up to 3.5 MPa (508 psi), at discrete locations, to densify soft loose or disturbed soil. Unlike hydrofracture grouting, the grout is intended to form a dense and coherent bulb without fracturing the surrounding soil. As a result of the pressure applied by the "concrete mushroom" on the surrounding soil, compaction takes place, and the compressibility of the compacted soil is considerably reduced.

Quite often, this type of compaction grouting is used to lift structures. The contractor forms one bulb on top of the other, until the "new foundation system" starts to push up the structure. In the earlier stages, at least in the United States, this process was used exclusively on shallow foundations. Nowadays it is used for liquefaction control below substantial buildings and dams.

Although compaction grouting does have practical and technical limitations, its popularity continues to grow, mainly due to the active and professional promotion in the technical press and conferences by dedicated, specialty contractors. Note that other grouting or piling techniques are often equally, if not better, suited to obtain the desired results, but an increasing number of engineers resort to compaction grouting. Warner,[27] Graf,[28] and Bandimere[29] are some of the most vocal and dedicated specialists in this field. They have provided valuable contributions to mix designs, rheology, and mechanism of compaction grouting.

Greenwood and Hutchinson[30] elaborate on *squeeze grouting,* whereby the fluid grout initially induces localized stresses centered around the injection point but then acts on a larger ground volume as the hydrofracture mechanism progresses in a controlled fashion.

Compensation grouting as described by Essler and Linnay[31] is another similar and recent offshoot of this concept and is apparently gaining popularity in Western Europe and Canada.

A spinoff of the patented anchor and pin-pile technology developed by Heenan and Naudts pertains to compaction grouting. In order to transfer loads to deep strata, a series of high-strength, high-flow geotextile bags are strapped around sleeve pipes and inflated one at a time, locking in considerable horizontal stresses in the soil strata. The geotextile prevents the soil fracturing, causes the pressure filtration of the grout, and compacts the surrounding soils. This method is particularly interesting if the soil investigation is performed with penetrometer testing equipment.

In areas which are still consolidating and which are prone to cause negative friction, the columns are kept to the minimum size to transfer the load. The lateral bearing capacity generated in the competent soils can be calculated, and the compaction grouting program can be performed with standard grouting equipment, via sleeve pipes, from the bottom up, with a low-viscosity suspension grout, without causing any additional settlement. Contrary to the common belief (especially in North America) that the compaction grouting has to extend to rock, a compaction grouting program can be executed in the absence of a "rock-hard" surface providing excellent end bearing. Enough shaft friction can be mobilized to provide the necessary bearing capacity.

The upper bags in the various sleeve pipes can be grouted simultaneously to generate a controlled lifting of the structure, if this is one of the goals of the operation.

5B.2.4 Jet Grouting

5B.2.4.1 General. Jet grouting developed by the Japanese during the early 1970s is one of the youngest ground treatment methods. Jet grouting uses high-velocity jet streams under very high pressure to cut, remove, and cement the soil in situ with a grout (usually a cement-based suspension grout, but occasionally solution grouts are used) (Fig. 5B.13). Depending on its grain size and consistency, the soil is removed to a certain extent by air-water jetting and simultaneously mixed and partly replaced by grout injection. The ASCE Geotechnical Engineering Division Committee on Grouting defined jet grouting as a "technique utilizing a special drill bit with horizontal and vertical high speed water jets to excavate alluvial soils and produce hard impervious columns by pumping grout through the horizontal nozzles that jets and mixes with foundation material as the drill bit is withdrawn."

Only three basic methods of jet grouting are being used, and these may be generically classified as follows:

- J1—one-fluid system, i.e., grout, in which the jet simultaneously erodes and injects. Involves only partial replacement of the soil.
- J2—two-fluid system, using a cement jet inside a compressed air cone. J2 gives a larger column diameter than J1 and gives a higher degree of soil replacement.

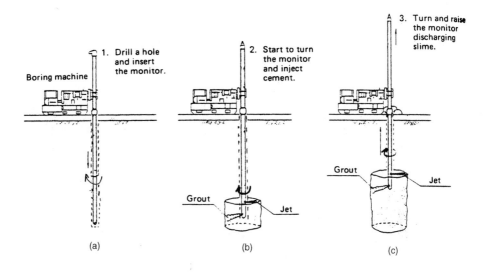

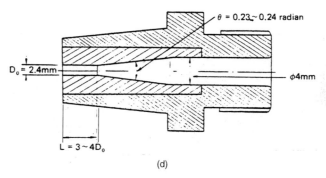

FIGURE 5B.13 Jet grouting: (*a*) boring the hole; (*b*) rotating grout pipe and injection; (*c*) rotating and raising injection pipe; (*d*) grout jet (typical).

- J3—three-fluid system, upper ejection of water inside an air envelope for excavation, with lower jet emitting grout for replacement of jetted soil.

The single-, double-, and triple-fluid systems of jet grouting all require two essential construction steps: First a vertical guide hole or grout hole must be drilled accurately down to the required depth. Then jet grouting follows, usually from the bottom to the top (Fig. 5B.14)

	Jet-Grouting Systems	
One-fluid system	Two-fluid system	Three-fluid system
Grout	Air	Air
	Grout	Water
	Air	Air

Source: From J. P. Welsh (1991).

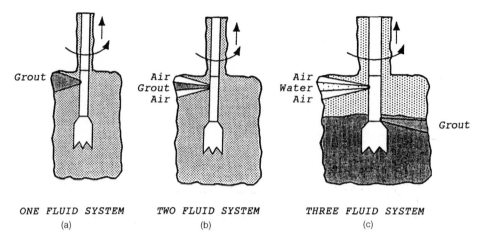

FIGURE 5B.14 Variations to jet grouting: (*a*) one-fluid system; (*b*) two-fluid system; (*c*) three-fluid system. (*From Joseph P. Welsh, 1991.*)

Major applications of jet grouting are:

- Underpinning of existing structures
- Formation of water cutoff walls or diaphragms [to 10^{-6} to 10^{-9} cm/s (0.4×10^{-6} to 0.4×10^{-9} in/s)]
- Foundation soil consolidation for new structures, embankments, and retaining walls
- Soil consolidation for the excavation of shallow tunnels
- Excavation support for open cuts and shafts
- Landslide stabilization

5B.2.4.2 General Operational Features of the Three Generic Methods. Table 5B.1 provides a general summary of operational parameters and grouted soil strengths. However, note that these data represent a typical range, and the details on any one project are dictated by a number of variables. For the J1 system, six jetting parameters must be determined: grout pressure, number and size of nozzles, w:c ratio, drill-rod lifting rate, and rotational speed. For J2, 2 additional variables are air pressure and delivery rate, and for J3 there are 10 parameters (the 7 for J2 plus water pressure, water jet nozzle size, and number).

The simplest and most straightforward of the three forms of jet grouting is the *single-fluid system*. This system relies on the high-velocity jet stream to cut, remove, and mix the in situ soil. It is more of a mix-in-place technique rather than replacement grouting method. The single-fluid system usually ejects the least amount of soil from the ground to form an in situ "soil-crete" cylinder. The diameters of the grout columns for the single-fluid system range from 40 to 80 cm (15.75 to 31.5 in) in cohesive soils, and 50 to 120 cm (20 to 47 in) in granular soils. The technical limitations of the single-fluid system are strongly governed by equipment. Pumps must be capable of delivering cement grout flow rates between 150 to 475 L/min (40 to 125.5 gal/min), at about 60 MPa (8702 psi), for one to eight hours of continuous operation. In J2 and J3, the air also aids evacuation of the debris, but as the jet is rotated from vertical to horizontal, the air lifting efficiency decreases. Therefore, J1 is used almost exclusively for horizontal jet grouting.

The most distinctive feature of the *double-fluid system* (J2) is the injection of high-velocity grout with a cone of compressed air. By introducing compressed air around the cutting jet, the double-fluid system is capable of creating columns about twice as large as those made using the single-fluid system, mainly because the dislodged soil is more efficiently removed from the area of cutting by the bubbling action of the compressed air lifting soil debris to the surface.

TABLE 5B.1 Typical Range of Jet-Grouting Parameters and Soilcrete Formed Using the Single-, Double-, and Triple-Fluid Systems

Jetting parameter		J1	J2	J3
Injection pressure				
Water jet	(bars)	PW	PW	300–550
Grout jet	(bars)	300–550	300–550	10–40
Compressed air	(bars)	Not used	7–17	7–17
Flow rates				
Water jet	(L/min)	PW	PW	70–100
Grout jet	(L/min)	60–150	100–150	150–250
Compressed air	(m^3/min)	Not used	1–3	1–3
Nozzle sizes				
Water jet	(mm)	PW	PW	1.8–2.6
Grout jet	(mm)	1.8–3.0	2.4–3.4	3.5–6
Number of water jets		PW	PW	1–2
Number of grout jets		2–6	1–2	1
Cement grout $W:C$ is 0.80:1 to 2:1				
Cement consumption	(kg/m)	200–500	300–1000	500–2000
	(kg/m^3)	400–1000	150–550	150–650
Rod rotation speed	(rpm)	10–30	10–30	3–8
Lifting speed	(min/m)	3–8	3–10	10–25
Column diameter				
Course-grained soil	(m)	0.5–1	1–2	1.5–3
Fine-grained soil	(m)	0.4–0.8	1–1.5	1–2
Soilcrete strength				
Sandy soil	(MPa)	10–30	7.5–15	10–20
Clayey soil	(MPa)	1.5–10	1.5–5	1.5–7.5

Note: PW = Water jets only used during prewashing.

The major drawback of the double-fluid system is that the cemented soil has a high air content. The soilcrete strength is usually the lowest in comparison to each of the two other jet-grouting systems. J3 is the most involved and sophisticated jet-grouting system due to the simultaneous injection of three different "fluids"—air, water, and grout (Fig. 5B.15). The injection of the three fluids causes more soil to be removed, and, therefore, the triple system constitutes almost a full replacement of the in situ soil with grout. The column sizes formed using the triple system are usually the largest in diameter due to the separate injection of air, water, and grout. Typical column sizes in sands vary from 50 to 300 cm (20 to 120 in) in diameter and from 50 to 150 cm (20 to 60 in) in clays.

5B.2.4.3 General Advantages and Limitations of Jet Grouting. One of the primary advantages of jet grouting over the other grouting techniques is that all jet-grouting methods can be used to treat a wide range of soils from gravels, sands, and heavy clays. Large-diameter columns (50 to 300 cm or 20 to 120 in) can be created starting from a relatively small diameter drill hole (9 cm or 3.5 in). Obstructions such as timber piles or small boulders can be bypassed or incorporated into the soilcrete. Jet grouting can be conducted from any suitable access point and can be terminated at any elevation, providing treatment only in the target zones. It can be conducted vertically or subhorizontally, above or below the water table. General disadvantages of all three jet-grout-

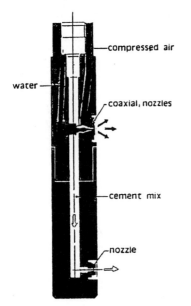

FIGURE 5B.15 J3 (three-fluid) grouting system. (*From Tornaghi and Perelli Cippo, 1985.*)

ing methods relate to the fluid mechanics requirement that there must be continuous flow of cuttings from the point of injection (nozzle) up to the surface, where atmospheric conditions exist. If a continuous material flow exists, the pressure near the point of injection is only slightly higher than the hydrostatic pressure caused by the slurry of cement and cuttings. If the path is restricted, then the pressure at the point of injection can rise to the pump pressure (up to 60 MPa or 8702 psi). These large pressures hydrofracture the soil, causing severe lateral soil movements and ground heave. If a continuous slurry flow can be maintained, then the jet-grouting method is a safe and high-velocity erosion process. Jet grouting is messy and requires a relatively large volume of spoil to be handled at the surface. Especially when polluted soils are to be treated, the treatment of the spoils can become an expensive exercise. Also, if the ground-water flow velocities are high, the fluid soilcrete may experience local removal of the cement prior to its stiffening and thus unevenness in quality.

5B.2.4.4 Selection of Jet-Grouting Parameters. The influence of nozzle size, pressure, grout flow rate, jet stream velocity, type of grout, monitor rotation, mechanical power requirements, and withdrawal speed have been widely investigated for various soils and hydrological conditions.

In general, the radius of influence is mainly related to the waiting time between the withdrawing of the drill bit, i.e., to the length of time spent treating the same level. An increase in pressure enhances the fracturing effects of the jet, reducing the time necessary to inject a given quantity of grout.

5B.2.4.5 Factors Influencing Soilcrete Strength. These factors play a role in soilcrete strength development:

1. *W:C ratio.* As the water content of the soil increases, the soilcrete strength decreases.
2. *Cement factor* (*CF*). Strengths increase with an increase in the cement factor (weight of cement per volume of soilcrete).
3. *Soil type and grain-size distribution.* Sands and gravel tend to produce stronger soilcrete, whereas clay and silt particles tend to produce weaker material.
4. *Age.* As soilcrete cures, the strength increases, but usually at a slower rate than concrete.
5. *Type of jet-grouting system used* (*single-, double-, or triple-fluid system*). The type of system can have a major influence on soilcrete strength. However, the appropriate jet grouting system can be selected for the project, which is a decision made by the owner or the engineer.

5B.2.4.6 Uncertainties Related to Soilcrete Strength. The difficulty in predicting the in situ soilcrete composition is due in part to the process. Predicting the in situ makeup is difficult, especially cement factor and W:C ratio of the soilcrete, because the grout jet stream is used to both excavate and strengthen the in situ soil as well as to mix the cement into the soil. The major uncertainties in predicting the in situ soilcrete composition relate to the following items:

1. The soil component of soilcrete is dictated by the in situ soil conditions, which may be variable. Also the grain-size distribution of the soilcrete usually is *not* the same as the initial soil

distribution, because fine sand and silt particles are more easily eroded from the matrix during jet grouting. However, removal of fines from the soilcrete is beneficial for increase in strength.
2. The final water content of the soilcrete in the column is influenced by four major factors, which include:
   a. In situ water content of the soil; the higher the soil water content, the higher the W:C ratio of the soilcrete. High soil-water content can be partially offset by using a grout with a higher cement content.
   b. W:C ratio of injected grout specified.
   c. Permeation of water out of the column, due to a gradient setup by residual jetting pressure in the column. This factor is especially important in sandy soils.
   d. Self-weight consolidation of the wet soilcrete, which causes a reduction in the W:C of the soilcrete.
3. The W:C ratio of the soilcrete and the in situ cement factor are a function of the amount of cement and water injected and proportion of soil, water, and cement ejected in the cuttings. The volume (column diameter) of the in situ soilcrete has a strong influence on both parameters. Only the amount of cement injected is under direct control. *Mass balance approaches* have been developed (Kauschinger[32]) for estimating the amount of soil, water, and cement left in situ, given measurements for the injected grout and cement content of the cuttings.
4. Soilcrete is a nonhomogeneous mixture that can be caused by several factors:
   a. Poor control over jetting parameters such as variable jetting pressure, worn nozzles, lift rate, and step size suggests the need for real-time data acquisition.
   b. Insufficient mixing of in situ soil with grout will cause the soilcrete strength to be the highest along the central axis and lower along the outer perimeter of the column.
   c. Soil conditions vary with depth

5B.2.4.7 Estimating Soilcrete Composition. Presently, there are no direct methods to verify the size of each soilcrete column formed during production. The most common method of measuring the size and strength of the jet grout column is by field trials. In a different approach to direct measurements, the size and composition of the soilcrete column can be estimated using principles of mass conservation. The analysis of the cutting composition can be entered into the mass balance equations (Kauschinger[32]) for fast calculation of the equilibrium between injected grout, ejected soil, water, cement, and the material deposited in the column. This information can then be used to estimate if the correct amount of cement, soil, water has stayed in the column.

The major component of the cuttings which has to be determined is the amount of cement. The heat of neutralization test was determined as being the most simple, quick, and accurate technique for estimating the amount of cement. The test requires reagents (sodium acetate, glacial acetic acid) in certain proportions to be added to the cuttings. A digital thermometer is used to measure the temperature rise as the cement becomes neutralized. The rise in temperature is related to the percentage of cement in the cuttings through the use of a calibration curve.

The major assumptions relied upon when developing the mass balance formulation include:

1. Soilcrete in the jet grout column is homogeneous.
2. All mass deposited in situ creates a uniform, cylindrical body, i.e., no hydrofracture or permeation through the soil mass.
3. Drainage of water from jet grout column during self-weight consolidation is not theoretically taken into account.
4. Cuttings sampled at the surface are representative of cuttings produced at point of injection.

5B.2.4.8 Theoretical Verification of the Size of Column. A rather theoretical approach has been developed by Kauschinger[32] based on the mass-balance approach. The total unit weight of the column, γ, is defined as

$$\gamma = \frac{W}{V} \tag{5B.6}$$

For cylindrical columns, the diameter of the column, D, equals

$$\left(\frac{4V}{\pi h}\right)^{1/2} \qquad (5B.7)$$

In which h equals the height of a cylindrical element of a soilcrete column. The total weight of a segment of a soilcrete column W equals

$$W = W_{cem} + W_{wat} + W_{soil} \qquad (5B.8)$$

D can be calculated from Eqs. (5B.6), (5B.7), and (5B.8). The unit weight of the hardened soilcrete can easily be measured. Only W needs to be determined:

W_{cem} = weight of the cement injected minus the weight of the cement ejected, or

$$W_{cem-inj} - W_{cem-ej} \qquad (5B.9)$$

$$W_{wat} = W_{wat\ in\ situ} + W_{wat-inj} - W_{wat-ej} \qquad (5B.10)$$

$$W_{soil} = W_{soil\ in\ situ} - W_{soil-ej} \qquad (5B.11)$$

Evaluation of In Situ and Injected Quantities. $W_{cem-inj}$ is easily derived from the total grout weight injected in a particular interval with height h. The total grout weight equals the volume of cement grout injected, divided by the specific gravity of the grout (as determined by the Baroid mud balance test). The water:cement ratio in turn determines the amount of cement ($W_{cem-inj}$) and the amount of water injected ($W_{wat-inj}$).

The soil in situ, affected by the jet-grouting process, is given by

$$W_{soil\ in\ situ} = V(\pi D^2 h/4)\, \gamma_{t^{in\ situ}} \left(\frac{1}{1+w}\right)$$

where ω = the in situ water content and
$\gamma_{t^{in\ situ}}$ = the in situ unit weight of the soil
$W_{wat\ in\ situ} = \omega\, W_{soil\ in\ situ}$

Evaluation of Ejected Quantities. The cement content can be established with the *heat of neutralization method*. The volume of the ejected spoils W_{cem-ej} to form a column, with a height h known, is easily determined. By drying out a sample of the spoils (microwave oven), the water content can be determined, provided a correction factor to compensate for the chemically bound water to the cement is added. Knowing the cement and the water content of the sample, the soil content can easily be calculated based on specific gravity, and W_{wat-ej} as well as $W_{soil-ej}$ are established. All the established data, using absolute numbers or one variable, D, only, are substituted in Eq. (5B.8) from which the theoretical diameter D of the section grout column with height h can be determined.

5B.2.4.9 *Characteristics of Jet-Grouted Soils.* The result of a treatment, in terms of uniformity and mechanical properties, depends on a number of interdependent factors concerning the soil properties and the jet-grouting parameters. In any case, experience has shown that for a given type of cement mixed with a given type of soil, the strength may be correlated statistically to the final cement:water ratio C/W, on condition that a fairly homogeneous soil-grout mixture could be obtained (Gallavresi[6]).

For saturated cohesive soils treated by the single-fluid procedure, Fig. 5B.16 represents the ranges of final C/W as a function of grout content, with four different cement:water contents (C_m/W_m) of the grout and three different water contents of the native soil. The C/W ratio of the treated soil is defined through indirect and statistical method, by measuring the final strength of the column proportional to the cement content.

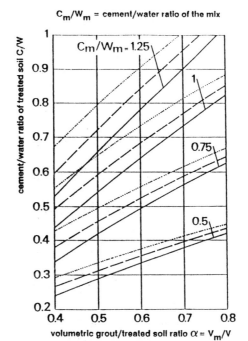

type	volumetric characteristics of native soil		
	saturated bulk density γ_t (t/m³)	dry density γ_a (t/m³)	water content W_t (%)
T1 ········	2.037	1.630	25
T2 — —	1.833	1.309	40
T3 ———	1.618	0.971	67

Saturated cohesive soils influence of treatment on final cement/water ratio

FIGURE 5B.16 Saturated, cohesive soils influence of treatment on final cement:water (C:W) ratio.

Figure 5B.16 shows the importance of cement content when appreciable drainage is unlikely to occur, that is, when soil permeability is low. In contrast, in permeable granular soils, the generally lower water content and drainage effect during the treatment are factors involving final C/W ratios even higher than C_m/W_m.

Figure 5B.17 also relates to fine cohesive soils, representing unconfined compressive strength R with two scales limiting the statistical range of maximum frequency of the strength index

$$R_o = \frac{R}{(C/W)^n}$$

The exponent n may range from 1.5 to 3, but in most cases of inorganic soil it may be assumed $n = 2$ and hence

$$R_o = \frac{R}{(C/W)^2}$$

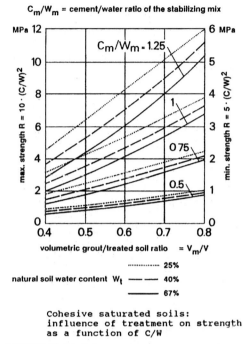

FIGURE 5B.17 Influence of treatment on strength as a function of C:W (cohesive, saturated soils).

The strength index R_o depends mainly on the type of cement, the physical properties of soil (mineralogy, grain-size distribution, plasticity) and, obviously, curing age.

Figure 5B.17 refers to treated cohesive soils after 1 to 2 months' curing and shows the strength ranges as a function of cement content in the grout, natural water content of the soil, and volumetric ratio of grout to treated soil; this ratio may be evaluated on samples of treated soil and therefore estimated for design according to case records.

Figure 5B.18 shows some average statistical data of strength as a function of cement content resulting from analytical evaluations. The involved soils, treated by single-fluid system, range from silty clays to silty sands; the three-fluid data refer to groups of columns in peaty soil.

5B.2.4.10 General Field Quality Control and Assurance Procedures. Most engineer-designed projects addressing difficult and challenging problems require that field quality controls be exercised during the preconstruction phase and during all phases of construction to ensure that the implementation is in accordance to the design criteria.

Preconstruction QC/QA. During preconstruction, the following items should be evaluated:

1. Subsurface conditions
   a. Soil stratigraphy (continuous sampling in critical areas to be jet-grouted)
   b. Soil/water chemistry to ensure that the grout and in situ conditions are compatible
   c. Elevation of ground water, and presence of hydraulic gradient; if seepage velocity is in excess of 6 cm/s, then a "heavy" grout should be used, or additives should be added to ensure that grout is not washed out
2. Laboratory tests
   a. Unconfined strength testing may be needed to determine the amount of injected cement needed for producing a prescribed soilcrete strength. The values of unconfined strength

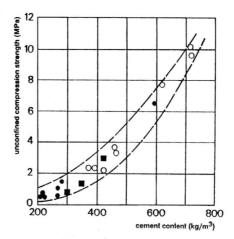

FIGURE 5B.18 Strength as a function of cement content: Mean experimental data.

versus W/C ratio, age, and cement factor can be used to help determine the composition (soil, water, cement) of the cuttings and of the column.
  b. Calibration curves should be established to estimate the amount of cement, soil, and water in the cuttings and the column.
3. Field trials. Presently no reliable means exists for calculating the size of a jet-grouted column over its entire length. Most often field trials are needed to determine the optimum jet-grouting parameters: jetting pressure, lifting speed, grout flow rate to produce the required column diameter, and soilcrete strength. However, if grouted columns are below the water table, then the submerged column characteristics must be determined. This may require drilling observation wells at various distances from the jet-grouting holes to determine the time required for the jet stream to cut the distance from the jet-grout hole to the observation well.

5B.2.4.11 Field Monitoring of Jet-Grouting Operation. The PAPERJET System has been developed for controlling the field operations when the jet grouting technique is utilized. It permits a continuous monitoring of the operations, enabling a real-time definition of the effectiveness of the treatment. The system executes the following operations:

- *Real-time* monitoring of drilling and jet-grouting parameters and their visual display in graphical form
- Printing in real time of all acquired numerical values
- Real-time storage of the same data in magnetic support
- Possibility of deferred plotting (and printing) of the variation of the parameters with depth

The real-time monitoring is useful, since, if some mechanical malfunctioning takes place (e.g., a drop of pressure or of flow rate, or clogging of the nozzles), the rig operator can immediately detect it on the display and take the necessary steps. Using this system, the site engineer is informed of any change in the soil condition and can accordingly modify the jet-grouting process. For example, if the system during drilling detects the presence of a cavity at a certain depth, the site engineer could decide to increase the treatment in that zone. In addition, the PAPERJET system permits the use of the data stored on magnetic support for the preparation of graphical outputs showing the variation of the measured quantities with depth.

5B.2.4.12 Some Final Thoughts on Jet Grouting.   Jet grouting has matured to a predictable and reliable soil modification technique. Independent developments in various places around the world have provided sufficient fundamental information. Much has been published that demonstrates that the result from excellent field trials can be duplicated in the actual projects. Field trials remain necessary to establish a number of parameters and to provide the necessary data for project design and subsequent performance. The emphasis is changing to quality control and quality assurance on an industrial scale, and this will feature, as far as is practical, the use of automated drilling, grouting, and recording systems.

Stimulated by the positive developments, and an increasingly impressive list of successful case histories, it appears that this technology will continue to enjoy growth around the world for decades to come.

5B.2.5 Unusual Approaches

5B.2.5.1 Precipitation Grouting.   It is amazing that the largest grouting project ever to take place on this planet predominantly involved an "alternative" grouting process called *precipitation grouting*. Precipitation grouting pertains to the injection of a "grout" which reacts with the existing ground water, triggering a precipitation of flocks or crystals, filling the channels in the medium to be treated.

None of the "reaction partners" (ground water nor grout) is curing on its own, but a precipitation of particles, forming a conglomerate, takes place under given pressure and temperature conditions.

Circumstances must be right to engage into a precipitation-grouting program. First, the ground water must be flowing, otherwise the mixing between the reaction partners cannot take place. Second, the ground water must contain certain minerals and/or metal ions from which crystals and complex mineral conglomerates can form.

It was observed that in the same potash mine, two different brines reacted with each other and precipitated salt and gypsum crystals. This precipitation phenomenon was discovered by accident, when pump lines clogged because of crystal growth. This amazing natural phenomenon was utilized to our benefit to control a major brine in-rush into a large potash mine in Saskatchewan. Inflows peaked at 12,000 gal/min (45 m^3/min). A calcium chloride brine was injected into the fast-flowing predominantly sodium chloride brine, triggering the growth of competent salt and gypsum crystals in the fast flowing brine, reducing the flow channels and curtailing the in-flow. The crystal growth choking off wide flow paths can be compared with the growth of zebra mussels in pipe lines.

The "grout" that causes the precipitation must not necessarily be a fluid. It was noted in water treatment plants that aeration of industrial wastewater (to increase the chemical oxygen demand) triggered substantial precipitation of metal flocs, causing clogging of the filters. Hence the Niers process was developed to prevent this from happening. The author experimented with the introduction of air into the polluted ground-water flow, below tailings dams of a mine. Iron oxide flocs clogged the soil matrix, turning the soil into an iron-rich sandstone. This patented method should be suitable to be executed, via sleeve pipes, in the same fashion as a regular precipitation-grouting project.

Field Application: The In-Flow into a Large Potash Mine in Saskatchewan

HISTORY   The general geology of the Saskatchewan potash basins was described by various authors during the first international symposium on potash mining in Saskatoon in 1983.[34] The prairie evaporites (salt formation) hosting the potash-rich horizons are located approximately 1000 m (1094 yd) below surface. Above the prairie evaporites, the Second Red Beds (a watertight mudstone formation, in which the cracks are sealed with salt), separate the water-bearing Dawson Bay formation from the salt horizon. The Dawson Bay formation contains several horizons with fissured and vuggy limestone, which in turn host the brine aquifer.

For two decades, potash had been mined very successfully in a very dry environment. Suddenly, a wild in-flow into the abandoned portion of a mine was detected. Specialists in a number of fields from all over the world pooled their expertise and knowledge to control this in-flow. Via a sophisticated three-dimensional analysis, the in-flow was pinpointed, and an access was mined toward the area. In the meantime, a deep-drilling program from surface both near and away from the infiltration point was carried out. The goal was to create a hydrogeological model of the in-flow.

Flow channels for the brine, from the Dawson Bay formation (limestone) through the prairie evaporites salt formation, down into the mine, were several meters in diameter, after erosion had taken place. The pressures in the producing aquifer fell from a virgin formation pressure of approximately 1250 psi (8.7 MPa) to a low of 450 psi (3.15 MPa) and were felt over an area of several square kilometers around the in-flow point.

The injection of a different type of brine (calcium chloride solution) into the fast-flowing brine triggered a major crystal growth, slowly choking off the in-flow into the mine. Formation pressures gradually recovered in the general area. Grout holes (vertical, from surface) were almost 1000 m (1094 yd) deep. Grouting specialists watched in disbelief as precipitation grouting reduced the in-flow to less than 10% of the original value, noticing how an aquifer-limited in-flow turned into an aperture-limited in-flow.

According to some of the most renowned specialists in rock mechanics, a natural hydrofracturing phenomenon resulting from a reduction in horizontal stresses in the aquifer formation had taken place and had caused the initial in-flow. The claim was made that the pressure in the aquifer had exceeded the (reduced) in situ horizontal stresses, causing (vertical) hydrofracturing. This reduction in the horizontal stresses was attributed to mining activities 20 to 50 m (66 to 164 ft) below the aquifer formation.

As a result of the continuous precipitation-grouting program, the formation pressures continued to climb and the in-flow continued to decrease, until the pressure in the formation equalled the horizontal stresses.

At this point, a new in-rush occurred in the same general area, but several hundred meters away from the first problem area. The second in-rush was tamed in the same fashion as the first, but in turn a third in-rush was triggered several months later, once the formation pressures were approaching the original values. The periodic injection of brine to control and stabilize the in-flow into the largest potash mine of the world has been an ongoing process for over 6 years.

5B.2.5.2 Reforestation and Erosion Control. During the late 1970s, several companies focused on reforestation and soil fixation processes, using solution grouts to accomplish their goal.

Polyacrylamide has been used in hydroseeding projects. The very weak gel (with long gel time) permeated the upper layer of soil and acted as a weak glue holding the soil particles together and preventing the grass seeds from being blown or washed away.

Hydrophillic polyurethanes were developed for the same purpose. Togo (Japan) and the N.V. Denys (Belgium) developed products called Uresol and Deci 162 and successfully carried out pilot reforestation projects in the Sahara desert as well in other arid, sensitive, sandy areas.

The role of the various solution grouts in reforestation applications is fixation of seeds and soil until the grass or plants have taken root, gradually taking over the protective role of the grouted soil crust. Initially, the grout prevents evaporation but allows water and moisture to percolate through the grouted crust, creating the appropriate circumstances for germination of the plant and grass seeds.

Because various solution grouts provide a variety of mechanical characteristics for grouted soil conglomerates and for surface treatment, the grout selection plays an important role. Because they have fast and adjustable reaction times and they create hard soil conglomerates, hydrophobic polyurethane prepolymers are used for military purposes to make helicopter landing pads in the desert. Maybe grouting *was* a factor in Operation Desert Storm....

5B.3 TYPES OF GROUT

5B.3.1 Cement-Based Suspension Grouts

5B.3.1.1 General Information. The days of permeation grouting, using a neat suspension grout containing only water and cement, are over. It will however take several years before the generation of the grouting dinosaurs will have died out.

Both laboratory testing, followed by mathematic modeling, as well as extensive, carefully monitored field applications have provided irrefutable evidence that the field application for unstable, cement-based suspension grouts is virtually nonexistent. Laboratory testing (e.g., Stripa, WIPP, AECL, Rodio, engineering contractors and independent researchers or universities, including our own testing and field monitoring programs) confirmed what the front runners in professional grouting experienced in the field: A properly balanced cementitious grout permeates considerably better than a thin, neat water-cement suspension grout. Moreover, balanced, stable grouts cause less secondary permeability and are more durable (i.e., lower permeability, higher resistance against erosion, chemical attack, and leachability).

The terms *stable* and *unstable* refer to decantation by gravity of particles from the suspension grout. If more than 5% bleed-water is formed, the suspension is considered unstable. *Stable grout* refers to a suspension grout in which the bleeding has been limited to less than 5% (often just using bentonite) without any further considerations about the impact on cohesion or viscosity. The term *balanced stable grouts* refers to suspension grouts with several additives, providing the desired rheologic characteristics.

Note that there is a fundamental difference between soil grouting using suspension grouts and rock and structural grouting. Soil grouting using regular cement-based suspension grouts historically had limited applications, since the pores in the soil are usually too "fine" to allow particles of a suspension grout to pass through. Microfine-based suspension grouts have enlarged the field application dramatically and have created a booming industry, more than 10 years after being introduced.

Various approaches evaluate the suitability and effectiveness for cement-based suspension grouts for soil and rock grouting. Regardless of the approach, which largely depends on the grouting background of the researcher or grouting specialist, the same conclusions are reached:

- Additives are necessary to create a balanced stable suspension grout.
- Penetrability of a grout with a given grain distribution is enhanced by reducing the cohesion and the viscosity of the mix and by increasing the resistance against pressure filtration.
- The finer the particle size in a properly balanced suspension grout, the better its amenability for treatment of media with lower permeability.
- Unstable grouts have inferior injectability and create a network of bleed-paths; their penetrability is unpredictable and erratic, and the durability of the cured grout is questionable.

The *amenability theory,* developed by this author, provides in situ evaluation of the suitability of selected suspension grouts for a given formation during the grouting operation. This information coupled to in-depth knowledge about the role of the various additives, concentrations, mixing time, and order, form the basis for in situ engineering of a grouting project. Grouting is complex and requires the full and continuous attention and skills of the experienced grouting engineer. It is most unfortunate that most engineering companies in North America resort to a "routine monitoring approach" (at best) to supervise a grouting project. It should come as no surprise that so many lousy projects are performed. (I once saw a young engineer, working for a large consulting firm, in charge of a critical environmental cutoff grouting project, study his professional engineering exams during the grouting operation.)

5B.3.1.2 Penetrability of Cement-Based Suspension Grouts for Rock and Structural Grouting.
The penetrability of a cement-based suspension grout into pores and fissures depends on four main factors: the hydraulic conductivity of the formation, the grain size of the particles in the suspension, the rheologic properties of the mix, and the evolution of the rheologic characteristics in the environment in which it is injected, under the given circumstances of pressure, temperature and chemistry.

The rheology of a cement-based suspension grout is an extremely complex matter. Researchers provide often confusing and even conflicting stories. The main reason is that most researchers are too isolated and are not in tune with the real grouting world and make or duplicate mistakes in formulating and mixing procedures or in selection of the test setup, or they simply go off on some theoretical tangent which cannot be correlated to any real-life field condition.

The research work performed by engineering contractors, is therefore often more credible and valuable. Most authors (e.g., Littlejohn,[35] Lombardi,[36] DePaoli and coworkers,[37] and Hakansson and coworkers[38] agree that the Bingham model, although rather basic, is, for all intents and purposes, the most appropriate rheological model to characterize the flow of a suspension grout.

Interparticle forces between the solids result in a yield stress that must be exceeded in order to initiate a flow. The yield stress is to be regarded as the material property that presents the transition between solidlike and fluidlike behavior.

The Bingham model states (Fig. 5B.19) that

$$\tau = C + \eta_B \left(\frac{dv}{dx}\right) \text{ or } \eta'\left(\frac{dv}{dx}\right)$$

where τ = shear stress or yield stress
C = yield value or cohesion
η_B = plastic viscosity
dv/dx = shear rate
η = apparent viscosity

It is clear that both the plastic viscosity and the cohesion influence the penetrability of a suspension grout.

Note that a stable, cement-based suspension grout does not behave like a Newtonian fluid. (Water and solution grouts are typical Newtonian fluids.) If a suspension grout remains stable (i.e., water does not separate from the cement), it behaves like a Binghamian fluid. If, however, the mix is unstable, it will separate into components, and its behavior will become unpredictable, being alternatively a Newtonian or a Binghamian fluid with internal friction. Basic experimental evidence shows that as soon as internal friction appears in the mix, the flow channels plug. The latter is the result of segregation due to gravity forces or applied pressure (pressure filtration).

Loss of water, during its travel through the medium to be injected, causes a rapid thickening of the suspension, with a consequential increase in viscosity and rigidity and formation of a *filtercake*.

To improve mix penetrability, it is necessary to increase mix stability under pressure. Lombardi[36] and Hakansson and coworkers[38] concluded that the cohesion determines the maximum distance the grout can travel in a uniform flow channel. For a cylindrical flow channel with radius r, the travel distance L is given by

$$L = \frac{pr}{2C}$$

where p = pressure applied
C = cohesion of the grout

They also concluded that the viscosity determines the pump flow.

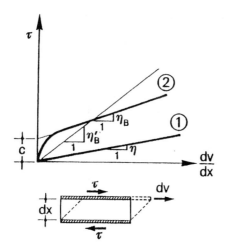

FIGURE 5B.19 Bingham model (plastic viscosity versus shear).

Circumstances permitting, the penetrability and the filling ratio should be optimized by reducing the initial cohesion of the mix and increasing the resistance against pressure filtration.

Some resistance against pressure filtration can be obtained, primarily by introducing clay-phyllo-silicates (i.e., bentonite) to the mix and to a lesser extent by using silica fumes or pozzolanic fillers. This inevitably increases the viscosity and the cohesion of the suspension. This effect has to be offset by introducing defloculators and dispersing agents.

Rodio developed the Mistra grouts, which, in essence, are bentonite cement-based suspension grouts stabilized with methacrylates. Substantial research work by this leading engineering contractor demonstrates the low cohesion and excellent resistance against pressure filtration of these types of formulations.

Other engineering contractors, although they have not performed such extensive research work, use dispersing agents, silica fume and, circumstances permitting, small percentages of fly ash, as well as polymers to obtain better resistance against pressure filtration and hence better injectability.

5B.3.1.3 Injectability of Suspension Grouts (Especially Microfine Cement-Based Grouts) into the Pores of the Soil. One of the earliest theoretical approaches to determine the injectability of suspension grouts into soil was done by the "father" of professional grouting, Cambefort.[39] Cambefort compared the *size of the particles* in a suspension grout to the *dimension of the channels between the pores*. His theoretical approach is based on the Poiseuille-Hagen law, the Darcy and Kozeny equation, and the assumption that the average diameter of the particles has to be smaller than one-third of the diameter of the capillary channels.

Thus Cambefort did not take the stability of the grout, its solid content, nor its resistance against pressure filtration into account. But he provided a valuable contribution about general injectability. He came to the conclusion that the average size of the particles d_i has to conform to the following

$$d_i < C \times \left(\frac{k}{n}\right)^{1/2} \quad (5B.12)$$

where C = constant
k = permeability coefficient
n = pore volume

More recent research work by De Paoli and coworkers confirmed that Cambefort's theoretical model is still the most reliable approach to injectability of soils with suspension grouts. De Paoli suggests replacing d_{50} of the grout with d_{85} or even the d_{95} of the particles in the grout to obtain a better correlation with the actual injectability of soils. In other words, the "tail end" of the grain-size distribution of the particles constituting the suspension grout is the most significant part.

By making the appropriate changes, to reflect the above, Eq. (5B.12) can be written as

$$d_{95} < C \times \left(\frac{k}{n}\right)^{1/2} \quad (5B.13)$$

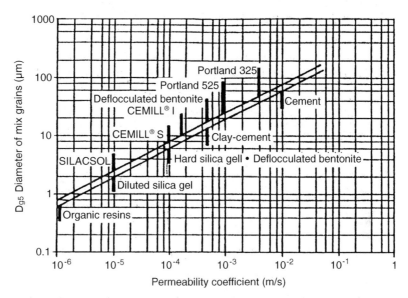

FIGURE 5B.20 Permeability limits of grouts—Cambefort (1967), lower line and DePaoli et al., upper line.

We will further refer to this as the equation of Cambefort-DePaoli. The injectability is even better than can be derived from Cambefort-DePaoli's equation if the stability is enhanced. The equation of Cambefort-DePaoli is depicted in Fig. 5B.20.

Krizek and colleagues[40–43] in their research work with microfine cement-based suspension grouts, correlated *the grain size of the soils with the grain-size of the particles* in the suspension grout. Quite a number of Krizek's findings do not agree with our own research and field experience. However, a practical but conservative number derived from his research work which generally holds true is that the $d_{10\text{soil}}$ has to be greater than 150 μm in order to be injectable with microfine cement-based suspension-grouts.

Krizek predominantly focused on one type of microfine cement, MC 500, manufactured by Onoda Cement in Japan. His team studied the influence of the mixing time on the bleeding and compressive strength of the cured grout as well as on the penetrability permeability and compressive strength of the grouted soil mass. Krizek's team provided valuable insight into the anisotropy of the grouted soils and did pioneering work in regard to the pore structure of soil injected with a cement-based suspension grout as a function of the W:C ratio as well as the sedimentation behavior of the suspended particles.

As far as injectability of soils is concerned, Krizek's team came to rather inconclusive results regarding *injectability limits*. They concluded that fine sands with more than 5% silt were not injectable with MC 500–based suspension grouts. Our field experience has repeatedly proved that a uniform soil conglomerate can be obtained, with microfine cement-based suspension grouts, when the silt concentration was near or slightly more than 10%. Thus our findings do not agree with theirs; we have suggested to Krizek that the boundary conditions of the soil test cylinders are the main cause for nonrepresentative results. Cambefort's *layer-grouting theory* provides the theoretical explanation for the rather ill-conceived testing setup.

We conclude with a quote from DePaoli: "The penetrability of the microfine cement mixes appears to be governed more by the fineness of the solid component than by the rheological and stability characteristics. These do nevertheless retain an important influence, as does the concentration of the solids in the water especially near the limit conditions of injectability".

5B.3.1.4 Resistance against Pressure Filtration of a Suspension Grout. A suspension grout should be characterized by its K_{pf} coefficient, its solid contents, and its grain-size distribution. Refer to Fig. 5B.21. The W:C factor alone does not provide relevant information. The K_{pf} coefficient is the pressure filtration resistance coefficient. It is defined as the volume of water lost in the pressure filtration test (using the API filter press) divided by the initial volume of water in the cup of the apparatus, divided by the square root of the filtration time (in minutes) under a pressure of 100 psi.

Figure 5B.20 illustrates the superior penetrability of the balanced, stable grouts. The unstable mixes have practically no cohesion nor resistance against pressure filtration. This is indicated by their high K_{pf}. Thus during a pressure grouting operation the unstable mixes will be subjected to substantial and almost immediate water loss, resulting in a quick refusal, especially if the medium to be grouted has a rather low permeability.

The stable cement-based grouts, have a considerably lower K_{pf}, but their higher cohesion limits the grout spread. By introducing dispersants, similar resistance against pressure filtration can be maintained, increasing the penetrability. (See Fig. 5B.21.)

The Mistra grout, using methacrylates to stabilize the bentonite cement suspension, have a low cohesion (which is maintained for a couple of hours) and a K_{pf} which is three to six times lower than the regular stable cement grouts. As a result, the penetrability and the grout spread obtained with balanced stable cement-based suspension grouts is substantially better than the regular stable grouts and simply cannot be compared with the unstable suspension grouts.

5B.3.1.5 Empirical Approaches to Penetrability and Injectability. It was discovered during the Stripa research project[44] in Scandinavia that a properly balanced, stable grout with viscosities between 32 and 45 s, containing water, naphthalene sulphonate, silica fume, fly ash, and cement prepared in a high shear mixer, permeated a 2-m (6.6-ft) long aperture with a width of 100 μm. A thin, neat suspension grout would not permeate apertures finer than 160 μm. Figure 5B.22 illustrates the Stripa test apparatus setup. Contrary to common (but fundamentally flawed) belief, it was demonstrated that the thick, balanced suspension grouts permeated finer apertures than the thin unstable suspension grouts. The above considerations provide the logic behind this phenomenon.

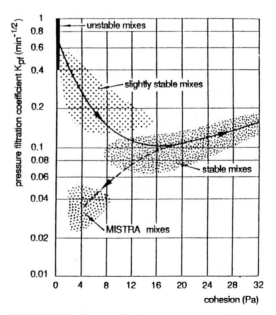

FIGURE 5B.21 Pressure filtration resistance versus cohesion.

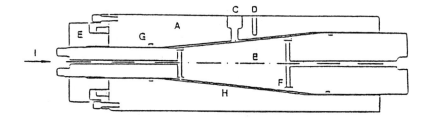

Cone-in-Cone apparatus
A   Outer cone, goldcoated brass
B   Inner cone, goldcoated brass
C   O-ring
D   Thermocouple
E   Slot expansion mechanism
F   Filter (bronze)
G   Evacuation hole
H   Sealing material
I   Water inlet

FIGURE 5B.22 Stripa test apparatus (cone in cone).

The AECL research program (Keil and coworkers[45]) unfortunately only duplicated the Stripa research findings, leaving bentonite out of the mix. The reasons for this was the concern about durability, which is a nonissue at low bentonite concentrations. The program however did some full-scale field testing in a fractured granite formation at an underground research facility. Microfine cement-based suspension grouts were injected into the granite formation at the underground facility. This full-scale field work provided (and confirmed) significant information for the industry.

Well-engineered strain measurements revealed opening and closure of the fracture zone by as much as 100 μm during the grouting operation. The significance of Cambefort's recommendations pertaining to grouting pressures and the importance of reaching a proper refusal was confirmed again. Cambefort stated that during rock grouting, some fine fissures are not filled but are squeezed tight, and therefore the formation cannot be given the chance to "spring back." In other words, a grouting operation has to be brought to an *absolute* refusal (no take at the maximum allowable pressure).

The AECL research team also discovered via "thin section analysis" of grouted specimen, removed from the grouted formation, that cracks as fine as 20 μm were filled with a microfine cement grout. This microfine cement grout contained ground-down sulphate-resistant cement, silica fume, and naphthalene sulphonate. It is important to know that a stable, rather thick (44-s marsh) microfine cement-based suspension grout was used for the full-scale test grouting.

The mechanical characteristics of the cured, neat cementitious grouts with a high $W:C$ factor are inferior. Not only does the bleeding cause major residual permeability, but the low compressive strength and high porosity of the cured portion of the grout often result in early deterioration of the grout, because of the leaching of the $Ca(OH)_2$ from the cement matrix, eventually leading to the complete deterioration of the grout curtain. Houlsby[19] and Petrovsky[46] among others, documented failures in grout curtains using unstable mixes.

In a balanced suspension grout, however, the bulk of the free $Ca(OH)_2$ is chemically locked in (after curing) because of the reaction of the pozzolanic additives in the mix with the free lime. Moreover, other additives considerably reduce the porosity of the cured grout, providing additional strength and site-specific characteristics.

Isolated bleed pockets, limited in size, could, however, still form as a result of pressure filtration and the existence of water in the formation. Therefore, multiple-phase grouting, in which secondary and tertiary holes provide escape routes for formation water (especially when under artesian pressure), forms an integral part of good grouting practice.

When grouting in water-saturated media, the formulations should be made water repellant, to minimize dilution of the grout.

The AECL researchers concluded that, in their opinion, in a balanced, stable, cement-based suspension grout, all the particles are "organized" in a particular arrangement, due to electrostatic and chemical interaction. There is limited or no segregation of the particles. The grout moves "in block" through the accessible pores, cracks, and fissures, with no or limited rearrangement of the particles. As a result, permeation in fissures is limited almost to the actual particle size of the largest grains in the suspension.

Hooton and Konecny[47] performed some focused research work regarding the penetrability of neat water cement grouts and silica fume–enhanced cement suspension grouts into fissured granite and into artificial wedge-shaped apertures. Although the researchers displayed a lack of grouting experience, their research work demonstrated that silica fume–enhanced grouts result in a superior penetrability, lower secondary permeability, and better bond. In view of the above evidence, only *stable*, cement-based grouts should be considered for most grouting operations.

In most European publications, the durability issue is not emphasized enough. Quite a number of applications involve grouting projects whereby the long-term durability is not very critical, hence the focus is more on penetrability and rheologic characteristics. Unfortunately, the same formulations are used for permanent applications.

5B.3.1.6 Grouting Philosophy (Pertaining to Rock and Structural Grouting). Professional grouting philosophy (especially in rock and in structural grouting) is based on the following principles:

1. Use regular cement-based suspension grouts, circumstances permitting.
2. Utilize a particle-size reducer, once it becomes available on the market.
3. If the particle size of the balanced, regular cementitious grouts becomes the factor that prohibits permeation, microfine cement-based grouts should be used.
4. The holes should *all* be brought to a proper refusal (i.e., *no* take at the maximum tolerated grouting pressure), allowing the grout to cure under pressure. The latter is very important. According to Cambefort (and later confirmed by AECL research work), some cracks widen and others tighten up during a pressure grouting operation. By allowing the formation to relax, fine, ungrouted cracks (squeezed tight during the grouting) are bound to open up again and cause residual permeability. For this reason, multiple-hole grouting is necessary. Once a hole has an apparent Lugeon value in the order of 1 to 3 Lu, it is further pressurized (in conjunction with other holes which are effectively taking grout), for an extra 30 min.

The long-term performance of cured grout relates directly to its mechanical characteristics, especially its porosity and compressive strength. Even a properly mixed, neat cementitious grout is an inferior product to the balanced grouts.

5B.3.1.7 Components of a Balanced, Stable, Cement-Based Suspension Grout

Regular Cement-based Grouts. Most of the following components are used in a balanced, stable cementitious grout:

Water   The carrier of the particles.

Naphthalene Sulphonate   Deflocculator. Enrobes the particles with a film, overprinting the particles, with the same negative electric charge. The result is that the same particles reject each other, preventing the formation of macro-flocs. The product is readily available from a number of concrete additive suppliers.

Methacrylates and Stability-Enhancing Polymers   Additives to enhance the resistance against pressure filtration and hence the stability of the mix.

Bentonite   Clay-phyllo-silicate of the 2-1 structures. This is the first ingredient to enter the high shear mixer. This product electrostatically attracts particles because of the polarity of its molecules; as a result, it reduces segregation in the suspension grout. It further lubricates particles and enhances the penetrability of the grout and increases the resistance against pressure filtration. Note that there is a minimum percentage of high-yield bentonite (use only sodium montmorillonite) required to stabilize the mix effectively. The findings published by Dr. Don Deere (New Orleans, 1982) are very useful for the formulator. Every type of bentonite behaves differently from the other types, and hydration speeds differ on the type and rpm of the colloidal mixer. Therefore, test formulations should be made prior to grouting. Bentonite increases the cohesion of the grout, and hence its apparent viscosity.

Silica Fume   Microfine powder (particle size of 0.5 μm) which enhances the stability of the mix, enhances the penetrability, and reduces the permeability of the cured grout (and as a result its durability). Silica fume enhances the resistance against pressure filtration and makes the grout more water-repellant. The small particles act as small ball bearings, carrying the larger cement particles into the cracks and crevices. Typical concentrations vary between 4 and 10%.

Type C Fly Ash   Rather inexpensive filler with pozzolanic characteristics. This product reacts with the free lime, generated by the hydraulic reaction between the cement and the water. A slight postcuring expansion is an added benefit. High concentrations (over 20% of the cement weight) cause adverse long-term effects to the cured grout (i.e., the grout slowly expands and crumbles). Type C fly ash increases the water-repellant characteristics of the grout and its resistance against pressure filtration.

High Early-Strength Cement   The finer the cement, the better the penetrability of the suspension grout. When the grout is introduced in wet and cold environments, the use of a faster curing grout is often necessary. Flash setting (especially when grout is applied at higher temperatures) can be avoided by adding retarders to the mix.

Viscosity Modifiers   Additives, producing a high viscosity, cohesive, water repellant grout, even when only minute quantities are introduced into the mix. The flavored version of some of these products is used to produce soft ice cream.

Trass (clay-phyllo-silicate 1-1 structures)   This natural pozzolan has been used for over 2000 years. Trass-enhanced mortars have stood up for 20 centuries in marine environments. Trass is a powdered tuff stone mined in Andernach (Germany).

Microfine Cement-based Grouts.

Ingredients.   The following ingredients are used in classic microfine cement-based formulations:

Water

Naphthalene sulphonate

Bentonite (only small percentages)

Silica fume

Microfine cement is either a very fine, pure Portland cement (such as Microcem 900 from Blue Circle; Degerhamn from Cementa; Mikrodur from Dyckerhoff), or a slag cement (such as M.C. 500, Nittetsu S.F. Spinor A12). These cements are characterized by their sieve curve and their specific surface or *Blaine fineness*. The Blaine fineness should be over 8000 cm^2/g to have a true microfine cement, according to the Europeans. The maximum grain size is a more representative characteristic. In order to have a microfine cement, the diameter of all particles should be smaller than 10 μm. Some of the Portland cement-based microfine cements contain citric acid as a retarder. According to our latest research, the final strength of a regular, stable, cementitious grout is drastically reduced when concentrations of citric acid over 0.1% are utilized. Some microfine cements contain in the order of 0.5% citric acid. As a result, the ultimate unconfined compressive strength of these microfine cements never exceeds 5 MPa (750 psi), provided a W:C factor of 1 or higher (practical application range) is utilized. We have experienced on *all* our projects an inconsistent performance of one particular type of microfine cement. (Flash "stiffening" occurred at high as well as low ambient temperatures as well as slow curing at normal temperatures; we experienced initial curing times, as long as 7 days. The use of accelerators provided some improvement but caused other problems, including fast increase in the apparent cohesion.)

With the other family of microfine cements, slag acts as a retarder. Over the past decade we have used and tested slag-based microfine cement (MC 500) formulations and never encountered a problem except in hydrogen sulfide environments.

CEMILL PROCESS   The Cemill process produces microfine cement during the grouting operation. It was developed by Rodio (Italy) for the following reasons (according to Rodio):

- To evolve an on-site manufacturing process to provide microfine grouts
- To use, as the starting material, conventional cement-based suspension grouts, in order to be independent of the local market availability of microfine cements (at present available only in a limited number of industrialized countries, often at fairly high cost)
- To overcome the limits of pure decantation, which permit only the manufacture of very diluted grouts and make difficult the disposal of the coarse, rejected portions
- To produce not only *unstable* mixes (water, cement, and a possibly dispersive agent, further referred to as Cemill-i) but also *stable* grouts using bentonite(further referred to as Cemill-S) and therefore unobtainable by decantation

The achievement of these goals has been made possible by the development of a special colloidal wet refiner which has a dual function:

- It imparts a very strong dispersive action to the cement used in the grout. In this way the effectiveness of the microfine portion already contained is safeguarded, even if dispersive agents are not used.
- It carries out a progressive refining process of the coarse portion until the desired fineness is reached, without requiring its physical elimination as in the case of decantation.

PARTICLE-SIZE REDUCTION PROCESS   A recent development may be an unprecedented breakthrough in the grouting industry—a (presently) rather complex particle reduction mill. This development is mainly the result of research conducted by Sandia National Laboratories, under the direction of Ahrens.[33]

The original test objectives of a pilot grouting plan to seal fractures and microfissures at the WIPP repository horizon included:

- Demonstrate the ability to practically and consistently produce microfine cement-based suspension grout at the site
- Demonstrate the penetration of microfine cement-based suspension grouts
- Obtain data on the ability of microfine cement-based suspension grouts to permeate microfractures (finer than 100 μm)
- Evaluate techniques that can assess the effectiveness of microfine cement-based suspension grout in reducing the gas transmissivity of fractured rock

Durability, injectability, and competence were three major targets for a "made in North America" microfine cement grout. Because of the significance and importance of the performance of the suspension grout, neither money nor effort were spared to adjust existing particle-reducing equipment, used in other industries, to produce an in situ microfine cement. Ahrens's team resorted to a batch-per-batch approach to produce a stable suspension grout containing pumice, cement, naphthalene sulphonate, and water. After milling the suspension grout for 84 min, the particle size of the stable suspension grout had been reduced to less than 5 μm. Quality control was assured by laser particle size and rheological analysis.

The grouting was conducted at the underground facilities, and the reduction in gas transmissivity was reduced by a factor of 3. Thorough instrumentation of the formation provided most valuable data and information which could benefit the grouting industry as a whole. Furthermore, cracks as narrow as 3 μm were discovered to be filled with grout. The methodology to establish that the material in these virtually invisible fractures was indeed grout was most credible and state-of-the-art. Mass spectroscopy provided irrefutable evidence that microfine stable cement-based suspension grouts have penetrability several orders of magnitude better than what can be accomplished with unstable, regular cement-based suspension grouts.

Ahrens's team elected to go with the batch-per-batch approach, which in its present form is only of scientific or academic interest. However, the approach sets the stage for further developments that could eventually benefit everyone in the business. The grouting industry is particularly interested in an economic continuous wet particle-reduction mill.

Further research work will be conducted to produce dry microfine cement-based grouts. One of the goals of Ahrens's team is to develop a prototype that can dry-mill the powders and that can be afforded by the average grouting contractor. Unfortunately, the research program went off on a tangent and only produced a microfine cement powder available in bags.

From a practical standpoint, the immediate benefit of the particle-size reducing mill is likely to be in the following approach:

- A properly balanced regular cement-based suspension grout is prepared in a colloidal mixer.
- Before it is dumped in the holding tank, the grout is fed through a continuous particle-size reducing mill. This mill, in essence, is a high, small-diameter, paddle mixer completely filled with small ball bearings. These balls provide the grinding mechanism and perform as a wet ball mill.

The first indications are that the maximum grain size produced by a basic particle-size reducing mill, is approximately 12 μm (Fig. 5B.23). The continuity of the grouting process should not be adversely affected by this type of continuous particle-size reducing mill.

Average cement particle size in a regular cement-based suspension grout is in the order of 50 to 80 μm. A reduction in maximum size to 12 μm enhances the penetrability of the grout to soils with permeability coefficient of 0.01 cm/s (4×10^{-3} in/s) or even lower (see Fig. 5B.20).

5B.3.1.8 The Amenability Theory (Rock and Structural Grouting). Most people in the grouting industry make poor or no use of the data provided by in situ hydraulic conductivity tests and of the grouting data. Usually, Lugeon values are calculated, and, in the best-case scenario, a trend analysis of the permeability tests is conducted.

When a grout selection is made to start the grouting of a particular zone or hole with a given permeability, the response of the formation has to be evaluated when the selected grout is introduced into the crack system. The apparent Lugeon value of the formation should also be evaluated on a regular basis during the grouting operation. Grouting is some kind of continuous permeability test, whereby the Newtonian test fluid (water) is replaced by a Binghamian fluid, with a given apparent viscosity.

For field applications, the following equation is used to calculate the apparent Lugeon value:

$$Lu_{apparent} = \frac{flow\ (L/min)}{1\ L/min} \frac{1\ m}{L_{zone}\ (m)} \frac{143\ psi}{P_{effect}\ (psi)} \frac{V_{marsh\ gr}(s)}{28\ s}$$

with $V_{marsh\ gr}$ = efflux time for 1 L of grout, via the marsh cone.

The apparent Lugeon value is further referred to as Lu_{gr}. For the calculation of the effective pressure, the head losses through the system for the given flow have to be taken into account. Note that the only way to monitor the apparent Lugeon values is with the help of a pressure/flow recorder providing continuous printout.

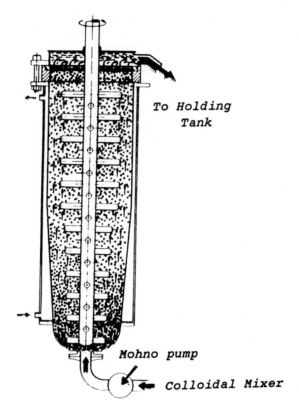

FIGURE 5B.23 Particle-size reduction mill.

Since water also permeates in fissures and pores not accessible to a particulate grout, it is to be expected that the initial grout take could be less than what can be expected, based on the Lugeon value derived from the permeability test (Lu_{wa}).

The amenability coefficient (A_c) of a particular grout, injected in a given formation is defined as

$$A_c = \frac{Lu_{gr}}{Lu_{wa}}$$

(Lu_{gr} refers to the initial stages of the grouting operation). The amenability is a measure of the suitability of a given suspension grout for permeating fissures and apertures accessible to water in the grout zone. Amenability provides a good indication of the secondary permeability to be encountered within the grout spread. If the amenability coefficient is approximately 75%, a residual permeability of at least 25% of the original hydraulic conductivity measured in the grout hole can be expected (within the grout radius) and even more a distance away from the grout hole. The amenability coefficient signals formulation design flaws immediately. Even microfine-based suspension grouts can provide a less than satisfactory amenability. At this point the decision has to be made to resort to appropriate solution grouts to achieve the design targets.

Expensive and useless grouting operations can be avoided if the amenability of the formation is too low. Unless the client is in a position to provide the funding for a grout with appropriate amenability, there is *absolutely no point* in trying to carry out the operation. Unpleasant surprises pertaining to poor injectability because of a lack of amenability can be avoided by executing a pilot grouting program.

5B.3.1.9 Analysis of Situations and Guidelines for the Sequence of Grout Mixes. It is not possible to provide a script for a grouting project that can be blindly followed by an engineer; flexibility to adjust the program to deal with specific circumstances, on a minute-to-minute basis, is necessary. This flexibility requires experience and a desire to perform on the part of the engineer. However, it is possible to provide a set of general guidelines developed through numerous grouting projects and based on the author's experience and the experience of other specialists.

It is good grouting practice to follow a grouting philosophy from which one might have to divert from time to time, as dictated by the field circumstances. It is obvious that there is a different scenario for holes with high hydraulic conductivity than for holes intersecting zones with low permeability.

During the grouting operation, one of the key elements to monitor is the amenability of the formation to accept a particular grout formulation and the evolution of the apparent hydraulic conductivity during grouting. This in turn will dictate whether a particular formulation is suitable or if changes are necessary. Therefore the *X-Y* recorder will have to be watched continuously and trend interpretations made on a continuous basis. A *trend interpretation* is related to the angle of the "take line" at constant (or slightly changing) grouting pressures. The take line signals if the solids content can or should be increased or decreased, if a runaway operation is in the making, if the grouting operation is consistent with void filling, or if a true permeation grouting with a suitable formulation is in progress.

A few typical situations which might occur during a structural grouting project follow.

Lugeon Values over 150 Lu, with No Signs of Mud Filling. Start the grouting operation with a low-viscosity, stable, regular cement-based suspension grout, with average solids content. If the apparent Lugeon value is more or less steady for an arbitrarily selected grout take (i.e., one to two times the theoretical volume) and if there are no signs of mud washing, gradually increase the solids content of the suspension grout to obtain a more durable final product.

If the apparent Lugeon value continues to remain steady and again exceeds one or two times the theoretical volume within a theoretical radius around the grout hole, the solids content should be continually increased, increasing the cohesion of the grout as little as possible while maintaining a good resistance against pressure filtration.

The friction losses in the grout should bring the operation to a natural refusal. If this is not the case, the decision should be made to artificially obtain a proper refusal by increasing the cohesion and consequently the apparent viscosity of the grout.

Initially, the solids content should be gradually increased, followed by the addition of minute quantities of a viscosity modifier to increase cohesion and thixotropy. Thus the hole can be brought to a proper refusal.

If, however, the apparent Lugeon value decreases at any point in the operation, the total take should be assessed:

- If there is insufficient grout being injected to form the theoretical spread, the resistance against pressure filtration is increased, the viscosity is lowered, or microfine cement-based formulations are introduced.
- If the total grout quantities are sufficient, the hole is brought to refusal without further changing the formulation.

If grout is leaking from cracks, compromises might have to be made, to stem the grout seepage. Typically, the grout flow in the zone causing the problem is reduced and the leak sealed with fast-curing (hydraulic) cement, oakum, or urethane drenched grout pads and wedges. (This can only be done if a multiple-hole grouting program is in progress).

Under no circumstances shall the grout be overdisplaced with bentonite or water. Another approach is to increase the thixotropy of the grout and, circumstances permitting, the viscosity too. Should this be not feasible or practical, the grout should be displaced with the same grout to which a retarder is added. Once the zone has been given "a rest" of a few minutes, the operation should be resumed at the particular hole.

Lugeon Values between 30 and 150 with No Signs of Mud Filling. Start the grouting operation with a low-viscosity regular cement-based suspension grout with low solids content. It might be necessary to start the operation with a microfine cement-based grout, if the amenability for regular, low-viscosity grouts is not acceptable. In any event, the starting mix has to be assessed for its amenability, as described above.

Increase the solids content gradually, circumstances permitting. (Keep the batch size small to maintain flexibility in the adjustment of formulations.) Follow the same steps as outlined above.

Lugeon Values over 30 and Signs of Mud Filling. Under these circumstances, only regular suspension grouts with low solids contents and good resistance against pressure filtration should be used, or alternatively, stable microfine cement-based formulations. Under no circumstances can the grouting be interrupted, because silt particles (foreign to the mix) will settle and clog the travel path for the grout, causing premature refusal.

Lugeon Values of less than 30 Lugeon and No Mud Washing. Start the grouting operation with a microfine cement-based formulation. Depending on the outflow rate and the Lugeon value, the solids content may be varied. Calculate and assess the amenability of the formation for the various, selected type of microfine cement grout formulations.

If the amenability is low, an attempt (only if practical) should be made to switch to the lowest solids content. In most cases, the take will be very low for the holes with low Lugeon value, and the hole will likely be refused once the batch with the thinner grout is ready to be introduced. Note that the apparent viscosity between mixes with a C:W ratio between 0.35 and 0.50 is negligible, and the benefit of low microfine cement contents is often more perceived than real. Of more importance is to maintain a pump flow to the hole. A "difficult" hole could immediately refuse, if the pressure suddenly drops. Since it is good grouting practice to grout holes with low permeability simultaneously (multiple-hole grouting), the inspector has to watch that any new hole is only gradually brought on stream, in order to prevent a sudden pressure drop in the other holes.

If the amenability is high, the hole should be fed the same formulation. If the apparent Lugeon value does not change, gradually increase the solids content of the microfine cement formulation. Only as a guideline, action should be taken when the theoretical quantity (based on the aperture theory) is exceeded by a factor 4 and the take line on the *X-Y* recorder remains vertical.

In the latter case, regular, stable cement-based suspension grouts with low viscosity and low solids contents should be introduced. It is always possible that fine fissures, intersected by the borehole, may lead to a larger porous zone which is amenable to regular cement-based grouts.

Depending on the "reservoir size" of the formation, the zone surrounding the borehole will be refused with either a microfine cement-based grout, with higher solids content, or with a low-viscosity (low solids content), regular cement-based suspension grouts.

If the amenability is low, solution grouts are more appropriate to saturate the formation. Even microfine cement-based formulations won't do more than just a "squeeze grouting" job. If the grout hole appears to connect with leaking cracks, causing dilution, hydrophobic water-reactive polyurethane prepolymer grout should be injected simultaneously via a resident tube.

Especially in secondary grout holes, the reservoir capacity could be limited; consequently, primarily microfine cement-based formulations will be used in the "second pass." In order to evaluate borehole spacing, the reduction in hydraulic conductivity in the secondary holes should be evaluated and assessed considering the amenability values of the adjacent primary holes.

Lugeon Values less than 30 Lugeon and Mud Washing. If silt, gouge, or clay filling fissures within the formation are eroded or extruded as a result of the grouting pressure, the grout take will gradually increase at a given grouting pressure. The pressure might even drop. At this point the amenability is (theoretically) greater than 100%, and the apparent Lugeon value is higher than the in situ hydraulic conductivity established with water. Formulations will have to be adjusted according to the same guidelines as explained above.

5B.3.1.10 Cement-foam-based Grouts. Especially when very pervious formations (such as rubble backfill) need to be stabilized, cement foam can be successfully used. Recent developments have made it possible to produce almost any rheology and strength to provide a custom-tailored solution to a given problem.

In essence, the following operations take place to produce a state-of-the-art cement foam:

- A cement-based suspension grout with the desired rheological characteristics is produced in a colloidal mixer.
- A stable "foam" is produced in a separate unit.
- Foam and cement-based suspension grout are mixed in a paddle mixer with horizontal axle. The mixing ratio depends on the desired mechanical characteristics for the cured grout. The lower the amount of foam introduced in the mix, the higher the unconfined compressive strength of the grout and the higher the cost of the cured product. High foam content produces a cement foam with low density, low cost, and good impact-absorbing characteristics.

Applications include the production of compressible fire and ventilation walls around shafts and ducts; cemented backfill in mining applications (room and pillar). Especially the latter application produces major cost savings to the mines. The impact absorption properties of cement foam reduce sloughing and dilution in the recovery of pillars during the blasting.

It is advisable to introduce the thixotropic cement foam together with the rock in the open stopes during the backfilling operation to obtain a homogeneous, high-quality backfill product.

In civil engineering projects, the nonsag characteristics of a thixotropic cohesive light foam is ideal for stabilization of rubble fill. It is an ideal and cheap filler around and between pipe bundles and ducts. Cement foam is also gaining popularity as an economic alternative for sand-cement grout as backfill material in caverns and abandoned mines to prevent subsidence to roads and structures.

5B.3.2 Solution Grouts

5B.3.2.1 Introduction. The term *solution grouts* (formerly known as *chemical grouts*) pertains to grouts which behave like Newtonian fluids when injected through cracks and soil channels. Contrary to suspension grouts, which behave like Binghamian fluids, solution grouts do not contain particles. As a result, the grouts are injectable even into very fine pores not accessible to (even microfine) suspension grouts.

The term *chemical grouts* is misleading, since cement-based suspension grouts are probably the most complex chemical grouts and are not classified (according to the ASCE committee on grouting) in this category.

Most solution grouts are characterized by a rather straightforward reaction pattern between stoichiometric quantities of the various components constituting the solution grout. The sensitivity (i.e., factors influencing the reaction or curing pattern during the reaction) of most solution grouts is lower than the sensitivity of cement-based suspension grouts.

Some authors distinguish between *evolutive solution grouts* and *true solution grouts.*

An evolutive solution grout is characterized by a steady increase in viscosity, as soon as the various reagents are mixed. Typical examples are the sodium silicate–based grouts.

A true solution grout is characterized by a flat viscosity curve followed by a sudden increase in viscosity immediately prior to gelation or curing. Acrylamides and most polyurethane-based grouts are typical examples of true solutions.

Solution grouts have made it possible to treat media with a measurable permeability, whether wood, glass, masonry, concrete, rock, or soils. Because of irresponsible use (especially from an environmental standpoint) and ill-conceived applications, especially during the pioneering days of solution grouting, solution grouts in general often have generated bad publicity. Since environmental issues are usually very complex and often quite controversial, solution grouts have developed a bad reputation during the "greener" years. Although some so-called chemical grouts are very durable, more predictable, and have better long-term stability than cement-based suspension grouts, they suffer from the bad reputation caused by unrelated other solution grouts.

It would be beneficial to the grouting industry if a competent (but nonpartisan) committee would be formed to provide leadership and develop guidelines for the commonly used solution grouts. A similar organization should look into testing and approving all existing and new grouts (before they are brought on the market) and should assign a range of application fields to the various products and ban others. All sides of the grouting industry should be represented in such a committee. Otherwise, the battles will continue to rage between economically biased parties advocating or opposing bans on particular acrylamide solution grouts, T.D.I.-based polyurethane grouts, classic sodium silicate hardeners, certain epoxy hardeners, and so on.

North America has been a trendsetter, increasing the awareness level of potential hazards with some of the solution grouts. The litigious environment has frightened some manufacturers, and production of certain toxic (under some circumstances) solution grouts was officially terminated. Unfortunately, several potentially more hazardous solution grouts are still manufactured and applied under circumstances in which they should not be used. Here we will clarify some of the mystique woven around solution grouts. The following product families are discussed:

1. Polyurethane grouts
2. Sodium silicate–based grouts
3. Acrylamide-based grouts
4. Epoxy grouts
5. Chrome-lignin-based grouts
6. Phenolic grouts
7. Formaldehyde-based grouts
8. Resorcin Phormol and aminoplast grouts

Polyurethane-based formulations are the most widely used solution grouts. Formulations based on sodium silicate and acrylamide are next, in order of usage. Silicate-based grouts offer relatively low cost and high strength, but the classic sodium silicate formulations only provide temporary soil modification. These materials also have a relatively high viscosity and may suffer deleterious syneresis and dissolution. Low-viscosity silicate solutions can be used at the expense of strength, gel time control, and permanence.

Acrylamide grouts offer low viscosity and excellent gel time control but are low in strength and pose some handling problems. Polyurethanes revolutionized the grouting industry and diversified and broadened the applications, opening untapped markets. Today, commercial chemical grouts cover a wide range of materials, properties, and costs, and give the grouter the opportunity to select a grout for specific job requirements.

The mechanical properties of any grout are important factors in its selection for a specific job; these properties include mechanical permanence, penetrability, permeability, and strength. Other factors influencing grout selection are chemical permanence, gel time control, suitability, sensitivity, and toxicity, as well as the economic factors including availability and cost. The following discussion is based partly on a publication by Karol.[48]

Permanence. All grouts which contain water not chemically bound to the grout particles are subject to mechanical deterioration if subjected to alternate freeze-thaw and/or wet-dry cycles. Mechanical deterioration of grouts can occur under a number of site-specific and environmental conditions.

Chemical deterioration of grouts can occur if the grouts react with the soil or ground water or grouted medium to form soluble reaction products, if the grout itself is soluble in ground water, or if the reaction products which form the grout are inherently unstable. In general, one would expect that materials with any of those unfavorable characteristics would not be proposed or used as a grout. However, well-oiled marketing machines promote far too many cure-all products, with no consideration for durability nor handling problems. There is virtually no publicity given to the failures. The numerous problems on small projects are indicative of the long-term failure rate of grouts. If North America would adopt the 10-year warranty program used in most countries in continental Europe, a number of products would rapidly disappear from the market.

One of the few market segments where long-term performance is vigorously monitored is the residential market. One small grouting contractor from the Toronto area stated that he had over

250 "call-backs" (virtually 90%) on projects (once accepted by the client) using a particular solution grout. Failures are usually the result of improper grouting techniques (or methodology), lack of durability of the grout, or a combination of both.

Penetrability.   The comparative ability of grouts to penetrate a medium is predominantly a function of their relative viscosities. Although there is some truth to the contention that lubricating fluids pass through pipes with less friction loss than nonlubricating fluids, the difference is small and cannot make up for significant viscosity differences. In general, viscosity alone could be used as a comparative guide to penetrability of solution grouts.

Penetrability can be related to permeability. A conservative criterion is that grouts with viscosities less than 2 cp (such as acrylamide-based materials) can usually be pumped without trouble into soils with permeabilities as low as 10^{-4} cm/s (0.4×10^{-4} in/s). All solution grouts may have trouble penetrating soils (by permeation only) when the silt fraction exceeds 20% of the total soil matrix.

Strength.   The strength of soil formations stabilized with solution grout is of very practical interest for soil grouting. In most structural grouting applications, the tensile strength, bond, elongations at break, and unconfined compressive strength play an important role.

Grouts which completely displace the fluid in the soil pores form a continuous but open and nonuniform latticework which binds the grains together. In so doing, the grout increases the resistance to relative motion between soil grains and thus adds shear strength to the soil mass and drastically reduces the compressibility of the soil.

Most applications, regardless of purpose, result in placing grout below the water table, and drying of the grouted mass will never or only periodically occur. Thus, the most significant strength factor is the "wet" strength: the strength of the stabilized soil formed and remaining immersed in a saturated medium.

In contrast, much of the literature devoted to strength data reports "dry" strength: strength of stabilized soils in which some of or all the contained water has been removed by air or oven drying. When water held by a grout is lost through desiccation, the grout matrix shrinks. Shrinkage of the grout between soil grains sets up forces analogous to capillary tensile forces but often much stronger. This increases the resistance to relative motion between grains and adds shear strength to the soil mass. Thus desiccated grouted soils will show strengths higher than "wet" strengths, often by as much as ten times.

The strength of a fully permeated grouted soil depends not only on the specific grout used but also on other factors such as density, average grain size, and grain distribution of the soil. In general, strength increases with increasing density and with decreasing effective grain size. Well-graded soils give higher strength than narrowly graded soils with the same effective grain size. Because of these variables, general discussion of the anticipated strength of grouted soils should present ranges of values rather than a specific number.

Cohesive soils almost always have permeabilities too low for effective grouting. Thus, virtually all soil grouting is in granular soils.

Almost all strength values reported in the literature for grouted soils are the results of unconfined compression tests. Unconfined compression tests measure the strength of the test sample at zero lateral pressure. Most soil grouting work is performed where a significant lateral pressure exists. Therefore, triaxial tests are a better replication in the laboratory of actual field stress conditions. The unconfined compression test is still very useful for comparative purposes, for example, checking the effects of additives to a grout or comparing different grouts.

The most important mechanical characteristic for soil grouting is to measure the increase of the compressibility coefficient of the grouted soil compared with the ungrouted horizon. The compressibility factor is determined by the static penetrometer test. For a given soil layer (at sufficient depth), the compressibility factor C equals $\frac{3}{2}$ of the cone resistance divided by the original vertical stress in the soil at that location.

Gel Time Control.   The ability to change and control the set time of a grout is an important factor in the successful execution of a grouting job. If all other factors are held constant, this time lag is a function of the concentration of the activator or inhibitor and/or the catalyst added to the formulation. With most grouts, it is possible to change the "pot life" by varying the concentration of any of the two or three (depending on the type of solution grouts) components. For some grouts a wide range of gel times can be obtained and accurately duplicated. For other grouts, gel times can be adjusted not at all or only within a narrow range.

The grouts generating an exotherm while reacting display a different behavior in very fine soils or cracks in comparison with coarse soils or wide cracks. The exotherm is more readily dissipated in fine pore channels, and the induction period is considerably longer in the medium than the pot life (pure grout) would indicate.

Sensitivity. For any given grout formulation, the gel time is a function of temperature. However, other factors affect the gel time or even prevent a complete reaction. The most important factor is the dilution by (running) ground water, which will adversely affect the reaction of water-incompatible grouts and might dilute other grouts to the extent that the concentration of active reagents drops below the point where no reaction takes place.

The water's chemistry (dissolved salts, gases) and its pH can change the reaction pattern. Even the chemistry of the medium in which the injection takes place (without the presence of water) can affect the curing of some grouts.

Toxicity. For too long, professionals have ignored occupational health and safety issues when dealing with solution grouts. Solution grouts unsuitable for particular applications have been washed out and have polluted aquifers and rivers. Other black pages in the grouting history books pertain to the stories of grouts which continue to pose a health hazard, even in their "cured stage," decades after they have been applied.

The following labels should accompany some products still in use: neurotoxic, carcinogenic, toxic, corrosive, skin irritating. All chemical grouts should be handled with care in the field, and workers should be properly trained in safe handling and application procedures.

The waste generated by any solution-grouting program, should be properly disposed of. Pump cleaner, solvents, residues of grout, as well as empty packaging require appropriate attention. In North America, some manufacturers provide good and practical guidelines to deal with the various types of waste generated and their proper disposal, whereas others provide insufficient or even misleading information.

Economic Factors. Cost is a most important factor in selecting the construction method best suited to solving a specific field problem. When comparing two different grouts, however, the mistake is often made of comparing raw material costs rather than job costs. Chemical grouts are available commercially at prices ranging from $1 to $50 per gallon. However, the ease of placing and the effectiveness of these materials vary, and the cost of placement is almost always a major part of the in-place cost. Thus, although raw material costs may vary over a 50:1 ratio, in-place costs generally only fluctuate from 10:1.

5B.3.2.2 Polyurethane Grouts

Water-Reactive Polyurethanes. Polyurethane prepolymers have one thing in common: They react with the in situ available (ground) water to create a foam or a gel which is either hydrophobic or hydrophillic. They are one-component products using "the enemy"—the water—as a reaction partner to create a gel- or foam-type product.

Water-reactive or water-activated polyurethane resins are classified into two different categories:

1. *Hydrophobic polyurethane resins* react with water but repel it after the final (cured) product has been formed.
2. *Hydrophillic polyurethane resins* also react with water but continue to physically absorb it after the chemical reaction has been completed.

During the exothermic reaction, the hydrophobic polyurethanes expand and penetrate fine cracks and even low-permeability soils, independent from the grouting pressure. They are "active" grouts. Clearly, the hydrophobic polyurethane prepolymers are not a replacement for acrylamides but are appropriate for straightforward applications in mining and geotechnical engineering.

Not all hydrophillic prepolymers expand. Depending on the type and/or the amount of water they are mixed with, they form either a foam or a gel. The hydrophillic polyurethanes, especially in North America, have often been used to fill part of the gap left by AM9, when production of this product was officially halted in 1978.

During the late 1970s, the prepolymers were considered the *deus ex machina* (cure-all) for any type of seepage problem. As the new technology matured, applicators learned the hard way that there were limits to each family of prepolymers.

Water-reactive prepolymers are high molecular grouting materials. They are primarily produced by mixing a polyol with an excessive amount of polyisocyanate to form a low prepolymeric compound, containing some free OCN groups. The injection resin is composed of this prepolymer, plasticizer, diluting agents, surface stabilizers, and the amine catalyst.

The mechanism of reaction among the isocyanate, polyol, and other components is rather complicated. In simple terms the following happens:

The reaction between the isocyanate and the polyol yields a prepolyurethane.

The reaction of polyisocyanate with water liberates carbon dioxide and urea derivatives.

The reaction of polyisocyanate with ureido develops molecular links and high molecular formation.

These reactions occur because free OCN groups in the grout can react with the compounds containing active hydrogen atoms, such as hydroxy, water, amino, and ureido. The hydrogen atoms move to link with the nitrogen atoms of the polyisocyanate and form high molecular polymers.

Hydrophobic Polyurethane Resins—"Rigid" One-Component Prepolymers The main application of these prepolymers is blocking water in-flows and soil grouting. Products used for water cutoff include: Tacss 20, Aquapreps 15, Deci 16, SK1, MME Universal, Rhone Poulenq P.U., Mountaingrout formulations, Adhesive Engineering 4058, Webac 151, Resicast GH67, DeNeef formulations, Spetec formulations, Stratathane 530, BASF & BAYER formulations. Products that are used in soil stabilization include Tacss 25, Aquapreps 05, Deci 161, SK3, MME Ultrafine, DeNeef formulations. These products react with the in situ water and expand during the exothermic chemical reaction, releasing carbon dioxide. They are totally inert, nontoxic, and stable after reaction but have very limited flexibility. The reaction time can be adjusted from 45 s up to 1 h by adding a tertiary amine-based catalyst. Figure 5B.24(*a*) illustrates a typical reaction curve for water-reactive hydrophobic polyurethane prepolymers.

Heating the products (to a maximum of 60°C or 140°F—never directly apply the heat to the grout containers) speeds up the reaction. If shorter set times are required, a resident tube is used: water or cement-based suspension grout is injected via separate lines into a collector tube, in conjunction with the prepolymer. This way the grout has already passed the "induction phase" of the chemical reaction when introduced into the formation or the water flow. By providing a grouting setup in which the water entry point into the resident pipe can be continually altered (mobile feeder pipe through stuffing box), the resident time can be continually changed, and washout of grout can be avoided.

The water cutoff series of prepolymers has been developed for stopping major inflows and seepage control. They are most suitable for the rather crude applications when water has to be stopped but not necessarily to the last drop. As a result, these prepolymers are ideal troubleshooting products for stopping or controlling major in-flows in geotechnical applications. Grouted soil or cured grout has a residual permeability in the 10^{-4} to 10^{-5} cm/s (0.4×10^{-4} to 0.4×10^{-5} in/s) range, which will allow minor seepage to take place.

Products within the same family differ due to:

- Viscosity
- Mixability with water
- Reactivity at elevated pressures
- Toxicity (solvent content and type of isocyanate)

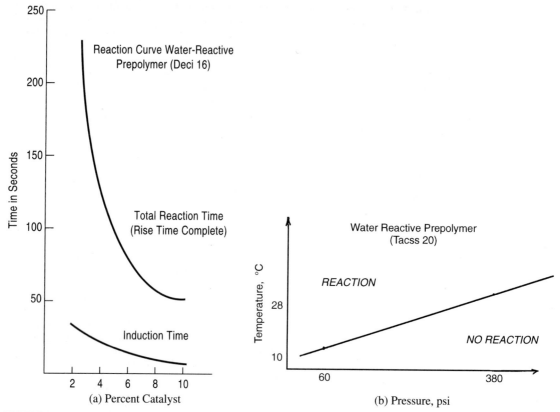

FIGURE 5B.24 (a) Reaction curve for water-reactive hydrophobic polyurethane prepolymers; (b) reaction diagram, temperature versus pressure (hydrophobic polyurethanes).

- Cellular structure of the free foam
- Hydraulic conductivity of grouted soils

Several manufacturers have obtained potable water approvals for their water-reactive hydrophobic polyurethane prepolymers.

In order to start the reaction, there is a minimum enthalpy required, which is higher as the pressure is higher. This means that if the enthalpy is too low (temperature too low), the reaction does not start unless the products are mixed thoroughly with the water. Some researchers have established a reaction-diagram for some of the hydrophobic polyurethane grouts [Fig. 5B.24(b)]. Above the "reaction line" the reaction takes place, and below this line reaction does not occur unless mechanical agitation of in situ water and the polyurethane grout is provided. The (turbulent) flow of the urethane-water mixture through cracks and channels often provides for enough in situ mixing. Some of the more recently developed prepolymers react regardless of pressure and ground-water temperature.

During grouting, the carbon dioxide generated during the chemical reaction will generate additional pressure, as the grout flows through the cracks and pore channels, pushing the grout into very fine cracks and crevices. The reaction pressure in totally confined circumstances is over 2.5 MPa (350 psi).

The best way to control the reaction times is by injecting the products as a two-component grout, with water or cement-based suspension grout, the second component, separately introduced into the manifold. By using the header pipe as resident pipe and selecting an induction period a little longer than the resident time, it is possible to control an in-flow. This is a fairly sensitive operation, as the flow pattern in the structure to be sealed is continually changing during the grouting operation.

Only MDI-based prepolymers should be used. The vapor pressure (at ambient temperature) of the solvent-free MDI-based products is low. However, ventilation is still required when working in enclosed spaces in order not to exceed the maximum threshold limit value limits of 100 ppb (8-h period), as determined by the International Institute for Isocyanate.

SOME TYPICAL APPLICATIONS

Curtain Grouting in Rock and Soils in Environmentally Sensitive Areas or in the Presence of a Considerable Ground-Water Flow   A rather well documented example is the curtain grouting around and below the temporary piers supporting the railway bridge over the Ochise River (near Toyohashi, Japan)[49] (Fig. 5B.25).

A grout curtain in permeable sand was installed around and below the temporary piers in the swiftly flowing Ochise River to prevent scour and to provide a uniform support to the precast footing. The project was successfully carried out in the spring of 1969. Simple strainer pipes (the Japanese have never really accepted sleeve pipes) were driven around and below the footing, and a single grout curtain with overlapping cylinders was first installed followed by the soil injection below the pier footing. The (now obsolete) Tacss 009 (similar to the current Tacss 25 version) was used to carry out this highly successful project.

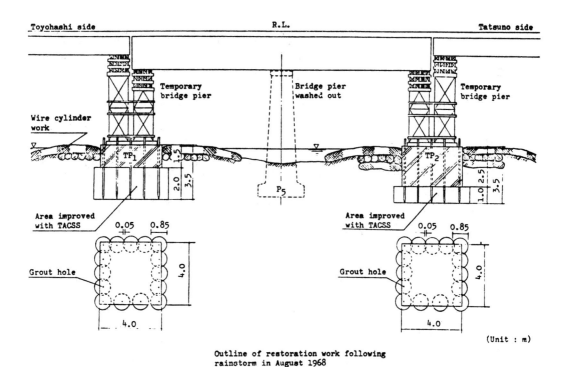

FIGURE 5B.25   Curtain-grouting application.

Sealing "Renards" (Soil-Piping Conditions) through Joints in Deep Tunnels Well Below the Water Table Most subways in Western Europe consist of multiple-level tunnels, as deep as 30 m (100 ft) below the ground-water table in fine granular soils. Failures in joints have devastating consequences. An example of a joint failure occurred between slurry-wall panels near the Midi Tower in Brussels (Central station) in 1982 (Fig. 5B.26). The Brussels subway job involved the construction of several tunnels excavated under the protection of heavily reinforced slurry walls. The excavation for the various levels of the tunnels was done, from the underground workings, after the "roof" of the tunnel had been cast, in order not to interrupt the traffic.

During the excavation of the second underground level, a barely leaking joint suddenly gave away at night, causing a major in-flow of water and fine sand into the tunnel, in turn resulting in a massive sinkhole on the Midi Square. Bentonite-soil lumps are often formed in the joints between various panels of slurry walls, especially in fine cohesionless soils.

Soil drifting into the subway disturbed a large area, endangering the foundation system of the Midi Tower. A swift night intervention involving prepolymer drenched pads and wedges blocked the soil in-flow within the hour and the water within 8 h after the intervention started. A soil-grouting program (using sleeve pipes) involving cement-based suspension grouts, sodium silicates, and prepolymers created a soil conglomerate on the outside of the joint, to full depth, successfully protecting it against future failures and allowing a safe rehabilitation program from inside the tunnel. The polyurethane grouts used in this project were Deci 16 and Deci 161.

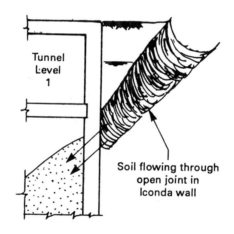

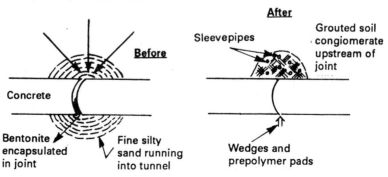

FIGURE 5B.26 Sealing joints in deep tunnels.

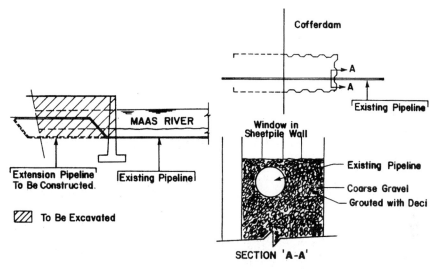

FIGURE 5B.27 Shutoff of major water flows in pipelines.

Sodium silicate–based grouts were used in conjunction with cement-based grouts, to stabilize the disturbed zone, away from the in-flow area, to prevent future subsidence. Sodium silicates (used with hardener 600 B) surrounded by cement-based grouts tend to remain stable in time.

Cutting off Major Water Flows and Soil-Piping Conditions through Open Joints or Openings in Retaining Walls in Deep Excavations (Cofferdams, Sheet-Pile Walls)   Although there are a variety of very interesting and quite unique case histories, the following case is typical for the effectiveness of prepolymer grouting (Fig. 5B.27).

When the Maas River (with heavy boat traffic) near Liege (Belgium) was widened, an existing natural gas pipeline had to be extended below the level of the future river bottom. A deep sheet-pile cofferdam was installed in the river bed, leaving a "window" for the pipeline. The sheet piles in the window reached only to the obvert of the pipeline. The new cofferdam could not be dewatered because of the ground water flow through the window, via the very permeable gravel horizon below the river bottom.

Because of environmental concerns and the presence of a ground-water flow in the gravel below the riverbed, water-reactive prepolymers were used to create a seal around and below the pipeline in the "window" area. Sleeve pipes were installed via a hollow pipe vibrated into the bottom of the river and a "knock-off shoe." The entire grouting operation (around the clock) lasted only one weekend.

Preventive Seepage Control in Rock Tunnels (Prior to Blasting)—Grouting Prepolymers in Conjunction with Cementitious Grouts in Leaking Blast Holes to Maintain Advance Rate   During the driving of an exploration decline at the Moss Lake gold exploration project (near Thunder Bay, Ontario), major in-flow and consequent flooding stopped the exploration work. Old exploration drill holes connected overlying aquifers with the exploration horizon. After the in-flows within the tunnel were grouted with prepolymers, tunneling activities resumed. Blasting was only allowed when all blast holes were dry. Leaking blast holes were grouted with prepolymers or a two-component prepolymer cementitious mixture. Blasting could then be carried out within 30 min after grouting. This approach saved the operation a 24-h waiting period that would have been necessary with the classic cement-grouting method (allowing the grout to cure) (Fig. 5B.28).

Tunnel Grouting in Hardrock Tunnels   In 1987, during the installation of a deep tunnel in Geneva (Switzerland) for CERN (Centre European de Recherche Nucleaire), major infiltration problems were encountered. The 27-km (17-mi) long rock tunnel is located 160 m (525 ft) below

FIGURE 5B.28 Seepage prevention.

the water table. The tunnel was excavated with a WIRT TBM and had a diameter of 4 m (13 ft) (Fig. 5B.29). At one point, a chalk horizon with interbedded clay seams was intersected. The clay was extruded from the seams, which resulted in a major in-flow. A cement grouting program was launched. After injecting 600 metric tons of cementitious grout, holding up further progress of the tunnel for more than 60 days, no progress could be measured. The total in-flow (aquifer limited) remained steady at 400 L/s (6300 gal/min). A decision was made to try water-reactive hydrophobic polyurethane prepolymers. The total in-flow was reduced to 37 L/s (9.8 gal/s) in only 2 days of grouting, meeting design criteria. Spetec H100 was used for this project. Total quantity injected was 3500 kg (7716 lb).

Dam and Dike Grouting to Prevent Seepage through the Abutment or the Embankment An extensive rock grouting program was performed in Montezic (French Alps) in the Digue de Monnes (owned by Electricité Défence) in 1988 to stop further deterioration (leaching) of an old grout curtain (sodium silicates and cementitious grouts).

A sleeve-pipe grouting program was launched to restore the integrity of the old dam. The casing grout surrounding the sleeve pipes clogged some of the minor fractures, which reduced access of the grout to the formation. (It is not good grouting practice to use casing grout around sleeve pipes for rock grouting.) Towel packers (or MPSP barriers) would have been much more appropriate (in lieu of the sleeve pipes) to create a grout curtain along the center line of the dike.

Rock hydrofracturing was required to access and grout the open cracks (with flowing water) and to eliminate seepage through this dike. The set time of the water-reactive grout was selected at 75% of the flow-through time. Twenty-five tons of Spetec H30 water-reactive prepolymer were used on this project.

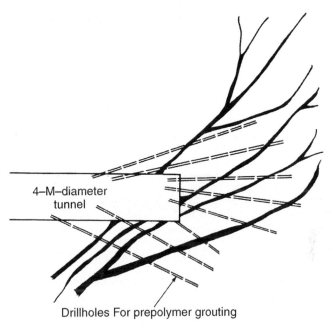

FIGURE 5B.29 Grouting clay seams in chalk (tunneling).

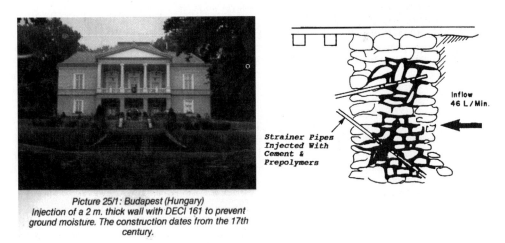

Picture 25/1: Budapest (Hungary)
Injection of a 2 m. thick wall with DECI 161 to prevent ground moisture. The construction dates from the 17th century.

FIGURE 5B.30 (*left*) Dutch embassy in Budapest, Hungary; (*right*) grouting brick and stone walls for water seepage control.

Structural Repairwork in Leaking Brick and Natural Stone Structures The 2-m thick, badly leaking stone walls in the basement of the Dutch embassy in Budapest (Hungary) were successfully sealed and rehabilitated by injecting Deci 16 in conjunction with a stable cementitious grout in the walls of this seventeenth-century building. Horizontal holes in a 40-cm-spaced pattern were installed in all exterior walls and pressure grouted until refusal (Fig. 5B.30). Quite often, reinforcing steel is placed in the drill holes to "stitch" the wall together.

Pin Piles for Temporary or Permanent Support to Tunnels or Structures—Especially in Emergencies
A typical example of a minipile that can be installed from confined spaces is the *combined root pile,* developed by this author. The pile consists of two concentric steel pipes: the outside pipe is a steel sleeve pipe to carry out a hydrofracturing/compaction cement grouting program; the inside pipe makes it possible to inject a solution grout through the pile point, into a competent pervious layer, to create superior end bearing.

Combined root piles are used to transfer load to deeper bearing strata when classic piles cannot be installed because of space and other constraints. Typical applications include renovations of older buildings or underpinning applications (preventing settlement because of construction adjacent and below the level of the footings).

Occasionally pin piles are installed to create a foundation system for new construction. This method is only economical when the rock formation is deep and a sufficiently thick sandy horizon with sufficient bearing capacity is present to form the base to provide support for the pile. The combined root pile works predominantly on end bearing.

A field application involved the combined root piles supporting the busy road tunnel in Ruisbroek, Belgium, in 1982. To solve traffic jams at the intersection between a six-track railway and a busy street, decision was made to construct an underpass for the road-traffic. The tunnel was to go, directly below the tracks, with interruption of the train traffic limited to 1 day only. Static penetrometer testing showed that the soil was very compressible, to a depth of approximately 13 m below the tracks.

The 39-m (128-ft) long, 14-m (46-ft) wide, 4.8-m (16-ft) high road tunnel was constructed in a dry dock near the intersection and jacked in place directly beneath the tracks. The installation, including the excavation, took less than 20 h.

While the tunnel was being constructed, two parallel horizontal minitunnels were jacked at an elevation corresponding with the elevation of the floor-slab of the future tunnel. These tunnels were to serve as "foundation beams" for the tunnel, transferring the load via a number of combined root piles to a competent sand layer 8 m (26 ft) deeper.

From the inside of the two 1200-mm (47-in) diameter pilot tunnels, 40 combined root piles were driven down in 800-mm (31-in) long segments (using a VW-jack) to a depth of 8 m (26 ft). A simple one-way valve system prevented soil and water from entering into the inner pipe (through the pile point). The outside sleeve-pipe was grouted first (multiple stages), to compact the soil around the pile and to create some hydrofracturing wings, providing corrosion protection and some lateral bearing to the root piles (Fig. 5B.31). The inner pipe was used to create a hard soil conglomerate below the pile point. A low viscosity hydrophobic polyurethane grout (Deci 161) was used for this purpose.

The "grout bulb" provided the end bearing of the pile. All piles were loaded to 40 metric tons (36.4 US tons) and passed the test. After the piles were constructed from within the two pilot minitunnels, a teflon-coated track was installed in each one of the small tunnels. Concrete (reinforced) was placed into the tunnels, to provide support for the tracks and make them monolithic with the pin piles. The minitunnels supported by the combined root piles formed the foundation system for the road tunnel.

The road tunnel was jacked over the teflon-coated tracks, and the crown of the pilot tunnels was continuously demolished as the tunnel advanced, to allow the tunnel to slide over the teflon coated tracks.

The existing railway tracks were directly fastened to the roof of the road tunnel. The train traffic was interrupted for only one day. More than 10 years after the installation of this "world premiere," no settlement has occurred.

Soil Grouting to Prevent Loss of Soil and Subsidence near Retaining Walls in Harbor Loading Docks
Every harbor commission is dealing with the problem of loss of soil near retaining structures. At the Texaco loading dock near Zeebrugge (Belgium), ground was lost through the horizontal joints of an ingeniously designed cantilever retaining wall (Fig. 5B.32)

Grout holes were drilled and sleeve pipes installed through the retaining wall intersecting the horizontal plane about 2 m from the face of the retaining structure joint. At low tide, a "custom-tailored," low-viscosity, water-reactive, hydrophobic prepolymer was injected, successfully stop-

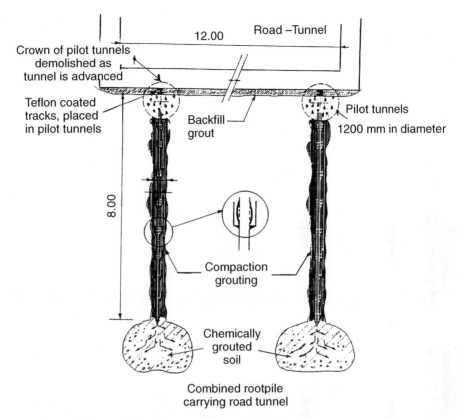

FIGURE 5B.31 Root-pile installation to stabilize road tunnel.

ping further internal erosion through this joint. A similar project has been executed in 1990 at the Pugsley Terminal in St. John (New Brunswick).

Renovation of Human-Accessible Sewers This is by far the most important application (volume-wise) for water-reactive prepolymers. The first rehabilitation of a human-accessible sewer, subject to internal erosion (even soil piping conditions, involving the grouting with water reactive prepolymers) occurred in Mechelen (Belgium) in 1977. The technique, now applied all over the world, only slowly found its way to North America. The first job of this kind was executed at Niagara Falls (Canada) (Fig. 5B.33).

During 1988 and 1989 the pavement of this very busy road collapsed a few times causing some spectacular accidents (vehicles driving into substantial sink holes. A 25-year-old sewer, located at a depth of about 7 m below the pavement, appeared to be the cause of the problem.

Loss of fine soil particles through leaking joints triggered an "internal erosion process" which in turn created voids above the water table. These voids became bigger and bigger, until the cohesion of the soil was no longer able to bridge the voids: the voids collapsed, resulting in major sinkholes. These sinkholes were usually formed after a rainfall or in the spring when the soil thawed.

The remedial measures consisted of the following activities:

- Installation of a proper ventilation system in the sewer line
- Installation of bulkheads and pumping the incoming sewage to the next manhole
- Cleaning of the section to be repaired

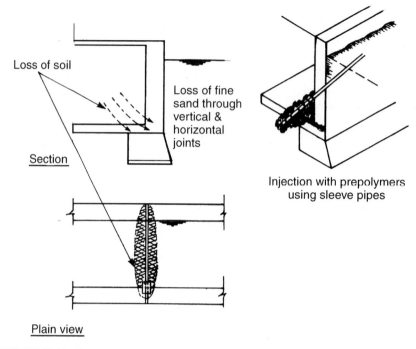

FIGURE 5B.32 Grout placement to prevent soil erosion at retaining walls (harbor loading decks).

- Drilling of at least four injection packers in each joint as pressure-relief holes and sealing the zone between the packers with hydraulic cement or with pads drenched in prepolymer
- Grouting of the joints (from packer to packer) with a water-reactive prepolymer until the grout travels through the entire joint, sealing it completely
- Installation of sleeve pipes (anchored in place, using prepolymer collar pads) from the inside of the sewer line followed by a cement suspension grouting program, embedding the sewer line in cement-based suspension grout and providing the required lateral uniform support to the sewer line to prevent ovalization and cracking.

More than 1.5 km (0.93 mi) of sewer grouting has been carried out in Niagara Falls since September 1989, successfully rehabilitating the old sewer for a fraction of the cost of a new line and without interrupting the traffic.

Hydrophobic "Flexible" Prepolymers. These products (such as Flex 44, MME Flex, Mountain Grout Flex, Webac 157, and DeNeef's Hydroactive Grout Flex L.V) remain flexible in time and are *form-stable*. The products have a high viscosity. The foam contains a certain percentage of open cells, when curing under atmospheric pressure. As a result, grout has to cure under pressure to form an effective elastic seal.

To evaluate a flexible prepolymer, a sample should be made in a closed bottle and the cell structure examined. To be acceptable, the foam should consist of very small, regular closed cells.

The flexible prepolymers should be used only for joint and crack injection and seepage control and not for soil grouting. The high viscosity of the prepolymer is an advantage when in-flows have to be sealed. The high viscosity increases the resident time of the prepolymer in the formation. Note, however, that the flexible hydrophobic prepolymers are less reactive than most other water-reactive prepolymers.

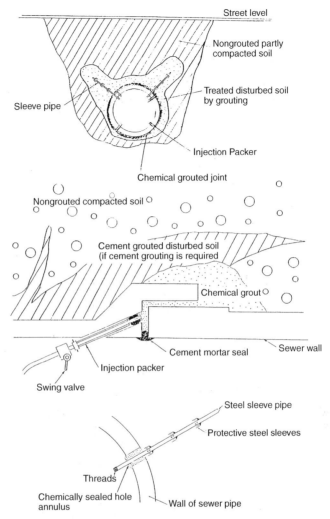

FIGURE 5B.33 Sewer rehabilitation (accessible).

Some applications are sealing leaking expansion joints and sealing leaking active cracks. As an example, the N.V. Nooren Diving Services from Stadskanaal (The Netherlands) developed a unique grouting method to seal open joints in water-retaining structures, below the water level. Recently, applications in the harbor of Nieuw Amsterdam, retaining walls in Smilde, Vissershaven Ymuiden, retaining walls in Seacanal Ghent—Terneuzen have successfully used this technique.

The patented method consists of the following procedures:

- The bad joint is cleaned.
- A sleigh is lowered under the water table and positioned over the crack.
- Via vacuum pads, the sleigh is anchored on the wall and the grouting "chamber" inserted in the sleigh frame. The chamber fits over the joint.

- This chamber is injected with a mixture of water and Spetec F 1000. The grouting pressure and the expansion of the grout create an elastic seal in the joint.
- The sleigh is moved up, and the operation is repeated at the next location.
- The fast set times allow quick and efficient execution of the job.

Hydrophilic One-Component Polyurethane Grouts. These products manufactured by 3M (US), Spetec (Belgium), Dai Ichi Kogyo Seyaku Co. (Japan), Togo (Japan), De Neef (Belgium), Denys (Belgium), Prime Resins (US), and other companies and are distributed under a variety of trade names such as CR grouts & Scotchseal, AV-Grouts, Superseal, Polygrout, MME Multigel, Penegrout, Strata Tech Series, and so on.

These products are grouted in conjunction with water and form a hydrophilic gel or a hydrophilic foam (depending on type and mixing ratio). The products often are more efficient than the hydrophobic prepolymers at stopping major in-flows, because they are more reactive (i.e., they react much faster, and the duration of the reaction is much shorter).

Hydrophilic gels appear not to be stable in time. The stability differs from product to product. The cured gels tend to physically absorb water in certain quantities, thereby increasing their porosity and losing their bond and some of or all their mechanical characteristics. Some products do not survive accelerated aging tests. Since these products absorb water after gelation, they exert a swell pressure that sometimes cannot be sustained by the structure, which might result in structural damage.

Utmost care is recommended when selecting these products for an application. When the products are mechanically confined, the postswelling is rather a positive characteristic, since a tighter "gasket" is created. It is the author's experience, however, that these products are very questionable for long-term solutions. Several projects using hydrophilic resins have been evaluated, and the long-term (over 3 years) performance is poor (author's experience and personal conversations with Todd Edmunds).

Most hydrophilic resins contain solvents (acetone) and are TDI-based (toluene di-isocyanate). Both the solvent factor and the presence of TDI make these products hazardous to the applicators unless substantial precautions are taken.

Products based on MDI (diphenylmethane-4-4-di-isocyanate) are considered significantly less toxic to the operators than TDI-based resins, and exposure-related problems are not anticipated under normal conditions. The higher volatility of TDI-based resins causes potential exposure-related problems during all phases of the product application. TDI-based isocyanate resins are known to be possible carcinogens. TDI-based resins frequently contain acetone, or other solvents, which increases volatility and exacerbates potential toxicity.

Due to the greater potential health hazard, use of TDI-based resins should be avoided whenever possible. The people in charge of projects where these products are indeed used should monitor the concentration of isocyanate vapors at the working site and ensure that the maximum threshold limit values are not exceeded. This limit has been established at 100 ppb, which is easily exceeded when working in enclosed areas, even when proper ventilation is in place.

For the sake of the health of the people employed in the grouting industry and its credibility to the public, a growing number of grouting contractors and engineers have urged the manufacturers of TDI-based products to stop production.

Examples of typical applications follow.

Sealing Expansion Joints in Submerged Structures During the construction of the Nipawin Hydro Electric Power Dam in Northern Saskatchewan, a gap between the dam and the structures upstream was noted. The gap was expected to open to twice its size, when the water in the headpond was raised to expected production levels. According to the engineers' lab tests, AV202 was the only grout that appeared to be able to deal with the widening of the expansion of the joints. No accelerated aging tests were performed. The AV202 was grouted (by divers) in vertical and horizontal expansion joints in 1985. Subsequent drilling through the joint revealed that the product was fully saturated, had become very porous, had lost its bond to the concrete, and—at best—was reducing the in-flow into the drainage blanket.

FIGURE 5B.34 Cutoff wall installation (sleeve-pipe grouting).

Soil Grouting in Certain Small-Scale Projects (Fig. 5B.34) In the Jackson Bluff dam, near Talahassee (Florida), a grout curtain was specified. The intent was to create a cutoff wall in the earth dam adjacent to the spillway. The in situ hydraulic conductivity was determined by injecting water at various soil horizons. Sleeve pipes, properly embedded in casing grout, were used for the testing and grouting. It was discovered (using equations of Cambefort-Naudts) that the spacing between the grout pipes could only be 60 cm (24 in) in order to create overlapping grout cylinders, instead of the 10 ft (3 m) specified by the engineer. The contractor successfully completed this project using a 3M product (CR356). This product had to be cooled down to 40°F (4.4°C) (using ice), and the maximum amount of retarder was used to slow down the reaction and obtain a gel time of 8 min. The selection of CR356 for soil grouting in this project was ill-conceived, because of its short reaction time, rather high viscosity, and questionable long-term stability.

Grouting Joints in Nonaccessible Sewers (Fig. 5B.35) Hydrophilic polyurethanes have been used to replace the acrylamide used before AM9 obtained its bad reputation. If grouting is properly carried out, the joints of leaky sewers can be successfully sealed with hydrophilic polyurethane grouts. Most sewer grouters, however, still prefer to use acrylamide grouts because of price and excellent track record, dating back to the fifties. Acrylamides, if properly handled, constitute less health hazards than the TDI-based hydrophilic polyurethanes, according to independent applicators.

Polyurethane Elastomers. These products consist of two components:

- The polyol (usually a poly-ether polyol) on which a catalyst is added to select the gel time
- The isocyanate, preferably an MDI (diphenyl methane di-isocyanate) type, since TDI (toluene di-isocyanate) types pose severe health hazards at ambient temperatures.

5.354 SOIL IMPROVEMENT AND STABILIZATION

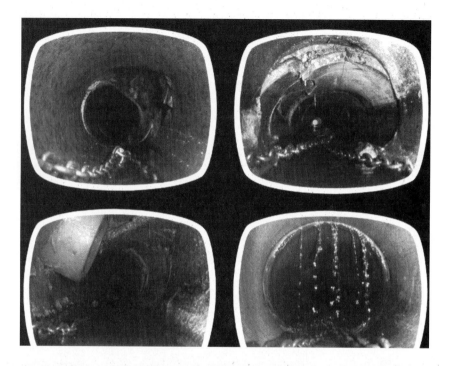

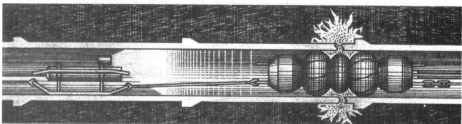

FIGURE 5B.35 Sewer grouting (nonaccessible).

The first generation of polyurethane elastomers has been used in Germany since the early 1960s under the name *polytixon*. The polytixon products are TDI-based; oils were used to lower the viscosity and to dilute the polyol.

Since 1984, a new generation of polyurethane elastomers has been introduced to the grouting industry resulting in a considerable improvement to rehabilitation and seepage control grouting.

HYDROPHOBIC POLYURETHANE ELASTOMERS Some examples are Resicast GH40 & 65, MME Polycast Std., MME Polycast WTR, MME Polycast LV, and DeNeef Hydroactive elastic. In cured form these grouts are totally inert and hydrophobic, and most of the above products remain flexible in time, even at low temperatures. However, it is the author's experience that the technical literature of some of these products is technically flawed and misleading.

The set time can be adjusted within narrow limits, by adding a catalyst (third component). These products have an excellent penetrability in cracks and are more suitable than the classic epoxy grouts for structural repairwork in concrete, because of their flexibility. An engineered approach toward continuous monitoring of cracks has been long overdue. Wiechmann[50] developed a continuous monitoring device which allows evaluation of movements of joints and cracks, pinpoints the cause, and determines in a meaningful and engineered fashion the appropriate remedial measures (methodology and product selection).

The tensile strength and the bond of the urethane has to exceed the tensile strength of the concrete, to be suitable to seal active cracks in concrete structures. It is our experience that cracks and joints fluctuating more than 40% in width are very difficult to seal, unless a cap on the upstream side of the joint or crack can be created.

The major reasons for structural grouting are:

- Sealing the cracks against penetration of air, water, soil, or steam
- Protecting reinforcing steel against corrosion
- Making the structure monolithic again

Contrary to what is accepted by a number of practitioners in the United States, a good bond of the grout on both sides of the joint is required to obtain a proper seal. Acidization (introduction of 5% diluted phosphoric acid), followed by "flush-grouting" of cracks with water, is absolutely necessary prior to grouting. After acidization, the cracks should be flushed extensively with water to remove residues of calcium and aluminum phosphates completely. These residues are not hygroscopic and remnants are not detrimental to the performance of the grout.

Few applications require the high compressive and tensile strengths that can be obtained with epoxy injection resins. Epoxies have far superior characteristics than concrete but are not able to absorb any deformation or movement of the structure. As a result, new cracks adjacent to the old ones often occur if the old cracks are active. In projects in which the structural integrity has to be restored, use stitch grouting rather than relying on the bond of the grout on the sides of the cracks.

Multiple-hole grouting is becoming increasingly popular. Rather high pressures are applied to saturate porous concrete or rock, or to fill fine fissures. The longer a borehole is exposed to a given pressure, the higher the spread-out radius of the grout. To obtain an "economical" take, expose a number of boreholes simultaneously to the same injection pressure (multiple-hole grouting). The flow through the porous medium is governed by the equations applicable for soil grouting.

The adjustability of set times of two component polyurethane elastomers is a major advantage over epoxies. Moreover, polyurethane grouts (MDI-based) are far less hazardous to the applicators than epoxies.

Sealing Cracks and Joints in Concrete and Masonry Several hundred meters of active cracks were discovered in the suspended slab and walls of the underground parking garage of a poorly constructed condominium complex near Toronto. First the cause of the movement was eliminated by reinforcing the foundation system, using pin piles. Afterward, the cracks in the suspended slabs were successfully injected with Resicast GH40 (1989–1990). It was discovered that this product becomes very brittle at low temperatures. In the walls, Polycast WTR was used. The latter product remains flexible, even at $-25°C$. A recent inspection revealed that the grouting still provides a 100% water-tight underground parking garage.

Injectable Tube Applications (Contact Grouting) Grouting via preplaced injectable tubes is gradually becoming an integral part of designs to create watertight construction joints, to "glue" old concrete to new concrete, or as an alternative to water stops.

Air and water leaks often occur through cold joints in structures. To prevent these types of problems, injectable tubes were invented in Germany in 1982 and are now used worldwide. This is one of the best applications of grouting as a construction tool rather than as a remedial measure.

The *injectable tube* is a flexible grout pipe usually consisting of a (steel) spiral-protected perforated pipe, including one or two layers of geotextile to prevent the passage of solids into the injectable tube (Fig. 5B.36).

The tubes are usually installed tight against the old concrete in construction joints, before the next pour of concrete is made (Fig. 5B.37). The tube will drain excess water from the fresh concrete, but the cement particles cannot enter the tube. In a later phase, after all shrinkage in the concrete and settlement (of the structure) has occurred, the most suitable solution grout is injected via the injectable tube into the joint, to create either rigid, flexible, or expansive seal between the construction elements. Verstraeten[51] is one of the first authors to publish about these tubes. He pro-

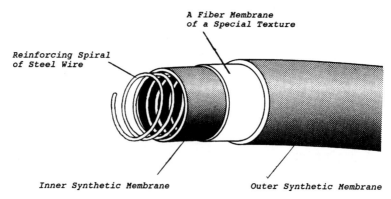

FIGURE 5B.36 Injection tube. (*From DeNeef brochure.*)

vides a historical overview of the evolution of the injectable tubes in Europe and reports on the various types available on the world market.

Because the injectable tube can be strategically installed in a construction, expansion, or cold joint (also between prefabricated elements), the "pay zone" can be injected in a later stage, without time-consuming drilling. The injectable tube is accessible via feeder tubes, installed at regular spacing. Some research was conducted in Germany in 1990 to determine the practical spacing of the feeder tubes for various types of injectable tubes. The conclusion was reached that 2 to 3 m (6.6 to 9.8 ft) was an appropriate spacing. Multiple-feeder-tube grouting allows for fast completion of an almost foolproof grouting project.

It should be emphasized that it is absolutely necessary to "acidize" the injectable tubes prior to grouting; ideally, a 5% concentrated phosphoric acid solution should be used to dissolve the calcium salts in the pores of the geotextiles surrounding the injectable tubes. After acidization, the tubes should be water- and air-flushed to detect any seepage through the joints. Pointing of seepage paths is necessary prior to grouting to create the proper confinement for the grout and complete filling (under pressure) of the joints.

Product selection is vital to the success of the operation. If dry or wet passive cold joints between structural elements have to be filled, a low-viscosity epoxy injection resin (water-compatible) should be selected. This is often used to glue prefab elements together in high-rise buildings not subject to thermal effects. A remarkable project in which the epoxy and injectable tubes system was applied successfully was in the huge LNG (liquid natural gas) concrete tank in Zeebrugge (Belgium) during the late 1980s.

To seal leaking active joints, flexible, hydrophobic polyurethane prepolymers are mostly used, although water-repellant hydrophobic polyurethane elastomers should be the products of choice if only minor movements are anticipated. Especially in tunnels and underground parking lots, the above described situation is encountered. More specifically, in the cold joints between walls and suspended slabs, leaks often occur. The injectable tubes should be located upstream of the water stop. The feeder tubes should run through or around the water stop and should be properly fastened to the concrete walls to prevent plugging by dirt.

A major (three-levels-deep) underground parking lot in downtown Toronto was one of the first applications for injectable tubes in Canada. The tubes were placed in *all* the cold joints. Injectable tubes are routinely installed in cold joints in most underground structures (tunnels and buildings) in continental Europe.

Leaking expansion joints containing an injectable tube should preferably be injected with a flexible, hydrophobic, polyurethane prepolymer. In North America, often hydrophilic polyurethane prepolymers often are used. Since the durability of these products is questionable, selecting this type of product involves considerable risk. Expansion joints in retaining walls near rivers and canals are routinely treated as described.

FIGURE 5B.37 Application of injectable tubes to drain excess water from fresh concrete (cement particles cannot permeate tube). (*From DeNeef brochure.*)

The ideal chemical grout to be installed via injectable tubes is the two-component hydro-expanding hydrophilic polyurethane elastomer (with hydrophobic matrix). Since it can only be installed in damp or dry injectable tubes, the injectable tubes should be air-flushed prior to grouting. Once the elastomer is cured, the grout expands (a predictable and well-determined percentage) upon contact with the water to form a tight gasket. This expansion is a 100% reversible process that does not affect the characteristics of the elastomer. This system was successfully installed, amongst others, in the following major projects: the Liefkenshoek tunnel in Antwerp (Belgium), various dams (lift joints) in France (EDF), and the Fjellinjen tunnel in Oslo (Norway).

Sealing Techniques around Bulkheads   The installation of a flood bulkhead in a gypsum mine in Shoals, Indiana, provided for contact grouting between the rock and a newly constructed bulkhead via seven rings of injectable tubes. Resicast GH40, a hydrophobic polyurethane elastomer, was injected via the feeder pipes, creating a perfect seal between the rock and the flood bulkhead. In the paper of DesRochers and coworkers,[52] the details of this application are described (Figs. 5B.38 and 5B.39). A similar injectable tube approach was designed by the author for a massive concrete plug in the Sungai Piah hydroelectrical project in Malaysia (1992), to provide a tight seal between the grouted rock formation and the concrete plug. The products used in this project were Spetec AL 1302 and Spetec Flex. This first grout is a hydrophobic water-repellant polyurethane elastomer with increased shore-hardness (and less flexibility), in comparison to the other product in this family available on the market. The grout was specially formulated for the project. The "flex" is a hydrophobic, one-component, flexible polyurethane prepolymer.

The oldest injectable tube application around a flood bulkhead—constantly subjected to 1250 psi (9 MPa) brine pressure since 1985—was executed in the potash mine in Rocanville. The product used for this application was the CSR sealring, a hydrophilic polyurethane elastomer with a hydrophobic matrix. The seal is still performing perfectly.

Water-compatible hydrophobic grouts (MME Polycast WTR, Spetec AL1302)   This product displaces water in cracks and cures, without foaming or reacting with the water, to form an hydrophobic elastomer with acceptable adhesion to the medium. The elastomer is not affected by wet-dry cycles and is stable in time.

This product is suitable for grouting into water-bearing formations, cracks, or joints (not running) and for grouting into injectable tubes filled with water to create an elastic "gasket" between two members. It is also very suitable for repair of wet cracks in concrete structures.

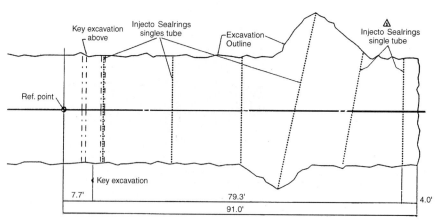

FIGURE 5B.38   Sealing rocks—bulkhead contact-injectable tube layout. (*From "Flood Protection for a Deep Gypsum Mine," by DesRochers, McKinstry, Naudts.*)

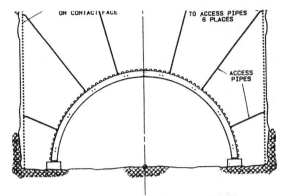

KEY SECTION WITH INJECTO TUBES

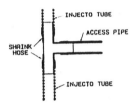

DETAIL - TEE CONNECTION

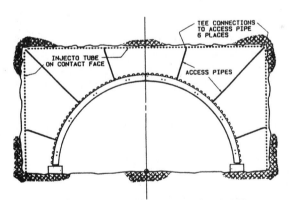

DRIFT SECTION WITH INJECTO TUBES

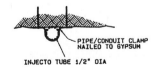

DETAIL - INJECTORING ATTACHMENT

FIGURE 5B.39 Detail of grouting procedure. (*From "Flood Protection for a Deep Gypsum Mine," by DesRochers, McKinstry, Naudts.*)

5.360 SOIL IMPROVEMENT AND STABILIZATION

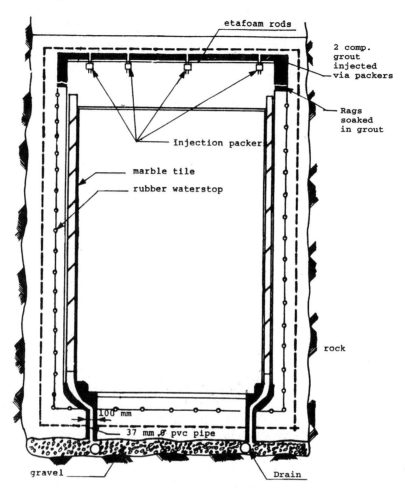

FIGURE 5B.40 Application of water-compatible, hydrophobic grouts (to shut off persistent seepage).

Preferably, this hydrophobic grout is injected as a one-component system whereby the A and B components are thoroughly mixed, prior to grouting. A catalyst can be added to adjust the gel time.

One of these products was used in solving the problem of persistent seepage, a serious nuisance in the pedestrian tunnel between the central block and west block on Parliament Hill in Ottawa (Fig. 5B.40). Water bypassed several waterproofing membranes (copper, rubber) and entered the tunnel via the expansion joints. During the rehabilitation program, carried out in December 1989, a water-repellant, two-component polyurethane grout was injected, forcing the water out of the joint and resulting in a flexible completely inert two-component polyurethane elastomer. The product selected for the project was Polycast WTR.

HYDROPHILIC POLYURETHANE ELASTOMERS (DOWEL CSR SEALRING, MME POLYCAST EXP) These grouts swell after they have cured in contact with water. They are predominantly used either for crack injection in dry or damp joints or for injectable tubes. Once cured, the grouts swell in a predictable way when in contact with water, forming a tight gasket.

The major difference with the pure one-component hydrophilic prepolymers is that the elastomers have a hydrophobic matrix, with "active antennas," attracting water. The percentage of swelling is completely predictable (in the new-generation elastomers) and depends on the formulation. These elastomers do not lose their mechanical characteristics, because of their hydrophobic matrix.

The old CSR sealring has been used extensively for "pickotage rings" in shafts and for sealing around flood bulkheads. The product is the most hydrophilic in this class, and, when given space to swell, it will expand up to 75% of its initial volume. CSR tend to lose most of its mechanical characteristics when swelling freely, which has caused secondary seepage in some cases. As a result, most sealrings had to be regrouted to tighten up the first ring.

Products such as Polycast EXP are only partly hydrophilic. A predetermined amount (custom-tailored formulation) of expansion will occur if the grout is given the room to expand. As a result, a tight seal can be created by properly designing the required expansion.

Typical applications include:

- Pickotage ring in shafts
- Sealing around flood bulkheads
- Prefabrication of "swell seals" of any size or dimension for joint sealing or prefab applications
- Injectable tube grouting (tunnels, parking lots, swimming pools)
- Cast-in-place expansion joints

Case history 1: E18-Fjellinjen Tunnel in Oslo (Norway) (Fig. 5B.41) This recently constructed tunnel incorporated in its design the installation of injectable tubes in every cold, construction, and expansion joint. Even for the construction, systematic grouting of the bedrock took place via 20-m (66-ft) long holes bored in a fan shape ahead of the tunnel face.

The injectable tubes were all grouted, after the structure had settled, with a two-component polyurethane elastomer, Polycast EXP. This grout cures to form an elastic rubber. When water penetrates the joint, the grout swells out to a maximum predetermined amount. In this project, the Polycast EXP was formulated to swell 20% after being submerged in water. (Extensive lab testing has confirmed the manufacturer's claim about the controlled and reversible expansion of the cured urethane elastomer.) The Polycast EXP has been one of the most fascinating developments of the late eighties.

Case history 2: In-flow during construction of Orly VAL Tunnel In 1990, during the tunneling at the Orly VAL project (France), a major in-flow (100 L/s or 26 gal/s) under considerable pressure was blocked. A defective tail joint in the tunneling shield (compressed air) caused the problem. Since the tunneling machine could not be "grouted in place," grouting was possible only from distinctive locations. The dye tests revealed resident times of less than 2 s. A two-component (cement-urethane) grout was developed by Verstraeten (personal conversation): The A-compo-

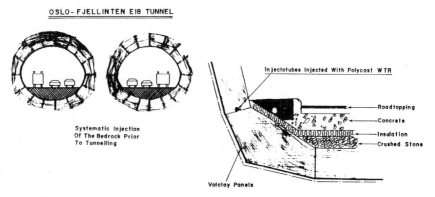

FIGURE 5B.41 Polycast EXP-injectable tube system grouts bedrock.

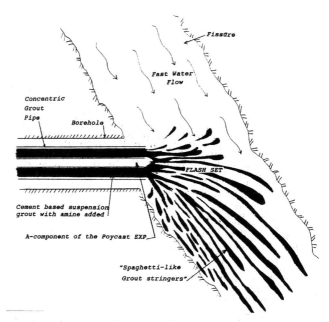

FIGURE 5B.42 Two-component grout system application for major water in-flow.

nent consisted of a partly stable cementitious grout on which an amine was added. The B-component consisted of a specially formulated polyol (A-component of the Polycast EXP). The mixing ratio (A:B) was approximately 10:15/1. Both products were injected via concentric pipes, whereby the B-component was introduced only at the exit point of the grout line (Fig. 5B.42). No static mixer was required. The reaction time (in spite of the water flow) of the grout was less than a second. The cement grout turned into an instant gel, forming spaghettilike stringers, clogging up the seepage path. Within hours, the entire flow was cut off. The soft-grout "spaghetti" continues to cure like a regular cement-based suspension grout, providing a permanent cutoff.

Two-Component Polyurethane Foam Grouts for Mining Applications. Some examples are Rockgrip, ICI-Foam, and Bevidol/Bevidam. The result of the reaction between a polyol (R-OH) and an isocyanate (R1-NCO) is the creation of a polyurethane. Depending on the type of polyol, blowing agents, catalysts, and a wide variety of foams with different characteristics can be formed, differing only in:

- Density
- Cellular structure
- Compressive strength
- Reaction pattern (cream time—tackfree time)
- Water absorption
- Fire resistance

The isocyanate has a high affinity for water and thus has a tendency to "steal" the isocyanate, leaving not enough isocyanate for the polyol to form a complete reaction.

These products are mainly injected to stabilize loose and fissured rock. They are extensively used in European, South Africa, and, to a lesser extent, in U.S. coal mines. A number of types deal with various ground conditions, including slow setting for deep grouting, fast setting for shallow depth grouting, and wet conditions. The applications are very crude and are mostly carried out with small two-component TURMAG pumps. If this foam catches fire, the gases liberated are lethal. During the last 15 years some tragic accidents in various mines in the United Kingdom, South Africa, United States, France, and Belgium have taken place, killing large numbers of miners. Only in United Kingdom and in some regions of France has the product been banned.

Typical applications (other than insulation applications) include:

- Stabilization of unstable rock in mines
- Sealing previous formations in front of flood bulkheads
- Sealing gaps and joints around ventilation doors
- Filling damaged air caissons (locks)
- Pipe line applications: (application as temporary floating blocks for river crossing, erosion blocks, and so on).

5B.3.2.3 Sodium Silicates

Classic Sodium Silicate–based Solution Grouts. Sodium silicates are usually prepared on site by mixing (static in-line mixer or small batches) a diluted sodium silicate solution with a hardener. Sodium silicate–based grouts are evolutive solution grouts: A gradual increase in viscosity takes place as soon as the silicate solution and the hardener have been mixed. The concentration of the hardener and the degree of dilution of the sodium silicate determine the pump time (and hence the gel time in the formation) of the mixed solution.

The key issue that determines the set time is the dilution factor. Classic sodium silicate solutions are, based on the authors' extensive experience (in several hundreds of projects), *not* permanent. Especially in the United States these solutions are still used for permanent soil modification purposes. Also in Europe, most major grouting contractors fail to admit the longevity problems.

In 1984, when the author asked the chief chemical engineer of the manufacturer of the most advanced hardener at that time—Durcisseur 1000—about the manufacturer's guarantee regarding longevity of a properly neutralized sodium silicate Durcisseur 1000 formulation, to be injected in fine sand, the following answer was received: "1 week." (This answer is in line with the author's experience.)

Only in particular environments (high pH soil) is the neutralization of the sodium silicate not reversed. Synersis is the major factor in this context. *Syneresis* is the phenomenon whereby water is exuded after gelation, causing shrinkage of the gel.

Creep effects of sodium silicate–grouted soils subjected to loading is a major problem documented by several authors. Although most grouting specialists claim that fine sands injected with sodium silicate–based suspension grouts are not subjected to syneresis, the author's opinion is that the so-called creep of sodium silicate–grouted soils is directly the result of the syneresis of the sodium silicates in the pores between the grains. The following observation is of particular importance: Eleven years after a massive grouting program had been executed for the construction of a subway in Europe, "grouted soil" samples were obtained within the grouted soil several meters below the water table. In the same locations, cores of grouted soil had been obtained during the execution of the project. The unconfined compressive strengths on the grouted soil, initially, were all in the 3 to 9 MPa (435 to 1305 psi) range (very hard gel). Eleven years later, none of the samples had a measurable compressive strength. Only the smell (of the hardener) of the sodium silicate–based grout remained. At the time this publication is written, a new catastrophe likely directly linked to the use of classic sodium silicate solutions, occurred on a grouting project in the docks of A. Sodium silicate grout was used by one of the major European grouting contractors.

For the sake of the credibility of the grouting industry, grouting contractors should pay far more attention to the durability problems associated with this family of products.

SODIUM SILICATE SOLUTION The density of the sodium silicate is expressed in degrees Beaume (°B). Typically, industrial sodium silicate solutions are supplied in the 36° to 42°B range. Sodium silicate is manufactured by burning a mixture of silica sand and sodium carbonate at 1400°C, which in turn is dissolved in water, using an autoclave at 140°C (284°F) under a pressure of 0.6 MPa (87 psi).

The higher the density, the higher the viscosity and the stronger the gel that will be formed. Usually the sodium silicate is diluted on site to 35° to 36°B, for use in the field.

SODIUM SILICATE HARDENERS In the past, the following hardeners have been used:

Formamide: creates a weak gel—very toxic (still used in the United States).

Ethyl acetate: very flammable, toxic hardener—causes synerisis fairly quickly. This was the hardener used in the Nurnberg catastrophe. Still extensively used under code names. Creates a hard, nonpermanent gel. Even contaminated ethyl acetate–based waste products from the photo industry are used as hardeners.

Calcium chloride: creates an instant gelation. Was used in the Joosten grouting process; has no practical use and has only historical importance.

Sodium bicarbonate and sodium aluminate: create weak gels—used for water-tightening purposes (in conjunction with diluted sodium silicate solutions). In Japan, hard (flash-setting) gels are obtained using these type of hardeners, and their grouting systems and methodology (i.e., *limited area grouting*—LAG—and double-tube drilling—DDS) make it possible to inject these flash setting grouts.

Inorganic acids (HCl, H_2SO_4, etc):—used in the darker days of our industry. Unfortunately still taken out of the old box (once in a while) by people trying to reinvent the wheel.

Nowadays, the following hardeners are used:

Glyoxal: mainly in Japan, producing debatable end products.

Mixtures of di-basic esters and organic acids: Durcisseur 600 series—developed in France during the 1960s; most commonly used commercially available hardeners during two decades. Durcisseur 1000 is produced by distilling the Durcisseur 600 to obtain a purer product. Good neutralization of the sodium silicates is obtained if the concentration of the sodium silicate (39°B concentration) in the grout is over 60% (by volume) and over 10% hardener is used.

The latter hardeners have been well researched, predominantly by Peruchon.[53] His graphs and tables serve as practical guidelines for the applicators. Note, however, that the soil chemistry, as well as external factors, can influence the data and prevent duplication for the project at hand.

CHEMICAL REACTION BETWEEN SODIUM SILICATE AND "DURCISSEURS" This reaction is shown below:

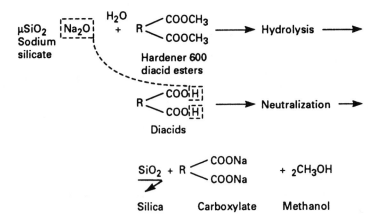

The Durcisseur has to be diluted with the dilution water, to prevent the formation of flocs, when introduced into the sodium silicate solution. The Durcisseur hydrolyzes slowly to produce diacids, which in turn neutralize the sodium silicate and release the silica.

The rate of neutralization is determined by the quantity of Durcisseur necessary to neutralize the soda of the silicate. Insufficient neutralization results in nonneutralized soda in the "water glass," slowly reversing the neutralization process and causing the deterioration of the gel.

The manufacturer recommends a neutralization of over 70% to obtain a "durable" gel and suggests (in its literature) that a neutralization of 65% will provide a life expectancy of "up to 24 months."

Note that the most expensive component of the sodium silicate solution is the Durcisseur. The higher the concentration of Durcisseur in the formulation, the shorter the gel time. Workable gel times can be obtained by diluting the sodium silicate solution, which unfortunately results in a weaker gel.

A number of conflicting factors make the selection of the appropriate sodium silicate solution quite a task for the grouting engineer. Careful analysis of Figs. 5B.43 through 5B.46 make it possible to establish formulations for particular applications. Weak gels can be formed using grouts with viscosities, less than 10 centipoise, whereas the harder gels will be formed from solution grouts with a higher viscosity. The higher viscosity and the shorter gel time restrict the grout spread of sodium silicate solutions, which form a strong gel.

The classic sodium silicate–based solution grouts are sensitive grouts and should be used only for temporary soil modification projects (mainly temporary stabilization of soils below or near structures—when excavation, or tunneling is taking place within the influence zone of the footings—or stabilization of tunnel faces).

Classic sodium silicates in conjunction with hardeners containing organic components (such as Durcisseur 1000) are "grundsetzlich verboten" (forbidden), unless a special permit can be obtained from three independent, official organizations (one pertaining to environment, one pertaining to ground-water supply) and the municipality. After a major grouting project in Berlin, independent research teams discovered a tremendous increase of bacteria, feeding on the decomposing sodium silicate gel. Research work in the universities of Gelchenkirchen and Berlin linked the increase of bacteria in the ground water to the leaching of the organic component in the hardener.

Newer Developments with Sodium Silicate–Based Grouts

SILACSOL   This type of sodium silicate–based grout appears to have a much more permanent character than the classic sodium silicate–based formulations. Silacsol is a true solution grout,

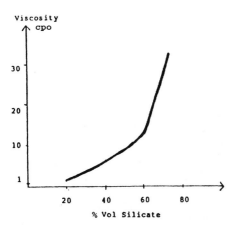

FIGURE 5B.43 Viscosity in terms of silicate concentration in the grout.

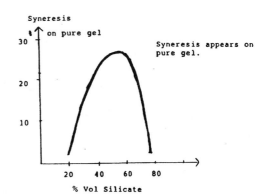

FIGURE 5B.44 Syneresis in terms of silicate concentration in the grout.

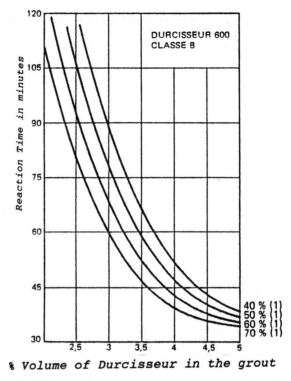

FIGURE 5B.45 Volume of Durcisseur versus reaction time. (1) % volume of sodium silicate in the grout. (*After Peruchon, Grout Injection of Ground and Concrete Structures.*)

characterized by a nonevolutive reaction or gelation. It has been developed in France by one of the leading grouting contractors.

Silacsol is made of an activated silica liquor and an inorganic calcium-based reagent. Contrary to commercial alkaline sodium silicates, the liquor is a true solution of activated silica. The activated, dissolved silica, when mixed with the reagent, produces calcium hydrosilicates with a crystalline structure which is quite similar to the one obtained by the hydration and setting of cement. The resulting product is a complex of permanently stable crystals.

The degree of neutralization is a nonissue when using Silacsol, because there is a direct reaction on a molecular scale, without the presence of silicized water. Hence, syneresis is not a consideration. The activated silica mix has the stability of a cement grout, owing to the nature of the resulting products (insoluble crystals of calcium silicate) and to the absence of polluting by-products.

The very low permeability of pure, cured silacsol is in the order of 10^{-10} cm/s (0.4×10^{-10} in/s) remaining constant over prolonged test periods. Unfortunately, this product is not readily available on the open (grouting) market.

Carbo-rock (sodium silicate carbon dioxide) This is a Japanese-developed product in which the hardener of the sodium silicate is CO_2. The CO_2 is almost completely absorbed by the solution, when introduced, via the double-tube rod injection technique. Shimada and coworkers[54] claim they can obtain either a short gel time or a long gel time, depending on the type of mixer, the amount of CO_2 introduced, the permeability, and the relative location of the soil horizons. The publication does not provide any details about the exact procedures nor does it report on the long-term stability of the gel. Shimada claims that the carbo-rock is environmentally friendly (nonpolluting) and that the reaction mechanism is easy to control and adjust (gas as hardener).

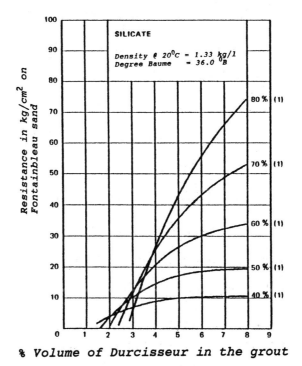

FIGURE 5B.46 Volume of Durcisseur versus resistance. (1) % volume of sodium silicate in the grout. (*After Peruchon, Grout Injection of Ground and Concrete Structures.*)

5B.3.2.4 Acrylic and Acrylamide-based Grouts

Classic Acrylamides. Acrylamide-based grouts are true solution grouts. For the longest time, acrylamide-based grouts have been associated with AM-9, which was developed in the United States in the early 1950s and backed by competent chemists. During almost 30 years it was used in a variety of applications, mainly to prevent or stop seepage through a variety of media with measurable permeability. AM-9 consists of a mixture of monomers which are polymerized at ambient temperature.

"Officially" the production of this type of monoacrylamide was interrupted in 1978, in the United States. Insiders in the grouting industry know better. A number of acrylamide-based products, commercially available at present, have been manufactured in Japan, but manufacturers in Europe and North America also produce these products.

Monoacrylamide-based grouts consist of a mixture of two organic monomers: acrylamide and a cross-linker, such as methylene-bis-acrylamide. Usually there is about 20 times more acrylamide than cross-linker. If more cross-linker is used, a harder or stiffer gel is formed (but still classified as a weak gel). Usually both products are mixed by the manufacturer in a 19:1 ratio. The handling of the powders requires utmost care and precautionary measures. Therefore it is strongly recommended to bring the grout to the site as a 40 to 48% solution, which is further diluted on site, usually to create a 10% solution. At this point, its viscosity is roughly the same as water (1.5 centipoise). As long as the concentration of the monomers is above 3%, a properly catalyzed acrylamide will form a gel. If even the slightest chance for dilution below this critical concentration exists, the original concentration should be increased to avoid the problem. If there is still a chance that excessive dilution might take place, acrylamide-based grouts should not be used, to avoid serious pollution problems. For those applications, one-component water-reactive hydrophobic polyurethane prepolymers or hot melts should be used instead.

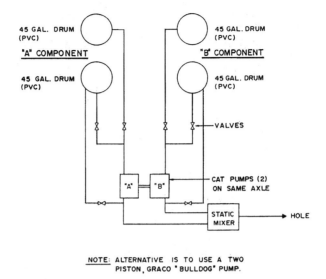

FIGURE 5B.47 Two-component grout mixing (acrylamide).

Acrylamide-based grouts are usually injected as a two-component system (Fig. 5B.47). On the A side, the concentrated acrylamide solution, water, and the activator (usually tri-ethanol-amine) are mixed. The B side consists predominantly of water, catalyst (ammonium persulphate), inhibitor (potassium ferricyanide)—used only when extended gel times are needed—buffer (to bring the pH to 8—baking soda usually does the trick), and additives to make the acrylamide compatible with the formation.

With acrylamide-based grouts, gel times from a few seconds to 10 to 12 h can be obtained in a predictable fashion and can easily be duplicated. Acrylamide produces an exothermic reaction. As a result, the gelation in the field will be a multiple of the pot life obtained in the laboratory. For critical soil- or rock-grouting applications, this correlation has to be established, via pilot-grouting programs. The temperature of the acrylamide solution (as well as the ground water) plays a very important role. The temperature curves supplied by the manufacturer should be consulted, prior to the execution of a grouting program.

Cured acrylamide consists predominantly of water, trapped in the "brush heap" structure of the gel. As a result, mechanical pressures can force water into the cured gel and cause it to swell. When the grout is placed in a dry or not completely saturated environment, the gel will gradually shrink. When saturation is reinstated, the grout will absorb the same amount of water it has previously lost. This is a reversible reaction, which has been tested through thousands of cycles.

Cracking of the gel occurs, resulting in an increased permeability, during the process. As a result, acrylamides should be used only for grouting in structures not subjected to wet-dry or freeze-thaw cycles. Ions can migrate through the water in a gel. As a result, acrylamide gel will steal salt ions from the formation when injected in salt and potash mines, causing secondary permeability in the formation.

Acrylamide still is the ideal grout to saturate formations (both rock and structures) in which the permeability is governed by matrix porosity. Hydraulic conductivities as low as 0.01 Lu have been obtained on a routine basis, when grouting in limestone formations (Naudts[55]). Because acrylamide forms rather weak gels, it should not be injected in wide cracks subjected to considerable hydraulic gradients. Selective pregrouting of the formation with cement-based suspension grouts is necessary and is good grouting practice anyway.

Applicators should be aware that a number of metals as well as sunlight cause flocculation of the acrylamide. No direct contact between acrylamide and steel or brass couplings, drums, valves, or lines can take place. Usually stainless steel or PVC materials are used instead. When it is not always possible to avoid contact between acrylamide and steel pipes (e.g., very deep drill holes such as used in the oil industry), the pipes are flushed with boric acid solutions to reduce the formation of flocs.

Acrylamide should be considered a very durable grout. In very harsh environments, acrylamides are far more resistant than cement-based suspension grouts.

Acrylamides in powder form or in solution are neurotoxic. If properly handled and utilized in applications where no excessive dilution can take place, acrylamides are *safe*.

Acrylamide-based grouts are still extensively used for automatic sewer-grouting applications in nonaccessible sewers and for seepage control where excessive dilution is not a problem. Also, remarkable results have been obtained in creating grout curtains under extremely high water pressures (potash mines).

Newest developments. A number of products were developed to fill the gap left by AM-9. The majority of these products came on the market and disappeared. The most commonly used classic acrylamide-based product in North America is probably the AV-100, which is not substantially different from the old AM-9. In Europe, Rockagil BT_2 is predominantly used for grouting in nonaccessible sewers. This product, manufactured in France, contains urea-formaldehyde, and performs more or less like the old AM-9. It is however more difficult to create a tailor-made formulation for particular applications with this product.

Siprogel is a sodium silicate acrylamide combination, used for seepage control in structures. The A component contains the active acrylamide components and the hardener for the sodium silicates. The B-component contains the sodium silicate solution and the catalyst for the acrylamide. A rather tough flexible gel can be obtained with this product. Gel times can be regulated within rather narrow limits. In Europe, this product has been overshadowed by the water-reactive prepolymers.

5B.3.2.5 *Epoxy Grouts.* Epoxy injection for structural repairwork is a valid and important technique in the rehabilitation business. For soil and rock grouting, epoxy injection resins should *not* be considered. The epoxy grouting industry has lost much of its old appeal since epoxy resins have been used too often for inappropriate applications, resulting in failures. The days that epoxy repair technology was franchised like fast-food restaurants are not completely over, but the dust has settled. Treating symptoms does not cure problems, and an engineered approach to structural problems often stemming from foundation-related matters is usually in order.

One major manufacturer of raw materials in Europe, Rutgers Epoxy (Bakelite), should be given credit for democratizing the industry. Their engineers taught engineering contractors how to formulate epoxy-based products, bringing prices down to affordable levels. More important, appropriate information about the chemical resistance, mechanical characteristics, performance, and toxicity of resin-hardener combinations was brought into the public domain. Test results and progress were shared by the industry.

Nowadays, epoxy injection resins are used no longer as curealls but as distinctive applications, where their superior strength characteristics are necessary to restore structural integrity.

Epoxy resins consist of bisphenol A&F resins. They are characterized by their epoxy number. Epoxy injection resins are reactively diluted resins with much shorter chains.

The hardeners are either amine- or amide-based resins. The more toxic hardeners are often partly reacted with an insufficient amount of bisphenol resin to form a less toxic "adduct hardener." The hardeners are characterized by their "amine equivalent." Once epoxy number and amine equivalent are known, the stoichiometric proportions of resin and hardener can easily be calculated.

The essential part of an epoxy resin is the reactive epoxide group and, in the amine hardener, the reactive amine hydrogen. The use of epoxy equivalent and amine equivalent values has been created because the hardening process takes place by the reaction of one epoxide group with one amine hydrogen. Epoxy resins with different oxygen content have different epoxide equivalents. Similar polyamines, which are used for the hardening process, have different amine equivalents, calculated on the amine hydrogen content.

It is common to calculate the amount of hardener required to react with 100 g (3.5 oz) of the epoxy resin. Therefore, the epoxy value of the resin has to be known as well as the amine equivalent of the hardener. The amount of hardener per 100 g (3.5 oz) of resin equals the amine equivalent times the epoxide value. Often the epoxide value is provided by the supplier of raw materials. The epoxy value equals 100 divided by the epoxide equivalent. With this information it is easy to verify the amount of resin various hardeners react with. The latter is necessary when hardener combinations are made.

Hardeners can be mixed to obtain particular characteristics. Formulating epoxies has similarities with dog breeding: Different characteristics (especially chemical resistance) are obtained by making different resin and hardener combinations. Additives mixed with either resin and/or hardener further change mechanical and rheological characteristics and keep the exotherm of the reaction under control.

Epoxy injection resins do not cure if the medium in which they are injected is too cold. The manufacturers' information should be carefully checked prior to the application. A thin-layer test, whereby a specimen, coated with the injection resin, is placed in a similar or identical environment as the one in which it will be grouted, is recommended to evaluate curing time and other mechanical characteristics.

The injection resins can be categorized into three categories:

1. The water-compatible epoxy resins, which repel and cure under water
2. The regular epoxy injection resins, which in turn can be divided into several subcategories, according to their viscosities (the very low viscosity resins require a fairly high minimum curing temperature or the resin never cures)
3. The semiflexible epoxies with low mechanical strength

The first two items are the most commonly used types of epoxy injection resins. Urethane elastomers are mostly used in lieu of the flexible epoxies.

Epoxy injection resins should not be injected into voids or wide cracks, because they can develop an exotherm which cannot readily be absorbed by the medium. Should this occur, the epoxy loses a great deal of its mechanical strength. It is better to inject an epoxy-based suspension grout or to resort to the preplaced aggregate technique when grouting larger voids, wide joints, or wide cracks. Only in rare situations (high tensile strength and/or chemical resistance requirements) is it necessary to use epoxy-based grouts in wide cracks. Cement-based grouts can be utilized instead on most occasions.

Epoxy does not bond to dusty, dirty, or calcium-salt-covered concrete. A chain is only as strong as the weakest link. Pressure flushing prior to grouting will likely remove some debris, but proper acidization (with diluted phosphoric acid) is required to remove calcium salts from fissures.

A lot of very debatable epoxy injection projects are performed in a wet environment or even under water. Most often, the surface preparation of the formation is ignored, and the epoxy formulations are not tailored for the specific needs of the project. Usually, other injection products are more economical or suitable for those applications.

5B.3.2.6 Phenolic Grouts: Phenoplasts and Aminoplasts. Phenolic foaming grouts were developed in Alsace (France) for the mining industry during the late 1970s to replace the two-component polyurethane foams for rock stabilization and ground-control purposes.

Especially in the French coal mines and in the deep South African mines, products like Mariflex are used in great quantities. The A component consists of formulated phenolic and urea formaldehyde–based resins. The B-component consists of concentrated organic and inorganic acids.

The various Mariflex formulations are injected with two-component pumps. The products can be injected in wet formations but are not truly water compatible. During the exothermic reaction, a competent foam is formed. Contrary to the polyurethane foams, this phenolic foam does not burn (M1 fire classification).

In the mines, Mariflex is also used to create firewalls and ventilation barriers and to prevent loss of air (spray-on applications) or to control methane gas. A variety of Mariflex products have been developed for the various types of ground control problems. Mariflex has occasionally been used in North American mines and has been used in European hard-rock tunnels but only sporadically in civil engineering applications.

Resorcinol Formaldehyde. These types of solutions were developed in the 1960s and are—most unfortunately—still used. Phenoplasts always contain a phenol, a formaldehyde, and an alkaline base. Resorcinol is toxic and caustic. Formaldehyde is well known to the general population, because its "fumes" cause chronic respiratory illnesses—even at ambient temperatures. Sodium hydroxide is a caustic material.

If the three ingredients are properly proportioned and mixed, the environmental hazard is minimal, although the catalyst may be gradually leached out. As is the case with any sensitive materials, things tend to go slightly wrong in the field, which will result in unreacted excess of one of the components, creating an environmental hazard. *Resorcin-Phormol* (as the French call it) is still actively promoted by a French grouting consultant.

The initial viscosity of resorcinol formaldehyde is very low (similar to acrylamides), and it remains low until gelation occurs. The degree of dilution with water determines the gel strength. The higher the concentration of the active components, the faster the reaction and the higher the gel strength.

Geoseal and Rockagil, two other products that are still used in the industry, are also considered phenoplasts.

For the sake of the survival of the grouting industry, manufacturers and distributors of these types of products should seriously reconsider further distribution and promotion in view of the potential human and environmental harm they might be causing.

Aminoplasts. Like the phenoplasts, aminoplasts require an acid environment to complete the reaction between a urea and a formaldehyde at ambient temperatures. Grouts using urea solutions are toxic and corrosive due to the formaldehyde and acid (from the catalyst) content. Prepolymers contain less formaldehyde than the monomers and therefore are less hazardous than the monomers. The foam, or gel formed at the end of the reaction is inert but generally contains small amounts of unreacted formaldehyde. This caused severe problems to homeowners who had their house insulated with urea-formaldehyde-based foams.

Urea-formaldehyde-based foams are still extensively used in the German mines (Iso-Schaum) and in the French mines (Iglo-Neige), to form fire barriers and temporary ventilation walls and to control or prevent the flow of methane.

Several urea-formaldehyde grouts have been used in North America, including Diarock, Cyanalock 62, and Herculox, but these are either gone or only sporadically used now.

Chrome Ligno Sulfonates. During the 1960s, some experimental solution grouts (chemical grouts) were developed for soil-grouting purposes. At the time there was very little or no concern about environmental matters, and some very toxic materials, potentially causing permanent damage to the ground-water quality, were injected in the soil.

Ligno-sulfonate-based grouts have only historical importance and are a reminder of the dark, pioneering days of solution grouting. During the production process of wood pulp, ligno sulfites are present in the wash water, which, through the addition of potassium bichromate ($K_2Cr_2O_7$) or sodium bichromate ($Na_2Cr_2O_7$), create gelation. The dry extract of this bisulfitic solution contains 76% sulfonated lignite, 19% reductive sugar, and 4% mineral ashes. A chemical reaction can be triggered by adding bichromate to the bisulfite solution. The amount of dilution water, as well as the amount of bichromate versus the dry extract of bisulfite, determine the gel time and gel strength.

The sugar is removed from the bisulfitic solution through a fermentation process (prior to grouting). As a result, the bichromate will only react with the ligno sulfonates and will not further oxidize the sugars, which speeds up the reaction.

Grouted sands can obtain an unconfined compressive strength of 0.5 MPa. This is considered to be a rigid gel. Weak gels can be obtained by adding more water and reducing the amount of bichromate added.

Usually the cured grout (or grouted soil conglomerate) is yellow as a result of the nonreacted bichromate. This bichromate is soluble in water and will color the ground water and surrounding soils. This makes it easy to track and identify the worst polluted areas in the vicinity of the grouted soils.

Especially when rigid gels had to be created, a relatively high amount of bichromate was introduced in the solution, which resulted in "extra" bichromate and the formation of Cr^{6+} ion solutions, which are, in contrast with the Cr^{3+} ion solutions, very toxic. Permanent environmental damage has been caused by grouting these solution grouts.

An old text by a French (unidentified) author (in the late 1960s) states: "It is necessary to take the proper precautions to avoid pollution of the ground water by the inevitably incomplete reaction between the ligno-sulphonates and the bichromates." The author, however, stopped short of explaining which and how these precautions should be implemented in the field.

5B.3.2.7 Hot Melts. Hot melts are not as widely used as the other main families of grouts (i.e., the suspension grouts and the solution grouts), although the hot melts have been used for sealing purposes, since the days of the Crusades. Hot bitumen grouting was successfully applied to stop major leaks below several dams in the United States during the first 30 years of the century. This old-fashioned but most effective grouting technique was adjusted to the technology of the 1980s in a few remarkable projects:

- Grouting of Lower Baker Dam in Concrete (Washington)
- Grouting of Stewartville Dam near Arnprior (Ontario)
- Grouting of major in-flow for paper mill near Sault-St-Marie (Ontario)
- Grouting of dike at Deferiet hydro project (New York)

All these projects had the following in common: A substantial water in-flow below a water-retaining structure had to be cut off or controlled.

Hot bitumen is injected in flowing water. Upon contact with water, the bitumen cools off immediately and forms stringers. As the grouting operation progresses, a grout flow through the cracks is established. The bitumen strings act as lava, slowly migrating through the cracks (Fig. 5B.48). The interior of the bitumen bulb remains warm and provides the medium to transfer pressure from the injection point to the fringes. Analyzing the flow-pressure data during a hot bitumen grouting program makes it very obvious that the grouting mode is compatible with permeation grouting. The wider the cracks, the more important the heat-sink provided by the bitumen becomes. Since bitumen cools off very slowly, the grouting operation can be interrupted for considerable periods without loss of access to the grout hole.

The placement technique requires a lot of attention. Insulated steel pipes with internal and external heating cables work very well. The grout spread can be monitored, via thermo couples installed in observation holes.

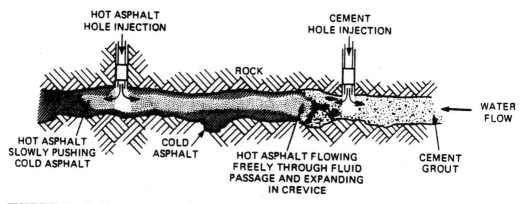

FIGURE 5B.48 Hot bitumen-cement grouting.

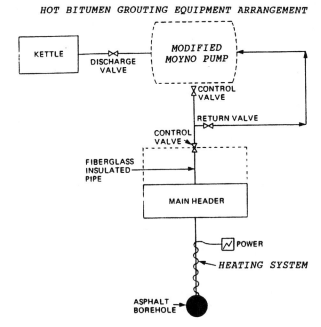

FIGURE 5B.49 Equipment arrangement for grouting.

Hot bitumen is not washed out, even in the wildest flows. It should however be pointed out that the thermal shrinkage of bitumen is substantial. Moreover, bitumen remains a fluid (with a viscosity of several hundred thousands of centipoise), even at very low temperatures. It is therefore mandatory to back up the bitumen with a cement-based grout to create a permanent cutoff system. The cement-based suspension grout takes advantage of the heat provided by the bitumen and causes a flash set.

The selection of the type of bitumen, as well as the details of the grouting design, are covered in a paper by Lukajic and coworkers[56] (Fig. 5B.49). The author has been credited with reviving this antiquated technology and molding it into a professional grouting methodology.

The worst leak was Lower Baker Dam—approximately 2000 L/s. The resident time for a grout in the formation was only 4 s, as established during a dye test.

5B.4 DESIGN AND SPECIFICATIONS FOR A GROUTING PROGRAM

5B.4.1 Introduction

Most grouting specifications are not properly written. Some grouting contractors have specialized in taking advantage of the lack of knowledge of the principal* to claim (often completely inappropriately) (and collect) large amounts of money for extras.

---

*In this section we will use the word *principal* to indicate either the owner or the engineer (referring to the person in charge of the operation on behalf of the owner).

This is one of the main reasons why grouting as an industry is on the decline. If the main grouting contractors would devote the same amount of effort to training their staff in engineering aspects and professional grouting procedures as they do on building frivolous claims, there would be hope for the industry.

Unfortunately, it is the author's experience that the site supervisors and the site engineers of most specialized grouting contractors are, at best, good operators with shallow knowledge of the applied science of grouting.

The principals, on the other hand can be classified in one of the four following categories:

- The useless spectators, who find grouting very boring. They are usually harmless. Contractors tend to accidentally open a bypass or sampling spout, when these principals become a nuisance.
- The administrators of activities and pay items, trying to keep the contractor honest. They are not interested in technical matters and only look after administrative issues.
- The competent and interested grouting engineers (assessing the program continuously), totally focused on the operation. They tend to be sucked into a technology vacuum often deliberately created by the sly or inexperienced or incompetent grouting contractor. This type enjoys working with a good grouting contractor and can easily relate to the practical and technical problems and will work with the contractor to find solutions.
- The pragmatic and arrogant types, exercising authority, justified or unjustified. Unless they are more knowledgeable than the contractor and fully comprehend the operation and the situation, they are a considerable liability to the owner. The "know-it-alls" with some shallow and/or pure theoretical knowledge, lacking patience and understanding to analyze a situation, especially can create serious problems. A little knowledge is always dangerous....

It is very important to define in the contract documents which type of contract should govern the execution of a project. Houlsby[19] provides very interesting information pertaining to this issue.

5B.4.2 Contractual Arrangements

Many claims and disputes could be avoided if the contract documents clearly stipulated which type of contract will govern the execution of the work. Moreover, an appropriate breakdown of pay items can considerably reduce the "lottery" aspect in the bidding process.

Some types of contract are types A, B, C, and D.

5B.4.2.1 Type A. The contractor provides the specified equipment, operators, and materials to perform the project. The work is performed under the direction of the "principal." Under these circumstances, contractors can claim to be "just the dumb contractors who do what they are told." Since most contractors do not provide anything more than just that anyway, it is a concept that should be used more often. It is the contractor's duty to provide only skilled and experienced operators; the principal takes care of administrative and technical matters.

Specifications should clearly state the type of contract. The operation can still be carried out as a unit rate contract, but prices should be lower than the ones obtained from the so-called specialist grouting contractors.

5B.4.2.2 Type B. In addition to what is required under a type A contract, in a type B, the contractor also provides the key *technical* people—experienced engineers and supervisors—to design or modify some of the practical aspects and details, further providing vital technical input and direction to the operation. The assessment and direction of the grouting operation is in the hands of the specialist grouting contractor, and her or his work only needs to be quietly checked by the principal, or eventually the principal's inspectors. The role of the principal is merely administrative, looking after the book work and public relations and verifying if the work is performed according to specifications. The principal does not interfere but relies on the expertise of the specialist contractor.

Most contracts in North America are type B contracts or at least are intended to be type B contracts. Quite often the contractor displays such incompetence that the principal is sucked into a technology "vacuum." At this point, the principal either interferes personally or brings in a specialist. This is exactly what the sly contractor wants: now he or she can play the "dumb contractor," declining all responsibility regarding the quality of the work. The type B contract thus becomes a type A contract. The worst part is that the contractor hereby shifts the entire responsibility for the outcome of the operation to the principal. Courts and arbitrators unfortunately have sided with contractors in this type of situation because contract documents were not clear enough.

Thus the true, professional grouting contractors, able to perform a type B contract and bidding accordingly, are undercut by the would-be specialists. The author has seen this scenario all too often. The worst part is that the principal ends up with all kinds of claims for extras and a marginally acceptable job.

Our suggestion is to provide a two-phase contract. The first phase, which should form an integral part of the overall tender package, should be to perform a very small portion of the grouting work, at the tendered unit rates for phase 1, to demonstrate the contractor's ability to perform a genuine type B contract. The specifications should stipulate that the contractor can be terminated without cost to the owner, and without any further explanation, at the end of the first phase. The contractual obligations of the specialist contractor to the owner should include keeping the same personnel on site for the execution of the second phase of the contract. Expensive litigation can be avoided this way.

5B.4.2.3   Type C.   The contractor provides a turnkey solution. Only the target is outlined by the specifications. The contractor provides all or most of the investigation, evaluation, engineering, bid items, estimated quantities, procedures, and complete execution of the job. The contractor further has to guarantee the performance of her or his work for a certain period outlined in the "target chapter" of the agreement.

The principal evaluates the various tender packages provided by the various groups before selecting the specialist contractor. The principal should evaluate which package appears closest to reality, especially where quantities are estimated.

Usually contractors provide their own quality control work and often have a peer review board with specialists scrutinizing approach, methodology, progress, workmanship, and quality of the execution.

5B.4.2.4   Type D.   *Negotiated contracts* are similar to type C contracts, but in them the owner commits to a particular organization, to buy its proposed solution. The owner can still entertain other concepts, without breaching the contractual obligation, with any other party proposing different solutions.

Usually, preengineering to establish the budgets and methodology is executed at cost price. Various approaches are possible from this point on: The project can be performed at unit rates, based on a cost plus a certain percentage for profit and overhead or using a cost and lump sum approach. An open-book accounting system, recording all costs and expenses during the execution of the project, is recommended. In a rather new approach, the owner creates a joint venture with the specialist contractor.

5B.4.3   Pay Items

Providing a list of bid items, for which the contractors submit unit rates or lump sum prices, allows remuneration to the contractor as fairly as possible for the actual work performed. The owner wants the best possible contractor for the best possible price. Therefore, it is in the owner's interest, to eliminate the "lotto" element as much as possible during the bidding process. This way, the most suitable contractor usually prevails. Exact quantities are seldom known in grouting projects, and therefore, the bid items should be selected to reduce the risk to both owner and contractor.

Contractors tend to "front-end" load the project in their mobilization and demobilization. Some distort their bid even further by overpricing items that will be—in their opinion—executed for sure, and they underestimate money needed to perform other bid items. Thus if the quantities for which the contractor has provided a below-cost price are exceeded, some contractors will become "inventive" to avoid losing their shirts. Nobody wins in such a situation.

The following tips, based on the author's experience, can help to provide fair bid items:

1. *Mobilization and demobilization.* The amount should be determined by the principal and should be filled in on the bid sheets. This way, all bidders have the same price for this item. The amount should be fair and should compensate the contractor to perform the preparation of the project, move equipment to the site, dispose of the waste generated during the program, and remove equipment from site.

2. *Drilling of holes.* State amounts per-unit length (meter, foot)

3. *Preparation of grout holes.* State costs in both a lump sum per hole (includes flushing and/or acidizing and/or setup—whichever applies) and a per-unit length (to install sleeve pipes, stand pipes, MPSP pipes, or packers—whichever applies).

4. *Grouting of holes.* State amounts per hole (brought to absolute refusal) plus per volume unit (placed grout) plus per-unit weight (kg, pound) of the various ingredients and additives constituting the various formulations. All ingredients going into the mixer are paid for, even if grout is dumped, unless dumping results because of mistakes made by the contractor. This way, the contractor gets compensated for all the additives and components of the various mixes, regardless of the formulations selected by the principal.

By specifying an orderly multiple-hole grouting program and the number of grout-phases in each zone or hole, the contractor has a good handle on how many holes will be finished, regardless of their permeability and grout-absorption, in a given time frame. It should be made clear to the contractor that all holes are brought to absolute refusal.

The guidelines for the grouting program (see Sec. 5B.3.1.7), applicable and appropriate for the particular grouting project, should be explained in the specifications. This way there are no surprises to the contractor.

5. In crack-injection projects, unit rates per linear meter (or foot) of crack should be a bid item. The unit rates should include all surface sealing, installation of grout ports, and acid and water flushing. The unit rates should also include an estimated average grout quantity per linear meter (or foot). The additional grout consumption should be compensated for per unit of volume. A reasonable estimate should be made regarding the additional quantities required.

6. The specifications (especially in type B contracts) should contain a clause to disqualify contractors from submitting unbalanced bids. Contract documents could even contain a statement to the extent that providing an unbalanced bid is an indication of technical incompetence, demonstrating a lack of knowledge about what is involved to perform particular tasks for which the unit rates do not cover the bare costs. This in turn forms the basis for rejection of the unbalanced bid. Even under type A contracts, provisions can be made to reject unbalanced bids.

5B.4.4 Design of a Soil-Grouting Program

5B.4.4.1 Definition of the Target. The reason for grouting has to be clearly defined. The problem at hand has to be fully understood. There is no point in treating symptoms unless the cause is tackled.

The type and extent of the field investigation prior to the execution of any work has to be decided on. Initially, assumptions have to be made to establish parameters from which a few potentially feasible grouting approaches can be derived.

Decisions have to be made when to perform field investigation—either during a design-build grouting program or prior to the tender for the work, independent from the grouting contract. The desired product has to be compared with what reasonably can be accomplished with various grouting approaches. Often, a pilot grouting program is in order to modify and adjust the original design and develop a fair tender package. Cost estimates have to be compared, and alternate solutions have to be evaluated.

5B.4.4.2 Field Investigation. Especially when the reduction of the compressibility of the soil is one of the prime or sole targets of the operations, a few vital soil characteristics have to be established:

- *Nature of the various soil layers.* This can be obtained with a number of representative boreholes. Samples from the various layers should be obtained and a sieve analysis established.
- *Compressibility coefficient of the various soil layers.* Ideally this is established with the penetrometer (Dutch cone). The compressibility coefficient can also (less precisely) be deduced from a standard penetration test. Usually the SPT values are deducted from the blow count driving the split spoon in the various horizons, during the execution of the boreholes.
- *Permeability coefficient of the various soil horizons.* These data are required first to evaluate the type of grouting to be performed, second to determine the type of grout formulations to be used, and third to calculate the grout spread that can be obtained with the selected grout formulations (in the case of permeation grouting).

The permeability coefficient can be estimated with Hazen's equation, with the American equation, or with the equation of Hooghout. These equations use the d_{10} of the various soil samples to estimate the in situ average hydraulic conductivity. This is a theoretical approach. It is our experience that for dense soils, these equations provide a good estimate.

The best way to establish the in situ hydraulic conductivity is by performing a water test using sleeve pipes properly embedded in casing grout. By injecting water at a constant pressure for a given time in a given soil stratum and monitoring the take, the in situ hydraulic conductivity can be calculated. Note that the line losses as well as the hydrostatic head should be taken into account to establish the effective grouting pressure. Also, the line losses through the entire testing system have to be determined prior to grouting, for different flows. Therefore, a sleeve pipe of the same type used in the field should be installed in a geotextile sock with the same diameter as the borehole. The geotextile sock should be filled with casing grout and allowed to cure. A double packer would be placed in the sleeve pipe. The pressures, recorded on the *X-Y* recorder, for different pump flows, indicate the line losses through the system, including the sleeves and the casing grout. A line losses versus flow curve is established to be used during the real testing program. Section 5B.1 discusses some real-life examples in detail.

Presence of ground-water flow or hydraulic gradient should be evaluated. Dye tests usually provide sufficient information to estimate the gradient.

The water chemistry, the presence of hydrogen sulphide or any other pollutants, must be evaluated. The presence of minerals, salts, gases, and other chemicals has a major impact on the reaction mechanism and the durability of a number of grouts. Especially in Europe, in the major towns, the subsoil often contains a variety of chemicals. At least once a year grouting projects fail because the presence of chemicals in the subsoil went undetected or was ignored.

5B.4.4.3 Actual Design (Permeation Grouting). Once all the relevant data are available, the design can be completed. The types of grouts have to be selected, and this is not too difficult. (See Sec. 5B.1.) The general guidelines for grout selection are:

- Regular, stable, cement-based suspensions should be used where the hydraulic conductivity is over 0.06 cm/s (0.024 in/s). The additives should be carefully selected to obtain optimum penetration, low cohesion, and good resistance against pressure filtration. Ideally, the grout should be treated in a particle size reducer (when and if it becomes available) prior to grouting, to further enhance its penetrability.

- Microfine cement-based suspension grouts should be selected to treat soils and formations with hydraulic conductivity greater than 0.001 cm/s (4.0×10^{-4} in/s) (and less than 10% silt).
- Water-reactive hydrophobic polyurethane prepolymers should be selected for soil grouting where there is a chance of washout.
- Classic sodium silicate solutions (in combination with Durcisseur 1000) should be selected for temporary soil modifications in soils only marginally injectable with microfine cement. This grout could also be considered for the grouting of secondary holes in which the primary holes have been grouted with microfine cement-based suspension grouts, for particular applications.
- Acrylamides should be selected where there is no chance for dilution nor washout during the curing and in soils for which a very low residual permeability has to be obtained. Acrylamide can be used after the soil has been grouted with microfine cement-based suspensions, even using the same sleeve pipes. Acrylamide is still the product of choice to permeate rock in which the permeability is governed by matrix porosity.

5B.4.4.4 Grout-Hole Spacing. Based on the actual in situ hydraulic conductivity of the soils, the grout spread is calculated for the various soil horizons. The equation of Cambefort should be used for soil horizons which are very thick and uniform. The equation of Naudts-Cambefort is more appropriate for layered soils.

Once the grout spread is determined, the hole spacing and location are established. The ultimate goal of the grouting operation plays an important role at this point: If a continuous cutoff curtain has to be established, the grout cylinders have to overlap. The spacing is then governed by the layers with the lowest permeability. If only a soil conglomerate with very low extra-compressibility (under the design load) has to be formed, the continuity between the grout cylinders is less critical.

The grouting pressures required to maintain a certain grout flow in a particular horizon can be estimated with the layer equation of Cambefort (Sec. 5B.2.1). This has to be compared with the in situ hydrofracturing pressure. Microfracture-induced permeation grouting is an acceptable technique, since the first type of fracturing occurs along semivertical planes, not lifting the soil.

5B.4.4.5 Depth of a Grout Curtain. If internal erosion below a water-retaining structure has to be prevented, the length of the flow lines can be determined with the criteria of Bligh and Lane. After examining numerous failures, Bligh and Lane came to the following empirical equation:

$$L_{\text{flowline}} > C_{\text{Lane}} (H - h)$$

where
C_{Lane} for gravel = 2
for coarse sand = 5
for dense undisturbed fine sand = 8
for disturbed loose fine sand and silt = 25

If the bearing capacity of the soil is the issue, then the method, as explained in detail in Sec. 5B.1, should be followed to determine the grouting depth.

5B.4.4.6 Required Grouting Gear for Suspension-Grouting Operations. Only manufactured sleeve pipes, properly embedded in casing grout, should be used for soil-grouting applications. The double packers should be leakfree, and enough spare parts should be kept on site. The use of approved high-shear mixers is mandatory. Holding tanks with slow paddle mixer should be in good working order and should have no grout buildup.

Grout pumps should be helical screw pumps (mohno pumps) or modern double-acting ram pumps with continuous output and adjustable speed. Dual acting, triple piston pumps are also suitable for particular applications. The pumps should not be leaking.

Flow-pressure recorders (*X-Y* recorders) with easy-to-read real-time printouts should be mandatory for most operations. Multiple-hole grouting devices are necessary in most cases. A particle-reducing mill should be requested as soon as the production models become available for full-size projects. A grouting shed, to protect the equipment and the crew against the elements, is also very important.

Special attention should be paid to the dry storage of products and materials on site. The environmental aspects and restrictions should be clearly spelled out. The removal of empty bags of the various ingredients, the handling of wash water and waste grout which are site-specific should also be addressed.

5B.4.4.7 Required Grouting Gear for Solution Grouts. The equipment suitable for the operation should be described, leaving room for minor deviations. The type of hoses, packers, multiple grouting devices, should be spelled out, but there should be always room for logical changes and improvements and suggestions by the contractor.

We do not want to go into details. Each type of solution grout has its particular most suitable equipment and setup. The principal should *never* provide specifications unless he or she has lived through a project in which the desired and particular family of grout(s) has been used.

Special attention should be given to the environmental aspects of the operation:

- Disposal of packaging
- Containment of spills
- Disposal of cured grout
- Disposal of cleaning agents, solvents, wash water
- Restoration of the work area

These should not be left up to the discretion of the contractor.

5B.4.5 Drilling for Soil-Grouting Projects

Bruce[57] has provided an excellent publication pertaining to this topic. His paper contains detailed information about the various ways to execute overburden drilling and points out the advantages and limitations of each system. The following is a summary of Bruce's paper.

5.B.4.5.1 Drive Drilling of Lancing. In principle, this is a percussive drilling system, in which a tube is drilled with the leading end terminating in either a knock-off shoe, bit, or crown. In particular for addressing soil conditions, and used predominantly for shallow applications, this technique is economical. Some companies use this technique for the installation of minipiles.

5.B.4.5.2 Rotary Duplex. With this method, the rod (with bit) and casing (with shoe) are simultaneously advanced. The basic method is rotary drilling.

5.B.4.5.3 Rotary Percussive Concentric Duplex. With this method both rods and casings are simultaneously percussed and rotated. The casings are very heavy, high-quality steel. In sensitive grounds, the bit can be retracted behind the casing shoe to minimize cavitation of ground and promote good flush return. In competent ground, flushing is introduced via an external flushing device, and high flushing rates at high pressures (2 to 3 MPa or 290 to 435 psi) are needed. Its drilling depth is limited (for the most robust types) to 60 m (197 ft).

5.B.4.5.4 Rotary Percussive Eccentric Duplex. Better known as the *ODEX system,* this drilling system allows drilling through soils containing "shifting boulders." This percussive duplex variant features a pilot bit with eccentric reamer that cuts a hole slightly larger than the casing.

5.B.4.5.5 Rotary or Driven Duplex Underreaming. The Acker drill is a typical example of this system. In principle, an oversized hole is cut by a bit, and the following casing is either driven or rotated. The oversize hole is not created by an eccentric bit but by activating-outward cutting blades above the pilot bit.

5.B.4.5.6 Double-Head Duplex Drilling. The two best-known makes are the Krupp and the Klemm drills. This is a very reliable drilling system to obtain very straight holes to a depth of 80 m (262 ft). This system is distinguished from conventional rotary duplex by the fact that rods and casings are simultaneously rotated but in opposite directions. The inner drill string, with right-hand rotation, carries either a down-the-hole hammer for hard ground conditions or a rotary bit for soft ground.

5B.4.5.7 Auger Drilling. This type of drilling is very popular for shallow soil investigation. The hollow-stem auger drill, especially, makes it possible to perform SPT tests, soil sampling, and sleeve-pipe installation. The continuous flight auger drill features the rotation of a "big screw" into the ground. This method of drilling is the most popular for small grouting projects.

5B.5 FIELD PROCEDURES

5B.5.1 Grouting Equipment

5B.5.1.1 Cement-based Suspension Grout. Very good publications by Houlsby[19] and Weaver[20] describe the various types of high-shear mixers and the importance of high-shear mixing in detail. They also elaborate on the holding tanks, the various setups, the return-line system, the recording of the grouting data (*X-Y* recorders), and a number of grouting and quality control devices. The following is a brief summary of the various, most commonly used components.

High-Shear Mixer. The high-shear mixer is used to establish a suspension grout in which the particles are separated from each other. The mixing usually occurs in a chamber below the drum, in which the ingredients are dumped.

The mixing chamber contains either a turbine, running in reverse, at very high speed (1200 to 2000 rpm), causing the high-shear action separating and wetting the particles, as in the Colcrete mixers, or a rotor with deep vanes, projecting down, creating great turbulence as in the Hany-type mixers. Other popular high-shear mixers are the Bachy-type and the Cemix. Figures 5B.50 and 5B.51 depict various types of high-shear mixers.

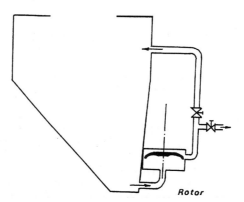

FIGURE 5B.50 High-shear (Hany) mixer. (*From Ref. 19.*)

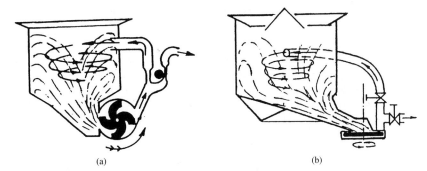

FIGURE 5B.51 High-shear mixers: (*a*) Colcrete, (*b*) Cernix. (*From Ref. 19.*)

It is our experience that there is a significant difference between the mixing action in the various types of high-shear mixers. Especially when the bentonite content of the mix is more than 4% (versus water), the difference in mixing quality becomes apparent. On some projects, we had the opportunity to compare various types of high-shear mixers. By executing mixing processes, containing the same ingredients, introduced in the same order, mixed for identical amounts of time, we noticed substantial difference in "apparent" viscosity (marsh funnel), cohesion, and resistance against pressure filtration. In other words, the wetting action in some mixers was not nearly as good as in others. The Colcrete-type mixers were far superior than any other type tested (advertising not paid for). The viscosity and the cohesion of the grout mixed with the Colcrete-type mixers was consistently higher than the grout prepared in the other type of mixers, indicating a better wetting of the particles.

The ideal mixing time depends on the type of mixer and the type of formulation. It is necessary to introduce the bentonite into the mix first and allow it to hydrate for at least 30 s, before inert or semi-inert fillers are introduced. The cement is the last component to be introduced into the mix.

If viscosity modifiers are necessary to control a runaway operation, they should be introduced in the holding tank.

In the oil-well industry, powerful jet mixers were and are still used. They are very suitable to place substantial quantities of suspension grout in a very short time. On one of our projects, 400,000 kg (441 US tons) of cement were injected in less than 6 h with this system. The mixing is very crude, and the rheology and consistency of the mix fluctuates considerably during the operation.

Some people still use paddle mixers for soil-grouting projects. This is a deplorable practice. Paddle mixers (with horizontal axle) are, however, suitable for mixing cement-based suspension grouts with a stable foam, to produce cellular cement foam.

Holding Tank. This is not a very critical component of the grout-placement operation. A low-rpm paddle mixer is sufficient to prevent flocculation of especially cohesive and thixotropic mixes. The holding tank should be located where both the quality control person as well as the operator can easily verify the grout level.

Flow Meter. A flow meter (pulse-type flow meter or magnetic type) should be installed on the main feeder line to the various grout holes. The meter should be positioned so that it is easily visible to the quality control person as well as the mixing crew.

X-Y Recorder. This recorder provides real-time monitoring of the grouting operation—a continuous display and printout of flow and pressure as a function of time. The graph immediately informs the experienced grouting engineer about the mode of the grouting operation and the suitability of the formulation for the formation at that particular stage of the operation. The data allow the principal to calculate the reduction rate in the apparent Lugeon value and compare it with the desired rate, which in turn dictates the changes in the formulation for the next batches. Some recorders made in the United States are considerably less expensive than the types provided by Craelius or Hany but provide the same data (if not more) as the European types.

Grout Hoses. The hoses are ideally approximately 25 mm in diameter. Prior to grouting they should be checked for grout buildup. It is good practice to flush the hoses from time to time with a 5% concentrated phosphoric acid solution. Quick couplings allow easy connections to packer-rods or grout-pipes.

Gauges and Gauge Savers. Gauges should be appropriate for the pressure range at which the operation will be conducted. Gauge savers should be used to prevent access of grout into or toward the gauge membrane. The oil and grease level should be topped up from time to time to provide accurate readings. The gauges should be placed immediately adjacent to the pressure transducers leading to the *X-Y* recorder.

Grout Pumps (Fig. 5B.52). Ideally, helical screw pumps (mohno pumps) should be used. The spaces containing the grout are moved by the steel rotor through the rubber stator chamber. This type of pump provides a virtually constant pressure output, and the flow is easily adjustable. Helical screw pumps are the most popular pumps for low-pressure grouting (up to 1.5 MPa or 218 psi) and are the most popular type in geotechnical engineering applications.

Double-acting duplex piston pumps provide a somewhat more pulsating action, and hence a less continuous output, but are also suitable for a lot of applications. The principle of the pumps is the following: One piston moves in one direction, pressurizing grout out of the cylinder house; grout is pulled in behind it, ready for the expulsion stroke. The other piston in the meantime runs in the other direction, doing the same thing. A ball-and-seat arrangement "directs" the flow inside the pump chamber. The Craelius pumps are especially popular in Europe.

Triple-cylinder, double-acting piston pumps are even more suitable than the duplex pumps and provide a more continuous output. Especially in deep mines, where high pressures are often necessary to overcome the pressure in the aquifer, the triple cylinder pumps work very well.

Ram pumps have been used for decades, especially in the mining industry. The single-acting Clivio pumps used to be very popular and were extensively used in France by the leading grouting contractors, but they provide a pulsating action. The double-acting ram pumps from Hany provide an almost continuous grout flow and are gaining in popularity.

Packer Rods and Packers. The double packers have to be suitable for use in the sleeve pipe in which they will be used. Since the packers have to fit snugly against the pipe wall, the packer cups or packer seals have to be inspected every day and replaced on a regular basis.

The packer rods are usually 12.5 mm (0.5 in) in diameter and preferably are color-coded. One color band should be identical to the distance between two sleeves on the sleeve pipe.

Especially when sleeve pipes with barrier bags are used (see Fig. 5B.10), it is important to position the double packer exactly over the sleeve around which the barrier bags are strapped. Bringing the double packer to the bottom of the grout pipe (straddling sleeve #1) and pulling it up (n-1) color bands, will allow the packer to straddle the nth sleeve.

There are single-fold, double-fold, and triple-fold double packers, named after the number of sleeves they straddle at a time. Tests should be conducted prior to grouting to check grout seepage past the packer seals. It is also advisable to grease the zone between the packer seals. Usually, the perforations of the packer rods between the upper and lower set of packer seals are too small and tend to clog. It is advisable to provide wider slots.

Multiple-hole Grouting Headers. Zones with similar hydraulic conductivity should be grouted simultaneously. Not only will this speed up the operation, but it allows marginally injectable soils to be fed at a slower pace. Especially during the second and third pass, when higher pressures are used, this technique can establish an overall economical pump flow while each grout zone is only marginally accepting grout. More important, the medium to be grouted can be brought to a proper refusal (zero take at the maximum allowable pressure).

It is not necessary to have an external flow-measuring device to each grout line. An extra mobile monitoring device is perfectly suited to check the individual lines from time to time. When monitoring and engineering a grouting operation, return lines cannot be used beyond the grout header where the flow and pressure are recorded.

Particle-Size Reducing Mill. If and when more test data are established on the consistency and reduction factor, this type of equipment will become standard on suspension-grouting projects. Affordable models are being developed that could benefit the specialist contractors.

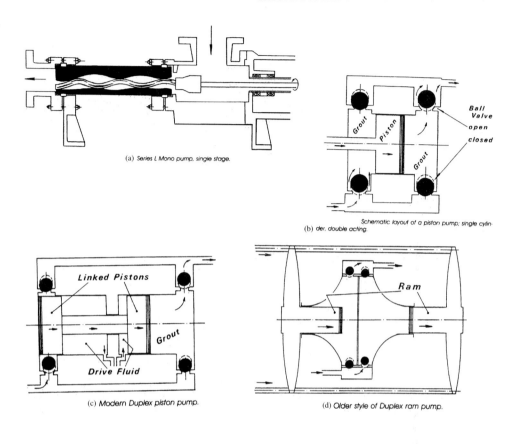
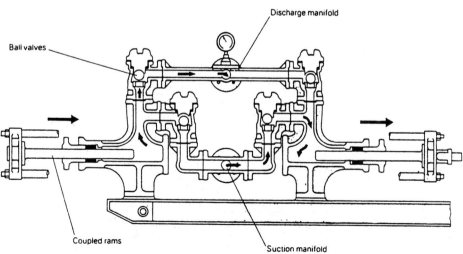

FIGURE 5B.52 Grout pumps: (*a*) moyno; (*b*) piston—single-cylinder, double acting; (*c*) piston—duplex; (*d*) ram—duplex; (*e*) ram—Evans. (*From Ref. 19.*)

Measuring Devices for the Various Ingredients. A water meter should be used to introduce the exact amount of water into the mixer. For the fluids in the mix (water reducers and deflocculators): measuring cups, large enough to hold the required range of fluidifiers, can be used in the various formulations. Graduated beakers or pails should be used for the various other additives, using a different-color pail for each additive. The bulk specific gravity for each of the additives should be established prior to the operation, and a table indicating the volumes of the additives for each formulation established. Several pails for each of the ingredients should be provided to allow the mixing crew to prepare the requested dosage for each additive in a continuous fashion. A properly organized grouting contractor has no problem continuously producing the requested formulations (changing as often as 5 to 8 times per grout hole). The principal should however tailor the formulations so that they contain an exact number of "units" for each one of the additives and an exact number for the cement bags.

A book with the various formulations should be at the grout plant, and the requested formulation should be posted above the mixer (plastified sheets).

5B.5.1.2 Solution Grouting Equipment. Because each type of product requires a different setup, it is quite a task to make a meaningful summary about what is used, what should be used, and rate suitability and performance.

5B.5.2 Engineering Supervision of a Grouting Program

5B.5.2.1 General Aspects. Since there is such a variety of grouting applications, no general guidelines can be given. The following is just an overview of the various tasks at hand.

The principal should verify if the contractor is supplying the necessary and appropriate equipment as has been stipulated in the contract documents. This should be verified prior to the contractor's mobilization to the site. The principal should be reasonable in assessing alternative equipment and approaches proposed by the contractor. The principal should make all possible efforts to fully understand what is proposed or offered by the contractor. This includes reading the technical literature supplied by the manufacturers. The principal, if in doubt, should seek the advice of specialists in the industry who might be more familiar with particular types of equipment.

The principal should verify the references provided by the contractor and those for key people. It is useful to contact other grouting specialists and obtain an opinion about the organization and the key people who run the day-to-day operation. Quite often, the key people are only good administrators or operators without any knowledge of the real engineering aspects of grouting.

The principal should interview the contractor prior to awarding a contract to make sure that the contractor has a clear understanding of all the aspects of the grouting project and to verify that the contractor agrees about every aspect of the proposed procedures and methodology. In this meeting, the contractor has to provide information about any changes proposed. Minutes of this meeting should form an integral part of the contract documents. It is important that the exact role of the principal and the contractor, as stipulated in the contract documents, be discussed and agreed upon.

On site, the principal and the contractor should sign daily reports pertaining to both technical and contractual matters as well as to quantities and pay items.

The principal's main task is to assess the grouting operation. One of the first things is to evaluate the amenability of grouts and grout formulations for the medium to be grouted. A principal who directs the program has to have hands on and direct and evaluate the in situ hydraulic conductivity tests. During the grouting operation the principal will evaluate the amenability of the selected grout formulation and continuously calculate apparent Lugeon values. The principal should oversee and evaluate the quality-control tests. For suspension grouts, the following tests are conducted:

- Quality control on mixing procedures
- Specific gravity of the grout (Baroid mud balance)
- Apparent viscosity immediately after mixing (marsh funnel)
- Temperature of mixing water and grout (especially during cold or very hot weather)

- Bleed-test (graduated cylinder)
- Initial gelation of the grout (i.e., pump time)
- Initial cure of the grout (Vicat needle)
- Resistance against pressure filtration of the various formulations (Baroid filter press)

For solution grouts, the following tests are necessary: pot life of the mixed products and quality control on the mixing procedures and quantities of the various ingredients introduced.

In a hands-on operation, the principal should remain completely focused on the operation and make trips, on a regular basis, from the mixing plant and recorders to where the grouting takes place.

Especially in structural grouting projects, the principal has to observe the "signs given by the structure," calling for additional instrumentation where and if required.

The principal will order, assist with, or execute particular tasks to verify spread, check movements (laser, extensometer) and seepage at particular locations, adjacent to the site, or other places.

The principal makes the call about which zones can be grouted simultaneously and when to bring other holes (or zones) on line, directs the changes in formulations, and makes decisions about how and when to deal with unforeseen circumstances. Usually, new zones are brought on line when the apparent Lugeon value in the grout holes on line decreased to a value compatible with the Lugeon value of the zones to be brought on line. This also means that the principal has to change formulations gradually back to the starter mix of the new zones to be brought on line.

Unforeseen circumstances may necessitate providing continuity in an operation despite mixing mistakes (human error), seepage of grout (especially in structural grouting and rock grouting), preferential grout migration, and washout. The principal has to act as the coach of the team, calling the mixes, based on the evolution of the apparent Lugeon value of the formation, its amenability, the grouting mode, the volumes injected, and particular circumstances at hand.

The principal cannot lose sight of the "big picture" and must continuously assess if the grouting program as executed will deliver the desired product. Modifications to the design are almost inevitable, especially during solution-grouting programs dealing with seepage control. The principal has to be able to adapt a strategy swiftly "to pull it off." This requires in-depth knowledge, heart, and a great deal of experience.

5B.5.2.2 Recording and Interpretation of the Grouting Data.

This section focuses on structural and soil grouting using suspension grouts. When using solution grouts in soil-grouting programs, some of the principles outlined in this section still apply. Because each type of solution grout behaves differently, no general approach can be outlined.

Figure 5B.53 displays a copy of a typical grout form (sheet 2) maintained by the principal during the permeation grouting of washed-out tremie concrete below the Detroit-Windsor railway tunnel. Figure 5B.54 displays a copy of a typical X-Y recorder sheet. Figures 5B.55–5B.58 display a copy of a daily report prepared on a daily basis by the principal. By analyzing all the data at the end of a shift and focusing on what transpired, the principal should discover a number of things every day and modify the grouting program to obtain the best possible result. Under no circumstances can the principal fall behind with the reporting, lest he or she discover major blunders far too late.

The following lists what is recorded on a grouting sheet:

1. *Crew members, date, type of shift, name of the inspector, and quality control person.* These data do not require any further explanation.

2. *Hole numbers and the time they are brought on stream.* The running time total ("time cum" on the sheets) as well as the duration of each time frame is recorded. A spot readout might not always reflect what the exact flow is, therefore it is better to take averages, and hence the duration of each test period has to be recorded.

3. *Type of mix—ID/batch number.* Each formulation has a code: A, B, C, D, etc. The characteristics of those mixes have been established prior to grouting, with the contractor's mixing equipment. Each batch of a particular mix type is recorded as type of mix followed by a number indicating the number of batches of that particular mix used during that shift (A1, A2, B1, B2, B3, B4, P1, P2, etc.).

4. *Viscosity.* This is the apparent viscosity measured with the marsh funnel, taken while the grout is dumped into the holding tank. This value in turn is used to calculate v/v_i and hence the apparent Lugeon value of the formation. The apparent viscosity is one of the fingerprints of a particular formulation. It detects mistakes made during the mixing. The variation in marsh funnel value for one particular mix for a well-defined mixing time should not vary more than 1.5 s. The quality control person performs this test for each batch.

5. *Specific gravity.* This is another fingerprint of a particular formulation. Errors in mixing are reflected by fluctuations in the specific gravity. The average specific gravity per type of formulation is used to calculate the yield per batch of that formulation during that day and compared with the average of previous days. With a well-organized contractor, fluctuations are in the order of 0.02 kg/l. The specific gravity is measured with the Baroid mud balance.

6. *Pressure.* The gauge pressure is also recorded manually at regular times. The gauge should be located where the pressure transducer is located. One of the reasons is to compare the gauge pressure with the numbers on the *X-Y* recorder chart. The head losses are derived from the head-losses graph. Note that the head losses have been established with water. To obtain the head losses established with grout, for a given flow, the values on the graph have to be multiplied by ($visc_{grout}/28$ s).

The effective pressure is calculated by deducting the head losses from the gauge pressure and adding the hydrostatic head. When grouting against artesian pressure, the hydrostatic head is a negative value (often called *closed-valve pressure*). The effective pressure allows the calculation of the apparent Lugeon value at any point during the operation.

7. *Lugeon value* (*in situ hydraulic conductivity*). The Lugeon value for a zone, or the average Lugeon value for a number of zones, has to be established prior to grouting and is again recorded on the grouting sheet (this is the so-called Lu_{wa}—see Sec. 5B.3.1). This value should be on the sheet to compare with the apparent Lugeon values established with grout (Lu_{gr}), which in turn makes it possible to calculate the amenability of a particular formulation. Apparent Lugeon value for the zones grouted is to be calculated continuously to establish the suitability and sequence of particular formulations (see Sec. 5B.3.1). The amenability for the starting formulation is calculated, based on the grout flow, usually after about 1 to 2 min in the operation. This process evaluates the suitability of the starter formulation.

The reduction rate in apparent Lugeon value is the most valuable tool in the hands of the grouting engineer, to call the mixes:

- Fast reduction in apparent Lugeon values indicates a need for finer suspension grouts and/or lower-viscosity grouts, with or without increased resistance against pressure filtration, with or without decreasing solids contents.
- Gradual reduction in apparent Lugeon values usually indicates that the formulation is suitable for the circumstances.
- Very slow or no reduction in apparent Lugeon value usually indicates that the grouting mode is compatible with void filling and that both cohesion and solids content of the grout can be increased.
- Increase in apparent Lugeon value indicates either void filling, in conjunction with mud washing, or grouting in a very loose coarse material. The solids contents in the grout and especially cohesion and thixotropy should be gradually increased.

The logic and the background for these decisions are explained in Sec. 5B.3.1.

By now, the reader should understand that the engineering of a grouting operation is not a boring task. It requires full concentration, fast data crunching, and dedication by the principal. The principal's task is not over when the grouting operation is halted for the day. The principal's task continues for several hours, to process the paperwork. By writing a daily report, the principal must analyze all the data and will thus understand the signals the formation has sent back during the operation.

GROUTING REPO. Date: JAN / 27 / 93

Crew: GAVIN, BILLY, KEVIN Inspector: L. NARDUZZO, B. FREDERICKS Sheet No. 2 of 5

Hole Number	GROUT DATE Di Day Sl Slewing Ht Height	TIME From	TIME To	TIME mins Past	TIME mins Cum	TYPE OF MIX ID/Batch	TYPE OF MIX Vol. (Litres)	TYPE OF MIX SGr (gm/cm³)	PRESSURE (psi) 1 Gage	PRESSURE (psi) 2 Line Loss	PRESSURE (psi) 3 Closed Valve	PRESSURE (psi) 1-2-3 Eff. Press.	AVERAGE TAKE LPM	AVERAGE TAKE Cum Grout Volume (litres)	AVERAGE TAKE Ave. Lugeon Value Line	AVERAGE TAKE Apparent Lugeon Value LUV	Amenability %	Comments
40+71.5	D	14:30		—	—	P1	38	1.39	—	—	—	—	—	—	—	—	FIRST SET OF HOLES GROUTED.
40+69.5		14:33		—	—				—	—	—	—	—	—	—	—	14:30 MAKE 1st BATCH
40+51 N		14:35		—	—				—	—	—	—	—	—	—	—	FILLING LINES: 45 litres
		14:37		—	—	P2	36	1.40	—	—	—	—	45	—	—	—	MAKE ANOTHER BATCH BEFORE
		14:40		0	0				50	—	—	—	12	0	11	—	—	STARTING: COUNTER RESET TO ZERO
		14:41		1	1				51	—	—	—	8	14	7.9	—	71%	
		14:42		1	2				65	—	—	—	3	21	3.7	—		Poor amenability for microfine
		14:43		1	3				135	—	—	—	5	25	1.8	—		suspension grout: 71%
		14:45		2	5				150	—	—	—	1	27	3.1	—		
		14:46		1	6				148	—	—	—	0	28	0	—		Pressurization to avoid the
		14:48		2	8				150	—	—	—	0	28	/	—		formation from springing back and
		14:52		4	12				150	—	—	—	0	29	/	—		to exceed pressure filtration time.
		14:58		6	18				150	—	—	—	0	30	Large value	—		
		15:02		4	22				150	—	—	—	0	30	/	—		App L.V. = $\frac{1}{4.8} \times \frac{143}{(6.5)} \times 17 \times \frac{38}{28} = 687$
40+75N		15:05		0	0				5	—	—	—	5	17	30	—		
40+63N		15:06		1	1				6	—	—	—	17	47	976	687	70%	
40+39N		15:08		2	3	E1	39	1.44	10	—	—	—	16	72		130		
40+27N		15:14		3	6				11	—	—	—	15	120		101		Spilled ~ 10 litres DURING
		15:14		3	9				17	—	—	—	13	154		44		TRANSFER
		15:21		7	16	E1	38	1.39	13	—	—	—	8	256				OUT OF MIX!
		15:24		3	19				15	—	—	—	12	271				
		15:26		2	21				20	—	—	—	20	305		54		
		15:26		2	23													

FIGURE 5B.53 Typical grout form (sheet 2).

5.388 SOIL IMPROVEMENT AND STABILIZATION

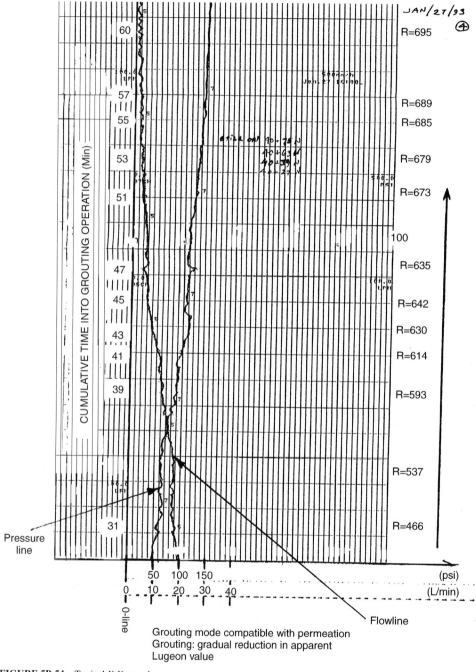

FIGURE 5B.54 Typical *X-Y* recorder.

GROUTING REPORT FOR JANUARY 27, 1993

LOCATION OF WORK PERFORMED

Stage II, Section A Experimental Area
Secondary Holes, 41+15N to 41+75N

STATISTICAL SUMMARY

Hours of Grouting	:14:30 - 18:30, 4 hours
Volume Grouted	:54 cubic feet
Holes Grouted	:12 holes
Distance Grouted	:72 feet
Volume per Hole	:4.5 cubic feet per hole
Volume per Foot	:0.75 cubic feet per hole
Volume per Hour	:13.5 cubic feet per hour
Holes per Hour	:3.0 holes per hour
Distance per Hour	:18 feet per hour

DAILY REPORT CONTENTS

a)	Technical Commentary	4 pages
b)	Grouting Maps	3 pages
c)	Quantity Reconciliation	3 pages
d)	Field Notes	10 pages
e)	X-Y Recorder Charts	9 pages

TECHNICAL COMMENTARY

1. Three holes on the south side of the tracks (40+33S, 40+45S, & 40+57S) intersected an area with permeability of zero Lugeon and were tremie grouted and subsequently pressurized.

2. The next three holes (40+21S (3 Lu), 40+51N (3 Lu), 40+68S (27 Lu)) consumed only 28 litres of microfine cement grout. Grouting pressures were elevated to 150 psi but no significant take was recorded.

   If the area being grouted had been a fissured structure, the take should have been at least 50 litres, based on the aperture theory. It is therefore concluded that the formation behaved like a porous medium (such as soil) with permeabilities of 3.0×10^{-3} cm/sec, 3.0×10^{-3} cm/sec, and 2.7×10^{-4} cm/sec for the three holes involved.

   Formations, such as this area, with permeabilities of less than 8.0×10^{-4} cm/sec are not injectable with suspension grouts, not even microfine suspension grouts. The active part of the grouting operation lasted for only 5 minutes, not including the holding time for the refusal process.

3. The next batch of holes (40+27N, 40+39N, 40+63N, & 40+75N) were all characterized by high Lugeon values. These holes were allowed to flow freely for 30 minutes prior to grouting, thereby depressurizing the formation from 28 psi. to 6 psi.

   After injecting the residual grout from the preceding operation, a "Mistral" grout (mix "E1") was used, being a stable cement grout, capable of permeating soil with permeability greater than 5.0×10^{-2} cm/sec.

   It took 15 minutes of grout injection before the equilibrium pressure was rebuilt, at which point the contractor's mixing crew ran out of grout. During the ensuing delay, the formation pressure dropped to only 3 psi. Ie. well below the original formation pressure.

   Based on this observation, it is concluded that the main feeder pathways connected to the porous formation being grouted must have been cutoff by grouting over a large area, due to four holes being on stream at the same time. This is considered to have been a significant finding from the day's activities.

   After the Contractor's crew had recharged their equipment, the recovery of equilibrium pressure was again a slow occurrence. After the equilibrium pressure was exceeded, the operation resumed a grouting mode which was consistent with permeation grouting (ie. gradual increase in pressure with decreasing flow).

   It is our opinion that the Mistral grout performed satisfactorily in dealing with the gravels and sands, or washed out tremie concrete.

   The grouting operation, as reflected on the X-Y charts, worked like a text book example. Pressure was gradually increasing as grout flow gradually decreased in response to increasing head losses due to grout migration through the porous formation.

   The quantity of grout consumed in those four holes was approximately 180 litres per hole.

   Grouting pressures higher than originally anticipated were used to obtain the desired grout spread, but without hydrofracturing the formation or creating otherwise adverse effects. It should be noted that there was no void system available to be pressurized, since the permeability was due to residual porosity in the tremie concrete.

4. It is recommended that the influence of the grouting operation be verified by drilling holes along the ditch.

5. Tremie grouting and pressurizing of the three zero Lugeon holes was undertaken at the same time as grouting of hole 40+15N.

   Grouting of hole 40+15N was considered as primary grouting as only one other hole in that diaphragm had previously been grouted. This hole was allowed to run freely for over two hours and was then treated with a Mistral grout.

   When this hole was near refusal, an adjacent Stage I hole was opened, which resulted in resumed grout consumption and a drop in grouting pressure. This demonstrates the limited feeder capacity of the formation after grouting cuts off the principal pathways.

   This hole consumed over 600 litres of grout and effective grouting pressures of up to 125 psi were applied in order to properly saturate the formation. This hole should be considered a primary grout hole and should be assessed independently of the other secondary grout holes which were previously grouted.

   The effectiveness of the grouting is therefore more pronounced on the south side:
   Primary Holes   70 Lugeon average permeability
   Secondary Holes   6 Lugeon average permeability

   A significant reduction in permeability was achieved in only one hole on the north side. This was in hole 40+51N (3 Lugeon), which was located within a diaphragm in which both the north and south primary holes were very pervious, and where a pure grout connection was made between the north and south holes during primary grouting (ie. grout of same S.G. flowing from connected hole as S.G. of grout being injected).

   The average grout consumption in the secondary holes was 73 litres, as compared with 110 litres in the primary hole.

7. Grouting pressures in secondary grout holes (even in soil grouting) are typically considerably higher since hydrofracturing of soil does not occur as fast during secondary grouting as during primary grouting.

   Of greater relevance to this project, the primary grouting filled all of the major voids which could have contributed to stress build-up in the liner, in the event that predetermined grouting pressures would be exceeded.

   From a structural analysis, it was determined that if an effective grouting of 100 psi would be applied to the liner, no adverse stress build-up would occur in the steel liner.

   In view of this analysis, and the absence of available voids during secondary grouting operations, it has been concluded that higher pressures which quickly dissipate through the porous formation can be safely applied. This was also confirmed by the recorded low pressures in adjacent holes.

8. One of the attached grout maps shows the total grout consumption for each diaphragm, as well as a breakdown showing primary and secondary grout consumption.

   The diaphragm with the highest primary consumption shows the lowest secondary consumption.

   There are two diaphragms which consumed close to 700 litres of grout per diaphragm. Assuming a "triangular void model" (going form south to north), a total aperture of almost 150 mm (6") wide, in the strata below the liner (north side) has been filled with grout.

9. Based on the (limited) experience gained during the grouting of the first set of holes, it can be concluded that application of only the specified pressures during the first pass grouting through Zone A is responsible for the negligible improvement in the formation encountered below the north side of the tracks.

FIGURE 5B.55 Typical grouting report.

DETROIT RIVER TUNNEL ENLARGEMENT PROJECT - GROUTING RECORDS

DAILY GROUTING REPORT FOR MONDAY, FEBRUARY 01, 1993   PAGE 1

SUMMARY OF SECTION 02330 PAYMENT ITEMS:

ITEM	DESCRIPTION	UNITS	QTY TODAY	ADJUSTE QTY	PREVIOUS QTY	TOTAL QTY
.8	GROUT (0 - 30)	CU FT	2	0	4	6
.9	GROUT (30 - 150)	CU FT	10	0	28	38
.10	GROUT (150+)	CU FT	33	0	225	259
.11	TYPE 10 CEMENT	LB	0	0	0	0
.12	TYPE 30 CEMENT	LB	1234	0	8199	9433
.13	MICROCEMENT	LB	441	0	1984	2424
.14	TYPE C FLYASH	LB	53	0	1217	1270
.15	N SULPHONATE	LB	12	0	83	95
.16	THIXO AGENT	LB	0.0	0.0	0	0.1
.17	SILICA FUME	LB	121	0	705	827
.18	BENTONITE	LB	62	0	242	304
.19	RESEALING	EACH	0	0	0	0

NOTES: GROUTED STAGE IID - SECOND PASS

ADJUSTMENTS:

UNIT RATE PAYABLE QUANTITIES FOR THIS DAY HAVE BEEN RECONCILED:

--------------------   --------------------
CONTRACTOR   ENGINEER

DATE PRINTED: 01-Feb-93

FIGURE 5B.56 Grouting report (*Continued*).

DETROIT RIVER TUNNEL ENLARGEMENT PROJECT - GROUTING RECORDS

DAILY GROUTING REPORT FOR MONDAY, FEBRUARY 01, 1993   PAGE 2

RECONCILIATION OF BATCHES MIXED AND MATERIALS USED:

GROUTING FORMULATIONS: MATERIAL USED

INGREDIENTS	UNITS	P	E1	A	KG	LB
WATER	KG	65	120	90	-	-
TYPE 10 CEMENT	KG	0	0	0	0	0
TYPE 30 CEMENT	KG	0	80	80	560	1234
MICROFINE CEMENT	KG	50	0	0	200	441
TYPE C FLYASH	KG	0	0	12	24	53
N SULPHONATE	KG	0.5	0.5	0.5	6	12
THIXO AGENT	KG	0	0	0	0.0	0.0
SILICA FUME	KG	5	5	5	55	121
BENTONITE	KG	1	4	2	28	62
TOTAL WEIGHT	KG	121.5	209.5	189.5		
SPECIFIC GRAVITY	KG/L	1.38	1.35	1.46		
YIELD PER BATCH	LITRES	88	155	130		
BATCHES MIXED	EACH	4	5	2		
VOLUME MIXED	LITRES	352	776	260		
VOLUME DUMPED	LITRES	0	100	0		
VOLUME GROUTED	LITRES	352	676	260		

TOTAL VOLUME GROUTED: 1288 LITRES

NOTES:

DATE PRINTED: 01-Feb-93

FIGURE 5B.56   (*Continued*)

SOIL IMPROVEMENT AND STABILIZATION

DETROIT RIVER TUNNEL ENLARGEMENT PROJECT - GROUTING RECORDS

DAILY GROUTING REPORT FOR MONDAY, FEBRUARY 01, 1993   PAGE 3

RECONCILIATION OF GROUTED VOLUMES:

VOLUMES GROUTED IN LITRES

CHAINAGE	OUTFLOW LUGEON	LUGEON <30	LUGEON 30+	LUGEON 150+	TOTAL
54+58 N	104		66		
54+64 S	20	16			
54+70 N	118		74		
54+76 S	32		23		
54+82 N	1578			945	
54+88 S	13	12			
54+94 N	114		72		
55+00 S	0	4			
55+06 N	15	13			
55+12 S	4.5	7			
55+18 N	87		56		
VOLUME GROUTED IN LITRES		52	291	945	1288
VOLUME GROUTED IN CUBIC FEE		2	10	33	46

NOTES: ALLOCATION OF VOLUME GROUTED HAS BEEN DETER

BASED ON FIELD NOTES RECORDED BY L. NARDUZZO

DATE PRINTED: 01-Feb-93

FIGURE 5B.56   *(Continued)*

GROUTING TO IMPROVE FOUNDATION SOIL 5.393

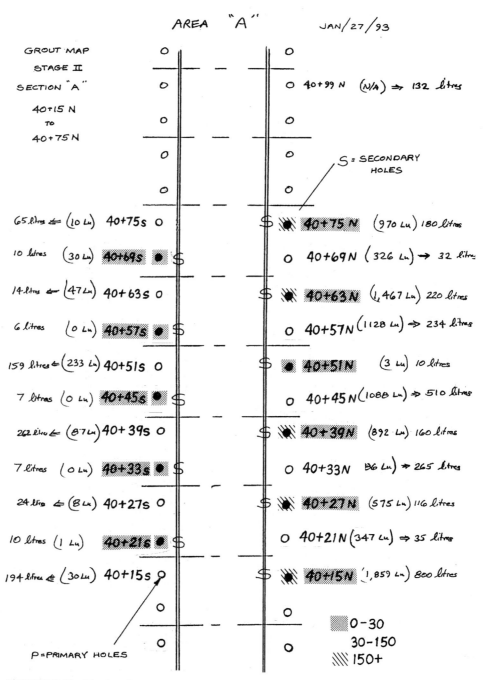

FIGURE 5B.56 (*Continued*)

GROUTING REPO... AFTER TESTING   Date: JAN / 27 / 93

Crew: KEVIN, BILLY, GAVIN   Inspector: L. NARDUZZO, B. FREDERICKS   Sheet No. 1 of 5

40+21S	0)	13:14	13:18	4	4	–	–	–	52	0	20	33	0.1	–	.1	–	–	GAME GAVIN BILLY, KEVIN
40-33S	0)	13:20	13:21	1	1	–	–	–	54	0	22	32	0	–	0	–	–	GONE LEFT TO GET DRAWINGS
40-45S	0)	13:22	13:23	1	1	–	–	–	55	0	0	55	0	–	0	–	–	STAMPED (13:15)
40-57S	0)	13.25	13:26	1	1	–	–	–	62	0	10	52	6	–	0	–	–	
40+69S	0)	13.28	13.32	4	4	–	–	–	48	0	28	21	5.4	–	30	–	–	
40+75N	0)	13:33	13:37	4	4	–	–	–	32	1	28	3	24.6	–	970	–	–	Lugeon Tests:
40-63N	0)	13:38	13:42	4	4	–	–	–	32	1	30	2	24.8	–	1467	–	–	Closed valve pressure refers to
40-51N	0)	13:45	13:49	4	4	–	–	–	55	0	12	43	1.0	–	3	–	–	the recorded hydrostatic pressure
40-39N	0)	13:57	14:01	4	4	–	–	–	40	4.5	30	6.5	49	–	892	–	–	in the standpipe prior to the
40+27N	0)	14:05	14:09	4	4	–	–	–	45	4.5	31	9.5	51	–	575	–	–	performance of the hydraulic
40+15N	0)	14:11	14:15	4	4	–	–	–	40	5.5	32	3.5	55	–	1051	–	–	conductivity tests.
																		Note: 1 psi is added to obtain
																		the effective pressure to
																		compensate for the
																		hydrostatic head (test equipment
																		above tracks)

GROUTING REPO.   Date: JAN / 27 / 93

Crew: GAVIN, BILLY, KEVIN   Inspector: L. NARDUZZO, B. FREDERICKS   Sheet No. 2 of 5

40+21S	0)	14:30		–	–	P1	39	1.39	–	–	–	–	–	–		–	–	FIRST SET OF HOLES GROUTED.
40-69S		14:33		–	–				–	–	–	–	–	–		–	–	14:30 MAKE 1st BATCH
40-51N		14:35		–	–				–	–	–	–	–	45		–	–	FILLING LINES: 45 litres.
		14:37		–	–	P2	36	1.40	–	–	–	–	–	45		–	–	MAKE ANOTHER BATCH BEFORE
		14:40		0	0				50				12	0	11			STARTING; COUNTER RESET TO ZERO
		14:41		1	1				51				8	14		7.9	71%	
		14:42		1	2				65				3	21		3.7		Poor amenability for microfine
		14:43		1	3				135				5	25		1.8		suspension grout: 71%
		14:45		2	5				150				1	27		3.1		
		14:46		1	6				140				0	28		0		Pressurization to avoid the
		14:48		2	8				150				0	28		/		formation from springing back and
		14:52		4	12				150				0	29		/		to exceed pressure filtration time.
		14:58		6	18				150				0	30		/		
		15:02		4	22				150				0	30				App. L.V. = 1/10 × 183/(6·3) × 17 × 38/28 = 687
40+75N	and	15:15		0	0				5			5	17	30				
40-63N	and	15:06		1	1				6				17	47	976	687	70%	
40+39N	and	15:08		2	3	E1	39	1.44	10				16	72		130		
40+27N		15:14		3	6				11				15	120		101		SPILLED ~10 litres DURING
		15:14		3	9				17				13	154		44		TRANSFER.
		15:21		7	16	E'1	38	1.39										OUT OF MIX!
		15:24		3	19				13				8	256				
		15:26		2	21				15				12	271				
		15:28		2	23				20				20	305		54		

GROUTING REPO.   Date: JAN / 27/93

Crew: GAVIN, KEVIN, BILLY   Inspector: L.N. & B.F.   Sheet No. 3 of 5

40+75N	and	15:30		2	25				35				22	348		29		
40-63N	and	15:32		2	27				45				20	388		18		
40-39N	and	15:34		2	29				47				19	430		16		
40-27N		15:36		2	31				63				17	466		11		FAST REDUCTION IN APPARENT
		15:40		4	35	E'2	38	1.37	70				16	537		9		LUGEON VALUE
		15:42		2	37				80				13	567		6.6		DROPS FROM 30 TO 5 IN
		15:44		2	39				90				11	593		5.2		15 MINUTES
		15:46		2	41	E'3	40	1.38	100				9	614		3.8		
		15:48		2	43				109				6	630		2.5		IS THE GROUT PICKING UP
		15:50		2	45				107				4	642		1.6		IMPURITIES IN THE FORMATION?
		15:52		2	47				110				4	655		1.5		
		15:54		2	49	E'4	40	1.37	121				4	664		1.4		BRIAN GARROD PRESENT TO INSPECT
		15:56		2	51				127				4	673		1.3		WORK (15:50 to 16:30)
		15:58		2	53				132				3	679		0.9		
		16:00		2	55				137				3	685		0.9		
		16:02		2	57				139				2	689		0.6		
		16:05		3	1:00				142				1.3	695		0.4		
		16:07		2	1:02				143				1	698		0.3		
		16:10		3	1:05				149				0.8	704		0.2		ADJUSTING PUMP SPEED
		16:13		3	1:08				150				0	705		0		REFUSAL PERIOD
		16:15		2	1:10				150				0	706		0		
		16:17		2	1:12				150				0	706		0		
		16:23		6	1:18				150				0	706		0		

FIGURE 5B.57 Grouting report (*Continued*).

GROUTING REPORT. Date: JAN / 27 / 93

Crew: GAVIN, KEVIN, BILLY Inspector: L. NARDUZZO, B. FREDERICKS Sheet No. 4 of 5

Hole Numbers	Grout Date D) Day S) Swing N) Night	Time From	Time To	Time (min) Part	Time (min) Cum	Type of Mix ID/Batch	Visc. (sec)	Bo/Gr (gm/cm3)	Pressure (psi) 1 Gage	2 Line Loss	3 Closed Valve*	1-2-3 Eff. Press.	Average Take LPM	Cum Grout Volume (litres)	Lugeon Value Lum	Apparent Lugeon Value Lugr	Amenability %	Comments
40+15N	S)	16:30		0	0				15		15		6	726				TREMIE GROUT (THEN PRESSURIZE)
		16:32		2	2				19		15		11	738	1859	468	25%	HOLES 40+575, 40+455
		16:34		2	4				25				20	770		340		and 40+335 → R = 700 TO 726
		16:36		2	6	E'5	40	1.38	38				15	793		110		Low amenability: influence from other areas.
		16:38		2	8				41				17	823		93		App.L.V. = $11 \times \frac{40}{28} \times \frac{143}{19-15} \times \frac{1}{1.2}$
		16:40		2	10				65				15	858		51		
		16:42		2	12				68				16	883		51		SHARP DROP IN APP. LU VALUE
		16:44		2	14				80				6	907		16		
		16:46		2	16				89				4	919		9.3		
		16:50		4	20				105				5	932		9.5		ALEX LEFT 16:50
		16:52		2	22				109				5	944		9		OPERATION EDGING TOWARDS
		16:54		2	24				107				4	951		7.4		REFUSAL
		16:58		4	28				121				5	979		8		
		17:02		4	32				123				5	998		7.8		
		17:05		3	35	E'6	40	1.37	126				4.5	1011		6.9		
		17:08		3	38				134				7	1026		10		OPERATION STARTS TO RUN AWAY.
		17:10		2	40				123				16	1065		25		MINOR FRACTURING? NO,
		17:13		3	43				124				14	1109		22		SOMEONE OPENED VALVE 30' AWAY
		17:16		3	46				124				11	1144		17*		* RELIEF VALVE CLOSED TEMPORARILY
		17:18		2	48				122				19	1180		30		
		17:19		1	49				119				15	1195		24		
		17:22		3	52				121				13	1237		21		
		17:25		3	55				121				10	1271		16		

GROUTING REPO Date: JAN / 27 / 93

GAVIN, KEVIN, BILLY Inspector: L. NARDUZZO, B. FREDERICKS Sheet No. 5 of 5

Hole Numbers	Grout Date D) Day S) Swing N) Night	Time From	Time To	Time (min) Part	Time (min) Cum	Type of Mix ID/Batch	Visc. (sec)	Bo/Gr (gm/cm3)	Pressure (psi) 1 Gage	2 Line Loss	3 Closed Valve*	1-2-3 Eff. Press.	Average Take LPM	Cum Grout Volume (litres)	Lugeon Value Lum	Apparent Lugeon Value Lugr	Amenability %	Comments
40+15N		17:26		1	56				122				8	1282		12.7		
		17:29		3	59				123				8	1306		12.7		
		17:30		1	60				126				6	1313		9.2		
		17:32		2	1:02				130				7	1328		10.3		
		17:36		4	1:06				124				5.5	1354		8.6		
		17:38		2	1:08				124				3.8	1366		6		
		17:40		2	1:10				125				4.0	1372		6.1		
		17:42		2	1:12				130				5	1380		7.3		
		17:45		3	1:15				135				3	1387		4.2		NO
		17:48		3	1:18				121				1	1396		1.6		"LEAK" APPEARED IN NEXT DIAPHRAM
		17:53		5	1:23				123				0.4	1403		0.6		VALUE OPEN ON 39+91 N
		17:55		2	1:25				135				0.5	1407		0.7		AND PRESSURE DROPPED
		17:59		4	1:29				131				1	1409		1.4		
		18:02		3	1:32				133				0.9	1412		1.3		
		18:06		4	1:36				138				0.3	1414		0.4		
		18:08		2	1:38				130				0.0	1414		0		
		18:10		2	1:40				135				0.0	1414		0		
		18:12		2	1:42				133					1414		0		TURN MIXER ON TO CLEAN - LOST
		18:14		2	1:44				130				•	1422*		0		PRESSURE.
		18:18		4	1:48				130					1422		0		
		18:22		4	1:52				130									20 litres DUMPED

FIGURE 5B.58 Grouting report (*Continued*).

5B.5.2.3 Quality Control and Testing

Testing of Apparent Viscosity. The marsh funnel is still the most practical device in the field to obtain valuable information about the flow characteristics of a particular suspension grout formulation. Only one type of marsh funnel should be used to avoid mass-confusion about data produced at different locations at this planet. The *marsh-funnel viscosity* should be defined as the afflux time of 1 L (0.26 gal) of test fluid from a true marsh funnel filled to the screen. Water has a marsh viscosity of 28 s. Note that the marsh funnel should be washed out thoroughly after every test.

The marsh viscosity reflects rheological characteristics in combination with friction losses due to wall roughness of the spout. Therefore, the term *apparent viscosity* is appropriate to describe the characteristics of the grout determined with this device. Deere[58] and Lombardi[36] provided valuable information to correlate marsh funnel viscosity with plastic viscosity.

Some other flow cones are still used, and they prepare relative comparison between the apparent viscosities of the formulations. The finer afflux spout offers a more relevant comparison between the low-viscosity grouts and provides a more meaningful insight in comparing flow characteristics through fine channels. The Brooklyn viscosity meter (ASTM D4016) can be used in a site lab but is too sensitive to be used at the "front line."

Density or Specific Gravity of the Grout. The Baroid mud balance (API specification 13A) provides a fast and practical way to assess the specific gravity of a specific grout formulation. Every mix should be checked to verify if all the ingredients in the appropriate proportions have been introduced in the mix. The specific gravity makes it possible to calculate the yield of each batch and is useful to determine pay quantities. The quality control person should wash out the balance after each mix.

Bleed Test. On a regular basis (once a week), bleed tests should be performed in accordance to ASTM C191 standards, for each of the formulations used. The formulations should be virtually bleedfree.

Resistance against Pressure Filtration. Each formulation should be subjected to a *pressure filtration test*. This test provides most valuable information about what actually occurs when grout is subjected to pressure while flowing through a porous medium. Good resistance against pressure filtration is necessary to obtain good permeation. (see also Fig. 5B.21). The resistance against pressure filtration tests should be conducted on each formulation, prior to the execution of the grouting program, and should be repeated on a routine basis throughout the grouting program.

In essence, the pressure filtration test involves filling a standard API filter press and subjecting it to 100 psi (0.7 MPa) air pressure, provided by CO_2 cartridges. The volume of water squeezed out with time is monitored until all water is squeezed out (apparatus produces a big "puff"). The pressure filtration coefficient can be calculated (as explained in Sec. 5B.3.1) and compared with the values in Fig. 5B.21.

The resistance against pressure filtration should not be obtained at the expense of increased cohesion (unless under specific conditions or circumstances). Polymers and additives are available to obtain excellent resistance against pressure filtration without increasing the cohesion or the apparent viscosity of the mix. This type of information is crucial in designing and changing formulations to meet specific targets.

Initial Cure of Grout (Gravity Cure). Especially for applications at low temperatures, it is necessary to do *initial cure tests* (Vicat needle—ASTM C191), at similar conditions and temperatures to what the grout is subjected. Also for regular applications this type of information is very useful.

Unconfined Compressive Strength. It is good practice to cast grout cubes of the various formulations at regular times throughout the grouting operation. Especially the disposable plastic cubes are very practical and inexpensive. Test cubes should be broken at appropriate times, which could mean 7 days but also 56 or even 128 days. Formulations containing a considerable amount of F fly ash cure very slowly.

Cubes should also be prepared with paste obtained from the pressure filtration apparatus. The paste will have to be squeezed into the mold, but the difference in unconfined compressive strength with the regular grout will become apparent.

5B.5.3 Evaluation of the Grouting Operation and Nondestructive Quality Control

5B.5.3.1 Core Sampling. It is rather difficult to obtain cores from grouted soils unless the unconfined compressive strength of the grouted soil is reasonably high. Bruce[59] suggests that the core barrel should be at least 100 mm in diameter and that a triple-tube system be used for the probing.

5B.5.3.2 Borehole Camera Investigation. Usually, this test is rather valuable, especially with small boreholes. It often requires some referencing with ungrouted materials, but it can provide a good understanding about what has been accomplished.

5B.5.3.3 Comparison with the Data from Production Grout-Hole Drilling. When large grouting projects are conducted, specific drilling parameters should be continuously recorded and correlated to data obtained in pilot holes. The Enpasol method was developed by Soletanche and the Papero method by Rodio during the mid-1980s. Sensors, fitted to the drill, record the penetration rate, rotational speed, torque, and flush pressure as a function of depth. The computer relates these data to ground type, and a geological map of the formation is obtained. Each production hole hereby becomes an investigation hole.

Drilling the grouted soils and continuously recording the drilling parameters and correlating them with a thoroughly investigated pilot hole drilled in grouted soils can provide a profile of the quality and continuity of the grouted soils, especially if the drilling is done in conjunction with other tests (e.g., in situ hydraulic conductivity, standard penetration test).

5B.5.3.4 Penetrometer Testing and SPT Testing. Although penetrometer testing is a far better way to verify the characteristics of grouted soils (especially the compressibility coefficient), the Dutch cone testing apparatus is not available in most places in North America. This test allows direct determination of the compressibility of soils and hence their bearing capacity. If this type of test is performed in conjunction with in situ hydraulic conductivity tests, the results for the various grouted layers can be easily evaluated.

5B.5.3.5 Pressure Meter Testing. This test provides a good evaluation of the reduction in compressibility of the grouted soils and should be executed before and after grouting to obtain data about the local improvement (reduction in compressibility) due to the grouting.

5B.5.3.6 In Situ Hydraulic Conductivity Testing. In this testing, the same methodology as during the water testing prior to the execution of the soil grouting program should be followed.

5B.5.3.7 Acoustic Emission Monitoring (AEM) of Injection Pressures. This method detects hydraulic fracturing during grouting. Indications of fracturing are bursts of microseismic noises heard by the system, denoted by increased acoustic emission count rates.

The sensor is placed in an adjacent pipe at the approximate level of the zone in which grouting takes place. It can filter out frequencies below 1000 Hz, which include most construction noises. After calibrating the system, the pressure can be increased until a fracture is induced. The actual grouting pressure is then limited for a comfortable safety margin. Some claim that AEM can track the flow of grout through founding soils. Koerner and coworkers[60] claim that AEM is "a likely candidate" to track and detect subsurface flows. The U.S. Army Corps of Engineers Waterways Experiment Station at Vicksburgh, Pennsylvania, is pursuing this avenue.

5B.5.3.8 Geophysical Quality Assurance Tools. According to Baker,[61] the most useful geophysical tests for evaluating grouted soils include cross-hole seismic profiling and ground-probing radar. These are well suited to defining decreased compressibility and grout presence, respectively.

Borehole Radar. In the preferred method of transillumination profiling, a transmitter is lowered down one borehole and a receiver down an adjacent hole, to the same level. They are then raised simultaneously to provide a radar profile, by taking profiles before and after grouting. The effects of the grouting can be seen in the comparison of the profiles. This system is best suited to determining grout location and to quantifying the grout presence. Its use should however be restricted to granular soils, since its degree of resolution and range in cohesive soils are rather low.

Cross-Hole Acoustic Velocity. Cross-hole acoustic transmissions measure the acoustic velocity and spectra of received signals. Profiles are obtained between boreholes as in the radar method, except that the signal is mechanical rather than electromagnetic. This system is set to determine if the transmitted spectrum indicates an improved acoustic medium after impregnation with grout. Attenuation of acoustic energy in soil is highly dependant upon the stiffness of the ground. For example, grouted sands are known to increase in low-strain stiffness and thus show increased velocity. Such surveys demonstrate qualitatively the strength of the grouted zone, and relative changes in acoustic velocity are of significance. Baker[61] notes that velocities in grouted soils may be as high as 2000 m/s—up to ten times that of ungrouted soil, depending on the water table, diagnostic of a change from soil to weak rock—and therefore indicate the existence of significantly treated soils.

A third category of testing was described by Komine[62] who used electrical resistivity methods to identify void filling efficiency. He found this promising only if the electrical resistivity of the ground, ground water, grout, grouted soil, void ratio, and grain-size distribution were known in advance.

Typically, these geophysical techniques involve exceptional skill and expertise, and are therefore costly.

5B.6 SOME FINAL COMMENTS

It has never been the intent to provide a complete document on soil grouting. This subject is too complex, and there are so many aspects, methodologies, approaches, theories, publications, products, and methods, that it is simply impossible to even try and summarize it. The reader is also reminded that this section barely touches the subjects of rock, structural, and mine grouting.

Those disappointed or offended by what has or has *not* been written should understand that it was never the intent to cover everything about the many complex and evolving branches of science that constitute the applied science of grouting. Furthermore, a lot of excellent or useful publications have not even been mentioned; in no way should one interpret this that they are less valuable. There are numerous good publications, particularly the focused papers discussing only one aspect of grouting, especially if written by a true grouting specialist. The reader should be encouraged to explore the literature and find out more about grouting and grouting related matters. In this publication, I have tried to put some more emphasis on the less well-known and less understood engineering aspects of grouting.

The author wishes to thank the many people who have assisted in providing and exchanging technical information over the years, especially Don Bruce and Doug Heenan. All the "clients" who trusted my expertise and gave me the opportunity to further develop my skills in challenging projects, indirectly, provided the most important contribution to this publication. Also many thanks to my loyal, right-hand man, Luigi Narduzzo, for managing to always tie up the loose ends.

5B.7 REFERENCES

1. D. A. Bruce and R. A. Jewell, "Soil-Nailing: Parts 1 & 2." *Ground Engineering,* May 1986.
2. D. A. Bruce, "The Construction and Performance of Prestressed Ground Anchors in Soils and Weak Rocks: A Personal Overview," *Proceedings 16th Annual Meeting,* DFI, Chicago, October 1991.
3. F. Gularte, "Soil Grouting for Tunnelling and Excavation. Short Course on Grouting," University of Wisconsin, November 1990.

4. M. Potoschnik, "Settlement Reduction by Soil Fracture Grouting." International Grouting Conference, ASCE, New Orleans, 1992.
5. D. A. Bruce, *Handbook on Soil-Grouting,* New York, John Wiley and Sons, 1994.
6. F. Gallavresi, "Grouting Improvement on Foundation Soils," International Grouting Conference, ASCE, New Orleans, 1992.
7. J. Kauschinger and J. Welsh, "Jet Grouting for Urban Construction Geotechnical lecture series, Boston Society of Civil Engineering Design Construction and Performance of Earth Support Systems, Massachusetts Institute of Technology, Cambridge, Mass, November 1989.
8. J. Welsh, "Jet Grouting for Support of Structures," ASCE Conference on Grouting for Support of Structures, Seattle, 1986; J. Welsh and Burke, "Jet Grouting—Uses for Soil Improvement," Proceedings ASCE Conference, Evanston, Ill, 2 vols., June 25–29, 1991.
9. J. Wolosick, "Construction of Anchored Retaining Walls, Underpinning and Slope Stabilization," University of Wisconsin Short Course on Grouting, 1991.
10. J. A. Yeats and N. J. O'Riordan, "The Design and Construction of Large Diameter Base Grouted Piles in Thanet Sand at Blackwall Yard London," Piling Conference, London, 1989.
11. Sherwood and Mitchell, "Toe grouting of caissons in London Thanet Sands," Piling Conference, London, 1989.
12. Gouvenot and Gabaix, "A New Foundation Technique Using Piles, Sealed by Cement Grouting under High Pressure, Proceedings of the Seventh Annual Offshore Technique Conference, Houston, 1975.
12a. Oosterbos, "Installation of Tubular Piles in Oude Schelde Dordrecht," Piling Conference, London, 1989.
13. S. Bandimere, "Grouting: Engineers and Contractors Working Together," University of Wisconsin Short Course on Grouting, 1991.
14. D. A. Bruce, "Enhancing the Performance of Large Diameter Piles by Grouting, Parts 1 & 2," *Ground Engineering,* nos. 4–5, 1986.
15. Stocker, "The Influence of Grouting on the Load Bearing Capacity of Bored Piles," Eighth European Conference on Soil Mechanics and Foundation Engineering, Helsinki, 1983.
16. G. S. Littlejohn and D. A. Bruce, "Improvement in Base Resistance of Large Diameter Piles Founded in Silty Sand," Eighth European Conference on Soil Mechanics and Foundation Engineering, Helsinki, 1983.
17. Endo, "Relation below Design and Construction in Soil Engineering—Deep Foundation: Caisson and Pile Systems, Proc. 9th ICSMFE, Tokyo, 1977.
18. A. Naudts, "Enhancing the Performance of Cast in Situ Piles by Grouting—Some Practical and Theoretical Considerations, Second International Grouting Conference, Toronto, May 1990.
19. A. C. Houlsby, *Construction and Design of Cement Grouting,* New York, John Wiley & Sons, 1990.
20. K. Weaver, *Damfoundation Grouting,* New York, 1991.
21. B. Garrod, A. Naudts, and G. Pozzobon, "The Detroit River Tunnel Remedial Grouting Program," Eleventh Annual Canadian Tunnelling Conference, Toronto, October 1993.
22. M. Jones, "Consolidation Grouting of Heritage Masonry Walls and Foundations at Christ's Church Cathedral, Hamilton Ontario," Second International Grouting Conference, Toronto, 1990.
23. D. Heenan and A. Naudts, "Case History: Rehabilitation Ritson Road Culvert, Second International Grouting Conference, Toronto, 1990.
24. A. Naudts, "Engineering Aspects of Grouting—Grouting Projects Are Not Found, They Are Created," Short Course on Grouting, Technical University of Nova Scotia, June 1991.
25. Abidi, Barnes, and Young, "Investigation and Remedial Grouting of a Dam Undergoing Piping in Its Foundation," Canadian Dam Safety Conference, Toronto, 1990.
26. Z. Aziz and J. Levay, *Grouting Practices,* Quebec, Quebec Hydro Yearbook Canadian Tunnelling Association, 1990.
27. J. Warner, "Compaction Grouting—The First Thirty Years," International Grouting Conference, ASCE, New Orleans, 1982.
28. E. Graf, "Compaction Grout," International Grouting Conference, ASCE, New Orleans, 1992.
29. S. Bandimere, "Compaction Grouting to Stabilize Collapsing Silty Soil, Second International Grouting Conference, Toronto, 1990.
30. D. Greenwood and T. Hutchinson, "Squeeze Grouting Unstable Ground in Deep Tunnels," International Grouting Conference, ASCE, New Orleans, 1982.
31. Essler and Linnay, "Compensation Grouting Trial Works at Redcross Way, London," ICE Conference about Grouting in the Ground, London, November 1992.
32. J. Kauschinger, "State of the Practice and Methods to Estimate Composition of Jet-Grouted Bodies," International Grouting Conference, New Orleans, ASCE, 1992.
33. E. Ahrens, "Test Plan—Sealing of the Disturbed Rock Zone including Marker Bed 139 and the Overlying Halite below the Repository Horizon at the Waste Isolation Pilot Plant," internal publications, Sandia National Laboratories, Albuquerque, New Mexico, 1992, 1994.

34. "Potash 83," Proceedings of the International Conference on Potash Mining, Saskatoon, Saskatchewan, October 1983.
35. G. S. Littlejohn, "Design of Cement Based Grout," International Grouting Conference, ASCE, New Orleans, 1982.
36. Lombardi, "The Role of the Cohesion in Cement Grouting of Rock," Fifteenth Congress de grandes Barrages, Lausanne, 1985.
37. B. DePaoli, B. Bosco, R. Granata, and D. A. Bruce, "Fundamental Observations on Cement Based Grouts: Microfine Cements and the Cemill Process. Fundamental Observations on Cement Based Grouts: Traditional Materials," International Grouting Conference, ASCE, New Orleans, 1992.
38. U. Hakansson, L. Hassler, and H. Stille, "Properties of Microfine Cement Grouts with Additives," International Grouting Conference, New Orleans, 1992.
39. Cambefort, *Injections des Sols,* Paris, Eyrolles, 1964; "The Principles and Applications of Grouting," *Quarterly Journal of Engineering Geology,* vol. 10, 1977, Geological Society of London.
40. R. Krizek, "Effects of Mixing on Rheological Properties of Microfine Cement Grout," International Grouting Conference, New Orleans, 1992.
41. R. Krizek, "Preferred Orientation of Pore Structure in Cement Grouted Sand," International Grouting Conference, ASCE, New Orleans, 1992.
42. R. Krizek, "Injection of Fine Sands with Very Fine Cement Grout," Second International Grouting Conference, Toronto, 1990.
43. R. Krizek, "Preferred Orientation of Pore Structure in Cement-Grouted Sand," International Grouting Conference, New Orleans, 1992.
44. Stripa Report 88-04.
45. D. Keil, "Some New Initiatives in Cement Grouts and Grouting," Forty-Second Canadian Geotechnical Conference, Winnipeg, October 1989.
46. M. Petrovsky, "Monitoring of Grout Leaching at Three Dam Curtains in Crystalline Rock Foundations," International Grouting Conference, ASCE, New Orleans, 1982.
47. D. Hooton and L. Konecny, "Permeability of Grouted Fractures in Granite," *Concrete International Magazine,* 1990.
48. R. Karol, "Chemical Grouts and Their Properties," International Grouting Conference, ASCE, New Orleans, 1982.
49. Tacss Know-How Books, Internal documents Takenaka Comuten Inc., Japan, 1975.
50. M. Wiechmann, "The Need for Continuous Monitoring by Accurate Instrumentation in Analyzing Structural Movements," Second International Grouting Conference, Toronto, 1990.
51. P. Verstraeten, "The European Approach for Preventing Seepage in Tunnelling Design—the Installation and Injection of Injectotubes in Construction and Expansion Joints," Second International Grouting Conference, Toronto, May 1990.
52. E. DesRochers, B. McKinstry, and A. Naudts, "Flood Protection for a Deep Gypsum Mine, Second International Grouting Conference, Toronto, May 1990.
53. R. Peruchon, "Grout Injection of Ground and Concrete Structures," First International Grouting Conference, Toronto, 1989.
54. S. Shimada, "Development of a Gas-Liquid Reaction Injection System," International Grouting Conference, ASCE, New Orleans, 1992.
55. A. Naudts, "Curtain Grouting from a 900 Metres Deep Donut Shaped Tunnel to Dry Up Leaking Shaft at IMC's K2 Mine," *Il Consolidamento Del Suolo e delle Rocce nelle Realizzazioni in Soterraneo,* Milan, March 1991.
56. B. Lukajic, G. Smith, and J. Deans, "Use of Hot Asphalt in Treatment of Dam Foundation Leakage, Stewartville Dam," First International Grouting Conference, Toronto, April 1989.
57. D. A. Bruce, "Contemporary Practice in Geotechnical Drilling and Grouting," First International Grouting Conference, Toronto, 1989.
58. D. Deere, "Cement-Bentonite Grouting for Dams," International Grouting Conference, ASCE, New Orleans, 1982.
59. D. A. Bruce, "Permeation Grouting," *Handbook on Soil Grouting,* New York, J. Wiley & Sons, 1994.
60. Koerner, "Acoustic Emission Monitoring of Grout Movement—ASCE Conference Issues on Dam Grouting," Denver, Colo, April 1985.
61. W. H. Baker, "Planning and Performing Structural Chemical Grouting," International Grouting Conference, ASCE, New Orleans, 1982.
62. H. Komine, "Estimation of Chemical Grout Void Filling by Electrical Resistivity," International Grouting Conference, New Orleans, 1992.

P · A · R · T · 6

FOUNDATION FAILURES AND REPAIR: RESIDENTIAL AND LIGHT COMMERCIAL BUILDINGS

As with almost every undertaking, even the best made plans occasionally go awry. This is certainly true where foundation design and construction are concerned, and especially with light foundations. This section will address those instances where the problems do arise.

Section 6A defines the common causes for foundation failures, how to identify the problems, and it describes some of the more common misconceptions associated with differential foundation movement.

Section 6B presents the spectrum of foundation repair options, special conditions which resist restoration, and it describes actual field practices.

Section 6C provides maintenance procedures which, if implemented, could prevent or minimize foundation problems.

Section 6D suggests steps helpful to prospective buyers for evaluating the structural conditions of real estate properties. In geographical areas with widespread foundation problems it has become routine for lenders and insurers to require structural inspections before mortgages or certain insurance policies will be issued. Information in this section can also be useful for building inspectors (see Sec. 8A).

Much of this information, and more, was presented in *Foundation Repair and Behavior: Residential and Light Commercial,* published by McGraw-Hill in 1992. This prior work is a complete reference to foundation behavior, failure, repair, and prevention of damage. Similar information included herein is intended to help separate this handbook from other handbooks.

SECTION 6A
FOUNDATION FAILURES

Robert Wade Brown

6A.1 CAUSES FOR FOUNDATION FAILURES 6.3
   6A.1.1 Nonexpansive Soils 6.3
   6A.1.2 Expansive Soils 6.6
   6A.1.3 Summary 6.10
6A.2 SETTLEMENT 6.11
6A.3 UPHEAVAL 6.12
6A.4 OCCURRENCE OF SETTLEMENT VERSUS UPHEAVAL 6.12
6A.5 DIAGNOSIS OF SETTLEMENT VERSUS UPHEAVAL 6.16
6A.6 SLIDING 6.17
6A.7 THE CASE FOR AND AGAINST USING GRADE ELEVATIONS FOR EVALUATING FOUNDATION DISTRESS 6.18
6A.8 SOIL SWELL (UPHEAVAL) VERSUS MOISTURE CHANGES 6.20
6A.9 HOW IS THE NEED FOR FOUNDATION REPAIR ESTABLISHED? 6.23
6A.10 REFERENCES 6.28

6A.1 CAUSES FOR FOUNDATION FAILURES

In general, people tend to associate foundation failures with expansive soils. Basically, the association is justified, especially with residential properties. However, problems are also encountered in areas with nonexpansive soils. Furthermore, foundation failures can occur which are not directly related to bearing soil behavior. Examples of the latter could involve those instances where the foundation is designed or constructed improperly. This could include such flaws as placement or sizing of reinforcement, stressing or placement of cables (posttension), design and placement of beams or piers, or selection, quality, and placement of concrete. The following sections will discuss situations most commonly linked to differential foundation movement, not normally considered as design-related. In general, the statements are applicable whether the foundation design is residential or commercial, slab or pier and beam. For simplicity, the presentation is subdivided into nonexpansive soils and expansive soils. In all cases it is assumed that the foundation was designed appropriately and of concrete construction.

6A.1.1 Nonexpansive Soils

Often the nonexpansive soils are also referred to as *noncohesive*. This is generally true, except that some cohesive soils can be relatively nonexpansive. An example is kaolinite. Typical nonexpansive soils are sands, gravels, coral, and sometimes, though to a lesser extent, certain peats, silts, and clays. Soils with a swell potential of less than 1.0% are often considered nonexpansive. Foundations situated on these types of soils can suffer distress failure, usually as a result of one or more of the following:

   1. *Frost heave and permafrost*   Frost heave occurs as a result of soil water freezing with sufficient expansion to cause the foundation member to heave or, in some cases, as with basement walls, to collapse inward. Soils in subfreezing climates which are porous and permeable are the most susceptible to this problem. Permafrost represents a problem of a different nature. Here the principal defect involves melting of the bearing soil, which results in subsidence of the foundation.

   2. *Undercompaction*   Bearing soils consisting of fill or disturbed native material often must be compacted to provide the competent, consistent bearing surface required to carry structural loads. Well-graded sands and gravels (dry and relatively pure) will consolidate almost immediately upon application of load and, accordingly, would offer few problems. (An exception could be collapsible soils.[1,2]) On the other hand, heterogeneous, low-density, expansive, or wet soils could represent a principal concern. Since undercompaction is followed by consolidation, the differential foundation movement, if any, would be that of settlement. The introduction of water will often exacerbate the process.

3. *Overcompaction*   Overcompaction can be a natural consequence when overburden is removed from a soil and not replaced by an equivalent or greater load. Alternately, fill could be mechanically overcompacted. In either event, the net result could be soil rebound and an upward thrust of the structure (upheaval).

4. *Removal of water* (*drain*)   Removal of excessive water from the cell structure of specific soils often results in consolidation. Sometimes this loss is such that structural subsidence follows. This could be the result of pumping water from aquifers (Mexico City) or simply of lowering the level of a nearby lake (Florida). This phenomenon is often described as a *consolidating soil*.

5. *Addition of water* (*collapsing soils*)   Under appropriate conditions, specific soils are susceptible to a substantial reduction in void ratio upon addition of water. These soils might exist naturally at low moisture contents with apparent high strengths largely attributable to particle bonding. Water destroys this structural bond and allows local compression. The majority of soils which tend to experience this collapse are aeolian (wind) deposits such as sands, silts, sandy silts, or clayey sands of low plasticity in which there is a loose structure with relatively low density. As opposed to time-dependent consolidation (reduction of void ratio permitted by removal of pore water), settlement of these collapsing soils occurs somewhat rapidly upon the influx of water.[1-5]

6. *Decay of organic material in fills*   Organic materials decay. In the process, methane gas is produced and a void is created. If the volume of the organic material (void) is sufficient, structural subsidence is likely to follow. (The exception is when the foundation bearing surface is below the level of the organic material.) As a rule, when the process of decay is ongoing, methane gas can be detected.

7. *Construction on a sloping site with down-dip soil bedding planes*   In this instance a downslope stress can be transmitted to the foundation. If the sliding stress is sufficient, downslope movement of the structure could occur unless, of course, the foundation support is constructed through (beneath) the problem soil-soil interface. Horizontal parallel bedding planes would not suggest this problem [Fig. 6A.1(*a*)].

8. *Cut and fill site*   Anytime a foundation is placed on cut and fill, the inherent tendency toward instability exists. The cut soil may lose desirable properties, and the fill area represents a soil of altered or different properties. The latter tends to be true even if the fill is native material, supposedly compacted to the original density and moisture. The disturbed, reconstituted soil seldom, if ever, exhibits properties identical to those of the natural undisturbed soil.

9. *Marginal bearing capacity*   Some soils, such as silts and loess, have a low bearing capacity, which tends to make them unsuitable for foundation support. In this case, as with items 7 and 8, special foundation design procedures are required.

10. *Sulfide or sulfate attack on concrete or steel*   Sulfate corrosion occurs when high sulfate levels contact exposed concrete. Sulfide corrosion occurs when sulfates are converted into sulfides, or when free sulfide (H_2S) is naturally available. Bacteria can convert hydrogen sulfide into sulfuric acid (H_2SO_4), which attacks concrete and metals alike. In other instances, hydrogen sulfide (H_2S) can react directly with metals. The best prevention is not to allow the sulfate or sulfide access to the foundation or to use sulfate-resistant concrete (where sulfates are a known problem). See Sec. 2A and Kienow and Kienow.[6] As an aside, the introduction of soluble sulfates into soils containing lime can result in the formation of ettringite accompanied by soil heave. Refer to Petry and Little,[7] and to Sec. 6B.10.5.

11. *Earthquakes or slides*   These problems normally result in total destruction of the property and are not addressed as being remediable repairs. Refer to Sec. 3B.

12. *Sand or gravel cushion beneath slabs-on-grade*   Where the cushion is not separated from the bearing soil, water entering the porous fill material can cause the sand or gravel to penetrate the

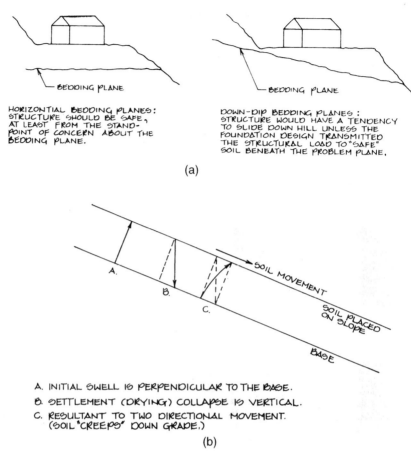

FIGURE 6A.1 (*a*) Construction on natural slope. (*b*) Soil creep.

wet soil. A cushion 6 in (15 cm) thick can allow a slab settlement of 2 in (5 cm). This problem is more pronounced with soils that are water-sensitive, that is, expansive. Alleviation of this condition can involve either the use of impermeable moisture barriers (6-mil polyethylene sheets) to separate the fill cushion from the soil or elimination of the granular fill altogether. One advantage of the cushion is to permit a higher finished floor grade without requiring additional concrete.

13. *Soil creep*   Anytime soils are placed on slopes without sealing in the soil moisture, the soil has a tendency to creep, or move down the slope. Structures built on soils subject to this condition can move as well. This phenomenon is enhanced by wetting and drying cycles as well as the presence of moisture-sensitive soils. (An example is the movement of flatwork laterally away from a foundation.) See Fig. 6A.1(*b*).

Note in particular that of the foregoing potential causes of water-related distress, none are relative to changes in *combined* soil moisture, such as surface-absorbed or crystalline interlayer water. Further, in no instance was there a significant physical change noted in the soil particles.

On occasion such factors as lightning and vibrations (usually from blasting) have been alleged to be the cause for foundation damage. While this is possible, the author has not documented such a cause in over 35,000 inspections spread over 30 years.

6A.1.2 Expansive Soils

The subject of foundation failure relative to differential movement takes on a new significance with the introduction of a new parameter—expansive soils. In addition to the problems described in Sec. 6A.1.1, the soil itself now becomes a principal contributor to the movement. Expansive soils will shrink upon loss of moisture and, more importantly, will swell upon introduction of water. The expansive clay constituent within these soils represents the basic problem. One such clay, montmorillonite, can swell up to 20 times its volume. Soils with montmorillonite present to the extent of 40% total volume of solids have been known to show swell potentials in excess of 30%, with expansive pressures in excess of 20,000 lb/ft^2 (97,600 kg/m^2).

In addition to the percentage and the type of clay, the swell potential of a soil is dependent on such factors as initial and final moisture contents. A soil with a natural moisture content of 1.0% or so above its plastic limit may already have experienced most of its swell. As a rule of thumb, residential foundations are designed to tolerate swell potentials up to about 3%. Above that, damage may occur. (Some engineers suggest even lower "safe" swell potentials, often 2% or less.) With the new trouble factor in mind, the frequent causes of foundation failure include:

1. *Abnormal soil moisture at the time the foundation is poured* As a rule, this problem is mostly relegated to slab construction, more particularly that of residential or light commercial design.

If the natural soil moisture is unusually low at the time the slab is poured, two possibilities for differential foundation movement exist—perimeter uplift or center heave. The presence of the slab itself tends to cause the confined soil moisture content to increase somewhat, creating a wetter condition beneath the slab as compared to the dryer external soil. Consistent with this, several points should be acknowledged:

- The dry exterior soil has access to greater water availability from periodic rains or watering, either of which could equalize or reverse the natural soil moisture differential. In this instance, should the condition persist sufficiently long, perimeter uplift could occur.
- Water naturally existing beneath the slab tends over time to be of rather uniform distribution and a few percentage points above that of the original in situ moisture (Fig. 6A.2). The final level of soil moisture will be determined by the availability of water from the lower soils within the active zone. Obviously, these potential "donors" must have in situ moisture above that of the shallow soils.
- The principal threat comes from the introduction of excess water into the relatively dry soil beneath the foundation. The source may be either domestic or various forms of precipitation. The latter is usually accompanied by some drainage deficiency. In either instance, the problem is center slab uplift (upheaval), which is the most preponderant cause of slab foundation failure.

Figure 6A.3 depicts the relative conditions of center heave, perimeter settlement, and perimeter uplift.

If the soil moisture is relatively high when the slab foundation is placed, the general tendency could be the eventual appearance of a center slab heave (doming) or perimeter settlement. This is prompted by the relatively wetter interior soils as compared to the dry exterior soil. The heave could be uniform (or avoided) if the interior moisture were maintained constant over the entire slab area and soil moisture migration toward the exterior from the interior were prevented. Attempts to do this artificially have involved the use of both moisture barriers (see Sec. 6C.3.2) and some form of supplemental external watering system (see Sec. 6C.2). The latter would be helpful to ensure perimeter stability as well. Remember, moisture in the soil beneath a slab foundation is essentially "bottled up." Capillary or soil suction pressures are restricted to a large extent by the very low permeability of clay. Hence any situation that tends to move soil moisture will be quite slow in happening and can be arrested or reversed by proper availability of water. Furthermore, in these instances where the soil's vertical and horizontal permeabilities are roughly equal, natural forces would favor the condition of uniform moisture over the confined area.

Generally speaking, it seems that the foundation distress induced by natural soil moisture contents at the time of construction poses more of a theoretical concern than a real problem. These soil conditions should be addressed *before* the foundation is placed.

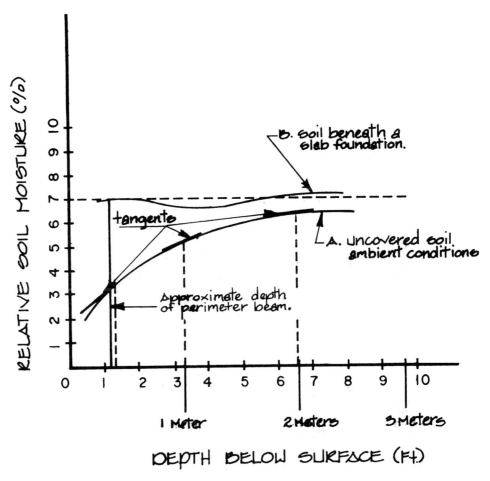

FIGURE 6A.2 Natural soil moisture versus depth.

Note: An in-depth study of the moisture profile beneath slab foundations can be found in two recent research articles.[8] One important point that these papers bring out is the absence of the sometimes assumed "central doming" effect. The soil confined by the foundations does heave, but not in the pattern associated with the so-called doming. Furthermore, their data downplay the overall effectiveness of vertical soil barriers. Several articles describing similar findings can be found in the *Proceedings of the 7th Conference on Expansive Soils.*[9]

    2. *Undercompaction*   The problem is basically the same as for nonexpansive soils (see Sec. 6A.1.1, item 2).

    3. *Overcompaction*   The problem is somewhat the same as in Sec. 6A.1.1, item 3, except that the new parameter of soil swell is introduced. Refer also to Lawton et al.[4]

    4. *Frost heave*   This is often not a serious problem in areas with expansive soil due to the extremely low permeability and normally low porosity. Water must be present within the soil matrix at a depth sufficiently shallow to be affected by freezing temperatures. In some cold-climate wet areas the expansive soil might have sufficient root channels, intrinsic fractures, or slickensides to provide the necessary permeability and porosity to become the exception. As the name implies, since the movement is one of expansion, upheaval is the mode of failure (see also Sec. 6A.1.1, item 1).

6.8 FOUNDATION FAILURES AND REPAIR

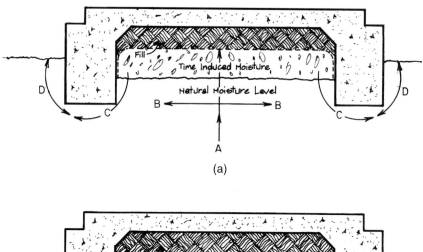

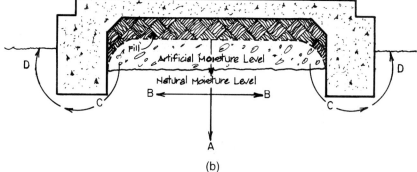

FIGURE 6A.3 (*a*) Soil dry when foundation poured. 1. Prior to placement of foundation—major moisture loss (evaporation) in *A* direction. 2. After placement—evapotranspiration restricted, moisture tends to increase in *A* and *B* directions. Removal of moisture would be from *C* to *D*, provided *D* existed in a dryer state. 3. Natural conditions—Moderate moisture addition at *D* would tend to avoid moisture migration and differential foundation movement. Alternately, without moisture, some perimeter settlement could be possible. Excess moisture at *D* might tend to produce perimeter heave. 4. Preponderant threat—any accumulation of extraneous moisture in *A–B* area which could cause serious upheaval. (*b*) Soil wet when foundation poured. 1. Moisture levels at *A*, *B*, and *C* are much higher than at *D*. Major soil moisture flow would be in *A–B* direction and to a variable extent *C* dependent upon moisture level at *D*. (Direction of flow is always from wet to dry.) 2. Major moisture loss from *C* to *D* would, over time, produce appearance of a center high and/or perimeter low. 3. Time required for excessive movement or equilibrium of soil moisture to develop would be dependent on such factors as in situ soil moisture at lower levels (*A* direction), relative permeabilities, and amount and rate of transfer from *C* to *D*. In many cases equilibrium might not be attained during useful life of structure. 4. Introduction of extraneous water beneath slab might not produce serious upheaval due to already expanded state of the soil.

    5. *Decay of organic material*   See Sec. 6A.1.1, item 6.

    6. *Plant roots*   Plants tend to sap more soil moisture through transpiration than wind and heat do through evaporation. As long as this moisture loss is confined to capillary water (sometimes referred to as pore or interstitial water), the soil itself is not particularly affected, as would often be the case with nonexpansive soils. However, at some point, if other moisture is pulled from the clay, the results can be soil shrinkage with potential foundation settlement. In general, foundation distress relative to transpiration is probably much less than normally assumed. This is supported by the following discussion.

    Vegetation feeder roots tend to be both shallow and limited in radial extent from the plant trunk. Most botanists suggest that the feeder roots of many trees and plants exist within the top 1 to 2 ft (0.3 to 0.6 m) below the ground surface.[10-14] If the perimeter beam is over 2 ft (0.6 m) deep, intruding roots are not likely to cause any serious concern with respect to the interior foundation areas. Roots already in place might be another story, as might be trees located quite close to a shallow foundation. The latter conditions could give rise to mechanical upheaval of the foundation beam as the roots grow in diameter. Under these conditions, similar effects could be ascribed to foundation problems on nonexpansive soils.

On the other hand, tree roots can be beneficial to foundation stability since they tend to act as a soil reinforcement and increase the soil's shearing resistance.[13] The practical result of this effect could be to impede foundation settlement and reduce soil cracks, even though the roots do remove soil moisture.

Figure 6A.4 shows the relationship between tree roots and foundation. The foregoing representations were taken from leading botanists such as Dr. Don Smith at North Texas State University and the references.[10-14] Stevens[14] reconfirms the shallow depth of feeder roots.

7. *Loss of soil moisture* Any condition that causes an expansive soil to lose moisture has the potential to cause foundation settlement. As a rule, settlement represents the least worrisome foundation problem, since generally the extent of movement is relatively limited and the condition is often reversed easily by the introduction of proper watering procedures. Deep foundations rarely suffer from this problem because moisture loss is quite limited in depth.[15-20] Generally, the overall consensus places the depth of the active zone at somewhere between 1 and 11.5 ft (0.3 and 3.5 m). However, all seem to agree that at least 80% of total moisture change occurs within the top 3.2 ft (1.0 m) or less (see Fig. 6A.2).

8. *Gain of soil moisture* With limited exceptions an expansive soil will swell when subjected to excess water. Conversely, some clay soils will lose structure (cohesion) and creep at extremely high moisture contents, usually those bordering the liquid limit. In this relatively rare instance, the expansive soil might give the impression of shrinking, or creeping.

Soil swell and foundation heave represent the most serious problem to foundation stability when dealing with an expansive soil. The source of water could result from such causes as surface water (improper drainage), utility leaks (principally slab foundations), wet-weather aquifers, or subsurface water.[21,22]

9. *Construction on sloping site with down-dip bedding planes* The soil-soil interface at the bedding plane has a lower shear resistance because of reduced contact friction. The soil is therefore more subject to slipping or sliding. This is particularly true with expansive soils. In sliding, the mode of failure is not up or down but horizontal. See also Sec. 6A.1.1, item 7.

10. *Cut and fill site* The problems are the same as in Sec. 6A.1.1, item 8, only more so. When rather permeable fill materials are used which extend outside the building areas, surface water can penetrate the fill and wet the expansive soil beneath the foundation, resulting in foundation heave. Only well-compacted clay fill should extend outside or beyond the building lines.

11. *Marginal bearing capacity* Problems are the same as in Sec. 6A.1.1, item 9.

12. *Sulfate or sulfide attack* Problems are the same as in Sec. 6A.1.1, item 10.

13. *Earthquakes or slides* Problems are the same as in Sec. 6A.1.1, item 11.

14. *Sand or gravel cushion beneath slabs-on-grade* Refer to Sec. 6A.1.1, item 12. This problem seems to suggest an instance where moisture causes an expansive soil to appear to settle rather than swell. (See also item 8, this section.) In this instance the amount of moisture or the swell capacity of the clay are such that any expansive swell is less than the volumetric loss due to the porosity provided by the granular fill. Theoretically a 6-in (15-cm)-thick granular fill could produce a 2-in (5-cm) void. In any event, water must have access to the cushion fill. In flatwork this is rather straightforward. However, foundation beams create a substantial obstruction to the invasion of surface moisture. Domestic leaks are not subject to this restriction.

15. *Soil creep* See Sec. 6A.1.1, item 13.

The foregoing presented the basic factors that threaten the safety and stability of foundations in general. The following discussions will be more specific.

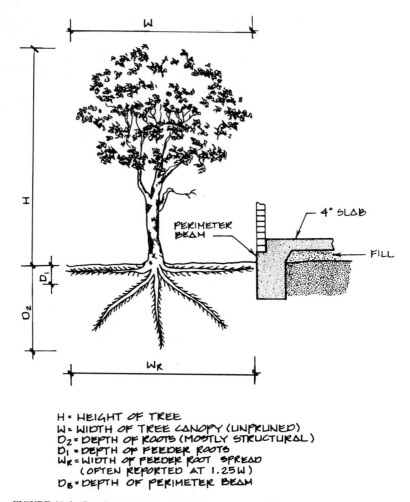

FIGURE 6A.4 Development of tree root system versus foundation.

6A.1.3 Summary

To date much concern has been devoted to avoid foundation failures resulting from distress—or differential movement—brought about by design deficiencies or environmental extremes. Annual rainfall, temperature ranges, clay content of the supporting soil, soil composition, and so on, combine to help produce the design and construction requirements for foundations. In highly plastic clay soils such conditions as the relative moisture content in the soil immediately prior to construction can affect the future stability or behavior of the foundation.

Each of these factors can have a direct influence on settlement (soil shrinkage) as well as on upheaval (soil swelling). The following sections will introduce an argument, supported by field results, that upheaval, not settlement, is in fact the most serious and prevalent problem for slab-on-grade foundations within many areas featuring expansive soils. Holland and Lawrence[19] report[19,20] on Australia, whereas Frelund et al.[23] suggest the preponderance of upheaval in Canada. In the Dallas, Texas, metropolis upheaval accounts for as much as 70% of all failures (see Sec. 6A.4). These investigations assume an expansive soil with an ambient moisture no higher than perhaps the plastic limit. The lower the natural moisture, the greater the swell potential (Fig. 6A.5).

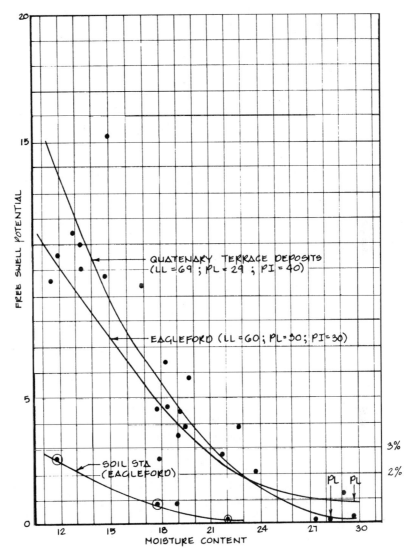

FIGURE 6A.5 Free swell versus soil moisture content.

To avoid any problems of semantics, it is advisable to define the terms *settlement* and *upheaval*. In general this chapter focuses more on slab foundations. Similar results and problems are encountered with pier and beam foundations. However, slab construction has recently become numerically dominant over piers and beams.

6A.2 SETTLEMENT

Settlement is the instance when some portion of the foundation drops below the original as-built grade. This occurs as a result of loss of bearing, such as compaction of fill, erosion of supporting soil, or dehydration (shrinkage of supporting soil). As an example of settlement brought about by loss of soil moisture, Tucker and Poor[15] reported a change of ½ in (1.27 cm) in vertical displacement for a slab foundation over a period of about 6 months from the winter of 1971 to the summer of 1972. The conditions of exposure (and loss of soil moisture) were more severe than one would expect for normal residential conditions, since the slab was completely exposed (the house had been removed) and the perimeter had not been watered.[15]

In general settlement originates and is more pronounced at the perimeter of the slab, and more particularly at a corner. Settlement progresses relatively slowly and can often be reversed or arrested by proper watering. Figure 6A.6 shows typical examples of foundation settlement. Figure 6A.6(a) is an artist's rendering of settlement and Fig. 6A.6(b) shows an actual occurrence. Cracks generally appear first at structural weak points, such as openings or joints, but may develop in other locations, depending on specific tension forces. (Failures normally develop in a direction perpendicular to tension.) Table 6A.1 presents a generalization of exterior crack or failure patterns in masonry.

6A.3 UPHEAVAL

Upheaval relates to that situation in which the internal and, on rare occasions, external areas of the foundation raise above the as-built position. Figure 6A.7 shows typical examples of foundation upheaval. Refer also to Table 6A.1. In highly plastic soils this phenomenon results almost without exception from the introduction of moisture under the foundation. Upheaval can develop rather rapidly, depending on the quantity of available water, the starting or in situ water content, and the clay constituent within the soil. The author has recorded instances where 3 to 4 in (7.5 to 10 cm) vertical deflection occurred in a slab foundation over a time interval of less than 1 month. Admittedly this is unusual, but it has been documented. Once the slab heaves, the reinforcing steel is stressed into what amounts to a permanent elongation. Even if the expanded soil were to shrink, the domed slab would tend to remain in the deformed contour. As a matter of interest, a 4-in (10-cm) heave over a 12-ft (3.6-m) distance would cause the rebars to be stretched about ¼ in (0.6 cm) (Fig. 6A.8). The ¼ in (0.6 cm) does not seem to be very much, but as a point of interest, ¼ in (0.6 cm) linear elongation (stretch) is often used as one criterion for ultimate failure of steel piles in tension.[24]

The moisture may originate from normal precipitation (improper grade), underground water, or, more frequently, from domestic sources such as leaks in supply or waste systems.[21-25] Visual observations will usually reveal any grade problems. An understanding of the particular locale will help predict or evaluate the prevalence of underground water. A qualified plumbing check will verify or sometimes eliminate the possibility of any domestic leak.

Watering around the perimeter can sometimes retard the noticeable progress of upheaval temporarily, but will not normally reverse the problem. The latter is true for two reasons:

1. In the case of slab foundations, the weakest area (interior) is domed and stressed into a deformed contour. Watering may tend to raise the perimeter, but the interior area is likely to merely move upward as well. When the availability of water ceases, the exterior soil moisture decreases and the beam may settle back.
2. In the case of pier and beam foundations, surface watering is not apt to influence the grade of the perimeter beam because the piers tend to act as anchors.

6A.4 OCCURRENCE OF SETTLEMENT VERSUS UPHEAVAL

The author reviewed 502 foundation inspections during a particular study period.[21,26,27] The basic soils within the subject area exhibit representative plastic limits (PL) of about 18, liquid limits

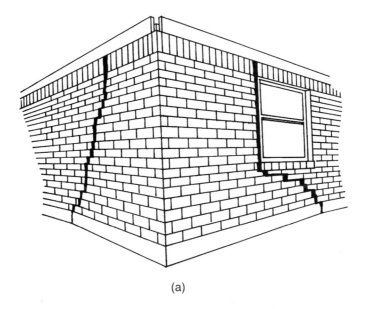

(a)

(b)

FIGURE 6A.6 (*a*) Typical foundation settlement. (*b*) Foundation settlement at corner. Foundation settlement at the corner of the perimeter beam caused the very noticeable separation in the brick mortar. In this instance, the crack is in excess of 2 in (5 cm). Note that the brick has also slipped laterally.

TABLE 6A.1 Evidence of Exterior Tension (Movement)

1. A *vertical crack* in the center of the wall extending from the grade beam upward, with the crack wider at the top, generally indicates either upward movement in the central region of the grade beam (upheaval) or downward movement of the corner(s) (settlement).
2. *Diagonal cracks* in brick often indicate rotational movement of the grade beam. Depending on specific conditions, these may reflect either settlement or upheaval, or sometimes even a combination of both.
3. *Horizontal cracks* in brick mortar often suggest upheaval.
4. *Cracks larger at the bottom* than at the top indicate either upward movement of the corners or downward movement of the central portion of the wall.
5. *Separations between door or window frames and brick veneer* can also be indicative of either settlement or upheaval.
6. *Fills* placed in a manner inconsistent with good engineering practice can complicate the stress pattern.

Source: Foundation Seminar at UTA Arlington, Tex., 1985.

(LL) of 60, and plasticity indexes (PI) of 42. The natural moisture was about 18% and montmorillonite content was around 40%. The study area was primarily confined to Dallas County, Tex.

Of the cases reported, 216 (43%) were no bids, that is, there was no problem that required attention other than adequate maintenance. Most structures exhibited the result of some settlement, but often the movement was not considered sufficiently serious to warrant repairs. Identifying upheaval tendencies early often allows the cause to be identified and eliminated, thus avoiding serious problems. Virtually all structures within the study area showed signs of some distress.

The remaining 286 cases were considered to have problems sufficiently serious to warrant foundation repairs; 87 of those foundation failures were obvious, unquestioned conditions of settlement and 199 cases exhibited some indication of various degrees of upheaval. From these data it appears that settlement versus upheaval would occur on the basis of 87 to 199, or 30% as compared to 70%. Other studies have developed similar data for comparable soils. (See Sec. 8A.)

Many individuals have been reluctant to accept the preponderance and seriousness of upheaval. However, the available data give proof that upheaval is indeed the most serious deterrent to slab-on-grade construction. It is also interesting to note that in most cases the movement arrests sometime after the source of water has been eliminated. What does all this mean? Water under a foundation is truly a serious problem and must be avoided. This suggests measures to control subsurface water and surface drainage as well as preventing deficiencies or leaks in the plumbing systems. Failing to do so will almost certainly invite problems.

Surface grade problems can be handled easily, and construction in areas subject to problem subsurface water can be avoided or the condition remedied prior to construction. Simple quality control would probably eliminate a majority of the plumbing-induced problems. According to the plumbing contractors, whose findings and data helped provide the basis of this chapter,[27] the principal or most common plumbing problems have the following origins:

1. Faulty or omitted shower pans
2. Faulty installation of water closets (collapsed lead sleeves)
3. Faulty or ill-designed drain waste valves
4. Ill-fitted or careless installation of connection joints
5. Substandard sewer line materials
6. Broken pipes

Early signs that might forewarn the existence of any of these problems would include fixtures which begin to drain slowly, backups, or the appearance of wet spots. In other words, added attention to the plumbing installations might reduce foundation failures in slab-on-grade construction

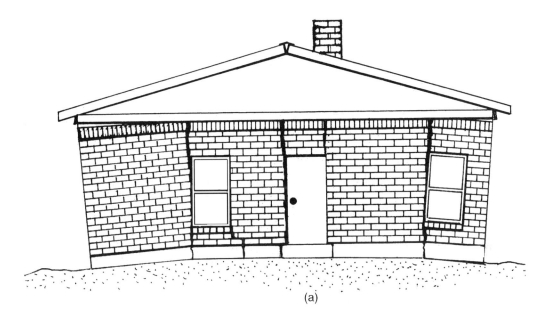

FIGURE 6A.7 (*a*) Typical foundation upheaval. (*b*) Upheaval in interior slab floor.

by as much as 50 to 60% within areas that contain soils similar to those studied. Along these lines a warning was published in the June 3, 1990, *Dallas Morning News,* Section H, "Flexible Plastic Plumbing Causing Problems, Lawsuits," pointing out the particular hazards associated with the use of polybutyl plastic pipes and fittings. Special concern should be given to the use of this material beneath the slab.

Bear in mind two facts: (1) water from leaks beneath the slab is accumulative and any water under the slab, regardless of source, tends to accumulate in the plumbing ditch; (2) sewer lines are

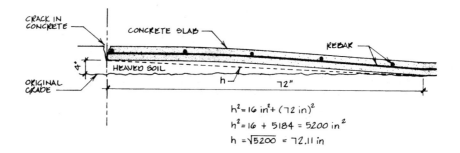

FIGURE 6A.8 Rebar stretch versus heave.

graded for gravity drainage away from the foundation, but the ditches in which they are situated may not be. Anything that could be done to alleviate the accumulation of water in the plumbing ditch would obviously be beneficial to slab-on-grade construction.

Also be advised that seldom does the water accumulate at the source of the leak and very little water is needed to cause serious upheaval. In fact, assuming:

1. A 100 ft² (9.3 m²) affected slab area
2. A soil with a PI of 42 and W_s = 81.96 lb/ft³ (1313 kg/m³)
3. An initial 22% moisture
4. Occurrence of 86% of principal soil moisture change within the top 3.5 ft (1.1 m)

Under these conditions, a 3% change in moisture would likely cause upheaval in the foundation of approximately 1.25 in (3.2 cm) and would require less than 0.5 oz/h (14.6 mL/h) of water over the period of a year. (The foundation area most often affected is the weakest point—the interior slab.) (See also Sec 6A.8.) Since water is accumulative, nearly any sort of persistent leak or drainage deficiency would handily account for the stated amount of water. In some drier soils, a 3% change in moisture would account for substantially higher swells. Conversely, a soil with a natural moisture content near or in excess of the plastic limit will normally exhibit little retentive swell potential.

6A.5 DIAGNOSIS OF SETTLEMENT VERSUS UPHEAVAL

Upheaval, as stated previously, is virtually always the result of some water accumulation beneath the slab. Settlement describes the condition in which some area of the foundation falls below the original grade. The differentiation between the two is most critical and difficult, requiring exten-

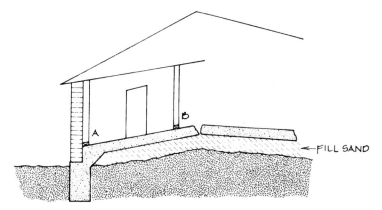

FIGURE 6A.9 Example of foundation deflection.

sive experience and knowledge (Fig. 6A.9). Did the slab foundation heave near the interior partition, or did the perimeter settle? Consider the evidence:

1. The door frame is off 1 in (2.5 cm) across the 30-in (0.75-m) header with the inside jamb being high. The grade change from point B to point A is 4.75 in (12 cm). The distance from B to A is 12 lin ft (3.6 m).
2. The brick mortar joint shows a high area near the midpoint of the wall A.
3. The cornice trim at both front and rear corners is out less than $\frac{1}{2}$ in (1.27 cm).
4. At both corners the mortar joints down adjoining walls are reasonably straight, except for the single high area previously cited.
5. There is no appreciable separation of brick veneer from window frames in the end walls.

Analysis of the data gives rise to only one conclusion: the center has heaved. Alternately, had the perimeter settled, the greatest differential movement [4.75 in (12 cm)] would show at the corners, and observations 3, 4, and 5 would have been impossible. To further complicate the matter, both settlement and upheaval often exist in the same structure. Before proper repair procedures can be established, the cause of the problem must first be diagnosed.

Failures which are diagnosed as interior slab settlement can sometimes be the result of a slab having been installed during a period of prolonged dry weather when the soil moisture might be abnormally low. When rains or domestic watering supply the perimeter areas with moisture, natural swelling occurs, which raises the perimeter, giving the overall appearance of interior settlement. The opposite situation, that is, construction of a slab after extended moist weather, can occur, which gives the impression of central slab upheaval when the perimeter area bearing soil returns to normal moisture content and leaves the central slab area high. In either event the vertical displacement is brought about by variations in moisture content within the expansive soil, and the magnitude of this differential is normally limited.

6A.6 SLIDING

A third type of movement, sliding, sometimes occurs when a structure is erected on a slope. Here the movement is not limited to up or down (vertical) but possesses a lateral or horizontal component. In expansive soils, cut-and-fill operations can be particularly precarious. The clay constituents in the cut tend to become unusually expansive, even though recompacted to (or near) their

6.18 FOUNDATION FAILURES AND REPAIR

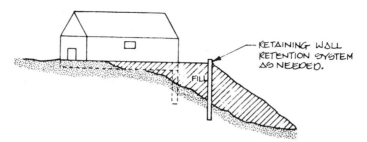

FIGURE 6A.10 Construction on filled slope.

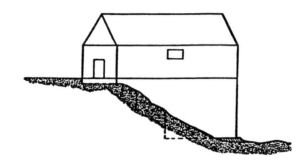

FIGURE 6A.11 Step-beam construction on slope.

original density. This is due in large part to the breakup of cementation within the cut soil, poor compaction of fill, or, to a lesser extent, a change in the permeability or infiltration of the fill.

Figure 6A.10 illustrates a construction on a filled slope and Fig. 6A.11 shows the use of a step beam to conform to the contour of the land. The applied forces are basically the same in both examples, and foundation failure generally results in movement down the slope. As a rule, restoration of this problem is handled as settlement. The lower areas can be raised and restored to a vertical position, but rarely is the horizontal component reversed. Fortunately this type of movement is relatively rare.

If slope stability is questionable, retaining walls, piers, or pilings are often required. To prevent the future movement or sliding of fill and, consequently, potential foundation failure, the retention system must be designed to withstand the downhill stress. Refer to Sec. 3. Figure 6A.10 shows construction on a filled slope. Piers can be drilled through the fill into undisturbed soil and used to support the foundation, or a step beam can be designed which conforms to the contour of the land, as illustrated in Fig. 6A.11. Either approach will reduce or neutralize the downhill stress.

6A.7 THE CASE FOR AND AGAINST USING GRADE ELEVATIONS FOR EVALUATING FOUNDATION DISTRESS

Grade elevations *can* be a useful tool in evaluating the scope and extent of differential movement. The information can also be quite misleading. Instrument readings show the grade elevations of the foundation at the time the data are taken. Since virtually no foundation is constructed truly level, these data will reflect uneven elevations due to as-built conditions, but they do not differentiate between this condition and differential movement. (The latter is the only concern influencing foundation repair.) Consistent with this, Fig. 6A.12 is an overly simplified depiction of the problems inherent in grade elevations.

The elevations in Fig. 6A.12(A) indicate a grade differential between points *a* and *b*, $\Delta_{ab} = 3$ in (7.5 cm). This represents the as-built condition and does not reflect differential movement. A second set of elevations is reflected in Fig. 6A.12(B). Has the house actually improved? By no means. Point *b* has subsided or settled $1\frac{1}{2}$ in (3.7 cm), which is manifested in the usual signs of distress within the structure, that is, doors out of plumb, sheetrock cracks, cracks in concrete or mortar, and so on. This is true although the elevations might indicate a reduced movement between points *a* and *b*. The same analysis holds for the condition simulated in Fig. 6A.12(C). The difference is that point *b* remained stationary whereas point *a* heaved $1\frac{1}{2}$ in (3.7 cm). Figure 6A.12(D) represents a rather unique situation. Point *b* subsided $1\frac{1}{2}$ in (3.7 cm) and point *a* heaved

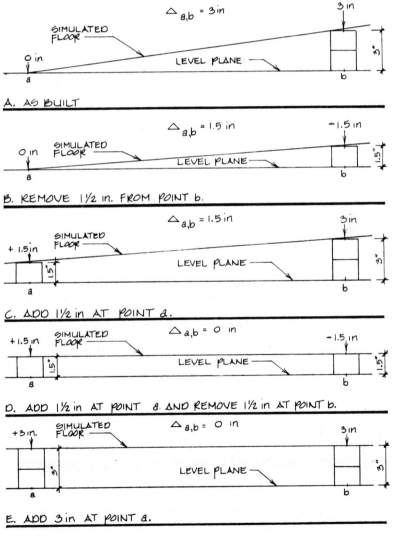

FIGURE 6A.12 Floor elevations (A)–(F).

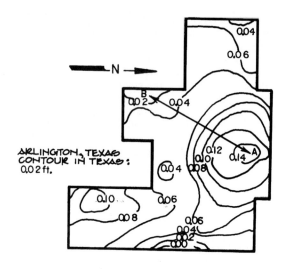

F. DIFFERENTIAL LEVEL OF NEW SLAB-ON-GROUND, ARLINGTON, TEXAS.

FIGURE 6A.12 (*Continued*)

1½ in (3.7 cm). Elevations show a zero differential between the two locations, suggesting a level floor and perhaps concealing the problems. In fact, the distress apparent in the structure is proof that the differential movement occurred although the elevations might suggest level. Figure 6A.12(*E*) is similar, although a more significant differential movement has occurred. This obviously has caused more extensive structural damage, even though, once again, the elevations would suggest a level floor.

Exception might be taken that it is unusual to construct a foundation as much as 3 in (7.5 cm) out of level. Agreed, it would be unusual, but it can happen.[27–29] More important, the analogy is true regardless of the numbers. Reduce all the numbers by a factor of 3, and the points are both true and clear.

Figure 6A.12(*F*) depicts actual elevations taken on a newly poured conventional, deformed-bar, monolithic slab-on-grade foundation. Note that the elevation differential between points *A* and *B* is 1.44 in (3.66 cm) over 28 ft (8.4 m).[15] The source for this drawing contained similar findings for posttensioned slabs, where the differential was even more pronounced, namely, 0.96 in (2.4 cm) over 12 ft (3.6 m). Expressed over a similar distance of 28 ft, this differential would equal 2.33×0.96, or 2.24 in (5.77 cm).

The point taken is: do not rely too heavily on grade elevations when analyzing foundation problems. If you feel compelled to use this type of approach, at least use a series of elevations taken some time apart. These can serve to pinpoint differential movement which occurs over the time interval between the readings.

6A.8 SOIL SWELL (UPHEAVAL) VERSUS MOISTURE CHANGES

Some discussion has already been devoted to the issue of soil swell. Tucker and Davis[30] (Fig. 6A.2) presented data for a particular soil where a 4% increase in soil moisture was sufficient to cause a 1¼-in (3.2-cm) vertical rise in a type *B* FHA slab foundation with a resulting potential swell force of 9000 lb/ft^2 (43,900 kg/m^2). Note also that the plotted data suggest a soil active depth of about 3 ft (0.9 m) (exterior to the foundation).

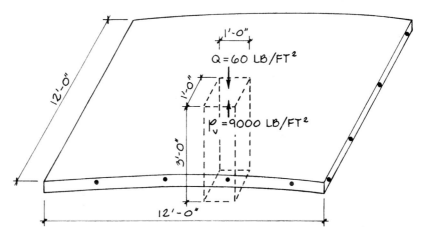

FIGURE 6A.13 Heave of expansive soil on concrete slab.

What can all these seemingly unimportant factors tell about soil swell? First, refer to Fig. 6A.13. Here a cubic foot of soil is shown isolated from the surrounding soil. Assume an imaginary glass box 1 ft (0.3 m) square by 3 ft (0.9 m) deep filled with the soil described by Tucker and existing at an initial moisture content of some 20% with a final moisture content of 24%. The 9000 lb/ft² (43,900 kg/m²) swell potential of the "confined" soil will tend to raise the lightly loaded slab [approximately 60 lb/ft² (293 kg/m²)] over an area equivalent to counter the upward thrust. In this case 9000 lb/60 lb/ft² = 150 ft² (13.5 m²), or roughly an area of 12 by 12 ft (3.6 by 3.6 m). This analysis admittedly takes certain liberties, but it is nonetheless technically valid. The resistance to heave would be materially influenced and enhanced by such factors as steel reinforcement, cross beams, or load-bearing walls. However, the intent here is not to confuse but to illuminate. The resisting moment of a beam is a square function of its depth (steel reinforcement not considered). For example, under identical conditions a piece of wood 2 × 4 in (5 × 10 cm) on edge will support about four times the load that it would if laid flat. When steel rebar within a concrete beam is considered, the moment of inertia (or stiffness) varies as the cube of the depth. In this case, doubling the depth of the beam would increase its rigidity by a factor of about 8.[29–32]

Staying with this train of thought, actually how much water is required to produce the 4% increase? Assume

$$W_s + W_w = 100 \text{ lb/ft (1602 kg/m3)}$$

$$W_s = 86 \text{ lb/ft}^3 \text{ (1376 kg/m}^3\text{)}$$

Initial moisture $W_\% = 16$ $W_w = 14 \text{ lb/ft}^3 \text{ (224 kg/m}^3\text{)}$

Final moisture, $W_\% = 20$ $W_w = 17 \text{ lb/ft}^3 \text{ (272 kg/m}^3\text{)}$.

where W_w = weight of water
W_s = weight of soil
$W_\% = W_w/W_s$

All values are based on a 1-ft³ (16-kg/m³) sample. Based on this, the added weight of water would be

$$\Delta W_w = 17 - 14 = 3 \text{ lb/ft}^3 \text{ (48 kg/m}^3\text{), or 0.36 gal (1.4 L) per cubic foot}$$

Assume the constraints set forth by Fig. 6A.13. In the case at hand this would approximate only 1.2 gal (0.4 gal/ft³×3 ft³). Again for simplicity assume that the source for water had preexisted for 12 months. Then the daily input of water (and the amount required to produce the 4% increase) would be only 143 drops per day.[27] For example,

$$1.2 \text{ gal}/12 \text{ mo} = 0.10 \text{ gal/mo} = 0.0033 \text{ gal/day} = 0.43 \text{ oz/day} (14.3 \text{ ml/day})$$

$$= 143 \text{ drops/day, or } 0.10 \text{ drop/min}$$

Admitted this example isolates the cubicle of wetted soil. In real life this could not occur. However, the relative quantities show a clear picture. Also, as a broader area is wetted, the potential soil heaved expands proportionately. In other words, if the wetted area expands laterally, the heaved area of the slab expands almost in direct proportion, although the magnitude of the heave could be lessened.

As the soil expands, what happens to the slab? First, assuming a conventional monolithic deformed-bar slab, the steel is stressed into a distorted length, which is not going to recover without some form of reverse stress. Along these lines consider Fig. 6A.8. If this slab is heaved by 4 in (10 cm), each steel rebar within the 12-ft (3.6-m) heaved area will be stretched by ¼ in (0.6 cm). Note as well that the elongation of the rebar will be anything but uniform. (With post-tension slabs, where the cables are sleeved, this would not be the absolute case.) Even after the cause of the heave is alleviated, the domed area is not likely to return to a level, or near-level, condition. In fact, experience dictates that the distressed area will not improve materially unless appropriate remedial actions are initiated.

Where can the water come from to cause the swell? This subject has been discussed earlier to some extent. Figure 6A.14 depicts a sewer line with a separation. The normal, minimum-gravity fall in the sewer pipe is ⅛ in (0.3 cm) per linear foot [approximately 1 ft (0.3 m) per 100 ft (30 m)]. Waste directed into the sewer vortices (turbulent flow) creates a centrifugal force, tending to throw liquid from the pipe if any separation exists. (As shown in Sec. 6A.8, very little water is required to cause a potentially serious problem.)

Eventually the flow will settle down to the laminar regime, with the major velocity down the pipe centerline. In this instance conditions might lessen the amount of water leaking from the pipe. However, as shown in the preceding, very little water is required to cause a potentially serious threat.

Will all expansive soils swell when subjected to available water? The simple answer is *no* (see Fig. 6A.5). If the existing moisture in these particular soils is above 24%, virtually all capacity for swell has been lost.

Will slab heave always appear to be fairly uniform, as depicted by Fig. 6A.13? The answer to this question is a resounding *NO*. Figure 6A.13 is intended to illustrate a principle of force versus resistance and borders on an ideal condition. For example, the wetted area is assumed to be at the surface. Often the affected area is at the bottom of the plumbing ditch and could be several feet below surface. Also the presence of porous fill or contact with foundation beams can influence

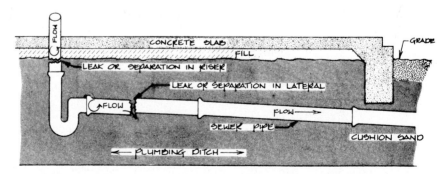

FIGURE 6A.14 Residential sewer.

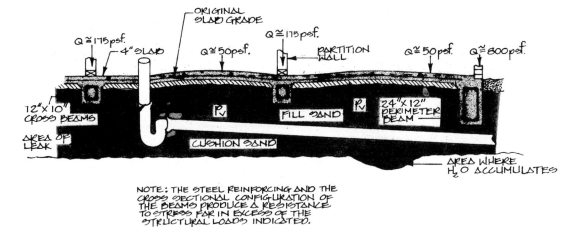

FIGURE 6A.15 Slab displacement due to upheaval from soil swell.

the pattern of water movement. In many cases, particularly with older properties, proper mudjacking eliminates these flow channels.

Realistically the deformity of a concrete slab foundation would appear more as Fig. 6A.15 suggests than as shown in Fig. 6A.13. The added structural load on the beam coupled with the increased resistance to deflection provided by the beams distort the doming effect. In effect, the slab resembles a "quilted" surface, except that the individual cells need not be of organized dimension. A topographic view of such a slab might resemble Fig. 6A.16 or Fig. 6A.17.

Figure 6A.16 depicts actual elevations with indications of minor settlement at the northwest, west of entry, and possibly the southeast corners. The major movement is center slab heave in the shaded area. Figure 6A.17 presents a foundation inspection field drawing intended to provide information sufficient for a repair estimate. The heaved area is generalized based on observations of differential movement.

Reference to Sec. 6A.7 will provide additional discussion regarding the difference between grade elevations and differential movement. Foundation repair should address only differential movement. Little, if anything, can be done to improve variations in grade elevation caused by initial construction. In fact attempts to do so are likely to cause serious additional distress (see Fig. 6A.12).

For general information regarding upheaval refer also to Secs. 6A.3 through 6A.5, 6B.3.2, 6B.5, and 6B.9.1.

6A.9 HOW IS THE NEED FOR FOUNDATION REPAIR ESTABLISHED?

What extent of movement is sufficiently serious to warrant or demand repairs? There is no uniformly established rule. Several factors further complicate this issue. First, few if any residential foundations are constructed totally level. Second, the foundation being out of level does not normally render the property uninhabitable or unsafe. Third, due at least in part to the foregoing, virtually all approaches to foundation repair (leveling) include some degree of compromise. The net goal is a foundation of tolerable appearance and behavior. The problem lies in the conceptual definition of "tolerable."

In attempting to establish a reasonable approach to defining the need and scope of foundation repair, the preferred approach is to provide the most acceptable appearance along with a foundation that is more stable at the least possible cost.

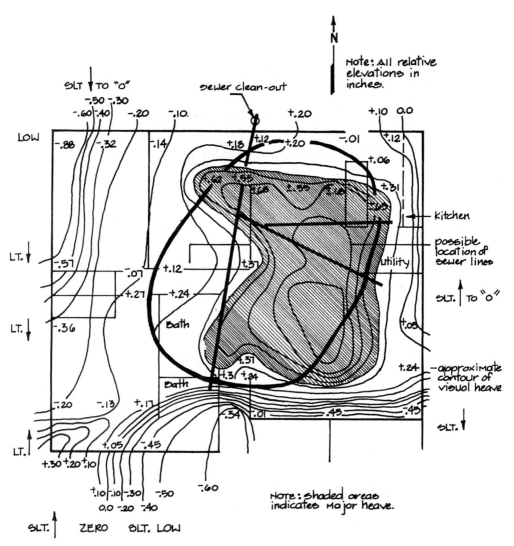

FIGURE 6A.16 Differential elevations in slab foundation.

Determining the point of movement at which foundation repair is demanded is equally elusive.[33–37] Since most foundation repairs are performed through aesthetic choice as opposed to true structural necessity, the mental attitude of the property owner becomes a prime factor. Other factors which play a part are property age and value, spendable income of the owner, peer pressure, cost of foundation repair versus continued cosmetic repairs, cause of distress (settlement versus upheaval), and the likelihood that proper maintenance could arrest the movement. Consistent with the latter, before any foundation repair is performed, the cause of the differential movement must be identified and subsequently corrected. Otherwise, recurrent problems are likely, regardless of the foundation repairs.

A semitechnical approach for establishing "tolerable" and "intolerable" foundation movement is developed in the following. Tensile strain for a typical mortared masonry wall is 0.0005 in/in.

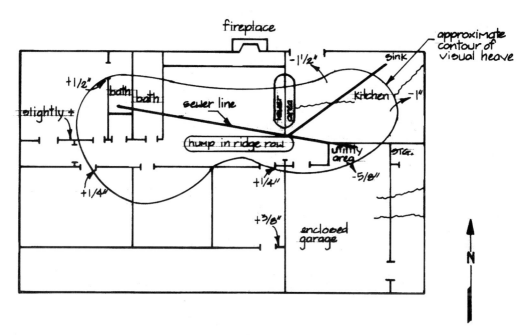

FIGURE 6A.17 Contour in slab foundation as evidenced by structural distress.

(Strain is determined by measuring deflection divided by distance.) Thus such a wall 8 ft (2.4 m) in height could resist movements up to about 1 in (2.5 cm) over 65 ft (20 m). On the other hand a 1-in (2.5-cm) movement over 15 ft (4.6 m) could produce a single-crack separation of about $\frac{1}{2}$ in (1.25 cm) (Fig. 6A.18). The same deflection applied over 30 ft (10 m) could produce a single-crack separation of about $\frac{1}{4}$ in (0.6 cm). Note the emphasis on *single-crack* separation. Multiple cracks reduce the separation distance proportionately. Lambe and Whitman present the relationship between strain and distance somewhat differently[32] (Fig. 6A.19). If L is replaced by x in Fig. 6A.19(*b*) and by $x/2$ in Fig. 6A.19(*c*), the analysis is similar to that shown in Fig. 6A.18.

Consistent with this analogy, interior features such as door frames can tolerate movement on the order of 1 in (2.5 cm) over 12.5 ft (3.8 m). This is borderline. Increased movement usually causes the doors to become nonfunctional.

This discussion brings into focus the problems inherent to defining tolerable deflection. Various conditions suggest different values. The distance over which the deflection occurs is paramount, followed by pertinent structural construction.

The arbitrary value for acceptable movement recognized by some repair contractors is crack separations in excess of $\frac{1}{4}$ to $\frac{1}{8}$ in (0.6 to 1 cm). These roughly equate to a differential movement of 1 in (2.5 cm) over 30 ft (10 m), or 25 ft (8 m). Normal construction often accepts a tolerance of as much as 1 in (2.5 cm) in 20 ft (6 m).[33] This again emphasizes the difference between possible as-built grade variation and differential movement.

Meyer suggests the values presented in Table 6A.2 as the criteria for foundation failure threshold.[28] In other words, movements which exceed the numbers indicated in Table 6A.2 suggest a failure condition that should be considered serious and warranting remedial attention.

Kaminetzky approaches the situation from a different perspective.[37] Table 6A.3 presents his classification of distress based on visible damage. He defines moderate damage as occurring at approximate crack widths of $\frac{1}{8}$ to $\frac{1}{2}$ in (3.2 to 12.7 mm). This is the point at which remedial action is suggested.

Home Owners' Warranty (HOW) uses a still different approach. Their booklet describes major structural defects as those which meet two criteria:

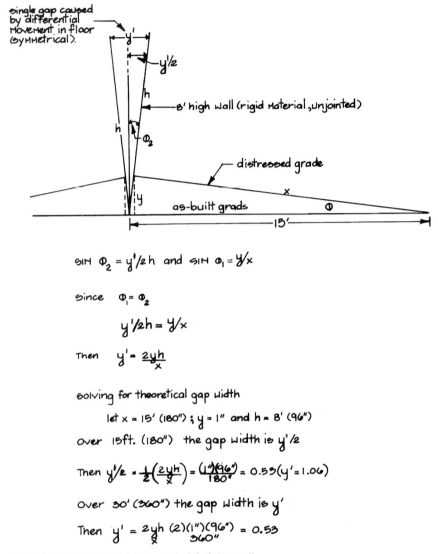

FIGURE 6A.18 Theoretical single gap in 8-ft (2.4-m) wall.

1. "It must represent actual damage to the load-bearing portion of the home that affects its load-bearing function, and
2. The damage must vitally affect or is imminently likely to produce a vital effect on the use of the home for residential purposes."

In a slightly different vein, under the HOW section 3, "Concrete," the following are cited as excessive cracks: foundation or basement walls—greater than $\frac{1}{8}$ in (0.3 cm), basement floors—in excess of $\frac{3}{16}$ in (0.45 cm), garage slabs—wider than $\frac{1}{4}$ in (0.6 cm), foundation slab floor—wider than $\frac{1}{8}$ in (0.3 cm). Section 4, "Masonry," defines cracks in veneer as excessive when they exceed $\frac{3}{8}$ in (0.9 cm) in width. Section 6, "Wood and Plastics," defines as intolerable floors that

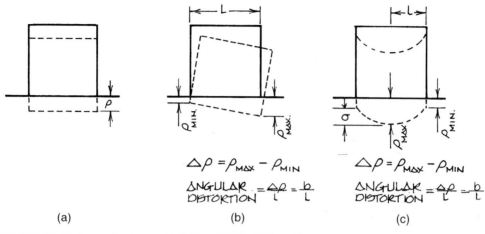

FIGURE 6A.19 Types of settlement. (*a*) Uniform. (*b*) Tilt. (*c*) Nonuniform.

TABLE 6A.2 Recommended Foundation Failure Criteria for Selected Building Conditions (Maximum Allowable Values)

Condition	Differential across two points (y/x)*	Overall slope, % (y/x)*	Total vertical movement, in
Single- or multifamily wood-frame dwellings up to three stories:		0.3 (1 in/30 ft)	4
Exterior masonry 8 ft high	1 in/40 ft		
Exterior masonry >8 ft high	1 in/50 ft		
Exterior plaster 8 ft high	1 in/40 ft		
Exterior nonmasonry	1 in/16.7 ft		
Interior sheetrock walls 8 ft high	1 in/30 ft		
Interior sheetrock walls >8 ft high	1 in/40 ft		
Interior wood paneling walls	1 in/76.7 ft		
Small-span structures with cross roof trusses	1 in /30 ft		
Steel-framed buildings with metallic skin; no sensitive equipment; average drainage	1 in/16.7 ft	1.0	6
(if in isolated location with soft adjacent improvements and maximum drainage)			12
Concrete-framed industrial buildings with CMU or masonry walls	1 in/30 ft		
High-exposure low-rise retail or office buildings with glass or architectural masonry walls	1 in/50 ft	0.3	2 (but check entry transitions)

  *x is lateral distance over which vertical deflection y occurs. In case of overall slope, x would generally be the length or width of the foundation and y the overall deflection.
  Source: After Meyer.[28] See also elsewhere.[36,37]

TABLE 6A.3 Classification of Visible Damage

Damage	Description of typical damage*	Crack width, in (mm)†
1. Negligible	Hairline cracks‡	<0.005 (0.13)
2. Very light	Isolated light fracture in building. Cracks in exterior walls§ visible on close inspection.	1/64 to 1/32 (0.40 to 0.80)
3. Light	Light fracture visible on building inside (floors, partitions) and exterior. Doors and windows may stick slightly.	1/32 to 1/8 (0.80 to 3.2)
4. Moderate	Moderate cracks visible on building inside and exterior. Doors and windows stick. Utility pipes and glass may fracture. Water and air penetration through exterior.	1/8 to 1/2 (3.2 to 12.7)
5. Extensive	Severe cracks visible on building inside and exterior. Windows and door frames are skewed and "locked in"; floors slope noticeably. Walls lean or bulge noticeably; some loss of bearing in beams. Utility pipes and glass broken. Water and air penetration through exterior.	1/2 to 1 (12.7 to 25.4)
6. Very extensive	Very extensive cracks visible on building inside and exterior. Broken pipes and glass. Full loss of beam bearing. Walls lean or bulge dangerously. Structure requires shoring. Danger of instability.	1 or wider (25.4 or wider), but depends on number and location of cracks

*In evaluating the degree of damage, consideration must be given to its location in the building and structure (e.g., points of maximum stress).
†Crack width is only one aspect of damage and should not be used alone as a direct measure of damage.
‡Refers to existing old cracks. New cracks must be monitored for possible increase in width.
§A criterion related to visible cracking is useful since tensile cracking is so often associated with settlement or movement damage. Therefore it may be assumed that the state of visible cracking in a given material is associated with its limit of tensile strain.
Source: From Kaminetzky.[37]

are out of level more than 1/4 in (0.6 cm) over a linear distance of 32 in (0.8 m). General floor slope within any room should not exceed 1 in (2.5 cm) over 20 ft (6 m). On the other hand, HUD (at least some regional offices) often considers a property uninsurable if the floor is off level more than 11/16 in (1.7 cm) over 20 ft (6 m). [This translates to 1 in (2.5 cm) over 30 ft (10 m).] Both HOW and HUD offer the homeowner some measure of insurance. HUD insures homeowner mortgage loans and HOW provides them with insurance protection against major structural defects. Warranty Underwriters Insurance Company and Home Buyers Warranty are examples of other organizations that offer home buyers some degree of protection in the form of insurance against major household defects such as foundation failure. Each entity offers a somewhat different policy regarding insured losses.

Basic homeowners' insurance policies generally cover foundation repair only in cases where foundation movements are caused by the accidental discharge of water from a household plumbing system (sewer or supply).

If you have insurance and suspect a foundation problem, the best bet is to contact your insurer first. If doubt or dispute arises you might then want to contact a qualified engineer or foundation repair expert.

6A.10 REFERENCES

1. S. L. Houston, W. N. Houston, and D. J. Spadola, "Prediction of Field Collapse of Soils Due to Wetting," *J. Geotech. Eng., ASCE,* vol. 114, Jan. 1988.
2. S. L. Houston and M. el Ehwany, "Settlement and Moisture Movement in Collapsible Soils," *J. Geotech. Eng., ASCE,* vol. 116, Oct. 1990.

3. C. D. Cozart and J. W. Burke, "Collapsible Soils in Texas," ASCE, Austin, Tex., Section, Apr. 1990.
4. E. C. Lawton, R. J. Fragaszy, and J. H. Hardcastle, "Collapse of Compacted, Clayey Sand," *J. Geotech. Eng., ASCE,* vol. 115, Sept. 1989.
5. R. W. Day, "Sample Disturbance of Collapsible Soil," *J. Geotech. Eng., ASCE,* vol. 116, Jan. 1990.
6. K. Kienow and K. Kienow, "Corrosion below Sewer Structures," *Civ. Eng.,* Sept. 1991.
7. T. Petry and D. L. Little, "Update of Sulfate Induced in Lime and Portland Cement Treated Soils," presented at the 1992 Annual Meeting of the Transportation Research Board, Aug. 1991.
8. J. D. Nelson et al., "Moisture Content and Heave beneath Slabs on Grade" and "Prediction of Floor Heave," presented at the 5th Int. Conf. on Expansive Soils, Adelaide, Australia, May 1984.
9. *Proc. 7th Int. Conf. on Expansive Soils,* ASCE, Dallas, Tex., 1992.
10. N. Sperry, *Complete Guide to Texas Gardening,* Taylor Publishing, Dallas, Tex., 1982.
11. J. Haller, *Tree Care,* Macmillan, New York, and Collier Macmillan, London, 1986.
12. G. Hall, "Garden Questions: How to Get a Fruitful Apple Tree," *Dallas Times Herald,* Mar. 24, 1989.
13. T. H. Wu, R. M. McOmber, R. T. Erb, and P. E. Beal, "Study of Soil-Root Interaction," *J. Geotech. Eng., ASCE,* vol. 114, Dec. 1988.
14. H. S. Stevens, "The Top Five Mistakes We Make with Trees," *Dallas Morning News,* Apr. 23, 1993.
15. R. L. Tucker and A. Poor, "Field Study of Moisture Effects on Slab Movements," *J. Geotech. Eng., ASCE,* Apr. 1978.
16. A. Komornik and J. G. Zeitlen, "Effects of Swelling Clay on Piles," Israel Institute of Technology, Haifa, Israel, 1973.
17. V. A. Sowa, "Influence of Construction Conditions on Heave of Slab-on-Grade Floors Constructed on Swelling Clays," in *Theory and Practice in Foundation Engineering, Proc. 38th Can. Geotech. Conf.,* Sept. 1985.
18. R. G. McKeen and L. D. Johnson, "Climate Controlled Soil Design Parameters for Mat Foundations," *J. Geotech. Eng.,* vol. 116, July 1990.
19. J. E. Holland and C. Lawrence, "Seasonal Heave of Australian Clay Soils," presented at the 4th Int. Conf. on Expansive Soils, ASCE, Denver, Colo., 1980.
20. J. E. Holland, C. Lawrence, et al., "The Behavior and Design of Housing Slabs on Expansive Soils," presented at the 4th Int. Conf. on Expansive Soils, ASCE, Denver, Colo., 1980.
21. R. W. Brown, "A Field Evaluation of Current Foundation Design vs. Failure," *Texas Contractor,* July 6, 1976.
22. J. P. Krohn and J. E. Slosson, "Assessment of Expansive Soils in the United States," presented at the 4th Int. Conf. on Expansive Soils, ASCE, Denver, Colo., 1980.
23. D. C. Frelund et al., "Variability of an Expansive Clay Deposit," presented at the 4th Int. Conf. on Expansive Soils, ASCE, Denver, Colo., 1980.
24. S. Prakash and H. D. Sarma, *Pile Foundations in Engineering Practice,* Wiley, New York, p. 659.
25. T. Petry, C. Armstrong, and A. Poor, "Foundation Seminar," University of Texas at Arlington, Dec. 13, 1985.
26. R. W. Brown and C. H. Smith, Jr., "The Effects of Soil Moisture on the Behavior of Residential Foundations in Active Soils," *Texas Contractor,* May 1, 1980.
27. R. W. Brown, *Foundation Repair and Behavior: Residential and Light Commercial,* McGraw-Hill, New York, 1992.
28. K. T. Meyer, "Defining Foundation Failure," Texas Section, ASCE, South Padre, Tex., Fall 1991.
29. National Research Council, "Criteria for Selection and Design of Residential Slabs-on-Ground," Report 33, Publ. 1571, National Academy of Science, Washington, D.C., 1968.
30. R. Tucker and R. C. Davis, "Soil Moisture and Temperature Variation Beneath a Slab Barrier on Expansive Soils," Report No. TR-3-73, Construction Research Center, Univ. of Texas at Arlington, Arlington, Tex., May 1, 1973.
31. P. M. Ferguson, *Reinforced Concrete Fundamentals,* 3d ed., Wiley, New York, 1973.
32. E. E. Seeyle, *Data Book for Civil Engineers, Design,* vol. 1, 2d ed., Wiley, New York, 1945.
33. T. W. Lambe and R. V. Whitman, *Soil Mechanics,* Wiley, New York, 1969.
34. D. E. Bobbitt, "Theory and Application of Helical Anchors for Underpinning and Tiebacks," Helical Piers and Tieback Seminar, Atlanta, Ga., Mar., 1990.
35. D. Kaminetzky, *Design and Construction Failures: Lessons from Forensic Investigations,* McGraw-Hill, New York, 1991.
36. ACI Reports 302.1R-80, 318-83, and 435.3R-84, American Concrete Institute, Detroit, Mich.
37. D. Kaminetzky, "Rehabilitation and Renovation of Concrete Buildings," in *Proc. Workshop, National Science Foundation and ASCE,* New York, Feb. 1985, pp. 68–78.

SECTION 6B
FOUNDATION REPAIR PROCEDURES

Robert Wade Brown

6B.1 INTRODUCTION 6.31
   6B.1.1 Foundation Failure: Cause versus Repair Option 6.32
6B.2 PIER AND BEAM FOUNDATIONS 6.32
   6B.2.1 Underpinning 6.32
   6B.2.2 Interior Floors (Shimming) 6.45
6B.3 SLAB FOUNDATIONS 6.48
   6B.3.1 Mud Jacking 6.48
   6B.3.2 Upheaval 6.50
   6B.3.3 Settlement 6.51
   6B.3.4 Posttension Retrofit 6.55
6B.4 SLIM PIERS OR MINIPILES 6.55
   6B.4.1 Design Concerns for Minipiles 6.56
   6B.4.2 Deterioration of Steel Piles and Piers 6.58
   6B.4.3 Action of Soils around a Driven Pile 6.58
   6B.4.4 Screw or Earth Anchors 6.59
   6B.4.5 Spread Footings with Multiple Drilled Piers 6.59
   6B.4.6 Comparative Bearing Capacities 6.59
6B.5 REVIEW OF REPAIR LONGEVITY 6.60
6B.6 GROUTING 6.62
   6B.6.1 Introduction 6.62
   6B.6.2 Mechanics of Grouting Applications 6.63
   6B.6.3 Placement of Injection Pipes 6.64
   6B.6.4 Subgrade Waterproofing 6.66
   6B.6.5 Soil Consolidation 6.66
6B.7 SPECIAL PROBLEMS ADVERSELY AFFECTING LEVELING 6.67
6B.7.1 Upheaval 6.67
6B.7.2 Warped Substructure 6.67
6B.7.3 Construction Interferences 6.67
6B.8 CONCLUSIONS 6.71
6B.9 INTERESTING EXAMPLES OF FOUNDATION REPAIR TECHNIQUES 6.72
   6B.9.1 Florida 6.72
   6B.9.2 London, England 6.72
   6B.9.3 Piers of Driven Steel Pipe 6.74
   6B.9.4 Lowering a Foundation 6.76
   6B.9.5 Apartment Building 6.79
   6B.9.6 Basements and Foundation Walls 6.79
   6B.9.7 Chemical Stabilization 6.85
   6B.9.8 Uretek—An Alternative to Mud Jacking? 6.86
   6B.9.9 Conclusion 6.86
6B.10 FAILURES IN FOUNDATION REPAIR 6.86
   6B.10.1 Substandard Pier Construction 6.87
   6B.10.2 Failed Minipile 6.87
   6B.10.3 Failure to Mud Jack a Slab Foundation 6.89
   6B.10.4 Improper Placement of Haunch beneath Beam 6.89
   6B.10.5 Sulfate Reactions with Certain Soils 6.91
   6B.10.6 Ultraslim Concrete Piers 6.92
   6B.10.7 Hydraulically Driven Concrete Cylinders 6.92
6B.11 REFERENCES 6.92

6B.1 INTRODUCTION

When foundations fail for one reason or another, prompt and competent repairs are demanded; otherwise the problems will increase and jeopardize the value and safety of the structure. The repair procedures discussed in this chapter have been developed over the last 30 years basically through a process of trial and error. When a technique proves successful over several thousand field evaluations, the conscientious contractor or engineer should be reluctant to accept unproven changes. In foundation repair, results far outweigh theory. As time goes on, new methods (or variations of existing ones) will be introduced and, results warranting, perhaps accepted.

Moisture variations within expansive soils cause the majority of all foundation failures. The soil swells when wet (upheaval) and shrinks when dry (settlement). This volumetric change in the bearing soil causes, in turn, differential movement in the supporting foundation. According to Jones and Holtz,[1] 60% of all residential foundations constructed on expansive soils will experience some distress caused by differential foundation movement, and 10% will experience problems sufficient to demand repairs. Specific areas suffer even higher failure rates.

Remedial procedures depend on both the nature of the problem and the type of foundation. Settlement is corrected by raising or restoring the affected area to the approximate original grade. Upheaval is corrected by raising the lowermost areas around the raised crown to a new grade sufficient to "feather" the differential.

All foundation failures are generally manifested in one or more of the following signs: interior or exterior wall cracks, ceiling cracks, sticking doors or windows, pulled roof trusses, and broken windows (Fig. 6B.1). A common misconception is that foundation movement occurs instantaneously. The mistake is promoted by the fact that some of the signs of distress do appear instantaneous. However, the walls, ceilings, and other structural features are somewhat elastic in nature, and ultimate failure (cracks, for example) occurs when the applied stress exceeds the elasticity of the particular surface. The behavior is somewhat analogous to that of a stretched rubber band. Stress is not evident until the band parts.

On occasion signs of distress are apparent which do not suggest the need for foundation repair [Fig. 6B.1(*H*)]. Light foundations on expansive soils are virtually assured of some movement. The question arises as to when or at what point repairs are considered necessary. This decision is somewhat arbitrary and was discussed in some detail in Sec. 6A.9.

In correcting any failure, it is first necessary to isolate and recognize the particular problem. For example, it is often difficult, and always critical, to differentiate between settlement and upheaval. Many of the outward signs of differential foundation movement are similar, whether the actual distress is relative to settlement or to upheaval. (For example, refer to Figs. 6A.6 and 6A.7.) Both have cracks, and both show lateral displacement of the perimeter walls. However, the causes are opposite. Many foundation problems have been aggravated rather than solved by errors in recognizing the cause of distress. However, once the problem has been identified, proper restitution can commence. The general delineation for the following discussions will be based on the type of foundation—either slab or pier and beam. Note also that all leveling techniques generally involve raising some portion of the foundation to a higher elevation. None of the repair methods are intended to tie the foundation down once it has been raised.

6B.1.1 Foundation Failure: Cause versus Repair Option

Table 6B.1 presents an overview of the various causes for foundation problems combined with specific repair options. The subsequent sections will detail selected repair procedures.

6B.2 PIER AND BEAM FOUNDATIONS

Pier and beam foundation movement is generally confined to settlement, although, in limited cases, upheaval is observed. For practical purposes, the solution for upheaval is again to treat the lowermost sections of the foundation as though settlement had occurred and raise them to the higher grade. Settlement is generally alleviated by raising the beam mechanically and sustaining the desired position by installing new supporting structures. Interior floors are leveled by shimming on piers caps.

6B.2.1 Underpinning

For our purposes the discussion on underpinning will focus generally on the two most commonly used designs:

1. Spread footings
2. Drilled concrete piers

Limited discussion will be devoted to other methods and variations (see also Sec. 6B.4). Underpinning is generally restricted to applications concerning the perimeter beam.

FIGURE 6B.1 Evidence of differential foundation movement. (*a*) Brick veneer tilted inward; often an indication of interior slab settlement. (*b*) Separation of brick veneer from garage door jamb. (*c*) Brick veneer separated from window frame.

(d)

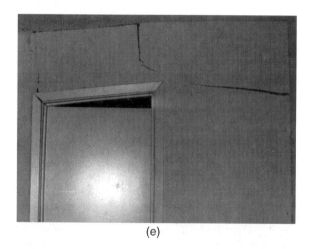

(e)

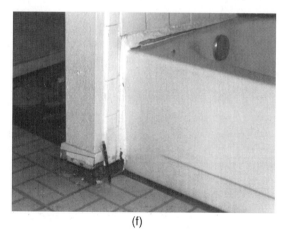

(f)

FIGURE 6B.1 (*Continued*). (*d*) Horizontal separation in brick mortar. (*e*) Interior door frame out of plumb, causing sheet rock cracks. This illustrates the effect of upheaval in an interior slab. The high point is to the right side of the door. (*f*) Interior slab settlement. This shows separation of interior slab from the wall partition. Note also that the bathtub has settled away from the ceramic tie.

(g)

(h)

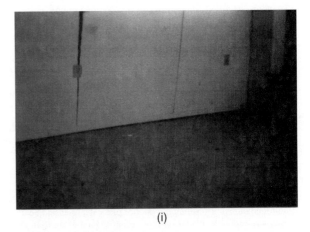

(i)

FIGURE 6B.1 (*Continued*). (*g*) Crack in perimeter beam. This illustrates a major crack in the perimeter beam accompanied by secondary cracks in brick veneer and brick mortar. (*h*) Stair step separation in brick mortar. This shows an obvious crack in the brick mortar but does it represent a problem of concern? First, the crack width is less than $\frac{1}{4}$ in (0.6 cm). Second, the mortar joints are straight. Third, the cornice trim at the corner is tight. Fourth, no interior damage was noted. Thus, the damage is not one of concern. (*i*) Slab upheaval. A fairly typical representation of slab upheaval. Note the very obvious crown in the slab surface (approx. 4 in or 10 cm) accompanied by vertical separation in wallboard, cracks in floor slab, and separation of shoe mold from slab.

TABLE 6B.1 Foundation Repair Options for Residential and Light Commercial Foundations

Nonexpansive soils	Expansive soils
\multicolumn{2}{c}{a. Frost heave (permafrost)}	
Problem is best addressed in original foundation design. Foundation base support should be below level of frost line, and subgrade walls should be reinforced to resist inward stress. Steps to prevent water accumulation in shallow soils will also help, both as a component of original design and as remedial approach. (Refer also to Sec. 3B.)	Not often a serious problem, due to inherent lack of soil permeability and porosity. Could be approached as suggested for nonexpansive soils if need should arise.
b. Undercompaction	
Generally shows as problem relatively rapidly following construction, if at all.	Settlement is corrected, as a rule, by underpinning the perimeter and either mud jacking interior slab (slab foundation) or shimming interior pier caps to level interior floors (pier and beam construction).
c. Overcompaction	
Not normally a serious problem for shallow foundations.	Same as nonexpansive soils. When exceptions occur, repair procedures normally associated with upheaval could be applicable.
d. Water drain	
Restitution or stabilization would require replacement of lost water or filling void space by pressure grouting. Often, after voids are filled, structure is leveled by underpinning the perimeter and mud-jacking slab or reshimming interior pier caps on pier and beam foundation.	Not often a problem within the definition.
e. Decay of organics	
Consolidating or collapsing soils are treated the same regardless of classification. Cure is to fill voids and create a cemented volume sufficient to carry applied structural load. In case of compressible materials, such as peat, material will not likely be cemented, but compression to a pressure above intended load is adequate. Once deep-seated problem has been cured, surface appearance can be remedied by conventional leveling procedures.	
f. Construction on slope	
Slopes with down-dip bedding planes create similar problems regardless of soil type. While expansive soils do present a greater propensity to this failure, the cure is still almost identical. Remedial choices generally require underpinning the foundation with added support members extending safely below problem slip plane into solid bearing. Most often underpinning technique requires installation of drilled concrete piers or driven pilings. Releveling the stabilized structure would most often require mud jacking (slab foundations) or reshimming interior pier caps (pier and beam foundations). In rare instances interface at slip surface might be pressure-grouted with cementitious grout to retard slip. (See Figs. 6A.10 and 6A.11.)	
g. Abnormal soil moisture at time of construction	
Not normally a point of particular concern.	Natural moisture differentials between foundation bearing soils and exterior create shrink-swell tendencies, especially where slab-on-grade foundations are concerned. Dry soil at the perimeter can often be watered to the extent that differential movement is minimal. Dry interior soil might allow slab to settle to the extent that mud jacking would be required. In extreme cases any repair approach might include chemical stabilization of clay to control affinity to water.

TABLE 6B.1 Foundation Repair Options for Residential and Light Commercial Foundations (*Cont.*)

Nonexpansive soils	Expansive soils
h. Plant roots	
Seldom a problem.	Plant roots remove moisture from soil, sometimes to the extent that foundation stability is at risk, particularly with respect to slab foundations. Prevention is preferred to cure. However, when a situation develops, the cure is to either (1) create a situation that restores soil moisture, (2) construct a root barrier at the foundation perimeter to exclude root intrusion, or (3) if situation demands, raise perimeter beam by underpinning and mud jack interior slab.
i. Loss of soil water	
Seldom a serious concern. However, any remedial approach would be consistent with that for expansive soils.	Remedial approach similar to item h, except that with pier and beam foundations interior floors would be releveled by shimming.
j. Gain of soil water	
Seldom a serious problem. The exception could be a collapsing soil when silts or sandy silts are involved. [See item 5 in Sec. 6A.1.1.]	Excess water beneath foundation causes upheaval, especially with slab foundations. First eliminate source of water, which could require plumbing repairs, drainage corrections, installation of French drains, and so on, depending on source of water. Then foundation is leveled by conventional means as described. The principal exception would be the desirability to treat soil with a chemical stabilizer, such as Soil Sta.
k. Marginal bearing capacity	
Solution is the same for both expansive or nonexpansive soils. However, if need be, soil bearing capacity can often be increased by introducing chemicals, often of a cementitious nature. Nonexpansive soils show a greater propensity to suffer from this problem than expansive soils.[2,3]	
l. Sand or gravel cushion beneath slabs-on-grade	
Loss in bearing support can normally be corrected by mud jacking.	
m. Soil creep	
Refer to item f.	

6B.2.1.1 Spread Footings. Typically, the supports or spread footings consist of:

1. Steel-reinforced footings of sufficient size to distribute the beam load adequately, and poured to a depth relatively independent of seasonal soil moisture variations
2. A steel-reinforced pier tied to the footing with steel and poured to the bottom of the foundation beam (Fig. 6B.2)

Design and placement of these spread footings are critical if future beam movement is to be averted. The footing design must consider the possible problems of both future settlement and upheaval and should be of sufficient area to develop adequate bearing by the soil. The pier should be of sufficient diameter or size to carry the foundation load. It also should be poured to intimate contact with the irregular configuration of the undersurface of the beam. Precast masonry, steel,

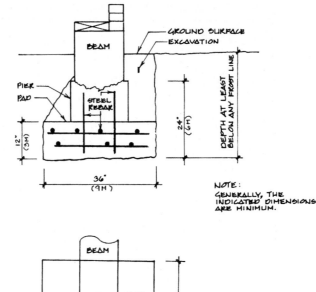

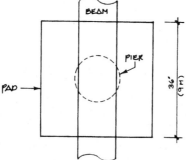

FIGURE 6B.2A Underpinning—typical design of spread footing.

or wood shims should not be considered as the pier cap material. Figure 6B.2B depicts actual field development of a typical spread footing. Figure 6B.2B(top) shows a spread footing base pad poured in place. Figure 6B.2B(bottom) illustrates the pier as poured. The jack used to raise the beam is still in place to the right of the pier, and the steel pipe used to support the beam temporarily until the concrete pier cures is evident to the left.

The principle of footing design is to distribute the foundation load over an extended area at a stable depth and thus provide increased support capacity on even substandard bearing soil. The typical design represented by Fig. 6B.2 provides a bearing area of 9 ft² (0.8 m²). Effectively, a load of 300 lb/ft² (1465 kg/m²) applied over 1 ft² (0.09 m²) would require a soil bearing strength of 300 lb/ft² (1465 kg/m²). The same load distributed over 9 ft² (0.8 m²) would require a soil bearing strength of only 33 lb/ft² (161 kg/m²). Expressed another way, a 9-ft² (0.8-m²) spread footing on a soil with an unconfined compressive strength q_u of 1500 lb/ft² (7320 kg/m²) would provide a load resistance of 13,500 lb (6136 kg). ($Q_r = q_u A$.) This capacity would exceed the structural loads imposed by residential construction by a wide margin. The spread footings are predominantly used when the overburdened soil is thick and the soil bearing strengths are low or marginal. In general the diameter of the pier is greater than or at least equal to the width of the existing beam. Since the pier is essentially in compression, utilizing the principal strength of concrete, its design features are not as critical as those for the footings.

As stated previously, the spread footing should be located at a depth sufficient to be relatively independent of soil moisture variations due to climate conditions. In London, England, this depth

FIGURE 6B.2B Spread footing. Top: Concrete pad is poured in place. Bottom: The concrete pier cap has been poured on the pad to support the foundation beam. The jack used to raise the beam is still in place and the steel pipe used as a temporary support is in place.

is reportedly in the range of 3 to 3.5 ft (0.9 to 1.1 m).[3] In the United States this depth is reportedly in the range of 2 to 3.5 ft (0.6 to 1.1 m).[4] In Australia the depth is considered to be less than about 4 ft (1.25 m).[5,6] In Canada this depth has been reported to be as shallow as 1 ft (0.3 m).[7] Along this line of thought, Chen[8] questions the reliability of theoretical approaches to predict heave. The heave prediction methods are based on an assumed depth of wetting, which varies considerably among investigators. Chen also suggests that heave predictions are generally much greater than the heave actually measured in the field. Does this mean that the seasonal depths of soil moisture change are, in fact, considerably less than the values normally assumed?

Proper soil moisture maintenance would ensure the stability of the footing with minimal concern for the effective active depth, or the depth of ambient soil moisture variation.

6B.2.1.2 Drilled Concrete Piers.

An alternate to the spread footing is the pier or piling. Figure 6B.3 depicts equipment used to drill piers. The truck-mounted unit represents the quickest and least expensive means for drilling, but it requires access and head room. (The expense advantage becomes even more pronounced as the pier depth and diameter increase.) This equipment can drill soft rock. The truck-mounted drill is also used to provide shafts for deep grouting, sometimes to a depth of 100 ft (30 m). The tractor rig is effective when access is limited, shaft diameters are 24 in (0.6 m) or less, and depths are less than about 30 ft (9 m). The latest equipment is the limited-access rig, which is used in those cases where access is critically limited. This equipment can drill with 7-ft (2.1-m) head clearance and no more than 4 or 5 ft (1.2 or 1.5 m) of lateral or surrounding access. The rig pictured is capable of drilling a 24-in (0.6-m) shaft to a depth of 30 ft (9 m) and providing a 42-in (1.4-m) bell. The problem is the time required to drill such a pier and the inherent inability to penetrate rock to any degree. The machine can, however, drill smaller-diameter shafts to shallower depths at relatively reasonable costs. For example, a 12-in (0.3-m)-diameter pier can be drilled to 12 ft (3.6 m) for a cost of about $130 more than that for the truck rig and $90 more than that for the tractor-mounted machine (1995 dollars). By the same token, the cost to drill this pier is roughly equivalent to that associated with the excavation for a spread footing. The same pier can be drilled with either of the other machines at a cost below that required to excavate the spread footing. Unless the truck rig remains on paved surfaces, considerable yard and landscaping damage could result. The tractor rig is susceptible to creating similar surface

(1)   (2)

FIGURE 6B.3(a) Preparation of concrete piers for underpinning. (A) Drilling equipment used. (1) Truck mounted rig. (2) Tractor drill.

(3)

(4)

FIGURE 6B.3(a) (*Continued*). (3) Limited access drill. (4) Hydraulic power source limited access drill.

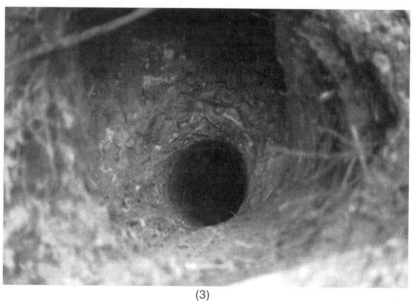

FIGURE 6B.3(*b*) Drilling sequence used for preparation of concrete piers for underpinning. (1) Drilling the pilot hole for haunch with 24-in bit. (2) Drilling the 12-in-diameter shaft to 15 ft depth. The man at right is enlarging and shaping the haunch. (3) Haunch and pier ready for reinforcing steel and concrete.

(1)

(2)

FIGURE 6B.3(c) Constructing the concrete pier. (1) Rebar caged, ready to place into pier hole (commercial application). (2) Pouring the concrete.

(1)

(2)

FIGURE 6B.3(*d*) Underpinning and raising the foundation. (1) Haunch cured and jack in place for raising the perimeter beam. (2) The final step—The beam has been raised and the pier cap poured. The steel pipe beside the pier cap is intended only to provide temporary support to the beam until the concrete pier cap cures.

disturbances. This is expressly true in instances where the yards are less than dry. Figure 6B.3(*b*) illustrates the drilling sequence for pier shaft preparation, Fig. 6B.3(*c*) depicts the pier construction, and Fig. 6B.3(*d*) shows the haunch and final raising.

In general the piers or pilings depend on endbearing and skin friction (except in high-clay soils) for their support capacity. Piers or pilings are normally extended through the marginal soils to either rock or other competent bearing strata. This obviously enhances and satisfies the support requirement. The use of piers or pilings, however, is generally restricted to instances where adequate bearing materials can be found at reasonably shallow depths. When that is possible, the installation of the piers or pilings is sometimes less costly than that of the spread footing. Generally speaking, for residential repairs a competent stratum depth below 15 ft (4.5 m) resists the use of the pier or piling technique.

Greenfield and Shen[9] suggest a minimum concrete pier diameter of 12 in (0.3 m). The reasoning is to minimize pier heave, provide adequate bearing, and facilitate proper steel reinforcing. [It is impractical, if not impossible, to utilize caged rebar in a pier diameter of 10 in (25 cm) or less.] The pier can be belled whenever the soil bearing capacity is low or questionable. The same reference also notes that pier spacing should be a maximum consistent with beam design, load requirements, and site characteristics. The logic is to provide maximum safe load on the pier to counter the experience of expansive soil heave. While Greenfield and Shen principally address original construction, the same design concerns would be applicable to remedial procedures.

6B.2.1.3 Combination Spread Footing and Drilled Pier.
On occasion the piers or pilings are used in conjunction with the spread footing (haunch). Here the theory is to utilize the best support features of each design in the hope of achieving a synergistic effect. (Also, as a practical matter, the haunch is necessary as a base from which to raise the beam.) In practice the goal is not always attained. When soil conditions dictate the spread footing, the pier provides little, if any, added benefit. However, the integration of the deep pier as part of the spread footing has no deleterious features, provided the cross-sectional area of the footing is not diminished, and the pier does not penetrate a highly expansive substratum which has access to water. Along these lines, consider Fig. 6B.4. Water in contact with the CH clay at a 5- to 7½-ft (1.5- to 2.3-m) depth could cause the pier to heave, whereas the spread footing would be stable.[10] Certain steps are available to avoid or minimize friction (upheaval) in the design of shallow piers. These efforts include: (1) the needle or slim pier or piling (reduced surface area), (2) belling the pier bottoms (usually effective with conventional-diameter shafts), and (3) placing a friction-reducing membrane between the pier and the sidewall of the hole.

6B.2.1.4 Summary: Underpinning Pier and Beam Foundations.
Regardless of the type of underpinning support, two precautions must be exercised. First, no steel or wood should be exposed below grade either as shim or as pier materials. Second, the pier or pier cap should be poured in place to ensure intimate contact between the pier and the irregular bottom of the perimeter beam. In the first instance, the exposure to the soil and water will corrode the steel and rot the wood within an unusually short period of time. In the latter instance, flat surfaces in contact with the irregular bottom surface of the beam will result in either immediate damage to the concrete beam or subsequent resettlement over a period of time due to the same effect, such as crushing of concrete protrusions.

Also, bear in mind that none of the underpinning techniques attempts to "fix" the foundation to prevent upward movement. This is by design to avoid subsequent uncontrolled damage to the foundation and emphasizes the fact that, if shallow expansive soils are subjected to sufficient water, the soil expansion will raise the foundation off the supports. (Refer to Part 3 for detailed information on piers and pilings.)

6B.2.2 Interior Floors (Shimming)

For practical purposes, the residence interior is leveled by shimming on existing pier caps. This shimming will not guarantee against future recurrence of interior settlement since the true prob-

6.46 FOUNDATION FAILURES AND REPAIR

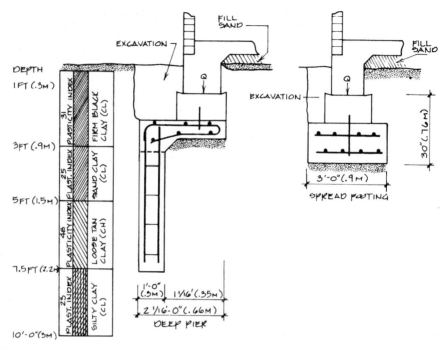

FIGURE 6B.4 Deep piers versus spread footing.

lem has not been addressed. However, reshimming is relatively inexpensive, and the rate of settlement decreases with time because the bearing soil beneath the pier is being continually compacted. True correction of the problem would normally require the installation of new piers and pier caps at considerable expense, since the flooring and perhaps even the wall partitions would need to be removed to provide clearance for drilling.

In some instances, existing interior pier supports are unacceptable. This could relate either to concrete piers which, for one reason or another, have lost their structural competence, or to piers constructed from deficient material. Again, the ideal solution would be the installation of new piers drilled into a competent bearing stratum. As noted earlier, this approach is prohibitively expensive for existing structures. Hence the usual approach is to provide supplemental pier caps supported essentially on the surface soil. The best compromise has been a design represented by Fig. 6B.5.

The support depicted in Fig. 6B.5 involves a concrete base pad, a concrete pier cap, suitable hardwood spacers, and a tapered shim for final adjustment. The base pad can be either poured in place or precast. The choice and size depend on the anticipated load and accessibility. For single-story frame construction the pad can be precast, at least $18 \times 18 \times 4$ in ($46 \times 46 \times 10$ cm) thick, with or without steel reinforcing.

For single-story and normal two-story brick construction load conditions, the pad should be steel-reinforced and at least $24 \times 24 \times 4$ in ($61 \times 61 \times 10$ cm) thick. For unusually heavy load areas, such as a multiple-story stairwell, the pad should be larger, thicker, and reinforced with more steel. In the latter case, the pad is normally poured in place. (The added weight creates severe handling problems.) In any event, the pad is leveled on or into the soil surface to produce a solid bearing. Conditions rarely warrant any attempt to place the pads materially below grade.

When the structure is supported on wood piers or stiff legs, it is rarely justified to replace only a portion of the wood supports with the superior concrete design. Normally such a practice would represent little, if any, benefit and be a waste of money.

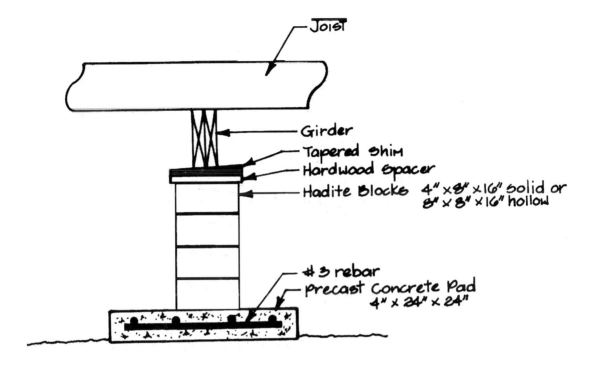
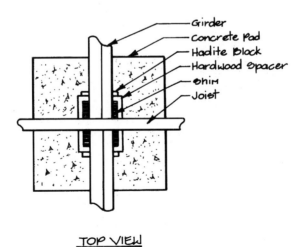

FIGURE 6B.5 Interior pier cap: end view and top view.

Once the pad is prepared, the pier cap can be either poured in place or precast. In most cases the limited work space demands the precast cap. Figure 6B.5 shows a precast design. Choices for the precast design include concrete cylinders, hadite (lightweight concrete) blocks, and other square masonry blocks. (Ideally, the head of the pier cap should be as wide as the girder to be supported.) Either material is acceptable for selected loads. The hadite blocks, for example, are occasionally inadequate for multiple-story concentrated loads unless the voids within the blocks are filled with concrete or solid blocks are utilized. The precast concrete can be designed to support a load comparable to that of a poured-in-place pier cap. [As an aside, the crushing strength of a hollow 8- × 16-in (20- × 40-cm) hadite block is about 110,000 lb (49,900 kg)].

The wood spacers are hardwood and of suitable size and thickness to fill most of the space or gap between the pier cap and the girder to be supported. In instances where the pier-cap head is not as wide as the supported girder, the hardwood spacers should be. If the grain of the wood in the spacer is perpendicular to the plane of the girder, the width and length can be the same. If the wood blocking is placed with the grain in line with the beam, the length should be at least twice the width. The final leveling is accomplished by the thin shim or wedge (usually a shingle). The combined thickness of the shims should normally not exceed about $\frac{3}{8}$ in (0.9 cm).

6B.3 SLAB FOUNDATIONS

Slab foundations suffer the problems of both settlement and upheaval. The general approach to the perimeter is the same as that described in Sec. 6B.2, except that mud jacking is required to raise or stabilize the foundation. Refer also to Sec. 6B.3.3 and the references.[2,10–14] For all practical purposes, conventional deformed-bar and posttensioned slab foundations are treated as being the same. In the latter case it sometimes becomes also necessary to repair or retension defective cables.

6B.3.1 Mud Jacking

Mud jacking is a process whereby a water and soil-cement or soil-lime cement grout is pumped beneath the slab, under pressure, to produce a lifting force which literally floats the slab to the desired position.[15,16] Figure 6B.6 depicts the normal equipment required to mix and pump the grout. The mud jack shown behind the truck is a Koehring model 50, theoretically capable of pump rates up to 10 yd^3/h (7.86 m^3/h) and pressures to 250 psi (1725 kPa). The grout is introduced through small holes drilled through the concrete. The pattern for holes drilled to facilitate grout injection is dictated by the specific job condition and guided by experience.

Seldom is any uniform grid applicable except in routine slab raising such as pavements, walks, floating slabs, such as warehouse floors, or some aspects of pressure grouting (See Sec. 6B.6). In fact, often during slab foundation leveling the hole pattern is adjusted to provide the desired control and results.

The next consideration is control of grout consistency. This is handily accomplished under normal conditions by adjusting the solids-to-water ratio. A typical grout could consist of four sacks of cement [376 lb (170 kg)], 1800 lb (820 kg) siliceous soil, and 70 gal (265 L) of water per cubic yard (0.786 m^3) of dry mix. A lower-consistency grout (higher water-to-solids content) is useful when concentrated loads resist the attempt to raise and localized force is required. A thinner grout will migrate over a larger area, sometimes allowing a greater lifting force at the same pump pressure ($F = P \times A$). A thicker grout (less flow) will be restricted to a smaller area, but allows a greater lifting force through an increased confined pressure. In some instances the thicker grout still tends to escape from the work area. Consider, for example, that situation where the grout flows beyond the foundation or intended work area. This could represent a project involving an interior fireplace (concentrated load) surrounded by a normal slab (low weight and

FIGURE 6B.6 Mud-jacking equipment.

strength), a shallow or no perimeter beam, and so on. If the thickened grout does not serve the purpose, other options in order of increasing difficulty (and expense) are:

1. *Stage pumping.* This practice involves the placement of a volume of grout followed by a period of time sufficient for the grout to reach initial or intermediate set. The process is then repeated.
2. *Shoring for containment.* Sheet piling (plywood or suitable material) is driven or buried into the soil at the slab perimeter and shored by suitable bracing. Under most conditions the seal between the sheet piling and the concrete perimeter is sufficient to contain the grout.
3. *Underpinning.* Underpinninng can be used to raise the slab perimeter and, followed by mud jacking, to fill voids. On occasion sheet piling might again be utilized to contain the grout.

Grout placement pressure is another concern. This aspect is most often overrated because foundation slabs represent very minimal loads. For example, the weight of a 4-in (10-cm)-thick slab is about 50 lb/ft^2 or 0.35 psi (2.4 kPa). In most residential construction the added live load plus dead load on the interior slab is minimal, often assumed to be 60 lb/ft^2 (2.9 kPa). The perimeter loads are substantially greater [perhaps 600 to 800 lb/lin ft (890 to 1190 kg/m)]. However, the perimeters are seldom intended to be raised solely by mud jacking. (This was not always the rule, but expediency and lack of gifted operators has prompted the change.) Multiple-story construction increases the weight, but still the loads are relatively minimal. Hence most mud jacking is accomplished at very low pressure. Significant pump pressure surges are generally the result of various forms of friction increase—plugged lines, improper grout mixture, or, on occasion, that event where the member to be raised suffers some form of mechanical binding or resistance. A sudden pressure drop most often suggests a parted hose or blowout. Once the grout leaves the pump hose there are few instances which could account for a measurable pressure drop. This is best comprehended by understanding the minimal resistance offered by the slab during mud-jacking procedures.

More significant than placement pressure is the placed volume since this aspect has a direct bearing on project cost. In normal slab-raising operations an operator who is capable of mixing and pumping on the fly can place up to about 16 yd^3 (12.8 m^3) per 8-h day. Under near ideal conditions the complexities imposed by foundation leveling reduce this capacity to a maximum of about 10 yd^3/day (8 m^3/day). For a matter of perspective 10 yd^3 (8 m^3) corresponds to an area of about 800 ft² (74 m²), 4 in (10 cm) thick. Void filling (amounting to open-ended pumping) can sometimes reach a placement volume approximating 20 yd³ (16 m³) per 8-h day if material supply and handling can support the quantity. If placement volumes are required in excess of these, the best solution is multiple pumps. For other grouting operations, greater capacities are possible. The material can sometimes be batched, and alternative pumps can be utilized which are capable of pumping 25 to 30 yd³ (19 to 23 m³) in less than 1 h, or 200 to 240 yd³ (160 to 190 m³) per 8-h day. (Refer again to Sec. 6B.6.)

Neglecting slowdown imposed by changing injection sites, allowance for set times, materials supply, and so on, the placement limitation imposed solely by pressure can be explained by the relationship

$$\text{HHP} = k\text{BHP} = \frac{Q \times P}{7.2}$$

where HHP = hydraulic horse power
k = efficiency constant (frequently 0.75 or less)
BHP = brake horse power
Q = ft³/min
P = placement pressure, psi

At maximum conditions this equation shows that as the placement pressure increases, the volume pumped decreases proportionately. Note that the placement pressure is the resistance at the pump cylinder head and not the pressure resistance offered by the member being jacked. Friction pressures developed during normal mud-jacking operations far exceed the load resistance required in slab raising. In fact something in excess of about 150 to 200 ft (45 to 60 m) of 2-in (5-cm) inner-diameter pump hose and a thick grout will approach the capacity of the average Koehring machine. Figure 6B.7 illustrates the mud-jacking process. The captions describe the various activities.

Mud jacking has the attendant feature of *some* chemical stabilization of the soil, which helps preclude future differential movement produced by soil moisture variations (previously discussed as a preventive treatment).

In leveling a slab foundation the primary concerns are:

1. Unnecessary damage to floor covering. Generally every attempt is made to limit or eliminate the need for drilling holes through ceramic tile or sheet linoleum
2. Raising the areas that are below the desired grade.
3. Ascertaining that all voids are properly grouted.
4. Rendering the foundation to as near as-built as practical without creating undue additional damage to the structure.

Note: Opinions have been expressed which tend to minimize the effectiveness of mud jacking. For example, in Winterkorn and Fang[17] it is stated: "This method (mud jacking) is rarely successful unless the entire consolidation of the ground has taken place." Such a statement is, at best, misleading, particularly as far as residential or light commercial foundations are concerned. Mud jacking is more permanent than the unconsolidated soil beneath. Recurrent problems would be the result of continued loss of soil bearing capacity and not the failure of mud jacking. (Reference to this statement was deleted from the revised edition.)

6B.3.2 Upheaval

Upheaval of a slab foundation presents the most serious restoration problem. If the upward movement is pronounced, the only true cure is to break out the slab, excavate the affected base to proper grade, prepare the base, and repour the slab section. This approach is, however, inordinately expensive. If the movement is nominal, the lowermost slab areas can be raised to the grade of the

FIGURE 6B.7 Mud jacking a slab. (*a*) Injection hole drilled through perimeter beam. (*b*) Nozzle in place, and mud jacking is in progress. (*c*) Interior pumping. (*d*) Exterior pumping of patio slab.

heaved section. This operation is extremely sensitive since the grade of the entire structure is being altered. At the same time, particular care must be exercised not to aggravate the heaved area. The leveling is accomplished by mud jacking, often in conjunction with the supplementary use of spread footings or other underpinning, as described earlier.

6B.3.3 Settlement

Slab settlement repair is a relatively straightforward problem. Here the lower sections are merely raised to meet the original grade, thus completely and truly restoring the foundation. The raising is accomplished by the mud-jack method as described. In some instances where concentrated loads are located on an outside beam, mud jacking may be augmented by mechanical jacking (installation of spread footings or other underpinning). In no instance should an attempt be made to level a slab foundation by mechanical means alone. Rather than providing interior support (and stabilizing the subsoil), mechanical raising creates voids which, if neglected, may ultimately cause more problems than had existed originally. As a rule, foundation slabs are not designed as structural bridging members and should not exist unsupported.[2,10–14,18] Mechanical techniques normally make no contribution toward correcting interior slab settlement. Raising the slab beam mechanically and backfilling with grout represents a certain improvement over mechanical methods alone, but still leaves much to be desired. This approach loses the benefits of pressure injection normally associated with the true mud-jack method. In other words, the voids are not adequately filled, which thus prompts resettlement. ("High-pressure injection" is a bit of a misnomer when slab foundation raising is concerned, as discussed.)

Figure 6B.8 illustrates the effect of leveling. In instances involving interior slabs, leveling was accomplished by mud jacking. Perimeters were generally leveled by underpinning techniques. In the "before" picture [Fig. 6B.8(a)] the separations under the wall partition in the brick mortar,

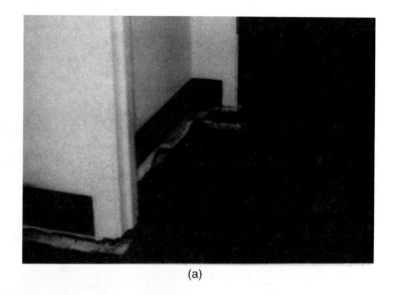

(a)

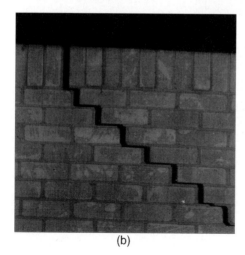

(b)

FIGURE 6B.8 Foundation leveling. (*a*) The floor is separated from the wall partition by about 4 in (10 cm). (*b*) The separation in brick mortar is in excess of 2 in (5 cm).

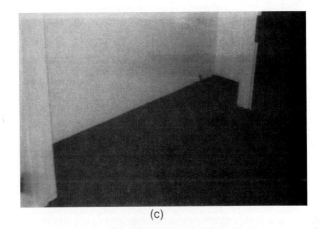

FIGURE 6B.8 (*Continued*). (*c*) and (*d*) are results from foundation levelings of (*a*) and (*b*), respectively. The separations are closed. The end results of foundation leveling are not always so impressive.

between brick and door frame, and the settlement of the floor slab are obvious indications of foundation distress. In the "after" picture [Fig. 6B.8(*b*)] the separations are completely closed, illustrating that the movement has been reversed or corrected. In this particular example, the leveling operation produced nearly perfect restoration. This is not always the case, as discussed in various sections in this book. The remaining illustrations depict the mud-jacking process in action. The captions are self-explanatory.

Mud jacking is not intended to be a cure-all; it is intended to provide specific benefits. In truth, the "rap" against mud jacking per se is ill-founded and based on a lack of understanding or misapplication, or both. As a rule, mud jacking is essential to the proper repair of conventional residential slab foundations.[9–14,18] Foundations can be designed to cantilever or bridge void areas.

(e)

(f)

FIGURE 6B.8 (*Continued*). (*e*) Mud jacking in progress. The shovel and scribed marks monitor the raise of the slab. At this point, the raise has been more than the width of a brick. ($3\frac{1}{2}$ in or 9 cm). This is shown by the distance between the shovel point and the original scribed mark. (The shovel handle is on a stationary surface.) (*f*) In this example the shovel point raises as the floating slab is mud jacked. (The shovel point is on the stationary surface.) The floor slab is down about 4 in (10 cm).

FIGURE 6B.8 (*Continued*). (*g*) The floor is restored to its original grade.

However, such design practices are not common to residential or light commercial foundations. To further complicate the situation, the actual foundation design is most often not known to the foundation repair engineer or contractor.

6B.3.4 Posttension Retrofit

On occasion posttensioned slabs suffer differential movement due to either improper cable tension or failure of cable tension. In this instance the problem can sometimes be remedied by retensioning the cables. This effort most likely will require subsequent mud jacking, but may well avoid the need for underpinning. A new approach to slab repair has been suggested by Barchas.[19] This method suggests the use of external posttensioning (EPT) as a remedial procedure for sagging, settled, cracked, or spalled concrete slabs or beams. The approach is interesting and may prove to bear merit as its use is documented.

6B.4 SLIM PIERS OR MINIPILES

In 1976 a new, patented process was introduced to correct failed foundations. This technique, labeled Perma-Jack,[20] utilizes a hydraulic ram to drive jointed sections of 3-in (7.5-cm)-diameter, 3-ft (0.9-m)-long steel pipe supposedly to rock or suitable bearing. When a solid base or other resistance is reached, continued exertion of the hydraulic ram will raise the foundation. The structural weight provides the force used to drive the pipe. Accordingly, the system allows for no safety margin. (The safety factor F is 1.0 where most foundation design systems suggest values between 2 and 3; see Sec. 6B.4.6.) Once at grade, the pipe is pinned through the jack bracket to hold the foundation in place. The jack brackets can be positioned from the interior (Fig. 6B.9) or exterior. In the case of slab foundations, the void beneath the slab must then be filled by mud jacking.[2,11,12]

The Perma-Jack (or similar) process is more expensive than conventional methods and in most areas did not seem to provide comparable results. It would appear that the technique might be of some use to the foundation repair contractor in areas where the more conventional methods demonstrate inadequacy.[2] Five basic problems with the slim pier or the minipile techniques are

(1) uncertainty of alignment,[11,12] (2) susceptibility of the slim piers or piles to fail when subjected to lateral stress (soil swell, for example),[11,21] (3) dependence of support capacity almost solely on skin friction,[11,21] (4) concern for margin of safety in load capacity, and (5) imminent corrosion of the steel pipe exposed to the wet soil[2,11] (Fig. 6B.9). Note that none of the underpinning techniques prevents the foundation from raising off the support.

In several alternatives the slip joints are replaced by threaded or welded (usually with epoxy) couplings. One variation is to fill the driven pipe with concrete. This provides little, if any, improvement.[11,12] Some methods drive the pipe from the point as opposed to the top. This should provide both better alignment and greater driving force, allowing a more favorable safety factor. Yet another technique professes to drive sections of precast concrete cylinders stacked one on top of the other. This approach would seem to have little or no use in expansive soils due to the basic problems discussed earlier for minipiles. Problems of failure in lateral stress, negative skin friction, and uncertain alignment would be even more pronounced. Another technique is to drill the pier holes, case them if necessary, and place steel reinforcing and concrete. This approach would offer the advantages of both a properly aligned pier and elimination of the corrosion concern, but it could enhance the more serious problem of pier uplift. Admittedly, with slim piers this threat is somewhat reduced as compared to the conventional design because of reduced surface area. A technique sometimes used to further reduce upheaval is to sleeve the uppermost several feet of the hole with a so-called friction reducer, such as cardboard or polyethylene. Belling the pier shaft is most often the best preventive against heave. (See Part 3 for more detailed information.) However, for this approach to be effective, the pier shaft must be larger than about 6 or 8 in (15 to 20 cm) in diameter.

Neely[23] presents an interesting study on yet another option—continuous-flight augered piers, which are, in turn, pressure-grouted bottom to top. These piers normally have a bearing capacity less than that of a conventional driven or cast-in-place pile of similar dimensions, but for moderate loads they can be effective and less expensive. (See also Secs. 6B.6 and 3B.)

6B.4.1 Design Concerns for Minipiles

As a matter of interest, the following facts were taken from an authoritative British publication.[11]

1. Minipiles are ineffective where:
   - Lateral soil movements can be reasonably anticipated, that is, for expansive soils, sloping sites, soil consolidation, collapsing soils.
   - Appropriate bearing stratum is not readily accessible. [In the instance of a 3-in (7.5-cm) minipile this depth must be no more that about 18 ft (5.4 m).]
   - Structural loads are high [over about 6.5 tons (60 kN)].

2. In soft clay or loose granular soils no contribution to load design is assigned to shaft friction. In some cases the shaft friction factor is assigned a negative number, that is, soil consolidation or surcharge load.

3. In silty or sandy soils the load-bearing capacity of the pile can diminish with time due to movement of water between soil particles (collapsing soil).

4. Raked piles are more susceptible to failure due to lateral stress (heave in clay soils) than those truly vertical. A rake angle of 10° is most common, but it must not exceed 15°. (See also Fig. 6B.15.)

5. Minipiles (high length-to-diameter ratio) often fail more in strength as a column as opposed to failure in soil support.

6. Straightness, true vertical alignment, and concentric loading are useful or necessary to prevent buckling of the pile.

7. Pile lengths should be limited to no more than 75 times their diameter. [For a 3-in (7.5-cm)-diameter pile, the maximum effective length should be 18.75 ft (5.6 m).]

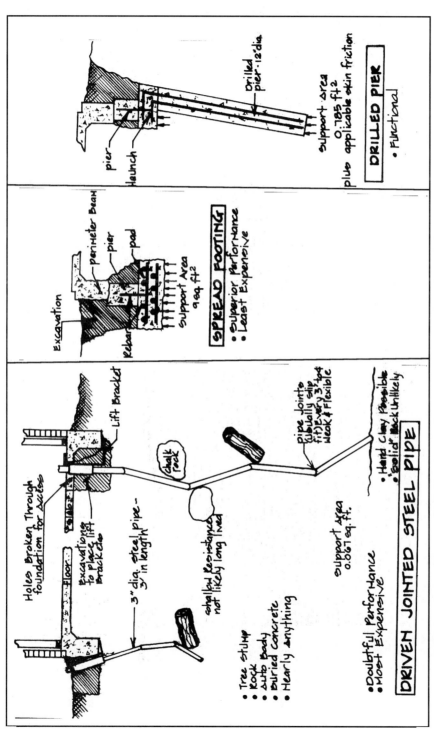

FIGURE 6B.9 Comparison of underpinning techniques.

8. A single, central rebar is useless for resisting lateral forces (bending). It does, however, provide additional resistance to tensile forces.
9. Casing provides only minimal resistance to bending of minipiles. Caged rebar is the best solution, but the pile or pier diameter must be such as to permit proper placement of concrete.
10. Prior to underpinning, all voids beneath the floor slab must be grouted (mud jacked). (When underpinning is performed in conjunction with raising the foundation, the author prefers that mud jacking be performed subsequent to raising.)
11. Minipile spacing is normally limited to about 5 ft (1.5 m) maximum.
12. Piles larger than approximately 6 in (15 cm) are normally augered.
13. A safety factor of between 2½ and 3 is suggested.

Bruce and Nicholson presented an interesting study.[24] The diameters of the piles in their study were somewhat larger than those generally considered in Digest 313.[11] Bruce and Nicholson used drilled shafts generally in the range of 8 to 12 in (20 to 30 cm) with depths ranging to 200 ft (60 m). The shafts were either grouted full with a short section of surface casing or filled with a pin pile which was cemented bottom to top. The reported capacities for piles installed wholly in soil suggested safe working loads approaching 100 tons (890 kN). When founded in rock, the safe working loads might approach 300 tons (2670 kN). The most significant disadvantage to these approaches was the inordinately high cost. For example, the long pin pile was reported to cost in excess of $100/ft (0.3 m).

Prakash and Sharma[25] provide valuable information regarding the use of piles. Two areas which they discuss and which impact upon repair procedures are presented in this section.

6B.4.2 Deterioration of Steel Piles and Piers

Deterioration of steel piles may occur either when they are damaged during driving (deflected, welds, wrench scars, and so on) or when they become corroded. Mechanical damage can be controlled to a large extent by the methods and techniques used to drive the pile. When it does occur, mechanical damage further increases the susceptibility to corrosion damage. The severity of corrosion damage will depend on the composition of the piles and the nature of the environment into which the pile is placed. The rate of corrosion of metal varies drastically with such factors as soil composition and texture, pH, ionized elements, depth of embedment, moisture, and availability of oxygen. For example, even in moist clays, an oxygen deficiency may preclude extensive corrosion at significant depths. On the other hand, in a coarse-grained soil the corrosion at some depth may be equivalent to that at atmospheric (near surface) conditions. Pile composition, the use of noncorrosive coating, and, in extreme cases, cathodic or anodic protection can control chemical corrosion.

6B.4.3 Action of Soils around a Driven Pile

The process of pile driving is reflected in a remolding of the soil around the pile. First, the volume of the pile or pipe must be displaced. Second, in cohesive soils the effects in the adjacent matrix include altering the internal stress, buildup of hydrostatic excess, and a long-term phenomenon of strength regain. For noncohesive soils the major effects are densification and increase in the effective lateral pressure around the pile. Meyerhof[26] reports, based on field analysis, that in dense sand the effective lateral stress is equivalent to the effective vertical stress ($K = 1.0$). The effective value for loose sand was suggested as $K = 0.5$ times the effective vertical stress. Fourie[27] discusses values of lateral stress which measured as much as twice that for corresponding vertical stress. (See also Sec. 3D.) When available, specific geotechnical information should be used to determine the actual K value. This is particularly significant due to the propensity for minipiles to fail in external stress.

6B.4.4 Screw or Earth Anchors

Screw or earth anchors have been used for many years, principally as a means to provide resistance to uplift forces and lateral moments. In general this involves retention systems for retaining walls, excavations, transmission towers, hydraulic structures, and the like. Anchors are manufactured in a variety of configurations such as plate anchors, pile anchors, grouted anchors, tension anchors (dywidag bars or posttension cables grouted into drilled holes), prestressed concrete anchors, and single- or multiple-screw helical anchors. The chief advantages of screw anchors over the alternatives are that they can be loaded immediately after installation, are easy to install with lightweight equipment, and are cost-effective in cases of high underground water tables because dewatering is not necessary.

A December 1990 issue of *Concrete Repair Digest* contained a short article (advertisement actually) addressing the use of helical screw anchors as a means for underpinning residential foundations. A screw helical anchor is often constructed by welding one or more screw helices onto a steel shaft. The anchor is installed by simply screwing it into the ground. Once placed, axial compression can be applied to the shaft which is sufficient to lift the weight of the residential foundation. Once raised, the anchor can be attached to the foundation in such a manner as to sustain the beam elevation. If the foundation is of slab design, the raising must be accompanied by mud jacking in order to fill voids and restore the support for the slab. Screw anchors suffer similar problems and limitations as minipiles and piers. Since placement is not dependent on structural dead weight, screw systems can provide some margin of safety between installation and design or actual load. The screw anchors are also more expensive for the consumer than conventional underpinning methods, that is, drilled concrete piers or spread footings.[27,28] Ghaly and Hanna have authored or coauthored several articles dealing with screw anchors.[29]

6B.4.5 Spread Footings with Multiple Drilled Piers

Pod piers beneath spread footings represent yet another variation. Shallow piers, on a bipod or tripod pattern, are installed beneath the pads used with the conventional spread footings (see Fig. 6B.2). As a rule, the piers are 6 to 8 in (15 to 20 cm) in diameter and extend to depths of 3 to 9 ft (0.9 to 2.7 m) below the pad. This technique was carefully evaluated in the mid to late 1960s and discarded. Field tests indicated that the inclusion of the shallow piers added nothing to the stability of the pad (conventional spread footing). This technique is similar to the conventional drilled pier with haunch, except for the following:

1. As a rule the regular-size spread footing base [3 × 3 ft (0.9 × 0.9 m)] is used as opposed to the smaller haunch [24 × 24 in (0.6 × 0.6 m)].
2. The multipiers are smaller in diameter than the conventional drilled pier. This could introduce an area of weakness, particularly in applications where lateral stress is a factor.
3. The placement of steel reinforcement is seriously restricted.
4. Two or three piers could provide a load capacity greater than a single pier of identical size. (However, as noted, the conventional drilled pier is significantly larger in diameter than the pod pier. As a matter of fact, a single 12-in (0.3-m) pier will have a bearing capacity greater than that of two 8-in (20-cm)-diameter piers of equal length.
5. The drilled pier can be installed with little or no rake (angle off vertical), whereas the pod piers are slanted (see Fig. 6B.9).

6B.4.6 Comparative Bearing Capacities

For purposes of discussion, the perimeter load of a single-story brick-veneer dwelling would be in the range of 0.375 ton (3.5 kN) per linear foot (0.3 m). Assuming a safety factor of 3 and a spacing of 8 ft (2.4 m), the required underpinning load capacity would be about 9 tons (83 kN). In remedial applications the load capacity requirement for the underpin is greater than that indicated until upward movement is established, at which point the compressive load becomes strictly the weight of the structure.

By comparison, consider the following ultimate load capacity data for various underpins as reported by Petry[30] and by O'Neill and Mofor.[22] Petry's tests were conducted in northwest Dallas, where the soil is the Eagle Ford Terrace deposit. The O'Neill and Mofor tests were performed at the University of Houston test facility utilizing soil designated as Beaumont clay. The general trend of these data can be presented as follows:

1. *Petry tests* (Installations were performed by a single foundation repair contractor who prefers the bipod design.)
   a. 3- × 3-ft (1- × 1-m) spread footing: 50 tons (460 kN)
   b. 8-in × 12-ft (20-cm × 3.6-m) concrete pier with 2- × 2-ft (0.6- × 0.6-m) haunch: 40 tons (275 kN)
   c. 2- × 2-ft (0.6- × 0.6-m) footing with two 8-in × 12-ft (20-cm × 3.6-m) piers: 60 tons (550 kN)
2. *O'Neill tests*
   a. 4-in (10-cm)-diameter steel pipe driven to 21 ft (6.6 m), filled with concrete and reinforced with a single deformed bar: 6.5 tons (60 kN)
   b. 4-in (10-cm)-diameter open shaft to 23.5 ft (7 m) filled with concrete and single rebar: 17 tons (155 kN)

Note: The bearing capacities reported in tests 2a and 2b were attributed almost solely to skin friction, a design feature somewhat variant in clay soils. Pier depths for Petry tests were measured from the ground surface. The actual pier lengths would be about 9 ft (2.7 m). The maximum load capacity values recorded from the Petry tests likely represent failures in bearing soil rather than underpin.

The reduced bearing capacity reported in test 2a was alleged to be the result of reduced skin friction resulting from the collars' lateral thrust of soil away from the pile during placement. (A uniform-outside-diameter pipe without a friction collar or enlarged point would help reduce this problem.) Both researchers allegedly utilized similar test methods, namely, the Texas quick test method or some simple version thereof.[22] (See Part 3.) The effects of lateral stress were not considered.

These data suggest that either of the methods except 2a could be adequate for underpinning normal residential structures when bearing capacity is the only concern. However, method 2b would be clearly suspect for use in clay soils due both to its dependence on skin friction for its bearing capacity and to its susceptibility to failure in lateral stress. All the other methods are clear overkill with respect to bearing capacity.

Important note: None of these capacity data should be used indiscriminately for design calculations. The information is intended for comparative purposes only and would vary with soils as well as with load or test conditions.

Other interesting studies on bearing capacities were presented first by Neely,[31] who develops an empirical equation suitable for calculating the ultimate load of expanded basic piles (Franki piles or pressure-injected footings) at depths of up to 10 base diameters. The second was a feature article,[32] and the others were by Fellenius[33] and Ismad et al.[34] These tests deal with sandy soils. Therefore skin friction becomes a viable design criterion. For more detail on pile or pier behavior, see Sec. 3B.

6B.5 REVIEW OF REPAIR LONGEVITY (COMMON REPAIR METHODS)

For a comprehensive evaluation of the relative success of foundation repairs, refer to Table 6B.2. These data were collected from repair records compiled essentially from Dallas, Tex.[9,10,27] However, similar results could be anticipated from any area if one assumes competent repair procedures were utilized. These data also include random results from Louisiana, Arkansas,

TABLE 6B.2 Success of Foundation Repairs

Test period	Number of jobs	Total redos*	Cause of failure		
			Upheaval	Settlement	Indeterminate
7/90–6/91	423	12 (2.8%)	3 (0.7%)	9 (2.1%)	
6/89–6/90	401	11 (2.7%)	4 (1%)	7 (1.7%)	—
Total	824	23 (2.78%)	7 (0.8%)	16 (1.9%)	
1/82–6/82	166	7 (4.2%)	6 (3.6%)	1 (0.4%)	—
1981	336	29 (8.6%)	23 (6.8%)	6 (1.7%)	—
1980	302	36 (11.9%)	22 (7.2%)	14 (4.6%)	—
Total	804	72 (8.9%)	51 (6.3%)	21 (2.7%)	
12/78–12/79	243	20 (8.2%)	12 (4.0%)	5 (2.0%)	3
12/77–12/78	291	32 (10.9%)	18 (6.2%)	8 (2.7%)	6
Total	534	52 (9.7%)	30 (5.6%)	13 (2.4%)	9
7/64–1/69	380	18 (4.7%)	4 (1.1%)	4 (1.1%)	10

*Excludes simple reshim of pier and beam interior floors.

Mississippi, Florida, and virtually all other parts of Texas. No form of soil stabilization was incorporated into the example repairs until the 1989–1991 period, at which time some chemical stabilization was used on occasion.

Table 6B.2 also emphasizes the relative occurrence of upheaval versus settlement as a reason for redo. Also it is obvious that the redo rate is quite favorable, with an overall success ratio of better than 90% (upheaval and settlement combined) under the most adverse conditions. These data do not distinguish between slab or pier and beam foundations. However, overall the ratio of slab repairs to pier and beam repairs was about 3:1.

For our purposes the term *redo* refers to any rework of the foundation during the initial warranty period, usually 12 months. It might be interesting to note, however, that less than 1% of the foundations repaired developed recurrent problems after a year. At this point a careful analysis of the data presented in Table 6B.2 would be in order.

At the time the 1964–1969 data were compiled, the slab repairs were performed by mud jacking alone—spread footings or piers were not utilized to supplement the raising. Further, the prevalence of upheaval was not fully recognized.

The 1977–1979 data cover a span containing an inordinately wet cycle. In fact, the period of December 1977 through about May 1978 represented one of the wettest periods recorded in Texas history. The heavy precipitation gave rise to an unusually high proportion of redos, particularly upheaval. Along these lines, the overall redo rate expressed for this period is 9.7%, as compared to 4.7% for the earlier period, a nearly twofold increase. Further, the relative rates of recurrent settlement vary by about the same ratio, 2.4% versus 1.1%. The significant variation involves the reported comparative incidence of recurrent upheaval, namely, 5.6% versus 1.1%. This disproportionate difference is likely due to the following facts:

1. Until the 1970s foundation repair contractors were not properly aware of the tendencies toward upheaval and probably mislabeled some of the causes.
2. The 1977–1979 data were accumulated during a cycle of excessive and unusual moisture, which prompted increase in upheaval.

The third set of data covers the 30-month period immediately prior to July 1, 1982. Again the influence of climate conditions is clearly reflected. The 1980 data show the results of a record July heat wave following the extremely wet years 1978 and 1979. The year 1981 was again a near record wet year following the unusually dry 1980. The first 6 months of 1982 represent a "return to normal" trend.

The fourth set of data reflects a period where Soil Sta was sometimes used (note the obvious reduction in upheaval), and drilled concrete piers were used as often as spread footings. The 1989–1991 data were cross-checked to see whether any relationship existed between redo rate and method of underpinning (spread footing as opposed to drilled concrete piers). No such correlation was noted, and the relative success of either method was indicated at better than 97%.

The data in Table 6B.2 were developed by two separate firms, the 1964–1969 data by one and the other three sets by another. Still, all factors considered, there is amazing consistency. Regrettably, efforts to accumulate additional reliable data from other sources were unsuccessful.

The incidence of upheaval (as opposed to settlement) as a cause for redo occurred at the ratio of a little over 2:1 until the 1989–1990 data, at which time the ratios were reversed. This is largely due to the use of Soil Sta, which impedes or prevents upheaval. As a matter of interest, the ratio of slab to pier and beam failure was about 3:1. The incidence of redo after 1 year was reflected as 0.2% for the 1989–1990 period and 1.6% for the period of 1990–1991. These data reflect the introduction of an extended limited warranty, which extended coverage from 1 year to 10 years. The extended warranty was introduced fully in 1984. The 1991 data reflect a $7\frac{1}{2}$-year history.

At least 50%, and perhaps as many as 75%, of the redo cases were brought about by the refusal or neglect of the owner to properly care for or maintain the property. Under normal climatic conditions, the anticipated chance for any type of redo should be less than approximately 4 out of 100 jobs performed (96% success). This obviously assumes current U.S. construction practices and competent repair procedures, and it excludes the use of effective soil stabilization, which should reduce the incidence of failures even further. Since upheaval is caused by excessive accumulation of water beneath the foundation, any measures that prevent this will minimize the chance for redo. For that matter, the same precautions will help avoid the initial problem as well. On the other hand, proper watering will help to arrest movement relative to settlement.

6B.6 GROUTING

If the soil problem is one of poor consolidation, this condition should be corrected if the foundation is to remain stable. Mud jacking or a variation thereof—pressure grouting—is most often quite successful in cementing and consolidating or compacting loose soils. The cementitious aspect depends on the design of the grout and the nature of the soil being grouted. As an aside, the 7-day compressive strength of typical soil-cement grout is 230 lb/in^2 [33,120 lb/ft^2 (161,626 kg/m^2)].[35] Specific mixes can be selected to provide a bearing capacity highly superior to that of even the most favorable soils.

This section discusses grouting as it relates to foundation repair. Section 5B addresses overall grouting operations.

6B.6.1 Introduction

For purposes of the following discussion, deep-grouting operations include any activity whereby the success of the operation depends on the ability of an injected material to penetrate and permeate a relatively deep soil bed. Once in place, the grout must either set into a strong, competent mass or intersperse with and into the soil to create a strong, competent mass. Specifically, these operations include: (1) stabilization, (2) subgrade water control or waterproofing, (3) soil consolidation for control of bearing strength or sloughing, (4) pile or pier enhancement (Franki pile), (5) waste containment, (6) grouting slip liners, and (7) certain backfilling operations. Often grouting is coupled with mechanical leveling or hydraulic stabilization to correct foundation problems originating deep below the soil surface. Similarly, grouting can be used to "freeze in" existing foundation piers, improve the bearing capacity of piers used in remedial operations, or stabilize fill beneath problem foundations. The physical properties of the soil matrix and the specific problem to be addressed dictate the specific grout to be used. For example, the particle size and permeability of the soil particles tend to suggest the nature of the grout. For soils with an average grain size greater than about 10.0 to 1.0 mm, cement grouts can be effective. Soils with smaller grain sizes suggest chemical grouts. For example, grain sizes between 1.0 and 0.1 mm can be amenable to the silicates, lignins, and perhaps specific bentonitic gels. Particle sizes between 0.1 and 0.01 mm would require the less viscous phenols or resins.[36] The two basic grout types are discussed in the following paragraphs.

6B.6.1.1 Cement Grouts. For optimum penetration, the consistency (viscosity) of the invading grout should be controlled. For maximum penetration, the consistency should be minimal. In other cases it may be desirable to restrict penetration by increasing the consistency. In the latter case it is normally sufficient to merely decrease the water-to-solids ratio; in the former, the solution is not so simple. Although grout consistency can be decreased by increasing the water-to-solids ratio, this approach is highly restrictive because increasing the water content will drastically reduce the compressive strength of the set grout, and the additional water can create severe handling problems brought about by separation between solids and liquid during transportation (pumping). Perhaps one of the best solutions to high strength–low consistency grout involves the use of a very fine cement.[37] Such a product, MC500, has a specific surface of about 8200 cm^2/g whereas that of other cements falls in the range of 3200 to 6300 cm^2/g. (Specific surface values are determined by the Blaine method.) Due to this fineness, MC500 grouts have been used successfully to permeate soils with a coefficient of permeability k substantially below 0.1 cm/s. The unconfined compressive strength of the hardened MC500 paste taken after 28 days curing averages about 1.5 times that for normal portland cement pastes.

Grout consistency is approximated in the field by timing a measured volume to flow from a cone or funnel. Refer to U.S. Corps of Engineers.[38] Using these data, the mix composition can be varied to produce the desired grout property.

6B.6.1.2 Chemical Grouts. Chemical grouts and chemically treated cement grouts can also be useful for consistency-controlled grouts. Several additives useful in cement grouts were discussed in Brown.[35] New developments on chemical grouting were discussed by Nandts.[39] He addresses such procedures as precipitation, jet and sewer grouting, water shutoff, bearing capacity enhancement, flexible form grouting (geotextile mattresses filled with concrete or grout), root piles (prefab piles consisting of two concentric steel pipes, the outer pipe being used for compaction grouting and the inner pipe allowing chemical grouting through the pile point), and injectotubes (flexible grout pipe installed in cold or expansion joints before concrete is poured). Nandts also touches on custom grouts, principally hydrophobic (polyurethanes), hydrophyllic (urethanes and acrylamides), and cementitious grouts with controlled consistencies.

6B.6.2 Mechanics of Grouting Applications

The specific mechanics required for a grouting operation depend entirely on the type of operation involved. For certain kinds of grouting it is necessary to form a continuous consolidated boundary or area of sufficient strength, often referred to as a *grout curtain,* to permit excavation or provide adequate bearing strength. The hole pattern used to introduce the grout will depend on the particular problem, and each project should be carefully designed and engineered. Generally three or more rows of holes are split spaced 4 to 8 ft (1.2 to 2.4 m) apart. Figure 6B.10 illustrates such a pattern. This example utilizes 6-ft (1.8-m) spacing and would be most applicable for subsurface water shutoff from the northeastern direction, or consolidation of fill at the northeast corner. Typically row *A* would be grouted to refusal, then row *C* followed by row *B*. (Rather than refusal, grouting can terminate at some predetermined injection pressure.) The same general pattern is used for either chemical injection or soil stabilization at the perimeter. Often, especially with chemical injection or foundation leveling, the pattern extends to all four sides and includes a uniform pattern for the interior. The depth of injection is designated by the particular problem. With shallow foundations this depth is often 12 to 20 ft (3.6 to 6 m).

6.64 FOUNDATION FAILURES AND REPAIR

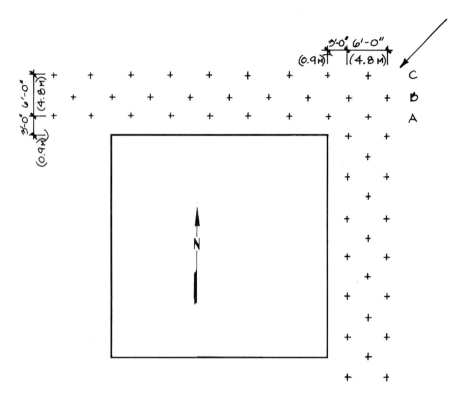

FIGURE 6B.10 Pressure grouting layout.

In general, however, the fewer the holes and the wider the spacing, the more economical the operation. For this to be practical, the grout must flow efficiently, once again pointing out the need for a consistency control. Along these same lines, the diameter of the grout injection pipes is influenced by the grout consistency as well as the rate of grout placement. Typically the grout pipes vary between $1\frac{1}{4}$ and 2 in (3 and 5 cm) in diameter.

Other interesting grout applications are discussed in the literature.[40–45] Weaver et al.[43] deal with ground-water recovery at a waste disposal site.

In the late 1950s and early 1960s the author dealt with the problem of permanent disposal of intermediate-level atomic wastes. This project utilized a mixture of bentonite (montmorillonite), cement, and contaminated water to produce a grout which was then injected into a deep subgrade stratum. Upon setting up, the grout permanently contained the atomic waste. The bentonite served to base-exchange the radioactive cations, and the cement caused the grout to become a solid mass.

Many other applications for pressure grouting are described in detail elsewhere, in particular, in Secs. 5B and 6B. Another application of interest is mud jacking (low-pressure grouting) to protect and extend the life of aboveground storage tanks.

The mechanics of pressure grouting for applications relative to this section will be discussed in the following paragraphs.

6B.6.3 Placement of Injection Pipes

The placement of the grout injection pipes can be quite simple but also quite complicated. The easiest solution is to either drive or push the pipes to depth. Pneumatic paving breakers equipped with a drive head can be used to drive the pipes. Front-end loaders, forklifts, or the equivalent can often be used to push the pipes. In either case, care must be taken to keep the pipes from being plugged or damaged during placement. More often than not the side-slotted design shown in Fig. 6B.11(a) is adequate. Any intruding soil or foreign materials can often be either blown out with air or washed out with water. Another approach would be the use of a pointed drive cap or plug which is loose fitting and is removed by injection pressure or by pulling the pipe upward slightly.

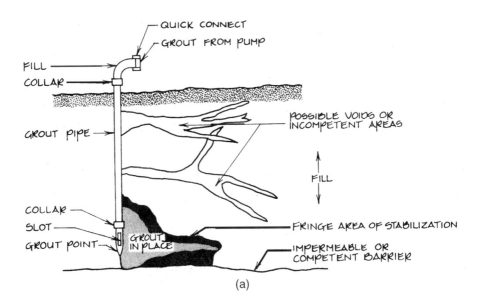

(a)

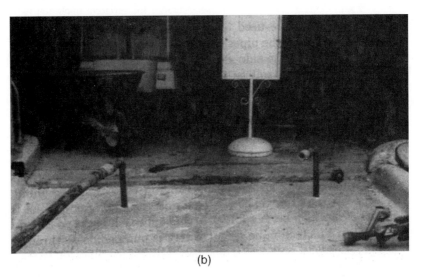

(b)

FIGURE 6B.11 (a) Pressure grouting to consolidate fill. (b) Pressure grouting in progress. The pipe to the left is being pumped. The pipe at right has been set and is ready. The air hammer, lower right, is fitted with the pipe-driving head used to set pipes.

6.66 FOUNDATION FAILURES AND REPAIR

In either event, the cap or plug is left in the hole. On occasion the pipes can be merely washed down, a forerunner to jet grouting. Figure 6B.11(*b*) depicts pressure grouting in progress.

The more complicated (and expensive) alternative involves drilling the holes, with or without casing. In this technique the injection pipes normally require some method to pack off or seal the annular space between the grout injection pipe and the casing or wall of the hole, as the case may be. A spinoff to this approach is the so-called jet grouting.[45] In this technique, grout slurry is pumped down through a small-diameter drill pipe and forced laterally through small-diameter orifices or ports. The high-velocity grout impacts the soil matrix. The pipe is rotated and withdrawn slowly to create an enlarged-diameter column of grout and grouted soil. Pump pressures are often in the range of 4000 lb/in^2 (192 kPa). The withdrawal rate is on the order of 1 ft/min (0.3 m/min), and the grout composition is often a 1:1 mix of cement and water.[46] The grout injection spacing often involves a multiple series of holes, split spaced, from 3 to 8 ft (0.9 to 2.4 m) apart.

6B.6.4 Subgrade Waterproofing

Since the processes of deep grouting are more common to repair procedures for commercial foundations, the following presentation will be brief and limited to those applications most likely to involve residential foundations. One example of deep grouting would be in waterproofing subgrade basements and foundation walls. Another example would be deep grouting to fill voids and increase bearing qualities of deep fills, particularly those of organic composition.

Waterproofing subgrade walls is a rather routine, straightforward operation. The real difficulty lies in the fact that the stoppage of water intrusion at one point often results in the reappearance of water at unpredictable new locations. Technically, subgrade waterproofing is not materially different from regular mud jacking, the differences being that in subgrade waterproofing, (1) rather than being pumped through a floor slab, the grout is often injected through the walls, (2) working or injection pressures are higher, and (3) the grout composition is often modified, principally by increasing the cement content. Since each situation is different, the appropriate repair technique must be designed for the specific problem.

The results obtained from this process are generally satisfactory. On occasion, as noted, the intrusion of water may reappear at a new location, or the water flow may be reduced to an intolerable seepage. In either of these instances the grouting operation is normally repeated. From cases within the author's personal knowledge, the initial procedure was successful in over 80% of the cases. Less than 5% of the cases required a third application, and none required a fourth. Success varies directly with technique, nature of the water problem, and, on occasion, the particular job or site conditions.

6B.6.5 Soil Consolidation

Consolidation of deep fill, particularly organic material, by pressure grouting represents a more involved operation. The grout must be placed to fill voids as well as permeate the soil matrix to a sufficient extent to develop the desired bearing strength of the fill segment. The quality and the nature of the fill material dictate the grout composition. Often a cement grout, with or without chemical additives, is acceptable. In any event, the selected grout is mixed and pumped into the target soil. Generally the placement commences at the lowermost elevation and proceeds upward to the surface or to some predetermined level. Consistent with this, the grout pipe is placed at the bottom of the zone to be grouted and progressively raised in lifts to accommodate the upper levels.

Pumping normally continues at each level until either some selected resistance pressure is encountered or undesired movement is detected at the surface. Figure 6B.11 illustrates typical pressure-grouting operations. The represented void areas are not meant to imply that either voids or incompetent sections follow any defined pattern. The grout placed also may develop into a bulb configuration around the grout pipe, with little or no penetration into the soil. This might well be satisfactory, provided the placement pressure is sufficiently high either to cause mechanical consolidation of the surrounding fill or that the grout permeates and penetrates the matrix soil to the extent that sufficient cementitious consolidation occurs.

The so-called fringe area of stabilization [Fig. 6B.11(a)] may or may not occur, depending on the particular properties of the specific grout. In general the aqueous phase of the grout is responsible for this benefit, which often results in a significant improvement of the competence of the soil matrix.

On occasion, stage grouting is required to provide the desired results. Here the grouting operation is merely repeated until acceptable consolidation is achieved. In general no more than two to three stages are required within a given area to produce the desired results.

6B.7 SPECIAL PROBLEMS ADVERSELY AFFECTING LEVELING

This section will touch on the inherent problems that impede the intended foundation leveling or some related operation. In most cases specific and marked improvement of the condition is experienced, although the overall results might be less than desired. Factors to be considered include upheaval, warped wood members, and various construction or as-built conditions.

6B.7.1 Upheaval

As noted in previous sections, upheaval is the condition when some area of the foundation is forced above the original grade. Upheaval is caused by water accumulation and subsequent swell of the clay soil, which exerts an upward force on the foundation. When this condition occurs, it is rare that the foundation can be completely restored to a level condition. As a rule, the lowermost surrounding areas are raised to approximate or "feather" the heaved crown, and the foundation is then stabilized to prevent future distress. The result is normally acceptable as far as utilization is concerned.

6B.7.2 Warped Substructure

Warped wood members are another antagonist to foundation leveling. This problem normally involves pier and beam construction, but it can also include slab construction where the floors are laid on a wood screed system. In either event, a wood member warps into an arc because of the externally applied stress, which is often the result of differential foundation movement. In the stressed condition, the wood is in tension and thus is often tight against the foundation supports. When the extremities (ends) of the wood are raised in the attempt to level, the stress may be relieved, but the member may retain the warped or distorted state (Fig. 6B.12). The floor surfaces at points A and B are level after the raising operation, but the midsection at C is above the desired grade.

From a foundation repair standpoint, nothing further can be done to improve the situation. Often, given time, the wood will relax and the crown will depress to approximately normal. In an effort to encourage this event, the appropriate space is left above the shim at point C. In cases of slab construction with wood screed, the analogy is identical except that the screed rather than the girder is warped.

6B.7.3 Construction Interferences

Often specific conditions of construction will restrict or prevent the desired foundation repair results. A few specific examples are discussed in the following.

6B.7.3.1 Add-on or Remodeling without Correcting Preexisting Foundation Distress. In one condition that often exists, an add-on is constructed abutting an existing structure, which has

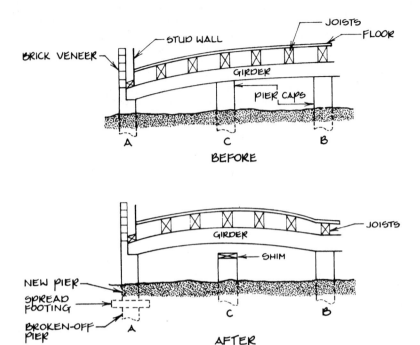

FIGURE 6B.12 Warped substructures.

previous and ongoing differential movement. In time the condition progresses to the extent that foundation leveling is desired. With a pier and beam construction, the only technique is to underpin (mechanically raise) the affected beam section. To accomplish this, proper access must be provided. In general this would require removal of floor sections in either the add-on or the original structure and doing the work from within. Obviously this could be quite expensive. For slab construction, the problem area might be restored to an intermediate grade by mud jacking inside the original structure. In general this is not a serious problem, and when applicable, end results are satisfactory.

In either event, whether slab or pier and beam foundation, the original structure is restored to that grade or to the level existing when the add-on was constructed, though not necessarily to level. To attain a nearly true level would require raising the entire add-on to proper established grade consistent with the original structure. Again this approach would involve considerable expense, and generally the results would not justify or warrant the cost.

Remodeling operations can create a similar problem. Suppose a wall partition is added in an area where the floor is not level. If the partition is built rigid, one of two things will occur: either some wall studs will be longer than others, or one or both plates will be shimmed. Leveling the floor will then push the longer stud area through the ceiling. If this condition is intolerable, the leveling must be curtailed short of the desired results. The plates are wood members placed horizontally at each end of the studs. The ceiling plate forms the top header, and the floor or sole plate the bottom. (See Fig. 6B.13, for example.)

6B.7.3.2 Creation of New Distress. Most structures are not constructed perfectly level at the outset. Any attempt to correct an as-built problem will result in the creation of new distress. For that reason a foundation is normally raised to a level appearance rather than a true level. Even then some compromise is often required to equate the degree of levelness with practical job conditions.

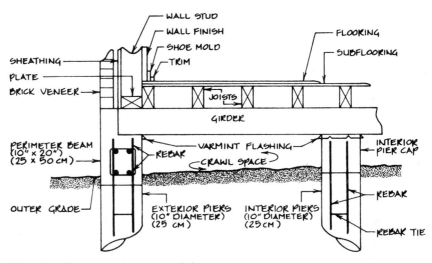

FIGURE 6B.13 Typical pier and beam design.

For example, if a leveling operation intended to close a ¼-in (0.6-cm) crack develops an offsetting ½-in (1.27-cm) crack, nothing is being gained and a decision or compromise must be made. Usually all attempts to raise are stopped.

The same is true in the instance of offsetting doors. Assume two doors are opening in opposite directions, on either 180 or 90° planes. If door A is plumb and door B is not, little or nothing constructive can be gained by attempting to level door B. Efforts to improve door B will most often cause an equivalent, offsetting movement for door A.

Occasionally an upper floor will develop a sag in an area not directly supported by the lower floor or foundation. Since there is no vertical connection between the problem area and the lower floor, foundation leveling will neither affect nor improve the sag. One solution has been to install wood beams across the lower floor ceiling, properly anchored, to raise and support the area. Alternatively, where practical, columns might be installed to provide the support or to provide contact with the foundation. This latter solution is often not acceptable, particularly with residential construction. The columns would be obstructive to movement and generally unsightly.

6B.7.3.3 Incompetent Foundations. The competence of the foundation materials is sometimes below standard and hence prevents or deters proper foundation repairs. In general this condition involves older concrete foundations, either poured with faulty concrete or deteriorated by unusual chemical or weather attack. Occasionally the problem is merely poor construction of concrete floors, such as (1) improper or no reinforcement, (2) exceptionally thin concrete, or (3) badly cracked concrete. In other instances the problem might involve rotted or deficient wood. If the foundation will not withstand the stress required for leveling, little can be accomplished.

In some marginal conditions, long steel plates or beams can be placed under the foundation member to distribute the load and facilitate leveling. This process is relatively expensive and does not replace the deficient materials. Faulty wood can generally be replaced economically if proper access or work space is available. Replacing faulty concrete is generally quite expensive, and since older properties are most often involved, not economically feasible.

6B.7.3.4 Interior Fireplaces. When interior fireplaces settle, expensive problems also develop. Even with slab foundations, the usual repair technique requires the installation of spread footings (typical) and mechanical raising. Access then becomes a serious and costly obstacle. If the movement is not substantial, an acceptable approach is either to cut the surrounding areas loose from the fireplace and level the floors by normal shimming operation (pier and beam foundations) or, in the case of slabs, to stabilize the foundation in order to, hopefully, prevent future progressive movement.

6B.7.3.5 Crossbeams, Interior Piers, or Intensive Loads (Slab). In slab construction the condition that sometimes develops involves a problem area where a stiffened slab (crossbeams) or intensified load abuts a normal slab floor section. An attempt to raise the weighted area by mud jacking alone could crown the weaker slab. In this instance hydraulic leveling attempts (mud jacking) must be aborted to protect the remaining slab. The alternative would be to break out the adjacent slab for access and underpin the problem area mechanically. As a rule this is not advisable from both a cost and a structural viewpoint. The identical problem is encountered where interior piers (belled or otherwise) are tied into the slab. Again, the general decision would be to level to a practical extent and stabilize to prevent future progressive movement.

6B.7.3.6 Lateral Movement. Under certain conditions one foundation member may move laterally with respect to another. Specific examples are a patio slab moving away from the perimeter beam and a perimeter beam moving horizontally from an interior floating slab floor. In either instance the problem member can be raised vertically quite easily. However, it is an altogether different problem to move that structure laterally to restore the original position. As a rule this is beyond reasonable expectation. Prevention is therefore the best solution. The next best is to stabilize the moving member to preclude continued displacement.

While there are other conditions that interfere with proper or desired foundation leveling, those presented in the preceding sections are the most common.

6B.7.3.7 Frame Structures on Wood Foundations: Remedial Approaches. While most foundation repair concerns are directed toward concrete design, there exists the often neglected issue of frame structures on wood foundations. The discussion about to unfold will address several aspects of foundation repair inherent to frame design. It will become apparent that the foundation repair may involve:

1. Leveling of floor systems basically "as is"
2. Leveling the floors and creating minimal clearance between the wood substructure and ground
3. Providing a crawl space with concurrent leveling

6B.7.3.7.1 Frame Foundations with Limited Crawl Space. Wood members near or in contact with the ground are influenced more by moisture than those further removed and protected by air circulation. Contact with moisture encourages the wood framing to warp and rot. Differential movement over a prolonged period of time acerbates the same warp. Further, the limited access to interior floor joists and girders complicates remedial activities. In the end, each condition represents serious deterrents to foundation repair. In other words:

1. The limited or nonexistent crawl space must be addressed. Access is required for wood replacement, access to utilities, or foundation leveling. In some instances a crawl space of 18 in (0.45 m) is specified. This represents yet another problem, as discussed in following paragraphs.
2. Preconditioned warp in the substructure must be understood and compromised. Many times leveling is neither practical nor possible.
3. Relative costs for various alternatives are sometimes prohibitive.

1. *Providing Access* This situation offers several alternatives, none of which is easy or inexpensive. For example, the interior floors can be removed as required to provide access to the floor substructure. As necessary, the skirting can be removed to provide access to the perimeter sill beam. At this point the structure can be leveled or even raised to create a desired crawl space. Once the access is provided, masonry piers can be installed. Figure 6B.5 shows a typical pier design. If the foundation is to be raised to provide the 18-in (0.45-m) crawl space, concern must be given to utility connections in advance in order to circumvent unnecessary damage to water, sewer, or electrical connections. Care must be taken in the use of the word *level*. This term does not apply to foundation repair procedures in general, even more so when old frame substructures exhibit warp.

In other cases steel beams can be used to raise and support the structure independently of the foundations. (Placement of the beams is generally consistent with the existing girder or perimeter support system.) As a rule house movers are better equipped to handle this operation. Once the structure is elevated, two options become available. First, if lateral space permits, the structure can be moved aside to provide access to drill piers and set forms for the installation of new concrete piers, pier caps, and perimeter beam. Second, if space is not available, precast masonry piers and pads can be located beneath the elevated structure to support the lightly loaded interior floors. Conventional spread footings, concrete piers with haunches and pier caps, or continuous concrete perimeter beams can be poured in place to support the exterior. Access problems add considerably to the costs for implementing the concrete perimeter foundation.

2. *Summary*   In summary, in most cases the most favorable option is to remove sections of the floor and excavate soil to provide the desired access and minimal clearance between the wood substructure and the ground. This approach is often the least expensive. The most expensive solution normally involves those instances where excessive excavation is required for the creation of a crawl space. Many problems are resolved once the foundation is raised by whatever means and supported by a masonry or concrete foundation. Access, ventilation, and insulation from the ground are handily resolved. The final goal is to level the floors. This ambition is often met with compromise since warped wood subjected to no appreciable load is not likely to relax or straighten. A high spot left unsupported might, over time, settle back somewhat to project a degree of leveling. However, this event is by no means predictable. Thus the end results are clearly a degree of compromise.

6B.7.3.7.2   Frame Foundations with Adequate Access.   This situation provides a much better, less expensive set of approaches.

1. *All-Frame Foundations*   If all foundation support members are frame, two options exist. (1) The floor system, including the perimeter, can be reshimmed on the existing wood piers. (2) The wood piers can be completely replaced by masonry pier caps on concrete pads (see Fig. 6B.5).

2. *Concrete Perimeter with Frame Interior Piers*   For foundations with concrete perimeter beams and wood interior piers the interior wood piers can be either shimmed "as is" or replaced, as before, with masonry pier caps on pads. Problems with the perimeter beam are classically addressed by conventional underpinning (see Fig. 6B.9).

6B.8   CONCLUSIONS

Based on the foregoing, the following becomes apparent:

1. Foundation repairs tend to protect principally against recurrent settlement.
2. Recurrent upheaval is a concern, especially with slab foundations, unless proper maintenance procedures are instituted concurrent with or prior to proper foundation repairs.
3. Spread footings appear equal or, in some cases perhaps, superior to deep piers as a safeguard against resettlement.[10]
4. Deep piers may, in some instances, be conducive to upheaval.[10]
5. Upheaval accounts for more foundation failures than settlement.
6. Moisture changes which influence the foundation occur within relatively shallow depths.

7. The effects of upheaval distress occur more rapidly and to a potentially greater extent than do those of settlement distress.
8. The cause of foundation problems must be diagnosed and eliminated if recurrent distress is to be avoided.
9. Weather influences foundation behavior, both prior to and after construction.

Note: The term *deep piers* is a colloquialism used to refer to piers constructed to a minimum depth of 10 ft (3 m).[28]

6B.9 INTERESTING EXAMPLES OF FOUNDATION REPAIR TECHNIQUES

Four examples of interesting and unusual foundation repair exercises are described in the following sections. This will be followed by field examples which illustrate foundation repair failures.

6B.9.1 Florida

This problem involved extreme subsidence brought about principally when the water level in an adjacent lake was lowered several feet. The bearing soil consisted of a top layer of silty sand, a midlayer of decayed organics (peat), and a base of coral sand.

The two-story brick veneer dwelling suffered from differential foundation settlement reaching 6 in (15 cm) in magnitude. The objectives were (1) to underpin the perimeter beam to facilitate leveling (as well as provide future stability), (2) to consolidate the peat stratum by pressure grouting to provide a solid base for the repaired structure, and (3) to mud jack the entire foundation area to create a level or more nearly level structure. Figure 6B.14(*a*) and (*b*) depicts the repair process.

First, excavations were made at strategic locations to provide the base for the spread footings (underpinning). Next, grout pipes were driven through the base of the excavations into the peat identified for consolidation. Steel-reinforced concrete was then poured into the excavations and allowed to cure. While the footing pads were curing, the entire slab was drilled for mud jacking and interior deep grouting. Deep grouting was then initiated, starting with the permanent grout pipes set through the footings and continuing to the interior.

At each site, grouting was continued either to refusal, a breakthrough of grout to the surface, or start of an unwanted raise. Interior grout pipes were removed during the grouting process. The permanent exterior pipes were cleared of grout and capped. After the grout had set, the top section of the perimeter pipes was removed and capped below grade. This procedure would permit regrouting at some future date should subsidence recur. Next the perimeter beam was raised to the desired grade and pinned by the installation of the poured concrete pier. The final step was to mud jack the entire foundation for the final grading.

Repairs were completely successful and have remained so since 1980. Prior to the work, the dwelling could be neither inhabited nor sold.

6B.9.2 London, England

The soil in the United Kingdom is not materially different from that found in parts of the United States. The plasticity indexes classically run in the range of 40 to 50 and the problem clay (montmorillonite) in the vicinity of 20 to 40%. The annual rainfall is not particularly high, only about 30 in (76 cm) per year, but the rain is well distributed over the year (150 days), with seldom more than 3 in (7.5 cm) during one month. The Dallas Metropolis, for example, will have about 30 in (76 cm) of rain per year, but perhaps 75 to 80% of the total will fall in less than 15 days. (The latter experiences something like 90% runoff.) Further, London's high temperatures are generally in

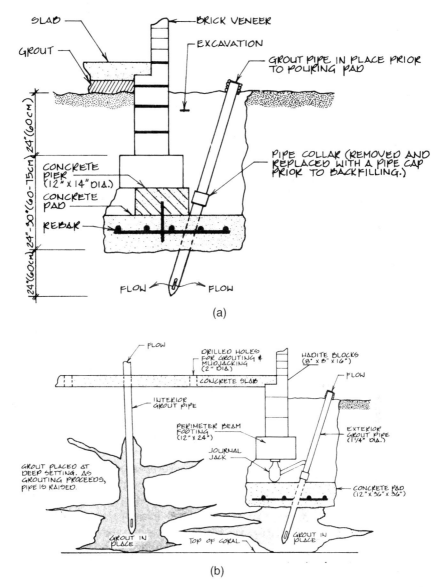

FIGURE 6B.14 Excavation for spread footing. (*a*) Grout pipe in place. (*b*) Typical deep grouting procedure.

the 70°F (21°C) range, whereas the Dallas highs exceed 105°F (40.5°C). London's climate produces a high, fairly consistent moisture content within the soil.

Often the soil moisture tends to persistently approach or exceed the plastic limit, indicating little, if any, normal residual swell potential. It is also interesting to note that the moisture content taken from soil borings in close proximity to trees often shows little, if any, reduction in percent water between the depth of approximately 3.3 ft (1 m) and perhaps 49 ft (15 m). Obviously this indicates that, over the centuries, the soil has attained a level of unique moisture balance.

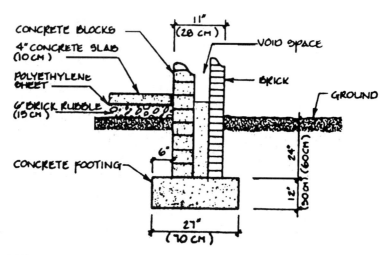

FIGURE 6B.15 Variation—slab foundation, United Kingdom.

Occasionally a prolonged change in climate does come along in London, which tends to disturb the balance temporarily, such as the drought of 1976. During that period the soil moisture within 6.6 ft (2 m) or so was significantly reduced, reportedly causing severe and extensive problems of subsidence. Later, upon return of the normal moisture, the problems became even more severe due to soil swell and upheaval.

The London projects involved restoring the foundation of four banks of garages to the extent that the repairs could be expected to alleviate future distress for a period of at least 20 years. The repair procedure included the pressure injection of Soil Sta into the bearing soil beneath the foundations to a depth of 6 ft (1.8 m). The chemical was injected on the basis of ¼ gal/ft^2 (1 mL/cm^2) of the surface treated. Soil Sta was used principally to preclude the recurrence of the effects of drought conditions such as those of 1976, specifically the upheaval phase. Next, spread footings were installed beneath the load-bearing perimeter to permit mechanical raising and underpinning. Soil Sta was injected through the base of each footing excavation prior to pouring concrete. The chemical volume and purpose were the same as specified earlier. Figure 6B.15 gives a detail of the foundation design. The final stage involved mud jacking to fill voids, raise and level the slab foundation, and fill any voids beneath the perimeter beam which resulted from the underpinning.

The jobs were considered successful. The procedure was substantially less expensive than methods of deep pilings or needle piers considered previously as the conventional approach. In addition, the foregoing methods cause less damage to the landscaping and are quicker and less involved to perform.

6B.9.3 Piers of Driven Steel Pipe

Figure 6B.16 shows steel pipe placed by a hydraulic driver and pinned by bolts through the lift bracket. In Fig. 6B.16(a) the pipe is slanted inward at over 30° (proving nonalignment). In this instance, if one assumes an axial load on each steel pile of 6000 lb (27 kN), the lateral vector F_x would be 3000 lb (13.35 kN). This force plus any lateral force created by the soil is responsible for pile failure in lateral stress. (See Sec. 6B.4.1 and Fig. 6B.17.) The lateral component of soil stress is also discussed in Secs. 3D.1.7.1 and 3D.1.7.2. For other detailed information refer to Prakash and Sharma.[25]

Figure 6B.16(b) shows the lift bracket not in contact with the beam. The settlement of the pipe could be the result of soil dilatency, clay bearing failure, or ultimate failure in whatever material

(a)

(b)

FIGURE 6B.16 Failure in steel minipiles. (*a*) Pipe slanted inward, proving nonalignment. (*b*) Lift bracket out of control with the beam.

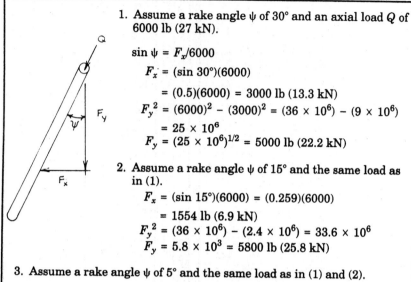

1. Assume a rake angle ψ of 30° and an axial load Q of 6000 lb (27 kN).

$$\sin \psi = F_x/6000$$
$$F_x = (\sin 30°)(6000)$$
$$= (0.5)(6000) = 3000 \text{ lb } (13.3 \text{ kN})$$
$$F_y^2 = (6000)^2 - (3000)^2 = (36 \times 10^6) - (9 \times 10^6)$$
$$= 25 \times 10^6$$
$$F_y = (25 \times 10^6)^{1/2} = 5000 \text{ lb } (22.2 \text{ kN})$$

2. Assume a rake angle ψ of 15° and the same load as in (1).

$$F_x = (\sin 15°)(6000) = (0.259)(6000)$$
$$= 1554 \text{ lb } (6.9 \text{ kN})$$
$$F_y^2 = (36 \times 10^6) - (2.4 \times 10^6) = 33.6 \times 10^6$$
$$F_y = 5.8 \times 10^3 = 5800 \text{ lb } (25.8 \text{ kN})$$

3. Assume a rake angle ψ of 5° and the same load as in (1) and (2).

$$F_x = (\sin 5°)(6000) = (0.087)(6000)$$
$$= 522 \text{ lb } (2.3 \text{ kN})$$

The lateral component of the axial force Q reduces to zero as the rake angle ψ approaches 0°. This is an overly simplified example, but it serves to illustrate a point.

FIGURE 6B.17 Calculation of lateral stress on raked steel minipile.

or object into which the pier tip is embedded. The shiny spots on both pipes represent a prior attempt to adjust the pipes in order to reraise the beam. Although Fig. 6B.16 represents only one example, this type of performance appears to accompany the current driven steel pipe procedures.

Other conditions of pile or pier reactions to load exist. For example, a vertical pile subjected to an inclined load Q at angle ψ is equivalent in behavior to a batter (rake) pile or pier inclined at an angle ψ and subject to a vertical load Q. The simplest and preferred condition occurs when the pile or pier is vertical with the load applied concentrically.

It certainly seems safe to say that this nonperformance represents the rule rather than the exception, at least within certain areas. Another problem, limited to slab foundations, has been the failure of contractors to follow the piling process with competent mud jacking of the slab. Since the slab is not designed to be a bridging member, the voids, whether already existing or created by raising the perimeter, encourage interior settlement of the floors. This must be circumvented by mud jacking. Proper mud jacking could, in fact, eliminate or minimize some of the other inherent deficiencies of the driven pipe process. Mud jacking alone will normally hold a raised slab foundation, provided proper maintenance procedures are instituted and followed (see Sec. 6D).

6B.9.4 Lowering a Foundation

On rare occasions it becomes desirable to both raise and lower sections of a foundation. Figure 6B.18 represents such a situation. As the elevations indicate [Fig. 6B.18(A)] the south two-thirds of the foundation has settled significantly [up to 3 in (7.5 cm)] whereas the garage, north, north-

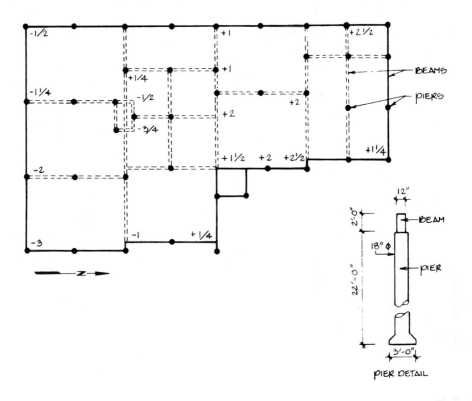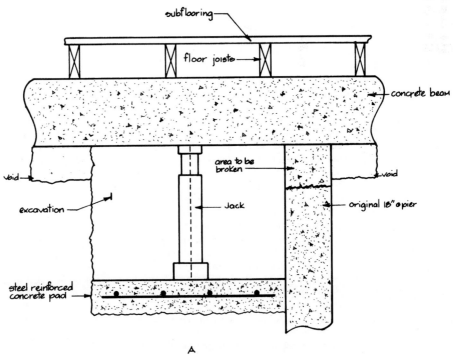

FIGURE 6B.18 (A) Foundation plan and elevation and schematic drawing of mechanical setup.

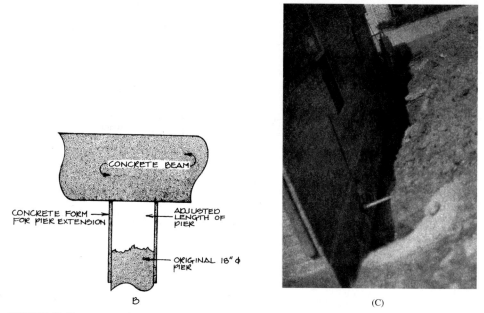

FIGURE 6B.18 (*Continued*). (*B*) Foundation lowering process. (*C*) North wall of garage undermined.

west, and north central beams have heaved [at least 2½ in (6.25 cm)]. The elevations are considered accurate to the extent that relative grade positions are shown. The question lies in defining the true differential movement. The floors were not in contact with the soil and the beams were originally underlain with 12-in (0.3-m) void boxes. The void had been lost in many areas due to either beam settlement, filling in by soil, or some combination of both. The void area was to be restored by excavating beneath the beams as required or raising the beam, or both.

Conventional methods were used to raise the settled areas of the beam [Fig. 6B.18(*B*)]. Next the existing piers were adjusted and extended to resupport the beam, as opposed to adding new drilled piers or spread footings. Spread footing pads were used as a base from which to raise the beam sections after the existing belled piers had been broken free from the beams.

The heaved segments of the beam were lowered by excavating the soil beneath the affected length of the beam, supporting the beam on jacks (resting on spread footing pads as previously noted), removing any above-grade lengths of the pier, pouring appropriate additions to the existing pier at each location, then reloading the adjusted piers by lowering the beam onto the "new" piers, and, finally, removing the jacks.

At the completion of the project, grade elevation suggested that the maximum remaining vertical deflection in the foundation was less than ½ in (1.27 cm), whereas at the beginning the maximum deflection was 5½ in (14 cm).

This particular job presents an interesting question. What is the merit of the lowering procedures? In undermining the approximate 250 lin ft (75 m) of beam it was necessary to remove approximately 300 ft² (27 m²) of floor and subflooring to gain access to the interior beams. This area, plus the perimeter beams, was then excavated [Fig. 6B.18(*C*)]. The combined cost of undermining, floor replacement, and the attendant lowering operation represented about one-half of the entire foundation repair bill. This approach was necessary to give the customer what he wanted, but could some compromise have been more cost-effective? Perhaps not in this particular instance, since the home was vacant and on the market to sell.

The author tends to disfavor lowering operations in general. Raising lower areas to meet the high is almost always the most practical solution, particularly when dealing with normal residential or light commercial construction. Nonetheless, this procedure does give the repair contractor another option.

6B.9.5 Apartment Building

This project involved a more complicated problem. The foundation distress was such that the masonry exterior walls were forced outward to the point where the second-story precast-concrete floor slabs were pulled almost off their bases. The foundation problem was addressed as a typical slab repair. The perimeter was underpinned (spread footings in this case), and the interior floor slab was mud jacked. Concurrently the interior was crisscrossed with dywidag bars at the first-floor ceiling, extending through the drywall to the exterior (Fig. 6B.19). The walls are plumbed as tension is applied by tightening the nuts on the dywidag bars against the steel beams or plates. All beams were later replaced with steel plates. A ceiling furr down was used to conceal the bars on the inside. The exterior bars were cut flush with the nuts, plastered over, and painted to match the exterior walls.

6B.9.6 Basements and Foundation Walls

Failure in basement walls occurs in areas with or without expansive soil and is generally limited to colder-climate regions. Where concerns for frost lines exist, the propensity toward construction with basements is enhanced. As far as expansive soils are concerned (see Fig. 6C.5), Colorado is the principal area where basements are prevalent.

Basement wall construction can vary from wood to steel-reinforced concrete. The former is relatively limited. Therefore the following will concentrate on masonry and concrete construction.

Basement construction, and hence repair, is not common in areas where the author has developed 30 years of experience. The following examples constitute a compilation of repair procedures utilized by the author or, as the case may be, by other consulting engineers. In each instance only a typical design is depicted. In practice, specific project conditions must be taken into account. Such considerations would include condition, design, and strength of existing foundation and floor members, structural loads, soil characteristics, drain and hydrostatic conditions of subsurface structural members and soils, waterproofing and shutoff, goal to accomplish, and costs. Obviously all these issues cannot be factored into a simple sketch.

It might be interesting to note that above-grade foundation walls (with built-up floor systems, that is, dock high) suffer failures similar to those described in this section, with the exception that the foundation failure (rotation) is toward the exterior. Repair or restoration procedures are often much the same as those described in the following examples. As a rule, the retaining or plumbing procedures are installed on the external side of the foundation wall where property lines permit.

6B.9.6.1 Basement or Foundation Wall Repair. The first repair example involves the construction of a new member inside the original wall, sometimes referred to as a sister wall. The structural load is ultimately transferred to this addition. No effort is made to plumb or reinforce the existing basement foundation wall [Fig. 6B.20(*a*)]. The sequence of construction is essentially as follows:

1. Shore existing floor joists to lessen or remove the structural load from the defective wall.
2. Sandblast or scarify surface of existing wall to remove laitance (loose concrete material) and provide an improved bonding surface.
3. Break out floor slab and protruding strip footing to prepare for the installation of supporting concrete piers and haunches. [Typically, for a wall height less than about 10 ft (3.0 m) with construction loads less than about 1500 lb/in ft (2268 kg/m), the piers would be 12 in (30 cm) in diameter (spaced on approximately 8-ft (2.4-m) centers.]

6.80 FOUNDATION FAILURES AND REPAIR

(a)

(b)

FIGURE 6B.19 Apartment building repair. (*a*) Dywidag bars crisscross the interior space immediately below the first floor ceiling. (*b*) Steel beams distribute tension and allow walls to to be pulled inward by tightening a nut on the Dywidag, threaded bar.

4. Place steel lintels across keyways to hold fascia brick. (An alternative would be to remove brick.)
5. Break out concrete or remove sufficient concrete block to create keyways of approximately 1×2 ft (0.3×0.6 m), on 6-ft (1.8-m) centers.
6. Drill dowel holes into existing foundation or basement wall. Often the rebars would be no. 7s ($7/8$ in or 2.2 cm) spaced to create a pattern on the order of 12 to 18 ft^2 (1.0 to 1.5 m^2).

(c)

FIGURE 6B.19 (*Continued*). (*c*) Steel beams are replaced with smaller plates once the wall has been plumbed. The excess bar is cut off flush with the nut and the assembly stuccoed over and painted.

7. Place steel and pour piers and haunches.
8. Place steel and pour sister wall. Often the concrete is placed through the keyways by the use of concrete pumps.

Note: This and the foregoing examples are meant to be general. Specific loads and job conditions will dictate the size of reinforcement and the spacing of the support members. Also, virtually every example will benefit from some type of waterproofing.

Figure 6B.20(*B*) presents yet another example. This approach permits some plumbing of the defective wall and is more effective with masonry wall construction. The wall materials should suffer little or no deterioration. The general procedure for this installation is as follows:

1. Excavate fill adjacent to wall exterior.
2. Drill holes through basement or foundation wall to facilitate take-up bolts. Generally, the bolts would be 1 to 1½ in (2.5 to 3.8 cm) in diameter, located at least three to a tier and on 3- to 6-ft (0.9- to 1.8-m) centers.
3. Place channel iron on each side of the wall. Install bolts and tighten.
4. Waterproof wall as necessary and backfill with gravel hydrostatic drain system.
5. As with virtually all attempts to plumb a failed basement or foundation wall, it is often desirable to supplement the system with additional force to push the wall into plumb [Fig. 6B.20(*c*)].

Another technique, more suited to actually plumb a concrete basement or foundation wall, is depicted in Fig. 6B.20(*c*). This technique uses an opposing wall to secure the jacking system that is utilized to plumb or align the basement or foundation wall. Depending on such factors as existing wall design, load conditions, and degree of rotation, a second battery of jacks may be required. Typically each battery of jacks would be placed 4 to 8 ft (1.2 to 2.4 m) apart. A typical sequence for the installation of this system would be as follows:

1. Drill holes for dywidag or similar bars, 1½ to 2 in (3.8 to 5.0 cm) in diameter, through the basement or foundation wall. Locate the bar in the drilled pier shaft. Holes are typically three to each pier, on 4- to 8-ft (1.2- to 2.4-m) centers.

2. Drill and pour the external concrete piers.
3. Excavate behind the existing wall. Alternately the backfill could be excavated prior to placement of the pier. This would necessitate forming the piers, but it would allow the strip footing to be broken out at pier locations. Removal of the strip footing would allow contact of the pier to existing wall with straight alignment.
4. Place channel iron and commence the jacking operation.
5. Waterproof wall as required and backfill with suitable gravel hydrostatic drain.

The knee brace is yet another approach to retrofit basement or foundation walls. This method is fairly simple and adequate to sustain a wall rotation. It is not intended to accomplish any degree of vertical alignment to the existing wall. Fig. 6B.20(d) depicts the design of a typical knee brace. Depending on specific conditions, a second pier may be required immediately adjacent to the existing wall. However, for lightly loaded conditions a schedule of dowels will prevent any movement between the wall and the brace. Considering normal basement heights [less than 10 ft (3 m)] and lightly loaded conditions, the placement of the braces might be on 6- to 10-ft (1.8- to 3-m) centers.

The knee brace is also frequently used to control outward rotation of foundation walls in deck-high construction. The primary limitation would be instances where the defective wall is situated on a zero lot line. When the foundation walls are 5 ft (1.5 m) high or less and the wall structure permits, the placement of the supports might be as far apart as 10 to 20 ft (3 to 6 m).

For certain minor problems, particularly those associated with light load conditions, an adequate restoration approach could be to utilize the jacking system illustrated in Fig. 6B.20(a) to raise and level the floor joist system. Permanent supports would consist of supplemental beams (wood as a rule) supported atop lally columns. In general the columns [often 4- to 6-in (10- to 15-cm) steel pipe] would be fitted with steel plates at the top and bottom. Frame basement walls are particularly susceptible to this option. Job conditions control the number, design, and placement of these supports.

6B.9.6.2 Hadite Block Walls. Other relatively minor problems sometimes manifest themselves with hollow concrete block (hadite) walls. If the intent is to strengthen the wall or shut off minor water seepage, the problem might be addressed by filling the blocks with a concrete mix. Normally at least two rows of injection holes are drilled through the inside of a concrete block wall. Typically the lowest row would be about 4 ft (1.2 m) off the floor and a second row near the top of the wall. Adjacent (lateral) holes are used to ensure complete penetration of all voids. Initially the holes might be approximately on 4 ft (1.2 m) centers. If any question arises concerning filling all voids, intermediate holes can be drilled. The lowest row of holes is normally injected first and may or may not be allowed to attain initial set before the top row is pumped. Stage pumping reduces the hydrostatic load on the lower courses of block.

There are many other options to correct problems with basement or foundation walls. However, the foregoing should provide a sense of direction. Other ideas can be found in Secs. 4B and 6B.4.4.

6B.9.6.3 Basement Water Infiltration. With persistent water infiltration the solution shown in Fig. 6B.20(e) might be an acceptable solution. Here the water that penetrates the wall is collected in the trough at the perimeter of the floor slab. Water is then transported to an adequate sump or drain. A slight variation with this is to (1) construct a sump in the floor slab and (2) build a screed floor over the concrete slab. The sump accumulates the water and pumps it to a drain. The wood screed floor remains dry. The screed members should be rot resistant, such as redwood, cypress, cedar, or chemically treated pine. Neither of these options addresses structural concerns per se. The expressed intent is to control unwanted water. A much more in-depth study of water shutoff can be found in Secs. 5B and 6B.6.

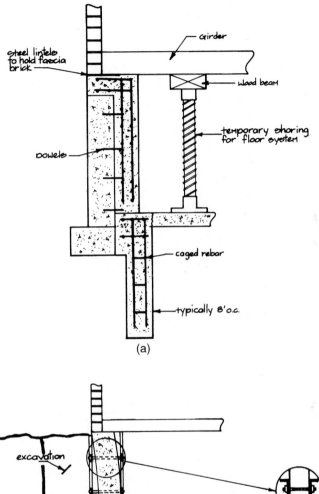

FIGURE 6B.20 Basement and foundation wall repair.

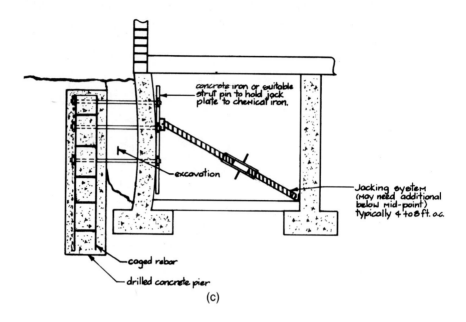

(c)

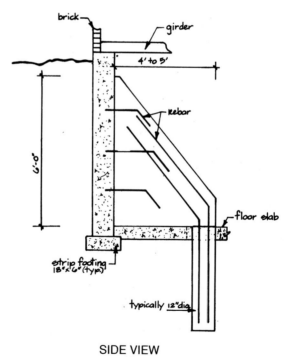

SIDE VIEW
(d)

FIGURE 6B.20 (*Continued*)

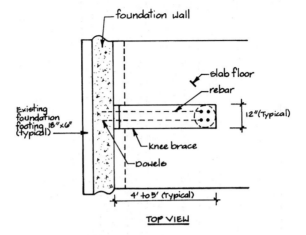

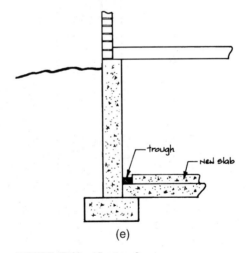

FIGURE 6B.20 *(Continued)*

6B.9.7 Chemical Stabilization

Chemical stabilization of the expansive constituents within a soil represents yet another method for dealing with certain foundation problems. This approach is generally one of preventive rather than remedial nature. However, on occasion chemical stabilization is used in conjunction with normal repair procedures. One such example was discussed in Sec. 6B.9.2. For complete details on the subject of chemical stabilization refer to Part 5.

6B.9.8 URETEK—An Alternative to Mud Jacking?

In 1976 Uretek introduced a method and product designed to be an alternate solution to slab settlement problems. This procedure is specifically intended to compete with conventional mud jacking. The Uretek method injects a polyurethane compound beneath the slab through strategically placed ½-in (1.27-cm)-diameter drilled holes. Once placed, the polyurethane expands and hardens. The expansion represents the force available to raise the slab. Supposedly this expansion exerts a lifting force of 8000 lb/ft^2 (380 kPa). (Conventional mud jacking relies on a hydraulic force to raise the slab. Refer to Sec. 6B.3.1.) The Uretek 486 is reported to harden to 90% of its full strength in about 15 min. The company claims that the expansion will develop a some 30-fold volume increase. The ultimate compressive strength of the material is reported to be in the range of 50 psi (345 kPa), as compared to about 400 to 500 psi (2760 to 3450 kPa) for a normal soil-cement grout containing four sacks of cement per cubic yard of dry mix. (1 psi is equivalent to 144 psf.)

The disadvantages which seem inherent to the Uretek method would include such concerns as cost, toxic nature of polyurethane, shortage of data and history dealing with occupied structures, and, despite the company's claim, the apparent problems in such areas as the control, accuracy, and efficiency of the basic lifting process.[46] Also, based on company literature, it would seem that Uretek has a much more restricted tolerance to ambient temperature ranges than would the typical soil-cement grout. Soil-cement-water grouts are difficult to place at subfreezing temperatures, but once placed they have little reaction to normal temperature ranges.

6B.9.9 Conclusion

The foregoing examples illustrate a few of the many tools available to correct distressed foundations (structures). One famous problem has thus far eluded our best efforts—the leaning Tower of Pisa. Many approaches have been considered, but as of this writing no corrective action has proved successful.

6B.10 FAILURES IN FOUNDATION REPAIR

The instances of foundation repair failure can generally be grouped in one or more of the following:

1. The cause of the original foundation distress was not identified and remedied.
2. Repair techniques were utilized that were inadequate to address the specific distress [see Fig. 6B.21(a) to (c)].
3. Some condition was subsequently introduced which the repair methods were not intended to address. An example would be where shallow surface soils are subjected to excess water, resulting in soil swell and foundation heave (see Secs. 6A.3, 6B.1, and 6B.2.1.4).
4. The full scope of the foundation problem was not addressed. This could represent a situation where a sanitary fill has not completely reacted, where fill has not completely consolidated, or perhaps where a slip plane in an unstable soil was not adequately addressed. Often cost concerns suggest a compromise rather than a total solution. This choice can permit an ongoing problem to exist, although one that is a calculated risk.
5. Occasionally preexisting conditions not only impede the leveling operation but might also result in recurrent problems (see Sec. 6B.7).

The property owner in consultation with the foundation contractor can basically resolve issues 3, 4, and 5. A competent repair contractor can offer protection against issues 1 and 2. However, the consumer must heed the advice of the contractor (issue 1) and be selective in choosing a repair method. Figure 6B.21 documents recurrent foundation failures brought about by ineffective repair techniques.

6B.10.1 Substandard Pier Construction

Figure 6B.21(*a*) illustrates a failure due to an inadequate pier. It depicts a second repair of a residential foundation. The first attempt utilized solid hadite (lightweight concrete) blocks stacked on a small concrete haunch or pad. Wood and steel were used to shim on top of the blocks. The method failed as illustrated. Note the space created above the makeshift pier as the beam was raised and leveled by the jack. This space represents settlement of the hadite pier due to lateral failure or deterioration of the pier materials. Below-grade wood rots and steel corrodes (especially at shallow depths). Further, the stacked hadite blocks do not offer stability. The problem was corrected by replacing the precast blocks and steel shim with a poured concrete pier in full contact with the perimeter beam.

6B.10.2 Failed Minipile

Figure 6B.21(*b*) shows the underpinning failure of a driven steel pile or pipe. The pile failure in this figure shows multiple problems. First, the lift bracket has pulled away from the beam and no longer provides support. Second, the steel pipe is raked at an angle in excess of 20°. These obser-

(*a*)

FIGURE 6B.21 Foundation repair failures. (*a*) Substandard pier cap construction.

(b)

(c)

FIGURE 6B.21 (*Continued*). (*b*) Failure of a driven steel minipile. (*c*) Interior slab failure due to improper mud jacking.

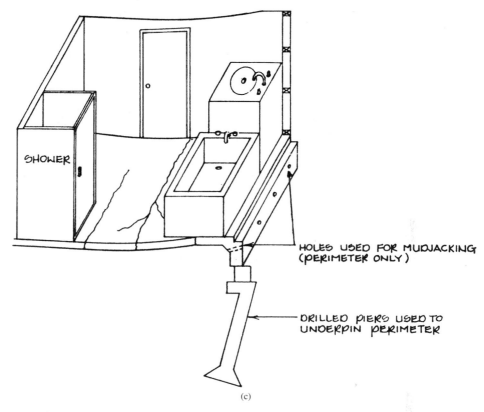

(c)

FIGURE 6B.21 (*Continued*). (*c*) Interior slab failure due to improper mud jacking (artist's rendering).

vations are not uncommon, particularly where pipe is driven into the cohesive expansive soils. The remedial action was to replace the driven steel pipes by concrete spread footings. (Minipiles are generally not recommended for expansive soils.)[2,11,12]

6B.10.3 Failure to Mud Jack a Slab Foundation

Figure 6B.21(*c*) depicts interior slab failure caused by incomplete mud jacking after underpinning. The problem was corrected by properly mud jacking the depressed area of the slab foundation. Any time a slab foundation (except a structural slab) is underpinned, mud jacking is necessary.[2,11–14] Refer also to Sec. 6B.3 for details about proper technique.

6B.10.4 Improper Placement of Haunch beneath Beam

Figure 6B.21(*d*) depicts a foundation repair failure yet to happen. The artist's rendition of the situation illustrates, in some detail, the fallacies of this approach. First, the greatest load (on the edge of the haunch) occurs during jacking. This could cause the haunch to crack and fail. Second, the eccentric load on the perimeter beam, as well as that transferred to the pier cap, could cause the beam either to rotate inward or to fail in shear (compression), or both. The 8-in (20-cm)-diameter pier is subjected to a bending movement, assuming that it is tied into the haunch. The precast con-

6.90 FOUNDATION FAILURES AND REPAIR

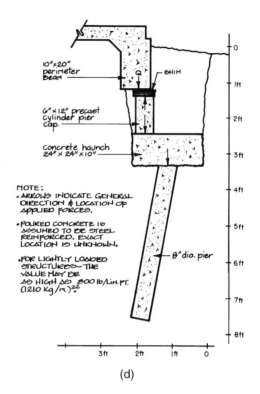

(d)

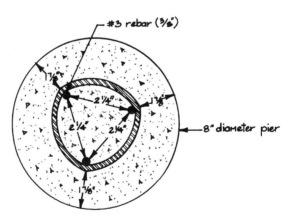

(e)

FIGURE 6B.21 (*d*) Artist's rendering of part (*c*). (*e*) Substandard pier diameter.

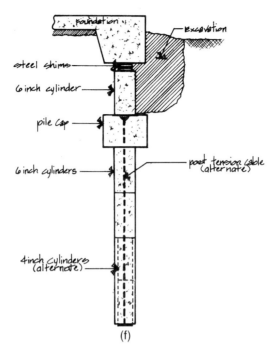

FIGURE 6B.21 *(Continued).* (*f*) Hydraulically driven concrete pilings.

crete cylinder used as a pier cap could fail in shear. Either of these occurrences could result in a failed foundation repair attempt. The thickness of the haunch, the depth of the pier, as well as the existence and placement of rebars, if any, are unknown values.

6B.10.5 Sulfate Reactions with Certain Soils

In 1990 the detrimental effects of sulfates on clay soils became an issue.[2] Grubbs[47] and Perrin[48] reported significant, uncontrolled heave in pavements upon the introduction of lime into expansive soils containing sulfates.[47,48] It was reported that the chemical reaction between lime and sulfate produced ettringite, a calcium aluminum hydroxide sulfate hydrate. Petry and Little[49] discussed a similar problem with similar results. This paper dealt with the heave of lime-containing soils following the introduction of sulfates ($SO_4^=$).

Wrench and Geldenhuis[50] present yet another aspect of the same general problem. They describe the effects of sulfuric acid attack on soils as being "considerable heave (up to 1 m) of lightly loaded structures or settlement of moderately loaded structures." One general conclusion from this work was that, in many cases, remedial measures were not available which could satisfactorily address the foundation problems. Thus it was often decided to simply relocate the structure.

Three other problems seem possible. (1) Specific bacteria (*Desulfovibrio desulfricus*) can reduce sulfate to produce hydrogen sulfide (H_2S), which is a poisonous gas. (2) Sulfuric acid (H_2SO_4) will chemically attack steel and concrete. This is true whether the sulfuric acid is introduced as H_2SO_4 or the result of a chemical reaction of H_2S to produce the acid. (3) The Environmental Protection Agency might take steps to require cleanup of soil contaminated by H_2SO_4 or condemn any structure situated on said soil. The practice of introducing sulfates as

6B.10.6 Ultraslim Concrete Piers

Other researchers have established the optimum concrete pier diameter to be 12 in (0.3 m) for applications involving lightly loaded foundation systems on expansive soils.[21] The practice of using small-diameter concrete piers [less than 10 in (0.25 m)] for underpinning is questionable for the reasons of possible bearing-strength deficiency, difficulty in terms of steel and concrete placement, possible failure under lateral stress,[11] and, perhaps to some extent, nonuniformity of pier diameter to beam width (pier diameter does not span full width of beam).

Figure 6B.21(e) illustrates a typical 8-in (0.2-m) pier. With the steel centralized as shown, the various clearance distances are indicated. It is generally recognized that the minimum safe clearance for the passage of solid objects through a restriction is three times the largest diameter of those objects. Hence the minimum safe clearance for the passage of $1\frac{1}{2}$ in (3.8 cm) minus aggregate would be 4.5 in (11.4 cm). For 1 in (2.5 cm) the safe clearance would be 3.0 in (7.5 cm). Further, the concrete cover between rebar and soil is frequently suggested to be at least 3 in (7.5 cm).

Referring to Fig. 6B.21(e), the options for placing concrete in the 8-in (0.2-m) pier would be to move the steel to the side of the shaft wall, reduce the amount and placement of steel, or use a pea gravel mix concrete. Neither of these choices is structurally viable. Couple this with the other concerns, and it becomes apparent that the use of small-diameter piers for underpinning is not a defensible choice.

6B.10.7 Hydraulically Driven Concrete Cylinders

In the late 1980s or early 1990s the process of driving steel minipiles hydraulically was given a novel twist. Rather than using short sections of small-diameter steel pipe, the process uses concrete cylinders approximately 6 in (15 cm) in diameter by 12 in (30 cm) in length [Fig. 6B.21(f)]. The claims for this method include such statements as "the weight of the building locks the sections of the pilings (pier) together with friction" and "the soil through which the pilings have been driven, and which was compacted by the driving process, firmly locks the pilings (piers) in vertical position." In the absence of strong proof, either allegation would seem most suspect.

From a technical view it would appear that this technique would suffer the same deficiencies as those described for the minipiles (Sec. 6B.4). These would include concerns involving alignment, absence of any margin of safety, failure in lateral stress, use of steel (shims) below grade, and so on.

An attempt to improve the alignment problem has been introduced. This involves threading the cylinders on a posttension cable. Once the desired numbers of cylinders are placed, the cable is tensioned, and it is hoped that it will both assure alignment and develop some tension resistance within the column. The benefits actually realized from this approach are most questionable. The cable might improve the piling's (pier) resistance to tensile forces but would offer little in resistance to lateral stress. Further the cable might tend to force the cylinders into closer and tighter contact but not likely improve overall alignment.

Another variation has been to taper the piling by initially placing 4-in (10-cm)-diameter cylinders and, at some point, switching to 6-in (15-cm) diameter. This technique would seem to make even less sense than utilizing the 6-in (15-cm)-diameter cylinders throughout, due, in part, to reduced strength and support factors.

6B.11 REFERENCES

1. E. Jones and W. Holtz, *Civil Eng.*, Aug. 1973.
2. R. W. Brown, *Foundation Behavior and Repair: Residential and Light Commercial Foundations*, McGraw-Hill, New York, 1992.

3. "Building Near Trees," Practice Note 3, House Building Council, London, England, 1985.
4. R. Tucker and A. Poor, "Field Study of Moisture Effects on Slab Movement," *J. Geotech. Eng., ASCE,* vol. 104, Apr. 1978.
5. J. E. Holland and C. E. Lawrence, "Seasonal Heave of Australian Clay Soils," presented at the 4th Int. Conf. on Expansive Soils, ASCE, June 1980.
6. J. E. Holland and C. Lawrence, "The Behavior and Design of Housing Slabs on Expansive Soils," presented at the 4th Int. Conf. on Expansive Soils, ASCE, June 1980.
7. V. A. Sowa, "Influence of Construction Conditions on Heave of Slab-on-Grade Floors Constructed on Swelling Clays," *Theory and Practice in Foundation Engineering,* Proc. 38th Canadian Geotech. Conf., Sept. 1985.
8. F. Chen, "Practical Approach on Heave Prediction," presented at the 7th Int. Conf. on Expansive Soils, ASCE, Dallas, Tex., 1992.
9. S. J. Greenfield and C. K. Shen, *Foundations in Problem Soils,* Prentice-Hall, Englewood Cliffs, N.J., 1992.
10. R. W. Brown and C. H. Smith, Jr., "The Effects of Soil Moisture on the Behavior of Residential Foundations in Active Soil," *Tex. Contractor,* May 1, 1980.
11. Digest 313, "Mini-Piling for Low Rise Building," Building Research Establishment, Garston, Watford, England, Sept. 1986.
12. T. Petry et al., Foundation Seminar, University of Texas at Arlington, May 20, 1983, and Dec. 13, 1985.
13. D. Kaminetzky, *Design and Construction Failure Lessons from Forensic Investigations,* McGraw-Hill, New York, 1991.
14. G. P. Tschebotarioff, *Foundations, Retaining and Earth Structures,* 2d ed., McGraw-Hill, New York, 1973.
15. R. W. Brown, "A Series on Stabilization of Soils by Pressure Grouting," *Tex. Contractor,* pt. 1A, Jan. 19, 1965; pt. 1B, Feb. 2, 1965; pt. 2B, Mar. 16, 1965.
16. R. W. Brown, "Concrete Foundation Failures," *Concrete Construction,* Mar. 1968.
17. Winterkorn and Fang, *Foundation Engineering Handbook,* Van Nostrand Reinhold, 1975.
18. Federal Housing Administration, "Criteria for Selection and Design of Residential Slab on Ground," Report 33, National Academy of Science, 1968.
19. K. Barchas, "Repair and Retrofit Using External Posttensioning," *Concrete Repair Dig.,* Feb. 1991.
20. G. F. Langenback, "Apparatus for a Method of Shoring a Foundation," Patent 3,902,326, Sept. 2, 1975.
21. D. E. Jones and W. G. Halz, "Expansive Soils—The Hidden Disaster," Civil Engineer ASCE, vol. 43, Aug. 1973.
22. M. O'Neill and D. Mofor, "Results of Concentric and Eccentric Axial Loading Tests on RB Piles (Mini-Pile)," University of Houston, Houston, Tex., 1988.
23. W. J. Neely, "Bearing Capacity of Auger-Cast Pilings in Sand," *J. Geotech. Eng., ASCE,* vol. 117, Feb. 1991.
24. D. A. Bruce and P. J. Nicholson, "Minipiles Mature in America," *Civil Eng.,* Dec. 1988.
25. S. Prakash and H. D. Sharma, *Pile Foundations in Engineering Practice,* Wiley, New York, 1990.
26. Meyerhof, "The Ultimate Bearing Capacity of Foundations," *Geotechnique,* vol. 2, 1951.
27. A. Fourie, "Laboratory Evaluation of Lateral Swelling Pressures," *J. Geotech. Eng., ASCE,* vol. 115, Oct. 1989.
28. R. W. Brown, "A Field Evaluation of Current Foundation Design vs. Failure," *Tex. Contractor,* July 6, 1976.
29. A. Ghaly and A. Hanna, "Uplift Behavior of Screw Anchors, I and II," *J. Geotech. Eng., ASCE,* vol. 117, May 1991.
30. T. Petry, Unpublished research report, University of Texas at Arlington, 1989.
31. W. J. Neely, "Bearing Capacity of Extended-Base Piles in Sand," *J. Geotech. Eng., ASCE,* vol. 116, Jan. 1990.
32. "Piles and Dynamic Pile Testing," *Geotech. News,* Dec. 1990.
33. B. H. Fellenius, "Introduction to the Dynamics of Pile Testing," *Geotech. News,* Dec. 1990.
34. N. Ismad, A. Khalidi, and M. A. Mollay, "Upgrading Footings in Sand with Bored Piles," *J. Geotech. Eng., ASCE,* vol. 115, Dec. 1989.
35. R. W. Brown, *Residential Foundations: Design, Behavior and Repair,* Van Nostrand Reinhold, Princeton, N.J., 1979 and 1984.
36. T. L. John and G. S. Lawrence, "Improvement of Disturbed Foundation Bearing Soils Adjacent to a Sheet Pile Supported Excavation by Grouting Methods," *Texas Civil Eng.,* Jan. 1986.
37. S. Zebovitz, R. J. Krizek, and D. K. Amatzidis, "Injection of Fine Sands with Very Fine Cements," *J. Geotech. Eng., ASCE,* vol. 115, Dec. 1989.
38. U.S. Corps of Engineers, "Flow Cone Method for Grout Consistency," CRD-C611.
39. A. Nandts, "Grouting Trends," *Civil Eng.,* Oct. 1989.
40. L. Lee, "Grouting Slip Liners," *Civil Eng.,* Sept. 1990.

41. D. C. Sego and K. W. Biggar, "Grouts for Use in Permafrost Regions," *Geotech. News,* Sept. 1990.
42. G. K. Burke and D. A. Metee, "Fixing Foundations," *Civil Eng.,* Mar. 1991.
43. K. Weaver et al., "Grouting against Hazwaste," *Civil Eng.,* May 1992.
44. R. H. Borden, R. D. Holz, and I. Juran, *Grouting, Soil Improvement and Geosynthetics,* vols. 1 and 2, American Society of Civil Engineers, New York, 1992.
45. P. Pettit and C. E. Wooden, "Jet Grouting: The Pace Quickens," *Civil Eng.,* 1988.
46. J. D. Nelson and D. J. Miller, *Expansive Soils,* Wiley Interscience, New York, 1992.
47. B. Grubbs, "Lime Induces Heave of Eagle Ford Clays in Southeast Dallas County," presented at the ASCE Fall Meeting, El Paso, Tex., 1990.
48. L. L. Perrin, "Expansive of Lime-Treated Clays Containing Sulfates," presented at the 7th Int. Conf. on Expansive Soils, ASCE, Dallas, Tex., 1992.
49. T. M. Petry and D. N. Little, "Update of Sulfate Induced Heave in Lime and Portland Cement Treated Clays," presented at the Annual Meeting of the Transportation Research Board, Aug. 1991.
50. B. P. Wrench and J. J. Geldenhuis, "Heave and Settlements of Soils Due to Acid Attack," presented at the 7th Int. Conf. on Expansive Soils, ASCE, Dallas, Tex., 1992.
51. M. Carter and S. P. Bentley, *Correlations of Soil Properties,* Pentech Press, London, 1991.
52. D. P. Conduto, *Foundation Design,* Prentice-Hall, Englewood Cliffs, N. J., 1994.
53. T. J. Freeman et al., "Has Your House Got Cracks?," Building Research Establishment and Institute of Civil Engineers, London, 1994.

SECTION 6C
PREVENTIVE MAINTENANCE

Robert Wade Brown

6C.1 INTRODUCTION   6.95
6C.2 WATERING   6.95
6C.3 DRAINAGE   6.97
   6C.3.1 French Drains and Subsurface Water   6.98
6C.3.2 Water or Capillary Barriers   6.98
6C.4 VEGETATION   6.101
6C.5 REFERENCES   6.102

6C.1 INTRODUCTION

Preventing a problem is always more desirable than curing one. Certain maintenance procedures can help prevent or arrest foundation problems if initiated at the proper time and carried out diligently. The following are specific suggestions on how to encourage foundation stability.

6C.2 WATERING

In dry periods, summer or winter, water the soil adjacent to the foundation to help maintain a constant moisture. Proper watering is the key. Also, be sure drainage is away from the foundation prior to watering.

When cracks appear between soil and foundation, the soil moisture is low and watering is in order. Conversely, water should not be allowed to stand in pools against the foundation. Watering should be uniform and preferably cover long areas at each setting, ideally 50 to 100 linear feet (15 to 30 m). Too little moisture causes the soil to shrink and the foundation to settle. Too much water—an excessive moisture differential—causes the soil to swell and heave the foundation. Never attempt to water the foundation with a root feeder or by placing a running garden hose adjacent to the beam. Either of these represents uncontrolled watering. Sprinkler systems often create a sense of false security because the shrub heads, normally in close proximity to the perimeter beam, are set to spray away from the structure. The design can be altered to put water at the perimeter and thereby serve the purpose.

The use of a soaker hose is normally the best solution. From previous studies of water infiltration and runoff it became evident that watering should be close to the foundation, within 6 to 18 in (15 to 45 cm), and timed to prevent excessive watering. Proper grading around the foundation will also serve to prevent unwanted accumulation of standing water.

More sophisticated watering systems which utilize a subsurface weep hose with electrically activated control valves and automatic moisture monitoring and control devices are now available (Figs. 6C.1 and 6C.2). The multiple control devices are situated to afford adequate soil moisture control automatically and evenly around the foundation perimeter. Reportedly the control can be set to limit moisture variations to $\pm 1\%$. Within this tolerance, little if any differential foundation movement would be expected in even the most volatile or expansive clay soils. The key to this system is the control of water output and placement.

Avoid watering procedures that make outlandish claims. They can often cause more problems than they cure. One example is the so-called water or hydro pier. The claim is that a weep hose placed vertically into the soil beneath the perimeter beam on convenient centers [often 6 to 8 ft

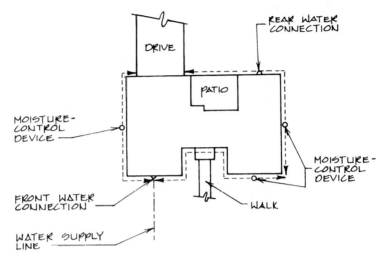

FIGURE 6C.1 Schematic design of watering system.

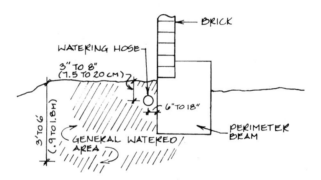

FIGURE 6C.2 Location for watering hose.

(1.8 to 2.4 m)] will develop a "pier" structure by virtue of expansion of the wetted clay soil. Supposedly this "pier" will then support and stabilize the foundation. Some contractors even claim the system will raise and level foundations. This system usually lacks water control in relation to the in situ moisture content. Such an oversight introduces several problems when dealing with expansive soils.

First, moisture distribution within the soil is seldom, if ever, uniform. Second, a soil moisture content approaching the liquid level can cause the heaving soil to actually *lose* strength. (Too much water presents a more serious consequence than does too little.) Third, although consistent moisture content in the expansive soil will normally prevent differential deflection of the foundation, the subject process or processes will not meet the requirements. Fourth, water replenishment into an expansive soil will seldom, if ever, singularly accomplish any acceptable degree of leveling. Minor settlement, limited in scope, could be the exception.

Where large plants or trees are located near the foundation, it could be advisable to water these as well, at least in areas with C_w factors below about 25 (see Fig. 6C.3 and Tschebotarioff.[1]). As far as foundation stability is concerned, the trees or plants most likely to require additional water

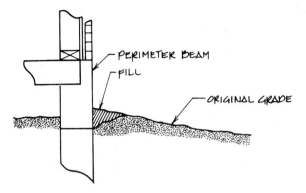

FIGURE 6C.3 Correct drainage.

would be those that (1) are immature, (2) develop root systems which tend to remove water from shallow soils, and (3) are situated within a few feet from the foundation. (Refer to Part 1 and Sec. 6C.4 for additional information.)

6C.3 DRAINAGE

For the reasons noted, it is important that the ground surface water drain away from the foundation. Where grade improvement is required, the fill should be a low-clay or sandy loam soil (Fig. 6C.4). The slope of the fill need not be exaggerated, but merely sufficient to cause the water to

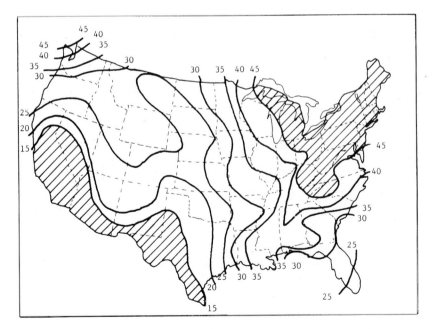

FIGURE 6C.4 Climatic ratings for continental United States.

flow outward from the structure. Figure 6C.5 presents minimum drainage swales as proposed by the Council of American Building Officials.[2] As a matter of clarification, a 1% slope is equivalent to a drop of 1 in (2.5 cm) over approximately 8 ft (2.4 m).

The surface of the fill must be below the air vent for pier and beam foundations and below the brick ledge for slabs. Surface water, whether from rainfall or watering, should never be allowed to collect and stand in areas adjacent to the foundation wall. Consistent with this, guttering and proper discharge of downspouts are quite important. Flower-bed curbing and planter boxes should drain freely and preclude trapped water at the perimeter. In essence, any procedure that controls and removes excess surface water is beneficial to foundation stability. (See also Part 1.)

Domestic plumbing leaks (supply and sewer) can be another source for unwanted water.[3,4] Refer also to Secs. 1B, 6A, and 6B. Extra care should be taken to prevent and/or correct this problem. Water accumulation beneath a slab foundation is responsible for perhaps 70% of all slab repairs.

Prior to construction the engineers are responsible for drainage design of the entire development. While this topic is beyond the scope of this section, Fig. 6C.6 presents some of the more common approaches. Part 1 covers this issue in detail.

6C.3.1 French Drains and Subsurface Water

French drains are required on occasion, when subsurface aquifers permit the migration of water beneath the foundation. When the foundation is supported by a volatile (high-clay) soil, this intrusion of unwanted water must be stopped. The installation of a French drain to intercept and divert the water is a useful approach. As shown in Fig. 6C.6, the drain consists of a suitable ditch cut to some depth below the level of the intruding water. The lowermost part of the ditch is filled with gravel surrounding a perforated pipe. The top of the gravel is continued to at least above the water level and often to the surface.

Figure 6C.7 illustrates the location of the French drain with respect to the foundation. Provisions are incorporated to remove the water from the drain either by a gravity pipe drain or by a suitable pump system. Simply stated, the French drain creates a more permeable route for flow and carries the water to a safe disposal point. If the slope of the terrain is not sufficient to afford gravity drainage, the use of a catch basin or sump pump system is required. The subsurface water commonly handled by the French drain is perched ground water or lateral flow from wet-weather springs or shallow aquifers.

Where the conditions warrant, the design of the drain can be modified to drain also excessive surface water. This is readily accomplished by either adding surface drains (risers) connected to the French drain system or carrying the gravel to the surface level. Water from downspouts should not be tied directly into the French drain; a separate pipe drain system is preferred. A solid pipe could, however, be placed in the French drain trench. A French drain is of little or no use in relieving water problems resulting from a spring within the confines of the foundation, as it is almost impossible to locate and tap a spring beneath a foundation. A proper drain intercepts and disposes of the water before it invades the foundation.

When a foundation problem exists prior to the installation of a French drain, foundation repairs are also often required. In that event, the repairs should be delayed to give ample time for the French drain to develop a condition of moisture equilibrium under the foundation. Otherwise, recurrent distress can be anticipated due to the influence of soil moisture introduced by the drain.

6C.3.2 Water or Capillary Barriers

In an effort to maintain constant soil moisture, measures which impede the transfer of soil moisture can be taken. The barriers may be either horizontal or vertical.

6C.3.2.1 Horizontal Barriers. Horizontal impermeable barriers can be as simple as asphalt or concrete paving or polyethylene film. These materials are used to cover the soil surface adjacent to the foundation and inhibit evaporation. Coincidentally, the covers could also restrict soil

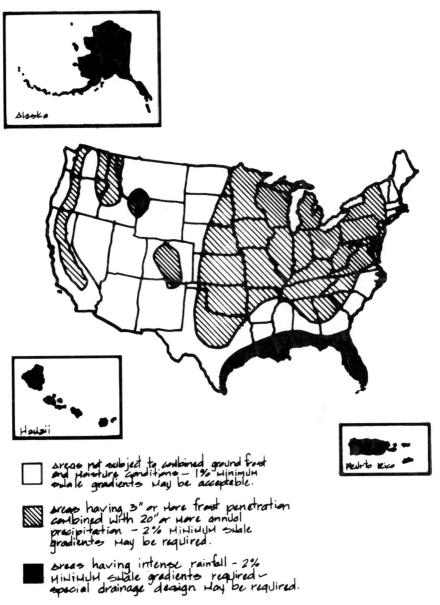

FIGURE 6C.5 General guidelines for areas of combined ground frost and moisture conditions.

moisture loss to transpiration since vegetation could neither grow nor be cultivated in the protected area. Permeable horizontal barriers usually consist of little more than landscaping gravel or granular fill placed on the soil surface previously graded for drainage. Moisture within the porous material cannot develop a surface tension and no adhesive forces will exist, both of which are required to create capillary, or pore, pressures. Gravity then becomes the factor dictating free water movement. The granular capillary barrier shown beneath the foundation slab in Fig. 6C.8 is one such example.

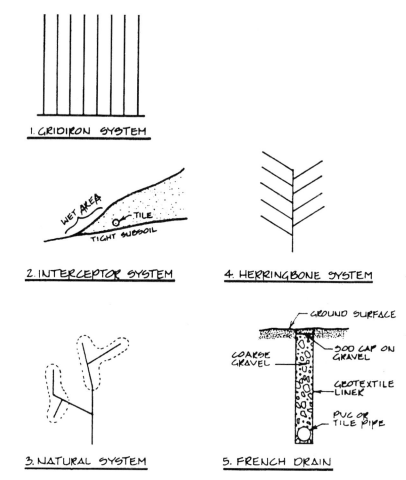

FIGURE 6C.6 Different forms of site drainage.

6C.3.2.2 Vertical Barriers. Vertical permeable barriers (VPCB) generally consist of a slit trench filled with a permeable material. Figure 6C.8 shows a typical permeable water or capillary barrier.[5] The VPCB will accept surface water and distribute it into the exposed soil matrix. At the same time it tends to block lateral capillary movement of water within the soil itself.

The vertical impermeable capillary or water barrier (VICB) is intended to block the transfer of water laterally within the affected soil matrix. Again refer to Fig. 6C.8.[5] Placed adjacent to a foundation, the VICB will hopefully maintain the soil moisture at a constant level within the foundation soil contained by the barrier. As an added benefit this approach will also prevent transpiration to the soil, since tree, plant, or shrub roots would be prevented from crossing the barrier. In some cases the soil moisture within the confines of the VICB is increased to 1% or so above the soil's plastic limit prior to construction of the foundation, particularly in the instances of slab foundations. Here the theory is that the soil is preswelled to a point that increased water is not likely to cause intolerable swell, and, at the same time, the barrier will prevent the decrease in soil moisture beneath the foundation. Hence a stable condition is created.

6C.3.2.3 Conclusion. The use of water or capillary barriers offers possible benefits, but field data made public to date leave much to be desired.[5–7]

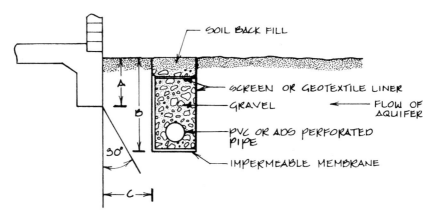

FIGURE 6C.7 Typical french drain. Generally C is equal to or greater than A and B is greater than $A + 2$ ft (0.6 m). The drain should be located outside the load surcharge area.

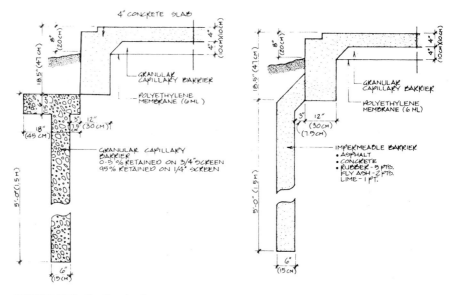

FIGURE 6C.8 Capillary barriers.

6C.4 VEGETATION

Certain trees, such as the weeping willow, cottonwood, and mesquite, have extensive shallow root systems which remove water from the soil. These plants can cause foundation and sewer problems even if located some distance from the structure. Many other plants and trees can cause different foundation problems if planted too close to the foundation. Plants with large, shallow root systems can grow under a shallow foundation, and as roots grow in diameter, they produce an upheaval in the foundation beam. Surfaces most susceptible to this include flat work such as sidewalks, driveways, and patios as well as some pier and beam foundations. The FHA (now HUD) suggests that trees be planted no closer than their ultimate height. In older properties this is often not possible. With proper care, the adverse effects to the foundation can be minimized or circumvented. Pruning the trees and plants will limit the root development. Watering, as discussed earlier, will also help.

Plants and trees can also remove water from the foundation soil (transpiration) and cause a drying effect, which in turn can produce foundation settlement. However, transpiration can often remove moisture to within a few percentage points above the plastic limit without creating excessive soil shrinkage.[3] Certainly, if the process occurred at a moisture content at or below the shrinkage limit, no shrinkage would occur. Further, Don Smith, a professor of botany at the University of North Texas, Denton, expresses the opinion that tree roots or other plant roots are not likely to grow beneath the foundation. This is due to several factors, the most important being:

1. Feeder roots tend to grow laterally within the top 24 in (0.6 m). The perimeter beam often extends to below that depth and would block root intrusion (see Fig. 6A.2).
2. Soil moisture (long range) and oxygen availability, both necessary for plant growth, are often lower beneath the foundation. More important, these confined areas have no normal access to a replenishing source for water. (Roots tend to "grow to water.")

For these reasons it would appear that trees pose no real threat to foundation stability other than that noted in the first paragraph. Along these same lines, even in a semiarid area with highly expansive soil, it is seldom that a significant earth crack is noted beneath the tree canopy. This is particularly true with trees exhibiting low canopies, which suggests an actual conservation of soil moisture.[8] Also if trees pose the problems that some seem to believe, why don't *all* foundations with like trees in close proximity show the same relative distress. Furthermore, in the thousands of instances where foundation repairs are made without removal of trees, why doesn't the problem always recur? In the latter instance the data indicate the contrary. Any extended differential in soil moisture can produce a corresponding movement in the foundation. If the differential movement is extensive, foundation failure will likely result. Of the two types of failures, settlement and upheaval, the latter is by far the more critical.

Even with proper care, foundation problems can develop. However, consideration and implementation of the foregoing procedures will afford a large measure of protection. It is possible that adherence to proper maintenance could eliminate perhaps 40% of all serious foundation problems. Anyone who can grow a flower bed can handle the maintenance requirements.

Remember one of the basic laws of physics—nothing moves unless forced to do so. The foundation is no different. This has been emphasized in prior chapters. All that foundation repair accomplishes is to restore the structural appearance. If the initial cause of the problem is not identified and eliminated, the problem is likely to recur.

6C.5 REFERENCES

1. G. P. Tschebotarioff, *Foundation Retaining and Earth Structures,* 2d ed., McGraw-Hill, 1973.
2. Council of American Building Officials, 5203 Leesburg Pike, Suite 708, Falls Church, VA 22041.
3. T. H. Petry and J. C. Armstrong, "Geotechnical Engineering Considerations for Design of Slabs on Active Clay Soils," presented at the ACI Convention, Dallas, Tex., Feb. 1981.
4. *Canadian Foundation Engineering Manual,* 2d ed., Canadian Geotechnical Society, 1985, p. 238.
5. A. Poor, "Experimental Residential Foundations on Expansive Clay Soils," HUD Contract H-224OR, University of Texas at Arlington, 1975–1979.
6. D. E. Jones and K. A. Jones, "Treating Expansive Soils," *Civil Eng.,* Aug. 1987.
7. J. D. Nelson et al., "Moisture Content and Heave beneath Slabs on Grade" and "Prediction of Floor Heave," presented at the 5th Int. Conf. on Expansive Soils, ASCE, Adelaide, Australia, May 1984.
8. T. H. Wu, "Study of Soil Root Interaction," *J. Geotech. Eng., ASCE,* Dec. 1988.

SECTION 6D
FOUNDATION INSPECTION AND EVALUATION FOR THE BUYER

Robert Wade Brown

6D.1 INTRODUCTION   6.103
6D.2 CHECKLIST FOR FOUNDATION INSPECTION   6.104
6D.3 IMPACT OF FOUNDATION REPAIR ON PROPERTY EVALUATION   6.105
6D.4 REFERENCES   6.106

6D.1 INTRODUCTION

If you are house shopping in a geographical locale with a known propensity for differential foundation movement, it is always wise to engage the assistance of a qualified foundation inspection service. The word *qualified* cannot be overemphasized. It is necessary both to evaluate the existence of any foundation-related problems and also to determine the cause of those problems. The latter is particularly important. Repairs will be futile if the original cause of the distress is not recognized and eliminated.

Several of the more obvious manifestations of distress relative to differential foundation movement were shown in Fig. 6B.1(*a*)–(*e*). With only limited experience one can learn to detect these signs. Often the problems of detection become more difficult when cosmetic attempts have been made to conceal the evidence. These activities, which commonly involve painting, patching, tuck-pointing, addition of trim, or installation of wall cover, require greater expertise for evaluation. The important issue in all cases is to decide whether the degree of distress is sufficient to demand foundation repair. This decision requires a great degree of experience since several factors influence the judgment:

1. The extent of vertical and lateral deflection. Does any structural threat exist or appear imminent?
2. Is the stress ongoing or arrested?
3. The age of the property.
4. The likelihood that the initiation of adequate maintenance would arrest continued movement. For example, in cases of upheaval, elimination of the source of water will often arrest the movement.
5. The value of the property as compared to the repair costs. Most foundation repair procedures require some degree of compromise. In order to arrive at a reasonable or practical cost, the usual primary concern is to render the foundation stable and the appearance tolerable. In most cases the cost to truly "level" a foundation, if it were possible, would be prohibitive.
6. The age, type, and condition of the existing foundation.
7. The possibility that, if the movement appears arrested, cosmetic approaches would produce an acceptable appearance.

It is difficult, if not impossible, to properly evaluate the foregoing without extensive first-hand experience with actual foundation repairs. To quote Terzaghi, the founder of soil mechanics: "In our field (engineering and geology) theoretical reasoning alone does not suffice to solve the problems which we are called upon to tackle. As a matter of fact it (engineering and geology) can even be misleading unless every drop of it is diluted by a pint of intelligently digested experience."[1] One good example of the importance of on-the-job experience is the proper determination of upheaval as opposed to settlement (see Sec. 6A.5). If upheaval were evaluated as settlement, the existence of water beneath the foundation would likely be overlooked. In that case future, more serious consequences would be nearly certain. However, there are no similar, serious consequences in making the error of labeling settlement as upheaval. Essentially this is true because all foundation repair techniques are designed to raise a lowermost area to some higher elevation. In cases of settlement, one merely raises the distressed area to as-built. In those instances of upheaval it becomes necessary to raise the as-built to approach the elevation of the distorted area. The latter is obviously far more difficult. Refer to Secs. 6A.2, 6A.3, and 6A.4 for greater detail.

The existence of foundation problems need not be particularly distressing so long as the buyer is aware of potential problems before the purchase. Refer to Sec. 6A.9 for additional background information. As a rule, the costs for foundation repair are not comparatively excessive, the results are most often satisfactory, and the net results are a stronger, more nearly stable foundation. It is the surprise that a buyer cannot afford. The cost for a residential foundation inspection service is quite nominal; it varies from about $50 to $175 (1995), depending on the time involved and the locale. The following section gives a simple checklist for evaluating the stability of a foundation. If you have questions or uncertainty regarding any of the items, consult a qualified authority. Another source for helpful information is the *Builders and Engineers Manual* provided by the Home Owners Warranty (HOW) Quality Control Program, revised August 1, 1990. (HOW began efforts to discontinue structural insurance coverage in 1994.)

6D.2 CHECKLIST FOR FOUNDATION INSPECTION

1. Check the exterior foundation and masonry surfaces for cracks, evidence of patching, irregularities in siding lines or brick mortar joints, separation of brick veneer from window and door frames, trim added along door jamb or window frames, separation or gaps in cornice trim, separation of brick from frieze or fascia trim (look for original paint lines on brick), separation of chimney from outside wall, masonry fireplace distress on interior surfaces, and so on.
2. Sight ridge rafter, roof line, and eaves for irregularities.
3. Check interior doors for fit and operation. Check for evidence of prior repairs and adjustment such as shims behind hinges, latches or keepers relocated, tops of doors shaved.
4. Check the plumb and square of door and window frames.
5. Note grade of floors. A simple method for checking the level of a floor (without carpet), window sill, counter top, and so on, is to place a marble or small ball bearing on the surface and observe its behavior. A rolling action would indicate a down-hill grade. (A hard surface such as a board or book placed on the floor will allow the test to be made on carpet.)
6. Inspect wall and ceiling surfaces for cracks or evidence of patching. *Note:* Any cracking should be evaluated on the basis of both extent and cause. Most hard construction surfaces tend to crack. Often this can be the result of thermal or moisture changes and not foundation movement. However, if the cracks approach or exceed $\frac{1}{4}$ in (0.6 cm) in width, the problem is likely to be structural. On the other hand, if a crack noticed is, for example, $\frac{1}{8}$ in (0.3 cm) wide, is it a sign of impending problems? A simple check to determine whether a crack is "growing" is to scribe a pencil mark at the apex of the existing crack and, using a straightedge, make two marks along the crack, one horizontal and one vertical (Fig. 6D.1). If the crack changes even slightly, one or more of the marks will no longer match in a straight line along the crack, or the crack will extend past the apex mark (Fig. 6D.2). A slight variation of this technique is to mark a straight line across a door and matching door frame. In a structure older than about 12 months, continued growth of the crack would be a strong indication of

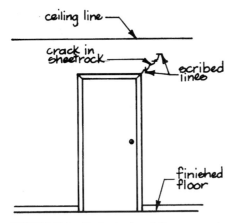
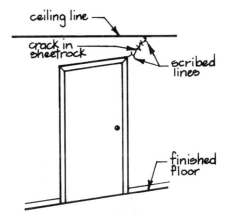

FIGURE 6D.1 Monitoring foundation movement.   **FIGURE 6D.2** Confirming movement.

foundation movement.

7. On pier-and-beam foundations, check floors for firmness, inspect the crawl space for evidence of deficient or deteriorated framing or support, and ascertain adequate ventilation.
8. Check exterior drainage adjacent to foundation beams. Any surface water should quickly drain away from the foundation and not pond or pool within 8 to 10 ft (2.4 to 3 m). Give attention to planter boxes, flower bed curbing, and downspouts on gutter systems.
9. Look for trees that might be located too close to the foundation. Many authorities feel that the safe planting distance from the foundation is 1 or preferably 1.5 times the anticipated ultimate height of the tree. More correctly, the distance of concern should be 1 to 1.5 times the canopy width. Consideration should be given to the type of tree. (Refer also to Part 1 and Sec. 6C.)
10. Are exposed concrete surfaces cracked? Hairline cracks can be expected in areas with expansive soils. However, larger cracks, approaching or exceeding $\frac{1}{8}$ in (0.3 cm) in width, warrant closer consideration.[2]

6D.3 IMPACT OF FOUNDATION REPAIR ON PROPERTY EVALUATION

Do foundation repairs impact resale values? This is a question often heard, particularly from appraisers or attorneys. Generally speaking, competent foundation repairs should produce an end product at least equivalent and, more often, superior to the original. Further, the repaired foundation will have no greater susceptibility to future problems. In fact, the properly repaired foundation might be considerably more resistant to such distress.

A foundation is no different than any other inanimate object—it refuses to move unless forced to do otherwise. Eliminate or prevent this force, and no movement occurs.

In the case of foundations built on expansive soils, the primary force is provided by variations in soil moisture. Add water and the soil swells. Remove water and the soil shrinks. Do neither and no movement occurs. It makes no difference whether the foundation is new, old, or previously repaired.

Proper maintenance and alert observation to the early warning signs of impending distress will prevent or minimize all foundation movement. Drainage and proper watering can be easily controlled by common sense. The unknown in the formula involves the accumulation of water beneath slab foundations due to some form of utility leak, the most serious being a sewer leak. As a rule the latter is usually detected from the various signs of differential foundation movement as manifested by sheetrock cracks, mortar cracks, ill-fitted doors, windows, or cabinets, or deviated floors. Catch the problem sufficiently early, eliminate the cause, and, in most cases, avoid the need for foundation repairs. Notice that no differentiation is given as to age or history of the subject foundation. It makes no difference.

The principal foundation concern in purchasing a property should be whether the foundation was designed properly for the site conditions and then competently constructed. The occurrence of foundation repair would have no negative impact, provided the repairs were properly executed.

6D.4 REFERENCES

1. H. H. Einstein, "Observations, Quantification and Judgment: Terzaghi and Engineering Geology," *J. Geotech. Eng.*, Nov. 1991.
2. D. Kaminetzky, "Rehabilitation and Renovation of Concrete Buildings," in *Proc. Workshop, National Science Foundation and ASCE,* New York, Feb. 1985.

PART 7

FOUNDATION FAILURES AND REPAIR: HIGH-RISE AND HEAVY CONSTRUCTION

PART 7

FOUNDATION FAILURES AND REPAIR: HIGH-RISE AND HEAVY CONSTRUCTION

Dov Kaminetzky

7.1 INTRODUCTION 7.3
   7.1.1 High-Rise and Heavy Construction 7.3
   7.1.2 Foundation Failures 7.5
7.2 UNDERMINING OF SAFE SUPPORT 7.7
   7.2.1 New York Suburb Sewers 7.8
   7.2.2 Electronics Plant, Upstate New York 7.11
   7.2.3 East Side Hospital, New York City 7.15
   7.2.4 Manhattan Hospital Complex, New York City 7.16
7.3 LOAD-TRANSFER FAILURE 7.21
7.4 LATERAL MOVEMENT 7.22
   7.4.1 New Jersey Postal Facility—Cracked Pile Caps 7.23
   7.4.2 The Collapse of the Hartford, Connecticut, Foundation Walls 7.24
   7.4.3 Long Island Water Pollution Control Plant Outfall 7.27
7.5 UNEQUAL SUPPORT 7.28
7.6 DRAG-DOWN AND HEAVE 7.31
   7.6.1 Queens Apartments, New York 7.32
   7.6.2 Edgewater, New Jersey, High Rise 7.34
7.7 DESIGN ERROR 7.35
   7.7.1 The Alaska Pier Collapse 7.36
   7.7.2 Washington, D.C., Foundation Mat 7.40
   7.7.3 Boston Housing Slab Settlement 7.43

7.8 CONSTRUCTION ERRORS 7.44
   7.8.1 Lower East Side Collapses 7.44
   7.8.2 Apple Juice Factory—Drilled-in Piers 7.47
   7.8.3 Marble Hill 7.52
7.9 FLOATATION AND WATER–LEVEL CHANGE 7.52
   7.9.1 Brooklyn Hospital—Settlement of Delivery Room 7.53
   7.9.2 Industrial Building, Hackensack, New Jersey 7.55
   7.9.3 West 41st Street Structure—Wood Piles 7.57
7.10 VIBRATION EFFECTS 7.57
   7.10.1 Lower Manhattan Federal Office Building 7.58
7.11 COFFERDAMS—THE 14TH STREET COFFERDAM 7.59
7.12 CAISSONS AND PILES 7.60
   7.12.1 The North River Project—Caisson Deficiencies 7.60
   7.12.2 Philadelphia Federal Office Building—Caisson Settlement 7.67
7.13 COMMON PITFALLS 7.69
7.14 CHECKLIST 7.69
7.15 REFERENCES 7.70

7.1 INTRODUCTION

7.1.1 High-Rise and Heavy Construction

Foundation failures of high-rise and heavy construction generally do not differ from those of low-rise structures and single residences. Foundation failures are all caused by *human errors* and are the result of either one or a combination of these three types of errors:

1. Errors of knowledge (ignorance)
2. Errors of performance (carelessness and negligence)
3. Errors of intent (greed)

This section will help to minimize future failures resulting from the first two types of errors. This famous quote says it well: "A wise man learns from the mistakes of others. Nobody lives long enough to make them all himself."

7.4 FOUNDATION FAILURES AND REPAIR: HIGH-RISE AND HEAVY CONSTRUCTION

The consequences of high-rise failures are often very serious and most devastating because of the large scale and size of such structures. It is also self-evident that the potential for high loss of life is also great.

The pattern of foundation failures in high-rise structures is unique and differs from patterns of failures caused by forces such as lateral wind pressures, earthquakes, or punching shear of slabs. Since the foundations transfer the structural loads such as vertical dead and live load and lateral wind load to the ground at the bottom of the structural frame, foundation failures will typically *telegraph* their effect *upward* for the full height of the structure. Such failure also may affect the stability of the high-rise structure and often may result in a noticeable tilt of the structure. This is especially so in cases of unequal support. (See Figs. 7.1 and 7.2.)

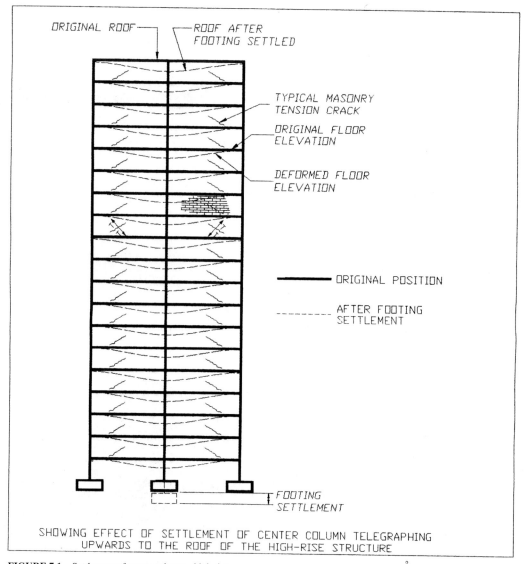

FIGURE 7.1 Settlement of center column—high rise.

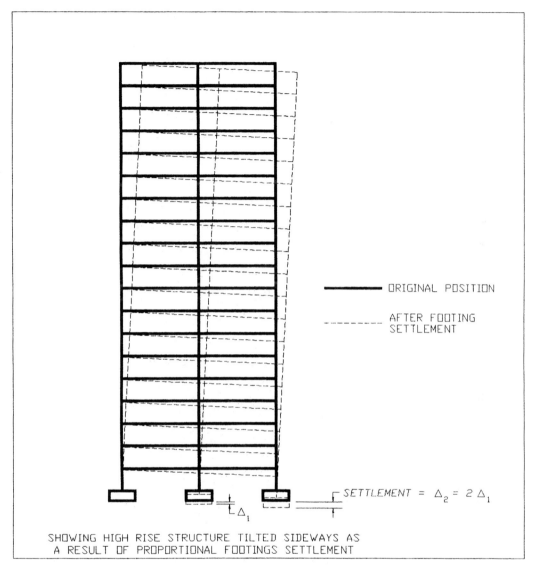

FIGURE 7.2 Structural tilt—a result of footings settlement.

7.1.2 Foundation Failures

The foundation is the structural element providing support for the various loads acting on the structure. It is the link between the structure and its eventual support, which is the soil itself. The actual transfer of load may be by direct bearing on soil or rock or by intermediary elements such as piles or caissons.

When we speak about foundation failures, we often refer to both the failure of the structural elements of the foundation such as footings or piles and the failure of the soil itself. Whereas the first type of failure may be the result of overloads on the foundation or its understrength, the second type results from overconfidence in the borings or other subsurface information or loss of bearing value because of adjacent work. Foundation failures resulting from failure of the footing itself is a rare occurrence. One such case was the fracture of the post-tensioned concrete foundation mat in Washington, D.C., which was caused by a design error. (See Figs. 7.49 to 7.51).

The foundation is usually a human-made structural element, whereas the soil is a material found in its natural state, disturbed or undisturbed by humans. Artificially made soils are also used in construction (these are mechanically compacted soils or soils made totally from nonnatural materials). The construction of a foundation introduces new conditions into the soil. This happens as the subgrade is exposed and, therefore, unloaded, as changes in the internal friction of the soil are effected by blasting or by the densification of soil by pile intrusion.

Ground conditions often vary considerably from one location to another within the confines of a single construction site. Sometimes the variation is so great that, when two borings do show identical soil layers, the validity of all borings will be questioned. Rock exposures may range from fine gray syenite, hard seamy limestones, granites, to schists of various hardnesses. These vary to such an extent that they are ringing hard or so soft that a pipe pile will penetrate over 10 ft (3.0 m) before indicating a 30-ton (267-kN) resistance. In some areas even seams of serpentine and asbestos are found.

However, granular materials will range from hardpans to uniformly sized, running fine sands. Soils also often contain clay deposits and organic silts. And, of course, there are the exposed and buried mud deposits with organic layers intermixed in a series of lenses to considerable depths. Because natural soils are often so varied and inconsistent, difficulties in foundation design and construction are expected and are encountered, often with associated trouble.

Actual *water level* in rock excavations sometimes has no relation to the level indicated by borings or to conditions next door. A typical example was the completed work for the New York Lincoln Center underground parking facilities and mechanical plant, where groundwater was first encountered at quite a high level. This water later disappeared after some anchorage drill holes were completed; it was later discovered almost 20 ft (6.1 m) lower than the groundwater encountered in the Philharmonic Hall, adjacent to it.

Bed rock surfaces are often very erratic and irregular. During the excavation for the Sixth Avenue subway in New York City during the 1930s, it was found that a strip only 100 ft (30 m) wide was 40 ft (12 m) lower on one side than on the other.

During the construction of the new Bellevue Hospital in Manhattan in 1960, the main hospital building was supported on piles as long as 100 ft (30 m). The adjacent parking facility, which was located no farther than 50 ft (15 m) away, was founded on concrete piers bearing directly on rock.

It is not uncommon to find large vertical mud holes or wide gaps filled with silt within a rock structure.

With such erratic rock strata, rock blasting has a serious effect on the behavior of rock. Rock seams must be found, traced along their length with the seams cleaned and packed and tied across with grouted dowels and rock anchors. Rock as a supporting subgrade can be made safe and sound, but only with careful planning and special work.

Granular soils are found in different varieties. There are some very good and some not-so-good granular soils. Some dense consolidated glacial sandy gravel deposits can safely sustain 10 tons per square foot (957 kPa), whereas uniformly fine grained and very loose silty sands are treacherous and extremely difficult to control. Glacial deposits also vary. Sometimes old sand fills are taken to be natural deposits. Standard borings contribute little information to distinguish the man-made from the natural geological sand layers. Only actual load tests are effective indicators, which often will show surprising looseness of these fills.

Layers of varied *silts and clays* cover many rock troughs and old shore lines. These layers are quite stable when left alone but become liquid when disturbed. Dewatering must be done slowly, to permit the seams to drain out, otherwise "boils" and collapse of braced sheeting excavation will result. Pile driving also generates local liquid conditions that flow readily. The plastic flow of the shore lines continuously pull the river structures outward. Even structures supported on piles carried to rock are affected and rotate with time.

Buried layers of *peat* are sometimes interspersed with silt or sand deposits and are found at great depths. Where piles receive their support at soil layers overlying peat deposits, serious settlements are common. Also, where pumping occurs in areas where footings are bearing above peat layers, settlements are common.

Pumping of water out of weak soils is known to cause consolidation of these soils and result in settlement damage to existing structures. Several precautionary methods are presently being employed to avoid this damage. Water-tight sheeting enclosures with deep cutoffs and internal pumping has been often found to be an adequate solution.

It is well recognized that all loads must be transferred to the underlying soils so that the resulting settlements can be tolerated by the structure without distress. At the same time, stability must be maintained over the life of the structure.

The ten most common categories of foundation failures are:

1. Undermining of safe support
2. Load transfer failure
3. Lateral movement
4. Unequal support
5. Drag down and heave
6. Design error
7. Construction error
8. Floatation and water-level change
9. Vibration effects
10. Earthquake effects

A foundation failure obviously is a serious event, since such a failure may trigger the collapse of the entire structure. This is critically so because most structures are based on structural systems whereby the support of the upper levels always depends on the structural integrity of the lower elements. Ironically, some foundation failures have been economical successes. One such example is the Leaning Tower of Pisa in Italy (Fig. 7.3). A close look at this tower will reveal that the tower started to lean during its construction. The reason for this conclusion is a slight change in the slope of the tower, near its midpoint. It is an indication that the builders attempted to correct the foundation-settling problem (Fig. 7.4).

7.2 UNDERMINING OF SAFE SUPPORT

Robert Frost in his poem "Mending Wall" (published in *North of Boston*) describes the troubles with stone walls in New England and concludes, "Before I built a wall I'd ask to know what I was walling in or walling out."[1] To paraphrase this warning, before a foundation is designed, one should know *what loads are to be carried and what soils are expected to carry the loads.*

The need for a thorough soil investigation prior to the undertaking of a construction project cannot be overemphasized. In addition to the careful study of the soil strata directly below the proposed structure, existing adjacent structures must be reviewed with care. The need for temporary supports must be evaluated. A well-designed bracing and shoring system is often needed to prevent a lateral shift. A permanent support structure such as underpinning should be installed where the new construction will undermine an existing present support system. Where these provisions are ignored or entirely omitted, serious distress follows (Fig. 7.5), and sometimes tragic consequences result (Fig. 7.6). Another example is the total collapse of a five-story building on 34th Street in New York City that occurred when the vertical support was lost as a result of loss of lat-

7.8 FOUNDATION FAILURES AND REPAIR: HIGH-RISE AND HEAVY CONSTRUCTION

FIGURE 7.3 Leaning Tower of Pisa.

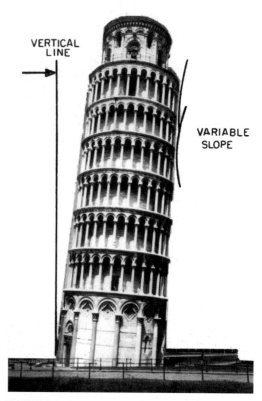

FIGURE 7.4 Leaning Tower of Pisa—variable tilt.

eral restraint. This happened when the excavation for a new high-rise building came too close to the foundation of the existing old building. Fortunately, the collapsed building had been evacuated a short time earlier because of unsanitary conditions. As a result of the collapse, the rubble filled the cellar completely and piled up almost to the second floor level.

Excavations for new sewer trenches adjacent to existing buildings have frequently caused undermining of footings resulting in distress and even total collapse. In the President Street collapse in Brooklyn, and similar other collapses, several occupants of existing buildings were killed or injured when their buildings were undermined. When these excavations were close to the existing footings, and within the influence line (Fig. 7.7), the footings were undermined, often with disastrous results.

7.2.1 New York Suburb Sewers

Construction for a new sewer main project in Brooklyn required the excavation of deep cuts adjacent to existing old residential buildings. The sewer lines were designed to run too close to existing footings. Not only were the new cuts within the influence lines of existing footings, but the contractor used vibratory pile drivers to install the piles. There was no monitoring of the

FIGURE 7.5 Structural damage caused by adjacent excavation.

FIGURE 7.6 Building collapse caused by adjacent excavation.

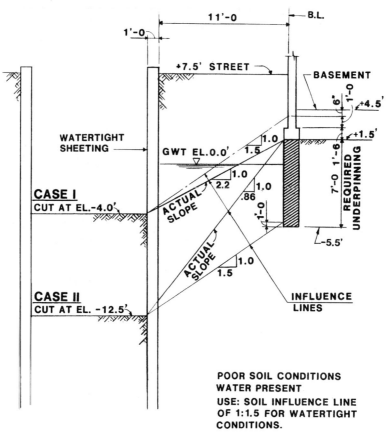

FIGURE 7.7 Underpinning existing structures adjacent to excavations.

vibration levels during pile driving. As a result, many nearby buildings settled, cracked, tilted, and shifted laterally. After the damage occurred, diagonal braces were added for lateral support (Fig. 7.8).

One building collapsed, killing one of the tenants (Fig. 7.9). The investigation revealed that the building was constructed using timber footings, a construction method common at the turn of century. The footings were found to be in good condition but had been disturbed by the sewer activity. Once the excavation was within the influence line, vertical support was lost, causing footing settlements. The high-grade steel bracing system (Fig. 7.10) could not save this building, because soil support was lost *below* the footing level.

To restore the integrity of the vertical support it was initially decided to underpin the buildings. The bid prices obtained for underpinning were so high, however, that it was more cost effective to buy the damaged buildings and demolish them.

Lessons to be learned:

1. Careful *preconstruction* study to determine the need for protection and underpinning of adjacent structures is required. This study should review existing plans, taking sufficient soil borings and evaluating the condition and strength of existing buildings and other structures.

FIGURE 7.8 Bracing structures for lateral support.

2. Excavations for sewers or other utilities should be moved away from existing buildings and so located that the excavation cut is outside footing influence lines.
3. Use of vibratory equipment for pile driving should be avoided in loose sands and similar soils.
4. Vibration readings, using seismographs, should be taken as often as needed to establish the suitability and proper energy of the driving hammers.

7.2.2 Electronics Plant, Upstate New York

An electronics plant was under construction when the entire building started to sink vertically and slide laterally toward the low adjacent valley. Instrument survey readings confirmed these visual observations. The footing movements, as is usual in these cases, were accompanied by distress in the building in the form of cracking of slabs and concrete grade beams (Fig. 7.11).

The main question that had to be answered was what was the effect of large movements of the order of 4 in (102 mm) on the structural steel frame and its connections (steel bolts)? To answer this question, we had to determine the level of stress in the loaded and deformed steel beams and the actual loads in the connection bolts. For that purpose we used the innovative stress-relief method (Fig. 7.12). This method, which our firm pioneered, proved that the level of stress actually present in the steel and its connections was low and acceptable.

This procedure requires the attachment of electrical strain gages to the steel flanges of the beams. Through each strain gage, a hole is then drilled directly into the steel. The reduction of stress at the hole is measured by the strain gage connected to a Wheatstone Bridge. The relieved stress thus measured gives the approximate stress level then existing in the loaded steel beam. The only structural corrective measure that was executed was to enlarge the existing footings, so as to enable the structure to support the additional live loads without considerable settlements.

The method of repair is shown in Figs. 7.13 and 7.14.

7.12 FOUNDATION FAILURES AND REPAIR: HIGH-RISE AND HEAVY CONSTRUCTION

FIGURE 7.9 Building collapse—ineffective bracing.

FIGURE 7.10 Bracing excavation cut.

FOUNDATION FAILURES AND REPAIR: HIGH-RISE AND HEAVY CONSTRUCTION 7.13

FIGURE 7.11 Structural failures (cracking) in deep-grade beams.

FIGURE 7.12 Stress-relief test.

7.14 FOUNDATION FAILURES AND REPAIR: HIGH-RISE AND HEAVY CONSTRUCTION

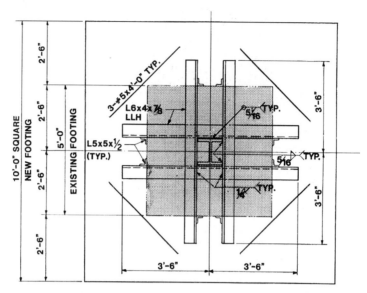

FIGURE 7.13 Increased footing size to accommodate additional live loads.

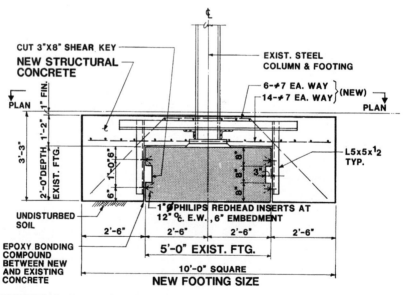

FIGURE 7.14 Cross-sectional view of Fig. 7.13.

Lessons to be learned:

1. High footings adjacent to severe soil slopes should be designed using low allowable bearing pressures to reduce possible settlements.
2. In extreme cases, the use of retaining walls should be considered.
3. The actual stress level in a deformed structure is the true yardstick for evaluating the adequacy of a distressed structure.
4. The stress-relief method proved to be an effective tool in strength evaluation of distressed structures.

7.2.3 East Side Hospital, New York City

During excavation for a new high-rise hospital structure, an alarm went out to everybody connected with the project. It was discovered that, as a result of rock removal by the foundation contractor, the adjacent high-rise apartment building to the west was precariously sitting on a "sliver" of rock (Fig 7.15). Construction work came to a halt, and serious consideration was given to evacuating the fully occupied apartment building.

Probes into the wall revealed the following conditions (Fig. 7.16):

1. The width of the rock "sliver," of unknown strength, was approximately 14 in (355 mm). Incredibly, it was the only structural element supporting the high-rise building.
2. Even more remarkably, the footings called for in the original design for the existing apartment building were not built at all.
3. The easterly wall of the building cellar was not a standard concrete cellar wall but was actually a rock face with stucco finish.
4. The plans that had been filed with the Department of Buildings did not show the "as-built" conditions.

An emergency plan for underpinning was immediately launched. This plan included new concrete underpinning piers (braced laterally) cut into the rock "sliver" and bearing on the rock below the existing cellar level (Fig. 7.17). The plan was executed perfectly, and the alarm was turned off.

As usual, the responsibility for the unexpected additional cost was in dispute. The questions asked by everybody were:

1. Was the original 1964 construction, which deviated from the design, bearing the apartment building columns on the neighbor's rock, legal?
2. Did it follow accepted good construction practice?
3. What responsibility, if any, does the Structural Engineer of Record have for the contractor's variation from the plans?

Regarding the first question, our research into the building code showed that the code was completely silent on this issue. Regardless, we found no requirement forbidding the practice. Our opinion was that it is poor construction practice for one building owner to utilize a neighbor's rock for bearing for the owner's building, especially since he or she has no control over nor any right to limit the neighbor's construction activity.

The Engineer of Record had no "Controlled Inspection" responsibility for this project, neither was he required to perform such an inspection of the construction according to the then-pertinent code.

According to many building codes, when neighboring construction goes *below* the bottom of an existing footing, it is the responsibility of the neighbor doing the new construction to underpin the existing structure, when his or her footings are more than 10 ft (3.0 m) below curb level.

7.16 FOUNDATION FAILURES AND REPAIR: HIGH-RISE AND HEAVY CONSTRUCTION

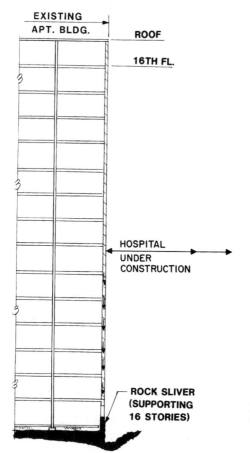

FIGURE 7.15 Rock sliver precariously supports 16-story brick wall.

The New York City Building Code, for example, states[2]:

(b) Support of adjoining structures.—
(1) Excavation Depth More Than 10 Ft. (3.0 m)—When an excavation is carried to depth more than 10 ft. (3.0 m) below the legally established curb level, the person who causes such excavation to be made shall preserve and protect from injury any adjoining structures....

Lessons to be learned:

1. Do not use the rock outside your property for support of your structure, regardless of the apparent immediate savings.
2. You may support your structure only on the rock totally within your own side of the property line.
3. Rock "slivers" should not be relied upon for foundation support.

7.2.4 Manhattan Hospital Complex, New York City

When the structures adjacent to a deep excavation begin cracking, it slowly starts to sink in that something is wrong. An excavation extending an entire site was progressing toward a total depth

FOUNDATION FAILURES AND REPAIR: HIGH-RISE AND HEAVY CONSTRUCTION   7.17

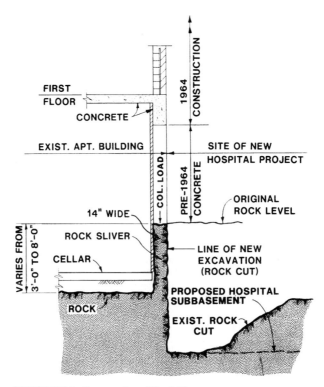

FIGURE 7.16 Cross section of Fig. 7.15.

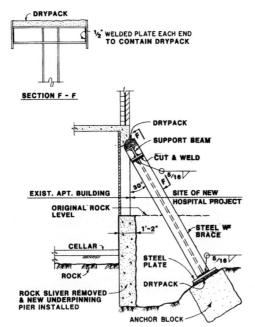

FIGURE 7.17 Emergency plan for underpinning existing building to create proper bearing.

of 45 ft (14 m) below grade. The existing structures contained only single-level basements. Initially, it was expected that much of the excavation would be into sound rock, which was identified as gneiss. The ground-water level within the buildings was controlled by a sump pump located below an existing basement slab.

During the design phase it was obvious that the adjacent structures would have to be underpinned. The New York City Building Code requires that underpinning will be designed by a professional engineer. This requirement was met, and a design using underpinning pits with post-tensioning cables to resist lateral earth pressures was devised.

As excavation for the pits was proceeding, movement was detected in the building directly east of the site. This movement was recorded by the tell-tales (movement gages) which were properly installed for monitoring such movements. As water was found several feet below the pits, attempts were made to dewater the soil surrounding the pit excavation. But, as recorded by the inspection engineer, "…this proved futile because of the impermeable nature of the soil existing at this location."

Dewatering by well points was then tried. This also proved ineffective, because the silts in the soil clogged the well point screens. A third alternative using jack-piles was then considered. This method of pushing pipe segments into the ground requires the use of the building weight to act as a reaction for the jacking forces. In this project, unfortunately, it readily became apparent that the footings of the existing building, which consisted of decomposed rubble, did not offer sufficient resistance to the jacking loads.

Even an attempt to use a steel beam to jack against failed, and the scheme was then abandoned. While these efforts were going on, a further adverse movement was detected in the adjacent building. At this time the safety of the structure was questioned, and temporary rakers (sloped braces) were installed. For all practical purposes, construction operations came to a stop.

The two buildings on the east exhibited serious cracking throughout (Figs. 7.18 and 7.19). The one nearest the excavation settled, causing the entire building to pivot about the foundation walls. This pivoting effect in turn caused the entire party wall to move westward, leaving a 6-in

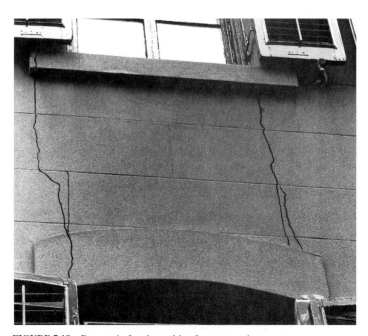

FIGURE 7.18 Damage in facade resulting from excavation.

FIGURE 7.19 Cracking of partition wall and settlement of stairs due to footing movement.

(152-mm) gap at the top (Fig. 7.20). This movement caused enormous distress on the interior of the building. Floors settled, ceilings and walls cracked, and even the elevator got stuck as its shaft deformed and would not permit free movement of the cab. At the fourth floor there was actually a separation between the wood joists and the party wall, causing the floor to settle by as much as 1-1/8 in (28 mm) at one point.

The foundation contractor, realizing the danger, immediately installed vertical shores under the west end of the wood joists, from the first floor all the way up to the roof.

It was at this point that our firm was retained. Subsequently it was recommended that the contractor take weekly movement readings using surveying instruments and continue doing so until the settlement stopped. The contractor was also instructed not to proceed with any restoration work until all settlement had stopped. *The New York Times* of July 3, 1977, printed the following article regarding this project under the headline "When the Earth Opens and Walls Move[3]":

> George N. and Barbara K. don't live here any more. Their former apartment, in a brownstone in the Gramercy Park area, once looked like the home of many a professional couple—antique furniture, luxuriant plants and exposed wooden beams, lots of beams.
>
> ...Their apartment is one of the occasional casualties of excavation work, the bulldozing, blasting and drilling that create the cavernous holes meant to hold a new building's foundation.
>
> Though New Yorkers frequently sue or arrange out-of-court settlements with excavators (claims range from apartment damage to sexual inadequacy caused by noise), George and Barbara just decided not to renew their lease.
>
> ...The whole experience was very unnerving, "Mr. N. recalls. "I was really concerned how much physical damage would be done to my home by the end of a day. *It started out as a hair-line crack and got bigger every day* (author's italics). I saw the baseboard move further away from its original place on the floor. Then we had to take down a shelf in the closet because it was no longer wide enough. *The wall had moved four inches (102 mm) toward New Jersey.*" (author's italics).

7.20 FOUNDATION FAILURES AND REPAIR: HIGH-RISE AND HEAVY CONSTRUCTION

FIGURE 7.20 Lateral displacement of existing building toward excavation.

...A side of the ceiling was now without support and, just as Mr. N. had feared, it soon gave way. Fortunately, their apartment was saved by beams, which had been installed by workmen a few days earlier. [The vertical shores referred to above (author's comment).]

...Dr. F., a physician, and her husband, resident-owners of the building next to the excavation, never contemplated selling their property.

Although the roar of the machinery can reduce telephone conversations to the F. residence to frustrating shouts, Dr. F. says that somehow her practice at home (psychiatry) "has not been disturbed."

...Mr. R., project engineer for the excavation company, says that the unknown factor in that site was an unexpectedly high underground water elevation...."We had what you call test borings....As it turned out, these test borings were greatly different from what we encountered."

When all efforts to support the building failed, the *method of last resort* was used, very costly soil solidification by freezing the soil. The saturated soil proved to be an advantage for the freezing process. This process with all its pipes and equipment took about 3 weeks to install (Fig. 7.21). It proved to be a total success, as practically all movements within the adjacent "ailing" structure stopped. Once the soil under the building was consolidated, the original conventional underpinning design was implemented. Just prior to freezing the soil, a cross-lot bracing system was installed (Fig. 7.22). The purpose of this system was to eliminate any further movement in the building on the east and to stabilize it during the subsequent construction operations. The giant cross-lot braces which spanned the entire lot permitted construction to proceed without interruption. After completion of the underpinning, the adjacent buildings were restored.

Lessons to be learned:

1. Accurate preconstruction subsoil study is essential to the success of foundation work. Inadequate test data can cause serious delays, cost overruns, damage, and even collapses. In this case, had accurate information been obtained, design changes and soil solidification could have been planned prior to start of construction.
2. Adjacent structures must be carefully monitored during underpinning operations, so if and when movements are detected, changes in methods or procedures can be implemented.

FIGURE 7.21 Soil solidification by freezing.

FIGURE 7.22 Cross-view of Fig. 7.21.

7.3 LOAD-TRANSFER FAILURE

A well-designed and constructed rigid-frame structure will tolerate even substantial foundation movements. When an assembly of walls, floors, frame, and partitions is rigidly connected, the system will adequately adjust itself to differential foundation movements. The load transfer is made by frame action through the support offered by the foundation. When this interconnected

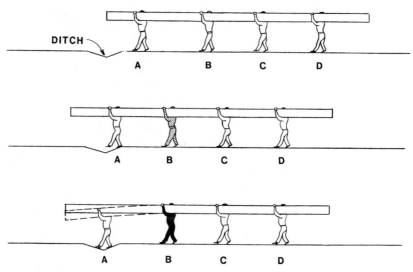

FIGURE 7.23 Diagram of log deflection.

rigidity is absent, the load transfer will be through a single support so that the load will go directly to the soil vertically. Where this single support in the soil is missing, the structure will fail, unless the structural rigidity will transfer it horizontally to other footings and then, in turn, to the soil.

Where the adjacent rigidity is present but lacks sufficient strength, the adjacent structure will fail. Figure 7.23 shows a diagrammatic description of this action. Four men are carrying a log, its weight uniformly distributed. When A steps into a ditch (an inadequate foundation), his portion of the load is suddenly transferred to B, who may be unable to support the additional load.

One such typical example was a 16-story warehouse building in lower Manhattan, supported by four continuous pile caps and timber piles. Pumping operations at a neighboring 20-ft (6-m) deep excavation for a depressed roadway lowered the water table and caused one line of piles to settle and pull the pile cap down.

This released the exterior support so that the load transfer to the interior line of columns overloaded their steel beam grillages. The 24-in (610-mm) beams collapsed down to a 12-in (305-mm) depth with consequent 12-in (305-mm) settlement of every floor in the building. The floor beams resting on steel girders between columns had only "soft" support so that they took the new trough shape without failure.

The entire building was shored, and new shallow grillages were installed to replace the failed ones. Rather than jacking the entire structure, which entailed definite risks, especially for an old structure, each floor was releveled, and the building was put back into use.

7.4 LATERAL MOVEMENT

It is well known that 1 inch (25 mm) of lateral movement of a foundation causes more damage than 1 in (25 mm) of vertical settlement. Lateral movements are caused from either the elimination of existing lateral resistances or from the addition of active lateral pressures and loads. The changes in the active pressures and passive resistances as a result of variable water conditions must be considered. Saturation of the soil often increases the active pressures and reduces the passive resistances. Many collapses have occurred as a result of demolition of adjacent buildings. In these instances the backfilling of adjoining cellar space generated lateral soil pressures on the old cellar partitions. These walls, in turn, failed because they were not designed and constructed as retaining walls.

Lateral flow of soil under buildings is also known to cause collapses of buildings. In Sao Paulo, Brazil, such lateral soil flow caused the total collapse of a 24-story office building in 1943.

A similar case occurred in 1957 in Rio de Janeiro, also in Brazil, where an 11-story residential building collapsed completely after the excavation of an adjoining wall removed the lateral restraint of the building piers. The attempted underpinning, which was started too late, could not save the rotating structure.

There also are numerous cases of wall failures from broken drains alongside the footings with washout of the soil during heavy storms.

Change in pressure intensity against walls often causes failure, especially in the unreinforced concrete basement walls for residential buildings. These walls are unfortunately seldom investigated for high soil pressures. Surcharging soil on land adjacent to structures often causes large lateral pressures. Mounds of debris from demolitions are frequently piled adjacent to basement walls which had not been designed to resist such loads. These bowed basement walls often cave in and cause the total collapse of the structure.

The stability of retaining walls is often similarly affected, resulting in failure by overturning, either because of inadequate base width or because of high soil pressure caused by defective drainage or adjacent load surcharge.

Failure of buried sanitary structures are also common, especially when they are emptied for cleaning or repairs.

7.4.1 New Jersey Postal Facility—Cracked Pile Caps

Shortly after completion of the construction of this mail bag repair shop, the entire structure started to shift sideways, causing many of the piles to bend and the pile caps to crack (Fig. 7.24). In addition, an interior block wall parallel to the direction of the shift also cracked (Fig. 7.25).

FIGURE 7.24 Lateral failure of H piles and concrete caps.

Our firm was called to investigate and to determine the structural adequacy of the entire facility. What was immediately evident was the very heavy vertical gravity loads imposed on the piles. Figure 7.26 shows heavy bales of paper and cloth sitting directly on the already shifted and distressed piles.

The area directly east of the plant was found to be surcharged with mounds of heavy garbage dumping (Fig. 7.27). It quickly became clear that the soft clay soil directly underneath this extreme garbage weight was surcharging the soil, causing lateral pressures on the long, unbraced steel piles. Previous attempts to "hold the line" by welding steel cross-ties had been unsuccessful, and rather than stopping the lateral shift, they only transferred the lateral loads to the adjacent piles, which eventually failed, too. The additional lateral cross ties may be seen in Fig. 7.28.

We immediately recommended that the owners reduce the actual live loads on the piles, by moving the heavy bales to unaffected areas. A complete underpinning program involving drilling through the existing slabs and jackpiling down through the soil, was then recommended. The owners dragged their feet for a while, but finally a program similar to one proposed by our firm was implemented several years later.

Lesson to be learned:

The placement of high surcharge adjacent to structures generates high lateral loads that can cause shifting of piles and bowing of foundation walls, and should be avoided.

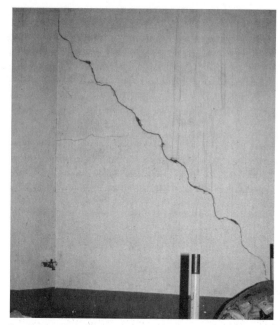

FIGURE 7.25 Diagonal masonry cracks due to lateral and downward settlement.

FIGURE 7.26 Unsafe loading on damaged piles (Fig. 7.24).

7.4.2 The Collapse of the Hartford, Connecticut, Foundation Walls

This large housing development is located on a sloping terrain, using concrete foundation walls enclosing the basements. The walls were unreinforced, common for low-rise residential buildings. Rather than using controlled granular fill, the contractor used clay backfill during a dry period of weather. This clay backfill shrank away from the concrete walls, leaving narrow gaps along the

FIGURE 7.27 Lateral soil shift causes pile failure.

FIGURE 7.28 Steel cross-ties stabilize piles.

full height of the walls. After a heavy rainstorm, the runoff filled these gaps, causing very high hydrostatic lateral pressures. Some of the walls which were not yet braced at the top toppled over, and some bent, cracked, and completely failed [Figs. 7.29(a), 7.29(b), and 7.30].

Lessons to be learned:

1. Foundation walls must be braced at top and bottom prior to any backfilling.
2. Only well-controlled and drainable, porous backfill should be placed against foundations and other retaining walls to avoid shrinking soils and subsequent development of high lateral soil and water pressures.

FIGURE 7.29 (*a*) Hydrostatic loads cause severe cracks in wall. (*b*) Buckling of foundation walls.

FIGURE 7.30 Total collapse of foundation walls [Fig. 7.29(a) and (b)].

7.4.3 Long Island Water Pollution Control Plant Outfall

During 1972 and 1973, precast concrete sections were installed for the Long Island Water Pollution Control Plant Outfall. Later, in April of 1974, during a dye test conducted to check the continuity of the ocean outfall, it was discovered that there were open joints along the bay length. Further investigation disclosed that 2300 ft (710 m) of pipe sections out of the overall length of 22,000 ft (6706 m) were either displaced, damaged, or completed separated.

The outfall consisted of an 84-in diameter, prestressed concrete pipeline pipe sections, 20-ft (6.1-m) long, which were generally assembled into 100-ft (30.5-m) strings and installed on pile supports spaced at intervals of 20 ft (6.1 m). The outfall consisted of three sections: a 17,000-ft (5182-m) bay section which was essentially horizontal; a 5000-ft (1524-m) section sloping at a 3% grade; and the remaining horizontal section. The failure occurred in the sloping portion, which deformed into a lateral S-shaped form.

This S shape was consistent with a buckling failure of a compression member under high stress. These types of failures may occur under several basic conditions:

1. Structural continuity
2. Axial compressive force
3. End restraints
4. A slight eccentricity with/or a lateral load adequate to trigger the initiation of the failure

Our review of all the data indicated that all the above-listed conditions were met in the summer of 1973.

1. The pipeline was completely closed, and all the sections were assembled and laid into the water under winter conditions.
2. The change in water temperature within and without the pipeline, which was estimated to be at least 30°F (-1°C) between February and August, induced the expansion of the concrete pipe.

3. End restraint was provided by the pile caps.
4. A small lateral soil pressure was considered to be the trigger most likely to have started the movement. The S bending shape was governed by lateral loading conditions. Movement downward was resisted by the pile support, and movement upward was counteracted by gravitational force. Therefore lateral movement was met with the least resistance. Additionally, uneven backfill will result in horizontal pressure component resulting in an unbalance sufficient to trigger the movement.

Lessons to be learned:

1. Special care should be taken to provide balanced lateral pressure by providing continuous quality control of backfill operations at underwater structures.
2. Although underwater structures are subjected to a rather small temperature differential, some relief must be provided to permit linear expansions.

7.5 UNEQUAL SUPPORT

A basic rule of engineering is that there is no load transfer without deformation. When loads are transferred to the soil through the foundation, this transfer is associated with a deformation of the soil. In other words, all footings settle when they are loaded. The amount of settlement is equal for different footings when soil resistances are identical and load distributions are equal. When they are not, differential settlements occur, and load transfer or tipping of the structure will result. The portion of the structure founded on the weaker soil will tip away. Where the framework is not continuous, the brittle masonry enclosure will crack (usually in a diagonal pattern) during the shear transfer. There are a great number of such structures, with foundations partly on rock and partly on soil, partly on rock and partly on friction piles, partly on stiff soil and partly on soft soil. There are no doubt many structures with footings bearing on different soils with different and unequal soil bearing resistances. The results are invariably distress and severe cracking (Fig. 7.31).

All these soil support deficiencies may be corrected and are often so rectified. However, these corrections are usually quite costly and involve underpinning the weaker soil or the weaker support. The use of additional piles or the installation of jack piles are some of the most commonly used and most successful methods. The famous Transcona grain elevator in Manitoba, Canada (Fig. 7.32) showed unequal settlement and tipped. This structure consisting of 65 concrete bins 93-ft (28-m) high on a concrete mat 77×175 ft (24×53 m), settled 12 in (305 mm) and then tipped

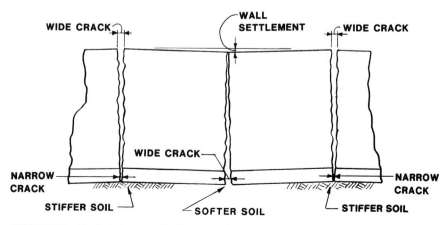

FIGURE 7.31 Cracks indicate rotation from vertical (heave) or settlement movement.

FIGURE 7.32 Tilt in Canadian grain elevator.

27° when about 85% full of grain. The structure weighed 20,000 tons (178 MN) and contained at the time grain weighing 22,000 tons (195 MN). Already when repair work was begun in 1914 the massive structure, which was 34 ft (10 m) out of level, was jacked back to use with underpinning to rock—a remarkable engineering feat.

Sometimes, other methods of soil stabilization by cement or chemical injection are used. One other correction procedure utilizes the addition of a subsurface enclosure (usually a tight sheet-pile cell) that increases the bearing value of the soil. This was done on a grain elevator in Portland, Oregon, in 1919, where 2600 pinch piles were added to enclose the foundations after the structure showed progressive unequal settlement and was tipping.

Not so fortunate was a smaller grain elevator in Fargo, North Dakota, which tilted in 1955 when it was only 1 year old. This elevator, which contained 20 bins 120 ft (37 m) high, was supported on a concrete mat measuring 52×216 feet (16×66 m). As the mat tilted, the bins crumbled and the structure was wrecked and had to be abandoned.

There are many classical examples of tilting buildings as a result of unequal settlements. The Tower of Pisa is probably the best known and researched example and is still in use, as is the Palace of Fine Arts in Mexico City (Fig. 7.33) with its settlements measured in meters. The Guadalupe Shrine and Cathedral, also in Mexico City (Fig. 7.34), is so distressed, cracked, and tilted that it makes visiting tourists nervous.

Buildings partially on earth and partially on rock have too often been designed on the incorrect assumption that the mere use of local building code allowable bearing values will ensure equal settlement. Where such variable bearing conditions exist, a separation joint must be provided so that each section of the building can then act as an individual separate structure. These buildings, especially when they utilize bearing walls, often crack almost immediately where such a joint is missing. Thus, nature provides the omitted joint (Fig. 7.35). Underpinning and other repair costs together with the additional economic loss caused by delayed or vacated use far exceed the cost of properly designed and constructed separated foundations.

Our firm has also investigated several instances where soil shrinkage due to local dewatering triggered differential settlements even though the original construction was proper and followed faultless design.

FIGURE 7.33 Settlement of Palace of Fine Arts, Mexico City.

FIGURE 7.34 Foundation failure of Guadalupe shrine.

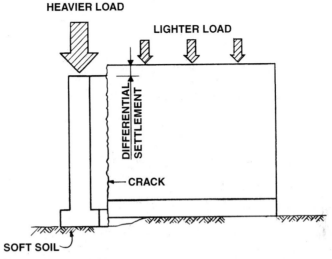

FIGURE 7.35 Differential settlement.

7.6 DRAG-DOWN AND HEAVE

As a footing is loaded, the supporting soil reacts by yielding and compressing to provide resistance. The actual compressing of the soil takes place rapidly in the case of granular soils but much more slowly for clays. Once this compression occurs, the structure remains stable, since the foundation no longer settles. This stability depends on fringe areas as well as the soil directly below the footing, or, in the case of piles, on the soil near the pile tip. If the soil below the footing is removed or disturbed, settlement or lateral movement is induced. Similarly, if the fringe area is removed, the soil reaction pattern must change. Even if sufficient resistance still remains without measurable additional settlement, the center of the resistance is changed, thus affecting the stability of the footing. When the fringe area is loaded by a new structure, causing new compression in the affected soil volume, the old area can be required to carry some of the new load by internal shear strength. In such a case, there will be an additional unexpected new settlement of the previously stable building. In the event the new building is not separated from the existing construction, the settlement caused by the new load will cause a partial load transfer to the existing wall, with possible overloading of the previously stable footing.

In plastic soils, these new settlements often are accompanied by upward movements and heaves (Fig. 7.36) some distance away. The liquid in the soils cannot change volume, and every new settlement must produce an equal volume heave.

When an entire structure settles at a slow rate for a long period, these settlements may be acceptable as long as they are *uniform. Large differential settlements* will result in damage.

To disregard such elementary considerations is often the cause of much litigation stemming from cases where new foundation construction in urban areas suddenly places neighboring existing construction in danger of serious damage or even complete failure. Under several building codes (The New York City Building Code is one such example), if new construction requires excavation more than 10 ft (3.0 m) in depth adjacent to an existing foundation, all precautions to avoid damage to the existing conditions are the obligation of the new project. If the new excavation depth is 10 ft (3.0 m) or less, the obligation for protection falls on the owner of the existing building should its foundations be higher than the new footings.

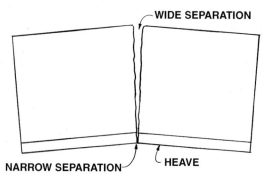

FIGURE 7.36 Upheaval.

In medium- to well-compacted sands, pile driving demands heavy power, and the vibration impulses can shake adjacent structures to the point of serious damage.

Drag-down from soil shrinkage when water tables recede or from desiccation by tree growth, with consequent differential settlements, is a phenomenon observed in many localities. The failure of a theater wall in London in 1942 was correlated with the growth of a line of poplar trees. In Kansas City, Missouri, 65% of the homes in one residential area were found to be affected by soil desiccation from vegetation growth in foundation plantings.

Piles embedded in soil layers that will consolidate from dewatering or from load surcharge are subject to overloading when the greater soil density increases surface friction and the new soil loads "hang" on the piles. This phenomenon is sometimes referred to as *negative friction*. The added load may cause considerable increase in settlement and may even pull the pile out of the pile cap or pull the pile cap free of the column or wall.

Dewatering operations are commonly caused by new construction or pumping by water supply companies.

In some instances, as consolidation of organic clay strata takes place, piles continue to sink with additional settlement of the structure. This was the case at the Queens Apartments in New York.

7.6.1 Queens Apartments, New York

Settlements and masonry cracks developed in many levels of the buildings of this project (Fig. 7.37). The cracks appeared in both interior partitions and the exterior brick walls.

These structures, located in Flushing, New York, were constructed over marsh land and were supported by timber piles. An earlier subsurface investigation of this project indicated that it was likely that at least some of the piles in the three-pile group supporting the distressed corner of the building had broken. This fact could not be confirmed without excavation or probes. The breakage may have occurred when the piles were driven or subsequently when the loads resulting from consolidation of the organic clay stratum were imposed on them.

In an effort to assess the potential for additional settlement and their likely magnitude, a laboratory and a field program were performed. In the laboratory, the rate of settlement due to a secondary compression of samples of the organic clay was measured. In the field, a sensitive settlement plate (Fig. 7.38) was monitored for movement every 2 weeks for about 6 months, using a depth caliper accurate to 0.001 in (0.025 mm).

The area selected for installation of the plate was the interior corner directly outside of the laundry room. This area was chosen due to its close proximity to settlement cracks; it also provided sufficient overhead clearance for drilling.

FIGURE 7.37 Masonry cracks due to settlement.

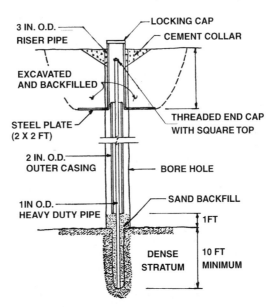

FIGURE 7.38 Instrumentation for measuring soil settlement.

7.34 FOUNDATION FAILURES AND REPAIR: HIGH-RISE AND HEAVY CONSTRUCTION

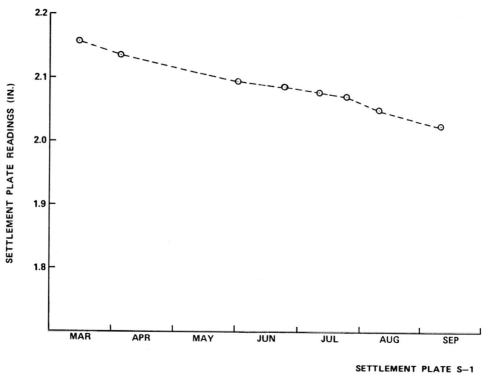

FIGURE 7.39 Chart of measured settlement.

A 1-in (25-mm) diameter heavy-duty pipe was driven into the lower sand stratum to provide a fixed reference point. A 2-in (51-mm) diameter outer casing protected this pipe from down-drag effects from the overlying fill and clay.

Analysis of laboratory data has suggested that settlement of the organic clay, as a result of the fill above it, can be expected to be about 1 in (25 mm) every 10 years. The field measurements from the settlement plate indicated even greater settlement.

Figure 7.39 presents the data obtained from the settlement plate device, with settlement rates of up to about ¼ in (6.4 mm) per year. Thus, the field data suggested settlements about twice those indicated by the laboratory data.

Lessons to be learned:

1. Do not guess whether a structure that has settled will continue to do so. Instrument the foundations and the soil to assess potential future settlements.
2. Placing a structure on piles bearing in marsh land always involves risks of settlements.

7.6.2 Edgewater, New Jersey, High Rise

The piles supporting the first-floor slabs of this complex settled and pulled the slab downward. This caused settlements and cracks of the slabs and tilting and cracking of partitions (Fig. 7.40). The first signs of distress were "sticking" doors and windows.

An in-depth structural investigation was carried out, revealing that although the main supports for the columns were bearing on steel piles driven to resistance in rock, the piles supporting the

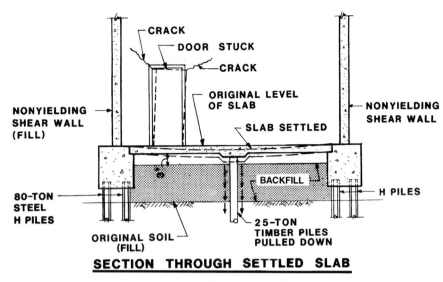

FIGURE 7.40 Structural distress relative to differential slab deflections.

slabs were only wooden friction piles. The pattern of slab deflection was confirmed by instrument survey and the corresponding contour map (Fig. 7.41).

Probes below the slab showed that the backfill had settled approximately 3 in (76 mm) (Fig. 7.42). The negative friction effect of the downward settlement of the backfill and consolidation of the soil pulled the piles down. Consequently, the concrete slab attached at the top of the piles was dragged down.

The repair required cutting into existing slabs and adding new reinforced-concrete beams which transferred the slab loads into the reliable column piles (Fig. 7.43).

Lesson to be learned:

When piles are driven into compressible soils, negative friction effects should be taken into consideration. Additional soil surcharge will increase the drag-down forces.

Heaving of footing contact surface is known to neutralize soil bearing by frost action. We have investigated several cases where during extremely cold winter, frost penetrated below slabs on grade and cracked buried water and sprinkler lines, causing ice buildup which heaved the structure above and even sheared some of the supporting columns.

7.7 DESIGN ERROR

Unfortunately, many foundations are designed with insufficient prior subsurface investigation or with disregard of the true soil conditions. This results in inadequate support for the structure, often requiring later expensive corrective work.

Standard structural design procedures for foundations will result in an adequate factor of safety; most errors are in judgment, leading to assumptions or decisions which are not consistent with the actual behavior of the soil later.

A common design error is often made, usually in an effort to save initial construction costs. In these cases the designers provide for pile support for the walls and the roof, while the main floor is placed on "more or less" compacted sand over fills, rubbish, peats, organic silts, and other com-

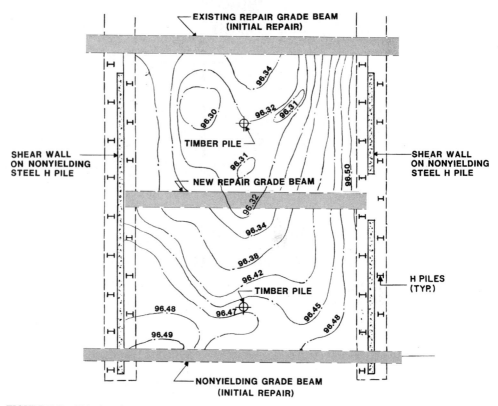

FIGURE 7.41 Slab elevations suggesting settlement of up to 2½ in (64 mm).

pressible layers. Our own firm has seen many industrial plants where the roof was supported on piles while the floor supporting expensive equipment was placed over soil fill. It seems more logical to support expensive equipment on piles and let the roof settle. Heavy factory machinery generally requires installation on precisely level floors, and small differential settlements can become cause for concern.

Even in warehouse use, floor deformations become objectionable, making efficient use of forklifts impractical. Placing floor slabs on inadequate soil fills obviously is a false economy. True cost efficiency is measured by the total of the initial and in-life service costs.

7.7.1 The Alaska Pier Collapse

This unusual failure was caused by a design error resulting from inadequate preinvestigation study of local data and conditions. The formation of huge ice blocks atop the pier pilings was totally unexpected. The pier designers had not taken into account loads of such magnitude being imposed on the structure. The total collapse of the pier occurred at low tide when the total weight of the ice was suddenly suspended from the sloped piles (Fig. 7.44).

There were unique conditions that contributed to this collapse. These were the extremely large variations between low and high tide, which were approximately 20 ft (6.1 m). The blocks of ice that had formed during the winter suddenly lost their support when the thaw came in the spring.

FIGURE 7.42 Settlement of fill leaving 3-in (76-mm) void.

FIGURE 7.43 New concrete beams shift slab loads to steel piles.

FIGURE 7.44 Ice-encrusted piles.

The enormous weight of the ice was shifted to the piles, which bent, causing fracture of the concrete pile caps. The thawing of the ice at the contact surface with the piles also caused the ice blocks to slide down diagonally while being wedged between the piles (Fig. 7.45). The bending stresses at the ends caused by the wedging forces (see force and moment diagrams shown in Figs. 7.46 and 7.47) severely cracked the concrete pile caps (Fig. 7.48) leading to the eventual collapse of the pier.

The author's files show that the partially completed pier was supported on pile bents spaced 20 ft (6.1 m) on center. The bents consisted of both vertical and diagonal piles. Twenty-inch octagonal prestressed vertical piles and 20-in (508-mm) diameter batter steel piles were arranged in pairs forming quasi A-frames. The length of the piles was between 70 and 80 ft (21 to 24 m). The piles were driven into 15 ft (5.0 m) of silt, then 55 ft (16.8 m) of silty gravel and sand underlain by layers of clay. By midwinter, huge blocks of ice, approximately 20-ft thick (6.1 m), had formed on the piles. The blocks weighing 50 tons (445 kN) or more slid down the sloped piles, damaging the piles and pile caps.

The recorded telephone conversation between the design engineers on the project went like this:

> D: We've got great big problems up here...I went down the dock...a good half of our pile caps where the brace pile...are cracked.
> They're cracked bad. They've got an inch and a half (38 mm) crack in some of these. We know it's because of the tidal action and the ice.
>
> A: It looks like...the weight of the ice is causing such a moment up there that it is cracking the heck out of it. We're going to have to take the panels off, go in there and remove the caps and redesign the whole thing....
>
> D: We've got a good connection up there,...it's very, very, rigid, but it's bending that's doing it. Would have saved us I think, if we'd had a lot more of those stirrups in there, but there are only No. 5 bars at about 18 in (457 mm) on center and it's just not enough.

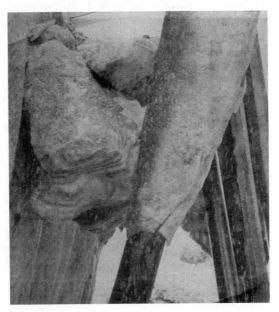

FIGURE 7.45 Ice melts, and blocks become wedged between piles (Fig. 7.44).

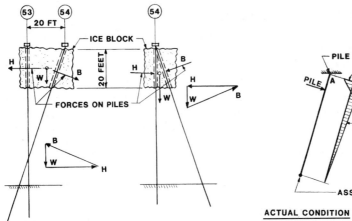

FIGURE 7.46 Lateral forces on battered and vertical piles due to ice action (Fig. 7.45).

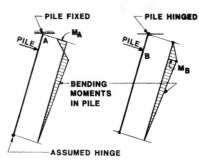

FIGURE 7.47 Lateral forces on battered piles due to ice melting (Fig. 7.46).

7.40 FOUNDATION FAILURES AND REPAIR: HIGH-RISE AND HEAVY CONSTRUCTION

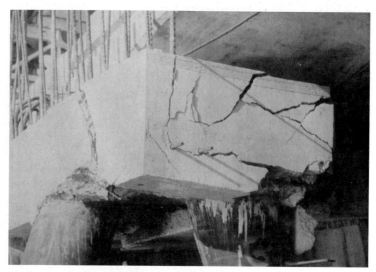

FIGURE 7.48 Failure of concrete pile caps due to ice action.

While this failure was considered by many investigators to be caused principally by a design error, it was also established that damage was caused to the vertical piles during installation with overjetting and pounding with the diesel hammer. Insufficient penetration of the batter piles was also confirmed in several cases.

Lessons to be learned:

1. The effect of ice formation and resulting forces must be considered in the design of structures in northern exposures.
2. Piles subjected to lateral loading and bending must be sufficiently embedded into the concrete pile caps.
3. The consequences of extreme variations of tidal rise and fall should always be taken into consideration in design of marine structures.

7.7.2 Washington, D.C., Foundation Mat

A 3-ft thick (0.92-m) foundation mat was cast with draped tendons (Fig. 7.49) in accordance with the design requirements, and shortly afterward was post-tensioned. Unfortunately, as soon as the high prestress loads were applied, the slab lifted upward and cracked (Fig. 7.50).

The error was discovered immediately because it was so obvious. It was amazing that the incorrect procedure eluded everybody involved with the project. The fact that the designer of the mat was a prestigious engineer only proves that no one is immune to mistakes.

As can be seen in Fig. 7.51, there could be no equilibrium of forces in the foundation mat without the presence of the building columns. Realistically, the design should have planned for the prestressing to be executed in stages, with the prestress forces increasing as more and more weight of the upper levels of the structure was imposed. This procedure was not followed, and the failure ensued.

The repair was carried out using rock anchors to provide the necessary vertical reactions to the prestressing forces (Fig. 7.52). The mat was repaired, and the prestressing was successful the second time around.

FIGURE 7.49 Concrete reinforcing prior to placement of concrete.

FIGURE 7.50 Prestress loads cause slab uplift.

7.42 FOUNDATION FAILURES AND REPAIR: HIGH-RISE AND HEAVY CONSTRUCTION

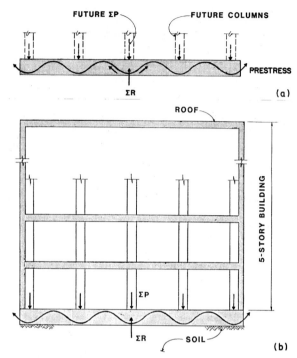

FIGURE 7.51 (*a*) Distress caused by design error in assumption of prestressing forces. (*b*) Lack of equilibrium of prestress forces is obvious.

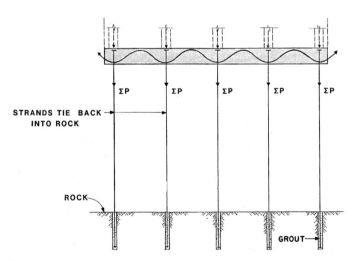

FIGURE 7.52 Rock anchors provide vertical reactions to prestressing forces [Fig. 7.49(*a*) and (*b*)].

Lessons to be learned:

1. Posttension sequence must be carefully reviewed on a step-by-step basis.
2. Total equilibrium of forces must be ensured.

7.7.3 Boston Housing Slab Settlement

Where slabs on grade are expected to be at the same level as adjacent structurally supported slabs, total reliance on well-compacted backfill is merely wishful thinking (see Fig. 7.53). Unless a detailed compaction control program is set in advance of construction, an undesirable step will ensue as a result of differential settlement.

At the Boston project, the entrance to the 10-story building resulted in a hazardous step [painted red (Fig. 7.54, see arrows)]. The proper design detail (Fig. 7.53) will support the slab on grade (or fill) directly on a seat prepared in the foundation wall, thus preventing downward settlement.

Lesson to be learned:

To avoid steps caused by differential settlement, concrete slab on grade must be either cast on well-compacted fill or designed to bear on adjacent walls.

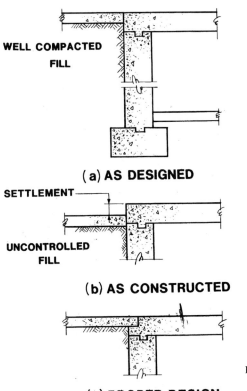

FIGURE 7.53 Slab design on fill.

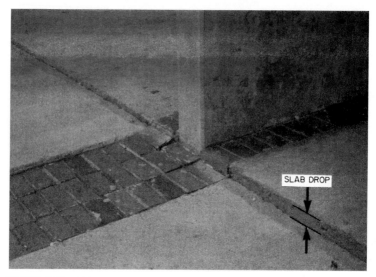

FIGURE 7.54 Slab settlement.

7.8 CONSTRUCTION ERRORS

There are two common types of construction errors:

1. Temporary protection measures during the construction phase
2. Foundation work itself

Most foundation failures involve the first type of error, relating to temporary shoring and bracings and temporary cofferdams for lateral protection and pumping operations. Because of the temporary nature of these structures, safety measures are often kept to a minimum for economy reasons.

Many construction errors on foundation work also are a result of the same deficiencies that exist in superstructure concrete work, such as improper concrete quality and improper rebar placement. Where deficient material is provided, in general there is absolutely no difference between superstructure and foundation work. The exceptions are foundation elements such as piles and caissons where "blind work" is involved. Cavities within cast-in-place piles are often found where quality control was lacking. The omission of verification methods early in the construction progress is partially responsible for the enormous losses caused by such errors.

The unique cases involve the placement of foundation elements such as footings, piers, or piles on inadequate bearing soil strata—footings bearing on soft soils, or piles driven "short" so as to hang high in improper and insufficient bearing soil.

In 1958, a newly constructed 11-story building in Rio de Janeiro toppled and overturned. It was reported that the driven piles were about 20 ft (6.1 m) shorter than piles used for adjacent structures. Initially the settlement of the building was considered to be "natural." After several days of attempts to save the leaning building, the building was evacuated. Hours after the evacuation, the structure fell flat in *just 20 seconds*.

7.8.1 Lower East Side Collapses

On April 4, 1984, two structures located at Delancey Street in lower Manhattan collapsed during construction (Fig. 7.55) about 1 h after a 3½-in (89-mm) concrete floor slab had been

FIGURE 7.55 Collapse of structure.

poured. Two workers were killed by the collapse, including the general superintendent for construction.

On arrival on the scene a day later, I found that the debris was still being removed from the site. There was severe damage to the two adjoining structures, both of which were in precarious condition. One structure was subsequently demolished, and the other was braced and shored. Our investigation determined several improper construction procedures including inadequate field supervision. We also identified a number of design deficiencies.

Upon exposure after the collapse, several of the footings were found to have been cast as less than half (46%) the size required by the design drawings. Out of three interior footings, two footings had been cast eccentrically to the center line location [one 7½ in (191 mm) and the other by as much as 11½ in (292 mm)]. The eccentrically positioned footings were visibly rotated and pitched from one end to the other by as much as 5 in (127 mm). The general layout of the building may be seen in Fig. 7.56.

The structural analysis indicated that the bearing capacity of these footings was considerably reduced. Using the actual loads computed to be present at the time of collapse, we estimated that even using the optimum possible bearing capacity of these footings, they would have been loaded at 25% over ultimate capacity *and had to fail.* See Figs. 7.57 and 7.58 for footing diagrams and load table.

Another serious deficiency we uncovered related to the method of temporary support provided for the central bearing walls. These walls were subjected to a twisting rotational couple due to the weight of the two unequal bearing walls (Fig. 7.59). The fact that the second floor was missing (it had previously been removed) compounded this effect through the lack of lateral support. The method of temporary support used by the contractor had a considerable inherent potential for instability. Thus, when the footings settled vertically, dislodging wedges and shims, the resultant movements initiated the total collapse.

One of the striking omissions in this failure was the lack of a construction procedure, which is essential to the successful execution of a difficult foundation reconstruction of an old structure.

The design deficiencies included main girders which were calculated to be stressed between 33 to 39 ksi (228 to 268 MPa) when subjected to full dead and live loads. Even the review of the design of the footings indicated overstresses in the range of 27 to 44%. Needless to say, the lack of shop drawings did not help matters.

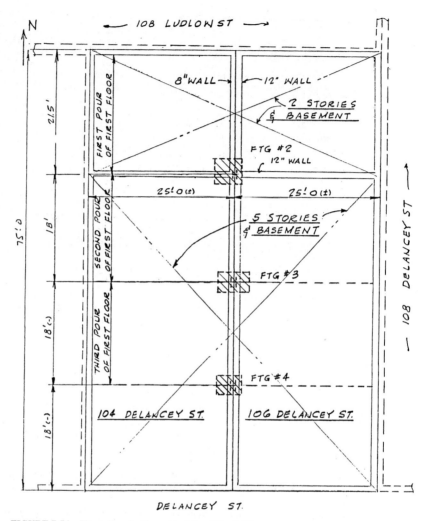

FIGURE 7.56 Floor plan of collapsed building (Fig. 7.55).

Lessons to be learned:

1. Eccentrically loaded footings should be avoided at all costs, unless provided for in the design by oversized footings.
2. A construction procedure outlining step by step the method of temporary support, underpinning, shoring, and bracings is a must in construction of new foundations for old and deteriorated structures.
3. Lateral stability and eccentric effects on critical bearing walls must be considered and accounted for.
4. There is no substitute for good control of a project, including the preparation of shop drawings and thorough field inspection.

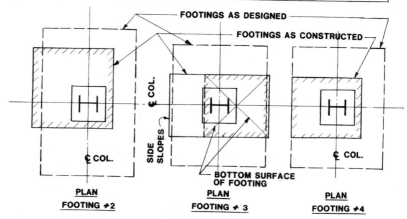

FIGURE 7.57 Substandard footings responsible for collapse (Fig. 7.55).

7.8.2 Apple Juice Factory—Drilled-in Piers

When I first saw this project, which had come to a halt, both owners and contractors were aware of the serious problem they were facing. The exposed sides of several concrete piers could be seen to contain concrete contaminated with large quantities of brown mud (Fig. 7.60). Windsor probe tests had already been taken, and these confirmed the nonuniformity of the strength of the concrete. It was absolutely clear to everybody that expensive corrective work was required. The nature and extent of this work and the causes for the defects were sure to be in contention.

The foundation was being constructed to support an apple juice plant. The foundation system consisted of concrete drilled-in piers (caissons), which were cast in place into big holes augured into the ground. These piers ranged in size from 30 to 42 in (762 to 1067 mm) diameter, with "bells" of 42 to 90 in (1067 to 2286 mm) diameter, and were 15 to 55 ft (5 to 17 m) long. The piers were designed to bear on top of bedrock with a bearing value of 12 tons per ft^2 (1149 kPa). Unfortunately, *all the piers had been completed* at the time the defects were discovered.

The design of drilled piers with bells in clayey or silty sand is feasible provided the soil in which the bells are formed is cohesive. That condition may be realized where the bells are above the ground-water level. Where the shafts are flooded, as the case often was at this particular construction site, the success of belling is highly questionable. Especially doubtful is the possibility of installing the bells and maintaining them intact without the sides caving in and collapsing (Fig. 7.61). The failed caissons confirmed this opinion. The presence of groundwater at several of the caissons should have been expected from the 23 borings taken at the site prior to construction.

What was more stunning was the interference of the design Engineer of Record with the contractor's methods and operations. When it became clear that it was impossible to pump the holes

7.48 FOUNDATION FAILURES AND REPAIR: HIGH-RISE AND HEAVY CONSTRUCTION

	1*	2*	3†	4†	5	6	7	8
	ALLOWABLE LOADS		DESIGN LOADS		DES.LD. ÷ ALLOW.LD.		DES.LD. ÷ ALLOW. LD.	
	N.Y.C. BLDG. CODE	ONE-THIRD OF ULTIMATE	WITH 1-12" & 1-8" WALLS	1-12" WALL ONLY	3 ÷ 1	3 ÷ 2	4 ÷ 1	4 ÷ 2
FOOTING #2	128^K	67^K	336^K	315^K	2.63 (163% OVER)	5.03 (403% OVER)	2.46 (146% OVER)	4.70 (370% OVER)
FOOTING #3 (FULL AREA)	121^K	216^K	358^K	313^K	2.95 (195% OVER)	1.66 (66% OVER)	2.59 (159% OVER)	1.45 (45% OVER)
FOOTING #3 (BOTTOM AREA)	79^K	272^K	358^K	313^K	4.53 (353% OVER)	1.32 (32% OVER)	3.96 (296% OVER)	1.15 (15% OVER)
FOOTING #4	83^K	114^K	358^K	313^K	4.30 (330% OVER)	3.14 (214% OVER)	3.77 (277% OVER)	2.74 (174% OVER)

	9	10	11**	12	13
	ESTIMATED LOAD AT TIME OF COLLAPSE		ULTIMATE LOAD	COLLAPSE LD. ÷ ULTIMATE LD.	
	WITH 1-12" & 1-8" WALLS	WITH 1-12" WALL ONLY	(PER WOODWARD-CLYDE)	9 ÷ 11	10 ÷ 11
FOOTING #2	251^K	217^K	200^K	1.26 (26% OVER)	1.09 (9.0% OVER)
FOOTING #3 (FULL AREA)	210^K	142^K	648^K	0.32 (32%)	0.22 (22%)
FOOTING #3 (BOTTOM AREA)	210^K	142^K	167^K	1.26 (26% OVER)	0.85 (85%)
FOOTING #4	210^K	142^K	342^K	0.61 (61%)	0.42 (42%)

ALLOWABLE BEARING DETERMINED BY WOODWARD-CLYDE CONSULTANTS INC.
*BASED ON ACTUAL FOOTING SIZES; N.Y.C. BLDG. CODE ALLOWABLE, 3 TONS PER SQ.FT.; EFFECT OF ECCENTRICITY NOT INCLUDED.
**EFFECT OF ECCENTRICITY CONSIDERED.
†DESIGN LOADS USED BY OUR FIRM.

LIVE LOADS
1ST FLOOR: 100 PSF (RETAIL)
2ND FLOOR: 75 PSF (RETAIL)
3RD, 4TH, 5TH FLOORS: 100 PSF (LIGHT STORAGE)
ROOF: 30 PSF

FIGURE 7.58 Load table illustrates design flaws (Fig. 7.55).

dry, the engineer directed the subcontractor to use larger pumps. [The ground-water infiltration was greater than the capacities of the pumps by more than 1000 gal/min (6308 L/s).] When the contractor did not comply, the engineer *bought large pumps using his own funds* and delivered them to the contractor.

Thus the Engineer of Record committed three serious errors:

1. *A first major technical error.* It should be common knowledge that it is improper to cast concrete into holes while simultaneously pumping the holes. Such pumping only serves to disturb the wet concrete and mix it with soil, resulting in contaminated and defective concrete as per Fig. 7.61 (properly filled caisson is shown in Fig. 7.62).

2. *A second technical error.* There was no way for the engineer's inexperienced inspector to verify in the flooded shafts the conditions of the bearing rock and whether the rock was level or sloped (Fig. 7.63).

3. *A policy error.* An engineer should never interfere with the contractor's methods and means and definitely should not supply the contractor with the tools to perform the engineer's instructions, which also happened to be completely wrong. Needless to say, the subcontractor for the

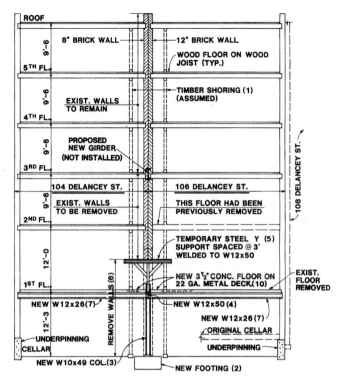

FIGURE 7.59 Construction alternatives contribute to collapse (Fig. 7.56).

drilled piers should not have knowingly followed the erroneous instructions or should have known as a supposed expert in his field that they were flawed.

An experienced caisson contractor would also have known to continuously keep the bottom of the tremie (underwater concrete) pipe inserted into the fresh concrete (Fig. 7.64). Whereas the selected foundation system for this structure was totally inappropriate (a system of H piles or even a continuous concrete mat should have been used), the construction methods used failed to produce a reliable foundation. Not only was the concrete defective in many instances, but even the quality and levelness of the rock bearing strata could not be confirmed.

The sad ending to this story was that the foundation system as constructed had to be completely abandoned, and lengthy litigation ensued.

Lessons to be learned:

1. Designers should carefully check the feasibility of bell excavations in soil prior to concrete casting and especially in water.
2. Tremie placing of concrete should always be in still (static) water. Pumping during concrete placement operations should not be attempted.
3. The installation of drilled concrete piers should be performed only by experienced contractors.
4. Full-depth concrete test cores should be extracted from the first five piers, to verify the quality and strength of the concrete. The *early detection* of deficiencies is essential.
5. Engineers should not direct contractors' field operations, should not supply them with tools to perform the work, and should not interfere with their methods and means unless safety is involved.

7.50 FOUNDATION FAILURES AND REPAIR: HIGH-RISE AND HEAVY CONSTRUCTION

FIGURE 7.60 Defective concrete piers.

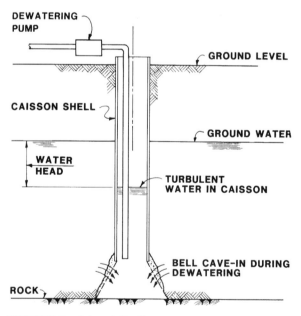

FIGURE 7.61 Caisson bell collapse.

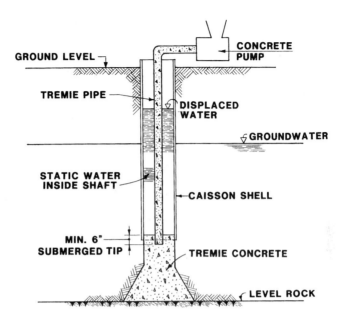

FIGURE 7.62 Proper concrete placement in caisson.

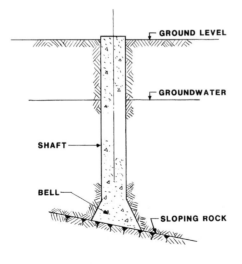

FIGURE 7.63 Pier construction on sloping rock.

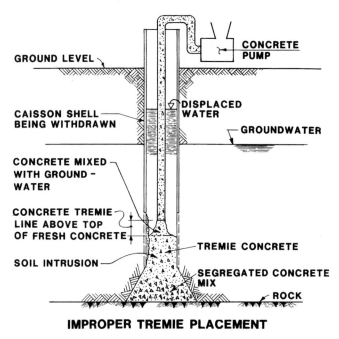

FIGURE 7.64 Improper use of tremie.

7.8.3 Marble Hill

When cracks in the brick wall of the Marble Hill one-story structure started to widen, it became apparent that a serious condition existed below the concrete grade beam supporting the wall. After the soil adjacent to the grade beam was removed by excavation, a long diagonal crack was uncovered in the beam. The crack had been previously crudely patched with mortar, which itself was badly cracked and served no purpose whatsoever. There was no attempt to find the culprit! The failure was a classical shear crack with a vertical shift on both sides of the crack. A close look at this brick wall told the story of a settlement that had taken place *prior* to its construction. There was an additional course of brick on the left side of the crack (Fig. 7.65). The solution to this mystery was found when, after further excavation, it was discovered that a pile designed to support the grade beam was missing. It was clearly a construction error. The attempt to patch the crack obviously was not adequate to correct the serious omission.

Lesson to be learned:

When cracking occurs, determine the cause of the distress. Do not guess! Only when the true causes are found can an efficient, long-term correction be designed and executed.

7.9 FLOATATION AND WATER-LEVEL CHANGE

Except in well-consolidated granular soils, a change in water content will modify the dimensions and structure of the supporting soil, whether from flooding or from dewatering. Many cases have been recorded where pumping by occupants for cooling water or by water companies to increase the capacity of their water supply resulted in receding of ground levels and in turn caused settlements with severe damage.

FIGURE 7.65 Foundation wall failure during construction.

Pumping from adjacent construction excavations has also affected the stability of existing spread footings and even caused drag-down on short piles. This fact is in part responsible for the insertion of the requirement of recharging of ground-water levels in some foundation contracts. The effectiveness of these recharging pools has been and continues to be a touchy subject among foundation designers and geotechnical engineers.

Construction of new dams has also been found to be responsible for lowering of river levels, thereby causing severe damage and cracking of adjacent structures.

Clay heaves from oversaturation must also be expected and, in such soils, the structures must either be designed to tolerate upward displacement or else the supporting soil must be protected against flooding. Many parts of the world have trouble from heaving bentonitic clays, for which a perfect solution is yet to be found. Where water infiltration into the soil is effectively prevented, and there is no change in water level, the soil volume should stay stable.

7.9.1 Brooklyn Hospital—Settlement of Delivery Room

This three-story hospital had performed perfectly for many years after its construction. Suddenly, masonry walls started to crack, "for no apparent reason" (Fig. 7.66). The most severely affected wing of the building contained both the newborn delivery room and the general surgery room, whose continuous operation could not be stopped. I was actually measuring cracks inside the delivery room (dressed in scrubs) while a newborn was delivered. It was then that I found it necessary to temporarily support this wing, accomplished by the use of diagonal steel bracing system (Fig. 7.67).

Once the structure was secured, we started the investigation of the sudden cracking and distress. To our amazement we found that the water supply company operating in the area had recently abandoned a great number of existing wells.

As a result, the ground-water level had risen sharply in the entire area. The immediate result was water penetration through foundation walls of the hospital which had not been waterproofed to such a high level. To solve the new problem of water infiltration into basement areas, pumps

7.54 FOUNDATION FAILURES AND REPAIR: HIGH-RISE AND HEAVY CONSTRUCTION

FIGURE 7.66 Vertical stress crack, hospital.

FIGURE 7.67 Shoring to abate settlement (Fig. 7.66).

FIGURE 7.68 Underpinning and stabilizing building corner (Fig. 7.66).

had then been installed by the hospital management. However, an unforeseen side effect was that the continuous pumping caused the removal of most of the fines in the soil directly below the footings supporting the affected wing, thereby causing the cracks and settlements.

There was no easy solution short of underpinning the building. This was accomplished by the use of jack-piles, which were pushed down into the ground, with the weight of the building serving as counterweights (Fig. 7.68). The steel pipes which had been pretested for the required load were then filled with concrete, and then the jacks were removed.

Lessons to be learned:

1. Lowering or raising of ground-water level affects the bearing capacity of soils.
2. Pumping for new excavations may cause settlements of buildings. Therefore, water-level readings should be monitored, and protective measures taken.

7.9.2 Industrial Building, Hackensack, New Jersey

This 1-story factory and office building was located in the bottom of a geological lake, atop deep glacial deposits of silts and clays in successive layers, forming what is known technically as *varved silt deposit*. In this very loose material, areas of soft shore land become marsh land. Considerable areas of such marsh land have been filled in to form usable development property. The existence of the soft underlying material should signal potential difficulties and must be taken into account in all construction operations. This type of soil is very susceptible to drainage and is associated with a considerable volume change.

The local sewer authority had issued a contract for the construction of a new sewer. A deep trench was excavated using steel "soldier beams" 10 ft (3.0 m) on center and timber sheeting. The excavation was braced at two levels. As the excavation approached the factory, various signs of damage started to appear in the form of cave-ins and vertical and diagonal cracking of walls and partitions (Figs. 7.69 and 7.70).

FIGURE 7.69 Masonry distress caused by soil shrinkage (loss of bearing support).

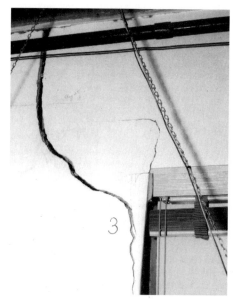

FIGURE 7.70 Interior distress (Fig. 7.69).

The damage was reviewed by our firm, and it was evident that the undermining was caused by the drainage operations in the sewer trench causing a shrinkage of the subsoil under the building's foundation. This was followed by the lateral movement of soil from under the building toward the deep hole at the sewer trench. The nearest edge of the sewer excavation was approximately 220 ft (67 m) from the building. Yet the effect of the drainage of the sewer excavation extended to great distances on each side of the sewer, and soil settlement was even visible in the parking area in addition to along the walls of the building itself.

Fortunately, the main structure of the building was supported on piles which had been carried into materials deep enough and dense enough not to be affected by this drainage. There was, however, still the effect of "negative friction" on the piles, with the additional soil weight adding to the loads which were already present.

Once the pumping operations ceased, no additional damage was recorded. Slabs which had settled by as much as 3 to 4 in (76 to 101 mm) had to be removed and rebuilt. Slabs which settled to a lesser degree, leaving a hollow space between the underside of the slab and the ground, were restored by pressure grouting. Wall cracks were repaired using ordinary cement mortar and refinished.

Lessons to be learned:

1. The ground-water level outside a deep excavation should be monitored using piezometer pipes installed at the adjacent structures which may be affected. Criteria for the maximum permissible water-level drop should be set.
2. In the event of a drop exceeding the criteria and/or the initiation of damage, the pumping methods and procedures should be reviewed to minimize the damage.
3. The services of competent geotechnical and structural engineers should be solicited.

7.9.3 West 41st Street Structure—Wood Piles

Damage to the exterior brick masonry of this building was caused by settlement of timber piles resulting from lowering of ground-water level at an adjacent railroad construction. Probes below the building showed that the timber piles were rotted and could not accommodate the additional loads generated by the drop of the water level. The damage to exterior and interior masonry is shown in Figs. 7.71 and 7.72.

7.10 VIBRATION EFFECTS

Earth masses which are not fully consolidated will change volume when exposed to vibration impulses. The vibration source can be construction equipment, especially pile drivers, mechanical equipment in a completed building, or even traffic on rough or potholed pavement, as well as blasting shock.

We have investigated many cases of damage as a result of pile driving. Heavy damage can be caused when both impact hammers and vibratory hammers are used. In several cases, entire rows of buildings have had to be condemned as unsafe and were subsequently demolished.

FIGURE 7.71 Brick distress resulting from soil water removal.

7.58 FOUNDATION FAILURES AND REPAIR: HIGH-RISE AND HEAVY CONSTRUCTION

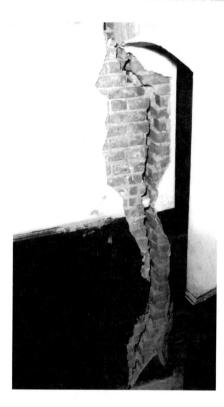

FIGURE 7.72 Interior distress (Fig. 7.71).

Blasting operations must be carefully programmed, to avoid serious damage to adjacent structures. A criterion for maximum particle velocity, which is a measure of the intensity of vibration, has to be established, and constant readings by the use of seismographs have to be taken. Damage from blasting is often so potentially severe that the size of explosive charges have to be limited to keep the vibrations to tolerable levels. Factors such as the condition, rigidity, material brittleness, and age of nearby structures must be considered. Where adjacent structures are founded on rock, the seam structure of the rock itself also must be taken into account. A complete study of vibration transmission and correlation among intensity, wave length, and possible damage to various types of structures is given in a Liberty Mutual Insurance Company report, "Ground Vibration Due to Blasting," by F. J. Crandell.[4]

7.10.1 Lower Manhattan Federal Office Building

After heavy impact hammers were used to install the piles for a prestigious building, the lower Manhattan Federal Office Building, the adjacent structures were seriously shaken. Street pavements settled (Fig. 7.73) as much as 16 in (406 mm) after only one-fourth of the piles had been driven. The installation of steel sheetpiling was then attempted in an effort to protect the streets against further subsidence. This attempt was successful against further settlement and damage to sewer and water lines, but even after the contractor shifted to vibratory hammers, the damage continued. High-rise buildings as tall as 19 stories were also damaged. The damage was so heavy that eventually all buildings within a radius of 400 ft (122 m) were condemned as unsafe and had to be demolished. In a similar construction some 1000 ft (305 m) away, when first signs of trouble became evident, the adjacent buildings were underpinned on piles down to the tip level of the new

FIGURE 7.73 Pavement settlement due to pile-driving vibrations.

building piles. From here on, the consolidation of sand layers as a result of pile driving had no ill-effect, as these sand layers were no longer supporting building loads.

Lessons to be learned:

1. The use of vibratory pile-driving equipment must be avoided where possible. Such equipment should be used only after careful evaluation of potential damage.
2. Underpinning of adjacent structures must be seriously considered for support to minimize damage.
3. Where possible the use of augured piles should be considered.

7.11 COFFERDAMS—THE 14TH STREET COFFERDAM

In order to construct a deep pumping station picking up flow from a 9-ft (2.74-m) sewer on the east side of New York City, a temporary steel cofferdam was constructed. The size of the pit was 100 ft × 250 ft × 37 ft (30.48 m × 76.20 m × 11.28 m). Adjacent was a high-rise public housing building on short piles so that the excavation was 30 ft (9.14 m) below the pile tips.

The site was filled-in land overlying some organic silt and with sand 80 ft (24.38 m) down, showing full hydrostatic pressure. It was not possible to drive interlocking sheetpiling through the sand to impervious till or rock, about 95 ft (28.96 m), since major removal of the rough fill might affect the stability of the apartment building. The solution was to core through the fills and place 30-in (762-mm) pipe sleeves and drive soldier beams on 5-ft (1.52-m) spacing, first welding sheet-pile interlocks to two edges of the wide flange (see Fig. 7.74). The soldiers were driven to bedrock. The space between the soldiers was then filled in with three units of flat steel sheeting, draped into a catenary arc and designed to act as such. Sheeting was driven into the till layer. The horizontal cross-bracings were all installed by jacking design loads into all members in each direction. Bracing levels were installed as the cofferdam was excavated and pumped out. Water was controlled by two deep wells carried into the sand. Some of the sheeting engaged an old buried timber crib, and some of the interlocks were broken so that water cutoff was not accomplished.

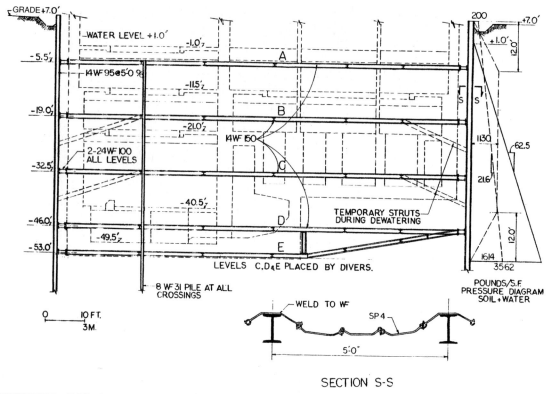

FIGURE 7.74 Soldier beams used to construct cofferdam.

Although the basic design of the interlocking sheet piles was proper, neither the 5-ft (1.52-m) spacing nor the verticality of the piles could be accurately controlled. Thus, the cofferdam walls failed at certain locations as the sheet-pile fingers tore, leaving the wall with open gaps for inflow of soil.

After several blows through the bent sheeting stopped progress, a slurry trench cutoff to rock was installed along the distorted and damaged section, which fortunately was located furthest from the apartment building. The open area near the apartment building was recharged with pumped water to maintain the original water table level, and no damage resulted either to the building or to the trees.

Lesson to be learned:

Driving of sheetpiling should be closely controlled so that extremely hard driving with potential damage to the sheets can be avoided.

7.12 CAISSONS AND PILES

7.12.1 The North River Project—Caisson Deficiencies

Manhattan's enormous North River water pollution project was constructed on a 30-acre (121,400 m^2) concrete platform supported on 2500 drilled caissons (Fig. 7.75). The caissons extend to the Hudson River bottom and are socketed into the rock underlying the lower sand layers. They were

FIGURE 7.75 2500 caissons installed in Hudson River.

from 80 to 250 ft (24 to 76 m) long and were constructed of steel pipe shells in diameters varying from 36 to 42 in (914 to 1067 mm).

For economy reasons, the caissons were designed to carry very heavy loads of up to a maximum of 5000 tons (45 MN) each. During the construction of the platform, two major problems arose:

1. Defective concrete was found to be cast in the caisson shells.
2. Caissons shifted laterally at the top and bent. These problems came to light after a substantial number of the caissons had been completed.

A full-scale study was launched, and it was determined that the problem of the *defective concrete* was caused by improper tremie concrete placement. It became clear early during the installation of the caissons that about one-quarter of the caissons had low-strength concrete and discontinuities of different varieties: uncemented aggregates, unhydrated cement, and sand layers intermixed with the concrete.

After driving the steel shells into the river, the shells were seated into the rock. Sockets were then drilled in the rock, and the shells were emptied of river mud and sand. The contractor made several attempts at casting the caissons in the dry cement after sealing the bottom and pumping the water out of the shell. When these attempts failed, a tremie method (casting concrete in water) was used. This method, which unfortunately did not utilize pressure pumping of concrete, also failed, a fact that was brought to light only later. This was so because the accidental lateral shift of the caissons caused sufficient concern, necessitating full-scale in-place load tests. The first of the load tests (described later) resulted in failure of the tested caisson, which sank vertically, a failure which confirmed the concrete defects within the caissons.

The concrete subsequently was repaired by new high-strength rebar [75-ksi (517-MPa)] bundles installed in vertical holes drilled in the caissons. These holes were then pressure-grouted with high-strength grout.

The *shifting of the tops* was determined to be the result of one-sided soil dredging of a deep trench. The creation of the trench caused unbalanced lateral soil pressure on the caissons, resulting in horizontal movement and excessive bowing. This horizontal movement of the tops

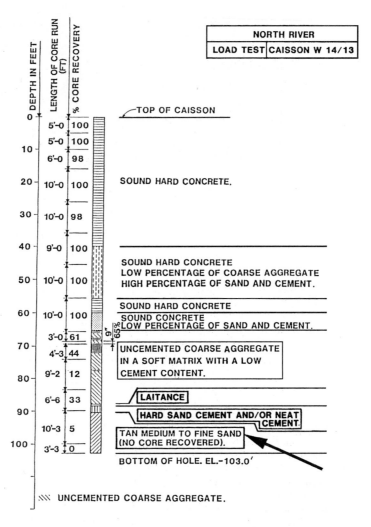

FIGURE 7.76 Coring log (Fig. 7.75).

exceeded 2 ft (610 mm) in many cases. The caissons were later forced back into position by a "pulling program" involving the application of high lateral loads coupled with jetting of the caissons. It was only after the lateral shift of several hundred caissons occurred that the concrete problem was discovered, as it was decided to core-drill several of the caissons. The coring revealed many deficiencies within the caissons, where some "concrete" near the bottom of the caissons was nothing but sand (Fig. 7.76). Before any remedial measures were considered and proposed, it was decided to load-test one of the affected caissons. To determine the load capacity of the shifted and bent caissons, a full-scale load test was ordered. The tested caisson was 36 in (914 mm) in diameter, 98 ft (30 m) long, socketed 5 ft (1.5 m) into rock, with a design capacity of 680 tons (6.1 MN). The top of this caisson had been displaced laterally as much as 2.35 ft (0.71 m) after it was installed.

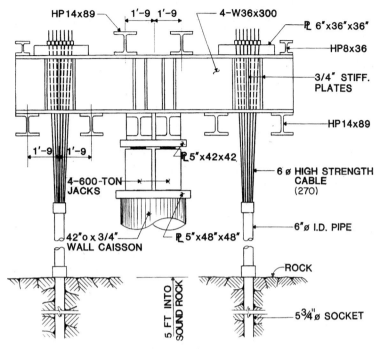

FIGURE 7.77 Test-loading questionable caissons (Fig. 7.76).

The vertical load was applied to the test caisson by means of four hydraulic jacks; the jacks reacted against a loading frame held in place by four vertical cable anchorages grouted 45 ft (13.7 m) into sound rock (Fig. 7.77).

Four 0.001-in (0.025-mm) dial extensometer gages were mounted on an independent common frame to measure vertical displacements at the top of the caisson. The top of the test caisson was braced at the top to adjacent caissons to prevent lateral translation during the test. The braces were pin-connected to eliminate undesired transferral of the vertical load through the bracing system. The load was applied in two phases.

During the first phase, a load of 850 tons (7.5 MN) was applied. The load was cycled in small increments. All loads were maintained until the vertical displacements were stabilized. Only the 765- and 850-ton (6.8- and 7.5-MN) loads were maintained for 15 and 18 h, respectively. Later the load was *cycled five times*. The vertical displacement under a load of 850 tons (7.5 MN) (125% of the design load) after 17 h was 2.6 in (66 mm) with a permanent set of 2.2 in (56 mm).

During the second phase, the load was increased and cycled further in increments to a maximum of 1360 tons (12 MN).

This maximum load was held for 60 hours and then recycled again ten times. To everybody's shock, the caisson failed to support the load and sank. An additional deflection of 3.6 in (92 mm) was measured for a total vertical displacement of 5.8 in (147 mm). The total permanent set was 4.5 in (114 mm) (Fig. 7.78).

The excessive vertical displacement and permanent set were caused by either one or a combination of two factors:

1. The existence of uncemented zones within the lower 40 ft (12.2 m) of the caisson resulted in the transfer of the entire caisson load to the steel shell which was subjected to a stress of 50 ksi (345 MPa).

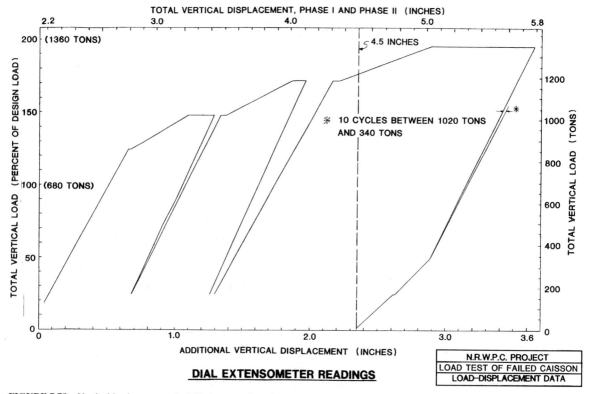

FIGURE 7.78 Vertical load versus vertical displacement for caissons.

2. Lacking sound concrete in the socket itself (there was actually uncemented sand in the socket), the cutting edge of the shell was punched further into the socket.

It was not possible to determine whether the lateral shift of the caisson was, in fact, a significant factor causing the vertical displacement of the caisson and its eventual failure. For this reason an additional caisson was tested 2 months later. This later test proved that the lateral displacement of 2.5 ft (0.76 m) at the top of the caisson had no discernible effect on the elastic behavior and load-carrying capacity of the structurally sound portion of the caisson. Therefore, all caissons with displacements of less than 2.5 ft (0.76 m) were incorporated into the structure. The unbalanced lateral loads resulting from the lateral shift were subsequently balanced by batter caissons.

A second defective caisson was then tested similarly. This caisson was also cored with an NX-size double tube core barrel prior to load testing. What we found was that the upper 80 ft (25 m) contained generally sound concrete, contrary to the first tested caisson. To measure the vertical displacement at the bottom of the caisson, measuring devices were inserted into 1-in (25-mm) diameter pipes installed in holes drilled in the caisson. The devices were ½-in (13-mm) diameter smooth steel rod tell-tales lowered into the pipes. This caisson held the load for the required 60 hours, with a total displacement of only 0.35 in (9 mm). This led us to conclude that the caissons were repairable and could be incorporated into the structure without any reduction of their load capacity. We also determined that the presence of the powdery cement layer was apparently the result of the concreting procedures employed, whereby up to three bags of cement were deposited to absorb the water which was not removed by pumping out of the shell.

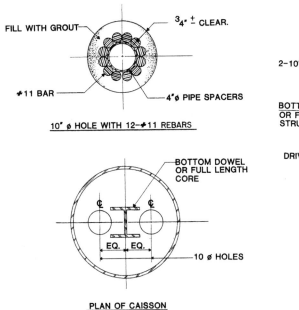
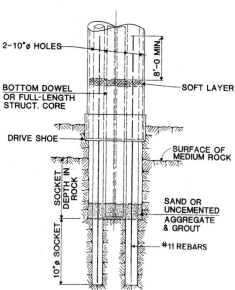

FIGURE 7.79 Repair of defective caissons.

FIGURE 7.80 Schematic diagram for grouted rebar extended 12 ft (3.7 m) into rock.

Remedial Work. The caissons were repaired by bridging the discontinuities with grouted rebars. Two 10-in (254-mm) holes were drilled within the concrete for the full depth of the caissons, including the socket, into rock. After each hole was cleaned of loose concrete, rebar bundles consisting of twelve no. 11 high-strength [75 ksi yield (517 MPa)] bars tied together around pipe spacers were lowered into the holes and grouted (Figs. 7.79 and 7.80).

Lateral Shifting of the Top of the Caissons. As soon as it was discovered that a large group of caissons had shifted laterally after installation, an immediate investigation was launched. Survey teams using laser-based instruments took readings of 180 individual caissons to establish their accurate locations. The top of the caissons were surveyed for a year. (See typical movement chart in Fig. 7.81). Lateral displacements of as much as 2.5 ft (0.76 m) were measured. Temperature variations were found (by analytical computation) to account for only a small portion of the movement. Examination of the horizontal steel 8HP36 staylath members (Fig. 7.82) and their connections showed them to be in good condition and without any signs of distress. Analytical computations established that very high loads were, indeed, required to be generated to cause such a shift.

It was also determined that a one-sided soil pressure is capable of generating such high loads. The actual behavior of the group of caissons which were tied together by means of steel bracings followed the pattern of lateral shift resulting from an unbalanced lateral load. It was soon discovered that such unbalance did, in fact, occur as a result of a sizable dredging operation performed to remove existing boulders which were obstructing caisson-driving operations. It was established that, although most caissons moved west (toward the dredged trench, which was excavated parallel to the shoreline), several caissons located west of the trench did, indeed, move toward the trench in the easterly direction.

At this time we had to determine whether the movements in the southeast section of the foundation systems were continuing and/or could be expected to continue after the completion of the concrete deck supported by the caissons. We also needed to assess the effect of such possible movements on the superstructure.

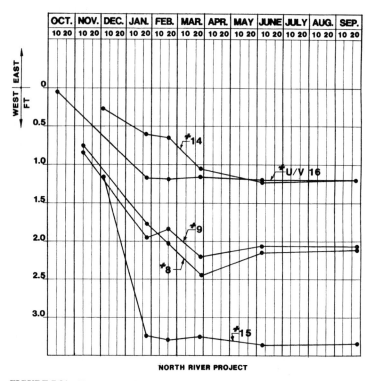

FIGURE 7.81 Test survey of caisson movement before and after stabilization.

To answer these questions, we released a number of caissons from their staylath bracings and installed inclinometers in drilled holes within the caissons. The inclinometers, which are devices that measure *tilt,* or the angular deviation from the vertical, were the slope indicator type.

The results of the surveys (Fig. 7.83) indicated that, except for the initial adjustment of the casing after installation, no movement was occurring. We therefore concluded that the westward movements of the caissons and/or soil in the southeast area of the project had ceased.

The main concern at this juncture was to establish whether the structural integrity of the caissons was still preserved. The stress analysis that followed did confirm this fact, and we then looked for a method of moving the displaced caissons back to their original vertical positions.

"Pulling" procedures were developed, using "come-alongs" to move the caissons laterally. The loads (measured by dynamometers) were applied in small increments, so as not to damage the caissons. Jetting of the river bottom was utilized to assist in pulling the caissons to the desired locations. It should be pointed out that the enormous construction difficulties encountered in this project were effectively resolved due to the excellent cooperation and combined efforts of the owner and both the design and construction teams.

Lessons to be learned:

1. Tremie casting of concrete into long caissons should be performed under pressure by the use of pumps.
2. The bottom of the tremie pipe must at all times be embedded into the fresh concrete.
3. The quality and strength of concrete in caissons, and the reliability of the tremie concrete procedures, *must be confirmed by core drilling at the beginning of the operation.*

FIGURE 7.82 Bracing caissons (Fig. 7.75).

4. Cutting deep trenches adjacent to caissons or similar pile foundations will result in a lateral shift and should be avoided.
5. Good cooperation among the various parties involved in a construction failure is essential to a speedy, successful, and litigation-free resolution.

7.12.2 Philadelphia Federal Office Building—Caisson Settlement

In 1974, several piers of this 23-story steel frame high-rise building settled with one caisson settling approximately 2 in (50.8 mm). The base of the building was 274 ft×240 ft (83.52 m×73.15 m) with a narrow tower approximately 120 ft×150 ft (36.58 m×45.72 m) in size. Wind frames were constructed in two perpendicular directions by the use of welded and high-strength bolt connections.

After stress analysis was performed, it was decided to jack one of those columns that settled; however, no action was taken from 1974 to 1976 (2 years).

In 1977, four 400-ton (3559-kN) jacks were placed between the structural steel column and the caisson. The structure was to be raised 1 in (25 m) in two stages. The relative movement of the column in respect to the caisson was measured by dial gages. At this time accurate survey revealed that 44 caissons had settled.

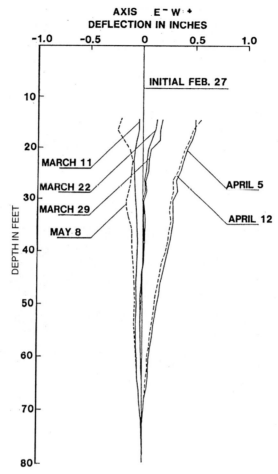

FIGURE 7.83 Inclinometer readings of caissons.

The original plans called for the caissons to bear on rock at an approximate depth of 60 to 70 ft (18.29 m to 21.34 m), but later (1969) a change order was issued (with a $70,000 savings) which required the caisson to go down only 50 to 55 ft (15.24 m to 16.76 m).

During construction, there were reports of sudden drop of the top of the caisson and of water accumulation at the top of the caissons. There were two possible causes for the settlements of the caissons:

1. Inadequate bearing strata at the bottom of the caisson
2. Discontinuities and voids within the concrete caisson itself

The load test that was conducted was inconclusive, since the monitoring setup was not properly instrumented. Whereas only the top of the caisson was monitored, it was not possible to determine whether the bottom of the caisson dropped.

What was confirmed was that the jacking of the caisson actually resulted in a drop of the caisson instead of the lifting of the column above.

The final repair utilized grout which was injected into, around, and below the settled caissons.
Lessons to be learned:

1. Caisson construction must include coring of typical caissons to confirm the quality and continuity of the concrete. Coring program must be performed at the start of construction.
2. Bearing surface must be confirmed by 5-ft-long (1.52-m) rock cores extracted prior to casting of the concrete.

Recommendations for design and construction improvement include avoiding common pitfalls and review of checklists.

7.13 COMMON PITFALLS

1. One structure supported on two different foundation systems, i.e., piles and spread footings
2. Inadequate drainage for retaining and basement walls, leading to buildup of hydrostatic pressure
3. Inadequate lateral stability of structures with one-sided soil pressure
4. Corrosion of piles and sheet piles in the splash zone
5. Lack of waterstops in foundation walls and pressure slabs
6. Ignoring negative friction effect on piles in unconsolidated soils
7. Lack of piles to support slabs on grade where soils are of poor bearing capacity
8. Insufficient or no borings
9. Borings not taken deep enough
10. No consideration of ground-water level
11. Failure to design for uniform settlement
12. Backfilling before foundation walls are adequately braced by floor framing
13. Inadequate depth of foundations subjected to scouring
14. Compacting upper 10-in (25.4-mm) layer of loose, unconsolidated soil and fills

7.14 CHECKLIST

1. Visit site to observe existing structures which may be affected by new construction. Find out type of foundation used for nearby existing buildings.
2. Take proper borings of sufficient number and adequate depth. Verify ground-water table.
3. Choose appropriate type of foundation. Consult geotechnical engineer.
4. Design for uniform bearing and settlement or provide for any differential settlements which may occur between parts of the structure, by use of expansion joints.
5. Brace foundation walls before backfilling. Do not load unbraced foundations with material or equipment surcharge.
6. Provide drainage at retaining or basement walls.
7. Verify that actual soil is as assumed in design, both for vertical support and for lateral load against walls below grade.
8. Check for effect that pumping and/or vibratory equipment may have on existing structures.
9. Design foundation for all directional loads and within allowable stresses and settlement tolerances.

10. Support slabs on piles when slabs are bearing on compressible or otherwise inadequate soils.
11. Provide step-by-step construction procedure when temporary support, underpinning, shoring, or bracing is required during installation of new foundation.
12. Tremie placing of concrete should always be performed in still water. Do not pump water out of the shell during placement operations.
13. For drilled pier and caisson installations, extract full-depth concrete test cores from the first five to verify integrity of the concrete and adequacy of bottom bearing.
14. Monitor ground-water level during construction, especially if pumping is required, since it may affect existing buildings and the bearing capacity for new structures.
15. Establish vibration criteria for effect on existing structures during pile-driving operations. Special care is required for old and historical landmark buildings.
16. Provide batter piles to resist lateral loads.
17. Avoid bending moments and eccentricities on footings.
18. Check pile caps for eccentricities of as-driven pile locations.

7.15 REFERENCES

1. Robert Frost, *North of Boston,* Holt, New York, 1979.
2. Building Code of the City of New York, Sec. 26-1903.1(b), Department of General Services, New York, May 1990.
3. A. Koral, "When the Earth Opens and Walls Move," *The New York Times,* July 3, 1977.
4. F. J. Crandell, "Ground Vibration Due to Blasting," Liberty Mutual Insurance Company.

PART 8

MISCELLANEOUS CONCERNS

SECTION 8A
THE PROFESSIONAL ENGINEER'S (P.E.'s) ROLE IN FOUNDATION ANALYSIS

Jim Irwin

8A.1 INTRODUCTION   8.3
8A.2 ANALYSIS AND EVALUATION OF DISTRESS   8.3
8A.3 OTHER FACTORS THAT MIGHT BE ASSUMED TO CAUSE FOUNDATION DISTRESS   8.4

8A.3.1 Effect of Trees on Foundation Stability   8.4
8A.4 RECOMMENDED REPAIR OR OTHER REMEDIAL ACTION   8.5
8A.5 CONCLUSION   8.5
8A.6 REFERENCES   8.5

8A.1 INTRODUCTION

Parts 2 and 3 established the engineer's role in foundation design. Over the last 10 years or so, a new opportunity has developed for the alert engineer. This new discipline involves the structural analysis of existing properties to establish the overall condition of a property, principally the foundation. (Licensed property inspectors generally handle the evaluation of such features as roof, insulation, framing, electrical, mechanical, plumbing, and appliances.[1] These individuals are not generally registered professional engineers.) The demand for the P.E.'s service to inspect and evaluate the structural condition of a foundation is generally initiated by such entities as lenders, purchasers, realtors, or insurers, including those companies specializing in structural insurance, such as HUD and the Veterans Administration. Refer also to Secs. 8C and 8D.

The role of the professional engineer in foundation appraisal involves his or her use or development of expertise in such areas as soil analysis, hydrology, foundation design and construction, as well as structural analysis. Thorough training in these areas supported by hands-on experience is necessary for competent performance. Incompetent performance often results in litigation (see Sec. 8B).

Parts 6 and 7 presented a rather thorough discussion of foundation failures and repair, including preventive measures and evaluation. There is little need to reiterate in this section. It might, however, be worthwhile to review from the engineer's point of view certain areas of remedial commitment.

8A.2 ANALYSIS AND EVALUATION OF DISTRESS

A diagnosis of foundation problems is accomplished by an analysis of distress symptoms and by considering the site and type of soil. Although distress symptoms can be disguised or hidden by tuck-pointing, patching, wall covering, painting, or adding wood trim, an experienced inspector can usually detect the problems.

A comparison of the construction trades will reveal basic data about construction practices. This involves comparing slopes of carpenters' interior work (framing of doors, windows, and floors) in relation to drywall and ceilings and to the exterior masonry and foundation. New construction is virtually never level. In fact, new construction tolerances can be in the range of ½ in per 10 lin ft (1.27 cm per 3 m). Significant foundation concerns involve differential movement, not inconsistencies in original construction. (Refer to Sec. 6A.7.)

Once foundation problems are identified, they are generally classified as either upheaval or settlement. According to statistics compiled by the author between 1984 and 1992, settlement, as opposed to upheaval, was recognized as the preponderant cause of foundation distress on a basis of something like 2:1. However, these same statistics, relative to slab foundations on expansive soils, show upheaval as the cause for necessary repairs to be 2 out of 3. Settlement is often arrested by proper preventive measures and therefore does not always present the degree of movement necessary to demand repairs.

8A.3 OTHER FACTORS THAT MIGHT BE ASSUMED TO CAUSE FOUNDATION DISTRESS

Wind loading on roofs can cause torsional stress on the walls and ceiling of a light residential or commercial structure, sometimes resulting in cracks in the interior walls. Obviously this damage is not soil-related. Where expansive soils and slab foundations are involved, some inspectors with clients (insurers, for example) who, for self-serving reasons, do not wish to accept responsibilities for foundation repairs tend to misidentify the causes. This prompts conflicts of interest leading the inspectors/engineers to suggest such factors as wind loading, lightning, or blasting to be the cause for damage when in fact the foundation experienced problems due to expansive soils. Similarly, the same engineers, for the same reasons, tend to label settlement as the cause for virtually all foundation problems. Under most home owners' insurance, foundation problems caused by acts of God are excluded. However, damage caused by domestic leaks (upheaval) might well be covered. Slopes in the floor, accompanied by distorted door frames and cracks in the gypsum walls and ceilings that concentrate in areas where plumbing lines would normally run (kitchen, utility room, baths), provide the preponderance of evidence indicating heave as the most likely cause. Contracting a competent plumber to conduct a plumbing leak test cannot be overemphasized. Cosmetic symptoms from heave due to plumbing leaks, although "repaired," will continue to reappear as long as the primary cause is neglected. Heave normally ceases some time after the source of the water is eliminated. Upheaval can occur in the perimeter beams of either slab or pier and beam foundations, but the incidence is relatively limited. The structurally weak, lightly loaded interior slab is most generally affected.

Factors such as earthquakes, hurricanes, floods, mud slides, or tornadoes cause serious structural damage to property. These calamities are not considered here because restitution usually requires complete reconstruction.

8A.3.1 Effect of Trees on Foundation Stability

Trees planted close to the foundation can absorb moisture from soils beneath the foundation when feeder roots grow under the foundation. The types of trees most damaging include fruitless mulberry, willow, cottonwood, and pecan trees, which typically grow in bottom lands in their native state. Several opinions exist concerning the minimum distance a tree should be planted from a foundation. Some experts recommend planting the tree a distance from the foundation greater than the drip line of its mature canopy. Generally the detrimental effects of tree roots are overemphasized. Feeder roots tend to be quite shallow (1 to 1.5 ft or 0.45 m maximum) and would not likely penetrate a normal perimeter beam.[2] Refer to Sec. 1A.

8A.4 RECOMMENDED REPAIR OR OTHER REMEDIAL ACTION

Section 6B quite adequately covers the subject of foundation repair for residential and other light construction. However, the following statements offer a concise review of a few of the more salient issues.

Piers are often used in foundation repairs to raise perimeter beams and are spaced from 6 to 9 ft center to center (1.8 to 2.7 m), depending on soil conditions and the requirements for support of the damaged foundation. Generally 12-in-diameter (0.3 ft) concrete piers should be used in expansive clay soils. Piers should be poured within 8 h after shafts are drilled. Piers may be either straight or belled. The belled portions have sloping sides, formed typically at 45° angles or less. The end of the belled portion of a pier transmits a structural load over an expanded area, reducing allowable soil bearing pressure. This in effect increases the pier's load-bearing capacity. Belled piers also resist upheaval in expansive soils. Typically, straight or belled piers are made of reinforced concrete, with steel extending the length of the pier. (See Secs. 6B.2, 6B.3, as well as 3E.)

Pressure grouting is necessary to fill voids under concrete slabs. Pressure grouting, or mud jacking, to fill voids under the slab after raising is very important. Otherwise the slab will be unsupported in some areas and will not have the structural ability to support live loads.

8A.5 CONCLUSION

There is a certain amount of teamwork involved in real estate inspections. The licensed real estate inspector often works with the electrical, mechanical, and plumbing systems, whereas the engineer conducts the structural foundation inspection. Often the real estate inspector is the first on the scene. His or her alertness is most important. If no structural problems exist, there is no need for engineer involvement. However, if any question exists, the inspector should recommend the services of an engineer. It is therefore most important that the inspector be able to recognize the symptoms of differential foundation movement. Should he or she fail to recognize an existing problem, the engineer may never be consulted.

The engineer's and the real estate inspector's roles have been increased by reactions of the FDIC to the failure of lending agencies. New government agencies such as the Resolution Trust Corporation have been created to dispose of property from failed bank and savings and loan companies. Policies of existing agencies such as Fannie Mae, Fannie Mac, and HUD have been tightened and made more strict, requiring additional and more exacting work by the inspector. One positive advantage for the engineer or inspector is that work opportunities are available. Lending institutions require inspections of real estate with symptoms that were previously considered inconsequential. In addition, appraisers and FHA-licensed inspectors are referring many more jobs for engineering reports.

8A.6 REFERENCES

1. J. Wren, *Home Buyer's Guide,* Barnes and Noble, New York, 1970.
2. R. W. Brown, *Foundation Behavior and Repair,* McGraw-Hill, New York, 1992.

SECTION 8B
BUSINESS RISK MANAGEMENT

Thomas M. Smith

8B.1 INTRODUCTION 8.7
8B.2 THE RISKS 8.7
8B.3 RISK MANAGEMENT 8.7
8B.4 LIMITATION OF LIABILITY 8.8
8B.5 BUSINESS ENTITIES 8.9

8B.6 BUSINESS AND PROFESSIONAL LIABILITY INSURANCE 8.9
8B.7 ALTERNATIVE DISPUTE RESOLUTION 8.10
8B.8 CONCLUSION 8.11

8B.1 INTRODUCTION

It is assumed that those reading this text are either engineers, working toward becoming an engineer, or otherwise employed in the business of foundation construction or repair. Like the engineering complexities of foundation design presented in the previous sections, the business of foundation construction is no less complex. Technological and regulatory complexities, the vagaries of human endeavor, environmental issues, and the day in–day out operation of a business make it easy to understand that construction projects are not for the risk averse. The following is intended to provide a short overview of some of the issues the reader may face.

8B.2 THE RISKS

Lawsuits seem to come from new sources each year and appear to be limited only by the imagination of those who bring them. Theoretically, the engineering professional (individual or business entity) owes a duty of care to any party who could possibly be injured, either physically or economically, from his or her acts or omissions. Many claims have no basis in fact and are added by litigants either to gross up the potential damages as an initial intimidation effort or, should litigation reach a jury, to more easily convince the jury that there must be some damage to the allegedly injured party. Whether the claim is legitimate or not, the defendant or his or her insurance carrier faces the inevitable cost of defending.

In recent years owners have increasingly attempted to shift liability, first from themselves to the general contractors, who then shift as much liability as possible to the subcontractor, and finally from both the owner and the general contractor to the engineer. No matter what category you fall into, the concept of risk management will affect both the cost of services and ultimately the survival of the business.

8B.3 RISK MANAGEMENT

The concept of risk management has been in existence for years. However, in recent years there has been an increase in construction litigation and a resulting interested focus on what risks each party to the project should be responsible for. The practice of risk management entails all parties to the project, identifying the project's risks and allocating those risks in a fair manner. This process is known as risk allocation. The basic underlying force in any project is the cost. Typically, the greater the risk accepted by a knowledgeable party to a project, the higher the fee. This may result from, among other factors, increased cost of casualty insurance as well as the cost of unanticipated time delays and the possible resulting financial losses.

When these risks can be addressed in a collaborative manner, it generates a team spirit which by itself can lower the risk. In the construction industry the term *partnering* is used to describe this relationship. Partnering seeks to change the working environment on a construction project from its traditional adversarial nature to one of trust and cooperation. Savings come in the form of reduced costs, such as lower insurance premiums. Insurance rates are based on the claims history of the firm or company. The fewer the claims, the lower the rate.

8B.4 LIMITATION OF LIABILITY

Unfortunately, cooperative risk allocation is not the norm in today's business world. The more common term used to describe the current business climate is *limitation of liability*. In the engineering and construction business, indemnity provisions frequently are used to transfer risk from the owner to a contractor or from the contractor to a subcontractor. Since the risks inherent in this business can be significant, bargaining for an allocation of risks associated with personal and bodily injury or property damage is critical. For example, one entity may agree to indemnify several entities on a site. This places the risk of loss on one entity who can acquire insurance to cover the risk.

Clauses limiting professional liability are as numerous as there are drafters. The following indemnity provision is a typical example of the allocation of risk between parties:

> (a) Subcontractor shall fully protect, indemnify and defend Contractor and hold it harmless from and against any and all claims, demands, liens, damages, causes of action and liabilities of any and every nature whatsoever arising in any manner, directly or indirectly, out of or in connection with or in the course of or incidental to any of Subcontractor's work or operations hereunder or in connection herewith (regardless of cause or of any concurrent or contributing fault or negligence of Contractor) or any breach of or failure to comply with any of the provisions of this Subcontract or the Contract Document by Subcontractor.
>
> (b) Subcontractor shall fully protect, indemnify and defend Contractor and hold it harmless from and against any and all claims, demands, causes of action, damages and liabilities for injury to or death of Subcontractor or any one or more of Subcontractor's employees or agents, or any Subcontractor or supplies of Subcontractor, or any employee or agent of any such Subcontractor or supplier, arising in any manner, directly or indirectly, out of or in connection with or in the course of or incidental to any work or operation or operations of Subcontractor or Contractor or any other Contractor or Subcontractor or party, or otherwise in the course and scope of their employment, and regardless of cause or of any fault or negligence of Contractor.

In drafting limitation clauses, professional assistance may be obtained from attorneys specializing in construction law or from publications of professional engineers' or contractors' organizations. Whatever the source, avoid the use of "legalese." Plain English, simply written, will provide the clarity needed to explain the contractual limitation to all parties to an agreement and, fortunately, if necessary, to the jury. Judges and jurors are more likely to enforce a clause if it is clear and unambiguous, and written in a manner which can be easily understood by all parties to the agreement.

The limitation language should be given emphasis by the use of such techniques as bold print, capital lettering, providing a place to initial the clause, large type size, underlining, or highlighting. Some drafters would use all of these techniques; others feel that one or two are sufficient. The intention is to draw the parties' attention to the limitation and avoid a later challenge that it was concealed in the contract as "boilerplate" language minor.

One may well ask: "Why would anyone sign a contract which substantially limits the contractor's or engineer's professional liability?" Surprisingly there is not as much resistance as one might expect. In these days of expanding liability, most owners or contractors realize the need to limit one's contracting liability by accepting only those risks that the contractor or engineer specifically chooses to accept.

8B.5 BUSINESS ENTITIES

The type of business entity under which the design professional or contractor chooses to operate can significantly impact the liability of its owners. Unlike most business professionals in Texas, the engineering professional cannot limit his or her liability by incorporating. Engineers are personally liable for their mistakes from the moment they place their seals on reports or plans. The use of limited liability partnerships or corporations will insulate the individual engineering professional from other engineering professionals in a business. The primary protection for a mistake by the individual engineering professional is insurance.

Contractors, however, have the ability to structure their business in a manner so as to protect themselves from personal liability. Typically contractors operate their businesses as corporations. If properly formed and maintained, the corporation protects the contractor's personal assets from judgment creditors. There are significant tax and liability differences among the various forms of business entities. These differences should be discussed thoroughly with both a certified public accountant and an attorney who represents clients in the construction business.

8B.6 BUSINESS AND PROFESSIONAL LIABILITY INSURANCE

In the event of a claim, coverage is available for foundation construction or repair companies under a commercial general liability policy and for design professionals under an errors and omissions policy. Typically the design professional's policy provides that the insurance company will pay on behalf of the insured all sums which the insured shall become legally obligated to pay as damages if legal liability arises out of the performance of professional services for others in the insured's capacity as an architect or engineer and if such legal liability is caused by an error, omission, or negligent act of the insured for which the insured is legally liable.

Virtually all of these policies are now claims made policies, which means that the claim must be reported during the term of the policy or a renewal thereof. If a design professional is changing companies or has been uninsured for a period of time, prior acts coverage is available, so that the design professional will have coverage for acts and omissions which occurred prior to the inception date of coverage. Design professionals' policies frequently contain large deductibles and provide that defense costs are to be applied against the limit of liability.

The general liability policy for a contractor provides that the insurance company will pay on behalf of the insured all sums which the insured shall become legally obligated to pay as damages because of (1) bodily injury or (2) property damage to which the insurance applies, caused by an occurrence, and the insurance company shall have the right and duty to defend any suit against the insured seeking damages on account of such bodily injury or property damage, even if any of the allegations of the suit are groundless, false, or fraudulent, and may make such investigation and settlement of any claim or suit as it deems expedient. But the insurance company shall not be obligated to pay any claim or judgment or to defend any suit after the applicable limit of the insurance company's liability has been exhausted by payment of judgments or settlements. Most policies exclude coverage for repair or replacement of the work performed by the contractor and provide coverage only for consequential damages and liability to third parties.

8.10 MISCELLANEOUS CONCERNS

Special policies are available to provide coverage for the design-and-build contractor and for construction managers. Project liability coverage is also available for both design professionals and contractors. Such a policy ensures insurance coverage to uniform limits and allows the cost of coverage to be charged as a reimbursable expense on the project.

As with all insurance policies, there are conditions which must be met by the insured in order to be covered. For example, the insured must give timely notice of a claim so as not to prejudice the rights of the insurance company to provide a defense. The policy will also contain an aggregate limit of liability in any one year, and it will provide the insurance company a right to settle a claim.

Upon receiving notice of a claim, the insured should notify the insurance company by certified mail, forward to the company any documents received with regard to the claim, and request that the insurance company provide a defense and indemnify the insured for the claim. Unless otherwise negotiated, a standard insurance policy will normally give the insurance company the right to choose legal counsel. However, in Texas, if the insurance company takes the position that there may not be coverage, the insurance company may lose the right to choose counsel. The proper handling of the defense of a claim may lower the claim's impact on future insurance rates.

Apart from the foregoing, homeowners have access to insurance which provides coverage against specific perils. Similarly, specialty insurance programs provide alternative or additional structural insurance policies to protect homeowners and building contractors (see Sec. 8C). Coverage disputes sometimes arise between these insurers and the insured. The problems and resolutions common to these forms of insurance are comparable to those described in the foregoing paragraphs.

8B.7 ALTERNATIVE DISPUTE RESOLUTION

When a claim is brought to the attention of a contractor or engineering professional, the use of a dispute resolution technique other than the traditional adversarial trial system should be considered. The construction business has been at the forefront of alternative dispute resolution (ADR) due to the unique nature of the business. Construction litigation involves fact-intensive, heavily documented cases. Job records, change orders, plans, specifications, job meeting minutes, daily, weekly, and monthly reports, and financial records can fill a small room. All of these items must be sorted and summarized into the documents and information relevant to a dispute. Just as a construction project involves owners, suppliers, design professionals, subcontractors, lenders, and insurance sureties, the typical construction-related lawsuit will also involve multiple parties. The technical issues involved in most construction litigation will invariably lead to the use of expert witnesses (see Sec. 8A). These inherent characteristics make construction litigation an inordinately expensive and debilitating process.

The construction industry has utilized virtually all forms of ADR. Binding arbitration has long been the common ADR method of choice and is virtually the only form for resolving international disputes. Both federal and state courts favor arbitration. A policy favoring arbitration takes statutory form in the Federal Arbitration Act and in the various state versions of the Uniform Arbitration Act. Both federal and state statutes permit contracting parties to choose arbitration as a remedy and to enforce in court a motion to compel its use.

In recent years binding arbitration has come under criticism in the construction industry, and as a result there is an increased interest in mediation. Parties at risk in large claims are reluctant to arbitrate without advice of counsel. As lawyers have become a component of the arbitration process, they have sought to change the informal nature of arbitration to one of more formal rules for presentation of evidence and stricter procedure.

As a result, nonbinding mediation has been used increasingly as an alternative to arbitration. Mediation is a process of assisted negotiation in a form which provides an impartial person, the mediator, who facilitates communication between the opposing parties so as to achieve reconciliation and settlement of the dispute. The mediator focuses the parties by assisting in defining the relevant issues, exploring alternative solutions, and providing a framework in which the opposing parties may reach a settlement. Whether the parties resolve their dispute with arbitration or mediation, the use of ADR offers an attractive alternative to our adversarial trial-based system of law.

8B.8 CONCLUSION

Operating an engineering or construction business in today's litigious environment may appear daunting; however, with careful planning, risks can be minimized. The use of a proper business entity along with use of partnering, risk allocation, insurance, and effective claims management will substantially reduce risks and their associated expense.

SECTION 8C
STRUCTURAL INSURED PROTECTION

Charles A. Windham

8C.1 SCOPE 8.13
  8C.1.1 New Construction 8.13
  8C.1.2 Existing Construction 8.13
  8C.1.3 Summary 8.13
8C.2 POLICIES AND WARRANTIES 8.14
  8C.2.1 Major Structural Defects 8.14
  8C.2.2 Coverage 8.15
8C.3 EXAMPLES OF MAJOR STRUCTURAL DEFECTS 8.16

8C.3.1 Arbitration 8.16
8C.3.2 Transfer of Coverage 8.17
8C.4 POLICY/WARRANTY COVERAGES 8.17
  8C.4.1 Inspections 8.19
8C.5 TYPICAL COSTS (1992 DOLLARS) 8.19
8C.6 SETTLEMENT OF CLAIMS 8.20

8C.1 SCOPE

First it is appropriate to mention at the outset that the subject policies/warranties issued by the various insuring companies are *not* a maintenance or repair agreement. There is no coverage allowed for normal and natural wear and tear (whether or not induced under normal living conditions by individuals), the contained environment, or natural elements. As expected, there is no coverage for abuse of any kind or type to the home. Most of all, the policies/warranties *do not* entirely take the place of the homebuilder's responsibilities, either under the law or contractually.

8C.1.1 New Construction

In light of the foregoing, structural protection relative to newly constructed residential homes is available only through those homebuilders approved by the various insured protection programs throughout the United States. The coverage under a plan is specified by the policy/warranty. These programs will insure residential properties *only*. Apartments, for example, are considered commercial properties and would not be insured. In order for a home to be considered for insurance under these programs, the individual owners of record (by title transfer) must occupy the home. Once a home is leased or rented to others, that home, under most circumstances, becomes commercial property and is therefore excluded from coverage subject to the exclusions section of the policy/warranty of the available programs.

8C.1.2 Existing Construction

Structural insured protection for existing homes is also offered by some insurers in some localities. The commercial aspect under this type of coverage is virtually the same as that for the newly constructed home. Under the existing home coverage programs, the only coverage normally available is relative to the load-bearing structural elements, as defined by the particular policy.

8C.1.3 Summary

Homebuilders who are approved to participate in the "new" construction programs are required to attest that they will construct all homes (which they propose to insure) in accordance with the acceptable prevailing model building code and applicable standards in the locality where the home that is to be insured is constructed. Otherwise the homebuilders must attest that they will abide by the Council of American Building Officials (CABO) One- and Two-Family Dwelling Building Code and applicable standards. Some programs have more stringent building requirements than those promulgated by the building codes and standards mentioned. However, while residential building plans and specifications may meet or exceed the minimum criteria of the insured protection program or the various codes and standards, installation compliance in accordance with the codes is the key. If compliance of the installation at each job site cannot be achieved through the on-site inspection process and procedures, then building code standard requirements are to no avail, no matter how stringent they might be.

The design of residential foundations by professional structural or civil engineers is either done at the behest of the individual homebuilder, or it is required by the local building authorities or the individual insuring program. However, where the design of residential foundations is less than that prescribed by the governing building code and applicable standards, the foundation design must be an engineered design. Other structural aspects of a residence above the foundation may also require the input of a professional structural or civil engineer.

8C.2 POLICIES AND WARRANTIES

The insured protection programs came onto the scene around 1978. The coverage criteria of these programs, for the most part, where new homes are concerned, were derived from the minimum criteria required by HUD/FHA (Department of Housing and Urban Development/Federal Housing Administration) for approval under their mortgage insurance programs. Other agencies of the U.S. government, namely, the Veterans Administration (VA) and the Farmers Home Administration (FmHA), substantially required the same criteria for approval under their mortgage insurance programs.

8C.2.1 Major Structural Defects

The definition of a major structural defect, for example, is essentially the same for all of the HUD/FHA approved insured protection programs. In fact, that definition, which is outlined in every approved insuring company's document, is virtually verbatim the definition outlined in the statutes of the Federal Housing Act governing the required coverage criteria of these various insured protection programs in order to achieve approval by HUD/FHA, VA, and FmHA.

Coverage is very limited under these insured protection programs, thus the reason for the very low cost of such protection over a 10-year period. A major structural defect as defined, for example, would have to be a very serious or catastrophic type failure or loss in order to be covered by these programs. Minor structural impairment due to normal and natural settlement, or minor differential movement, the result of the natural and expected expansion and contraction of the supporting soils, would not be covered. Minor latent design structural impairment would also not be covered.

A major structural defect as defined, for example, is a defect to the defined load-bearing elements to the extent that the home becomes unsafe, unsanitary, or otherwise unlivable from a safety or health standpoint. Unlivable from the standpoint of impaired aesthetics, as determined by others, is not covered by these programs. Any other structural coverage greater than that stipulated would be cost-prohibitive. The author is not aware of any insuring company that offers coverage beyond that outlined, although there may be those whose definitions of major structural defect may be less restrictive.

What does that definition really say? Those who are not versed in the vernacular and meaning of the industry jargon, particularly attorneys who endeavor to "build" a claim, say that such a definition is too vague. Perhaps this might be true where the inexperienced or unprofessional is concerned. However, where experienced individuals or professionals are concerned, this is not believed to be the case, particularly since the definition is derived from the federal statutes, namely, the Federal Housing Act.

8C.2.2 Coverage

It would be most unusual to find any lengthy written document that is not vague to some degree. Nonetheless, insuring companies are continually rewriting various aspects of their policies/warranties in an effort to better clarify the intent of their coverage. The written law at any level of government is no exception. It has been said by every court in the land, for example, that ignorance of the law is no excuse. The dockets of all the courts attest to that fact.

Why, then, should ignorance of coverage under a policy or warranty be an excuse? Consider the case where the young son of a homebuyer, twice over a very short period of time, drove the family car into the corner of their attached garage. Not only did he knock down the brick veneer both times, but the overhead garage door and the door jamb and trim also required repair. The mother of the young boy was, simply put, livid because the builder, without recompense, would not take responsibility for either instance. She actually believed, and still does to this day, that because these incidents (accidents) happened during their first year of possession of the home, the builder was obligated to correct the damages. This mother, to make matters worse, was a college graduate with supposedly better than average powers of understanding and reasoning. No amount of "clarity" with respect to a written document of any type can deal with or pacify that kind of mentality.

When one has issue where the law is concerned, one seeks assistance from a professional, that is, an attorney. Why, then, when one has a technical problem or concern involving one's home, should one not also seek assistance from a professional, that is, a structural or civil engineer rather than going to an attorney? Where structural coverage is concerned, this alternative, with the input of a qualified arbiter, as stipulated under the policy/warranty terms, would not only lessen the dockets of the courts considerably, but it could achieve the settlement of such claims much more swiftly and at lower costs to all parties concerned.

However, don't be misled! Engineers are not always the solution. Sometimes they become part of the problem. A segment of the engineering community, perhaps referred to as "company men," give the insurance companies less than professional in-depth input, which often prompts unnecessary, unsuccessful, and expensive litigation. These engineers often own their companies but derive perhaps 90% of their income from a very limited number of insurance companies. Through ignorance or design, they virtually always tend to report in favor of their client, oblivious to facts. This causes insurance companies to initially deny legitimate claims. Ultimately such claims are often paid along with astronomical attorney, expert witness (often engineers), and court costs.

As an aside, one of the claim areas most commonly abused involves determination and assessment of utility leaks beneath residential slab foundations. Comments attributed to professional engineers include:

1. Center doming (or upheaval) is characteristic of all slab foundations and is not the result of utility leaks.
2. Sewer leaks beneath a slab foundation cannot cause upheaval.
3. Sewer leaks sufficient to cause significant slab upheaval would require thousands of gallons of water.
4. The perimeter beam settled (uniformly) as opposed to the center raising.
5. The leak could not be responsible since the major heaving was not in the immediate location of the leak.

8.16 MISCELLANEOUS CONCERNS

The list goes on, but the point is clear. Each of these allegations is both false and absurd.

The insurance companies would fare much better, given the truth. Legitimate claims would be settled without unnecessary litigation costs and consumer ill wills.

8C.3 EXAMPLES OF MAJOR STRUCTURAL DEFECTS

Some examples, not all inclusive, of a major structural defect as defined, and where proper and normal maintenance has evidently been performed, are as follows:

1. A failure of the defined structural elements (not the brick veneer or other finish materials) to the extent that collapse or other danger of a home is imminent. That would be an unsafe condition and therefore otherwise unlivable.

2. A failure of the defined structural elements to the extent that doors or windows that are classified as either primary or secondary means of ingress or egress to and from a home are sticking and inoperable by emergency personnel, or by one who is old enough, or has the capacity to utilize these means of egress from a room or area in a home that is classified as a habitable room during an emergency. This would be an unsafe condition, and therefore otherwise unlivable.

3. A failure of the defined structural elements to the extent that moisture under normal weather conditions penetrates the roof or exterior walls into the attic or wall cavity and/or into the home. This would be an unsanitary condition and therefore otherwise unlivable.

4. A failure of the defined structural elements to the extent that plumbing waste, drain, vent, or water distribution lines are severed or leak into the house. This would represent an unsanitary condition and therefore represent an unlivable situation.

5. A failure of the structural elements to the extent that moisture, mud, insects (termites), or other vermin (mice) enter a house through cracks in the floors, walls, or roof caused by such failure. That would be an unsanitary condition and therefore otherwise unlivable.

Every impaired structural condition claimed would have to be physically judged on its own merits by an experienced individual or professional person to determine whether such structural impairment is covered, as defined.

None of these insuring companies will cover a defect caused by the homeowner, either willfully or negligently. A major structural defect claim of record by one of the insuring companies, for example, was denied due to the homeowner's negligence. In this case the homeowner constructed a swimming pool in the backyard of his home after he had taken possession of the home. The pool was installed in such a manner that positive drainage away from a U-shaped area in the rear of the home was severely impaired. The excess moisture trapped in the area could not readily drain away, and the excess moisture in this isolated or confined area caused the soil (a highly expansive type) to expand excessively (heave), to the extent that structural impairment occurred in this confined area of the home. This also caused undue cracking of the finish materials in some of the other areas of the home. While the aesthetics of the home were a "mess," and even though the problems were induced by the homeowner's negligence, it was also determined by the insurance company that a major structural defect, as defined, did not exist.

8C.3.1 Arbitration

One of the good features about these insured protection programs is that once a determination has been made relative to a major structural defect, or other as defined, either party, that is, the builder, buyer, or insurance/warranty company, can request that the matter be arbitrated by a recognized arbitrator. This arbitration provision and procedure is required where HUD/FHA, VA, or FmHA approved coverage is concerned. The arbitration procedure (and the arbiter's decision is considered binding and final) was designed, hopefully, to avoid litigation. Unfortunately the first notice to the insurer of a pending claim is often through some homebuyer's attorney. There is sometimes a place for the attorney to intervene eventually, particularly if a claim is disputed after review by both the insuring company and an arbiter. Insuring company experts, and even arbiters, sometimes err in their judgments. It has been known that arbiters sometimes even address conditions or areas of concern other than those which they were commissioned to arbitrate. These insuring companies by and large do an exceptional job in judging the coverages under their policies/warranties. Claims, on the other hand, that are initially submitted through attorneys most generally tend to both hamper and prolong settlements. As the old saying goes, only the attorneys win such claims or litigation when all is said and done.

8C.3.2 Transfer of Coverage

Another good benefit is that once a policy/warranty has been issued, it usually is transferable without cost to whoever owns the home during the 10-year insured period. This procedure is required by federal statute where the mortgage is insured by either HUD/FHA, VA, or FmHA.

Where new homes are concerned, some insuring companies offer a structural *only* insured protection program for homes sold with conventional mortgages, or homes that are sold without mortgages. New homes that are sold where the mortgages are insured by either HUD/FHA, VA, or FmHA *must* have three levels of coverage, which is discussed in Sec. 8C.4. Those companies that offer insuring programs for existing homes insure *only* the defined structural elements against major structural defects, as defined. The defined aspect is substantially the same as that already discussed. Before an existing home can be insured by these companies, a structural condition report by a licensed professional structural or civil engineer, or an otherwise approved individual of the insuring company, is required to be submitted to the insuring company by the individuals wanting the insurance. On the basis of the conditions addressed in that report, the insuring company will decide if the home is a viable risk to insure. The programs for existing homes, usually, can be purchased by anyone without having to be an approved registrant in some program.

8C.4 POLICY/WARRANTY COVERAGES

An insurance/warranty company policy or warranty document usually outlines the following:

1. The declaration's page, which shows the homebuyer's name and address, type mortgage, original selling price of the home, and effective dates of the policy/warranty insured period
2. A foreword or prelude
3. The insuring agreements
4. An aggregate limit of loss, usually the original selling price of the home
5. Deductibles, usually a one-time $250 deductible (the amounts are governed by the federal statute for HUD insured mortgages), for first- or second-year claims, and a $250 deductible for each major structural defect claim. Special deductibles apply to town house and condominium-type homes. The deductibles may vary with other insuring companies where conventionally mortgaged or sold homes are concerned
6. A mortgagee clause
7. Claim notice requirements, conditions precedent to claim, claim settlements, and repetitive claims
8. Conditions
9. Exclusions
10. Definitions of the various terms used.

Where newly constructed homes are concerned, particularly homes with mortgages insured by either HUD/FHA, VA, or FmHA, three levels of coverage are required from any insurance/warranty program approved by these agencies. To reiterate, the homebuilders must also be approved by the insuring company. Usually there is a first-year insuring agreement, a second-year insuring agreement, and an extended-coverage insuring agreement that is effective through the 10th year from the date of closing, where HUD/FHA, VA, or FmHA is concerned, or from the date of possession if that date occurred prior to closing, where conventional mortgages or sales are concerned.

The first-year insuring agreement is relative to the workmanship and materials furnished and installed by the homebuilder or the homebuilder's subcontractors and suppliers. While the homebuilder is directly responsible for his or her workmanship and materials under the first-year insuring agreement, the homebuyer is still protected by the insurance/warranty company, less a deductible, should the homebuilder fail or refuse to perform. The insuring company, however, must be notified within the prescribed time frames for such claims to be honored.

Except for at least one insuring company, which operates only in Texas, the second-year insuring agreement is relative to the MEP systems (mechanical, electrical, and plumbing) and major structural defects, as defined. Except for the company in Texas, the homebuilder is again responsible under this agreement for any defects occurring only to the wiring, piping, and ducts where the MEP systems are concerned, and where any major structural defect, as defined, is concerned. As with the first-year insuring agreement, if the builder fails or refuses to perform, the homebuyer is still protected by the insuring company, less a deductible, where such claims are filed within the prescribed time frames.

The extended coverage insuring agreement, again with the exception of one Texas company, protects the homebuyer against major structural defects, as defined, from the third through tenth years. The insuring company, under the policy/warranty coverage, is solely responsible for the coverage under this agreement, less the applicable deductible.

The coverages of one Texas company are a little different from other programs. The marketing and administrative entity for their builder program approves the house for their insured coverage program. Next the insurance is placed through an insurance agency, and an insurance company issues an insurance policy (not a warrranty with an insurance certificate) to each individual homebuyer. A list of HUD/FHA, VA, and FmHA approved builder programs is published by those agencies.

The program coverage outlined in the policy from the Texas insurance company is as follows:

The first-year insuring agreement is identical to that outlined earlier.

The second-year insuring agreement differs somewhat. Under the Texas insurance company program the homebuilder's responsibility ends after the first-year coverage agreement has expired. During the second year under this agreement the insuring company takes full responsibility, less a deductible, for the MEP systems. Unlike all the other companies, the Texas insurance company, with the exception of two components, insures the MEP systems virtually complete, less a deductible, instead of just the wiring, piping, and ducts. This program does not insure, neither does any of the other companies, anything that carries a manufacturer's warranty. The air-conditioning compressor on the mechanical system usually carries a 5-year manufacturer's warranty. Therefore it is not covered under the program. The hot-water heater on the plumbing system usually carries a 5- to 10-year manufacturer's warranty. Therefore it is not covered. Except for those two components, however, the program covers the MEP systems virtually 100%, less the deductible. There are some other components that are either plugged or tied into the covered systems, which are not considered to be a part of the systems, and are not covered under any of the insuring programs in the second year. These other components are, but are not limited to, the range, oven, dishwasher, garbage disposal, trash compactor, vent-a-hood, microwave, clothes washer, clothes dryer, garage door openers, and other similar appliance components. They each carry a manufacturer's warranty at least during the first year.

The extended coverage insuring agreement differs somewhat in that the Texas insurance company takes full responsibility, less a deductible, for major structural defects, as defined, from the second through tenth years.

The insuring/warranty company has printed construction standards governing the first- and second-year coverage claims, which are attached to and made a part of the policy/warranty. These are both HUD approved and substantially nationally recognized standards. While the standards do not outline everything with respect to the construction of a home, they do outline those conditions which claims most commonly reflect. The standards outline the tolerances and limitations under which certain conditions are considered either acceptable or unacceptable. The construction standards usually are not relevant to the extended coverage agreements, that is, major structural defects.

8C.4.1 Inspections

All insuring companies require certain compliance inspections to be performed by qualified and approved inspectors at various stages during construction of a new home. Most programs require at least three inspections, or four if drilled piers are specified for the foundation construction or repair. These would involve:

1. A drilled pier hole inspection
2. A first-stage inspection prior to placement of the concrete for a foundation
3. A second-stage inspection before insulation and wallboarding inside and bricking outside
4. A final inspection when the house is completely finished and ready for occupancy

Once a final inspection is 60 days old or older, most insurers will require another inspection prior to closing or final settlement.

There are some things that an inspector never gets to inspect, however, because the items are either not installed at the time one inspection is made, or they are completed or covered up by the time that inspector returns for a subsequent inspection. The concrete for a foundation, for example, is usually never observed being placed by a compliance inspector. The inspector inspects the foundation installation (everything that is installed under the concrete) before the concrete is placed, and does not see the concrete until after it is placed when he or she returns for the second-stage inspection. Inspections performed are verified either by an approved municipality green tag and building permit procedure, or by individual compliance inspection reports by approved independent inspectors.

Coverage for existing homes is usually for major structural defects only, as addressed previously. Few companies known to the author will consider insuring an existing home, and then only if the property is considered a viable risk based on a structural condition report as addressed earlier. The coverage for existing homes is usually for a 3-year period. The coverage is renewable for an additional 3 years, but another structural condition report is necessary to assess the continued risk viability.

Under HUD/FHA guidelines relative to mortgage insurance for an existing home that has had recent (less than 1 year) foundation repairs, or is about to have such repairs, HUD will accept certain programs' insured structural protection policies/warranties in lieu of a bonded constructor's warranty.

8C.5 TYPICAL COSTS (1992 DOLLARS)

Where the program for newly constructed homes is concerned, homebuilders that are approved to participate in that program pay an annual registration fee, usually $200 for the first year and $100 for each year thereafter. For each home enrolled in the program, the homebuilder usually pays a $25 home enrollment fee. The homebuilder pays for the compliance inspections that are required by the insuring company or the approved municipal authorities. The cost of municipal building permits normally cover all municipal compliance inspections and are paid for by the homebuilder. When a new home goes to closing or final settlement, the homebuilder then pays a closing fee of at least $2.50 per each $1000 of the selling price to cover placement of the insurance. The cost is for the full 10 years of coverage.

Where the mortgage on a new home is insured by either HUD/FHA, VA, or FmHA, the homebuilder *must* pay for the insured protection, which is included in the selling price of the home. HUD and the other agencies, by federal statute, will not allow the homebuyers to pay directly for such insured coverage.

Other insuring programs charge similarly for their insured protection, although costs or procedures may vary somewhat.

The cost of the programs' coverage on an existing home, in addition to a structural condition report already addressed, is $3.00 or more for each $1000 of the selling price or appraised value (since this coverage is sometimes purchased without transfer of title, that is, without selling the home). This cost is for the full term of coverage, usually 3 years. With regard to this coverage, anyone can purchase the coverage, that is, the builder, buyer, lender, seller, foundation repair company, or purchaser. Also, as stated earlier, one does not have to belong to a program in order to be able to purchase this type coverage.

8C.6 SETTLEMENT OF CLAIMS

The program's insurance company handles the settlement of claims similar to procedures established by many insurance companies, particularly auto insurance companies. Once a claim has been determined to be covered under the policy, the insurer requires the homebuyer to solicit at least three detailed, itemized bids to correct the defects, subsequent to the policy terms and conditions. The insurer then would pay the lowest bid less the $250 deductible. The company would pay only for the work necessary to rectify the covered defect. As applicable, the insurer will advise the homeowner to verify that the procedure for repairing a covered defect (particularly major structural defects) is done under the auspices of a licensed professional structural or civil engineer or competent repair contractor, or both.

Where a mortgage is involved with a home, the insurer will issue a joint check in settlement of a claim in the name of both the mortgagee and the homeowner. This procedure is required by federal statute where a mortgage is insured by HUD/FHA.

Other insuring companies may use either a similar or a totally different procedure for settling such claims. However, the federal statute applies to all such companies where the joint claim check payment on HUD/FHA insured mortgages is concerned.

SECTION 8D
HUD

Richard J. Sazinski

8D.1 INTRODUCTION 8.21
8D.2 SCOPE 8.21
   8D.2.1 Single-Family Units 8.23
   8D.2.2 Multifamily Property 8.25
8D.3 POLICY DEVELOPMENT AND RESEARCH 8.26
   8D.3.1 Remedial Measures for Houses Damaged by Expansive Clay 8.27
   8D.3.2 Experimental Residential Foundation Designs on Expansive Clay Soils 8.27
8D.4 TECHNICAL SUITABILITY OF PRODUCTS PROGRAM 8.27
8D.5 LOWER-COST FOUNDATIONS 8.29
   8D.5.1 Concrete Footings, Foundation Walls, and Slabs 8.29
   8D.5.2 Frost-Protected Shallow Foundations 8.29
   8D.5.3 Permanent Wood Foundation (PWF) 8.29
   8D.5.4 Posttensioned Slab-on-Grade 8.29
8D.6 MANUFACTURED HOUSING (MOBILE HOME) FOUNDATIONS 8.31
8D.7 CLOSURE 8.40
8D.8 ACKNOWLEDGMENT 8.40
8D.9 REFERENCES 8.41

8D.1 INTRODUCTION

The Department of Housing and Urban Development (HUD) covers the United States with numerous HUD field offices (Fig. 8D.1). One or more of these offices may be located in each state. The various HUD field offices apply, monitor, or oversee the varied HUD programs, pursuant to Congress's allocation of funds, through appropriate HUD handbooks and procedures.

The United States Congress has mandated that HUD create conditions for every family to have decent and affordable housing. Numerous HUD programs[1] and activities may, in part, involve new or rehabilitation design and construction of several types of foundation systems for varied types of structures and projects. The current and future long-term stability of any new or rehabilitated foundation system, involved in any HUD program area, is critical to the economic interest of the American taxpayer and necessary to assure occupant safety.

Several HUD programs discussed in this part that are involved with foundation design, construction, repair, or rehabilitation include single-family (SF) homes, multifamily (MF) projects (new or rehabilitated), existing SF properties/MF projects, policy development and research, HUD's Technical Suitability of Products Program, affordable housing (via lower-cost foundations), and manufactured housing (mobile homes).

The views presented in the HUD part of this handbook are those of the author as an individual and do not necessarily represent the views of HUD.

8D.2 SCOPE

Numerous HUD programs assist borrowers in acquiring SF homes or are used for developing MF projects.[1] Such homes and projects are planned, located, and constructed to serve the present and predictable needs of tenants and to encourage owners in the maintenance and preservation of their investment. HUD may provide mortgage insurance for a loan obtained from a financial institution to purchase an SF property or finance an MF project. Acceptable risk to HUD's insurance fund (mortgage insurance) can be defined where a mortgage is secured by a property or project which meets the requirements of HUD's Minimum Property Standards (MPS).[2] The strength of an owner's desire to retain ownership will depend largely on continuing satisfaction with a property or

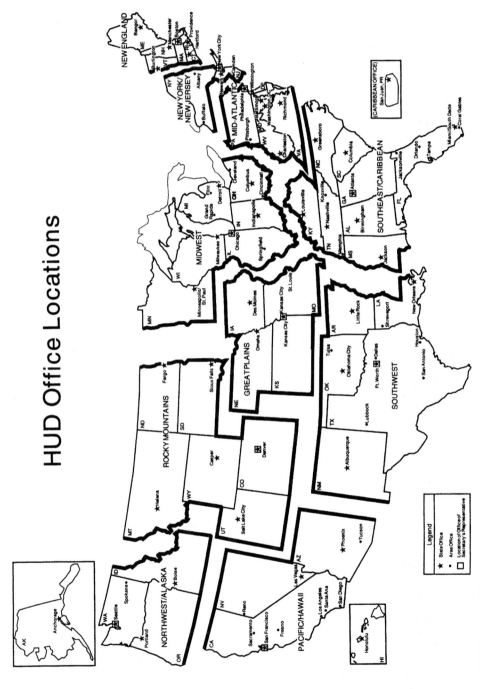

FIGURE 8D.1 HUD field offices.

project. Dissatisfaction arising from any cause, including foundation-related problems, lessens motivating interest in ownership and increases mortgage risk.

HUD relies on local codes or national model codes, in addition to standards listed in HUD's MPS for Housing,[2] as the basis for defining acceptable foundation standards and foundation design and construction practices. National model codes include the Council of American Building Officials (CABO) One and Two Family Dwelling Code, the Uniform Building Code (UBC), the Building Officials and Code Administrators (BOCA) Basic/National Building Code, and the Standard Building Code (SBC). HUD typically requires that foundation design and construction, as well as reports or repair methods related to foundations, be designed and completed by qualified professionals, that is, licensed architects, engineers, or contractors. Certifications from qualified professionals may also be required by HUD as a condition of mortgage reinsurance. Usually, completed foundation construction carries a 1-year workmanship and materials warranty. For any foundation-related repairs, corrections should be permanent in nature and commensurate with the value and type of property or project.

HUD SF properties or MF projects are involved in nearly the full range of foundation types, design, construction, inspections, and repairs which have been covered in Parts 2, 3, and 5–7, as well as this part of this handbook. HUD SF and MF areas involving foundations are further discussed in the sections that follow and cover new (proposed) construction, rehabilitation construction, or existing construction.

8D.2.1 Single-Family Units

SF units can be detached, semidetached, duplex, or row houses, either site-built or fully or partially factory manufactured. For example purposes, this section deals with typical site-built, detached SF homes covered by FHA mortgage insurance.

8D.2.1.1 New (Proposed) SF Construction. New (proposed) construction of SF units must comply with HUD's MPS for One- and Two-Family Dwellings (Appendix K of Ref. 2). HUD currently utilizes, wherever possible, acceptable state or local codes as a portion of its MPS for any particular property. Should no acceptable state or local code be available in an area, HUD will utilize the CABO One and Two Family Dwelling Code. HUD field offices maintain lists of jurisdictions and their appropriate codes which HUD utilizes as part of its MPS. HUD enforces and interprets codes for its own purposes and not for any local entity.

The CABO code, or any other acceptable state or local code, will address foundation systems. Any foundation system section of a code, used by HUD as part of its MPS, must address foundation depths, footings, and foundation materials criteria. Such codes also typically reference acceptable engineering practices and standards relative to foundations constructed of varying types of materials and methods, such as brick, masonry, treated wood, and concrete. HUD also defines acceptable material standards and engineering practices related to foundations (Appendices C, E, and F of Ref. 2).

HUD SF processing procedures for new proposed construction are defined in HUD Handbook 4145.1 REV-1.[3] These procedures define exhibits which must be submitted for each unit. These exhibits include a separate foundation plan and technical reports and other exhibits when the mortgage risk could be affected by unstable soil or other differential ground movement. Such reports or exhibits might include, but are not limited to, an engineer's report on soil exploration and testing along with special foundation designs for conditions found.

At times HUD may receive complaints from homeowners relative to foundation problems. If the property is covered by one of the typical 10-year structural warranty policies offered by various entities, HUD will rely on said warranty company as the first avenue of resolution. (Refer to Sec. 8C and to Appendix 10 of Ref. 3.) If no assistance is available from one of these warranty companies, HUD does have a process available to assist homeowners in correcting structural defects.[4] Under Section 518(a) of the National Housing Act, HUD can assist in the repair of structural defects reported within the first 4 years of the issuance of mortgage insurance. Problems with foundations can possibly qualify for assistance if deemed to be the builder's responsibility and the builder refuses to repair or rectify the situation. HUD may contract for repairs, reimburse the homeowner for completed repairs, or buy back the property from the homeowner.

8D.2.1.2 *Rehabilitation of SF Construction.* HUD promotes and facilitates the restoration and preservation of the nation's existing SF housing stock. One avenue provided by HUD in which a property can be acquired and rehabilitated is via Section 203(k) of the National Housing Act. There is a minimum $5000 requirement for the cost estimate of eligible rehabilitation or improvement items. As in other SF mortgage insurance programs, a Section 203(k) mortgage is funded by a HUD approved lender, and the mortgage is insured by HUD.

The procedures of the 203(k) program are detailed in HUD Handbook 4240.4 REV-2.[5] Under this program, one of the architectural exhibits required by HUD to be submitted is an inspection report from a qualified architectural, engineering, or home inspection service defining the adequacy or required upgrading of existing systems, such as heating, plumbing, and foundation. (Refer to Sec. 8A for further discussions of home inspections.) Homes that have been demolished, or will be razed as part of the rehabilitation work, are eligible provided the existing foundation system is not affected and will still be used. The complete foundation system must remain in place. A report from a licensed structural engineer is required stating that the existing foundation is structurally sound and capable of supporting the proposed dwelling construction.

Other foundation structural alterations and reconstruction, if necessary, could also be covered under a 203(k) mortgage in the form of foundation repairs or new foundation elements for new additions. (Refer to Parts 2, 3, 5, and 6.) Any new foundation construction must conform with local or state codes and HUD's MPS (Appendix K of Ref. 2). HUD Handbook 4905.1 REV-1,[6] which addresses acceptable conditions of existing units, applies in part to this program when defective conditions, including foundation problems, must be rectified (see Sec. 8D.2.1.3 for further discussion).

8D.2.1.3 *Existing SF Construction.* HUD requirements for the quality of existing SF properties, including technical acceptability, are defined in HUD Handbook 4905.1 REV-1.[6] Existing SF units with defective conditions are unacceptable for involvement in normal HUD program areas. Such conditions include defective construction, evidence of continuing settlement, or other conditions impairing the structural soundness of the dwelling. Such defective conditions shall render the property unacceptable until the defects or conditions have been remedied and the probability of further damage has been eliminated. Also, properties shall be free of hazards that may adversely affect the structural soundness of the improvements.

HUD's treatment of existing properties is also outlined in HUD Handbook 4150.1 REV-1.[7] HUD technical personnel, either HUD staff or fee staff (architects, engineers, construction analysts, and so on), are usually not involved in an inspection of an existing SF property for determining acceptability, unless specifically requested. Usually a HUD fee appraiser will determine the value of a property for FHA mortgage insurance purposes. The appraisal is performed for the benefit of HUD, not for the public. In addition to providing an estimate of value, the appraiser inspects the property for any visible deficiencies which may affect the health and safety of the occupants or the continued marketability of the property. HUD makes no warranties as to the value or condition of the house. Therefore the borrower must determine that the price of the property is reasonable and that its condition is acceptable. The borrower is also urged to hire a home inspection service to check out a property for adequacy. (See Sec. 8A for a discussion on inspections.)

The condition of existing building improvements is examined at the time of appraisal to determine whether repairs, alterations, or additions are necessary. When an examination of existing construction reveals noncompliance with the general acceptability criteria defined in HUD Handbook 4905.1 REV-1,[6] the appraiser must define an appropriate specific correction of the deficiency, if correction is feasible. A typical condition requiring repair, which would be noted by the appraiser, might involve visible foundation damage or damage related to foundation problems. Usually the appraiser may either define the required repairs, or, more commonly, the appraiser will request an inspection by a licensed or registered structural engineer defining the cause and cure of any foundation-related problem. Any cures defined by the structural engineer may also be required to be inspected and cleared by the same structural engineer as to adequacy of completion. (Refer to Parts 6 and 7 for typical foundation repairs.)

HUD acquired SF properties are usually the responsibility of HUD's Real Estate Owned Branch located in HUD field offices. When requested, HUD technical staff (architect, engineer, construction analyst, and so on) will determine the best way to restore a property to a good condition. Such restoration may entail foundation repairs prior to offering the unit as eligible for future mortgage insurance involvement.

8D.2.2 Multifamily Property

MF property can involve buildings designed and used for normal MF occupancy, including both unsubsidized and subsidized insured housing. Buildings with three or more living units are typically involved.

8D.2.2.1 New (Proposed) MF Construction. MF projects involved in HUD programs[1] cover a wide range of foundations and structures (single-story, stick-built to multistory, highrise concrete). Such structures typically must comply with HUD's MPS for Housing[2] along with a partially or fully acceptable state or local building code. If no partially or fully acceptable state or local code is in force or acceptable, a recognized model code must be utilized (such as UBC, BOCA, or SBC). HUD field offices maintain lists of jurisdictions with typically utilized or accepted codes for HUD MF project development. HUD enforces and interprets codes for its own purposes and not for any local entity.

Any acceptable state or local code, as well as any of the recognized model codes, covers foundation systems. Foundation system code sections must cover soil tests, foundation depths, footings, foundation materials criteria, piles (materials, allowable stresses, design), and excavations. Codes also will reference acceptable engineering practices and standards relative to foundations which are constructed of varying types of materials and methods (brick, masonry, treated wood, concrete, and so on). HUD also provides similar or additional listings of foundation-related acceptable material standards and engineering practices (Appendices C, E, and F of Ref. 2).

HUD requires that reliable subsurface exploration information be available (soils report, test boring logs, test pit data, soil bearing values, geotechnical study, and so on).[8] Usually soils reports will recommend foundation types along with design parameters for HUD MF projects. Also, HUD requires any MF project structural foundation plans to be stamped by a registered architect or professional engineer.[8]

8D.2.2.2 Rehabilitation of MF Construction. MF-type dwelling structures have always been and probably will continue to be, for the foreseeable future, a major source of housing for lower- and moderate-income families. Over the years, because of the economic conditions and sometimes the result of higher operating costs, many of these dwelling units have been neglected, and some are seriously deteriorated. Typical HUD procedures for MF rehabilitation are defined in HUD Handbook 4460.1 REV-1 (refer to Chap. 4 of Ref. 8).

A building should undergo a preliminary examination in order to clarify whether or not it is generally suitable for rehabilitation. After said preliminary examination, HUD may require an engineering report to examine major building components, such as the foundation system, to define any upgrading or replacement to assure structural soundness.

The services of a state licensed or registered engineer will be required by HUD when the design and construction of an MF rehabilitation project necessitates plans and specifications to properly define the scope and concept of the rehabilitation if structural changes are necessary or an addition is proposed to an existing building. Rehabilitation must comply with applicable local codes and ordinances. All new construction or additions that enlarge existing buildings must meet applicable codes and HUD's MPS for new construction. (Refer to Sec. 8D.2.2.1 and Ref. 2.) For a discussion of typical types of foundation repairs or new foundation elements involved, refer to Parts 2, 3, and 5–7.

Examples of HUD MF projects involving foundation rehabilitation are briefly summarized as follows[9]:

1. *Minerva Place Apartments* (St. Louis, MO./FHA #085-25-339)   Constructed in 1942 as a two-story YMCA with a swimming pool, it was converted to a 56-unit apartment complex with a new third floor added. A subsurface soil investigation, which determined the soil bearing capacity, required installation of new larger footings.

2. *Douglas Manor Apartments* (Webster Groves, MO./FHA #085-35-350)   Constructed in 1945 as a two-story public elementary school with classrooms and gymnasium, it was converted to a 41-unit apartment building. The critical element in the structure was found to be the allowable soil bearing under existing column spread footings. Soil analysis indicated that the shear strengths of the soil were low and erratic, and the allowable bearing capacity would be less than the load applied to the soil due to the installation of additional floors. Therefore the 12 columns supporting the gymnasium were shored so that the spread footings could be underpinned and enlarged for an allowable soil pressure of 2700 lb/ft^2 (13,200 kg/m^2).

3. *Patterson Place* (Bismarck, N.D./Proj. #094-35040-LD-SR-L8)   Constructed in 1910 as a six-story hotel with approximately 100 rooms, over the years (around the 1930s) four more floors were added to extend the structure to a 10-story height. The proposed rehabilitation consisted of converting the hotel into 117 apartments for the elderly. Because of the addition of four more floors, the foundation and columns became underdesigned. To reduce the loads on column footing pads, the rehabilitation required the removal of 2- to 4-in (0.8- to 1.6-cm)-thick concrete floor toppings, nonbearing concrete partitions, and heavy plastered ceilings. Steel stud drywall partition walls were installed, and overloaded columns from the basement to the fifth floor were upgraded.

8D.2.2.3 Existing MF Construction.   HUD offices nationwide are involved with a large inventory of varying types of existing MF structures (wood frame, reinforced concrete, steel, and so on) originally constructed under various HUD programs [such as 221(d)(4) insured apartments, 202 elderly, or public housing].[1] At any time, an MF structure might develop a foundation problem necessitating action. Such foundation problems may surface during one of HUD's periodic maintenance inspections, or, they may be reported by project personnel or by tenants themselves. A report from a registered or licensed structural engineer, defining causes and cures, may be requested. Any repair (cures) or retrofit projects can be funded by HUD under various programs.

As an example, a project known as Jackson Heights Elderly Highrise (Proj. #SD-045-001W), constructed in 1978 in Rapid City, S.D., started to exhibit stress conditions on several interior wall surfaces in 1984/1985 (cracked drywall, door frame racking, and so on). A structural engineer and a soils consultant were employed to define causes and cures. The structure was six stories in height, consisted of 106 units, and was constructed of standard concrete columns bearing on isolated concrete footing pads. A drastic change in the moisture content of a subsurface bearing stratum caused an approximate 2-in (5-cm) differential settlement between adjacent reinforced concrete columns and pads spaced from 17 ft to 20 ft (5.2 to 6.1 m) apart. 12 isolated concrete pad footings of one wing were stabilized via compaction grouting. An experienced contractor completed the necessary repairs. For a discussion of compaction grouting, see Sec. 5B.

8D.3 POLICY DEVELOPMENT AND RESEARCH

To carry out presidential and congressional mandates in the areas of housing, community development, and fair housing efficiently and effectively, HUD is structured so that research, economic and policy analyses, and program evaluations are the responsibility of the Assistant Secretary for Policy Development and Research (PD&R) in Washington, D.C. All research activities are centralized in PD&R. The research data HUD uses in policy development are made available to interested parties such as state and local governments, financial institutions, builders, developers, universities, and colleges.

The research program of PD&R has focused, in part, on granting research contracts to various groups to investigate and report on hazards to housing (expansive soils, lead-based paint, radon gas, and so on). Relative to foundation systems, work carried out under research contracts is usually based on developments in applied structural and geotechnical theory. Cooperative efforts of research institutions (colleges, universities, and so on), contractors, practicing engineers, and others involved in such research contracts have produced a better understanding of the factors that contribute to foundation performance. Examples of HUD funded research involving foundation systems includes a study of remedial measures for houses damaged by expansive clay,[10] as well as a study of experimental residential foundation designs for use on expansive clay soils.[11]

8D.3.1 Remedial Measures for Houses Damaged by Expansive Clay

One such research contract investigated certain remedial techniques which could be used to restore houses with damaged foundations caused by expansive clay soils.[10] All techniques were directed toward stabilizing a mass of soil beneath house floor slabs to a depth of 5 ft (1.5 m) or more. The 10 houses included in the investigation were HUD-owned properties identified by the HUD Dallas office, and all were severely distressed by vertical movement and volume change of the soil. Data acquisition continued over approximately 2 years, or four seasonal climatic cycles.

The remedial techniques included lime slurry pressure injection, a subsurface irrigation system, and three types of moisture migration barriers (granular, recycled rubber, and lean concrete). The lime slurry injection process, utilized around the foundation perimeters to stabilize the soil mass and inhibit moisture migration, was very costly and not very effective. The subsurface irrigation system was not completely effective due to additional climatic effects and difficulty in keeping the mechanical components of the system active. All three of the moisture migration barriers (granular, recycled rubber, and lean concrete) appeared to be viable techniques when accompanied by increasing the foundation soil moisture to 2 to 3% above the plastic limit. These barriers, approximately 5 ft (1.5 m) deep and the width of trenching equipment, isolated a mass of soil beneath the floor slab and caused the slab and soil to interact as a layered system. Once a stable equilibrium condition was realized after water infusion, cosmetic repairs could be effected both inside and outside the damaged structure and the foundation restored to a usable condition.

8D.3.2 Experimental Residential Foundation Designs on Expansive Clay Soils

Preventive measures, applied at the time of construction, were also studied under research contract.[11] These have been discussed partly in Sec. 6C. The study involved 11 experimental homes, each with a different type of foundation system, constructed over expansive clay soils in Grand Prairie, Tex. The performance of the homes was monitored over a 3-year period, or six seasonal cycles. Foundation systems utilized various reinforced concrete, posttensioning, moisture barrier stabilization techniques, and subsurface irrigation systems.

The typical reinforced concrete systems (BRAB waffle slab design) performed adequately, yet exhibited a prohibitively high cost. Posttensioned systems exhibited reasonably effective performance. (See Sec. 8D.5.4 and Refs. 12–14 for further discussion.) Active subsurface irrigation systems proved not to be justified because of indeterminate effects of interruptions of water service, tampering with controls, and concerns for the longevity and durability of any buried pipe. Perimeter passive vertical moisture barrier stabilization techniques were deemed viable, provided the soils encompassed by any barrier were preswelled and moisture was contained.

8D.4 TECHNICAL SUITABILITY OF PRODUCTS PROGRAM

Section 521 of the National Housing Act of 1965 directs HUD to adopt a uniform procedure for the acceptance of materials and products to be used under HUD housing programs. To comply with this mandate, HUD developed the Technical Suitability of Products Program for review and acceptance of building systems, components, products, and materials. The objectives of this program are the acceptance of new and innovative building materials and systems, and to encourage improvements in, and development of, technological advances in home building.

The program and processing procedures are outlined in HUD Handbook 4950.1 REV-2.[15] Under this program, one such type of acceptance document that can be issued by HUD is a structural engineering bulletin (SEB). An SEB can be issued to accept complex components which have structural features not addressed by, or not in compliance with, HUD requirements for new construction. A floor, wall, or roof system can be covered by an SEB issued by HUD headquarters.

Wall systems can, in part, cover foundation systems. An example is SEB #1117 issued to Superior Walls of America, Ltd., for their precast concrete stud wall panels utilized in SF home basement construction. The wall panels, in themselves, are designed to act as a reinforced concrete monolithic concrete panel with a footing (Fig. 8D.2). In the past 9 years, over 10,000 Superior Walls foundations have been installed throughout the northeastern United States. The precast reinforced concrete insulated foundation basement wall panels can be used for dwellings up to 2½ stories (plus basement). Wall panels can be precast in 4-ft (1.2-m), 8-ft (2.4-m), and 10-ft (3-m) heights and up to 20 ft (6 m) wide. The panels consist of reinforced concrete exterior skins, interior studs, and interior insulation. Minimum concrete strength is 5000 psi (34.5 MPa).

The concrete exterior skins are 1¾ in (4.5 cm) thick, monolithic with 10½-in (26.7-cm) top ledger and bottom footing reinforced with 2 no. 3 horizontal bars. The reinforced concrete interior studs are 2¼ in (5.7 cm) by 6¾ in (17.1 cm) with a no. 4 vertical bar. Studs are spaced 24 in (60 cm) center to center and are mechanically anchored to the concrete skin, top ledger, and bot-

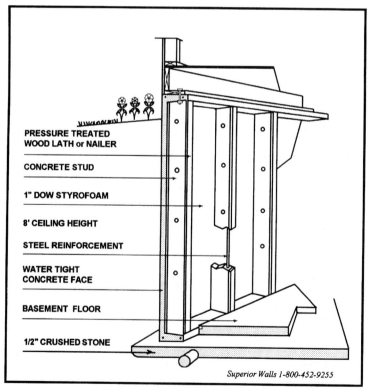

FIGURE 8D.2 Superior Wall and Foundation System.

tom footing. Treated wood nailers are attached to the interior face of the studs. Basement foundation wall panels have a maximum allowable load (earth pressure) of 60 lb/ft^3 (960 kg/m^3) equivalent fluid pressure and a maximum allowable vertical load (building structure plus live loads) of 4200 lb/lin ft (6250 kg/m).

8D.5 LOWER-COST FOUNDATIONS

Homes which are safe, durable, marketable, and affordable must be provided to meet national housing goals. In January, 1982, HUD announced the formation of the Joint Venture for Affordable Housing (JVAH) as a public-private partnership to reduce housing costs. Over the years, several HUD JVAH demonstration projects have utilized lower-cost structural foundation systems to assist in reducing housing costs.[16,17] Lower-cost foundation systems used in the JVAH effort, as well as other cost-saving foundation systems, are discussed in this section.

8D.5.1 Concrete Footings, Foundation Walls, and Slabs

Footing widths could be reduced if accurate soil data were utilized rather than providing local practice minimum widths. Footing reinforcing can be deleted for footings placed on undisturbed stable soil. Reinforcement in foundation walls is usually not required in nonexpansive soil and in areas of low seismic risk. Under stable base conditions, concrete slab floors do not require typical welded-wire fabric (WWF), concrete slab thickness can be reduced to 2½ in (6 cm), and concrete slab strength can also be reduced to 2000 psi (13.8 MPa). Highly expansive soils may necessitate specialized construction.

8D.5.2 Frost-Protected Shallow Foundations

Perimeter insulation is added along with a shallow foundation system in lieu of a full-frost-depth foundation (Fig. 8D.3).[18] Insulation dimensioning is according to design guides of the Norwegian Building Research Institute along with some extrapolation and adjustment for U.S. energy code minima. Corner insulation is typically extrawide due to three-dimensional heat loss. Cost savings have been projected from $1239 to $2875 for insulated shallow foundation systems versus full-frost-depth foundations used in several areas of the country [1612-ft^2 (150-m^2) home, block and concrete foundations].[18]

8D.5.3 Permanent Wood Foundation (PWF)

The PWF system (Fig. 8D.4),[19] previously identified as the AWWF (All-Weather Wood Foundation), is basically a below-grade stud wall built of pressure-preservative-treated lumber and American Plywood Association (APA) trademarked plywood. Basic advantages of the PWF are installation in nearly any weather condition, speed of construction, ease of interior finishing, and versatility. HUD JVAH demonstration projects in Tulsa, Okla., and Fairbanks, Alaska, utilized the PWF system, and cost savings amounted to $1470 and $1035 per unit, respectively.[17]

8D.5.4 Posttensioned Slab-on-Grade

Certain foundation configurations are designed to be relatively insensitive to the fluctuations in soil moisture and the consequent variations in supporting soil volume. Typical foundations

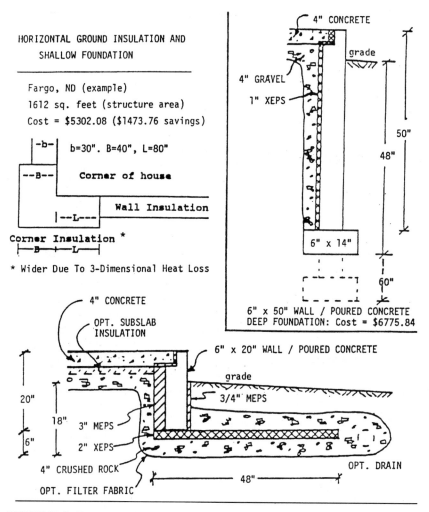

FIGURE 8D.3 Frost-protected shallow foundation and full-frost-depth foundation.

depend on stable supporting soil. The use of posttensioned slabs-on-ground is most often specified in locations with highly expansive soils or in other situations where the foundation-supporting material provides uncertain or variable supporting qualities.[12,13]

The system can be utilized for sites with highly expansive or compressible soils where the use of conventional foundations might require extensive site soil modifications (soil removal and replacement). More extensive deep foundation systems otherwise might be required. Costs vary with site soil conditions and building configurations. Part 3 further discusses this type of foundation system.

Some HUD concerns were recently expressed relative to these types of systems and their utilization in Texas for units involving FHA mortgage insurance.[14] However, HUD concerns could be acceptably mitigated with assured certifications as to defining proper soil design parameters, following current design procedures, and providing for adequate supervision over construction.[14]

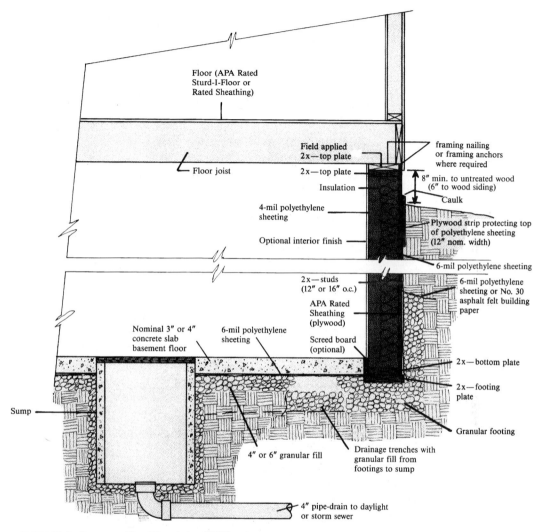

FIGURE 8D.4 Permanent Wood Foundation (PWF) system.

8D.6 MANUFACTURED HOUSING (MOBILE HOME) FOUNDATIONS

In 1983 HUD extended eligibility for typical FHA SF mortgage insurance to manufactured homes constructed in conformance with the Manufactured Home Construction and Safety Standards (MHCSS), when permanently attached to a site-built foundation. Neither HUD's MPS nor the Manual of Acceptable Practices (MAP) adequately covered special requirements for permanent foundations for manufactured-housing. Due to this inadequacy, HUD headquarters realized an urgent need to furnish HUD field offices with technical guidance concerning acceptance of permanent site-built foundations for manufactured housing if they are to be covered by typical 30-year FHA SF mortgage insurance.

8.32 MISCELLANEOUS CONCERNS

To fulfill this need, HUD entered into a contract (#H-5640) with the Small Homes Council of the University of Illinois to develop guidelines, procedures, and details to be included in a HUD Handbook to contain foundation concepts that are adequate in design, and for use in permanent foundation systems for manufactured housing. This contract work resulted in the issuance of HUD Handbook 4930.3.[20]

Companies building manufactured homes assisted in the preparation of HUD Handbook 4930.3 by providing over 40 foundation systems appropriate for manufactured housing. This information was assembled and used as the basis for the foundation design concepts presented in Appendix A of the handbook.[20] The three basic types of foundation systems defined are as follows (Fig. 8D.5):

1. *Type C systems*   Supported and anchored at chassis only
2. *Type E systems*   Supported at both the chassis and the exterior wall, but tied down or anchored at the exterior wall only

TYPE C (Choose for foundation design concepts C2, C3, C4)

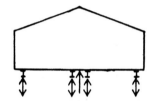

TYPE E (Choose for foundation design concepts E1, E2, & E8)

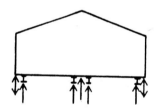

TYPE I (Concepts are included here as a possible future design concept. No Type I foundations were submitted.)

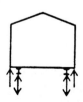
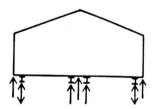

FIGURE 8D.5   Manufactured housing foundation system types.

3. *Type I systems* Supported at both the chassis and the exterior wall, but tied down or anchored at the chassis only

Type C1 and E1 systems are depicted in Figs. 8D.6 and 8D.7. Once the system is defined, Appendix B of the handbook[20] presents foundation design tables for use in determining foundation anchorage and footing sizes for all foundation types. All structural design was based on generally accepted engineering practice. Design loads (wind, snow, and so on), except for cases of higher local code requirements, are based on ANSI A58.1-1982, "Minimum Design Loads for Buildings and Other Structures." (*Note:* ANSI A58.1 is now designated ASCE 7-88.) Normal ranges of allowable footing bearing pressures are also defined in the handbook.[20] Foundation capacity tables are presented in Appendix C of the handbook[20] to determine the sizes and spacing of anchors, required size and depth of footings, and necessary reinforcement. Horizontal and vertical anchorage and withdrawal resistance capacities are defined for foundation systems of reinforced concrete, grouted masonry, isolated piers, and all-weather wood systems (on concrete or gravel) (Figs. 8D.8 through 8D.12).

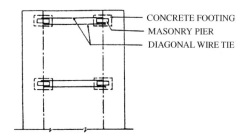

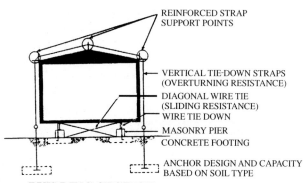

FIGURE 8D.6 Type C1 foundation type.

8.34 MISCELLANEOUS CONCERNS

FOUNDATION TYPE	SYSTEM NUMBER
Reinforced perimeter wall, unreinforced piers at chassis	
SUPERSTRUCTURE TYPE	E1
Exterior anchored, chassis supported single or multi-wide	

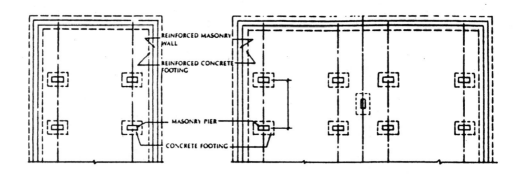

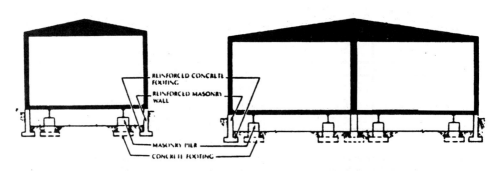

FIGURE 8D.7 Type E1 foundation type.

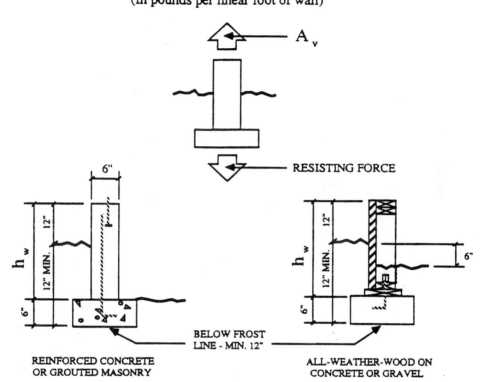

TABLE 1
WITHDRAWAL RESISTANCE*
CONTINUOUS FOUNDATIONS[1,2]
(In pounds per linear foot of wall)

h_w	Reinforced Concrete		Masonry-Fully Grouted		Masonry-Grouted @ 48" o.c.		All-Weather Wood w/ Conc. Footing		All-Weather Wood w/ Gravel Base
	12"	16"	12"	16"	12"	16"	12"	16"	12" & 16"
2'	285	350	245	310	220	285	170	225	35
3'	420	525	360	465	320	425	290	385	75
4'	555	700	475	620	420	565	340	475	120

* Potential resistance to withdrawl is the maximum uplift resistance which can be provided by the foundations shown. It is computed by adding the weights of building materials and soil over the top of the footing. To fully develop this potential, adequate connections to the footing and superstructure must be provided.

[1] Foundations must be designed for bearing pressure, gravity loads, and uplift loads in addition to meeting the anchorage requirements tabulated in the Foundation Design Tables.

[2] Values shown in this table could be increased by widening the footing, provided the system is designed for the increased load, or by a more detailed analysis of the shearing strength of the soil overburden.

FIGURE 8D.8 Withdrawal resistance.[20]

TABLE 2
WITHDRAWAL* RESISTANCE FOR PIERS[1,2]
(In pounds per pier)

h_p Depth	Width of Square Footing			
	1' - 0"	2' - 0"	3' - 0"	4' - 0"
2' - 0"	260	990	2065	3560
2' - 8"	340	1325	2780	4785
3' - 4"	410	1655	3485	5990
4' - 0"	490	1990	4200	7220
4' - 8"	565	2320	4910	8445

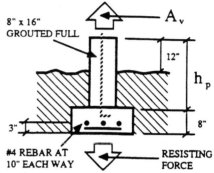

* Potential withdrawal resistance is the maximum uplift resistance which can be provided by the foundations shown. It is computed by adding the weights of building materials and soil over the top of the footing. To fully develop this potential, adequate connections to the footing and superstructure must be provided.

[1] Foundations must be designed for lateral soil pressure, bearing pressure, gravity loads, and uplift loads, in addition to meeting the anchorage requirements tabulated in the Foundation Design Tables. The bottom of the footing must also be below the maximum depth of frost penetration.

[2] Values shown in this table could be increased by widening the footing, providing the wall system is designed for the increased load, or by a more detailed analysis of the shear strength of the soil overburden.

TABLE 3
VERTICAL ANCHOR CAPACITY FOR PIERS*
(In pounds)

Anchor bolt dia	Capacity per number of bolts	
	1	2
1/2"	3930	7860
5/8"	6140	12280

TABLE 3A[1]

Anchor bolt dia.	Vertical rebar	Minimum lap splice	Rebar hook
1/2"	# 4	20"	6"
5/8"	# 5	25"	7"

* The vertical anchor capacity is based upon the working capacity of ASTM A - 307 anchor bolts in 2500 psi concrete or grout. To fully develop this capacity, anchor bolts must be properly connected to the foundation and superstructure.

[1] The capacity is based on $F_c = 2500$ psi; $f_y = 40{,}000$ psi

FIGURE 8D.9 Withdrawal resistance and anchor capacity.[20]

TABLE 4A
VERTICAL ANCHOR CAPACITY FOR LONG FOUNDATION WALL*
(A_v in pounds per linear foot of wall.)

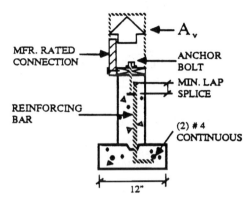

CONCRETE OR MASONRY WALL

A_v lbs./ft.	Required Anchorage[1]		
	Anchor Bolt[4]	Rebar	Spacing[5]
650	1/2"	#4	72"
785	1/2"	#4	60"
See Table 3A For Rebar Details			

* Anchor capacity is based on the capacity of ASTM A-307 anchor bolts properly embedded, enclosed, and spaced. The foundation wall must be adequate to develop the full capacity of the bolts.

[1] Values are based on horizontal load per foot of wall. Select A_h for pier spacing of 4-feet for use with this table.

[2] It is assumed that a reinforcing bar of the same diameter as the anchor is adequately embedded in the footing and lapped with the anchor. In the case of a treated wood foundation wall, the wood wall and its connections must be designed to transfer the anchor load to a concrete footing. This table does not apply to treated wood foundation walls on gravel bases.

[3] Spacing based on capacity of anchor bolt in grout / concrete. (:

FIGURE 8D.10 Vertical anchor capacity for concrete or masonry wall.[20]

TABLE 4B
VERTICAL ANCHOR CAPACITY FOR LONG FOUNDATION WALL*
(A_v in pounds per linear foot of wall.)

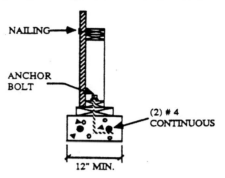

TREATED WOOD WALL

A_v lbs./ft.	Required Nailing[2,3] (Edge Spacing, in.)	Min. Plywood Thickness	Required Anchorage[1]	
			Anchor Bolt Diameter	Bolt Spacing[6]
75	6d @ 6" o.c.	3/8"	1/2"	72"
90	↓	↓	↓	60"
110	↓	↓	↓	48"
150	↓	↓	↓	36"
170	↓	↓	↓	32"
225	8d @ 6" o.c.	7/16"	3/4"	42"
340	8d @ 4" o.c.	↓	↓	28"
450	10d @ 4" o.c.	15/32"	↓	20"
***				

*** For Av greater than 450 lb./ft., consider using a different foundation material or utilize an engineered design with a higher capacity.

* Anchor capacity is based on the capacity of ASTM A-307 anchor bolts properly embedded, enclosed, and spaced. The foundation wall must be adequate to develop the full capacity of the bolts.

[1] Values are based on horizontal load per foot of wall. Select A_h for pier spacing of 4-feet for use with this table.

[2] Assuming 1 1/2" thick sill plate, 3/4" edge distance for wood or composite nailer plates or 20 diameter end distance for plywood sheathing; C-C exterior, properly seasoned wood, 33% increase for wind loads; Group III woods, not permanently loaded, and a 10% reduction for fasteners in treated wood.

[3] Nailing scheduled in this table is intended to secure the superstructure to the foundation only, and not to provide required edge fastening for plywood siding or sheathing

[4] Spacing based on allowable compression of wood perpendicular to grain for F_c = 565 psi and washer with total diameter equal to bolt diameter plus 1/2". (*special design required.)

FIGURE 8D.11 Vertical anchor capacity for treated wood wall.[20]

TABLE 5A
HORIZONTAL ANCHOR CAPACITY FOR SHEAR WALLS*
(In pounds per foot of wall)

CONCRETE OR MASONRY

A_h[1] lbs./ft.	Required Anchorage		
	Anchor Bolt[4]	Rebar	Spacing[5]
600	1/2"	#4	36" o.c.
675	↓	↓	32" o.c.
900	↓	↓	24" o.c.
1350	↓	↓	16" o.c.
1800	↓	↓	12" o.c.
***	See Table 3A For Rebar Details		

*** For greater than 1800 lb./ft. consider using an engineered design with a higher capacity.

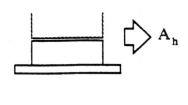

TABLE 5B

TREATED WOOD

A_h[1] lbs./ft.	Required Nailing[2,3] (Edge Spacing, in.)	Min. Plywood[3] Nailer Thickness	Required Anchorage	
			Anchor Bolt Diameter	Bolt Spacing[6]
330	8d @ 4" o.c.	7/16"	1/2"	36"
370	8d @ 4" o.c.	15/32"	1/2"	32"
495	10d @ 4" o.c.	19/32"	1/2"	24"

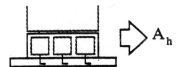

* Anchor capacity is based on the capacity of ASTM A-307 anchor bolts properly embedded, enclosed, and spaced. The foundation wall must be adequate to develop the full capacity of the bolts.
[1] Values are based on horizontal load per foot of wall. Select A_h for pier spacing of 4-feet for use with this table.
[2] Assuming 1 1/2" thick sill plate, 3/4" edge distance for wood or composite nailer plates or 20 diameter end distance for plywood sheathing; C-C exterior, properly seasoned wood, 33% increase for wind loads; Group III woods, not permanently loaded, and a 10% reduction for fasteners in treated wood.
[3] Nailing scheduled in this table is intended to secure the superstructure to the foundation only, and not to provide required edge fastening for plywood siding or sheathing
[4] It is assumed that a reinforcing bar of the same diameter as the anchor is adequately embedded in the footing and lapped with the anchor. In the case of a treated wood foundation wall, the wood wall and its connections must be designed to transfer the anchor load to a concrete footing. This table does not apply to treated wood foundation walls on gravel bases.
[5] Spacing based on capacity of anchor bolt in grout / concrete. (see also #4.)
[6] Spacing based on allowable compression of wood perpendicular to grain for F_c = 565 psi and washer with total diameter equal to bolt diameter plus 1/2". (*special design required.)

FIGURE 8D.12 Horizontal anchor capacity.[20]

The handbook's format is arranged for a licensed professional (engineer or architect) to progress through a series of logical steps designated to lead to approval. In developing the handbook,[20] seismic loads based on ANSI A58.1 were computed for all foundation systems of each type of structure and were found to be less than the forces produced by the minimum wind load of 70 mi/h (113 km/h). In seismic zones 3 and 4 (as defined in ANSI A58.1) the handbook[20] requires foundations to be designed by a professional engineer licensed in the applicable state. Screw-in-ground anchors are not considered permanent foundations.

8D.7 CLOSURE

The previous sections have described several HUD program areas involved with foundation design, construction, repair, and rehabilitation. Numerous types of foundation systems and repair techniques are and have been utilized in various HUD program areas. HUD typically relies on foundation design and construction procedures defined in national model codes, acceptable state and local codes, or defined in its MPS for Housing. Policy development and research activities have been sponsored by HUD to study remedial measures for damaged foundations as well as acceptable foundation systems for use over expansive clay soils. HUD's Technical Suitability of Products Program is an avenue which has been used by sponsors to obtain a review and acceptance of foundation-related housing components or systems. Various types of economically built foundation systems have been shown to contribute to affordable housing in demonstration projects developed under HUD's Joint Venture for Affordable Housing (JVAH). Manufactured housing (mobile home) foundations, considered permanent in nature, have been defined by HUD through cooperative efforts of the industry and the Small Homes Council of the University of Illinois.

It has not been possible in the course of these brief sections to cover all the types of HUD program areas that may involve foundation design and construction. Omitted, for example, are discussions of public and Indian housing projects, community planning and development grants, and HUD-held properties or projects. However, in such omitted program areas, similar procedures are utilized and are based on the use of acceptable foundation standards, complying with building codes, involving qualified design professionals, and utilizing experienced contractors. Key goals are to assure structural integrity and longevity, occupant safety, durability, protection of HUD's mortgage insurance fund, and the efficient and economical use of congressionally allocated funds.

Some important programs, projects, or properties may regrettably have gone unmentioned because they have not come to the attention of the author. Nevertheless, it is hoped that the preceding review, which has attempted to emphasize several HUD programs, projects, and property types, will provide an informative guide as to how HUD handles the critical housing subject of foundations. For more detailed information on HUD programs, including available publications or handbooks, contact the HUD field office in your area.

8D.8 ACKNOWLEDGMENTS

Grateful acknowledgment should be expressed to many colleagues, especially to the limited nationwide staff of HUD technical professionals (architects, engineers, and construction analysts) who strive to assure that properties and projects involved in HUD program areas have adequately designed and constructed foundations. A special acknowledgment is also expressed to G. Robert Fuller (retired, HUD headquarters), Kenneth L. Crandall (HUD headquarters), and Robert Wade Brown, whose helpful suggestions and cooperation have contributed significantly to the preparation of this section.

8D.9 REFERENCES

1. U.S. Department of Housing and Urban Development (HUD), "Programs of HUD," HUD-214-PA(18), Washington, D.C., May 1992.
2. U.S. Department of Housing and Urban Development (HUD), "Minimum Property Standards for Housing," HUD Handbook 4910.1, Washington, D.C., 1994.
3. U.S. Department of Housing and Urban Development (HUD), "Architectural Processing and Inspections for Home Mortgage Insurance," HUD Handbook 4145.1 REV-1, Washington, D.C., Mar. 1990.
4. U.S. Department of Housing and Urban Development (HUD), "The Construction Complaints and Section 518(a) and (b) Handbook," HUD Handbook 4070.1 REV-2, Washington, D.C., Aug. 1981.
5. U.S. Department of Housing and Urban Development (HUD), "203(k) Handbook—Rehabilitation Home Mortgage Insurance," HUD Handbook 4240.4 REV-2, Washington, D.C., Dec. 1991.
6. U.S. Department of Housing and Urban Development (HUD), "Requirements for Existing One to Four Family Units," HUD Handbook 4905.1 REV-1, Washington, D.C., Aug. 1991.
7. U.S. Department of Housing and Urban Development (HUD), "Valuation Analysis for Home Mortgage Insurance," HUD Handbook 4150.1 REV-1, Washington, D.C., Feb. 1990.
8. U.S. Department of Housing and Urban Development (HUD), "Architectural Analysis and Inspections for Project Mortgage Insurance," HUD Handbook 4460.1 REV-1, Washington, D.C., Aug. 1991.
9. G. R. Fuller and S. Asar, "Structural Evaluation and Rehabilitation of Multistory Residential Buildings," presented at the ASCE Spring Convention, May 1983.
10. A. Poor, "Remedial Measures for Houses Damaged by Expansive Clay," HUD Contract H-2240R, HUD-PDR-415-2, University of Texas at Arlington, Apr. 1979.
11. A. Poor, "Experimental Residential Foundation Designs on Expansive Clay Soils. Data Supplement," HUD Contract H-2240R, University of Texas at Arlington, Dec. 1979.
12. Post-Tensioning Institute (PTI), "Design and Construction of Post-Tensioned Slabs-on-Ground," PTI, Phoenix, Ariz., 1980.
13. Post-Tensioning Institute (PTI), "Tentative Recommendations for Prestressed Slabs-on-Ground Used on Expansive or Compressible Soil," PTI, Phoenix, Ariz., Sept. 1975.
14. R. J. Sazinski, "HUD Concerns: Post-Tensioned Foundations in Texas," in *Proc. 7th Int. Conf.* on Expansive Soils, Dallas, Tex., Aug. 1992.
15. U.S. Department of Housing and Urban Development (HUD), "Technical Suitability of Products Program Processing Procedures," HUD Handbook 4950.1 REV-2, Washington, D.C., Mar. 1988.
16. R. J. Sazinski, "Lower Cost Structural Techniques for Housing," in *Proc. ASCE Specialty Conf.* on Affordable Housing, Miami, Fla., Dec. 1989.
17. U.S. Department of Housing and Urban Development (HUD), "Affordable Housing, Challenge and Response, vol. II: Affordable Residential Construction; a Guide for Home Builders," HUD-1128-PDR(vol. 2) prepared for HUD by National Association of Home Builders (NAHB)/National Research Center (NRC), Upper Marlboro, Md., July 1987.
18. R. A. Morris, "Frost-Protected Shallow Foundations: Current State-of-the-Art and Potential Application in the U.S.," prepared for the Society of the Plastics Industry by NAHB National Research Center (NAHB/NRC Proj. 2070), Upper Marlboro, Md., Aug. 1988.
19. American Forest and Paper Association (AFPA), American Wood Council, "Permanent Wood Foundation, Guide to Design and Construction," Form A400Q, Washington, D.C., Revised Jan. 1990.
20. U.S. Department of Housing and Urban Development (HUD), "Permanent Foundations Guide for Manufactured Housing," HUD Handbook 4930.3, Washington, D.C., Feb. 1992.

INDEX

Accelerated strength test, **3**.43–**3**.44
   autogenous method, **3**.44
   boiling-water method, **3**.43
   high-temperature method, **3**.44
   pressure method, **3**.44
   warm-water method, **3**.43
ACI code, **3**.19–**3**.20, **3**.117
Acrylamide grout, **5**.367–**5**.369
Activity, values of, **2**.11
Activity number, **2**.11
Admixtures (concrete):
   accelerating, **3**.13
   air-detraining, **3**.16
   air-entraining, **3**.13–**3**.15
   blast-furnace slag, **3**.18
   bonding, **3**.16
   calcium chloride, **3**.15
   chemical, **3**.13–**3**.16
   coloring, **3**.16
   corrosion inhibiting, **3**.16
   dampproofing, **3**.16
   expansion producing, **3**.16
   flocculating, **3**.16
   fly ash, **3**.16–**3**.18
   fungicidal, germicidal, and insecticidal, **3**.16
   gas forming, **3**.16
   mineral, **3**.16–**3**.18
   pozzolans, **3**.18
   reducing alkali-aggregate expansion, **3**.16
   rice-husk ash, **3**.18
   set-controlling, **3**.16
   shrinkage and, **3**.59–**3**.60
   silica fume, **3**.18
   water-reducing, **3**.12–**3**.13
Aeration zone, **1**.3
   vs. soil moisture, **1**.5–**1**.6
Aggregates, **3**.6–**3**.11
   cement ratio and, **3**.51
   characteristics, **3**.10
   coarse, **3**.6–**3**.8, **3**.32, **3**.33, **3**.53
   fine, **3**.32, **3**.33
   grading requirements, **3**.8–**3**.10
   heavyweight, **3**.7
   lightweight, **3**.7

Aggregates (*Cont.*):
   moisture adjustments, **3**.31
   production of, **3**.8–**3**.9
   recycled concrete and other products, **3**.8
   selection, **3**.10–**3**.11
   shrinkage and, **3**.59
   size, **3**.9–**3**.10
Alternative dispute resolution (ADR), **8**.10–**8**.11
American Association of State Highway and
   Transportation Officials (AASHTO), **2**.14
American Plywood Association (APA), **8**.29
Anchors:
   earth, **6**.59
   mobile homes, **8**.36–**8**.39
      vertical anchor capacity for long foundation
         wall, **8**.37–**8**.38
      vertical anchor capacity for piers, **8**.36
      vertical anchor capacity for shear walls, **8**.39
Anisotropic fabric, **5**.78
Apartment building, foundation repair and, **6**.79, **6**.80
Apparent cohesion, **5**.81
Applied stress, **2**.22–**2**.26
Aspdin, Joseph, **3**.4
Atterberg limits, **2**.10
   uses of, **2**.11

Backfill:
   classifications, **4**.24
   stress on walls, **4**.24–**4**.29
   types of, **4**.24
Balaguru, P., **3**.3
Bars:
   compression development length, **3**.120
   reinforced concrete, **3**.117–**3**.120
   tension development length, **3**.118–**3**.120
Basements, **3**.184–**3**.185
   flooding, **6**.82
   foundation repair, **6**.79–**6**.84
Batter, **1**.22
Beam foundations:
   with crawl space, **3**.184
   repairing, **6**.32–**6**.48**

Beam foundations, repairing (*Cont.*)
    shimming interior floors, **6**.45–**6**.48
    spread footings, **6**.37–**6**.40
    underpinning, **6**.32, **6**.44, **6**.45
Beams:
  shear strength of reinforced concrete, **3**.94–**3**.99
  torsion strength of reinforced concrete, **3**.100–**3**.102
Bearing capacity:
  absolute, **5**.278
  cohesive soil, **5**.127
  comparative foundation, **6**.59–**6**.60
  deformation, **5**.278
  equation, **2**.48–**2**.49, **2**.51–**2**.54
  factors surrounding, **2**.51–**2**.54
    depth, **2**.52–**2**.53
    inclination, **2**.53–**2**.54
    safety, **2**.54
    shape, **2**.52
  failure of retaining walls, **4**.31–**4**.32
  field application:
    PI Case, **5**.284–**5**.289
    Rogers' Reservoir, **5**.279–**5**.284
    Ryerson Polytechnical Institute, **5**.289
  foundation failures and, **6**.4, **6**.9, **6**.37
  gravity walls, **4**.36–**4**.37
  mat foundations, **2**.54–**2**.55
  overexcavation and replacement practices, **5**.5–**5**.14, **5**.31–**5**.33
  shallow foundations, **2**.48–**2**.54
  soil grouting, **5**.278–**5**.289
  soils under shallow foundations, **3**.122–**3**.123
  unit weight, **2**.50
  water table influences, **2**.50–**2**.51
Bed rock, **7**.6
Borehole log, **1**.10
Bowle's settlement method, **2**.61–**2**.62
Building Research Advisory Board (BRAB), **3**.183
Brown, Robert Wade, **1**.3, **1**.21, **5**.229, **6**.3, **6**.31, **6**.95, **6**.103, **8**.40
Bruce, Donald A., **5**.398
Building codes, **7**.31, **8**.23
Building Officials and Code Administrators (BOCA), **8**.23
Business:
  alternative dispute resolution, **8**.10–**8**.11
  entities, **8**.9
  insurance, **8**.9–**8**.10
  lawsuits, **8**.7
  liability limitations, **8**.8–**8**.9
  risk management, **8**.7–**8**.11
  (*See also* Contracts; Insurance; Warranties)
Buttressed walls, **4**.19–**4**.21, **4**.33–**4**.35

Caissons, deficient, **7**.60–**7**.69
California bearing ratio (CBR), **5**.178, **5**.202
Cantilever walls, **4**.19–**4**.21, **4**.33–**4**.35
Capillary barriers, **6**.98–**6**.101
  horizontal, **6**.98–**6**.99
  vertical, **6**.100
Capillary pressure, **1**.4, **1**.6
  (*See also* Matrix suction)
Capillary rise, **1**.4
  (*See also* Vertical permeability)
Carbo-rock, **5**.366
Carbonate, **2**.6–**2**.7
Carbonation, **3**.60
Casagrande's method, **2**.29–**2**.32
Catch basins, **1**.26
Cement, **3**.4–**3**.6
  aggregate-cement ratio, **3**.51
  Cemill process, **5**.332
  chemical stabilization of soil using, **5**.211–**5**.214
  composition, **3**.4–**3**.5
  fineness, **3**.5
  high early-strength, **5**.331
  manufacture of, **3**.4
  microfine, **5**.331–**5**.332
  Portland, **5**.211–**5**.212
  shrinkage and, **3**.59
  testing methods, **3**.6
  type and age factors, **3**.51–**3**.53
  types of, **3**.5
  water-cement ratio, **3**.50–**3**.51
Cementation, **5**.171–**5**.173
  effects of, **5**.98–**5**.100
Cemill process, **5**.332
Center doming, **1**.4, **1**.18
Center lift, **1**.4
Chemical admixtures, **3**.13–**3**.16
  accelerating, **3**.13
  air-entraining, **3**.13–**3**.15
  bonding, **3**.16
  calcium chloride, **3**.15
  coloring, **3**.16
  corrosion inhibiting, **3**.16
  dampproofing, **3**.16
  expansion producing, **3**.16
  flocculating, **3**.16
  fungicidal, germicidal, and insecticidal, **3**.16
  gas forming, **3**.16
  reducing alkali-aggregate expansion, **3**.16
  set-controlling, **3**.16
Chemical stabilization, **5**.205–**5**.264
  cation exchange:
    chemical bonding in, **5**.207–**5**.210
    chemistry of, **5**.207
  cement, **5**.211–**5**.214
  chemicals and reactions, **5**.205–**5**.224

Chemical stabilization (*Cont.*):
  clay:
    cation and base exchange with, **5.**206–**5.**211
    organic modification of, **5.**210–**5.**211
  curing conditions, **5.**263
  emulsions, **5.**222–**5.**224
  fly ash, **5.**219–**5.**222
  foundation repair and, **6.**85
  injection methods, **5.**253–**5.**256
  lime, **5.**214–**5.**219
  mixing methods, **5.**253–**5.**263
    deep soil, **5.**261–**5.**263
    off-site, **5.**253–**5.**259
    on-site, **5.**260–**5.**261
  quality assurance, **5.**263–**5.**264
  sodium silicates, **5.**222–**5.**224
  soil:
    collapse potential, **5.**231–**5.**234
    compaction moisture-density relationship, **5.**244–**5.**248
    engineering properties and behavior, **5.**224–**5.**252
    liquefaction potential, **5.**250–**5.**252
    permeability, **5.**248–**5.**250
    plasticity and workability, **5.**224–**5.**225
    shrinkage, **5.**231
    stress-strain-strength behavior, **5.**234–**5.**244
    swelling, **5.**225–**5.**231
Chemical weathering, **2.**6
Clay, **1.**7, **1.**8, **2.**50, **7.**6
  chemical stabilization of, **5.**206–**5.**211
  composite, **2.**13
  consistency, **2.**13
  consolidated drained strength of saturated, **2.**37–**2.**38
  consolidated undrained strength of consolidated, **2.**38–**2.**41
  consolidation settlement of foundations on, **2.**67
  expansive, **8.**27
  fat, **1.**9
  London, **1.**8, **1.**11
  mineralogy, **2.**7, **2.**11
  organic modification of, **5.**210–**5.**211
  particle of, **2.**10
  sensitivity, **2.**13
  shear strength of, **2.**34–**2.**44
  shear strength of saturated overconsolidated, **2.**42–**2.**44
  unconfined compression test of consolidated, **2.**41–**2.**42
  unconsolidated-undrained test, **2.**41–**2.**42
  wetting-induced volume change, **5.**135–**5.**136
Coarse-grained soil, **2.**9, **2.**14
Coefficient of consolidation, **2.**30–**2.**32
Colley, Barbara, **1.**27

Columns:
  axially loaded, **3.**106
  biaxial bending, **3.**116–**3.**117
  buckling effect, **3.**110, **3.**115–**3.**116
  general information, **3.**105–**3.**106
  jet grouting, **5.**317–**5.**318
  load-moment interaction diagrams, **3.**109–**3.**115
  reinforced concrete, **3.**105–**3.**117
  slender, **3.**110, **3.**115–**3.**116
  uniaxial bending and compression, **3.**106–**3.**109
  verifying size, **5.**317–**5.**318
Compaction, **5.**36–**5.**205
  cohesive soil and, **5.**111–**5.**118
  critical prewetting relative, **5.**138
  degree of wetting, **5.**141–**5.**143
  embankments, **5.**36–**5.**37
  engineering properties of soil affected by, **5.**38
  equipment, **5.**40–**5.**54
    classification of, **5.**44
    grid rollers, **5.**48
    impact roller, **5.**50–**5.**51
    rammers and tampers, **5.**53–**5.**54
    rubber-tire and pneumatic rollers, **5.**41
    sheepsfoot rollers, **5.**42, **5.**43
    smooth-drum and smooth-wheel rollers, **5.**41, **5.**44
    tamping foot and padfoot rollers, **5.**42, **5.**46
    vibratory plate compactor, **5.**51–**5.**53
    vibratory rollers, **5.**48–**5.**50
    wobble-wheel rollers, **5.**41–**5.**42
  impact, **5.**63–**5.**66
  kneading, **5.**67–**5.**70
  laboratory tests, **5.**59–**5.**72
    confined-ring specimens, **5.**71–**5.**72
    impact, **5.**63–**5.**66
    kneading, **5.**67–**5.**70
    static, **5.**63
    vibratory, **5.**70–**5.**71
  method of, **5.**139–**5.**141
  moisture-density relationship, **5.**54–**5.**59
  moisture-density relationship after chemical stabilization, **5.**244–**5.**248
  near-surface, **5.**36
  overcompaction, **6.**4, **6.**7, **6.**36
  oversize particles, **5.**143–**5.**147
  permeability, **5.**153–**5.**163
  quality assurance, **5.**173–**5.**205
  shrinkage, **5.**150–**5.**153
  soil:
    aging, **5.**163, **5.**165, **5.**167, **5.**172, **5.**173
    cementation, **5.**98–**5.**100, **5.**171–**5.**173
    cohesive, **5.**134–**5.**163
    thixotropy, **5.**163–**5.**173
    (*See also* Soil, cohesive; Soil, noncohesive)

Compaction (*Cont.*):
  specifications, **5.**173–**5.**182
    density, **5.**174–**5.**175
    end product, **5.**173–**5.**181
    moisture condition, **5.**175–**5.**178
  static, **5.**63
  swelling pressure, **5.**149–**5.**150
  tests:
    apparent hydraulic conductivity method, **5.**204
    California bearing ratio, **5.**178, **5.**202
    cone penetration, **5.**179, **5.**202
    drive-cylinder method, **5.**186
    dynamic penetration, **5.**178
    Hilf's rapid method, **5.**189–**5.**194
    laboratory permeability test, **5.**179
    moisture-density, **5.**183
    nuclear method, **5.**186–**5.**189
    one-dimensional compression, **5.**179
    one-dimensional swelling and collapse, **5.**179
    plate load, **5.**178, **5.**196–**5.**202
    Proctor penetrometer, **5.**178, **5.**194–**5.**195
    rubber balloon, **5.**183–**5.**186
    sand cone, **5.**183–**5.**186
    sand replacement method, **5.**186
    sealed double ring infiltration, **5.**179, **5.**202–**5.**204
    sleeve method, **5.**186
    standard penetration, **5.**179, **5.**202
    suction head method, **5.**204
    suction measured using tensiometer, **5.**179
    test pit method, **5.**186
    triaxial compression, **5.**179
    unconfined compression, **5.**179
    variability and statistical specifications, **5.**180–**5.**182
    water replacement method, **5.**186
    wetting front method, **5.**205
  types of, **5.**36–**5.**37
  undercompaction, **6.**3–**6.**4, **6.**7, **6.**36
  vibratory, **5.**70–**5.**71
  wetting induced volume change:
    control of, **5.**148–**5.**149
    time rate, **5.**147
Compaction grouting, **5.**311–**5.**312
Compensation grouting, **5.**312
Compression:
  cohesive soil, **5.**114
  one-dimensional, **5.**115
  primary, **5.**121–**5.**122
  secondary, **5.**121
Compressive strength test, **3.**36–**3.**40
  capping cylindrical specimens, **3.**38
  mixing, molding and curing specimens, **3.**36–**3.**38

Compressive strength test (*Cont.*):
  procedure, **3.**39–**3.**40
  specimen preparation, **3.**39
  test apparatus, **3.**38
  unconfined, **2.**41–**2.**42
Concrete, **3.**3–**3.**73
  admixtures:
    accelerating, **3.**13
    air-detraining, **3.**16
    air-entraining, **3.**13–**3.**15
    blast-furnace slag, **3.**18
    bonding, **3.**16
    calcium chloride, **3.**15
    chemical, **3.**13–**3.**16
    coloring, **3.**16
    corrosion inhibiting, **3.**16
    dampproofing, **3.**16
    expansion producing, **3.**16
    flocculating, **3.**16
    fly ash, **3.**16–**3.**18
    fungicidal, germicidal, and insecticidal, **3.**16
    gas forming, **3.**16
    mineral, **3.**16–**3.**18
    pozzolans, **3.**18
    reducing alkali-aggregate expansion, **3.**16
    rice-husk ash, **3.**18
    set-controlling, **3.**16
    shrinkage and, **3.**59–**3.**60
    silica fume, **3.**18
    water-reducing, **3.**12–**3.**13
  composition, **3.**77
    aggregates, **3.**6–**3.**11, **3.**53
    aggregate-cement ratio, **3.**51
    air content, **3.**54–**3.**55
    cement, **3.**4–**3.**6
    cement type and age, **3.**51–**3.**53
    chemical admixtures, **3.**13–**3.**16
    mineral admixtures, **3.**16–**3.**18
    water, **3.**11
    water-cement ratio, **3.**3–**3.**18, **3.**50–**3.**51
    water-reducing admixtures, **3.**12–**3.**13
  creep, **3.**60–**3.**62
  curing conditions, **3.**55–**3.**56
  mechanical properties, **3.**49–**3.**65
  pumping, **3.**71–**3.**73
  quality assurance, **3.**33–**3.**49
    charts, **3.**42
    by using accelerated strength tests, **3.**44–**3.**46
  reinforced (*see* Reinforced concrete)
  shrinkage, **3.**58–**3.**60
  strength:
    compressive, **3.**50–**3.**56
    computing mix proportions, **3.**26–**3.**33
    computing required average, **3.**19–**3.**24

Concrete, strength (*Cont.*):
   flexural, **3.**57
   flowchart for selecting proportions, **3.**21
   maximum strain values, **3.**40
   modulus of elasticity, **3.**58
   modulus of rupture, **3.**57
   proportioning based on field-strength records, **3.**25–**3.**26
   proportioning based on maximum water-cement ratio, **3.**26
   proportioning based on trial mixtures, **3.**26
   relations between compressive, flexural, and tensile, **3.**57
   selecting proportions, **3.**25
   shear, **3.**57–**3.**58
   slump recommendations, **3.**27
   tensile, **3.**56–**3.**57
   testing, **3.**36–**3.**40, **3.**43–**3.**44, **3.**47–**3.**48
   variations in, **3.**41–**3.**43
   volume of dry-rodded coarse aggregate per unit, **3.**30
   water-cement ratio and compressive strength, **3.**30
   water-cement ratio for sever exposures, **3.**30
   water mixing and air content requirements, **3.**29
 stress:
   behavior under multiaxial, **3.**62–**3.**63
   creep and, **3.**60–**3.**62
   fatigue loading, **3.**63–**3.**65
   shrinkage and, **3.**58–**3.**60, **3.**62
 tests, **3.**34–**3.**40
   accelerated strength, **3.**43–**3.**44
   acoustical method, **3.**49
   air void parameter, **3.**36
   compaction factor, **3.**35
   compressive strength, **3.**36–**3.**40
   conditions for, **3.**56
   core, **3.**49
   creep under compression, **3.**36
   electromagnetic method, **3.**49
   flexural strength, **3.**36
   fresh concrete, **3.**34–**3.**35
   hardened concrete, **3.**36
   length change of hardened paste or mortar, **3.**36
   maturity concepts, **3.**49
   modulus of elasticity and Poisson's ratio, **3.**36
   nondestructive, **3.**47–**3.**49
   penetration resistance method, **3.**48
   pull-out method, **3.**48
   resistance to rapid freezing/thawing, **3.**36
   resistance to scaling, **3.**36
   slump, **3.**34–**3.**35
   splitting tensile strength, **3.**36
   surface hardness method, **3.**47–**3.**48

Concrete, tests (*Cont.*):
   ultrasonic pulse velocity method, **3.**48–**3.**49
   V-B, **3.**35
   weather conditions:
     cold, **3.**65–**3.**67
     hot, **3.**67–**3.**71
   weight of fresh, **3.**31
Concrete piles:
   cast-in-place, **3.**156–**3.**157
   mandrel-driven, **3.**156–**3.**157
   precast-reinforced, **3.**156
   prestressed, **3.**154–**3.**155
Conductivity, hydraulic, **5.**306–**5.**307, **5.**309, **5.**386
   (*See also* Permeability)
Cone penetration test (CPT), **5.**98
Consolidation test, **2.**26–**2.**32
   calculation of coefficient of consolidation, **2.**30–**2.**32
   calculation of laboratory consolidation curve, **2.**29
   performing, **2.**27–**2.**29
   reconstruction of field consolidation curve, **2.**29–**2.**30
Contractile skin, **5.**79
Contracts:
   grouting projects, **5.**374–**5.**375
   pay items, **5.**375–**5.**376
   Type A, **5.**374
   Type B, **5.**374–**5.**375
   Type C, **5.**375
   Type D, **5.**375
Controlled hydrofracturing, **5.**310
Coral, **2.**6
Coulomb theory, **4.**13–**4.**17
Council of American Building Officials (CABO), **8.**14, **8.**23
Counterfort walls, **4.**19–**4.**21, **4.**33–**4.**35
Crandall, Kenneth L., **8.**40
Creep, **3.**60–**3.**62, **6.**5, **6.**9, **6.**37
Cross slope, **1.**22
Crown, **1.**22
Cut, **1.**22
Cyclic mobility, **5.**89

Darcy's law, **2.**17–**2.**18
Das's settlement method, **2.**64–**2.**66
Deep grouting, **6.**62–**6.**67, **6.**72
   (*See also* Mud jacking; Pressure grouting)
Drainage, **1.**24–**1.**29, **6.**97–**6.**100
   catch basins, **1.**26
   drop inlets, **1.**26
   foundation failures and, **6.**4, **6.**36
   manholes, **1.**26
   pipes, **1.**27

Drainage (*Cont.*):
  sanitary sewer systems, **1.**27–**1.**28
  subsurface water and, **1.**29, **6.**98
  surface, **1.**25–**1.**26
  water and capillary barriers, **6.**98–**6.**100
Drainage systems, **1.**29–**1.**30
  foundation, **1.**31–**1.**32
  French drains, **1.**29–**1.**30, **6.**98
  gridiron, **1.**29–**1.**30
  herringbone, **1.**29–**1.**30
  interceptor, **1.**29–**1.**30
  natural, **1.**29–**1.**30
Drilling:
  auger, **5.**380
  concrete piers, **6.**42–**6.**45
  double-head duplex, **5.**380
  drive drilling of lancing, **5.**379
  driven duplex underreaming, **5.**380
  equipment, **6.**40–**6.**41
  rotary duplex, **5.**379
  rotary duplex underreaming, **5.**380
  rotary percussive concentric duplex, **5.**379
  rotary percussive eccentric duplex, **5.**379
  soil-grouting projects, **5.**379–**5.**380
Drop inlets, **1.**26

Earth pressure, **4.**3–**4.**37
  active:
    on buttressed walls, **4.**33–**4.**35
    on cantilever walls, **4.**33–**4.**35
    on counterfort walls, **4.**33–**4.**35
    on solid gravity walls, **4.**33
  at rest, **4.**3–**4.**5
  Coulomb theory, **4.**13–**4.**17
    active, **4.**13–**4.**15
    point of application, **4.**16
    trial wedge method, **4.**15–**4.**16
  deformation and boundary conditions, **4.**12–**4.**13
  movement required to generate stresses, **4.**13
  passive, **4.**35
  Rankine theory, **4.**5–**4.**13
    active, **4.**6–**4.**8
    passive, **4.**9–**4.**12
  walls and, **4.**21
Earthquake regions, **3.**147–**3.**152
  dynamic properties of soil in, **3.**148–**3.**150
  foundation design considerations, **3.**150–**3.**152
  foundation failures and, **6.**4, **6.**9
Effective stress, **2.**21
Elastic theory, **2.**62–**2.**64
Embankments, **5.**36–**5.**37
Engineers, foundation analysis, **8.**3–**8.**5
Epoxy grout, **5.**369–**5.**370

Equipment, **5.**39–**5.**54
  backhoe, **5.**39, **5.**41
  bulldozers, **5.**39, **5.**43
  compaction, **5.**40–**5.**54
  excavation, hauling, and spreading, **5.**39–**5.**40
  graders, **5.**39, **5.**43
  grid rollers, **5.**48
  grouting, **5.**378–**5.**379, **5.**380–**5.**384
  haul trucks, **5.**39, **5.**40
  hydraulic power souce limited access drill, **6.**41
  impact roller, **5.**50–**5.**51
  limited access drill, **6.**41
  loaders, **5.**39, **5.**40
  rammers and tampers, **5.**53–**5.**54
  rubber-tire and pneumatic rollers, **5.**41, **5.**45
  scrapers, **5.**39
  sheepsfoot rollers, **5.**42, **5.**43, **5.**46
  smooth-drum and smooth-wheel rollers, **5.**41, **5.**44
  tamping foot and padfoot rollers, **5.**42, **5.**46–**5.**47
  tractor drill, **6.**40
  truck mounted rig, **6.**40
  vibratory plate compactor, **5.**51–**5.**53
  vibratory rollers, **5.**48–**5.**50
  wobble-wheel rollers, **5.**41–**5.**42
Erosion control, **5.**323–**5.**324
  grouting techiques, **5.**299–**5.**303
  plans, **1.**24
Evaporation, **1.**5, **1.**16
Excavation:
  equipment, **5.**39–**5.**40
  overexcavation and replacement (*see* Overexcavation)
  stabilizing techniques, **5.**289–**5.**295
External posttensioning (EPT), **6.**55
Ezeldin, A. Samer, **3.**75, **3.**153

Fat clay, **1.**9
Federal Housing Act, **8.**14
Feldspar, **2.**6
Field capacity, **1.**4
Fill, **1.**22
Filtercake, **5.**325
Fine-grained soil, **2.**9, **2.**14
Finite element method (FEM), **5.**30
Fireplaces, interior, **6.**69–**6.**70
Flexural strength, **3.**57
  reinforced concrete, **3.**77–**3.**93
  testing concrete, **3.**36
Floors, shimming, **6.**45–**6.**48
Fly ash:
  chemical composition, **5.**220
  chemical reactions, **5.**221–**5.**222

Fly ash (*Cont.*):
  chemical stabilization of soil using, **5.**219–**5.**222
  concrete admixture, **3.**16–**3.**18
  physical properties, **5.**220
  type C, **5.**331
Footings, **2.**45–**2.**46
  column, **2.**45
  combined, **2.**45, **3.**121–**3.**122, **3.**135–**3.**141
  concrete, **8.**29
  continuous, **2.**45–**2.**46
  designing, **3.**126–**3.**130
    combined, **3.**135–**3.**141
    pile caps, **3.**174–**3.**180
    rectangular, **3.**131–**3.**135
  failure, **3.**124–**3.**130
    diagonal tension, **3.**124
    flexure, **3.**124–**3.**125
    one-way shear, **3.**124
  isolated, **2.**45
  load transfer from column to, **3.**126–**3.**130
  mat, **3.**121–**3.**122
  pile caps, **3.**121–**3.**122, **3.**174–**3.**180
  proportioning for equal settlement, **2.**67
  rectangular, **3.**131–**3.**135
  spread, **2.**45, **3.**121–**3.**122, **3.**135, **6.**32, **6.**37–**6.**40, **6.**59
  strip, **2.**45–**2.**46, **3.**121–**3.**122
  types of, **3.**121–**3.**122
Foundation failures, **6.**3–**6.**29, **6.**86–**6.**92
  causes, **6.**3–**6.**11
    bearing capacity marginal, **6.**4, **6.**9, **6.**37
    caisson and pile deficiencies, **7.**60–**7.**69
    cofferdams, **7.**59–**7.**60
    cohesive soil, **6.**6–**6.**9
    common pitfalls, **7.**69
    construction errors, **7.**44–**7.**52
    cut and fill site, **6.**4, **6.**9
    design error, **7.**35–**7.**43
    drag-down and heave, **7.**31–**7.**35
    earth slides, **6.**4, **6.**9
    earthquakes, **6.**4, **6.**9
    expansive clay, **8.**27
    floatation and water-level change, **7.**52–**7.**57
    frost heave, **6.**3, **6.**7, **6.**36
    human error, **7.**3
    lateral movement, **7.**22–**7.**26
    load-transfer, **7.**21–**7.**22
    noncohesive soils, **6.**3–**6.**5
    organic material decaying, **6.**4, **6.**8, **6.**36
    overcompaction, **6.**4, **6.**7, **6.**36
    permafrost, **6.**3
    plant roots, **6.**8–**6.**9, **6.**37
    sand and gravel cushion beneath slabs-on-grade, **6.**4, **6.**9, **6.**37

Foundation failures, causes (*Cont.*):
    settlement, **7.**4, **7.**5
    sloping site construction, **6.**4, **6.**9, **6.**36
    soil creep, **6.**5, **6.**9, **6.**37
    soil moisture:
      high, **6.**6–**6.**7, **6.**9, **6.**36
      low, **6.**9, **6.**37
    sulfide and sulfate attack on concrete and steel, **6.**4, **6.**9
    ten most common categories, **7.**7
    undercompaction, **6.**3–**6.**4, **6.**7, **6.**36
    undermining of safe support, **7.**7–**7.**20
    unequal support, **7.**28–**7.**31
    vibration, **7.**57–**7.**59
    water, **6.**4
    water drainage, **6.**4, **6.**36
  checklist to avoid, **7.**69–**7.**70
  classification of visible damage, **6.**28
  criteria for, **6.**27
  evidence of exterior tension, **6.**14
  examples:
    Alaska pier collapse, **7.**36–**7.**40
    apple juice plant, **7.**47–**7.**52
    Canadian grain elevator, **7.**29
    Connecticut, Hartford foundation walls, **7.**24–**7.**27
    Guadalupe shrine, **7.**30
    Leaning Tower of Pisa, **7.**8
    Long Island water pollution control plant, **7.**27–**7.**28
    Marble Hill, **7.**52
    Massachusettes, Boston housing slab settlement, **7.**43
    Mexico City Palace of Fine Arts, **7.**30
    New Jersey:
      Edgewater high rise, **7.**34–**7.**35
      Hackensack industrial building, **7.**55–**7.**57
      postal facility, **7.**23–**7.**24
    New York:
      14th street, **7.**59–**7.**60
      41st street structure, **7.**57
      Brooklyn hospital, **7.**53–**7.**55
      collapsed building, **7.**7, **7.**9, **7.**44–**7.**46
      east side hospital, **7.**15–**7.**16
      electronics plant, **7.**11–**7.**15
      Manhattan hospital complex, **7.**16–**7.**21
      Manhattan office building, **7.**58–**7.**59
      North River project, **7.**60–**7.**67
      Queens apartments, **7.**32–**7.**34
      suburb sewers, **7.**8, **7.**10–**7.**11
      Washington DC foundation mat, **7.**40–**7.**43
    Pennsylvania, Philadelphia office building, **7.**67–**7.**69

Foundation failures (*Cont.*):
  grade elevations, **6.**18–**6.**20
  high-rise and heavy construction, **7.**3–**7.**70
  hydraulically driven concrete cylinders, **6.**92
  improper placement of haunch beneath beam, **6.**89–**6.**91
  minipiles, **6.**87–**6.**89
  mud jacking, **6.**89
  settlement, **6.**11–**6.**12
  settlement vs. upheaval, **6.**12–**6.**17
  sliding, **6.**17–**6.**18
  soil swell vs. moisture changes, **6.**20–**6.**23
  substandard pier construction, **6.**87
  sulfate reactions with certain soils, **6.**91–**6.**92
  ultraslim concrete piers, **6.**92
  upheaval, **6.**12
Foundation repair, **6.**31–**6.**94, **8.**5
  adding on creates new distress problem, **6.**68–**6.**69
  adding on without correcting preexisting distress, **6.**67–**6.**68
  apartment buildings, **6.**79, **6.**80
  basements, **6.**79–**6.**84
    water infiltration, **6.**82
  beams, **6.**32–**6.**48
  bearing capacity and, **6.**59–**6.**60
  chemical stabilization, **6.**85
  construction interferences, **6.**67–**6.**71
  distress signs, **6.**32–**6.**35
  establishing need for, **6.**23–**6.**28
  example techniques:
    Florida, **6.**72
    London, **6.**72–**6.**74
  failures, **6.**86–**6.**67
    hydraulically driven concrete cylinders, **6.**92
    improper placement of haunch beneath beam, **6.**89–**6.**91
    minipiles, **6.**87–**6.**89
    mud jacking, **6.**89
    substandard pier construction, **6.**87
    sulfate reactions with certain soils, **6.**91–**6.**92
    ultraslim concrete piers, **6.**92
  frame structures on wood, **6.**70–**6.**71
  grouting, **6.**62–**6.**67
  home resale value and, **6.**105–**6.**106
  interior fireplaces and, **6.**69–**6.**70
  interior floors, **6.**45–**6.**48
  lateral movement, **6.**70
  leveling problems, **6.**67–**6.**71
  longevity, **6.**60–**6.**62
  lowering a foundation, **6.**76–**6.**79
  material incompatibility, **6.**69
  minipiles, **6.**55–**6.**60
  mud jacking, **6.**48–**6.**50, **6.**53–**6.**54, **6.**62
  options, **6.**36–**6.**37

Foundation repair (*Cont.*):
  pier, **6.**32–**6.**48
  posttensioning, **6.**55
  settlement problems, **6.**51–**6.**55
  slab, **6.**48–**6.**55
  slim piers, **6.**55–**6.**60
  underpinning, **6.**32, **6.**44, **6.**45, **6.**49, **6.**57
  upheaval problems, **6.**50–**6.**51, **6.**67
  Uretek method, **6.**86
  walls, **6.**79–**6.**84
    hadite block, **6.**82
  warped substructure, **6.**67
Foundations:
  analyzing, **8.**3–**8.**5
  basements, **3.**184–**3.**185
  beam:
    with crawl space, **3.**184
    repairing, **6.**32–**6.**48
  designing, **3.**121–**3.**152
    footings (*see* Footings)
    mat, **3.**142–**3.**145
  distress factors, **8.**3–**8.**4
  drain systems, **1.**31–**1.**32
  earth anchors, **6.**59
  footings (*see* Footings)
  fundamentals of construction and design, **3.**1–**3.**181
  geographical areas:
    cold-weather regions, **3.**145–**3.**147
    earthquake regions, **3.**147–**3.**152
  inspecting, **6.**103–**6.**106
  lightly loaded, **3.**183–**3.**185
  lower-cost, **8.**29–**8.**30
  lowering, **6.**76–**6.**79
  mat, **2.**46–**2.**47, **2.**54–**2.**55, **2.**56, **3.**141–**3.**145
  minipiles, **6.**55–**6.**60
  mobile home, **8.**31–**8.**40
  on filled slopes, **1.**24
  permanent wood, **8.**29
  pier:
    with crawl space, **3.**184
    repairing, **6.**32–**6.**48
  pile, **3.**153–**3.**181, **5.**295–**5.**298, **6.**55–**6.**60
  preventive maintenance, **6.**95–**6.**102
  shallow (*see* Shallow foundations)
  slab, **6.**12, **6.**20–**6.**23, **6.**48–**6.**55, **8.**29
    differential elevations in, **6.**24
    leveling, **6.**50, **6.**52, **6.**53
    nonstructural, **3.**183
    repairing, **6.**48–**6.**55
    ribbed slab with void space, **3.**184
    ribbed slabs-on-ground, **3.**184
    soil swell vs. moisture changes, **6.**20–**6.**23
    structural, **3.**183–**3.**184
    upheaval, **6.**12

Foundations (*Cont.*):
  slim piers, **6.**55–**6.**60
  soil moisture and stability of, **1.**18
  spread footing and drilled pier combination, **6.**45
  subgrade waterproofing, **6.**66
  vegetation surrounding, **6.**101–**6.**102
Foundation walls, **4.**19–**4.**21
Friction angle, **2.**33
Frost:
  penetration map of USA, **3.**147
  soil susceptible to, **3.**146
Frost heave, **6.**3, **6.**7, **6.**36
Fuller, G. Robert, **8.**40

Geostatic stress, **2.**21–**2.**22
Grade, **1.**22
Grade elevations, **6.**18–**6.**20
Grading:
  erosion control plans, **1.**24
  finished, **1.**23
  plans for, **1.**23–**1.**24
  rough, **1.**23
  terms used in, **1.**22
Gravel, **3.**7
Gravity, soil moisture and, **1.**5–**1.**6
Gravity walls, **4.**19–**4.**21, **4.**33
  bearing capacity, **4.**36–**4.**37
  overturning, **4.**35–**4.**36
  sliding, **4.**37
Ground modification techniques (*see* Soil stabilization)
Ground water, **1.**3
  grouting techniques, **5.**299–**5.**303
  recharge, **1.**8
Grout, **5.**324–**5.**373
  cement-based, **5.**324–**5.**337, **6.**63
    components, **5.**330–**5.**333
    description, **5.**324–**5.**325
    empirical approaches to penetrability and injectability, **5.**328–**5.**330
    injecting into pores of soil, **5.**326–**5.**327
    pressure filtration resistance, **5.**328
    rock and structural grouting, **5.**325–**5.**326, **5.**330
  cement-foam-based, **5.**336–**5.**337
  evolutive solution, **5.**337
  guidelines for sequence of mixes, **5.**335–**5.**336
  hot melts, **5.**372–**5.**373
  Lugeon values, **5.**333–**5.**334, **5.**336, **5.**386
  selection guidelines, **5.**377–**5.**378
  solution, **5.**337–**5.**373, **6.**63
    acrylamide, **5.**338, **5.**367–**5.**369
    carbo-rock, **5.**366
    chrome-lignin, **5.**338, **5.**371

Grout, solution (*Cont.*):
  economic factors, **5.**340
  epoxy, **5.**338, **5.**369–**5.**370
  formaldehyde, **5.**338, **5.**371
  gel time, **5.**339–**5.**40
  penetrability, **5.**339
  permanence, **5.**338–**5.**339
  phenolic, **5.**338, **5.**370–**5.**372
  polyurethane, **5.**338, **5.**340–**5.**363
  resorcin phormol and aminoplast, **5.**338, **5.**370–**5.**371
  Rockagil, **5.**369
  sensitivity, **5.**340
  silacsol, **5.**365–**5.**366
  siprogel, **5.**369
  sodium silicate, **5.**338, **5.**363–**5.**366
  strength, **5.**339
  toxicity, **5.**340
  true solution, **5.**337
Grouting, **5.**277–**5.**400
  applications:
    curtain in rock and soil, **5.**343
    cutting off major water flows, **5.**345
    hardrock tunnels, **5.**345–**5.**346
    injectable tube (contact), **5.**355–**5.**358
    pin piles for support to tunnels and structures, **5.**348
    pressure, **6.**62–**6.**67
    renovation of human-accessible sewers, **5.**349–**5.**350
    repairing leaking brick and stone structures, **5.**347
    retaining walls in harbor loading docks, **5.**348–**5.**349
    sealing around bulkheads, **5.**358
    sealing cracks and joints in concrete/masonry, **5.**355
    sealing expansion joints in submerged structures, **5.**352
    sealing renards in tunnels, **5.**344–**5.**345
    seepage control in rock tunnels, **5.**345
    seepage prevention in dams and dikes, **5.**346
    soil grouting small-scale projects, **5.**353
  bearing capacity and, **5.**278–**5.**289
  curtain, **6.**63
  drilling, **5.**379–**5.**380
    auger, **5.**380
    double-head duplex, **5.**380
    drive drilling of lancing, **5.**379
    driven duplex underreaming, **5.**380
    rotary duplex underreaming, **5.**380
    rotary duplex, **5.**379
    rotary percussive concentric duplex, **5.**379
    rotary percussive eccentric duplex, **5.**379

Grouting (*Cont.*):
  dual-phase toe, **5.**298
  equipment, **5.**380–**5.**384
    flow meter, **5.**381
    gauges, **5.**382
    grout hoses, **5.**382
    grout pumps, **5.**382, **5.**383
    high-shear mixer, **5.**380–**5.**381
    holding tank, **5.**381
    measuring devices, **5.**384
    multiple-hole grouting headers, **5.**382
    packer rods, **5.**382
    packers, **5.**382
    particle-size reducing mill, **5.**382
    X-Y recorder, **5.**381
  field application:
    Credit River internal erosion, **5.**302
    deep sewer network in France, **5.**291–**5.**293
    E-18-Fjellinjen tunnel, **5.**361
    European subways, **5.**290–**5.**291
    flood prevention for high-rise complex, **5.**300–**5.**302
    foundation underpinning, **5.**293–**5.**294
    Orly VAL tunnel, **5.**361–**5.**362
    PI case, **5.**284–**5.**289
    pipeline from Zeebruges to Antwerp, **5.**294–**5.**295
    potash mine in Saskatchewan, **5.**322–**5.**323
    restoration of historical locks, **5.**305–**5.**306
    Rogers' reservoir, **5.**279–**5.**284
    Roosevelt Square, **5.**294
    Ryerson Polytechnical Institute, **5.**289
    tunnel for SIAAP in Paris, **5.**303
    tunneling below railway embankment, **5.**299–**5.**300
    underpinning of historical building, **5.**304
  field procedures, **5.**380–**5.**398
  foundations, **6.**62–**6.**67
  ground water and erosion control, **5.**299–**5.**303
  hazardous chemical and waste control, **5.**298–**5.**299
  injection pipe placement, **6.**64–**6.**66
  mechanics, **6.**63–**6.**64
  multiple-hole, **5.**355
  multiple-stage permeation, **5.**298
  pile and pier enhancement, **5.**295–**5.**298
  program design, **5.**376–**5.**379
    curtain depth, **5.**378
    equipment and materials needed, **5.**378–**5.**379
    field investigation, **5.**377
    grout selection guidelines, **5.**377–**5.**378
    hole spacing, **5.**378
    target definition, **5.**376–**5.**377

Grouting (*Cont.*):
  program supervision, **5.**384–**5.**396
    general aspects, **5.**384–**5.**385
    quality control and testing, **5.**396
    recording and interpreting data, **5.**385–**5.**395
  projects, **5.**373–**5.**376
    contractural agreements, **5.**374–**5.**375
    pay items, **5.**375–**5.**376
  reforestation and erosion control, **5.**323–**5.**324
  remedial, **5.**303–**5.**306
  rocks, **5.**298, **5.**325–**5.**326, **5.**330, **5.**333–**5.**334
  sciences of, **5.**277–**5.**278
  slope or excavation stabilizing technique, **5.**289–**5.**295
  slurry walls, **5.**299
  soil consolidation, **6.**66–**6.**67
  structural, **5.**325–**5.**326, **5.**330
  subgrade waterproofing, **6.**66
  techniques, **5.**306–**5.**324
    jet, **5.**293–**5.**294
    compaction, **5.**311–**5.**312
    compensation, **5.**312
    hydrofracture, **5.**310–**5.**311
    jet, **5.**299, **5.**312–**5.**322
    permeation, **5.**306–**5.**310, **5.**377–**5.**378
    precipitation, **5.**322–**5.**323
  tests:
    acoustic emission monitoring, **5.**397
    Baroid mud balance, **5.**396
    bleed, **5.**396
    borehole camera investigation, **5.**397
    borehole radar, **5.**397–**5.**398
    compressive strength, **5.**396
    core sampling, **5.**397
    cross-hole acoustic velocity, **5.**398
    grout-hole drilling, **5.**397
    hydraulic conductivity, **5.**397
    initial cure, **5.**396
    penetrometer, **5.**397
    pressure filtration, **5.**396
    pressure meter, **5.**397
    SPT, **5.**397
    viscosity, **5.**396
  uses, **5.**278

Heaving, **6.**3, **6.**7, **6.**12, **6.**36, **7.**31–**7.**35
Heenan, Doug, **5.**398
Home Owners' Warranty (HOW), **6.**25–**6.**26
Homes:
  HUD regulations and, **8.**21–**8.**41
  mobile, **8.**31–**8.**40

Homes (*Cont.*):
    multifamily units, **8.**25–**8.**26
        existing construction, **8.**26
        new construction, **8.**25
        rehabilitation construction, **8.**25–**8.**26
    single-family units, **8.**23–**8.**25
        existing construction, **8.**24–**8.**25
        new construction, **8.**23–**8.**24
        rehabilitation construction, **8.**24
Housing and Urban Development (HUD), **8.**21–**8.**41
    building codes, **8.**23
    description, **8.**21
    lower-cost foundations, **8.**29–**8.**30
    minimum property standards, **8.**21
    mobile home foundations, **8.**31–**8.**40
    multifamily property, **8.**25–**8.**26
    office locations and jurisdictions, **8.**22
    policy development, **8.**26–**8.**27
    scope of programs, **8.**21–**8.**26
    single-family units, **8.**23–**8.**25
    technical suitability of products, **8.**27–**8.**29
Hydration, **2.**6
Hydraulic conductivity, **5.**306–**5.**307, **5.**309, **5.**386
    (*See also* Permeability)
Hydrofracture grouting, **5.**310–**5.**311
Hydrofracturing:
    controlled, **5.**310
    vertical, **5.**310

Illite, **2.**7
Impact compaction, **5.**63–**5.**66
Infiltration, water, **1.**6–**1.**7
Inspections:
    foundation analysis, **8.**3–**8.**5
    insurance, **8.**19
Insurance:
    business, **8.**9–**8.**10
    claims settlement, **8.**20
    costs, **8.**19–**8.**20
    coverage, **8.**15–**8.**16
    examples of main structural defects, **8.**16
    existing construction, **8.**13
    home owner's policies, **8.**4
    new construction, **8.**13
    policies and warranties, **8.**14–**8.**19
    property inspections, **8.**19
    structural, **8.**13–**8.**20
    transfer of coverage, **8.**17
Interception, **1.**5
Interclod pores, **5.**101–**5.**102
Irwin, Jim, **8.**3

Jet grouting, **5.**293–**5.**294, **5.**299, **5.**312–**5.**322
    advantages, **5.**315
    applications, **5.**314
    column size verification, **5.**317–**5.**318
    description, **5.**312
    double-fluid system, **5.**314–**5.**315
    field monitoring during, **5.**321–**5.**322
    limitations, **5.**316
    methods, **5.**312–**5.**313
    operational features, **5.**314–**5.**315
    parameter selection, **5.**316
    quality assurance, **5.**320–**5.**321
    single-fluid system, **5.**314
    soil characteristics, **5.**318–**5.**320
    soilcrete composition, **5.**317
    soilcrete strength, **5.**316–**5.**317
    triple-fluid system, **5.**315
Joint Venture for Affordable Housing (JVAH), **8.**29, **8.**40

Kaminetzky, Dov, **7.**3
Kaolinite, **2.**7
kneading compaction, **5.**67–**5.**70
Kunkel, Jerry, **3.**183

Land planning, **1.**30–**1.**31
Landscaping, **1.**30
Lawsuits, **8.**10–**8.**11, **8.**16–**8.**17, **8.**20
Lawton, Evert, **5.**3
Leaching, **2.**6
Lifts, **5.**36
Lime:
    agricultural, **5.**215
    chemical stabilization of soil using, **5.**214–**5.**219
    construction, **5.**214
    hydrated, **5.**214
    mixing with water, **5.**216
    quicklime, **5.**214
    sulfate reactions with soil, **6.**91
Liquefaction, **5.**89–**5.**98
    characteristics, **5.**91
    complete, **5.**90
    CPT vs. SPT, **5.**98
    initial, **5.**90
    partial, **5.**90
    potential after chemical stabilization, **5.**250–**5.**252
    SPT procedure, **5.**97
    tests, **5.**97–**5.**98
Liquid limit (LL), **2.**10
Liquidity index, **2.**11
London clay, **1.**8, **1.**11
Lugeon values, **5.**333–**5.**334, **5.**336, **5.**386

Manholes, **1.**26
Manufactured Home Construction Safety Standards (MHCSS), **8.**31
Mat foundations, **2.**46–**2.**47, **3.**141–**3.**145
  allowable settlement, **2.**56
  bearing capacity, **2.**54–**2.**55
  designing, **3.**142–**3.**145
    elastic method, **3.**143–**3.**145
    numerical method, **3.**145
    rigid method, **3.**142–**3.**143
  types of, **3.**142
Matric suction, **5.**80, **5.**81
  (*See also* Capillary pressure; Osmotic suction; Soil suction)
Mechanical weathering, **2.**5–**2.**6
Mica, **2.**6
Mineral admixtures, **3.**16–**3.**18
  blast-furnace slag, **3.**18
  fly ash, **3.**16–**3.**18
  pozzolans, **3.**18
  rice-husk ash, **3.**18
  silica fume, **3.**18
Mineralogy, **2.**6–**2.**7, **2.**11, **5.**154, **5.**206
Minipiles, **6.**55–**6.**60, **6.**87–**6.**89
  application failures, **6.**87–**6.**92
  design concerns, **6.**56
  deterioration, **6.**58
Mobile homes:
  foundations, **8.**31–**8.**40
    type C systems, **8.**32
    type C1 systems, **8.**33
    type E systems, **8.**32
    type E1 systems, **8.**33–**8.**34
    type I systems, **8.**32
    vertical anchor capacity for long foundation wall, **8.**37–**8.**38
    vertical anchor capacity for piers, **8.**36
    vertical anchor capacity for shear walls, **8.**39
    withdrawal resistance for continuous, **8.**35
    withdrawal resistance for piers, **8.**36
Modulus of elasticity, **3.**58
Modulus of rupture, **3.**57
Mohr's theory, **2.**35–**2.**44
Moisture (*see* Soil moisture; Water)
Montmorillonite, **2.**7
Mud jacking, **6.**48–**6.**50, **6.**53–**6.**54, **6.**62, **6.**89, **7.**97
  (*See also* Pressure grouting)

Narduzzo, Luigi, **5.**398
National Housing Act of 1965, **8.**27
Naudts, Alex A., **5.**277
Negative skin friction, **3.**172

O'Neill test, **6.**60
Organic decomposition, **2.**6
Osmotic suction, **5.**79
  (*See also* Matric suction; Pore stress)
Overexcavation, **5.**4–**5.**36
  bearing capacity failures and, **5.**5–**5.**14
  finite element method, **5.**30
  limitations, **5.**30–**5.**36
  settlement factors, **5.**14–**5.**29

PAPERJET System, **5.**321
Peat, **2.**6, **7.**7
Perma-Jack, **6.**55
Permanent wood foundation (PWF), **8.**29
Permeability, **1.**7, **2.**19–**2.**20, **5.**84–**5.**86
  chemical stabilization changes in, **5.**248–**5.**250
  compacted cohesive soils and, **5.**153–**5.**163
  horizontal barriers, **5.**264–**5.**265
    (*See also* Soil moisture)
  intrinsic, **5.**84
  laboratory methods of determination, **2.**19
  tests:
    constant head, **2.**19
    falling head, **2.**19
    sources of error in, **2.**20
    triaxial flexible wall, **2.**20
  vertical barriers, **5.**265–**5.**267
    (*See also* Capillary rise)
Permeation grouting, **5.**306–**5.**310, **5.**377–**5.**378
  maximum pressures and limitations, **5.**310
  solution grouts, **5.**306–**5.**309
  suspension grouts, **5.**309
Petry, Dr. Tom, **5.**229
Petry test, **6.**60
Phenolic grout, **5.**370–**5.**372
Phreatic boundary, **1.**4
Phreatophytes, **1.**5
Pier caps, **6.**47
Pier foundations:
  with crawl space, **3.**184
  repairing, **6.**32–**6.**48
    drilled concrete piers, **6.**40–**6.**45
    shimming interior floors, **6.**45–**6.**48
    underpinning, **6.**32, **6.**44, **6.**45
Piers:
  concrete, **6.**42–**6.**45
  deterioration, **6.**58
  driven steel pipe, **6.**74–**6.**76
  multiple drilled, **6.**59
  ultraslim concrete, **6.**92
Pile caps, **3.**121–**3.**122
  designing, **3.**174–**3.**180

Pile foundations, **3.**153–**3.**181, **6.**55–**6.**60
  concrete, **3.**154–**3.**157
  designing, pile caps, **3.**174–**3.**180
  displacement, **3.**153
  enhancing through grouting techniques, **5.**295–**5.**298
  Franki, **3.**153
  laterally loaded, **3.**172–**3.**174
  loading capacity, **3.**160–**3.**169
    axial single-pile, **3.**160–**3.**167
    cohesionless soil, **3.**160–**3.**163
    cohesive soil, **3.**163–**3.**165
    dynamics equations, **3.**169–**3.**170
    groups of piles, **3.**167–**3.**169
    silt, **3.**165–**3.**167
    testing, **3.**170–**3.**171
  negative skin friction, **3.**172
  nondisplacement, **3.**153
  Raymond, **3.**153
  soil action around driven, **6.**58
  steel, **3.**157–**3.**159
  stress, **3.**154–**3.**159
    concrete, **3.**154–**3.**157
    steel, **3.**157–**3.**159
    timber, **3.**159
  tests, **3.**170–**3.**171
  timber, **3.**159
  types of, **3.**153
Pipes:
  ADS plastics, **1.**27, **1.**29
  drainage, **1.**27
  PVC plastics, **1.**27, **1.**29
Pitch, **1.**22
Plastic limit, **2.**10
Plasticity, **2.**9–**2.**10, **2.**17
Plasticity index, **2.**11
Poiseuille-Hagen law, **5.**326
Polyurethane grout, **5.**338, **5.**340–**5.**363
  applications, **5.**343–**5.**350, **5.**352–**5.**353, **5.**355–**5.**359, **5.**361–**5.**362
    curtain grouting in rock and soil, **5.**343
    cutting off major water flows, **5.**345
    hardrock tunnels, **5.**345–**5.**346
    injectible tube and contact grouting, **5.**355–**5.**358
    joints in nonaccessible sewers, **5.**353
    pin piles for support to tunnels and structures, **5.**348
    renovation of human-accessible sewers, **5.**349–**5.**350
    repairing leaking brick and stone structures, **5.**347
    retaining walls in harbor loading docks, **5.**348–**5.**349

Polyurethane grout, applications (*Cont.*):
    sealing around bulkheads, **5.**358
    sealing cracks and joints in concrete/masonry, **5.**35
    sealing expansion joints in submerged structure, **5.**352
    sealing renards in tunnels, **5.**344–**5.**345
    seepage control in rock tunnels, **5.**345
    seepage prevention in dams and dikes, **5.**346
    soil grouting small-scale projects, **5.**353
  elastomers, **5.**353–**5.**355
    hydrophillic, **5.**360–**5.**361, **5.**354–**5.**355
  hydrophillic, **5.**340–**5.**341, **5.**352
  hydrophobic, **5.**340, **5.**341–**5.**350
    flexible prepolymers, **5.**350–**5.**352
    water-compatible, **5.**358, **5.**360
  two-component foam for mining applications, **5.**362–**5.**363
  water-reactive, **5.**340–**5.**341
Pore stress, **2.**21
  (*See also* Osmotic suction; Matric suction)
Posttensioning, **8.**29–**8.**30
  external, **6.**55
Post-Tensioning Institute, **3.**183
Pozzolans, **3.**18
Precipitation grouting, **5.**322–**5.**323
Pressure grouting, **6.**62–**6.**67, **7.**97
  (*See also* Mud jacking; Grouting)
Pumps:
  concrete, **3.**71–**3.**73
  sump, **1.**31

Quartz, **2.**6

Rankine theory, **4.**5–**4.**13
Reeves, Dr. Malcom, **5.**229
Reforestation, **5.**323–**5.**324
Reinforced concrete, **3.**77–**3.**120
  bars, **3.**117–**3.**120
    compression development length, **3.**120
    tension development length, **3.**118–**3.**120
  beams, **3.**94–**3.**102
    shear strength, **3.**94–**3.**99
    torsion strength, **3.**100–**3.**102
  columns, **3.**105–**3.**117
    axially loaded, **3.**106
    biaxial bending, **3.**116–**3.**117
    buckling effect, **3.**110, **3.**115–**3.**116
    general, **3.**105–**3.**106
    load-moment interaction diagrams, **3.**109–**3.**115

Reinforced concrete (*Cont.*):
  slender, **3.**110, **3.**115–**3.**116
  uniaxial bending and compression, **3.**106–**3.**109
  flexure behavior, **3.**77–**3.**93
    cracked and elastic range, **3.**81–**3.**82
    cracked and inelastic range, **3.**82–**3.**83
    doubly reinforced sections, **3.**86–**3.**90
    flanged section, **3.**90–**3.**93
    rectangular sections with tension reinforcement, **3.**83–**3.**85
    uncracked and elastic range, **3.**78–**3.**80
  overview, **3.**77
  shear behavior, **3.**93–**3.**99
    beams, **3.**94–**3.**99
  strength:
    combined shear, bending, and torsion, **3.**102–**3.**105
    flexural, **3.**77–**3.**93
    load factors, **3.**78
    reduction factor, **3.**78
    shear, **3.**93–**3.**99
    torsion, **3.**100–**3.**105
  torsion behavior, **3.**100–**3.**105
    beams, **3.**100–**3.**102
Repairing foundations, **6.**31–**6.**94, **8.**5
  establishing need for, **6.**23–**6.**28
  drilled concrete piers, **6.**40–**6.**45
  failure criteria, **6.**27
  slab, **6.**48–**6.**55
  spread footings, **6.**37–**6.**40
Retaining walls, **4.**19–**4.**37
  backfilling, **4.**24–**4.**29
  bearing capacity failure, **4.**31–**4.**32, **4.**36–**4.**37
  buttressed, **4.**19–**4.**21
    active earth pressure on, **4.**33–**4.**35
  cantilever, **4.**19–**4.**21
    active earth pressure on, **4.**33–**4.**35
  counterfort, **4.**19–**4.**21
    active earth pressure on, **4.**33–**4.**35
  depth of base, **4.**22
  design limits, **4.**22
  designing:
    simplified method, **4.**23–**4.**32
    theoretical method, **4.**32–**4.**37
  earth pressures applied to, **4.**21
  foundation, **4.**19–**4.**21
  gravity, **4.**19–**4.**21
    active earth pressure on, **4.**33
    bearing capacity, **4.**36–**4.**37
    overturning, **4.**35–**4.**36
    sliding, **4.**37
  movement limits, **4.**21–**4.**22

Retaining walls (*Cont.*):
  overturning, **4.**29–**4.**32, **4.**35–**4.**36
  sliding, **4.**32, **4.**37
  stability, **4.**29–**4.**32, **4.**35–**4.**37
  stress:
    estimating applied forces under active conditions, **4.**24–**4.**29
    estimating on nonyielding walls, **4.**23
  types of, **4.**19–**4.**21
  unyielding, **4.**32–**4.**33
Risk management, **8.**7–**8.**11
Rockagil, **5.**369
Rocks, **2.**3, **5.**88
  grouting, **5.**298, **5.**325–**5.**326, **5.**330, **5.**333–**5.**334
  minerals of, **2.**6–**2.**7
  weathering, **2.**5
Roots, **1.**5
  depth of tree, **1.**11
  foundation failures and, **6.**8–**6.**9, **6.**37
  soil moisture vs. development of, **1.**8–**1.**18
  system diagram, **1.**11
Root zone, **1.**3
Runoff, **1.**7–**1.**8, **1.**25

Safety, bearing capacity and, **2.**54
Sand, shear strength of, **2.**33–**2.**34
Saturation, **5.**36
Saturation zone, **1.**3
Saylor, David, **3.**4
Sazinski, Richard J., **8.**21
Schmertmann's settlement method, **2.**64–**2.**66
Sedimentation, **2.**26
Settlement:
  allowable, **2.**55–**2.**56
  blowcount adjustment factors, **2.**59
  Bowles's method, **2.**61–**2.**62
  computing methods, **2.**60–**2.**67
  consolidation of foundations on clay, **2.**67
  Das's method, **2.**64
  differential, **2.**56
  distortion, **2.**56
  elastic, **2.**56–**2.**67
    key variables in, **2.**56–**2.**57
  elastic soil, **2.**57–**2.**58
  elastic theory, **2.**62–**2.**64
  embedment correction factors, **2.**60
  factors concerning overexcavation and replacement, **5.**14–**5.**29, **5.**34–**5.**35
  foundation failures and, **6.**11–**6.**12
  high rise foundations, **7.**4, **7.**5
  immediate, **5.**123
  mat foundations, **2.**56

Settlement (*Cont.*):
　maximum allowable, **2**.56
　models, **2**.57–**2**.60
　overburden correction factors, **2**.58
　Peck-Hanson-Thronburn method, **2**.61
　proportioning footings for equal, **2**.67
　Schmertmann's method, **2**.64–**2**.66
　shallow foundations, **2**.55–**2**.67
　slab foundations, **6**.51–**6**.55
　standard penetration test, **2**.58–**2**.60
　Teng's method, **2**.61
　Terzaghi's method, **2**.57, **2**.60
　three-dimensional, **5**.126–**5**.127
　tilt, **2**.56
　time rate for cohesive soil, **5**.118–**5**.123
　total, **5**.123–**5**.126
　types of foundation, **2**.55
　uniform, **2**.55–**2**.56
　vs. upheaval, **6**.12–**6**.17
Sewer systems, **1**.27–**1**.28, **7**.8, **7**.10–**7**.11
　peak flow accomodation, **1**.27
Shallow foundations:
　bearing capacity, **2**.48–**2**.54
　　of soils under, **3**.122–**3**.123
　design parameters, **2**.47–**2**.48
　　geotechnical, **2**.47–**2**.48
　　structural, **2**.47
　footings, **2**.45–**2**.46
　frost-protected, **8**.29
　settlement, **2**.55–**2**.67
　　allowable, **2**.55–**2**.56
Shear strength:
　cohesive soil, **5**.114
　concrete, **3**.57–**3**.58
　reinforced concrete, **3**.93–**3**.99
　saturated overconsolidated clay, **2**.42–**2**.44
　triaxial tests, **2**.36–**2**.37
　unsaturated soil, **5**.106–**5**.109
Shear strength test, **2**.32–**2**.44
　clay, **2**.34–**2**.44
　sand, **2**.33–**2**.34
Shrinkage limit, **2**.10
Silacsol, **5**.365–**5**.366
Silt, **7**.6
　piles in, **3**.165–**3**.167
Siprogel, **5**.369
Site development, **1**.1–**1**.33
Site hydrology, **1**.21
Site preparation, **1**.21–**1**.33
　drainage, **1**.24–**1**.29
　drainage systems, **1**.29–**1**.30
　foundation drain systems, **1**.31–**1**.32
　foundation on filled slopes, **1**.24

Site preparation (*Cont.*):
　grading plans, **1**.23–**1**.24
　land planning, **1**.30–**1**.31
　landscaping, **1**.30
Skin friction, **3**.172
Skin resistance, **3**.160
Slab foundations, **8**.29
　differential elevations in, **6**.24
　leveling, **6**.50, **6**.52, **6**.53
　nonstructural, **3**.183
　posttension, **3**.183–**3**.184
　repairing, **6**.48–**6**.55
　　mud jacking, **6**.48–**6**.50, **6**.53–**6**.54
　　posttensioning, **6**.55, **8**.29–**8**.30
　　settlement, **6**.51–**6**.55
　　underpinning, **6**.49
　　upheaval, **6**.50–**6**.51
　ribbed slab with void space, **3**.184
　ribbed slabs-on-ground, **3**.184
　　pier supported, **3**.184
　soil swell vs. moisture changes, **6**.20–**6**.23
　structural, **3**.183–**3**.184
　upheaval, **6**.12
Slopes, **1**.22
Smith, Dr. Cecil, **5**.229
Smith, Dr. Don, **1**.9, **6**.102
Smith, Thomas M., **8**.7
Sodium silicate grout, **5**.363–**5**.366
　chemical reaction to Durcisseurs, **5**.364–**5**.365
　hardeners, **5**.364–**5**.365
　newer developments with, **5**.365–**5**.366
Soil, **2**.3–**2**.44
　aggregate properties, **2**.12
　aging, **5**.163, **5**.165, **5**.167, **5**.172, **5**.173
　backfill (*see* Backfill)
　bearing capacity (*see* Bearing capacity)
　cementation, **5**.171–**5**.173
　characteristics of jet grouted, **5**.318–**5**.320
　classifying, **2**.13–**2**.17
　clay (*see* Clay)
　coarse-grained, **2**.7–**2**.8, **2**.9, **2**.14
　cohesionless (*see* Soil, noncohesive)
　cohesive, **2**.12–**2**.13, **2**.50, **5**.100–**5**.163, **6**.6–**6**.9
　　bearing capacity, **5**.127
　　compaction method, **5**.134–**5**.163
　　compactive prestress, **5**.104–**5**.105
　　compressibility, **5**.127–**5**.128
　　effects of density, **5**.109–**5**.111
　　effects of fabric, **5**.129–**5**.131
　　effects of moisture in, **5**.105–**5**.109
　　fabric and structure, **5**.100–**5**.103**
　　macrofabric, **5**.101
　　method of compaction, **5**.111–**5**.118

Soil, cohesive (*Cont.*):
    microfabric, **5.**100
    minifabric, **5.**101
    permeability, **5.**153–**5.**163
    piles in, **3.**163–**3.**165
    settlement, **5.**118–**5.**127
    shrinkage potential, **5.**150–**5.**153
    strength, **5.**128–**5.**129
    stress-strain-strength behavior, **5.**103–**5.**163
    void space, **5.**101
    wetting-induced volume change, **5.**131–**5.**135
collapse potential, **5.**231–**5.**234
compacted (*see* Compaction)
components, **2.**3
compressibility, **2.**11
consolidated, **2.**29
consolidation test, **2.**26–**2.**32
deposits, **2.**14
dispersed, **2.**12
earth pressure and (*see* Earth pressure)
earthquake regions, **3.**147–**3.**150
effects of cementation, **5.**98–**5.**100
elastic, **2.**57–**2.**58
engineering properties of compacted, **5.**72–**5.**173
erosion control plans, **1.**24
expansive, **1.**4, **1.**15
fine-grained, **2.**8, **2.**14
flocculated, **2.**12
formation of, **2.**5–**2.**6
frost-penetration map of USA, **3.**147
frost-susceptible, **3.**146
grain shape, **2.**9
grain size, **2.**7–**2.**9
granular, **7.**6
grouting (*see* Grouting)
heaving (*see* Heaving)
homogeneous, **5.**19
honeycombed, **2.**12
hydraulic conductivity, **5.**306–**5.**307, **5.**309, **5.**386
identifying, **2.**5–**2.**13
liquefaction, **5.**250–**5.**252
liquefaction potential, **5.**89–**5.**98
movement, **1.**16
noncohesive, **1.**5, **1.**15, **2.**12, **2.**50, **6.**3–**6.**5
    compacted, **5.**72–**5.**100
    effect of moisture on compressibility and strength, **5.**78–**5.**86
    piles in, **3.**160–**3.**163
    static stress-strain-strength behavior, **5.**72–**5.**78
    wetting-induced volume changes, **5.**86–**5.**100
overconsolidated, **2.**29
plasticity, **2.**9–**2.**10, **2.**17, **5.**224–**5.**225

Soil (*Cont.*):
    residual, **2.**13–**2.**14
    sand, **2.**33–**2.**34
    saturated, **5.**36
    shear strength, **2.**32–**2.**44
    shrinkage, **1.**15, **5.**89, **5.**150–**5.**153, **5.**231
    silt, piles in, **3.**165–**3.**167
    single grained, **2.**12
    strength, cohesive, **5.**128–**5.**129
    stress:
        applied, **2.**22–**2.**26
        geostatic, **2.**21–**2.**22
    suction, **1.**16
        (*See also* Capillary pressure; Osmotic suction; Pore stress)
    sulfate reactions with, **6.**91–**6.**92
    swelling, **5.**149–**5.**150, **5.**225–**5.**231, **6.**20–**6.**23
    tests:
        compaction, **5.**59–**5.**72
        consolidated-drained, **2.**36–**2.**37
        consolidated-undrained, **2.**37
        consolidation, **2.**26–**2.**32
        cyclic triaxial, **3.**148–**3.**150
        in earthquake regions, **3.**148–**3.**150
        relative density, **3.**148
        shear strength, **2.**32–**2.**44
        triaxial, **2.**36–**2.**37
        unconfined compression, **2.**41–**2.**42
        unconsolidated-undrained, **2.**37, **2.**41–**2.**42
    Texas, **1.**15
    thixotropy, **5.**163–**5.**173
    three-phase, **2.**5
    topsoil, **1.**23–**1.**24
    transported, **2.**14
    two-phase, **2.**3–**2.**5
    water behavior in, **1.**3–**1.**19
    water flow in, **2.**17–**2.**20
    well-graded, **2.**8
Soil active zone (SAZ), **1.**17
Soil classification, **2.**14–**2.**17
    AASHTO system, **2.**14–**2.**15
    unified system, **2.**14, **2.**16
Soil improvement, near-surface compaction, **5.**36–**5.**205
    (*See also* Chemical stabilization; Soil stabilization)
Soil moisture, **1.**3–**1.**4
    vs. aeration zone, **1.**5–**1.**6
    barriers, **5.**264–**5.**267
        horizontal, **5.**264–**5.**265
        vertical, **5.**265–**5.**267
    compaction and, **5.**54–**5.**59, **5.**175–**5.**178
    conditioning period, **5.**61

Soil moisture (*Cont.*):
  evaporation, **1.**5–**1.**6
  factors influencing foundation stability, **1.**18
  foundation failures caused by, **6.**6–**6.**7, **6.**36
  gravity and, **1.**5–**1.**6
  London clay, **1.**11
  permeability vs. infiltration, **1.**6–**1.**7
  vs. root development, **1.**8–**1.**18
  saturated, **5.**36
  transpiration, **1.**5, **1.**14, **1.**16, **1.**17
  typical loss of, **1.**13
  vs. water table, **1.**4–**1.**5
Soil Sta, **5.**229–**5.**231
Soil stabilization, **5.**1–**5.**400
  grouting techniques, **5.**277–**5.**400
  moisture barriers, **5.**264–**5.**267
    horizontal, **5.**264–**5.**265
    vertical, **5.**265–**5.**267
  nongrouting techniques, **5.**3–**5.**276
  overexcavation and replacement, **5.**4–**5.**36
  using chemicals (*see* Chemical stabilization)
Soil water belt, **1.**3
Soilcrete, **5.**316–**5.**317
Standard Building Code (SBC), **8.**23
Standard penetration test (SPT), **2.**58–**2.**60
Static compaction, **5.**63
Steel piles, **3.**157–**3.**159
  (*See also* Minipiles)
Stephenson, Richard W., **4.**3, **4.**19
Stone, crushed, **3.**7
Storm sewers, **1.**25, **1.**31–**1.**32
Strain, calculating, **6.**25
Stress:
  allowable in pile foundations, **3.**154–**3.**159
  applied in soil, **2.**22–**2.**26
  average stress influence chart, **2.**26
  biaxial interaction curves, **3.**65
  concrete, **3.**58–**3.**65
  creep and, **3.**60–**3.**62
  effective, **2.**22
  factors concerning overexcavation and replacement, **5.**35–**5.**36
  geostatic in soil, **2.**21–**2.**22
  Mohr's theory, **2.**35–**2.**44
  point load, **2.**22, **2.**35
  pore, **2.**21
  total, **2.**21
  triaxial, **2.**36–**2.**37
  uniformly loaded circular area, **2.**23–**2.**24
  uniformly loaded rectangular area, **2.**25
  uniformly loaded strip, **2.**22
  vertical vs. depth, **2.**23
  vertical vs. radius, **2.**23

Structural engineering bulletin (SEB), **8.**28
Subsurface water, **1.**3, **6.**98
Sump pumps, **1.**31
Superplasticizers, **3.**12
Surface drainage, **1.**25–**1.**26
Surface tension, **5.**79
Swell potential, **2.**11
Syneresis, **5.**363

Taylor's method, **2.**30–**2.**32
Temperature:
  cementitious reactions and, **5.**238
  climatic ratings for USA, **6.**97
  concrete pouring and, **3.**65–**3.**71
  curing conditions for chemicals added to soil, **5.**263
  influence of temperature on moisture-density relationship, **5.**62
  pozzolanic reactions and, **5.**238
Teng's settlement method, **2.**61
Tensile strength:
  concrete, **3.**56–**3.**57
  testing concrete, **3.**36
Terzaghi's settlement method, **2.**57, **2.**60
Tests:
  accelerated strength test, **3.**43–**3.**44
    autogenous method, **3.**44
    boiling-water method, **3.**43
    high-temperature method, **3.**44
    pressure method, **3.**44
    warm-water method, **3.**43
  cement, **3.**6
  compaction, **5.**59–**5.**72, **5.**178–**5.**205
    apparent hydraulic conductivity method, **5.**204
    California bearing ratio, **5.**178, **5.**202
    cone penetration, **5.**179, **5.**202
    confined-ring specimens, **5.**71–**5.**72
    drive-cylinder method, **5.**186
    dynamic penetration, **5.**178
    Hilf's rapid method, **5.**189–**5.**194
    impact, **5.**63–**5.**66
    kneading, **5.**67–**5.**70
    laboratory permeability test, **5.**179
    moisture-density, **5.**183
    nuclear method, **5.**186–**5.**189
    one-dimensional compression, **5.**179
    one-dimensional swelling and collapse, **5.**179
    plate load, **5.**178, **5.**196–**5.**202
    Proctor penetrometer, **5.**178, **5.**194–**5.**195
    rubber balloon, **5.**183–**5.**186
    sand cone, **5.**183–**5.**186

Tests, compaction (*Cont.*):
    sand replacement method, **5.**186
    sealed double ring infiltration, **5.**179, **5.**202–**5.**204
    sleeve method, **5.**186
    standard penetration, **5.**179, **5.**202
    static, **5.**63
    suction head method, **5.**204
    suction measured using tensiometer, **5.**179
    test pit method, **5.**186
    triaxial compression, **5.**179
    unconfined compression, **5.**179
    variability and statistical specifications, **5.**180–**5.**182
    vibratory, **5.**70–**5.**71
    water replacement method, **5.**186
    wetting front method, **5.**205
  concrete, **3.**34–**3.**40, **3.**43–**3.**44, **3.**47–**3.**49
    accelerated strength, **3.**43–**3.**44
    acoustical method, **3.**49
    air void parameter, **3.**36
    compaction factor, **3.**35
    compressive strength, **3.**36–**3.**40
    conditions for, **3.**56
    core, **3.**49
    creep under compression, **3.**36
    electromagnetic method, **3.**49
    flexural strength, **3.**36
    fresh concrete, **3.**34–**3.**35
    hardened concrete, **3.**36
    length change of hardened paste or mortar, **3.**36
    maturity concepts, **3.**49
    modulus of elasticity and Poisson's ratio, **3.**36
    nondestructive, **3.**47–**3.**49
    penetration resistance method, **3.**48
    pull-out method, **3.**48
    resistance to rapid freezing/thawing, **3.**36
    resistance to scaling, **3.**36
    slump, **3.**34–**3.**35
    splitting tensile strength, **3.**36
    surface hardness method, **3.**47–**3.**48
    ultrasonic pulse velocity method, **3.**48–**3.**49
    V-B, **3.**35
  grouting, **5.**396–**5.**398
    acoustic emission monitoring, **5.**397
    Baroid mud balance, **5.**396
    bleed, **5.**396
    borehole camera investigation, **5.**397
    borehole radar, **5.**397–**5.**398
    compressive strength, **5.**396
    core sampling, **5.**397
    cross-hole acoustic velocity, **5.**398

Tests, grouting (*Cont.*):
    grout-hole drilling, **5.**397
    hydraulic conductivity, **5.**397
    initial cure, **5.**396
    penetrometer, **5.**397
    pressure filtration, **5.**396
    pressure meter, **5.**397
    SPT, **5.**397
    viscosity, **5.**396
  liquefaction, **5.**97–**5.**98
  O'Neill, **6.**60
  permeability, **2.**19–**2.**20
    constant head, **2.**19
    falling head, **2.**19
    trixial flexible wall, **2.**20
  Petry, **6.**60
  pile foundation load capacity, **3.**170–**3.**171
  soil:
    compaction, **5.**59–**5.**72
    consolidated-drained, **2.**36–**2.**37
    consolidated-undrained, **2.**37
    consolidation, **2.**26–**2.**32
    in earthquake regions, **3.**148–**3.**150
    shear strength, **2.**32–**2.**44
    triaxial, **2.**36–**2.**37
    unconfined compression, **2.**41–**2.**42
    unconsolidated-undrained, **2.**37, **2.**41–**2.**42
  sources of error in permeability, **2.**20
  standard penetration settlement, **2.**58–**2.**60
Thixotropy, **5.**163–**5.**173
  definition, **5.**163–**5.**164
  factors that affect, **5.**168–**5.**171
Timber piles, **3.**159
Topsoil, **1.**23–**1.**24
Torsion strength, reinforced concrete beams, **3.**100–**3.**102
Total stress, **2.**21
Total stress envelope, **2.**41
Transpiration, **1.**5, **1.**14, **1.**16
Trees:
  depth of roots, **1.**11
  foundations and, **6.**101–**6.**102
  foundation stability and, **8.**4
  timber piles made from, **3.**159
Triaxial shear apparatus, **2.**35
Triaxial stress, **2.**36–**2.**37

Unconfined compression test, **2.**41–**2.**42
Underpinning, **6.**32, **6.**44, **6.**45, **6.**49, **6.**57
Upheaval, **6.**12, **6.**67
  vs. settlement, **6.**12–**6.**17
  slab foundations, **6.**50–**6.**51

Uretek method, **6.**86
(*See also* Polyurethane grout)

Vegetation:
    foundations and, **6.**101–**6.**102
    roots and foundation failures, **6.**8–**6.**9, **6.**37
    soil moisture removal from, **1.**5, **1.**17
Vertical hydrofracturing, **5.**310
Vertical impermeable capillary barrier (VICB), **6.**100
Vertical permeable capillary barrier (VPCB), **6.**100
Vibratory compaction, **5.**70–**5.**71
Volume change potential, **2.**11–**2.**12

Walls:
    buttressed, **4.**19–**4.**21, **4.**33–**4.**35
    cantilever, **4.**19–**4.**21, **4.**33–**4.**35
    counterfort, **4.**19–**4.**21, **4.**33–**4.**35
    foundation, **4.**19–**4.**21
    gravity, **4.**19–**4.**21, **4.**33, **4.**35–**4.**37
        bearing capacity, **4.**36–**4.**37
        overturning, **4.**35–**4.**36
        sliding, **4.**37
    hadite block, **6.**82
    repairing foundation, **6.**79–**6.**84
    retaining (*see* Retaining walls)
    trixial flexible test, **2.**20
Warranties, **8.**14–**8.**19
    Home Owners' Warranty (HOW), **6.**25–**6.**26
Wash, **1.**22

Water:
    barriers for, **6.**98–**6.**100
    behavior in soil, **1.**3–**1.**19
    capillary, **1.**4
    cement ratio and, **3.**50–**3.**51
    evaporation, **1.**5, **1.**16
    flow of in soil, **2.**17–**2.**20
    flow velocity in soil, **2.**18–**2.**19
    foundation failures and, **6.**4
    freezing, **3.**146
    ground, **1.**3
        recharge, **1.**8
    impurities, **3.**11
    level in rock excavations, **7.**6
    moisture regimes, **1.**3–**1.**4
    permeability (*see* Permeability)
    runoff, **1.**7–**1.**8, **1.**25
    soil moisture vs. aeration zone, **1.**5–**1.**6
    soil moisture vs. root development, **1.**8–**1.**18
    soil moisture vs. water table, **1.**4–**1.**5
    subsurface, **1.**3, **6.**98
    using to make concrete, **3.**11–**3.**13
Water table, **1.**3, **2.**50–**2.**51
    vs. soil moisture, **1.**4–**1.**5
Water-reducing admixtures, **3.**12–**3.**13
Watering, **6.**95–**6.**97
Waterproofing, subgrade foundation, **6.**66
Wier, John, **1.**26
Williams, Jim, **3.**183
Windham, Charles A., **8.**13

Zero air voids line (ZAVL), **5.**55, **5.**58

ABOUT THE EDITOR

Robert Wade Brown is the president of Brown Foundation Repair and Consulting, Inc. in Dallas, Texas. He is also the author of *Foundation Repair and Behavior: Residential and Light Commercial* and *Design and Repair of Residential and Light Commercial Foundations*, both available from McGraw-Hill, as well as more than 50 articles in the field of foundation engineering.